DIAGNOSTIC MOLECULAR PATHOLOGY

D1587479

ELSEVIER

science & technology books

Companion Web Site:

http://booksite.elsevier.com/9780128008867

Diagnostic Molecular Pathology: A Guide to Applied Molecular Testing
William B. Coleman and Gregory J. Tsongalis, Editors

Available Resources:

• All figures and tables from the book available for download

• Case studies

• Question and Answers

ELSEVIER

ACADEMIC
PRESS

DIAGNOSTIC MOLECULAR PATHOLOGY

A Guide to Applied Molecular Testing

Edited by

WILLIAM B. COLEMAN, PhD

Department of Pathology and Laboratory Medicine, Program in Translational Medicine,
UNC Lineberger Comprehensive Cancer Center, University of North Carolina School of Medicine,
Chapel Hill, NC, United States

GREGORY J. TSONGALIS, PhD, HCLD, CC

Laboratory for Clinical Genomics and Advanced Technology (CGAT),
Department of Pathology and Laboratory Medicine,
Dartmouth-Hitchcock Medical Center and Norris Cotton Cancer Center, Lebanon, NH, United States;
Geisel School of Medicine at Dartmouth, Hanover, NH, United States

MEDICAL LIBRARY
QUEENS MEDICAL CENTRE

AMSTERDAM • BOSTON • HEIDELBERG • LONDON
NEW YORK • OXFORD • PARIS • SAN DIEGO
SAN FRANCISCO • SINGAPORE • SYDNEY • TOKYO

Academic Press is an imprint of Elsevier

Academic Press is an imprint of Elsevier
125 London Wall, London EC2Y 5AS, United Kingdom
525 B Street, Suite 1800, San Diego, CA 92101-4495, United States
50 Hampshire Street, 5th Floor, Cambridge, MA 02139, United States
The Boulevard, Langford Lane, Kidlington, Oxford OX5 1GB, United Kingdom

Copyright © 2017 Elsevier Inc. All rights reserved.

No part of this publication may be reproduced or transmitted in any form or by any means, electronic
or mechanical, including photocopying, recording, or any information storage and retrieval system,
without permission in writing from the publisher. Details on how to seek permission, further information about
the Publisher's permissions policies and our arrangements with organizations such as the Copyright Clearance
Center and the Copyright Licensing Agency, can be found at our website: www.elsevier.com/permissions.

This book and the individual contributions contained in it are protected under copyright by the Publisher
(other than as may be noted herein).

Notices
Knowledge and best practice in this field are constantly changing. As new research and experience broaden
our understanding, changes in research methods, professional practices, or medical treatment may become
necessary.

Practitioners and researchers must always rely on their own experience and knowledge in evaluating and
using any information, methods, compounds, or experiments described herein. In using such information
or methods they should be mindful of their own safety and the safety of others, including parties for whom
they have a professional responsibility.

To the fullest extent of the law, neither the Publisher nor the authors, contributors, or editors, assume any
liability for any injury and/or damage to persons or property as a matter of products liability, negligence
or otherwise, or from any use or operation of any methods, products, instructions, or ideas contained in the
material herein.

British Library Cataloguing-in-Publication Data
A catalogue record for this book is available from the British Library

Library of Congress Cataloging-in-Publication Data
A catalog record for this book is available from the Library of Congress

ISBN: 978-0-12-800886-7

For Information on all Academic Press publications
visit our website at https://www.elsevier.com

1008018316

 **Working together
to grow libraries in
developing countries**

www.elsevier.com • www.bookaid.org

Publisher: Mica Haley
Acquisition Editor: Tari Broderick
Editorial Project Manager: Jeff Rosetti
Production Project Manager: Julia Haynes
Designer: Inês Cruz

Typeset by MPS Limited, Chennai, India

Endorsements

Diagnostic Molecular Pathology is destined to become an important cornerstone and go-to volume for pathologists, researchers, and clinicians—indeed anyone who wants to understand the approach and application of modern molecular techniques in disease detection and diagnosis. Drs. Coleman and Tsongalis—and their impressive stable of contributors—are to be congratulated for making the exciting and rapidly expanding science of molecular diagnosis accessible to the novice, while also providing the expert with important details on the nuances. Beyond just a compendium of useful information, it is a well-curated journey through basic concepts, infectious diseases, malignancy, hematopathology, and genetic diseases—and even includes access to a website with lab test videos and decision-making exercises.

Richard N. Mitchell, MD, PhD (Brigham and Women's Hospital and Harvard Medical School)

With much fanfare, molecular techniques have revolutionized clinical medicine, from diagnosis to personalized medical therapeutics. It is with this background that *Diagnostic Molecular Pathology* approaches the daunting task of summarizing the advances in the field from infectious diseases, heritable and acquired genetic diseases, hematological malignancies, to pharmacogenomics. In a comprehensive publication, the editors perform a yeoman's effort in covering this dynamic and exciting field.

Lawrence M. Silverman, PhD (University of Virginia School of Medicine)

Diagnostic Molecular Pathology provides a panoramic view of diagnostic molecular pathology testing, written by leaders in the field. It is a great reference and learning tool.

Dani S. Zander, MD (University of Cincinnati Medical Center)

Dedication

This textbook describes the emerging field of *diagnostic molecular pathology* and its application to various forms of human disease. Despite the relative youthfulness of this field, *diagnostic molecular pathology* has become critically important in the contemporary practice of personalized medicine and is built upon the collective knowledgebase that reflects our understanding of the pathology, pathogenesis, and pathophysiology of human disease. As such, the information contained in this textbook represents the culmination of innumerable small successes that emerged from the ceaseless pursuit of new knowledge by countless clinical and experimental pathologists working around the world on all aspects of human disease. Their ingenuity and hard work have dramatically advanced the field of molecular pathology over time, and particularly during the last 25 years. This book is a tribute to the dedication, diligence, and perseverance of individual scientists who contributed to the advancement of our understanding of the molecular basis of human disease, especially graduate students, laboratory technicians, and postdoctoral fellows, whose efforts are so frequently taken for granted, whose accomplishments are so often unrecognized, and whose contributions are so quickly forgotten.

Diagnostic Molecular Pathology: A Guide to Applied Molecular Testing is dedicated to the memory of Dr. Kathleen Rao who passed away on March 24, 2016, following a brief battle with cancer. Dr. Rao earned a PhD in genetics from the University of North Carolina at Chapel Hill and was a member of the faculty in the UNC School of Medicine from 1984 until the time of her death. Dr. Rao was a Professor of Pediatrics, Genetics, and Pathology and Laboratory Medicine, and served as the Director of the Cytogenetics Laboratory for UNC Hospitals. Dr. Rao made numerous contributions to the field of cytogenetics, was a Founding Fellow of the American College of Medical Genetics and Genomics, and was the recipient of the 2016 Distinguished Cytogeneticist Award. She served on the International Standing Committee on Cytogenetic Nomenclature, the Children's Oncology Group Cytogenetics Committee, and the Cancer and Leukemia Group B Cytogenetics Review Committee. Dr. Rao was also well-recognized as an extraordinary medical educator at the University of North Carolina where she was a Founding member of the UNC School of Medicine's Academy of Educators. As Director of the Cytogenetics Laboratory Fellowship Training Program at UNC, Dr. Rao taught and mentored numerous students who now work in the field of cytogenetics throughout the United States. Dr. Rao was a dear friend and cherished colleague to many people at the University of North Carolina and across the country. We are proud to have known her and worked with her through the years. We are also extremely honored to have her as a contributor to this textbook (see chapter: Molecular Testing in Pediatric Cancers) and regret that we will not have another chance to work with her on a project like this one. This book is dictated to the example Dr. Rao provides all of us—as a distinguished educator, an accomplished molecular pathologist, and a genuinely good person.

We also dedicate *Diagnostic Molecular Pathology: A Guide to Applied Molecular Testing* to the many people that have played crucial roles in our successes. We thank our many scientific colleagues, past and present, for their camaraderie, collegiality, and support. We especially thank our scientific mentors for their example of dedication to research excellence. We are truly thankful for the positive working relationships and friendships that we have with our faculty colleagues, for the mentoring we received from our elders, and for the opportunity to mentor those that follow us. We also thank our undergraduate students, graduate students, and postdoctoral fellows for teaching us more than we might have taught them. We thank our parents for believing in higher education, for encouragement through the years, and for helping make dreams into reality. We thank our brothers and sisters, and extended families, for the many years of love, friendship, and tolerance. We thank our wives, Monty and Nancy, for their unqualified love, unselfish support of our endeavors, understanding of our work ethic, and appreciation for what we do. Lastly, we give special thanks to our children, Tess, Sophie, Pete, and Zoe. Their achievements and successes as young adults are a greater source of pride for us than our own accomplishments. As when they were children, we thank them for providing an unwavering bright spot in our lives, for their unbridled enthusiasm and boundless energy, and for giving us a million reasons to take an occasional day off from work just to have fun.

William B. Coleman
Gregory J. Tsongalis

Contents

I
INTRODUCTION TO MOLECULAR TESTING

1. Basic Concepts in Molecular Pathology— Introduction to Molecular Testing in Human Disease

W.B. COLEMAN AND G.J. TSONGALIS

2. Laboratory Approaches in Molecular Pathology—The Polymerase Chain Reaction

W.B. COLEMAN AND G.J. TSONGALIS

3. Next-Generation Sequencing in the Clinical Laboratory

D.L. DUNCAN AND N.M. PATEL

4. Automation of the Molecular Diagnostic Laboratory

S.A. TURNER AND G.J. TSONGALIS

II
MOLECULAR TESTING IN INFECTIOUS DISEASE

5. Molecular Testing for Human Imunodeficiency Virus

M. MEMMI, T. BOURLET AND B. POZZETTO

6. Molecular Testing in Hepatitis Virus Related Disease

P.M. MULROONEY-COUSINS AND T.I. MICHALAK

III

MOLECULAR TESTING IN GENETIC DISEASE

IV

MOLECULAR TESTING IN ONCOLOGY

V

MOLECULAR TESTING
IN HEMATOPATHOLOGY

List of Contributors

Kimberly H. Allison, MD Department of Pathology, Stanford University School of Medicine, Stanford, CA, United States

Megan A. Allyse, PhD Department of Health Sciences Research, Mayo Clinic School of Medicine, Rochester, MN, United States

Rodney C. Arcenas, PhD, D(ABMM) Molecular Microbiology and Immunology, Memorial Healthcare System, Pathology Consultants of South Broward, Hollywood, FL, United States

Michael J. Bartel, MD Division of Gastroenterology & Hepatology, Mayo Clinic, Jacksonville, FL, United States

Amir Behdad, MD Division of Hematopathology, Northwestern University, Feinberg School of Medicine, Northwestern Memorial Hospital, Chicago, IL, United States

Katie M. Bennett, PhD, MB (ASCP)CM, NRCC-CC Texas Tech University Health Sciences Center, School of Health Professions, Molecular Pathology Program, Lubbock, TX, United States

Jonathan S. Berg, MD, PhD Department of Genetics, University of North Carolina School of Medicine, Chapel Hill, NC, United States

D. Hunter Best, PhD Department of Pathology, University of Utah School of Medicine, Salt Lake City, UT, United States; Molecular Genetics and Genomics, ARUP Laboratories, University of Utah School of Medicine, Salt Lake City, UT, United States

Bryan L. Betz, PhD Department of Pathology, University of Michigan, Ann Arbor, MI, United States

Jessica K. Booker, PhD Department of Pathology and Laboratory Medicine; Department of Genetics, University of North Carolina at Chapel Hill, Chapel Hill, NC, United States

Kristi S. Borowski, MD Departments of Medical Genetics and Obstetrics and Gynecology, Mayo Clinic School of Medicine, Rochester, MN, United States

Thomas Bourlet, PharmD, PhD GIMAP EA3064, University of Lyon, Saint-Etienne, France; Laboratory of Infectious Agents and Hygiene, University Hospital of Saint-Etienne, Saint-Etienne, France

Pierre Brissot, MD, PhD National Center of Reference for Rare Genetic Iron Overload Diseases, Pontchaillou University Hospital, Rennes, France; Inserm-UMR 991, University of Rennes 1, Rennes, France

Noah A. Brown, MD Department of Pathology, University of Michigan, Ann Arbor, MI, United States

Marcin Bula, PhD The Wolfson Centre for Personalised Medicine, Institute of Translational Medicine, University of Liverpool, Liverpool, United Kingdom

Richard M. Caprioli, PhD Mass Spectrometry Research Center, and Department of Biochemistry, Vanderbilt University School of Medicine, Nashville, TN, United States

Subhankar Chakraborty, MD Division of Gastroenterology & Hepatology, Mayo Clinic, Rochester, MN, United States

William B. Coleman, PhD Department of Pathology and Laboratory Medicine, Program in Translational Medicine, UNC Lineberger Comprehensive Cancer Center, University of North Carolina School of Medicine, Chapel Hill, NC, United States

Kristy R. Crooks, PhD Department of Pathology, University of Colorado, Anschutz Medical Campus, Aurora, CO, United States

Jianli Dong, MD Department of Pathology, University of Texas Medical Branch, Galveston, TX, United States

Harry A. Drabkin, MD Department of Medicine, Division of Hematology/Oncology, Medical University of South Carolina, Charleston, SC, United States

Daniel L. Duncan, MD Department of Pathology and Laboratory Medicine, University of North Carolina School of Medicine, Chapel Hill, NC, United States

Jawed Fareed, PhD Department of Pathology, Loyola University Health System, Maywood, IL, United States

Andrea Ferreira-Gonzalez, PhD Division of Molecular Diagnostics, Department of Pathology, Virginia Commonwealth University, Richmond, VA, United States

Birgit H. Funke, PhD, FACMG Laboratory for Molecular Medicine, Partners Personalized Medicine, Boston, MA, United States; Department of Pathology, Harvard Medical School, Boston, MA, United States; Department of Pathology, Massachusetts General Hospital, Boston, MA, United States

Larissa V. Furtado, MD Department of Pathology, University of Chicago, Chicago, IL, United States

Giorgio Gallinella, MD, PhD Department of Pharmacy and Biotechnology, S. Orsola-Malpighi Hospital — Microbiology, University of Bologna, Bologna, Italy

Sonzalo Gonzalo, PharmD GIMAP EA3064, University of Lyon, Saint-Etienne, France; Laboratory of Infectious Agents and Hygiene, University Hospital of Saint-Etienne, Saint-Etienne, France

Florence Grattard, MD, PhD GIMAP EA3064, University of Lyon, Saint-Etienne, France; Laboratory of Infectious Agents and Hygiene, University Hospital of Saint-Etienne, Saint-Etienne, France

Danielle B. Gutierrez, PhD Mass Spectrometry Research Center, and Department of Biochemistry, Vanderbilt University School of Medicine, Nashville, TN, United States

Gloria T. Haskell, PhD Department of Genetics, Duke University, Durham, NC, United States

Amin A. Hedayat, MD Department of Pathology, Dartmouth-Hitchcock Medical Center, Lebanon, NH, United States

W. Edward Highsmith, Jr, PhD Departments of Laboratory Medicine and Pathology, and Medical Genetics, Mayo Clinic School of Medicine, Rochester, MN, United States

Susan J. Hsiao, MD, PhD Department of Pathology & Cell Biology, Columbia University Medical Center, New York, NY, United States

Omer Iqbal, MD Department of Pathology, Loyola University Health System, Maywood, IL, United States

Nahed Ismail, MD, PhD, D(ABMM), D(ABMLI) Department of Pathology, University of Pittsburgh, Pittsburgh, PA, United States

Anne-Marie Jouanolle, PharmD National Center of Reference for Rare Genetic Iron Overload Diseases, Laboratory of Molecular Genetics and Genomics, Pontchaillou University Hospital, Rennes, France

Sarah E. Kerr, MD Department of Laboratory Medicine and Pathology, College of Medicine, Mayo Clinic, Rochester, MN, United States

Olivier Loréal, MD, PhD National Center of Reference for Rare Genetic Iron Overload Diseases, Pontchaillou University Hospital, Rennes, France; Inserm-UMR 991, University of Rennes 1, Rennes, France

Heather M. McLaughlin, PhD Laboratory for Molecular Medicine, Partners Personalized Medicine, Department of Pathology, Harvard Medical School, Massachusetts General Hospital, Boston, MA, United States

Meriam Memmi, PhD GIMAP EA3064, University of Lyon, Saint-Etienne, France; Laboratory of Infectious Agents and Hygiene, University Hospital of Saint-Etienne, Saint-Etienne, France

Tomasz I. Michalak, MD, PhD, FAASLD, FCAHS Molecular Virology and Hepatology Research Group, Division of BioMedical Science, Faculty of Medicine, Health Sciences Centre, Memorial University, St. John's, Newfoundland, Canada

Melissa B. Miller, PhD Department of Pathology and Laboratory Medicine, UNC School of Medicine, Chapel Hill, NC, United States

Patricia M. Mulrooney-Cousins, PhD Molecular Virology and Hepatology Research Group, Division of BioMedical Science, Faculty of Medicine, Health Sciences Centre, Memorial University, St. John's, Newfoundland, Canada

Yuri E. Nikiforov, MD, PhD Division of Molecular & Genomic Pathology, Department of Pathology, University of Pittsburgh School of Medicine, Pittsburgh, PA, United States

Jeremy L. Norris, PhD Mass Spectrometry Research Center, and Department of Biochemistry, Vanderbilt University School of Medicine, Nashville, TN, United States

Nirali M. Patel, MD Department of Pathology and Laboratory Medicine, University of North Carolina School of Medicine, Chapel Hill, NC, United States

Peter L. Perrotta, MD Department of Pathology, West Virginia University, Morgantown, WV, United States

Benjamin A. Pinsky, MD, PhD Department of Medicine, Division of Infectious Diseases and Geographic Medicine, Department of Pathology, Stanford University School of Medicine, Stanford, CA, United States

Munir Pirmohamed, MBChB (Hons), PhD, FRCP, FRCP(E) The Wolfson Centre for Personalised Medicine, Institute of Translational Medicine, University of Liverpool, Liverpool, United Kingdom

Rongpong Plongla, MD, MSc Division of Infectious Diseases, Chulalongkorn University and King Chulalongkorn Memorial Hospital, Bangkok, Thailand; Department of Pathology and Laboratory Medicine, UNC School of Medicine, Chapel Hill, NC, United States

Bruno Pozzetto, MD, PhD GIMAP EA3064, University of Lyon, Saint-Etienne, France; Laboratory of Infectious Agents and Hygiene, University Hospital of Saint-Etienne, Saint-Etienne, France

Victoria M. Pratt, PhD, FACMG Pharmacogenomics Laboratory, Department of Medical and Molecular Genetics, Indiana University School of Medicine, Indianapolis, IN, United States

Gary W. Procop, MD, MS Section of Clinical Microbiology, Department of Laboratory Medicine, Cleveland Clinic, Cleveland, OH, United States

Massimo Raimondo, MD Division of Gastroenterology & Hepatology, Mayo Clinic, Jacksonville, FL, United States

⁺Kathleen W. Rao, PhD Departments of Pediatrics, Pathology and Laboratory Medicine, and Genetics, University of North Carolina School of Medicine, Chapel Hill, NC, United States; Cytogenetics Laboratory, McLendon Clinical Laboratories, UNC Hospitals, Chapel Hill, NC, United States

Stuart A. Scott, PhD, FACMG Department of Genetics and Genomic Sciences, Icahn School of Medicine at Mount Sinai, New York, NY, United States

Chanjuan Shi, MD, PhD Department of Pathology, Microbiology, and Immunology, Vanderbilt University Medical Center, Nashville, TN, United States

Carolyn J. Shiau, MD, FRCPC Department of Pathology, University Health Network, Toronto, ON, Canada

Yue Si, PhD Department of Pathology, University of Utah School of Medicine, Salt Lake City, UT, United States

Steven C. Smith, MD, PhD Department of Pathology, Virginia Commonwealth University School of Medicine, Richmond, VA, United States

Matthew B. Smolkin, MD Department of Pathology, West Virginia University, Morgantown, WV, United States

Kathleen A. Stellrecht, PhD Department of Pathology and Laboratory Medicine, Albany Medical College; Albany Medical Center Hospital, Albany, NY, United States

Susanna K. Tan, MD Department of Medicine, Division of Infectious Diseases and Geographic Medicine, Stanford University School of Medicine, Stanford, CA, United States

Jessica S. Thomas, MD, PhD, MPH Department of Pathology, Microbiology, and Immunology, Vanderbilt University Medical Center, Nashville, TN, United States

Scott A. Tomlins, MD, PhD Department of Pathology, Michigan Center for Translational Pathology, Department of Urology, Comprehensive Cancer Center, University of Michigan Medical School, Ann Arbor, MI, United States

Dimitri G. Trembath, MD, PhD Division of Neuropathology, Department of Pathology and Laboratory Medicine, The University of North Carolina at Chapel Hill, Chapel Hill, NC, United States

Ming-Sound Tsao, MD, FRCPC Department of Pathology, University Health Network, Toronto, ON, Canada

Gregory J. Tsongalis, PhD, HCLD, CC Laboratory for Clinical Genomics and Advanced Technology (CGAT), Department of Pathology and Laboratory Medicine, Dartmouth-Hitchcock Medical Center and Norris Cotton Cancer Center, Geisel School of Medicine at Dartmouth, Hanover, NH, United States

Richard M. Turner, MB, BChir, MA, MRCP The Wolfson Centre for Personalised Medicine, Institute of Translational Medicine, University of Liverpool, Liverpool, United Kingdom

Scott A. Turner, MD, PhD Laboratory for Clinical Genomics and Advanced Technology (CGAT), Department of Pathology and Laboratory Medicine, Dartmouth-Hitchcock Medical Center and Norris Cotton Cancer Center, Geisel School of Medicine at Dartmouth, Hanover, NH, United States

Aaron M. Udager, MD, PhD Department of Pathology, University of Michigan Medical School, Ann Arbor, MI, United States

Paul Verhoeven, MD, PhD GIMAP EA3064, University of Lyon, Saint-Etienne, France; Laboratory of Infectious Agents and Hygiene, University Hospital of Saint-Etienne, Saint-Etienne, France

David H. Walker, MD Department of Pathology, University of Texas Medical Branch, Galveston, TX, United States

Myra J. Wick, MD Departments of Medical Genetics and Obstetrics and Gynecology, Mayo Clinic School of Medicine, Rochester, MN, United States

Kathryn Willoughby, MD Department of Medicine, Division of Hematology/Oncology, Medical University of South Carolina, Charleston, SC, United States

Shaofeng Yan, MD, PhD Department of Pathology, Dartmouth-Hitchcock Medical Center, Lebanon, NH, United States

Belinda Yen-Lieberman, MS, PhD Section of Clinical Microbiology, Department of Laboratory Medicine, Cleveland Clinic, Cleveland, OH, United States

Preface

Pathology is the scientific study of the nature of disease and its causes, processes, development, and consequences. The field of pathology emerged from the application of the scientific method to the study of human disease. Thus, pathology as a discipline represents the complimentary intersection of medicine and basic science. Early pathologists were typically practicing physicians who described the various diseases that they treated and made observations related to factors that contributed to the development of these diseases. The description of disease evolved over time from gross observation to structural and ultrastructural inspection of diseased tissues based upon light and electron microscopy. As hospital-based and community-based registries of disease were developed, the ability of investigators to identify factors that cause disease and assign risk to specific types of exposures expanded to increase our knowledge of the epidemiology of disease. While descriptive pathology can be dated to the earliest written histories of medicine and the modern practice of diagnostic pathology dates back perhaps 200 years, the elucidation of mechanisms of disease and linkage of disease pathogenesis to specific causative factors occurred more recently from studies in experimental pathology. The field of experimental pathology embodies the conceptual foundation of early pathology—the application of the scientific method to the study of disease—and applies modern investigational tools of cell and molecular biology to advanced animal model systems and studies of human subjects. Whereas the molecular era of biological science began over 50 years ago, recent advances in our knowledge of molecular mechanisms of disease have propelled the field of molecular pathology. These advances were facilitated by significant improvements and new developments associated with the techniques and methodologies available to pose questions related to the molecular biology of normal and diseased states affecting cells, tissues, and organisms. Today, molecular pathology encompasses the investigation of the molecular mechanisms of disease and interfaces with translational medicine where new basic science discoveries form the basis for the development of new therapeutic approaches and targeted therapies for the new

strategies for prevention, and treatment of disease. Diagnostic molecular pathology is a new field that is focused on exploitation of molecular features and mechanisms of disease for the development of practical molecular diagnostic tools for disease detection, diagnosis, and prognostication. Diagnostic molecular pathology is essential for the realization of true personalized medicine. As this field continues to expand and mature, new molecular tests will emerge that will have utility in the sensitive and specific detection, diagnosis, and prognostication of human disease. Over time, the molecular technologies required will become increasingly economically practical and accessible to all patients whether treated in academic medical centers or community hospitals.

With the remarkable pace of scientific discovery in the field of *diagnostic molecular pathology*, basic scientists, clinical scientists, and physicians have a need for a source of information on the current state-of-the-art of our understanding of the molecular basis of human disease and how we harness the molecular features of disease for practical molecular testing. More importantly, the complete and effective training of today's graduate students, medical students, postdoctoral fellows, and others, for careers related to the investigation and treatment of human disease requires textbooks that have been designed to reflect our current knowledge of the molecular mechanisms of disease pathogenesis, as well as emerging concepts related to translational medicine. In this volume on *Diagnostic Molecular Pathology: A Guide to Applied Molecular Testing* we have assembled a group of experts to discuss the molecular basis and mechanisms of major human diseases and disease processes, presented in the context of traditional pathology, and how these molecular features of disease can be effectively harnessed to develop practical molecular tests for disease detection, diagnosis, and prognostication. This volume is intended to serve as a multiuse textbook that would be appropriate as a classroom teaching tool for medical students, biomedical graduate students, allied health students, and others (such as advanced undergraduates). Further, this textbook will be valuable for pathology residents and other postdoctoral fellows who desire to advance their understanding of molecular mechanisms of disease and practical applications related to these mechanisms, beyond what they learned in medical/graduate school. In addition, this textbook is useful as a reference book for practicing basic scientists and physician scientists who perform disease-related basic science and translational research, who require a ready information resource on the molecular basis of various human diseases and disease states and the molecular tests that are used during patient workup in a modern hospital laboratory. To be sure, our understanding of the many causes and molecular mechanisms that govern the development of human diseases is far from complete, and molecular testing has not yet become available for all human diseases. Nevertheless, the amount of information related to the practical exploitation of molecular mechanisms of human disease has increased tremendously in recent years and areas of thematic and conceptual consensus have emerged. We hope that *Diagnostic Molecular Pathology: A Guide to Applied Molecular Testing* will accomplish its purpose of providing students, researchers, and practitioners with in-depth coverage of the molecular basis of major human diseases and associated molecular testing so as to stimulate new research aimed at furthering our understanding of these molecular mechanisms of human disease and practice of molecular medicine through the development of new and novel molecular technologies and tests.

William B. Coleman
Gregory J. Tsongalis

INTRODUCTION TO
MOLECULAR TESTING

1

Basic Concepts in Molecular Pathology— Introduction to Molecular Testing in Human Disease

W.B. Coleman[1] and G.J. Tsongalis[2]

[1]Department of Pathology and Laboratory Medicine, Program in Translational Medicine, UNC Lineberger Comprehensive Cancer Center, University of North Carolina School of Medicine, Chapel Hill, NC, United States
[2]Laboratory for Clinical Genomics and Advanced Technology (CGAT), Department of Pathology and Laboratory Medicine, Dartmouth-Hitchcock Medical Center and Norris Cotton Cancer Center, Geisel School of Medicine at Dartmouth, Hanover, NH, United States

INTRODUCTION

Human diseases reflect a spectrum of pathologies and mechanisms of disease pathogenesis. The general categories of disease affecting humans include (1) hereditary diseases, (2) infectious diseases, (3) inflammatory diseases, and (4) neoplastic diseases. Pathologic conditions representing each of these general categories have been described for every tissue in the body. Despite the grouping of diseases by the common features of the general disease type, the pathogenesis of all of the various diseases is unique, and in some cases multiple distinct mechanisms can give rise to a similar pathology (disease manifestation). Disease causation may be related to intrinsic factors or extrinsic factors, but many/most diseases are multifactorial, involving a combination of intrinsic (genetic) and extrinsic factors (exposures). It is now well recognized that most major diseases are ultimately the result of aberrant gene expression, and that susceptibility to disease is significantly influenced by patterns of gene expression in target cells or tissues for a particular type of pathology. It follows that gene mutations and other genetic alterations are important in the pathogenesis of many human diseases. Hence, molecular diagnostic testing for genetic alterations may (1) facilitate disease detection, (2) aid in disease classification (diagnosis), (3) predict disease outcomes (prognostication), and/or (4) guide therapy

(Fig. 1.1). Likewise, nongenetic alterations affecting the expression of key genes (termed epimutations) may also contribute to the genesis of disease at many tissue sites. Molecular testing focused on epigenetic alterations in human disease is emerging and in development. Like genetic alterations, epigenetic changes may significantly impact on certain disease characteristics that confer diagnostic value. Epigenetic alterations can lead to gene silencing events (which are mechanistically equivalent to inactivating mutations or gene deletions) and may contribute to gene expression signatures that have predictive value with respect to clinical features of disease.

In this chapter we describe basic concepts in molecular pathology and molecular diagnostic testing for human disease. This is intended to be an introductory review of the field, rather than a comprehensive review of the field. Hence, when needed, examples are drawn preferentially from the cancer literature. Interested readers will find comparable literature in numerous other biomedical fields.

MUTATIONS AND EPIMUTATIONS

Mutation refers to changes in the genome that are characterized by alteration in the nucleotide sequence

© 2017 Elsevier Inc. All rights reserved.

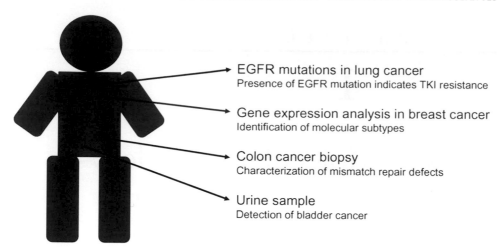

EGFR mutations in lung cancer
Presence of EGFR mutation indicates TKI resistance

Gene expression analysis in breast cancer
Identification of molecular subtypes

Colon cancer biopsy
Characterization of mismatch repair defects

Urine sample
Detection of bladder cancer

FIGURE 1.1 Utilization of DNA biomarkers in disease detection, diagnosis, classification, and guided treatment. This schematic provides examples from human cancer where DNA biomarkers obtained from non-invasive or invasive sources are used in the clinical workup of patients.

of a specific gene and/or other alterations at the level of the primary structure of DNA. Point mutations, insertions, deletions, and chromosomal abnormalities are all classified as mutations. In contrast, epimutation refers to alterations in the genome that do not involve changes in the primary sequence of the DNA. Aberrant DNA hypermethylation or hypomethylation and/or abnormal histone modifications resulting in alterations in chromatin structure are considered epimutations. Despite the differences between mutation and epimutation, the consequences of these molecular processes on the normal expression/function of critical genes/proteins may be the same—alteration of normal gene expression and/or normal protein function. These alterations may reflect (1) loss or reduction of normal levels of gene expression with consequent loss of protein function, (2) loss of function due to loss of protein or synthesis of defective protein, (3) increased levels of gene expression with consequent overexpression of protein, or (4) gain-of-function mutation with consequent altered protein function. Whereas many human diseases can be attributed to genetic alteration or epimutation affecting a single gene (or a few genes), the actual molecular consequence of these changes can be very dramatic, resulting in major alterations in gene expression patterns secondary to the primary genetic or epigenetic gene defect.

Genetic Alterations

Disease-related genetic alterations can be categorized into two major groups: nucleotide sequence abnormalities and chromosomal abnormalities. Examples of both of these forms of molecular lesion have been characterized in familial and acquired diseases affecting various human tissues.

Nucleotide sequence alterations include changes in individual genes involving single nucleotide changes (missense and nonsense), and small insertions or deletions (some of which result in frameshift mutations). Single nucleotide alterations that involve a change in the normal coding sequence of the gene (point mutations) can give rise to an alteration in the amino acid sequence of the encoded protein. Missense mutations alter the translation of the affected codon, while nonsense mutations alter codons that encode amino acids to produce stop codons. This results in premature termination of translation and the synthesis of a truncated protein product. Small deletions and insertions are typically classified as frameshift mutations because deletion or insertion of a single nucleotide (for instance) will alter the reading frame of the gene on the 3'-side of the affected site. This alteration can result in the synthesis of a protein that bears very little resemblance to the normal gene product or production of an abnormal/truncated protein due to the presence of a stop codon in the altered reading frame. In addition, deletion or insertion of one or more groups of three nucleotides will not alter the reading frame of the gene, but will alter the resulting polypeptide product, which will exhibit either loss of specific amino acids or the presence of additional amino acids within its primary structure.

Chromosomal alterations include the gain or loss of one or more chromosomes (aneuploidy), chromosomal rearrangements resulting from DNA strand breakage (translocations, inversions, and other rearrangements), and gain or loss of portions of chromosomes (amplification, large-scale deletion). The direct result of chromosomal translocation is the movement of a segment of DNA from its natural location into a new location within the genome, which can result in altered expression of the genes that are contained within the translocated region. If the chromosomal breakpoints utilized in a translocation are located within structural genes, then hybrid (chimeric) genes can be generated. The major consequence of a chromosomal deletion (involving a whole chromosome or a large chromosomal region) is the loss of specific genes that are localized to the deleted chromosomal segment, resulting in changes in the copy number of the affected genes. Likewise, gain of chromosome number or amplification of chromosomal regions results in an increase in the copy numbers of genes found in these chromosomal locations.

Epigenetic Alterations

In contemporary terms, epigenetics refers to modifications of the genome that are heritable during cell division, but do not involve a change in the DNA sequence. Therefore epigenetics describes heritable changes in gene expression that are not simply attributable to nucleotide sequence variation. It is now recognized that epigenetic regulation of gene expression reflects contributions from both DNA methylation as well as complex modifications of histone proteins and chromatin structure. Nonetheless, DNA methylation plays a central role in nongenomic inheritance and in the preservation of epigenetic states, and remains the most accessible epigenomic feature due to its inherent stability. Thus DNA methylation represents a target of fundamental importance in the characterization of the epigenome and for defining the role of epigenetics in disease pathogenesis.

SOURCES OF NUCLEIC ACIDS FOR MOLECULAR TESTING

Molecular diagnostic testing is now firmly engrained in the clinical testing menu of most/all hospital clinical laboratories. To conduct molecular testing in the workup of patients with known or suspected disease, sources of nucleic acids (primarily DNA) for use as biomarkers in molecular diagnostic assays are required. There are a large number of potential sources for patient-derived DNA (Fig. 1.2). These sources can be divided based upon the difficulty in sampling and/or the discomfort to the patient during sampling as (1) invasive sources of DNA biomarkers or (2) noninvasive sources of DNA biomarkers (Fig. 1.2). Tremendous research effort is now focused on utilization of noninvasive sources of biomarkers in the detection, diagnosis, prognostication, and classification of human disease. Noninvasive sources of DNA cause minimal to no discomfort to the patient. Collection of a urine sample represents a procedure with no discomfort, while collection of peripheral blood is a procedure with minimal discomfort to the patient. In contrast, invasive sources of DNA, while valuable for molecular testing, can require surgical procedures to obtain (such as in the case of a tissue biopsy) and may involve considerable discomfort to the patient (which is the case for Pap smears and spinal taps). In all cases DNA from diseased tissue or cells is the desired product. This DNA may be derived from cells collected in one of these procedures or through isolation of cell-free DNA (in some cases). No matter the source of biomarkers or the procedures used to collect the sample, it is critical that the intended use of the sample is kept in mind to ensure that samples are collected, stored, and processed in a manner that will not compromise the DNA. Numerous commercial sources provide kits for preparation of nucleic acids from various bodily fluids and tissue samples. Furthermore, this process has been automated in many cases through the use of commercial instrumentation.

CLASSIFICATION OF DISEASE

The classification of disease has historically been based upon (1) site of the pathological lesion (organ or

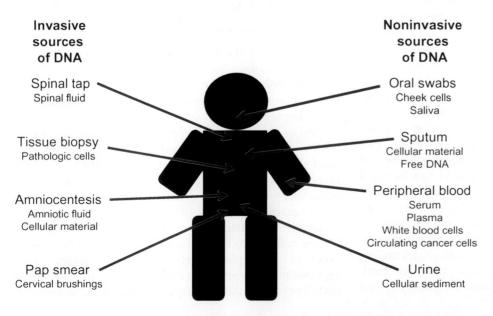

Invasive sources of DNA

Spinal tap
Spinal fluid

Tissue biopsy
Pathologic cells

Amniocentesis
Amniotic fluid
Cellular material

Pap smear
Cervical brushings

Noninvasive sources of DNA

Oral swabs
Cheek cells
Saliva

Sputum
Cellular material
Free DNA

Peripheral blood
Serum
Plasma
White blood cells
Circulating cancer cells

Urine
Cellular sediment

FIGURE 1.2 **Sources of DNA biomarkers for molecular testing.** Sources of DNA biomarkers for molecular testing can be grouped according to the relative difficulty in sampling and/or the relative discomfort to the patient during sample collection as (1) invasive sources, and (2) noninvasive sources. Among the invasive sources of DNA biomarkers, tissue biopsy may be used throughout the body to collect tissue for molecular testing (or routine pathologic examination). Some tissue biopsies can be obtained through simple surgical procedures (such as a skin biopsy), while others require a more extensive surgical procedure (bronchoscopy-based lung biopsy). In some cases the source of molecular biomarkers reflects an infectious agent (bacterium or virus) rather than host cells or cellular material.

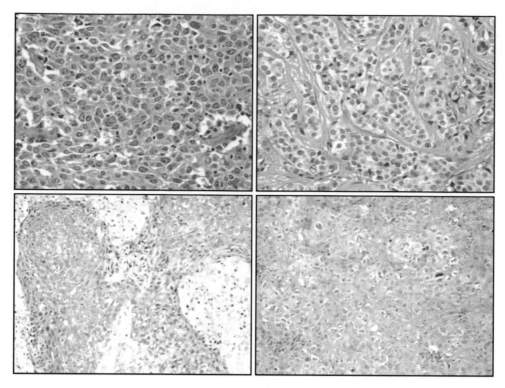

FIGURE 1.3 **Invasive ductal carcinoma of the breast.** Examples of invasive ductal carcinoma of the breast from four different patients are shown with hematoxylin and eosin staining. While some subtle histologic differences can be observed among these breast cancers, all of them are given the same clinical diagnosis—invasive ductal carcinoma.

tissue) and (2) nature of the pathological lesion (neoplasia, inflammation, etc.) [1]. Hence we are accustomed to diseases being classified into broad categories: for example, (1) organ-specific cancers, (2) cardiomyopathies, or (3) coagulopathies, among others. These broad disease classifications are typically associated with heterogeneity of disease presentation or response to therapy, suggesting multiple disease subtypes within these categories [1]. It is now well recognized that molecular subtypes exist within many of these broad disease classifications. For example, approximately 80% of all breast cancers are classified as invasive ductal carcinoma based upon routine pathologic evaluation (Fig. 1.3). However, breast cancer tends to be a heterogeneous disease based upon presentation, natural history, and responses to therapy, suggesting that breast cancer is not a single disease entity. In fact, examination of gene expression patterns among invasive ductal carcinomas identified five molecular subtypes that predict the aggressiveness of the disease in the individual patient and can be used to predict clinical course [2]. While the molecular subtypes of breast cancer were identified based upon complex gene expression patterns, comprehensive molecular analyses have revealed that breast cancers contain any number of molecular alterations, including chromosomal aberrations (structural and number), gene mutations, distinct gene expression patterns, and

changes in noncoding RNA expression (microRNAs and others) (Fig. 1.4) [3,4]. Epigenetic changes associated with histone modifications and DNA methylation directly affect gene expression patterns [5–8]. Likewise, changes in expression of noncoding RNAs may alter posttranscriptional regulation of gene expression patterns [6].

MOLECULAR CLASSIFICATION OF DISEASE

Evidence for the existence of molecular subtypes of disease includes the observation by clinicians that patients with the same disease diagnosis and diseases that apparently share many phenotypic characteristics will display widely varying clinical courses and responses to therapy. Hence, when a cohort of patients with a given disease are treated with a common standard therapy, only a subset of patients are expected to respond favorably (Fig. 1.5A). Upon the discovery of molecular subtypes of disease, the mechanistic basis for favorable outcomes of subsets of patients with a given disease becomes more apparent (Fig. 1.5B).

The molecular classification of breast cancer is an excellent example. Early microarray-based gene expression analyses of invasive breast cancers identified five molecular subtypes: luminal A, luminal B,

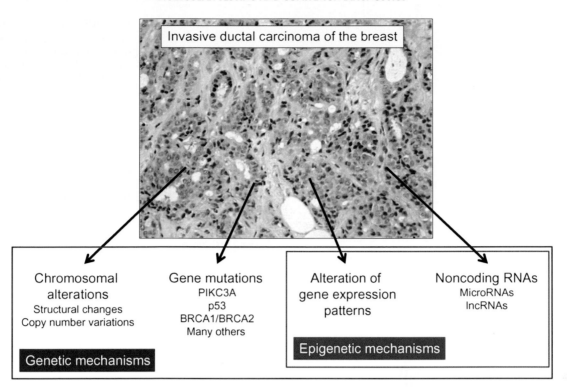

FIGURE 1.4 **Molecular alterations in breast cancer.** While the majority of breast cancers are of the same histologic type (invasive ductal carcinoma), each individual cancer contains numerous genetic and epigenetic alterations. The unique pattern of alterations found in individual cancers will influence its intrinsic characteristics, drive clinical behavior, and affect responses to therapeutic interventions. All of the genetic and epigenetic alterations found in an individual breast cancer represent potentially useful biomarkers for exploitation in molecular testing.

HER2-enriched, basal-like, and normal-like [2]. An example of an unsupervised cluster analysis of microarray-based gene expression data from 294 breast cancers is shown in Fig. 1.6. Since the early studies of gene expression patterns to determine molecular subtypes among breast cancers, new molecular assays based upon polymerase chain reaction (PCR) (PAM50) and RNA sequencing have emerged that faithfully classify breast cancers in a similar manner [3,9,10]. Of significance was the observation that the natural history of breast cancers differs with molecular subtype—luminal A breast cancers demonstrate excellent long-term survival, while basal-like breast cancers exhibit rapid progression of disease and poor long-term patient outcomes [11,12]. Hence, knowledge of the molecular subtype of breast cancer informs the clinician of the likely aggressiveness of the disease. However, these basic molecular subtypes of breast cancer are not homogeneous groupings. Rather, considerable heterogeneity has been observed [13,14]. In recent studies triple-negative breast cancers have been subclassified into groupings that predict response to specific chemotherapeutic agents [15–17]. These new data provide hope that advances in the molecular classification of various diseases will translate into practical molecular tests with utility in the clinical setting as an aid to clinicians in the management of individual patients. With the generation of large datasets related to gene expression, gene copy number variations, mutations, and molecular pathways in human disease, new challenges have emerged with respect to disease classification [1]. Certainly, achieving molecular classification of disease does not automatically equate to better patient management.

MOLECULAR TESTING AND COMPANION DIAGNOSTICS

Companion diagnostics are generally defined as a molecular test that produces a result that predicts the likely success of using a specific therapeutic agent in an individual patient with a given disease [18,19]. In many/most cases the molecular test is not based upon the mutation or expression status of a single gene. Rather, these tests tend to utilize multiple genes to improve sensitivity, specificity, and predictive value. The objective of the companion diagnostic is to identify patients that will benefit from a given therapeutic drug or drug combination, ideally with minimal toxicity. These two consequences (toxicity and benefit) of drug treatment are not mutually exclusive. When a

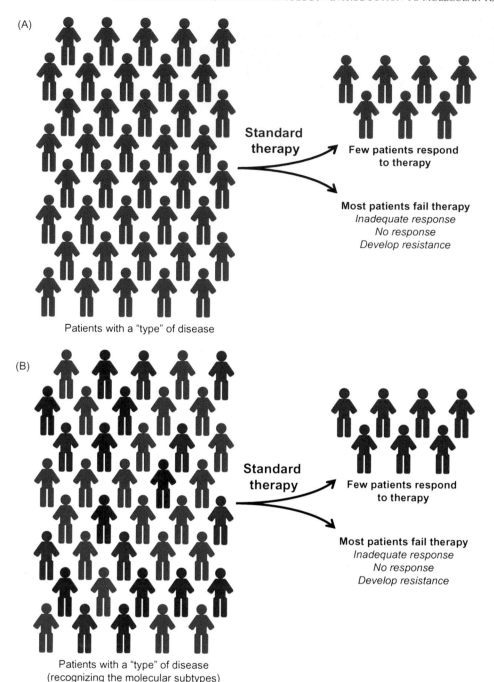

FIGURE 1.5 Molecular subtypes identify patient groupings with similar disease outcomes. (A) When a cohort of patients with the same disease diagnosis are treated with similar forms of standard therapy, only a subset of patients will respond favorably. This observation indicates that therapy cannot be effectively prescribed based upon a simple diagnosis. (B) Upon recognition of molecular subtypes of disease, it becomes apparent while certain subsets of patients with the same disease diagnosis respond differentially (favorably) to a given therapy. This observation suggests that therapeutic approaches should be governed by molecular profiling of patients.

drug is administered to a given patient, there are four possible outcomes: (1) the drug is toxic and not beneficial, (2) the drug is toxic, but beneficial, (3) the drug is not toxic, but not beneficial, and (4) the drug is not toxic, but is beneficial (Fig. 1.7). However, toxicity to the patient (reflecting effects on the patient's normal physiology) is most often an intrinsic property of the drug itself. Thus in the clinical setting, numerous drugs and drug regimens are employed on a routine basis that not only are toxic to patients but also provide benefit in the treatment of their disease.

Molecular testing includes companion diagnostics as well as other diagnostic assays that provide insights into the nature and characteristics of a disease in a given individual patient. There is little question that as we gain additional insights into molecular subtypes of specific diseases and how they impact on disease progression, outcomes, and responses to therapy, molecular testing to achieve subclassification of disease (and groupings of patients) will become very important (Fig. 1.8). In some cases subclasses of disease will be associated with patient outcomes

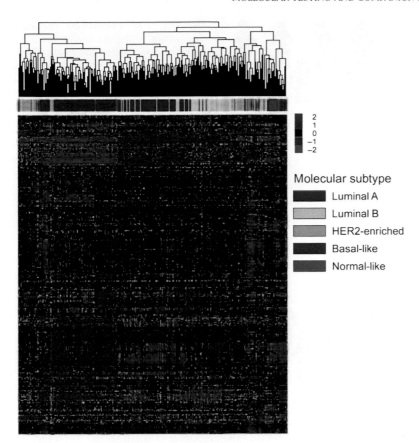

FIGURE 1.6 **Microarray analysis of breast cancer gene expression patterns.** Unsupervised cluster analysis of 294 primary breast cancers identifies five major molecular subtypes: (1) luminal A, (2) luminal B, (3) HER2-enriched, (4) basal-like, and (5) normal-like.

Molecular subtype

Luminal A
Luminal B
HER2-enriched
Basal-like
Normal-like

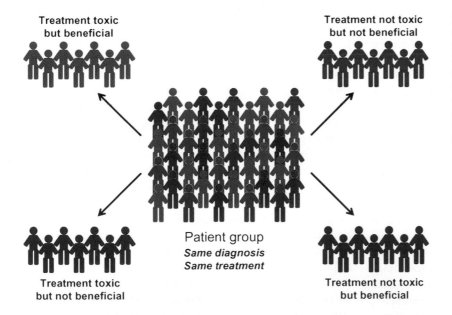

FIGURE 1.7 **Possible treatment outcomes for any given therapy.** For any cohort of patients that are treated with a given drug, there are four possible treatment outcomes. The ideal outcome corresponds to the treatment where there is no toxicity of the drug, but the drug is beneficial to the patient. In cancer chemotherapy the typical outcome corresponds to where the drug is toxic to the patient (and so administered in a dose-limiting fashion), but provides benefit (kills the cancer cells).

(short-term disease-free survival or long-term overall survival). Knowledge of the likelihood of aggressive disease versus more benign disease (Fig. 1.9) will be useful in decisions related to patient management, even if a specific treatment course is not prescribed in response to testing. By the same token, development of new biomarkers of disease will enable generation of practical molecular tests for selection of appropriate therapies (especially as new drugs and drug regimens emerge), expanding our menu of companion diagnostics for use in designing therapeutic strategies for individual patients (Fig. 1.9).

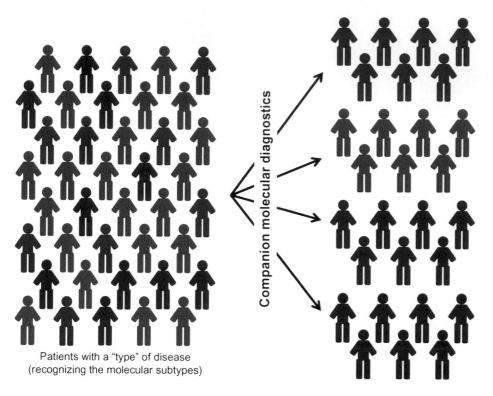

Companion molecular diagnostics

Patients with a "type" of disease
(recognizing the molecular subtypes)

FIGURE 1.8 **Companion molecular diagnostics.** Molecular testing through a companion molecular diagnostic enables classification of patients and their disease by identifying those with certain intrinsic characteristics. Hence, these molecular tests may be employed to (1) identify subsets of disease, (2) identify cohorts of patients that have better or worse prognosis, or (3) identify patients that according to likelihood of response to therapy.

MOLECULAR DETECTION OF DISEASE

For many years, investigators have explored the use of molecular biomarkers for prediction of disease susceptibility and for the detection of occult (subclinical) disease. Molecular testing is typically based upon PCR or molecular methods of equal sensitivity. Early experimental studies demonstrated the power of molecular testing for detection of lung cancer in high-risk individuals (cigarette smokers) [20]. Sputum samples were collected from high-risk individuals prior to their diagnosis with lung cancer and tested for the presence of mutations in **K-*ras* and *p53* [20]. These mutations occur frequently in lung cancer [21]. The investigators found that 10/15 lung cancers were positive for K-*ras* and/or *p53* mutations [20]. Remarkably, 8 of 10 sputum samples (which were cytologically negative for a cancer diagnosis) corresponding to patients with a lung cancer diagnosis were positive for K-*ras* or *p53* mutations [20]. In one case the gene mutation was detectable in the sputum >1 year prior to clinical diagnosis of lung cancer [20]. Hence, molecular testing has the power to detect driver mutations in rare neoplastic cells in patients that will eventually progress to form a clinical cancer, potentially enabling the employment of cancer preventative strategies or very early therapeutic intervention in patients following a positive test. However, caution must be used because there is the possibility that a positive molecular test will be obtained in an individual that does not progress to a

clinical cancer (Fig. 1.10). That is, a true positive molecular test result (K-*ras* or *p53* mutation detected in sputum) can be associated with a false-positive prediction of the development of lung cancer (when K-*ras* or *p53* mutant cells do not progress to a clinical cancer). This sort of prospective molecular testing is only practical when there is a high-risk population to screen. Population-based screening using molecular testing has only proved to be effective for certain forms of cancer that are relatively slow growing and homogeneous in presentation (like colon cancer), while screening for cancers that exhibit greater heterogeneity (like breast cancer) are not yet practical and effective [22]. Further, in the example of molecular testing for early lung cancer detection, former smokers will remain at high risk for many years after cessation of smoking. Hence, a true negative result early in testing may become a true positive result over time and/or an individual with a true negative result may develop lung cancer over time in the absence of conversion to a positive molecular test. Additional limitations to this approach for the early detection of cancer relates to the design of the molecular test. It is clear from studies performed to date that testing of one or two gene mutations will not be sufficient to capture all developing cancers and that multigene panels will be required to achieve the necessary sensitivity and specificity for a routine screening application. Early detection of cancer through molecular testing is likely to improve long-term outcomes for patients,

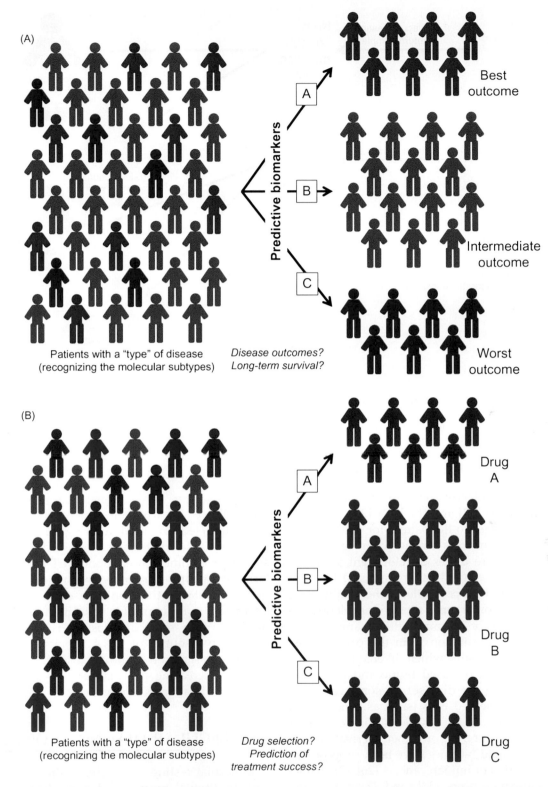

(A)

Patients with a "type" of disease
(recognizing the molecular subtypes)

Disease outcomes?
Long-term survival?

Predictive biomarkers

A → Best outcome

B → Intermediate outcome

C → Worst outcome

(B)

Patients with a "type" of disease
(recognizing the molecular subtypes)

Drug selection?
Prediction of
treatment success?

Predictive biomarkers

A → Drug A

B → Drug B

C → Drug C

FIGURE 1.9 **Utilization of molecular biomarkers for prediction of disease outcomes and drug selection.** (A) Molecular testing can be used to identify individual patients or patient subsets that have a better or worse prognosis based upon results with a group of informative biomarkers. Measures of prognosis may be short-term or long-term survival, relapse-free interval, or likelihood of disease progression. (B) Molecular testing can be utilized for drug selection (or therapeutic strategy) based upon results with a group of biomarkers.

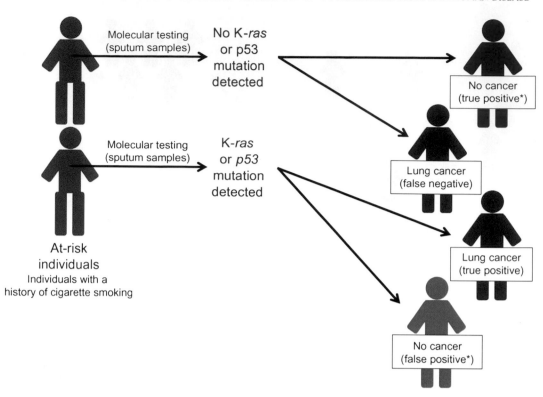

FIGURE 1.10 Molecular testing for detection of lung cancer. This schematic represents exploitation of gene mutations affecting *p53* or K-*ras* in the detection of lung cancer in high-risk patients using cytologically negative sputum samples. A true-positive molecular test (*) is achieved when no mutation is detected and the patient does not develop cancer (over time). However a false-positive molecular test (*) results when a mutation is detected, but the patient does not develop cancer.

especially for difficult-to-treat cancers like those affecting the pancreas [23] and the liver [24].

Given the advantages to the patient for using noninvasive sources of biomarkers for molecular testing, so-called *liquid biopsies* are under development for many forms of human cancer (and other diseases) [25–27]. The concept of liquid biopsy reflects the noninvasive source for disease biomarkers—most often blood or a blood-derived (serum or plasma) patient sample—and the exploitation of information from the primary disease (cancer in this case)—which likens to a tissue biopsy (taken from the diseased tissue and providing information regarding the disease). In some cases liquid biopsy targets diseased cells (such as cancer cells) in the blood, while in other cases the biomarkers reflect cell-free DNA, microRNAs, or other derivative cell products [28,29]. Liquid biopsies have been proposed as a tool for (1) detection of human cancers [26], (2) characterization of human cancers [30], and (3) monitoring cancer status (recurrence/relapse) [31–33].

In cancer molecular testing, detection of recurrent disease through screening of patients with a prior diagnosis and successful treatment of cancer represents another important application. For many major forms of cancer, conventional treatment involves surgery followed by adjuvant chemotherapy, radiation, or use of targeted drugs. Monitoring for cancer relapse or progression to metastatic disease involves imaging studies over time (based upon PET scanning or similar). Investigators are now examining the possibility that molecular testing can be used as a less expensive alternative to radiographic scanning to detect recurrent disease. When molecular testing is combined with a noninvasive source of DNA biomarkers, it becomes practical to screen individual patients more frequently, possibly increasing the chances of detecting recurrent disease early after its emergence.

MOLECULAR DIAGNOSIS OF DISEASE

Molecular testing has a great value when combined with traditional means of disease diagnosis. In some cases the results of a molecular test confirms the diagnosis based upon clinical symptoms and evaluation of other information and samples (such as a biopsy of tissue). In other cases the results of the molecular test provides information related to the intrinsic properties of the disease enabling the physician to make good treatment decisions. In addition, molecular profiling of gene mutations in individual patients with a given disease may identify a genetic cause of that disease and enable targeted screening for at-risk relatives of the affected individual.

The use of Tamoxifen in the treatment of estrogen receptor-positive (ER+) breast cancers is an example of a targeted therapy. In the clinical workup of breast cancer patients, cancer tissues (from biopsy or surgery) are evaluated by immunohistochemistry for expression of ER. In patients with ER+ disease, Tamoxifen (or other antiestrogenic agents) is typically prescribed, whereas ER-negative patients do not receive this treatment because lack of ER renders the cancer Tamoxifen-resistant. Likewise, breast cancers that overexpress human epidermal growth factor 2 (HER2) are treated with Herceptin (or similar drugs that target HER2). Immunostaining for HER2 identifies some patients who will benefit from Herceptin treatment, while others are confirmed with molecular testing based upon HER2 fluorescence in situ hybridization to detect *Her2* gene amplification [34,35].

Triple-negative breast cancer represents a difficult-to-treat subtype of invasive ductal carcinoma that is classified based upon results of immunostaining for ER, progesterone receptor (PR), and HER2. Triple-negative breast cancers are negative for ER and PR, and do not exhibit overexpression of HER2. This subset of invasive breast cancers is very heterogenous [14]. In recent studies molecular profiling of triple-negative breast cancers based upon gene expression profiles has identified a number of distinct subsets [15,36,37]. Refinement of the subtyping of triple-negative breast cancers [38] produced four molecular subtypes: Basal-like 1 (BL1), Basal-like 2 (BL2), mesenchymal-type (M), and luminal androgen receptor-type (LAR) [15]. Each of these molecular subtypes emerges from the clinical designation of triple-negative breast cancer based upon immunohistochemistry, so the identification of these subtypes does not modify the diagnosis of disease. However, the molecular subtypes identified in triple-negative breast cancer differentially respond to specific chemotherapeutic drug combinations [15,36,39].

MOLECULAR PROGNOSTICATION OF DISEASE

Molecular testing of tissue samples from patients undergoing surgery (or biopsy) in treatment (or diagnosis) of disease can be used to predict (or prognosticate) patient outcomes. Typical readouts for prediction of patient outcomes include (1) relapse-free survival, (2) probability of recurrence, (3) probability of progression to metastatic disease (in the example of cancer), and (4) overall survival.

An excellent example of a molecular test that provides information related to various measures of prognosis is the Oncotype DX assay [40]. Oncotype DX utilizes a 21-gene signature to predict the likelihood of disease recurrence among patients with node-negative ER+ breast cancer [40–42]. The recurrence score (RS) provided by Oncotype DX provides a measure of risk for recurrence during a 10-year follow-up period: low risk, intermediate risk, or high risk. When Oncotype DX results were evaluated with respect to benefit from adjuvant chemotherapy, it was found that the low-risk group does not benefit from chemotherapy [40]. Hence, a low RS is now used as a basis to spare patients with early stage ER+ breast cancers chemotherapy after surgical intervention. In these cases surgery alone probably eliminates all the cancer and so the use of adjuvant chemotherapy is unnecessary. In contrast, patients with a high-risk RS benefit from adjuvant chemotherapy. Oncotype DX has also been evaluated as a predictor of local regional failure (local recurrence). Patients with a high-risk RS have a greater chance of local recurrence versus those with a low-risk RS [40].

PERSPECTIVES

The technologies associated with molecular testing have advanced significantly in the last 25 years, and continue to improve with respect to sensitivity and accuracy, as well as flexibility of testing platforms and methodologies. Advances are also being made in our understanding of normal human biology, and the pathology, pathogenesis, and pathophysiology of disease. Emerging from the confluence of these scientific disciplines are molecular testing platforms and assays that are and will contribute to improved disease detection, diagnosis, and prediction of outcomes for individual patients. Likewise, as new molecular diagnostics are developed, application of molecular assays to biomarker detection will begin to guide therapeutic strategies for individual patients. With continued advancement in these areas, true personalized medicine for a variety of human diseases will become a reality. With true personalized medicine, individual patients will be evaluated and the results from molecular testing will determine the best therapeutic approaches and treatment regimens for that person's disease (with its unique intrinsic properties), and long-term patient outcomes will improve.

References

[1] Song Q, Merajver SD, Li JZ. Cancer classification in the genomic era: five contemporary problems. Hum Genomics 2015;9:27.

[2] Perou CM, Sorlie T, Eisen MB, et al. Molecular portraits of human breast tumours. Nature 2000;406:747—52.

[3] Cancer Genome Atlas Network. Comprehensive molecular portraits of human breast tumours. Nature 2012;490:61—70.

[4] Prat A, Perou CM. Deconstructing the molecular portraits of breast cancer. Mol Oncol 2011;5:5—23.

[5] Jovanovic J, Ronneberg JA, Tost J, Kristensen V. The epigenetics of breast cancer. Mol Oncol 2010;4:242—54.

[6] Veeck J, Esteller M. Breast cancer epigenetics: from DNA methylation to microRNAs. J Mammary Gland Biol Neoplasia 2010;15:5—17.

[7] Huang Y, Nayak S, Jankowitz R, Davidson NE, Oesterreich S. Epigenetics in breast cancer: what's new? Breast Cancer Res 2011;13:225.

[8] Atalay C. Epigenetics in breast cancer. Exp Oncol 2013;35:246—9.

[9] Parker JS, Mullins M, Cheang MC, et al. Supervised risk predictor of breast cancer based on intrinsic subtypes. J Clin Oncol 2009;27:1160—7.

[10] Prat A, Parker JS, Fan C, Perou CM. PAM50 assay and the three-gene model for identifying the major and clinically relevant molecular subtypes of breast cancer. Breast Cancer Res. Treat. 2012;135:301—6.

[11] Sorlie T. Molecular portraits of breast cancer: tumour subtypes as distinct disease entities. Eur J Cancer 2004;40:2667—75.

[12] Sorlie T, Perou CM, Tibshirani R, et al. Gene expression patterns of breast carcinomas distinguish tumor subclasses with clinical implications. Proc Natl Acad Sci USA 2001;98:10869—74.

[13] Parker JS, Perou CM. Tumor heterogeneity: focus on the leaves, the trees, or the forest? Cancer Cell 2015;28:149—50.

[14] Lehmann BD, Pietenpol JA. Clinical implications of molecular heterogeneity in triple negative breast cancer. Breast 2015;24 (Suppl. 2):S36—40.

[15] Lehmann BD, Jovanovic B, Chen X, et al. Refinement of triple-negative breast cancer molecular subtypes: implications for neoadjuvant chemotherapy selection. PLoS One 2016;11:e0157368.

[16] Lehmann BD, Pietenpol JA, Tan AR. Triple-negative breast cancer: molecular subtypes and new targets for therapy. Am Soc Clin Oncol Educ Book 2015;2015:e31—9.

[17] Abramson VG, Lehmann BD, Ballinger TJ, Pietenpol JA. Subtyping of triple-negative breast cancer: implications for therapy. Cancer 2015;121:8—16.

[18] Jorgensen JT. Clinical application of companion diagnostics. Trends Mol Med 2015;21:405—7.

[19] Jorgensen JT. Companion diagnostics: the key to personalized medicine. Foreword. Expert Rev Mol Diagn 2015;15:153—6.

[20] Mao L, Hruban RH, Boyle JO, Tockman M, Sidransky D. Detection of oncogene mutations in sputum precedes diagnosis of lung cancer. Cancer Res 1994;54:1634—7.

[21] Ahrendt SA, Decker PA, Alawi EA, et al. Cigarette smoking is strongly associated with mutation of the K-ras gene in patients with primary adenocarcinoma of the lung. Cancer 2001;92:1525—30.

[22] Shieh Y, Eklund M, Sawaya GF, et al. Population-based screening for cancer: hope and hype. Nat Rev Clin Oncol 2016. Available from: http://dx.doi.org/10.1038/nrclinonc.2016.50.

[23] Lennon AM, Wolfgang CL, Canto MI, et al. The early detection of pancreatic cancer: what will it take to diagnose and treat curable pancreatic neoplasia? Cancer Res 2014;74:3381—9.

[24] Tsuchiya N, Sawada Y, Endo I, et al. Biomarkers for the early diagnosis of hepatocellular carcinoma. World J Gastroenterol 2015;21:10573—83.

[25] Pantel K, Alix-Panabieres C. Liquid biopsy: potential and challenges. Mol Oncol 2016;10:371—3.

[26] Lancet Oncology. Liquid cancer biopsy: the future of cancer detection? Lancet Oncol 2016;17:123.

[27] Salinas Sanchez AS, Martinez Sanchis C, Gimenez Bachs JM, Garcia Olmo DC. Liquid biopsy in cancer. Actas Urol Esp 2016;40:1—2.

[28] Giallombardo M, Chacartegui Borras J, Castiglia M, et al. Exosomal miRNA analysis in non-small cell lung cancer (NSCLC) patients' plasma through qPCR: a feasible liquid biopsy tool. J Vis Exp 2016. Available from: http://dx.doi.org/10.3791/53900.

[29] Cheng F, Su L, Qian C. Circulating tumor DNA: a promising biomarker in the liquid biopsy of cancer. Oncotarget 2016;. Available from: http://dx.doi.org/10.18632/oncotarget.9453.

[30] Liquid biopsy holds its own in tumor profiling. Cancer Discov 2016;6:686.

[31] Li T, Zheng Y, Sun H, et al. K-Ras mutation detection in liquid biopsy and tumor tissue as prognostic biomarker in patients with pancreatic cancer: a systematic review with meta-analysis. Med Oncol 2016;33:61.

[32] Openshaw MR, Page K, Fernandez-Garcia D, Guttery D, Shaw JA. The role of ctDNA detection and the potential of the liquid biopsy for breast cancer monitoring. Expert Rev Mol Diagn 2016;16:751—5.

[33] Diaz LA. The promise of liquid biopsy in colorectal cancer. Clin Adv Hematol Oncol 2014;12:688—9.

[34] Prendeville S, Corrigan MA, Livingstone V, et al. Optimal scoring of brightfield dual-color in situ hybridization for evaluation of HER2 amplification in breast carcinoma: how many cells are enough? Am J Clin Pathol 2016;145:316—22.

[35] Onguru O, Zhang PJ. The relation between percentage of immunostained cells and amplification status in breast cancers with equivocal result for Her2 immunohistochemistry. Pathol Res Pract 2016;212:381—4.

[36] Lehmann BD, Bauer JA, Chen X, et al. Identification of human triple-negative breast cancer subtypes and preclinical models for selection of targeted therapies. J Clin Invest 2011;121:2750—67.

[37] Burstein MD, Tsimelzon A, Poage GM, et al. Comprehensive genomic analysis identifies novel subtypes and targets of triple-negative breast cancer. Clin Cancer Res 2015;21:1688—98.

[38] Chen X, Li J, Gray WH, et al. TNBCtype: a subtyping tool for triple-negative breast cancer. Cancer Inform 2012;11:147—56.

[39] Masuda H, Baggerly KA, Wang Y, et al. Differential response to neoadjuvant chemotherapy among 7 triple-negative breast cancer molecular subtypes. Clin Cancer Res 2013;19:5533—40.

[40] Kaklamani V. A genetic signature can predict prognosis and response to therapy in breast cancer: Oncotype DX. Expert Rev Mol Diagn 2006;6:803—9.

[41] Freitas MR, Simon SD. Comparison between Oncotype DX test and standard prognostic criteria in estrogen receptor positive early-stage breast cancer. Einstein (Sao Paulo) 2011;9:354—8.

[42] Conlin AK, Seidman AD. Use of the Oncotype DX 21-gene assay to guide adjuvant decision making in early-stage breast cancer. Mol Diagn Ther 2007;11:355—60.

2

Laboratory Approaches in Molecular Pathology—The Polymerase Chain Reaction

W.B. Coleman[1] and G.J. Tsongalis[2]

[1]Department of Pathology and Laboratory Medicine, Program in Translational Medicine, UNC Lineberger Comprehensive Cancer Center, University of North Carolina School of Medicine, Chapel Hill, NC, United States
[2]Laboratory for Clinical Genomics and Advanced Technology (CGAT), Department of Pathology and Laboratory Medicine, Dartmouth-Hitchcock Medical Center and Norris Cotton Cancer Center, Geisel School of Medicine at Dartmouth, Hanover, NH, United States

INTRODUCTION

The polymerase chain reaction (PCR) represents a rapid, sensitive, and specific method for in vitro amplification of nucleic acid sequences. Through utilization of specific oligodeoxynucleotide primers, the PCR is capable of identifying a target sequence and then using a DNA polymerase able to amplify millions of copies (amplicons) of the target. The concept of the PCR was first described in the mid-1980s [1−3]. This technology made its initial impact on the research laboratory, but once the power of this technique was realized it quickly became the basis for numerous applications in clinical testing. Since that time, developments related to methodological modifications and new forms of instrumentation have combined to enhance the technology, which has evolved into a reliable, affordable, user-friendly method that is performed in laboratories world-wide. There is little question that the PCR has had an extraordinary impact as a modern technology on the field of molecular diagnostics. PCR methodology has now become routine and the instrumentation required is common/available to most/all laboratories. Thus it is easy to underestimate the significant impact of the PCR on day-to-day operation of both clinical molecular diagnostics laboratories and basic science research laboratories.

As a molecular technology, the PCR facilitates the amplification of specific nucleic acid sequences to produce a quantity of amplified product that can be analyzed by other methods. Hence, the PCR offers a very sensitive and specific method to perform quantitative and qualitative analyses of target sequences. The development of various chemistries for primer and probe labeling has produced an extraordinary technology with respect to performance characteristics. Early PCR methods utilized the Klenow fragment of *Escherichia coli* DNA polymerase I for DNA synthesis during each amplification cycle [1]. However, Klenow fragment is not thermally stabile. Therefore, this method required the addition of fresh enzyme after each denaturation step as samples were quickly cooled to avoid heat denaturation of the enzyme. In addition, the primer annealing and DNA synthesis steps were carried out at 30°C to preserve the activity of the polymerase enzyme. This low-temperature annealing enabled hybridization of primers to nontarget sequences, contributing to nonspecific amplification [4]. The first major technological breakthrough in the development of the PCR was the introduction of a thermostable DNA polymerase [3]. *Thermus aquaticus* is a bacterium that is found in hot springs and is adapted to the variations in ambient temperature that accompany its environment. The DNA polymerase enzyme expressed in *T. aquaticus* (known as Taq polymerase) exhibits robust polymerase activity that is relatively unaffected by rapid fluctuations in temperature over a wide range [5]. Introduction of Taq polymerase to the PCR improved the practicality of this methodology because this polymerase enzyme can survive extended incubation at the elevated temperatures required for DNA denaturation (93−95°C) [5], eliminating the need for addition of fresh enzyme after each cycle. The second major technological breakthrough in

© 2017 Elsevier Inc. All rights reserved.

the development of the PCR was the introduction of the programmable heat block that automatically changes the reaction temperature during each amplification cycle—the thermocycler. The thermocycler instrument enabled automation of the PCR and the basic methodology has not changed significantly since the late 1980s [6].

THE POLYMERASE CHAIN REACTION

In a typical PCR, successive cycles are performed in which a DNA polymerase copies target DNA sequences from a template molecule in vitro. The amplification products produced during each cycle provide new templates for the successive rounds of amplification (Fig. 2.1). Hence, the concentration of the target DNA sequence increases exponentially over the course of the PCR. The typical PCR reaction mixture contains (1) a thermostable DNA polymerase (Taq polymerase), (2) target-specific forward and reverse oligodeoxynucleotide primers, (3) each of the four deoxynucleotide triphosphates (dNTPs), (4) reaction buffer, and (5) a source of template (genomic DNA, cDNA, or cell lysates). The target sequence is defined by the specificity of the oligodeoxynucleotide primers that anneal to complementary sequences on opposite template strands flanking the region of interest. During the PCR, thee primers are extended in the $5' \rightarrow 3'$ direction by the DNA polymerase enzyme to yield overlapping copies of the original template. Each cycle of the PCR proceeds through three distinct phases: (1) denaturation, (2) primer annealing, and (3) primer extension (Fig. 2.2). The denaturation step is typically accomplished by incubation of samples for up to 1 minute at 94°C to render the DNA containing the sequence of interest into a single-stranded template. The primer annealing step is accomplished at a temperature that is specific for the PCR primers and conditions used. During the annealing step oligodeoxynucleotide primers recognize and hybridize (hydrogen bond) to the target sequence contained in the single-stranded template. The primer extension step is accomplished at 72°C. During this step the polymerase enzyme catalyzes the polymerization of dNTPs in a $5' \rightarrow 3'$ DNA-directed DNA synthesis reaction. The actual times used for each cycle (and each step in the cycle) will vary from 15 seconds to 1−2 minutes depending upon the type of thermocycler used and its temperature ramping speed. Amplification of a target sequence is accomplished through repetition of these incubations for 25−30 (or more) cycles. The exact number of cycles necessary to produce sufficient amplicons for detection will depend upon the starting concentration of the target sequence. By the end of the third cycle of amplification (in a typical 25−30 cycle PCR), a new double-stranded template molecular is formed in which the 5' and 3' ends coincide exactly with the oligodeoxynucleotide primers used. Because the copy number theoretically doubles after each successive cycle of amplification, an exponential increase of 2^n (where n is the total

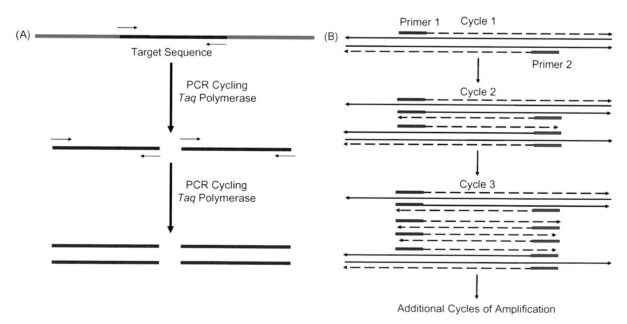

FIGURE 2.1   **Schematic representation of the polymerase chain reaction (PCR).** (A) This simplified schematic depicts two cycles of PCR amplification from a single target sequence (blue bars) found in a complex DNA template (green bar). Primers are depicted as arrows. (B) This schematic shows additional detail for the initial rounds of PCR amplification from a single target sequence. Forward primers are shown as red lines and reverse primers are depicted as yellow lines. Newly synthesized DNA in each cycle is depicted as a dashed line, while the original template DNA (and new templates in each cycle) are depicted as solid lines.

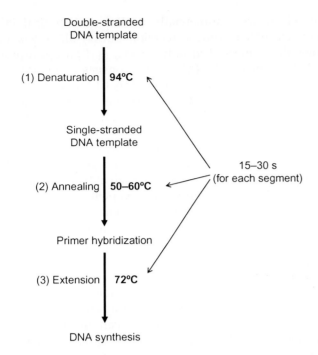

Double-stranded
DNA template

(1) Denaturation **94°C**

Single-stranded
DNA template

15–30 s
(for each segment)

(2) Annealing **50–60°C**

Primer hybridization

(3) Extension **72°C**

DNA synthesis

FIGURE 2.2 **Steps and temperatures for a single PCR cycle.** One cycle of PCR amplification is depicted. Each cycle consists of three segments: (1) template denaturation, (2) primer annealing, and (3) primer extension. Typical temperatures are indicated for each segment. Each segment will typically occur over 15–30 seconds (using state-of-the-art PCR instruments), but some instruments and applications call for longer segment times (up to 1–2 minutes each).

number of cycles of PCR performed) is accomplished during the complete reaction. Accumulation of amplicons corresponding to the target sequence eventually reaches a plateau. The initial number of target sequences contained within the template sample, the efficiency of primer extension, and the number of PCR cycles performed determine the upper limits of amplification.

COMPONENTS OF THE PCR

The PCR is dependent on successful synthesis of target sequences through a series of amplification cycles, beginning with the first cycle and continuing through the final elongation step. Similarly, each PCR is performed in the presence of reagents that are critical to the performance of the reaction.

DNA Template

The PCR amplifies specific sequences from DNA templates (genomic DNA or cDNA derived from RNA) that can be prepared from various sample sources. Clinical specimens may be derived from various bodily fluids (for instance blood, urine, or amniotic fluid) or surgical samples (for instance frozen cancer specimens) [7,8]. Forensic specimens may be derived from blood, semen, hair, or tissue (for instance skin cells). In addition to

fresh specimens, DNA derived from fixed tissues (formalin-fixed paraffin-embedded specimens) can be used routinely in PCR applications [9]. Most PCR reactions amplify small targets from the template sample (100–500 bp in size). Hence, high molecular weight DNA is not necessary, and highly fragmented DNA (like that obtained from formalin-fixed paraffin-embedded tissues) can be effectively utilized. However, certain tissue fixatives (such as Bouin solution which contains picric acid) and tissue treatments (such as tissue decalcification) damage DNA and can render tissue samples useless as a source of DNA for molecular analysis. Preparation of RNA can be accomplished from fresh tissues or from formalin-fixed paraffin-embedded samples.

DNA Polymerase Enzyme

The active component of the PCR is a DNA polymerase enzyme that is required for DNA synthesis during the primer extension step of the PCR. Contemporary PCR uses Taq polymerase (isolated from *T. aquaticus*) or similar [5,9]. Taq polymerase exhibits $5' \rightarrow 3'$ polymerase activity, $5' \rightarrow 3'$ exonuclease activity, thermostability, and optimum performance at 70–80°C. Temperature, pH, and ion concentrations (Mg^{2+}) can influence the activity of Taq polymerase. The half-life of Taq polymerase activity at 95°C is approximately 40–60 minutes. Extremely high denaturation temperatures (>97 °C) will significantly reduce the activity of Taq polymerase. Because time and temperature represent the critical variables for maintenance of Taq polymerase activity, lowering of the denaturation temperature or reduction of the denaturation time can prolong the activity of the enzyme during a PCR reaction. The optimum pH for a given PCR reaction is between 8.0 and 10.0, but must be determined empirically. The typical PCR reaction is carried out in a buffer (usually Tris-Cl) that is pH 8.3. Taq polymerase activity requires divalent cations in the form of Mg^{2+}. Lower divalent cation (Mg^{2+}) concentrations decrease the rate of dissociation of enzyme from template by stabilizing the enzyme-nucleic acid interaction. Most PCR mixtures contain at least 1.5 mM $MgCl_2$. However, $MgCl_2$ titration is recommended for any new template-primer combination. Although Taq polymerase is ideal for routine PCR applications, several other thermostable DNA polymerases with unique qualities [10] are available. The properties of these alternative thermostable polymerase enzymes make them useful for specialized applications such as amplification of long stretches of DNA sequence or high-fidelity amplification.

Oligodeoxynucleotide Primers

The oligodeoxynucleotide primers utilized in a PCR determine target specificity of the amplification reaction. Effective oligodeoxynucleotide primers for PCR are highly sequence-specific, free of secondary structure, and form stable duplexes with target sequences. Four variables need to be considered when designing oligodeoxynucleotide primers: (1) size of the target sequence to be amplified, (2) the location of the target sequence within the overall genomic DNA (or cDNA) sequence, (3) secondary structure within the target sequence and flanking regions, and (4) specificity of amplification. The size of the target sequence should be selected such that the PCR products produced range from 100 to 500 bp in length. Primer length can influence target specificity and efficiency of hybridization. A long oligodeoxynucleotide primer may be more specific for the target sequence, but is less efficient at hybridization. A short oligodeoxynucleotide primer is efficient at hybridization, but is less specific for the target sequence. As a general guideline, oligodeoxynucleotide primers should be 17−30 nucleotides in length. Whenever possible, both primers should be of the same length because primer length influences the calculated optimal annealing temperature for a specific primer. The base composition of the oligodeoxynucleotide primers is also important because annealing temperature is governed in part by the G + C content of the primers. Ideally, G + C content should be 50−60%, and the percent G + C should be the same or very similar for both oligodeoxynucleotide primers in any given primer pair. The 3′ terminus of an oligodeoxynucleotide primer should always contain a G, C, GC, or CG. Repetitive or palindromic sequences should be avoided and primer pairs should not contain sequences that are complimentary to one another. Likewise, oligodeoxynucleotide primers should not anneal elsewhere in the gene of interest or in other sequences contained in the genome.

The optimal annealing temperature for a given oligodeoxynucleotide primer set is critical for designing an effective PCR reaction with respect to amplification specificity. The melting temperature T_m of an oligodeoxynucleotide primer can be calculated using a simplified formula that is generally valid for primers that are 18−24 nucleotides in length [11]: $T_m = 69.3 + 0.41$ (%G + C) − (650/L). In this formula, L is the primer length in bases and the result is the theoretical annealing temperature in degree Celsius. Online tools are now available to assist with prediction of properties of oligodeoxynucleotides, including annealing temperatures (see http://biotools.nubic.northwestern.edu/OligoCalc.html).

PCR Reaction Buffer

All of the components of a PCR reaction can be and often are combined into a single reaction mixture, most of which are commercially available. A typical PCR mixture will include a reaction buffer, oligodeoxynucleotide primers, Taq polymerase, and an appropriate DNA template. The PCR buffer consists of 50 mM KCl, 1.5 mM MgCl$_2$, 10−50 mM Tris-Cl (pH 8.3), and 50−200 μM dNTPs. Concentrations of KCl that are >50 mM can inhibit the enzymatic activity of Taq polymerase and should be avoided. However, the presence of KCl is necessary to encourage oligodeoxynucleotide primer annealing to target sequences in the template DNA. Likewise, excessive NaCl concentrations in the PCR mixture can adversely affect the enzymatic activity of Taq polymerase. The amount of MgCl$_2$ that is optimal for a given PCR reaction must be empirically determined. However, most standard PCR amplification reactions can be accomplished using 1.5−2 mM MgCl$_2$. The final concentration of dNTPs is 200 μM for a typical PCR, but some applications can be accomplished using much lower concentrations. Higher concentrations of dNTPs (or MgCl$_2$) can encourage errors related to dNTP misincorporation by Taq polymerase and should be avoided. The concentration of oligodeoxynucleotide primers should not exceed 1 μM unless the primers used contain a high degree of degeneracy. Taq polymerase enzyme is provided by commercial suppliers at 5 U/μL. One unit (U) of enzymatic activity is defined as the amount of enzyme required to catalyze the incorporation of 10 nmol of dNTP into acid-insoluble material in 30 minutes under standard reaction conditions. A 50 μL reaction will typically require 2.5 U of enzyme activity, while a 10 μL reaction will only require 0.5 U of enzyme activity. The amount of DNA template included in a PCR reaction will vary with the nature of the template source and the target sequence. Amplification from genomic DNA may require as much as 100 ng of DNA for a 50 μL reaction, whereas amplification from a plasmid template (for example) may only require 5 ng of DNA. Likewise, amplification of a target sequence that corresponds to a single allele may require more template, while amplification of a repetitive sequence (Alu for example) will require substantially less template.

The inclusion of gelatin or bovine serum albumin (BSA) can enhance the efficiency of the PCR amplification reaction. Gelatin or BSA can be included in the PCR mixture at concentrations up to 100 μg/mL. These agents act to stabilize the polymerase enzyme. The addition of helix destabilizing chemicals may be necessary if the PCR target sequence is located in a DNA region that is known to be of high G + C content. For example, dimethylsulfoxide (DMSO), dimethyl formamide (DMF), formamide, or urea may be included for this purpose. In most cases, these additives are included in the reaction mixture at 10% (w/v or v/v). These additives are thought to lower the T_m of the target sequence. However, care must be exercised when additives of this type are included in PCR reaction mixtures as high concentrations of these chemicals can adversely affect polymerase activity.

OPTIMIZATION OF PCR AMPLIFICATION REACTIONS

Several difference factors can significantly affect PCR sensitivity and specificity, including (1) oligodeoxynucleotide primer design, (2) PCR cycling variables (number of cycles, cycle times, and temperatures), and (3) composition of the PCR reaction mixture (divalent cation concentration). For most PCR applications, the most critical variable is the annealing temperature for the oligodeoxynucleotide primers employed. The maximum annealing temperature is determined by the primer with the lowest T_m. Exceeding this T_m by more than a few degrees will diminish the ability of the oligodeoxynucleotide primers to anneal to the target sequence and may lead to failure to produce the amplification product of interest. If utilization of an annealing temperature that is equal to the T_m of the oligodeoxynucleotide primers fails to produce the desired amplification product, then it may be necessary to further lower the annealing temperature. If the desired amplicon is produced at a lower T_m, but the amount of background products is high, then the annealing temperature should be increased. Salt concentrations also affect several aspects of the PCR reaction. Mg^{2+} concentration can affect oligodeoxynucleotide primer annealing to the target sequence, the T_m of the oligodeoxynucleotide-template complexes, as well as enzyme activity and fidelity. Taq polymerase requires free Mg^{2+} for activity. Thus, sufficient $MgCl_2$ must be included in the PCR mixture to provide adequate Mg^{2+} for the enzyme after some of the cation is lost to chelation by the oligodeoxynucleotide primers and the template DNA. The concentration of other salts can also affect the PCR reaction (including KCl). However, optimization of most PCR applications can be achieved through modification of Mg^{2+} concentration. Complete optimization of the reaction conditions may require several adjustments to the annealing temperature, PCR cycle variables, and salt concentrations. Several commercial sources offer kits which provide a range of PCR reaction mixtures for simple and rapid optimization of specific PCR conditions for a specific target sequence and its primers. Likewise, several manufacturers offer gradient thermocyclers which feature heating blocks where temperatures can be varied across the samples, enabling optimization of temperatures in a single PCR run.

INCREASING PCR SPECIFICITY AND SENSITIVITY

Taq polymerase has substantial enzymatic activity at 37°C, although its optimal activity is expressed at a much higher temperature (approximately 72°C). This low-temperature polymerase activity is the basis for nonspecific amplification associated with mispriming events that occur during the initial phase of the PCR reaction. Extension can occur from oligodeoxynucleotide primers that anneal nonspecifically to template DNA before the first denaturation step at 93–95°C. Because of this, several modified polymerase enzymes have been created to avoid this nonspecific primer extension activity. One example of this is Platinum Taq Polymerase (from Invitrogen Life Technologies, http://www.thermofisher.com). By including a thermolabile inhibitor of Taq polymerase in the form of a monoclonal antibody, the enzyme does not become active until the inhibitor is heat inactivated. Hence, the Taq polymerase becomes active after the elevated temperature destroys the monoclonal antibody during the initial denaturation phase of the PCR reaction which results in release of the functional enzyme. The antibody-mediated inhibition of Taq polymerase allows for room temperature assembly of the PCR reaction mixture. Nonspecific amplification associated primer extension from mispriming events is eliminated or reduced by holding the Taq polymerase functionally inactive until the critical temperature is reached.

PCR CONTAMINANTS

When performing PCR amplification it is critical to be aware of potential sources of DNA contamination and to employ procedures to ensure contamination-free working conditions. The power of PCR to amplify very small quantities of DNA producing detectable amplification products demands that special care be taken to prevent cross-contamination between different samples. This is especially true for PCR targets expected to be present in low amounts because greater efforts are usually required to amplify those sequences. Sources of contamination include (1) genomic DNA contaminating RNA samples, (2) cross-contamination among different nucleic acid samples processed simultaneously, (3) laboratory contamination of cloned target sequences (genomic DNA or cDNA), and (4) carryover of PCR products. In general, working in a clean laboratory and using good laboratory practices (such as wearing clean gloves at all times) substantially reduces the likelihood of contamination. Carryover products from other PCR reactions can be effectively controlled by the use of aerosol-free pipette tips, by using dedicated pipetters and solutions, and by maintaining separate areas to handle pre-PCR and post-PCR solutions and samples. In all PCR applications, it is essential to include proper positive-control and negative-control reactions to guard against systemic contamination of PCR reagents and to ensure that the desired amplicon is produced in positive reactions.

INHIBITORS OF PCR

Organic and inorganic compounds that inhibit PCR amplification of nucleic acids are common contaminants in DNA samples from various origins. These contaminating substances can interfere with the PCR reaction at several levels, leading to different degrees of attenuation and to complete inhibition. Many PCR inhibitors have been reported, and they appear to be particularly abundant in complex samples such as bodily fluids and samples containing high numbers of bacteria. Most of these contaminants (polysaccharides, urea, humic acids, hemoglobin) exhibit similar solubility in aqueous solution as DNA. As a consequence, they are not completely removed when typical extraction procedures are used in the preparation of the template DNA (detergent, protease, and phenol—chloroform treatments). Several methods have been developed to avoid these contaminating substances. Some of these methods are simple but lead to loss of significant amounts of the original sample, whereas others are very specific methods directed against specific forms of contaminations and may require expensive reagents.

ANALYSIS OF PCR PRODUCTS

There are numerous methods for analysis of PCR products (Fig. 2.3). The method of choice for analysis of PCR products will depend on the type of information that is desired. Typical analysis of PCR products involves electrophoretic separation of amplicons and visualization with ethidium bromide staining (or similar DNA dye). In most cases, amplification products can be analyzed by standard agarose gel electrophoresis. Agarose gel electrophoresis effectively separates DNA products over a wide range of sizes (100 bp to >25 kbp). PCR products from 200 to 2000 bp can be separated quickly on a 1.6% agarose gel. When greater resolution or separation power is required, such as in the analysis of very small PCR products (<100 bp), polyacrylamide gel electrophoresis is the method of choice. DNA products are easily visualized by ultraviolet illumination after ethidium bromide staining. Another method often utilized to quantify products is

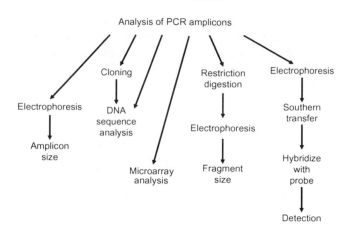

FIGURE 2.3 **Methodologies for analysis of PCR products.** PCR products (amplicons) generated through traditional PCR reactions can be analyzed in various ways using several methodologies. For some questions, the presence or absence of a PCR product answers the question, and a simple electrophoretic analysis with visualization of the product will suffice. For other questions the amplicon may require cloning or direct sequencing, restriction analysis, or blotting to obtain the desired information. PCR products can also be analyzed using various array-based techniques.

the incorporation of radioactive, fluorescent, or biotinylated markers. PCR products may be labeled by incorporating labeled nucleotides or through the use of labeled oligodeoxynucleotide primers. Labeled PCR products are separated by electrophoresis on agarose or polyacrylamide gels and visualized using appropriate techniques (for instance autoradiography for radiolabeled products). In the event that multiplex PCR analysis is being performed or for the most accurate sizing of amplicons within banding patterns, laboratories now employ capillary electrophoresis systems. In some cases, the desired information resulting from a PCR reaction can be obtained through a simple analytical gel separation (for instance in assays where the presence or absence of a PCR product answers the question), whereas in other instances additional information is required (like DNA sequencing to detect a gene mutation). Hence, PCR products can be cloned and used for sequence analysis, constriction of molecular probes, mutation analysis, in vitro mutagenesis, and studies of gene expression (Fig. 2.3).

VARIATIONS OF THE TYPICAL PCR AMPLIFICATION REACTION

Over the years, many modifications have been made to the standard PCR reaction. Some of the more significant modifications include: (1) hot-start PCR, (2) nested PCR, (3) reverse transcriptase PCR (RT-PCR), and (4) real-time PCR.

Hot-start and Nested PCR

Hot-start PCR was developed to reduce background from nonspecific amplification. Initial hot-start PCR was performed by limiting the Mg^{2+}, dNTP, or enzyme concentration. Alternatively, hot-start can be achieved by separating the reaction components with a wax bead barrier that melted as the mixture is heated during the initial denaturation step of the PCR. In either case, hot-start prevents polymerization of new DNA during the initial phase of the reaction when nonspecific binding may occur between primers and nonspecific DNA targets [12,13]. More recently chemical or antibody engineered polymerases have become commercially available that are activated once a specific temperature is reached. Nested PCR can increase both the sensitivity and specificity of amplification [14]. The amplification product(s) generated in the first PCR reaction are used as the template for a second PCR reaction, in which primers are used that are internal, or nested, within the first primer pair.

PCR Analysis of RNA

RT-PCR is an excellent method for analysis of RNA transcripts, especially for measuring low-abundance species or working with limited amounts of starting material. RT-PCR couples the tremendous DNA amplification powers of the PCR with the ability of reverse transcriptase (RT) to reverse transcribe small quantities of total RNA. RT-PCR is basically a four-step process: (1) RNA isolation, (2) reverse transcription, (3) PCR amplification, and (4) PCR product analysis. RNA is isolated from cells or tissue using various chemical-based extraction techniques or affinity-based (column) methods to eliminate contaminating DNA. This RNA is then used as a template in a reverse transcription reaction that produces cDNA, which serves as a template for the PCR reaction. RT (retroviral RNA-directed DNA polymerase) is the enzyme used to catalyze cDNA synthesis. The RT reaction consists of (1) cDNA synthesis primer, (2) an appropriate RT reaction buffer, (3) dNTPs, (4) RNA template (total RNA or mRNA), and (5) RT enzyme. There are several commercially available RT enzyme preparations that an be used in standard RT-PCR reactions. These include the Moloney murine leukemia virus (MMLV) RT and the avian myeloblastosis virus (AMV) RT. More recently, recombinant derivatives of these RT enzymes have become available that offer advantages over the native enzymes. One example of these recombinant enzymes is SuperScript III Reverse Transcriptase from Invitrogen (http://www.thermofisher.com). Advanced enzyme preparations such as these produce the highest yields and confer high specificity when gene-specific primers are employed.

RT-PCR is an excellent method for analysis of RNA transcripts, especially for measuring low-abundance species or working with limited amounts of starting material (such as those obtained from formalin-fixed paraffin-embedded samples). Traditional blotting and solution hybridization assays require much more RNA for analysis, and lack the speed and ease of technique afforded by PCR-based applications. Some of these traditional methods (such as northern blotting) require high-quality intact RNA species, whereas RT-PCR approaches can tolerate some RNA degradation. RT-PCR combines the tremendous DNA amplification power of PCR with the ability of RT to reverse transcribe very small quantities of total RNA (<1 ng) into cDNA. The use of total RNA preparations rather than poly(A)-purified RNA reduces the possibility of losing specific rare or low-abundance mRNAs during the purification process and allows for the use of very small quantities of starting material (cells or tissues). Additional advantages of RT-PCR include versatility, sensitivity, rapid turnaround time, and the ability to simultaneously compare multiple samples.

Real-Time PCR

Real-time PCR combines the amplification steps of traditional PCR with simultaneous detection steps that do not require post-PCR manipulation or interrogation of amplified products (Fig. 2.4). The PCR product is directly examined within the reaction tube. During real-time PCR, the exponential phase of PCR is monitored as it occurs using fluorescently-labeled probes [15]. During the exponential phase the amount of PCR product present in the reaction tube is directly proportional to the amount of emitted fluorescence and the amount of the initial target sequence [16,17]. Thus, these reactions can also be quantitative. There are two types of detection chemistries for real-time PCR: (1) those which use intercalating DNA binding dyes such as ethidium bromide or SYBR green, and (2) those that use various types of fluorescently-labeled probes.

Intercalating DNA binding dyes allow for the simple determination of the presence or absence of an amplicon. SYBR green, like ethidium bromide, is a dye that emits fluorescence when it is bound to double-stranded DNA. During the PCR reaction, there is an increase in the copy number of the amplicon and a simultaneous increase in the amount of intercalated SYBR green dye. This will then increase the amount of emitted fluorescent signal in direct proportion to the copy number [18]. One disadvantage associated with DNA binding dyes of this type is that they are nonspecific and will bind to any double-stranded DNA.

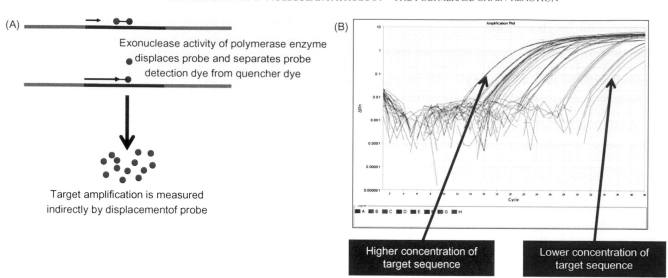

FIGURE 2.4 **Real-time PCR amplification of target sequences.** (A) The target sequence is shown as a blue bar within the genomic DNA (green bar). The real-time primer is depicted as a black arrow, and the real-time probe is depicted as a black line with a fluorescent tag (green ball) and quencher (red ball). With primer extension and exonuclease activity, the fluorescent tag is released from the quencher. Accumulation of the fluorescent signal provides a measure of abundance of the target sequence. (B) Results from a real-time amplification run with multiple samples analyzed in triplicate. The red arrow indicates an amplification curve corresponding to a target sequence of high abundance. The blue arrow indicates amplification of a target sequence with lower (but detectable) abundance.

Detection of real-time PCR products can also be accomplished using fluorescently-labeled probes of various types. There are three main detection chemistries for these probes: (1) cleavage-based (5′ exonuclease) probes, (2) molecular beacons, and (3) fluorescence resonance energy transfer (FRET) probes. Cleavage-based probes depend upon the 5′→3′ exonuclease activity of the Taq polymerase. These assays are commercially available as Taqman assays (Fig. 2.4). Molecular beacons are self-complimentary single-stranded oligonucleotides that form a hairpin loop structure and consist of a probe homologous to the target sequence flanked by sequences that are homologous to each other. Attached to one end is a reporter dye (such as FAM or TAMRA) and to the other end a quencher (such as DABCYL). When the beacon binds to the target sequence, the quencher and reporter are separated and fluorescence is emitted. FRET probes are composed of two separate fluorescently-labeled oligonucleotides, one with a 5′ donor molecule and the other with a 3′ acceptor molecule attached. When these probes hybridize with close proximity, energy can be transferred from the donor to the acceptor, resulting in fluorescence emission.

Real-time PCR is rapidly becoming the method of choice for most molecular diagnostics laboratories because of its increased sensitivity/specificity and turnaround times. This technology can be used for quantitative and qualitative assessment of target sequences and for distinguishing mutant from wild-type sequences. For single-nucleotide polymorphism (SNP) genotyping

and small mutation testing, two different labeled probes are designed—one for the wild-type allele and one for the mutant allele. The mismatch between the wild-type allele and the mutant probe facilitates competitive hybridization. Therefore, fluorescence will only be detected when the correct probe binds the correct target sequence. If binding-dye chemistries are used, another powerful feature of most real-time PCR instruments is the ability to perform melting curve analyses [18]. The T_m of a specific amplicon can be identified by an additional thermal step on the PCR product in the same reaction tube. Real-time PCR can also be used to determine the copy number of specific target sequences for infectious disease and oncology applications [19]. By multiplexing the primers and probes for the target sequence with primers and probes for a control sequence, accurate assessment of target copy number can be made in a relative quantification reaction. In contrast, using external standards of known concentration to create a standard curve and enables determination of absolute quantities of target copy number.

The main advantages of real-time PCR are the speed with which samples can be analyzed (because no post-PCR processing steps are required), and the closed single-tube nature of the technology. The analysis of results via amplification curve and melting curve analysis is very simple and contributes to overall increases in the speed of PCR analysis. With respect to potential contamination issues, another major advantage of have no post-PCR steps is that real-time PCR is a closed tube method of analysis which greatly reduces (1) the chance that a sample will be contaminated, (2) errors from mistakes in tube transfers, or (3) the possibility of amplicons aerosolizing into the laboratory environment.

Some investigators argue that a limitation of real-time PCR is the initial capital investment for instrumentation. However, since the first real-time PCR instruments were introduced on the market, a number of new real-time platforms have become available that are very cost effective. In fact, pricing of some instruments may be less than a traditional thermocycler and separate detection system. One technical limitation that should be noted is when using DNA binding dyes for the detection of real-time PCR products, nonspecific amplicons may also be detected. Such nonspecific amplicons would be observed through traditional post-PCR product analysis using electrophoresis. Hence, optimization of reaction conditions and melting curve analysis should be performed to distinguish between the desired (correct) amplicon and those corresponding to nonspecific amplification products.

References

[1] Saiki RK, Scharf S, Faloona F, et al. Enzymatic amplification of beta-globin genomic sequences and restriction site analysis for diagnosis of sickle cell anemia. Science 1985;230:1350–4.

[2] Mullis K, Faloona F, Scharf S, et al. Specific enzymatic amplification of DNA in vitro: the polymerase chain reaction. Cold Spring Harb Symp Quant Biol 1986;51(Pt 1):263–73.

[3] Saiki RK, Gelfand DH, Stoffel S, et al. Primer-directed enzymatic amplification of DNA with a thermostable DNA polymerase. Science 1988;239:487–91.

[4] Mullis KB, Faloona FA. Specific synthesis of DNA in vitro via a polymerase-catalyzed chain reaction. Methods Enzymol 1987; 155:335–50.

[5] Lawyer FC, Stoffel S, Saiki RK, et al. Isolation, characterization, and expression in *Escherichia coli* of the DNA polymerase gene from *Thermus aquaticus*. J Biol Chem 1989;264:6427–37.

[6] Vosberg HP. The polymerase chain reaction: an improved method for the analysis of nucleic acids. Hum Genet 1989; 83:1–15.

[7] Rogers BB. Application of the polymerase chain reaction to archival material. Perspect Pediatr Pathol 1992;16:99–119.

[8] Rogers BB, Josephson SL, Mak SK, Sweeney PJ. Polymerase chain reaction amplification of herpes simplex virus DNA from clinical samples. Obstet Gynecol 1992;79:464–9.

[9] Vieille C, Zeikus GJ. Hyperthermophilic enzymes: sources, uses, and molecular mechanisms for thermostability. Microbiol Mol Biol Rev 2001;65:1–43.

[10] Perler FB, Kumar S, Kong H. Thermostable DNA polymerases. Adv Protein Chem 1996;48:377–435.

[11] Wallace RB, Shaffer J, Murphy RF, et al. Hybridization of synthetic oligodeoxyribonucleotides to phi chi 174 DNA: the effect of single base pair mismatch. Nucleic Acids Res 1979;6: 3543–57.

[12] Kaijalainen S, Karhunen PJ, Lalu K, Lindstrom K. An alternative hot start technique for PCR in small volumes using beads of wax-embedded reaction components dried in trehalose. Nucleic Acids Res 1993;21:2959–60.

[13] Bassam BJ, Caetano-Anolles G. Automated "hot start" PCR using mineral oil and paraffin wax. Biotechniques 1993;14: 30–4.

[14] Porter-Jordan K, Rosenberg EI, Keiser JF, et al. Nested polymerase chain reaction assay for the detection of cytomegalovirus overcomes false positives caused by contamination with fragmented DNA. J Med Virol 1990;30:85–91.

[15] Wilhelm J, Pingoud A. Real-time polymerase chain reaction. Chembiochem 2003;4:1120–8.

[16] Wittwer CT, Herrmann MG, Moss AA, Rasmussen RP. Continuous fluorescence monitoring of rapid cycle DNA amplification. Biotechniques 1997;22:130–1, 134–8.

[17] Parks SB, Popovich BW, Press RD. Real-time polymerase chain reaction with fluorescent hybridization probes for the detection of prevalent mutations causing common thrombophilic and iron overload phenotypes. Am J Clin Pathol 2001;115:439–47.

[18] Wittwer CT, Reed GH, Gundry CN, Vandersteen JG, Pryor RJ. High-resolution genotyping by amplicon melting analysis using LCGreen. Clin Chem 2003;49:853–60.

[19] Heid CA, Stevens J, Livak KJ, Williams PM. Real time quantitative PCR. Genome Res 1996;6:986–94.

3

Next-Generation Sequencing in the Clinical Laboratory

D.L. Duncan and N.M. Patel

Department of Pathology and Laboratory Medicine, University of North Carolina School of Medicine, Chapel Hill, NC, United States

INTRODUCTION

Sequencing is the general term for any technique designed to determine the order of nucleotides in a nucleic acid molecule. Because nucleotide sequences are ultimately translated into cellular processes, sequencing technology has played a central role in biological research for decades. Similarly, nucleotide sequences and sequence variants contribute greatly to medical practice, making sequencing a mainstay technology in the clinical realm. Clinical genetics relies on identifying germline DNA variants to diagnose heritable disease and explain phenotypes of likely genetic origin. Increasingly, medical oncology utilizes sequencing to identify somatic variants or sequence variants specific to tumors. Presence or absence of somatic variants can offer prognostic information to clinicians, and even assist in diagnosis. Furthermore, identifying somatic variants impacts treatment options as many drugs are targeted to specific variants, genes or metabolic pathways.

Until recently, most routine sequencing was performed using the Sanger method (first-generation sequencing; Fig. 3.1). Sanger sequencing utilizes the natural properties of DNA polymerase as well as modified di-deoxynucleotides to generate sequence data [1]. Sanger sequencing and derivative techniques can generate sequencing data from DNA and RNA, as well as provide epigenetic information about virtually any genomic target. However, Sanger sequencing has fundamental limitations that hinder its application in modern clinical practice. Sanger sequencing can only be performed on one target per reaction, and that target has a maximum size of a few hundred nucleotides. Generating large amounts of sequence information over a broad range of targets is extremely time-consuming. Furthermore, the sensitivity of Sanger sequencing is often insufficient to identify somatic variants in tumor samples as these variants often exist at a low level in only a subpopulation of tumor cells [2]. These obstacles led to the development of new technologies to address the need for faster, more sensitive, and more comprehensive sequencing. These so-called "next-generation" sequencing methodologies rely on the combination of innovative laboratory techniques and massive computational power.

Why Perform Clinical Next-Generation Sequencing?

Clinical laboratories have rapidly adopted next-generation sequencing (which will be referred to as "massively parallel sequencing" because of its ability to sequence multiple nucleic acid molecules at the same time) as their technique of choice for performing sequence analysis [3]. Next-generation sequencing offers numerous technical advantages (improved cost-effectiveness, scalability, resolution) over Sanger sequencing. However, the true driving pressure behind the move toward next-generation sequencing is the clinical demand for more sequencing data.

As our understanding of both heritable and cancer genetics grows, more genes and individual mutations become relevant to everyday clinical practice.

Diagnostic Molecular Pathology
DOI: http://dx.doi.org/10.1016/B978-0-12-800886-7.00003-0

© 2017 Elsevier Inc. All rights reserved.

Cycle sequencing

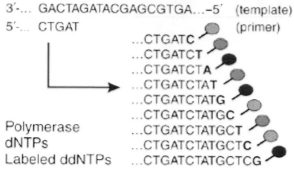

```
3'-... GACTAGATACGAGCGTGA...-5'   (template)
5'-... CTGAT                       (primer)
                    ...CTGATC
                    ...CTGATCT
                    ...CTGATCTA
                    ...CTGATCTAT
                    ...CTGATCTATG
Polymerase          ...CTGATCTATGC
dNTPs               ...CTGATCTATGCT
Labeled ddNTPs      ...CTGATCTATGCTC
                    ...CTGATCTATGCTCG
```

Electrophoresis
(1 read/capillary)

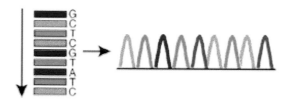

FIGURE 3.1 Sanger sequencing. Each Sanger-based sequencing reaction takes place within a microliter-scale volume, generating a ladder of ddNTP-terminated, dye-labeled products, which are subjected to high-resolution electrophoretic separation within capillaries of a sequencing instrument. As fluorescently labeled fragments of discrete sizes pass a detector, the four-channel emission spectrum is used to generate a sequencing trace. *Source: This image is reproduced with permission from: Shendure J, Ji H. Next-generation DNA sequencing. Nat Biotechnol 2008;26:1135–45.*

Particularly, some drugs have gene or pathway specific activity making them efficacious in patients who harbor certain mutations, such as the usage of vemurafenib in patients whose tumors harbor the *BRAF* V600E mutation. Therefore, sequencing is absolutely necessary for making appropriate treatment decisions in these patients. Professional guidelines and recommendations have been published to establish sequencing as standard of care in many cancer types including lung adenocarcinoma and melanoma [4,5]. Similarly, there are characteristic, prognostic, and even diagnostic genetic variants that warrant sequence analysis as part of standard of care treatment [6]. So, a large number of cancer patients require sequencing analysis as routine care, and that number will increase as new drugs are developed and new informative mutations are discovered. In order to sequence all of the important genomic regions in all of these patients efficiently, the throughput offered by next-generation sequencing makes this technique almost essential to modern molecular laboratories.

How Are Assays Performed?

The power of massively parallel sequencing lies in the confluence of three disciplines: chemistry, computer science, and clinical knowledge. These components may seem discrete from one another, but generating useful clinical information from such sequencing is possible only through the interdependency of these fields. Unique laboratory methodologies are required to generate the quantity of sequencing data on the scale that is desired for clinical use. Such a massive amount of data requires sophisticated computational processing before it can be interpreted by a human. Finally, there must be application of broad clinical knowledge to analyze the sequencing data and generate a report that is relevant to patient care. In the following sections, we will discuss these three components (chemistry, computer science, and clinical knowledge) separately, while acknowledging their reliance upon each other.

DNA SEQUENCING CHEMISTRY

The first step toward generating a massively parallel sequencing report is generating sequence data. This can be done using a variety of laboratory techniques, though all of them share certain qualities that allow for the large-scale and high accuracy required of clinical testing. First, sample DNA must go through a selection and modification process known as library preparation. Library preparation fragments DNA molecules and isolates regions of DNA that are to be sequenced. The area of the region to be sequenced can range from a single gene to the entire human genome. Then, target DNA is modified with adapters making it compatible with the sequencing platform, as well as with indexes or barcodes that are specific to the sample. This pool of modified target DNA is known as the library. Once generated, the library is loaded into the sequencing instrument. Libraries from multiple samples may be loaded and sequenced simultaneously. To enable this simultaneous, parallel sequencing, the instrument must separate and immobilize individual target DNA molecules. Physically separating molecules allows them to be sequenced individually, and the indexes added to the molecules during the library preparation step can be used to tie these individual sequences to the original sample. In this way, sequence information can be generated about all of the genomic targets within multiple samples in a single reaction [7].

Research groups and biotechnology companies alike have taken differing approaches to this general principle of massively parallel sequencing, and new laboratory methodologies are constantly being developed.

However, at this time there are three major platforms for performing massively parallel sequencing that have proven to be reliable for clinical use: (1) the MiSeq platform from Illumina, (2) Ion Torrent from Life Technologies, and (3) Single Molecule Real Time (SMRT) sequencing from Pacific Biosciences. Each of these platforms leverages basic principles of DNA replication in a unique way to generate sequence data.

Illumina Sequencing

Illumina manufactures a number of sequencing instruments including the MiSeq, HiSeq, and NextSeq. The MiSeq is designed specifically for targeted sequencing and is therefore the most popular Illumina instrument in clinical use. Despite individual differences, all of the Illumina instruments utilize the same proprietary technique of sequencing by synthesis.

Sequencing by synthesis takes place on a solid state flow cell [8]. The flow cell is coated with a lawn of oligonucleotides that have a generic sequence. During the library preparation step, sample DNA is modified to add adapter sequences to sequencing targets. These adapter sequences are designed to bind complimentarily to the oligonucleotides on the flow cell. The flow cell is flooded with library material, and target molecules are physically separated and immobilized by binding to the oligonucleotide lawn.

After the target material is immobilized on the flow cell, there is a short amplification step that generates up to 1000 copies of each target molecule. These identical molecules are also bound to the flow cell in close proximity to the original and are cumulatively known as a cluster (Fig. 3.2). Millions of clusters are generated on the flow cell, each representing one target molecule from one sample [9].

After cluster generation, sequencing proceeds by a series of repeated cycles. During a cycle, fluorescently labeled nucleotides flood the flow cell. These nucleotides compete for incorporation into growing complementary nucleic acids associated with each target molecule. The fluorescent label on each nucleotide serves as a polymerization terminator, so that only one nucleotide may be added to the growing nucleic acid

during each cycle [10]. After the nucleotides bind, a camera images the fluorescence on the flow cell to identify the base that was incorporated at each cluster. The dye is then enzymatically cleaved so that a new labeled nucleotide can be added during the next cycle.

Because the camera can image the entire flow cell, each cycle generates information about every cluster at once. And, because bases are added sequentially, the position of each base is known by the cycle number. In this manner, all of the clusters on the flow cell (corresponding to the original parent target molecules) are interrogated in parallel. The sequencing process continues for a number of cycles corresponding to the size of the target molecules so that the entire target is sequenced. So, by the end of the sequencing process, the raw fluorescence intensity data is recorded for every cluster position on the cell across every cycle. Additionally, the index (or bar code) that was added to each sample during library preparation is sequenced. By sequencing these indexes, the target molecule is linked to its originating sample. The raw data for each sequence is stored and is now ready for bioinformatic processing [7].

Ion Torrent Sequencing

The Life Technologies Ion Torrent platform uses the same principles of complementary base pairing and physically separated target molecules as the MiSeq with a few technical differences.

Rather than sequencing taking place on a flow cell, the Ion Torrent utilizes a semiconductor chip with microwells. Individual micro-beads are each covered with a clonally amplified target molecule (Fig. 3.3) and then separated and identified by their microwell location with one target molecule per well, instead of by their position on a flow cell. To perform sequencing, the microwell chip is successively flooded with unmodified single nucleotides. Only one species of nucleotide floods the chip at a time. When the available nucleotide species in the well is complementary to the target template molecule at the leading position, it will be incorporated into a growing nucleic acid [11].

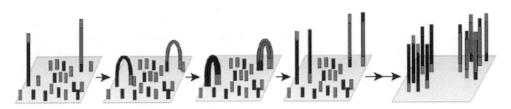

FIGURE 3.2 Illumina sequencing. The Illumina platform immobilizes individual product molecules on a flow cell and then uses bridge PCR to form clonally amplified colonies. *Source: This image is reproduced with permission from: Shendure J, Ji H. Next-generation DNA sequencing. Nat Biotechnol 2008;26:1135—45.*

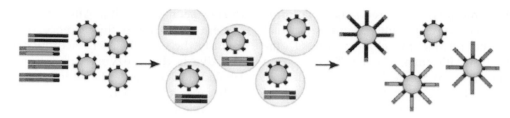

FIGURE 3.3   Ion Torrent sequencing. The Ion Torrent platform physically separates individual product molecules and micro-beads using emulsion PCR. Primers on the micro-beads allow for tethering and clonal amplification of the molecules prior to sequencing. *Source: This image is reproduced with permission from: Shendure J, Ji H. Next-generation DNA sequencing. Nat Biotechnol 2008;26:1135—45.*

The incorporation of a nucleotide into a nucleic acid causes the release of a hydrogen ion. Below each microwell, embedded in the semiconductor chip is an ion sensor. These sensors are sensitive enough to detect the hydrogen ion release each time a complementary nucleotide is added to the target molecule template. The sequence for a given target molecule, then, is determined by evaluating the electrical signal intensity during each sequential exposure to different nucleotide species. If there is a repeated sequence of the same nucleotide in the target molecule, then the signal intensity will be high when the complementary nucleotide floods the chip. Conversely, there will be no signal at time points when the target molecule is exposed to noncomplementary nucleotides. The size of the semiconductor chip and the sensitivity of the ion sensors allow for raw electrical data to be recorded simultaneously for thousands of target molecules as the chip is sequentially flooded with nucleotides [12]. The raw ion sensor data from each microwell over the time course of the run is stored for bioinformatic processing. So, similar to the MiSeq platform, sequencing occurs in parallel across multiple targets and samples.

SMRT Sequencing

SMRT sequencing, developed by Pacific Biosciences, separates target molecules by location in microwells similar to the Ion Torrent. However, raw data collection utilizes a distinctive technique. There is a single DNA polymerase enzyme locked in the bottom of each microwell [13]. Also, rather than amplified target DNA, each microwell holds a single copy of target molecule. The microwells are flooded with nucleotides. Each nucleotide species is labeled with a unique fluorescent dye. The nucleotides are incorporated in a complementary manner to the template target molecule by the polymerase. A light detector at the base of the microwell detects signal from the polymerase. As a nucleotide is incorporated into the growing nucleic acid, the fluorescence is measured by the detector. The dye is cleaved during the incorporation, and the active

site of the polymerase is free for the next nucleotide to be added [14].

Measuring raw data as fluorescence with an optical detector is similar to the MiSeq platform. However, unlike the MiSeq instrument sequencing does not proceed in cycles. The SMRT sequencer adds an excess of nucleotides that are allowed to be incorporated by the polymerase in real time. The sequencing reaction proceeds quickly as fluorescence signal is recorded from all microwells simultaneously and for each nucleotide of the single template molecule in each microwell.

Advantages and Disadvantages of Sequencing Technologies

There are myriad technical differences in this brief cross section of three popular massively parallel sequencing platforms. With these differences come advantages and disadvantages for each technique. Generating amplified clusters on the MiSeq flow cell yields high sequencing accuracy on a base-by-base basis. However, the cyclical nature of the sequencing process leads to longer sequencing times on the MiSeq instrument. Conversely, the Ion Torrent can complete a sequencing run in a fraction of the time that the MiSeq requires. But, this speed comes at the cost of accuracy. Specifically, the Ion Torrent has difficulties quantifying single nucleotide repeats with its reliance on quantifying an electrical impulse for each unique nucleotide that is incorporated, and this contributes to the higher raw error rate of the Ion Torrent over the Illumina instrument [15]. The SMRT sequencer is an even more extreme example. Sequencing is extremely rapid, but observing fluorescence at the single molecule level predisposes to errors in signal detection. Furthermore, individual polymerase molecules have an inherent error rate when incorporating nucleotides into nucleic acids [16]. By anchoring a single polymerase and template molecule, the SMRT sequencer can sequence much longer target molecules (roughly 1500 bases per read) than the MiSeq or Ion Torrent which have average read lengths of 150—200 bases [17]. This increased read length is advantageous in the bioinformatic processing stages of analysis. So, no sequencing technique is perfect, but each has enough advantages to warrant use in clinical sequence data generation.

Perhaps as striking as the specific unique features of these sequencing platforms are the similarities. The central axiom of separating target molecules and sequencing in a massively parallel manner is preserved in all methodologies. Separating targets allows for molecule specific sequencing which is essential for multiplexing. A massively parallel sequencing reaction generates raw data for all copies of all target molecules simultaneously. These two capabilities enable the massively parallel sequencers to keep up with the clinical demand for higher sample throughput and more sequencing targets. But, generating the raw data alone is only the first step toward creating a clinical sequencing report. A multistep bioinformatic process is necessary before the sequencing data is ready for interpretation.

COMPUTER SCIENCE

Following the sequencing reaction and the generation of raw data, the next integral step in creating a massively parallel sequencing report is bioinformatic processing. A single massively parallel sequencing reaction will generate millions of individual molecules, each with a specific sequence and unique identifier. This is far too much information for a person or group of people to analyze and interpret by hand. Fortunately, there are powerful computational tools that are designed to compile the most salient sequencing data into a format that is understandable to a molecular pathologist. Briefly, raw data from the instrument must be translated into individual sequences known as "reads." Then, each read must be "mapped" by identifying its targeted region in the genome. Finally, the aligned reads are used to determine differences in nucleotide sequence between the sample DNA and a standardized reference sequence. These basic steps—base calling, sequence alignment, and variant calling—are explained in further detail below.

Base Calling

Base calling is the process by which raw data from the sequencing instrument is converted to nucleotide sequences. This is performed by base calling software that is usually run from the instrument itself. As we have seen, sequencing instruments generate raw data in different forms (fluorescence intensity, electrical impulse, etc.), and therefore base calling is specific to the particular platform being used. However, the general concepts and goals of base calling are similar across all platforms.

The primary goal of the base calling software is to assign a nucleotide species to each position in the target DNA sequence. One target molecule's base sequence is called a read. The nature of massively parallel sequencing leads to the generation of data at the level of individual target molecules, so the base caller transforms raw data at the scale of one read per originating molecule. Thus, millions of reads are generated from a single sequencing reaction.

To assign a nucleotide to a position in a read, the base caller must evaluate the raw signal generated for that position. The intensity of signal relative to the background noise or artifact will allow the base caller to generate a confidence score associated with each nucleotide. The confidence score is the likelihood that the chosen nucleotide is the correct call for that position [18].

For example, using the MiSeq there will be a target molecule at a given position on the flow cell. The fluorescent intensity generated at that position during each cycle is recorded as raw data. Stronger fluorescence gives greater confidence in the nucleotide species that was incorporated during that cycle. However, if other target clusters are in close physical proximity there can be overlapping fluorescent signal resulting in background noise. The presence of this noise reduces the confidence of the nucleotide call [19].

Ultimately, the base caller uses proprietary algorithms that are specific to the sequencing platform to separate signal from noise for each nucleotide. This generates a specific nucleotide along with a confidence score at every position in every read. The read data is recorded and saved in an electronic format, the most popular of these being FASTQ.

The FASTQ format is a standardized text based format for storing specific sequencing read data. The format has three components: (1) a read identifier, (2) nucleotide sequence, and (3) confidence scores. The read identifier links the sequence data to the original individual target molecule from which the raw data was generated. The nucleotide sequence is simply the order of nucleotides in that read identified by the base caller. Finally, the confidence score is a representation of the confidence in the nucleotide choice generated by the base caller for each nucleotide in the sequence [20]. The data is ready for the alignment step once the base caller generates FASTQ formatted read for each target molecule in the sequencing reaction.

Sequence Alignment

The FASTQ output from the base caller contains all of the DNA sequences from all of the target molecules in the reaction. However, the read data alone is not useful because it gives no indication of what genomic regions have been sequenced. So, a step is needed to map all of the individual reads to their locations in the genome. This mapping process is known as sequence alignment.

Sequence alignment is performed by software that utilizes one or more alignment algorithms to map the read data from the base caller. Unlike base calling, alignment is not platform specific and can be performed using any available aligning software given the appropriate sequence read input. However, most clinical sequencing platforms are sold with proprietary software that performs alignment. Regardless of software choice, the basic principles of alignment are the same.

To perform read mapping, the aligning software requires the read data and a reference sequence. The reference sequence is necessary to serve as a standard against which individual reads can be compared. Generally the reference sequence is a standardized whole human genome, though it can be customized or changed to serve different purposes. For example, known viral or bacterial genomic data can be added to the reference sequence if infectious agents are included in the targets for sequencing [21].

The aligning software will take each read that was generated by the base caller and compare it individually to the reference sequence with the goal of placing each read at the most appropriate location in the reference. To perform this goal, the aligner takes many factors into account. The fidelity of the read sequence to the reference is the most obvious and significant factor in alignment. If a read is identical to a portion of the reference sequence, then it most likely corresponds to that genomic position. However, not all reads will perfectly match the reference. Many biologic factors such as benign variants and deleterious mutations result in deviations from reference. Additionally, technical factors such as errors during sequencing or inaccurate base calling contribute to imperfect read alignment [22].

The alignment software allows for these occurrences by generating penalties for inexact alignment. This is similar to the confidence score made by the base caller and creates room for reads to be aligned despite some differences from the reference. Different factors can result in different penalties. For instance, a read with low base confidence that otherwise matches the reference will not be penalized as heavily as a read with numerous bases that differ from the reference entirely. At a certain penalty threshold, a read is considered unalignable and cannot be mapped to any region in the reference sequence. These unaligned reads are discarded and not analyzed. The penalty scheme can be customized to allow for more stringent or more lenient alignments. The stricter the penalties, the fewer reads

can be aligned and included in analysis. The more lenient the penalties, the more incorrectly mapped reads will be incorporated into the final data. This penalty scheme is absolutely crucial to alignment. If only perfectly aligned reads were kept for analysis, then no mutations could ever be identified as a mutation is by definition a variant from the expected sequence [23].

The process of performing sequence alignment on the read data generates large matrices of alignment data that can be stored in multiple file formats. The most popular of these alignment formats is the Sequence Alignment/Map (SAM) format. The SAM format, and its binary counterpart known as BAM, stores alignment data indexed by reference sequence location. These SAM and BAM files can be opened using a number of alignment viewing programs to generate a graphical display of the aligned reads. The graphical interface typically shows the reference sequence with parallel rows of reads corresponding to where each read has been mapped against the reference. The total number of reference sequence bases for which reads from a sample have been aligned is known as the "coverage" of that sample. Most reference sequence bases in the target regions will be covered multiple times, as multiple reads are typically generated for the same genomic target. The number of reads that align over a single base in the reference sequence is known as the "depth of coverage" at that base. Depth of coverage is important as it leads to certainty about the true biological sequence in a sample. Each read that is aligned over a given position acts to independently verify the base call at that position with the other aligned reads. So, greater depth of coverage leads to greater confidence in the sequencing data. Typically, 20-fold coverage (a depth of coverage of 20 reads over the target region) is required to be confident of the base called in a pure or germline specimen, whereas 1000-fold coverage is needed to identify variants in mixed samples such as tumor specimens [24].

Manually viewing the alignment data is interesting and can be helpful in verifying sequence variants and interpreting complicated variants. However, the large number of aligned reads would be impossible to review by hand in order to identify mutations in the sequenced samples. Therefore, once the alignment data is generated it enters a final bioinformatic processing step—variant calling.

Variant Calling

The sequence alignment process yields useful coverage and depth of coverage data about the sequencing run. But the true value of sequencing is the ability to identify sequence differences between the sample nucleic acid and the expected, typical sequence. The alignment alone does not identify these differences or variants. So, a final bioinformatic step is required to decide what positions in the alignment represent true variants in the sequenced sample.

This variant calling step, similar to alignment, is performed by software that is independent of the sequencing platform used to generate the raw data. There are multiple free and third-party variant callers in addition to the proprietary variant calling software usually included with the sequencing instrument. Different variant callers utilize unique algorithms or combinations of algorithms to tackle the same question—given the alignment data, what is the true biological nucleotide sequence [25]?

To identify sequence variants the variant caller must consider a number of factors provided by the alignment. Depth of coverage and the variant frequency affect the confidence in a variant's validity. In heterogeneous samples, such as tumor specimens, variants are only expected to occur in a subset of reads, so a minimum variant frequency must be established as a "true variant" threshold. Additionally, the alignment score generated by the penalty scheme impacts the variant caller. Generally, variants in reads with stronger alignment are favored. However, this strategy is problematic as complicated insertions, deletions, and duplications will necessarily be poorly aligned. Thus, single nucleotide variants are rather simple for variant calling algorithms to identify whereas more complicated variants can be misidentified or missed entirely [26].

The data generated by the variant calling process can be stored in multiple file formats, though the most commonly used is the Variant Call Format (VCF). The VCF stores only the location and type of variants identified in each sample, making it much smaller than the aggregate read data or sequence alignment data [27]. VCF formatted data can be accessed using a variety of software packages that will display identified variants by sample and genomic location. This information can be manually reviewed and interpreted to create a clinical report.

SEQUENCE INTERPRETATION AND CLINICAL REPORTING

After sequencing is complete and bioinformatic processing has been performed, the resultant data is ready for analysis. The ultimate goal of clinical sequencing is generating a report that is useful to clinicians and patients in helping to guide therapy, making diagnoses, or providing prognostic information. Furthermore, the report must be constructed in a manner that provides relevant information in a way that is easy for clinicians and patients to understand. So, the sequencing analyst—generally a molecular genetic pathologist or clinical molecular geneticist—must be versed in both clinical medicine and genetic interpretation and reporting.

Of primary importance to generating a clinically useful report is an understanding of current clinicogenetic principles and practice. In general, the sequencing report will address variants that are established in clinical practice as significant to the patient's clinical care. A good report will explain the relevance of each variant (diagnostic, prognostic, or therapy related) in terms of the specific patient's current condition. Even targeted sequencing panels will often times identify multiple variants in a single patient, presenting an interpretive challenge to the analyst [28]. Thus, sequencing offers a great opportunity for individualizing patient care based on a host of information that would not otherwise be attainable.

Beyond clinical variant interpretation, the analyst's job is to organize the reportable data into a concise but comprehensive report. Professional guidelines have been published to detail best practices for variant reporting. Generally, variants are reported in tiers that suggest their level of clinical actionability. Variants known to have prognostic or therapeutic implications are reported first, and variants of unknown clinical significance are reported last [29]. There can be other intermediate reporting categories for likely benign or likely pathogenic variants, and some institutions create a category for variants that are likely to impact clinical trial eligibility. The category system serves to highlight the most relevant variants foremost, while also documenting all findings. While variants of unknown significance offer little immediate clinical assistance, they may become informative at a later time, so thorough recordkeeping is paramount. To that end, variants must be reported in a way that can be easily searched in the future. Therefore, there are additional reporting guidelines for variant nomenclature and even gene name and coordinate recording [30]. The location of all variants is established during sequence alignment based on a standard reference sequence. So if a different reference is used for sequence alignment of the same sequencing data, the variants will all be different. The recommended standard reference sequence is the Genome Reference Consortium standard human genome. However, this reference sequence is periodically updated and changed. Therefore, careful documentation of the reference sequence used during alignment is necessary in every sequencing report [31].

A combination of broad medical and genomic knowledge allows the sequence analyst to interpret each patient's variants in terms of the patient's clinical presentation. Then, an understanding of reporting guidelines and best practices is necessary to generate a report that provides immediately actionable information as well as serves as a useful record. The final clinical sequencing report represents the laborious convergence of multiple scientific disciplines and technologies.

CONCLUSION

Generating a massively parallel sequencing report from an initial DNA sample is an extremely complicated process. We have presented a brief overview of the technology in its present state, but these techniques are constantly changing and improving. However, the fundamental, multidisciplinary approach of the sequencing process is unlikely to become obsolete. The sequencing chemistry is dependent upon bioinformatic processing which, in turn, relies upon an analyst's interpretive capabilities. The interdependence of these elements provides opportunities for broadening the clinical utility of sequencing in multiple areas. New sequencing technologies promise higher throughput and more accurate sequencing. Novel alignment and variant calling approaches will inevitably overcome current challenges to mapping. In the clinical realm, drug trials and improved variant databases will offer new actionable targets for analysis. So, while massively parallel sequencing has already established its utility in the clinical lab, the reliance on this technology will only grow with the growing ability to offer patient and disease-specific, personalized therapy.

References

[1] Hutchison CA. DNA sequencing: bench to bedside and beyond. Nucleic Acids Res 2007;35:6227–37.

[2] Altimari A, de Biase D, De Maglio G, Gruppioni E, Capizzi E, Degiovanni A, et al. 454 next generation-sequencing outperforms allele-specific PCR, Sanger sequencing, and pyrosequencing for routine KRAS mutation analysis of formalin-fixed, paraffin-embedded samples. Onco Targets Ther 2013;6:1057–64.

[3] Mardis ER. A decade's perspective on DNA sequencing technology. Nature 2011;470:198–203.

[4] Lindeman NI, Cagle PT, Beasley MB, Chitale DA, Dacic S, Giaccone G, et al. Molecular testing guideline for selection of lung cancer patients for EGFR and ALK tyrosine kinase inhibitors: guideline from the College of American Pathologists, International Association for the Study of Lung Cancer, and Association for Molecular Pathology. J Mol Diagn 2013;15:415–53.

[5] Gonzalez D, Fearfield L, Nathan P, Tanière P, Wallace A, Brown E, et al. BRAF mutation testing algorithm for vemurafenib treatment in melanoma: recommendations from an expert panel: recommendations for BRAF testing for vemurafenib. Br J Dermatol 2013;168:700–7.

[6] Thomas L, Di Stefano AL, Ducray F. Predictive biomarkers in adult gliomas: the present and the future. Curr Opin Oncol 2013;25:689–94.

[7] Metzker ML. Sequencing technologies—the next generation. Nat Rev Genet 2010;11:31–46.

[8] Fedurco M, Romieu A, Williams S, Lawrence I, Turcatti G. BTA, a novel reagent for DNA attachment on glass and efficient generation of solid-phase amplified DNA colonies. Nucleic Acids Res 2006;34 e22-e22.

[9] Ju J, Kim DH, Bi L, Meng Q, Bai X, Li Z, et al. Four-color DNA sequencing by synthesis using cleavable fluorescent nucleotide reversible terminators. Proc Natl Acad Sci USA 2006;103: 19635–40.

[10] Chen C-Y. DNA polymerases drive DNA sequencing-by-synthesis technologies: both past and present. Front Microbiol 2014;5:305.

[11] Bragg LM, Stone G, Butler MK, Hugenholtz P, Tyson GW. Shining a light on dark sequencing: characterising errors in Ion Torrent PGM data. PLoS Comput Biol 2013;9:e1003031.

[12] Rothberg JM, Hinz W, Rearick TM, Schultz J, Mileski W, Davey M, et al. An integrated semiconductor device enabling non-optical genome sequencing. Nature 2011;475:348–52.

[13] Levene MJ, Korlach J, Turner SW, Foquet M, Craighead HG, Webb WW. Zero-mode waveguides for single-molecule analysis at high concentrations. Science 2003;299:682–6.

[14] Eid J, Fehr A, Gray J, Luong K, Lyle J, Otto G, et al. Real-time DNA sequencing from single polymerase molecules. Science 2009;323:133–8.

[15] Loman NJ, Misra RV, Dallman TJ, Constantinidou C, Gharbia SE, Wain J, et al. Performance comparison of benchtop high-throughput sequencing platforms. Nat Biotech 2012;30:434–9.

[16] Carneiro MO, Russ C, Ross MG, Gabriel SB, Nusbaum C, DePristo MA. Pacific biosciences sequencing technology for genotyping and variation discovery in human data. BMC Genomics 2012;13:375.

[17] Quail MA, Smith M, Coupland P, Otto TD, Harris SR, Connor TR, et al. A tale of three next generation sequencing platforms: comparison of Ion Torrent, Pacific Biosciences and Illumina MiSeq sequencers. BMC Genomics 2012;13:341.

[18] Ledergerber C, Dessimoz C. Base-calling for next-generation sequencing platforms. Brief Bioinform 2011;12:489–97.

[19] Ewing B, Green P. Base-calling of automated sequencer traces using Phred. II. Error probabilities. Genome Res 1998;8:186–94.

[20] Deorowicz S, Grabowski S. Compression of DNA sequence reads in FASTQ format. Bioinformatics 2011;27:860–2.

[21] Flicek P, Birney E. Sense from sequence reads: methods for alignment and assembly. Nat Methods 2009;6:S6–12.

[22] Bao R, Huang L, Andrade J, Tan W, Kibbe WA, Jiang H, et al. Review of current methods, applications, and data management for the bioinformatics analysis of whole exome sequencing. Cancer Inform 2014;13:67–82.

[23] Li H, Homer N. A survey of sequence alignment algorithms for next-generation sequencing. Brief Bioinform 2010;11:473–83.

[24] Schrijver I, Aziz N, Farkas DH, Furtado M, Gonzalez AF, Greiner TC, et al. Opportunities and challenges associated with clinical diagnostic genome sequencing: a report of the Association for Molecular Pathology. J Mol Diagn 2012;14:525–40.

[25] Liu X, Han S, Wang Z, Gelernter J, Yang B-Z. Variant callers for next-generation sequencing data: a comparison study. PLoS ONE 2013;8:e75619.

[26] Pabinger S, Dander A, Fischer M, Snajder R, Sperk M, Efremova M, et al. A survey of tools for variant analysis of next-generation genome sequencing data. Brief Bioinform 2014;15:256–78.

[27] Dolled-Filhart MP, Lee M, Ou-yang C, Haraksingh RR, Lin JC-H. Computational and bioinformatics frameworks for next-generation whole exome and genome sequencing. Scientific World J 2013;2013:730210.

[28] Johnston JJ, Rubinstein WS, Facio FM, Ng D, Singh LN, Teer JK, et al. Secondary variants in individuals undergoing exome sequencing: screening of 572 individuals identifies high-penetrance mutations in cancer-susceptibility genes. Am J Hum Genet 2012;91:97−108.

[29] Green RC, Berg JS, Grody WW, Kalia SS, Korf BR, Martin CL, et al. ACMG recommendations for reporting of incidental findings in clinical exome and genome sequencing. Genet Med 2013;15:565−74.

[30] Richards CS, Bale S, Bellissimo DB, Das S, Grody WW, Hegde MR, et al. ACMG recommendations for standards for interpretation and reporting of sequence variations: revisions 2007. Genet Med 2008;10:294−300.

[31] Rehm HL, Bale SJ, Bayrak-Toydemir P, Berg JS, Brown KK, Deignan JL, et al. ACMG clinical laboratory standards for next-generation sequencing. Genet Med 2013;15:733−47.

4

Automation of the Molecular Diagnostic Laboratory

S.A. Turner[1,2] and G.J. Tsongalis[1,2]

[1]Laboratory for Clinical Genomics and Advanced Technology (CGAT), Department of Pathology and Laboratory Medicine, Dartmouth-Hitchcock Medical Center and Norris Cotton Cancer Center, Lebanon, NH, United States
[2]Geisel School of Medicine at Dartmouth, Hanover, NH, United States

INTRODUCTION

The increasing demand for highly reliable, accurate, cost-effective, and expedited diagnostics in today's health care system has resulted in the evolution of the modern clinical laboratory. This evolution has resulted in the adoption of *molecular diagnostics* into many clinical settings. Molecular diagnostics is a term used to describe a family of techniques used to analyze biological markers in an individual's genetic code (genome) and to analyze how their cells express their genes as proteins (proteome). This form of diagnostics applies molecular biology techniques to clinical testing in order to diagnose and monitor disease, detect risk, and personalize therapies by determining which treatments will work best for an individual patient.

Research into the molecular basis of disease is driving the growing demand for molecular diagnostics across many different clinical indications, including infectious disease, medical genetics, and molecular oncology, by demonstrating the usefulness or utility that these tests have in diagnosing and treating the patient. In fact, molecular diagnostics is one of the fastest growing segments of laboratory medicine (Fig. 4.1). In turn, the expanding molecular testing catalog and increased testing volumes are often constrained by available resources, including cost and the availability of highly trained personnel. In many cases, these constraints have resulted in the centralization of diagnostic testing, causing a reduction in point-of-care testing and an increase in samples being sent out to larger reference laboratories for analysis. The need to overcome these constraints for molecular diagnostics laboratories of all sizes has driven the advancement of technologies to decrease demands of personnel by increasing automation.

Automation or *automatic control* is defined as the use of various control systems for operating equipment such as machinery in a factory, or processes such as network switching requiring no, minimal, or reduced human intervention. In the molecular diagnostic laboratory what was once a labor-intensive process requiring substantial time investment by a highly trained clinical laboratory scientist can now be automated, resulting in reduced technician time and cost, faster testing turnaround times, and in many cases improved accuracy and reliability of highly complex molecular diagnostic testing.

One of the earliest and most profound examples of automation in the molecular diagnosis of disease is the invention of the thermal cycler (or thermocycler), a device responsible for controlling temperature during polymerase chain reaction (PCR) amplification of gene targets. PCR requires template DNA, dideoxynucleotides (ddNTPs), oligonucleotide primers, and a polymerase enzyme to synthesize copies of the target region. The reaction mixture undergoes successive rounds of heating and cooling to denature the double-stranded DNA and produces an exact copy of the targeted sequence. After 30–40 successive temperature cycles, the gene target is amplified hundreds of millions of times. Upon invention of PCR in 1983, this technique was largely manual, requiring a scientist to place each reaction into a heated or cooled water bath to complete each cycle. This led to the invention of the first automated thermal cycler known as "Mr. Cycle" by a PerkinElmer Cetus Instruments

© 2017 Elsevier Inc. All rights reserved.

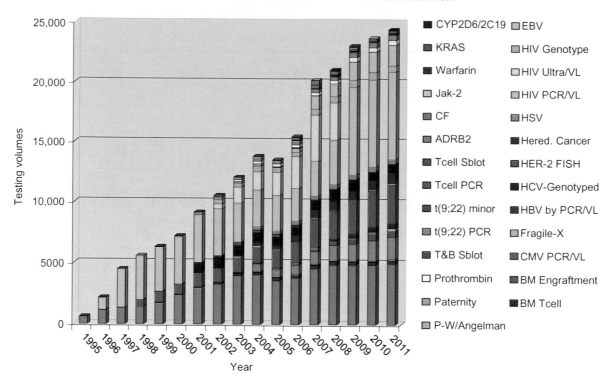

FIGURE 4.1   Representative growth in test volumes and testing menu of an average academic medical center molecular diagnostics laboratory over the past 20 years. Continued expansion of testing in infectious disease, molecular oncology, and hereditary disease screening, including cystic fibrosis screening, will drive this expansion in the years to come.

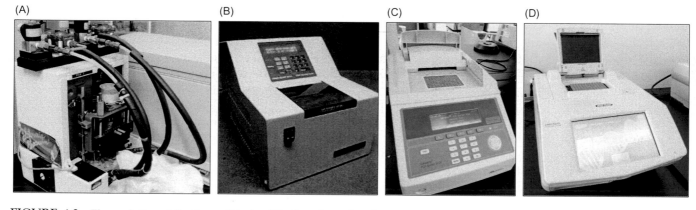

FIGURE 4.2   The evolution of the thermal cycler. The first thermal cycler prototype named (A) "Mr. Cycle" was introduced in 1985 by PerkinElmer Cetus Instruments (PECI). Only three were ever built and currently reside in the National Museum of Natural History (Smithsonian), the Science Museum of London, and University of Southern California. The first commercially available thermal cycler was the (B) Thermal Cycler 480 by PECI released in 1987. Over the years, additional advancements have led to the mass adoption of the instruments like the (C) GeneAmp 9700 (Applied Biosystems) introduced in 2007, and popular touch-screen models like the (D) C1000 (Bio-Rad) introduced in 2011.

(PECI) collaboration. This early automation device consisted of a four-unit prototype containing an automatic pipettor, a simple robot, and temperature-controlled water baths. This thermal cycler would control cycling temperature by placing each sample in the appropriate water bath and replenish the nonstable polymerase after each successive temperature cycle, allowing for dramatically less human intervention. The first published

application using this type of process was a test for diagnosis of sickle cell disease [1]. The subsequent use of a thermostable polymerase, Taq polymerase [2,3], allowed for further advancements in automation and the creation of the first fully automated commercially available PECI, called the *DNA Thermal Cycler 480 system*, allowing for the wide adoption of PCR as the basis of many molecular techniques today (Fig. 4.2).

In this chapter, we will describe how automation has helped meet the high demands facing today's molecular diagnostic laboratory. Automation has impacted every aspect of the diagnostic laboratory from extraction of nucleic acids to the massively parallel sequencing of tumors. While many of the technologies presented herein can be applied to many applications, this chapter will describe the role of automation in terms of diagnosis of infection disease, genetic disorders, and molecular characterization of neoplasms routinely performed in today's molecular diagnostics laboratory.

AUTOMATION OF NUCLEIC ACID EXTRACTION

Almost all molecular diagnostic applications start with the efficient extraction of high-quality nucleic acids, DNA and RNA (including mRNA, miRNA), from biological samples. Manual extraction methods have benefited from advancements in nucleic acid chemistry over the years. Such methods typically employed by a molecular diagnostic laboratory range from chemical-based extractions relying on manual phenol—chloroform purification, to the use of manufactured extraction kits, which utilize silica-based column or magnetic bead nucleic acid purifications. While these methods vary in their chemical makeup, extraction efficiency, and nucleic acid purification techniques, most are capable of providing high-purity nucleic acids required for quality molecular diagnostic results [4]. However, the ability to recover nucleic acids and remove inhibitors and contaminants is not the only important aspect of an extraction method.

Manual extraction methods of all kinds are prone to highly variable results [5,6]. While variation in extraction efficiencies and nucleic acid quality exists between methodologies, significant variation is also observed with the same methodologies within a single laboratory. This variation often warrants the need for a highly trained technologist with significant hands-on expertise with a specific extraction method. Since no single extraction method or kit is adequate for extraction of all nucleic acids from a growing variety of biological samples found in the molecular diagnostics laboratory including whole blood, urine [7], feces [8], saliva [9], buccal [10] and nasopharyngeal swabs [11], bone marrow biopsies [12], fresh and frozen tissues, and formalin-fixed paraffin-embedded tissues [13], this amount of hands-on expertise is challenging for most diagnostics laboratories. To ensure consistent, high-quality, manual nucleic acids extraction, today's molecular technologist must possess expertise in an expanding number of extraction methods for a large number of sample types, using a variety of kits and methodologies.

The expanding test menu, diversity of sample types, and increased demand for molecular diagnostics result in a significant increase in the volume of samples requiring nucleic acid extraction. While these manual extraction methods may be adequate for lower volume sample types, most moderately sized molecular diagnostics laboratories can no longer solely rely on such methods for nucleic acid extraction. By automating nucleic acid extraction, the molecular diagnostics laboratory can (1) increase throughput by reducing manual manipulation, (2) reduce turnaround times by increasing capacity, and (3) improve and standardize extraction efficiency and quality from a growing number of biological samples.

A wide variety of automated extraction systems are used in today's molecular diagnostics laboratory. They range from small stand-alone extraction instruments designed for low to moderate volumes [14] to large robotic liquid handlers often incorporated into highly automated workflows capable of high-volume molecular testing [15].

The stand-alone extraction instrument is by far the most popular choice of many small- to medium-sized molecular diagnostics laboratories found in academic centers and hospitals. Instruments like the EZ1 Advanced XL (Qiagen) are capable of extracting between 1 and 16 samples in a single 20- to 40-min run. While these instruments are considerably cheaper than larger automated extraction systems, they do not offer the same extraction capacity. However, one of the primary benefits that these smaller stand-alone systems offer is the flexibility they provide to a diagnostics laboratory with moderate volumes of diverse sample types. These instruments are designed as "closed" systems, meaning they require manufacturer-provided reagents to perform preprogramed extraction protocols. Switching between sample types and extraction methods typically requires the insertion of a new reagent kit and the running of a new, preprogrammed protocol. This means that by training a technologist to perform extractions on this instrument for one sample type allows for minimal additional training in order to produce high-quality and consistent results for a variety of additional sample types for both DNA and RNA extractions.

The high-capacity automated extraction instruments come in both "closed" and "open" formats. As with smaller instrumentation, the closed format systems, such as the COBAS *AmpliPrep* (Roche), use dedicated reagent kits and protocols to perform extraction for specific sample types [16,17]. Rather than 16 samples per run, these instruments are capable of extracting up to 96 samples in a single run and typically can process over 384 samples in a working day. While these systems are technologically more advanced, they remain user-friendly with simplified user interfaces allowing for easier technologist training and regulatory compliance. These types of closed systems are common in laboratories with only a few higher volume tests, which require dedicated equipment to meet daily testing requirements. Many of these closed systems are used to extract nucleic acids as part of a testing platform for screening and diagnosis of infectious disease.

The high-capacity "open" format instruments provide the greatest flexibility in high-volume testing. Systems like the BioMek FX^P (Beckman Coulter) are designed as a completely customizable automated workstation. These systems at minimum consist of an advanced robotic liquid handler, but also typically contain heating elements, magnets, shakers, etc., to extract from 8 to 384 samples in a single run. The customizable nature of these instruments results in the capability to adapt a wide variety of extraction methods and kits to suit the laboratory's needs. However, one significant drawback to open extraction systems is the requirement for a high level of instrument and workflow expertise. Such expertise is required to develop protocols in order for the utility of these systems to be harnessed. These limitations result in open systems being better suited for automating either complex molecular workflows, such as next-generation sequencing (NGS) nucleic acid extraction and library preparation, or novel techniques or methods where no commercially available high-throughput technology is available.

AUTOMATION IN INFECTIOUS DISEASE MOLECULAR DIAGNOSTICS

Automation of the molecular diagnostics laboratory has had a profound effect on the diagnosis of infectious disease. Infectious disease diagnostics once solely relied upon the use of highly manual multiday microbial culture [18] and viral plaque assays [19] to diagnose and monitor infectious disease. While these methods are still considered gold standard for many indications [20], there is a growing number of PCR-based applications being employed in molecular diagnostics laboratories for the rapid detection of infectious agents [17,21–23].

The specific molecular techniques applied largely depend on the question being asked. For example, real-time PCR-based approaches may be adequate in diagnosis of species-specific infections, while postamplification sequencing may be required to identify and distinguish among a variety of coinfections in a sample. While many of these methods have benefited from various advancements in automation, those most commonly implemented into the molecular diagnostic laboratory are integrated sample-to-result diagnostic platforms that allow for the input of raw biological material and result in a final diagnostic report (Table 4.1). While these systems range in footprint and sample capacity, all of these automated systems allow for nucleic acid extraction and PCR detection of viral or bacterial genomic targets with minimal to no technologist intervention.

TABLE 4.1 List of Common Fully Automated Devices and FDA-Approved Assay for Infectious Disease Molecular Testing

Platform	Manufacturer	FDA-approved assays	Sample throughput
Cobas Liat	Roche	Flu, Strep A	Only one sample/instrument
FilmArray	BioFire	Respiratory Panel, GI Panel, Meningitis Panel, Blood Culture ID Panel	Only one sample/instrument
Verigene SP	NanoSphere	Respiratory Panel, SA/SE, Enteric Pathogen Panel, GC-GN, GC-GP	Only one sample/instrument (expandable)
ESensor XT-8	GenMarkDx	Respiratory Panel	Up to eight samples/instrument (expandable)
Aries	Luminex	HSV 1 & 2	Up to 12 samples/run
BDMax	Becton Dickson	GBS, MRSA, C. diff, Enteric Bacterial Panel, Enteric Parasite Panel	Up to 24 samples/run
GeneXpert/ GeneXpert Infinity	Cepheid	Flu/RSV/EV, GBS, MRSA, MTB/RIF, C. diff, Norovirus, CT/NG, TV, GBS	Up to 80 samples/run
m2000 RealTime	Abbott Molecular	HBV, HCV, HIV, HSV 1 & 2, Flu, C. diff	96 samples/run (128 in 8 h)
COBAS Ampliprep	Roche	HBV, HCV, CMV, HIV-1	48 samples/run (168 in 8 h)
Panther	Hologic/Gen-Probe	CT/NG, TV, HPV	120 samples/run (275 in 8 h)
Cobas 4800	Roche	CT/NG, HPV	96 samples/run (288 in 8 h)
Cobas 6800/8800	Roche	HBV, HCV, HIV	96 samples/run (384/960 in 8 h)
Tigris DTS	Hologic Gen-Probe	CT/NG, TV, HPV	182 samples/run (450 in 8 h)

The system adopted by a laboratory largely depends on the volumes of samples being tested and the time required to obtain a result. For low to moderate volume testing, there are single sample cartridge or reagent strip based automated systems, such as the GeneXpert (Cepheid), BDMax (BD Molecular Diagnostics), ESensor XT8 (GenmarkDx), and the Aries (Luminex) systems that allow for rapid detection of a variety of different infectious agents. These automated instruments can accommodate from 1 to 80 samples to be tested simultaneously. Each biological sample is dispensed directly into a single assay cartridge or reagent strip, which contains all the necessary reagents to extract nucleic acids and amplify the chosen genomic target(s). Currently, these platforms have over a dozen FDA-approved in vitro diagnostics (IVD) assays for the diagnosis of infectious disease including: multi-drug resistant TB [24–26], influenza (types A and B) [27]; health care related infections including: methicillin-resistant *Staphylococcus aureus* [28,29], *Clostridium difficile* [30,31]; and sexually transmitted infections including *Chlamydia trachomatis*, *Neisseria gonorrhoeae*, and *Trichomonas vaginalis* [32,33]. These automated cartridge and strip-based systems can typically be completed with as little as a few minutes of technical hands-on time per sample allowing for diagnosis within 60–90 min for low-volume testing.

While these smaller automated systems are excellent for low-volume testing, many molecular diagnostic laboratories are becoming inundated with samples requiring screening and/or monitoring of viral loads in patients with human immunodeficiency virus (HIV), hepatitis B virus (HBV), hepatitis C virus (HCV), cytomegalovirus (CMV), human papilloma virus (HPV), and others. This has resulted in the need for adequate high-volume testing capabilities that will accommodate this increasing demand with little to no change in the molecular technologist workload. A prime example of this increasing demand is found in recent reports identifying the clinical utility of high-risk HPV testing as the primary screening method for cervical cancer. Current screening recommendations suggest the use of cotesting (cytology and HPV screening) in women between the ages of 30 and 65 [34]. However, in the spring of 2014, the FDA approved the Cobas HPV test for primary screening of cervical cancer [35], allowing for HPV-positive samples to be followed up with cytology screening [36]. While this information has yet to be included in screening recommendation, there is support for expanding HPV molecular testing potentially resulting in increased HPV testing volumes [36]. Often the only way to achieve the turnaround times required for this testing is to incorporate fully automated high-volume molecular testing platforms into the diagnostic workflow.

There are a number of commercially available platforms for high-volume infectious disease diagnostics including the Panther and Tigris DTS systems (Hologic), the Cobas 4800/6800/8800 systems (Roche), the COBAS *AmpliPrep*/TaqMan instrument (Roche), and the m2000 RealTime System (Abbott). The manufacturers of these platforms have developed assay kits for virus detection and quantification of viral loads for a number of different infectious diseases. Many of these kits have achieved FDA-approved IVD status, which only requires a verification for adopting a highly automated system into the molecular laboratory.

While each of these platforms varies in component technology, assay chemistry, and sample capacity, all are designed to allow for barcoded and/or radio frequency sample and reagent tracking, nucleic acid extraction, PCR assay setup, and amplicon detection on a massive scale. Some fully integrated platforms allow for complete process automation, including direct sample tube input, eliminating the laborious process of transferring samples into platform-specific disposables. This level of automation affords the molecular diagnostics laboratory the ability to process up to 960 tests in an 8-h shift, and over 3000 tests in a 24-h period, without having to drastically increase the amount of technologist time required to achieve testing requirements.

Early versions of many of these platforms limited versatility by requiring all samples be run for a single test, and assays had to be completed before subsequent samples could be loaded. More recent versions of these platforms can now accommodate running up to three assays on the platform at one time, as well as allow for loading of additional samples while other samples are being processed. These further advancements allow for an even greater flexibility in high-volume infectious disease testing.

Testing that once took multiple days and required significant technical expertise can now be accomplished quickly with minimal technical training. The described automation has resulted in rapid diagnosis of infectious disease, allowing for the potential reduction of disease transmission and outbreak prevention [37] and better critical care patient management [38,39]. As the clinical utility of rapid molecular infectious disease testing grows, increased automation will allow the molecular diagnostics laboratories to continue to meet the growing demand.

AUTOMATION IN GENETICS AND MOLECULAR ONCOLOGY DIAGNOSTICS

In the fields of genetics and molecular oncology, there is an expanding need for automated diagnostics. The increased utility of noninvasive prenatal testing from circulating fetal DNA [40,41], expanded carrier screening for conditions such as cystic fibrosis [42], the characterization of how a patient's genotype affects drug response (pharmocogenetics) [43], and the use of large multigene panels for genetic diagnosis [44] are largely driving this expansion. In addition, the routine care of today's cancer patient has been transformed by molecular diagnostics. It is now commonplace for the molecular diagnostics laboratory to provide detailed molecular characterization of a patient's cancer. Successful identification of these molecular alterations has changed the treatment paradigm for a number of different cancer types. To determine the most appropriate treatment for lung adenocarcinomas, today's oncologist requires detailed information on the presence or absence of variants in the epidermal growth factor receptor (EGFR), for example [45]. These variants are known to either sensitize (EGFR exon 19 deletions) or make a tumor cell resistant (EGFR p. T790M) to treatment with common first-line EGFR tyrosine kinase inhibitors [46]. With the FDA approval of additional targeted therapeutics for personalized cancer treatment and expanded use of genetic testing, the diagnostic laboratory must be prepared to meet this increasing demand.

Advancements in postamplification detection of inherited single nucleotide polymorphisms (SNPs), acquired single nucleotide variants (SNVs), small insertion/deletions (INDELS), large copy number variants (CNVs), and other chromosomal aberrations have allowed the molecular diagnostics laboratory to vastly expand the testing menu while still being able to offer high-quality, timely results.

One of the first PCR-based molecular diagnostic technologies used restriction enzymes to digest a PCR-amplified product in order to determine the presence of a pathogenic variant [1]. This early version of the restriction fragment length polymorphism (RFLP) assay required the use of gel electrophoresis, in which the digested amplicon was electrophoresed through an agarose gel using an electric current to separate fragments by size. The size of these restriction fragments could then be compared to samples known to be positive for the pathogenic variant and the appropriate diagnosis could be made. This groundbreaking technique proved successful and has been repeated for genotyping of variants for numerous diagnostic applications. However, the use of RFLP for detection of variants has a number of limitations, including (1) limited sequence recognition by restriction enzymes, (2) the ability to detect only one variant per reaction, and (3) the inability to quantitatively determine the amount of variation in a sample. In part, these limitations have led to the advancement of numerous additional molecular techniques including refractory mutation system

PCR (ARMS-PCR) [47], 5'-nuclease "TaqMan" PCR [48], primer extension assays [49], oligonucleotide ligation assays [50], chromosomal microarrays [51], and various sequencing techniques [52].

Real-Time Quantitative PCR

At their core, all of these techniques continue to utilize PCR amplification. However, advancements in automated amplification detection has allowed for expanded applications of this methodology. The advent of real-time quantitative PCR (qPCR) has provided the ability to detect the relative quantity of specific PCR products during the PCR reaction through the use of fluorometry [53]. Molecular methods employing qPCR vary in design, but by detecting various flourophore-labeled PCR products or probes, these techniques are capable of simultaneous detection of up to seven targets in a single reaction. In reality, it can be a challenge to achieve this level of multiplexing due to oligonucleotide primer competition and cross-reactivity [54], but it is not uncommon to routinely detect two to three targets per reaction. This reduces the variation introduced by manually pipetting one sample into numerous separate reactions, allowing for more precise target quantification and normalization across the assay. In addition, the qPCR instruments found in today's molecular diagnostics laboratories, including the 7500 Fast (Thermo Fisher) or CFX384 (Bio-Rad), are capable of simultaneously testing from 96 to 384 samples in a single run.

qPCR methods employing the use of fluorophore-labeled oligonucleotide probes designed to pair only with a single variant, allow for the rapid detection of multiple variants in a single PCR run. This increases the molecular diagnostics laboratory's ability to detect all kinds of SNVs, including those used to determine which drug treatments are best suited for a particular patient (pharmacogenetics).

Digital Droplet PCR

qPCR instruments have been commercially available since 1996, and although they have become more refined over time, they still rely on the basic principles of traditional PCR. In 2006, a new form of PCR, termed digital droplet PCR (ddPCR) [55,56], was made commercially available. In brief, ddPCR uses the same chemical components to amplify nucleic acid targets, however, instead of relying on exponential amplification to calculate the amount of product at the end of each PCR cycle, ddPCR uses various technologies to isolate each individual target reaction, resulting in thousands of discrete measurements, allowing for more accurate and quantitative analysis of initial target amounts [57]. A ddPCR instrument, such as the QX200 (Bio-Rad), uses microfluidics to create over 20,000 droplets each containing the components for a single PCR reaction. These droplet PCR reactions are run on a traditional thermal cycler, and postamplification fluorescence detection is performed on individual droplets to determine the number of droplets positive for the target being amplified. In the molecular diagnostics laboratory, automated ddPCR allows for the accurate detection of CNVs for diagnosis of genetic disease, identification of low frequencies variants in heterogeneous tumor samples, as well as allows for a higher level of multiplexing due to the elimination of competing primers within a reaction. By automating the PCR amplification and target detection through qPCR and ddPCR, there is a reduction in technologist workload per sample and an increase in sensitivity and accuracy of PCR-based diagnostics.

Capillary Electrophoresis

Many clinical PCR-based applications still require postamplification detection of end point PCR products. With the introduction of the capillary array electrophoresis instrument into the molecular diagnostics laboratory, the days of having to rely on gels for determining fragment size have quickly come to an end. Capillary electrophoresis instruments, like the AppliedBiosystems 3500 series (Thermo Fisher), use the same principles as traditionally gel electrophoresis. An electrical current is applied to a matrix containing the amplified PCR fragments. The fragments are separated by size as the larger fragments move more slowly through the matrix. In capillary electrophoresis, each sample is injected into a single reusable capillary containing a replenishable gel-like matrix eliminating the need to use disposable gel cassettes. These instruments commonly consist of arrays containing between 8 and 16 capillaries, allowing for up to 128 samples to be loaded on a run, resulting in the capability of analyzing a large number of samples with no additional hands-on time. However, the most valuable addition to capillary electrophoresis is the automated detection and analysis of PCR fragments as they flow through the capillary. These instruments are equipped with a fluorescence detector capable of identifying fluorescently labeled fragments and converting that signal into a read out of relative fluorescence resulting in a fragment peak. When run alongside a fragment ladder, the fragment length of the product can be calculated. This allows for the accurate determination of fragments differing by as little as a single base pair. With the addition of multiple fluorophores, similarly sized fragments can be multiplexed into one reaction.

In the molecular diagnostics laboratory, capillary electrophoresis is commonly used for a variety of diagnostic assays, including the detection of microsatellites to monitor bone marrow engraftment after transplantation [12] and the diagnosis of expansion repeat diseases such as Fragile X syndrome [58] and Huntington's disease [59]. However, the capillary electrophoresis instrument is also responsible for automating what is still considered the gold standard in molecular diagnostics, the sequencing of genetic material [60]. Prior to the adoption of these instruments in the laboratory, Sanger sequencing required the running of large format gels to identify radioactively labeled ddNTPs present at each nucleotide position [61]. By modifying these ddNTPs to include various fluorescent labels, the detector on a capillary electrophoresis instrument can distinguish between each ddNTPs and produce an easy-to-read sequencing trace [62]. It once took technologists hours to painstakingly read and reread a single sequencing reaction but now this process is automated, allowing for numerous sequencing reactions to be performed concurrently. It is this level of automation that made the sequencing of the human genome possible [63,64].

These sequencing methods are now used in the molecular diagnostics laboratory to identify pathogenic variants in genetic disease. Many genetic diseases are the result of variants found throughout the coding region of a gene. To make these diagnoses, every base pair within an exon must be examined, which makes the use of a qPCR genotyping assay unfeasible. This same technology has been adopted to perform primer extension assays, commonly referred to as SNaPshot (Thermo Fisher), to identify variants present one base pair upstream of a designed primer. This technique is used to identify actionable hotspot mutations, defined as common cancer mutations which have identified therapeutics for a given tumor type. Primer extension assays are considered more sensitive than qPCR methods for the detection of low-level variation often present in a heterogeneous population of tumor cells.

Chromosomal Microarray Analysis

The analysis of chromosomes is used to identify large chromosomal deletion and amplifications across a patient's entire genome. These chromosomal aberrations alter the normal copy number of genes within the affected regions. In the molecular diagnostics laboratory, detection of these CNVs can result in diagnosis of various genetic disorders [65] or alter the treatment selection for certain tumor types [66]. This cytogenetic testing once solely relied on the use of karyotyping, a process requiring the multiday culture of live cells and microscopic analysis of chromosomal banding patterns, and fluorescence in situ hybridization (FISH), a technique by which cells are treated with very large fluorescently labeled probes, which pair with target chromosomes and allow for microscopic detection of fluorescent spots. These highly manual techniques are often very challenging and limited by the ability of cells to grow in culture and the availability of FISH probes for a given chromosome target. The introduction of the chromosomal microarray analysis allowed for automated molecular analysis of chromosomes resulting in a virtual karyotype with higher sensitivity and increased diagnostic yield [67].

The most common chromosomal microarrays used in molecular diagnostics laboratories include the comparative genomic hybridization (CGH) array, the SNP array, or a combination of the two (CGH + SNP). These arrays consist of hundreds of thousands to upward of a few million oligonucleotide probes bound to the surface of a solid array chip. While these array technologies vary, when a patient's DNA binds to a probe with matching sequence, a fluorescent signal can be detected. Optical array scanners detect signals from millions of individual probes on a single chip for each patient. Algorithms within propriety software packages then automatically interpret that information to determine regions of the chromosome where higher (amplification) or lower (deletion) signal is observed when compared to a normal reference. The more probes present on the chip the higher the resolution, as there is less genomic space between each probe. When these arrays contain probes that cover common SNPs, additional chromosomal information can be gained, including copy-neutral loss of heterozygosity resulting in the diagnosis of uniparental disomy disorders such as Prader-Willi and Angelman syndromes [68].

The automation of molecular chromosome analysis has resulted in higher resolution detection of CNVs. Manual cytogenetic methods are capable of detecting CNVs down to 5–10 million base pairs while commercially available microarrays can accurately detect variants as small as 10,000–20,000 base pairs. This increased resolution has resulted in finer mapping of CNV breakpoints, resulting in the discovery and characterization of a number of new microdeletion disorders [69].

Next-Generation Sequencing

The most recent technological advancement to reach the molecular diagnostics laboratory is the use of NGS technologies to diagnose genetic disease and identify large numbers of actionable variants in human cancer. The term *next-generation sequencing* applies to all sequencing technologies that allow for massively parallel or deep sequencing of oligonucleotide strands. This means that instead of being able to sequence one fragment of DNA per reaction, as is the case with Sanger sequencing, the laboratory can now sequence tens of thousands of fragments thousands of times during a single sequencing run. With this new technology, the entire human genome can be sequenced and analyzed within a matter of days for as little as $1000 (Veritas Genetics, Press Release—September 29, 2015).

Historically, it takes approximately 10 years of clinical research before new technologies are adopted for use in clinical diagnosis. However, due to its disruptive nature, the adoption of NGS into the molecular diagnostics laboratory has been expedited. The potential exists for NGS to allow for test consolidation, permitting one assay to be run for a large number of patients for a variety of different indications. For example, instead of running dozens of Sanger sequencing assays to identify breast cancer associated variants in *BRCA1*, we can now examine all variations across a number of different genes with known breast cancer associations. With that same assay, we can also examine variants in *CFTR* and determine the risk a child has of inheriting cystic fibrosis. In yet another patient, a pediatric geneticist is looking to identify the cause of intellectual disability by determining the presence of pathogenic CNVs. While currently there remains a number of technical, financial, and ethical hurdles to be cleared the diagnostic potential of NGS is clear.

Currently, the clinical implication for much of the variation observed across the genome remains unknown [70]. Therefore, the majority of molecular diagnostics laboratories employing NGS methods are using either targeted sequencing panels or whole exome sequencing. These directed methods of analysis help narrow the scope of the diagnostic assay allowing for easier adoption into a clinical setting. Targeted sequencing panels contain anywhere from a few genes to hundreds of gene targets. Predesigned gene panels for a variety of indications are commercially available from companies such as Qiagen, Illumina, and Thermo Fisher. If predesigned panels are not available, custom panels can be created either in-house or through commercial services. While custom panels typically offer the most flexibility, commercially available panels have typically undergone rigorous performance verification prior to release, potentially allowing for less in-house development.

There are a number of sequencing platforms currently being used in the research setting. However, the most common NGS platforms found in the molecular diagnostics laboratory are the IonTorrent PGM (Thermo Fisher), the Illumina MiSeq, the Illumina NextSeq 500, and the Illumina HiSeq sequencers. The

(A)   (B)   (C)

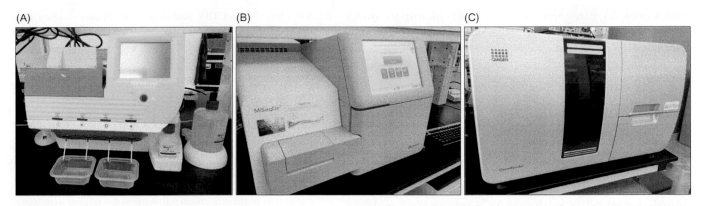

FIGURE 4.3 The desktop DNA sequencer. Desktop sequencers are a favorite of many molecular diagnostic laboratories due to the small footprint, assay adaptability, and sequencing capacity. These sequencers have the capacity to screen and diagnosis genetic disease through gene specific or gene panel tests, identify low-level variation in solid tumors through "hotspot" gene panel testing, and sequence the clinical exome. The most common instruments found in today's molecular diagnostics laboratory's are (A) the IonTorrent PGM (Thermo Fisher), (B) the MiSeqDx (Illumina) which is currently the only FDA-approved sequencer, and (C) the GeneReader NGS system (Qiagen) which is the only fully automated desktop sequencer released in late 2015.

platform selected largely depends on the volume of testing, the size of the gene panel, and the suspected frequency of variation being identified. These platforms also vary in their sequencing chemistries and sequencing capacity.

The smaller desktop sequencers (Fig. 4.3), including the IonTorrent PGM and MiSeq, have the lowest sequencing capacity of the platforms in the clinical laboratory. These platforms are therefore best suited for smaller targeted sequencing panels. This reduced sequencing capacity makes appropriate target selection imperative. Cancers are known to consist of a heterogeneous population of normal and neoplastic cells. Therefore, the cancer-causing variants may be underrepresented, resulting in only a small number of gene targets containing the variant. In order to ensure that these low-level variants are detected, each gene region targeted must be sequenced 500–1000 times. This depth consistently allows variants with allelic frequencies as low as 5% to be identified [71]. The only way to get this kind of coverage on a desktop sequencer is to limit the number of targets to be sequenced. Cancer panels are typically designed to cover only hotspot regions of genes known to be mutated in various cancers.

These desktop sequencers are also capable of sequencing full genes for the diagnosis of genetic disease. Unlike somatic variants, inherited variation is typically present at allelic frequency of at least 50%. This means that each targeted base only needs to be sequenced 20 times, allowing for much higher gene content per panel. These hereditary panels can range in size depending on indication, but larger panels typically range from dozens to hundreds of genes.

Currently, the desktop sequencers on the market are not capable of sequencing the entire human exome, the protein-coding region of the genome, at a minimum depth of $20 \times$. The exome contains upward of 20,000 genes, so in order to sequence that many targets sequencing capacity must be increased. The Illumina HiSeq and the newly introduced Illumina NextSeq provide enough capacity to sequence the entire exome of a number of patients in a single run. However, some companies are offering panels for what has been termed the *clinical exome*. These targeted panels contain 6000–7000 genes identified to have high clinical significance. Desktop sequencers like the Illumina MiSeq has the capacity to run one of these modified exome panels on a single patient per run, limiting the throughput of such testing.

The NGS wet-bench and analytical pipeline varies depending on sequencing application. Preparing libraries for sequencing typically takes multiple days and requires highly trained and skilled technologists to ensure high-quality sequencing data. Recent efforts have been made to automate these largely manual processes through the use of both platform-specific or open automation systems. The IonTorrent Chef system (Thermo Fisher) and the NeoPrep system (Illumina) are designed to automate platform-specific libraries. While the larger, open systems as previously described for infection disease automation, such as the BioMek FX^P (Beckman Coulter), contain robot liquid handlers that can be adapted to a variety of different library preparation protocols.

The analytical pipelines used to align sequencing reads, identify variants present, and determine the clinical significance are still largely laboratory specific. While portions of these pipelines can be performed directly on the sequencing platform, quite often the need for additional bioinformatics analysis exists. Similarly, variant interpretation largely remains a manual process, requiring the need to follow issued guidelines to categorize each variant identified into benign, variant of unknown significance, or pathogenic classifications. In cancer, the growing number of targeted therapies in clinical trials adds an additional wrinkle to this interpretation. This is an area ripe for innovation in automation. Recently, the Qiagen Clinical Insight and CLC Clinical Workbench software applications were released, that combine variant analysis and clinical interpretation pipelines, some of which reside within the cloud (Press Release, November 2, 2015). The standardization and implementation of integrated NGS analytical workflows remains the biggest hurdle for introducing clinical NGS into today's molecular diagnostics laboratory.

Clinical Laboratory Improvement Amendments (CLIA) and Centers for Medicare and Medicaid Services (CMS) regulation and guidelines require thorough and rigorous validation processes prior to offering any laboratory-developed procedure including all NGS diagnostic services. These large-scale validations are costly and therefore limit the flexibility of a molecular diagnostics laboratory to make improvements to validated sequencing assays. On the other hand, the MiSeqDX and Cystic Fibrosis 139-Variant and Diagnostic Sequencing assays are currently the only FDA-approved IVDs for clinical sequencing. However, it is likely that more will be approved in the coming years as technologies and analytical pipelines improve. With these improvements, clinical NGS is sure to revolutionize tomorrow's molecular diagnostics laboratory.

CONCLUSIONS

As technological developments continue to improve on existing instrumentation, there is a growing need to decentralize molecular diagnostic testing to near-patient or at-home testing. To this end, we anticipate many molecular diagnostics platforms to continue to become smaller with single-use cartridge type assays that can be run by anyone, including the patient. The improvement in throughput and cost of NGS will be a major driver for the routine assessment of the clinical exome, whole exome, and even whole genome by many molecular diagnostics laboratories. Improvement of analytical pipelines, through the cloud or by other mechanisms, will expedite the interpretation of data and make the new age of molecular diagnostic testing seamless.

References

[1] Saiki RK, Scharf S, Faloona F, et al. Enzymatic amplification of beta-globin genomic sequences and restriction site analysis for diagnosis of sickle cell anemia. Science 1985;230:1350–4.

[2] Saiki RK, Gelfand DH, Stoffel S, et al. Primer-directed enzymatic amplification of DNA with a thermostable DNA polymerase. Science 1988;239:487–91.

[3] Vosberg HP. The polymerase chain reaction: an improved method for the analysis of nucleic acids. Hum Genet 1989;83:1–15.

[4] Thatcher SA. DNA/RNA preparation for molecular detection. Clin Chem 2015;61:89–99.

[5] Riemann K, Adamzik M, Frauenrath S, et al. Comparison of manual and automated nucleic acid extraction from whole-blood samples. J Clin Lab Anal 2007;21:244–8.

[6] Dundas N, Leos NK, Mitui M, Revell P, Rogers BB. Comparison of automated nucleic acid extraction methods with manual extraction. J Mol Diagn 2008;10:311–16.

[7] Tang YW, Sefers SE, Li H, Kohn DJ, Procop GW. Comparative evaluation of three commercial systems for nucleic acid extraction from urine specimens. J Clin Microbiol 2005;43:4830–3.

[8] Claassen S, du Toit E, Kaba M, et al. A comparison of the efficiency of five different commercial DNA extraction kits for extraction of DNA from faecal samples. J Microbiol Methods 2013;94:103–10.

[9] Goode MR, Cheong SY, Li N, Ray WC, Bartlett CW. Collection and extraction of saliva DNA for next generation sequencing. J Vis Exp 2014;. Available from: http://dx.doi.org/10.3791/51697.

[10] Saab YB, Kabbara W, Chbib C, Gard PR. Buccal cell DNA extraction: yield, purity, and cost: a comparison of two methods. Genet Test 2007;11:413–16.

[11] Hajia M, Rahbar M, Fallah F, Safadel N. Detection of *Bordetella pertussis* in infants suspected to have whooping cough. Open Respir Med J 2012;6:34–6.

[12] Odriozola A, Riancho JA, Colorado M, Zarrabeitia MT. Evaluation of the sensitivity of two recently developed STR multiplexes for the analysis of chimerism after haematopoietic stem cell transplantation. Int J Immunogenet 2013;40:88–92.

[13] Spencer DH, Sehn JK, Abel HJ, et al. Comparison of clinical targeted next-generation sequence data from formalin-fixed and fresh-frozen tissue specimens. J Mol Diagn 2013;15:623–33.

[14] Davis CP, King JL, Budowle B, Eisenberg AJ, Turnbough MA. Extraction platform evaluations: a comparison of AutoMate Express, EZ1(R) Advanced XL, and Maxwell(R) 16 Bench-top DNA extraction systems. Leg Med 2012;14:36–9.

[15] Micalessi IM, Boulet GA, Bogers JJ, Benoy IH, Depuydt CE. High-throughput detection, genotyping and quantification of the human papillomavirus using real-time PCR. Clin Chem Lab Med 2012;50:655–61.

[16] Pyne MT, Mallory M, Hillyard DR HCV. RNA measurement in samples with diverse genotypes using versions 1 and 2 of the Roche COBAS(R) AmpliPrep/COBAS(R) TaqMan(R) HCV test. J Clin Virol 2015;65:54–7.

[17] Margariti A, Chatzidimitriou D, Metallidis S, et al. Comparing Abbott m2000 RealTime HIV test and Roche COBAS Ampliprep/COBAS Taqman HIV test, v2.0 in treated HIV-1 B and non-B subjects with low viraemia. J Med Virol 2015;88:724–7.

[18] Lin D, Lehmann PF, Hamory BH, et al. Comparison of three typing methods for clinical and environmental isolates of *Aspergillus fumigatus*. J Clin Microbiol 1995;33:1596–601.

[19] Boeckh M, Boivin G. Quantitation of cytomegalovirus: methodologic aspects and clinical applications. Clin Microbiol Rev 1998;11:533–54.

[20] Bloomfield MG, Balm MN, Blackmore TK. Molecular testing for viral and bacterial enteric pathogens: gold standard for viruses, but don't let culture go just yet? Pathology 2015;47:227–33.

[21] Derache A, Wallis CL, Vardhanabhuti S, et al. Phenotype, genotype, and drug resistance in subtype C HIV-1 infection. J Infect Dis 2016;213:250–6.

[22] Somerville LK, Ratnamohan VM, Dwyer DE, Kok J. Molecular diagnosis of respiratory viruses. Pathology 2015;47:243—9.

[23] Sloots TP, Nissen MD, Ginn AN, Iredell JR. Rapid identification of pathogens using molecular techniques. Pathology 2015;47: 191—8.

[24] Division of Microbiology Devices, Office of In Vitro Diagnostics and Radiological Health, Center for Devices and Radiological Health, Food and Drug Administration; Centers for Disease Control and Prevention (CDC). Revised device labeling for the Cepheid Xpert MTB/RIF assay for detecting *Mycobacterium tuberculosis*. Morb Mortal Wkly Rep 2015;64:193.

[25] Shinnick TM, Starks AM, Alexander HL, Castro KG. Evaluation of the Cepheid Xpert MTB/RIF assay. Exp Rev Mol Diagn 2015;15:9—22.

[26] Lepainteur M, Delattre S, Cozza S, et al. Comparative evaluation of two PCR-based methods for detection of methicillin-resistant *Staphylococcus aureus* (MRSA): Xpert MRSA Gen 3 and BD-Max MRSA XT. J Clin Microbiol 2015;53:1955—8.

[27] Sambol AR, Iwen PC, Pieretti M, et al. Validation of the Cepheid Xpert Flu A real time RT-PCR detection panel for emergency use authorization. J Clin Virol 2010;48:234—8.

[28] Coombs GW, Morgan JP, Tan HL, Pearson JC, Robinson JO. Evaluation of the BD GeneOhm MRSA ACP Assay and the Cepheid GeneXpert MRSA Assay to detect genetically diverse CA-MRSA. Pathology 2013;45:713—15.

[29] Rossney AS, Herra CM, Brennan GI, Morgan PM, O'Connell B. Evaluation of the Xpert methicillin-resistant *Staphylococcus aureus* (MRSA) assay using the GeneXpert real-time PCR platform for rapid detection of MRSA from screening specimens. J Clin Microbiol 2008;46:3285—90.

[30] Yoo J, Lee H, Park KG, et al. Evaluation of 3 automated real-time PCR (Xpert *C. difficile* assay, BD MAX Cdiff, and IMDx *C. difficile* for Abbott m2000 assay) for detecting *Clostridium difficile* toxin gene compared to toxigenic culture in stool specimens. Diagn Microbiol Infect Dis 2015;83:7—10.

[31] Verhoeven PO, Carricajo A, Pillet S, et al. Evaluation of the new CE-IVD marked BD MAX Cdiff Assay for the detection of toxigenic *Clostridium difficile* harboring the tcdB gene from clinical stool samples. J Microbiol Methods 2013;94:58—60.

[32] Badman SG, Causer LM, Guy R, et al. A preliminary evaluation of a new GeneXpert (Gx) molecular point-of-care test for the detection of *Trichomonas vaginalis*. Sex Transm Infect 2015. Available from: http://dx.doi.org/10.1136/sextrans-2015-052384.

[33] Gaydos CA. Review of use of a new rapid real-time PCR, the Cepheid GeneXpert(R) (Xpert) CT/NG assay, for *Chlamydia trachomatis* and *Neisseria gonorrhoeae*: results for patients while in a clinical setting. Exp Rev Mol Diagn 2014;14:135—7.

[34] Smith RA, Manassaram-Baptiste D, Brooks D, et al. Cancer screening in the United States, 2015: a review of current American Cancer Society guidelines and current issues in cancer screening. CA Cancer J Clin 2015;65:30—54.

[35] Food and Drug Administration. Cobas HPV Test—P10020/ S008. 2014 [cited 2015 September 1st, 2015]. Available from: http://www.fda.gov/MedicalDevices/ProductsandMedical Procedures/DeviceApprovalsandClearances/Recently-Approved Devices/ucm395694.htm.

[36] Huh WK, Ault KA, Chelmow D, et al. Use of primary high-risk human papillomavirus testing for cervical cancer screening: interim clinical guidance. Obstet Gynecol 2015;125:330—7.

[37] Kumar S, Henrickson KJ. Update on influenza diagnostics: lessons from the novel H1N1 influenza A pandemic. Clin Microbiol Rev 2012;25:344—61.

[38] Bissonnette L, Bergeron MG. Infectious disease management through point-of-care personalized medicine molecular diagnostic technologies. J Pers Med 2012;2:50—70.

[39] Krishna NK, Cunnion KM. Role of molecular diagnostics in the management of infectious disease emergencies. Med Clin North Am 2012;96:1067—78.

[40] Cuckle H, Benn P, Pergament E. Cell-free DNA screening for fetal aneuploidy as a clinical service. Clin Biochem 2015;48: 932—41.

[41] Brady P, Brison N, Van Den Bogaert K, et al. Clinical implementation of NIPT—technical and biological challenges. Clin Genet 2015;. Available from: http://dx.doi.org/10.1111/cge.12598.

[42] Brennan ML, Schrijver I. Cystic fibrosis: a review of associated phenotypes, use of molecular diagnostic approaches, genetic characteristics, progress, and dilemmas. J Mol Diagn 2016;18:3—14.

[43] Yip VL, Hawcutt DB, Pirmohamed M. Pharmacogenetic markers of drug efficacy and toxicity. Clin Pharmacol Ther 2015;98:61—70.

[44] Hoffman JD, Park JJ, Schreiber-Agus N, et al. The Ashkenazi Jewish carrier screening panel: evolution, status quo, and disparities. Prenat Diagn 2014;34:1161—7.

[45] Travis WD, Brambilla E, Noguchi M, et al. International Association for the Study of Lung Cancer/American Thoracic Society/ European Respiratory Society international multidisciplinary classification of lung adenocarcinoma. J Thorac Oncol 2011;6:244—85.

[46] Ellis PM, Coakley N, Feld R, Kuruvilla S, Ung YC. Use of the epidermal growth factor receptor inhibitors gefitinib, erlotinib, afatinib, dacomitinib, and icotinib in the treatment of non-small-cell lung cancer: a systematic review. Curr Oncol 2015;22:e183—215.

[47] Nanfack AJ, Agyingi L, Noubiap JJ, et al. Use of amplification refractory mutation system PCR assay as a simple and effective tool to detect HIV-1 drug resistance mutations. J Clin Microbiol 2015;53:1662—71.

[48] Kho SL, Chua KH, George E, Tan JA. High throughput molecular confirmation of beta-thalassemia mutations using novel TaqMan probes. Sensors 2013;13:2506—14.

[49] Abramson VG, Cooper Lloyd M, Ballinger T, et al. Characterization of breast cancers with PI3K mutations in an academic practice setting using SNaPshot profiling. Breast Cancer Res Treat 2014;145:389—99.

[50] Macdonald SJ. Genotyping by oligonucleotide ligation assay (OLA). Cold Spring Harb Protoc 2007;2007 pdb.prot4843.

[51] Batzir NA, Shohat M, Maya I. Chromosomal microarray analysis (CMA): a clinical diagnostic tool in the prenatal and postnatal settings. Pediatr Endocrinol Rev 2015;13:448—54.

[52] Yang Y, Muzny DM, Reid JG, et al. Clinical whole-exome sequencing for the diagnosis of Mendelian disorders. N Engl J Med 2013;369:1502—11.

[53] Heid CA, Stevens J, Livak KJ, Williams PM. Real time quantitative PCR. Genome Res 1996;6:986—94.

[54] Henegariu O, Heerema NA, Dlouhy SR, Vance GH, Vogt PH. Multiplex PCR: critical parameters and step-by-step protocol. BioTechniques 1997;23:504—11.

[55] Ottesen EA, Hong JW, Quake SR, Leadbetter JR. Microfluidic digital PCR enables multigene analysis of individual environmental bacteria. Science 2006;314:1464—7.

[56] Warren L, Bryder D, Weissman IL, Quake SR. Transcription factor profiling in individual hematopoietic progenitors by digital RT-PCR. Proc Natl Acad Sci USA 2006;103:17807—12.

[57] Drandi D, Kubiczkova-Besse L, Ferrero S, et al. Minimal residual disease detection by droplet digital PCR in multiple myeloma, mantle cell lymphoma, and follicular lymphoma: a comparison with real-time PCR. J Mol Diagn 2015;17:652—60.

[58] Lyon E, Laver T, Yu P, et al. A simple, high-throughput assay for Fragile X expanded alleles using triple repeat primed PCR and capillary electrophoresis. J Mol Diagn 2010;12:505—11.

[59] Toth T, Findlay I, Nagy B, Papp Z. Accurate sizing of (CAG)n repeats causing Huntington disease by fluorescent PCR. Clin Chem 1997;43:2422—3.

I. INTRODUCTION TO MOLECULAR TESTING

[60] Bakker E. Is the DNA sequence the gold standard in genetic testing? Quality of molecular genetic tests assessed. Clin Chem 2006;52:557—8.

[61] Sanger F, Nicklen S, Coulson AR. DNA sequencing with chain-terminating inhibitors. Proc Natl Acad Sci USA 1977;74: 5463—7.

[62] Gocayne J, Robinson DA, FitzGerald MG, et al. Primary structure of rat cardiac beta-adrenergic and muscarinic cholinergic receptors obtained by automated DNA sequence analysis: further evidence for a multigene family. Proc Natl Acad Sci USA 1987;84:8296—300.

[63] Venter JC, Adams MD, Myers EW, et al. The sequence of the human genome. Science 2001;291:1304—51.

[64] Lander ES, Linton LM, Birren B, et al. Initial sequencing and analysis of the human genome. Nature 2001;409:860—921.

[65] Girirajan S, Campbell CD, Eichler EE. Human copy number variation and complex genetic disease. Annu Rev Genet 2011;45: 203—26.

[66] Piccart-Gebhart MJ, Procter M, Leyland-Jones B, et al. Trastuzumab after adjuvant chemotherapy in HER2-positive breast cancer. N Engl J Med 2005;353:1659—72.

[67] Miller DT, Adam MP, Aradhya S, et al. Consensus statement: chromosomal microarray is a first-tier clinical diagnostic test for individuals with developmental disabilities or congenital anomalies. Am J Hum Genet 2010;86:749—64.

[68] Schaaf CP, Wiszniewska J, Beaudet AL. Copy number and SNP arrays in clinical diagnostics. Annu Rev Genomics Hum Genet 2011;12:25—51.

[69] Slavotinek AM. Novel microdeletion syndromes detected by chromosome microarrays. Hum Genet 2008;124:1—17.

[70] Chrystoja CC, Diamandis EP. Whole genome sequencing as a diagnostic test: challenges and opportunities. Clin Chem 2014;60:724—33.

[71] D'Haene N, Le Mercier M, De Neve N, et al. Clinical validation of targeted next generation sequencing for colon and lung cancers. PLoS One 2015;10:e0138245.

MOLECULAR TESTING IN INFECTIOUS DISEASE

MOLECULAR TESTING IN INFECTIOUS DISEASE

5

Molecular Testing for Human Imunodeficiency Virus

M. Memmi[1,2], T. Bourlet[1,2] and B. Pozzetto[1,2]

[1]GIMAP EA3064, University of Lyon, Saint-Etienne, France [2]Laboratory of Infectious Agents and Hygiene, University Hospital of Saint-Etienne, Saint-Etienne, France

INTRODUCTION

The emergence of human immunodeficiency virus (HIV) infection in the 1980s [1] was contemporary of the discovery of polymerase chain reaction (PCR) [2], the most popular method used for the molecular diagnosis of viral diseases. Consequently, molecular testing occupies a large place in the screening and follow-up of HIV-infected subjects. The HIV species, a member of the *Retroviridae* family and of the *Lentivirus* genus, includes two serotypes named HIV-1 and HIV-2—both of them are responsible for severe immunodeficiency in humans (AIDS) but the first one is distributed worldwide whereas the other is rather limited to West Africa. Given this major difference in terms of Public Health, most of the attention will be dedicated to HIV-1 in this review.

BACKGROUND ON HIV INFECTION AND AIDS

Overall Epidemiology

Since the beginning of the HIV pandemic, almost 78 million people have been infected with the HIV-1 virus and about half of them died. At the end of 2013, according to the World Health Organization (WHO), 35 million (33.1−37.2 million) people were living with HIV [3]. That same year, some 2.1 million people became newly infected, and 1.5 million died of AIDS-related causes. Approximately 0.8% of individuals aged 15−49 years worldwide are living with HIV. The burden of the pandemic varies dramatically between countries and regions, Sub-Saharan Africa remaining the most severely affected, with nearly 1 in every 20 adults living with HIV and accounting for nearly 71% of the people living with HIV worldwide (women comprised 59% of infected people in this area).

Life Cycle of HIV

HIV is a single-stranded, positive-sense RNA enveloped virus of about 120 nm in diameter. Due to the glycolipids constituting its envelope, the virion is relatively fragile in the environment. The entry of the virus through the envelope glycoproteins into the competent cells (mainly immune cells and notably T cells, monocytes−macrophages, and dendritic cells) requires the presence of receptors (mainly CD4 molecule) and chemokine co-receptors (CCR5 or CXCR4) at the surface of the cell. Once the viral RNA is delivered into the cell, it is reverse transcribed into DNA by an RNA-dependent DNA polymerase encoded by the viral genome. The newly generated double-stranded DNA is exported to the nucleus where it is integrated within the cellular DNA by a viral integrase. This phase, which is mandatory for the continuation of the viral cycle, is also crucial for the constitution of viral reservoirs that will persist definitively, despite the further use of antiviral or immunomodulating treatments (at least at this stage of our knowledge). The viral DNA is then transcribed into different RNAs that are used for generating both genomic RNA and messengers coding for viral proteins that are further processed by viral proteases. The virion is then released from the cell by budding through the cytoplasmic membrane.

Diagnostic Molecular Pathology
DOI: http://dx.doi.org/10.1016/B978-0-12-800886-7.00005-4

© 2017 Elsevier Inc. All rights reserved.

Transmission of HIV

The virus is mostly transmitted via sexual route, mainly by vaginal and anal intercourse, although oral sex can also be incriminated. The transmission via contaminated products of human origin is another common way of infection, notably in intravenous drug users sharing syringes and in patients receiving unsafe blood products. The latter mode of transmission explains the need for screening donors of human products (blood, semen, other tissues, organs) with sensitive techniques. The third way of HIV transmission is from mother to child (MTC) during pregnancy, delivery, and breastfeeding. The detection of infected babies is also a major goal of HIV diagnosis.

Natural History of HIV Infection

After a stage of primary infection that can be clinically symptomatic or asymptomatic, the infected subject experiences a long phase of clinical latency during which the viral replication is ongoing in most cases, but at variable levels. A minority of individuals, called long-term nonprogressors, remain asymptomatic for years without developing immunodeficiency. By contrast, most of the infected subjects, if not diagnosed and treated, develop in less than 10 years a progressive loss of their immune functions, affecting principally the T-cell repertoire. AIDS is characterized by an acquired immunodeficiency that results in the occurrence of opportunistic infections and/or cancers that are responsible for the death of patients.

HIV Evolution During Treatment

There is presently no preventive vaccine against HIV infection and also no definite cure. However, lifelong effective treatment with antiretroviral (ARV) drugs can control the virus so that HIV infection can now be considered a chronic disease. Current typical treatments include a combination of three drugs in order to avoid the emergence of resistant strains. More than 25 approved molecules belonging to different classes of ARV are available, some of them being combined in the same pill for facilitating the daily observance of treatment. In most cases, this regimen is able to maintain the plasmatic viral load at an undetectable level and to preserve the immune defenses. In 2013, 12.9 million people living with HIV were receiving ART, of which 11.7 million from low-income or middle-income countries [4]. In the latter areas and during the same year, one-third of the total number of infected adults had access to ART whereas this proportion was only of one in four for children.

MOLECULAR TOOLS IN THE DIAGNOSIS AND FOLLOW-UP OF HIV INFECTION

The first role of the clinical laboratory in the management of HIV infection is to identify the subjects who are not already recognized as infected by HIV, whatever the stage of HIV infection is (from primary infection to AIDS) at the time of the first diagnosis. In other words, the laboratory result constitutes the concrete event that classifies the subject in the category of HIV-infected people. The further mission of the clinical laboratory toward HIV-infected individuals consists in the lifelong surveillance of the direct and indirect biological parameters that reflect the control of viral replication, whatever the patient is treated or not by ARV. A third goal of the laboratory would be, as for hepatitis C virus (HCV) infection, to predict the treated individuals who are definitively cured from infection. However, this objective remains purely elusive before solutions for curing HIV from the reservoirs are found. For accomplishing the first two missions of the clinical laboratory, different strategies and methods have been proposed. This chapter aims to identify the place of molecular tools among other laboratory tests used in the management of HIV infection.

Diagnosis of HIV Infection

Historically, the detection of antibodies specific for HIV in serum was the first test used for determining the status of HIV-seropositive patient. Up to now, the serological tools remain at the basis of the identification of HIV-infected people. Fig. 5.1 summarizes the sequence of appearance of laboratory markers in the course of HIV-1 infection [5]. The eclipse period corresponds to the short phase following HIV infection without any detectable circulating marker. The seroconversion window period is the interval between HIV infection and the appearance of HIV antibodies using the most sensitive immunoassays—it corresponds to phases I (positivity of HIV RNA only) and II (positivity of both HIV RNA and p24 capsid antigen) in the Fiebig staging system [6]. Acute HIV infection is the period separating the detection of HIV RNA from that of the first detection of HIV antibodies—it corresponds to phases III (positive immunoassay but negative western blot profile), IV (positive immunoassay and indeterminate western blot profile), and V (positive immunoassay and incomplete western blot profile) in the Fiebig staging system [6]. Finally, the established HIV infection corresponds to a fully developed antibody response, classically assessed by positive reactivity to all the major bands of HIV proteins by using western blotting (phase VI in the Fiebig staging system).

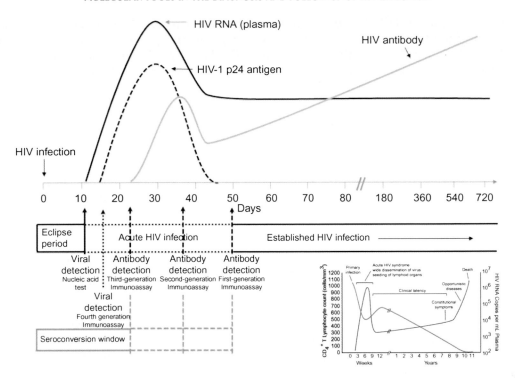

FIGURE 5.1 Kinetics of virological and serological markers during the first steps of HIV-1 infection [5]. The pictogram in the bottom right (originating from *Wikimedia commons*) illustrates the inverse kinetics of CD4 + T cell count (blue curve) and HIV-1 RNA load (red curve) during the natural history of HIV-1 infection.

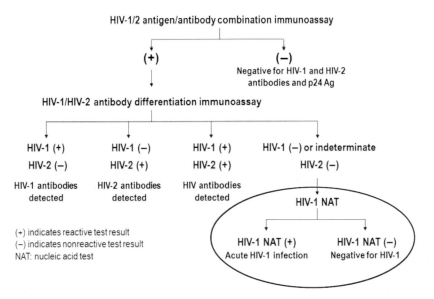

FIGURE 5.2 Algorithm proposed in 2014 by the CDCs for the diagnosis of HIV-1 infection [5]. The red circle illustrates the introduction for the first time of a molecular test for resolving some indeterminate serological results.

The screening of HIV-infected individuals must be performed with sensitive fourth-generation reagents able to detect antibodies against HIV-1 (including strains of group O that circulate in Central Africa), antibodies against HIV-2, and p24 capsid antigen from HIV-1. According to the recent algorithm recommended by the CDCs [5] illustrated in Fig. 5.2, a negative result does not require further testing (except if the subject is suspected to be in the Fiebig I stage).

When the first test is not negative, it is recommended to test a new serum specimen in order to exclude any error of identity between the subject and the blood sample or a contamination from a tube to another. For confirming a nonnegative screening test, the CDCs algorithm requires a second serological test allowing differentiating HIV-1 from HIV-2 infection (Fig. 5.2). In case of discrepant results between the initial and the differentiation tests, an HIV-1 molecular test is required for segregating acute HIV-1 infection from a false-positive result of the first test (Fig. 5.2). This new strategy, which involves for the first time molecular tests in HIV screening, combines several advantages: (1) it is cost-effective, (2) it is very sensitive, even at the early phase of HIV-1 infection, (3) it allows a clear differentiation of HIV-1 and HIV-2 infections (which was not the case with previous strategies using western blot testing because of cross-reactivity between the antibodies directed against the two serotypes), and (4) it permits to exclude false-positive results of HIV-1 serological tests. The nucleic acid tests (NATs) used at this stage requires a qualitative detection of HIV-1 RNA, although a sensitive quantitative HIV test may also be used for this purpose.

Follow-Up of HIV-Infected Subjects

Once a subject is detected seropositive for HIV, whatever the stage of HIV infection, it is crucial to initiate a strict follow-up aimed at preserving or restoring the immune system of the infected individual. This close surveillance relies on two kinds of biological markers: (1) the determination of CD4 + T lymphocyte cell count (typically through flow cytometry analysis) and (2) the measure of the viral load (through quantitative NAT based either on qPCR or alternative methods of quantification of HIV nucleic acids) [7−9]. The tests that are used on a routine basis for determining the viral load target circulating HIV RNA [10−12]. More recently, it has been proposed to explore in parallel the DNA viral load with the aim of quantifying the HIV reservoirs, both in blood and in different tissues, including the intestinal and the semen reservoirs [13,14]. The aim of the follow-up of these different markers is to determine the optimal time for starting an HIV ART triple therapy.

In the course of HIV primary infection, the current recommendations are to start a potent anti-HIV as soon as possible according to the test-and-treat concept in order to reduce the size of the viral reservoir, to decrease the rate of viral mutation by suppressing viral replication, and to preserve immune function [15−19]. According to this objective, it is important to differentiate accurately HIV-1 from HIV-2 infection since some ARV drugs active against HIV-1 are not effective against HIV-2 [20,21].

For HIV infection discovered at a later stage, the markers described help to distinguish long-term non-progressors who only need a surveillance of immune and viral markers from the majority of other HIV-infected subjects who require an ART triple (or more) therapy for preserving or restoring immune function, according to the stage of infection. Indeed, it is important to note that the initiation of a potent ART treatment is able to stop viral replication, even at the stage of AIDS, and to improve the immune deficiency. The WHO guidelines issued in 2013 recommend ART initiation when the CD4 + T lymphocyte cell count drops below 500 cells/μL [22]. Through the combination of biological surveillance and lifelong effective drug regimens, most HIV infections whose natural evolution would lead to death in a few years have become a chronic disease compatible with a life expectancy close to that of noninfected people, at least in the world areas where there is no economic limitation to surveillance and treatment access. In treated patients, the molecular tests are also very useful to monitor the development of HIV resistance to viral drugs.

Summary of HIV Molecular Testing in the Course of HIV Infection

Table 5.1 lists a number of molecular tests that are currently available or in development for the screening and follow-up of HIV-infected people. This list summarizes the panel of the different HIV tests that can be proposed for the diagnosis and surveillance of HIV infection according to their indications. Some of these tests are qualified of point-of-care (POC) NAT, which means that they can be used as near-patient testing in different clinical settings when a rapid answer is required or when high-tech laboratory facilities are unavailable [23,24]. According to UNITAID, a POC test must be ASSURED, which means affordable, sensitive, specific, user-friendly, rapid and robust, equipment-free, and deliverable to end users [25].

Table 5.2 recapitulates the different indications of HIV molecular testing for the screening and follow-up of HIV-infected subjects.

SPECIAL EMPHASIS ON THE USE OF HIV MOLECULAR TOOLS IN SPECIFIC CLINICAL SITUATIONS

Screening of Blood Products

In addition to serological testing, NAT is presently indicated for HIV-1 screening in blood donors, either by minipools (MP-NAT) or by individual donation (ID-NAT) in most developed countries. HIV-1 NAT was implemented in 1999 in the United States and in 2001 in France. This strategy is usually associated to the detection of other viral genomes, including hepatitis B virus (HBV) and HCV. It resulted in a significant reduction of the HIV-1 window period, from 22 days without NAT to 9−12 days with NAT [26−28].

Two systems are mainly used around the world for HIV NAT of blood products—the Procleix Tigris system (Grifols Engineering) based on transcription-

TABLE 5.1 Examples of Molecular Tests for Qualitative or Quantitative Detection of HIV-1 Genome That Are Either Commercially-Available in 2015 or in Development

Marker category	Molecular technology	Commercial kit or platform (company)
HIV qualitative nucleic acid assays used for HIV screening in patients	rtPCR	Cobas Taqman (Roche Diagnostics)
	rtPCR	HIV-1 qualitative assay (Abbott Molecular)
	rtPCR	Generic HIV DNA cell (Biocentric)
	TMA	Aptima HIV-1 RNA qualitative assay (Hologic)
HIV qualitative nucleic acid assays used for HIV screening of donors of blood products (in combination with HBV and HCV)	rtPCR	Cobas s 201 system (Roche Diagnostics)
	TMA	Procleix Tigris system (Grifols Engineering)
HIV RNA viral load	rtPCR	Cobas TaqMan HIV-1 (Roche Diagnostics)
	rtPCR	RealTime HIV-1 m2000rt (Abbott Molecular)
	rtPCR	VERSANT HIV-1 RNA (kPCR) (Siemens)
	rtPCR	Generic HIV RNA viral load (Biocentric)
	rtPCR	*artus* HI Virus-1 QS-RGQ (Qiagen)
	rtPCR	ExaVir Load3 (Cavidi)
	TMA	Aptima HIV-1 Quant Dx (Hologic)
	rtNASBA	NucliSens EasyQ HIV-1 (bioMérieux)
	Kinetic PCR	VERSANT kPCR Molecular System (Siemens Healthcare)
	bDNA	VERSANT 440 Molecular System (Siemens Healthcare)
POC testing for HIV	rtPCR	Liat HIV Quant Assay (IQuum Inc)
	rtPCR	Xpert HIV-1 Viral Load (Cepheid)
	rtPCR rtPCR	Truelab Real Time micro PCR system (Molbio Diagnostics Pvt Ltd)
	Isothermal amplification	Savanna HIV viral load test (NWGHF)
	Isothermal amplification	SAMBA platform (Diagnostics for the Real World Ltd)
		EOSCAPE-HIV (Wave 80 Biosciences)
	Isothermal amplification	Alere Q (Alere)
	Isothermal amplification	RT CPA HIV-1 Viral Load test (Ustar Biotechnologies)
	Isothermal amplification	Bioluminescent Assay in Real-Time or BART (Lumora Ltd)
	Chip-based system	Gene-RADAR platform (Nanobiosym Diagnostics)
	RT activity measurement	ZIVA Automated RT Viral Load (Cavidi)

HBV, hepatitis B virus; HCV, hepatitis C virus; rtPCR, real-time PCR; TMA, transcription-mediated amplification; rtNASBA, real-time nucleic acid sequence based amplification; bDNA, branched DNA; RT, reverse transcriptase.

mediated amplification (TMA) technology and the Cobas s 201 system (Roche Diagnostics) based on real-time PCR (rtPCR) technology. The first system is a fully-automated, closed machine allowing the simultaneous detection of HIV, HBV, and HCV genomes on individual specimens. The second system combines several modules adapted to the distribution of specimens, extraction of nucleic acids, and amplification of HIV, HBV, and HCV genomes. It can be used on individual sera but is optimized for six-donor minipools. Additional targets can be tested if needed, such as West Nile virus or dengue virus genomes. The sensitivity of the two systems is very similar [29,30].

In the era of NAT screening, the residual risk for HIV transmission by labile blood products was estimated to be 1:1,800,000 (MP-NAT) and 1:2,800,000 (SD-NAT) in the United States [31], 1:4,300,000 in Germany [32], and 1:2,400,000 in France [33].

TABLE 5.2 Panel of Molecular Tools Currently Available for the Diagnosis and Follow-Up of HIV-1 Infection According to the Clinical Context and the Level of Resource

Clinical context	Level of resource	Matrix	Technology used for HIV testing	Major quality/ies required
Screening				
• Acute HIV infection	High-income areas	Blood	RNA viral load	High sensitivity
• Mother-to-child transmission	High-income areas	Infant blood	DNA or RNA viral load	Sensitivity
	Low-income areas	Infant blood	Quantitative molecular POC testing	ASSURED[a]
• Blood products and other products of human origin	High-income areas	Blood	DNA or RNA qualitative NAT (ideally multiplexed with other agents as hepatitis B and C viruses)	Rapid and easy testing
• Medical assisted procreation	High-income areas	Semen	Quantitative RNA viral load	Robust testing for avoiding PCR inhibitors
Follow-up				
• ART monitoring	High-income areas	Blood	RNA viral load	High sensitivity
	Low-income areas	Blood	Quantitative molecular POC testing	ASSURED
• Drug-resistance testing	High-income areas	Blood	Genotype or phenotype molecular testing	Good predictive value for ART efficiency
• Exploration of reservoirs	High-income areas	Blood/tissues	Single-cell qualitative assays	For research only

[a]*ASSURED, affordable, sensitive, specific, user-friendly, rapid and robust, equipment-free, and deliverable to end users; POC, point of care; NAT, nucleic acid testing; ART, antiretroviral treatment.*

Screening of Organs and Tissues from Human Donors

By contrast to blood donors who represent a well-defined population that can be secured by regular screening, organ and tissue donors are occasional donors for whom blood safety is much more difficult to assess. Schematically, the situations encountered in clinical practice involve three categories of donors: (1) beating heart cadavers from whom solid organs (heart, liver, lungs, kidneys, and pancreas) and different tissues (bones, tendons, vessels, skin, cornea, and others) can be sampled, (2) non-beating heart cadavers from whom tissues can be sampled, and (3) living donors from whom some organs (mainly kidney and bone marrow) and tissues can be sampled. HIV transmission to recipients was occasionally observed in all these situations [34–40]. The more at-risk situation for the transmission of blood-borne viruses is that involving beating heart cadavers: (1) the decision of organ sampling must be taken within a few hours due to the short cold-ischemic time of the grafts, (2) the demographics of this population exposed to sudden death and the frequent absence of medical and behavioral risk assessment history increase the probability of encountering subjects with primary viral infection, and (3) the organ penury with regard to medical needs may represent an additional risk of being less vigilant in terms of viral safety.

In addition to serological and antigen testing that is currently performed for the most at-risk pathogens, including HIV, HBV, and HCV, the use of NAT testing is possible and recommended for not-beating heart cadavers and living donors because there is enough time for implementing sensitive genome detection of blood-borne viruses (Table 5.1 for HIV-1). For beating heart cadaver, the situation is much more difficult since no molecular test can be currently performed on-demand, within 3–5 h. This subject was extensively debated through a conference held in 2010 under the auspices of North-American Societies of transplantation [41]. In 2015, new commercially-available tests (including the Hologic and geneXpert tests listed in Table 5.1) could permit detecting HIV-1 genome in emergency conditions, even if studies are required for evaluating their performances in this context. Additional tests detecting simultaneously HBV and HCV genomes would be needed to offer full viral safety in graft donors, even if the small size of the market from an economic point of view does not incite companies trained in molecular technologies to invest in this domain.

Screening HIV-1 Genome in Semen in Medically-Assisted Procreation

In addition to solving problems of female infertility, medically-assisted procreation (MAP) techniques are recommended in fertile women who want to be pregnant and whose male partner is infected by HIV-1 [42,43]. In this context, it is useful to control that the semen fraction used for MAP procedures is free of HIV RNA. Unfortunately, PCR-based techniques may be inhibited by the presence of compounds (notably zinc ions) naturally present in some semen specimens. To our knowledge, none of the commercial tests that are licensed for blood specimens are validated for the use of semen samples. Laboratories involved in MAP viral safety must develop adapted protocols of commercial tests for checking that the amplification step was not inhibited [44–46]. Due to these precautions, no case of HIV infection has been reported in babies born after using MAP procedures [47,48].

Screening of Neonates Born to HIV-Infected Mothers

In the absence of any intervention, the combined risk of MTC transmission of HIV-1 in utero and intrapartum is 15–30%, and breast-feeding increases the risk to 20–45%. Factors of risk are age, maternal viral load, clinical stage of HIV-1 infection, and presence or absence of therapeutic and/or prophylactic ART in mother and infant. Since infected children have a high morbidity and mortality in the first 2 years of life, an early diagnosis is essential for establishing the infectious status of the neonate, and in case of infection, for initiating appropriate ART [49,50]. This objective implies the need of sensitive tests for the diagnosis of HIV infection in newborns. Serological tests are useless because of the detection of maternal HIV antibodies in the blood of newborns up to 18 months of age. Therefore, early qualitative or quantitative detection of HIV-1 DNA in peripheral blood mononuclear cells (PBMCs) and RNA in plasma have become the methods of choice for evaluating HIV infection in neonates whose mother is positive for HIV-1 [51–56]. The WHO recommends a systematic HIV-1 screening in all exposed newborns at 4–6 weeks of age by using DNA or RNA molecular test [57], both approaches having equal sensitivity rates [58]. If negative, an additional testing must be performed 2 months after breast-feeding cessation. This strategy is aimed to contribute to the rise of an HIV-free generation, notably in those areas where the rate of HIV-infected pregnant women is very high.

Use of HIV-1 Molecular Tools in Low-Income and Middle-Income Countries

Most HIV-infected people worldwide reside in low-income and middle-income countries. During the last 10 years, the availability of generic drugs and price reductions of patented medications have allowed a significant scale-up of ART in subjects living with HIV-1 in developing countries. It is estimated that approximately one-third of HIV-infected subjects had received ART by the end of 2013 in those areas [59]. Therefore, laboratory HIV-1 testing has become a major challenge both for identifying newly-infected individuals and monitoring the efficacy of treatments in those who benefit of them. In this context, HIV-1 molecular tools can schematically be indicated in three main situations [60]: (1) a qualitative approach is needed for identifying infected infants exposed to HIV-1 during pregnancy or breast-feeding according to the strategies summarized above; (2) a semiquantitative approach is recommended, in combination with a simplified determination of CD4 cell count [61], on the basis of at least one measure per year for monitoring HIV-infected subjects on a routine basis either for deciding of the opportunity to start ART or for evaluating the efficacy of HIV treatments in those subjects who are already treated; and (3) a quantitative approach is indicated in clinical trials conducted in ART-treated patients on the basis of at least four measures per year for research purpose. From a practical point of view, two strategies can be proposed according to the laboratory facilities available in each country from low-income areas. The first one consists in implementing simple POC tests in remote rural areas in order to circumvent the absence of well-equipped laboratories and trained operators. The alternative approach consists in centralizing the tests in laboratories located in cities and benefiting of molecular facilities close to those available in developed countries. In order to facilitate the transport of specimens from remote areas to reference laboratories, it is possible to use dried blood or dried plasma spots (DBS/DPS). The blood or plasma, from venous puncture or finger-prick or heel-prick, is dried onto a filter paper, placed in gas-impermeable zipper storage bag with desiccant, and sent to the central laboratory at ambient temperature. Previous studies have demonstrated the benefit of the use of DBS/DPS for detecting and quantifying HIV-1 nucleic acids with sensitivity performances close to those using fresh blood [62–76]. However, a strict standardization of the procedures is needed prior recommending large-scale use of DBS/DPS in clinical settings [77,78].

NEW INSIGHTS IN MOLECULAR HIV TESTING

New molecular tools that could be used for the diagnosis and follow-up of HIV infection are discussed in the context of the following topics: (1) improvement of the sensitivity of quantitative HIV-1 viral load tests, (2) development of new POC molecular tests for HIV testing, and (3) present and future of ART resistance determination.

Improvement of the Sensitivity of Quantitative HIV-1 Viral Load Tests

At the era of potent ART, there is a need for developing highly sensitive tests able to measure the HIV-1 viral load at the cellular level. In effect, if the first objective of ART is to make the RNA viral load undetectable by using current quantitative tests, the final aim of these treatments would be to eradicate the integrated DNA provirus from the circulating and tissue reservoirs.

The decay dynamics of plasma HIV-1 RNA under potent ART can be summarized in four successive phases (Fig. 5.3) [14]. The first phase with a half-life of approximately 1.5 days corresponds to the elimination of free virus and productively infected T cells. The second phase with a half-life of approximately 28 days corresponds to the destruction of more resistant cells represented by activated T cells and monocytes–macrophages. These two phases can be investigated by sensitive RNA quantitative commercially-available tests exhibiting a threshold of 20–50 copies/mL, as depicted in Table 5.3. The third phase corresponds to a slow decrease (half-life of about 1 year) of the viral load under the threshold of

50 copies/mL. Beyond this phase that may last for 5–7 years, there is a stabilization of the viral load at very low level (fourth phase with a half-life of infinity) that reflects the persistence of viral replication in long-lived cellular reservoirs. The exploration of these two last phases needs new ultrasensitive molecular tools able to detect residual integrated DNA [79].

Sensitive in-house rtPCR tests targeting various HIV-1 genes [14,79–81] were developed for exploring low-level HIV viremia observed in HIV-1-infected subjects successfully treated with potent ART. However, these tests are technologically limited by the high signal-to-noise ratio that is observed near the limit of detection. To circumvent this problem, a new molecular approach called droplet digital PCR (ddPCR) was recently proposed. ddPCR uses a PCR mixture including the template and fluorescent hydrolysis probes that is emulsified into droplet generation oil containing stabilizing surfactants. Each droplet is then transferred to a 96-well plate and submitted to PCR amplification. The number of fluorescent droplets is counted in a droplet fluorescence reader using Poisson modeling [82]. This test was shown to be well correlated to DNA rtPCR assays for the measure of viral DNA in treated patients with undetectable plasmatic viral load [83]. This technique can be coupled to single-cell analysis by microengraving. The latter technique consists in the simultaneous culture of single lymphocytes in an array of subnanoliter wells for exploring cytokine production kinetics. The corresponding RNA can be amplified by using single-cell digital PCR within the culture array. The combination of these techniques offers an opportunity for exploring the latent cellular reservoirs in blood and tissues (intestine, genital tract, central nervous system) with the objective of curing them through HIV eradication.

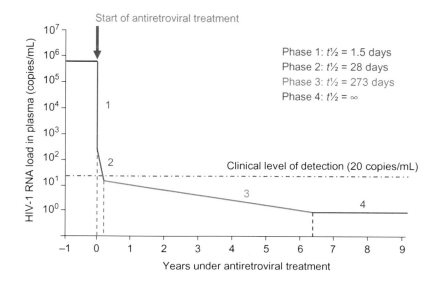

FIGURE 5.3 Theoretical decay kinetics of plasmatic HIV-1 RNA load in patients receiving a potent ARV treatment. The half-life of each phase was recorded from the model of Hilldorfer et al. [14].

Phase 1: $t\frac{1}{2}$ = 1.5 days
Phase 2: $t\frac{1}{2}$ = 28 days
Phase 3: $t\frac{1}{2}$ = 273 days
Phase 4: $t\frac{1}{2}$ = ∞

TABLE 5.3 Examples of SBS (Sequencing by Synthesis) Platforms Used for HIV-1 Resistance Testing

Synthesis strategy	Generation sequencing	Platform	Enzyme	Amplification step	Principle
Stepwise (base-by-base)	First	Sanger sequencing	Polymerase	Yes (clones or PCR)	Polymerization of fluorescent ddNTPs generating chain termination and fragment analysis by gel
	NGS/Second	Illumina	Polymerase	Yes (bridge amplification)	Polymerization of fluorescent ddNTPs generating reversible chain termination and fragment analysis by capillary electrophoresis
Sequential	NGS/Second	Life technologies; SOLiD	Ligase	Yes (PCR)	Ligation of labeled oligonucleotides
	NGS/Second	Life technologies; Ion Torrent	Polymerase	Yes (PCR)	Polymerization of natural dNTPs and measure of H$^+$ release using a semiconductor chip
	NGS/Second	Roche; 454 Life Sciences	Polymerase	Yes (PCR)	Polymerization of natural dNTPs and luminescence detection of pyrophosphate ("pyrosequencing")
Single molecule real time (SMRT)	NGS/Third	Pacific Biosciences; PACBIO RS II	Polymerase	No	Continuous polymerization of labeled dNTPs and real-time recording of light pulses emitted as a by-product of nucleotide incorporation by DNA single molecules

NGS, next-generation sequencing; dNTP, deoxribonucleotide triphosphate; ddNTP, di deoxribonucleotide triphosphate.

Development of New POC Molecular Assays for HIV Testing

At the same time that ultrasensitive molecular tools are developed for anticipating the control of the hypothetical HIV cure in patients from high-income countries, there is an urgent need for implementing simple platforms able to either detect the presence of HIV, notably in children born to infected mothers, or monitor the viral load of treated patients from low-resource areas where high-technology laboratories are scarce. A number of recent review articles offer an overview of the techniques that are presently available or in development to reach these goals [24,61,77]. Fig. 5.4 illustrates the POC molecular tests that are readily commercialized or that will soon be available for determination of HIV vial load in infected subjects or for qualitative NAT in early infant diagnosis [25]. These tests offer the advantage of combining extraction of nucleic acids, amplification of specific targets, and detection of amplicons through a single platform. As an alternative to classical technologies that require a sophisticated laboratory environment, most of these POC tests can be performed by operators not trained to NAT and some of them can be implemented in low-resource facilities that have no access to electricity (replaced by batteries), plasma samples (replaced by capillary blood collected by heel-prick or finger-prick), or air-conditioning (tests can be performed without constraints of controlled ambient temperature or humidity). The cost of these tests is another limitation. It must be as low as possible in order to avoid facing the paradox of having a system to monitor ART treatment that is more expensive than the treatment itself [60].

Present and Future of ART Resistance Determination

ART resistance determination has become a key element for monitoring treatment efficacy and transmission of drug-resistant mutants, despite the fact that the benefit of resistance testing is not easy to assess [84,85]. Conceptually, three categories of tests can be used [86]. The first one, designed as actual phenotype resistance assay, evaluates directly the susceptibility of the virus strain contained in the PBMCs of the subject, to different antiviral drugs. It is fastidious, time-consuming, and subject to the ability to cultivate the patient's viral strain. An alternative strategy, also based on cell culture and designed as virtual phenotype resistance assay (*VirtualPhenotype*, Tibotec-Virco), consists of using an easy-cultivatable HIV reference strain in which the genes that constitute the targets of ARV drugs have been replaced by those of the patient's strain. This technique is easier and more rapid to perform than the former one and well-standardized. In practice, because of its cost, this method is limited to clinical trials. The most popular assays that are used for ART resistance determination are the genotypic tests based on sequencing the target genes of ARV

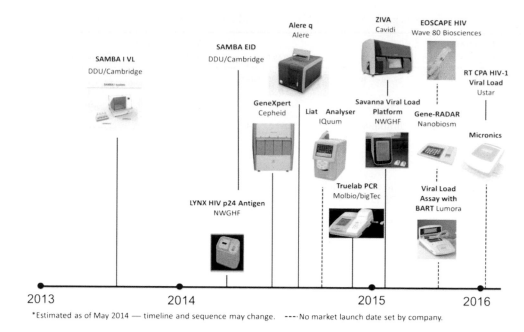

FIGURE 5.4 POC systems (enzyme immunoassays and molecular tests) that are either already commercially-available or in development for measuring the HIV-1 load according to UNITAID [25]. The molecular tools depicted in this figure are briefly described in the bottom of Table 5.1.

drugs, namely protease, reverse transcriptase, integrase, and envelope glycoproteins. The presence of specific mutations in these genes has been shown to predict resistance to the corresponding drugs. In addition, they are able to determine the tropism of HIV strains, which is useful when entry inhibitors such as maraviroc targeting the CCR5 co-receptor are used.

As given in Table 5.3, different generations of sequencing techniques are available. Those based on the Sanger method are currently used on a routine basis, at least in high-resource areas. These include in-house techniques and commercially-available tests (such as the ViroSeq HIV-1 Genotyping System from Celera Diagnostics). The main limitation is their lack of sensitivity for resistant minority variants representing less than 20% of the whole quasi-species, thus risking selection through use of drugs active in the major population. Next-generation sequencing tests (Table 5.3) can overcome this problem by allowing the individual analysis of all the components of the quasi-species, even those representing as little as 1% of the total. However, the use of PCR for amplifying these variants may lead to unequal amplification of some of them or nucleotide mis-incorporation due to polymerase infidelity, leading to the over- or under-estimation of some mutants. In addition, the high quantity of sequences generated by these approaches makes difficult the interpretation of the data on a routine basis. Third-generation sequencing techniques able to amplify single DNA molecules (SMRT in Table 5.3),

which are still at an experimental stage, do not require an amplification step. Although lacking sensitivity in samples with low viral loads, this method could provide the true picture of the breadth and depth of individual circulating viruses [87]. Detailed recent reviews on NGS techniques are available in Refs. [88–90].

In practice, sequencing techniques are recommended in treatment-naïve patients for whom ART is indicated, especially when low genetic barrier drugs such as nonnucleoside reverse transcriptase inhibitors (Nevirapine, Effavirenz) are intended to be used. Sequencing can also be used to monitor drug changes in patients experiencing treatment failure [91], notably for predicting HIV-1 tropism when changing to maraviroc is required [92,93]. However, due to their high cost and to the need of high-technology platforms, sequencing approaches are reserved to high-income countries, at least on a routine basis. To overcome this disadvantage in low-resource areas, the WHO encourages alternative strategies based on the setting of early warning indicators of treatment failure [94,95].

CONCLUSIONS

This overview of molecular HIV testing highlights the central part that NAT occupies in the diagnosis and monitoring of all the stages of HIV infection, from primary infection to treatment management. Sensitive tests are readily available in high- and middle-resource countries and tests based on new technologies are in continuous development for improving the life expectancy and quality of life of HIV-infected people. In low-income areas, the major focus is enhancement of the recognition of HIV-infected subjects and provision of access to effective ART. Molecular tools are of great help for reaching these objectives as illustrated by molecular POC-available tests for HIV-1 that combine high technology, excellent sensitivity, and simple/rapid use on the field. The development of molecular tests combining the detection of different pathogens associated with HIV infection, including hepatitis viruses or tuberculosis agents, is another technological challenge that will be faced in the coming years. Thus, the worldwide fight against HIV infection appears to be, at least in part, driven by the performance of molecular testing, with the final hope that solidarity will help to make these tools available to people that are the most in need of them.

References

[1] Barré-Sinoussi F, Chermann JC, Rey F, et al. Isolation of a T-lymphotropic retrovirus from a patient at risk for acquired immune deficiency syndrome (AIDS). Science 1983;220:868–71.

[2] Saiki RK, Scharf S, Faloona F, et al. Enzymatic amplification of beta-globin genomic sequences and restriction site analysis for diagnosis of sickle cell anemia. Science 1985;230:1350–4.

[3] World Health Organization. Available from: <http://www.who.int/hiv/data/epi_core_dec2014.png?ua=1> [accessed March 2015].

[4] World Health Organization. Available from: <http://www.who.int/mediacentre/factsheets/fs360/en/> [accessed March 2015].

[5] Centers for Disease Control and Prevention and Association of Public Health Laboratories. Laboratory testing for the diagnosis of HIV infection: updated recommendations. Published June 27, 2014. Available from: <http://stacks.cdc.gov/view/cdc/23447> [accessed March 2015].

[6] Fiebig EW, Wright DJ, Rawal BD, et al. Dynamics of HIV viremia and antibody seroconversion in plasma donors: implications for diagnosis and staging of primary HIV infection. AIDS 2003;17:1871–9.

[7] Ho DD, Neumann AU, Perelson AS, Chen W, Leonard JM, Markowitz M. Rapid turnover of plasma virions and CD4 lymphocytes in HIV-1 infection. Nature 1995;373:123–6.

[8] Mellors JW, Rinaldo Jr CR, Gupta P, White RM, Todd JA, Kingsley LA. Prognosis in HIV-1 infection predicted by the quantity of virus in plasma. Science 1996;272:1167–70.

[9] Mellors JW, Muñoz A, Giorgi JV, et al. Plasma viral load and CD4 + lymphocytes as prognostic markers of HIV-1 infection. Ann Intern Med 1997;126:946–54.

[10] Peter JB, Sevall JS. Molecular-based methods for quantifying HIV viral load. AIDS Patient Care STDS 2004;18:75–9.

[11] Wittek M1, Stürmer M, Doerr HW, Berger A. Molecular assays for monitoring HIV infection and antiretroviral therapy. Expert Rev Mol Diagn 2007;7:237–46.

[12] Rouet F, Ménan H, Viljoen J, et al. In-house HIV-1 RNA real-time RT-PCR assays: principle, available tests and usefulness in developing countries. Expert Rev Mol Diagn 2008;8:635–50.

[13] Graf EH, O'Doherty U. Quantitation of integrated proviral DNA in viral reservoirs. Curr Opin HIV AIDS 2013;8:100–5.

[14] Hilldorfer BB, Cillo AR, Besson GJ, Bedison MA, Mellors JW. New tools for quantifying HIV-1 reservoirs: plasma RNA single copy assays and beyond. Curr HIV/AIDS Rep 2012;9:91–100.

[15] Panel on Antiretroviral Guidelines for Adults and Adolescents. Guidelines for the use of antiretroviral agents in HIV-1-infected adults and adolescents. 2013. Available from: <http://www.aidsinfo.nih.gov/ContentFiles/AdultandAdolescentGL.pdf> [accessed March 2015].

[16] Hogan CM, Degruttola V, Sun X, et al. The setpoint study (ACTG A5217): effect of immediate versus deferred antiretroviral therapy on virologic set point in recently HIV-1-infected individuals. J Infect Dis 2012;205:87–96.

[17] Saez-Cirion A, Bacchus C, Hocqueloux L, et al. Post-treatment HIV-1 controllers with a long-term virological remission after the interruption of early initiated antiretroviral therapy ANRS VISCONTI Study. PLoS Pathog 2013;9:e1003211.

[18] Hocqueloux L, Saez-Cirion A, Rouzioux C. Immunovirologic control 24 months after interruption of antiretroviral therapy initiated close to HIV seroconversion. JAMA Intern Med 2013;173:475–6.

[19] Smith MK, Rutstein SE, Powers KA, et al. The detection and management of early HIV infection: a clinical and public health emergency. J Acquir Immune Defic Syndr 2013;63:S187–99.

[20] Ntemgwa ML, d'Aquin Toni T, Brenner BG, Camacho RJ, Wainberg MA. Antiretroviral drug resistance in human immunodeficiency virus type 2. Antimicrob Agents Chemother 2009;53:3611–19.

[21] Hizi A, Tal R, Shaharabany M, et al. Specific inhibition of the reverse transcriptase of human immunodeficiency virus type 1 and the chimeric enzymes of human immunodeficiency virus type 1 and type 2 by nonnucleoside inhibitors. Antimicrob Agents Chemother 1993;37:1037–42.

[22] World Health Organization. Global update on HIV treatment 2013: results, impact and opportunities. Available from: <http://apps.who.int/iris/bitstream/10665/85326/1/9789241505734_eng.pdf> [accessed March 2015].

[23] Clerc O, Greub G. Routine use of point-of-care tests: usefulness and application in clinical microbiology. Clin Microbiol Infect 2010;16:1054–61.

[24] Niemz A, Ferguson TM, Boyle DS. Point-of-care nucleic acid testing for infectious diseases. Trends Biotechnol 2011;29:240–50.

[25] UNITAID. 2014. HIV/AIDS diagnostic technology landscape: 4th ed. Available from: <http://www.unitaid.eu/images/marketdynamics/publications/UNITAID-HIV_Diagnostic_Landscape-4th_edition.pdf> [accessed March 2015].

[26] Busch MP, Glynn SA, Stramer SL, et al. A new strategy for estimating risks of transfusion-transmitted viral infections based on rates of detection of recently infected donors. Transfusion 2005;45:254–64.

[27] Pillonel J, Laperche S, Saura C, Desenglos JC, Courouce AM. Transfusion-Transmissible Agents Working Group of the French Society of Blood Transfusion. Trends in residual risk of transfusion-transmitted viral infections in France between 1992 and 2000. Transfusion 2002;42:980–98.

[28] Assal A, Coste J, Barlet V, et al. Application de la biologie moléculaire à la sécurité virale transfusionnelle: le dépistage génomique viral. Transfus Clin Biol 2003;10:217–26.

[29] Margaritis AR, Brown SM, Seed CR, Kiely P, d'Agostino B, Keller AJ. Comparison of two automated nucleic acid testing systems for simultaneous detection of human

immunodeficiency virus and hepatitis C virus RNA and hepatitis B virus DNA. Transfusion 2007;47:1783–93.

[30] Assal A, Barlet V, Deschaseaux M, et al. Sensitivity of two hepatitis B virus, hepatitis C virus (HCV), and human immunodeficiency virus (HIV) nucleic acid test systems relative to hepatitis B surface antigen, anti-HCV, anti-HIV, and p24/anti-HIV combination assays in seroconversion panels. Transfusion 2009; 49:301–10.

[31] Jackson BR, Busch MP, Stramer SL, AuBuchon JP. The cost-effectiveness of NAT for HIV, HCV, and HBV in whole-blood donations. Transfusion 2003;43:721–9.

[32] Hourfar MK, Jork C, Schottstedt V, et al. Experience of German Red Cross blood donor services with nucleic acid testing: results of screening more than 30 million blood donations for human immunodeficiency virus-1, hepatitis C virus, and hepatitis B virus. Transfusion 2008;48:1558–66.

[33] Barlet V. Évolutions technologiques en qualification biologique du don et leur impact sur le risque résiduel transfusionnel. Trans Clin Biol 2011;18:292–301.

[34] Ahn J, Cohen SM. Transmission of human immunodeficiency virus and hepatitis C virus through liver transplantation. Liver Transpl 2008;14:1603–8.

[35] Ison MG, Llata E, Conover CS, et al. Transmission of human immunodeficiency virus and hepatitis C virus from an organ donor to four transplant recipients. Am J Transplant 2011;11:1218–25.

[36] Schratt HE, Regel G, Kiesewetter B, Tscherne H. HIV-Infektion durch kältekonservierte Knochentransplantate. Unfallchirurg 1996;9:679–84.

[37] Centers for Disease Control and Prevention (CDC). Transmission of HIV through bone transplantation: case report and public health recommendations. Morbid Mortal Wkly Rep 1988;37:597–9.

[38] Simonds RJ, Holmberg SD, Hurwitz RL. Transmission of human immunodeficiency virus type 1 from a seronegative organ and tissue donor. N Engl J Med 1992;326:726–32.

[39] Clarke JA. HIV transmission and skin grafts. Lancet 1987;i:983.

[40] Li CM, Ho YR, Liu YC. Transmission of human immunodeficiency virus through bone transplantation: a case report. J Formosan Med Assoc 2001;100:350–1.

[41] Humar A, Morris M, Blumberg E. Nucleic acid testing (NAT) of organ donors: is the "best" test the right test? A consensus conference report. Am J Transplant 2010;10:889–99.

[42] Wingfield M, Cottell E. Viral screening of couples undergoing partner donation in assisted reproduction with regard to EU Directives 2004/23/EC, 2006/17/EC and 2006/86/EC: what is the evidence for repeated screening? Hum Reprod 2010;25:3058–65.

[43] Practice Committee of American Society for Reproductive Medicine. Recommendations for reducing the risk of viral transmission during fertility treatment with the use of autologous gametes: a committee opinion. Fertil Steril 2013;99:340–6.

[44] Garrido N, Meseguer M, Simon C, Pellicer A, Remohi J. Assisted reproduction in HIV and HCV infected men of serodiscordant couples. Arch Androl 2004;50:105–11.

[45] Pasquier C, Anderson D, Andreutti-Zaugg C, et al. Multicenter quality control of the detection of HIV-1 genome in semen before medically assisted procreation. J Med Virol 2006;78: 877–82.

[46] Pasquier C, Andreutti C, Bertrand E, et al. Multicenter assessment of HIV-1 RNA quantitation in semen in the CREAThE network. J Med Virol 2012;84:183–7.

[47] Vitorino RL, Grinsztejn BG, de Andrade CA, et al. Systematic review of the effectiveness and safety of assisted reproduction techniques in couples serodiscordant for human

immunodeficiency virus where the man is positive. Fertil Steril 2011;95:1684–90.

[48] Bujan L, Hollander L, Coudert M, et al. Safety and efficacy of sperm washing in HIV-1-serodiscordant couples where the male is infected: results from the European CREAThE network. AIDS 2007;21:1909–14.

[49] van Rossum AM, Fraaij PL, de Groot R. Efficacy of highly active antiretroviral therapy in HIV-1 infected children. Lancet Infect Dis 2002;2:93–102.

[50] Resino S, Resino R, Maria Bellon J, et al. Clinical outcomes improve with highly active antiretroviral therapy in vertically HIV type-1-infected children. Clin Infect Dis 2006;43:243–52.

[51] Rouet F, Montcho C, Rouzioux C, et al. Early diagnosis of paediatric HIV-1 infection among African breast-fed children using a quantitative plasma HIV RNA assay. AIDS 2001;15:1849–56.

[52] Sherman GG, Cooper PA, Coovadia AH, et al. Polymerase chain reaction for diagnosis of human immunodeficiency virus infection in infancy in low resource settings. Pediatr Infect Dis J 2005;24:993–7.

[53] Stevens WS, Noble L, Berrie L, Sarang S, Scott LE. Ultra-high-throughput, automated nucleic acid detection of human immunodeficiency virus (HIV) for infant infection diagnosis using the Gen-Probe Aptima HIV-1 screening assay. J Clin Microbiol 2009;47:2465–9.

[54] Owen SM, Yang C, Spira T, et al. Alternative algorithms for human immunodeficiency virus infection diagnosis using tests that are licensed in the United States. J Clin Microbiol 2008;46: 1588–95.

[55] Anoje C, Aiyenigba B, Suzuki C, et al. Reducing mother-to-child transmission of HIV: findings from an early infant diagnosis program in south-south region of Nigeria. BMC Public Health 2012;12:184.

[56] Torpey K, Mandala J, Kasonde P, et al. Analysis of HIV early infant diagnosis data to estimate rates of perinatal HIV transmission in Zambia. PLoS One 2012;7:e42859.

[57] World Health Organization. Consolidated guidelines on the use of antiretroviral drugs for treating and preventing HIV infection: recommendations for a public health approach, June 2013. Available from: <http://apps.who.int/iris/bitstream/10665/85321/1/9789241505727_eng.pdf> [accessed March 2015].

[58] Lilian RR, Kalk E, Bhowan K, et al. Early diagnosis of in utero and intrapartum HIV infection in infants prior to 6 weeks of age. J Clin Microbiol 2012;50:2373–7.

[59] World Health Organization. 2014. Global update on the health sector response to HIV. Available from: <http://www.who.int/hiv/pub/progressreports/update2014/en/> [accessed March 2015].

[60] Rouet F, Rouzioux C. HIV-1 viral load testing cost in developing countries: what's new? Expert Rev Mol Diagn 2007;7:703–7.

[61] Shafiee H, Wang S, Inci F, et al. Emerging technologies for point-of-care management of HIV infection. Annu Rev Med 2015;66:387–405.

[62] Cassol SA, Gill MJ, Pilon R, et al. Quantification of human immunodeficiency virus type 1 RNA from dried plasma spots collected on filter paper. J Clin Microbiol 1997;35:2795–801.

[63] Fiscus SA, Brambilla D, Grosso L, Schock J, Cronin M. Quantitation of human immunodeficiency virus type 1 RNA in plasma by using blood dried on filter paper. J Clin Microbiol 1998;36:258–60.

[64] Biggar RJ, Broadhead R, Janes M, Kumwenda N, Taha TET, Cassol S. Viral levels in newborn African infants undergoing primary HIV-1 infection. AIDS 2001;15:1311–13.

[65] Brambilla D, Jennings C, Aldrovandi G, et al. Multicenter evaluation of use of dried blood and plasma spot specimens in quantitative assays for human immunodeficiency virus

RNA: measurement, precision, and RNA stability. J Clin Microbiol 2003;41:1888–93.

[66] Mwaba P, Cassol S, Nunn A, et al. Whole blood versus plasma spots for measurement of HIV-1 viral load in HIV-infected African patients. Lancet 2003;362:2067–8.

[67] de Baar MP, Timmeremans EC, Buitelaar M, et al. Evaluation of the HIV-1 RNA RetinaTM Rainbow assay on plasma and dried plasma spots: correlation with the Roche Amplicor HIV-1 Monitor v 1.5 assay. Antivir Ther 2003;8:S533.

[68] Uttayamakul S, LikAnonsakul S, Sunthornkachit R, et al. Usage of dried blood spots for molecular diagnosis and monitoring HIV-1 infection. J Virol Methods 2005;128:128–34.

[69] Luo W, Yang H, Rathbun K, Pau CP, Ou CY. Detection of human immunodeficiency virus type 1 DNA in dried blood spots by a duplex real-time PCR assay. J Clin Microbiol 2005; 43:1851–7.

[70] Sherman GG, Stevens G, Jones SA, Horsfield P, Stevens WS. Dried blood spots improve access to HIV diagnosis and care for infants in low-resource settings. J Acquir Immune Defic Syndr 2005;38:615–17.

[71] Lofgren SM, Morrissey AB, Chevallier CC, et al. Evaluation of a dried blood spot HIV-1 RNA program for early infant diagnosis and viral load monitoring at rural and remote healthcare facilities. AIDS 2009;23:2459–66.

[72] Kerr RJ, Player G, Fiscus SA, Nelson JA. Qualitative human immunodeficiency virus RNA analysis of dried blood spots for diagnosis of infections in infants. J Clin Microbiol 2009;47: 220–2.

[73] Viljoen J, Gampini S, Danaviah S, et al. Dried blood spot HIV-1 RNA quantification using open real-time systems in South Africa and Burkina Faso. J Acquir Immune Defic Syndr 2010; 55:290–8.

[74] Huang S, Erickson B, Mak WB, Salituro J, Abravaya K. A novel realtime HIV-1 qualitative assay for the detection of HIV-1 nucleic acids in dried blood spots and plasma. J Virol Methods 2011;178:216–24.

[75] Nkenfou CN, Lobé EE, Ouwe-Missi-Oukem-Boyer O, et al. Implementation of HIV early infant diagnosis and HIV type 1 RNA viral load determination on dried blood spots in Cameroon: challenges and propositions. AIDS Res Hum Retroviruses 2012;28:176–81.

[76] Okonji JA, Basavaraju SV, Mwangi J, et al. Comparison of HIV-1 detection in plasma specimens and dried blood spots using the Roche COBAS Ampliscreen HIV-1 test in Kisumu, Kenya. J Virol Methods 2012;179:21–5.

[77] Rouet F, Rouzioux C. The measurement of HIV-1 viral load in resource-limited settings: how and where? Clin Lab 2007; 53:135–48.

[78] Bourlet T, Memmi M, Saoudin H, Pozzetto B. Molecular HIV screening. Expert Rev Mol Diagn 2013;13:693–705.

[79] Alidjinou EK, Bocket L, Hober D. Quantification of viral DNA during HIV-1 infection: a review of relevant clinical uses and laboratory methods. Pathol Biol (Paris) 2015;63:53–9.

[80] Palmer S. Advances in detection and monitoring of plasma viremia in HIV-infected individuals receiving antiretroviral therapy. Curr Opin HIV AIDS 2013;8:87–92.

[81] Strain MC, Richman DD. New assays for monitoring residual HIV burden in effectively treated individuals. Curr Opin HIV AIDS 2013;8:106–10.

[82] Hindson BJ, Ness KD, Masquelier DA, et al. High-throughput droplet digital PCR system for absolute quantitation of DNA copy number. Anal Chem 2011;83:8604–10.

[83] Eriksson S, Graf EH, Dahl V, et al. Comparative analysis of measures of viral reservoirs in HIV-1 eradication studies. PLoS Pathog 2013;9:e1003174.

[84] Torre D, Tambini R. Antiretroviral drug resistance testing in patients with HIV-1 infection: a meta-analysis study. HIV Clin Trials 2002;3:1–8.

[85] Panidou ET, Trikalinos TA, Ioannidis JP. Limited benefit of antiretroviral resistance testing in treatment-experienced patients: a meta-analysis. AIDS 2004;18:2153–61.

[86] Schutten M. Resistance assays. In: Geretti AM, editor. Antiretroviral resistance in clinical practice. London: Mediscript; 2006 [chapter 5]. Available from: <http://www.ncbi.nlm.nih.gov/books/NBK2252/> [accessed March 2015]

[87] Smit E. Antiviral resistance testing. Curr Opin Infect Dis 2014;27:566–72.

[88] Fuller CW, Middendorf LR, Benner SA, et al. The challenges of sequencing by synthesis. Nat Biotechnol 2009;27:1013–23.

[89] Barzon L, Lavezzo E, Costanzi G, Franchin E, Toppo S, Palù G. Next-generation sequencing technologies in diagnostic virology. J Clin Virol 2013;58:346–50.

[90] Chen CY. DNA polymerases drive DNA sequencing-by-synthesis technologies: both past and present. Front Microbiol 2014;5:305.

[91] Pou C, Noguera-Julian M, Pérez-Álvarez S, et al. Improved prediction of salvage antiretroviral therapy outcomes using ultrasensitive HIV-1 drug resistance testing. Clin Infect Dis 2014;59:578–88.

[92] Swenson LC, Mo T, Dong WW, et al. Deep sequencing to infer HIV-1 co-receptor usage: application to three clinical trials of maraviroc in treatment-experienced patients. J Infect Dis 2011;203:237–45.

[93] Swenson LC, Mo T, Dong WW, et al. Deep V3 sequencing for HIV type 1 tropism in treatment-naive patients: a reanalysis of the MERIT trial of maraviroc. Clin Infect Dis 2011;53:732–42.

[94] Fokam J, Billong SC, Bissek AC, et al. Declining trends in early warning indicators for HIV drug resistance in Cameroon from 2008–2010: lessons and challenges for low-resource settings. BMC Public Health 2013;13:308.

[95] World Health Organization. Meeting report on assessment of World Health Organization HIV drug resistance early warning indicators: report of the Early Advisory Indicator Panel meeting, 11–12 August 2011, Geneva, Switzerland. Available from <http://apps.who.int/iris/bitstream/10665/75186/1/9789241503945_eng.pdf> [accessed March 2015].

6

Molecular Testing in Hepatitis Virus Related Disease

P.M. Mulrooney-Cousins and T.I. Michalak

Molecular Virology and Hepatology Research Group, Division of BioMedical Science, Faculty of Medicine, Health Sciences Centre, Memorial University, St. John's, Newfoundland, Canada

INTRODUCTION

It should be noted that all information regarding the sensitivity or lowest limit of detection (LLD) of viral genomes by the assays discussed in this chapter was based on the data reported for analysis of individual serum or plasma samples, but not their pools or nucleic acid preparations from infected cells or tissues. In this context, it should be indicated that the blood bank screening schemes globally utilize plasma or serum minipools (MP) and this tends to decrease the overall sensitivity of virus detection in individual samples. It also should be noticed that in cases where nucleic acids are subjected to multistep preparation and amplification processes, thoughtful standardization of the assay protocols, especially with regard to collection and storage of samples, nucleic acid isolation procedures, and reliability of the specificity and quantitative controls used are of paramount importance. Minor deviations in this regard may result in inconsistent or uninterpretable results, particularly when samples with low levels of viral genomes are examined. In addition, the results of many tests for quantitative detection of hepatitis viruses are currently presented in international units (IU) per mL of serum or plasma but not in actual virus genome copy numbers (also called virus genome equivalents, vge) per mL. This can be of concern and a confusing factor when comparing loads for some hepatitis viruses, as it is in the case of hepatitis C virus (HCV), where for some assays arbitrary designated IU by individual producers are used. For these reasons, interpretation of the data from molecular assays applied for quantitative detection of

viral pathogens causing hepatitis requires basic understanding of the principles of the assays utilized by the local clinical laboratory and their relevance to the results from other laboratories or those reported in literature. This is particularly important when diagnosis of virus clearance, based solely on testing of the circulating virus, and termination of antiviral therapy are considered.

HAV AND HAV RNA DETECTION TESTS

Hepatitis A virus (HAV) is a nonenveloped virus that belongs to the *Hepatovirus* genus of the *Picornaviridae* family. Its genome is comprised of a linear, single-stranded RNA approximately 7.5 kb in length. There is some genetic variability across HAV isolates, mainly in the junction between virus proteins 1 and 2A, resulting in the recognition of six genotypes (I–VI). HAV causes acute hepatitis that does not result in chronic illness or apparent long-term persistence of the virus. However, fulminant hepatitis is observed in about 1% of those infected. The World Health Organization (WHO) estimates that there are about 1.4 million cases globally, despite the availability of very effective prophylactic vaccines [1,2]. The infection is predominately transmitted by fecal–oral route, but it can also be passed by transfusion of contaminated blood.

Serological diagnosis of hepatitis A is based on the detection of anti-HAV antibodies of the IgM class, indicating acute infection. Detection of anti-HAV IgG antibodies is indicative of the past exposure to HAV or clinical convalescence from the recent acute infection.

© 2017 Elsevier Inc. All rights reserved.

Molecular tests designed to identify HAV RNA are rarely applied for clinical diagnostic purposes. However, polymerase chain reaction (PCR) with reverse transcription (RT) step (RT-PCR) has demonstrated that viremia can be detected prior to antibody development (window period infection) and that virus can persist into the clinically defined convalescence for months after onset of icterus [3–5]. Nested RT-PCR is rarely utilized for HAV RNA identification, mainly to increase sensitivity of virus detection in research setting or in cases where there are no anti-HAV antibodies detected but exposure to HAV is suspected [3]. HAV RNA testing is not incorporated in blood bank screening.

Real-time RT-PCR with SYBR-green, Taqman probes, and molecular beacons, mainly targeting the 5'-noncoding region (5'-NCR), have been used for detection of HAV genome in research laboratory settings and in outbreak situations [6–8]. These tests utilize RNA extracted from serum, saliva, fecal, and/or environmental samples.

There are no tests for HAV RNA detection approved for clinical use by the FDA. However, there are a number of commercial assays and those most frequently used are listed in Table 6.1. There is a WHO international standard for HAV RNA (identification code: 00/560) and a CE (Conformité Européene)-certified HAV RNA working reagent (identification code: 01/488) from the National Institute for Standards and Controls (NIBSC; Hertfordshire, UK) that can be used to validate and determine sensitivity of in-house developed nucleic acid tests (NATs). The availability of these reference standards significantly improves consistency of the data reporting and interpretation from HAV RNA detection tests generated in various laboratories.

HBV INFECTION

Infection with hepatitis B virus (HBV), despite the availability during the last three decades of effective prophylactic vaccines, remains the major cause of life-threatening liver diseases, such as chronic or fulminant hepatitis culminating in liver failure, cirrhosis, and primary hepatocellular carcinoma (HCC) [9]. HBV is one of the smallest enveloped DNA viruses known and its highly compact genome is composed of four overlapping open reading frames. HBV has a very unique replication strategy, whereby the virus partially double-stranded DNA is fully repaired in the nucleus of infected cells to form a mini-chromosome, referred to as circular covalently closed DNA (cccDNA), which serves as the template for the virus mRNA transcripts. Detection of HBV cccDNA and mRNA are considered to be reliable indicators of active virus replication [10]. HBV is classified into at least eight distinctive genotypes (A–H), with inter-genotypic divergence of 8% or more within the complete genome sequence [11]. HBV infection occurs as serologically apparent, that is, HBV surface antigen (HBsAg)-positive, chronic infection in at least 370 million people worldwide (Fig. 6.1) [12]. Recent WHO estimates imply that up to 2 billion people have been exposed to HBV, which further emphasizes a global epidemiological and potential pathogenic significance of this virus [12,13]. Serological assays for the identification of HBV antigens and respective antibodies have been widely used for decades. However, with the relatively recent advent of commercial assays for detection of HBV DNA and highly sensitive research assays [12,14,15], it became apparent that the virus replicates from the earliest

TABLE 6.1 Assays for HAV RNA Detection and Quantification

Assay	Manufacturer	Method used	Sensitivity		Dynamic range	
			IU/mL	copies/mL	IU/mL	copies/mL
RealStar HAV RT-PCR Kit 1.0	Altona Diagnostics	RT-PCR	~12	n.p.	n.a.	n.a.
hepatitis A@ceeramTools.health	CEERAM S.A.S.	qRT-PCR	n.p.	5–50/reaction[b]	n.p.	n.p
HAV Real time RT-PCR Kit	Liferiver, Gentaur	RT-PCR	n.p.	n.p.	n.a.	n.a.
UltraQual-100, 2X HAV RT-PCR	National Genetics Institute	RT-PCR	2.08	n.p.	n.a.	n.a.
artus HAV LightCycler RT-PCR Kit	QiAgen	qRT-PCR	50	n.p.	1×10^4–1×10^8	n.p.
COBAS TaqScreen DPX Test[a]	Roche Diagnostics	RT-PCR	1.1	n.p.	n.a.	n.a.
LightCycler HAV Quantification Kit	Roche Diagnostics	qRT-PCR	n.p.	500	n.p.	2×10^4–2×10^8

[a]*Can also quantify parvovirus B1.9.*
[b]*Amount of input RNA not given.*
IU, international units; qRT-PCR, quantitative RT-PCR; n.p., not provided; n.a., not applicable.

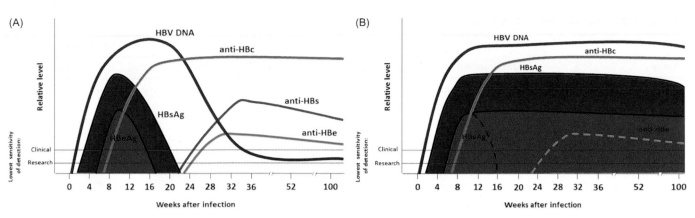

FIGURE 6.1 Serological (immunovirological) and molecular markers detectable during the course of HBV infection. (A) Self-limited acute HBV infection with serologically evident recovery followed by a low-level HBV DNA persistence reflecting existence of asymptomatic (occult) HBV infection (OBI). (B) Serum HBsAg-positive chronic HBV infection typically accompanied by chronic hepatitis B. Colored lines and areas under the lines indicate relative levels of HBsAg (blue), HBeAg (orange), anti-HBc (green), anti-HBs (yellow), anti-HBe (purple), and HBV DNA (red) detected in serum or plasma. The lowest sensitivity limits of HBV DNA detection by standard clinical or research assays are indicated by dashed black lines. HBeAg may or may not be detectable during the course of chronic HBV infection, as indicated by the dashed orange line, as well as anti-HBe may develop or not, as marked by dashed purple line.

stages after exposure and persists at low levels as occult HBV infection for decades, if not for lifetime, after clinical resolution of either self-limited acute hepatitis or therapy-induced recovery from chronic hepatitis B (Fig. 6.1A) [14,16]. In addition to serum HBsAg, it is now evident that detection of antibodies to HBV core (nucleocapsid) antigen (anti-HBc) and/or HBV DNA alone, that is, in the absence of detectable serum HBsAg, is indicative of prior exposure and persistence of biologically competent virus at low levels [16–20]. The availability of molecular tests for detection of HBV DNA in clinic has drastically improved identification of HBV infection, evaluation of its responsiveness to antiviral therapy, and enhanced safety of blood and organ donations [21]. However, the persistence of HBV at low levels, frequently not detectable by the current clinical laboratory tests, in a significant portion of individuals with asymptomatic infection and the findings of occult HBV infection relevance to the pathogenesis of HCC [15,22,23] highlights the need for continual improvements to the sensitivity of molecular assays detecting HBV nucleic acids and to their employment in clinic, and for population, blood, and organ donation screenings.

HBV DNA Detection Assays

Molecular assays for HBV DNA identification are of increasing importance in the diagnosis and therapeutic management of chronic hepatitis B, as well as in the identification of window period infections and occult HBV persistence occurring in the absence of clinical symptoms and serological markers of infection [24].

Many commercial assays are now available for HBV DNA qualitative and quantitative identification, but they are limited to detection of virus genome in serum or plasma only. There are no currently commercial assays capable of identifying active HBV replication by analyzing virus genome replication intermediates in infected tissues or cells. There are also limitations in regard to the sensitivity of HBV DNA detection, the amount of sample material required for testing, and if testing is of an individual sample or MP [25]. In regard to testing of sample MP, it has recently been acknowledged that the sensitivity of the currently available NAT is related to the number of individual donations within the MP tested [26,27]. Also, HBV DNA reactive MP, where individual samples were found to be HBV DNA nonreactive, is currently identified as one of the most puzzling findings which impedes full identification of potentially infected donor units [28,29].

A WHO international standard, which is available from the NIBSC (identification code: NIBSC-10/264) and a CE-certified HBV DNA working reagent (identification code: 11/182-001) are available as references for determination of LLD and the linear ranges of HBV DNA detection assays. It is accepted that approximately five copies of HBV DNA are equivalent to 1 IU [30].

Qualitative detection of HBV DNA in blood donors is considered as sufficient [31], but for the evaluation of viral load in plasma for monitoring of the effectiveness of anti-HBV therapy quantitative assessment is required [32]. Table 6.2 presents some of the commonly used HBV DNA tests currently available. It is important to note that the sensitivities of these assays and their linear detection ranges differ, and that not all of them

TABLE 6.2　Assays for HBV DNA Detection and Quantification

Assay	Manufacturer	Method used	Sensitivity		Dynamic range	
			IU/mL	copies/mL	IU/mL	copies/mL
HBV SuperQuant PCR Assay	National Genetics Institute	qPCR	n.p.	100	n.p.	$1 \times 10^2 - 5 \times 10^9$
artus HBV LC PCR Kit	QiAgen	qPCR	~6	n.p.	$31.6 - 1 \times 10^{10}$	n.p.
artus HBV TM PCR Kit	QiAgen	qPCR	~4	n.p.	$8.8 - 5.6 \times 10^9$	n.p.
COBAS Ampliprep/COBAS TaqMan HBV Test v2.0	Roche Diagnostics	Automated qPCR	20	n.p.	$20 - 1.7 \times 10^8$	n.p.
COBAS TaqMan HBV Test	Roche Diagnostics	qPCR	6	n.p.	$29 - 1.1 \times 10^8$	n.p.
COBAS TaqMan MPX[a]	Roche Diagnostics	PCR	3	n.p.	n.a.	n.a.
HBV Ampliscreen	Roche Diagnostics	PCR	5	~50	n.a.	n.a.
Versant HBV bDNA 3.0	Seimens	Branched DNA	n.p.	2×10^3	n.p.	$2 \times 10^3 - 1 \times 10^8$

[a]*Multiplex, can also detect HCV, HIV-1, and HIV-2.*
IU, international units; qPCR, quantitative PCR; n.p., not provided; n.a., not applicable.

are approved for clinical purposes. Therefore, whenever possible, it is important to use one of the approved tests to determine HBV loads in a patient prior to and on antiviral therapy, especially if the available assay does not include quantitative standards.

Since most of the current HBV DNA tests are based on targeted PCR amplification, they are highly specific with less than 3% of seronegative subjects showing false-positive results [33]. As mentioned, there have been problems with reproducibility at the LLD that may occur in samples obtained during window period infections and those during occult HBV infection continuing after clinically resolved or primary silent infection [33,34]. Real-time PCR-based assays have somewhat circumvented this problem, such as COBAS (Roche Molecular Diagnostics, Pleasanton, CA) and Artus HBV DNA real-time PCR assay (Qiagen, Valencia, CA). However, the underlying problems with interpretation of results from real-time PCR are the variety of methods applied for serum or plasma acquisition, preparation and storage, and methods of DNA extraction. A more standardized approach has come with the fully automated extractions and amplifications offered by Roche Molecular Diagnostics, known as the COBAS-Ampliprep technology. The automated nucleic acid extractions enable "samples in/results out" and it eliminates any manual intervention by laboratory staff. Recently, there also became available multiplex assays capable of simultaneously testing for HIV type 1, HBV, and HCV in serum or plasma, for example, Procleix Ultrio Plus test (Grifols, Barcelona, Spain). Some of these tests are FDA approved and they are becoming applied for blood safety testing and in screening of donors of tissues and organs for transplantation. Some of the multiplex

systems are fully automated, for example, the Procleix Tigris instrument (Grifols via Novarits Diagnostics) and the COBAS TaqScreen MPX Test (Roche Molecular Diagnostics), decreasing likelihood for user-based errors.

HBV Genotyping

It has become evident that HBV genotype is an important factor in determining the outcomes of therapeutic strategies against HBV, especially in endemic areas where infections with genotypes B and C are most prevalent [35]. For example, interferon alpha (IFN-α) therapy induces HBV seroconversion more often in patients with genotype A or B than in those with genotype C or D [36–38]. On the other hand, genotypes C, D, and F have been associated with faster progressing and more severe liver disease and HCC development than other HBV genotypes [39,40]. Although HBV genotyping is not yet commonly performed, it progressively becomes an important clinical tool since, as mentioned, different HBV genotypes respond to therapy differently. Certain laboratories now offer HBV genotyping of clinical samples by employing such techniques as direct sequencing, for example, Mayo Medical Laboratories (Andover, MA) and ARUP laboratories (Salt Lake City, UT). Further, sequencing of the whole HBV genome or the S gene and genotype-specific PCR have also been shown to give reliable results. However, these approaches are time-consuming and costly. At the present time, only one assay, INNO-LiPA (line probe assay), developed by Innogenetics (now Fujirebio Europe N.V., Ghent, Belgium) is commercially available [41]. This assay employs a line-probe hybridization to the pre-S1 sequence of the S gene to determine HBV genotype. The accuracy of this assay to identify a singular genotype is as high as 99% and it offers an alternative for samples which are difficult to sequence. For direct sequencing, a kit TRUGENE from Seimens Medical Solutions (Cary, NC) is available. It can be used in samples with relatively low HBV loads (near as 1000 genome copies per mL) [42]. However, plasma HBV load levels in patients with occult HBV infection are normally lower than 100–200 genome copies per mL and are below this detection level. HBV Sequencing Assay from Abbott Molecular Inc. is a CE-certified test that requires 500 µL of patient serum or plasma and uses a sequencer from Applied Biosystems Inc. (Grand Island, NY) to identify all known HBV genotypes. However, this system is not FDA approved as screening or diagnostic tool.

Molecular Testing for HBV Antiviral Resistance

Monitoring for development or breakthrough of antiviral drug resistant mutants is of central importance in the management of patients on nucleotide/nucleoside-analogue antiviral therapy. These classes of drugs lead to virus-adaptive mutations, mainly in HBV reverse transcriptase region of the DNA polymerase gene, which culminates in virus breakthrough, rendering therapy essentially useless. In general, when patients who are on anti-HBV therapy are found to have rising serum HBV loads, the emergence of antiviral resistant mutants should be assumed. HBV rebound is defined as a recovery of virus replication due to compensatory mutations that help restore viral fitness and, thereby, increase virus plasma loads by more than 1-log [43]. There are very few commercial assays capable of identifying such escape mutants. For example, the reverse hybridization test INNO-LiPA DR (Fujirebio Europe N.V.) has an expanded lamivudine (LMV, also known as 3TC) resistance panel and also can detect HBV adefovir resistant mutations [44]. This assay offers an advantage over direct sequencing in that mixed viral populations are not often detected by the previously mentioned TRUGENE test. Another commercial assay is the Affigene HBV DE/3TC test from Sangtec Molecular Diagnostics AB (Bromma, Sweden). This test combines hybridization and direct sequencing to detect LMV resistant mutants. It works comparably well but has higher undeterminate rates than the INNO-LiPA DR assay [45]. These two commercial kits however identify only mutant sequences if they comprise 5% (detectable by hybridization) or 20% (detectable by direct sequencing) of the total virus population.

More recent approaches to HBV genotyping, identification of variants, and drug escape mutants in mixed viral populations are those utilizing ultra-deep pyrosequencing [46]. They have a significant advantage over the current assays because they also identify novel mutations and mixed genotypes at once [47]. The current hybridization-based assays are capable of highly specific discrimination, but only of known nucleotide mutations. Therefore, if new mutations arise, the line probe assays must be updated [48]. Recently published studies identifying hepatotropic viruses from plasma using deep sequencing have provided much new information [47]. At this stage, however, this likely next-generation diagnostic tool is too complex and expensive to be employed for routine clinical testing.

Other complex yet highly sensitive techniques, such as restriction fragment mass polymorphism and oligonucleotide microarray chips, can detect mutations of less than 1% of the total virus population [40]. The associated costs make these techniques unfeasible for diagnostic use at the present time [46]. It is also becoming increasingly evident that very minor variant populations, called quasi-species, with antiviral resistance may exist prior to the initiation of therapy [49,50]. Currently, antiviral resistance testing is not performed prior to treatment, except in some research trial settings. These naturally occurring drug resistant variants constitute well below 5% of the wild-type virus pool, which is the current limit of mutant genotypic identification in patients with overt HBV infection.

HCV AND HCV RNA DETECTION ASSAYS

HCV is a positive-strand RNA virus that belongs to the *Hepacivirus* genus of the family *Flaviviridae*. Its 9.6-kb-long genome encodes a polyprotein of almost 3000 amino acids flanked by two untranslated regions, that is, 3′- and 5′-UTR. HCV isolates are categorized into six genotypes with 30–35% sequence variability with further variation within subtypes. HCV genotypes are essentially geographically distributed. Approximately 150 million people worldwide suffer from chronic HCV infection, and about 30% of those will develop cirrhosis and potentially HCC [51]. There is no prophylaxic vaccine yet available, which is one of main reasons why HCV infection remains of a global concern.

Testing of serum or plasma for HCV RNA by NAT is now accepted as the standard for diagnosis of ongoing hepatitis C and its resolution (Fig. 6.2) [52–55], and for the screening of blood, blood products, and organ and tissue donations [53]. Different molecular approaches, such as RT-PCR, branched DNA (bDNA) amplification, and transcription-mediated amplification, were and are

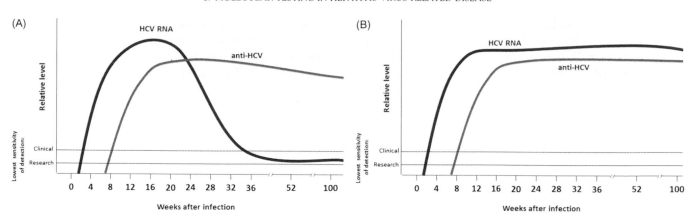

FIGURE 6.2 Typical serological and molecular profiles of HCV infection. (A) Acute HCV infection with recovery. After resolution of clinical evident acute hepatitis C, HCV RNA can remain detectable at very low levels for the prolonged period of time. (B) Chronic HCV infection. Colored lines indicate relative levels of anti-HCV antibodies (green) and HCV RNA (red) detected in serum or plasma. The lowest sensitivity limits of HCV RNA detection based on standard clinical or research assays are indicated by dashed black lines.

TABLE 6.3 Assays for HCV RNA Detection and Quantification

Assay	Manufacturer	Method used	Sensitivity		Dynamic range	
			IU/mL	copies/mL	IU/mL	copies/mL
RealTime HCV/m2000sp/m2000rt	Abbott Molecular	Automated qRT-PCR	12	n.p.	$12-1 \times 10^8$	n.p.
Hepatitis C Virus RT-PCR Assay	Biolife Plasma Services	RT-PCR	n.p.	n.p.	n.a.	n.a.
HCV SuperQuant	National Genetics Institute	qRT-PCR	n.p.	100	n.p.	$1 \times 10^2-5 \times 10^6$
artus HCV RG RT-PCR	QiAgen	qRT-PCR	33.6	n.p.	$65-1 \times 10^6$	n.p.
COBAS Amplicor HCV v2.0	Roche Diagnostics	Automated RT-PCR	50–60	n.p.	n.a.	n.a.
COBAS Ampliprep/TaqMan HCV v2.0	Roche Diagnostics	Automated qRT-PCR	15	n.p.	$15-1 \times 10^8$	n.p.
COBAS Ampliscreen HCV v2.0	Roche Diagnostics	Automated RT-PCR	21	57	n.a.	n.a.
COBAS TaqMan HCV v2.0	Roche Diagnostics	qRT-PCR	9.3	n.p.	$25-3.9 \times 10^8$	n.p.
COBAS TaqMan MPX[a]	Roche Diagnostics	RT-PCR	11	n.p.	n.a.	n.a.
Versant HCV RNA Test 1.0 (kPCR)	Siemens	qRT-PCR	37	n.p.	$37-1.1 \times 10^7$	n.p.
Versant HCV RNA Test 3.0	Siemens	Branched DNA	615	n.p.	$6.2 \times 10^2-7.7 \times 10^6$	n.p.

[a]*Multiplex, can also detect HBV, HIV-1, and HIV-2.*
IU, international units; qRT-PCR, quantitative RT-PCR; n.p., not provided; n.a., not applicable.

utilized to develop HCV RNA detection assays. Several of the commercially available tests are listed in Table 6.3. It should be noted that not all of them have FDA approval for diagnosis and clinical monitoring purposes. However, most of them are quantitative and demonstrate acceptable specificity and sensitivity. Many of the NAT currently available utilize as a quantitative reference the WHO HCV 4th International Standard for HCV RNA amplification techniques (identification code: 06/102). Also CE-certified HCV RNA working reagent (identification code: 02/264-003) is available from the NIBSC. Although there is now somewhat less discrepancy in reporting on the detectable viral amounts by individual assays, there are still issues with conversion of the LLD values to virus genome copy equivalents for some assays and this information is not always easily available from the supplier. Also, in many of the assays that are not fully automated, it is up to the end user which methodology for recovery of template HCV RNA from patients' samples is applied. In general, as in the case of HBV DNA detection, preferably the same assay should be used when determining viral loads in patients before, during, and after therapy to avoid potential discrepancies. It should also be noted that many in-house assays, which use more efficient methods of RNA extraction, testing of different amounts of RNA template, and applying customized PCR amplification conditions and amplicon detection methods, have better LLD than clinical assays. However, they are time-consuming and require expertise that is not often available in the clinical laboratory setting. Nonetheless, these in-house assays significantly contributed to our better understanding of the natural history and initially unapparent consequences of HCV infection. For example, they played a central role in the identification of long-term occult persistence of small amounts of infectious HCV after either spontaneous or therapy-induced resolution of hepatitis C [56–59], and the discovery that immune cells are commonly, if not invariably, infected by HCV regardless of whether the infection is symptomatic or clinically unapparent [56,59–62].

HCV Genotyping

HCV genotyping is considered essential for determining the type and duration of anti-HCV therapy, since different genotypes and even subtypes respond differently to both IFN-α/ribavirin and directly acting antivirals (DAAs) [63,64]. In addition to standards for HCV quantitation, there is now a HCV RNA genotype performance panel (Sera Care Life Sciences, Milford, MA) that can be used to validate assays designed to delineate genotypic and major subtypes of HCV sequences. Some commercially available assays use direct cDNA sequencing (eg, TRUGENE HCV Genotyping Assay, Siemens Medical Solutions) and a line probe hybridization for 5′-UTR and core gene (eg, INNO-LiPA HCV II Genotype Test, Fujirebio Europe N.V.) for differentiation of HCV genotypes in serum from patients with chronic hepatitis C [63]. The only FDA-approved assay is Abbott RealTime HCV Genotype II which can differentiate genotypes 1, 1a, 1b, 2, 3, 4, and 5 and has been found to be applicable in the clinical setting [65]. This is a significant breakthrough. In the era of DAAs against HCV, access to highly sensitive and specific assays which are reliable in identifying HCV genotypes while monitoring patients' viral load in responses to treatment will remain of a very high priority.

HDV AND HDV RNA DETECTION

Hepatitis D virus (HDV) is considered a subviral agent that requires the lipoprotein envelope of HBV to enable a 1.7 kb RNA genome entry to hepatocyte. The HDV genome encodes a single protein (HDAg) that can exist in two isoforms with corresponding molecular weights of 24 and 27 kDa. Because of the need of the HBV envelope protein to permit HDV infection, the existence of serum HBsAg-positive HBV infection in concert with antibodies to HDV (anti-HDV) is usually required for the diagnosis of hepatitis D. HDV infection is considered to be increasing globally, and some countries, like France and Italy, have noted greater numbers of HDV cases during the past two decades [66]. Current estimates suggest that 15–20 million people are infected globally, despite the availability of the HBV prophylactic vaccine which protects against HDV infection [67]. Hepatitis D is a clinically significant illness that leads to fulminant hepatitis 10 times more often and with 2–20% greater mortality rates than other types of viral hepatitis. HDV coinfection or superinfection tends to exacerbate chronic hepatitis B leading to higher rate of liver cirrhosis and HCC than in those infected with HBV alone [67].

Molecular testing for HDV RNA in serum has been attempted using full-length cDNA probes, RNA hybridization, and RT-PCR. In-house qualitative and/ or semiquantitative RT-PCR tests have been widely used for the study of HDV, but considering that there are eight genotypes that can be up to 40% divergent, designing of primers to detect different genotypes is highly challenging. Most consistent results were obtained using primers corresponding to the genome fragment covering the ribozyme, due to the conserved sequence of this region. It has been reported, using an in-house test with primers enabling detection of all genotypes, that the sensitivity of 100 genome copies per mL was achieved for detection of all HDV genotypes [68]. Such assays are not yet commercially available, despite having proved useful for monitoring serum HDV loads over the course of IFN-α therapy. Two assays are commercially available for research purposes only: LightMix Kit HDV from TIB MOLBIOL GmbH (Berlin, Germany) for use on the Roche LightCycler, which detects genotype 1 (linear range $10–10^6$ copies/ reaction) and HDV RNA Quantitation DRNA from Dia. Pro Ltd. (Milan, Italy), which has LLD of 300 copies of HDV RNA per mL with a linear range of $10^3–10^{12}$ copies per reaction. This assay has CE mark designation, but is not used for diagnostic purposes. It has been noted that the commercial assay may give underestimation of true HDV RNA loads [69], when compared to in-house assays. It is also notable that there is no international quantitative standard for HDV.

HEV AND DETECTION OF HEV RNA

Hepatitis E virus (HEV) is a nonenveloped, RNA-positive strand virus of the family *Hepeviridae* that infects estimated 20 million people globally, mainly in Eastern and Southern Asia. Hepatitis E is considered to be a self-limiting disease, but HEV infection can cause fulminant hepatitis, particularly in pregnant woman [70]. The virus is usually spread by fecal—oral route, but also can be transmitted through blood products and vertically from mother to fetus. HEV causes a high risk infection in pregnant women and fulminant hepatitis much more frequently than infections with other hepatitis viruses leading to 20—25% mortality rate among third-trimester women. There are four genotypes of HEV with up to 27.7% sequence divergence among them and several minor subtypes [71,72]. Genotypes 1 and 2 are pathogenic to humans, while genotypes 3 and 4 are considered to be animal strains that are rarely transmitted to people. A prophylactic vaccine against HEV was developed in China and was registered in 2011 [73], but it is not globally available.

In practice, the specific diagnosis of HEV is based on the detection of virus-specific IgM and/or IgG antibodies, and molecular NATs are not routinely used. However, strides in the standardization of HEV RNA detection and monitoring HEV infection by RT-PCR have exceeded that for HDV [74]. As of October 2011, there has been a WHO international standard for HEV RNA genotype 3A diluted in plasma and lyophylized to yield a standard of 2.5×10^5 IU/mL. There are several RT-PCR tests available for detection of HEV RNA, many with CE mark designation, but none have FDA approval (Table 6.4). Some of these tests are able to detect all four genotypes at levels as low as 50 genome copies per mL.

GB VIRUS C AND MOLECULAR DETECTION OF GB VIRUS C GENOME

GB virus C (GBV-C), also known as hepatitis G virus (HGV), is a member of the *Flaviviridae* family. Its genome consists of a positive-sense-stranded RNA encoding a single polyprotein of approximately 3000 amino acids. Currently, GBV-C is assumed to be a lymphotropic virus that does not cause hepatitis [75—77]. However, since historically HGV was considered to be hepatitis virus, we briefly discussed molecular detection of its genome.

Progressing GBV-C infection can only be diagnosed by identifying genomic RNA. The exact sensitivity and specificity of testing for HGV RNA using in-house assays are not well established. Antibodies to GBV-C appear to be only detectable when GBV-C RNA is no longer detected, and the true seroconversion rates are not known. It has been postulated that GBV-C is highly prevalent in the general global population (1.7—2.0%) [78], as determined by limited screening of blood donations. Most recently, it has been suggested that GBV-C infection may actually be beneficial in slowing the progression of disease in HIV/HCV coinfected patients [79,80]. HIV-infected patients who are reactive for GBV-C RNA may also have more positive outcomes than those who are not due to the maintenance of an intact Th1 cytokine profile [81], or modulation of entry receptors [82]. In addition, there have been reports of GBV-C presence in the lymphoid cells of patients with hematological malignancies [83,84]. In this regard, cells of the bone marrow were found to be infected with GBV-C in almost 20% of patients with hematological malignancies tested [83]. Patients with cirrhosis and HCC have occasionally tested positive for GBV-C RNA [85—87]. However, the causal role for

TABLE 6.4 Assays for HEV RNA Detection and Quantification

Assay	Manufacturer	Method used	Sensitivity		Dynamic range	
			IU/mL	copies/mL	IU/mL	copies/mL
RealStar HEV RT-PCR Kit 1.0	Altona Diagnostics	RT-PCR	20—100	n.p.	n.a.	n.a.
hepatitis E@ceeramTool.health	CEERAM S.A.S.	qRT-PCR	n.p.	5—50/reaction[a]	n.p.	n.p.
Geno-Sen's HEV Real Time PCR Kit	Genome Diagnostics	qRT-PCR	n.p.	80	n.p.	$1 \times 10^2 - 1 \times 10^6$
HEV Real Time RT-PCR Kit	Liferiver, Gentaur	RT-PCR	n.p.	n.p.	n.p.	n.p
ampliCUBE HEV	MIKROGEN	RT-PCR	n.p.	$<10^4$	n.a.	n.a.
Path-HEV	PrimerDesign	qRT-PCR	n.p.	100	n.p.	$1 \times 10^2 - 1 \times 10^7$
COBAS HEV Test	Roche Diagnostics	Automated RT-PCR	18.6	n.p.	n.a.	n.a.

[a]*Amount of input RNA not given.*

IU, international units; qRT-PCR, quantitative RT-PCR; n.p., not provided; n.a., not applicable.

GBV-C infection in these pathologies has never been shown. The current opinion assumes that the virus is not directly linked to any disease and has no apparent effect on patients after liver transplantation regardless of the cause of the end-stage liver disease.

There are no longer any commercial assays available for detection of GBV-C RNA. However, since the virus appears to be highly prevalent, several research laboratories have established testing of clinical samples using highly sensitive RT-PCR assays [83,88]. It is likely, as potentially stronger linkages to human pathology are uncovered, that providers of diagnostic tools may once again show interest in developing molecular tests for GBV-C detection.

References

[1] van Damme P, van Herck K. A review of the long-term protection after hepatitis A and B vaccination. Travel Med Infect Dis 2007;5:79—84.

[2] World Health Organization. Hepatitis A. Fact sheet no. 328. Geneva, Switzerland: World Health Organization. www.who.int/mediacentre/factsheets/fs328/en [accessed 13.11.14].

[3] dePaula VS, Villar LM, Morais LM, Lewis-Ximenez LL, Neil C, Gaspar AM. Detection of hepatitis A virus RNA in serum during the window period of infection. J Clin Virol 2004;29:245—9.

[4] Lanford RE, Feng Z, Chavez D, et al. Acute hepatitis A virus infection is associated with a limited type I interferon response and persistence of intrahepatic viral RNA. Proc Natl Acad Sci USA 2011;108:11223—8.

[5] Normann A, Jung C, Vallbracht A, Flehmig B. Time course of hepatitis A viremia and viral load in the blood of human hepatitis A patients. J Med Virol 2004;72:10—16.

[6] Roque-Afonso AM, Desbois D, Dussaix E. Hepatitis A virus: serology and molecular diagnostics. Future Virol 2010;5:233—42.

[7] de Paula VS. Laboratory diagnosis of hepatitis A. Future Virol 2012;7:461—72.

[8] Lemon S, Jansen RW, Brown EA. Genetic, antigenic and biological differences between strains of hepatitis A virus. Vaccine 1992;10:S40—4.

[9] Lavanchy D. Hepatitis B virus epidemiology, disease burden, treatment, and current and emerging prevention and control measures. J Viral Hep 2004;11:97—107.

[10] Locarnini S. Molecular virology of hepatitis B. Semin Liver Dis 2004;24S1:3—10.

[11] Kay A, Zoulim F. Hepatitis B virus genetic variability and evolution. Virus Res 2007;127:164—76.

[12] Mulrooney-Cousins PM, Michalak TI. Diagnostic assays for hepatitis B virus. Hot Topics Viral Hep 2009;15:7—13.

[13] World Health Organization. Hepatitis B. Fact sheet no. 204. Geneva, Switzerland: World Health Organization. www.who.int/mediacentre/factsheets/fs204/en [accessed 13.11.14].

[14] Michalak TI, Pasquinelli C, Guilhot S, Chisari FV. Hepatitis B virus persistence after recovery from acute viral hepatitis. J Clin Invest 1994;93:230—9.

[15] Michalak TI, Pham TNQ, Mulrooney-Cousins PM. Molecular diagnosis of occult hepatitis C and hepatitis B virus infections. Future Virol 2007;2:451—65.

[16] Reherman B, Ferrari C, Pasquinelli C, Chisari FV. The hepatitis B virus persists for decades after patients' recovery from acute viral hepatitis despite active maintenance of a cytotoxic T-lymphocyte response. Nature Med 1996;2:1104—8.

[17] Satoh K, Iwata-Takakura A, Yoshikawa A, et al. A new method of concentrating hepatitis B virus (HBV) DNA and HBV surface antigen: an application of the method to the detection of occult HBV infection. Vox Sang 2008;95:173—80.

[18] Coffin CS, Pham TNQ, Mulrooney PM, Churchill ND, Michalak TI. Persistence of isolated antibodies to woodchuck hepatitis virus core antigen is indicative of occult infection. Hepatology 2004;40:1053—61.

[19] Michalak TI. Occult persistence and lymphotropism of hepadnaviral infection: insights from the woodchuck viral hepatitis model. Immunol Rev 2000;174:98—111.

[20] Raimondo G, Navarra G, Mondello S, et al. Occult hepatitis B virus in liver tissue of individuals without hepatic disease. J Hepatol 2008;48:743—6.

[21] Galli C, Orlandini E, Penzo L, et al. What is the role of serology for the study of chronic hepatitis B virus infection in the age of molecular biology? J Med Virol 2008;80:974—9.

[22] Michalak TI. Immunology of hepatitis B virus. In: Colacino JM, Heinz BA, editors. Hepatitis prevention and treatment. Basel, Switzerland: Birkhauser Verlag; 2004. p. 87—105.

[23] Raimondo G, Allain JP, Brunetto MR, et al. Statements from the Taormina expert meeting on occult hepatitis B virus infection. J Hepatol 2008;49:652—7.

[24] Kao JH. Diagnosis of hepatitis B virus infection through serological and virological markers. Expert Rev Gastroenterol Hepatol 2008;2:553—62.

[25] Allain JP, Candotti D. Diagnostic algorithm for HBV safe transfusion. Blood Transf 2009;7:174—82.

[26] Vermeulen M, Coleman C, Mitchel J, et al. Sensitivity of individual-donation and minipool nucleic acid amplification test options in detecting window period and occult hepatitis B virus infections. Transfusion 2013;53:2459—66.

[27] Vermeulen M, van Drimmelen H, Coleman C, Mitchel J, Reddy R, Lelie N. A mathematical approach to estimate the efficacy of individual-donation and minipool nucleic acid amplification test options in preventing transmission risk by window period and occult hepatitis B virus infections. Transfusion 2014;54:2496—504.

[28] Taira R, Satake M, Momose S, et al. Residual risk of transfusion-transmitted hepatitis B virus (HBV) infection caused by blood components derived from donors with occult HBV infection in Japan. Transfusion 2013;53:1393—404.

[29] Wang L, Chang L, Xie Y, et al. What is the meaning of a nonresolved viral nucleic acid test-reactive minipool? Transfusion 2014;. Available from: http://dx.doi.org/10.1111/trf.12818.

[30] Shah SM, Singh SP. Hepatitis B virus serology: use and interpretation. Hep B Annual 2007;4:39—54.

[31] Kuhns MC, Busch MP. New strategies for blood donor screening for hepatitis B virus: nucleic acid testing versus immunoassay methods. Mol Diagn Ther 2006;10:77—91.

[32] Hatzakis A, Magiorkinis E, Haida C. HBV virological assessment. J Hepatol 2006;44:S71—6.

[33] Germer JJ, Qutub MO, Mandrekar JN, Mitchell PS, Yao JD. Quantification of hepatitis B virus (HBV) DNA with a TaqMan HBV analyte-specific reagent following sample processing with the MagNAPure LC instrument. J Clin Microbiol 2006;44 1490—4.

[34] Laperche S, Thibault V, Bouchardeau F, et al. Expertise of laboratories in viral load quantification, genotyping, and precore mutant determination for hepatitis B virus in a multicenter study. J Clin Microbiol 2006;44:3600—7.

[35] Valsamakis A. Molecular testing in the diagnosis and management of chronic hepatitis B. Clin Microbiol Rev 2007;20:426—39.

[36] Flink HJ, van Zonneveld M, Hansen BE, et al. Treatment with Peg-interferon alpha-2b for HBeAg-positive chronic hepatitis B: HBsAg loss is associated with HBV genotype. Am J Gastroenterol 2009;101:297—303.

[37] Lau GK, Piratvisuth T, Luo KX, et al. Peginterferon alfa-2a, lamivudine, and the combination for HBsAg-positive chronic hepatitis B. N Engl J Med 2005;352:2682–95.

[38] Buster EH, Flink HJ, Cakaloglu Y, et al. Sustained HBeAg and HBsAg loss after long-term follow-up of HBeAg-positive patients treated wih peg-interferon alpha-2b. Gastroenterology 2008;135:459–67.

[39] Allain JP, Candotti D, ISBT HBV Safety Collaborative Group. Hepatitis B virus in transfusion medicine: still a problem? Biologicals 2012;40:180–6.

[40] Guirgis BS, Abbas RO, Azzazy HM. Hepatitis B virus genotyping: current methods and clinical implications. Int J Infect Dis 2010;14:e941–53.

[41] Aberle SW, Kletzmayer B, Watschinher B, Schmied B, Vetter N, Puchhammer-Stockl E. Comparison of sequence analysis and the INNO-LiPA HBV DR line probe assay for detection of lamivudine-resistant hepatitis B virus strains in patients under various clinical conditions. J Clin Microbiol 2001;39:1972–4.

[42] Gintowt AA, Germer JJ, Mitchell PS, Yao JD. Evaluation of the MagNA Pure LC used with the TRUGENE HBV genotyping kit. J Clin Virol 2005;34:155–7.

[43] Ghany MG, Doo EC. Antiviral resistance and hepatitis B therapy. Hepatology 2009;49:S174–84.

[44] Osiowy C, Villeneuve JP, Heathcote EJ, Giles E, Borlang J. Detection of rtN236T and rtA181V/T mutations associated with resistance to adefovir dipivoxil in samples from patients with chronic hepatitis B virus infection by the INNO-LiPA HBV DR line probe assay (version 2). J Clin Microbiol 2006;44:1994–7.

[45] Olivero A, Ciancio A, Abate ML, Gaia S, Smedile A, Rizzetto M. Performance of sequence analysis, INNO-LiPA line probe assays and AFFIGENE assays in the detection of hepatitis B virus polymerase and precore/core promoter mutations. J Viral Hepat 2006;13:355–62.

[46] Kim JH, Park YK, Park ES, Kim KH. Molecular diagnosis and treatment of drug-resistant hepatitis B virus. World J Gastroenterol 2014;20:5708–20.

[47] Law J, Jovel J, Patterson J, et al. Identification of hepatotropic viruses from plasma using deep sequencing: a next generation diagnostic tool. PLoS One 2013;8:e60595.

[48] Sablon E, Shapiro F. Advances in molecular diagnosis of HBV infection and drug resistance. Int J Med Sci 2005;2:8–16.

[49] Dupouey J, Gerolami R, Solas C, Colson P. Hepatitis B virus variant with the a194t substitution within reverse transcriptase before and under adefovir and tenofovir therapy. Clin Res Hepatol Gastroenterol 2012;36:e26–8.

[50] Coffin CS, Mulrooney-Cousins PM, Peters MG, et al. Molecular characterization of intrahepatic and extrahepatic hepatitis B virus (HBV) reservoirs in patients on suppressive antiviral therapy. J Viral Hepat 2011;18:415–23.

[51] World Health Organization. Hepatitis C. Fact sheet no. 164. Geneva, Switzerland: World Health Organization. www.who.int/mediacentre/factsheets/fs164/en [accessed 13.11.14].

[52] Chakravarty R. Diagnosis and monitoring of chronic viral hepatitis: serologic and molecular markers. Front Biosci (Schol Ed) 2011;3:156–67.

[53] Gupta E, Bajpai M, Choudhary A. Hepatitis C virus: screening, diagnosis, and interpretation of laboratory assays. Asian J Transfus Sci 2014;8:19–25.

[54] Kamili S, Drobeniuc J, Araujo AC, Hayden TM. Laboratory diagnostics for hepatitis C virus infection. Clin Infect Dis 2012;55(Suppl 1):S43–8.

[55] Scott JD, Gretch DR. Molecular diagnostics of hepatitis C virus infection: a systematic review. JAMA 2007;297:724–32.

[56] Pham TN, MacParland SA, Mulrooney PM, Cooksley H, Naoumov NV, Michalak TI. Hepatitis C virus persistence after spontaneous or treatment-induced resolution of hepatitis C. J Virol 2004;78:5867–74.

[57] Radkowski M, Gallegos-Orozco JF, Jablonska J, et al. Persistence of hepatitis C virus in patients successfully treated for chronic hepatitis C. Hepatology 2005;41:106–14.

[58] MacParland SA, Pham TN, Guy CS, Michalak TI. Hepatitis C virus persisting after clinically apparent sustained virological response to antiviral therapy retains infectivity in vitro. Hepatology 2009;49:1431–41.

[59] Chen AY, Zeremski M, Chauhan R, Jacobson IM, Talal AH, Michalak TI. Persistence of hepatitis C virus during and after otherwise clinically successful treatment of chronic hepatitis C with standard pegylated interferon α-2b and ribavirin therapy. PLoS One 2013;8:e80078.

[60] Pham TN, King D, Macparland SA, et al. Hepatitis C virus replicates in the same immune cell subsets in chronic hepatitis C and occult infection. Gastroenterology 2008;134:812–22.

[61] Carreño V. Seronegative occult hepatitis C virus infection: clinical implications. J Clin Virol 2014;61:315–20.

[62] Pham TN, Mulrooney-Cousins PM, Mercer SE, et al. Antagonistic expression of hepatitis C virus and alpha interferon in lymphoid cells during persistent occult infection. J Viral Hepat 2007;14:537–48.

[63] Cobb B, Heilek G, Vilchez RA. Molecular diagnostics in the management of chronic hepatitis C: key considerations in the era of new antiviral therapies. BMC Infect Dis 2014;14:S8.

[64] Cobb B, Pockros PJ, Vilchez RA, Vierling JM. HCV RNA viral load assessments in the era of direct-acting antivirals. Am J Gastroenterol 2013;108:471–5.

[65] González V, Gomes-Fernandes M, Bascuñana E, et al. Accuracy of a commercially available assay for HCV genotyping and subtyping in the clinical practice. J Clin Virol 2013;58:249–53.

[66] Niro GA, Fontana R, Ippolito AM, Andriulli A. Epidemiology and diagnosis of hepatitis D virus. Future Virol 2012;7:709–17.

[67] World Health Organization. Hepatitis D. Geneva, Switzerland: World Health Organization. http://www.who.int/csr/disease/hepatitis/whocdscsrncs20011/en/ [accessed 13.11.14].

[68] Le Gal F, Gordien E, Affolabi D, et al. Quantification of hepatitis delta virus RNA in serum by consensus real-time PCR indicates different patterns of virological response to interferon therapy in chronically infected patients. J Clin Microbiol 2005;43:2363–9.

[69] Brichler S, Le Gal F, Butt A, Chevret S, Gordien E. Commercial real-time reverse transcriptase PCR assays can underestimate or fail to quantify hepatitis delta virus viremia. Clin Gastroenterol Hepatol 2013;11:734–40.

[70] World Health Organization. Hepatitis E. Fact sheet no. 280. Geneva, Switzerland: World Health Organization. www.who.int/mediacentre/factsheets/fs280/en [accessed 13.11.14].

[71] Lu L, Li C, Hagedorn CH. Phylogenetic analysis of global hepatitis E virus sequences: genetic diversity, subtypes, and zoonosis. Rev Med Virol 2006;16:5–36.

[72] Inoue J, Nishizawa T, Takahashi M, et al. Analysis of the full-length genome of genotype 4 hepatitis E virus isolates from patients with fulminant or acute self-limited hepatitis E. J Med Virol 2006;78:476–84.

[73] Pischke S, Wedemeyer H. Hepatitis E virus infection: multiple faces of an underestimated problem. J Hepatol 2013;58:1045–6.

[74] La Rosa G, Fratini M, Muscillo M, et al. Molecular characterization of human hepatitis E virus from Italy: comparative analysis of five reverse transcription-PCR assays. Virol J 2014;11:72.

[75] Theodore E, Lemon SM. GB virus C, hepatitis G virus, or human orphan flavivirus? Hepatology 1997;25:1285–6.

[76] Feucht HH, Zöllner B, Polywka S. Distribution of hepatitis G viremia and antibody response to recombinant proteins with special regard to risk factors in 709 patients. Hepatology 1997;26:491–4.

[77] Bhattarai N, Stapleton JT. GB virus C: the good boy virus? Trends Microbiol 2012;20:124–30.

[78] Reshetnyak VI, Karlovich TI, Ilchenko LU. Hepatitis G virus. World J Gastroenterol 2008;14:4725–34.

[79] Lefrère JJ, Roudot-Thoraval F, Morand-Joubert L, et al. Carriage of GB virus C/hepatitis G virus RNA is associated with a slower immunologic, virologic, and clinical progression of human immunodeficiency virus disease in coinfected persons. J Infect Dis 1999;179:783–9.

[80] Berzsenyi MD, Bowden DS, Kelly HA, et al. Reduction in hepatitis C-related liver disease associated with GB virus C in human immunodeficiency virus coinfection. Gastroenterology 2007;133:1821–30.

[81] Xiang J, George SL, Wünschmann S, Chang Q, Klinzman D, Stapleton JT. Inhibition of HIV-1 replication by GB virus C infection through increases in RANTES, MIP-1alpha, MIP-1beta, and SDF-1. Lancet 2004;363:2040–6.

[82] Schwarze-Zander C, Neibecker M, Othman S, et al. GB virus C coinfection in advanced HIV type-1 disease is associated with low CCR5 and CXCR4 surface expression on CD4(+) T-cells. Antivir Ther 2010;15:745–52.

[83] Kisiel E, Cortez KC, Pawełczyk A, et al. Hepatitis G virus/GBV-C in serum, peripheral blood mononuclear cells and bone marrow in patients with hematological malignancies. Infect Genet Evol 2013;19:195–9.

[84] Chang CM, Stapleton JT, Klinzman D, et al. GBV-C infection and risk of NHL among U.S. adults. Cancer Res 2014;74:5553–60.

[85] Di Bisceglie AM. Hepatitis G virus: a work in progress. Ann Intern Med 1996;125:772–3.

[86] Kanda T, Yokosuka O, Imazeki F, et al. GB virus-C RNA in Japanese patients with hepatocellular carcinoma and cirrhosis. J Hepatol 1997;27:464–9.

[87] Yoshiba M, Okamoto H, Mishiro S, et al. Detection of the GBV-C hepatitis virus genome in serum from patients with fulminant hepatitis of unknown aetiology. Lancet 1995;346:1131–2.

[88] Chivero ET, Bhattarai N, Rydze RT, Winters MA, Holodniy M, Stapleton JT. Human pegivirus RNA is found in multiple blood mononuclear cells in vivo and serum-derived viral RNA-containing particles are infectious in vitro. J Gen Virol 2014;95:1307–19.

7

Molecular Testing for Human Papillomaviruses

K.M. Bennett

Texas Tech University Health Sciences Center, School of Health Professions,
Molecular Pathology Program, Lubbock, TX, United States

INTRODUCTION

Human papillomavirus (HPV) is the most common sexually-transmitted infection in the United States. The overall prevalence of HPV infection is estimated at about 27–43% in US females between the ages of 14 and 59 [1,2]. HPV is also very common in men, with an estimated prevalence of 52–65% [3,4]. The Centers for Disease Control and Prevention (CDC) estimates that there are over 79 million new and existing HPV infections in the United States, which makes up over 70% of all sexually-transmitted infections [5]. HPV infection is typically transient, with over 90% of new infections naturally cleared within 6 months to 2 years after infection [6].

HPV infection is often asymptomatic, but has two major clinical impacts: genital warts and cancer. Genital warts (condyloma acuminatum) are growths on the cervical or vulval mucosa in females, or on the glans or prepuce in males. HPV-associated cancers include cancer of the oropharynx, cervix, vulva, vagina, anus, and penis [7,8]. Approximately 70% of oropharyngeal cancer is associated with HPV infection [9], but the most notable clinical manifestation of HPV infection is cervical cancer. In 2011, there were an estimated 249,632 women living with cervical cancer in the United States, with 7.8 new cases per 100,000 women per year between 2007 and 2011 [10].

There are currently 170 different types of HPV recognized [11], with over 40 of those types considered sexually transmitted [6]. HPV types are classified as low risk or high risk, depending on the potential for causing cervical dysplasia and cancer [12–14]. The low-risk strains tend to be associated with genital warts, with types 6 and 11 causing about 90% of all genital warts cases, while the high-risk strains are associated with cervical cancer development [13,15]. The low- and high-risk HPV types are summarized in Table 7.1.

Molecular testing for HPV is thus focused on the detection of high-risk HPV types, with the purpose of identifying cervical cancer or its precursors. This chapter will highlight the characteristics of the HPV and the mechanisms of cervical lesion development caused by HPV infection. The diagnostic strategies for HPV and cervical cancer detection will be outlined, with a discussion of the common molecular diagnostic methodologies used in the clinical laboratory.

MOLECULAR TARGET

Pathophysiology of HPVs

HPVs are nonenveloped viruses belonging to the Papillomaviridae family. The virus has a double-stranded, circular DNA genome that is approximately 7900 base pairs in size. The HPV genome is divided into three sections: (1) early expressed genes, (2) late expressed genes, and (3) the long control region (LCR). The expressed sections of the genome include eight overlapping open reading frames. The early genes include six open reading frames that are translated into proteins that play a role in genome replication and transcription. The E6 and E7 oncogenes can form complexes with tumor suppressors, induce mitotic spindle malformations and abnormal centrosome numbers during mitosis [16,17]. The late genes, downstream of the early region, include the L1 and L2 genes which are translated into the major and minor capsid proteins, respectively. The last 10% of the HPV genome is the LCR, which holds no protein-coding

© 2017 Elsevier Inc. All rights reserved.

TABLE 7.1 HPV Types

High-risk HPV types	16	18	31	33	35	39	45	51	52	56	58	59	66[a]	68	73	82	83
Low-risk HPV types	6	11	26	40	42	43	44	53	54	55	66[a]	84					

[a]Type 66 has been categorized as both low- and high risk.

function. The LCR includes the origin of replication and a number of regulatory sites that are recognized by transcription factors [18]. The HPV genome is diagrammed in Fig. 7.1.

HPVs have a progressive life cycle that is dependent on the differentiation of epithelial cells. HPVs are transmitted by contact with the skin or mucosa of an infected individual. The target cell is the basal epithelium of the skin or mucosa, likely accessed through small tears that expose the basal layer. The cutaneous varieties of HPV are considered epidermitrophic, thus typically infecting the hands and feet. Mucosal types of HPV target the mouth, throat, or anogenital region [19]. The productive stage of the HPV life cycle follows the differentiation of the basal cells into mature keratinocytes, where the viral DNA is replicated, capsid proteins are synthesized, and viral release occurs. The oncogenic E6 gene product interacts with the tumor suppressor p53, resulting in degradation. The HPV E7 protein disrupts the cellular retinoblastoma protein, pRB, causing the stimulation of cellular DNA synthesis and cell division. This manipulation by HPV results in

the transformation of cells from a terminal differentiated state to an active state that allows for viral replication. Depending on the type of HPV infection, this aberrant cell division can result in warts or mucosal lesions, such as cervical dysplasia. While in the benign lesion state, HPV is found in the nucleus, but outside of the cellular chromosomes. This is called the episomal state of infection. In advanced stages, the infection progresses to cancer, and HPV becomes integrated into the host genome. Once the HPV genome is integrated into the host cell, the E6 and E7 gene expression is upregulated and further binding of p53 and pRB occurs [19]. This results in a cascade of cellular disruptions such as an uncoupling of centrosome duplication from the cell cycle, which contributes to genomic instability and increases the production of abnormal cells [17]. HPVs play a role in the etiology of a variety of benign conditions, including common warts, plantar warts, anogenital warts (condylomata acuminata), respiratory papillomatosis, and others. Molecular HPV testing is rarely utilized for these conditions, as they can be usually diagnosed by visual examination, biopsy, and histology. However, infection with high-risk types of HPV can result in cell dysplasia, premalignancies, and eventually cancer. Cancer can occur in any of the anogenital regions of women or men, including the vulva, vagina, cervix, anus, or penis [20]. The most clinically relevant application of molecular detection of HPV infection is to diagnose cervical cancer or its precursors.

HPVs and Cervical Cancer

HPV infection has been established as a necessary, but not sufficient, cause of cervical cancer. HPV is found in over 99% of cervical carcinomas worldwide, indicating that HPV infection serves as a trigger for development of cervical cancer [21]. However, it is important to recognize that HPV infection alone is not sufficient to cause cancer, as infection can be transient and cleared naturally. Persistent infection of the cervical epithelium increases the risk of malignancy due to additional genomic deletions and chromosomal aberrations that lead to inactivation of tumor suppressor genes [22]. The HPV types 16 and 18 have been determined to be the most prevalent strains found in cervical malignancies and so top the list of the 14 high-risk HPV types [23]. Due to the strong

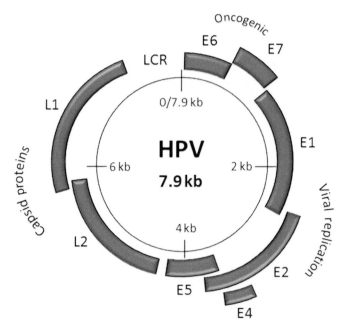

FIGURE 7.1 The HPV genome. The diagram shows a general HPV genome of 7.9 kb of circular DNA. The overlapping open reading frames are shown in blue. The E6 and E7 genes are oncogenic. The E1–E5 gene products participate in viral replication, while the L1 and L2 are capsid proteins. The L1 gene is a common target of HPV molecular assays.

association of high-risk HPV infection with the development of cervical cancer, HPV testing has been integrated into the traditional cervical screening algorithms.

The current gold standard for detection of abnormal cervical epithelial cells is the Papanicolaou (Pap) stain. This entails collection of cervical epithelial cells and spreading them onto a slide for microscopic analysis of cell morphology. Pap smears may be done conventionally by smearing cells directly onto the slide, or cells can be collected into liquid cytology medium, which is centrifuged and filtered prior to monolayer preparation. Liquid cytology methods, such as ThinPrep (Hologic, Inc.), tend to show better sensitivity and specificity for the detection of cervical dysplasia, due to standardized preparation and decreased presence of obscuring material such as blood and mucus [24]. Morphological findings from a cytology examination define the level of risk for developing cervical malignancy.

Classification of cervical abnormalities is established by several nomenclature standards. The spectrum for all nomenclatures begins with normal or negative cells, spanning through increasing levels of cervical dysplasia, and ending with cervical adenocarcinoma. The cervical intraepithelial neoplasia (CIN) nomenclature is widely used, although the most recent classification is the Bethesda 2001 system. A comparison of the four cytological classifications is reviewed in detail by Sherman [25]. The general progression of cervical dysplasia to carcinoma is illustrated in Fig. 7.2. Cervical epithelial cells determined to show a low level of atypical or abnormal morphology are termed "atypical squamous cells of undetermined significance" (ASCUS) or "cannot exclude high-grade lesions" (ASC-H). ASCUS is the equivocal gate-keeper diagnosis, meaning that its presence is often the stage at which further examination may be warranted. ASCUS may indicate low-grade squamous intraepithelial lesions (LSIL) or

CIN1, which is indicative of HPV infection. HPV infection in this state may be transient or self-resolving, and cervical dysplasia can resolve back to normal. Dysplasia classifications include CIN2 and progress to CIN3 or high-grade squamous intraepithelial lesions (HSIL). If HPV infection does not resolve or if there is recurrent infection, then a woman is at higher risk of HSIL/CIN3, followed by cervical carcinoma, particularly after age 30 [26]. The widespread use of cervical cytology exams has led to a significant decline in cervical cancer, although the mortality rate has essentially leveled off in the last 20 years. It is estimated that in 2014, 4020 deaths will be attributed to cervical cancer in the United States [10].

Primary prevention of HPV infection is a promising measure in reducing the incidence of cervical cancer. In 2006, the Gardasil (Merck) prophylactic HPV vaccine was approved by the Food and Drug Administration (FDA). The vaccine is a series of three shots given over a 6-month interval in males and females aged 9–26 years. The quadrivalent vaccine is indicated to prevent infections from HPV types 6, 11, 16, and 18, to prevent anogenital warts, cervical cancer, and other neoplasias [27,28]. The vaccine can be given before or after the patient becomes sexually active and is recommended as part of routine vaccination for boys and girls at the age of 11 or 12 [29]. A 53% decline in the prevalence of vaccine-covered HPV types was observed within 4 years of vaccine introduction [30]. Recent evaluations of the prevalence of other high-risk HPV types in cervical cancer provide evidence that a nine-valent HPV vaccine (HPV 6, 11, 16, 18, 31, 33, 45, 52, 58) could prevent the majority of cases of CIN2 and higher [31]. Despite widespread recommendation by several medical organizations, the HPV vaccine is underutilized, with only 37.6% of adolescent girls and 13.9% of adolescent boys completing the total vaccine series [32]. There is data showing that combined vaccination and cervical cancer screening could prevent almost 93% of new cancer

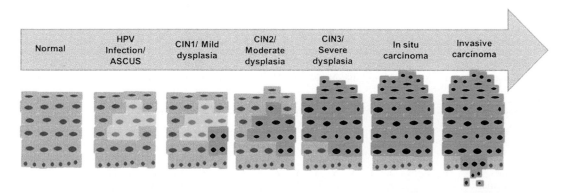

FIGURE 7.2 Progression of cervical dysplasia. The figure illustrates the progression from normal cervical epithelium to invasive carcinoma after infection with HPV. Infection with HPV triggers cellular transformation, which may be transient and regress or may progress into more severe stages. The goal of HPV testing is to detect CIN3 or more severe cervical disease.

cases [33]. Continued education of both the public and medical providers on the benefits of HPV vaccination and HPV testing is pivotal in reaching this goal. Despite well-established screening algorithms for detection of cervical cancer, approximately 8.2 million women in the United States have not been screened for cervical cancer in the past 5 years [34]. Molecular HPV testing is a critical component of cervical cancer screening that can significantly improve the early detection of cancer and its precursors.

MOLECULAR TECHNOLOGIES

Clinical testing for HPVs is exclusively molecular in methodology, as culture of the virus is extremely difficult. Methods have been established to culture HPV by recreating the three-dimensional epithelium that the virus requires for its life cycle [35]. However, this is not a practical approach in the diagnostic environment. Thus, molecular detection of HPV nucleic acid is the gold standard for clinical testing. It is important to note that the purpose of HPV testing is not to detect HPV itself, but to identify patients at risk for developing severe cervical dysplasia or cervical cancer. Today, there are a number of options that the medical provider has to choose from when deciding the best diagnostic method for HPV testing. This section will focus on the FDA-approved methods for HPV testing [36], although a multitude of research-use only methodologies also exist that have been reviewed in the literature [37–39].

The principles for detection of the HPV genome vary greatly between methods. The initial assays were signal amplification methods, which increase DNA-proportional signals to detectable levels. These signals can be detected by color change, fluorescence, or chemiluminescence. Target amplification, which includes polymerase chain reaction (PCR), amplifies fragments of DNA from targeted sequences using the DNA polymerase enzyme. PCR products can be detected either using end point analysis, such as gel electrophoresis, or in real time by using fluorescence-detection chemistries. Another approach is transcription-mediated amplification (TMA), which is an isothermal reaction that uses RNA polymerase to amplify a nucleic acid target. Summaries of the FDA-approved HPV assays are provided in Tables 7.2 and 7.3.

Digene Hybrid Capture 2

The Digene Hybrid Capture 2 (HC2), now manufactured by Qiagen, Inc., was the first FDA-approved method for detection of HPV DNA. The first-generation assay was approved in 1995, followed by the second version (HC2) in 1999, which has since been refined to detect high-risk genotypes only [40]. The Digene HC2 assay is a signal amplification technique, so the chemiluminescent signal is amplified rather than the target HPV DNA, in contrast to PCR. The HC2 assay remains the most widely used HPV detection method, still being utilized by most of the high-volume clinical reference laboratories in the United States. The HC2 assay can

TABLE 7.2 FDA-Approved HPV Assays

Assay name	Manufacturer	HPV types detected (Yes/No)															Specimen type/minimum volume	Automation
		16	18	31	33	35	39	45	51	52	56	58	59	66	68		
Aptima HPV Assay	Hologic, Inc.	Y	Y	Y	Y	Y	Y	Y	Y	Y	Y	Y	Y	Y	Y	ThinPrep/PreservCyt/ 1 mL	Panther/Tigris Systems
Aptima HPV 16 18/45 Genotype Assay	Hologic, Inc.	Y	Y	N	N	N	N	Y	N	N	N	N	N	N	N	ThinPrep/PreservCyt/ 1 mL	Panther/Tigris Systems
Cervista HPV HR	Hologic, Inc.	Y	Y	Y	Y	Y	Y	Y	Y	Y	Y	Y	Y	Y	Y	ThinPrep/PreservCyt/ 2 mL	Cervista HTA System
Cervista HPV 16/18	Hologic, Inc.	Y	Y	N	N	N	N	N	N	N	N	N	N	N	N	ThinPrep/PreservCyt/ 2 mL	Cervista HTA System
Digene Hybrid Capture 2 (HC2) High-Risk HPV DNA	Qiagen, Inc.	Y	Y	Y	Y	Y	Y	Y	Y	Y	Y	Y	Y	N	Y	PreservCyt/4 mL	Rapid Capture System
cobas HPV Test	Roche Molecular Systems, Inc.	Y	Y	Y	Y	Y	Y	Y	Y	Y	Y	Y	Y	Y	Y	ThinPrep T2000 & 3000/ PreservCyt/cobas PCR Cell Collection Media/ 1 mL	cobas 4800 System

TABLE 7.3 Characteristics of HPV Assays

Assay	HPV target	Principle	Internal control	Indications for use		
				ASCUS	Co-testing	Primary screening
Aptima HPV Assay	E6/E7 mRNA	TMA	Exogenous[a]	✓	✓	
Aptima HPV 16 18/45	E6/E7 mRNA	TMA	Exogenous[a]	✓[b]	✓	
Cervista HPV HR	Mixed genomic	Invader probe	Histone 2	✓	✓	
Cervista HPV 16/18	Mixed genomic	Invader probe	Histone 2	✓[b]	✓	
Digene HC2	Mixed genomic	Hybrid capture	None	✓	✓	
cobas HPV Test	L1 DNA	Real-time PCR	β-globin	✓	✓	✓

[a]Proprietary exogenous internal control is spiked into the assay to monitor nucleic acid capture, amplification, and detection. This does not monitor cellularity of the specimen.
[b]Genotyping assay is indicated as a reflex test after positive result on the corresponding high-risk screening test.

qualitatively detect 13 high-risk HPV genotypes. It is currently indicated as a follow-up after ASCUS diagnosis, to determine need for colposcopy, and for women greater than 30 years of age to be used adjunctively with the Pap screen.

Cervical specimens are collected using one of three methods: (1) the HC2 DNA Collection device, (2) biopsies collected in Specimen Transport Medium (STM), or (3) using a broom-type collection device placed in Cytyc ThinPrep PreservCyt solution. The assay is approved for use after preparation of ThinPrep slides, but requires at least 4 mL of the PreservCyt Solution remaining for the HPV test. Denatured specimens are mixed with a HPV RNA probe cocktail. If HPV DNA is present, the DNA will anneal to the RNA probe(s), creating a DNA:RNA hybrid. The hybrids are then captured to the surface of a microplate well that is coated with antibodies that bind DNA:RNA hybrids. The now immobilized hybrids are detected by binding with a second antibody that is conjugated with alkaline phosphatase. Substrate is added to the well, which is then cleaved by bound alkaline phosphatase, creating a chemiluminescent light signal. The hybrids can bind multiple conjugated antibodies and each antibody has multiple conjugated alkaline phosphatase molecules. This results in amplification of the signal, increasing the signal-to-target ratio. The microplate is then placed into a luminometer, which measures the light intensity of each well. If the intensity crosses a defined threshold, the result is recorded as "detected" for high-risk HPV genotypes [41]. The Qiagen Rapid Capture automated platform can be used in high-throughput laboratories. This instrument is a pipetting and dilution system that can process up to 352 samples in about 8 h. This automated option reduces the laborious sample transfer and washing steps.

The HC2 assay is qualitative only and cannot distinguish between genotypes. Thus, a positive result means that at least one of the 13 detectable genotypes is present. However, because the risk level of each of the HPV genotypes is not equal, this is a notable limitation of the HC2 assay. Other limitations include documented cross-reactivity of the assay to low-risk genotypes 6, 42, and 70, and to possible high-risk types 53 and 66 [41,42], causing false-positive results. The consequence of false-positive results in HPV detection is unnecessary colposcopy, which may have uncommon but significant effects on fertility, obstetric outcomes, and psychological anxiety [43]. The HC2 assay is further lacking an internal control, which means the specimen is not monitored for cellularity. Newer assays have implemented various internal controls to monitor various stages of the processing and analysis (Table 7.3). While the hybrid capture assay is the most widely used methodology, it may eventually be replaced by more sensitive, specific, and better refined clinical assays. Nonetheless, it remains the gold standard against which every new assay is compared, because of its well-characterized status.

Cervista HPV HR and Cervista HPV 16/18

The Digene HC2 assay remained the only FDA-approved test for approximately 10 years, until approval in 2009 of the Cervista HPV HR qualitative screening test and the Cervista HPV 16/18 genotyping test. The original Cervista high-risk assay was developed by Third Wave Technologies, and the Cervista line is now marketed by Hologic, Inc. The Cervista HPV HR test screens for 14 high-risk HPV types. This includes HPV type 66, which is not a direct target of the HC2 test. The Cervista HPV 16/18 assay is a genotyping test that can determine the presence of types 16 and 18, which indicate the highest risk of developing cervical cancer [44]. The genotyping assay is intended for use as a reflex test after a positive result on the Cervista HPV HR test and should not be used as a stand-alone test. The capability to genotype HPV types

16 and 18 resulted in amended recommendations by American Society for Colposcopy and Cervical Pathology (ASCCP) to triage women who are positive for one or both types to immediate colposcopy. Women with high-risk HPV, but not types 16/18, are to be monitored more frequently, but less aggressively in the absence of cytological abnormalities.

Both Cervista assays are unique signal amplification methods using the proprietary Invader cleavage-based technology. Fig. 7.3 is a summary of the assay steps [38]. The Invader chemistry includes two isothermal reactions. In the first reaction, a sequence-specific probe oligonucleotide binds to the HPV DNA target. At the same time, an Invader probe also binds. The Invader probe inserts itself between the target and the

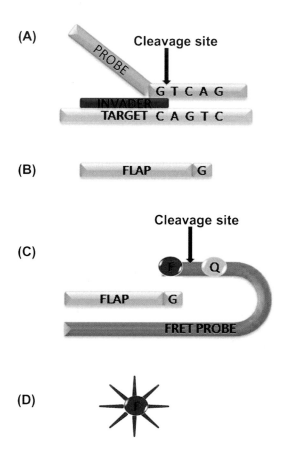

FIGURE 7.3 Invader assay. The schematic shows the major steps of a signal amplification assay using Invader technology. The target DNA is denatured, and a sequence-specific probe binds in conjunction with an Invader oligo. (A) The formation of this complex results in cleavage at the indicated site. (B) Cleavage releases a free flap of the probe molecule. Multiple flaps are generated from each target DNA molecule. (C) The flap then associates with a FRET probe, resulting in cleavage at the indicated site. (D) The fluorophore, F, is released from the quencher, Q, causing a fluorescent signal proportionate to the amount of target DNA. *Source: This figure was reproduced with permission from Arney A, Bennett KM. Molecular diagnostics of human papillomavirus. Lab Medicine 2010;41:523–30.*

sequence-specific probe, causing a brief secondary structure to form. This is recognized by a proprietary enzyme that cleaves the sequence-specific probe at the overlapping site, releasing a 5′ section of the probe called a flap. The 5′ flap then participates in the secondary reaction, in which the flap now behaves as an Invader oligo by binding to a hairpin oligonucleotide. The hairpin probe utilizes fluorescence resonance energy transfer (FRET) between attached fluorophore and quencher molecules. When the 5′ flap binds to the FRET hairpin probe, the structure is cleaved at a site between the fluorophore and the quencher, producing a fluorescent signal. The signal amplification occurs because there is a large molar excess of the sequence-specific probes so that many 5′ flaps are produced per target HPV DNA sequence. The 5′ flaps cycle on and off of the hairpin FRET probes, serving to produce further signal amplification from the original target (up to 10^7-fold signal amplification per hour) [45].

One of the key benefits of the Cervista assays over HC2 is the use of an internal control. The human histone 2 gene (H2be) is used to monitor appropriate test processing, inhibition, and to detect insufficient sample (cellularity). This is important to rule out false-negative results. The internal control signal is differentiated by a red fluorophore, which is spectrally distinct from the green signal produced by the target reaction. Other controls include a positive control for each of the three probe mixtures (assorted by gene homology of the HPV types) and a negative control (Yeast tRNA). Results are reported as positive or negative for high-risk HPV, with further testing required to determine if the signal was caused by types 16 or 18 [45].

The Cervista 16/18 genotyping assay is based on the same assay principle as the high-risk screening test. It utilizes two probe mixtures, one for HPV 16 and one for HPV 18, in separate reactions. The H2be internal control is integrated with each probe mixture. Results are reported as HPV 16 positive/negative, HPV 18 positive/negative, or HPV 16 and 18 positive/negative. The genotyping test is only run if the screening test showed a positive result and is not indicated as a stand-alone assay [46].

Both Cervista assays are approved for use with the ThinPrep Pap Test PreservCyt solution (Hologic, Inc.) and a number of broom-type collection devices. The assays have been evaluated for post-cytology analysis from the automated ThinPrep 2000 system only, but have not yet been established for use with other cytology processors. Of note, the minimum sample volume required for the Cervista assays is half that of the HC2 assay (Table 7.2). The high-risk screen does not exhibit as much cross-reactivity with nontarget HPV types as does the HC2 assay, but the Cervista HPV HR test does warn of cross-reactivity to HPV types 67 and 70. The Cervista HPV 16/18 assay further shows cross-reactivity to high levels of high-risk type 31, which can mimic an HPV 16 positive result [45,46].

In comparison to the ALTS clinical trial, the Cervista assay showed substantially similar performance characteristics to the established Digene HC2 assay, per FDA approval. Some studies indicate that specificity is equivalent or even increased in the Cervista assay [47–49]. Another study suggests that clinical specificity can be further increased for the Cervista assay by changing the cutoff value for positivity. This was shown to reduce the number of false-positive results that are due to interfering factors unrelated to HPV [50].

Aptima HPV and Aptima HPV 16 18/45

In October 2011, Gen-Probe, Inc. received FDA approval for a new assay called the Aptima HPV assay. Now marketed by Hologic, Inc., the Aptima HPV assay uses a very different principle of HPV detection as compared to its predecessors. The Aptima assay detects 14 high-risk HPV types using a messenger RNA (mRNA)-based detection system called TMA. It targets the two major oncogenes, E6 and E7 (Fig. 7.1). A year later, the Aptima HPV 16 18/45 genotyping assay was approved by the FDA. This genotyping assay can distinguish HPV types 16, 18, and/or 45. HPV type 45, although relatively uncommon overall, has been implicated as the third most common genotype in invasive cervical cancer [51]. The Aptima genotyping assay, like the Cervista HPV 16/18 assay, is designed to serve as a reflex test after a positive result in the screening test.

The principle of the Aptima test includes the capture of mRNA target, amplification using MMLV reverse transcriptase and T7 RNA polymerase, and detection of amplicon by a hybridization protection assay (HPA) (Fig. 7.4). Like the other FDA-approved assays, the Aptima assays are approved for use with ThinPrep liquid cytology specimens, including postcytology analysis on the ThinPrep 2000 system.

The mRNA target capture step is accomplished using capture oligonucleotides that have two sections: one section is complementary to the target HPV mRNA and another section is a string of deoxyadenosine residues (polyA). The capture oligomers and the target mRNA is mixed with magnetic microparticles that have poly-deoxythymidine (polyT) molecules covalently attached. The sequence-specific section of the capture oligo binds to the target HPV mRNA, and then after a cooling step, the polyA section of the capture oligo binds to the polyT section of the magnetic bead. The magnetic microparticles, now with target mRNA attached, are pulled to the side of the tube using magnets. This allows for nonbound nucleic acid and other cellular components to be aspirated, followed by washing to remove residual specimen matrix. Amplification by TMA follows target capture. TMA begins with production of a DNA copy of the target HPV mRNA sequences, using MMLV reverse transcriptase. Then T7 RNA polymerase produces multiple copies of RNA amplicon from the DNA copy template. In the third step, detection is achieved using a HPA. The HPA uses single-stranded labeled probes that are complementary to the target RNA amplicon. A selection reagent is added that inactivates the labels on probes that are not hybridized to amplicon. Thus, the probes that are hybridized to the target are protected from label inactivation. The protected hybrids can then be detected using a luminometer. Relative light units that exceed the cutoff are considered positive reactions. Both the Aptima HPV and the Aptima HPV 16 18/45 assays use the same principle [52,53].

The Aptima assays do include an internal control, although it is an exogenous control that is spiked into samples during the assay. This internal control will monitor target capture, amplification, and detection, but cannot provide any information about specimen cellularity (specimen adequacy). The internal control nucleic acid participates in the TMA reaction in a similar manner as the HPV target RNA. During the HPA detection, the internal control can be differentiated from the HPV target by use of a probe with varying light emission kinetics, known as a Dual Kinetic Assay (DKA). In the genotyping assay, the HPV 16 signal can be further distinguished from HPV 18/45 using the DKA principle. The assay does not differentiate between signal from HPV 18 and 45. Thus a genotyping result will be reported as positive/negative for HPV 16 and then positive/negative for HPV 18 and/or 45 together [52].

An advantage of the Aptima test is that the detection of the E6/E7 mRNA may increase clinical specificity for cervical cancer, due to the overexpression of E6/E7 from genome integration during persistent HPV infection. A meta-analysis showed comparable clinical sensitivities between Aptima HPV and HC2, but a superior clinical specificity for the Aptima assay [54]. The increased clinical specificity may also be a factor of the lack of cross-reactivity of the Aptima assay to low-risk HPV types.

Since the Aptima and Cervista assays are both marketed by Hologic, Inc., practitioners are likely to question which assay is better. The answer is not simple, as both laboratory testing volume and desired assay characteristics play a role in the decision. The Aptima tests are generally marketed to the higher volume laboratories because of the automated testing options on the Tigris and Panther systems. Cervista can also be

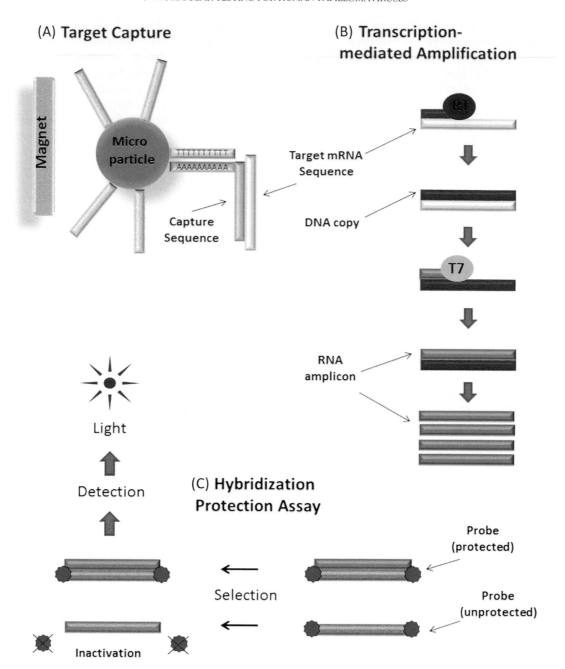

FIGURE 7.4 Aptima TMA assay. The figure shows the major steps of the Aptima TMA assay. (A) First, target mRNA sequences are captured using magnetic microparticles. A capture sequence containing a poly-adenosine sequence hybridizes to a poly-thymidine oligo that is bound to the microparticle. The sequence-specific portion of the capture oligo then binds to the target mRNA. (B) After mRNA sequences are captured, they are amplified using TMA. Reverse transcriptase synthesizes a complementary DNA sequence from the mRNA target. T7 RNA polymerase creates many RNA copies of the DNA template. (C) A hybridization protection assay is used for amplicon detection. The RNA amplicon binds to a probe containing several chemiluminescent molecules. The hybridization of the probe to target protects the probe from degradation during a chemical selection step. Intact probes are then subjected to a detection reaction that releases a light signal.

automated on the Cervista HTA system, but is more flexible for a smaller volume laboratory. The true internal control in the Cervista assay is important if specimen adequacy is a concern in causing false-negative results. A study by Nolte and Ribeiro-Nesbitt in 2014 compared both of the Hologic tests using 208 patients and found only 88% agreement between the two assays [55]. Analysis showed that 18 specimens were determined to be false positives by the Cervista test. This was highly correlated with a triple-positive occurrence in which all three Cervista probe sets were above the cutoff value, similar to a previous study that suggested the cutoff should be increased for the Cervista assay to prevent such false positives [50]. The Nolte study suggested that the Aptima assay has a significantly higher specificity than the Cervista assay and recommended that triple-positive Cervista results should be confirmed by another test or reported as indeterminate [55].

Roche cobas HPV Test

In April 2011, the Roche cobas HPV test was approved by the FDA for HPV screening in patients 21 years and older with abnormal cytology (ASCUS) and for co-testing with cervical cytology in women 30 and older. The Roche cobas HPV assay was the first HPV test to include screening for 12 high-risk HPV types (Table 7.2) while simultaneously genotyping for types 16 and 18, all in the same reaction. The clinical performance of the cobas HPV assay was well established by the ATHENA study, which validated the assay for risk stratification of patients with cervical disease [56]. Unlike its counterparts, the cobas HPV assay is based on target amplification-real-time PCR. The assay is performed using the cobas 4800 system, which is a highly automated platform that includes DNA isolation, PCR reaction setup, amplification, and detection with minimal hands-on time. The Roche cobas HPV test is a multiplex real-time PCR assay using uniquely labeled 5' nuclease probes. The assay begins with specimen preparation on the cobas x480 instrument. Approved specimen types are specimens collected in the ThinPrep medium, including post-cytology processing using either the ThinPrep 2000 system or the ThinPrep 3000 system [57,58]. Specimens are lysed, and DNA is purified by adsorption to magnetic glass particles. The HPV DNA is selected by primers and probes specific for the HPV L1 gene. The human β-globin gene serves as an internal control to control for processing, detection, and specimen cellularity (specimen adequacy). During real-time PCR, the target sequences are amplified by DNA polymerase, creating an exponential increase in the amount of amplicon. The sequence-specific probes include a 5' reporter dye and a 3' quencher dye. When the probe binds to the amplicon, DNA polymerase cleaves the probe by 5' to 3' nuclease activity, releasing the reporter dye from the quencher. This produces a fluorescent signal unique to the probe. The fluorescence is detected by the instrument during each PCR cycle, creating an accumulating signal pattern by the end of the reaction. The characteristic wavelength of each probe allows the differentiation of signal from HPV-16, HPV-18, other high-risk HPV (pooled), and the internal control. Results are reported as HPV 16 positive/negative, HPV 18 positive/negative, and other high-risk HPV positive/negative [59].

The Roche cobas HPV test offers some advantages over some of the other offerings on the market. The cobas assay does not require a reflex test for genotyping as does the Aptima and Cervista assays. Instead the cobas HPV assay includes high-risk screening and genotyping together in a multiplexed reaction. Since PCR-based assays have a high level of analytical sensitivity, the cobas assay includes a chemical reaction to prevent contamination from previous amplification reactions, called AmpErase uracil-N-glycosylase. The integration of an endogenous internal control (β-globin) increases confidence in the quality of the specimen and reduces false negatives. The Roche cobas assay does not show cross-reactivity with low-risk HPV types [59]. Comparisons between the cobas assay and other assays have indicated a high sensitivity and specificity in detection and genotyping of HPV. As compared to HC2 and the Aptima assay, one study determined that clinical sensitivity was equivalent in the Roche cobas assay, but the clinical specificity of the Aptima assay for detecting CIN2 or greater remained higher than the other tests [60]. In another comparison, the Roche cobas HPV assay out-performed the Cervista assay, revealing a concerning number of false positives using the Invader technology [61].

Three years after initial FDA approval (April 2014), the FDA announced that the Roche cobas HPV assay could be used as a primary test for cervical cancer screening in women 25 years and older [62]. This was the first time an HPV test had been approved as a replacement to the Pap test for primary screening. The intended use of the Roche cobas HPV test now includes options for primary HPV screening, along with the currently recommended uses for screening women with ASCUS cytology results or co-testing with cytology [63]. The approval of HPV testing as a front-line screen for cervical cancer will likely cause a paradigm shift in routine patient management in women's health.

CLINICAL UTILITY

The clearly established link of HPV infection to the progression of cervical dysplasia to carcinoma has prompted numerous studies into the management of women with abnormal cytology screening tests. A large study was conducted by the National Cancer Institute named the ASCUS-LSIL Triage Study (ALTS) in order to compare different methods of patient management after an ASCUS (equivocal) diagnosis, including triage by HPV DNA testing [64]. The findings led to the 2006 American Society for Colposcopy and Cervical Pathology (ASCCP) consensus guidelines for the management of women with cervical neoplasia or abnormal cervical screening tests. The 2006 ASCCP guidelines recommended that a woman over age 30 with a negative cytology report should be screened for high-risk HPV infection. When genotyping assays became available to distinguish HPV 16 and 18 from other high-risk types, the guidelines were amended in 2009 to state that colposcopy should be done immediately if HPV types 16 or 18 were present in women over 30. Soon, the co-testing with cytology and HPV testing became a widely accepted strategy for cervical cancer screening for women aged 30–64 years [26,65].

The 2012 ASCCP recommendations are quite complex and a practitioner can benefit from published algorithms in flow-chart format [66] or even from smart phone apps. In sum, women who are 21–65 years in age may be screened with cytology every 3 years, or for women ages 30–65, a combination of cytology and HPV testing (co-testing) may be performed every 5 years. In the case of ASCUS cytology results, HPV testing is recommended to determine triage to colposcopy in women 25 and older [65]. The prevailing guidelines for management of abnormal cervical screens are currently changing due to emerging data that proposes the HPV test can even replace front-line Pap stain screening.

Large-scale clinical studies have shown the benefits of using the more sensitive HPV test first to determine potential risk for development of high-grade cervical dysplasia or cancer. The largest such study was at Kaiser-Permanente Northern California, in which more than 1 million women 30–64 years old were tested for both HPV (by HC2 assay) and by cervical cytology (Pap) [67]. This study expanded beyond the data from the ALTS study and allowed for a more detailed stratification of recommendations based on age, cytology classification, and HPV status. The most recent update from the study found that the risk of developing CIN3 or greater was significantly lower after an HPV-negative result than after a Pap-negative result [68]. Other studies have shown similar results when comparing HPV testing to cervical cytology alone. An analysis of four European clinical trials showed that HPV screening provided a 60–70% greater protection against invasive cervical cancer as compared to cytology [69]. The approval of the Roche cobas HPV assay was largely based on the results of the ATHENA study, which examined women with ASCUS, with normal cervical cytology, and a broad population of women 25 years and older [56,70,71]. The 2012 ASCCP guidelines did not recommend HPV co-testing until age 30, and only recommend HPV testing in women 25–29 if first given an ASCUS diagnosis by cytology [66]. However, cervical cancer incidence begins to increase significantly between ages 25 and 35 [10], and although HPV infection can be transient, this age group still has a significant risk of developing cervical dysplasia that can lead to cancer. There is also evidence that HPV screening is still necessary through age 65 [72]. A new testing algorithm proposed by the ATHENA study, approved by the FDA, and now recommended by the ASCCP is significantly different from the current guidelines.

The newest clinical guidance was released in January 2015 by the Society of Gynecologic Oncology and the ASCCP [73]. The guidance states that primary HPV screening can be used as an alternative to the current cytology-based cancer screening approaches. Primary HPV screening may begin for women 25 years and older, by an FDA-approved HPV assay that includes an approved indication for primary screening. If positive for high-risk HPV types other than 16 and 18, the patient should be triaged to cytology testing, followed by colposcopy in the case of ASCUS result or worse. If positive for HPV types 16 or 18, the patient should be referred immediately to colposcopy. Patients with negative HPV results should remain under routine screening schedules, in which rescreening for HPV should occur no sooner than every 3 years [62,73]. A negative high-risk HPV test has been shown to provide greater reassurance of low risk to develop CIN3 or greater, as compared to a negative cytology result [73]. It is important to note that in the case of HPV primary screening, the Pap test is not necessarily eliminated, but instead shifted to serve as the second-line test. The exception would be in the case of HPV 16- or 18-positive patients, where the Pap test would be skipped for triage to immediate colposcopy. This interim guidance did not eliminate the option for co-testing or primary cytology screening, but does offer the practitioner additional screening options in light of recent research.

LIMITATIONS

As expected for any laboratory test, molecular HPV testing has limitations, both from clinical and technical perspectives. One clinical limitation is the age restriction for the use of HPV testing. HPV screening is not recommended below the age of 25, since infection is common in the younger age group and is often cleared naturally without causing cervical dysplasia. HPV testing cannot distinguish between transient and persistent infection. A positive result must be considered carefully in the clinical context of the patient. A positive result for HPV is not a guaranteed indicator of cancer or even cervical abnormality. Instead it should be viewed as a heightened risk factor that warrants further investigation, either by cervical cytology or colposcopy, depending on the screening algorithm used. Practitioners and patients can be more reassured from a negative HPV result, as a negative result indicates a 3-year cumulative incidence rate of CIN3 + of only 0.3−0.5% [73]. Still, even with such a high negative predictive value, there is always the possibility of a false negative, either due to extremely rare cases of HPV-negative cervical cancer or because of technical limitations of the assay. During the visual inspection of a cervical cytology specimen, occasional incidental findings can be discovered, such as other metastatic tumors or non-HPV infections. Such discoveries would not occur in the case of primary HPV screening, if the patient is HPV negative. There is little data on this particular aspect of reduction or elimination of front-line cervical cytology, so further research may be needed.

Technical limitations of HPV testing vary depending on the assay utilized. One universal limitation appears to be the variability of assay performance using the SurePath (Becton Dickinson) liquid cytology medium. While widely used for cervical cytology, the SurePath medium is not cleared for use in any of the current FDA-approved HPV assays. Many laboratories have performed independent validations of the SurePath medium for HPV testing, essentially operating off-label. However, there are concerns that the stability of the HPV DNA is compromised in the medium, causing an elevated false-negative rate [74]. If a practitioner chooses to order an HPV test using SurePath sample medium, it is important to inquire with the laboratory regarding the validations performed and the limitations of the test.

Other technical limitations of some HPV assays include lack of internal controls or use of controls that cannot challenge the cellularity of the specimen (Table 7.3). Preanalytical error, such as inadequate sampling, can cause a false-negative result. Cross-reactivity between HPV types is another concern in some assays. Cross-reactivity of an assay to a low-risk strain may falsely indicate the presence of a high-risk strain, potentially causing patient harm in the form of unnecessary invasive follow-up testing. Even if cross-reactivity is not published for a particular assay, it is important to note that new HPV strains are still being classified with the help of metagenomics [11], with unknown clinical correlations.

Despite the limitations, testing for HPV in the prevention of cervical cancer is firmly established in medical practice and will continue to expand. As molecular technology evolves, more HPV assays will likely be offered on the market, giving the medical provider an array of choices for cervical cancer screening. HPV testing will continue to be critical in monitoring the effectiveness of HPV vaccination [75]. Further clinical studies in other HPV-triggered cancers, such as head, neck, anal, and penile cancers, will determine whether HPV molecular testing may eventually play a clinically significant role in the management of those patients as well.

References

[1] Dunne EF, Unger ER, Sternberg M, et al. Prevalence of HPV infection among females in the United States. JAMA 2007;297:813−19.

[2] Hariri S, Unger ER, Sternberg M, et al. Prevalence of genital human papillomavirus among females in the United States, the National Health And Nutrition Examination Survey, 2003−2006. J Infect Dis 2011;204:566−73.

[3] Giuliano AR, Lu B, Nielson CM, et al. Age-specific prevalence, incidence, and duration of human papillomavirus infections in a cohort of 290 US men. J Infect Dis 2008;198:827−35.

[4] Giuliano AR, Anic G, Nyitray AG. Epidemiology and pathology of HPV disease in males. Gynecol Oncol 2010;117(2 Suppl): S15−19.

[5] Centers for Disease Control and Prevention: Incidence, prevalence, and cost of sexually transmitted infections in the United States, <http://www.cdc.gov/std/stats/sti-estimates-fact-sheet-feb-2013.pdf/>; 2013 [accessed 01.11.14].

[6] Hariri S, Dunne, E., Saraiya, M., Unger, E., Markowitz, L. Manual for the surveillance of vaccine-preventable diseases. Chapter 5: Human Papillomavirus, <http://www.cdc.gov/vaccines/pubs/surv-manual/chpt05-hpv.html/>; 2011 [accessed 01.11.14].

[7] Centers for Disease Control and Prevention: Human papillomavirus (HPV)-associated cancers, <http://www.cdc.gov/cancer/hpv/statistics/cases.htm/>; 2014 [accessed 01.11 14].

[8] Moscicki AB, Schiffman M, Burchell A, et al. Updating the natural history of human papillomavirus and anogenital cancers. Vaccine 2012;30(Suppl 5):F24−33.

[9] Chaturvedi AK, Engels EA, Pfeiffer RM, et al. Human papillomavirus and rising oropharyngeal cancer incidence in the United States. J Clin Oncol 2011;29:4294−301.

[10] Howlader N, Noone AM, Krapcho M, et al., ediotrs. SEER Cancer Statistics Review, 1975−2011, <http://seer.cancer.gov/csr/1975_2011/>; 2014 [accessed 01.11.14].

[11] de Villiers EM. Cross-roads in the classification of papillomaviruses. Virology 2013;445:2−10.

[12] Zuna RE, Allen RA, Moore WE, Lu Y, Mattu R, Dunn ST. Distribution of HPV genotypes in 282 women with cervical lesions: evidence for three categories of intraepithelial lesions based on morphology and HPV type. Mod Pathol 2007;20:167−74.

[13] Muñoz N, Bosch FX, de Sanjosé S, et al. Epidemiologic classification of human papillomavirus types associated with cervical cancer. N Engl J Med 2003;348:518−27.

[14] Cogliano V, Baan R, Straif K, et al. Carcinogenicity of human papillomaviruses. Lancet Oncol 2005;6:204.

[15] Greer CE, Wheeler CM, Ladner MB, et al. Human papillomavirus (HPV) type distribution and serological response to HPV type 6 virus-like particles in patients with genital warts. J Clin Microbiol 1995;33:2058−63.

[16] Schiffman M, Castle PE, Jeronimo J, Rodriguez AC, Wacholder S. Human papillomavirus and cervical cancer. Lancet 2007;370:890−907.

[17] Duensing S, Lee LY, Duensing A, et al. The human papillomavirus type 16 E6 and E7 oncoproteins cooperate to induce mitotic defects and genomic instability by uncoupling centrosome duplication from the cell division cycle. Proc Natl Acad Sci USA 2000;97:10002−7.

[18] Zheng ZM, Baker CC. Papillomavirus genome structure, expression, and post-transcriptional regulation. Front Biosci 2006;11:2286−302.

[19] Burd EM. Human papillomavirus and cervical cancer. Clin Microbiol Rev 2003;16:1−17.

[20] Steben M, Duarte-Franco E. Human papillomavirus infection: epidemiology and pathophysiology. Gynecol Oncol 2007;107 (2 Suppl. 1):S2–5.

[21] Walboomers JM, Jacobs MV, Manos MM, et al. Human papillomavirus is a necessary cause of invasive cervical cancer worldwide. J Pathol 1999;189:12–19.

[22] Steenbergen RD, de Wilde J, Wilting SM, Brink AA, Snijders PJ, Meijer CJ. HPV-mediated transformation of the anogenital tract. J Clin Virol 2005;32(Suppl 1):S25–33.

[23] Bosch FX, Manos MM, Muñoz N, et al. Prevalence of human papillomavirus in cervical cancer: a worldwide perspective. International biological study on cervical cancer (IBSCC) Study Group. J Natl Cancer Inst 1995;87:796–802.

[24] Abulafia O, Pezzullo JC, Sherer DM. Performance of ThinPrep liquid-based cervical cytology in comparison with conventionally prepared Papanicolaou smears: a quantitative survey. Gynecol Oncol 2003;90:137–44.

[25] Sherman ME. Chapter 11: Future directions in cervical pathology. J Natl Cancer Inst Monogr 2003;31:72–9.

[26] Massad LS, Einstein MH, Huh WK, et al. Updated consensus guidelines for the management of abnormal cervical cancer screening tests and cancer precursors. Obstet Gynecol 2012;2013 (121):829–46.

[27] GARDASIL Package Insert. Whitehouse Station, NJ: Merck & Co., Inc.; 2011.

[28] Harrison C, Britt H, Garland S, et al. Decreased management of genital warts in young women in australian general practice post introduction of national HPV vaccination program: results from a nationally representative cross-sectional general practice study. PLoS One 2014;9:e105967.

[29] Markowitz LE, Dunne EF, Saraiya M, et al. Human papillomavirus vaccination: recommendations of the Advisory Committee on Immunization Practices (ACIP). MMWR Recomm Rep 2014;63:1–30.

[30] Markowitz LE, Hariri S, Lin C, et al. Reduction in human papillomavirus (HPV) prevalence among young women following HPV vaccine introduction in the United States, National Health and Nutrition Examination Surveys, 2003–2010. J Infect Dis 2013;208:385–93.

[31] Joura EA, Ault KA, Bosch FX, et al. Attribution of 12 high-risk human papillomavirus genotypes to infection and cervical disease. Cancer Epidemiol Biomarkers Prev 2014;23:1997–2008.

[32] Elam-Evans LD, Yankey D, Jeyarajah J, et al. National, regional, state, and selected local area vaccination coverage among adolescents aged 13–17 years—United States, 2013. MMWR Morb Mortal Wkly Rep 2014;63:625–33.

[33] Goldhaber-Fiebert JD, Stout NK, Salomon JA, Kuntz KM, Goldie SJ. Cost-effectiveness of cervical cancer screening with human papillomavirus DNA testing and HPV-16,18 vaccination. J Natl Cancer Inst 2008;100:308–20.

[34] Benard VB, Thomas CC, King J, Massetti GM, Doria-Rose VP, Saraiya M. Vital signs: cervical cancer incidence, mortality, and screening—United States, 2007–2012. MMWR Morb Mortal Wkly Rep 2014;63:1004–9.

[35] Flores ER, Allen-Hoffmann BL, Lee D, Sattler CA, Lambert PF. Establishment of the human papillomavirus type 16 (HPV-16) life cycle in an immortalized human foreskin keratinocyte cell line. Virology 1999;262:344–54.

[36] Food and Drug Administration: Nucleic Acid Based Tests. <http://www.fda.gov/MedicalDevices/ ProductsandMedicalProcedures/InVitroDiagnostics/ ucm330711.htm/>; 2014 [accessed 13.10.14].

[37] Abreu AL, Souza RP, Gimenes F, Consolaro ME. A review of methods for detect human papillomavirus infection. Virol J 2012;9:262.

[38] Arney A, Bennett KM. Molecular diagnostics of human papillomavirus. Lab Medicine 2010;41:523–30.

[39] Molijn A, Kleter B, Quint W, van Doorn LJ. Molecular diagnosis of human papillomavirus (HPV) infections. J Clin Virol 2005;32 (Suppl. 1):S43–51.

[40] Clavel C, Masure M, Putaud I, et al. Hybrid capture II, a new sensitive test for human papillomavirus detection. Comparison with hybrid capture I and PCR results in cervical lesions. J Clin Pathol 1998;51:737–40.

[41] Qiagen Digene Hybrid Capture 2 High-Risk HPV DNA Test [package insert]. Gaithersburg, MD: Digene Corporation; 2007.

[42] Gillio-Tos A, De Marco L, Carozzi FM, et al. Clinical impact of the analytical specificity of the hybrid capture 2 test: data from the New Technologies for Cervical Cancer (NTCC) study. J Clin Microbiol 2013;51:2901–7.

[43] Flanagan SM, Wilson S, Luesley D, Damery SL, Greenfield SM. Adverse outcomes after colposcopy. BMC Womens Health 2011;11:2.

[44] Khan MJ, Castle PE, Lorincz AT, et al. The elevated 10-year risk of cervical precancer and cancer in women with human papillomavirus (HPV) type 16 or 18 and the possible utility of type-specific HPV testing in clinical practice. J Natl Cancer Inst 2005;97:1072–9.

[45] Cervista HPV HR [package insert]. San Diego, CA: Hologic, Inc; 2012.

[46] Cervista HPV 16/18 [package insert]. San Diego, CA: Hologic, Inc.; 2010.

[47] Ginocchio CC, Barth D, Zhang F. Comparison of the Third Wave Invader human papillomavirus (HPV) assay and the digene HPV hybrid capture 2 assay for detection of high-risk HPV DNA. J Clin Microbiol 2008;46:1641–6.

[48] Stillman MJ, Day SP, Schutzbank TE. A comparative review of laboratory-developed tests utilizing Invader HPV analyte-specific reagents for the detection of high-risk human papillomavirus. J Clin Virol 2009;45(Suppl. 1):S73–7.

[49] Belinson JL, Wu R, Belinson SE, et al. A population-based clinical trial comparing endocervical high-risk HPV testing using hybrid capture 2 and Cervista from the SHENCCAST II Study. Am J Clin Pathol 2011;135:790–5.

[50] Boers A, Slagter-Menkema L, van Hemel BM, et al. Comparing the Cervista HPV HR test and Hybrid Capture 2 assay in a Dutch screening population: improved specificity of the Cervista HPV HR test by changing the cut-off. PLoS One 2014;9:e101930.

[51] de Sanjose S, Quint WG, Alemany L, et al. Human papillomavirus genotype attribution in invasive cervical cancer: a retrospective cross-sectional worldwide study. Lancet Oncol 2010;11:1048–56.

[52] Aptima HPV 16 18/45 Genotype Assay [package insert]. San Diego, CA: Gen-Probe Inc.; 2013.

[53] Aptima HPV Assay [package insert]. San Diego, CA: Gen-Probe Inc.; 2013.

[54] Arbyn M, Roelens J, Cuschieri K, et al. The APTIMA HPV assay versus the Hybrid Capture 2 test in triage of women with ASC-US or LSIL cervical cytology: a meta-analysis of the diagnostic accuracy. Int J Cancer 2013;132:101–8.

[55] Nolte FS, Ribeiro-Nesbitt DG. Comparison of the Aptima and Cervista tests for detection of high-risk human papillomavirus in cervical cytology specimens. Am J Clin Pathol 2014;142:561–6.

[56] Stoler MH, Wright TC, Sharma A, et al. High-risk human papillomavirus testing in women with ASC-US cytology: results from the ATHENA HPV study. Am J Clin Pathol 2011;135:468–75.

[57] Rao A, Young S, Krevolin M, et al. Comparison of cobas human papillomavirus test results from primary versus secondary vials

of PreservCyt specimens and evaluation of potential cross-contamination. Cancer Cytopathol 2012;120:380−9.

[58] Use the cobas HPV Test pre-cytology or post-cytology. <https://www.hpv16and18.com/labs/lab-efficiencies/testing-flexibility.html/>; 2014 [accessed 22.11.14].

[59] Roche cobas HPV Test [package insert]. Pleasanton, CA: Roche Molecular Systems, Inc.; 2010.

[60] Cuzick J, Cadman L, Mesher D, et al. Comparing the performance of six human papillomavirus tests in a screening population. Br J Cancer 2013;108:908−13.

[61] Martin IW, Steinmetz HB, Lefferts CL, Dumont LJ, Tafe LJ, Tsongalis GJ. Evaluation of the cobas 4800 HPV test for detecting high-risk human papilloma-virus in cervical cytology specimens. Pathogens 2012;1:30−6.

[62] FDA approves first human papillomavirus test for primary cervical cancer screening, <http://www.fda.gov/newsevents/newsroom/pressannouncements/ucm394773.htm/>; 2014 [accessed 22.11.14].

[63] The cobas HPV Test. Intended use, <https://www.hpv16and18.com/hcp/cobas-hpv-test/intended-use.html/>; 2014 [accessed 22.11.14].

[64] Schiffman M, Solomon D. Findings to date from the ASCUS-LSIL Triage Study (ALTS). Arch Pathol Lab Med 2003;127:946−9.

[65] Saslow D, Solomon D, Lawson HW, et al. American Cancer Society, American Society for Colposcopy and Cervical Pathology, and American Society for Clinical Pathology screening guidelines for the prevention and early detection of cervical cancer. Am J Clin Pathol 2012;137:516−42.

[66] American Society for Colposcopy and Cervical Pathology (ASCCP) Management Guidelines. <http://www.asccp.org/Guidelines-2/Management-Guidelines-2/>; 2013 [accessed 09.11.14].

[67] Katki HA, Kinney WK, Fetterman B, et al. Cervical cancer risk for women undergoing concurrent testing for human papillomavirus and cervical cytology: a population-based study in routine clinical practice. Lancet Oncol 2011;12:663−72.

[68] Gage JC, Schiffman M, Katki HA, et al. Reassurance against future risk of precancer and cancer conferred by a negative human papillomavirus test. J Natl Cancer Inst 2014;106 dju153.

[69] Ronco G, Dillner J, Elfström KM, et al. Efficacy of HPV-based screening for prevention of invasive cervical cancer: follow-up of four European randomised controlled trials. Lancet 2014; 383:524−32.

[70] Castle PE, Stoler MH, Wright TC, Sharma A, Wright TL, Behrens CM. Performance of carcinogenic human papillomavirus (HPV) testing and HPV16 or HPV18 genotyping for cervical cancer screening of women aged 25 years and older: a subanalysis of the ATHENA study. Lancet Oncol 2011;12:880−90.

[71] Wright TC, Stoler MH, Behrens CM, Apple R, Derion T, Wright TL. The ATHENA human papillomavirus study: design, methods, and baseline results. Am J Obstet Gynecol 2012;206:46. e1−46.e11.

[72] Gage JC, Katki HA, Schiffman M, et al. Age-stratified 5-year risks of cervical precancer among women with enrollment and newly detected HPV infection. Int J Cancer 2014;136:1665−71.

[73] Huh WK, Ault KA, Chelmow D, et al. Use of primary high-risk human papillomavirus testing for cervical cancer screening: Interim clinical guidance. Gynecol Oncol 2015;136:178−82.

[74] Naryshkin S, Austin RM. Limitations of widely used high-risk human papillomavirus laboratory-developed testing in cervical cancer screening. Drug Healthc Patient Saf 2012;4:167−72.

[75] Meites E, Lin C, Unger ER, et al. Can clinical tests help monitor human papillomavirus vaccine impact? Int J Cancer 2013;133: 1101−6.

8

Molecular Testing for Herpes Viruses

S.K. Tan[1] and B.A. Pinsky[1,2]

[1]Department of Medicine, Division of Infectious Diseases and Geographic Medicine,
Stanford University School of Medicine, Stanford, CA, United States [2]Department of Pathology,
Stanford University School of Medicine, Stanford, CA, United States

INTRODUCTION

The family *Herpesviridae* comprises over a hundred viruses which include eight human herpes viruses: (1) herpes simplex-1 (HSV-1), (2) herpes simplex-2 (HSV-2), (3) varicella-zoster virus (VZV), (4) Epstein−Barr virus (EBV), (5) cytomegalovirus (CMV), (6) human herpesvirus-6 (HHV-6), (7) human herpesvirus-7 (HHV-7), and (8) human herpesvirus-8 (HHV-8). Herpes viruses possess a double-stranded DNA genome surrounded by an icosahedral nucleocapsid encased by an envelope with viral glycoproteins. All herpes viruses encode a core set of proteins involved in nucleic acid synthesis, nucleic acid metabolism, protein modification, and virion structure [1]. The viral life cycle consists of a lytic phase with active viral replication and resultant cell destruction and a latent phase with dormant virus within neurons or lymphocytes that can later reactivate and produce infectious virus. Reactivation is more likely to occur under stress and in immunocompromised hosts with decreased T-cell immunity [2].

Based on serological surveillance, nearly all adults have been infected with one or more human herpesviruses. Primary infection usually occurs through direct contact with infected oral or genital secretions, although VZV is an exception and can be spread by airborne transmission. Intrauterine infection and transmission via organ transplantation can also occur. Several human herpes viruses cause similar clinical syndromes including vesicular skin lesions, mononucleosis, hepatitis, encephalitis, retinitis, and mononucleosis (Tables 8.1−8.4). Importantly, human herpes viruses cause more severe disease in persons with impaired cellular immunity, such as solid organ transplant (SOT) recipients receiving immunosuppression therapy, hematopoietic cell transplant (HCT) recipients, and patients with acquired immunodeficiency syndrome (AIDS).

The diagnosis and monitoring of human herpes virus infections increasingly relies upon the detection and quantitation of human herpes virus DNA in clinical specimens [3,4]. These nucleic acid amplification techniques have in many laboratories replaced conventional viral diagnostic methods, such as viral culture and antigen detection, and have redefined the manner in which human herpes virus infections are diagnosed and managed. This chapter describes each human herpesvirus in turn and highlights the application and utility of molecular diagnostics for specific clinical syndromes.

HERPES SIMPLEX VIRUS

HSV-1 and HSV-2 are a common cause of dermal, oral, and genital infections worldwide (Table 8.1). HSV-1 has a seroprevalence of 60% and is acquired at a younger age, while the seroprevalence of HSV-2 is 20% and correlates with the onset of sexual activity [5,6]. Both viruses are transmitted through direct contact. After primary infection, HSV establishes latency in neurons and can later reactivate and produce recurrent symptoms. Primary infections are more severe than recurrent ones, but reactivation of latent virus can result in frequent clinical manifestations. HSV infection most commonly results in painful vesicles on an erythematous base that subsequently ulcerate. Oral lesions are most often due to HSV-1 and genital lesions due to HSV-2, although genital lesions can be caused by either HSV-1 or HSV-2. HSV infection can also lead to serious disease such as aseptic meningitis,

DOI: http://dx.doi.org/10.1016/B978-0-12-800886-7.00008-X
© 2017 Elsevier Inc. All rights reserved.

TABLE 8.1 Clinical Syndromes and Application of Molecular Diagnostics for HSV-1, HSV-2, and VZV

	Clinical syndrome	Method	Specimen type(s)
HSV-1 AND HSV-2			
Immunocompetent persons	Gingivostomatitis	NAAT	Lesion swab in VTM
	Genital herpes	NAAT	Lesion swab in VTM
	Cutaneous herpes	NAAT	Lesion swab in VTM
	Keratoconjunctivitis	NAAT	Conjunctival swab in VTM
	Meningitis, Encephalitis	NAAT	CSF
Immunocompromised persons[a]	Dissemination, visceral involvement	NAAT	Plasma, tissue, CSF
	Drug resistance	Sequencing	Plasma, tissue, CSF, swab in VTM
VZV			
Immunocompetent persons	Chickenpox	Not performed	Not applicable
	Herpes zoster	NAAT	Lesion swab in VTM
Immunocompromised persons	Dissemination, visceral involvement	NAAT	Plasma, tissue, CSF

[a]*Includes clinical syndromes seen in immunocompetent persons.*
HSV, herpes simplex virus; VZV, varicella-zoster virus; NAAT, nucleic acid amplification test; VTM, viral transport media; CSF, cerebrospinal fluid.

TABLE 8.2 Clinical Syndromes and Application of Molecular Diagnostics for EBV

	Clinical syndrome	Method	Specimen type(s)
EBV			
Immunocompetent persons	Mononucleosis	Not performed	Not applicable
	Nasopharyngeal carcinoma	NAAT	Plasma, nasopharyngeal swab in VTM
Immunocompromised persons	PTLD	NAAT	Plasma, whole blood, PBMC

EBV, Epstein–Barr virus; NAAT, nucleic acid amplification testing; VTM, viral transport media; PTLD, posttransplant lymphoproliferative disorder; PBMC, peripheral blood mononuclear cells.

TABLE 8.3 Clinical Syndromes and Application of Molecular Diagnostics for CMV

	Clinical syndrome	Method	Specimen type(s)
CMV			
Immunocompetent persons	Mononucleosis	Not performed	Not applicable
	Congenital infection	NAAT	Urine, saliva in the neonate
			Amniotic fluid in pregnancy
Immunocompromised persons	GI disease	NAAT	Tissue
	Pneumonitis	NAAT	BAL fluid
	Hepatitis	NAAT	Tissue
	Retinitis	NAAT	Ocular fluid
	Encephalitis	NAAT	CSF
	Drug resistance	Sequencing	Plasma, tissue, BAL fluid, ocular fluid
Special consideration in the immunocompromised	Disease prevention	NAAT	Plasma

CMV, cytomegalovirus; NAAT, nucleic acid amplification testing; GI, gastrointestinal disease; BAL, bronchoalveolar lavage; CSF, cerebrospinal fluid.

encephalitis, and ocular involvement with keratitis, blepharitis, conjunctivitis, and retinitis [7]. In immunocompromised patients, HSV may disseminate to visceral organs involving the gastrointestinal (GI) tract, liver, lungs, adrenal glands, and bone marrow with resultant high morbidity and mortality [8,9].

The preferred method for the diagnosis of HSV infection is the detection of HSV DNA by nucleic acid

TABLE 8.4 Clinical Syndromes and Application of Molecular Diagnostics for HHV-6, HHV-7, and HHV-8

	Clinical syndrome	Method	Specimen type(s)
HHV-6			
Immunocompetent persons	Roseola infantum	Not performed	Not applicable
Immunocompromised persons	Encephalitis	NAAT	CSF
	Dissemination, visceral involvement	NAAT	Plasma, tissue
HHV-7			
Immunocompetent persons	Roseola infantum	Not performed	Not applicable
Immunocompromised persons[a]	Encephalitis	NAAT	CSF
	Dissemination, visceral involvement	NAAT	Plasma, tissue
HHV-8			
Immunocompetent persons	Febrile exanthum	Not performed	Not applicable
Immunocompromised persons[b]	Kaposi sarcoma	NAAT	Plasma, PBMC, tissue
	Multicentric Castleman disease	NAAT	Plasma, PBMC, tissue
	Primary effusion lymphoma	NAAT	Plasma, PBMC, effusion fluid

[a]Routine testing for HHV-7 DNA in plasma is not recommended, as detection often does not correlate with disease.
[b]The diagnosis of HHV-8-associated malignancies requires histopathological evaluation.
HHV-6, human herpes virus-6; HHV-7, human herpes virus-7; HHV-8, human herpes virus-8; NAAT, nucleic acid amplification testing; CSF, cerebrospinal fluid; PBMC, peripheral blood mononuclear cell.

amplification testing (NAAT), most commonly polymerase chain reaction (PCR), of lesion scrapings, fluids, or tissue [10]. Isolation of virus and identification of cytopathic effect on tissue culture is three to four times less sensitive than PCR and has a longer turnaround time of 2—4 days compared to hours [8,11]. Wright or Tzanck straining of scrapings from the base of HSV lesions may demonstrate characteristic giant cells or intranuclear inclusions, but similarly has low sensitivity and cannot differentiate between HSV and VZV infections [7,10]. Compared to patients with biopsy proven and clinically correlated HSV encephalitis, detection of HSV DNA by PCR in cerebrospinal fluid (CSF) is highly sensitive and is the method of choice for the diagnosis of HSV encephalitis and meningitis [12,13]. One study evaluating the role of quantitative HSV PCR in CSF found that higher viral load, defined as more than 10^4 copies/mL, was associated with the presence of brain lesions on computed tomography (CT) and magnetic resonance imaging (MRI) and that these patients had poorer clinical outcomes than patients with lower HSV viral loads in the CSF [14]. Additional studies are needed to further understand the utility of quantitative HSV PCR in CSF as such testing is not currently routinely performed.

Acyclovir is the first-line treatment for HSV infections. Patients who receive long-term acyclovir and have profound immunosuppression are at increased risk for the development of acyclovir-resistant HSV [15,16]. Among immunocompromised patients, the prevalence of drug resistance reaches 3.5—10%, with

rates highest in HCT recipients [16]. Without acyclovir prophylaxis, 80% of HSV seropositive patients reactivate infection after receiving HCT [17]. As a result, prophylaxis is recommended in these recipients for at least the first several weeks to months after transplantation. Prophylaxis against HSV has prevented and shortened the course of mucocutaneous infection and reduced bacteremia with oral pathogens [18—20]. In immunocompetent hosts, HSV resistance to acyclovir is less than 1% [21]. Cases of resistance in immunocompetent patients are most often associated with recurrent genital herpes [16], but have also been reported in patients with HSV infection involving immune-privileged sites such as the eye and central nervous system (CNS) [21,22]. Drug resistance testing should be considered in patients who develop or do not resolve HSV lesions while on prolonged acyclovir therapy.

In clinical laboratories, HSV drug resistance testing is currently performed using phenotypic methods. However, genotypic drug resistance testing is an important application of molecular diagnostics that predicts drug resistance based on the identification of specific mutations. In particular, mutations in HSV thymidine kinase encoded by the *UL23* gene and DNA polymerase encoded by the *UL30* gene result in resistance to acyclovir [23]. As new mutations associated with drug resistance continue to be described and sequencing technologies become even more accessible, it is anticipated that HSV genotypic drug resistance testing will be more commonly used outside of the research setting [24—27].

VARICELLA-ZOSTER VIRUS

VZV infection results in two major clinical syndromes: chickenpox and herpes zoster (Table 8.1). Primary infection through respiratory or direct contact results in chickenpox, an infection presenting as diffuse, pruritic vesicles with eventual crusting of lesions [28]. Chickenpox most commonly manifests during childhood and is self-resolving, but can cause more serious manifestations in adults and life-threatening disease in immunocompromised hosts. In such patients, the duration of healing of cutaneous lesions can be prolonged and there is risk for VZV dissemination to visceral organs such as the lungs, liver, and CNS [29,30]. Furthermore, presentation of VZV can be atypical [29]. After primary infection, the virus establishes latency in the dorsal root ganglia, but can later reactivate and lead to herpes zoster. Zoster is characterized by painful, unilateral vesicular lesions in a dermatomal distribution. Herpes zoster occurs at all ages, but is more frequent in the elderly or the immunocompromised [31,32].

Detection of VZV DNA by PCR in vesicle fluid, blood, spinal fluid, and other tissues is the most sensitive test for diagnosing varicella disease [33–37]. In patients with zoster, the most specific and sensitive samples for the detection of VZV DNA by PCR are from crusts and vesicle fluid swabs. In one study, 97% of crusts, 94% of vesicle swabs, 90% of crust swabs, and 84% of papule swabs were VZV DNA positive ($\geq$10 DNA copies/sample). The probability of a false-negative result was 5% for crusts, 6% for vesicle swabs, 14% for papule swabs, and 24% for crust swabs [36]. Other methods for detecting VZV from lesions, such as direct fluorescent antibody (DFA) testing and viral culture, have lower sensitivity [33,38]. In plasma, detection of VZV DNA has been correlated with symptomatic VZV disease. One study demonstrated that 10/10 immunocompetent patients with zoster and 4/4 immunocompromised patients with visceral VZV disease had detectable VZV DNA in plasma, while none of the 108 asymptomatic SOT recipients who were greater than 1 year posttransplantation and off antiviral therapy had positive plasma VZV DNA [34]. Higher viral loads in immunocompromised patients with visceral VZV may also predict prolonged clinical courses [39].

EPSTEIN–BARR VIRUS

EBV is ubiquitous with a seroprevalence of 90–95% in adults (Table 8.2). Primary infection with EBV results from exposure to the oral secretions of seropositive individuals. The virus has a tropism for epithelial cells and establishes lifelong latency in B cells with periodic reactivation and viral shedding in saliva. Illness from EBV can include infectious mononucleosis, characterized by fever, sore throat, and lymphadenopathy, as well as hepatitis, pneumonitis, and leukopenia. In immunocompromised hosts, uncontrolled proliferation of EBV-infected B cells may occur resulting in lymphoproliferative diseases. Posttransplant lymphoproliferative disorder (PTLD) is seen in SOT and HCT recipients, but can occur in any patient who is on high-dose immunosuppression or with underlying dysfunction in T-cell immunity. EBV has also been implicated in the pathogenesis of several malignancies including nasopharyngeal carcinoma, Burkitt lymphoma, and Hodgkin lymphoma.

PTLD is a serious complication affecting transplant recipients. PTLD presents as a wide spectrum of disease ranging from indolent lymphoproliferation to malignant aggressive lymphoma with high mortality despite aggressive chemotherapy [40]. Overall mortality of PTLD is difficult to establish, but has been reported to be 40–70% after SOT and 90% after HCT [41]. The prevalence of PTLD ranges from 1% to 20%, but varies by presence of risk factors. Development of PTLD is greatest within the first few months of transplantation, although the risk is prolonged in SOT recipients given the ongoing need for immunosuppressive therapy to minimize graft rejection. PTLD has been reported to occur as far out as a decade from SOT [42]. In SOT, prevalence varies with type of organ transplanted and pretransplant EBV serostatus. Lung and intestinal transplants, which are lymphoid-rich organs, have the highest rates of PTLD. Similarly, seronegativity prior to transplant and primary EBV infection after transplant are significant risk factors for PTLD [40,42,43]. In HCT, risk factors include HLA-mismatching, T-cell depletion, use of anti-lymphocyte antibodies, and graft versus host disease [41].

Given the significance of PTLD, early recognition and initiation of therapy is important. EBV quantitation has been increasingly incorporated into management algorithms for PTLD, although the optimal way to perform, interpret, and utilize EBV PCR assays for surveillance, diagnostic, and disease monitoring requires ongoing evaluation [44]. The optimal blood compartment (whole blood vs plasma) to test for EBV testing remains controversial. Both whole blood and plasma have been found to be useful for EBV DNA measurements, although plasma is more specific for the diagnosis of PTLD in immunosuppressed patients [45]. Prior attempts at defining viral load thresholds to initiate action have been hampered by significant interlaboratory variability in qualitative and quantitative EBV results. To address this, the World Health Organization (WHO) approved the 1st International

Standard for EBV for Nucleic Acid Amplification Techniques in 2012. This reference standard defines an EBV international unit (IU) in order to facilitate the comparison of EBV measurements worldwide. The commutability, the ability of a reference material to have interassay properties comparable to that of authentic specimens, of the EBV WHO standard has also been evaluated for two commonly used assays with favorable results [46]. Efforts toward harmonizing viral load results across laboratories will allow for increased generalizability of findings from studies evaluating the role of EBV quantitation in the management of patients with PTLD.

Surveillance monitoring of EBV DNA levels has been useful in predicting risk for PTLD. Additionally, initiation of preemptive treatment with reduction of immunosuppression or use of the anti-CD20 antibody rituximab has resulted in decreased incidence of PTLD in transplant recipients [47–50]. High viral loads often antedate the clinical presentation of EBV-associated PTLD, although a strong correlation does not always exist [51,52]. The viral load can remain low if the site of PTLD is protected such as early in the graft itself or within GI tissues. Furthermore, patients with viremia do not always have or develop PTLD. In high-risk asymptomatic SOT recipients being serially monitored, EBV DNA testing had good sensitivity for detecting EBV-positive PTLD. However, it had poor specificity, resulting in good negative (>90%) but poor positive predictive value (as low as 28% and not >65%) in these populations [44,52,53]. Optimal monitoring frequency remains uncertain. Since EBV doubling times are as short as 49–56 h, weekly monitoring over the high-risk period has been recommended [44]. The suitability of EBV level as a predictive marker for PTLD relapse remains unclear. Discordance in viral load in whole blood and clinical course has been observed, with some EBV-associated PTLD patients in clinical remission having EBV levels as high as those recorded at the onset of PTLD and some patients with progressive disease having a decline in EBV [54].

EBV quantitation is also increasingly used in the diagnosis and management of nasopharyngeal carcinoma. Plasma EBV levels can identify high-risk patients at risk for relapse. Compared to patients who sustained clinical remission, patients with relapsed disease had significantly higher pretreatment EBV DNA and persistently detectable posttreatment EBV DNA levels in plasma. Additionally, decreased survival was associated with higher pretreatment and persistently positive posttreatment plasma EBV DNA [46,55,56]. EBV DNA has also been evaluated in salivary and nasopharyngeal biopsy specimens. EBV DNA levels in salivary specimens were higher posttreatment than pretreatment with a trend toward higher levels in

patients with advanced stage compared to early-stage nasopharyngeal carcinoma [57]. Nasopharyngeal brush sampling was found to correlate with nasopharyngeal carcinoma and may be a less invasive adjunct in diagnosis and posttreatment monitoring [58]. The role of EBV DNA in the management of patients with PTLD and nasopharyngeal carcinoma is significant and would benefit from continued study.

CYTOMEGALOVIRUS

CMV seroprevalence ranges from 45% to 100% depending on the population, with infection most often occurring within the first two decades of life [59,60]. Most CMV infections are mild or asymptomatic. Primary infection with CMV can produce an infectious mononucleosis-like syndrome with fever, lymphadenopathy, and relative lymphocytosis (Table 8.3). After initial infection, the virus establishes life-long latency in peripheral blood mononuclear cells (PBMCs) [60,61]. While individuals with intact immune responses are able to suppress viral replication, immunocompromised and immunologically immature hosts are most at risk for the development of severe disease after primary infection or reactivation. Symptomatic CMV disease can affect any organ in the body, resulting in interstitial pneumonitis, hepatitis, esophagitis, colitis, retinitis, and myocarditis [62].

In transplantation, CMV is a major cause of morbidity and preventable cause of mortality. Without a prevention strategy, CMV disease typically occurs within the first 3 months after SOT and first 120 days in HCT recipients [63,64]. Donor and recipient CMV serostatus is perhaps the most important predictor of CMV infection and disease, with serostatus mismatch conferring the highest risk [65–68]. The risk for CMV disease is also increased in transplant recipients requiring high levels of immunosuppression such as lung transplants [69,70]. The strategies aimed at CMV disease prevention involve the use of universal prophylactic and preemptive therapies. Universal prophylaxis entails the administration of antivirals directed to all predisposed transplant recipients for a defined high-risk period. Preemptive therapy involves the use of antivirals guided by the early detection of CMV viremia prior to end-organ involvement, as high CMV levels and rate of rise of viremia are risk factors for the development of CMV disease [71–74]. A prophylaxis strategy is easier to coordinate but has higher drug costs and increased monitoring for drug toxicity. A preemptive strategy requires weekly CMV surveillance, with the benefit of potentially shorter courses of antivirals and less drug toxicity. In HCT recipients, a preemptive approach is most often recommended. The prophylactic use of ganciclovir has raised concern for prolonged neutropenia and bacterial infection [69,75,76]. In SOT recipients, the superiority of preemptive versus prophylactic therapy for CMV disease prevention remains unclear due to few randomized trials and significant heterogeneity between studies [77–81]. However, prophylactic antiviral therapy is generally recommended as the preferred prevention strategy in CMV donor positive/recipient negative (D + /R −) and recipient positive (R +) individuals.

The widespread implementation of quantitative CMV testing has significantly contributed to the success of preemptive strategies. Initially, determination of CMV levels in the blood was made via microscopy, utilizing indirect immunofluorescence to detect the CMV pp65 structural phosphoprotein antigen in peripheral blood leukocytes. However, antigenemia testing is laborious, requires rapid specimen processing, and is insensitive during neutropenic episodes. As such, nucleic acid amplification techniques, particularly real-time PCR methods, have become the preferred approach for CMV detection and quantitation. Significantly, studies in immunocompromised patients have shown that real-time PCR assays are more sensitive than pp65 antigenemia without loss of specificity [82,83]. In addition, real-time PCR is less affected by specimen transport conditions and can be automated to efficiently process large numbers of specimens [82,84]. Despite the advances in quantitative CMV testing, optimal viral thresholds at which to initiate preemptive treatment have yet to be identified. Previous attempts at establishing broadly applicable quantitative cutoff values have been hampered by significant interassay variability. While many factors play a role in the variability of quantitative results, selection of the quantitative calibrator, commercially prepared primers and probes, and amplification target gene were most prominently associated with variability in a multivariate analysis [85]. In an attempt to address this issue, the WHO created the 1st International Standard for Human CMV for Nucleic Acid Amplification Techniques in 2010. While the commutability of the WHO International Standard and secondary standards calibrated to the IU need further evaluation, standardization that delivers comparable quantitative data across different laboratories provides the opportunity to identify clinical viral load thresholds in future prospective trials.

In addition to the evaluation of blood samples, primarily plasma, PCR testing for CMV DNA in tissues and fluids is increasingly being used to aid in the diagnosis of CMV end-organ disease, such as CMV colitis and pneumonitis. Definitive diagnosis of CMV disease has required histologic demonstration of characteristic viral inclusions or detection of viral proteins by immunohistochemical (IHC) staining in tissue samples. However, a number of studies have shown that detection of CMV DNA by PCR in GI specimens, both fresh and formalin-fixed, paraffin-embedded tissues, may be a useful adjunct to histologic evaluation for the diagnosis of CMV GI disease [86,87]. Future studies will be required to evaluate the utility of nucleic acid amplification techniques for detection of CMV end-organ disease in other tissue types.

The utility of CMV DNA testing in bronchoalveolar lavage (BAL) fluid has also been evaluated in retrospective studies. Testing of CMV by PCR in BAL is believed to be most useful in patients with underlying immunosuppression, and in particular, the absence of CMV DNA has good negative predictive value [88,89]. However, the clinical significance of qualitative and quantitative detection of CMV in BAL fluid on disease detection and outcomes remains unclear. Though detection of CMV in BAL fluid is common 6 months post-lung transplant, correlation with the development of bronchiolitis obliterans syndrome is inconsistent [90–92]. Similarly, though one study of 27 lung transplant recipients noted that a BAL viral load of more than 500,000 copies/mL by quantitative hybrid capture assay correlated with CMV pneumonitis, other studies have found no such correlation of disease with BAL viral load [93]. More data is needed to better understand the application of CMV DNA positive BAL fluid in immunocompromised patients.

CMV genotypic drug resistance testing is another application of molecular diagnostics that has a critical role in the management of transplant recipients. In SOT recipients, the rate of CMV drug resistance varies from 2% to 17.6% [94–97]. CMV drug resistance in HCT recipients is infrequent. However, significant resistance (14.5%) has been noted in patients receiving haploidentical transplants, likely due to high viral levels, and prolonged duration of viremia and exposure to antivirals [98]. Clinical outcome in patients with resistance-associated disease is poor with higher rates of progressive viremia and CMV-associated mortality [97,99]. Guidelines suggest that CMV drug resistance testing should be considered in patients who have persistent or increasing viral loads despite 2–3 weeks of antiviral therapy [100].

CMV drug resistance mutations are located in the UL97 phosphotransferase and UL54 DNA polymerase genes. All current antivirals recommended for CMV treatment (ganciclovir, valganciclovir, cidofovir, and foscarnet) ultimately target viral DNA polymerase, although ganciclovir and its prodrug, valganciclovir, require phosphorylation by the viral UL97 kinase for antiviral activity. The most common antiviral resistance strains contain mutations in the UL97 gene, which confer resistance to ganciclovir and valganciclovir [101]. Mutations in the *UL54* gene can confer cross-resistance to more than one antiviral, particularly with ganciclovir and cidofovir, and less commonly with foscarnet [102]. Alternative mutations may become clinically relevant as new antivirals with different mechanisms of actions are approved.

Molecular diagnostics have also significantly contributed to the early diagnosis of CMV in nontransplant recipients. CMV is a leading cause of congenital infection worldwide, occurring in 0.2 − 2.2% of live births [62]. While the majority (85 − 90%) of infants with congenital CMV are asymptomatic, symptomatic infants can develop jaundice, hepatosplenomegaly, microcephaly, intrauterine growth retardation, and chorioretinitis [103]. Beyond acute manifestations, symptomatic infants are at increased risk for permanent disabilities such as sensorineural hearing loss and mental disabilities. Ten to fifteen percentage of asymptomatic infants can also develop these permanent disabilities [103]. The standard method for the diagnosis of congenital CMV infection in neonates has been viral culture of urine or saliva specimens obtained by the end of the third week of life. As an alternative to culture, PCR has been evaluated for the diagnosis of congenital CMV for its enhanced sensitivity and rapid turnaround. Several studies have shown that PCR performs as well as rapid shell vial culture of urine or saliva specimens for the diagnosis of congenital CMV infection [104−106]. Evaluation of amniotic fluid at 21 weeks gestation for CMV DNA by PCR has been found to have high specificity for congenital CMV, but false positives have been noted [107,108]. The utility of quantitative CMV testing in amniotic fluid is unclear, though there have been case reports suggesting a possible correlation of high CMV levels with disease [107,109].

HUMAN HERPESVIRUS-6

Like other herpes viruses, HHV-6 is ubiquitous with a seroprevalence of more than 90% in adults [110]. Two variants of HHV-6 have been identified, HHV-6a and HHV-6b, and are now considered separate species. The majority of primary infections and reactivation events are due to HHV-6b. Primary infection with HHV-6b occurs early in life and can present asymptomatically or with classic roseola infantum with fever and rash (Table 8.4). After primary infection, HHV-6 establishes latency in salivary glands and monocytes with potential for reactivation and subsequent transmission via saliva exposure. Uniquely, 1% of the population possesses a chromosomally-integrated version of HHV-6 (ciHHV-6) [111]. Both HHV-6a and HHV-6b can integrate into the genome of a germ cell and be inherited in a Mendelian manner by resultant offspring. In these offspring, at least one integrated copy of the HHV-6 genome is present in every nucleated cell. Consequently, these individuals have very high levels of HHV-6 DNA in blood and tissue samples.

In transplant recipients, HHV-6 DNA is detected in the blood in 30 − 50% of patients and has been associated with disease [112]. HHV-6 encephalitis is the most serious complication portending a poor prognosis [113]. HHV-6 mRNA and antigen have been found in brain lesions on autopsy of patients with HHV-6 encephalitis, with the hippocampus most commonly affected [114]. In HCT recipients, HHV-6 has also been associated with CMV reactivation, acute graft versus host disease (GVHD), bone marrow suppression, pneumonitis, and mortality [115−119]. Risk factors for the development of HHV-6 encephalitis in HCT recipients include myeloablative conditioning and umbilical cord blood transplantation [120]. In SOT recipients, encephalitis has been reported, and CMV reactivation, organ dysfunction, and rejection remain as concerns [121]. ciHHV-6 has been described in both HCT and SOT recipients. Identification of these individuals from those with actively replicating virus is important to minimize unnecessary antiviral treatment. Beyond this, the significance ciHHV-6 in transplant recipients remains uncertain. Some small studies have suggested an association with higher frequency of bacterial infection and indirect effects such as allograft rejection, while others have found no disease associations [122,123].

HHV-6 reactivation is diagnosed via detection of HHV-6 DNA in plasma by PCR. Quantitative real-time PCR methods are often performed, but clinically significant thresholds have yet to be determined. Patients with ciHHV-6 can be identified using fluorescent in situ hybridization (FISH) with a specific HHV-6 probe performed on metaphase chromosome preparations from peripheral blood. While FISH is considered the reference standard technique, alternative methods include detection of HHV-6 DNA by PCR of hair follicles and determining the ratio of HHV-6 to cellular DNA by droplet digital PCR in cellular and plasma samples [124,125].

The clinical significance of the presence of HHV-6 DNA in plasma has been evaluated in several studies. Higher levels of plasma HHV-6 DNA is associated with increased risk of HHV-6 encephalitis, although not all patients with high levels will develop encephalitis. In one study, HCT recipients were prospectively followed for CNS dysfunction and monitored with biweekly plasma HHV-6 PCR. None of the HCT patients with low level HHV-6 reactivation, defined as plasma HHV-6 DNA less than or equal to 10^4 copies/mL, developed encephalitis, while 8% of patients with high plasma HHV-6 levels developed HHV-6 encephalitis [126].

Although indirect effects of HHV-6 such as CMV reactivation and acute GVHD have been described, viremia more commonly presents asymptomatically in both HCT and SOT recipients [127–129]. In one study evaluating double-umbilical cord blood transplant recipients, HHV-6 DNA detection in plasma was frequent, although resolution was observed in untreated patients and correlated with absolute lymphocyte count recovery [130]. The clinical significance of HHV-6 DNA levels and the role of molecular testing for detection of HHV-6 in plasma remain to be determined.

Diagnosis of HHV-6 encephalitis requires the detection of HHV-6 DNA in CSF. However, the presence of detectable HHV-6 in CSF does not necessarily confirm the diagnosis of HHV-6 encephalitis. In one study, HHV-6 DNA in CSF was detected in both patients with HHV-6 encephalitis and those who had an alternate explanation for their neurologic symptoms. However, patients with HHV-6 related CNS dysfunction had significantly higher HHV-6 levels in CSF (median peak copies/mL, 9050; range, 54–450,000) compared to all patients without HHV-6 CNS dysfunction (median peak copies/mL, 655; range, 25–260,000; $p = 0.05$). Regardless, all patients had poor survival [131]. Resolution of HHV-6 DNA in CSF does not necessarily predict resolution of active HHV-6 infection, as noted in three patients who died of HHV-6 encephalitis and were found to have active infection in brain tissue even after HHV-6 DNA in CSF and serum had become undetectable [114]. Furthermore, detection of HHV-6 DNA in CSF in the absence of CNS dysfunction is believed to occur in 0–0.9% of immunocompromised patients [132].

Given the limited but growing understanding of HHV-6 in transplant recipients, there is no current role for preemptive HHV-6 monitoring and no specific treatment recommendations. A study comparing a preemptively monitored group for HHV-6 in liver transplant recipients found no symptomatic HHV-6 disease and no differences in adverse events indirectly attributable to viral reactivation such as opportunistic infection, graft rejection, and hepatitis C recurrence [133]. In vitro studies suggest that ganciclovir, foscarnet, and cidofovir have activity against HHV-6, although the clinical management of HHV-6 in practice is variable [134]. Additional studies on HHV-6 treatment in transplant recipients and other patients are warranted.

HUMAN HERPESVIRUS-7

Like HHV-6, HHV-7 is also highly seroprevalent and acquired at a young age from exposure to infected saliva. Primary infection of HHV-7 may be asymptomatic or associated with fever and rash (Table 8.4).

Detection of HHV-7 DNA in plasma is much less common than for HHV-6. While HHV-7 DNA can be detected in about 50% of HCT and 20% of SOT recipients, reactivation of HHV-7 is less common than HHV-6 in these immunocompromised patients [135,136]. Healthy control subjects have also been found to have similar median HHV-7 viral loads compared to HCT recipients [137]. Accordingly, the correlation of HHV-7 with disease in immunocompromised patients remains unclear. Levels of HHV-7 DNA in blood have not correlated with disease in immunocompromised patients [137]. However, there are case reports that suggest a role for HHV-7 in hepatitis, encephalitis, and graft dysfunction [138–140]. Other studies have evaluated the role of routine surveillance of HHV-7 DNA in blood samples in HCT and SOT recipients and have not found benefit in graft or patient outcome [133,136]. Thus, monitoring of HHV-7 by molecular methods is not currently recommended in transplant recipients.

HUMAN HERPESVIRUS-8

HHV-8 is transmitted through saliva, but infection may also be acquired through sexual intercourse, blood transfusion, and organ transplantation [141]. HHV-8 causes Kaposi sarcoma (KS) and is linked with primary effusion lymphoma and multicentric Castleman disease (Table 8.4) [142]. HHV-8 has also been reported to cause fever and other constitutional symptoms, bone marrow suppression, hemophagocytic syndrome, and clonal gammopathy after transplantation [143–145]. Diagnosis of KS and other HHV-8-associated malignancies requires histopathologic evaluation of biopsied tissue. HHV-8 testing with IHC or PCR of biopsied tissue is not essential, but can be helpful if the diagnosis is unclear.

Detection and quantitation of HHV-8 DNA in PBMCs and plasma can aid in predicting the development of KS and treatment outcomes. In one study, KS incidence was 10-fold higher in HIV-positive patients with detectable PBMC-associated HHV-8 DNA (30.3 per 100 person years vs 3.4 per 100 person years) compared to those without detectable HHV-8 DNA [146]. Pretreatment HHV-8 DNA levels in plasma in patients with KS have also been associated with decreased survival and poor clinical response to treatment, and thus may be useful in risk stratification, selection of treatment strategy, and monitoring treatment response [147–149]. A study comparing simultaneously collected plasma and PBMCs from patients with different HHV-8-related lymphoproliferative diseases found comparable plasma and PBMC HHV-8 levels, suggesting that either specimen type may be suitable for HHV-8 testing [150]. Regardless of which blood compartment is evaluated, more studies are needed to further determine the clinical utility of HHV-8 quantitation in patients at risk for KS, as well as those patients that have already developed disease.

CONCLUSIONS

This chapter highlights the diverse and expanding clinical applications of molecular techniques in the diagnosis and management of herpes viruses, with an emphasis on the immunocompromised host. Advances in molecular diagnostics have been critical to the determination and risk stratification of active disease, assessment of therapeutic response, preemptive treatment strategies, and genotypic antiviral resistance testing. In many instances, the utility of molecular diagnostics in the diagnosis and management of herpes virus infections deserves further exploration and study. Novel applications of molecular methods for the clinical evaluation and monitoring of herpes viruses can be anticipated in the future.

Acknowledgments

Salary support was provided by NIH training grant 5T32AI007502-19 (SKT) and a generous gift from Beta Sigma Phi (SKT), an international women's service organization.

References

[1] Cohen J. Introduction to herpesviridae. In: Mandell G, Bennett J, Dolin R, editors. Mandell, Douglas, and Bennett's principles and practice of infectious diseases. 7th ed. Philadelphia, PA: Elsevier; 2009. p. 1937–42.

[2] Grinde B. Herpesviruses: latency and reactivation—viral strategies and host response. J Oral Microbiol 2013;5:1–9.

[3] Madhavan HN, Priya K, Anand AR, Therese KL. Detection of Herpes simplex virus (HSV) genome using polymerase chain reaction (PCR) in clinical samples—comparison of PCR with standard laboratory methods for the detection of HSV. J Clin Virol 1999;14:145–51.

[4] Schmutzhard J, Merete Riedel H, Zweygberg Wirgart B, Grillner L. Detection of herpes simplex virus type 1, herpes simplex virus type 2 and varicella-zoster virus in skin lesions. Comparison of real-time PCR, nested PCR and virus isolation. J Clin Virol 2004;29:120–6.

[5] Schillinger JA, Xu F, Sternberg MR, et al. National seroprevalence and trends in herpes simplex virus type 1 in the United States, 1976-1994. Sex Transm Dis 2004;31:753–60.

[6] Xu F, Sternberg MR, Kottiri BJ, et al. Trends in herpes simplex virus type 1 and type 2 seroprevalence in the United States. JAMA 2006;296:964–73.

[7] Schiffer J, Corey L. Herpes simplex virus. In: Bennett J, Dolin R, Blaser M, editors. Mandell, Douglas, and Bennett's principles and practice of infectious diseases. Philadelphia, PA: Elsevier/Saunders; 2014. p. 1713–30.

[8] Ito J. Herpes simplex virus. In: Appelbaum F, Forman SJ, Negrin RS, Blume KG, editors. Thomas' hematopoietic cell transplantation: stem cell transplantation. 4th ed Hoboken, NJ: Blackwell Publishing Ltd; 2004. p. 1382–7.

[9] Wilck MB, Zuckerman RA. AST Infectious Diseases Community of Practice. Herpes simplex virus in solid organ transplantation. Am J Transplant 2013;13:121–7.

[10] LeGoff J, Pere H, Belec L. Diagnosis of genital herpes simplex virus infection in the clinical laboratory. Virol J 2014;11:83.

[11] Wald A, Huang ML, Carrell D, Selke S, Corey L. Polymerase chain reaction for detection of herpes simplex virus (HSV) DNA on mucosal surfaces: comparison with HSV isolation in cell culture. J Infect Dis 2003;188:1345–51.

[12] Lakeman FD, Whitley RJ. Diagnosis of herpes simplex encephalitis: application of polymerase chain reaction to cerebrospinal fluid from brain-biopsied patients and correlation with disease. National Institute of Allergy and Infectious Diseases Collaborative Antiviral Study Group. J Infect Dis 1995;171:857–63.

[13] Mitchell PS, Espy MJ, Smith TF, et al. Laboratory diagnosis of central nervous system infections with herpes simplex virus by PCR performed with cerebrospinal fluid specimens. J Clin Microbiol 1997;35:2873–7.

[14] Bhullar SS, Chandak NH, Purohit HJ, Taori GM, Daginawala HF, Kashyap RS. Determination of viral load by quantitative real-time PCR in herpes simplex encephalitis patients. Intervirology 2014;57:1–7.

[15] Danve-Szatanek C, Aymard M, Thouvenot D, et al. Surveillance network for herpes simplex virus resistance to antiviral drugs: 3-year follow-up. J Clin Microbiol 2004;42:242–9.

[16] Piret J, Boivin G. Resistance of herpes simplex viruses to nucleoside analogues: mechanisms, prevalence, and management. Antimicrob Agents Chemother 2011;55:459–72.

[17] Sandherr M, Einsele H, Hebart H, et al. Antiviral prophylaxis in patients with haematological malignancies and solid tumours: guidelines of the Infectious Diseases Working Party (AGIHO) of the German Society for Hematology and Oncology (DGHO). Ann Oncol 2006;17:1051–9.

[18] Saral R, Burns WH, Laskin OL, Santos GW, Lietman PS. Acyclovir prophylaxis of herpes-simplex-virus infections. N Engl J Med 1981;305:63–7.

[19] Ringden O, Heimdahl A, Lonnqvist B, Malmborg AS, Wilczek H. Decreased incidence of viridans streptococcal septicaemia in allogeneic bone marrow transplant recipients after the introduction of acyclovir. Lancet 1984;1:744.

[20] Wade JC, Newton B, Flournoy N, Meyers JD. Oral acyclovir for prevention of herpes simplex virus reactivation after marrow transplantation. Ann Intern Med 1984;100:823–8.

[21] Bacon TH, Boon RJ, Schultz M, Hodges-Savola C. Surveillance for antiviral-agent-resistant herpes simplex virus in the general population with recurrent herpes labialis. Antimicrob Agents Chemother 2002;46:3042–4.

[22] van Velzen M, van de Vijver DA, van Loenen FB, Osterhaus AD, Remeijer L, Verjans GM. Acyclovir prophylaxis predisposes to antiviral-resistant recurrent herpetic keratitis. J Infect Dis 2013;208:1359–65.

[23] Hussin A, Md Nor NS, Ibrahim N. Phenotypic and genotypic characterization of induced acyclovir-resistant clinical isolates of herpes simplex virus type 1. Antiviral Res 2013;100:306–13.

[24] Sauerbrei A, Liermann K, Bohn K, et al. Significance of amino acid substitutions in the thymidine kinase gene of herpes simplex virus type 1 for resistance. Antiviral Res 2012;96:105–7.

[25] Andrei G, Georgala A, Topalis D, et al. Heterogeneity and evolution of thymidine kinase and DNA polymerase mutants of herpes simplex virus type 1: implications for antiviral therapy. J Infect Dis 2013;207:1295–305.

[26] Burrel S, Aime C, Hermet L, Ait-Arkoub Z, Agut H, Boutolleau D. Surveillance of herpes simplex virus resistance to antivirals: a 4-year survey. Antiviral Res 2013;100:365–72.

[27] Sauerbrei A, Bohn K, Heim A, et al. Novel resistance-associated mutations of thymidine kinase and DNA polymerase genes of herpes simplex virus type 1 and type 2. Antivir Ther 2011;16:1297–308.

[28] Whitley RJ. Chickenpox and herpes zoster (varicella zoster virus). In: Bennett J, Dolin R, Blaser M, editors. Mandell, Douglas, and Bennett's principles and practice of infectious diseases. Philadelphia, PA: Elsevier/Saunders; 2014. p. 1731–7.

[29] Pergam SA, Limaye AP. AST Infectious Diseases Community of Practice. Varicella zoster virus in solid organ transplantation. Am J Transplant 2013;13:138–46.

[30] Ho D, Arvin A. Varicella-zoster virus infections. In: Appelbaum F, Forman SJ, Negrin RS, Blume KG, editors. Thomas' hematopoietic cell transplantation: stem cell transplantation. Hoboken, NJ: Blackwell Publishing Ltd; 2004. p. 1388–409.

[31] Weitzman D, Shavit O, Stein M, Cohen R, Chodick G, Shalev V. A population based study of the epidemiology of Herpes zoster and its complications. J Infect 2013;67:463–9.

[32] Choi WS, Kwon SS, Lee J, et al. Immunity and the burden of herpes zoster. J Med Virol 2014;86:525–30.

[33] Leung J, Harpaz R, Baughman AL, et al. Evaluation of laboratory methods for diagnosis of varicella. Clin Infect Dis 2010;51:23–32.

[34] Kronenberg A, Bossart W, Wuthrich RP, et al. Retrospective analysis of varicella zoster virus (VZV) copy DNA numbers in plasma of immunocompetent patients with herpes zoster, of immunocompromised patients with disseminated VZV disease, and of asymptomatic solid organ transplant recipients. Transpl Infect Dis 2005;7:116–21.

[35] Aberle SW, Aberle JH, Steininger C, Puchhammer-Stockl E. Quantitative real time PCR detection of varicella-zoster virus DNA in cerebrospinal fluid in patients with neurological disease. Med Microbiol Immunol 2005;194:7–12.

[36] Mols JF, Ledent E, Heineman TC. Sampling of herpes zoster skin lesion types and the impact on viral DNA detection. J Virol Methods 2013;188:145–7.

[37] Koropchak CM, Graham G, Palmer J, et al. Investigation of varicella-zoster virus infection by polymerase chain reaction in the immunocompetent host with acute varicella. J Infect Dis 1991;163:1016–22.

[38] Coffin SE, Hodinka RL. Utility of direct immunofluorescence and virus culture for detection of varicella-zoster virus in skin lesions. J Clin Microbiol 1995;33:2792–5.

[39] Ishizaki Y, Tezuka J, Ohga S, et al. Quantification of circulating varicella zoster virus-DNA for the early diagnosis of visceral varicella. J Infect 2003;47:133–8.

[40] San-Juan R, Commoli P, Caillard S, Moulin B, Hirsch H, Meylan P. EBV-related post transplant lymphoproliferative disorder (PTLD) in solid organ transplant (SOT) recipients. Clin Microbiol Infect 2014;20:109–18.

[41] Loren AW, Porter DL, Stadtmauer EA, Tsai DE. Post-transplant lymphoproliferative disorder: a review. Bone Marrow Transplant 2003;31:145–55.

[42] Green M, Michaels MG. Epstein-Barr virus infection and posttransplant lymphoproliferative disorder. Am J Transplant 2013;13:41–54.

[43] Kremer BE, Reshef R, Misleh JG, et al. Post-transplant lymphoproliferative disorder after lung transplantation: a review of 35 cases. J Heart Lung Transplant 2012;31:296–304.

[44] Allen UD, Preiksaitis JK. AST Infectious Diseases Community of Practice. Epstein-Barr virus and posttransplant lymphoproliferative disorder in solid organ transplantation. Am J Transplant 2013;13:107–20.

[45] Wagner HJ, Wessel M, Jabs W, et al. Patients at risk for development of posttransplant lymphoproliferative disorder: plasma versus peripheral blood mononuclear cells as material for quantification of Epstein-Barr viral load by using real-time quantitative polymerase chain reaction. Transplantation 2001;72:1012–19.

[46] Abeynayake J, Johnson R, Libiran P, et al. Commutability of the Epstein-Barr virus WHO international standard across two quantitative PCR methods. J Clin Microbiol 2014;52:3802–4.

[47] Choquet S, Varnous S, Deback C, Golmard JL, Leblond V. Adapted treatment of Epstein-Barr virus infection to prevent posttransplant lymphoproliferative disorder after heart transplantation. Am J Transplant 2014;14:857–66.

[48] Gulley ML, Tang WH. Using Epstein-Barr viral load assays to diagnose, monitor, and prevent posttransplant lymphoproliferative disorder. Clin Microbiol Rev 2010;23:350–66.

[49] Holman CJ, Karger AB, Mullan BD, Brundage RC, Balfour Jr. HH. Quantitative Epstein-Barr virus shedding and its correlation with the risk of post-transplant lymphoproliferative disorder. Clin Transplant 2012;26:741–7.

[50] van der Velden WJ, Mori T, Stevens WB, et al. Reduced PTLD-related mortality in patients experiencing EBV infection following allo-SCT after the introduction of a protocol incorporating preemptive rituximab. Bone Marrow Transplant 2013;48:1465–71.

[51] Gaeta A, Nazzari C, Verzaro S, et al. Early evidence of lymphoproliferative disorder: post-transplant monitoring of Epstein-Barr infection in adult and pediatric patients. New Microbiol 2006;29:231–41.

[52] Gärtner BC, Schäfer H, Marggraff K, et al. Evaluation of use of Epstein–Barr viral load in patients after allogeneic stem cell transplantation to diagnose and monitor posttransplant lymphoproliferative disease. J Clin Microbiol 2002;40:351–8.

[53] Stevens SJ, Verschuuren EA, Verkuijlen SA, Van Den Brule AJ, Meijer CJ, Middeldorp JM. Role of Epstein-Barr virus DNA load monitoring in prevention and early detection of posttransplant lymphoproliferative disease. Leuk Lymphoma 2002;43:831–40.

[54] Oertel S, Trappe RU, Zeidler K, et al. Epstein-Barr viral load in whole blood of adults with posttransplant lymphoproliferative disorder after solid organ transplantation does not correlate with clinical course. Ann Hematol 2006;85:478–84.

[55] Lin JC, Wang WY, Chen KY, et al. Quantification of plasma Epstein-Barr virus DNA in patients with advanced nasopharyngeal carcinoma. N Engl J Med 2004;350:2461–70.

[56] Chan JY, Wong ST. The role of plasma Epstein-Barr virus DNA in the management of recurrent nasopharyngeal carcinoma. Laryngoscope 2014;124:126–30.

[57] Pow E, Law MY, Tsang PC, Perera RA, Kwong DL. Salivary Epstein-Barr virus DNA level in patients with nasopharyngeal carcinoma following radiotherapy. Oral Oncol 2011;47:879–82.

[58] Adham M, Greijer AE, Verkuijlen SAWM, et al. Epstein-Barr virus DNA load in nasopharyngeal brushings and whole blood in nasopharyngeal carcinoma patients before and after treatment. Clin Cancer Res 2013;19:2175–86.

[59] Cannon MJ, Schmid DS, Hyde TB. Review of cytomegalovirus seroprevalence and demographic characteristics associated with infection. Rev Med Virol 2010;20:202–13.

[60] Bate SL, Dollard SC, Cannon MJ. Cytomegalovirus seroprevalence in the United States: the national health and nutrition examination surveys, 1988–2004. Clin Infect Dis 2010;50:1439–47.

[61] Hahn G, Jores R, Mocarski ES. Cytomegalovirus remains latent in a common precursor of dendritic and myeloid cells. Proc Natl Acad Sci USA 1998;95:3937–42.

[62] Crumpacker C. Cytomegalovirus. In: Bennett J, Dolin R, Blaser M, editors. Mandell, Douglas, and Bennett's principles and practice

of infectious diseases. Philadelphia, PA: Elsevier/Saunders; 2014. p. 1737−53.

[63] Ramana P, Razonable RR. Cytomegalovirus infections in solid organ transplantation: a review. Infect Chemother 2013;45: 260−71.

[64] Winston DJ, Ho WG, Bartoni K, et al. Ganciclovir prophylaxis of cytomegalovirus infection and disease in allogeneic bone marrow transplant recipients. Results of a placebo-controlled, double-blind trial. Ann Intern Med 1993;118:179−84.

[65] George B, Pati N, Gilroy N, et al. Pre-transplant cytomegalovirus (CMV) serostatus remains the most important determinant of CMV reactivation after allogeneic hematopoietic stem cell transplantation in the era of surveillance and preemptive therapy. Transpl Infect Dis 2010;12:322−9.

[66] Jaskula E, Bochenska J, Kocwin E, Tarnowska A, Lange A. CMV serostatus of donor-recipient pairs influences the risk of CMV infection/reactivation in HSCT patients. Bone Marrow Res 2012;2012:375075.

[67] Humar A, Mazzulli T, Moussa G, et al. Clinical utility of cytomegalovirus (CMV) serology testing in high-risk CMV D+/R− transplant recipients. Am J Transplant 2005;5:1065−70.

[68] Ljungman P, Brand R, Hoek J, et al. Donor cytomegalovirus status influences the outcome of allogeneic stem cell transplant: a study by the European group for blood and marrow transplantation. Clin Infect Dis 2014;59:473−81.

[69] Zaia JA. Prevention of cytomegalovirus disease in hematopoietic stem cell transplantation. Clin Infect Dis 2002;35:999−1004.

[70] Zamora MR, Nicolls MR, Hodges TN, et al. Following universal prophylaxis with intravenous ganciclovir and cytomegalovirus immune globulin, valganciclovir is safe and effective for prevention of CMV infection following lung transplantation. Am J Transplant 2004;4:1635−42.

[71] Mendez J, Espy M, Smith TF, Wilson J, Wiesner R, Paya CV. Clinical significance of viral load in the diagnosis of cytomegalovirus disease after liver transplantation. Transplantation 1998;65:1477−81.

[72] Sia IG, Wilson JA, Groettum CM, Espy MJ, Smith TF, Paya CV. Cytomegalovirus (CMV) DNA load predicts relapsing CMV infection after solid organ transplantation. J Infect Dis 2000;181:717−20.

[73] Humar A, Gregson D, Caliendo AM, et al. Clinical utility of quantitative cytomegalovirus viral load determination for predicting cytomegalovirus disease in liver transplant recipients. Transplantation 1999;68:1305−11.

[74] Meyers JD, Ljungman P, Fisher LD. Cytomegalovirus excretion as a predictor of cytomegalovirus disease after marrow transplantation—importance of cytomegalovirus viremia. J Infect Dis 1990;162:373−80.

[75] Goodrich JM, Bowden RA, Fisher L, Keller C, Schoch G, Meyers JD. Ganciclovir prophylaxis to prevent cytomegalovirus disease after allogeneic marrow transplant. Ann Intern Med 1993;118:173−8.

[76] Green ML, Leisenring W, Stachel D, et al. Efficacy of a viral load-based, risk-adapted, preemptive treatment strategy for prevention of cytomegalovirus disease after hematopoietic cell transplantation. Biol Blood Marrow Transplant 2012;18: 1687−99.

[77] Owers DS, Webster AC, Strippoli GF, Kable K, Hodson EM. Pre-emptive treatment for cytomegalovirus viraemia to prevent cytomegalovirus disease in solid organ transplant recipients. Cochrane Database Syst Rev 2013;2.

[78] Hodson EM, Ladhani M, Webster AC, Strippoli GF, Craig JC. Antiviral medications for preventing cytomegalovirus disease in solid organ transplant recipients. Cochrane Database Syst Rev 2013;2.

[79] Khoury JA, Storch GA, Bohl DL, et al. Prophylactic versus preemptive oral valganciclovir for the management of cytomegalovirus infection in adult renal transplant recipients. Am J Transplant 2006;6:2134−43.

[80] van der Beek MT, Berger SP, Vossen AC, et al. Preemptive versus sequential prophylactic-preemptive treatment regimens for cytomegalovirus in renal transplantation: comparison of treatment failure and antiviral resistance. Transplantation 2010;89: 320−6.

[81] Witzke O, Hauser IA, Bartels M, et al. Valganciclovir prophylaxis versus preemptive therapy in cytomegalovirus-positive renal allograft recipients: 1-year results of a randomized clinical trial. Transplantation 2012;93:61−8.

[82] Marchetti S, Santangelo R, Manzara S, D'onghia S, Fadda G, Cattani P. Comparison of real-time PCR and pp65 antigen assays for monitoring the development of Cytomegalovirus disease in recipients of solid organ and bone marrow transplant. New Microbiol 2011;34:157−64.

[83] Cariani E, Pollara CP, Valloncini B, Perandin F, Bonfanti C, Manca N. Relationship between pp65 antigenemia levels and real-time quantitative DNA PCR for Human Cytomegalovirus (HCMV) management in immunocompromised patients. BMC Infect Dis 2007;7:1−7.

[84] Flexman J, Kay I, Fonte R, Herrmann R, Gabbay E, Palladino S. Differences between the quantitative antigenemia assay and the COBAS Amplicor Monitor quantitative PCR assay for detecting CMV viraemia in bone marrow and solid organ transplant patients. J Med Virol 2001;64:275−82.

[85] Hayden RT, Yan X, Wick MT, et al. Factors contributing to variability of quantitative viral PCR results in proficiency testing samples: a multivariate analysis. J Clin Microbiol 2012;50: 337−45.

[86] Mills AM, Guo FP, Copland AP, Pai RK, Pinsky BA. A comparison of CMV detection in gastrointestinal mucosal biopsies using immunohistochemistry and PCR performed on formalin-fixed, paraffin-embedded tissue. Am J Surg Pathol 2013;37:995−1000.

[87] McCoy MH, Post K, Sen JD, et al. qPCR increases sensitivity to detect cytomegalovirus in formalin-fixed, paraffin-embedded tissue of gastrointestinal biopsies. Hum Pathol 2014;45:48−53.

[88] Tachikawa R, Tomii K, Seo R, et al. Detection of herpes viruses by multiplex and real-time polymerase chain reaction in bronchoalveolar lavage fluid of patients with acute lung injury or acute respiratory distress syndrome. Respiration 2014;87: 279−86.

[89] Jouneau S, Poineuf JS, Minjolle S, et al. Which patients should be tested for viruses on bronchoalveolar lavage fluid? Eur J Clin Microbiol Infect Dis 2013;32:671−7.

[90] Schlischewsky E, Fuehner T, Warnecke G, et al. Clinical significance of quantitative cytomegalovirus detection in bronchoalveolar lavage fluid in lung transplant recipient. Transpl Infect Dis 2013;15:60−9.

[91] Paraskeva M, Bailey M, Levvey BJ, et al. Cytomegalovirus replication within the lung allograft is associated with bronchiolitis obliterans syndrome. Am J Transplant 2011;11:2190−6.

[92] Costa C, Delsedime L, Solidoro P, et al. Herpesviruses detection by quantitative real-time polymerase chain reaction in bronchoalveolar lavage and transbronchial biopsy in lung transplant: viral infections and histopathological correlation. Transplant Proc 2010;42:1270−4.

[93] Chemaly RF, Yen-Lieberman B, Chapman J, et al. Clinical utility of cytomegalovirus viral load in bronchoalveolar lavage in lung transplant recipients. Am J Transplant 2005;5:544−8.

[94] Reddy AJ, Zaas AK, Hanson KE, Palmer SM. A single-center experience with ganciclovir-resistant cytomegalovirus in lung transplant recipients: treatment and outcome. J Heart Lung Transplant 2007;26:1286–92.

[95] Myhre HA, Haug Dorenberg D, Kristiansen KI, et al. Incidence and outcomes of ganciclovir-resistant cytomegalovirus infections in 1244 kidney transplant recipients. Transplantation 2011;92:217–23.

[96] Limaye AP, Corey L, Koelle DM, Davis CL, Boeckh M. Emergence of ganciclovir-resistant cytomegalovirus disease among recipients of solid-organ transplants. Lancet 2000;356:645–9.

[97] Boivin G, Goyette N, Rollag H, et al. Cytomegalovirus resistance in solid organ transplant recipients treated with intravenous ganciclovir or oral valganciclovir. Antivir Ther 2009;14:697–704.

[98] Shmueli E, Or R, Shapira MY, et al. High rate of cytomegalovirus drug resistance among patients receiving preemptive antiviral treatment after haploidentical stem cell transplantation. J Infect Dis 2014;209:557–61.

[99] Eid AJ, Arthurs SK, Deziel PJ, Wilhelm MP, Razonable RR. Emergence of drug-resistant cytomegalovirus in the era of valganciclovir prophylaxis: therapeutic implications and outcomes. Clin Transplant 2008;22:162–70.

[100] Le Page AK, Jager MM, Iwasenko JM, Scott GM, Alain S, Rawlinson WD. Clinical aspects of cytomegalovirus antiviral resistance in solid organ transplant recipients. Clin Infect Dis 2013;56:1018–29.

[101] Lurain NS, Chou S. Antiviral drug resistance of human cytomegalovirus. Clin Microbiol Rev 2010;23:689–712.

[102] Smith IL, Cherrington JM, Jiles RE, Fuller MD, Freeman WR, Spector SA. High-level resistance of cytomegalovirus to ganciclovir is associated with alterations in both the UL97 and DNA polymerase genes. J Infect Dis 1997;176:69–77.

[103] Boppana SB, Ross SA, Fowler KB. Congenital cytomegalovirus infection: clinical outcome. Clin Infect Dis 2013;57(Suppl. 4):S178–181.

[104] Ross SA, Ahmed A, Palmer AL, et al. Detection of congenital cytomegalovirus infection by real-time polymerase chain reaction analysis of saliva or urine specimens. J Infect Dis 2014;210:1415–18.

[105] Boppana SB, Ross SA, Shimamura M, et al. Saliva polymerase-chain-reaction assay for cytomegalovirus screening in newborns. N Engl J Med 2011;364:2111–18.

[106] de Vries J, van der Eijk AA, Wolthers KC, et al. Real-time PCR versus viral culture on urine as a gold standard in the diagnosis of congenital cytomegalovirus infection. J Clin Virol 2012;53:167–70.

[107] Goegebuer T, Van Meensel B, Beuselinck K, et al. Clinical predictive value of real-time PCR quantification of human cytomegalovirus DNA in amniotic fluid samples. J Clin Microbiol 2009;47:660–5.

[108] Liesnard C, Donner C, Brancart F, Gosselin F, Delforge ML, Rodesch F. Prenatal diagnosis of congenital cytomegalovirus infection: prospective study of 237 pregnancies at risk. Obstet Gynecol 2000;95:881–8.

[109] Lazzarotto T, Gabrielli L, Foschini MP, et al. Congenital cytomegalovirus infection in twin pregnancies: viral load in the amniotic fluid and pregnancy outcome. Pediatrics 2003;112:e153–157.

[110] Cohen J. Human herpesvirus types 6 and 7. In: Bennett J, Dolin R, Blaser M, editors. Mandell, Douglas, and Bennett's principles and practice of infectious diseases. 8th ed. Philadelphia, PA: Elsevier/Saunders; 2014. p. 1772–6.

[111] Pellett PE, Ablashi DV, Ambros PF, et al. Chromosomally integrated human herpesvirus 6: questions and answers. Rev Med Virol 2012;22:144–55.

[112] Zerr DM. Human herpesvirus 6 (HHV-6) disease in the setting of transplantation. Curr Opin Infect Dis 2012;25:438–44.

[113] Shimazu Y, Kondo T, Ishikawa T, Yamashita K, Takaori-Kondo A. Human herpesvirus-6 encephalitis during hematopoietic stem cell transplantation leads to poor prognosis. Transpl Infect Dis 2013;15:195–201.

[114] Fotheringham J, Akhyani N, Vortmeyer A, et al. Detection of active human herpesvirus-6 infection in the brain: correlation with polymerase chain reaction detection in cerebrospinal fluid. J Infect Dis 2007;195:450–4.

[115] Le Bourgeois A, Labopin M, Guillaume T, et al. Human herpesvirus 6 reactivation before engraftment is strongly predictive of graft failure after double umbilical cord blood allogeneic stem cell transplantation in adults. Exp Hematol 2014;42:945–54.

[116] Gotoh M, Yoshizawa S, Katagiri S, et al. Human herpesvirus 6 reactivation on the 30th day after allogeneic hematopoietic stem cell transplantation can predict grade 2-4 acute graft-versus-host disease. Transpl Infect Dis 2014;16:440–9.

[117] Van Leer-Buter CC, Sanders JS, Vroom HE, Riezebos-Brilman A, Niesters HG. Human herpesvirus-6 DNAemia is a sign of impending primary CMV infection in CMV sero-discordant renal transplantations. J Clin Virol 2013;58:422–6.

[118] de Pagter PJ, Schuurman R, Keukens L, et al. Human herpes virus 6 reactivation: important predictor for poor outcome after myeloablative, but not non-myeloablative allo-SCT. Bone Marrow Transplant 2013;48:1460–4.

[119] Zerr DM, Boeckh M, Delaney C, et al. HHV-6 reactivation and associated sequelae after hematopoietic cell transplantation. Biol Blood Marrow Transplant 2012;18:1700–8.

[120] Jeulin H, Agrinier N, Guery M, et al. Human herpesvirus 6 infection after allogeneic stem cell transplantation: incidence, outcome, and factors associated with HHV-6 reactivation. Transplantation 2013;95:1292–8.

[121] Lautenschlager I, Razonable RR. Human herpesvirus-6 infections in kidney, liver, lung, and heart transplantation: review. Transpl Int 2012;25:493–502.

[122] Lee SO, Brown RA, Razonable RR. Clinical significance of pre-transplant chromosomally integrated human herpesvirus-6 in liver transplant recipients. Transplantation 2011;92:224–9.

[123] Lee SO, Brown RA, Razonable RR. Chromosomally integrated human herpesvirus-6 in transplant recipients. Transpl Infect Dis 2012;14:346–54.

[124] Ward KN, Leong HN, Nacheva EP, et al. Human herpesvirus 6 chromosomal integration in immunocompetent patients results in high levels of viral DNA in blood, sera, and hair follicles. J Clin Microbiol 2006;44:1571–4.

[125] Sedlak RH, Cook L, Huang ML, et al. Identification of chromosomally integrated human herpesvirus 6 by droplet digital PCR. Clin Chem 2014;60:765–72.

[126] Ogata M, Satou T, Kadota J, et al. Human herpesvirus 6 (HHV-6) reactivation and HHV-6 encephalitis after allogeneic hematopoietic cell transplantation: a multicenter, prospective study. Clin Infect Dis 2013;57:671–81.

[127] Al Fawaz T, Ng V, Richardson SE, Barton M, Allen U. Clinical consequences of human herpesvirus-6 DNAemia in peripheral blood in pediatric liver transplant recipients. Pediatr Transplant 2014;18:47–51.

[128] Luiz CR, Machado CM, Canto CL, et al. Monitoring for HHV-6 infection after renal transplantation: evaluation of risk factors for sustained viral replication. Transplantation 2013;95:842–6.

[129] Illiaquer M, Malard F, Guillaume T, et al. Long-lasting HHV-6 reactivation in long-term adult survivors after double umbilical cord blood allogeneic stem cell transplantation. J Infect Dis 2014;210:567−70.

[130] Olson AL, Dahi PB, Zheng J, et al. Frequent human herpesvirus-6 viremia but low incidence of encephalitis in double-unit cord blood recipients transplanted without antithymocyte globulin. Biol Blood Marrow Transplant 2014;20:787−93.

[131] Hill JA, Boeckh MJ, Sedlak RH, Jerome KR, Zerr DM. Human herpesvirus 6 can be detected in cerebrospinal fluid without associated symptoms after allogeneic hematopoietic cell transplantation. J Clin Virol 2014;61:289−92.

[132] Wang F, Linde A, Hägglund H, Testa M, Locasciulli A, Ljungman P. Human herpesvirus 6 DNA in cerebrospinal fluid specimens from allogeneic bone marrow transplant patients: does it have clinical significance? Clin Infect Dis 1999;28:562−8.

[133] Fernandez-Ruiz M, Kumar D, Husain S, et al. Utility of a monitoring strategy for human herpesviruses 6 and 7 viremia after liver transplantation: a randomized clinical trial. Transplantation 2015;99:106−13.

[134] De Bolle L, Naesens L, De Clercq E. Update on human herpesvirus 6 biology, clinical features, and therapy. Clin Microbiol Rev 2005;18:217−45.

[135] Dockrell DH, Paya CV. Human herpesvirus-6 and -7 in transplantation. Rev Med Virol 2001;11:23−36.

[136] Chan PK, Li CK, Chik KW, et al. Risk factors and clinical consequences of human herpesvirus 7 infection in paediatric haematopoietic stem cell transplant recipients. J Med Virol 2004;72:668−74.

[137] Boutolleau D, Fernandez C, Andre E, et al. Human herpesvirus (HHV)-6 and HHV-7: two closely related viruses with different infection profiles in stem cell transplantation recipients. J Infect Dis 2003;187:179−86.

[138] Fule Robles JD, Cheuk DK, Ha SY, Chiang AK, Chan GC. Human herpesvirus types 6 and 7 infection in pediatric hematopoietic stem cell transplant recipients. Ann Transplant 2014;19:269−76.

[139] Holden SR, Vas AL. Severe encephalitis in a haematopoietic stem cell transplant recipient caused by reactivation of human herpesvirus 6 and 7. J Clin Virol 2007;40:245−7.

[140] Thomasini RL, Sampaio AM, Bonon SH, et al. Detection and monitoring of human herpesvirus 7 in adult liver transplant patients: impact on clinical course and association with cytomegalovirus. Transplant Proc 2007;39:1537−9.

[141] Le J, Gantt S. AST Infectious Diseases Community of Practice. Human herpesvirus 6, 7 and 8 in solid organ transplantation. Am J Transplant 2013;13:128−37.

[142] Kaye K. Kaposi's sarcoma−associated herpesvirus (human herpesvirus 8). In: Bennett J, Dolin R, Blaser M, editors. Mandell, Douglas, and Bennett's principles and practice of infectious diseases. Philadelphia, PA: Elsevier/Saunders; 2014. p. 1772−6.

[143] Luppi M, Barozzi P, Rasini V, et al. Severe pancytopenia and hemophagocytosis after HHV-8 primary infection in a renal transplant patient successfully treated with foscarnet. Transplantation 2002;74:131−2.

[144] Regamey N, Hess V, Passweg J, et al. Infection with human herpesvirus 8 and transplant-associated gammopathy. Transplantation 2004;77:1551−4.

[145] Pietrosi G, Vizzini G, Pipitone L, et al. Primary and reactivated HHV8 infection and disease after liver transplantation: a prospective study. Am J Transplant 2011;11:2715−23.

[146] Engels EA, Biggar RJ, Marshall VA, et al. Detection and quantification of Kaposi's sarcoma-associated herpesvirus to predict AIDS-associated Kaposi's sarcoma. AIDS 2003;17:1847−51.

[147] El Amari EB, Toutous-Trellu L, Gayet-Ageron A, et al. Predicting the evolution of Kaposi sarcoma, in the highly active antiretroviral therapy era. AIDS 2008;22:1019−28.

[148] Borok M, Fiorillo S, Gudza I, et al. Evaluation of plasma human herpesvirus 8 DNA as a marker of clinical outcomes during antiretroviral therapy for AIDS-related Kaposi sarcoma in Zimbabwe. Clin Infect Dis 2010;51:342−9.

[149] Simonelli C, Tedeschi R, Gloghini A, et al. Plasma HHV-8 viral load in HHV-8-related lymphoproliferative disorders associated with HIV infection. J Med Virol 2009;81:888−96.

[150] Tedeschi R, Marus A, Bidoli E, Simonelli C, De Paoli P. Human herpesvirus 8 DNA quantification in matched plasma and PBMCs samples of patients with HHV8-related lymphoproliferative diseases. J Clin Virol 2008;43:255−9.

[151] Luppi M, Barozzi P, Schulz TF, et al. Bone marrow failure associated with human herpesvirus 8 infection after transplantation. N Engl J Med 2000;343:1378−85.

9

Molecular Testing for Parvoviruses

G. Gallinella[1,2]

[1]Department of Pharmacy and Biotechnology, University of Bologna, Bologna, Italy
[2]S. Orsola-Malpighi Hospital — Microbiology, University of Bologna, Bologna, Italy

THE FAMILY PARVOVIRIDAE

The family Parvoviridae includes viruses with a single-stranded DNA genome, encapsidated in an icosahedral protein capsid, about 22—26 nm in diameter. Replication occurs in the nucleus of infected cells and is highly dependent on cellular environment, so that a productive cycle is usually achieved only in actively replicating cells, or in some cases when supported by complementation from helper viruses. The subfamily Parvovirinae includes viruses able to infect vertebrate hosts, within it the most recent taxonomical revision distinguishes eight viral genera, and within each genus individual virus species that collect the viral isolates normally recognized in laboratory or clinical settings [1].

The present formal classification is the result of an approach mainly relying on genome sequence information, rather than on phenotypic classification. Application of molecular biology techniques, in particular of the more advanced high-throughput sequencing techniques and related bioinformatics analysis, led to a classification scheme better suited to describe the biological diversity of viruses within the family, and in turn useful as a framework for defining molecular diagnostic approaches.

Viruses adapted to the human host are found in the genera *Dependoparvovirus* (adeno-associated viruses, AAV), *Erythroparvovirus* (B19V), *Bocaparvovirus* (HBoV1-4), *Tetraparvovirus* (PARV4). While AAV viruses are considered nonpathogenic and have been exploited as transduction vectors, the others possess a pathogenic potential that prompts for the development of diagnostic molecular testing in a clinical setting. Parvovirus B19 (B19V) can be considered the most relevant human pathogenic virus [2], while human bocaviruses have gained more recent interest as respiratory or enteric pathogenic viruses [3].

PARVOVIRUS B19

The Virus

B19V genome is a linear ssDNA molecule of 5.6 kb in length. Strands of either polarity are separately encapsidated at the same frequency and are functionally equivalent. The genome organization (Fig. 9.1) is composed of a unique internal region, containing all the coding sequences, flanked by repeated, inverted terminal regions that serve as origins of replication. The genome encodes for three major proteins, the nonstructural (NS) protein in the left side and the two colinear capsid proteins, VP1 and VP2, in the right side, and for additional minor NS proteins. The capsid, composed of 5—10% VP1 and 90—95% VP2 proteins, forms an icosahedral structure in $T = 1$ arrangement, about 25 nm in diameter.

B19V shows a selective tropism for erythroid progenitor cells in the bone marrow, linked to the presence of specific receptors, such as globoside and integrins [4,5], and functional internalization processes [6,7]. In a permissive cellular environment, a coordinated series of macromolecular syntheses occurs [8,9]. DNA repair synthesis generates a double-stranded DNA template, then first-phase transcription mainly produces mRNAs coding for the NS protein, followed by rolling hairpin replication of the genome and extended transcription, including mRNAs coding for structural proteins. Accumulation of VP proteins eventually leads to the assembly of capsids, encapsidation of single-stranded genomes, and release of virions

Diagnostic Molecular Pathology
DOI: http://dx.doi.org/10.1016/B978-0-12-800886-7.00009-1

© 2017 Elsevier Inc. All rights reserved.

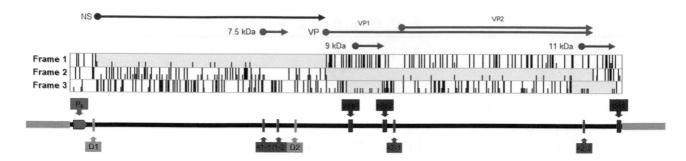

FIGURE 9.1 Schematic representation of B19V genome organization and functional mapping. (Top) Open reading frames (ORFs) identified in the positive strand of genome, arrows indicate the coding regions for viral proteins positioned on the ORF map. (Bottom) Genome organization, with distinct representation of the terminal and internal regions, and indication of the positions of promoter (P6), splice donor (D1, D2), splice acceptor (A1-1/2, A2-2/2), and cleavage-polyadenylation (pAp1, pAp2, pAd) sites. The internal coding region is flanked by two inverted terminal repeat regions that are extensively palindromic and can fold in hairpin, double-stranded structures that serve as priming sites for the second-strand synthesis and function as origins of replications via a rolling hairpin replication mechanism. The internal region encompasses all reading frames, divided in two main blocks. In the left side of the genome, a single ORF codes for the viral NS protein that operates many functions in viral replication and interaction with the cellular environment (cell cycle arrest, induction of apoptosis, induction of proinflammatory cytokines). In the right side of the genome, a single ORF encodes for the viral capsid proteins and depending on the mRNA processing two colinear proteins can be produced: (1) the longer VP1, whose N-terminus (VP1u) is unique with respect to VP2 and has a PLA2 phospholipase activity that is essential for viral infectivity and (2) the shorter VP2, colinear with the C terminus of VP1, that constitutes the core structure of capsid. Additional minor ORFs encode for small NS proteins (11, 9, and 7.5 kDa).

from infected cells. The permissive environment is restricted to cells in the erythroid lineage at differentiation stages ranging from CFU-E to erythroblasts [10,11]. In these cells, the virus exerts a complex series of effects, including arrest of the cell cycle [12—14] and induction of apoptosis [15], causing a temporary block in erythropoiesis that can manifest as a transient or persistent erythroid aplasia. The virus can infect other different cellular types in diverse tissues, including endothelial, stromal, or synovial cells [16—18]. Cellular environments other than erythroid progenitor cells are normally nonpermissive—infection is usually abortive and the presence of the viral genome is not associated with replication, transcription, or protein synthesis [19], although a productive replication can be sporadically documented and contribute to pathological processes.

Transmission

B19V is widely and worldwide diffuse. In most epidemiological settings, considering the presence of specific IgG as the marker of past infection, the highest rate of infection occurs before age of 20, reaching a prevalence of about 60% of population, but infection can occur until elder ages, reaching maximal prevalence values higher than 80% [20].

The main route of transmission of the virus is through the respiratory system, and close contacts in the household, school, or hospital settings are associated with rates of transmission up to 50%. In temperate climate countries, circulation of the virus is higher in the Spring/early Summer months, and epidemic cycles are reported to occur every 4—5 years. The virus can be transmitted from mother to fetus with possible fetal damage, and B19V should be included in the antenatal assessment of risk of fetal infections [21—23]. Finally, due to a viremic phase with high viral loads, there is a risk of iatrogenic transmission of the virus via blood and blood-derived products, producing blood and blood product safety issues [24].

Infection—Early Events

Following contact and the primary viremic phase of infection (normally undetected), the virus gains access to the bone marrow and infects erythroid progenitor cells, achieving a productive infection and exerting cytotoxic effects. In this phase, the bone marrow shows erythroid aplasia and the presence of characteristic giant erythroblasts. The pathogenic effects on bone marrow are derived from the ability of the virus to induce cell-cycle arrest, block of erythroid differentiation, and eventually apoptosis of susceptible and infected cells [25]. The clinical impact on the host reflects the depression of bone marrow activity, linked to the volume and turnover rate of the erythroid compartment, and the ability of immune system to mount an effective specific response [26,27].

In individuals with physiological erythropoiesis and normal immune system, infection is limited in extent and temporal frame and is controlled by the development of a specific neutralizing immune response. Levels of hemoglobin decrease only marginally and infection is usually asymptomatic from the hematological perspective. Production of antibodies with neutralizing activity (IgM followed by IgG) contributes to the progressive clearance of infection over 3–4 months with constantly decreasing viral load levels, even if very low levels can be detected for several months following primary infection [28–30]. When preexisting alterations in the erythropoiesis process, or defects in the immune response, alter the balance between viral replication and cellular turnover, infection can manifest as pure red cell aplasia (PRCA) and anemia [31]. In situations where the number of erythroid progenitors and their replication rate are expanded because of a reduced lifespan of erythrocytes, or increased need, infection can lead to an acute episode of profound anemia, presenting as classical aplastic crisis. In situations where the immune system lacks the capacity to control, neutralize, and clear viral infection, the infection may become persistent. This can occur in cases of congenital or acquired immunodeficiency, such as HIV infection [32], in cancer during the course of chemotherapy [33], or during immunosuppressive treatments in bone marrow or solid organ transplant recipients [34–36]. In all these subjects, depression of erythropoiesis can be persistent and manifest with anemia of different grades, but also anemia that is compensated and unapparent.

Infection—Late Events

Bone marrow supports a productive infection and release of progeny virus into the blood, leading to a secondary viremia characterized by high viral load levels ($>10^{12}$ virus/mL) in the acute phase, systemic distribution of the virus and possible late clinical manifestations of infection. In this phase, other cell types, including endothelial, stromal, or synovial cells, can be infected. In particular, endothelial cells constitute a diffuse target that can account for the wide distribution of virus and its detection in disparate tissues. Endothelial cells are normally nonpermissive and a possible site of persistence of the virus, but in some cases markers of viral activity have been precisely localized to endothelial cells and causally linked to pathological processes [18].

In this phase, the pathogenic mechanisms usually involve the induction of inflammatory responses [37–39], more rarely the induction of necrotic processes [40] or the development of autoimmune disorders [41]. The two classical late manifestations of B19V infection are erythema infectiosum (typical of children) and arthropathies (typical of adult patients), and with a tendency to chronicity. The virus has been recognized as cause of acute myocarditis [42,43], and possibly chronic inflammatory cardiomyopathies [44], and has been involved in the development of rheumatic [45,46] or autoimmune diseases [41]. The spectrum of clinical manifestations associated to B19V infection has been constantly increasing to involve almost all organs and tissues, and descriptions of clinical presentations have progressively stressed atypical aspects. Strict diagnostic criteria and sound methodologies should always be adopted to link B19V infection to atypical pathological processes.

Virus Persistence

Following primary infection, B19V can normally be detected in a wide range of tissues, probably lifelong [47]. The virus can be detected in bone marrow not only in cases of persistent infections with constant low-level viremia, but also in normal subjects without any evidence of active viral replication [48,49]. The presence of viral DNA has been reported in lymphoid tissue, including spleen, lymph nodes, and tonsils [49]. Viral DNA can be detected in liver [50], and in the heart the common presence of B19V DNA has been the focus of a debate on its potential role in the development of cardiomyopathies [51,52]. Viral DNA is commonly found in synovial tissues [53] and skin [54]. The initial picture of a virus capable of acute infections and rapidly cleared by the organism as a consequence of the immune response has given place to the picture of a virus able to establish long-term relationship with human hosts, and the current assumption is that persistence of viral DNA in tissues can be the normal outcome of infections, making B19V a relevant part of the human virome [47]. Whether this persistence might imply integration of viral DNA in the host genome, or be related to reactivation and productive infections, is still a matter of investigation.

Fetal Infection

A relevant property of B19V is its ability to cross the placental barrier and infect the fetus. The viral receptor, globoside, is present on the villous trophoblast layer of the placenta and its expression may facilitate transcytosis of virus to the fetal circulation [55]. Endothelial placental cells can be productively infected, facilitating the establishment of fetal infection and contributing to placental damage [56]. When in the fetal circulation, the virus can infect erythroid progenitor cells, in liver and/or bone marrow depending on the gestational age, and can be detected in cells circulating in the vessels of several tissues as well as in the amniotic fluid [57]. The virus can induce a block in fetal erythropoiesis whose effect will depend on fetal developmental stage, the rate of expansion of fetal erythroid compartment, and the maturity of the immune response [58]. Infections occurring at earlier stages of pregnancy carry a higher risk of fetal death, up to about 10%, while infections occurring in the central part of pregnancy more frequently lead to fetal hydrops. Hydrops may eventually cause fetal death, but frequently the fetus can recover without persistent developmental damage. In the third trimester the overall risk of fetal damage decreases to background values, although late intrauterine fetal death can occur [21,59]. Newborns may show transient presence of virus at birth [23,60], only sporadically associated with neonatal anemia or anomalies [61], while consequences of fetal anemia on the long-term neurological development are still under investigation [62].

MOLECULAR TARGET

Target and Samples

The low structural complexity of B19V restricts the range of molecular targets relevant for diagnostics mainly to viral DNA. Detection of viral DNA in serum/plasma from peripheral blood, other body fluids, fetal cord blood, or amniotic fluid is indicative of productive viral replication and active infection. Due to the characteristics of the viral replication and the dynamics of the infectious process, it is critical for molecular amplification methods to produce reliable quantitative assessment of the viral load in order to obtain useful diagnostic information. Viral DNA can also be sought in cellular samples, such as bone marrow aspirates, solid tissue biopsies, fetal biopsies, or placental biopsies. In these cases, to differentiate active infections from silent persistence of virus, both quantitative evaluation of the genome copy number and detection of viral mRNAs can be considered appropriate. In these instances, besides quantitative molecular amplification methods, viral nucleic acids may be efficiently detected by in situ hybridization, or viral NS or VP proteins can be detected by immunological methods.

B19V Genotypes

B19V as a species is subdivided into three genotypes: the prototype genotype 1 and two variant genotypes 2 and 3 [63,64]. At the nucleotide level, the diversity between genotype clusters is about 10% for genotype 1 and 5–6% for genotypes 2 and 3, while the diversity within each genotype cluster is normally lower than 2% for genotype 1 and in the range 3–10% for genotypes 2 and 3. All genotypes cocirculate, but with different frequencies and geographical distributions. The prototype genotype 1 is ubiquitous worldwide and includes the greatest part of circulating virus [65,66]. The majority of isolates are referred to as subtype 1a, while subtype 1b and other variants are rarer and confined to limited geographical areas. Genotype 2 appears to be an older variant with respect to genotypes 1 and 3, is commonly harbored in tissues of elderly populations [47], and only sporadically detected as circulating virus [67]. Genotype 3, divided in the two distinct subtypes 3a and 3b, circulates at relatively higher frequencies in western Africa [68] and at lower frequencies in other geographic areas. The three B19V genotypes are assumed to have similar biological properties, pathogenic capacity, transmission routes and pose a similar diagnostic challenge in the clinical setting [69]. Hence, nucleic acid amplification procedures should enable the detection and standardized quantification of all genotypes. The pursuit of these main requirements can take advantage of an ample amount of sequence information from nucleotide databases and can rely on the availability of international standards as reference materials for all genotypes [70,71].

For genotype 1, available sequence information indicates a continuous evolutionary process [72–74], but still yields a consistent consensus sequence coupled with a limited genetic diversity of the different isolates. The low genetic diversity of genotype 1 suggests that primers and probes used in molecular detection assays will recognize most targets with high expectation values. Genotypes 2 and 3 pose different problems. In both cases, fewer genomic sequences are presently available, inter- and intragenotypic diversity is higher, the consistency of consensus sequences is lower than for genotype 1, and it may be expected that addition of new sequences may demonstrate additional genetic diversity. This higher sequence divergence may pose diagnostic challenges, and primers and probes may recognize targets with lower expectation values.

MOLECULAR TECHNOLOGIES

Technical Developments

The fact that B19V is not readily adapted to grow in cell cultures prompted the development of direct molecular methods for its detection in clinical specimens. Over the years, the advancements in molecular analytic techniques have always found a complete paradigm in the applications to the detection of B19V, with a particular focus on the viral genome as diagnostic target. In the progress toward a rapid and accurate molecular diagnosis, a wide array of molecular hybridization and nucleic acid amplification techniques have been developed [75–80]. Standardization and inclusion of competitor or internal controls have been developed for PCR protocols to ensure accuracy and robustness [81–83]. Currently, quantitative and internally controlled real-time PCR techniques (qPCR) represent the standard analytical method for the molecular detection of B19V DNA [84–87]. There are two main requirements that should be met: (1) the capability of detection of all genotypes of B19V and (2) a calibrated and standardized quantification of target. Both of these requirements take advantage of international standards [70] and can be challenged by international proficiency panels [88,89].

qPCR Assay Design

Considerations on genotype distribution and sequence heterogeneity among isolates guide the design of primers and probes for a molecular amplification assay [90]. Sequence alignment permits the definition of consensus sequences for each genotype, and of a whole-species consensus that can be used to define positions of amplification primers and probes. As a strategy for the design of a molecular assay (Fig. 9.2), a target sequence within the internal region of B19V genome, conserved enough to be amplified by consensus primers with equal efficiency for all genotypes, but

encompassing specific signature sequences allowing distinction among genotypes by means of specific probes, can be chosen to ensure both the detection of all genotypes with similar analytical performances, and further genotyping. However, any choice of primers and probes does not exclude the possibility of mismatches to individual clinical isolates, causing impaired annealing and leading to underestimation or misdetection of targets. In well-designed assays, single-base mismatches can be present either on one of the primer binding sites, or on the probe binding site, at an expected frequency lower than 1%. Sequencing of amplification products might be finally carried mainly for epidemiological studies, especially to confirm the presence and identity of variant genotypes.

qPCR Assay Validation

Alternative qPCR protocols can be developed to allow choice with respect to operational systems and diagnostic requirements. Consensus genotype-independent detection of B19V DNA by means of intercalating dyes may be an alternative to genotype-specific detection by means of fluorescence probes, or the two detection formats may be successfully combined maintaining equal sensitivity and specificity of the assay. This latter scheme guarantees a specific identification of prototype or variant genotypes

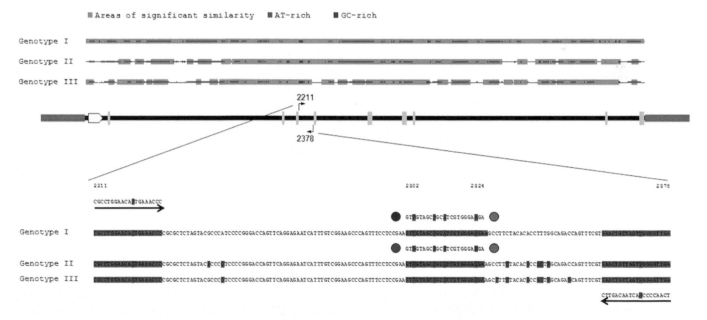

FIGURE 9.2 qPCR assay design. (Top) Sequence similarity among B19V genotypes. Consensus sequences for each genotype, each derived from the alignment of a dataset of complete sequences from NCBI, are aligned to show areas of significant (>90%) sequence similarity between genotypes. (Center) Schematic representation of B19V genome organization and functional mapping (as shown in Fig. 9.1). The position of two suitable, genotype consensus qPCR primer is indicated. (Bottom) qPCR target sequence, sequence alignment for the three genotypes. Sequences recognized by amplification primers and genotype-specific probes are highlighted in yellow, nucleotide differences are highlighted in green. Two alternative hydrolysis probes, recognizing genotype 1 or genotypes 2/3, are shown, with alternative reporter dyes [90].

coupled to a high flexibility in the detection of newly-emerging variants with possible additional sequence heterogeneity. An internal reaction control is essential for routine clinical applications. In the case of serum/plasma samples, an exogenous control target can be added to the sample during the nucleic acid purification steps and used as an in-process control. Analytical performance of the designed molecular assay in different detection formats need to be evaluated according to guidelines for harmonization in analytical assays. The availability of an international panel of standard reference material allows the setup of the assay and the construction of quantitative calibration curves for assay validation and operation. Analytical requirements need to be met in terms of limit of detection and quantification, and linear range of the assay. Results should be given in International Units (IU) rather than genome copies. A lower limit of detection or quantification can be set in the range 10^2-10^3 IU/mL, while the range of linearity of the assay should extend to at least up to 10^9 IU/mL. Diagnostic laboratories can rely on commercially-available validated diagnostic assays. However, diagnostic kits encounter problems due to target heterogeneity, and for B19V some reports and proficiency panel evaluations suggest that these qPCR assays may not always conform to analytical performance standards [91–93]. Conversely, the in-house development of molecular assays may allow a wider operational choice but requires a high degree of conformity to analytical guidelines and validation procedures to ensure reliable results [94].

In Situ Hybridization Techniques

A useful complement to qPCR is offered by in situ hybridization for the detection of viral nucleic acids within cells or tissues with preservation of cellular morphology. In the case of biopsies, the quantitative information offered by qPCR on purified nucleic acids should not be considered of primary relevance, even if standardized by a reference endogenous target, because of the variability inherent in the sampling of a nonuniform cell population with different degrees of permissiveness to viral replication or different spread of virus. By in situ hybridization techniques, infected cells can be easily identified in the sampled material and give indication on the distribution of infection within tissues, and to what cellular types are involved. Several methodologies have been developed [95–100], differing in the choice of probe (DNA, oligo-DNA, or oligo-PNA probes), labeling method and moiety, and detection method. Immunofluorescence or immunoenzymatic detection methods can be used. In the latter case, chromogenic substrates offer the advantage of an easy microscopic inspection of the analyzed tissue, while the use of chemiluminescent substrates may offer the advantage of a quantitative assessment of the abundance of targets within infected cells [101].

Immunological Techniques

Immunological detection of viral proteins is not advisable for detection of virus in blood [102], but can be considered as an useful tool for identification of productively infected cells in tissue samples. Commercially available monoclonal or polyclonal antibodies are directed against the viral capsid proteins, mostly toward the VP1/VP2 common epitopes [103]. In selected applications, such as in situ hybridization techniques, immunologic detection of viral proteins, and/or of cellular markers, might be combined to better characterize a productive viral infection and identify the phenotype of target cells [56,101].

CLINICAL UTILITY

Diagnostic Approach

A laboratory diagnosis is necessary to confirm or exclude B19V infection and to differentiate acute from persistent infections. Moreover, B19V detection can be part of screening procedures, especially considering antenatal screening, or part of prophylactic measures to prevent iatrogenic transmission of virus through blood or blood derivatives.

B19V is a virus capable of infections presenting with different courses depending on the interplay with host factors and the efficacy of the immune system response. An accurate laboratory diagnosis of B19V infection relies on a multiparametric approach, combining as much as possible the molecular detection of viral components, mainly viral DNA, to the immunological detection of virus-specific antibodies (Fig. 9.3) [104]. The acute phase of infection characterized by high viremic load and the emergence of specific IgM/IgG antibodies is normally followed by a progressive clearance of viremia still in the presence of specific IgM and IgG, or may lead to persistent infections, usually in the presence of IgG only, and finally may result in the silent persistence of virus in tissues.

The indirect, immunological approach to the diagnosis of B19V infection is commonly followed as a first-level investigation. In this case, coupled determination of IgG/IgM antibodies against VP proteins is required. The presence of IgM reactivity normally indicates an active or a recent infection, so that a repeated determination can be used as a confirmatory assay. In addition, assays able to discriminate a differential

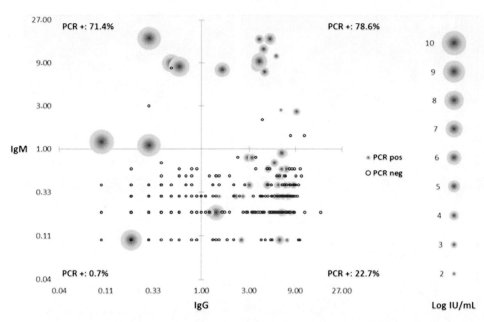

FIGURE 9.3 Combined qPCR and immunological detection of anti-B19V IgG and IgM. Multiparametric analysis on 354 consecutive serum samples, for combined detection of B19V-specific IgG and IgM antibodies (EIA, DiaSorin) and B19V DNA (qPCR) [90]. Of these, 61 were PCR positive, with viral load ranging from 10^2 to more than 10^{10} IU/mL, and 293 were PCR negative. Samples are plotted according to the respective index values for both IgG and IgM, in case of positive qPCR the diameter of the bubble is proportional to the log of viral load (IU/mL) according to scale.

reactivity against conformational or linear epitopes can be used to confirm or better characterize the immunological response [104–106]. However, the immunological approach presents several limitations that prompt molecular diagnostic testing.

The highest viremic levels are reached in the initial phase of infection, before or coupled to the development of a detectable immune response. In this case, a single immunological determination may yield unreliable results, while repeated testing will lead to a delay in diagnosis. In a later phase of infection, progressively lowering viral loads can be detected in the presence of both IgG and IgM. Clearance of viremia might be delayed and parallel the waning of IgM, but in persistent infections even sustained levels of viremia can be present when IgM antibodies are no longer detected.

Indications for molecular testing include the confirmation of suspected active infections, the follow-up of the course of documented infections, and the characterization of persistent infections. In particular, in immunodeficient, immunosuppressed, or in transfused patients, antibody detection is unreliable for diagnostic purposes and detection of viral DNA should be considered the only relevant diagnostic parameter. In antenatal screening or prenatal diagnosis, relying on maternal immune status only will lead to underestimation of active infections in the mother and correlated risk to fetus, so detection of viral DNA in maternal blood should be part of the diagnostic workup [23,107].

Molecular analysis is normally carried out from peripheral blood since productive infections are characterized by viremia, and viremic levels are correlated with the clinical course. Both serum and plasma are suitable samples, so that detection of both viral genomes and specific antibodies can be obtained from the same sample. In case of predominant hematological involvement, the virus can also be detected from bone marrow aspirates. Biopsy samples can be obtained from other tissues where an active B19V infection is suspected, for example, in the case of acute myocarditis, and in these cases the critical issue will be to differentiate an active infection from the persistence of viral genomes. Molecular analysis may be necessary to confirm infection of the fetus, and for this purpose it is not necessary to analyze cord blood, since the virus can also be detected in amniotic fluid, or in placenta at term, or in biopsies in the case of postmortem analysis (Fig. 9.4).

qPCR Result Interpretation and Relevance

As a main issue, molecular amplification methods leading to a reliable quantitative assessment of viral load in an ample range are critically required to obtain useful diagnostic information. In fact, considering peripheral blood as the sample of choice and the characteristics of a typical time course of infection, a high viral load in the acute phase will progressively decline to lower levels in the following months. Hence, qualitative assessment of the presence/absence of viral DNA within a sample would be of very little informative content, while a quantitative evaluation of the viral load is to be considered necessary for an accurate molecular diagnosis of B19V infection. Even in normal subjects a complete clearance of B19V DNA from

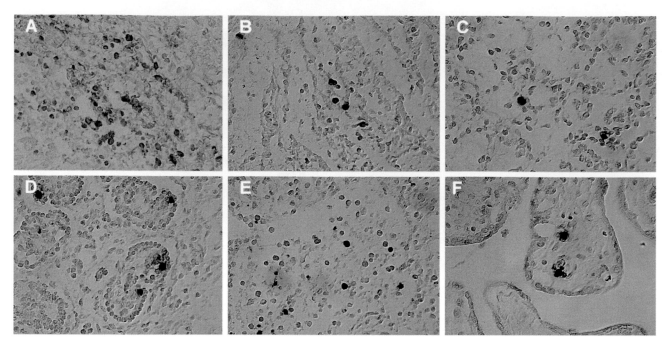

FIGURE 9.4 In situ hybridization detection of B19V DNA infected cells in fetal tissues. Biopsies were obtained postmortem from a miscarriage (19th week of gestation) with moderate fetal hydrops. Formalin-fixed, paraffin-embedded tissue sections have been processed for detection of B19V nucleic acids by means of a digoxigenin-labeled DNA probe and immunological detection of hybridized probes by anti-Dig alkaline phosphatase antibodies followed by BCIP-NBT substrate development [57]. Positive cells appear dark blue within tissues. Sections: (A) liver; (B) heart; (C) lung; (D) kidney; (E), mesentery; and (F) placenta. Positive cells have been identified as hemopoietic (liver) or circulating erythroid progenitor cells. Original magnification: 400 × .

peripheral blood may not be achieved in a short time [30,108]. Therefore, the diagnostic question will be to discriminate this situation from an active, persistent infection and determination of the viral load is the only relevant parameter for this purpose. In pregnant women, either in case of antenatal screening or in response to a specific diagnostic question, determination of B19V DNA is a more reliable marker of infection than determination of the immunological status [23], while confirmation of intrauterine or congenital infection needs to rely on the detection of B19V DNA in fetal or neonatal samples [60].

Genotyping is a secondary issue with respect to diagnosis, as all genotypes of B19V share the same pathogenetic characteristics [69], but genotype determination is relevant for epidemiological investigations. However, it is important that all viral genotypes are detected by a molecular assay with the same analytical sensitivity. Thus, the relevance to detect B19V with a qPCR method that is able to quantitate and possibly discriminate all different genotypes derives from two considerations. First, variant genotypes are known to circulate at higher frequencies in areas other than Europe or the United States, and in addition the changing demographic and epidemiological scenario will probably lead to a more global circulation of genotype 3 [65]. Second, a reliable

viral diagnosis should be able to detect variant genotypes even in particular sporadic clinical situations when assays focused on genotype 1 only may fail in B19V detection [93,109,110].

Limitations of Testing

B19V is underestimated from a clinical perspective. Its wide circulation and prevalent benign and self-limiting clinical course generally lead to a diminished appreciation of its pathogenic potential. However, B19V is a possible etiological agent in a large ensemble of diseases, encompassing practically all tissues and organs. An extended awareness and definition of the actual pathogenic role of B19V among human diseases will be fostered by the development of better diagnostic methods and algorithms. In this respect, integration of the common immunological diagnostic scheme with the application of molecular detection methods will be crucial. Molecular detection of B19V is easy and reliable. qPCR techniques for the detection of B19V can be carried out from biological specimens with standard preanalytical processing procedures. The choice of qPCR techniques ranges from commercially-available, to integrated in common-flow analytical platforms, to the tailored or lab-developed assays, requiring standard laboratory equipment. Development of even more versatile analytical platforms, and of alternative molecular assays with point-of-care characteristics [111], will widen in the future the opportunities for molecular testing for B19V, aimed at improving performance and informative content, and reducing time and costs. The pattern of B19V genetic evolution and diversity, its biological characteristics and complex relationship with the host, and its diverse clinical manifestations of infection are all topics that are far from completely understood and will benefit from information emerging from the wider use of molecular diagnostic approaches.

References

[1] Cotmore SF, Agbandje-McKenna M, Chiorini JA, et al. The family Parvoviridae. Arch Virol 2014;159:1239—47.

[2] Gallinella G. Parvovirus B19 achievements and challenges. ISRN Virology 2013;. Available from: http://dx.doi.org/10.5402/2013/898730.

[3] Jartti T, Hedman K, Jartti L, Ruuskanen O, Allander T, Soderlund-Venermo M. Human bocavirus-the first 5 years. Rev Med Virol 2012;22:46—64.

[4] Brown KE, Anderson SM, Young NS. Erythrocyte P antigen: cellular receptor for B19 parvovirus. Science 1993;262:114—17.

[5] Weigel-Kelley KA, Yoder MC, Srivastava A. Alpha5beta1 integrin as a cellular coreceptor for human parvovirus B19: requirement of functional activation of beta1 integrin for viral entry. Blood 2003;102:3927—33.

[6] Quattrocchi S, Ruprecht N, Bonsch C, et al. Characterization of the early steps of human parvovirus B19 infection. J Virol 2012;86:9274—84.

[7] Leisi R, Ruprecht N, Kempf C, Ros C. Parvovirus B19 uptake is a highly selective process controlled by VP1u, a novel determinant of viral tropism. J Virol 2013;87:13161—7.

[8] Bonvicini F, Filippone C, Delbarba S, et al. Parvovirus B19 genome as a single, two-state replicative and transcriptional unit. Virology 2006;347:447—54.

[9] Bonvicini F, Filippone C, Manaresi E, Zerbini M, Musiani M, Gallinella G. Functional analysis and quantitative determination of the expression profile of human parvovirus B19. Virology 2008;381:168—77.

[10] Wong S, Zhi N, Filippone C, et al. Ex vivo-generated CD36 + erythroid progenitors are highly permissive to human parvovirus B19 replication. J Virol 2008;82:2470—6.

[11] Filippone C, Franssila R, Kumar A, et al. Erythroid progenitor cells expanded from peripheral blood without mobilization or preselection: molecular characteristics and functional competence. PLoS One 2010;5:e9496.

[12] Morita E, Nakashima A, Asao H, Sato H, Sugamura K. Human parvovirus B19 nonstructural protein (NS1) induces cell cycle arrest at G(1) phase. J Virol 2003;77:2915—21.

[13] Wan Z, Zhi N, Wong S, et al. Human parvovirus B19 causes cell cycle arrest of human erythroid progenitors via deregulation of the E2F family of transcription factors. J Clin Invest 2010;120:3530—44.

[14] Luo Y, Kleiboeker S, Deng X, Qiu J. Human parvovirus B19 infection causes cell cycle arrest of human erythroid progenitors at late S phase that favors viral DNA replication. J Virol 2013;87:12766—75.

[15] Yaegashi N, Niinuma T, Chisaka H, et al. Parvovirus B19 infection induces apoptosis of erythroid cells in vitro and in vivo. J Infect 1999;39:68—76.

[16] Ray NB, Nieva DR, Seftor EA, Khalkhali-Ellis Z, Naides SJ. Induction of an invasive phenotype by human parvovirus B19 in normal human synovial fibroblasts. Arthritis Rheum 2001;44:1582—6.

[17] Zakrzewska K, Cortivo R, Tonello C, et al. Human parvovirus B19 experimental infection in human fibroblasts and endothelial cells cultures. Virus Res 2005;114:1—5.

[18] von Kietzell K, Pozzuto T, Heilbronn R, Grossl T, Fechner H, Weger S. Antibody-mediated enhancement of parvovirus B19 uptake into endothelial cells mediated by a receptor for complement factor C1q. J Virol 2014;88:8102—15.

[19] Bonvicini F, Manaresi E, Di Furio F, De Falco L, Gallinella G. Parvovirus B19 DNA CpG dinucleotide methylation and epigenetic regulation of viral expression. PLoS One 2012;7:e33316.

[20] Mossong J, Hens N, Friederichs V, et al. Parvovirus B19 infection in five European countries: seroepidemiology, force of infection and maternal risk of infection. Epidemiol Infect 2008;136:1059—68.

[21] Enders M, Weidner A, Zoellner I, Searle K, Enders G. Fetal morbidity and mortality after acute human parvovirus B19 infection in pregnancy: prospective evaluation of 1018 cases. Prenat Diagn 2004;24:513—18.

[22] Lamont RF, Sobel JD, Vaisbuch E, et al. Parvovirus B19 infection in human pregnancy. BJOG 2011;118:175—86.

[23] Bonvicini F, Puccetti C, Salfi NC, et al. Gestational and fetal outcomes in B19 maternal infection: a problem of diagnosis. J Clin Microbiol 2011;49:3514—18.

[24] Blumel J, Burger R, Drosten C, et al. Parvovirus B19—revised. Transfus Med Hemother 2010;37:339—50.

[25] Chisaka H, Morita E, Yaegashi N, Sugamura K. Parvovirus B19 and the pathogenesis of anaemia. Rev Med Virol 2003;13:347—59.

[26] Brown KE. Haematological consequences of parvovirus B19 infection. Baillieres Best Pract Res Clin Haematol 2000;13:245—59.

[27] Young NS, Brown KE. Parvovirus B19. N Engl J Med 2004;350:586—97.

[28] Musiani M, Zerbini M, Gentilomi G, Plazzi M, Gallinella G, Venturoli S. Parvovirus B19 clearance from peripheral blood after acute infection. J Infect Dis 1995;172:1360—3.

[29] Lindblom A, Isa A, Norbeck O, et al. Slow clearance of human parvovirus B19 viremia following acute infection. Clin Infect Dis 2005;41:1201—3.

[30] Lefrere JJ, Servant-Delmas A, Candotti D, et al. Persistent B19 infection in immunocompetent individuals: implications for transfusion safety. Blood 2005;106:2890—5.

[31] Young NS, Abkowitz JL, Luzzatto L. New insights into the pathophysiology of acquired cytopenias. Hematology (Am Soc Hematol Educ Program) 2000;18—38.

[32] Koduri PR. Parvovirus B19-related anemia in HIV-infected patients. AIDS Patient Care STDS 2000;14:7—11.

[33] Broliden K, Tolfvenstam T, Ohlsson S, Henter JI. Persistent B19 parvovirus infection in pediatric malignancies. Med Pediatr Oncol 1998;31:66—72.

[34] Gallinella G, Manaresi E, Venturoli S, Grazi GL, Musiani M, Zerbini M. Occurrence and clinical role of active parvovirus B19 infection in transplant recipients. Eur J Clin Microbiol Infect Dis 1999;18:811—13.

[35] Broliden K. Parvovirus B19 infection in pediatric solid-organ and bone marrow transplantation. Pediatr Transplant 2001;5:320—30.

[36] Eid AJ, Brown RA, Patel R, Razonable RR. Parvovirus B19 infection after transplantation: a review of 98 cases. Clin Infect Dis 2006;4:40—8.

[37] Moffatt S, Tanaka N, Tada K, et al. A cytotoxic nonstructural protein, NS1, of human parvovirus B19 induces activation of interleukin-6 gene expression. J Virol 1996;70:8485−91.

[38] Lu J, Zhi N, Wong S, Brown KE. Activation of synoviocytes by the secreted phospholipase A2 motif in the VP1-unique region of parvovirus B19 minor capsid protein. J Infect Dis 2006;193:582−90.

[39] Duechting A, Tschope C, Kaiser H, et al. Human parvovirus B19 NS1 protein modulates inflammatory signaling by activation of STAT3/PIAS3 in human endothelial cells. J Virol 2008;82:7942−52.

[40] Tsitsikas DA, Gallinella G, Patel S, Seligman H, Greaves P, Amos RJ. Bone marrow necrosis and fat embolism syndrome in sickle cell disease: increased susceptibility of patients with non-SS genotypes and a possible association with human parvovirus B19 infection. Blood Rev 2014;28:23−30.

[41] Lunardi C, Tinazzi E, Bason C, Dolcino M, Corrocher R, Puccetti A. Human parvovirus B19 infection and autoimmunity. Autoimmun Rev 2008;8:116−20.

[42] Bultmann BD, Klingel K, Sotlar K, et al. Fatal parvovirus B19-associated myocarditis clinically mimicking ischemic heart disease: an endothelial cell-mediated disease. Hum Pathol 2003;34:92−5.

[43] Andreoletti L, Leveque N, Boulagnon C, Brasselet C, Fornes P. Viral causes of human myocarditis. Arch Cardiovasc Dis 2009;102:559−68.

[44] Modrow S. Parvovirus B19: the causative agent of dilated cardiomyopathy or a harmless passenger of the human myocard? Ernst Schering Res Found Workshop 2006;55:63−82.

[45] Moore TL. Parvovirus-associated arthritis. Curr Opin Rheumatol 2000;12:289−94.

[46] Kerr JR. Pathogenesis of human parvovirus B19 in rheumatic disease. Ann Rheum Dis 2000;59:672−83.

[47] Norja P, Hokynar K, Aaltonen LM, et al. Bioportfolio: lifelong persistence of variant and prototypic erythrovirus DNA genomes in human tissue. Proc Natl Acad Sci USA 2006;103:7450−743.

[48] Cassinotti P, Burtonboy G, Fopp M, Siegl G. Evidence for persistence of human parvovirus B19 DNA in bone marrow. J Med Virol 1997;53:229−32.

[49] Manning A, Willey SJ, Bell JE, Simmonds P. Comparison of tissue distribution, persistence, and molecular epidemiology of parvovirus B19 and novel human parvoviruses PARV4 and human bocavirus. J Infect Dis 2007;195:1345−52.

[50] Schneider B, Hone A, Tolba RH, Fischer HP, Blumel J, Eis-Hubinger AM. Simultaneous persistence of multiple genome variants of human parvovirus B19. J Gen Virol 2008;89:164−76.

[51] Kuhl U, Pauschinger M, Seeberg B, et al. Viral persistence in the myocardium is associated with progressive cardiac dysfunction. Circulation 2005;112:1965−70.

[52] Lotze U, Egerer R, Gluck B, et al. Low level myocardial parvovirus B19 persistence is a frequent finding in patients with heart disease but unrelated to ongoing myocardial injury. J Med Virol 2010;82:1449−57.

[53] Soderlund M, von Essen R, Haapasaari J, Kiistala U, Kiviluoto O, Hedman K. Persistence of parvovirus B19 DNA in synovial membranes of young patients with and without chronic arthropathy. Lancet 1997;349:1063−5.

[54] Bonvicini F, La Placa M, Manaresi E, et al. Parvovirus B19 DNA is commonly harboured in human skin. Dermatology 2010;220:138−42.

[55] Wegner CC, Jordan JA. Human parvovirus B19 VP2 empty capsids bind to human villous trophoblast cells in vitro via the globoside receptor. Infect Dis Obstet Gynecol 2004;12:69−78.

[56] Pasquinelli G, Bonvicini F, Foroni L, Salfi N, Gallinella G. Placental endothelial cells can be productively infected by Parvovirus B19. J Clin Virol 2009;44:33−8.

[57] Bonvicini F, Manaresi E, Gallinella G, Gentilomi GA, Musiani M, Zerbini M. Diagnosis of fetal parvovirus B19 infection: value of virological assays in fetal specimens. BJOG 2009;116:813−17.

[58] de Jong EP, Walther FJ, Kroes AC, Oepkes D. Parvovirus B19 infection in pregnancy: new insights and management. Prenat Diagn 2011;31:419−25.

[59] Riipinen A, Vaisanen E, Nuutila M, et al. Parvovirus B19 infection in fetal deaths. Clin Infect Dis 2008;47:1519−25.

[60] Puccetti C, Contoli M, Bonvicini F, et al. Parvovirus B19 in pregnancy: possible consequences of vertical transmission. Prenat Diagn 2012;32:897−902.

[61] Ergaz Z, Ornoy A. Parvovirus B19 in pregnancy. Reprod Toxicol 2006;21:421−35.

[62] De Jong EP, Lindenburg IT, van Klink JM, et al. Intrauterine transfusion for parvovirus B19 infection: long-term neurodevelopmental outcome. Am J Obstet Gynecol 2012;206:204e1−205e.

[63] Servant A, Laperche S, Lallemand F, et al. Genetic diversity within human erythroviruses: identification of three genotypes. J Virol 2002;76:9124−34.

[64] Gallinella G, Venturoli S, Manaresi E, Musiani M, Zerbini M. B19 virus genome diversity: epidemiological and clinical correlations. J Clin Virol 2003;28:1−13.

[65] Hubschen JM, Mihneva Z, Mentis AF, et al. Phylogenetic analysis of human parvovirus b19 sequences from eleven different countries confirms the predominance of genotype 1 and suggests the spread of genotype 3b. J Clin Microbiol 2009;47:3735−8.

[66] Corcoran C, Hardie D, Yeats J, Smuts H. Genetic variants of human parvovirus B19 in South Africa: cocirculation of three genotypes and identification of a novel subtype of genotype 1. J Clin Microbiol 2010;48:137−42.

[67] Eis-Hubinger AM, Reber U, Edelmann A, Kalus U, Hofmann J. Parvovirus B19 genotype 2 in blood donations. Transfusion 2014;54:1682−4.

[68] Parsyan A, Szmaragd C, Allain JP, Candotti D. Identification and genetic diversity of two human parvovirus B19 genotype 3 subtypes. J Gen Virol 2007;88:428−31.

[69] Ekman A, Hokynar K, Kakkola L, et al. Biological and immunological relations among human parvovirus B19 genotypes 1 to 3. J Virol 2007;81:6927−35.

[70] Baylis SA, Ma L, Padley DJ, Heath AB, Yu MW, Collaborative Study Group. Collaborative study to establish a World Health Organization International genotype panel for parvovirus B19 DNA nucleic acid amplification technology (NAT)-based assays. Vox Sang 2012;102:204−11.

[71] Trosemeier JH, Branting A, Lukashov VV, Blumel J, Baylis SA. Genome sequences of parvovirus b19 reference strains. Genome Announc 2014;2 e00830-14.

[72] Shackelton LA, Holmes EC. Phylogenetic evidence for the rapid evolution of human B19 erythrovirus. J Virol 2006;80:3666−9.

[73] Norja P, Eis-Hubinger AM, Soderlund-Venermo M, Hedman K, Simmonds P. Rapid sequence change and geographical spread of human parvovirus B19: comparison of B19 virus evolution in acute and persistent infections. J Virol 2008;82:6427−33.

[74] Molenaar-de Backer MW, Lukashov VV, van Binnendijk RS, Boot HJ, Zaaijer HL. Global co-existence of two evolutionary lineages of parvovirus B19 1a, different in genome-wide synonymous positions. PLoS One 2012;7:e43206.

[75] Anderson MJ, Jones SE, Minson AC. Diagnosis of human parvovirus infection by dot-blot hybridization using cloned viral DNA. J Med Virol 1985;15:163−72.

[76] Clewley JP. Detection of human parvovirus using a molecularly cloned probe. J Med Virol 1985;15:173—81.

[77] Salimans MM, Holsappel S, van de Rijke FM, Jiwa NM, Raap AK, Weiland HT. Rapid detection of human parvovirus B19 DNA by dot-hybridization and the polymerase chain reaction. J Virol Methods 1989;23:19—28.

[78] Zerbini M, Musiani M, Venturoli S, et al. Rapid screening for B19 parvovirus DNA in clinical specimens with a digoxigenin-labeled DNA hybridization probe. J Clin Microbiol 1990;28:2496—9.

[79] Musiani M, Zerbini M, Gibellini D, et al. Chemiluminescence dot blot hybridization assay for detection of B19 parvovirus DNA in human sera. J Clin Microbiol 1991;29:2047—50.

[80] Durigon EL, Erdman DD, Gary GW, Pallansch MA, Torok TJ, Anderson LJ. Multiple primer pairs for polymerase chain reaction (PCR) amplification of human parvovirus B19 DNA. J Virol Methods 1993;44:155—65.

[81] Zerbini M, Gibellini D, Musiani M, Venturoli S, Gallinella G, Gentilomi G. Automated detection of digoxigenin-labelled B19 parvovirus amplicons by a capture hybridization assay. J Virol Methods 1995;55:1—9.

[82] Gallinella G, Zerbini M, Musiani M, Venturoli S, Gentilomi G, Manaresi E. Quantitation of parvovirus B19 DNA sequences by competitive PCR: differential hybridization of the amplicons and immunoenzymatic detection on microplate. Mol Cell Probes 1997;11:127—33.

[83] Musiani M, Gallinella G, Venturoli S, Zerbini M. Competitive PCR-ELISA protocols for the quantitative and the standardized detection of viral genomes. Nature Protoc 2007;2:2511—19.

[84] Aberham C, Pendl C, Gross P, Zerlauth G, Gessner M. A quantitative, internally controlled real-time PCR Assay for the detection of parvovirus B19 DNA. J Virol Methods 2001;92:183—91.

[85] Gruber F, Falkner FG, Dorner F, Hammerle T. Quantitation of viral DNA by real-time PCR applying duplex amplification, internal standardization, and two-color fluorescence detection. Appl Environ Microbiol 2001;67:2837—9.

[86] Manaresi E, Gallinella G, Zuffi E, Bonvicini F, Zerbini M, Musiani M. Diagnosis and quantitative evaluation of parvovirus B19 infections by real-time PCR in the clinical laboratory. J Med Virol 2002;67:275—81.

[87] Gallinella G, Bonvicini F, Filippone C, et al. Calibrated real-time PCR for evaluation of parvovirus B19 viral load. Clin Chem 2004;50:759—62.

[88] Baylis SA. Standardization of nucleic acid amplification technique (NAT)-based assays for different genotypes of parvovirus B19: a meeting summary. Vox Sang 2008;94:74—80.

[89] Baylis SA, Buchheit KH. A proficiency testing study to evaluate laboratory performance for the detection of different genotypes of parvovirus B19. Vox Sang 2009;97:13—20.

[90] Bonvicini F, Manaresi E, Bua G, Venturoli S, Gallinella G. Keeping pace with parvovirus B19 genetic variability: a multiplex genotype-specific quantitative PCR assay. J Clin Microbiol 2013;51:3753—9.

[91] Baylis SA, Shah N, Minor PD. Evaluation of different assays for the detection of parvovirus B19 DNA in human plasma. J Virol Methods 2004;121:7—16.

[92] Hokynar K, Norja P, Laitinen H, et al. Detection and differentiation of human parvovirus variants by commercial quantitative real-time PCR tests. J Clin Microbiol 2004;42:2013—19.

[93] Cohen BJ, Gandhi J, Clewley JP. Genetic variants of parvovirus B19 identified in the United Kingdom: implications for diagnostic testing. J Clin Virol 2006;36:152—5.

[94] Bustin SA, Benes V, Garson JA, et al. The MIQE guidelines: minimum information for publication of quantitative real-time PCR experiments. Clin Chem 2009;55:611—22.

[95] Salimans MM, van de Rijke FM, Raap AK, van Elsacker-Niele AM. Detection of parvovirus B19 DNA in fetal tissues by in situ hybridisation and polymerase chain reaction. J Clin Pathol 1989;42:525—30.

[96] Morey AL, Porter HJ, Keeling JW, Fleming KA. Non-isotopic in situ hybridisation and immunophenotyping of infected cells in the investigation of human fetal parvovirus infection. J Clin Pathol 1992;45:673—8.

[97] Gentilomi G, Zerbini M, Musiani M, et al. In situ detection of B19 DNA in bone marrow of immunodeficient patients using a digoxigenin-labelled probe. Mol Cell Probes 1993;7:19—24.

[98] Gallinella G, Young NS, Brown KE. In situ hybridisation and in situ polymerase chain reaction detection of parvovirus B19 DNA within cells. J Virol Methods 1994;50:67—74.

[99] Bonvicini F, Filippone C, Manaresi E, et al. Peptide nucleic acid-based in situ hybridization assay for detection of parvovirus B19 nucleic acids. Clin Chem 2006;52:973—8.

[100] Bonvicini F, Mirasoli M, Gallinella G, Zerbini M, Musiani M, Roda A. PNA-based probe for quantitative chemiluminescent in situ hybridisation imaging of cellular parvovirus B19 replication kinetics. Analyst 2007;132:519—23.

[101] Bonvicini F, Mirasoli M, Manaresi E, et al. Single-cell chemiluminescence imaging of parvovirus B19 life cycle. Virus Res 2013;178:517—21.

[102] Corcoran A, Kerr S, Elliott G, Koppelman M, Doyle S. Improved detection of acute parvovirus B19 infection by immunoglobulin M EIA in combination with a novel antigen EIA. Vox Sang 2007;93:216—22.

[103] Morey AL, O'Neill HJ, Coyle PV, Fleming KA. Immunohistological detection of human parvovirus B19 in formalin-fixed, paraffin-embedded tissues. J Pathol 1992;166:105—8.

[104] Gallinella G, Zuffi E, Gentilomi G, et al. Relevance of B19 markers in serum samples for a diagnosis of parvovirus B19-correlated diseases. J Med Virol 2003;71:135—9.

[105] Kerr S, O'Keeffe G, Kilty C, Doyle S. Undenatured parvovirus B19 antigens are essential for the accurate detection of parvovirus B19 IgG. J Med Virol 1999;57:179—85.

[106] Manaresi E, Gallinella G, Venturoli S, Zerbini M, Musiani M. Detection of parvovirus B19 IgG: choice of antigens and serological tests. J Clin Virol 2004;29:51—3.

[107] Enders M, Schalasta G, Baisch C, et al. Human parvovirus B19 infection during pregnancy—value of modern molecular and serological diagnostics. J Clin Virol 2006;35:400—6.

[108] Juhl D, Steppat D, Gorg S, Hennig H. Parvovirus b19 infections and blood counts in blood donors. Transfus Med Hemother 2014;41:52—9.

[109] Liefeldt L, Plentz A, Klempa B, et al. Recurrent high level parvovirus B19/genotype 2 viremia in a renal transplant recipient analyzed by real-time PCR for simultaneous detection of genotypes 1 to 3. J Med Virol 2005;75:161—9.

[110] Knoester M, von dem Borne PA, Vossen AC, Kroes AC, Claas EC. Human parvovirus B19 genotype 3 associated with chronic anemia after stem cell transplantation, missed by routine PCR testing. J Clin Virol 2012;54:368—70.

[111] Mirasoli M, Bonvicini F, Dolci LS, Zangheri M, Gallinella G, Roda A. Portable chemiluminescence multiplex biosensor for quantitative detection of three B19 DNA genotypes. Anal Bioanal Chem 2013;405:1139—43.

II. MOLECULAR TESTING IN INFECTIOUS DISEASE

10

Molecular Testing for Polyomaviruses

G.W. Procop and B. Yen-Lieberman

Section of Clinical Microbiology, Department of Laboratory Medicine, Cleveland Clinic, Cleveland, OH, United States

INTRODUCTION

Polyomaviruses are small (45 nm), nonenveloped viruses with icosahedral nucleocapsids. The genetic material is double-stranded DNA, with genomes that range 4.5—5.5 kb. The Polyomaviridae contains two groups: the *Orthopolyomaviruses* and the *Wukipolyomaviruses* [1,2]. The *Orthopolyomavirus* genus contains the three most important human pathogens: the BK virus, the JC virus, and the Merkel cell polyomavirus (MCPV) [2]. It also contains the tricho-dysplasia spinulosa-associated polyomavirus. The BK and JC viruses received their curious nomenclature from the initials of the patients from which they were originally isolated [3,4]. The *Wukipolyomavirus* genus contains the WU and KI viruses for which it was named, as well as the Human polyomaviruses 6, 7, 8, 9, and 10. Although the WU and KI polyomaviruses have been recovered from the respiratory, plasma, and/or urine specimens of transplant recipients, these have not been associated with disease [5]. The breadth of the Polyomaviridae is undetermined and the discovery of new polyomaviruses is expected. The most important etiologic agents of disease in this group, the BK and JC viruses, will command the attention of the majority of this chapter.

Human polyomaviruses viruses cause minimal to no disease in the immunocompetent host. Serologic studies suggest that the majority of humans are infected early in life. Infections are either asymptomatic or produce subclinical disease, possibly involving the respiratory tract. The precise mode of transmission is a matter of debate. Some have advocated a respiratory route of infection for the BK virus, since BK viral DNA has been detected in tonsillar tissues [6]. However, others have studied body fluids, such as saliva, and the respiratory secretions of children with upper respiratory tract infections for the presence of the BK and JC viruses, and these fluids did not harbor the viruses [7]. Uro—oral, fecal—oral, and transplacental transmission have also been postulated as modes of transmission, with the former seemingly very feasible given the permissive nature of urothelial cells for the replication of these viruses [8].

Regardless of the mode of transmission, these viruses are commonly found in the general population. The majority of children (ie, between 65% and 90%) are seropositive for the BK virus by the age of 10 [9]. Seropositivity rates for BK rise until around 40 years of age and then decrease slightly. The JC virus is also highly prevalent with 50—80% of the human population demonstrating serologic evidence of prior infection [10]. The seroprevalance of the other polyomaviruses has not been extensively studied. However, it has been demonstrated that up to 77% of the general population have been exposed to the MCPV [11].

After initial infection, a transient viremia likely ensues whereby the virus reaches the destination tissues for latency. It has been hypothesized that this dissemination could occur through the infection of mononuclear leukocytes in which BK virus DNA has been detected [12]. It is not known for certain whether these viruses become truly latent in the host cells or maintain subclinical (ie, minimal maintenance) replication once the viruses reach their destination tissues. It is known, however, that asymptomatic replication of both the BK and JC viruses occurs in the urothelium (ie, the lining epithelium of the bladder, ureters, and renal pelvis), with subsequent shedding of the virus into the urine [13]. Asymptomatic shedding of the BK virus has been documented in up to 10% of healthy adults [8]. Similarly, 25% of pregnant women have been shown to asymptomatically shed the BK virus in the urine [14]. Age-related shedding of the BK virus

© 2017 Elsevier Inc. All rights reserved.

has been reported, with shedding occurring less frequently in individuals less than 30 years old and gradually increasing in individuals greater than or equal to 30 years old [15]. This asymptomatic shedding of virus may contribute to the transmission of the virus to those who were previously unexposed and is supportive of a uro—oral route of infection.

MOLECULAR TARGETS

The molecular targets used for the detection of the BK and JC viruses include the *VP1* gene, the *VP2* gene, and the *T antigen* gene. Variations in the *VP1* sequence are associated with different subtypes [16]. Primer sets for the quantitative assessment of BK virus are commercially available (Table 10.1), as are numerous laboratory-developed tests. Similarly, laboratory-developed tests have been used for the detection of the JC virus and MCPV. The RealStar JC virus (Altona, Hamburg, Germany), which targets the large T antigen gene, is available. The linear range claimed is $1000-10^{12}$ copies/mL. Similarly, the GeneProof JC Virus assay (GeneProof, Brno, Czech Republic) that targets the junction of the *VP1* and *VP2* genes is also available. They claim a linear range from 528 to 10^{10} copies/mL. GeneProof also offers a combined BK/JC assay.

MOLECULAR TECHNOLOGIES

Quantitative PCR is the most commonly used method in the assessment of clinical specimens for the detection and quantitation of BK and JC polyomaviruses. This has been done in conjunction with a variety of probe types. Less commonly, qualitative PCR is used to determine the presence of these viruses. In situ hybridization and immunohistochemistry are used in the assessment of biopsy specimens for these viruses [17,18]. Electron microscopy was used in the distant past, but it has largely been replaced by these

technologies [19]. However, electron microscopy remains an important tool for discovery and in instances wherein these other technologies fail or inconsistent results are produced.

Quantitative PCR is used predominantly for the detection and quantitative monitoring of the BK virus in plasma. A wide variety of PCR assays have been described for the detection and/or quantitation of the BK virus. Many of these are laboratory-developed tests, but some are commercially available [20–25]. These assays usually employ some type of nucleic acid extraction method prior to PCR amplification. Although all aspects of tests need to undergo a thorough assessment during assay validation, most commercially-available extraction methods are robust in the removal of inhibitors and are largely comparable with respect to DNA yield [26]. The quantity of virus present in the plasma is important, since higher values are more predictive of BK nephropathy.

Qualitative PCR for the BK virus in urine may be used to identify (or screen out) patients for BK nephropathy, but this method is not commonly used in this manner. However, the qualitative detection of the JC virus in the cerebrospinal fluid (CSF) of patients with progressive multifocal leukoencephalopathy (PML) is sufficient to support this diagnosis in the appropriate clinical setting, which includes supportive radiological findings. Given the relative simplicity of converting a qualitative rapid cycle PCR assay into a quantitative assay, quantitative values are often reported when the JC virus is detected in CSF or other body fluids.

Beyond quantitative PCR assays, loop-mediated isothermal amplification (LAMP) techniques have been developed for the detection of the BK virus [27]. Although commercially-available LAMP methods for BK viruses are not available, this type of technology is attractive because it uses standard laboratory equipment, has an acceptable level of sensitivity, and may be used without DNA extraction and a thermocycler. Such a technology could conceivably be used to inexpensively screen for the presence of the BK virus in

TABLE 10.1 Commercially-Available BK Virus Assays

Vendor/Assay	Assay specifics	Target region	Dynamic range (copies/mL)
ELiTechMGB Alert BK Virus Primers	MGB probes	*Large T antigen*	12.5×10^1 to 12.5×10^6
Focus DiagnosticsSimplexa BKV Kit	Scorpion	*VP2*	5.1×10^2 to 1.0×10^8
Luminex (Eragen)MultiCode BK Virus Primers	MultiCode	*Large T antigen*	5.0×10^2 to 5.0×10^6
QiagenArtus BK Virus RG Kit	Real-time PCR	*Large T antigen*	5.0×10^2 to 5.0×10^6
GeneProof BK Virus Kit	Real-time PCR	*VP1/VP2* junction	5.96×10^2 to 1.0×10^{10}
RealStar BK virus (Altona)	Real-time PCR	Unknown	1.0×10^3 to 1.0×10^{12}

at-risk populations. In addition to monoplex assays, multiplex assays have been developed. Some target only the most common polyomaviruses (ie, BK and JC), whereas others target additional viruses that may cause disease in the immunocompromised host, such as adenovirus [21,25]. Fluorescence resonance energy transfer probe sets that utilize broad-range polyomavirus primers have also been used to detect and differentiate the BK and JC viruses in a single reaction (Fig. 10.1) [22]. Such an assay may be useful in institutions that screen urine specimens, since either virus may be present. In addition to these, qualitative seminested PCR assays have also been described [28].

In situ hybridization and immunohistochemistry are technologies largely used by anatomic pathologists for the detection of these and other viruses in histologic preparations. The availability of these stains is important to confirm the nature of intranuclear inclusions thought to likely represent polyomaviral inclusions. Other viruses, such as adenovirus and cytomegalovirus, can cause intranuclear inclusions in the same at-risk patient population and are represented in the differential diagnosis. These tools are particularly useful when the intranuclear inclusions are not typical. Although biopsy suffers from sampling error, the demonstration of BK viral inclusions associated with interstitial nephritis is the gold standard for the diagnosis of BK nephropathy. Fortunately, tissue is regularly available for assessment, since renal biopsy is necessary to monitor and/or assess for transplant rejection. However, the same is not always true regarding brain biopsy and the diagnosis of PML. Tissue is not readily available in these instances and most would like to avoid a brain biopsy if at all possible. Although the

demonstration of the JC virus in infected oligodendroglial cells in immunocompromised patients with a demyelinating disease may be the gold standard for the diagnosis of PML, most are satisfied with the demonstration of the JC virus in the CSF to support the diagnosis of PML in a patient with the appropriate clinical and radiological findings.

CLINICAL UTILITY

Although most individuals become infected with a polyomavirus sometime during life, serious symptomatic diseases caused by polyomaviruses occur in immunocompromised patients. The type of immunologic compromise and the degree to which the immune system is suppressed are directly related to the type of polyomavirus infection that is most likely to occur, and in some instances with the severity of infection. The main types of diseases caused by polyomaviruses, as well as associated risk factors and the clinical utility of testing, will be discussed here.

The BK Virus

The BK virus is responsible for two main types of disease. It causes nephropathy that can result in graft failure in renal transplant recipients and hemorrhagic cystitis in patients undergoing stem cell transplantation.

Early studies described the severe tubulointerstitial nephritis caused by the BK virus in renal allografts [29]. Serologic studies are supportive of the hypothesis that symptomatic disease is most commonly the result of reactivation of latent, endogenous virus rather than primary infection [30,31]. BK nephropathy most commonly occurs in the transplanted kidney, but infection of the native kidney has also been described [10,32]. Disease usually presents 10–13 months after transplantation [10]. Risk factors include a seropositive donor status and/or a seronegative recipient status, older patient age, male gender, ischemic or immunologic injury, and the degree of HLA mismatch. This latter factor is likely more an indicator of the degree of immunosuppression that will be necessary to avoid transplant rejection. Otherwise stated, the greater the HLA mismatch, the higher the number of rejection episodes that are likely to occur, which will necessitate antilymphocyte therapy to control rejection, which facilitates the emergence of the BK virus [33]. However, there is debate regarding the contribution of either cold ischemia or rejection episodes to BK virus reactivation [34].

The diagnosis and monitoring of patients at risk for BK nephropathy is multifactorial and has evolved with

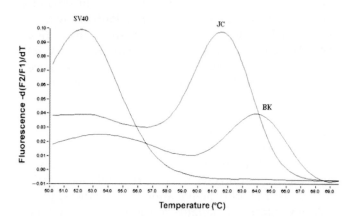

FIGURE 10.1 This post-amplification melt curve was produced following real-time PCR amplification using broad-range polyomavirus primers. Fluorescence resonance energy transfer probes that hybridized with complete complementarity with the BK virus amplicon had two mismatches with the JC virus amplicon and several mismatches with the SV40 virus amplicon. This demonstrates the qualitative detection and differentiation of the medically-important BK and JC polyomaviruses in a single assay.

the introduction of molecular methods. The BK virus can be grown in cell culture or shell vial assay, or directly detected in the urine by immunofluoroescent methods. These methods, which were used in the past, have been replaced by quantitative PCR, which is faster, more specific, and less labor intense [35,36]. Quantitative PCR for the BK polyomavirus is predominantly performed on extracts from plasma specimens from renal transplant recipients. This is usually performed periodically throughout the lives of these patients and the presence and quantity of BK virus are measured and trends are established. A baseline should be established relatively soon after transplantation. Testing should be done with the greatest frequency through the 10- to 13-month window after transplantation when the incidence of disease is greatest, then at a regular interval. A recommended protocol for the screening of renal transplant recipients is to screen the urine every 3 months and the plasma every 1–3 months for the first 2 years after transplantation or when graft dysfunction occurs [37]. An increasing viral load in the presence of decreasing renal function correlates with BK nephropathy [38]. The same correlation is not as clear for high BK viral loads in the urine, since the BK virus replicates in the urothelium and relatively high viral loads may be present in the absence of renal involvement. However, Pang et al. undertook a 1-year prospective study evaluating urine and plasma specimens from renal transplant recipients [39]. They found that as the viral load in the urine increased from 7.0 to 10.0 log(10) copies/mL the percentage of patients with viremia increased from 22% to 100%, respectively. These authors suggested that plasma viral load testing could be reserved for those patients who have greater than or equal to 7.0 log(10) copies/mL of virus in the urine. The absence of BK virus in the urine has a high negative predictive value for BK nephropathy.

The routine use of BK viral load testing represents an important advancement in the diagnosis and monitoring of patients for BK nephropathy. It is particularly important since it has direct implications on the immunosuppressive regimen given to the patient. Decreased kidney function may occur for a variety of reasons in the renal transplant recipient. In addition to BK nephropathy, decreased renal function may occur secondary to transplant allograft rejection. The differentiation of these conditions is critical, since transplant rejection is addressed by increasing the immunosuppression, whereas BK nephropathy is addressed by decreasing the immunosuppression.

The BK virus is the most common cause of hemorrhagic cystitis in hematopoietic stem cell transplant recipients [40,41]. Hemorrhagic cystitis usually occurs 2 weeks after transplant and may affect up to 25% of these patients [42]. Hemorrhagic cystitis ranges in severity from mild (only microscopic hematuria) to severe (clots of blood become obstructive to the flow of urine). There is a great deal of morbidity associated with moderate to severe disease. When the BK virus is the cause of hemorrhagic cystitis the BK viral loads in the urine are extremely high and may be as great as 10^4 copies/mL. The absence of significant quantities of the BK virus in the urine of a patient with hemorrhagic cystitis is suggestive of another etiology. Other etiologies include infections by adenovirus or cytomegalovirus, as well as the chemotherapeutic treatment, albeit this latter cause has become less common as an etiology due to preventive measures taken to prepare the patient for therapy [1].

The BK virus has been associated with ureteral stenosis, which is not surprising given the locus of viral replication. Additionally, it has only rarely been associated with other diseases, such as native kidney infection in recipients of other types of transplants (eg, heart and stem cell), pneumonia, and encephalitis [43–45]. Individuals have sought associations of these viruses with other diseases, but these searches have largely been unfruitful. For example, an examination of lung tissue extracts from 33 patients with documented idiopathic pulmonary fibrosis found no evidence for the presence of either the BK or JC virus by PCR [46].

The JC Virus

The JC virus causes PML [47]. This disease was associated with a variety of malignancies or sarcoidosis prior to the HIV epidemic [48,49]. Patients with progressive HIV infection (ie, AIDS) are at high risk of PML. The CD4 + T-cell reduction caused by HIV infection produces the perfect milieu for the emergence of the JC virus in the central nervous system. The lytic infection of myelin-producing oligodendroglial cells is the cause of this demyelinating disease. As with BK nephropathy, serologic studies are supportive of reactivation of endogenous virus rather than a primary infection [50,51]. This disease has become significantly less common in resource-rich countries wherein patients have access to highly active antiretroviral therapy.

A new group of patients are now recognized as at-risk for PML. This group consists of patients who undergo immunotherapy with biologics for a variety of autoimmune conditions. PML has been reported in patients receiving natalizumab, for multiple sclerosis, efalizumab for psoriasis, and infliximab for Crohn's disease and other autoimmune diseases [52–57]. In rare instances, PML may present in patients with minimal to occult immunosuppression, the possibility of this disease should remain in the differential diagnosis of patients with the appropriate clinical and radiological findings [58].

Quantitative or qualitative PCR for the JC virus on extracts from CSF provides the necessary supportive data to confirm the diagnosis of PML in the appropriate clinical setting [59]. PCR for the JC virus in this setting has largely replaced brain biopsy for the diagnosis of PML, just as temporal lobe biopsy for the diagnosis of herpes simplex virus (HSV) meningoencephalitis has been replaced by HSV PCR performed on CSF specimens.

The JC virus, like the BK virus, has also been noted to be a cause of urethral stenosis. There have been rare instances of JC nephropathy in renal transplant recipients [60]. Therefore, some have suggested that this condition should be termed polyomavirus nephropathy. JC nephropathy should be suspected in patients with decreasing renal function, but with undetectable to very low BK viral loads in the plasma. An infection with a BK variant that is not detected or not quantitated well by the BK virus PCR should also be suspected in this scenario. The histopathologic assessment of the kidney biopsy for evidence is critically important in this type of situation. If JC nephropathy is suspected, then quantitative PCR for the JC virus should be performed on plasma specimens. Similarly, the detection and/or quantitation of the JC virus in the urine may be used to disclose the presence and quantity of this virus in patients with ureteral stenosis.

Other Polyomaviruses

The Merkel cell carcinoma (MCC) polyomavirus has received a great deal of study recently, given the etiologic association of this virus with MCC. MCC is a neuroendocrine carcinoma that arises from the cell that bears the same name. These carcinomas occur most commonly in transplant recipients. However, even in this patient population the incidence of disease is low. The lesion is characterized by neoplastic cells with neuroendocrine features that in many ways resemble small cell carcinoma. The presence of immunoreactivity to cytokeratin 20 is useful in differentiating MCC from small cell carcinoma.

Although the MCPV has been strongly associated with MCC, there are MCCs that lack the presence of the MCPV and likely have arisen through other mechanisms [61]. Therefore, the absence of the MCPV does not exclude the diagnosis of MCC. Currently there is no role for testing for this virus in routine medical practice laboratory.

There is currently no documented clinical utility in testing the numerous other polyomaviruses that have been described.

LIMITATIONS OF TESTING

The presence of PCR inhibitors may occur in any clinical specimen, but these are minimal in specimens that are extracted using modern techniques for the recovery of DNA. The possibility of specimen-to-specimen contamination or extract-to-extract contamination predominantly concerns predominantly the processing and testing of urine specimen extracts where the viral load may be very high. These should not be processed in the same run as plasma specimens.

Hayden et al. used results from a national proficiency testing provider to study factors that contributed to variability in quantitative PCR results for viral assays [62]. They reiterate the importance of standardized quantitative control material, but explore other potential factors. They found that the selection of the quantitative calibrator, the use of commercially-prepared primers and probes, and the target gene selected for amplification all were associated with variability. The differences in these variables between different laboratories are in part responsible for interlaboratory variability of BK testing, which has been described [62]. Genomic sequence variability occurs in the BK virus and has an impact on the detection and reliable quantitation of this virus. Randhawa et al. studied the variability of the *VP1* gene sequence, which was their target for a hydrolysis probe PCR. Of 184 publically-available sequences for review, only 44% ($n = 81$) demonstrated a perfect match for either primers or probes [63]. Not surprisingly, they determined that BK genotypes with mismatches at the primers and probe sites would not be detected unless they were present in high concentrations (ie, the sensitivity of the assay was decreased for these subtypes). Furthermore, they reported that the calculated viral loads would be between 0.57% and 3.26% of the expected values for BK strains with greater than or equal to two mismatches [63]. Similarly, Hoffman et al. compared seven different BK virus quantitative PCR assays and reported substantial disagreement [64]. Like others, this was due to primer and probe mismatches. However, they noted that this primarily occurred with subtypes III and IV. The seven assays were described as typically uniform for the more common subtypes (ie, Ia, V, and VI). Others, similarly, have reported the failure to detect BK virus strains secondary to mismatches, which has necessitated assay redesign [23,24]. It has been a goal of many to design an assay that will detect all BK subtypes equally well [24,64]. Therefore, it is recommended that target sequences are annually reviewed against newly available sequences [24].

New subtypes of the BK virus are still being described [65]. Therefore, it remains important to continually assess the ability of the assays in routine use to detect and adequately quantitate these subtypes. It is important to consider the genomic variability that is known to occur among BK subtypes for particular genes and to consider this factor, if the BK virus is not detected or the viral load is inconsistent with the clinical findings [63].

The quantities of polyomavirus shed into the urine are often logarithmically greater than the quantities detected in the plasma. Therefore, an important practical consideration is to not mix specimen types in quantitative PCR runs. A contamination event from a specimen or a urine extract with a very high BK viral load could produce misleading, false-positive BK viral load results in nearby extracts from plasma. It is our recommendation that if both plasma and urine specimens are to be assessed for the presence and quantity of the BK virus, then these should be processed separately and the PCR assays run at different times.

The cross-reactivity of in situ hybridization probes and immunohistochemistry reagents has been described. This is expected for the immunohistochemistry product, since the commercially-available antibody is directed against antigens on the SV40 polyomavirus. Therefore, these reagents should be considered sensitive for the visual detection of polyomavirus in tissue preparations that have been appropriately processed, but nonspecific with respect to which type of polyomavirus is present. This is usually not an issue since the polyomavirus associated with a demyelinating disease in the brain would be the JC virus and the polyomavirus associated with renal histopathology is most likely, but not exclusively, the BK virus. Therefore, we recommend correlating the results of in situ hybridization or immunohistochemistry studies with the results of PCR assay that target the respective polyomaviruses.

References

[1] Bennett SM, Broekema NM, Imperiale MJ. BK polyomavirus: emerging pathogen. Microbes Infect 2012;14:672–83.

[2] Johne R, Buck CB, Allander T, Atwood WJ, Garcea RL, Imperiale MJ, et al. Taxonomical developments in the family Polyomaviridae. Arch Virol 2011;156:1627–34.

[3] Gardner SD, Field AM, Coleman DV, Hulme B. New human papovirus (B.K.) isolated from urine after renal transplantation. Lancet 1971;1:1253–7.

[4] Padgett BL, Walker DL, ZuRhein GM, Eckroade RJ, Dessel BH. Cultivation of papova-like virus from human brain with progressive multifocal leucoencephalopathy. Lancet 1971;1 1257–60.

[5] Csoma E, Meszaros B, Asztalos L, Gergely L. WU and KI polyomaviruses in respiratory, blood and urine samples from renal transplant patients. J Clin Virol 2015;64:28–33.

[6] Goudsmit J, Wertheim-van Dillen P, van Strien A, van der Noordaa J. The role of BK virus in acute respiratory tract disease and the presence of BKV DNA in tonsils. J Med Virol 1982;10:91–9.

[7] Sundsfjord A, Flaegstad T, Flo R, Spein AR, Pedersen M, Permin H, et al. BK and JC viruses in human immunodeficiency virus type 1-infected persons: prevalence, excretion, viremia, and viral regulatory regions. J Infect Dis 1994;169:485–90.

[8] Jiang M, Abend JR, Johnson SF, Imperiale MJ. The role of polyomaviruses in human disease. Virology 2009;384:266–73.

[9] Knowles WA. Discovery and epidemiology of the human polyomaviruses BK virus (BKV) and JC virus (JCV). Adv Exp Med Biol 2006;577:19–45.

[10] Boothpur R, Brennan DC. Human polyoma viruses and disease with emphasis on clinical BK and JC. J Clin Virol 2010;47:306–12.

[11] Touze A, Gaitan J, Arnold F, Cazal R, Fleury MJ, Combelas N, et al. Generation of Merkel cell polyomavirus (MCV)-like particles and their application to detection of MCV antibodies. J Clin Microbiol 2010;48:1767–70.

[12] Chatterjee M, Weyandt TB, Frisque RJ. Identification of archetype and rearranged forms of BK virus in leukocytes from healthy individuals. J Med Virol 2000;60:353–62.

[13] Heritage J, Chesters PM, McCance DJ. The persistence of papovavirus BK DNA sequences in normal human renal tissue. J Med Virol 1981;8:143–50.

[14] Jin L, Gibson PE, Booth JC, Clewley JP. Genomic typing of BK virus in clinical specimens by direct sequencing of polymerase chain reaction products. J Med Virol 1993;41:11–17.

[15] Zhong S, Zheng HY, Suzuki M, Chen Q, Ikegaya H, Aoki N, et al. Age-related urinary excretion of BK polyomavirus by nonimmunocompromised individuals. J Clin Microbiol 2007;45:193–8.

[16] Jin L, Gibson PE, Knowles WA, Clewley JP. BK virus antigenic variants: sequence analysis within the capsid VP1 epitope. J Med Virol 1993;39:50–6.

[17] Procop GW, Beck RC, Pettay JD, Kohn DJ, Tuohy MJ, Yen-Lieberman B, et al. JC virus chromogenic in situ hybridization in brain biopsies from patients with and without PML. Diagn Mol Pathol 2006;15:70–3.

[18] Wang Z, Portier BP, Hu B, Chiesa-Vottero A, Myles J, Procop GW, et al. Diagnosis of BK viral nephropathy in the renal allograft biopsy: role of fluorescence in situ hybridization. J Mol Diagn 2012;14:494–500.

[19] Singh HK, Madden V, Shen YJ, Thompson BD, Nickeleit V. Negative-staining electron microscopy of the urine for the detection of polyomavirus infections. Ultrastruct Pathol 2006;30:329–38.

[20] Moret H, Brodard V, Barranger C, Jovenin N, Joannes M, Andreoletti L. New commercially available PCR and microplate hybridization assay for detection and differentiation of human polyomaviruses JC and BK in cerebrospinal fluid, serum, and urine samples. J Clin Microbiol 2006;44:1305–9.

[21] Dumonceaux TJ, Mesa C, Severini A. Internally controlled triplex quantitative PCR assay for human polyomaviruses JC and BK. J Clin Microbiol 2008;46:2829–36.

[22] Whiley DM, Mackay IM, Sloots TP. Detection and differentiation of human polyomaviruses JC and BK by LightCycler PCR. J Clin Microbiol 2001;39:4357–61.

[23] Lamontagne B, Girard N, Boucher A, Labbe AC. Improved detection and quantitation of human BK polyomavirus by PCR assay. J Clin Microbiol 2011;49:2778.

[24] Dumoulin A, Hirsch HH. Reevaluating and optimizing polyomavirus BK and JC real-time PCR assays to detect rare sequence polymorphisms. J Clin Microbiol 2011;49:1382–8.

[25] Funahashi Y, Iwata S, Ito Y, Kojima S, Yoshikawa T, Hattori R, et al. Multiplex real-time PCR assay for simultaneous quantification of BK polyomavirus, JC polyomavirus, and adenovirus DNA. J Clin Microbiol 2010;48:825–30.

[26] Tang YW, Sefers SE, Li H, Kohn DJ, Procop GW. Comparative evaluation of three commercial systems for nucleic acid extraction from urine specimens. J Clin Microbiol 2005;43:4830–3.

[27] Bista BR, Ishwad C, Wadowsky RM, Manna P, Randhawa PS, Gupta G, et al. Development of a loop-mediated isothermal amplification assay for rapid detection of BK virus. J Clin Microbiol 2007;45:1581–7.

[28] Held TK, Biel SS, Nitsche A, Kurth A, Chen S, Gelderblom HR, et al. Treatment of BK virus-associated hemorrhagic cystitis and simultaneous CMV reactivation with cidofovir. Bone Marrow Transplant 2000;26:347–50.

[29] Smith RD, Galla JH, Skahan K, Anderson P, Linnemann Jr. CC, Ault GS, et al. Tubulointerstitial nephritis due to a mutant polyomavirus BK virus strain, BKV(Cin), causing end-stage renal disease. J Clin Microbiol 1998;36:1660–5.

[30] Hogan TF, Borden EC, McBain JA, Padgett BL, Walker DL. Human polyomavirus infections with JC virus and BK virus in renal transplant patients. Ann Intern Med 1980;92:373–8.

[31] Nickeleit V, Singh HK, Mihatsch MJ. Latent and productive polyomavirus infections of renal allografts: morphological, clinical, and pathophysiological aspects. Adv Exp Med Biol 2006;577:190–200.

[32] Nickeleit V, Mihatsch MJ. Polyomavirus nephropathy in native kidneys and renal allografts: an update on an escalating threat. Transplant Int 2006;19:960–73.

[33] Awadalla Y, Randhawa P, Ruppert K, Zeevi A, Duquesnoy RJ. HLA mismatching increases the risk of BK virus nephropathy in renal transplant recipients. Am J Transplant 2004;4(10):1691–6.

[34] Priftakis P, Bogdanovic G, Tyden G, Dalianis T. Polyomaviruria in renal transplant patients is not correlated to the cold ischemia period or to rejection episodes. J Clin Microbiol 2000;38:406–7.

[35] Marshall WF, Telenti A, Proper J, Aksamit AJ, Smith TF. Rapid detection of polyomavirus BK by a shell vial cell culture assay. J Clin Microbiol 1990;28:1613–15.

[36] Hogan TF, Padgett BL, Walker DL, Borden EC, McBain JA. Rapid detection and identification of JC virus and BK virus in human urine by using immunofluorescence microscopy. J Clin Microbiol 1980;11:178–83.

[37] Hirsch HH, Brennan DC, Drachenberg CB, Ginevri F, Gordon J, Limaye AP, et al. Polyomavirus-associated nephropathy in renal transplantation: interdisciplinary analyses and recommendations. Transplantation 2005;79:1277–86.

[38] Randhawa P, Ho A, Shapiro R, Vats A, Swalsky P, Finkelstein S, et al. Correlates of quantitative measurement of BK polyomavirus (BKV) DNA with clinical course of BKV infection in renal transplant patients. J Clin Microbiol 2004;42:1176–80.

[39] Pang XL, Doucette K, LeBlanc B, Cockfield SM, Preiksaitis JK. Monitoring of polyomavirus BK virus viruria and viremia in renal allograft recipients by use of a quantitative real-time PCR assay: one-year prospective study. J Clin Microbiol 2007;45:3568–73.

[40] Schechter T, Liebman M, Gassas A, Ngan BY, Navarro OM. BK virus-associated hemorrhagic cystitis presenting as mural nodules in the urinary bladder after hematopoietic stem cell transplantation. Pediatr Radiol 2010;40 1430-1423.

[41] Bogdanovic G, Priftakis P, Giraud G, Kuzniar M, Ferraldeschi R, Kokhaei P, et al. Association between a high BK virus load in urine samples of patients with graft-versus-host disease and development of hemorrhagic cystitis after hematopoietic stem cell transplantation. J Clin Microbiol 2004;42:5394–6.

[42] Dropulic LK, Jones RJ. Polyomavirus BK infection in blood and marrow transplant recipients. Bone Marrow Transplant 2008;41:11–18.

[43] Galan A, Rauch CA, Otis CN. Fatal BK polyoma viral pneumonia associated with immunosuppression. Hum Pathol 2005;36:1031–4.

[44] Sandler ES, Aquino VM, Goss-Shohet E, Hinrichs S, Krisher K. BK papova virus pneumonia following hematopoietic stem cell transplantation. Bone Marrow Transplant 1997;20:163–5.

[45] Cubukcu-Dimopulo O, Greco A, Kumar A, Karluk D, Mittal K, Jagirdar J. BK virus infection in AIDS. Am J Surg Pathol 2000;24:145–9.

[46] Procop GW, Kohn DJ, Johnson JE, Li HJ, Loyd JE, Yen-Lieberman B, et al. BK and JC polyomaviruses are not associated with idiopathic pulmonary fibrosis. J Clin Microbiol 2005;43:1385–6.

[47] Major EO, Amemiya K, Tornatore CS, Houff SA, Berger JR. Pathogenesis and molecular biology of progressive multifocal leukoencephalopathy, the JC virus-induced demyelinating disease of the human brain. Clin Microbiol Rev 1992;5 49–73.

[48] Astrom KE, Mancall EL, Richardson Jr. EP. Progressive multifocal leukoencephalopathy. Brain 1958;81:93–127.

[49] Richardson Jr. EP. Progressive multifocal leukoencephalopathy. N Engl J Med 1961;265:815–23.

[50] Padgett BL, Walker DL. Virologic and serologic studies of progressive multifocal leukoencephalopathy. Prog Clin Biol Res 1983;105:107–17.

[51] Brooks BR, Walker DL. Progressive multifocal leukoencephalopathy. Neurol Clin 1984;2:299–313.

[52] Di Lernia V. Progressive multifocal leukoencephalopathy and antipsoriatic drugs: assessing the risk of immunosuppressive treatments. Int J Dermatol 2010;49:631–5.

[53] Warnke C, Menge T, Hartung HP, Racke MK, Cravens PD, Bennett JL, et al. Natalizumab and progressive multifocal leukoencephalopathy: what are the causal factors and can it be avoided? Arch Neurol 2010;67:923–30.

[54] Lysandropoulos AP, Du Pasquier RA. Demyelination as a complication of new immunomodulatory treatments. Curr Opin Neurol 2010;23:226–33.

[55] Bellizzi A, Barucca V, Fioriti D, Colosimo MT, Mischitelli M, Anzivino E, et al. Early years of biological agents therapy in Crohn's disease and risk of the human polyomavirus JC reactivation. J Cell Physiol 2010;224:316–26.

[56] Clifford DB, De Luca A, Simpson DM, Arendt G, Giovannoni G, Nath A. Natalizumab-associated progressive multifocal leukoencephalopathy in patients with multiple sclerosis: lessons from 28 cases. Lancet Neurol 2010;9:438–46.

[57] Tan CS, Koralnik IJ. Progressive multifocal leukoencephalopathy and other disorders caused by JC virus: clinical features and pathogenesis. Lancet Neurol 2010;9:425–37.

[58] Gheuens S, Pierone G, Peeters P, Koralnik IJ. Progressive multifocal leukoencephalopathy in individuals with minimal or occult immunosuppression. J Neurol Neurosurg Psychiatry 2010;81:247–54.

[59] Hammarin AL, Bogdanovic G, Svedhem V, Pirskanen R, Morfeldt L, Grandien M. Analysis of PCR as a tool for detection of JC virus DNA in cerebrospinal fluid for diagnosis of progressive multifocal leukoencephalopathy. J Clin Microbiol 1996;34:2929–32.

[60] Drachenberg CB, Hirsch HH, Papadimitriou JC, Gosert R, Wali RK, Munivenkatappa R, et al. Polyomavirus BK versus JC replication and nephropathy in renal transplant recipients: a prospective evaluation. Transplantation 2007;84:323−30.

[61] Miner AG, Patel RM, Wilson DA, Procop GW, Minca EC, Fullen DR, et al. Cytokeratin 20-negative Merkel cell carcinoma is infrequently associated with the Merkel cell polyomavirus. Mod Pathol 2015;28:498−504.

[62] Hayden RT, Yan X, Wick MT, Rodriguez AB, Xiong X, Ginocchio CC, et al. Factors contributing to variability of quantitative viral PCR results in proficiency testing samples: a multivariate analysis. J Clin Microbiol 2012;50:337−45.

[63] Randhawa P, Kant J, Shapiro R, Tan H, Basu A, Luo C. Impact of genomic sequence variability on quantitative PCR assays for diagnosis of polyomavirus BK infection. J Clin Microbiol 2011;49:4072−6.

[64] Hoffman NG, Cook L, Atienza EE, Limaye AP, Jerome KR. Marked variability of BK virus load measurement using quantitative real-time PCR among commonly used assays. J Clin Microbiol 2008;46:2671−80.

[65] Kapusinszky B, Chen SF, Sahoo MK, Lefterova MI, Kjelson L, Grimm PC, et al. BK polyomavirus subtype III in a pediatric renal transplant patient with nephropathy. J Clin Microbiol 2013;51:4255−8.

11

Molecular Testing for Respiratory Viruses

K.A. Stellrecht

Department of Pathology and Laboratory Medicine, Albany Medical College;
Albany Medical Center Hospital, Albany, NY, United States

INTRODUCTION

Respiratory tract infections (RTIs) are common and are associated with significant health burden. For example, pneumonia is the fourth leading cause of death globally and the leading infectious cause [1]. Despite being generally mild and self-limiting, the common cold is associated with an enormous economic burden, both in lost productivity and in expenditures for treatment [2]. The major viral agents of RTIs include influenza viruses A and B, respiratory syncytial virus (RSV), human metapneumovirus (HMPV), parainfluenza virus (PIV), adenovirus (AdV), rhinoviruses (RVs), enteroviruses (EVs), and human coronavirus (HCoV). Common to these viruses are their ability to infect airway epithelial cells, co-opt host cell proteins to facilitate infection, modulate both innate and adaptive immune responses, and to mediate proinflammatory responses which contribute to disease pathogenesis (Table 11.1). Yet, some of the unique features of these viruses can lead to diagnostic limitations.

MOLECULAR TARGETS

Influenza

Influenza viruses are some of the most important human pathogens, infecting hundreds of millions of people annually with 250,000−500,000 deaths worldwide [3]. As members of the *Orthomyxoviridae* family, these viruses are classified into three distinct types, A, B, and C viruses based on major antigenic differences, subdivisions based on antigenic characterization of the surface glycoproteins hemagglutinin (HA) and neuraminidase (NA). Currently, among the type A viruses, there are 16 HA subtypes and 9 NA subtypes.

Influenza infections are usually acute, self-limited, febrile illness which manifest clinically as fever, malaise, and cough with attack rates as high as 10−40% [4]. Their occurrence is generally seasonal with outbreaks of varying severity observed almost every winter. Pandemics have occurred in 1918, 1957, 1968, and 2009 and were caused by different antigenic subtypes of influenza A: H1N1, H2N2, H3N2, and again H1N1 (Fig. 11.1).

Historically, H3N2 is associated with higher mortality [4]. Alternately, other stains are associated with more severe infection among individuals with certain high-risk factors such as obesity, pregnancy, and other comorbidities [5,6]. Furthermore, specific viral mutations are associated with higher virulence and cell receptor binding, which affects their predilection for the upper (URTI) or lower respiratory tract infection (LRTI). The Glu222Gly substitution in the *HA* gene can be found in strains of avian influenza, and to a lesser extent, in some strains of 2009 H1N1 [7,8]. Whereas most strains of influenza replicate in the URT where α-2,6-linked sialic acid receptors predominate on cell surfaces, this amino acid substitution is associated with a greater affinity for α-2,3-linked receptors which are more abundant in the LRT, resulting in a greater risk for viral pneumonia [9−12]. Despite the greater number of influenza A hospitalizations, there appears to be no significant difference between influenza A and B in rates of high-risk conditions, median length of stay, intensive care unit (ICU) admissions, or deaths [13].

Human infection with zoonotic strains is more concerning as these strains have the potential to be more pathogenic, as seen with the avian H5N1 strains, and

© 2017 Elsevier Inc. All rights reserved.

TABLE 11.1 Natural History, Pathogenesis, and Clinical Presentation of Common Respiratory Viruses

	Seasonality in temperate climates	Incubation period	Duration of illness	Period of shedding[a]	Replication site	Cell receptor	Mechanism of pathogenesis	Presentation	Respiratory disease syndrome(s)	Extrapulmonary manifestation	Optimal specimen
Flu	Sharp annual peak lasting ~6 to 8 weeks	1–4 days (average = 2 days)	3–7 days Cough, malaise more than 2 weeks	Day −1 to day 10	1° ciliated columnar, also alveolar and dendritic	Predominately α-2,6-linked sialic acids	H5N1 and to a lesser extent 2009 H1N1, predominately LRT where α-2,3-linked sialic acids predominate on cuboidal bronchiolar cells	Abrupt onset of flu symptoms (fever, myalgia, headache, malaise, dry cough, pharyngitis, and rhinitis). Otitis media, nausea, and vomiting may also be observed	Flu Pneumonia	Cytokine related encephalitis	Usually NP aspirate, washes, and swabs Influenza pneumonia: include BAL, sputum, or throat
RSV	RSV broad peak 15–20 weeks (Oct–May)	3–8 days (average = 5 days)	7–14 days	Day −3 to day 14	1° ciliated columnar, also alveolar and dendritic	Heparan sulfate and nucleolin	Accessory proteins interfere with IFN pathways. Rapid inhibition of Na + transport, resulting in apical fluid accumulation	Begins with rhinorrhea, pharyngitis, cough, headache, fatigue, and fever. Bronchiolitis ±	URI with or without LRI Bronchiolitis Pneumonia Tracheobronchitis Croup	Rare encephalitis, myocarditis	NP aspirate, washes, swabs
HMPV	Late winter and early spring biennial pattern	4–6 days	5–10 days	1–2 weeks	1° ciliated columnar, also alveolar	Heparan sulfate	Accessory proteins interfere with IFN pathways. Rapid inhibition of Na + transport, resulting in apical fluid accumulation	Begins with rhinorrhea, pharyngitis, cough, headache, fatigue, and fever. Bronchiolitis	URI with or without LRI Bronchiolitis Pneumonia Tracheobronchitis Croup	Rare encephalitis	NP aspirate, washes, swabs
PIV	PIV1: biennial autumn; PIV2: also autumn; PIV3: endemic with spring time	2–7 days	7–10 days	Day −3 to day 20	Ciliated columnar	PIV1: sialic acids with terminal NeuAcα2-3Gal PIV3: sialic acids with terminal NeuAcα2-6Gal or NeuGcα2-3Gal	Accessory proteins interfere with IFN pathways. Rapid inhibition of Na + transport, resulting in apical fluid accumulation	Begins with rhinitis, pharyngitis, cough (croupy), and hoarseness, usually with fever PIV1 and PIV2: Croup PIV3: Bronchiolitis	PIV1 and PIV2: URI with or without LRI Croup Pneumonia Bronchiolitis PIV3: URI with or without LRI Pneumonia Croup	Rare meningitis, hepatic infection	NP aspirate, washes, swabs

											Affected sites:
AdV	Endemic with winter or early spring epidemics	2–14 days		3–6 weeks, some months to years	Nonciliated epithelial persistence within lymphocytes	A, C, E, and F: CAR B and D: CD46	Destruction of respiratory epithelial cells Disruption of the integrity of cell–cell contact	Fever, pharyngitis, exudative tonsillitis, and cough with or without diarrhea, vomiting and abdominal pain and/or with or without conjunctivitis	URI with or without GI Conjunctivitis Pneumonia Croup Bronchiolitis	Conjunctivitis, GI, cystitis, rare meningitis, myocarditis, myositis	NP aspirate, washes, swabs, throat, BAL, urine stool, blood
RV	Endemic with peaks in fall (Aug–Sep) and spring (Apr–May)	1–7 days (average = 2 days)	10–14 days (average = 10 days)	Day –1 to day 14 some 3 weeks	1° ciliated epithelial, also nonciliated	RV A and B: ICAM-1 and LDLR RV C: other but unknown RV C: other but unknown	Nonspecific host inflammatory responses	Rhinorrhea, pharyngitis, cough, headache, malaise, mild fever	URI Asthma exacerbation Bronchiolitis Pneumonia	Rare meningitis, myocarditis	NP aspirate, washes, swabs, tracheal or bronchial aspirate, BAL
HCoV	Typical HCoV: endemic with peaks in winter to early spring. Peaks in 2–4 years SARS & MERS: zoonotic	Typical HCoV: 2–5 days (average = 3 days) SARS: 4–7 days	Typical HCoV: 3–18 days (average = 7 days) SARS: 7–21 days	Typical HCoV: day 1 to day 21 MERS: Day 1 to day 33	Typical HCoV: ciliated epithelial MERS: alveolar and blood vessel endothelium	229E: human aminopeptidase N (hAPN); OC43: carcinoembryonic antigen (CEA) NL63: ACE2 HKU1: HLA-C SARS: ACE2 MERS: dipeptidyl Peptidase IV (DPP4)	Typical HCoV: destruction of upper respiratory epithelial cells SARS: diffuse alveolar damage	Typical HCoV: rhinorrhea, pharyngitis, cough, headache, malaise, mild fever SARS and MERS: fever, cough, dyspnea, malaise, headache. Diarrhea in some	Typical HCoV: URI Pneumonia Bronchiolitis SARS and MERS: Pneumonia GI Renal failure (MERS)	Typical HCoV: None SARS: GI, kidney, liver	Typical HCoV: SARS and MERS: NP and throat, BAL, sputum serum, stool

aShedding in children is generally longer than adults. Immunocompromised can shed virus for weeks to months.

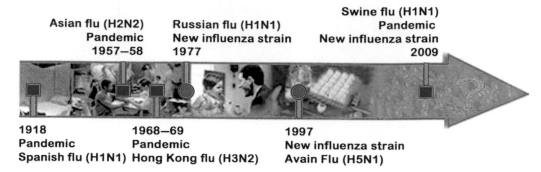

FIGURE 11.1   Timeline of human flu pandemics. ■ Major pandemic; ● The appearance of influenza strain in the human population. *Source: Adapted from http://www.niaid.nih.gov/topics/Flu/Research/Pandemic/Pages/TimelineHumanPandemics.aspx.*

they have the potential to be the source of the next pandemic due to low levels of immunity in the population. Human infection with many of these strains is associated with unique presentations, such as conjunctivitis with H7 stains, and atypical symptoms like nausea, vomiting, encephalopathy, and bleeding gums and nose with H5 strains [4], which may delay clinical diagnosis and recognition of zoonotic transmission. Interesting, single amino acid changes appear to be responsible for changes in host range [14].

Typically, influenza infections present with systemic symptoms, fever and myalgia, along with upper airway symptoms, such as pharyngitis and dry cough. They usually begin with an abrupt onset of symptoms after an incubation period of 1–2 days and last 4–5 days. However, prolonged infection with or without disease has been reported to last weeks to months in immunocompromised individuals. Less commonly, the virus infects the lung, either via contiguous spread from the URT or via inhalation, causing primary viral pneumonia. Influenza pneumonia frequently requires ICU admission and mortality is high [4]. Secondary bacterial pneumonia is a well-recognized complication of viral pneumonia and accounts for a large proportion of the morbidity and mortality of viral LRT disease, especially in adults. Bronchiolitis and croup may also occur with influenza infection, albeit much less frequently than RSV and PIV. Influenza can be associated with exacerbation of chronic pulmonary diseases such as chronic bronchitis, asthma and worsening pulmonary function in children with cystic fibrosis.

Nonpulmonary complications include myocarditis and pericarditis, as well as exacerbations of other underlying disease such as chronic heart failure and chronic renal disease [15]. Myocarditis is not highly uncommon during influenza infection and may present as asymptomatic myocardial involvement to fulminant myocarditis resulting in cardiogenic shock and death [16]. Central nervous system involvements

include the rare occurrence of transverse myelitis and encephalitis which appear to be immune rather than viral mediated [17]. Guillain-Barré syndrome is also associated with immune mechanisms following influenza infection [18].

Paramyxoviruses

RSV and HMPV are from the *Pneumovirinae* subfamily of the *Paramyxoviridae* family. RSV is the major cause of LRT illness in young children and is associated with an estimated 132,000–172,000 pediatric hospitalizations in the United States annually [19] and globally it is an important cause of death [20]. Most infants (50–69%) are infected during the first year of life and virtually all are infected by age 2 [21]. HMPV also causes a broad range of URTI/LRTI, which are clinically indistinguishable from RSV. It accounts for about 1–5% of childhood URTI and 10–15% of hospitalizations for LRTI, depending on age group and year of study [22–24]. Primary infection with HMPV tends to occur at a slightly older age than RSV and by age 5 most children have been infected [25,26].

PIVs also belong to the *Paramyxoviridae* family and are classified as four types and two subtypes (PIV1, 2, 3, 4a, and 4b). PIV1 and to a lesser extent PIV2 are the most significant cause of croup while PIV3 is a significant cause of bronchiolitis, bronchitis, and pneumonia. Indeed these viruses accounted for 6–8% of all hospitalizations for fever or acute respiratory illnesses in children less than 5 years of age [27]. By 5 years of age most children have antibodies against PIV3 and approximately 75% have antibodies against PIV1 and PIV2.

Primary infections with paramyxoviruses are usually symptomatic and present as URTI beginning 2–8 days after infection through the nose or eyes. Although all these viruses replicate in the ciliated columnar cells of the nasopharyngeal (NP) tract [28–30], it is believed that varying cell receptor usage, including sialic acid containing molecules usage by different PIV strains, likely contributes to the differences in pathogenesis [31,32]. The viruses may then spread to the LRT within 1–3 days as the result of viral impairment of the ciliary epithelium [33]. Paramyxovirus pathogenesis is then associated with necrosis and sloughing of the ciliated epithelial cells which along with edema and increased mucus secretion, obstructs airway, and leads to airway hyperresponsiveness [34,35].

LRTI with RSV and HMPV occurs in 25–40% of cases and manifests most commonly as bronchiolitis, followed by pneumonia and tracheobronchitis, and lastly croup [21,26,36]. Risk factors for bronchiolitis requiring hospitalization include young age, prematurity, male sex for RSV and female sex for HMPV, chronic illness, lower socioeconomic status, smoke exposure, and asthma [21,37–39]. PIV develops into LRTI in 15–25% of cases [40]. There is a tendency for PIV1 and PIV2 to involve the larynx and upper trachea, resulting in the croup syndrome, while PIV3 spreads to the small air passages with the development of bronchopneumonia, bronchiolitis, and/or bronchitis when it is associated with severe disease [27]. There is compelling evidence that the level of virus replication correlates to the disease severity, but innate immune responses also appear to be important [41–43]. HMPV infection appears somewhat milder than that of RSV, but dual HMPV and RSV infections have been reported as more severe than with either virus alone [44,45]. Among the two antigenic subgroups, RSV A is associated with more severe disease than subgroup B [46,47], while the severity of illness associated with HMPV A is similar to HMPV B infection [48].

Reinfection with paramyxoviruses occurs throughout life and is usually present as mild URTI in children and adults with RSV, HMPV, and PIV causing about 7%, 2%, and 5% of acute respiratory illnesses in adults, respectively [49,50]. Reinfection in immunocompromised individuals has a higher risk of more serious disease. Extrapulmonary manifestations from paramyxoviruses are rare and controversial. However, there have been a few reports of paramyxoviruses in CSF in cases of encephalitis or meningitis, as well as in myocardium and liver [43,51–53].

Adenovirus

Human AdVs, belonging to the genera *Mastadenovirus*, are further divided into seven species (A through G) and 57 types [54]. These viruses cause a broad range of clinical syndromes, with groups A, B, C, and E causing 5–10% of pediatric and 1–7% of adult URTI and LRTI [55]. Several group B AdVs, including serotypes 3, 7, 14, and 21, have caused outbreaks of acute respiratory disease (ARD). Although fatal AdV infections in immunocompetent adults are rare, ARD outbreaks due to a virulent strain of serotype 14 in 2006 and 2007 was associated with a significant number of ICU admissions and deaths in previously healthy young adults [56].

Approximately 50% of all AdV infections result in subclinical disease, and most symptomatic infections are mild and self-resolving within 2 weeks [57]. AdV infection begins with replication in nonciliated respiratory epithelium of the tonsils and adenoids [54]. A brief period of viremia ensues. URTI symptoms in children and young adults include fever, pharyngitis, tonsillitis, and cough, with or without GI symptoms or conjunctivitis [55]. Disruption of the integrity of cell–cell contact enables infection of other cells of the respiratory tract [54]. Worldwide, pneumonia occurs in up to 20% of young children with fatality rates for severe AdV pneumonia exceeding 50% [55]. AdVs utilize cell receptors that are abundantly expressed in epithelial cells in multiple organs or tissues (CAR for groups A, C, E, and F, and CD46 for groups B and D) [58,59]. Hence, extrapulmonary manifestations are common in normal host and include conjunctivitis, GI illness, and cystitis, as well as the more rare occurrences of meningitis, myocarditis, and myositis. AdV can persist as a latent infection for years after an acute initial infection and may reside in lymphoid tissue, renal parenchyma, or other tissues [55]. Reactivation may occur in severely immunosuppressed patients.

AdV causes considerable destruction of respiratory epithelial cells due to inhibition of cellular DNA, mRNA, and protein synthesis resulting in the formation of characteristic smudge cells with enlarged nuclei containing basophilic inclusion bodies surrounded by thin rims of cytoplasm [54]. The penton base structural protein, which causes cells to detachment in vitro, may be involved in pathogenesis in vivo.

EVs—Including RVs and Human Parechoviruses

EVs and human parechoviruses (HPeVs) of the *Picornaviridae* family are associated with RTI in addition to a wide array of other disease. In fact, EVs are responsible for up to approximately 19% of LRTI in hospitalized children [60]. Human infections are associated with four species of EV (EV A–D), three species of RVs (RV A–C) from the EV genus and one species from the HPeV genus (HPeV A). Although strains from all species may infect the respiratory tract, EV C (C104, C109, C117), EV D (D68), RV A, and RV C are associated with more serious respiratory disease.

RV is undoubtedly the most commonly detected respiratory virus in all age groups, accounting for 25% of all respiratory infections, with asymptomatic infection occurring in at least 20% of healthy individuals

[61]. RV preferentially infects the URT, primarily the paranasal sinuses and nasopharynx. One to three days after infection, URTI frequently begins as a sore or scratchy throat followed by nasal obstruction and rhinorrhea with cough, headache, malaise, and sometimes fever. Large- and medium-sized airways also maintain high-level RV replication [62]. As a result, RV is associated with bronchiolitis in infants and exacerbations in patients with chronic asthma. LRTI such as pneumonia, croup, and bronchitis also occur and result in a significant number of hospitalizations [63]. Cytopathogenicity of this virus is low and pathology is primarily due to nonspecific host inflammatory responses.

Coronaviruses

Most members of *Coronaviridae* family infecting humans (229E and OC43 from the alpha-CoV genus, NL63 and HKU1 from the beta-CoV genus) cause mild URT diseases. In fact, these typical HCoV cause up to 30% of all URTIs [64]. However, two novel beta-CoV, severe acute respiratory syndrome associated CoV (SARS-CoV), and Middle East respiratory syndrome CoV (MERS-CoV) cause serious viral pneumonitis, leading to hospitalization and death with overall mortality rates of 10% and 30%, respectively [65].

Infection with the typical HCoV, of which 30% are asymptomatic, begins with replication in the ciliated epithelial cells of the nasopharynx. Direct destruction of ciliated epithelial cells in conjunction with innate immune responses produces rhinorrhea, pharyngitis, cough, headache, malaise, and mild fever 2–5 days

after infection. These viruses have also been associated with severe pneumonia and bronchiolitis in neonates and the elderly, especially those with underlying illnesses. In addition, HCoV-NL63 is also an important cause of croup [66]. Infection frequently occurs in young children with seropositivity in 50% of school-age children [64]. Reinfection as well as coinfection is common.

SARS begins with fever, headache, malaise, or myalgia, followed by nonproductive cough and dyspnea in a few days to a week after onset of symptoms. Although the upper airway is also infected, there is little epithelial cell damage and URT disease is lacking. Virus rapidly spreads to the alveoli, causing diffuse alveolar damage leading to pneumonia and ARDS in 25% of cases [67]. Diarrhea is common. MERS is also associated with a biphasic illness strikingly similar to SARS except for more frequent renal failure [68]. Most patients who are hospitalized with SARS and MERS have chronic comorbidities. Interestingly, asymptomatic infections with both viruses have been reported [69,70].

CLINICAL UTILITY OF MOLECULAR DIAGNOSTICS FOR RESPIRATORY VIRUS INFECTION

Respiratory viruses can infect both the URT and the LRT (Fig. 11.2) and tend to cause distinct clinical syndromes based on their tropism for different sites of the respiratory tract. Most commonly these viruses only

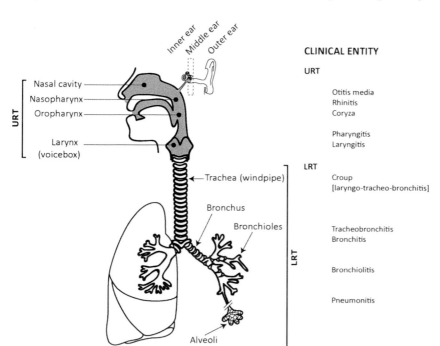

FIGURE 11.2   Schematic representation of the human respiratory tract. The upper (shaded pink) and lower respiratory tract (URT/LRT) and the components of the ear are indicated. The approximate locations of URT and LRT diseases associated with respiratory virus infection are indicated. *Source: Reprinted with permission from Caister Academic Press (from Mackay IM, Arden KE, Nissen MD, Sloots TP. Challenges facing real-time PCR characterization of acute respiratory tract infections. In: Mackay IM, editor. Real-time PCR in microbiology: from diagnosis to characterization. Caister Academic Press; 2007. pp. 269–318).*

infect the URT, and when LRT infection does occur, it is most often due to contiguous spread.

Upper Respiratory Tract Infections

The Common Cold

The common cold refers to a syndrome of upper respiratory symptoms that may be caused by a variety of viral pathogens. These symptoms include nasal blockage, runny nose, sneezing, cough, and sore throat, sometimes with headache or other body aches, and typically begin 1–3 days after infection. Fever and other constitutional symptoms are more often seen in URTIs associated with influenza, RSV, HMPV, and AdV. Colds usually last about 1 week, but virus shedding can persist for 2–3 weeks. Otitis media can develop from URTI with any of these viruses and can due to secondary bacterial infection or direct viral infection. Indeed, virus can be detected in middle ear fluids with RSV, influenza, HCoV, and RV being the most common [71].

The pathogens most frequently associated with common cold symptoms are the EV/RV, which cause approximately half of all colds in children and almost three-quarters of colds in adults, and HCoV (Table 11.2). It is often forgotten that influenza viruses can present with only mild URTI symptoms and is in fact a common cause of the cold. Other important pathogens that are also associated with cold symptoms include AdV, RSV, HMPV, and PIV. Coinfections are common.

Although generally mild and self-limited, these illnesses are associated with an enormous economic burden both in lost productivity and in expenditures for treatment. Hence, attempts have been made to create and market antiviral agents targeting causes of the common cold, particularly EV/RV [72]. Due to the lack of success in therapeutic interventions, diagnostic testing outside of epidemiological investigations is not warranted.

Influenza-Like Illness

Influenza-like illness (ILI) is on the other end of the spectrum of URTIs and is defined as the presence of fever of greater than or equal to 100°F, in addition to cough or sore throat, in the absence of an alternative cause. After an incubation period of 1–4 days, there is an abrupt onset of constitutional and respiratory signs and symptoms which generally lasts 5–7 days. The constitutional symptoms can include malaise, body aches, headache, loss of appetite, and nausea and are generally due to cytokines released by immune system activation. Interestingly, influenza only causes 35–45% of ILI cases during peak seasons. But many other viral infections can present as flu-like, particularly RV/EV and RSV (Table 11.2).

Appropriate treatment of patients with respiratory illness depends on accurate and timely diagnosis. Early diagnosis of influenza can reduce the inappropriate use of antibiotics, provide the option of using antiviral therapy and is an important infection prevention measure. The causative agent of ILI is difficult to determine on the basis of signs and symptoms alone. Sensitivity and predictive value of clinical definitions vary, depending on the prevalence of other respiratory pathogens and the level of influenza activity. Among generally healthy adults living in areas during the peak of influenza activity, the positive predictive value (PPV) of a simple clinical definition of influenza (acute onset of cough and fever) can be over 80%. However, the presentation in children, the elderly, and individuals with comorbidities is less likely to be typical, in

TABLE 11.2 Relative Rates of Respiratory Viruses Among Respiratory Tract Syndromes

Virus	Common cold (pediatric/adult)	ILI[a] (all ages)	Croup (pediatric)	Bronchiolitis (pediatric)	Pneumonia (pediatric/adult)
AdV	5–10/1	0.4–9	1	1–8	1–10/3–13
HCoV	10–15/11	0.2–10	2[b]	1–8	3–7/6–13
Influenza	25–30/8	8–52	9	1–10	4–22/21–31
HMPV	1–5/1	0.2–10	<1	3–12	1–13/3–22
PIV	1–5/5	0.4–11	42[c]	1–3	8–28/6–14
RSV	1–5/3	0.4–19	15	70–80	3–45/13–24
RV/EV	40–50/71	4–29	21	15–35	3–45/13–24
References	[74,124]	[79,138,139]	[140]	[141–143]	[77,144]

[a]*Influenza-like Illness = fever, myalgia, pharyngitis, and dry cough.*
[b]*Frequencies not yet well established.*
[c]*PIV1 = 31%, PIV2 = 5%, PIV3 = 6%.*

which case the PPV of clinical impression can be as low as 17—30% in these populations [73].

Dagnostic testing is not needed for all patients with ILI to make antiviral treatment decisions once high levels of influenza activity have been identified in the region. For most outpatient and emergency room settings, results for molecular assays are generally not available to assist in clinical decision making. Fortunately, that paradigm is changing with advent of rapid tests which provide a wide panel of results in approximately 1 h, or 20-min point-of-care devices for influenza. But generally, molecular testing has been considered to be most appropriate for hospitalized patients if a positive test would result in a change in clinical management, including infection control practices.

Lower Respiratory Tract Infections

Croup

Croup is a common childhood disease characterized by sudden onset of a distinctive barky cough that is usually accompanied by inhalation stridor, hoarse voice, and respiratory distress resulting from upper airway obstruction that worsens at night. Although the illness is generally mild and short-lived, the presentation in a child is alarming. In fact 85% of cases typically present mild croup and fewer than 5% are hospitalized [74].

Typically, this disease affects children between 3 months and 3 years of age. Frequently it begins with a nonspecific URTI 12—48 h prior to the development of classic symptoms. The barky cough resolves within 3—4 days for 60% of cases, but some patients will continue to have symptoms for up to 1 week [74].

Although present year-round, croup often presents with biannual peaks in late autumn and again in spring, particularly in odd-numbered years, correlating with the prevalence of PIV (Fig. 11.3). Of the PIV strains, type 1 is the primary cause of croup, followed by type 3, and then 2 [74]. This finding appears contradictory since type 3 is usually associated with bronchiolitis. However, this observation is easily explained by the greater prevalence of type 3 virus over type 2 virus. Other viruses implicated in the disorder include influenza, AdV, RSV, HMPV, and HCoV-NL63. In addition, measles remains an important cause of croup in nonimmunized children. RV coinfection is frequent.

Croup is a clinical diagnosis. Laboratory tests are not needed to confirm the diagnosis. Laboratory analysis generally should be limited to tests necessary for management of a more severely ill child. Viral identification may be warranted when specific antiviral therapy is being considered, such as for severely ill or high-risk children with influenza.

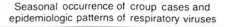

Seasonal occurrence of croup cases and epidemiologic patterns of respiratory viruses

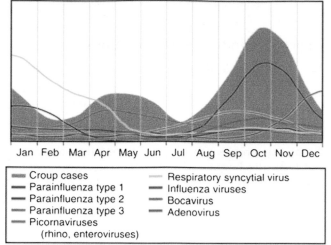

- Croup cases
- Parainfluenza type 1
- Parainfluenza type 2
- Parainfluenza type 3
- Picornaviruses (rhino, enteroviruses)
- Respiratory syncytial virus
- Influenza viruses
- Bocavirus
- Adenovirus

FIGURE 11.3 The seasonal occurrence of croup cases is shown in relation to the epidemiologic activity of the respiratory viruses associated with croup. *Source: Reprinted from Hall and McBride (Bower J, McBride JT. Croup in children (acute laryngotracheobronchitis). In: Bennett JE, Dolin R, and Blaser MJ, Editors. Mandell, Douglas, and Bennett's principles and practice of infectious diseases, 8th ed. Philadelphia, PA: Saunders/Elsevier; 2015, pp. 762—766).*

Bronchiolitis

Bronchiolitis is the most common acute viral LRT illness in children less than 2 years of age. Clinical signs and symptoms of bronchiolitis include rhinorrhea, cough, tachypnea, wheezing, rales, and increased respiratory effort which typically lasts 3—7 days. There is commonly a prodromal URTI (coryza, cough, and mild fever) which lasts for several days. Complications with bronchiolitis, such as apnea and aspiration, occur most frequently in infants within the first several months of life, in premature infants, and in children with chronic conditions. Indeed, it is the most common cause of hospitalization among infants during the first 12 months of life. Although the hospitalization rates for bronchiolitis have been increasing, mortality rates have declined [75].

Peak occurrence of bronchiolitis is during the winter to early spring, and usually correlates with the prevalence of RSV, which causes of about 70—80% of cases. RV/EV and HMPV are other leading cause of bronchiolitis, but all respiratory viruses have been associated with bronchiolitis (Table 11.2), and a considerable fraction of cases (30%) involve multiviral infections. Again, bronchiolitis is a clinical diagnosis and laboratory tests are not needed to confirm the diagnosis. In fact, the American Academy of Pediatrics recommends against radiographic or laboratory studies routinely [76].

Pneumonia

Pneumonia is a common illness with high morbidity and mortality, particularly in children less than 5 years old and in adults over 75. Viruses are more commonly associated with pneumonia in children, particularly influenza, RSV, RV, HMPV, and PIV (Table 11.2) [77]. The prevalence of the causative agents is age-dependent, with RSV and PIV being more common causes of pneumonia in children less than 2 years old than in older children. Dual viral infections are common, and a third of children have evidence of viral—bacterial coinfection, particularly with *Streptococcus pneumoniae* and *Staphylococcus aureus*. In adults, viral agents are an important cause of pneumonia in the elderly, although historically their role has been underestimated given the insensitivity of antigen assays and viral culture in this population. As the result of nucleic acid amplification testing, it is now evident that viruses, in particular influenza viruses, RVs, and coronaviruses, are the putative causative agents in a third of cases of community-acquired pneumonia [78,79].

Viral infection of the lung can be the result of either contiguous spread from the URT or by direct inhalation, with the former beginning with typical URTI symptoms followed by a rapid progression of fever, cough, dyspnea, and cyanosis. Nonrespiratory symptoms include fatigue, sweats, headache, nausea, and myalgia. With increasing age, both respiratory and nonrespiratory symptoms of pneumonia become less frequent. Primary viral pneumonia frequently requires ICU admission and mortality is high. The diagnosis of pneumonia is determined clinically and confirmed by radiographic imaging, but identification of the etiological agent is important and recommended by the Infectious Diseases Society of America and the American Thoracic Society [80]. Indeed, viral pneumonia cannot be differentiated from bacterial pneumonia clinically, particularly in the elderly. Furthermore, secondary bacterial infection with certain bacteria may be virus-specific, increasing the need to know the causative agent [81].

MOLECULAR TECHNOLOGIES AND LIMITATIONS OF TESTING

Since most cases of viral respiratory infection (VRI) are associated with mild, self-limiting illness, laboratory testing is not necessary. However, for more serious cases, such as those requiring hospitalization or therapy, rapid laboratory diagnosis of the etiological agent can be important. Viral diagnostics can guide therapy, potentially eliminating unnecessary use of antibiotics and enabling the use of antivirals when appropriate. In addition, knowledge of the causative agent is important for infection control interventions to minimize the risk of nosocomial spread. Nucleic acid amplification tests (NAATs) have become the test of choice for VRI because rapid antigen tests have low sensitivities [82—86] and viral culture, which can take 3—10 days, lacks utility for patient management. Furthermore, NAATs have superior sensitivity and specificity in both pediatrics and adults, and results can be obtained within minutes to hours [87—89]. Not only has NAAT revolutionize the detection of traditional respiratory viruses with its exquisite sensitivity, but it enabled the discovery of new respiratory viruses, such as HMPV, many HCoV, and RV C group.

More recently, multiplexed molecular assays have been developed in order to diagnose a large number of respiratory viruses in single assays. As an added benefit, viruses that could not be detected by conventional virology have been included which further increases the diagnostic yield. Refer to the review by Gaydos [90] for details regarding the performance and workflow of many of these systems. The principle differences among NAATs are the throughput, turnaround time, ease of use, automation, versatility, use of a closed system to reduce contamination and cost.

Early problems with NAAT included lack of sensitivity for specific subtypes of AdV, the inability to differentiate RV from EV, and contamination issues with open platforms [91]. As expected manufacturers made or are working to make improvements, such as enhancing the range of AdV strain detection and reducing to eliminating postamplification processing [92]. In addition, point-of-care tests are now available, some with 20-min turnaround times.

The advantage of NAAT for VRI in terms of cost reduction is still unclear. Whereas rapid antigen tests have been shown to reduce length of stay, performance of ancillary diagnostic tests, and antibiotic consumption, the same cannot be said for multiplex assays, despite their higher sensitivity and specificity, and capacity to detect an extended range of viruses [91,93]. Initially, available multiplex systems were geared toward batched workflow usually performed once or twice per day. Subsequently, on-demand amplification methods, with potential turnaround time of 1—2 h, have come to market and are replacing batched test systems. Small studies have begun to show that identifying viral pathogens within a few hours does impact antibiotic or antiviral use and reduces labor cost in the Emergency Department [94,95]. More studies are needed to see if these results hold true.

Specimen Collection

The general rule for optimal specimens for the diagnosis of viral infection dictates that the specimens originate from the site of viral replication. Respiratory viruses are no different and since site of replication for these viruses is primarily the URT, in particular the NP region, it is best to sample that region for diagnostic testing. Of the URT samples, NP aspirates have traditionally been considered the most sensitive specimen for the detection of respiratory viruses [96,97]. However, a recent review by Jartti et al. [98] indicates that at least in children all NP samples, aspirates, washes, swabs, or brushings, have statistically equal sensitivity for NAAT, particularly when flocked swabs are used [99]. Historically, virus recovery in adults is much more difficult than in children because virus titers tend to be much lower in adults. The analytical sensitivity of NAAT appears to negate this concern [100]. Similarly, when flocked nasal swabs are used the sensitivity is similar to NP specimens in both pediatric and adult patients [100,101]. In addition, self-collected (in adult patients) or parent-collected flock nasal swabs specimens also show equivalency, opening the door for point-of-care devices [102−104].

Respiratory viruses can also be isolated from throat swabs or washes. Although the viral yield is typically lower than that seen with NP specimens, combining a throat swab and an NP swab may improve virus detection [87,105,106], and the general consensus is that throat swab alone is not recommended for most viruses. Exceptions include AdV, which replicates in the tonsils, and avian influenza, which primarily replicates in the LRT.

Calcium alginate swabs and swabs with wood shafts should not be used for respiratory specimen collection because they may interfere with NAATs. Specimens should be placed in sterile viral transport medium and refrigerated until transported to the laboratory for testing as soon as possible. However, some NAAT assays are approved for room temperature transport provided it occurs within a few hours. Clinicians should be aware of the approved clinical specimens, as well as specimen storage and transport, for the molecular assay being ordered. Freezing and thawing should be avoided or minimized to avoid degradation of virus particles, exposing the viral RNA to nucleases. Also viral integrity is needed if viral culture is to be performed, for example, for influenza resistance testing.

For cases of LRTI, sputum, endotracheal aspirates, or bronchoalveolar lavage specimens can increase the PCR diagnostic yield and should certainly be considered when URT specimens yield negative results but the suspicion is high. This is particularly true for viral pneumonia due to influenza, especially for cases of LRTI due to inhalation rather than spread from the URT. In fact, false-negative NP test occur in 10−35% of patients with viral pneumonia [107,108]. However, LRT specimens are not recommended for routine use as the diagnostic yield will not significantly improve. Furthermore, specificity is not necessarily improved with LRT specimens, particularly with the use of NAAT, as virus can be detected in LRT samples from asymptomatic children [98]. Lastly, none of the commercially-available tests have been validated for use with LRT specimens.

Although some respiratory viruses are shed from other sites such as urine or stool, as seen with AdV, these specimen types are not recommended for the diagnosis of respiratory illness. The only exceptions are SARS- and MERS-CoV where stool specimens may provide additional diagnostic yield. Rare occurrences of extrapulmonary manifestations have been reported with some respiratory viruses. Indeed, there have been anecdotal reports of respiratory virus detected in the CNS, myocardium, liver, and other sites [109]. Most often extrapulmonary syndromes are not due to a direct viral effect, but rather due to cytokine release as seen with influenza-associated encephalitis or myocarditis. In such cases, influenza RNA is almost never detected in CSF or myocardium. Likewise, rare cases of HMPV-associated encephalitis have been reported, but viral RNA has not been detected in CSF [53]. In contrast, AdV replicates in multiple organs and tissues. For example, AdV is often detected in urine, CSF, or myocardium in cases of AdV-associated hemorrhagic cystitis, meningitis or encephalitis, and myocarditis [55,110]. AdV DNA can even be found in serum during respiratory illness [55].

Timing of Disease/Virus Shedding

Generally peak respiratory virus shedding occurs on the first or second day of acute illness and generally declines substantially after 4 days [111−115]. However, duration varies with virus, patient age, severity of illness, comorbidities, and immune status. Often hospitalized patients, particularly those with LRTI, have higher viral titers and shed virus longer [111−114,116,117]. NAATs detect viral targets for a longer duration than other test methods and it is not unusual to detect viral nucleic acid a couple of weeks after infection, albeit the mean duration is generally 6−14 days [111−114,118,119]. Of the Paramyxoviruses, the duration of shedding for HMPV may be relatively shorter while PIV3 maybe longer [115,120]. Some viruses, particularly AdVs and picornaviruses, exhibit prolonged shedding in both asymptomatic and symptomatic patients, which can be a diagnostic conundrum [119,121].

Prolonged shedding of all respiratory viruses is not uncommon in severely immunocompromised patients and viral nucleic acids have been detected months after infection [109,118,119,122]. Prolonged shedding of influenza can be observed in this population even in the presence of treatment with antiviral agents. This then has been associated with the development of drug resistance mutations and subsequent community spread of resistant strains [118].

The literature has been inconsistent about the correlation between viral loads and disease severity [91,93]. Part of this discrepancy may be due in part to variances in the viruses themselves. More studies are needed to interpret the significance of viral loads. Likewise, there are mixed reports concerning the associations between infections with multiple viruses and more severe disease [91,93]. Indeed, asymptomatic, prolonged shedding associated with AdVs and picornaviruses complicates the interpretation.

Seasonality

In regions with temperate climates, the seasonal incidence of respiratory viruses is as diverse as the number of species associated with RTI, but the majority of infections occur between fall and early spring. In tropical climates, infections occur year-round or with increased incidence during the rainy season. The seasonal diversity of respiratory viruses is most evident with epidemiologic patterns of respiratory viruses associated with croup (Fig. 11.3) [123]. Influenza tends to produce a sharp annual peak lasting 6–8 weeks, while RSV tends to have a longer duration on the order of 15–20 weeks [109]. HMPV typically appears in late winter through early spring with a biennial pattern of epidemics [24]. Classically, PIV1 causes autumn epidemics in odd-numbered years and is sometimes accompanied by PIV2. PIV3 is more endemic with peaks in spring to early summer. Seasonality of PIV4 has not been as well characterized [109]. AdV infections occur throughout the year, but most epidemics occur in the winter or early spring [55]. EV infections usually occur in late summer to early fall, but those associated with respiratory disease also tend to be associated with sporadic outbreaks which can occur year-round. RV infections also occur throughout the year, but distinct peaks of illness are seen in the fall and spring [124]. HCoV are more endemic, but a bit more common in the cooler months [109]. Factors affecting seasonality in temperate climates are most likely due to environmental factors such as low temperatures and humidity, as well as social factors associated with colder months such as crowding indoors [125,126].

Interpretation of NAAT Results

Despite the high sensitivities and specificities of NAAT for respiratory virus detection, false-negative results can occur due to improper specimen collection or handling. A negative result can also occur when the patient is no longer shedding detectable virus, or at least at the site of collection. For hospitalized patients with LRT disease, if no other etiology is identified and viral pneumonia is still clinically suspected, the CDC recommends collecting LRT specimens. Sequence deviations or mutations at the site of primer or probe binding are also a potential source of false-negative results. In 2015, the majority of circulating influenza virus in the United States was characterized as A/Switzerland-like H3N2 viruses with significant genetic drift, loss of vaccine protection, and reduced ability to culture in many cell lines. Indeed, matrix gene primer or probe mismatches affect the performance of some commercial NAATs [127]. On the other hand, sequence deviations in the *H1* gene affected typing of A(H1N1)pdm09, leading to the serendipitous discovery of a strain of influenza A that cannot be typed. Similarly, pan-detection NAATs may not adequately detect all subtypes within a family of virus as commonly seen with commercial assays for the detection of AdV [128–130].

False-positive results, although rare, can occur (eg, due to lab contamination or other factors). A positive result indicates detection of viral nucleic acid, confirming virus infection, but does not necessarily mean the virus is the causative agent. Furthermore, patients vaccinated by intranasal administration of live attenuated influenza virus will likely test positive for 7–10 days [131].

Antiviral Resistance Testing

Antiviral resistance among influenza strains is a public health concern because resistant strains have spread rapidly in the community, quickly becoming the predominant virus [132,133]. Currently, circulating influenza A (H3N2) and 2009 H1N1 viruses are primarily susceptible to oseltamivir and zanamivir, but are resistant to the adamantanes (amantadine and rimantadine) [133]. However, prior to the 2009 pandemic the seasonal H1N1 virus developed into a predominantly oseltamivir-resistant strain [132]. Luckily this virus was susceptible to the adamantanes. Although only sporadic cases of oseltamivir resistance have been observed in isolates of A(H1N1)pdm09, this virus does have the sample potential as the seasonal H1 strain to become universally resistant.

Phenotypic susceptibility testing remains the gold standard for the assessment of viral resistance, but some of the more common resistance mutations have been identified which are useful for more rapid identification. For example, a histidine to tyrosine amino acid substitution at residue 275 of the NA protein (H274Y in N2 numbering; H275Y in N1 numbering) is associated with oseltamivir resistance [132]. Similarly, a change in amino acid 31 in the M2 gene product is associated with resistance to adamantanes [134]. Other potentially important mutations include a D199E mutation which is associated with reduced susceptibility to oseltamivir in seasonal H1N1 [135], a D198G (universal numbering, equivalent to site 199) mutation in H5N1 is associated with reduced susceptibility to oseltamivir and zanamivir [136], and a D198N mutation in influenza B virus is associated with high oseltamivir resistance [137].

It is important to point out that such testing is not available in most clinical laboratories and is generally performed by some public health laboratories or at the CDC for epidemiological purposes. Detection of the H275Y mutation is usually determined by a pyrosequencing assay developed by the CDC, while Sanger sequencing is used to assess mutations in the *NA* and *M2* genes.

CONCLUSIONS

Molecular diagnostics has revolutionized our ability to detect RTIs by increasing sensitivity of virus detection, broadening the array of viruses detected, and enabling the detection of multiple infections. Furthermore, test results are available in a timeframe that can better impact patient management. These tools have also advanced our understanding of the epidemiology and pathogenesis of respiratory virus disease.

References

[1] Lozano R, Naghavi M, Foreman K, et al. Global and regional mortality from 235 causes of death for 20 age groups in 1990 and 2010: a systematic analysis for the Global Burden of Disease Study 2010. Lancet 2012;380:2095—128.

[2] Fendrick AM, Monto AS, Nightengale B, Sarnes M. The economic burden of non-influenza-related viral respiratory tract infection in the United States. Arch Intern Med 2003;163:487—94.

[3] WHO. Influenza (seasonal), <http://www.who.int/mediacentre/factsheets/fs211/en/>; 2014 [accessed 23.11.14].

[4] Treanor JJ. Influenza (including avian influenza and swine influenza). In: Bennett JE, Dolin R, Blaser MJ, editors. Mandell, Douglas, and Bennett's principles and practice of infectious diseases. 8th ed. Philadelphia, PA: Saunders/Elsevier; 2015. p. 2000—24.

[5] Jain S, Kamimoto L, Bramley AM, et al. Hospitalized patients with 2009 H1N1 influenza in the United States. N Engl J Med 2009;361:1935—44.

[6] Siston AM, Rasmussen SA, Honein MA, et al. Pandemic 2009 influenza A(H1N1) virus illness among pregnant women in the United States. JAMA 2010;303:1517—25.

[7] Chen H, Wen X, To KK, et al. Quasispecies of the D225G substitution in the hemagglutinin of pandemic influenza A(H1N1) 2009 virus from patients with severe disease in Hong Kong, China. J Infect Dis 2010;201:1517—21.

[8] Melidou A, Gioula G, Exindari M, Chatzidimitriou D, Diza E, Malisiovas N. Molecular and phylogenetic analysis of the haemagglutinin gene of pandemic influenza H1N1 2009 viruses associated with severe and fatal infections. Virus Res 2010;151(2):192—9.

[9] Shinya K, Ebina M, Yamada S, Ono M, Kasai N, Kawaoka Y. Avian flu: influenza virus receptors in the human airway. Nature 2006;440:435—6.

[10] van Riel D, Munster VJ, de Wit E, et al. H5N1 virus attachment to lower respiratory tract. Science 2006;312:399.

[11] Itoh Y, Shinya K, Kiso M, et al. In vitro and in vivo characterization of new swine-origin H1N1 influenza viruses. Nature 2009;460:1021—5.

[12] Maines TR, Jayaraman A, Belser JA, et al. Transmission and pathogenesis of swine-origin 2009 A(H1N1) influenza viruses in ferrets and mice. Science 2009;325:484—7.

[13] Su S, Chaves SS, Perez A, et al. Comparing clinical characteristics between hospitalized adults with laboratory-confirmed influenza A and B virus infection. Clin Infect Dis 2014;59:252—5.

[14] Song W, Wang P, Mok BW-Y, et al. The K526R substitution in viral protein PB2 enhances the effects of E627K on influenza virus replication. Nat Commun 2014;5:5509.

[15] Craver RD, Sorrells K, Gohd R. Myocarditis with influenza B infection. Pediatr Infect Dis J 1997;16:629—30.

[16] Estabragh ZR, Mamas MA. The cardiovascular manifestations of influenza: a systematic review. Int J Cardiol 2013;167:2397—403.

[17] Steininger C, Popow-Kraupp T, Laferl H, et al. Acute encephalopathy associated with influenza A virus infection. Clin Infect Dis 2003;36:567—74.

[18] Sivadon-Tardy V, Orlikowski D, Porcher R, et al. Guillain-Barre syndrome and influenza virus infection. Clin Infect Dis 2009;48:48—56.

[19] Stockman LJ, Curns AT, Anderson LJ, Fischer-Langley G. Respiratory syncytial virus-associated hospitalizations among infants and young children in the United States, 1997—2006. Pediatr Infect Dis J 2012;31:5—9.

[20] Nair H, Nokes DJ, Gessner BD, et al. Global burden of acute lower respiratory infections due to respiratory syncytial virus in young children: a systematic review and meta-analysis. Lancet 2010;375:1545—55.

[21] Glezen WP, Taber LH, Frank AL, Kasel JA. Risk of primary infection and reinfection with respiratory syncytial virus. Am J Dis Child 1986;140:543—6.

[22] Gray GC, Capuano AW, Setterquist SF, et al. Multi-year study of human metapneumovirus infection at a large US Midwestern Medical Referral Center. J Clin Virol 2006;37:269—76.

[23] Edwards KM, Zhu Y, Griffin MR, et al. Burden of human metapneumovirus infection in young children. N Engl J Med 2013;368:633—43.

[24] Reiche J, Jacobsen S, Neubauer K, et al. Human metapneumovirus: insights from a ten-year molecular and epidemiological analysis in Germany. PLoS One 2014;9:e88342.

[25] van den Hoogen BG, de Jong JC, Groen J, et al. A newly discovered human pneumovirus isolated from young children with respiratory tract disease. Nat Med 2001;7:719—24.

[26] Boivin G, De Serres G, Cote S, et al. Human metapneumovirus infections in hospitalized children. Emerg Infect Dis 2003;9:634—40.

[27] Weinberg GA, Hall CB, Iwane MK, et al. Parainfluenza virus infection of young children: estimates of the population-based burden of hospitalization. J Pediatr 2009;154:694−9.

[28] Johnson JE, Gonzales RA, Olson SJ, Wright PF, Graham BS. The histopathology of fatal untreated human respiratory syncytial virus infection. Mod Pathol 2007;20:108−19.

[29] Feuillet F, Lina B, Rosa-Calatrava M, Boivin G. Ten years of human metapneumovirus research. J Clin Virol 2012;53:97−105.

[30] Zhang L, Bukreyev A, Thompson CI, et al. Infection of ciliated cells by human parainfluenza virus type 3 in an in vitro model of human airway epithelium. J Virol 2005;79:1113−24.

[31] Villar E, Barroso IM. Role of sialic acid-containing molecules in paramyxovirus entry into the host cell: a minireview. Glycoconj J 2006;23:5−17.

[32] Chang A, Dutch RE. Paramyxovirus fusion and entry: multiple paths to a common end. Viruses 2012;4:613−36.

[33] DeVincenzo JP. Factors predicting childhood respiratory syncytial virus severity: what they indicate about pathogenesis. Pediatr Infect Dis J 2005;24:S177−83.

[34] Rogers DF. Airway mucus hypersecretion in asthma: an undervalued pathology? Curr Opin Pharmacol 2004;4:241−50.

[35] Kuiken T, van den Hoogen BG, van Riel DA, et al. Experimental human metapneumovirus infection of cynomolgus macaques (Macaca fascicularis) results in virus replication in ciliated epithelial cells and pneumocytes with associated lesions throughout the respiratory tract. Am J Pathol 2004;164:1893−900.

[36] Hall CB, Weinberg GA, Iwane MK, et al. The burden of respiratory syncytial virus infection in young children. N Engl J Med 2009;360:588−98.

[37] Holman RC, Shay DK, Curns AT, Lingappa JR, Anderson LJ. Risk factors for bronchiolitis-associated deaths among infants in the United States. Pediatr Infect Dis J 2003;22:483−90.

[38] Schildgen V, van den Hoogen B, Fouchier R, et al. Human metapneumovirus: lessons learned over the first decade. Clin Microbiol Rev 2011;24:734−54.

[39] Papenburg J, Hamelin ME, Ouhoummane N, et al. Comparison of risk factors for human metapneumovirus and respiratory syncytial virus disease severity in young children. J Infect Dis 2012;206:178−89.

[40] Reed G, Jewett PH, Thompson J, Tollefson S, Wright PF. Epidemiology and clinical impact of parainfluenza virus infections in otherwise healthy infants and young children < 5 years old. J Infect Dis 1997;175:807−13.

[41] Martin ET, Kuypers J, Heugel J, Englund JA. Clinical disease and viral load in children infected with respiratory syncytial virus or human metapneumovirus. Diagn Microbiol Infect Dis 2008;62:382−8.

[42] Drunen Van, Littel-van Den Hurk S, Watkiss ER. Pathogenesis of respiratory syncytial virus. Curr Opin Virol 2012;2:300−5.

[43] Henrickson KJ. Parainfluenza viruses. Clin Microbiol Rev 2003;16:242−64.

[44] Semple MG, Cowell A, Dove W, et al. Dual infection of infants by human metapneumovirus and human respiratory syncytial virus is strongly associated with severe bronchiolitis. J Infect Dis 2005;191:382−6.

[45] Foulongne V, Guyon G, Rodiere M, Segondy M. Human metapneumovirus infection in young children hospitalized with respiratory tract disease. Pediatr Infect Dis J 2006;25:354−9.

[46] Walsh EE, McConnochie KM, Long CE, Hall CB. Severity of respiratory syncytial virus infection is related to virus strain. J Infect Dis 1997;175:814−20.

[47] Tran DN, Pham TM, Ha MT, et al. Molecular epidemiology and disease severity of human respiratory syncytial virus in Vietnam. PLoS One 2013;8:e45436.

[48] Agapov E, Sumino KC, Gaudreault-Keener M, Storch GA, Holtzman MJ. Genetic variability of human metapneumovirus infection: evidence of a shift in viral genotype without a change in illness. J Infect Dis 2006;193:396−403.

[49] Walsh EE, Peterson DR, Falsey AR. Human metapneumovirus infections in adults: another piece of the puzzle. Arch Intern Med 2008;168:2489−96.

[50] Hall CB, Long CE, Schnabel KC. Respiratory syncytial virus infections in previously healthy working adults. Clin Infect Dis 2001;33:792−6.

[51] Schildgen O, Glatzel T, Geikowski T, et al. Human metapneumovirus RNA in encephalitis patient. Emerg Infect Dis 2005;11:467−70.

[52] Eisenhut M. Extrapulmonary manifestations of severe respiratory syncytial virus infection—a systematic review. Crit Care 2006;10:R107.

[53] Arnold JC, Singh KK, Milder E, et al. Human metapneumovirus associated with central nervous system infection in children. Pediatr Infect Dis J 2009;28:1057−60.

[54] Wold WSM, Ison MG. Adenoviruses. In: Knipe DM, Howley PM, editors. Fields virology. 6th ed. Philadelphia, PA: Lippincott Williams & Wilkins/Wolters Kluwer; 2013. p. 1732−67.

[55] Lynch 3rd JP, Fishbein M, Echavarria M. Adenovirus. Semin Respir Crit Care Med 2011;32:494−511.

[56] Binn LN, Sanchez JL, Gaydos JC. Emergence of adenovirus type 14 in US military recruits—a new challenge. J Infect Dis 2007;196:1436−7.

[57] Fox JP, Brandt CD, Wassermann FE, et al. The virus watch program: a continuing surveillance of viral infections in metropolitan New York families. VI. Observations of adenovirus infections: virus excretion patterns, antibody response, efficiency of surveillance, patterns of infections, and relation to illness. Am J Epidemiol 1969;89:25−50.

[58] Meier O, Greber UF. Adenovirus endocytosis. J Gene Med 2004;6:S152−63.

[59] Gaggar A, Shayakhmetov DM, Lieber A. CD46 is a cellular receptor for group B adenoviruses. Nat Med 2003;9:1408−12.

[60] Stellrecht KA, Lamson DM, Romero JR. Enteroviruses and parechoviruses. In: Jorgensen JH, Pfaller MA, Carroll KC, Funke G, Landry ML, Richter SS, Warnock DW, editors. Manual of clinical microbiology. 11th ed. Washington, DC: ASM Press; 2015. p. 1536−50.

[61] Wright PF, Deatly AM, Karron RA, et al. Comparison of results of detection of rhinovirus by PCR and viral culture in human nasal wash specimens from subjects with and without clinical symptoms of respiratory illness. J Clin Microbiol 2007;45:2126−9.

[62] Gern JE, Palmenberg AC. Rhinoviruses. In: Knipe DM, Howley PM, editors. Fields virology. 6th ed. Philadelphia, PA: Lippincott Williams & Wilkins/Wolters Kluwer; 2013. p. 531−49.

[63] Miller EK, Lu X, Erdman DD, et al. Rhinovirus-associated hospitalizations in young children. J Infect Dis 2007;195:773−81.

[64] Masters PS, Perlman S. Coronaviridae. In: Knipe DM, Howley PM, editors. Fields virology. 6th ed. Philadelphia, PA: Lippincott Williams & Wilkins/Wolters Kluwer; 2013. p. 825−58.

[65] Hilgenfeld R, Peiris M. From SARS to MERS: 10 years of research on highly pathogenic human coronaviruses. Antiviral Res 2013;100:286−95.

[66] Abdul-Rasool S, Fielding BC. Understanding human coronavirus HCoV-NL63. Open Virol J 2010;4:76−84.

[67] Lew TW, Kwek TK, Tai D, et al. Acute respiratory distress syndrome in critically ill patients with severe acute respiratory syndrome. JAMA 2003;290:374−80.

[68] Assiri A, McGeer A, Perl TM, et al. Hospital outbreak of Middle East respiratory syndrome coronavirus. N Engl J Med 2013;369:407—16.

[69] Wilder-Smith A, Teleman MD, Heng BH, Earnest A, Ling AE, Leo YS. Asymptomatic SARS coronavirus infection among healthcare workers, Singapore. Emerg Infect Dis 2005;11:1142—5.

[70] Centers for Disease Control and Prevention. MERS clinical features, http://www.cdc.gov/coronavirus/mers/clinical-features.html; 2015 [accessed 13.02.15].

[71] Chonmaitree T, Revai K, Grady JJ, et al. Viral upper respiratory tract infection and otitis media complication in young children. Clin Infect Dis 2008;46:815—23.

[72] Chen T-C, Weng K-F, Chang S-C, Lin J-Y, Huang P-N, Shih S-R. Development of antiviral agents for enteroviruses. J Antimicrob Chemother 2008;62:1169—73.

[73] Fiore AE, Shay DK, Broder K, et al. Prevention and control of influenza: recommendations of the Advisory Committee on Immunization Practices (ACIP). MMWR Recomm Rep 2008;57:1—60.

[74] Bjornson CL, Johnson DW. Croup. Lancet 2008;371:329—39.

[75] Borchers AT, Chang C, Gershwin ME, Gershwin LJ. Respiratory syncytial virus—a comprehensive review. Clin Rev Allergy Immunol 2013;45:331—79.

[76] Ralston SL, Lieberthal AS, Meissner HC, et al. Clinical practice guideline: the diagnosis, management, and prevention of bronchiolitis. Pediatrics 2014;134:e1474—502.

[77] Ruuskanen O, Lahti E, Jennings LC, Murdoch DR. Viral pneumonia. Lancet 2011;377:1264—75.

[78] Cesario TC. Viruses associated with pneumonia in adults. Clin Infect Dis 2012;55:107—13.

[79] Falsey AR, McElhaney JE, Beran J, et al. Respiratory syncytial virus and other respiratory viral infections in older adults with moderate to severe influenza-like illness. J Infect Dis 2014;209:1873—81.

[80] Mandell LA, Wunderink RG, Anzueto A, et al. Infectious Diseases Society of America/American Thoracic Society consensus guidelines on the management of community-acquired pneumonia in adults. Clin Infect Dis 2007;44:S27—72.

[81] Hohenthal U, Vainionpaa R, Nikoskelainen J, Kotilainen P. The role of rhinoviruses and enteroviruses in community acquired pneumonia in adults. Thorax 2008;63:658—9.

[82] Ginocchio CC, Lotlikar M, Falk L, et al. Clinical performance of the 3M Rapid Detection Flu A + B Test compared to R-Mix culture, DFA and BinaxNOW Influenza A&B Test. J Clin Virol 2009;45:146—9.

[83] Miernyk K, Bulkow L, DeByle C, et al. Performance of a rapid antigen test (Binax NOW(R) RSV) for diagnosis of respiratory syncytial virus compared with real-time polymerase chain reaction in a pediatric population. J Clin Virol 2011;50:240—3.

[84] Lucas PM, Morgan OW, Gibbons TF, et al. Diagnosis of 2009 pandemic influenza A (pH1N1) and seasonal influenza using rapid influenza antigen tests, San Antonio, Texas, April-June 2009. Clin Infect Dis 2011;52:S116—22.

[85] Uyeki TM, Prasad R, Vukotich C, et al. Low sensitivity of rapid diagnostic test for influenza. Clin Infect Dis 2009;48: e89—92.

[86] Chartrand C, Leeflang MM, Minion J, Brewer T, Pai M. Accuracy of rapid influenza diagnostic tests: a meta-analysis. Ann Intern Med 2012;156:500—11.

[87] Weinberg GA, Erdman DD, Edwards KM, et al. Superiority of reverse-transcription polymerase chain reaction to conventional viral culture in the diagnosis of acute respiratory tract infections in children. J Infect Dis 2004;189:706—10.

[88] She RC, Polage CR, Caram LB, et al. Performance of diagnostic tests to detect respiratory viruses in older adults. Diagn Microbiol Infect Dis 2010;67:246—50.

[89] Talbot HK, Falsey AR. The diagnosis of viral respiratory disease in older adults. Clin Infect Dis 2010;50:747—51.

[90] Gaydos CA. What is the role of newer molecular tests in the management of CAP? Infect Dis Clin North Am 2013;27:49—69.

[91] Vallieres E, Renaud C. Clinical and economical impact of multiplex respiratory virus assays. Diagn Microbiol Infect Dis 2013;76:255—61.

[92] Doern CD, Lacey D, Huang R, Haag C. Evaluation and implementation of FilmArray version 1.7 for improved detection of adenovirus respiratory tract infection. J Clin Microbiol 2013;51:4036—9.

[93] Wishaupt JO, Versteegh FG, Hartwig NG. PCR testing for paediatric acute respiratory tract infections. Paediatr Respir Rev 2015;16:43—8.

[94] Rogers BB, Shankar P, Jerris RC, et al. Impact of a rapid respiratory panel test on patient outcomes. Arch Pathol Lab Med 2015;139:636—41.

[95] Xu M, Qin X, Astion ML, et al. Implementation of FilmArray respiratory viral panel in a core laboratory improves testing turnaround time and patient care. Am J Clin Pathol 2013;139:118—23.

[96] Ahluwalia G, Embree J, McNicol P, Law B, Hammond GW. Comparison of nasopharyngeal aspirate and nasopharyngeal swab specimens for respiratory syncytial virus diagnosis by cell culture, indirect immunofluorescence assay, and enzyme-linked immunosorbent assay. J Clin Microbiol 1987;25:763—7.

[97] Macfarlane P, Denham J, Assous J, Hughes C. RSV testing in bronchiolitis: which nasal sampling method is best? Arch Dis Child 2005;90:634—5.

[98] Jartti T, Soderlund-Venermo M, Hedman K, Ruuskanen O, Makela MJ. New molecular virus detection methods and their clinical value in lower respiratory tract infections in children. Paediatr Respir Rev 2013;14:38—45.

[99] Chan KH, Peiris JS, Lim W, Nicholls JM, Chiu SS. Comparison of nasopharyngeal flocked swabs and aspirates for rapid diagnosis of respiratory viruses in children. J Clin Virol 2008;42:65—9.

[100] Irving SA, Vandermause MF, Shay DK, Belongia EA. Comparison of nasal and nasopharyngeal swabs for influenza detection in adults. Clin Med Res 2012;10:215—18.

[101] Abu-Diab A, Azzeh M, Ghneim R, et al. Comparison between pernasal flocked swabs and nasopharyngeal aspirates for detection of common respiratory viruses in samples from children. J Clin Microbiol 2008;46:2414—17.

[102] Smieja M, Castriciano S, Carruthers S, et al. Development and evaluation of a flocked nasal midturbinate swab for self-collection in respiratory virus infection diagnostic testing. J Clin Microbiol 2010;48:3340—2.

[103] Larios OE, Coleman BL, Drews SJ, et al. Self-collected mid-turbinate swabs for the detection of respiratory viruses in adults with acute respiratory illnesses. PLoS One 2011;6:e21335.

[104] Esposito S, Molteni CG, Daleno C, et al. Collection by trained pediatricians or parents of mid-turbinate nasal flocked swabs for the detection of influenza viruses in childhood. Virol J 2010;7:85.

[105] Lambert SB, Whiley DM, O'Neill NT, et al. Comparing nose-throat swabs and nasopharyngeal aspirates collected from children with symptoms for respiratory virus identification using real-time polymerase chain reaction. Pediatrics 2008;122: e615—20.

[106] Spencer S, Gaglani M, Naleway A, et al. Consistency of influenza A virus detection test results across respiratory specimen

collection methods using real-time reverse transcription-PCR. J Clin Microbiol 2013;51:3880–2.

[107] Writing Committee of the WHOCoCAoPI, Bautista E, Chotpitayasunondh T, et al. Clinical aspects of pandemic 2009 influenza A (H1N1) virus infection. N Engl J Med 2010;362:1708–19.

[108] Lopez Roa P, Rodriguez-Sanchez B, Catalan P, et al. Diagnosis of influenza in intensive care units: lower respiratory tract samples are better than nose-throat swabs. Am J Respir Crit Care Med 2012;186:929–30.

[109] Robinson CC. Respiratory viruses. In: Specter S, Hodinka RL, Young SA, Wiedbrauk DL, editors. Clinical virology manual. 4th ed. Washington, DC: ASM Press; 2009. p. 203–48.

[110] Pankuweit S, Klingel K. Viral myocarditis: from experimental models to molecular diagnosis in patients. Heart Fail Rev 2013;18:683–702.

[111] Lee N, Chan PK, Hui DS, et al. Viral loads and duration of viral shedding in adult patients hospitalized with influenza. J Infect Dis 2009;200:492–500.

[112] To KK, Chan KH, Li IW, et al. Viral load in patients infected with pandemic H1N1 2009 influenza A virus. J Med Virol 2010;82:1–7.

[113] DeVincenzo JP, Wilkinson T, Vaishnaw A, et al. Viral load drives disease in humans experimentally infected with respiratory syncytial virus. Am J Respir Crit Care Med 2010;182:1305–14.

[114] Walsh EE, Peterson DR, Kalkanoglu AE, Lee FE, Falsey AR. Viral shedding and immune responses to respiratory syncytial virus infection in older adults. J Infect Dis 2013;207:1424–32.

[115] von Linstow ML, Eugen-Olsen J, Koch A, Winther TN, Westh H, Hogh B. Excretion patterns of human metapneumovirus and respiratory syncytial virus among young children. Eur J Med Res 2006;11:329–35.

[116] Kaiser L, Fritz RS, Straus SE, Gubareva L, Hayden FG. Symptom pathogenesis during acute influenza: interleukin-6 and other cytokine responses. J Med Virol 2001;64:262–8.

[117] Memish ZA, Assiri AM, Al-Tawfiq JA. Middle East respiratory syndrome coronavirus (MERS-CoV) viral shedding in the respiratory tract: an observational analysis with infection control implications. Int J Infect Dis 2014;29:307–8.

[118] Memoli MJ, Athota R, Reed S, et al. The natural history of influenza infection in the severely immunocompromised vs nonimmunocompromised hosts. Clin Infect Dis 2014;58:214–24.

[119] Loeffelholz MJ, Trujillo R, Pyles RB, et al. Duration of rhinovirus shedding in the upper respiratory tract in the first year of life. Pediatrics 2014;134:1144–50.

[120] Frank AL, Taber LH, Wells CR, Wells JM, Glezen WP, Paredes A. Patterns of shedding of myxoviruses and paramyxoviruses in children. J Infect Dis 1981;144:433–41.

[121] Kalu SU, Loeffelholz M, Beck E, et al. Persistence of adenovirus nucleic acids in nasopharyngeal secretions: a diagnostic conundrum. Pediatr Infect Dis J 2010;29:746–50.

[122] de Lima CR, Mirandolli TB, Carneiro LC, et al. Prolonged respiratory viral shedding in transplant patients. Transpl Infect Dis 2014;16:165–9.

[123] Bower J, McBride JT. Croup in children (acute laryngotracheobronchitis). In: Bennett JE, Dolin R, Blaser MJ, editors. Mandell, Douglas, and Bennett's principles and practice of infectious diseases. 8th ed. Philadelphia, PA: Saunders/Elsevier; 2015. p. 762–6.

[124] Turner RB. Rhinovirus. In: Bennett JE, Dolin R, Blaser MJ, editors. Mandell, Douglas, and Bennett's principles and practice of infectious diseases. 8th ed. Philadelphia, PA: Saunders/Elsevier; 2015. p. 2113–21.

[125] Paynter S. Humidity and respiratory virus transmission in tropical and temperate settings. Epidemiol Infect 2015;143:1110–18.

[126] Axelsen JB, Yaari R, Grenfell BT, Stone L. Multiannual forecasting of seasonal influenza dynamics reveals climatic and evolutionary drivers. Proc Natl Acad Sci USA 2014;111:9538–42.

[127] Nattanmai SM, Butt SA, Butt J, et al. Comparison of the Cobas® Influenza A and B Test with the FilmArray Respiratory Panel and the Prodesse ProFlu/ProFAST Assays for the detection of H3 influenza viruses circulating in 2015. 31st clinical virology symposium. Daytona Beach, FL; 2015.

[128] Mahony JB. Detection of respiratory viruses by molecular methods. Clin Microbiol Rev 2008;21:716–47.

[129] Loeffelholz MJ, Pong DL, Pyles RB, et al. Comparison of the FilmArray Respiratory Panel and Prodesse real-time PCR assays for detection of respiratory pathogens. J Clin Microbiol 2011;49:4083–8.

[130] Pierce VM, Elkan M, Leet M, McGowan KL, Hodinka RL. Comparison of the Idaho Technology FilmArray system to real-time PCR for detection of respiratory pathogens in children. J Clin Microbiol 2012;50:364–71.

[131] Lumley S, Atkinson C, Haque T. Respiratory PCR detects influenza after intranasal live-attenuated influenza vaccination. Arch Dis Child 2014;99:301.

[132] Moscona A. Oseltamivir resistance—disabling our influenza defenses. N Engl J Med 2005;353:2633–6.

[133] Centers for Disease Control and Prevention. Antiviral drug resistance among influenza viruses, <http://www.cdc.gov/flu/professionals/antivirals/antiviral-drug-resistance.htm/>; 2015 [accessed 24.05.15].

[134] Bright RA, Shay DK, Shu B, Cox NJ, Klimov AI. Adamantane resistance among influenza A viruses isolated early during the 2005–2006 influenza season in the United States. JAMA 2006;295:891–4.

[135] Deyde VM, Sheu TG, Trujillo AA, et al. Detection of molecular markers of drug resistance in 2009 pandemic influenza A (H1N1) viruses by pyrosequencing. Antimicrob Agents Chemother 2010;54:1102–10.

[136] Hurt AC, Holien JK, Barr IG. In vitro generation of neuraminidase inhibitor resistance in A(H5N1) influenza viruses. Antimicrob Agents Chemother 2009;53:4433–40.

[137] Ison MG, Gubareva LV, Atmar RL, Treanor J, Hayden FG. Recovery of drug-resistant influenza virus from immunocompromised patients: a case series. J Infect Dis 2006;193:760–4.

[138] Fowlkes A, Giorgi A, Erdman D, et al. Viruses associated with acute respiratory infections and influenza-like illness among outpatients from the Influenza Incidence Surveillance Project, 2010–2011. J Infect Dis 2014;209:1715–25.

[139] Thomas RE. Is influenza-like illness a useful concept and an appropriate test of influenza vaccine effectiveness? Vaccine 2014;32:2143–9.

[140] Rihkanen H, Ronkko E, Nieminen T, et al. Respiratory viruses in laryngeal croup of young children. J Pediatr 2008;152:661–5.

[141] Mansbach JM, Piedra PA, Teach SJ, et al. Prospective multicenter study of viral etiology and hospital length of stay in children with severe bronchiolitis. Arch Pediatr Adolesc Med 2012;166:700–6.

[142] Miller EK, Gebretsadik T, Carroll KN, et al. Viral etiologies of infant bronchiolitis, croup and upper respiratory illness during 4 consecutive years. Pediatr Infect Dis J 2013;32:950–5.

[143] Oymar K, Skjerven HO, Mikalsen IB. Acute bronchiolitis in infants: a review. Scand J Trauma Resusc Emerg Med 2014;22:23.

[144] Pavia AT. Viral infections of the lower respiratory tract: old viruses, new viruses, and the role of diagnosis. Clin Infect Dis 2011;52:S284–9.

12

Molecular Testing for Diseases Associated with Bacterial Infections

R. Plongla[1,2] and M.B. Miller[2]

[1]Division of Infectious Diseases, Chulalongkorn University and King Chulalongkorn Memorial Hospital, Bangkok, Thailand [2]Department of Pathology and Laboratory Medicine, UNC School of Medicine, Chapel Hill, NC, United States

INTRODUCTION

As the complexity of diagnostic bacteriology has increased, so have the methods employed to detect potential pathogens. In some cases, molecular technology has augmented traditional methods that have historically been the gold standard for pathogen detection, such as culture and serology. In other situations, molecular detection has completely replaced traditional methodologies. For routine bacteriology (ie, blood cultures, urine cultures, and respiratory cultures), culture has remained the gold standard primarily based on cost accounting and the potential complex nature of associated infections. However, in instances where there may be low quantities of the pathogen present, the patient may have received antibiotics prior to specimen collection, the etiologic agent may require unusual culture conditions, or a more rapid turnaround time is needed, molecular approaches are particularly beneficial.

Currently, the optimal use of molecular techniques in bacteriology resides with specimens in which a limited number of pathogenic organisms are sought (ie, pertussis or tuberculosis diagnosis) and in cases where the enhanced sensitivity, decreased turnaround time, and/or patient impact of molecular methods outweighs the increased cost to the laboratory (ie, molecular identification of organisms directly from positive blood cultures). A particularly exciting advancement in clinical microbiology is the use of mass spectrometry for the identification of a wide spectrum of bacterial organisms. This chapter aims to discuss the most common molecular methodologies and applications being used in clinical bacteriology laboratories.

IDENTIFICATION OF BACTERIA

Traditional biochemical methods used for bacterial identification have allowed for the differentiation of bacterial pathogens for many years. However, biochemical methods may take considerable time to identify an organism or may not be able to identify particular groups of organisms accurately. The use of probes, sequencing, and mass spectrometry has greatly improved the accuracy in which organisms are identified and has decreased the time to identification for many slow-growing or difficult-to-identify organisms. The use of molecular methods for identification relies on the fact that different species of bacteria have distinct nucleic acid sequences (and therefore peptides) that can be interrogated for species-level identification.

Probes

Molecular Target(s) and Technologies

Nucleic acid hybridization against specific sequences without the need for target amplification is the principle of genetic probes. Briefly, the two strands of DNA are dissociated by heat denaturation. A synthetic probe is then introduced and complementarity between probe and target sequence allows probe-target base pairing to form a hybrid strand, which can be DNA–DNA, DNA–RNA, or RNA–RNA. The hybrid can be detected by radioisotopes, enzymes, chemiluminescence, or fluorescence reporters labeled on the probe.

Peptide nucleic acid (PNA) probes are DNA analogues but the sugar phosphate backbone of DNA is

Diagnostic Molecular Pathology
DOI: http://dx.doi.org/10.1016/B978-0-12-800886-7.00012-1

© 2017 Elsevier Inc. All rights reserved.

substituted with polyamide or peptide backbone; therefore, PNA probes are noncharged [1]. This property helps PNA probes to overcome electrostatic repulsion between two negatively-charged DNA strands and provides more stability of the PNA–RNA hybrid during hybridization [2]. In addition, PNA probes are relatively hydrophobic which facilitate entrance into the hydrophobic cell membrane [3,4].

Clinical Utility

Rapid and accurate identification of bacteria provides clinicians the opportunity to use targeted antimicrobial therapy based on the organism and, therefore, limit a patient's unnecessary antimicrobial exposure. Several genetic probes are commercially available for direct detection of organisms from clinical samples and identification of bacteria after growth or isolation from culture. These probes are particularly useful as the culture confirmation assays for slow-growing or difficult-to-identify bacteria, such as mycobacteria.

FDA-cleared DNA probes for direct detection of group A streptococci as well as culture identification of *Listeria monocytogenes*, *Staphylococcus aureus*, *Streptococcus pneumoniae*, *Mycobacterium tuberculosis* complex, *M. avium*, *M. intracellulare*, *M. avium* complex, *M. gordonae*, and *M. kansasii* are available from Hologic (Bedford, MA). Fluorescence in situ hybridization using PNA probes (PNA-FISH) for identification of *Enterococcus faecalis*/other enterococci, *Staphylococcus aureus*/coagulase-negative staphylococci, *Escherichia coli*/*Klebsiella pneumoniae*/*Pseudomonas aeruginosa* from positive blood culture bottles, and *Streptococcus agalactiae* from broth cultures are available from AdvanDx (Woburn, MA).

Limitations of Testing

Genetic probes generally have lower analytic sensitivity than other molecular assays due to the nonamplified nature of the method. Therefore, the application is usually limited to specimens with high numbers of bacteria such as cases of group A streptococcal pharyngitis, positive blood cultures bottles, or isolate confirmation.

Sequencing

Molecular Target(s) and Technologies

Ribosomal RNA (rRNA) genes are present in all prokaryotes and weakly affected by horizontal gene transfer and mutation. These genes, in particular the sequence of the 16S rRNA gene, contain some regions that are highly conserved and other regions that have variable nucleic acid sequences. Using primers targeting conserved sequences of the 16S rRNA gene that

amplify a region of the gene that is variable allows for broad-range bacterial identification [5,6]. Often only the first 500 bp need to be amplified and sequenced to obtain a reliable species identification. The 16S–23S rRNA intergenic spacer (ITS) region and the large subunit (23S) rRNA genes can also be used as molecular targets [7,8]. The 16S–23S rRNA ITS region has higher copy number and sequence variability than the 16S rRNA gene. However, these regions have not been as broadly used as the 16S rRNA gene in clinical microbiology laboratories because of relatively limited sequence databases and the lack of commercially-available test kits. Alternative gene targets such as the heat shock proteins, *recA*, *rpoB*, *tuf*, *gyrA*, *gyrB*, and the *cpn60* family of proteins can be amplified and sequenced to differentiate closely related species [5].

DNA sequencing for bacterial identification generally comprises extraction of nucleic acids, amplification of the target sequence by PCR, sequence determination, and comparison with sequences found in public or commercial databases. The precision of the identification (ie, whether an organism is identified to genus or species level) depends on sequence homology within the database. The Clinical and Laboratory Standards Institute provides guidance for nucleic acid sequencing and interpretation in clinical laboratories [5].

The Sanger dideoxynucleotide chain termination method is the most widely used technique to determine DNA sequence. The use of fluorescent dye terminators and capillary electrophoresis has made the use of Sanger sequencing by diagnostic laboratories feasible [9]. Pyrosequencing (Qiagen, Gaithersburg, MD), which is based on the luminometric detection of pyrophosphate released during DNA synthesis, is an alternative approach for DNA sequencing [10]. Pyrosequencing provides reliable data for sequences adjacent to the sequencing primer termini and is a simple-to-use, robust platform for short-read-length sequencing.

Instead of looking for particular genes, several platforms for massively parallel sequencing (next-generation sequencing) are currently available. Although these high-throughput systems have not been routinely used in the clinical microbiology laboratory, they have potential applications in the areas of identification of unknown pathogens, direct specimen sequencing, and microbiome profiling.

Clinical Utility

Sequence analysis of the 16S rRNA gene increases our understanding of the phylogenetic relationships among bacteria and improves the identification of difficult-to-identify, unrecognized, or slowly-growing bacteria regardless of phenotypic characteristics. Cook et al. showed the cost-effectiveness advantages for using 16S rRNA gene sequencing compared with conventional methods for identification of nontuberculous mycobacteria. The turnaround time was improved from 2−6 weeks to 1−2 days with decreased expenses [11]. Moreover, direct sequencing from clinical specimens or formalin-fixed paraffin-embedded tissues offers the opportunity to identify the causative agent when cultures are negative (ie, specimens taken post-antimicrobial therapy or organisms that require special growth requirements).

Limitations of Testing

The major limitations for using sequencing for bacterial identification include the high cost of sequencer instrumentation and the need of experienced personnel to review the result since public databases are not always curated. Some organisms are very closely related or genetically indistinguishable and require multiple genes to be sequenced. For instance, *Mycobacterium abscessus* group and *M. chelonae* share 100% identity of 16S rRNA sequence. Thus, alternative targets (ie, *rpoB* or *hsp65*) are needed to provide better resolution to species. However, most alternative gene targets are not universal and must be chosen cautiously. Like with other nucleic acid testing, contamination prevention, quality control, standardization, and validation of assay protocols are needed to provide quality results. Importantly, careful interpretation of results and correlation with clinical presentation are critical, particularly in the setting of direct specimen sequencing. Unlike culture, ribosomal DNA sequencing does not inform viability of bacteria nor does it predict antimicrobial susceptibility.

Matrix-Assisted Laser Desorption Ionization-Time of Flight Mass Spectrometry

Molecular Target(s) and Technologies

Matrix-assisted laser desorption ionization-time of flight mass spectrometry (MALDI-TOF MS) uses proteomics for bacterial identification. The range of proteins being analyzed is 2−20 kDa, which enriches for ribosomal proteins. For direct colony identification, the organism is spread on a target plate and then overlaid with a polymeric matrix. The matrix (eg, α-cyano-4-hydroxycinnamic acid (CHCA)) isolates analyzed molecules and protects them from fragmentation by the laser. After firing the laser, matrix and proteins are desorbed and the charge is transferred to the molecules. The ionized molecules enter a vacuum flight tube and accelerate to a detector. As a result a spectrum signature is created according to the mass/ charge ratio of the molecular fragments [12]. The spectrum is compared with a database to determine the identification of the organism. The addition of formic acid or a preextraction step may improve identification of yeasts and some Gram-positive bacteria [13,14]. Two US FDA-cleared commercial MALDI-TOF MS systems are available: BioTyper (Bruker Daltonics, Billerica, MA) and Vitek MS (bioMérieux, Durham, NC).

Clinical Utility

MALDI-TOF MS can be used for identification of a wide range of bacteria that are commonly found in clinical laboratories. It is a high-throughput system providing identification in as little as 5−20 min with minimal hands-on time. Despite of the expensive capital equipment, MALDI-TOF MS has a low per-test cost (<$1/organism). The overall performance between the two commercial systems is comparable and most errors are associated with incomplete databases rather than the instrument. In a study of 1129 isolates, the Bruker LT BioTyper and the Vitek MS databases correctly identified 92.7% and 93.2%, respectively [15]. The authors concluded that both systems have equal analytical efficiency. Further discussion of the performance of these systems can be found in the reviews by Patel [12] and Clark et al. [16]. Besides the rapid and accurate identification of bacteria, Tran et al. also showed that MALDI-TOF MS provided cost savings of US$73,646 or 51.7% in total costs (including technologist time and maintenance costs) annually [17].

Limitations of Testing

The initial instrumentation is expensive, the databases are proprietary, and the databases need improvement and expansion. There have been some reports of misidentification. For example, MALDI-TOF MS cannot differentiate *E. coli* from *Shigella* spp. because of their genetic similarity [18]. Laboratories should develop refined criteria for identification using MALDI-TOF MS in combination with sequencing or reference laboratory to most accurately identify closely related organisms. Mucoid or tiny colonies may fail identification and current procedures require growth of a pure isolate to be tested. Protocols for identification of bacteria directly from specimens or positive blood cultures have been developed, but these require validation and standardization.

RESPIRATORY INFECTIONS

While multiplexed molecular detection for respiratory viruses has largely supplanted culture for detection, this is not the case for bacterial respiratory infections, with a few exceptions. Bacterial infections that can mimic viral respiratory tract infections, such as those caused by *Mycoplasma pneumoniae* and *Chlamydophila pneumoniae*, are included with some of the viral respiratory molecular panels. Otherwise, molecular tests for the detection of bacterial respiratory pathogens are limited to those organisms that are fastidious (*Bordetella pertussis*, *Mycoplasma pneumoniae*), slow-growing (*Mycobacterium tuberculosis*), or require a rapid result for therapeutic decisions (Group A *Streptococcus* (GAS)).

Molecular Target(s) and Technologies

M. pneumoniae, *C. pneumoniae*, and *B. pertussis* are the bacterial components of the 20-target FilmArray Respiratory Panel (BioFire, Salt Lake City, UT), which is FDA-cleared for testing of nasopharyngeal swabs. The FilmArray is a self-contained pouch and instrument that processes the sample from nucleic acid extraction to nested PCR amplification and ultimately melt curve analysis. For this moderately complex test, the hands-on time is less than 2 min with results available in about an hour. Meridian Bioscience (Cincinnati, OH) has moderately complex stand-alone amplification tests for *M. pneumoniae*, *B. pertussis*, and GAS. These tests use loop-mediated isothermal amplification (LAMP) which requires less than 2 min of hands-on time, and results are available in less than an hour with minimal laboratory equipment. Additional molecular tests for the detection of group A streptococci include the Alere i Strep A Rapid Molecular Test (Waltham, MA) and the Focus Diagnostics Simplexa Group A Strep Direct Test (Cypress, CA). Both of these assays are moderately complex tests and provide results in 8 min and 1 h, respectively.

There are two FDA-approved/cleared tests for the detection of *Mycobacterium tuberculosis* directly from respiratory specimens—Amplified MTD test (Hologic) and Xpert MTB/RIF test (Cepheid, Sunnyvale, CA). The Hologic assay targets the rRNA of the *M. tuberculosis* complex and uses transcription-mediated amplification followed by a hybridization protection assay to detect the amplicons. The Xpert test relies on nested real-time PCR of the *rpoB* gene followed by hybridization with five distinct molecular beacon probes, each labeled with a different fluorophore. The 81-bp region targeted in *rpoB* is the rifampin resistance-determining region. Therefore, by analyzing the hybridization patterns of the five probes, both *M. tuberculosis* complex DNA and rifampin resistance can be detected simultaneously. Both assays are approved for smear-positive and smear-negative respiratory specimens. In addition, the Xpert test has been FDA-cleared to aid physicians in determining if patients with suspected tuberculosis can be removed from airborne isolation precautions.

Clinical Utility

Although recognized as potential agents of community-acquired pneumonia, *M. pneumoniae* and *C. pneumoniae* are thought to be under-diagnosed due to the difficulty in culturing them and the reliance on a serologic diagnosis. As molecular detection of these organisms increases, so will our knowledge of their epidemiology and clinical features [19]. However, the ability to distinguish between asymptomatic carriage and true infection needs to be addressed in carefully conducted outcome studies.

Application of molecular assays to the detection of *B. pertussis* has greatly increased the sensitivity of detection, but often at a cost of specificity. Pseudo-outbreaks have been reported based on false-positive nucleic acid amplification test (NAAT) results and DNA contamination from clinics administering pertussis vaccine [20–22]. Therefore, it is recommended to augment molecular detection with culture during an outbreak [23]. The sensitivity of NAAT allows for detection of *B. pertussis* DNA well into the course of disease (~3 weeks) when culture would be negative. Cross-reactivity has been observed with other *Bordetella* spp., depending on the targets used in the NAAT [24].

The diagnosis of GAS pharyngitis in children usually relies on rapid antigen detection, but due to the lower sensitivity of antigen-based assays, specimens negative by rapid antigen should also be cultured [25]. The increased sensitivity (~99%) of GAS molecular detection compared to antigen/culture (81.7%) and its similar specificity (98.5–99.6%) allows for the elimination of the reflexed culture, thereby reducing the time to final result [26–28].

The use of NAAT for the detection of *M. tuberculosis* directly from patient specimens has the potential to impact therapeutic interventions and infection control. Although smear results have historically been used to determine the burden of mycobacterial disease and the need for airborne isolation, their low sensitivity requires that three smears be performed to rule out tuberculosis. This often takes 24–72 h. Studies have demonstrated that the use of a negative Xpert test results in reduced time in airborne isolation and institutional cost savings [29,30]. Further, acid-fast smears are not specific to *M. tuberculosis*. The use of NAATs on smear-positive respiratory specimens can quickly determine the presence of *M. tuberculosis* complex versus other mycobacteria. The Centers for Disease Control and Prevention (CDC) recommends NAAT be performed on at least one respiratory specimen from patients with signs and symptoms consistent with pulmonary tuberculosis, particularly if a diagnosis of tuberculosis has not yet been made and the test result would alter patient care and/or TB control [31]. Although the preference is to test the first collected specimen by NAAT to lessen the time to result, a smear-positive specimen should take priority due to the increased sensitivity of NAAT for smear-positive respiratory specimens. The sensitivity of the Xpert test is 98% for smear-positive respiratory samples compared to 72% for smear-negative samples, which increases to 86% if a second smear-negative specimen is tested [32]. It is important to note that NAAT should not replace or delay routine microbiologic methods, including smear.

The Xpert assay also detects rifampin resistance. The CDC has published recommendations for the reporting these results [33]. Clinical trial data indicate 96.1% sensitivity and 98.6% specificity for rifampin resistance, but the sites in the clinical trial (Peru, Azerbaijan, South Africa, and India) have a higher prevalence of TB and multi-drug resistant TB than the United States. Therefore, in the United States the positive predictive value is likely to be lower. Of note, a positive rifampin result by Xpert may be obtained even for silent mutations [34]. For these reasons, it is critical that initial rifampin resistance results be confirmed by traditional susceptibility testing. Nonetheless, molecular screening for rifampin resistance could potentially identify resistant isolates weeks earlier, allowing clinical care to be significantly impacted.

Limitations of Testing

The ability to distinguish between colonization and infection for *M. pneumoniae* and *C. pneumoniae* is a challenge. This limitation is one of the primary reasons there are not molecular tests for other potential causes of community-acquired pneumonia, such as *Streptococcus pneumoniae*, *Haemophilus influenzae*, and *Moraxella catarrhalis*. The expansion of molecular techniques to detect a broader range of bacterial pathogens will need to be accompanied by data and analysis tools to determine clinical relevance. The specificity of pertussis molecular detection and ease of contamination should be kept in mind. As with all molecular tests, a positive NAAT result does not differentiate between live and dead organisms. This is particularly problematic for *M. tuberculosis* detection. Amplification technologies should not be used on specimens collected from patients who have received antitubercular drugs for more than 7 days or have been treated for *M. tuberculosis* within 2 months of collection [35]. In addition,

about 4% of pulmonary and 19% of extrapulmonary specimens contain inhibitory substances that may lead to false-negative results [35].

GASTROINTESTINAL INFECTIONS

Acute gastroenteritis is a significant cause of morbidity and mortality worldwide, and a substantial healthcare burden in the United States with approximately 179 million cases and almost 1.3 million hospitalizations occurring annually [36]. While viruses account for the majority of cases, *Salmonella* and *Campylobacter jejuni* are the primary causes of bacterial gastroenteritis in the developed world, with *Shigella* and pathogenic *E. coli* also contributing to the disease burden. Detection of bacterial pathogens has historically relied on culture, which includes nonselective, enrichment and selective media to enhance the isolation of potential pathogens from the complex matrix of stool. Stool cultures are time-consuming and relatively expensive due to the significant labor required to distinguish nonpathogens from pathogens. Results are often not available for several days. Even with MALDI-TOF MS, there are challenges since it cannot differentiate normal microbiota *E. coli* from pathogenic *E. coli* or *Shigella*. Other methods used include *Campylobacter* enzyme immunoassay (EIA) and shigatoxin EIA, both suffer from low sensitivity and cannot be used to replace culture. With a low positivity rate (usually <5%), low sensitivity, and high labor costs, stool cultures are an easy target for replacement by molecular methods.

Molecular Target(s) and Technologies

Manufacturers have taken one of two approaches when developing molecular tests for gastrointestinal (GI) pathogens: (1) a broad multiplex panel that includes bacterial, viral, and parasitic targets or (2) a tiered approach where multiple panels are designed, each comprised of different targets (ie, bacterial/viral panel and a separate parasitic panel or extended bacterial panel). A summary of the FDA-cleared molecular panels that include bacterial targets can be found in Table 12.1.

The BD MAX [37,38], BioFire [39,40], and Nanosphere assays are sample-to-result platforms that do not require preextraction of nucleic acids and are labeled as CLIA moderate complexity. The Luminex [41] and Hologic [42] tests require extraction and an additional pipetting step making them more suitable for high-volume, batched-based testing in a molecular laboratory. These assays are labeled as CLIA

TABLE 12.1   FDA-Cleared Multiplex Gastrointestinal Tests with Bacterial Targets

Specimen types	BD MAX Enteric Bacterial Panel Stool Cary-Blair Stool	BioFire FilmArray GI Panel Cary-Blair Stool	Luminex xTAG GPP Stool	Nanosphere Verigene Enteric Pathogens Stool	Prodesse/Hologic SSCS Cary-Blair or Para-Pak C&S Stool
Campylobacter	X	X	X	X	X
C. difficile		X	X		
E. coli O157	[X][a]	X	X	[X]	[X]
EAEC, EPEC		X			
ETEC		X	X		
Plesiomonas shigelloides		X			
STEC	X	X	X	X	X
Salmonella	X	X	X	X	X
Shigella [EIEC]	X	X	X	X	X
Vibrio		X		X	
Yersinia enterocolitica		X		X	
Adenovirus 40/41, astrovirus, sapovirus		X			
Norovirus GI/GII		X	X		
Rotavirus		X	X		
Giardia		X	X		
Cryptosporidium		X	X		
Entamoeba histolytica, Cyclospora		X			

[a]_The brackets [X] indicate that E. coli O157 is detected by detecting the shiga-toxin gene(s), but is not specifically identified as O157 in these assays._
EAEC, enteroaggregative _E. coli_; EPEC, enteropathogenic _E. coli_; ETEC, enterotoxigenic _E. coli_; STEC, shiga-toxin producing _E. coli_; EIEC, enteroinvasive _E. coli_.

high complexity. Regardless of the workflow implemented, the time to result is vastly reduced from several days to as little as 1 h. A study comparing the performance of the two most comprehensive panels (BioFire and Luminex) showed high sensitivities and specificities for both tests, though the number of positive samples was relatively low [43].

Clinical Utility

Although many patients with acute gastroenteritis do not seek medical attention, for those that do it is important to determine the cause of acute gastroenteritis to appropriately inform clinicians whether antibiotics might be warranted or if they are contraindicated. In addition, since much of acute GI disease in the United States is food-borne, it is prudent to diagnose patients so that appropriate public health investigations can occur to potentially prevent more infections. Antimicrobial susceptibility testing is rarely needed for bacterial causes of diarrhea, so isolating the organism

is often not necessary. Since there are a limited number of bacterial causes of acute, community-acquired gastroenteritis, a molecular panel is particularly attractive to provide a syndromic-based approach to diagnosis. It has been reported that for 65% of the positive results obtained using a comprehensive molecular panel, the physician did not order testing for the positive pathogen, arguing that multianalyte panels provide a more accurate diagnosis [41]. It has also been noted that the positivity rate using molecular detection is higher than that of traditional methods which could be due to increased sensitivity or decreased specificity. Discrepant analysis has shown that molecular-positive culture-negative do not always confirm, so confirmation using traditional methods may be needed for some targets [44,45].

Limitations of Testing

Although there are a limited number of bacterial pathogens associated with acute gastroenteritis, the incidence of certain infections may vary depending on geography, and not all potential bacterial agents are in the currently available commercial panels. For example, *Aeromonas* and *Plesiomonas* are not in any of the panels, and the availability of *Vibrio* spp. and *Yersinia enterocolitica* is variable. Even if these analytes are included in panels, it may be difficult for laboratories that rarely see these infections to appropriately validate them for clinical reporting. Since the focus of the molecular diarrheal panels has been on community-onset disease, the question of what to do with *Clostridium difficile* often arises [46]. *Clostridium difficile* causes both healthcare- and community-associated diarrhea, but can also be a colonizer, particularly in children less than 2 years old. Some of the panels include *C. difficile*, while others do not. Therefore, careful consideration should be given to determine the best institutional approach for the diagnosis of *C. difficile*-associated disease. Since most of the organisms in the panels are not transmitted within the hospital, testing should be limited to outpatients or the first 3 days of hospitalization. As with other applications, molecular GI panels should not be used as a test of cure since the microbial DNA may remain for much longer than a viable organism.

Arguably the most significant impact of molecular testing for bacterial gastroenteritis is on public health. Although for clinical purposes, an isolate is usually not needed, public health relies on isolates for epidemiologic typing to assist in the identification of outbreaks [46]. The transition away from traditional culture has limited the number of isolates available for public health investigations. Currently, it is advisable for laboratories to either submit positive stool samples to their public health laboratory for culture, or to culture molecular positive samples and send an isolate to their public health laboratory. In time, molecular epidemiologic tools will catch up with those used in clinical laboratories, but currently isolates continue to be required.

BLOODSTREAM INFECTIONS

Bloodstream infections are responsible for increased patient morbidity and mortality. In the United States, the number of cases has been increasing due to aging of the population, increased numbers of patients with chronic illnesses, increase in invasive procedures, immunosuppressive therapy, chemotherapy, and transplantation, as well as increasing antibiotic resistance [47]. Molecular methods have not yet replaced the need for blood cultures largely due to the low organism burden seen with septicemia. Therefore, efforts have focused on the rapid and accurate identification of organisms and resistance determinates from positive blood culture bottles, decreasing the time to identification from 24–48 to 1–3 h. Prompt therapy with appropriate antibiotics is one of the key factors for reducing mortality in patients with sepsis [47,48].

Peptide Nucleic Acid Florescence In Situ Hybridization

Molecular Target(s) and Technologies

PNA probes incorporated with fluorescence in situ hybridization (PNA-FISH) provide rapid identification and direct visualization of bacteria from positive blood cultures. The probes are normally based on oligonucleotides that are complementary to the organism-specific rRNA sequences in the intact cell. Different targets are labeled with different fluorescent dyes allowing discrimination of various organisms. The assay comprises the following steps: (1) smear fixation, (2) hybridization, (3) posthybridization wash, (4) mounting, and (5) visualization by fluorescence microscopy. Multiple commercial assays are available from AdvanDx including identification for *Enterococcus faecalis*/other enterococci (OE), Gram-negative bacilli (*P. aeruginosa*, *Escherichia coli*, and *K. pneumoniae*; GNR Traffic Light), and staphylococci.

Clinical Utility

PNA-FISH provides species identification at least one workday earlier than traditional or MALDI-TOF MS identifications with the turnaround time of about 2.5 h. In a study, PNA-FISH (*Staphylococcus aureus*/CNS, *Enterococcus faecalis*/OE, GNR Traffic Light) had 98.4% concordant results with MALDI-TOF MS in 921 positive blood samples [49]. For *S. aureus*, PNA-FISH showed sensitivity, specificity, positive predictive value, and negative predictive value of 99–100%, 96–100%, 99–100%, and 98–100% [50–52]. Utilization of multicolored fluorescence dyes in PNA-FISH allows qualitative identification of mixed growth, which is often not appreciated by Gram stain alone. Several studies have shown that patients receive earlier appropriate antibiotic therapy after PNA-FISH [53–55]. Forrest et al. also showed a significant decrease in 30-day mortality after implementation of PNA-FISH for discrimination of enterococci (25% vs 45%) [53]. However, such benefits cannot be obtained without active notification or antimicrobial stewardship intervention [56].

Limitations of Testing

PNA-FISH requires more hands-on time than other molecular assays, fluorescent microscopy is required, and interpretation of results is subjective. Nonspecific probe binding to nontarget bacteria and inadequate washing after hybridization may result in false-positivity. Stringency between probe and target sequences is very important for the specificity of these tests. Insufficient fixation, ineffective penetration of the probe, and photo-bleaching may lead to false-negative results. The limit of detection of PNA-FISH is about 10^5 colony-forming units/mL, so this method is only applicable in the setting of a positive culture. While PNA-FISH provides a rapid identification, antimicrobial susceptibility cannot always be inferred.

Multiplex Assays from Positive Blood Cultures

Molecular Target(s) and Technologies

The US FDA-cleared tests for the multiplexed identification of bacteria in positive blood cultures include the Verigene Gram-Positive Blood Culture Nucleic Acid Test (BC-GP), the Verigene Blood Culture Gram Negative (BC-GN) (Nanosphere, Northbrook, IL), and the BioFire FilmArray Blood Culture Identification (BCID) test. These platforms are highly multiplexed panels for detection of multiple targets. There are also multiple FDA-cleared assays for the identification of *Staphylococcus aureus*, including the differentiation of MRSA from susceptible *S. aureus*. For a complete list see: http://www.fda.gov/MedicalDevices/Productsand MedicalProcedures/InVitroDiagnostics/ucm330711. htm#microbial.

The Verigene platform utilizes nonamplified nucleic acid with microarray-based detection. The instrument extracts nucleic acid, which then hybridizes to capture oligonucleotides on the test cartridge. A gold nanoparticle labeled probe with complimentary sequence to the target (mediator oligonucleotide) is introduced and coated with silver, resulting in an enhanced optical signal. The BC-GP test detects 15 targets for *Staphylococcus aureus*, *S. epidermidis*, *S. lugdunensis*, *Streptococcus anginosus* group, *S. agalactiae*, *S. pneumoniae*, *S. pyogenes*, *Enterococcus faecalis*, *E. faecium*, *Staphylococcus* spp., *Streptococcus* spp., *Listeria* spp., *mecA* for methicillin resistance, and *vanA* and *vanB* for vancomycin resistance. The BC-GN test detects eight identification targets (*Escherichia coli*, *Klebsiella pneumoniae*, *K. oxytoca*, *P. aeruginosa*, *Acinetobacter* spp., *Citrobacter* spp., *Enterobacter* spp., and *Proteus* spp.) and six resistance markers (CTX-M, IMP, KPC, NDM, OXA, and VIM) to detect extended spectrum beta-lactamases and carbapenemases.

The FilmArray BCID test enables detection of 27 targets including yeasts (*Candida albicans*, *C. glabrata*, *C. parapsilosis*, *C. tropicalis*, and *C. krusei*), Gram-positive bacteria (*Enterococcus* spp., *L. monocytogenes*, *Staphylococcus* spp.,

S. aureus, *Streptococcus* spp., *S. agalactiae*, *S. pneumoniae*, and *S. pyogenes*), Gram-negative bacteria (*Acinetobacter baumannii*, *H. influenzae*, *Neisseria meningitidis*, *P. aeruginosa*, *Enterobacteriaceae*, *Enterobacter cloacae* complex, *Escherichia coli*, *Klebsiella oxytoca*, *K. pneumoniae*, *Proteus* spp., and *Serratia marcescens*), and antibiotic resistance genes (*mecA*, *vanA/B*, and KPC). Like other FilmArray assays, the entire process from extraction to amplification by nested multiplex PCR to detection by endpoint melting curve is contained within one reaction pouch.

Clinical Utility

The Verigene BC-GP showed an overall concordance for organism identification of 92−97% in less than 2.5 h with less than 5 min hands-on time [57,58]. In a recent study, implementation of the Verigene BC-GP for identification of Gram-positive cocci in pairs or chains using an institution-developed algorithm reduced time to acceptable antibiotic overall from 13.2 to 1.9 h, and time to appropriate antibiotic for patients with vancomycin-resistant *Enterococcus* from 43.7 to 4.2 h [59,60]. The Verigene BC-GN showed 97.4% and 92.3% agreement with routine methods for identification and detection of resistant markers in 125 isolates [61]. With an average of 24 h faster for identification compared to traditional methods in a study, BC-GN could allow modification of medical management for 31.8% of patients 33 h sooner [62]. Shorter time for effective (3.3 vs 7.0 h) and optimal therapy (23.5 vs 41.8 h) has been demonstrated with the BG-GN assay with a sensitivity of 97.1% and specificity of 99.5% [63].

The FilmArray BCID test requires 2 min of hands-on time and turnaround time of approximately 1 h. The assay gives identification of 88−95% of pathogens recovered from positive blood cultures with correct identifications of 98% and 100% to the genus and the species/complex level [64−66]. Implementation of the assay, combined with an institution's antimicrobial stewardship program, could impact appropriate therapy for 99.2% of positive blood cultures [67].

Limitations of Testing

Large panel assays are less accurate with polymicrobial infections. For the Verigene BC-GP, concordant results decreased from 94% in monomicrobial infections to 76% for identification in polymicrobial infections, and the FilmArray could detect all targets in only 71% of the polymicrobial growth samples [57,68]. Similarly, the BC-GN identified all organisms at 54.5% and at least one organism at 95.4% of polymicrobial specimens [69]. It is worth noting that the manufacturers have not evaluated all species. Moreover, false-negative results can be due to sequence variants in the target, inhibitors in the specimen, or inadequate concentration for detection. False-positive *S. pneumoniae* and *Streptococcus* spp. BC-GP results can occur due to cross-reactivity of probes to *Streptococcusmitis* and *Lactococcus* species. Although rapid results are provided, the institution must have an appropriate reporting scheme to allow pharmacists or physicians to rapidly respond to the results, thereby leading to benefits in patient care.

GROUP B *STREPTOCOCCUS* SCREENING

Group B *Streptococcus* (GBS) causes invasive neonatal infections associated with morbidity and mortality. A major risk factor for early-onset neonatal disease is maternal colonization with GBS in the genitourinary or GI tracts, which is preventable by intravenous intrapartum antibiotic prophylaxis. About 10–30% of pregnant women are colonized, therefore the CDC recommends screening of all pregnant women for vaginal and rectal GBS colonization between 35 and 37 weeks gestation [70]. Traditional specimen processing requires 36–72 h, including 18–24 h growth in selective enrichment broth (ie, Lim broth) prior to subculture. Molecular techniques have been developed to increase sensitivity and facilitate rapid detection of GBS colonization.

Molecular Target(s) and Technologies

Nonamplification probe-based methods target *S. agalactiae*-specific rRNA sequences: the GBS PNA-FISH (AdvanDx) is indicated for turbid Lim broth and the Gen-Probe Accuprobe GBS test (Hologic) is approved for turbid Lim broth and culture identification. Multiple NAATs for the detection of GBS are available (see: http://www.fda.gov/MedicalDevices/ProductsandMedicalProcedures/InVitroDiagnostics/ucm330711.htm#microbial). The Illumigene GBS Assay targets 213 bp sequence residing in the 593–805 bp region of *S. agalactiae* genome segment 3 and is based on LAMP. Cepheid and BD platforms are real-time PCR-based technology. BD assays target 124 bp region of *cfb* gene sequence, the gene that encodes the CAMP factor, whereas Cepheid assays detect a target within a 3′ DNA region adjacent to *cfb* gene. The GeneXpert and the BD Max are fully automated systems while the Cepheid Smart GBS Assay and the GeneOhm StrepB Assay are performed on the SmartCycler through PCR amplification of the target and fluorogenic target specific hybridization (TaqMan and Molecular beacons probes, respectively).

Clinical Utility

Molecular assays are attractive in terms of sensitivity and short test times (about 55–75 min). Although antenatal culture using enrichment broth is considered the gold standard for detection of GBS colonization, the sensitivity has been shown to be as low as 53.6% [71]. NAAT performed on enriched culture increases the sensitivity to 90.9–100% with a specificity of 92.5–99.3% [71–74]. NAATs increased GBS detection rates from 15–26.5% using antenatal cultures to 30–31.5%. Only the Cepheid Smart GBS Assay, the Xpert GBS, and the GeneOhm StrepB have been approved for direct vaginal/rectal swab testing. Therefore, they can be useful for intrapartum testing for pregnant women who have suboptimal prenatal care to diminish the use of antibiotics. Intrapartum NAATs demonstrated sensitivities of 90.7–95.8% (vs 54.3–84.3% for antenatal cultures) with specificities of 64.5–97.6% [75–78]. A cost-effective analysis model also showed a cost-saving benefit of $6–7 from the implementation of intrapartum PCR testing over the 35- to 37-week culture for maternal risk stratification [79].

Limitations of Testing

Though not frequently needed, susceptibility testing of GBS is not provided by molecular methods. Although GBS is uniformly susceptible to penicillin, testing for clindamycin susceptibility is needed for penicillin-allergic patients. Implementation of intrapartum testing is limited by availability of testing and turnaround time. Hence, intrapartum testing does not replace antenatal culture. The American College of Obstetricians and Gynecologists recommends giving antibiotic prophylaxis regardless of the NAAT results for pregnant women with increased intrapartum risk factors (ie, fever, prolonged rupture of membranes) [80].

FUTURE PERSPECTIVES

Initial challenges such as reagent costs, instrumentation, and technical expertise prevented many laboratories from offering molecular tests for bacterial pathogens. Now, many manufacturers have competitive pricing and moderately complex tests that can easily be performed by any laboratorian. The simplicity of many of these tests allows them to be offered 24 h a day and 7 days a week, thereby greatly decreasing time to result and increasing potential impact on patient care. The cost–benefit of transitioning from traditional methods to molecular methods for bacterial detection and identification is obvious, for example, such as MALDI-TOF MS, but less clear for the large syndromic panels. As more outcome-based studies are published, the clinical utility of the newer syndromic-based molecular panels will become clearer.

Through the human microbiome project, the importance of variances in microbiota is being realized. As these data are aligned with specific clinical predictions, we will need to determine how and when technologies such as next-generation sequencing will be implemented in the clinical microbiology laboratory. There are many challenges associated with clinical next-generation sequencing, including cost, accurate interpretation, and clinically-relevant reporting, that need to be carefully considered prior to clinical implementation of microbiome analyses.

As methodologies continue to decrease in cost and complexity, molecular testing for clinical microbiology laboratories will no longer be limited to large, tertiary, or academic medical centers. They will, undoubtedly, transition closer to patient care where their impact can be more immediate. In fact, the first molecular point-of-care device has now been FDA-approved (for influenza and GAS). Never before has clinical microbiology changed at the rapid pace we are currently experiencing. We must remember that the power of molecular technologies should be coupled with well-controlled and clinically-relevant diagnostic approaches to have the greatest impact on patient care.

References

[1] Nielsen PE, Egholm M, Berg RH, Buchardt O. Sequence-selective recognition of DNA by strand displacement with a thymine-substituted polyamide. Science 1991;254:1497–500.

[2] Egholm M, Buchardt O, Christensen L, et al. PNA hybridizes to complementary oligonucleotides obeying the Watson-Crick hydrogen-bonding rules. Nature 1993;365:566–8.

[3] Stender H. PNA FISH: an intelligent stain for rapid diagnosis of infectious diseases. Expert Rev Mol Diagn 2003;3:649–55.

[4] Stender H, Fiandaca M, Hyldig-Nielsen JJ, Coull J. PNA for rapid microbiology. J Microbiol Methods 2002;48:1–17.

[5] Clinical and Laboratory Standards Institute. Interpretive criteria for identification of bacteria and fungi by DNA target sequencing; approved guideline. Clinical and Laboratory Standards Institute; 2008MM18-A

[6] Woese CR. Bacterial evolution. Microbiol Rev 1987;51:221–71.

[7] Gurtler V, Stanisich VA. New approaches to typing and identification of bacteria using the 16S-23S rDNA spacer region. Microbiology 1996;142:3–16.

[8] Shang S, Fu J, Dong G, Hong W, Du L, Yu X. Establishment and analysis of specific DNA patterns in 16S-23S rRNA gene spacer regions for differentiating different bacteria. Chin Med J (Engl) 2003;116:129–33.

[9] Felmlee TA, Oda RP, Persing DA, Landers JP. Capillary electrophoresis of DNA potential utility for clinical diagnoses. J Chromatogr A 1995;717:127–37.

[10] Diggle MA, Clarke SC. Pyrosequencing: sequence typing at the speed of light. Mol Biotechnol 2004;28:129–37.

[11] Cook VJ, Turenne CY, Wolfe J, Pauls R, Kabani A. Conventional methods versus 16S ribosomal DNA sequencing for identification of nontuberculous mycobacteria: cost analysis. J Clin Microbiol 2003;41:1010–15.

[12] Patel R. MALDI-TOF MS for the diagnosis of infectious diseases. Clin Chem 2015;61:100–11.

[13] McElvania Tekippe E, Shuey S, Winkler DW, Butler MA, Burnham CA. Optimizing identification of clinically relevant Gram-positive organisms by use of the Bruker Biotyper matrix-assisted laser desorption ionization-time of flight mass spectrometry system. J Clin Microbiol 2013;51:1421–7.

[14] Westblade LF, Jennemann R, Branda JA, et al. Multicenter study evaluating the Vitek MS system for identification of medically important yeasts. J Clin Microbiol 2013;51:2267–72.

[15] Martiny D, Busson L, Wybo I, El Haj RA, Dediste A, Vandenberg O. Comparison of the Microflex LT and Vitek MS systems for routine identification of bacteria by matrix-assisted laser desorption ionization-time of flight mass spectrometry. J Clin Microbiol 2012;50:1313–25.

[16] Clark AE, Kaleta EJ, Arora A, Wolk DM. Matrix-assisted laser desorption ionization-time of flight mass spectrometry: a fundamental shift in the routine practice of clinical microbiology. Clin Microbiol Rev 2013;26:547–603.

[17] Tran A, Alby K, Kerr A, Jones M, Gilligan PH. Cost savings incurred by implementation of routine microbiological identification by matrix-assisted laser desorption/ionization-time of flight (MALDI-TOF) mass spectrometry. J Clin Microbiol 2015;53(8):2473–9.

[18] Bizzini A, Durussel C, Bille J, Greub G, Prod'hom G. Performance of matrix-assisted laser desorption ionization-time of flight mass spectrometry for identification of bacterial strains routinely isolated in a clinical microbiology laboratory. J Clin Microbiol 2010;48:1549–54.

[19] Basarab M, Macrae MB, Curtis CM. Atypical pneumonia. Curr Opin Pulm Med 2014;20:247–51.

[20] Centers for Disease Control and Prevention. Outbreaks of respiratory illness mistakely attributed to pertussis—New Hampshire, Massachusetts, and Tennessee, 2004–2006. MMWR Morb Mortal Wkly Rep 2007;56:837–42.

[21] Mandal S, Tatti KM, Woods-Stout D, et al. Pertussis pseudo-outbreak linked to specimens contaminated by Bordetella pertussis DNA from clinic surfaces. Pediatrics 2012;129:e424–30.

[22] Weber DJ, Miller MB, Brooks RH, Brown VM, Rutala WA. Healthcare worker with "pertussis": consequences of a false-positive polymerase chain reaction test result. Infect Control Hosp Epidemiol 2010;31:306–7.

[23] Centers for Disease Control and Prevention. Pertussis. In: Rousch S, Baldy, L, editors. Manual for the surveillance of vaccine-preventable diseases. Atlanta, GA; 2015.

[24] Loeffelholz M. Towards improved accuracy of Bordetella pertussis nucleic acid amplification tests. J Clin Microbiol 2012;50:2186–90.

[25] Shulman ST, Bisno AL, Clegg HW, et al. Clinical practice guideline for the diagnosis and management of group A streptococcal pharyngitis: 2012 update by the Infectious Diseases Society of America. Clin Infect Dis 2012;55:1279–82.

[26] Anderson NW, Buchan BW, Mayne D, Mortensen JE, Mackey TL, Ledeboer NA. Multicenter clinical evaluation of the illumigene group A Streptococcus DNA amplification assay for detection of group A Streptococcus from pharyngeal swabs. J Clin Microbiol 2013;51:1474–7.

[27] Cohen DM, Russo ME, Jaggi P, Kline J, Gluckman W, Parekh A. Multicenter clinical evaluation of the novel Alere i Strep A isothermal nucleic acid amplification test. J Clin Microbiol 2015;53:2258–61.

[28] Felsenstein S, Faddoul D, Sposto R, Batoon K, Polanco CM, Dien Bard J. Molecular and clinical diagnosis of group A

streptococcal pharyngitis in children. J Clin Microbiol 2014;52:3884—9.

[29] Lippincott CK, Miller MB, Popowitch EB, Hanrahan CF, Van Rie A. Xpert MTB/RIF assay shortens airborne isolation for hospitalized patients with presumptive tuberculosis in the United States. Clin Infect Dis 2014;59:186—92.

[30] Millman AJ, Dowdy DW, Miller CR, et al. Rapid molecular testing for TB to guide respiratory isolation in the U.S.: a cost-benefit analysis. PLoS One 2013;8:e79669.

[31] Centers for Disease Control and Prevention. Updated guidelines for the use of nucleic acid amplification tests in the diagnosis of tuberculosis. MMWR Morb Mortal Wkly Rep 2009;58:7—10.

[32] Boehme CC, Nabeta P, Hillemann D, et al. Rapid molecular detection of tuberculosis and rifampin resistance. N Engl J Med 2010;363:1005—15.

[33] Centers for Disease Control and Prevention. Availability of an assay for detecting *Mycobacterium tuberculosis*, including rifampin-resistant strains, and considerations for its use—United States, 2013. MMWR Morb Mortal Wkly Rep 2013;62:821—4.

[34] Lippincott CK, Miller MB, Van Rie A, Weber DJ, Sena AC, Stout JE. The complexities of Xpert(R) MTB/RIF interpretation. Int J Tuberc Lung Dis 2015;19:273—5.

[35] Cheng VC, Yew WW, Yuen KY. Molecular diagnostics in tuberculosis. Eur J Clin Microbiol Infect Dis 2005;24:711—20.

[36] Zhang H, Morrison S, Tang YW. Multiplex polymerase chain reaction tests for detection of pathogens associated with gastroenteritis. Clin Lab Med 2015;35:461—86.

[37] Anderson NW, Buchan BW, Ledeboer NA. Comparison of the BD MAX enteric bacterial panel to routine culture methods for detection of *Campylobacter*, enterohemorrhagic *Escherichia coli* (O157), *Salmonella*, and *Shigella* isolates in preserved stool specimens. J Clin Microbiol 2014;52:1222—4.

[38] Harrington SM, Buchan BW, Doern C, et al. Multicenter evaluation of the BD max enteric bacterial panel PCR assay for rapid detection of *Salmonella* spp., *Shigella* spp., *Campylobacter* spp. (*C. jejuni* and *C. coli*), and Shiga toxin 1 and 2 genes. J Clin Microbiol 2015;53:1639—47.

[39] Buss SN, Leber A, Chapin K, et al. Multicenter evaluation of the BioFire FilmArray gastrointestinal panel for etiologic diagnosis of infectious gastroenteritis. J Clin Microbiol 2015;53:915—25.

[40] Spina A, Kerr KG, Cormican M, et al. Spectrum of enteropathogens detected by the FilmArray GI Panel in a multicentre study of community-acquired gastroenteritis. Clin Microbiol Infect 2015;21:719—28.

[41] Claas EC, Burnham CA, Mazzulli T, Templeton K, Topin F. Performance of the xTAG(R) gastrointestinal pathogen panel, a multiplex molecular assay for simultaneous detection of bacterial, viral, and parasitic causes of infectious gastroenteritis. J Microbiol Biotechnol 2013;23:1041—5.

[42] Buchan BW, Olson WJ, Pezewski M, et al. Clinical evaluation of a real-time PCR assay for identification of *Salmonella*, *Shigella*, *Campylobacter* (*Campylobacter jejuni* and *C. coli*), and shiga toxin-producing *Escherichia coli* isolates in stool specimens. J Clin Microbiol 2013;51:4001—7.

[43] Khare R, Espy MJ, Cebelinski E, et al. Comparative evaluation of two commercial multiplex panels for detection of gastrointestinal pathogens by use of clinical stool specimens. J Clin Microbiol 2014;52:3667—73.

[44] Vocale C, Rimoldi SG, Pagani C, et al. Comparative evaluation of the new xTAG GPP multiplex assay in the laboratory diagnosis of acute gastroenteritis. Clinical assessment and potential application from a multicentre Italian study. Int J Infect Dis 2015;34:33—7.

[45] Wessels E, Rusman LG, van Bussel MJ, Claas EC. Added value of multiplex Luminex Gastrointestinal Pathogen Panel (xTAG (R) GPP) testing in the diagnosis of infectious gastroenteritis. Clin Microbiol Infect 2013;20:O182—7.

[46] Bloomfield MG, Balm MN, Blackmore TK. Molecular testing for viral and bacterial enteric pathogens: gold standard for viruses, but don't let culture go just yet? Pathology 2015;47:227—33.

[47] Dellinger RP, Levy MM, Rhodes A, et al. Surviving sepsis campaign: international guidelines for management of severe sepsis and septic shock: 2012. Crit Care Med 2003;41:580—637.

[48] Kumar A, Ellis P, Arabi Y, et al. Initiation of inappropriate antimicrobial therapy results in a fivefold reduction of survival in human septic shock. Chest 2009;136:1237—48.

[49] Calderaro A, Martinelli M, Motta F, et al. Comparison of peptide nucleic acid fluorescence in situ hybridization assays with culture-based matrix-assisted laser desorption/ionization-time of flight mass spectrometry for the identification of bacteria and yeasts from blood cultures and cerebrospinal fluid cultures. Clin Microbiol Infect 2014;20:O468—75.

[50] Chapin K, Musgnug M. Evaluation of three rapid methods for the direct identification of *Staphylococcus aureus* from positive blood cultures. J Clin Microbiol 2003;41:4324—7.

[51] Gonzalez V, Padilla E, Gimenez M, et al. Rapid diagnosis of *Staphylococcus aureus* bacteremia using *S. aureus* PNA FISH. Eur J Clin Microbiol Infect Dis 2004;23:396—8.

[52] Oliveira K, Procop GW, Wilson D, Coull J, Stender H. Rapid identification of *Staphylococcus aureus* directly from blood cultures by fluorescence in situ hybridization with peptide nucleic acid probes. J Clin Microbiol 2002;40:247—51.

[53] Forrest GN, Roghmann MC, Toombs LS, et al. Peptide nucleic acid fluorescent in situ hybridization for hospital-acquired enterococcal bacteremia: delivering earlier effective antimicrobial therapy. Antimicrob Agents Chemother 2008;52:3558—63.

[54] Laub RR, Knudsen JD. Clinical consequences of using PNA-FISH in staphylococcal bacteraemia. Eur J Clin Microbiol Infect Dis 2014;33:599—601.

[55] Parcell BJ, Orange GV. PNA-FISH assays for early targeted bacteraemia treatment. J Microbiol Methods 2013;95:253—5.

[56] Holtzman C, Whitney D, Barlam T, Miller NS. Assessment of impact of peptide nucleic acid fluorescence in situ hybridization for rapid identification of coagulase-negative staphylococci in the absence of antimicrobial stewardship intervention. J Clin Microbiol 2011;49:1581—2.

[57] Samuel LP, Tibbetts RJ, Agotesku A, Fey M, Hensley R, Meier FA. Evaluation of a microarray-based assay for rapid identification of Gram-positive organisms and resistance markers in positive blood cultures. J Clin Microbiol 2013;51:1188—92.

[58] Wojewoda CM, Sercia L, Navas M, et al. Evaluation of the Verigene Gram-positive blood culture nucleic acid test for rapid detection of bacteria and resistance determinants. J Clin Microbiol 2013;51:2072—6.

[59] Alby K, Daniels LM, Weber DJ, Miller MB. Development of a treatment algorithm for streptococci and enterococci from positive blood cultures identified with the Verigene Gram-positive blood culture assay. J Clin Microbiol 2013;51:3869—71.

[60] Roshdy DG, Tran A, LeCroy N, et al. Impact of a rapid microarray-based assay for identification of positive blood cultures for treatment optimization for patients with streptococcal and enterococcal bacteremia. J Clin Microbiol 2015;53:1411—14.

[61] Dodemont M, De Mendonca R, Nonhoff C, Roisin S, Denis O. Performance of the Verigene Gram-negative blood culture assay for rapid detection of bacteria and resistance determinants. J Clin Microbiol 2014;52:3085—7.

[62] Hill JT, Tran KD, Barton KL, Labreche MJ, Sharp SE. Evaluation of the nanosphere Verigene BC-GN assay for direct identification

of gram-negative bacilli and antibiotic resistance markers from positive blood cultures and potential impact for more-rapid antibiotic interventions. J Clin Microbiol 2014;52:3805−7.

[63] Bork JT, Leekha S, Heil EL, Zhao L, Badamas R, Johnson JK. Rapid testing using the Verigene Gram-negative blood culture nucleic acid test in combination with antimicrobial stewardship intervention against Gram-negative bacteremia. Antimicrob Agents Chemother 2015;59:1588−95.

[64] Bhatti MM, Boonlayangoor S, Beavis KG, Tesic V. Evaluation of FilmArray and Verigene systems for rapid identification of positive blood cultures. J Clin Microbiol 2014;52:3433−6.

[65] Blaschke AJ, Heyrend C, Byington CL, et al. Rapid identification of pathogens from positive blood cultures by multiplex polymerase chain reaction using the FilmArray system. Diagn Microbiol Infect Dis 2012;74:349−55.

[66] Rand KH, Delano JP. Direct identification of bacteria in positive blood cultures: comparison of two rapid methods, FilmArray and mass spectrometry. Diagn Microbiol Infect Dis 2014;79:293−7.

[67] Southern TR, VanSchooneveld TC, Bannister DL, et al. Implementation and performance of the BioFire FilmArray(R) Blood Culture Identification panel with antimicrobial treatment recommendations for bloodstream infections at a midwestern academic tertiary hospital. Diagn Microbiol Infect Dis 2015;81:96−101.

[68] Altun O, Almuhayawi M, Ullberg M, Ozenci V. Clinical evaluation of the FilmArray blood culture identification panel in identification of bacteria and yeasts from positive blood culture bottles. J Clin Microbiol 2013;51:4130−6.

[69] Ledeboer NA, Lopansri BK, Dhiman N, et al. Identification of gram-negative bacteria and genetic resistance determinants from positive blood culture broths using the Verigene gram-negative blood culture multiplex microarray-based molecular assay. J Clin Microbiol 2015;53(8):2460−72.

[70] Centers for Disease Control and Prevention. Prevention of perinatal group B streptococcal disease—revised guidelines from CDC, 2010. MMWR Morb Mortal Wkly Rep—Recomm Rep 2010;59:1−36.

[71] Miller SA, Deak E, Humphries R. Comparison of the AmpliVue, BD Max System, and illumigene molecular assays

for detection of Group B Streptococcus in antenatal screening specimens. J Clin Microbiol 2015;53:1938−41.

[72] Block T, Munson E, Culver A, Vaughan K, Hryciuk JE. Comparison of carrot broth- and selective Todd-Hewitt broth-enhanced PCR protocols for real-time detection of Streptococcus agalactiae in prenatal vaginal/anorectal specimens. J Clin Microbiol 2008;46:3615−20.

[73] Goodrich JS, Miller MB. Comparison of culture and 2 real-time polymerase chain reaction assays to detect group B Streptococcus during antepartum screening. Diagn Microbiol Infect Dis 2007;59:17−22.

[74] Scicchitano LM, Bourbeau PP. Comparative evaluation of the AccuProbe Group B Streptococcus Culture Test, the BD GeneOhm Strep B assay, and culture for detection of group B streptococci in pregnant women. J Clin Microbiol 2009; 47:3021−3.

[75] Davies HD, Miller MA, Faro S, Gregson D, Kehl SC, Jordan JA. Multicenter study of a rapid molecular-based assay for the diagnosis of group B Streptococcus colonization in pregnant women. Clin Infect Dis 2004;39:1129−35.

[76] Gavino M, Wang E. A comparison of a new rapid real-time polymerase chain reaction system to traditional culture in determining group B Streptococcus colonization. Am J Obstet Gynecol 2007;197:388, e1−4.

[77] Money D, Dobson S, Cole L, et al. An evaluation of a rapid real time polymerase chain reaction assay for detection of group B Streptococcus as part of a neonatal group B Streptococcus prevention strategy. J Obstet Gynaecol Can 2008;30:770−5.

[78] Young BC, Dodge LE, Gupta M, Rhee JS, Hacker MR. Evaluation of a rapid, real-time intrapartum group B Streptococcus assay. Am J Obstet Gynecol 2011;205 372, e1−6.

[79] Haberland CA, Benitz WE, Sanders GD, et al. Perinatal screening for group B streptococci: cost-benefit analysis of rapid polymerase chain reaction. Pediatrics 2002;110:471−80.

[80] American College of Obstetricians and Gynecologists. ACOG Committee Opinion No. 485: prevention of early-onset group B streptococcal disease in newborns. Obstet Gynecol 2011; 117:1019−27.

13

Agents Associated with Sexually Transmitted Infections

P. Verhoeven[1,2], F. Grattard[1,2], S. Gonzalo[1,2], M. Memmi[1,2] and B. Pozzetto[1,2]

[1]GIMAP EA3064, University of Lyon, Saint-Etienne, France [2]Laboratory of Infectious Agents and Hygiene, University Hospital of Saint-Etienne, Saint-Etienne, France

INTRODUCTION

Sexually transmitted infections (STIs) represent a major burden in terms of Public Health. It was estimated by the World Health Organization (WHO) that 500 million cases occurred in 2008 worldwide among the most common of STIs, namely, syphilis, gonorrhea, chlamydia, and trichomoniasis (Fig. 13.1) [1]. In addition to their immediate medical impact, notably from a psychological point of view, STIs are the major factor in male and female infertility, are associated with severe obstetric and neonatal morbidity, and are responsible for chronic life-threatening diseases including immunodeficiency, cancer, liver and heart failure, and neuropsychiatric disease. Their social and economic impact is increasing with the change in sexual behaviors, notably in low-resource areas [2]. Because most of them are asymptomatic or paucisymptomatic at the initial phase, the diagnosis is often missed, which favors their epidemiological spread. The frequent association of several STI agents in the genital tract of a same subject is another feature that increases their dissemination. The transmission of the human immunodeficiency virus type 1 (HIV-1) is particularly influenced by the presence of other STI agents [3–5].

This chapter provides an overview of STIs associated with bacterial agents, namely, syphilis, gonorrhea, chlamydia, and *Mycoplasma genitalium* infection, and a genital parasite responsible for trichomoniasis. Nucleic acid amplification tests (NAATs) have played a major role during the last decade for the screening and diagnosis of STIs. The aim of this review is to delineate the place of molecular testing in the diagnosis of STIs of bacterial and parasitic origin. In addition to specific sections dedicated to the role of NAATs in the diagnosis of each of these agents, a concluding paragraph will focus on future directions, including (1) the study of vaginal microbiome for enhancing the control of bacterial STIs, (2) the multiplex approach of STI diagnosis by using molecular tools, (3) the implementation of point-of-care (POC) testing for better managing STIs, and (4) the need for molecular tools dedicated to the detection of antimicrobial resistance of STI agents.

SYPHILIS

Overview of the Disease and its Epidemiology

Syphilis is a complex bacterial disease caused by a motile spiral-shaped spirochete, called *Treponema pallidum*. It evolves classically in three stages including a primary phase occurring after an incubation period of 10–90 days and characterized by a nonpainful ulcer or chancre that is located at the site of primary inoculation (genital tract, anus, skin, mouth, etc.). Four to ten weeks after chancre appearance, the secondary phase involves blood dissemination of the motile bacteria that infiltrate tissues, including the central nervous system, and is sometimes revealed by various

Diagnostic Molecular Pathology
DOI: http://dx.doi.org/10.1016/B978-0-12-800886-7.00013-3

© 2017 Elsevier Inc. All rights reserved.

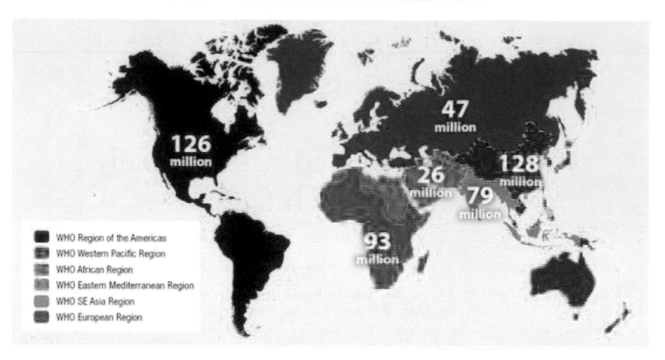

FIGURE 13.1   Estimated new cases of curable STIs (gonorrhea, chlamydia, syphilis, and trichomoniasis) by region from the WHO, 2008 [1].

nonspecific symptoms dominated by disseminated maculopapular rash, general malaise, and meningitis. When the infection is diagnosed during these two early phases, an antibiotic treatment, usually based on penicillin G (or macrolides in case of allergy to penicillin), is able to cure the infection definitively. If *T. pallidum* invasion is neglected, a latent infection is established with bacterium persistence in different tissues, which may lead after years or decades to noninfectious tertiary lesions including gumma, cardiovascular disease, and tertiary neurosyphilis with two clinical pictures named general paresis and tabes dorsalis [6]. If untreated, syphilis occurring during pregnancy may lead to stillbirth and congenital syphilis.

Syphilis is an old disease that was omnipresent worldwide during the 19th century and the first half of the 20th century. Whereas it remains highly prevalent in South and Southeast Asia and Sub-Saharan Africa, the introduction of antibiotics eradicated the disease in developed countries. However, its prevalence increased dramatically at the beginning of this century, notably in association with HIV-1 infection and within the networks of subjects with risky sexual practices, particularly in commercial sex workers and in men having sex with men (MSM), with outbreaks occurring in different parts of the developed world including the United States, Europe, Russia, and China [7].

Current Diagnosis of Syphilis

Despite tremendous efforts, *T. pallidum* remains nonculturable on inert media. At the primary phase, the bacterium can be detected in peripheral lesions by dark-field microscopy and direct immunofluorescence (Fig. 13.2A), although these methods are relatively insensitive and only accessible to specialized laboratories [8]. The diagnosis usually relies on serological testing that combines the detection of two categories of antibodies termed non-treponemal, which are directed against phospholipids, and treponemal, which recognize specific *T. pallidum* polypeptides. Non-treponemal antibodies are detected by agglutination methods whereas treponemal serology relies on immunofluorescence, hemagglutination, or, more recently, immunoassays. Different algorithms are used to date infection [6,9]. The non-treponemal antibodies are a good indicator of recent infection and their dynamics is also used for evaluating the efficacy of treatment. Both treponemal and non-treponemal tests may lead to false-positive results, notably in populations with low incidence of infection, such as pregnant women.

Place of Syphilis Molecular Testing

At the early stages of infection, the bacterium may be detected by PCR in swabs and biopsy specimens

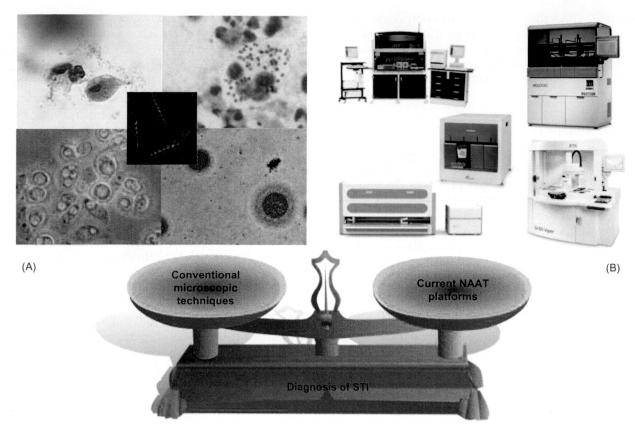

FIGURE 13.2 Schematic representation of some diagnosis tools used for the detection of nonviral agents responsible of STI. (A) Direct examination of a few agents using conventional microscopic methods (upper left: Giemsa staining of *Trichomonas vaginalis*; upper right: gram staining of purulent discharge with neutrophils and gram-negative cocci of *Neisseria gonorrhoeae*; lower right: "egg-fried" microcolonies of mycoplasma in culture; lower left: chlamydial inclusions in McCoy cells; center: fluorescent spirochetes of *Treponema pallidum*). (B) The most common commercial platforms that are currently used for the NAAT of STIs (upper left: Abbott m2000 platform; upper right: Hologic Panther platform; lower right: Becton Dickinson Visper platform; lower left: Roche Cobas 4800 platform; center: Cepheid GeneExpert platform).

from genital and mucosal ulcers, placental specimens, cerebrospinal fluid, and oral lesions, even if the presence of commensal treponemes may be responsible for false-positive results. Different targets of *T. pallidum* genome have been selected for PCR, including the *pol A* gene, the 47-kDa integral membrane lipoprotein gene, the *bmp* gene, and a 366 bp region of the 16S rRNA [8,10–12]. Molecular testing is also useful for typing strains of *T. pallidum* in order to better understand the spread of the disease within different communities. The genes that are targeted for epidemiological purposes are the *arp* gene and different *Tpr* subfamily II genes, using restriction fragment length polymorphism (RFLP) [13]. Typing is also useful for characterizing the epidemiology of macrolide-resistant strains [14,15]. As a whole, molecular tests are not readily used for the routine diagnosis of syphilis and remain the exclusivity of reference laboratories. However, their progressive incorporation to multiplex tools dedicated to the diagnosis of venereal diseases is in progress.

GONORRHEA

Overview of the Disease and its Epidemiology

Gonorrhea is a bacterial infection caused by *Neisseria gonorrhoeae*, a gram-negative coccus that is strictly restricted to humans and has been shown to occupy a particular environmental niche where it adapts rapidly to host influences, which is responsible for difficulties in terms of diagnosis and treatment. As for syphilis, gonorrhea burden, with an estimated 100 million cases worldwide each year [16], is high in low-income settings and notably in many developing countries.

Gonococcal genital infection consists mainly in urethritis in men that is symptomatic in most cases under the form of a purulent discharge (also termed "pissing glass"), and in vaginitis and cervicitis in women that is asymptomatic in approximately half of the cases. Rectal and pharyngeal localizations are also observed and exhibit a high rate of asymptomatic infections.

In women, gonorrhea may lead to complicated infection including pelvic inflammatory disease, tubal infertility, ectopic pregnancy, and chronic pelvic pain. It was also shown to favor the acquisition and transmission of HIV-1, especially in MSM [17].

Current Diagnosis of Gonorrhea

Bacterial culture remains the gold standard for the diagnosis and follow-up of gonococcal infection. In addition to its modest cost, it is presently the only way for managing the improving resistance of *N. gonorrhoeae* to antimicrobials, and notably to extended spectrum cephalosporins, which is an increasing challenge for the future infection [16,18]. Disadvantages of bacterial culture include the requirement of invasive specimens with high bacterial load, the need for rapid and appropriate transportation of these specimens in order to preserve germ viability, and the time required for getting the results that ranges from 2 to 5 days.

Direct examination after gram staining (Fig. 13.2A) is very useful in case of symptomatic urethritis with purulent discharge in males. In this situation, the sensitivity is similar to that of culture with specificity close to 100% and with results available within a few minutes. However, this test is insensitive in cases of asymptomatic genital infection or extragenital infection [19]. The same approach is valuable in the different antigen tests that were proposed as POC tests [9].

Nucleic Acid Tests Specific for Gonococcal Infection

Many nucleic acid tests have been proposed for the diagnosis of gonorrhea and are currently used in many laboratories due to their relative simplicity of use. They include hybridization methods and both in-house and commercial amplification methods. Concerning hybridization methods, two commercial tests were developed that used an oligonucleotide-specific probe, namely, GenProbe PACE II and Digene Hybrid Capture II assays. However, both were shown to have lower sensitivity and specificity than culture [20]. These assays have been supplanted by amplification-based assays that exhibit better analytical performance. In addition to in-house PCR assays targeting the *porA*, *opa*, or 16S genes of *N. gonorrhoeae* and mostly used as confirmatory tests [21], at least seven multiplex NAATs are commercially available for the simultaneous detection of *N. gonorrhoeae* and *Chlamydia trachomatis* in genital swabs and first-catch urine, with different amplification technologies (Fig. 13.2B). Three of them are based on PCR: (1) Cobas 4800 CT/NG Test (Roche Molecular Diagnostics, New Jersey, USA), (2)

GeneXpert CT/NG Assay (Cepheid, California, USA), and (3) RealTime CT/NG assay (Abbott Molecular, Illinois, USA). Two are based on transcription-mediated amplification (TMA): Aptima Combo 2 and Aptima CT assays (Hologic Gen-Probe Inc, California, USA). Two others are based on strand displacement amplification (SDA): ProbeTec ET CT/GC Amplified DNA and ProbeTec Qx Amplified DNA assays (Becton Dickinson, Maryland, USA).

The main advantages of these tests are (1) the use of specimens easy to collect (self-sampled vaginal swab, first-catch urine, tampon samples) and store (nonviable germs are sufficient), (2) their ability to be multiplexed and automated, and (3) their great sensitivity by comparison to culture. However, a series of disadvantages are also recognized [22,23], including (1) the inability to provide epidemiological data on antibiotic susceptibility, (2) the possibility of false-positive results leading to detrimental psychological consequences and useless antibiotic treatments, (3) the risk of selecting resistant strains to current antibiotics due to inappropriate treatment, (4) their nonvalidation for anal, pharyngeal, or semen specimens, and finally (5) their higher cost. Consequently, recent guidelines insisted on the need to avoid the systematic testing of *N. gonorrhoeae* by molecular assay in asymptomatic women belonging to low-prevalence population or to confirm systematically the positive results by a culture test [24,25]. By contrast, NAATs are well suited for the diagnosis of gonococcal genital symptomatic infections in at-risk populations (MSM, sex workers, indigenous communities with high prevalence of infection, and others). The development of new-generation NAATs, such as the Cepheid GenXpert assay that exhibits high sensitivity and specificity on genital samples and urine [26], may modulate these recommendations.

For epidemiological purposes and investigation of outbreaks, different molecular methods can be used, including RFLP, opa typing (based on the family of 11 *opa* genes), sequencing of hypervariable genes (*porB* or *tbpB*), and MultiLocus Sequence Typing [19,20]. Regarding antimicrobial resistance, PCR assays able to detect penicillinase-producing *N. gonorrhoeae* directly in clinical samples [27] and for predicting ciprofloxacin resistance [28,29] have been reported. Molecular tests targeting genes of antimicrobial resistance to ceftriaxone and azithromycin are also needed.

GENITAL CHLAMYDIA INFECTION

Overview of the Disease and its Epidemiology

Genital chlamydia infection is due to a spherical bacterium called *C. trachomatis*. Given its small size and its obligate intracellular parasitism, this organism was long confused with a virus. However, unlike viruses, it contains both types of nucleic acids and is susceptible to certain antibiotics, including macrolides, cyclines, and quinolones. With an estimated 100 million cases worldwide each year [1], chlamydia is (at equality with gonorrhea) the most common cause of curable bacterial STI and the most common STI in the United States and Europe. Early sexual exposure in young women is considered as a favoring factor, together with drug abuse, smoking, use of oral contraceptives, and poor socioeconomic condition.

In men, C. trachomatis infection is responsible for urethritis, epididymitis, proctitis, and pharyngitis, especially in MSM. In women, it is a common cause of asymptomatic endocervicitis. If untreated, C. trachomatis infection can lead to endometritis, salpingitis, pelvic inflammatory disease, ectopic pregnancy, and tubal infertility. Neonates born from infected mothers are at risk of developing conjunctivitis and severe pneumonitis. In both genders, it is associated with reactive arthritis. All these diseases are due to serovars D to K of C. trachomatis, whereas serovars L1, L2, and L3 are responsible for another STI called lymphogranuloma venereum (LGV), mostly in tropical settings, but also for proctitis in MSM. A recent review on this topic is available [30].

Nonmolecular Methods of Laboratory Diagnosis

Traditionally, tissue culture was considered the gold standard for the diagnosis. Due to technical difficulties inherent to cell culture that is found fastidious, long-lasting, expensive and limited to trained laboratories, alternative immunological techniques were developed, based on enzymatic or fluorescent assays using monoclonal antibodies. However, the sensitivity and specificity of these assays were relatively poor. Serological tests looking for IgA- or IgM-specific antibodies may also be useful in case of pelvic inflammatory disease.

Molecular Methods of Laboratory Diagnosis

Considering the limitations of traditional testing, molecular methods have progressively become the reference, notably those using amplification technologies (NAATs) [31,32]. These methods are easy to implement in laboratories trained to molecular biology, fast, sensitive (ranging from 86% to 100%), specific (>97%), and relatively inexpensive [20,21]. They can be performed on a wide range of clinical specimens including genital and rectal swabs, first-catch urine (considered as noninvasive specimens),

semen, pharyngeal specimens, conjunctival swabs, or peritoneal samples collected during pelvic surgery. As viable organisms are not required, the transportation of specimens is not a critical issue. The main genes targeted by NAATs for the diagnosis of C. trachomatis infection are the major outer membrane protein (MOMP) gene, the cryptic plasmid, the phospholipase gene, and the 16S and 23S rRNA gene. The choice of the sequence targeted by the test is critical for its sensitivity and specificity, as illustrated by the emergence of a mutant strain in 2006 in Sweden that was shown to be deleted of a 377 bp fragment in the cryptic plasmid [33,34]. This strain spread rapidly to other Nordic countries and led the manufacturers who had chosen this target for their molecular test to modify the design of their kit. The seven commercial NAATs currently used for the detection of C. trachomatis DNA are the same as those for the molecular diagnosis of gonorrhea since all these tests were multiplexed for detecting both organisms simultaneously (Fig. 13.2B). By contrast to N. gonorrhoeae infection, the limitations of NAATs used for the diagnosis of C. trachomatis are very few: (1) resistance to antibiotics is limited and cell culture cannot be used as an alternative for studying antimicrobial susceptibility, (2) false-positive results are limited to recent infections treated by antibiotics (NAATs become negative within 3 weeks of treatment), and (3) false-negative results due to amplification inhibitors can be detected by introducing an internal control targeting a human gene. The main limitation of most NAATS is the relative complexity of the platforms that are needed for performing the extraction and amplification steps of DNA. However, manufacturers are developing POC molecular tests that can be used by nontrained experimenters, as exemplified by the recent cartridge-based Cepheid GeneXpert CT/NG assay that can be performed within 90 min through a fully-automated platform, with excellent sensitivity and specificity on a wide range of clinical specimens [26]. Other POC molecular tests are in development for the diagnosis of chlamydia infection such as the Atlas Genetics platform based on PCR amplification and original electrochemical detection of amplicons [35] or the microwave-accelerated metal-enhanced fluorescent assay using silver metallic nanoparticles to amplify the fluorescent signal [36].

All the serovars of C. trachomatis, including those responsible for LGV, can be detected by the commercial NAATs described. From an epidemiological point of view, identification at the serovar level can be performed by direct omp 1 gene PCR-RFLP analysis or by nucleotide sequencing [20]. These techniques are reserved to reference laboratories.

MYCOPLASMA GENITALIUM INFECTION

Overview of the Disease and its Epidemiology

Mycoplasma genitalium is a motile flask-shaped bacterium of very small size (0.2–0.7 μm) that belongs to class *Mollicutes*, family *Mycoplasmataceae*, and genus *Mycoplasma*. It is the species with the smallest genome of any known living free agent, with a genome of only 580 bp. This genome has been fully sequenced [37]. The first two strains of *M. genitalium* were isolated in 1981 from the urethra of two men with non-gonococcal urethritis [38,39]. Unlike other members of the *Mycoplasmataceae* family naturally present in the human genital tract, including *M. hominis*, *Ureaplasma urealyticum*, and *U. parvum*, *M. genitalium* is not considered a commensal host of the genital tract but a sexually transmitted agent associated to different genital infections with transmission rate similar to that of *C. trachomatis* [40].

Several recent reviews [41–43] discuss in detail the involvement of *M. genitalium* in infections of the genital tract. Briefly, it is now well recognized that this bacterium is responsible for non-gonococcal urethritis in men, especially in persistent and recurrent ones. By contrast, its role in genital infection of women is still unclear. It has been associated to cervicitis and pelvic inflammatory disease. Its involvement in adverse pregnancy outcome and infertility needs to be documented by larger prospective clinical studies. Many studies on urethritis have pointed out the frequent association of *M. genitalium* to other agents of STI or to other *Mycoplasma* species. The place of *M. genitalium* in the epidemiology of STIs is difficult to delineate. From a series of studies involving a large number of patients of both sexes with different genital infections, the prevalence of the detection of the bacterium ranged from 4.0% to 38.2% [43].

Current Diagnosis

Mycoplasma genitalium is an emergent pathogen that was identified only recently as a significant agent of STIs. This is due in large part to the difficulty to obtain the bacterium in culture (Fig. 13.2B). In addition to the fact that all mycoplasmas are fastidious agents requiring specific conditions of culture for growing, the culture of *M. genitalium* is very slow (weeks to months), which is not practical for routine diagnosis.

Consequently, NAATs are the only tools that are available for detecting the presence of this agent in clinical samples. Appropriate specimens are urethral exudate, cervical exudate, and first-catch urine for both men and women. PCR assays have been implemented that target different genomic sequences including the *MgPa* adhesin gene [44], the 16S RNA gene [45], and the *gap* gene encoding glyceraldehyde 3-phosphate dehydrogenase [46]. The latter target is probably the less prone to false-positive results with other *Mycoplasma* species or to false-negative results consecutive to mutated sequences. Hologic (California, USA) commercialized a TMA real-time PCR assay that performs well when compared with other methods, but it is available in the United States only for research purposes [47–49].

The antibiotic treatment of infection relies on the same drugs as those used for chlamydia, including macrolides, quinolones, and cyclines. In order to identify the emergence of multiresistant strains, molecular tests are in development for targeting mutations involved in the resistance to macrolides [50,51] and quinolones [52].

TRICHOMONIASIS

Overview of the Disease and its Epidemiology

This parasitic venereal disease due to a flagellated motile protozoan named *Trichomonas vaginalis* is responsible for urethritis in men and vulvovaginitis in women. Although mostly asymptomatic, this infection may be associated to genital discharge both in females and males, pelvic inflammatory disease, and pregnancy-related complications. In addition, it was suspected to favor the acquisition or transmission of HIV-1 [53–55]. With an estimated annual incidence of 180 million cases worldwide [56], trichomoniasis is the most prevalent STI of nonviral origin.

Diagnosis Based on Conventional Techniques

Conventional techniques include (1) direct examination of urine in the absence (wet mount preparation) or presence of staining (Fig. 13.2A) that allows the observation of typical trophozoites and (2) culture from genital swabs or urine sediments using for instance the InPouch TV test (Biomed Diagnostics, Oregon, USA). The InPouch TV test consists of a pouch that contains a medium specifically adapted to the growth of the parasite. If positive, the device allows the observation within 2–5 days of typical living parasites by using a 10× lens. Despite their relative simplicity and the low cost of the first method, these tests are time-consuming, require experienced observers, very fresh samples, and are relatively insensitive. For instance, the sensitivity of wet mount microscopy and InPouch TV culture was shown to be 60% and 73%, respectively, by comparison to PCR [57]. Commercially available rapid antigen assays that detect *T. vaginalis* membrane proteins were also developed as POC tests. These include the immunochromatographic OSOM Trichomonas Rapid Test (Sekisui Diagnostics, California, USA) and the Tv latex agglutination test (Kalon Biological, Surrey, UK). The sensitivity of these assays is similar to that of culture [9].

Diagnosis Based on Molecular Testing

Molecular testing represents an interesting alternative to conventional techniques for the detection of T. vaginalis genome in genital specimens and first-catch urine. Affirm VPIII (Becton Dickinson, Maryland, USA) is a nonamplified nucleic acid probe hybridization test for the simultaneous detection of T. vaginalis, Gardnerella vaginalis, and Candida albicans. It can be completed within 45 min but its achievement is relatively complex. By comparison to amplification testing (NAATs), its sensitivity was shown to be of only 46.1% [58]. Currently, NAATs constitute the gold standard for the diagnosis of TV infection. They rely either on in-house PCR assays [59–63] or on commercial tests including the APTIMA T. vaginalis assay (Hologic Gen-Probe Inc, California, USA) based on TMA [64–66] and the ProbeTec T. vaginalis method (Becton-Dickinson, Maryland, USA) based on SDA [67]. These tests can be run on specific platforms (Fig. 13.2B) from a wide range of specimens (genital swabs, first-catch urine, throat and anal swabs, semen, etc.). Their sensitivity was shown to be between 88% and 100%. They can be combined with the search of other sexually transmitted pathogens on the same platform. However, they are expensive and require laboratory equipment and highly trained personnel. Persistent positive results following treatment are possible [54].

In conclusion, insensitive tests such as culture, antigen assays, and hybridization tests must be limited to genital samples from women with symptomatic infection, whereas NAATs can be used on a wider range of clinical specimens in populations with low prevalence of T. vaginalis infection, including asymptomatic infections. In case of recurrent infection in patients treated by nitroimidazoles, the strains need to be cultivated in order to detect T. vaginalis resistance to these drugs. For the phylogenetic analysis of T. vaginalis, techniques such as random amplified polymorphic DNA are interesting [20].

FUTURE DIRECTIONS

Vaginal Microbiome and STIs

According to R. M. Brotman [68], "...bacterial vaginosis (BV) is a gynecologic condition of unknown etiology and is traditionally characterized by a relatively low abundance of vaginal Lactobacillus sp. accompanied by polymicrobial anaerobic overgrowth...." In addition to representing a major cause of complaint in women, BV is recognized as a nonspecific marker associated to a higher risk of STI and its attendant complications including pelvic inflammatory disease, increased HIV transmission, and bad pregnancy outcome. From a clinical point of view, BV associates the following Amsel criteria [69]: vaginal pH of at least 4.5, gray-white malodorous fish-smelling discharge, and the presence of clue cells defined as epithelial vaginal cells constellated of attached bacteria. From a bacteriological point of view, the Nugent gram strain score takes into consideration the relative abundance of three kinds of bacteria according to their morphology: different gram-positive species of Lactobacillus, small gram-negative rods (Gardnerella, Bacteroides), and curved gram-negative bacilli (Mobiluncus) [70]. With an overall range from 0 to 10, a Nugent score of 7 or more is indicative of BV. The epidemiological factors that favor the disequilibrium of vaginal flora responsible for BV are numerous and include mainly menses, new sexual partners, vaginal douching, receptive oral sex, and lack of condom use [68].

The considerations noted indicate the importance of BV as a predisposing condition of STI. Its identification relies mainly on gram staining, which is greatly influenced by the subjective appreciation and expertise of the examiner. In this context, the new molecular tools available for exploring microbial flora, notably those based on deep sequencing methods, may be useful for exploring more objectively the bacterial diversity of vaginal flora. A recent study performed on 396 asymptomatic US women used pyrosequencing of the 16S RNA genes for characterizing the vaginal microbiome [71]. After phylogenic analysis, five different patterns were recognized—four of them were associated with a predominance of Lactobacillus species, whereas another one, present in approximately one-fifth of these asymptomatic women, was characterized by the predominance of other bacterial species. Interestingly, women from this latter group were characterized by elevated vaginal pH and Nugent score [71]. The development of these new sequencing methods represents a fascinating way of exploring more in depth the microbial diversity of vaginal flora, which may lead to the identification of women at risk of developing BV and, consequently, STIs. According to the concept of personalized medicine, these findings may open the way of new intervention strategies aimed at preventing STIs through the restoration of a normal vaginal flora.

Multiplex Approach of STI Diagnosis Using Molecular Tools

Two of the main concerns associated to the current diagnosis of STIs are the diversity of agents possibly involved in these pathologies, including bacteria, viruses, and parasites, together with the frequent association of several of them within the same infection. The asymptomatic character of some of these infections, notably in women, is an additional handicap to their easy recognition. Ideally, it would be very useful to produce a single multiplex molecular test, performed on a simple matrix such as self-sampled vaginal swab or first-catch urine, and able to detect simultaneously a wide range of STI agents.

During the past decade, numerous tests have been developed in order to reach, at least in part, this ambitious goal. Beyond the tests that are able to detect *C. trachomatis* and *N. gonorrhoeae* in the same assay, different studies reported the simultaneous detection of several sexually transmitted agents. For instance, highly specific and efficient primers were described for in-house multiplex PCR detection of *C. trachomatis*, *N. gonorrhoeae*, *M. hominis*, and *U. urealyticum* [72]. Another study reported the simultaneous identification of 14 genital microorganisms (namely, *T. vaginalis*, *Streptococcus pneumoniae*, *N. gonorrhoeae*, *C. trachomatis*, *U. parvum*, *U. urealyticum*, *G. vaginalis*, *Haemophilus influenzae*, herpes simplex virus (HSV) type 1 and type 2, *N. meningitidis*, *M. hominis*, *M. genitalium*, and adenovirus) in urine by using a multiplex PCR-based reverse line blot assay [73]. Other authors investigated the etiologies of cervicitis by using multiplex PCR testing targeting the following agents: cytomegalovirus, enterovirus, Epstein—Barr virus, varicella-zoster virus, HSV type 1 and type 2, *U. parvum*, *U. urealyticum*, *M. genitalium*, *M. hominis*, *C. trachomatis*, *T. vaginalis*, *T. pallidum*, group B streptococci, and adenovirus species A to E [74]. A recent study investigated the simultaneous detection of seven agents (*C. trachomatis*, *T. pallidum*, *M. genitalium*, *T. vaginalis*, *N. gonorrhoeae*, and HSV type 1 and type 2), together with human papillomaviruses, by single PCR in semen of men involved in infertility programs [75]. Fig. 13.3

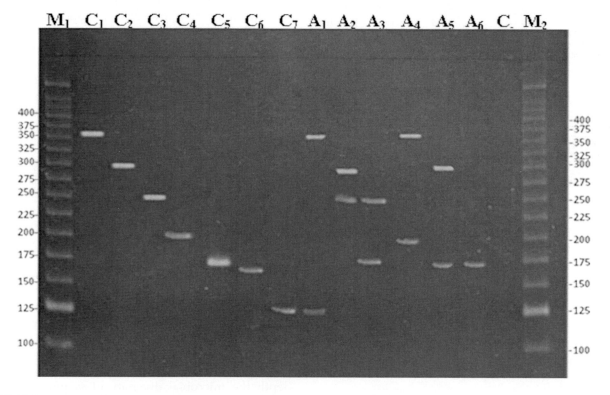

FIGURE 13.3   Example of in-house multiplex PCR assay detecting seven different pathogens involved in STI [75]. The picture illustrates the electrophoretic analysis of the amplified fragments in 8% polyacrylamide gel stained with ethidium bromide. Lanes C correspond to controls (C$_1$: 361 bp fragment of *Chlamydia trachomatis*; C$_2$: 291 bp fragment of *Treponema pallidum*; C$_3$: 249 bp fragment of HSV-2; C$_4$: 193 bp fragment of *Mycoplasma genitalium*; C$_5$: 170 bp fragment of *Trichomonas vaginalis*; C$_6$: 162 bp fragment of *Neisseria gonorrhoeae*; C$_7$: 123 bp fragment of HSV-1; C: negative control). Lanes M$_1$ and M$_2$ correspond to markers of molecular sizes (in bp). Lanes A correspond to infected semen specimens (A$_1$: double infection with *C. trachomatis* and HSV-1; A$_2$: double infection with *T. pallidum* and HSV-2; A$_3$: double infection with *T. vaginalis* and HSV-2; A$_4$: double infection with *C. trachomatis* and *M. genitalium*; A$_5$: double infection with *T. pallidum* and *T. vaginalis*; A$_6$: single infection with *T. vaginalis*).

illustrates the gel analysis of the PCR fragments obtained with the multiplex method used in the latter study [75]. All these in-house assays were shown to be sensitive and specific and exhibited a high proportion of coinfections by two or more microorganisms.

Different companies, including Bio-Rad (Hercules, California, USA), Amplex Biosystems (Giessen, Germany), PCR Diagnostics.eu (Bratislava, Slovak Republic), and Seegene (Seoul, Korea), market multiplex PCR-based kits using various formats and instrument platforms and are able to detect notably *C. trachomatis*, *N. gonorrhoeae*, *T. vaginalis*, *M. genitalium*, along with other urogenital mycoplasmas and ureaplasmas. None of these kits are yet registered by the US Food and Drug Administration. By comparison to other commercial tests, the Anyplex II multiplex real-time PCR from Seegene was found 97.8—100% sensitive and 99.3—100% specific according to the different pathogens on a large collection of urine and genital samples ($n = 897$) from subjects of both sexes [76].

As a whole, although further evaluation is needed for appreciating the performances and judicious selection of targets of these multiplex assays, it is clear that they represent a significant advance in the global management of STIs, notably for investigating the frequent association of multiple pathogens within a same patient. An increased choice of microorganism combination will be available during the next years and it will be important to edit recommendations relative to the list of pathogens that would need to be detected according to the different clinical and epidemiological situations that are most frequently encountered.

POC Diagnostics for STIs

Another goal of STI diagnosis is the immediate availability of microbiological investigations in order to propose an adequate treatment when possible, together with tailored recommendations regarding the sexual contacts. This need is particularly urgent when low-income or unstable populations are targeted, as it is often the case with STI subjects. The development of POC molecular tests is increasing rapidly [77], notably in the field of genital infections [9,78]. The PCR-based GeneXpert NG/CT assay for simultaneous detection of *C. trachomatis* and *N. gonorrhoeae* on genital swabs and urine is an illustration of the excellent performance that can be reached within 90 min of time with this kind of test [26]. Cheaper methods based on isothermal amplification, already optimized for viruses [79—81], and encompassing more pathogens are in development [77].

Molecular Tools Dedicated to Antimicrobial Susceptibility

The increased frequency of the worldwide prevalence of STIs is associated to a significant increase in antimicrobial treatments, which further leads to development of germ resistance. This is particularly true for *N. gonorrhoeae*, *C. trachomatis*, and *M. genitalium* and to a lesser extent for *T. pallidum* and *T. vaginalis*. Current phenotypic antimicrobial tests need the growth of the corresponding microorganism, which can be difficult due to its rapid inactivation (*N. gonorrhoeae*, *T. vaginalis*), its fastidious growth (*M. genitalium*), or its inability to grow in inert culture (*C. trachomatis*, *T. pallidum*). When feasible, the results are available within a few days, which can delay treatment or require adjustment.

To encompass these difficulties, it would be useful to develop molecular tests based on molecular tools, as it has been done for viruses or tuberculosis mycobacteria. Pulido et al. proposed recently an overview of the new techniques that are available for testing antimicrobial susceptibility, including those based on PCR assays, microarrays, microfluidics, cell lysis based approaches, or full genome sequencing [82]. Some molecular tests are in development for targeting mutations involved in the antibiotic resistance of *M. genitalium* [50—52] and *N. gonorrhoeae* [27—29]. Ideally, these tests of resistance would be associated to the multiplex assays described previously, either as second intention testing when the screening test is found positive for a target associated to antimicrobial resistance or even in the same test for frequent resistance patterns.

CONCLUDING REMARKS

This overview of the molecular tools used for STIs of nonviral origin illustrates that much progress has been made in implementing new diagnostic strategies in this field. Herein is not the place for discussing the cost—benefit balance of performing an etiological diagnosis of current STIs before treating them, especially in low-income areas, but it seems consensual that there is an urgent need in the future for low-cost molecular methods able to detect pathogens together with their resistance pattern in real time, in order to use adequate curative means. For this purpose, isothermal amplification methods could represent an interesting track from an economic point of view.

Technically, there are still a lot of issues that must be solved before NAATs become the reference for the diagnosis of all nonviral STIs. Except for chlamydia, the other agents reviewed in this chapter have shown some limitations when NAATs were used on a routine basis. This is particularly true when the positive predictive value is lower than 90%, as often recorded in low prevalence population, which justifies the implementation of confirmatory or complementary tests [20]. The use of standardized specimens and methods performed under rigorous norms (as ISO 15189) and strict quality controls is another exigency of the generalization of NAATs in microbiology settings. Nevertheless, molecular tools are becoming important in the management of STIs, for detecting the germs, evaluating their susceptibility to antimicrobials, and performing an epidemiological typing of strains. It is now clear that molecular assays are progressively supplanting conventional methods of diagnosis, at least in high-income areas.

The last words will concern prevention. In addition to education, vaccines [83,84], and prophylactic treatments such as local microbicides [85], a better knowledge of the genital microbiome is essential for understanding the complex interactions between agents of the normal flora and pathogens that are responsible for STIs. Revisiting this microbiological world in the light of the new generation sequencing techniques will probably help to identify those subjects that are the most at risk for developing complicated STIs and consequently to define new preventive strategies.

References

[1] World Health Organization. Global incidence and prevalence of selected curable sexually transmitted infections—2008. Geneva: World Health Organization; 2012. Available from: <http://www.who.int/reproductivehealth/publications/rtis/stisestimates/en/index.html> [accessed April 2015].

[2] Aral SO, Over M, Manhart L, Holmes KK. Disease control priorities in developing countries. In: Jamison DT, Breman JG, Measham AR, Alleyne G, Claeson M, Evans DB, et al., editors. Disease control priorities in developing countries. 2nd ed. Washington, DC: World Bank; 2006. p. 312–30.

[3] Simms I, Warburton F, Westrom L. Diagnosis of pelvic inflammatory disease: time for a rethink. Sex Transm Infect 2003; 2003;79:491–4.

[4] Gore-Felton C, Vosvick M, Bendel T, et al. Correlates of sexually transmitted disease infection among adults living with HIV. Int J STD AIDS 2003;14:539–46.

[5] Van der Pol B, Kwok C, Bosny PL, et al. *Trichomonas vaginalis* infection and human immunodeficiency virus acquisition in African women. J Infect Dis 2008;197:548–54.

[6] Ho EL, Lukehart SA. Syphilis: using modern approaches to understand an old disease. J Clin Invest 2011;121:4584–92.

[7] Stamm LV, Mudrak B. Old foes, new challenges: syphilis, cholera and TB. Future Microbiol 2013;8:177–89.

[8] Koek AG, Bruisten SM, Dierdorp M, van Dam AP, Templeton K. Specific and sensitive diagnosis of syphilis using a real-time PCR for *Treponema pallidum*. Clin Microbiol Infect 2006; 12:1233–6.

[9] Gaydos C, Hardick J. Point of care diagnostics for sexually transmitted infections: perspectives and advances. Expert Rev Anti Infect Ther 2014;12:657–72.

[10] Palmer HM, Higgins SP, Herring AJ, Kingston MA. Use of PCR in the diagnosis of early syphilis in the United Kingdom. Sex Transm Infect 2003;79:479–83.

[11] Bruisten SM, Cairo I, Fennema H, et al. Diagnosing genital ulcer disease in a clinic for sexually transmitted diseases in Amsterdam, The Netherlands. J Clin Microbiol 2001;39:601–5.

[12] Centurion-Lara A, Castro C, Shaffer JM, VanVoorhis WC, Marra CM, Lukehart SA. Detection of *Treponema pallidum* by a sensitive reverse transcriptase PCR. J Clin Microbiol 1997; 35:1348–52.

[13] Pillay A, Liu H, Chen CY, et al. Molecular subtyping of *Treponema pallidum* subspecies *pallidum*. Sex Transm Dis 1998;25:408–14.

[14] Martin IE, Gu W, Yang Y, Tsang RS. Macrolide resistance and molecular types of *Treponema pallidum* causing primary syphilis in Shanghai, China. Clin Infect Dis 2009;49:515–21.

[15] Marra CM, Colina AP, Godornes C, et al. Antibiotic selection may contribute to increases in macrolide-resistant *Treponema pallidum*. J Infect Dis 2006;194:1771–3.

[16] World Health Organization. Global action plan to control the spread and impact of antimicrobial resistance in *Neisseria gonorrhoeae*. Geneva: WHO; 2012. Available from: <whqlibdoc.who.int/publications/2012/9789241503501_eng.pdf> [accessed April 2015]

[17] Health Protection Agency/British Association for Sexual Health and HIV. Guidance for gonorrhoea testing in England and Wales. Available from: <https://www.gov.uk/government/uploads/system/uploads/attachment_data/file/405293/170215_Gonorrhoea_testing_guidance_REVISED__2_.pdf> [accessed April 2015].

[18] Tapsall JW, Ndowa F, Lewis DA, Unemo M. Meeting the public health challenge of multidrug- and extensively drug-resistant *Neisseria gonorrhoeae*. Expert Rev Anti Infect Ther 2009; 7:821–34.

[19] Whiley DM, Tapsall JW, Sloots TP. Nucleic acid amplification testing for *Neisseria gonorrhoeae*: an ongoing challenge. J Mol Diagn 2006;8:3–15.

[20] Vazquez F, Otero L, Melón S, de Oña M. Overview of molecular biological methods for the detection of pathogens causing sexually transmitted infections. Methods Mol Biol 2012; 903:1–20.

[21] Trembizki E, Costa AM, Tabrizi SN, Whiley DM, Twin J. Opportunities and pitfalls of molecular testing for detecting sexually transmitted pathogens. Pathology 2015;47:219–26.

[22] Low N, Unemo M, Skov Jensen J, Breuer J, Stephenson JM. Molecular diagnostics for gonorrhoea: implications for antimicrobial resistance and the threat of untreatable gonorrhoea. PLoS Med 2014;11:e1001598.

[23] Chow EP, Fehler G, Read TR, et al. Gonorrhoea notifications and nucleic acid amplification testing in a very low-prevalence Australian female population. Med J Aust 2015;202:321–3.

[24] Johnson RE, Newhall WJ, Papp JR, et al. Screening tests to detect *Chlamydia trachomatis* and *Neisseria gonorrhoeae* infections—2002. MMWR Recomm Rep 2002;51:1–38.

[25] Bell K, McCaffery KJ, Irwig L. Screening tests for gonorrhoea should first do no harm. Med J Aust 2015;202:281–2.

[26] Gaydos CA. Review of use of a new rapid real-time PCR, the Cepheid GeneXpert® (Xpert) CT/NG assay, for *Chlamydia trachomatis* and *Neisseria gonorrhoeae*: results for patients while in a clinical setting. Expert Rev Mol Diagn 2014;14:135–7.

[27] Goire N, Freeman K, Tapsall JW, et al. Enhancing gonococcal antimicrobial resistance surveillance: a real-time PCR assay for

detection of penicillinase-producing *Neisseria gonorrhoeae* by use of noncultured clinical samples. J Clin Microbiol 2011; 49:513–18.

[28] Siedner MJ, Pandori M, Castro L, et al. Real-time PCR assay for detection of quinolone-resistant *Neisseria gonorrhoeae* in urine samples. J Clin Microbiol 2007;45:1250–4.

[29] Magooa MP, Müller EE, Gumede L, Lewis DA. Determination of *Neisseria gonorrhoeae* susceptibility to ciprofloxacin in clinical specimens from men using a real-time PCR assay. Int J Antimicrob Agents 2013;42:63–7.

[30] Malhotra M, Sood S, Mukherjee A, Muralidhar S, Bala M. Genital *Chlamydia trachomatis*: an update. Indian J Med Res 2013;138:303–16.

[31] Horner P, Boag F. UK national guideline for the management of genital tract infection with *Chlamydia trachomatis*. BASHH; 2006. Available from: <http://www.bashh.org/documents/65. pdf> [accessed April 2015]

[32] Spersen DJ, Flatten KS, Jones MF, Smith TF. Prospective comparison of cell cultures and nucleic acid amplification tests for laboratory diagnosis of *Chlamydia trachomatis* infections. J Clin Microbiol 2005;43:5324–6.

[33] Ripa T, Nilsson P. A variant of *Chlamydia trachomatis* with deletion in cryptic plasmid: implications for use of PCR diagnostic tests. Euro Surveill 2006;11:e061109.2.

[34] Ripa T, Nilsson PA. A *Chlamydia trachomatis* strain with a 377-bp deletion in the cryptic plasmid causing false-negative nucleic acid amplification tests. Sex Transm Dis 2007;34:255–6.

[35] Pearce DM, Shenton DP, Holden J, Gaydos CA. Evaluation of a novel electrochemical detection method for *Chlamydia trachomatis*: application for point-of-care diagnostics. IEEE Trans Biomed Eng 2011;58:755–8.

[36] Melendez JH, Huppert JS, Jett-Goheen M, et al. Blind evaluation of the microwave-accelerated metal-enhanced fluorescence ultrarapid and sensitive *Chlamydia trachomatis* test by use of clinical samples. J Clin Microbiol 2013;51:2913–20.

[37] Fraser CM, Gocayne JD, White O, et al. The minimal gene complement of *Mycoplasma genitalium*. Science 1995;270:397–403.

[38] Tully JG, Taylor-Robinson D, Cole RM, Rose DL. A newly discovered mycoplasma in the human urogenital tract. Lancet 1981;1:1288–91.

[39] Taylor-Robinson D, Tully JG, Furr PM, Cole RM, Rose DL, Hanna NF. Urogenital mycoplasma infections of man: a review with observations on a recently discovered mycoplasma. Isr J Med Sci 1981;17:524–30.

[40] Jensen JS. *Mycoplasma genitalium* infections. Diagnosis, clinical aspects, and pathogenesis. Dan Med Bull 2006;53:1–27.

[41] Taylor-Robinson D, Jensen JS. *Mycoplasma genitalium*: from chrysalis to multicolored butterfly. Clin Microbiol Rev 2011;24:498–514.

[42] Waites KB, Xiao L, Paralanov V, Viscardi RM, Glass JI. Molecular methods for the detection of *Mycoplasma* and *Ureaplasma* infections in humans: a paper from the 2011 William Beaumont Hospital Symposium on molecular pathology. J Mol Diagn 2012;14:437–50.

[43] Sethi S, Singh G, Samanta P, Sharma M. *Mycoplasma genitalium*: an emerging sexually transmitted pathogen. Indian J Med Res 2012;136:942–55.

[44] Jensen JS, Uldum SA, Søndergård-Andersen J, Vuust J, Lind K. Polymerase chain reaction for detection of *Mycoplasma genitalium* in clinical samples. J Clin Microbiol 1991;29:46–50.

[45] Jensen JS, Borre MB, Dohn B. Detection of *Mycoplasma genitalium* by PCR amplification of the 16S rRNA gene. J Clin Microbiol 2003;41:261–6.

[46] Svenstrup HF, Jensen JS, Björnelius E, Lidbrink P, Birkelund S, Christiansen G. Development of a quantitative realtime PCR assay for detection of *Mycoplasma genitalium*. J Clin Microbiol 2005;43:3121–8.

[47] Hardick J, Giles J, Hardick A, Hsieh YH, Quinn T, Gaydos C. Performance of the gen-probe transcription-mediated amplification research assay compared to that of a multitarget real-time PCR for *Mycoplasma genitalium* detection. J Clin Microbiol 2006;44:1236–40.

[48] Wroblewski JK, Manhart LE, Dickey KA, Hudspeth MK, Totten PA. Comparison of transcription-mediated amplification and PCR assay results for various genital specimen types for detection of *Mycoplasma genitalium*. J Clin Microbiol 2006;44:3306–12.

[49] Huppert JS, Mortensen JE, Reed JL, Kahn JA, Rich KD, Hobbs MM. *Mycoplasma genitalium* detected by transcription-mediated amplification is associated with *Chlamydia trachomatis* in adolescent women. Sex Transm Dis 2008;35:250–4.

[50] Twin J, Jensen JS, Bradshaw CS, et al. Transmission and selection of macrolide resistant *Mycoplasma genitalium* infections detected by rapid high resolution melt analysis. PLoS One 2012;7:e35593.

[51] Touati A, Peuchant O, Jensen JS, Bébéar C, Pereyre S. Direct detection of macrolide resistance in *Mycoplasma genitalium* isolates from clinical specimens from France by use of real-time PCR and melting curve analysis. J Clin Microbiol 2014;52:1549–55.

[52] Yamaguchi Y, Takei M, Kishii R, Yasuda M, Deguchi T. Contribution of topoisomerase IV mutation to quinolone resistance in *Mycoplasma genitalium*. Antimicrob Agents Chemother 2013;57:1772–6.

[53] Laga M, Manoka A, Kivuvu M, et al. Non-ulcerative sexually transmitted diseases as risk factors for HIV-1 transmission in women: results from a cohort study. AIDS 1993;7:95–102.

[54] Hobbs MM, Kazembe P, Reed AW, et al. *Trichomonas vaginalis* as a cause of urethritis in Malawian men. Sex Transm Dis 1999;26:381–7.

[55] McClelland RS, Sangare L, Hassan WM, et al. Infection with *Trichomonas vaginalis* increases the risk of HIV-1 acquisition. J Infect Dis 2007;195:698–702.

[56] Weinstock H, Berman S, Cates Jr. W. Sexually transmitted diseases among American youth: incidence and prevalence estimates, 2000. Perspect Sex Reprod Health 2004;36:6–10.

[57] Patil MJ, Nagamoti JM, Metgud SC. Diagnosis of *Trichomonas vaginalis* from vaginal specimens by wet mount microscopy, In Pouch TV culture system, and PCR. J Glob Infect Dis 2012;4:22–5.

[58] Cartwright CP, Lembke BD, Ramachandran K, et al. Comparison of nucleic acid amplification assays with BD Affirm VPIII for diagnosis of vaginitis in symptomatic women. J Clin Microbiol 2013;51:3694–9.

[59] Caliendo AM, Jordan JA, Green AM, et al. Improves detection of *Trichomonas vaginalis* infection compared with culture using self-collected vaginal swabs. Infect Dis Obstet Gynecol 2005;13:145–50.

[60] Kengne P, Veas F, Vidal N, Rey JL, Cuny G. *Trichomonas vaginalis*: repeated DNA target for highly sensitive and specific polymerase chain reaction diagnosis. Cell Mol Biol 1994;40:819–31.

[61] Lawing LF, Hedges SR, Schwebke JR. Detection of trichomonosis in vaginal and urine specimens from women by culture and PCR. J Clin Microbiol 2000;38:3585–8.

[62] Madico G, Quinn TC, Rompalo A, McKee Jr KT, Gaydos CA. Diagnosis of *Trichomonas vaginalis* infection by PCR using vaginal swab samples. J Clin Microbiol 1998;36:3205–10.

[63] Mayta H, Gilman RH, Calderon MM, et al. 18S ribosomal DNA-based PCR for diagnosis of *Trichomonas vaginalis*. J Clin Microbiol 2000;38:2683–7.

[64] Munson E, Napierala M, Basile J, et al. *Trichomonas vaginalis* transcription-mediated amplification-based analyte-specific reagent and alternative target testing of primary clinical vaginal saline suspensions. Diagn Microbiol Infect Dis 2010;68:66–72.

[65] Munson E, Miller C, Napierala M, et al. Assessment of screening practices in a subacute clinical setting following introduction of *Trichomonas vaginalis* nucleic acid amplification testing. WMJ 2012;111:233−6.

[66] Munson KL, Napierala M, Munson E, et al. Screening of male patients for *Trichomonas vaginalis* with transcription-mediated amplification in a community with a high prevalence of sexually transmitted infection. J Clin Microbiol 2013;51:101−4.

[67] Van Der Pol B, Williams JA, Taylor SN, et al. Detection of *Trichomonas vaginalis* DNA by use of self-obtained vaginal swabs with the BD ProbeTec Qx assay on the BD Viper system. J Clin Microbiol 2014;52:885−9.

[68] Brotman RM. Vaginal microbiome and sexually transmitted infections: an epidemiologic perspective. J Clin Invest 2011;121:4610−17.

[69] Amsel R, Totten PA, Spiegel CA, Chen KC, Eschenbach D, Holmes KK. Nonspecific vaginitis. Diagnostic criteria and microbial and epidemiologic associations. Am J Med 1983; 74:14−22.

[70] Nugent RP, Krohn MA, Hillier SL. Reliability of diagnosing bacterial vaginosis is improved by a standardized method of gram stain interpretation. J Clin Microbiol 1991;29:297−301.

[71] Ravel J, Gajer P, Abdo Z, et al. Vaginal microbiome of reproductive-age women. Proc Natl Acad Sci USA 2011;108 (Suppl. 1):4680−7.

[72] Aguilera-Arreola MG, González-Cardel AM, Tenorio AM, Curiel-Quesada E, Castro-Escarpulli G. Highly specific and efficient primers for in-house multiplex PCR detection of *Chlamydia trachomatis, Neisseria gonorrhoeae, Mycoplasma hominis* and *Ureaplasma urealyticum*. BMC Res Notes 2014;7:433.

[73] Mckechnie ML, Hillman R, Couldwell D, et al. Simultaneous identification of 14 genital microorganisms in urine by use of a multiplex PCR-based reverse line blot assay. J Clin Microbiol 2009;47:1871−7.

[74] McIver CJ, Rismanto N, Smith C, et al. Multiplex PCR testing detection of higher-than-expected rates of cervical *Mycoplasma, Ureaplasma*, and *Trichomonas* and viral agent infections in sexually active Australian women. J Clin Microbiol 2009;47:1358−63.

[75] Gimenes F, Medina FS, de Abreu AL, et al. Sensitive simultaneous detection of seven sexually transmitted agents in semen by multiplex-PCR and of HPV by single PCR. PLoS One 2014;9: e98862.

[76] Choe HS, Lee DS, Lee SJ, et al. Performance of Anyplex II multiplex real-time PCR for the diagnosis of seven sexually transmitted infections: comparison with currently available methods. Int J Infect Dis 2013;17:e1134−40.

[77] Niemz A, Ferguson TM, Boyle DS. Point-of-care nucleic acid testing for infectious diseases. Trends Biotechnol 2011;29:240−50.

[78] Tucker JD, Bien CH, Peeling RW. Point-of-care testing for sexually transmitted infections: recent advances and implications for disease control. Curr Opin Infect Dis 2013;26:73−9.

[79] Curtis KA, Rudolph DL, Owen SM. Rapid detection of HIV-1 by reverse transcription, loop-mediated isothermal amplification (RT-LAMP). J Virol Methods 2008;151:264−70.

[80] Hagiwara M, Sasaki H, Matsuo K, Honda M, Kawase M, Nakagawa H. Loop-mediated isothermal amplification method for detection of human papillomavirus type 6, 11, 16, and 18. J Med Virol 2007;79:605−15.

[81] Enomoto Y, Yoshikawa T, Ihira M, et al. Rapid diagnosis of herpes simplex virus infection by a loop-mediated isothermal amplification method. J Clin Microbiol 2005;43:951−5.

[82] Pulido MR, García-Quintanilla M, Martín-Peña R, Cisneros JM, McConnell MJ. Progress on the development of rapid methods for antimicrobial susceptibility testing. J Antimicrob Chemother 2013;68:2710−17.

[83] Gottlieb SL, Low N, Newman LM, Bolan G, Kamb M, Broutet N. Toward global prevention of sexually transmitted infections (STIs): the need for STI vaccines. Vaccine 2014;32:1527−35.

[84] Broutet N, Fruth U, Deal C, Gottlieb SL, Rees H, Participants of the 2013 STI Vaccine Technical Consultation. Vaccines against sexually transmitted infections: the way forward. Vaccine 2014;32:1630−7.

[85] Pozzetto B, Delézay O, Brunon-Gagneux A, Hamzeh-Cognasse H, Lucht F, Bourlet T. Current and future microbicide approaches aimed at preventing HIV infection in women. Expert Rev Anti Infect Ther 2012;10:167−83.

14

Molecular Methods for Healthcare-Acquired Infections

R.C. Arcenas

Molecular Microbiology and Immunology, Memorial Healthcare System,
Pathology Consultants of South Broward, Hollywood, FL, United States

INTRODUCTION

Healthcare-acquired infections (HAIs) can put a significant physical and economic burden on hospitals and other healthcare institutions. A number of professional societies and organizations (CDC—Centers for Disease Control [1]; APIC—Association of Practitioners of Infection Control [2]; SHEA/IDSA—Society for Healthcare Epidemiology of America/Infectious Diseases Society of America [3]) have put forth recommendations toward the practice of preventing HAIs and/or monitoring HAIs within a particular setting or an institution. Additional efforts have also been put forth from the Centers for Medicare and Medicaid Services (CMS) to reduce preventable harm (which include HAIs) and improving patient safety. This CMS program essentially penalizes hospitals/healthcare institutions, in the form of reduced reimbursement, that rank low based on scoring criteria [4].

Methicillin-resistant *Staphylococcus aureus* (MRSA) and *Clostridium difficile* are two well-known pathogens that have traditionally been identified as hospital-acquired infections [5–7]. However, it is now recognized that these pathogens are not exclusively nosocomial pathogens [7,8]. Risk factors that have been identified for MRSA include: (1) prior antibiotic usage, (2) current colonization with MRSA, (3) admission to the intensive care unit (ICU), (4) presence of skin and soft tissue infection, and (5) prior history of MRSA infection/colonization [3,9–11]. Healthcare-acquired (HA) and community-acquired (CA) MRSA strains can be differentiated by molecular typing and epidemiological factors [12]. Differences have also been observed for HA-MRSA and CA-MRSA strain prevalence and geography [13–15]. Risk factors for CA-MRSA overlap with risk factors for HA-MRSA, but CA-MRSA strains are specifically associated with societal activities that involve some level of skin contact/exposure (ie, sports and athletics, intravenous drug abusers) [16,17].

In 2006 and 2010, Jarvis et al. conducted an MRSA prevalence survey to estimate the burden and impact of MRSA to healthcare institutions in the United States. They observed that the US MRSA colonization prevalence rate was higher compared to the last time the survey was conducted in 2006, 66.4 per 1000 inpatients versus 46.3 per 1000 inpatients, respectively [18,19]. Interestingly, they observed a decrease in the MRSA infection rate from 35 per 1000 inpatients (2006) to 25.3 per 1000 inpatients (2010). The increased colonization rate and decreased infection rate may reflect the observation that a higher proportion of healthcare institutions (76% in 2010 vs 29% in 2006) had an active surveillance program in place for MRSA and an improved ability to identify those patients that are colonized with MRSA. With improved ability to detect MRSA colonization comes a more focused approach to controlling colonization and infection.

In Canada, it was shown that the rate of MRSA infections increased from 0.46 to 5.90 per 1000 admissions from 1995 to 2004 [20]. That same study showed that patients with MRSA required prolonged hospitalization, special control measures, and more expensive treatments, compared to methicillin-sensitive *S. aureus* (MSSA) infections. The total MRSA-associated financial burden to the Canadian healthcare system was

© 2017 Elsevier Inc. All rights reserved.

estimated to be $82 million in 2004. Anderson et al. estimated that an MRSA surgical site infection (SSI) results in approximately $79,029 in hospital charges [21], compared to MSSA and uninfected controls where hospital charges were $55,667 and $38,735, respectively. Anderson and colleagues also showed a 5- to 6-day greater length of stay for MRSA SSIs versus MSSA SSIs [21]. Other studies show the significant economic and physical burden MRSA infections put on healthcare institutions [22–24].

Microbiology screening and testing have traditionally been performed using culture-based and other phenotypic methodologies. In general, clinical microbiology laboratories are now adopting molecular methods because results can be available to clinicians more rapidly versus what the traditional methods can offer. One of the bottlenecks in the time to result for the traditional microbiology testing is the incubation period required for culture growth. The shift to molecular testing has already happened in the clinical virology laboratories where it is now considered the new gold standard. This chapter will focus on MRSA and *C. difficile* as examples of HAIs and the molecular screening and diagnostic testing that are currently used in detecting these two pathogens.

MRSA MOLECULAR DIAGNOSTIC TESTING AND CLINICAL UTILITY

Molecular testing for MRSA can be divided into two broad categories: (1) diagnostic testing and (2) screening/epidemiology. FDA-approved assays are listed in Table 14.1. Molecular testing for diagnosis typically utilizes real-time PCR assays. However, there is an FDA approved method from AdvanDx/bioMerieux, Inc. that utilizes peptide nucleic acid—fluorescent in situ hybridization (PNA-FISH) and detection of a positive fluorescent signal directly from a positive blood culture. In this scenario, blood cultures are collected from a patient suspected to have bacteremia. The blood culture bottles are sent to the microbiology laboratory and are loaded onto the automated blood culture incubator. When the blood culture instrument flags a particular bottle as suspected of having growth, a gram-stained smear is prepared from the blood culture bottle to determine if bacteria (or yeast) are present. PNA-FISH offers four basic assays based on the gram stain: (1) *Staphylococcus* (gram positive), (2) *Enterococcus* (gram positive), (3) gram negative, and (4) *Candida*. Additional assays have the ability to separate *S. aureus*/coagulase-negative *Staphylococcus* utilizing a *mecA* probe for the identification of MRSA. The assay requires a fluorescent microscope to visualize the results. A number of studies have shown the clinical

benefits of PNA-FISH implementation as part of the blood culture workup—knowing methicillin resistance before confirmatory culture growth and before the full antibiotic susceptibilities can be performed [25–29]. This testing would not, at least currently, replace the traditional microbiology workup, but serves as another laboratory tool to more quickly accomplish pathogen identification so that clinical action can be taken earlier.

The FDA-approved PCR tests for MRSA are based upon detecting the *mecA* gene. Most of these assays are real-time PCR assays utilizing fluorescent probes with PCR cycling times taking approximately a few hours, depending on the vendor assay. The platforms available offer various levels of automation and hands-on-time (ie, nucleic acid extraction, post-PCR analyses) depending on the vendor. These are things to consider when assessing how the assay and instrumentation will fit based on a laboratory's volume and workflow. Clinical laboratories may choose to develop their own PCR tests for the detection of MRSA or modify FDA-approved assays by validating additional sample types. Doing so requires extensive validation testing to ensure analytical and clinical sensitivity/specificity performance meet certain standards in order to be used as a test for diagnostic purposes.

Similar to the PNA-FISH scenario, when utilizing Cepheid's Xpert MRSA/SA Blood Culture test the gram-stained smear prepared from the positive blood culture will determine whether Cepheid's assay should be run. Cepheid's methodology is real-time PCR based and does not require any subjective interpretation of the results by the laboratory staff. The cartridge for the Xpert MRSA/SA test houses all the reagents and is compartmentalized to accommodate the nucleic acid extraction, PCR amplification, and detection in a single cartridge device. The testing is automated once the sample is loaded into the test cartridge and the software analyzes the PCR amplification curves to determine if a patient's blood sample is positive or negative for the presence of MRSA or MSSA. Studies have shown the potential clinical benefits with the Xpert MRSA/SA test [30–33].

Nanosphere, Inc. and BioFire Diagnostics, Inc. approach *S. aureus*/MRSA testing from positive blood cultures as more of a syndrome of bacteremia or sepsis. These vendors are creating multiplex, molecular pathogen panels that target the most common bacteria and/or fungal pathogens occurring for a particular syndrome (ie, sepsis, gastrointestinal illness, respiratory infections, and encephalitis/meningitis). This is a unique approach that addresses the confounding overlap in patient symptoms for a specific syndrome or condition. When the symptoms overlap, this makes it difficult for clinicians to define the causative

TABLE 14.1 FDA-Approved Molecular Assays for the Identification of *S. aureus*

Vendor	Assay name	Method/intended use	MRSA gene target(s)
BD Diagnostics	BD Max MRSA	PCR/screening	*SCCmec*[a]/*orfX* junction, MREJ[b]
	BD GeneOhm StaphSR	PCR/screening	
	BD GeneOhm MRSA ACP	PCR/screening	
AdvanDx, Inc.	*Staphylococcus* QuickFISH BC	FISH[c]/diagnostic	*S. aureus* and coagulase-negative *Staphylococcus* ribosomal RNA sequences
	S. aureus PNA FISH	FISH/diagnostic	
bioMerieux, Inc.	NucliSENS EasyQ MRSA	NASBA[d]/screening	MRSA DNA (target(s) not specified)
Cepheid	Xpert MRSA/SA Blood Culture	PCR/diagnostic	*Spa*[e], *mecA*, *SCCmec*
	Xpert MRSA/SA Nasal Complete	PCR/screening	
	Xpert MRSA/SA SSTI	PCR/diagnostic	
	Xpert MRSA	PCR/screening	*attBsc*[f] of *SCCmec*
Roche	LightCycler MRSA Advanced	PCR/screening	*SCCmec*/*orfX* junction
Infectio Diagnostic, Inc.	IDI-MRSA	PCR/screening	*SCCmec*/*orfX* junction
Nanosphere, Inc.	Verigene Gram Positive Blood Culture Test	Multiplex PCR	*mecA*
BioFire Diagnostics, Inc.	Blood Culture Identification Panel	Multiplex PCR	*mecA*

[a]*Staphylococcal chromosomal cassette.*
[b]*mec right extremity junction.*
[c]*Fluorescent in situ hybridization.*
[d]*Nucleic acid sequence based amplification.*
[e]*Staphylococcus protein A.*
[f]*Sequence incorporating the insertion site (attBsc) of SCCmec.*
www.fda.gov/MedicalDevices/ProductsandMedicalProcedures/InVitroDiagnostics/ucm330711.htm (Web page updated January 27, 2015).

pathogen infecting their patient solely from the clinical picture, which makes it difficult to provide specific pathogen-based therapy. Historically, the microbiological diagnosis of these pathogens involves a combination of multiple cultures, antigen based, biochemical based, and maybe reference laboratory testing that piecemealed results to clinicians depending on how many tests were needed to rule-in (or rule-out) a particular pathogen as the cause of disease. Use of these syndromic multiplex molecular panels streamlines the testing process to more of a one-and-done approach.

Nanosphere offers several different panels: (1) gram positive (BC-GP panel), (2) gram negative (BC-GN panel), and (3) yeast (BC-Y panel). The organism(s) observed from the gram-stained smear determine which panel(s) to run. Additional testing with the gram-positive and gram-negative panels also includes testing for certain antibiotic resistance genes (*mecA*, *vanA*, and *vanB*) (Table 14.2). There are many studies showing overall good performance for the BC-GP panel [34—40]. However, Buchan et al. observed issues where a positive *mecA* target was not able to be

TABLE 14.2 Verigene BC-GP Targets

Targets
Staphylococcus spp.
Streptococcus spp.
Listeria spp.
Staphylococcus aureus
Staphylococcus epidermidis
Staphylococcus lugdunensis
Streptococcus anginosus group
Streptococcus agalactiae
Streptococcus pneumoniae
Streptococcus pyogenes
Enterococcus faecalis
Enterococcus faecium
mecA (methicillin resistance)
vanA (vancomycin resistance)
vanB (vancomycin resistance)

assigned due to the presence of a mixed infection [37]. In this situation, full antibiotic sensitivity testing is recommended because the traditional methods test each bacterial pathogen isolated individually. Beal et al. noted that when blood culture infections were caused by a single pathogen, there was generally good performance of the multiplex molecular assays [38]. When polymicrobial blood culture infections were noted, there was only 33% agreement with the routine cultures. Mestas et al. also noticed a lower percentage agreement for polymicrobial infections when compared to monomicrobial infections [39]. Polymicrobial bacteremia is relatively rare, but that does not mean that it cannot happen, and depending on the organisms present, can potentially have dire complications in the patient [41]. Again, cultures are required to determine the full antibiotic susceptibility results, and maybe for the detection of pathogens that are not included in the multiplex molecular panels.

The FilmArray Blood Culture Identification Panel (BCID) is a comprehensive panel that covers gram-positive, gram-negative, and yeast/fungal pathogens (Table 14.3), and so a positive blood culture from the automated instrument is still required. A gram-stained smear is not necessarily critical to have prior to running the FilmArray panel. However, it is good routine practice to perform the gram stain to correlate with results obtained from the molecular panel, and the eventual culture testing since full antibiotic susceptibility testing is required. Altun et al. observed that certain pathogens that are detected in routine cultures were not detected in the FilmArray panel, because the specific pathogens were not contained in the molecular panel [42]. So although the comprehensive panel covers the most common pathogens, clinical intuition is still ultimately needed especially when clinical symptoms and other laboratory data suggest a bacteremic process in the setting of a negative FilmArray panel. Most studies have demonstrated good performance of these assays, and the potential for a decreased turnaround time for obtaining a preliminary susceptibility profile based on the antibiotic resistance genes tested [40,42–44]. The ability to multiplex offers a greater advantage in knowing what may be present, but the negative predictive values of both tests offer clinical utility in knowing a particular set of pathogens are truly negative.

The ability to obtain a more rapid answer that is technically more sensitive and specific provides potentially positive downstream effects on patient care and antibiotic stewardship. Currently, few studies in the literature show a true analysis of the impact of these rapid PCR blood culture assays on the clinical end users (infection control, pharmacy, length of stay, and overall hospital costs). A study by Bauer et al. in 2010 was able to show clinical benefit after implementing the Cepheid Xpert MRSA/SA assay for positive blood cultures [31]. The investigators analyzed a 4-month pre-PCR period and then a 4-month post-PCR period and were able to show an overall shorter length of stay (6.2 days shorter) and the average hospital costs were $21,387 less than what was observed in the pre-PCR period. This study also showed that having a quicker result available enables infectious disease pharmacists to be more effective in deescalating or changing to more specific antibiotic therapies in a timelier manner when compared to the pre-PCR period [31]. So at least in this setting, there are benefits to be gained by having a more rapid and more sensitive test.

TABLE 14.3   FilmArray Blood Culture Identification Panel

Gram-positive bacteria	Gram-negative bacteria	Yeast	Antibiotic gene
Enterococcus	Acinetobacter baumannii	Candida albicans	mecA (methicillin resistance)
Listeria monocytogenes	Haemophilus influenzae	Candida glabrata	van A/B (vancomycin resistance)
Staphylococcus spp.	Neisseria meningitidis	Candida krusei	KPC (carbapenem resistance)
Staphylococcus aureus	Pseudomonas aeruginosa	Candida parapsilosis	
Streptococcus spp.	Enterobacteriacae	Candida tropicalis	
Streptococcus agalactiae	Enterobacter cloacae complex		
Streptococcus pyogenes	Escherichia coli		
Streptococcus pneumoniae	Klebsiella oxytoca		
	Klebsiella pneumonia		
	Proteus		
	Serratia marcescens		

Part of the difficulty in maintaining optimal performance of PCR assays is ensuring MRSA can be detected across the different MRSA genotypes. *mecA* is responsible for the resistance to methicillin and other beta-lactam antibiotics. *mecA* encodes penicillin-binding protein 2a (PBP2a) and this differs from the other penicillin-binding proteins as the active site is altered so that the beta-lactam antibiotic cannot bind [45]. The *mec* gene lies within the staphylococcal chromosomal cassette (SCC), which is a mobile genetic element [46,47]. There are 11 known SCC*mec* types along with different *mec* gene complexes (A, B, C1, C2, D, E) [46,48]. So as new SCC*mec* types and *mec* gene complexes are discovered, molecular assays need to be reassessed to determine if testing performance is compromised or affected. Ultimately, vendors may need to reformulate their assay(s) to maintain or improve test performance.

An example of this has been the observation of *mecA* dropout mutant isolates that are truly sensitive to methicillin by phenotypic testing methods, but will yield a false positive in some of the earlier generation of tests for MRSA detection [49,50]. Blanc et al. observed that 13% (28/217) patients screened for MRSA yielded false-positive results. The primers targeted regions that flanked the *mecA* gene and an amplicon was generated [51]. At the time, most of the commercially-available PCR tests developed used this same approach and the assays have improved their specificity. It was also noted by Blanc et al. that the MRSA assay from bioMerieux, Inc. does target specifically the *mecA* gene and should be more specific for MRSA. However, the problem lies in testing from a nonsterile site such as the anterior nares where methicillin resistance (ie, presence of *mecA*) does occur in other species of coagulase-negative *Staphylococcus* potentially yielding a false-positive PCR result.

MRSA MOLECULAR TYPING AND CLINICAL UTILITY

Understanding the epidemiology of MRSA cannot be completely appreciated without molecular strain typing. The use of molecular typing has helped in monitoring and limiting the spread of MRSA within healthcare facilities. These conventional molecular methods can also help in determining if a cluster of MRSA infections are the result of an outbreak (similar strains observed) or just a group of unrelated MRSA strains. Clinically, it may be important to know if separate episodes of an MRSA infection are due to a relapse of the initial infection or truly a second infection with a different MRSA strain.

TABLE 14.4 Molecular Typing Methodologies for MRSA

Molecular typing methods
Pulse field gel electrophoresis (PFGE)
Restriction endonuclease analysis (REA)
Restriction fragment length polymorphism (RFLP)
Ribotyping
SCC*mec* typing
agr typing
spa typing
Multilocus sequence typing (MLST)
Single-locus sequence typing (SLST)
DNA microarray hybridization
Binary typing
Arbitrarily primed PCR/randomly amplified polymorphic DNA (AP-PCR/RAPD)
Multiple-locus variable number tandem-repeat assay (MLVA)
PCR-RFLP
Sequencing (Sanger or next generation)

There are a number of conventional phenotypic and genotypic methods for determining the specific strain of MRSA. The phenotypic methods (antibiotic susceptibility profile, serotyping, phage, typing, and biotyping) [52] have limited ability to discriminate specific stains compared to genotypic methods (Table 14.4).

Pulsed field gel electrophoresis (PFGE) has been the traditional genotyping method for MRSA strain typing and is considered the gold standard [53]. The restriction enzyme *SmaI* is most commonly used in PFGE protocols. After the restriction enzyme digestions, the resulting fragments of MRSA DNA are subjected to agarose gel electrophoresis. In PFGE, the orientation of the electrical field is altered periodically to minimize the overlap of DNA fragments so that the resulting banding patterns can be interpreted more easily. Analysis of banding patterns is typically handled by a software package that determines the degree of relatedness between the various MRSA strains tested. This technique is highly reproducible and highly discriminatory. The major limitations of PFGE are reflected in the time to result (which can take at least 3 days) and the relative expense (compared to PCR-based typing) [52,54,55]. It is critical to impose standardization of the PFGE protocols used, analysis of the MRSA banding patterns, and nomenclature of the MRSA strain types because this allows a true comparison of MRSA strains from a global perspective [56].

There are other restriction enzyme based methodologies such as restriction endonuclease analysis (REA), REA with Southern blotting analysis, and ribotyping. Overall, these restriction enzyme methods do not necessarily provide a real-time result while an infection control investigation is ensuing, and this may be a scenario where the PCR-based typing methods can offer a quicker answer to better aid in the investigations to determine if an outbreak is truly occurring or it just happens to be a cluster of infections due to unrelated MRSA strains. However, compared to PFGE these other enzymatic methods do not provide the discriminatory power and technically more demanding [52].

PCR-based typing methods are generally not as time-consuming as restriction enzyme digestion methods. Hence, PCR-based typing potentially contributes to the real-time investigation of a suspected outbreak by providing preliminary results related to pathogen strain [57]. Although these methods exhibit quicker turnaround times for results, the PCR-based typing assays are not as reproducible and discriminatory compared to PFGE and thereby serve as a complement to the traditional gold standard [57].

There are a number of other molecular methods that can be used for MRSA typing and are utilized to determine molecular fingerprints [52,54,55,58–70]. The PCR-based methods focus on certain genes of interest such as *spa* (protein A) [70], *coa* (coagulase), SCC*mec* (SCC that has the *mecA* gene), *agr* (accessory gene regulator) [69], and then combinations of gene targets are employed to further discriminate MRSA strains [61]. Multilocus sequence typing (MLST) can be used to answer questions about the evolutionary and population biology of bacterial species and is based upon sequences of approximately 450-bp internal fragments of seven house-keeping genes amplified by PCR [55]. The PCR-amplified fragments are analyzed using a sequencer. The use and expense of a sequencer within the clinical microbiology laboratory may be prohibitive. Hence, this methodology may not be suitable for infection control outbreak investigations in real time. These molecular typing techniques have advantages and disadvantages, and varying levels of discriminatory power. There is no one best overall method for MRSA strain typing. However, the specific question(s) being asked in association with a single patient or group of patients will help to determine which molecular typing method(s) to choose.

Next-generation sequencing (NGS) (also known as massively parallel sequencing) is a newer methodology that has the potential for a quicker turnaround time to result while providing a greater ability to discriminate between different MRSA strains compared to PCR-based techniques. The delineation between CA-MRSA and HA-MRSA is not always clear from an epidemiological perspective or even from a genetic perspective, and NGS provides a more definitive result. Harris et al. utilized NGS to investigate an MRSA outbreak in a special care baby unit (SCBU) and compared the results to their traditional infection control/laboratory techniques [71]. They observed that NGS, in real time, was able to identify 26 additional related cases of MRSA carriage and showed that transmission occurred within the SCBU between mothers on a postnatal ward and in the community. NGS confirmed that the MRSA strain identified was in fact a new sequence type that was very closely related to the MRSA strain identified in the initial outbreak. Investigators traced the newly identified MRSA strain to a staff member that was colonized and provided the source of MRSA that supported the ongoing outbreak despite the deep cleaning that was performed in the SCBU.

Price et al. described that NGS does have the ability to discriminate down to single nucleotide difference that allows a more accurate typing analysis [72]. A number of other studies have shown the advantages of NGS as a stand-alone technique, as well as a complement to other molecular typing methodologies. NGS has also been utilized to look at antibiotic susceptibility genes and virulence [73–78]. As this technology becomes more mainstream and the ability to analyze the data becomes more standardized, clinical microbiology/molecular laboratories will adopt this as another infection control tool. With more distinct and discriminatory results for molecular typing, better inferences can be made during outbreak investigations for determining the most likely means of transmission. Coupled with a more specific result, infection control may be able to implement the appropriate countermeasures in a timelier manner. As the general costs for NGS decrease, the method will eventually become a viable option for clinical microbiology/molecular laboratories.

MRSA MOLECULAR SCREENING/SURVEILLANCE

The implementation of an active MRSA surveillance program is an infection control initiative for the purposes of identifying MRSA-colonized patients before they enter the hospital to be admitted or for a particular outpatient procedure. The ultimate goal is to reduce the incidence of MRSA HAIs among those patients, as well as prevent the horizontal transmission to other patients within the hospital. It has been shown that up to 33% of MRSA HAIs are a result of the patient's own colonizing MRSA strain [79]. Depending on the healthcare institution, patients that are identified as being colonized with MRSA may or may not be decolonized. Appropriate identification and implementation of infection control precautions of an MRSA-colonized patient also raises the awareness by the healthcare staff so that the proper measures (ie, hand hygiene compliance) can be taken to prevent horizontal transmission to other patients also under their care. Legislation has worked its way into shaping and developing MRSA screening programs, and several states have enacted laws mandating that surveillance and/or screening program be put in place for certain HAIs [3,18,80—85].

Many of the FDA-approved molecular screening methods (Table 14.1) at a minimum target the SCCmec. The same issues already mentioned with respect to MRSA specificity, different strain types, and *mecA* dropout mutants also apply to these tests intended for screening/surveillance.

The most common anatomical site sampled for MRSA screening is a swab of the anterior nares. However, it is well known that *S. aureus* can colonize extra-nasal sites such as oropharyngeal, axilla, groin, umbilicus, perineum, and less commonly the gastrointestinal tract and vagina [3,86—90]. Matheson et al. showed that combining anatomical sites increases the yield of finding an MRSA-colonized individual [91]. It has also been observed that pediatric patients have higher persistent *S. aureus* carriage when compared to adults [90].

Active MRSA screening or surveillance is a controversial topic. This debate is outside the scope of this chapter, and the reader is directed to a number of references that discuss the advantages and disadvantages of implementing an active surveillance program [64,83,85,91—100].

SHEA and IDSA provided a guidance document in 2014 that describes a basic framework of an active surveillance program for MRSA [3]. Depending on a myriad of institutional factors, culture or PCR may be the method of choice for screening patients. PCR does have the advantage of greater sensitivity and specificity, and also a shorter turnaround time. Being able to get a quicker result out to the healthcare staff/infection control allows them to implement infection control precautions as appropriate in a timelier manner versus waiting for culture results. Several studies compared the utility of culture-based methods versus molecular methods, and how these methods best support an active screening program [3,96—98,101—108]. It is important to recognize that MRSA DNA detected using a molecular assay from an individual could originate dead bacteria and might produce a clinical false positive despite the true presence of MRSA nucleic acid. It has also been shown that persons colonized with MRSA can be intermittent or persistent colonizers [90]. Thus, a person determined to be negative for MRSA colonization may be an intermittent carrier. This is a preanalytical factor to take into account when deciding how often to perform repeat testing on an individual if the institution develops an active surveillance program. Regardless of the methodology chosen, there should be a basic understanding of the molecular screening and its impact to infection control, hospital/laboratory finances, and HAI prevention.

CLOSTRIDIUM DIFFICILE

Clostridium difficile is an anaerobic, gram-positive, spore-forming bacterium that exists in the environment and in the gastrointestinal tract of humans and animals. This bacterium has long been recognized as the primary infectious cause of pseudomembranous colitis (PMC) and the principle cause of infectious diarrhea in hospitalized patients [109]. However, the manifestations of disease can vary from asymptomatic carrier to PMC, toxic megacolon, and death [110]. Infections outside the colon are rare. Clinical disease is mediated by the production of toxin: toxin A and toxin B, *tcdA* and *tcdB*, respectively [111]. Approximately 6—12.5% of *C. difficile* strains also produce a binary toxin (also called *C. difficile* transferase) [112].

A 2015 study by Lessa et al. showed that in the United States, *C. difficile* was responsible for approximately 500,000 infections and was associated with 29,000 deaths during 2011 [113]. The prevalence of hospital discharges with CDI (*C. difficile* infection) listed as a diagnosis increased from 3.82 per 1000 discharges in 2000 to 8.75 per 1000 discharges in 2008 [114]. Increases in disease incidence have been observed outside the United States as well [115,116]. Among pediatric patients, the incidence of CDI-related hospitalizations increased from 0.724 per 1000 hospital days in 1997 to 1.28 per 1000 hospital days in 2006 [117]. The highest incidence was observed in those 1—4 years of age and the lowest incidence occurring in infants less than 1 year of age.

Infected patients may experience a more severe clinical course that is caused by the more virulent *C. difficile* strain BI/NAP1/O27 [116,118]. This hypervirulent strain emerged as a result of several factors: polymorphisms in the down-regulatory gene (*tcdC*), presence of a gene encoding binary toxin (*ctdA* and *ctdB*), high-level fluoroquinolone resistance, and polymorphisms in the *tcdB* that could result in improved toxin binding [117]. It is now also recognized that this pathogen can be acquired from the community and is not exclusively a nosocomial pathogen. An ad hoc *C. difficile* surveillance group has developed three categories of infection related to the time of admission and discharge from the hospital/healthcare facility: (1) hospital onset (HO), (2) community-onset healthcare facility associated (CO-HCFA), and (3) CA [119].

C. DIFFICILE MOLECULAR TESTING AND CLINICAL UTILITY

The laboratory diagnosis of C. difficile is traditionally based on the detection of the toxin produced because nontoxin producing strains of C. difficile may be part of the normal gastrointestinal flora. The cytotoxicity neutralization assay (CCNA) was one of the first laboratory tests developed and is considered the gold standard. The diarrheal stool from an individual patient is processed and the stool filtrate is applied onto a monolayer of a particular cell line (Vero, McCoy, Hep2, human fibroblasts cell lines have been used). The cultures are incubated for 24–48 h and the cell lines are monitored for cytopathic effects (CPEs) induced by the presence of toxin. If CPE is observed, then a neutralization assay is performed to ensure that the CPE was mediated by the toxin and not a false-reacting substance. Toxigenic culture relies on the growth of C. difficile organism, and any culture growth is tested for the production of toxin.

The introduction of enzyme immunoassays (EIAs) for the detection of C. difficile toxin was widely accepted due to shorter turnaround time, elimination of the burden of maintaining cell lines, and higher throughput capabilities compared to CCNA and toxigenic culture. However, the EIAs suffer from a relatively low analytical sensitivity when used as the sole test. The subsequent generation of EIAs developed added a second analyte to be detected (glutamate dehydrogenase (GDH) antigen) along with testing for the presence of toxin in order to increase the sensitivity of the early-generation EIAs and improve the negative predictive value [120–124]. See Table 14.5 listing the methodologies available for C. difficile (molecular and nonmolecular).

Newer molecular-based methods offer comparable testing time, ease of technically performing the assays, and relatively better performance compared to EIA and culture-based assays [125–131]. Table 14.6 lists FDA-approved molecular assays available for C. difficile. The primary gene target is tcdB. However, some commercial assays target both the tcdB and tcdA genes. There are studies showing that clinical isolates of toxin-producing C. difficile express either tcdB or tcdB + tcdA, while strains that express tcdA only are rare [132]. Some assays also claim to detect the hypervirulent NAP1 strain by targeting the binary toxin (tcdC). Since there are nontoxin producing strains of C. difficile the toxin genes are the preferred gene targets. The strain type may factor into the performance of the molecular assays. Tenover et al. showed differences in the performance of molecular and EIA based with respect to the O27 (NAP1 hypervirulent strain) and non-O27 strains [133].

TABLE 14.5   Diagnostics Test for C. difficile

Diagnostic test	Advantages	Disadvantages
Cell culture cytotoxicity neutralization assay (CCCNA)	Good sensitivity	2-day TAT
		Requires cell culture capability
Toxigenic culture	Excellent sensitivity	Requires second-line test for toxin detection 3- to 4-day TAT
	Good specificity	Requires expertise in culturing C. difficile
Glutamate dehydrogenase (GDH)	Inexpensive	Very poor specificity
	Rapid	Requires second-line test for toxin detection
	Good sensitivity	
	Good negative predictive value	
Enzyme immunoassay (EIA) for toxin	Inexpensive	Poor sensitivity
	Rapid	Good specificity
GDH + EIA	Inexpensive	Poor sensitivity
	Rapid	Good specificity
Nucleic acid amplification/molecular	Excellent sensitivity	Relatively more expensive
	Excellent specificity	
	Rapid	

TABLE 14.6 Molecular Diagnostic Tests for C. *difficile*

Assay name	Vendor	Gene target	Method
ICEPlex C. *difficile* kit	PrimeraDx	*tcdB*	PCR with capillary electrophoresis
IMDx C. *difficile* for Abbott m2000	Intelligent Medical Devices, Inc.	*tcdB* and *tcdA*	Real-time PCR
Quidel Molecular Direct	Quidel Corp	*tcdB* and *tcdA*	Real-time PCR
Verigene C. *difficile* test	Nanosphere, Inc	*tcdB*, *tcdA*, *tcdC* and presumptive ID of O27/NAP1/BI strain	Real-time PCR/nanoparticle hybridization
Portrait Toxigenic C. *difficile* Assay	Great Basin Scientific, Inc.	*tcdB*	Helicase-dependent amplification/ microarray detection
Simplexa C. *difficile* Universal Direct Assay	Focus Diagnostics, Inc.	*tcdB*	Real-time PCR
Xpert C. *difficile*/Epi	Cepheid	*tcdB*, binary toxin gene (CDT), *tcdC*, also gives presumptive ID of O27/NAP1/BI strain	Real-time PCR
Xpert C. *difficile*	Cepheid	*tcdB*	Real-time PCR
Illumigene C *difficile*	Meridian Bioscience, Inc.	*tcdB*, *tcdA*	Loop-mediated isothermal amplification (LAMP)
ProGastro Cd	Prodesse/Hologic	*tcdB*	Real-time PCR
BD Max Cdiff	BD Diagnostics	*tcdB*	Real-time PCR
BD GeneOhm C diff	BD Diagnostics/GeneOhm	*tcdB*	Real-time PCR

http://www.fda.gov/MedicalDevices/ProductsandMedicalProcedures/InVitroDiagnostics/ucm330711.htm#microbial (Web page updated January 27, 2015).

Questions remain regarding whether an algorithmic process consisting of EIA + nucleic acid amplification test (NAAT) [121,134—138] or an NAAT-only test [127,139,140] provides better clinical benefit for patients. The clinical benefit is dependent upon a myriad of factors, one of which is the prevalence of C. *difficile* disease [141]. There are operational factors such as the total cost to the laboratory (instrumentation, reagents, and technical labor) to do the testing, getting results out in a timely manner, and having the laboratory staff expertise to perform molecular testing. These are important things to consider when a laboratory is considering adopting or changing the existing algorithm for C. *difficile*.

An important preanalytical factor is that only samples of diarrheal/liquid stool should be accepted, with the exception of a patient with an ileus where a swab may be acceptable. Normally, swabs of stool are not preferred and are rejected in most, if not all, clinical laboratories. There is very limited clinical utility when testing formed/solid stool from a patient that is not symptomatic for C. *difficile* disease. It is well established that individuals can be asymptomatically colonized with this bacteria and positive results confound clinical interpretation by the healthcare staff [110,112,115,142,143]. However, there is the possibility that the diarrheal stool is positive for C. *difficile* toxin and/or PCR, yet the cause of the gastrointestinal illness was due to some other cause/pathogen.

There is limited utility in testing neonates (<1 year old) for C. *difficile*. Interestingly, it has been observed that neonates in a special care nursery can be asymptomatically colonized with C. *difficile*, and potentially serve as a reservoir for transmission in the hospital setting [144,145]. It has been hypothesized that C. *difficile* cannot cause clinical disease in infants less than 1 year old because the receptor for toxin binding is not yet expressed [146]. Therefore, a positive C. *difficile* result in an infant has the potential to lead to unnecessary treatment and other causes of the patient's diarrhea should be investigated and/or ruled-out.

Surawicz et al. published guidelines on preanalytical and analytical aspects of testing [147]. These guidelines basically state that only diarrheal stools should be tested. These investigators also found NAAT to be superior to the toxin A/B EIA testing, but that algorithmic testing using a combination of EIA testing (GDH + Tox A/B) can be implemented with the caveat that the sensitivity is lower than the NAAT-only testing. Additionally, repeat testing is discouraged, especially if the test is being ordered as a test of cure. Testing for C. *difficile* " × 3" (ie, three consecutive sample collections for three separate tests) is a common order for microbiology laboratories, most likely due to the lack of confidence clinicians may have in the traditional EIA testing. Aichinger et al. showed that repeat testing for C. difficile after an initial negative result only increases the diagnostic yield by 1.9% and 1.7%, respectively, for repeat EIA and molecular testing [148]. Other studies have shown similar results indicating the lack of clinical utility for repeat testing [132,149−152].

Similar to the multiplex molecular panels for blood culture pathogen identification, C. difficile is part of two FDA-approved panels from Luminex Corp. and BioFire Diagnostics, Inc. Taking the approach of a one-and-done model these panels include a number of bacterial, viral, and parasitic pathogens that can be performed from a single diarrheal stool sample. The reader is directed to the respective vendor websites (https://www.luminexcorp.com/clinical/infectious-disease/gastrointestinal-pathogen-panel/ and http://filmarray.com/the-panels/) to see up-to-date lists of pathogens that are represented on their respective gastrointestinal pathogen panels.

C. DIFFICILE MOLECULAR TYPING AND CLINICAL UTILITY

Molecular typing is an important infection control tool to monitor the prevalence of certain strains within a healthcare institution or to investigate if a cluster of infections are unrelated or part of an outbreak [110,153]. These typing assays are not normally performed in the clinical laboratory because culture of this bacterium is required and are more likely to be performed in a public health or research laboratory setting. See Table 14.7 for some of the methods used

TABLE 14.7　Molecular Typing Methods for C. *difficile*

Methods
Amplified fragment length polymorphism (AFLP)
Multilocus sequence typing (MLST)
Multilocus variable number tandem repeat analysis (MLVA)
Pulsed field gel electrophoresis (PFGE)
Polymerase chain reaction ribotyping (PCR-RT)
Agarose-based PCR ribotyping
Sequence-based PCR ribotyping
Surface layer protein A gene sequence typing (slpAST)
Repetitive element PCR typing
Restriction endonuclease analysis (REA)
WGS—whole genome sequencing

for determining the genetic fingerprint of C. difficile strains.

PFGE is the most common method used in North America while PCR ribotyping is the most common method used in Europe [154]. PCR ribotyping analyzes the variability of the intergenic spacer region which is in between the 16S and 23S ribosomal RNA (rRNA). Primers are designed that target the conserved regions of the 16S and 23S rDNAs and variable PCR amplicons are produced that are separated by agarose gel electrophoresis. The different banding patterns observed for the C. difficile isolates are called PCR ribotypes. Several studies have compared a number of these methodologies from Table 14.7 and showed varying levels of comparability, advantages, and disadvantages [154−162]. It is important that despite the use of various molecular typing methods, that there is some level of standardization so that strains can be accurately compared from a more global perspective. Huber et al. noted that PCR ribotyping is becoming the method of choice and a move toward standardizing the protocol and interpretation is urgently needed [163].

References

[1] Siegel JD, Rhinehart E, Jackson M, Chiarello L, Committee HICPA. 2007 guideline for isolation precautions: preventing transmission of infectious agents in healthcare settings, <www.cdc.gov/ncidod/dhqp/pdf/isolation2007.pdf>; 2007.

[2] Aureden K, Arias K, Burns LA, et al. Guide to the elimination of Staphylococcus aureus (MRSA) transmission in hospital settings. 2nd ed.; 2010.

[3] Calfee DP, Salgado CD, Milstone AM, et al. Strategies to prevent methicillin-resistant Staphylococcus aureus transmission and infection in acute care hospitals: 2014 update. Infect Control Hosp Epidemiol 2014;35:772−96.

[4] Centers for Medicare & Medicaid Services. CMS to improve quality of care during hospital inpatient stays, <https://www.cms.gov/Newsroom/MediaReleaseDatabase/Fact-sheets/2014-Fact-sheets-items/2014-08-04-2.html>; 2015.

[5] Thompson R, Cabezudo I, Wenzel R. Epidemiology of nosocomial infections caused by methicillin-resistant Staphylococcus aureus. Ann Intern Med 1982;97:309−17.

[6] Boyce J. Methicillin-resistant Staphylococcus aureus. Detection, epidemiology, and control measures. Infect Dis Clin North Am 1989;3:901−13.

[7] Barbut F, Petit J. Epidemiology of Clostridium difficile-associated infections. Clin Microbiol Infect 2001;7:405−10.

[8] Tong SYC, Davis JS, Eichenberger E, Holland TL, Fowler VG. Staphylococcus aureus infections: epidemiology, pathophysiology, clinical manifestations, and management. Clin Microbiol Rev 2015;28:603−61.

[9] Klevens RM, Morrison MA, Nadle J, et al. Invasive methicillin-resistant Staphylococcus aureus infections in the United States. JAMA 2007;298:1763−71.

[10] Hidron AI, Kourbatova EV, Halvosa JS, et al. Risk factors for colonization with methicillin- resistant Staphylococcus aureus (MRSA) in patients admitted to an urban hospital: emergence of community-associated MRSA nasal carriage. Clin Infect Dis 2005;41:159−66.

[11] Moran GJ, Krishnadasan A, Gorwitz RJ, et al. S. aureus. N Engl J Med 2006;355:666−74.

[12] David MZ, Daum RS. Community-associated methicillin-resistant *Staphylococcus aureus*: epidemiology and clinical consequences of an emerging epidemic. Clin Microbiol Rev 2010;23:616−87.

[13] Chi C, Ho M, Ho C, Lin P, Wang J, Fung C. Molecular epidemiology of community-acquired methicillin-resistant *Staphylococcus aureus* bacteremia in a teaching hospital. J Microbiol Immunol Infect 2007;40:310−16.

[14] Adedeji A, Weller T, Gray J. MRSA in children presenting to hospitals in Birmingham, UK. J Hosp Infect 2007;65:29−34.

[15] Rollason J, Bastin L, Ac H, et al. Epidemiology of community-acquired methicillin-resistant *Staphylococcus aureus* obtained from the UK West Midlands region. J Hosp Infect 2008; 70:314−20.

[16] Beam JW, Buckley B. *Staphylococcus aureus*: prevalence and risk factors. J Athletic Train 2006;41:337−40.

[17] Sa F, Garbutt J, Elward A, Shannon W, Ga S. Prevalence of and risk factors for community-acquired methicillin-resistant and methicillin-sensitive *Staphylococcus aureus* colonization in children seen in a practice-based research network. Pediatrics 2008;121:1090−8.

[18] Jarvis WR, Jarvis AA, Chinn RY. National prevalence of methicillin-resistant *Staphylococcus aureus* in inpatients at United States health care facilities, 2010. Am J Infect Control 2012;40:194−200.

[19] Jarvis WR, Schlosser J, Chinn RY, Tweeten S, Jackson M. National prevalence of methicillin-resistant *Staphylococcus aureus* in inpatients at US health care facilities, 2006. Am J Infect Control 2007;35:631−7.

[20] Goetghebeur M, Landry P, Han D, Vicente C. Methicillin-resistant *Staphylococcus aureus*: a public health issue with economic consequences. Can J Infect Dis Med Microbiol 2007; 18:27−34.

[21] Anderson DJ, Kaye KS, Chen LF, et al. Clinical and financial outcomes due to methicillin resistant *Staphylococcus aureus* surgical site infection: a multi-center matched outcomes study. PLoS One 2009;4:1−8.

[22] Filice GA, Nyman JA, Lexau C, et al. Excess costs and utilization associated with methicillin resistance for patients with *Staphylococcus aureus* infection. Infect Control Hosp Epidemiol 2010;31:365−73.

[23] Kopp B, Nix D, Armstrong E. Clinical and economic analysis of methicillin-susceptible and -resistant *Staphylococcus aureus* infections. Ann Pharmacother 2004;38:1377−82.

[24] Lodise T, McKinnon P. Clinical and economic impact of methicillin resistance in patients with *Staphylococcus aureus* bacteremia. Diagn Microbiol Infect Dis 2005;52:113−22.

[25] Ly T, Gulia J, Pyrgos V, Waga M, Shoham S. Impact upon clinical outcomes of translation of PNA FISH-generated laboratory data from the clinical microbiology bench to bedside in real time. Ther Clin Risk Manag 2008;4:637−40.

[26] Gonzalez V, Padilla E, Gimenez M, et al. Rapid diagnosis of *Staphylococcus aureus* bacteremia using S. aureus PNA FISH. Eur J Clin Microbiol Infect Dis 2004;23:396−8.

[27] Hensley D, Tapia R, Encina Y. An evaluation of the AdvanDx *Staphylococcus aureus*/CNS PNA FISH assay. Clin Lab Sci 2009;22:30−3.

[28] Kothari A, Morgan M, Haake DA. Emerging technologies for rapid identification of bloodstream pathogens. Clin Infect Dis 2014;59:272−8.

[29] Laub R, Knudsen J. Clinical consequences of using PNA-FISH in staphylococcal bacteraemia. Eur J Clin Microbiol Infect Dis 2014;33:599−601.

[30] Parta M, Goebel M, Thomas J, Matloobi M, Stager C, Musher DM. Impact of an assay that enables rapid determination of *Staphylococcus* species and their drug susceptibility on the treatment of patients with positive blood culture results. Infect Control Hosp Epidemiol 2010;31:1043−8.

[31] Bauer KA, West JE, Pancholi P, Stevenson KB, Goff DA. An antimicrobial stewardship program's impact with rapid polymerase chain reaction methicillin-resistant *Staphylococcus aureus*/S. aureus blood culture test in patients with S. aureus bacteremia. Clin Infect Dis 2010;51:1074−80.

[32] Brown J, Paladino JA. Impact of rapid methicillin-resistant *Staphylococcus aureus* polymerase chain reaction testing on mortality and cost effectiveness in hospitalized patients with bacteraemia: a decision model. Pharmacoeconomics 2010;28:567−75.

[33] Scanvic A, Courdavault L, Sollet J, Le Turdu F. Interest of real-time PCR Xpert MRSA/SA on GeneXpert® DX System in the investigation of staphylococcal bacteremia. Pathol Biol (Paris) 2011;59:67−72.

[34] Wojewoda CM, Sercia L, Navas M, et al. Evaluation of the Verigene gram-positive blood culture nucleic acid test for rapid detection of bacteria and resistance determinants. J Clin Microbiol 2013;51:2072−6.

[35] Sullivan KV, Turner NN, Roundtree SS, et al. Rapid detection of gram-positive organisms by use of the Verigene gram-positive blood culture nucleic acid test and the BacT/Alert pediatric FAN system in a multicenter pediatric evaluation. J Clin Microbiol 2013;51:3579−84.

[36] Samuel LP, Tibbetts RJ, Agotesku A, Fey M, Hensley R, Meier FA. Evaluation of a microarray-based assay for rapid identification of gram-positive organisms and resistance markers in positive blood cultures. J Clin Microbiol 2013;51:1188−92.

[37] Buchan BW, Ginocchio CC, Manii R, et al. Multiplex identification of gram-positive bacteria and resistance determinants directly from positive blood culture broths: evaluation of an automated microarray-based nucleic acid test. PLoS Med 2013;10:1−13.

[38] Beal SG, Ciurca J, Smith G, et al. Evaluation of the nanosphere Verigene gram-positive blood culture assay with the VersaTREK blood culture system and assessment of possible impact on selected patients. J Clin Microbiol 2013;51:3988−92.

[39] Mestas J, Polanco CM, Felsenstein S, Bard D. Performance of the Verigene gram-positive blood culture assay for direct detection of gram-positive organisms and resistance markers in a pediatric hospital. J Clin Microbiol 2014;52:283−7.

[40] Bhatti MM, Boonlayangoor S, Beavis KG, Tesic V. Evaluation of FilmArray and Verigene systems for rapid identification of positive blood cultures. J Clin Microbiol 2014;52:3433−6.

[41] Park S, Park K, Bang K, et al. Clinical significance and outcome of polymicrobial *Staphylococcus aureus*. J Infect 2012;65:119−27.

[42] Altun O, Almuhayawi M, Ullberg M, Ozenci V. Clinical evaluation of the FilmArray blood culture identification panel in identification of bacteria and yeasts from positive blood culture bottles. J Clin Microbiol 2013;51:4130−6.

[43] Desoubeaux G, Bailly É, Le Brun C, et al. Prospective assessment of FilmArray technology for the rapid identification of yeast isolated from blood cultures. J Microbiol Methods 2014;106:119−22.

[44] Blaschke AJ, Heyrend C, Byington CL, et al. Rapid identification of pathogens from positive blood cultures by multiplex polymerase chain reaction using the FilmArray system. Diagn Microbiol Infect Dis 2012;74(4):349−55.

[45] Lowy FD. Antimicrobial resistance: the example of *Staphylococcus aureus*. J Clin Invest 2003;111:1265−73.

[46] Ito T, Hiramatsu K, Oliveira DC, et al. Classification of staphylococcal cassette chromosome mec(SCCmec): guidelines for

reporting novel SCCmec elements. Antimicrob Agents Chemother 2009;53:4961−7.

[47] Deurenberg R, Stobberingh E. The evolution of *Staphylococcus aureus*. Infect Genet Evol 2008;8:747−63.

[48] Ito T, Hiramatsu K, Oliveira DC, et al. International Working Group on the staphylococcal cassette chromosome elements, <http://www.sccmec.org/Pages/SCC_ClassificationEN.html> [accessed 24.06.15].

[49] Murray PR. Molecular diagnosis of methicillin-resistant *Staphylococcus aureus* colonization. J Clin Microbiol 2013;51:4284.

[50] Stamper PD, Louie L, Wong H, Simor AE, Farley JE, Carroll KC. Genotypic and phenotypic characterization of methicillin-susceptible *Staphylococcus aureus* isolates misidentified as methicillin-resistant *Staphylococcus aureus* by the BD GeneOhm MRSA assay. J Clin Microbiol 2011;49:1240−4.

[51] Blanc DS, Basset P, Nahimana-Tessemo I, Jaton K, Greub G, Zanetti G. High proportion of wrongly identified methicillin-resistant *Staphylococcus aureus* carriers by use of a rapid commercial PCR assay due to presence of staphylococcal cassette chromosome element lacking the *mecA* gene. J Clin Microbiol 2011;49:722−4.

[52] Mehndiratta P, Bhalla P. Typing of methicillin resistant *Staphylococcus aureus*: a technical review. Indian J Med Microbiol 2012;30:16−23.

[53] Struelens MJ, Deplano A, Godard C, Maes N, Serruys E. Epidemiologic typing and delineation of genetic relatedness of methicillin-resistant *Staphylococcus aureus* by macrorestriction analysis of genomic DNA by using pulsed-field gel electrophoresis. J Clin Microbiol 1992;30:2599−605.

[54] Tenover FC, Vaughn RR, Mcdougal LK, Fosheim GE, Mcgowan JE. Multiple-locus variable-number tandem-repeat assay analysis of methicillin-resistant *Staphylococcus aureus* strains. J Clin Microbiol 2007;45:2215−19.

[55] Szabo J. Molecular methods in epidemiology of methicillin resistant *Staphylococcus aureus* (MRSA): advantages, disadvantages of different techniques. J Med Microbiol Diagn 2014;3:147.

[56] He Y, Xie Y, Reed S. Pulsed-field gel electrophoresis typing of *Staphylococcus aureus* isolates. Methods Mol Biol 2014;1085:103−11.

[57] Strandén A, Frei R, Widmer AF. Molecular typing of methicillin-resistant *Staphylococcus aureus*: can PCR replace pulsed-field gel electrophoresis? J Clin Microbiol 2003;41:3181−6.

[58] Chung M, Lencastre H, de, Matthews P, et al. Molecular typing of MRSA by pulsed-field gel electrophoresis: comparison of results obtained in a multilaboratory effort using identical protocols and MRSA strains. Microb Drug Resist 2000;6:189−98.

[59] Chung S, Chung S, Yi J, et al. Comparison of modified multiple-locus variable-number tandem-repeat fingerprinting with pulsed-field gel electrophoresis for typing clinical isolates of *Staphylococcus aureus*. Ann Lab Med 2012;32:50−6.

[60] Elena VM, Rosario V, Dueñas H, Maria A, Eduardo RM. Pulsed field gel electrophoresis in molecular typing and epidemiological detection of methicillin resistant *Staphylococcus aureus* (MRSA). In: Magdeldin S, editor. Gel electrophoresis—advanced techniques. 2012. pp. 179−192. <www.intechopen.com/books/gel-electrophoresis-advanced-techniques/pulse-field-gel-electrophoresis-in-molecular-typing-and-epidemiological-detection-of-methicillin-re>.

[61] Omar NY, Ali HAS, Harfoush RAH, El Khayat EH. Molecular typing of methicillin resistant *Staphylococcus aureus* clinical isolates on the basis of protein A and coagulase gene polymorphisms. Int J Microbiol 2014;2014:1−11.

[62] Prosperi M, Veras N, Azarian T, et al. Molecular epidemiology of community-associated methicillin-resistant *Staphylococcus aureus* in the genomic era: a cross-sectional study. Sci Rep 2013;3:1902.

[63] Shopsin B, Kreiswirth BN. Molecular epidemiology of methicillin-resistant *Staphylococcus aureus*. Emerg Infect Dis 2001;7:323−6.

[64] Struelens MJ, Hawkey PM, French GL, Witte W, Tacconelli E. Laboratory tools and strategies for methicillin-resistant *Staphylococcus aureus* screening, surveillance and typing: state of the art and unmet needs. Clin Microbiol Infect 2009;15:112−19.

[65] Rajan V, Schoenfelder S, Ziebuhr W, Gopal S. Genotyping of community-associated methicillin resistant *Staphylococcus aureus* (CA-MRSA) in a tertiary care centre in Mysore, South India: ST2371-SCCmec IV emerges as the major clone. Infect Genet Evol 2015;34:230−5.

[66] Lee JH. Methicillin (oxacillin)-resistant *Staphylococcus aureus* strains isolated from major food animals and their potential transmission to humans. Appl Environ Microbiol 2003;69:6489−94.

[67] Machuca MA, Sosa LM, Gonza CI. Molecular typing and virulence characteristic of methicillin-resistant *Staphylococcus aureus* isolates from pediatric patients in Bucaramanga, Colombia. PLoS One 2013;8:1−8.

[68] Miller MB, Tang YW. Basic concepts of microarrays and potential applications in clinical microbiology. Clin Microbiol Rev 2009;22:611−33.

[69] Francois P, Koessler T, Huyghe A, et al. Rapid *Staphylococcus aureus* agr type determination by a novel multiplex real-time quantitative PCR assay. J Clin Microbiol 2006;44:1892−5.

[70] Narukawa M, Yasuoka A, Note R, Funada H. Sequence-based spa typing as a rapid screening method for the areal and nosocomial outbreaks of MRSA. Tohoku J Exp Med 2009; 218:207−13.

[71] Harris SR, Cartwright EJP, Török ME, et al. Whole-genome sequencing for analysis of an outbreak of methicillin-resistant *Staphylococcus aureus*: a descriptive study. Lancet Infect Dis 2013;13:130−6.

[72] Price J, Gordon NC, Crook D, Llewelyn M, Paul J. Whole genome sequencing in the prevention and control of *Staphylococcus aureus* infection. Clin Microbiol Infect 2013; 19:784−9.

[73] Bartels MD, Petersen A, Worning P, et al. Comparing whole-genome sequencing with Sanger sequencing for spa typing of methicillin-resistant *Staphylococcus aureus*. J Clin Microbiol 2014;52:4305−8.

[74] Gilchrist CA, Turner SD, Riley MF, Petri WA, Hewlett EL. Whole-genome sequencing in outbreak analysis. Clin Microbiol Rev 2015;28:541−63.

[75] Köser CU, Holden MTG, Ellington MJ, et al. Rapid whole-genome sequencing for investigation of a neonatal MRSA outbreak. N Engl J Med 2012;366:2267−75.

[76] Reuter S, Ellington MJ, Cartwright EJP, et al. Rapid bacterial whole-genome sequencing to enhance diagnostic and public health microbiology. JAMA Intern Med 2013;173:1397−404.

[77] Leopold SR, Goering RV, Witten A, Harmsen D, Mellmann A. Bacterial whole-genome sequencing revisited: portable, scalable, and standardized analysis for typing and detection of virulence and antibiotic resistance genes. J Clin Microbiol 2014;52:2365−70.

[78] Le VTM, Diep BA. Selected insights from application of whole genome sequencing for outbreak investigations. Curr Opin Crit Care 2013;19:432−9.

[79] Huang SS, Platt R. Risk of methicillin-resistant *Staphylococcus aureus* infection after previous infection or colonization. Clin Infect Dis 2003;36:281−5.

[80] State Legislation & Initiatives on Healthcare-Associated Infections. Committee to reduce infect deaths, <www.hospitalinfection.org/legislation.shtml>; 2011 [accessed 28.06.15].

[81] Wise ME, Weber SG, Schneider A, et al. Hospital staff perceptions of a legislative mandate for methicillin-resistant *Staphylococcus aureus* screening. Infect Control Hosp Epidemiol 2011;32:573—8.

[82] Weber SG, Huang SS, Oriola S, et al. Legislative mandates for use of active surveillance cultures to screen for methicillin-resistant *Staphylococcus aureus* and vancomycin resistant enterococci: position statement from the Joint SHEA and APIC Task Force. Infect Control Hosp Epidemiol 2007;28:249—60.

[83] Garcia R, Vonderheid S, McFarlin B, Djonlich M, Jang C, Maghirang J. Cost and health outcomes associated with mandatory MRSA screening in a special care nursery. Adv Neonatal Care 2011;11:200—7.

[84] Nelson RE, Jones M, Rubin MA. Decolonization with mupirocin and subsequent risk of methicillin-resistant *Staphylococcus aureus* carriage in veterans affairs hospitals. Infect Dis Ther 2012;1:1—7.

[85] Peterson LR, Diekema DJ, Doern GV. To screen or not to screen for methicillin-resistant *Staphylococcus aureus*. J Clin Microbiol 2010;48:683—9.

[86] Senn L, Basset P, Nahimana I, Zanetti G, Blanc DS. Which anatomical sites should be sampled for screening of methicillin-resistant *Staphylococcus aureus* carriage by culture or by rapid PCR test? Clin Microbiol Infect 2012;18:E31—3.

[87] El-Bouri K, El-Bouri W. Screening cultures for detection of methicillin-resistant *Staphylococcus aureus* in a population at high risk for MRSA colonisation: identification of optimal combinations of anatomical sites. Libyan J Med 2013;8:8—12.

[88] Lautenbach E, Nachamkin I, Hu B, et al. Surveillance cultures for detection of methicillin-resistant *Staphylococcus aureus*: diagnostic yield of anatomic sites and comparison of provider- and patient-collected samples. Infect Control Hosp Epidemiol 2009;30(4):380—2.

[89] Shurland SM, Stine OC, Venezia RA, et al. Colonization sites of USA300 methicillin-resistant *Staphylococcus aureus* in residents of extended care facilities. Infect Control Hosp Epidemiol 2009;30:313—18.

[90] Wertheim H, Melles D. The role of nasal carriage in *Staphylococcus aureus* infections. Lancet Infect Dis 2005;5:751—62.

[91] Matheson A, Christie P, Stari T, et al. Nasal swab screening for methicillin-resistant *Staphylococcus aureus*—how well does it perform? A cross-sectional study. Infect Control Hosp Epidemiol 2012;33:803—8.

[92] Tübbicke A, Hübner C, Hübner N-O, Wegner C, Kramer A, Fleßa S. Cost comparison of MRSA screening and management—a decision tree analysis. BMC Health Serv Res 2012; 12:438.

[93] Leonhardt KK, Yakusheva O, Phelan D, et al. Clinical effectiveness and cost benefit of universal versus targeted methicillin-resistant *Staphylococcus aureus* screening upon admission in hospitals. Infect Control Hosp Epidemiol 2011; 32:797—803.

[94] Brooks R. Screening for MRSA: an idea whose time has come? AAOS Now 2009; <http://www.aaos.org/news/aaosnow/mar09/clinical13.asp>.

[95] Deeny SR, Cooper BS, Cookson B, Hopkins S, Robotham JV. Targeted versus universal screening and decolonization to reduce healthcare-associated methicillin-resistant *Staphylococcus aureus* infection. J Hosp Infect 2013;85:33—44.

[96] Geiger K, Brown J. Rapid testing for methicillin-resistant *Staphylococcus aureus*: implications for antimicrobial stewardship. Am J Heal Pharm 2013;70:335—42.

[97] Olchanski N, Mathews C, Fusfeld L, Jarvis W. Assessment of the influence of test characteristics on the clinical and cost impacts of methicillin-resistant *Staphylococcus aureus* screening

programs in US hospitals. Infect Control Hosp Epidemiol 2011;32:250—7.

[98] Polisena J, Chen S, Cimon K, McGill S, Forward K, Gardam M. Clinical effectiveness of rapid tests for methicillin resistant *Staphylococcus aureus* (MRSA) in hospitalized patients: a systematic review. BMC Infect Dis 2011;11:336.

[99] Tacconelli E, De Angelis G, De Waure C, Cataldo MA, La Torre G, Cauda R. Rapid screening tests for methicillin-resistant *Staphylococcus aureus* at hospital admission: systematic review and meta-analysis. Lancet Infect Dis 2009; 9:546—54.

[100] Bode LGM, Kluytmans JAJW, Wertheim HFL, et al. Preventing surgical-site infections in nasal carriers of *Staphylococcus aureus*. N Engl J Med 2010;362:9-7.

[101] Arcenas RC, Spadoni S, Mohammad A, et al. Multicenter evaluation of the LightCycler MRSA advanced test, the Xpert MRSA assay, and MRSASelect directly plated culture with simulated workflow comparison for the detection of methicillin-resistant *Staphylococcus aureus* in nasal swabs. J Mol Diagn 2012;14:367—75.

[102] Lepainteur M, Delattre S, Cozza S, Lawrence C, Roux A-L, Rottman M. Comparative evaluation of two PCR-based methods for detection of methicillin-resistant *Staphylococcus aureus* (MRSA): Xpert MRSA Gen 3 and BD-Max MRSA XT. J Clin Microbiol 2015;53:1955—8.

[103] Parhizgar F, Colmer-Hamood J, Satterwhite J, Winn R, Nugent K. A comparative analysis of GeneXpert real-time PCR with culture for the detection of methicillin-resistant *Staphylococcus aureus* colonization in selected hospital admissions. ISRN Infect Dis 2013;2013:1—3.

[104] Patel P, Robicsek A, Grayes A, et al. Evaluation of multiple real-time PCR tests on nasal samples in a large MRSA surveillance program. Am J Clin Pathol 2015;143:652—8.

[105] Paule SM, Mehta M, Hacek DM, Gonzalzles TM, Robicsek A, Peterson LR. Chromogenic media vs real-time PCR for nasal surveillance of methicillin-resistant *Staphylococcus aureus* impact on detection of MRSA-positive persons. Am J Clin Pathol 2009;131:532—9.

[106] Peterson LR, Liesenfeld O, Woods CW, et al. Multicenter evaluation of the Lightcycler methicillin-resistant *Staphylococcus aureus* (MRSA) advanced test as a rapid method for detection of MRSA in nasal surveillance swabs. J Clin Microbiol 2010;48:1661—6.

[107] Wolk DM, Marx JL, Dominguez L, Driscoll D, Schifman RB. Comparison of MRSASelect agar, CHROMagar methicillin-resistant *Staphylococcus aureus* (MRSA) medium, and Xpert MRSA PCR for detection of MRSA in nares: diagnostic accuracy for surveillance samples with various bacterial densities. J Clin Microbiol 2009;47:3933—6.

[108] Yam WC, Siu GKH, Ho PL, et al. Evaluation of the Lightcycler methicillin-resistant *Staphylococcus aureus* (MRSA) advanced test for detection of MRSA nasal colonization. J Clin Microbiol 2013;51:2869—74.

[109] Bartlett JG, Wen Chang T, Gurwith M, Gorbach SL, Anderdonk AB. Antibiotic-associated pseudomembrane colitis due to toxin-producing clostridia. N Engl J Med 1978; 298:531—4.

[110] Burnham CAD, Carroll KC. Diagnosis of *Clostridium difficile* infection: an ongoing conundrum for clinicians and for clinical laboratories. Clin Microbiol Rev 2013;26:604—30.

[111] Voth D, Ballard J. *Clostridium difficile* toxins: mechanism of action and role in disease. Clin Microbiol Rev 2005;18:247—63.

[112] Carroll KC, Bartlett JG. Biology of *Clostridium difficile*: implications for epidemiology and diagnosis. Annu Rev Microbiol 2011;65:501—21.

[113] Lessa FC, Mu Y, Bamberg WM, et al. Burden of *Clostridium difficile* infection in the United States. N Engl J Med 2015;372:825–34.

[114] Miller BA, Chen LF, Sexton DJ, Anderson DJ. Comparison of the burdens of hospital-onset, healthcare facility-associated *Clostridium difficile* infection and of healthcare-associated infection due to methicillin-resistant *Staphylococcus aureus* in community hospitals. Infect Control Hosp Epidemiol 2011;32:387–90.

[115] Freeman J, Bauer MP, Baines SD, et al. The changing epidemiology of *Clostridium difficile* infections. Clin Microbiol Rev 2010;23:529–49.

[116] Kuijper EJ, Coignard B, Tu P. Emergence of *Clostridium difficile*-associated disease in North America and Europe. Clin Microbiol Infect 2006;12:2–18.

[117] Lessa FC, Gould CV, Clifford McDonald L. Current status of *Clostridium difficile* infection epidemiology. Clin Infect Dis 2012;55:65–70.

[118] Ghose C. *Clostridium difficile* infection in the twenty-first century. Emerg Microbes Infect 2013;2:e62.

[119] Cohen SH, Gerding D, Johnson S, et al. Clinical practice guidelines for *Clostridium difficile* infection in adults: 2010 update by the Society for Healthcare Epidemiology of America (SHEA) and the Infectious Diseases Society of America (IDSA). Infect Control Hosp Epidemiol 2011;31:431–55.

[120] Gilligan PH. Is a two-step glutamate dehydrogenase antigen-cytotoxicity neutralization assay algorithm superior to the premier toxin A and B enzyme immunoassay for laboratory detection of *Clostridium difficile*? J Clin Microbiol 2008; 46:1523–5.

[121] Kawada M, Annaka M, Kato H, et al. Evaluation of a simultaneous detection kit for the glutamate dehydrogenase antigen and toxin A/B in feces for diagnosis of *Clostridium difficile* infection. J Infect Chemother 2011;17:807–11.

[122] Shetty N, Wren MWD, Coen PG. The role of glutamate dehydrogenase for the detection of *Clostridium difficile* in faecal samples: a meta-analysis. J Hosp Infect 2011;77:1–6.

[123] Sharp SE, Ruden LO, Pohl JC, Hatcher PA, Jayne LM, Ivie WM. Evaluation of the C. Diff Quik Chek Complete Assay, a new glutamate dehydrogenase and A/B toxin combination lateral flow assay for use in rapid, simple diagnosis of *Clostridium difficile* disease. J Clin Microbiol 2010;48:2082–6.

[124] Quinn CD, Sefers SE, Babiker W, et al. C. Diff Quik Chek Complete Enzyme immunoassay provides a reliable first-line method for detection of *Clostridium difficile* in stool specimens. J Clin Microbiol 2010;48:603–5.

[125] Crobach MJT, Dekkers OM, Wilcox MH, Kuijper EJ. European Society of Clinical Microbiology and Infectious Diseases (ESCMID): data review and recommendations for diagnosing *Clostridium difficile*-infection (CDI). Clin Microbiol Infect 2009;15:1053–66.

[126] Huang H, Weintraub A, Fang H, Nord CE. Comparison of a commercial multiplex real-time PCR to the cell cytotoxicity neutralization assay for diagnosis of *Clostridium difficile* infections. J Clin Microbiol 2009;47:3729–31.

[127] Pancholi P, Kelly C, Raczkowski M, Balada-Llasat JM. Detection of toxigenic *Clostridium difficile*: comparison of the cell culture neutralization, Xpert C. *difficile*, Xpert C. *difficile*/Epi, and Illumigene C. *difficile* assays. J Clin Microbiol 2012;50:1331–5.

[128] Carroll KC, Buchan BW, Tan S, et al. Multicenter evaluation of the Verigene *Clostridium difficile* nucleic acid assay. J Clin Microbiol 2013;51:4120–5.

[129] Beck ET, Buchan BW, Riebe KM, et al. Multicenter evaluation of the Quidel Lyra Direct C. *difficile* nucleic acid amplification assay. J Clin Microbiol 2014;52:1998–2002.

[130] Deak E, Miller SA, Humphries RM. Comparison of Illumigene, Simplexa, and AmpliVue *Clostridium difficile* molecular assays for diagnosis of C. *difficile* infection. J Clin Microbiol 2014;52:960–3.

[131] Gilbreath J, Verma P, Abbitt A, Butler-Wu S. Comparison of the Verigene *Clostridium difficile*, Simplexa C. *difficile* Universal Direct, BD MAX Cdiff, and Xpert C. *difficile* assays for the detection of toxigenic C. *difficile*. Diagn Microbiol Infect Dis 2014;80:13–18.

[132] Kufelnicka AM, Kirn TJ. Effective utilization of evolving methods for the laboratory diagnosis of *Clostridium difficile* infection. Clin Infect Dis 2011;52:1451–7.

[133] Tenover FC, Novak-Weekley S, Woods CW, et al. Impact of strain type on detection of toxigenic *Clostridium difficile*: comparison of molecular diagnostic and enzyme immunoassay approaches. J Clin Microbiol 2010;48:3719–24.

[134] Culbreath K, Ager E, Nemeyer RJ, Kerr A, Gilligan PH. Evolution of testing algorithms at a university hospital for detection of *Clostridium difficile* infections. J Clin Microbiol 2012;50:3073–6.

[135] Peterson LR, Mehta MS, Patel PA, et al. Laboratory testing for *Clostridium difficile* infection: light at the end of the tunnel. Am J Clin Pathol 2011;136:372–80.

[136] Swindells J, Brenwald N, Reading N, Oppenheim B. Evaluation of diagnostic tests for *Clostridium difficile* infection. J Clin Microbiol 2010;48:606–8.

[137] Schmidt ML, Gilligan PH. *Clostridium difficile* testing algorithms: what is practical and feasible? Anaerobe 2009;15:270–3.

[138] Ota KV, McGowan KL. *Clostridium difficile* testing algorithms using glutamate dehydrogenase antigen and C. *difficile* toxin enzyme immunoassays with C. *difficile* nucleic acid amplification testing increase diagnostic yield in a tertiary pediatric population. J Clin Microbiol 2012;50:1185–8.

[139] Novak-Weekley SM, Marlowe EM, Miller JM, et al. *Clostridium difficile* testing in the clinical laboratory by use of multiple testing algorithms. J Clin Microbiol 2010;48:889–93.

[140] Larson AM, Fung AM, Fang FC. Evaluation of tcdB real-time PCR in a three-step diagnostic algorithm for detection of toxigenic *Clostridium difficile*. J Clin Microbiol 2010;48:124–30.

[141] Deshpande A, Pasupuleti V, Rolston DDK, et al. Diagnostic accuracy of real-time polymerase chain reaction in detection of *Clostridium difficile* in the stool samples of patients with suspected *Clostridium difficile* infection: a meta-analysis. Clin Infect Dis 2011;53:81–90.

[142] Delmée M. Laboratory diagnosis of *Clostridium difficile* disease. Clin Microbiol Infect 2001;7:411–16.

[143] McDonald LC, Coignard B, Dubberke E, Song X, Horan T, Kutty PK. Recommendations for surveillance of *Clostridium difficile*-associated disease. Infect Control Hosp Epidemiol 2007;28:140–5.

[144] Al-jumaili IJ, Shibley M, Lishman AH, Record C. Incidence and origin of *Clostridium difficile* in neonates. J Clin Microbiol 1984;19:77–8.

[145] Bolton RP, Tait SK, Dear PRF, Losowsky MS. Asymptomatic neonatal colonisation by *Clostridium difficile*. Arch Dis Child 1984;46:466–72.

[146] Eglow R, Pothoulakis C, Israel EJ, et al. Diminished *Clostridium difficile* toxin A sensitivity in newborn rabbit ileum is associated with decreased toxin A receptor. J Clin Invest 1992;90:822–9.

[147] Surawicz CM, Brandt LJ, Binion DG, et al. Guidelines for diagnosis, treatment, and prevention of *Clostridium difficile* infections. Am J Gastroenterol 2013;108:478–98.

[148] Aichinger E, Schleck CD, Harmsen WS, Nyre LM, Patel R. Nonutility of repeat laboratory testing for detection of

Clostridium difficile by use of PCR or enzyme immunoassay. J Clin Microbiol 2008;46:3795−7.

[149] Cardona DM, Rand KH. Evaluation of repeat *Clostridium difficile* enzyme immunoassay testing. J Clin Microbiol 2008; 46:3686−9.

[150] Khanna S, Pardi D, Rosenblatt J, Kammer P, Baddour L. An evaluation of repeat stool testing for *Clostridium difficile* infection by polymerase chain reaction. J Clin Gastroenterol 2012; 46:846−9.

[151] Drees M, Snydman D, O'Sullivan C. Repeated enzyme immunoassays have limited utility in diagnosing *Clostridium difficile*. Eur J Clin Microbiol Infect Dis 2008;27:397−9.

[152] Mohan S, McDermott B, Parchuri S, Cunha B. Lack of value of repeat stool testing for *Clostridium difficile* toxin. Am J Med 2006;119:7−8.

[153] Carrico RM, Bryant K, Lessa F, et al. Guide to preventing *Clostridium difficile* infections. Association for professionals in infection control and epidemiology. 2013.

[154] Knetsch CW, Lawley TD, Hensgens MP, Corver J, Wilcox MW, Kuijper EJ. Current application and future perspectives of molecular typing methods to study *Clostridium difficile* infections. Euro Surveill 2013;18:20381.

[155] Killgore G, Thompson A, Johnson S, et al. Comparison of seven techniques for typing international epidemic strains of *Clostridium difficile*: restriction endonuclease analysis, pulsed-field gel electrophoresis, PCR-ribotyping, multilocus sequence typing, multilocus variable-number tandem-repeat an. J Clin Microbiol 2008;46:431−7.

[156] Gurtler V, Grando D. New opportunities for improved ribotyping of C. *difficile* clinical isolates by exploring their genomes. J Micrbiol Methods 2013;93:257−72.

[157] Eckert C, Van Broeck J, Spigaglia P, et al. Comparison of a commercially available repetitive-element PCR system (DiversiLab) with PCR ribotyping for typing of *Clostridium difficile* strains. J Clin Microbiol 2011;49:3352−4.

[158] Church DL, Chow BL, Lloyd T, Gregson DB. Evaluation of automated repetitive-sequence-based PCR (DiversiLab) compared to PCR ribotyping for rapid molecular typing of community- and nosocomial-acquired *Clostridium difficile*. Diagn Microbiol Infect Dis 2011;70:183−90.

[159] Pasanen T, Kotila SM, Horsma J, et al. Comparison of repetitive extragenic palindromic sequence-based PCR with PCR ribotyping and pulsed-field gel electrophoresis in studying the clonality of *Clostridium difficile*. Clin Microbiol Infect 2011; 17:166−75.

[160] Manzoor SE, Tanner HE, Marriott CL, et al. Extended multilocus variable-number tandem-repeat analysis of *Clostridium difficile* correlates exactly with ribotyping and enables identification of hospital transmission. J Clin Microbiol 2011; 49:3523−30.

[161] Tenover FC, Åkerlund T, Gerding DN, et al. Comparison of strain typing results for *Clostridium difficile* isolates from North America. J Clin Microbiol 2011;49:1831−7.

[162] Hardy K, Manzoor S, Marriott C, et al. Utilizing rapid multiple-locus variable-number tandem-repeat analysis typing to aid control of hospital-acquired *Clostridium difficile* infection: a multicenter study. J Clin Microbiol 2012;50:3244−8.

[163] Huber CA, Foster NF, Riley TV, Paterson DL. Challenges for standardization of *Clostridium difficile* typing methods. J Clin Microbiol 2013;51:2810−14.

15

Molecular Testing in Emerging Infectious Diseases

J. Dong[1], N. Ismail[2] and D.H. Walker[1]

[1]Department of Pathology, University of Texas Medical Branch, Galveston, TX, United States [2]Department of Pathology, University of Pittsburgh, Pittsburgh, PA, United States

BACKGROUND AND CATALOGUE OF EMERGING INFECTIOUS AGENTS

By the late 1960s there was a widespread opinion that the era of infectious diseases was finished and that vaccines and antibiotics had controlled microbial pathogens. Indeed, it was commonly believed that we had discovered the important agents of infections and that there was little left to do in this scientific field. "...The war on infectious diseases is over and we have won..." was an often repeated conclusion. Yet in the quarter of a century between 1967 and 1992 more than 30 previously unrecognized pathogens were discovered as the etiologic agents of human infectious diseases (Table 15.1). Some of the diseases were well characterized, but the causes had been unknown. Other novel syndromes were recognized and the etiologic agents identified including acquired immunodeficiency syndrome (AIDS) and human immunodeficiency virus (HIV). Nevertheless, the general belief was that infectious diseases were less important than cardiovascular diseases and cancer, and they were not favored for research support and public health attention.

In 1992, the concept of emerging infectious diseases was defined and brought to the attention of physicians and scientists by a very widely distributed and read publication from the Institute of Medicine of the National Academies of Sciences, *Emerging Infections: Microbial Threats to Health in the United States*. The emergence of at least 16 novel infectious agents over the following 12 years (Table 15.2) emphasized that this phenomenon would be a continued series of events. The causes of awareness of the presence of an unknown pathogen are the abrupt onset of a cluster of severe illness (eg, *Legionella* pneumonia at a convention of the American Legion), recognition of distinct gross or microscopic pathologic lesions (eg, pseudomembranous colitis caused by *Clostridium difficile*), and clinical laboratory microscopy (eg, intramonocytic inclusions of *Ehrlichia chaffeensis* in patients with human monocytotropic ehrlichiosis). In numerous other instances application of an advanced technologic method identified the etiology of a well-defined syndrome (eg, noroviruses in Norwalk diarrheal illness; an outbreak had occurred and samples retained from years earlier).

DISCOVERY OF EMERGING INFECTIOUS AGENTS USING MOLECULAR METHODS

Many methods have been employed for the initial detection and identification of novel emerging pathogens including microscopy, bacterial culture, cell culture, animal inoculation, electron microscopy, archaic serologic tests, cross-reactive serologic tests, serendipitous serologic testing, and immunohistochemistry. However, currently molecular methods including probe hybridization, polymerase chain reaction (PCR) that amplifies the target or the signal, and nucleic acid sequencing are the most prominent methods for detection and characterization of newly emerging pathogens, both for discovering the agent and for determining that it is truly novel [1–11].

An example of the application of molecular methods to the identification of previously unidentified

© 2017 Elsevier Inc. All rights reserved.

TABLE 15.1 Chronological List of Emerged Infectious Agents/Diseases 1967–1992

Year	Agent	Agent characteristics	Disease	CDC molecular test name (test code)[a]	FDA-approved/cleared molecular test (manufacturer)[b]	Reference
1967	Marburg virus	Enveloped, single-stranded, negative sense RNA filovirus	Hemorrhagic fever	Marburg Identification (CDC-10349)	NA	1–5
1969	Lassa virus	Enveloped, single-stranded, bisegmented, ambisense RNA arenavirus	Hemorrhagic fever	Lassa Fever Identification (CDC-10343)	NA	1,6–8
1972	Norovirus	Nonenveloped, single-stranded RNA, viruses in the *Caliciviridae* family	Gastroenteritis	Norovirus Molecular Detection (CDC-10357), Norovirus Genotyping (CDC-10356), Norovirus Molecular Detection and Genotyping (CDC-10358)	NA	9
1973	Rotavirus	Double-stranded RNA virus. Five groups (A, B, C, D, and E): group A is the main human pathogen	Gastroenteritis	Rotavirus Molecular Detection and Genotyping (CDC-10410), Rotavirus Genotyping (CDC-10409)	NA	10–12
1975	Parvovirus B19	Nonenveloped, single-stranded DNA virus	Fifth disease or erythema infectiosum	Parvovirus B19 Molecular Detection (CDC-10363)	NA	13–15
1976	*Vibrio vulnificus*	Gram-negative, motile, curved, rod-shaped bacterium of the genus *Vibrio*	Vomiting, diarrhea, abdominal pain, and a blistering, cellulitis or septicemia	*Vibrio, Aeromonas,* and Related Organisms Study (CDC-10121), *Vibrio, Aeromonas,* and Related Organisms Identification (CDC-10120), *Vibrio* Subtyping (CDC-10122)	NA	16,17
1976	*Cryptosporidium parvum*	A protozoan	Cryptosporidiosis with symptoms including acute, watery, and nonbloody diarrhea	*Cryptosporidium* Special Study (CDC-10491)	NA	18–20
1977	Ebola virus	Enveloped, linear, single-stranded, negative-sense RNA filovirus	Hemorrhagic fever	Ebola Identification (CDC-10309)	FilmArray Biothreat-E test. Emergency Use Authorization (EUA) (Idaho Technology, Inc.)	2,5
1977	*Clostridium difficile*	A gram-positive bacterium	Colitis, diarrhea	*Clostridium difficile* Identification (CDC-10228), *Clostridium difficile* Outbreak Strain Typing (CDC-10229)	ICEPlex *C. difficile* Kit (PrimeraDx), IMDx *C. difficile* for Abbott m2000 (Intelligent Medical Devices, Inc.), BD Diagnostics BD MAX Cdiff Assay, (GeneOhm Sciences Canada Inc.), Quidel Molecular Direct *C. difficile* Assay, (Quidel Corporation), Verigene *C. difficile* Nucleic acid Test (Nanosphere, Inc.), Portrait Toxigenic *C. difficile* Assay (Great Basin Scientific, Inc.), Simplexa *C. difficile* Universal Direct Assay (Focus Diagnostics, Inc.), Xpert *C. difficile*/Epi (Cepheid),	21–23

Year	Organism	Description	Disease	Study	Illumigene C. difficile DNA Amplification Assay (Meridian Bioscience, Inc.), Ilumigene C. difficile Assay (Meridian Bioscience, Inc.), Xpert C. difficile (Cepheid), ProGastro Cd Assay (Prodesse, Inc.), BD GeneOhm Cdiff Assay (BD Diagnostics/GeneOhm Sciences, Inc.)	
1977	*Legionella pneumophila*	A thin, aerobic, pleomorphic, flagellated, non-spore forming, gram-negative bacterium of the genus *Legionella*	Legionnaires' disease	*Legionella* species Identification and Typing (CDC-10159), *Legionella* species Molecular Detection (CDC-10160), *Legionella* species Study (CDC-10161)	NA	24,25
1977	Hantaan virus	Single-stranded, enveloped, negative sense RNA viruses in the Bunyaviridae family	Hantavirus hemorrhagic fever with renal syndrome (HFRS) and hantavirus pulmonary syndrome (HPS)	Pathologic Evaluation of Unexplained Illness Due to Possible Infectious Etiology (CDC-10372)	NA	26,27
1977	Hepatitis delta virus	A small circular enveloped RNA virus	Superimposed on conditions of hepatitis with HBV	Hepatitis D Serology, NAT, and Genotyping (CDC-10328)	NA	28,29
1977	*Campylobacter* sp. (or *jejuni*)	Curved, helical-shaped, non-spore forming, gram-negative, and microaerophilic bacteria	Campylobacteriosis, Guillain-Barré syndrome (GBS)	*Campylobacter* and *Helicobacter* Study (CDC-10125), *Campylobacter*, *Helicobacter*, and Related Organisms Identification (CDC-10126), *Campylobacter*, *Helicobacter*, and Related Organisms Identification and Subtyping (CDC-10127)	NA	30,31
1979	*Cyclospora cayetanensis*	An apicomplexan, cyst-forming coccidian protozoan	Cyclosporiasis, gastroenteritis	Cyclospora Molecular Detection (CDC-10477)	NA	32,33
1980	HTLV-1	A retrovirus of the human T-lymphotropic virus (HTLV) family	Adult T-cell lymphoma (ATL), HTLV-1-associated myelopathy, uveitis, *Strongyloides stercoralis* hyper-infection	Pathologic Evaluation of Unexplained Illness Due to Possible Infectious Etiology (CDC-10372)	NA	34–37
1981	*Staphylococcus aureus* toxin	Exotoxins secreted by *S. aureus* that are compact, ellipsoidal proteins sharing a characteristic folding pattern with superantigen	Toxic shock syndrome	Staphylococcal Toxic Shock Syndrome Toxin (TSST-1) (CDC-10426)	NA	38–40
1982	*Borrelia burgdorferi*	A bacterial species of the spirochete class of the genus *Borrelia*	Lyme disease	*Borrelia* Culture and Identification (CDC-10299), *Borrelia* Special Study (CDC-10300)	NA	41,42

(Continued)

TABLE 15.1 (Continued)

Year	Agent	Agent characteristics	Disease	CDC molecular test name (test code)[a]	FDA-approved/cleared molecular test name (manufacturer)[b]	Reference
1982	*Escherichia coli* O157:H7	An enterohemorrhagic serotype of the bacterium *E. coli*	Hemolytic-uremic syndrome (HUS)	*Escherichia* and *Shigella* Identification, Serotyping, and Virulence Profiling (CDC-10114), Bacterial Select Agent Identification and AST (CDC-10224)	NA	43,44
1983	HIV-1	A lentivirus (a subgroup of retrovirus)	Acquired immune deficiency syndrome (AIDS)	HIV Molecular Surveillance Study (International Only) (CDC-10332), HIV-1 Drug Resistance Special Study (International Only) (CDC-10334), HIV-1 Genotype Drug Resistance (International Only) (CDC-10335), HIV-1 Nucleic Acid Amplification (Qualitative) (CDC-10275), HIV-1 Nucleic Acid Amplification (Viral Load) (CDC-10276), HIV-1 PCR (International Only) Qualitative (CDC-10336), HIV-1 PCR (International Only) Quantitative Viral Load (CDC-10337)	Abbott RealTime HIV-1 Assay (Abbott Molecular, Inc.), COBAS AmpliPrep/COBAS TaqMan HIV-1 Test (Roche Molecular Systems), APTIMA HIV-1 RNA Qualitative Assay (Gen-Probe, Inc.), ViroSeq HIV-1 Genotyping System (Abbott Molecular, Inc.), TRUGENEHIV-1 genotyping Kit and OpenGeneDNA Sequencing System (Siemens Healthcare Diagnostics)	45,46
1983	*Helicobacter pylori*	A gram-negative, microaerophilic bacterium	Peptic ulcer, MALT lymphoma, gastric cancer	*Helicobacter pylori* Special Study (CDC-10117)	NA	47,48
1984	*Haemophilus influenzae* biogroup *aegyptius*	Phylogenetically the same as *H. influenzae*, a gram-negative, coccobacillary, facultatively anaerobic bacterium belonging to the *Pasteurellaceae* family	Acute and often purulent conjunctivitis (pink eye)	*Haemophilus influenzae* Identification and Serotyping (CDC-10221), *Haemophilus influenzae* Study (CDC-10222), *Haemophilus* species (Not *H. influenzae*/*H. ducreyi*) ID (DC-10141)	NA	49,50
1985	*Enterocytozoon bieneusi*	A unicellular, obligate intracellular eukaryote, a species of the order microsporida	Diarrhea	Microsporidia Molecular Identification (CDC-10481), Enteric Isolation—Primary Specimen (CDC-10106)	NA	20,51
1986	*Chlamydophila pneumoniae*	An obligate intracellular bacterium in the species of *Chlamydophila*	Pneumonia	*Chlamydophila pneumoniae* Molecular Detection (CDC-10152)	FilmArray Respiratory Panel (RP) (Idaho Technology, Inc.)	52,53
1988	Human herpesvirus 6	Double-stranded DNA virus within the betaherpesvirinae subfamily and of the genus *Roseolovirus*	Neuroinflammatory diseases such as multiple sclerosis, exanthem subitum (also known as roseola infantum or sixth disease), and encephalitis, bone marrow suppression and pneumonitis in transplant recipients	Human Herpes Virus 6 (HHV6) Detection and Subtyping (CDC-10266)	NA	54,55

Year	Agent	Characteristics	Disease	CDC molecular test name (test code)[a]	FDA-approved/cleared molecular test (manufacturer)[b]	Ref.
1989	*Rickettsia japonica*	A genus of nonmotile, gram-negative, non-spore forming, highly pleomorphic bacteria	Japanese spotted fever	*Rickettsia* Molecular Detection (CDC-10402), *Rickettsia* Special Study (CDC-10405)	NA	56,57
1989	Hepatitis C virus	A small, enveloped, positive-sense single-stranded RNA virus of the family *Flaviviridae*	Hepatitis C	Hepatitis C Serology, NAT and Genotyping (CDC-10327)	Abbott RealTime HCV Genotype II (Abbott Molecular, Inc.), Abbott Realtime HCV Assay (Abbott Molecular, Inc.), COBAS AmpliPrep/ COBAS TaqMan HCV test (Roche Molecular Systems), Versant HCV 3.0 Assay (bDNA) (Siemens Healthcare Diagnostics), Versant HCV RNA Qualitative Assay (Gen-Probe, Inc.), COBAS AMPLICOR Hepatitis C Virus (HCV) Test (Roche Molecular Systems, Inc.), AMPLICOR HCV Test, v2.0 (Roche Molecular Systems, Inc.)	58,59
1990	Hepatitis E virus	A single-stranded positive-sense RNA, nonenveloped	Hepatitis	Hepatitis E Serology, NAT and Genotyping (CDC-10329)	NA	60,61
1990	*Balamuthia mandrillaris*	A free-living leptomyxid amoeba	Amoebiasis including granulomatous amoebic encephalitis (GAE)	*Balamuthia* Molecular Detection (CDC-10474), Ameba Identification (*Acanthamoeba, Balamuthia, Naegleria*) (CDC-10286)	NA	62,63
1990	Human herpesvirus 7	A member of Betaherpesviridae, a subfamily of the Herpesviridae	Exanthema subitum, acute febrile diseases	Human Herpes Virus 7 (HHV7) Detection (CDC-10267)	NA	64,65
1991	Guanarito virus	Enveloped, single-stranded, bisegmented RNA viruses with ambisense genomes	Venezuelan hemorrhagic fever	Pathologic Evaluation of Unexplained Illness Due to Possible Infectious Etiology (CDC-10372)	NA	66,67
1991	*Encephalitozoon hellem*	A unicellular, intracellular microsporidian species	Keratoconjunctivitis, infection of respiratory and genitourinary tract, and disseminated infection	Microsporidia Molecular Identification (CDC-10481)	NA	68,69
1991	*Ehrlichia chaffeensis*	An obligately intracellular gram-negative rickettsial bacterium	Human monocytotropic ehrlichiosis	*Anaplasma* and *Ehrlichia* Molecular Detection (CDC-10290), *Anaplasma* and *Ehrlichia* Special Study (CDC-10291)	NA	70–72

[a]CDC molecular test name (test code) are available from the Center for Disease Control and Prevention Test Directory, http://www.cdc.gov/laboratory/specimen-submission/list.html#M (last accessed 12/19/2014).
[b]FDA-approved/cleared molecular test (manufacturer) are available from the US Food and Drug Administration at http://www.fda.gov/MedicalDevices/ProductsandMedicalProcedures/InVitroDiagnostics/ucm330711.htm (last accessed 12/19/2014).
NA, not available.

References:
1. Drosten C, Gottig S, Schilling S, et al. Rapid detection and quantification of RNA of Ebola and Marburg viruses, Lassa virus, Crimean-Congo hemorrhagic fever virus, Rift Valley fever virus, dengue virus, and yellow fever virus by real-time reverse transcription-PCR. *J Clin Microbiol.* 2002;40(7):2323–2330.
2. Koehler JW, Hall AT, Rolfe PA, et al. Development and evaluation of a panel of filovirus sequence capture probes for pathogen detection by next-generation sequencing. *PLoS One.* 2014;9(9):e107007.
3. Muhlberger E, Trommer S, Funke C, Volchkov V, Klenk HD, Becker S. Termini of all mRNA species of Marburg virus: sequence and secondary structure. *Virology.* 1996;223(2):376–380.
4. Towner JS, Khristova ML, Sealy TK, et al. Marburgvirus genomics and association with a large hemorrhagic fever outbreak in Angola. *J Virol.* 2006;80(13):6497–6516.

(Continued)

TABLE 15.1 (Continued)

5. Euler M, Wang Y, Heidenreich D, et al. Development of a panel of recombinase polymerase amplification assays for detection of biothreat agents. *J Clin Microbiol.* 2013;51(4):1110–1117.

6. Asogun DA, Adomeh DI, Ehimuan J, et al. Molecular diagnostics for Lassa fever at Irrua specialist teaching hospital, Nigeria: lessons learnt from two years of laboratory operation. *PLoS Negl Trop Dis.* 2012;6(9):e1839.

7. Djavani M, Lukashevich IS, Sanchez A, Nichol ST, Salvato MS. Completion of the Lassa fever virus sequence and identification of a RING finger open reading frame at the L RNA 5' end. *Virology.* 1997;235 (2):414–418.

8. Ehichioya DU, Asogun DA, Ehimuan J, et al. Hospital-based surveillance for Lassa fever in Edo State, Nigeria, 2005–2008. *Trop Med Int Health.* 2012;17(8):1001–1004.

9. Cotten M, Petrova V, Phan MV, et al. Deep sequencing of norovirus genomes defines evolutionary patterns in an urban tropical setting. *J Virol.* 2014;88(19):11056–11069.

10. Martinez MA, Soto-Del Rio MD, Gutierrez RM, et al. DNA microarray for detection of gastrointestinal viruses. *J Clin Microbiol.* 2014.

11. Moore NE, Wang J, Hewitt J, et al. Metagenomic analysis of viruses in feces from unsolved outbreaks of gastroenteritis in humans. *J Clin Microbiol.* 2014.

12. Ye S, Lambert SB, Grimwood K, et al. Comparison of test specificity in commercial antigen and in-house PCR methods for rotavirus detection in stool specimens. *J Clin Microbiol.* 2014.

13. Bonvicini F, Manaresi E, Bua G, Venturoli S, Gallinella G. Keeping pace with parvovirus B19 genetic variability: a multiplex genotype-specific quantitative PCR assay. *J Clin Microbiol.* 2013;51 (11):3753–3759.

14. Maple PA, Hedman L, Dhanilall P, et al. Identification of past and recent parvovirus B19 infection in immunocompetent individuals by quantitative PCR and enzyme immunoassays: a dual-laboratory study. *J Clin Microbiol.* 2014;52(3):947–956.

15. Plentz A, Wurdinger M, Kudlich M, Modrow S. Low-level DNAemia of parvovirus B19 (genotypes 1-3) in adult transplant recipients is not associated with anaemia. *J Clin Virol: the official publication of the Pan American Society for Clinical Virology.* 2013;58(2):443–448.

16. Cruz CD, Win JK, Fletcher GC. An improved method for quantification of *Vibrio vulnificus* in oysters. *J Microbiol Methods.* 2013;95(3):397–399.

17. Wei S, Zhao H, Xian Y, Hussain MA, Wu X. Multiplex PCR assays for the detection of *Vibrio alginolyticus, Vibrio parahaemolyticus, Vibrio vulnificus,* and *Vibrio cholerae* with an internal amplification control. *Diagn Microbiol Infect Dis.* 2014;79(2):115–118.

18. Mazurie AJ, Alves JM, Ozaki LS, Zhou S, Schwartz DC, Buck GA. Comparative genomics of cryptosporidium. *Int J Genomics.* 2013;2013:832756.

19. Silva SO, Richtzenhain LJ, Barros IN, et al. A new set of primers directed to 18S rRNA gene for molecular identification of *Cryptosporidium* spp. and their performance in the detection and differentiation of oocysts shed by synanthropic rodents. *Exp Parasitol.* 2013;135(3):551–557.

20. Rubio JM, Lanza M, Fuentes I, Soliman RH. A novel nested multiplex PCR for the simultaneous detection and differentiation of *Cryptosporidium* spp., *Enterocytozoon bieneusi* and *Encephalitozoon intestinalis.* *Parasitol Int.* 2014;63(5):664–669.

21. Dingle KE, Elliott B, Robinson E, et al. Evolutionary history of the *Clostridium difficile* pathogenicity locus. *Genome Biol Evol.* 2014;6(1):36–52.

22. Eyre DW, Cule ML, Wilson DJ, et al. Diverse sources of *C. difficile* infection identified on whole-genome sequencing. *N Engl J Med.* 2013;369(13):1195–1205.

23. Leibowitz J, Soma VL, Rosen L, Ginocchio CC, Rubin LG. Similar proportions of stool specimens from hospitalized children with and without diarrhea test positive for *Clostridium difficile. Pediatr Infect Dis J.* 2014.

24. Gomez-Valero L, Rusniok C, Rolando M, et al. Comparative analyses of *Legionella* species identifies genetic features of strains causing Legionnaires inverted question mark disease. *Genome Biol.* 2014;15 (11):505.

25. Sanchez-Buso L, Comas I, Jorques G, Gonzalez-Candelas F. Recombination drives genome evolution in outbreak-related *Legionella pneumophila* isolates. *Nat Genet.* 2014;46(11):1205–1211.

26. Noh JY, Cheong HJ, Song JY, et al. Clinical and molecular epidemiological features of hemorrhagic fever with renal syndrome in Korea over a 10-year period. *J Clin Virol: the official publication of the Pan American Society for Clinical Virology.* 2013;58(1):11–17.

27. Wang ML, Lai JH, Zhu Y, et al. Genetic susceptibility to haemorrhagic fever with renal syndrome caused by Hantaan virus in Chinese Han population. *Int J Immunogenet.* 2009;36(4):227–229.

28. Karatayli E, Altunoglu YC, Karatayli SC, et al. A one step real time PCR method for the quantification of hepatitis delta virus RNA using an external armored RNA standard and intrinsic internal control. *J Clin Virol: the official publication of the Pan American Society for Clinical Virology.* 2014;60(1):11–15.

29. Le Gal F, Gault E, Ripault MP, et al. Eighth major clade for hepatitis delta virus. *Emerg Infect Dis.* 2006;12(9):1447–1450.

30. Taboada EN, Clark CG, Sproston EL, Carrillo CD. Current methods for molecular typing of *Campylobacter* species. *J Microbiol Methods.* 2013;95(1):24–31.

31. Vondrakova L, Pazlarova J, Demnerova K. Detection, identification and quantification of *Campylobacter jejuni, coli* and *lari* in food matrices all at once using multiplex qPCR. *Gut Pathog.* 2014;6:12.

32. Riner DK, Nichols T, Lucas SY, Mullin AS, Cross JH, Lindquist HD. Intragenomic sequence variation of the ITS-1 region within a single flow-cytometry-counted *Cyclospora cayetanensis* oocysts. *J Parasitol.* 2010;96(5):914–919.

33. Zhou Y, Lv B, Wang Q, et al. Prevalence and molecular characterization of *Cyclospora cayetanensis,* Henan, China. *Emerg Infect Dis.* 2011;17(10):1887–1890.

34. Brunetto GS, Massoud R, Leibovitch EC, et al. Digital droplet PCR (ddPCR) for the precise quantification of human T-lymphotropic virus 1 proviral loads in peripheral blood and cerebrospinal fluid of HAM/TSP patients and identification of viral mutations. *J Neurovirol.* 2014;20(4):341–351.

35. Firouzi S, Lopez Y, Suzuki Y, et al. Development and validation of a new high-throughput method to investigate the clonality of HTLV-1-infected cells based on provirus integration sites. *Genome Med.* 2014;6(6):46.

36. Pessoa R, Watanabe JT, Nukui Y, et al. Molecular characterization of human T-cell lymphotropic virus type 1 full and partial genomes by Illumina massively parallel sequencing technology. *PLoS One.* 2014;9(3):e93374.

37. Ratner L, Philpott T, Trowbridge DB. Nucleotide sequence analysis of isolates of human T-lymphotropic virus type 1 of diverse geographical origins. *AIDS Res Hum Retroviruses.* 1991;7(11):923–941.

38. Hait J, Tallent S, Melka D, Keys C, Bennett R. Prevalence of enterotoxins and toxin gene profiles of *Staphylococcus aureus* isolates recovered from a bakery involved in a second staphylococcal food poisoning occurrence. *J Appl Microbiol.* 2014;117(3):866–875.

39. Hait JM, Tallent SM, Bennett RW. Screening, detection, and serotyping methods for toxin genes and enterotoxins in *Staphylococcus* strains. *J AOAC Int.* 2014;97(4):1078–1083.

40. Leopold SR, Goering RV, Witten A, Harmsen D, Mellmann A. Bacterial whole-genome sequencing revisited: portable, scalable, and standardized analysis for typing and detection of virulence and antibiotic resistance genes. *J Clin Microbiol.* 2014;52(7):2365–2370.

41. Clark KL, Leydet BF, Threlkeld C. Geographical and genospecies distribution of *Borrelia burgdorferi sensu lato* DNA detected in humans in the USA. *J Med Microbiol.* 2014;63(Pt 5):674–684.

42. Jacquot M, Gonnet M, Ferquel E, et al. Comparative population genomics of the *Borrelia burgdorferi* species complex reveals high degree of genetic isolation among species and underscores benefits and constraints to studying intra-specific epidemiological processes. *PLoS One.* 2014;9(4):e94384.

43. Brewster JD, Paoli GC. DNA extraction protocol for rapid PCR detection of pathogenic bacteria. *Anal Biochem.* 2013;442(1):107–109.

44. Rump LV, Gonzalez-Escalona N, Ju W, et al. Genomic diversity and virulence characterization of historical *Escherichia coli* O157 strains isolated from clinical and environmental sources. *Appl Environ Microbiol.* 2014.

45. Casabianca A, Orlandi C, Canovari B, et al. A real time PCR platform for the simultaneous quantification of total and extrachromosomal HIV DNA forms in blood of HIV-1 infected patients. *PLoS One.* 2014;9(11):e111919.

46. Di Giallonardo F, Zagordi O, Duport Y, et al. Next-generation sequencing of HIV-1 RNA genomes: determination of error rates and minimizing artificial recombination. *PLoS One.* 2013;8(9):e74249.

47. Kao CY, Lee AY, Huang AH, et al. Heteroresistance of *Helicobacter pylori* from the same patient prior to antibiotic treatment. *Infect Genet Evol.* 2014;23:196–202.

48. Patel SK, Pratap CB, Jain AK, Gulati AK, Nath G. Diagnosis of *Helicobacter pylori*: What should be the gold standard? *World J Gastroenterol.* 2014;20(36):12847–12859.

49. Quentin R, Ruimy R, Rosenau A, Musser JM, Christen R. Genetic identification of cryptic genospecies of *Haemophilus* causing urogenital and neonatal infections by PCR using specific primers targeting genes coding for 16S rRNA. *J Clin Microbiol.* 1996;34(6):1380–1385.

50. Strouts FR, Power P, Croucher NJ, et al. Lineage-specific virulence determinants of *Haemophilus influenzae* biogroup aegyptius. *Emerg Infect Dis.* 2012;18(3):449–457.

51. Subrungruang I, Mungthin M, Chavalitshewinkoon-Petmitr P, Rangsin R, Naaglor T, Leelayoova S. Evaluation of DNA extraction and PCR methods for detection of *Enterocytozoon bieneusi* in stool specimens. *J Clin Microbiol.* 2004;42(8):3490–3494.

52. Benitez AJ, Thurman KA, Diaz MH, Conklin L, Kendig NE, Winchell JM. Comparison of real-time PCR and a microimmunofluorescence serological assay for detection of *Chlamydophila pneumoniae* infection in an outbreak investigation. *J Clin Microbiol.* 2012;50(1):151–153.

53. Ravindranath BS, Krishnamurthy V, Krishna V, C SK. In silico synteny based comparative genomics approach for identification and characterization of novel therapeutic targets in *Chlamydophila pneumoniae. Bioinformation.* 2013;9(10):506–510.

54. Debaugnies F, Busson L, Ferster A, et al. Detection of Herpesviridae in whole blood by multiplex PCR DNA-based microarray analysis after hematopoietic stem cell transplantation. *J Clin Microbiol.* 2014;52 (7):2552–2556.

55. Sedlak RH, Cook L, Huang ML, et al. Identification of chromosomally integrated human herpesvirus 6 by droplet digital PCR. *Clin Chem.* 2014;60(5):765–772.

56. Hanaoka N, Matsutani M, Kawabata H, et al. Diagnostic assay for *Rickettsia japonica. Emerg Infect Dis.* 2009;15(12):1994–1997.

57. Matsutani M, Ogawa M, Takaoka N, et al. Complete genomic DNA sequence of the East Asian spotted fever disease agent *Rickettsia japonica. PLoS One.* 2013;8(9):e71861.

58. Fevery B, Susser S, Lenz O, et al. HCV RNA quantification with different assays: implications for protease-inhibitor-based response-guided therapy. *Antivir Ther.* 2014.

59. Quer J, Gregori J, Rodriguez-Frias F, et al. High-resolution hepatitis C virus (HCV) subtyping, using NS5B deep sequencing and phylogeny, an alternative to current methods. *J Clin Microbiol.* 2014.

60. Vollmer T, Knabbe C, Dreier J. Comparison of real-time PCR and antigen assays for detection of hepatitis E virus in blood donors. *J Clin Microbiol.* 2014;52(6):2150–2156.

61. Zhou X, Wang Y, Metselaar HJ, Janssen HL, Peppelenbosch MP, Pan Q. Rapamycin and everolimus facilitate hepatitis E virus replication: revealing a basal defense mechanism of PI3K-PKB-mTOR pathway. *J Hepatol.* 2014;61(4):746–754.

62. da Rocha-Azevedo B, Tanowitz HB, Marciano-Cabral F. Diagnosis of infections caused by pathogenic free-living amoebae. *Interdiscip Perspect Infect Dis.* 2009;2009:251406.

63. Lares-Jimenez LF, Booton GC, Lares-Villa F, Velazquez-Contreras CA, Fuerst PA. Genetic analysis among environmental strains of *Balamuthia mandrillaris* recovered from an artificial lagoon and from soil in Sonora, Mexico. *Exp Parasitol.* 2014.

64. Donaldson CD, Clark DA, Kidd IM, Breuer J, Depledge DD. Genome sequence of human herpesvirus 7 strain UCL-1. *Genome Announc.* 2013;1(5).

65. Oakes B, Hoagland-Henefield M, Komaroff AL, Erickson JL, Huber BT. Human endogenous retrovirus-K18 superantigen expression and human herpesvirus-6 and human herpesvirus-7 viral loads in chronic fatigue patients. *Clin Infect Dis: an official publication of the Infectious Diseases Society of America.* 2013;56(10):1394–1400.

66. Fulhorst CF, Cajimat MN, Milazzo ML, et al. Genetic diversity between and within the arenavirus species indigenous to western Venezuela. *Virology.* 2008;378(2):205–213.

67. Vieth S, Drosten C, Charrel R, Feldmann H, Gunther S. Establishment of conventional and fluorescence resonance energy transfer-based real-time PCR assays for detection of pathogenic New World arenaviruses. *J Clin Virol: the official publication of the Pan American Society for Clinical Virology.* 2005;32(3):229–235.

68. Hester JD, Varma M, Bobst AM, Ware MW, Lindquist HD, Schaefer FW, 3rd. Species-specific detection of three human-pathogenic microsporidial species from the genus *Encephalitozoon* via fluorogenic 5' nuclease PCR assays. *Mol Cell Probes.* 2002;16(6):435–444.

69. Pombert JF, Selman M, Burki F, et al. Gain and loss of multiple functionally related, horizontally transferred genes in the reduced genomes of two microsporidian parasites. *Proc Natl Acad Sci USA.* 2012;109(31):12638–12643.

70. Breitschwerdt EB, Hegarty BC, Qurollo BA, et al. Intravascular persistence of *Anaplasma platys, Ehrlichia chaffeensis,* and *Ehrlichia ewingii* DNA in the blood of a dog and two family members. *Parasit Vectors.* 2014;7:298.

71. Doyle CK, Labruna MB, Breitschwerdt EB, et al. Detection of medically important *Ehrlichia* by quantitative multicolor TaqMan real-time polymerase chain reaction of the *dsb* gene. *J Mol Diagn.* 2005;7 (4):504–510.

72. Walker DH, Paddock CD, Dumler JS. Emerging and re-emerging tick-transmitted rickettsial and ehrlichial infections. *Med Clin North Am.* 2008;92(6):1345–1361, x.

TABLE 15.2 Chronological List of Emerged Infectious Agents/Diseases Since 1992

Year	Agent	Agent characteristics	Disease	CDC molecular test name (test code)[a]	FDA-approved/cleared molecular test (manufacturer)[b]	Reference
1992	Barmah Forest virus	An *Alphavirus* (small, spherical, enveloped viruses with a genome of a single-strand positive-sense RNA)	Epidemic polyarthritis (fever, malaise, rash, joint pain, and muscle tenderness)	Pathologic Evaluation of Unexplained Illness Due to Possible Infectious Etiology (CDC-10372)	NA	1,2
1992	*Vibrio cholerae* O139	A gram-negative, comma-shaped bacterium	Watery diarrhea and vomiting	*Vibrio cholerae* Identification (CDC-10119), *Vibrio* Subtyping (CDC-10122), *Vibrio, Aeromonas,* and Related Organisms Identification (CDC-10120), *Vibrio, Aeromonas,* and Related Organisms Study (CDC-10121)	NA	3,4
1992	*Bartonella henselae*	A proteobacterium	Cat-scratch disease, subacute regional lymphadenitis	*Bartonella* Molecular Identification (CDC-10295), *Bartonella* Special Study (CDC-10297)	NA	5,6
1992	*Rickettsia honei*	Nonmotile, obligately intracellular, gram-negative, non-spore forming bacteria	Flinders Island spotted fever	*Rickettsia* Molecular Detection (CDC-10402), *Rickettsia* Special Study (CDC-10405)	NA	7–9
1992	Sabia virus	An arenavirus (round, pleomorphic, and enveloped virus containing a beaded nucleocapsid with two single-stranded RNA segments)	Hemorrhagic fever	Pathologic Evaluation of Unexplained Illness Due to Possible Infectious Etiology (CDC-10372)	NA	10,11
1993	*Encephalitozoon intestinalis*	A parasite	Diarrhea	Microsporidia Molecular Identification (CDC-10481), Enteric Isolation—Primary Specimen (CDC-10106)	NA	12,13
1993	Sin Nombre virus	A single-stranded RNA negative-strand virus	Hantavirus cardiopulmonary syndrome (HCPS)	Pathologic Evaluation of Unexplained Illness Due to Possible Infectious Etiology (CDC-10372)	NA	14,15
1994	Human herpesvirus 8	A double-stranded DNA virus	Kaposi sarcoma	Human Herpes Virus 8 (HHV8) Detection (CDC-10268)	NA	16,17
1994	*Anaplasma phagocytophilum*	An obligately intracellular gram-negative bacterium	Human granulocytic anaplasmosis	*Anaplasma* and *Ehrlichia* Molecular Detection (CDC-10290), *Anaplasma* and *Ehrlichia* Special Study (CDC-10291)	NA	18,19
1994	*Rickettsia felis*	Nonmotile, obligately intracellular, gram-negative, non-spore forming bacteria	Flea-borne spotted fever	*Rickettsia* Molecular Detection (CDC-10402), *Rickettsia* Special Study (CDC-10405)	NA	7–9

Year	Agent	Characteristics	Disease	Test(s)	Commercial Assays	References
1994	*Rickettsia africae*	Nonmotile, obligately intracellular, gram-negative, non-spore forming bacteria	African tick bite fever	Rickettsia Molecular Detection (CDC-10402), Rickettsia Special Study (CDC-10405)	NA	7–9
1995	Hendra virus	Nonsegmented, single-stranded negative-sense RNA	Edema and hemorrhage of the lungs, encephalitis	Pathologic Evaluation of Unexplained Illness Due to Possible Infectious Etiology (CDC-10372)	NA	20,21
1995	Alkhumra virus	Enveloped virus with monopartite, linear, single-stranded RNA genomes	Tick-borne hemorrhagic fever	Alkhumra Identification (CDC-10274)	NA	22,23
1997	*Rickettsia slovaca*	Nonmotile, obligately intracellular, gram-negative, non-spore forming bacteria	Tick-borne lymphadenopathy	Rickettsia Molecular Detection (CDC-10402), Rickettsia Special Study (CDC-10405)	NA	7–9
1999	Nipah virus	Nonsegmented, single-stranded negative-sense RNA	Respiratory, gastrointestinal and neurologic symptoms, encephalitis	Nipah Virus Identification (CDC-10354)	NA	20,24
1999	West Nile virus	A positive-sense, single-stranded RNA virus	West Nile fever, encephalitis	Pathologic Evaluation of Unexplained Illness Due to Possible Infectious Etiology (CDC-10372)	NA	25,26
1999	*Ehrlichia ewingii*	An obligately intracellular gram-negative rickettsial bacteriaum	Ehrlichiosis ewingii infection	Anaplasma and Ehrlichia Molecular Detection (CDC-10290), Anaplasma and Ehrlichia Special Study (CDC-10291)	NA	9,27,28
2001	Human metapneumovirus	A negative-sense, single-stranded RNA virus	Pneumonia	Pathologic Evaluation of Unexplained Illness Due to Possible Infectious Etiology (CDC-10372)	Quidel Molecular RSV + hMPV Assay (Quidel Corporation), Quidel Molecular hMPV Assay (Quidel Corporation), Pro hMPV + Assay (Prodesse, Inc.), FilmArray Respiratory Panel (RP) (Idaho Technology, Inc.), xTAG Respiratory Viral Panel (RVP) (Luminex Molecular Diagnostics, Inc.), xTAG Respiratory Viral Panel Fast (RVP FAST) (Luminex Molecular Diagnostics, Inc.), eSensor Respiratory Viral Panel (RVP) (GenMark Diagnostic), ProFlu + Assay (Gen-Probe Prodesse, Inc.)	29,30

(Continued)

TABLE 15.2 (Continued)

Year	Agent	Agent characteristics	Disease	CDC molecular test name (test code)[a]	FDA-approved/cleared molecular test (manufacturer)[b]	Reference
2003	Monkeypox virus	A double-stranded DNA virus	Febrile enanthem	Pathologic Evaluation of Unexplained Illness Due to Possible Infectious Etiology (CDC-10372)	NA	31,32
2003	SARS coronavirus	A positive-sense and single-stranded RNA virus	Severe acute respiratory syndrome (SARS)	SARS Molecular Detection (CDC-10412)	NA	33–38
2004	*Rickettsia parkeri*	Nonmotile, obligately intracellular, gram-negative, non-spore forming bacteria	American tick bite fever	*Rickettsia* Molecular Detection (CDC-10402), *Rickettsia* Special Study (CDC-10405)	NA	7–9
2005	Human retroviruses (HTLV-3/4)	Human retroviruses	Unclear association with disease	Pathologic Evaluation of Unexplained Illness Due to Possible Infectious Etiology (CDC-10372)	NA	39,40
2005	Human bocavirus	A linear, nonsegmented single-stranded DNA viruses	Unclear association with disease	Pathologic Evaluation of Unexplained Illness Due to Possible Infectious Etiology (CDC-10372)	NA	41,42
2008	*Plasmodium knowlesi*	A primate malaria parasite	Malaria	Malaria Surveillance (CDC-10235)	NA	43,44
2008	Lujo virus	A bisegmented RNA arenavirus	Viral hemorrhagic fever (VHF)	Pathologic Evaluation of Unexplained Illness Due to Possible Infectious Etiology (CDC-10372)	NA	45,46
2008	Chapare virus	Enveloped, single-stranded, bisegmented, ambisense RNA arenavirus	Hemorrhagic fever	Pathologic Evaluation of Unexplained Illness Due to Possible Infectious Etiology (CDC-10372)	NA	47,48
2009	*Ehrlichia muris*–like	An obligate intracellular gram-negative rickettsial bacteriaum	Ehrlichiosis	*Anaplasma* and *Ehrlichia* Molecular Detection (CDC-10290), *Anaplasma* and *Ehrlichia* Special Study (CDC-10291), Bacterial ID of Unknown Isolate (Not Strict Anaerobe) (CDC-10145), Bacterial ID from Clinical Specimen (16S rRNA PCR) (CDC-10146)	NA	49,50
2009	Pandemic H1N1 influenza virus	A new influenza A subtype H1N1 RNA virus, having hemagglutinin (HA) of the H1 subtype and neuraminidase (NA) of the N1 subtype	Flu, pneumonia, acute respiratory distress syndrome (ARDS)	Pathologic Evaluation of Influenza and Other Viral Infections (CDC-10366)	Prodesse ProFAST Assat (Gen-Probe Prodesse, Inc.), Quidel Molecular Influenza A + B Assay (Quidel Corporation), IMDx Flu A/B and RSV for Abbott m2000 (Intelligent Medical Devices, Inc.), CDC Human Influenza Virus Real-Time RT-PCR Diagnostic Panel (CDC), Xpert Flu Assay (Cepheid),	51,52

Year	Agent	Description	Syndrome	CDC molecular test[a]	FDA-approved/cleared molecular test[b]	References
					Simplexa Flu A/B & RSV Direct (Focus Diagnostics, Inc.), FilmArray Respiratory Panel (RP) (Idaho Technology, Inc.), artus Infl A/B RG RT-PCR Kit (Qiagen GmbH), JBAIDS Influenza A Subtyping Kit (US Army Medical Materiel Development Activity), JBAIDS Influenza A&B Detection Kit (US Army Medical Materiel Development Activity), eSensor Respiratory Viral Panel (RVP) (GenMark Diagnostic), ProFlu + Assay (Gen-Probe Prodesse, Inc.), Verigene Respiratory Virus Plus Nucleic Acid Test (RV +) (Nanosphere, Inc.), Simplexa Flu A/B & RSV (Focus Diagnostics, Inc.), CDC Influenza 2009 A (H1N1) pdm Real-Time RT-PCR Panel (CDC), Simplexa Influenza A H1N1 (2009) (Focus Diagnostics, Inc.)	
2010	*Candidatus* Neoehrlichia mikurensis	An obligately intracellular gram-negative rickettsial bacteriaum	Ehrlichiosis-like syndrome	Pathologic Evaluation of Unexplained Illness Due to Possible Infectious Etiology (CDC-10372)	NA	53,54
2011	Severe fever with thrombocytopenia virus	Negative-stranded, enveloped RNA virus	Severe fever with thrombocytopenia syndrome (SFTS)	Pathologic Evaluation of Unexplained Illness Due to Possible Infectious Etiology (CDC-10372)	NA	55,56
2012	Middle East respiratory syndrome (MERS) coronavirus (MERS-CoV)	Positive-sense, single-stranded RNA coronavirus	Middle East respiratory syndrome	MERS-CoV PCR 9 (CDC-10488)	NA	57,58
2013	Novel H7N9 influenza virus (China)	A new influenza A subtype H7N9 RNA virus, having HA of the H7 subtype and NA of the N9 subtype	Flu, pneumonia, acute respiratory distress syndrome (ARDS)	Pathologic Evaluation of Influenza and Other Viral Infections (CDC-10366)	NA	59

[a] CDC molecular test name (test code) are available from the Center for Disease Control and Prevention Test Directory, http://www.cdc.gov/laboratory/specimen-submission/list.html#M (last accessed 12/19/2014).
[b] FDA-approved/cleared molecular test (manufacturer) are available from the US Food and Drug Administration at http://www.fda.gov/MedicalDevices/ProductsandMedicalProcedures/InVitroDiagnostics/ucm330711.htm (last accessed 12/19/2014).
NA, not available.

(Continued)

TABLE 15.2 (Continued)

References:

1. Lee E, Stocks C, Lobigs P, et al. Nucleotide sequence of the Barmah Forest virus genome. *Virology*. 1997;227(2):509–514.
2. Poidinger M, Roy S, Hall RA, et al. Genetic stability among temporally and geographically diverse isolates of Barmah Forest virus. *Am J Trop Med Hyg*. 1997;57(2):230–234.
3. Pang B, Zheng X, Diao B, et al. Whole genome PCR scanning reveals the syntenic genome structure of toxigenic *Vibrio cholerae* strains in the O1/O139 population. *PLoS One*. 2011;6(8):e24267.
4. Zhao J, Kang L, Hu R, et al. Rapid oligonucleotide suspension array-based multiplex detection of bacterial pathogens. *Foodborne Pathog Dis*. 2013;10(10):896–903.
5. Lantos PM, Maggi RG, Ferguson B, et al. Detection of *Bartonella* species in the blood of veterinarians and veterinary technicians: a newly recognized occupational hazard? *Vector Borne Zoonotic Dis*. 2014;14(8):563–570.
6. Psarros G, Riddell Jt, Gandhi T, Kauffman CA, Cinti SK. *Bartonella henselae* infections in solid organ transplant recipients: report of 5 cases and review of the literature. *Medicine (Baltimore)*. 2012;91(2):111–121.
7. Renvoise A, Rolain JM, Socolovschi C, Raoult D. Widespread use of real-time PCR for rickettsial diagnosis. *FEMS Immunol Med Microbiol*. 2012;64(1):126–129.
8. Sekeyova Z, Roux V, Raoult D. Phylogeny of *Rickettsia* spp. inferred by comparing sequences of 'gene D', which encodes an intracytoplasmic protein. *Int J Syst Evol Microbiol*. 2001;51(Pt 4):1353–1360.
9. Walker DH, Paddock CD, Dumler JS. Emerging and re-emerging tick-transmitted rickettsial and ehrlichial infections. *Med Clin North Am*. 2008;92(6):1345–1361, x.
10. Gonzalez JP, Bowen MD, Nichol ST, Rico-Hesse R. Genetic characterization and phylogeny of Sabia virus, an emergent pathogen in Brazil. *Virology*. 1996;221(2):318–324.
11. Vieth S, Drosten C, Charrel R, Feldmann H, Gunther S. Establishment of conventional and fluorescence resonance energy transfer-based real-time PCR assays for detection of pathogenic New World arenaviruses. *J Clin Virol: the official publication of the Pan American Society for Clinical Virology*. 2005;32(3):229–235.
12. Galvan A, Magnet A, Izquierdo F, Fenoy S, Henriques-Gil N, del Aguila C. Variability in minimal genomes: analysis of tandem repeats in the microsporidia *Encephalitozoon intestinalis*. *Infect Genet Evol*. 2013;20:26–33.
13. Rubio JM, Lanza M, Fuentes I, Soliman RH. A novel nested multiplex PCR for the simultaneous detection and differentiation of *Cryptosporidium* spp., *Enterocytozoon bieneusi* and *Encephalitozoon intestinalis*. *Parasitol Int*. 2014;63(5):664–669.
14. Black WCt, Doty JB, Hughes MT, Beaty BJ, Calisher CH. Temporal and geographic evidence for evolution of Sin Nombre virus using molecular analyses of viral RNA from Colorado, New Mexico and Montana. *Virol J*. 2009;6:102.
15. Henderson WW, Monroe MC, St Jeor SC, et al. Naturally occurring Sin Nombre virus genetic reassortants. *Virology*. 1995;214(2):602–610.
16. Dollard SC, Roback JD, Gunthel C, et al. Measurements of human herpesvirus 8 viral load in blood before and after leukoreduction filtration. *Transfusion*. 2013;53(10):2164–2167.
17. Speicher DJ, Johnson NW. Detection of human herpesvirus 8 by quantitative polymerase chain reaction: development and standardisation of methods. *BMC Infect Dis*. 2012;12:210.
18. Chan K, Marras SA, Parveen N. Sensitive multiplex PCR assay to differentiate Lyme spirochetes and emerging pathogens *Anaplasma phagocytophilum* and *Babesia microti*. *BMC Microbiol*. 2013;13:295.
19. Stuen S, Granquist EG, Silaghi C. *Anaplasma phagocytophilum*—a widespread multi-host pathogen with highly adaptive strategies. *Front Cell Infect Microbiol*. 2013;3:31.
20. Wang LF, Daniels P. Diagnosis of henipavirus infection: current capabilities and future directions. *Curr Top Microbiol Immunol*. 2012;359:179–196.
21. Yu M, Hansson E, Shiell B, Michalski W, Eaton BT, Wang LF. Sequence analysis of the Hendra virus nucleoprotein gene: comparison with other members of the subfamily Paramyxovirinae. *J Gen Virol*. 1998;79 (Pt 7):1775–1780.
22. Madani TA, Abuelzein el TM, Azhar EI, Al-Bar HM, Abu-Araki H, Ksiazek TG. Comparison of RT-PCR assay and virus isolation in cell culture for the detection of Alkhumra hemorrhagic fever virus. *J Med Virol*. 2014;86(7):1176–1180.
23. Madani TA, Azhar EI, Abuelzein el TM, et al. Complete genome sequencing and genetic characterization of Alkhumra hemorrhagic fever virus isolated from Najran, Saudi Arabia. *Intervirology*. 2014;57(5):300–310.
24. Harcourt BH, Tamin A, Ksiazek TG, et al. Molecular characterization of Nipah virus, a newly emergent paramyxovirus. *Virology*. 2000;271(2):334–349.
25. Lim SM, Koraka P, Osterhaus AD, Martina BE. Development of a strand-specific real-time qRT-PCR for the accurate detection and quantitation of West Nile virus RNA. *J Virol Methods*. 2013;194(1–2):146–153.
26. Pisani G, Pupella S, Cristiano K, et al. Detection of West Nile virus RNA (lineages 1 and 2) in an external quality assessment programme for laboratories screening blood and blood components for West Nile virus by nucleic acid amplification testing. *Blood Transfus*. 2012;10(4):515–520.
27. Breitschwerdt EB, Hegarty BC, Qurollo BA, et al. Intravascular persistence of *Anaplasma platys*, *Ehrlichia chaffeensis*, and *Ehrlichia ewingii* DNA in the blood of a dog and two family members. *Parasit Vectors*. 2014;7:298.
28. Doyle CK, Labruna MB, Breitschwerdt EB, et al. Detection of medically important *Ehrlichia* by quantitative multicolor TaqMan real-time polymerase chain reaction of the dsb gene. *J Mol Diagn*. 2005;7(4):504–510.
29. Klemenc J, Asad Ali S, Johnson M, et al. Real-time reverse transcriptase PCR assay for improved detection of human metapneumovirus. *J Clin Virol: the official publication of the Pan American Society for Clinical Virology*. 2012;54(4):371–375.
30. Roussy JF, Carbonneau J, Ouakki M, et al. Human metapneumovirus viral load is an important risk factor for disease severity in young children. *J Clin Virol: the official publication of the Pan American Society for Clinical Virology*. 2014;60(2):133–140.
31. Grant RJ, Baldwin CD, Nalca A, et al. Application of the Ibis-T5000 pan-Orthopoxvirus assay to quantitatively detect monkeypox viral loads in clinical specimens from macaques experimentally infected with aerosolized monkeypox virus. *Am J Trop Med Hyg*. 2010;82(2):318–323.
32. Li Y, Olson VA, Laue T, Laker MT, Damon IK. Detection of monkeypox virus with real-time PCR assays. *J Clin Virol: the official publication of the Pan American Society for Clinical Virology*. 2006;36(3):194–203.

33. Adachi D, Johnson G, Draker R, et al. Comprehensive detection and identification of human coronaviruses, including the SARS-associated coronavirus, with a single RT-PCR assay. *J Virol Methods.* 2004;122(1):29–36.

34. Huang JL, Lin HT, Wang YM, et al. Rapid and sensitive detection of multiple genes from the SARS-coronavirus using quantitative RT-PCR with dual systems. *J Med Virol.* 2005;77(2):151–158.

35. Drosten C, Gunther S, Preiser W, et al. Identification of a novel coronavirus in patients with severe acute respiratory syndrome. *N Engl J Med.* 2003;348(20):1967–1976.

36. Ksiazek TG, Erdman D, Goldsmith CS, et al. A novel coronavirus associated with severe acute respiratory syndrome. *N Engl J Med.* 2003;348(20):1953–1966.

37. Peiris JS, Lai ST, Poon LL, et al. Coronavirus as a possible cause of severe acute respiratory syndrome. *Lancet.* 2003;361(9366):1319–1325.

38. Rota PA, Oberste MS, Monroe SS, et al. Characterization of a novel coronavirus associated with severe acute respiratory syndrome. *Science.* 2003;300(5624):1394–1399.

39. Mahieux R, Gessain A. HTLV-3/STLV-3 and HTLV-4 viruses: discovery, epidemiology, serology and molecular aspects. *Viruses.* 2011;3(7):1074–1090.

40. Moens B, Lopez G, Adaui V, et al. Development and validation of a multiplex real-time PCR assay for simultaneous genotyping and human T-lymphotropic virus type 1, 2, and 3 proviral load determination. *J Clin Microbiol.* 2009;47(11):3682–3691.

41. Christensen A, Dollner H, Skanke LH, Krokstad S, Moe N, Nordbo SA. Detection of spliced mRNA from human bocavirus 1 in clinical samples from children with respiratory tract infections. *Emerg Infect Dis.* 2013;19(4):574–580.

42. Proenca-Modena JL, Gagliardi TB, Paula FE, et al. Detection of human bocavirus mRNA in respiratory secretions correlates with high viral load and concurrent diarrhea. *PLoS One.* 2011;6(6):e21083.

43. Foster D, Cox-Singh J, Mohamad DS, Krishna S, Chin PP, Singh B. Evaluation of three rapid diagnostic tests for the detection of human infections with *Plasmodium knowlesi*. *Malar J.* 2014;13:60.

44. Lucchi NW, Poorak M, Oberstaller J, et al. A new single-step PCR assay for the detection of the zoonotic malaria parasite *Plasmodium knowlesi*. *PLoS One.* 2012;7(2):e31848.

45. Atkinson B, Chamberlain J, Dowall SD, Cook N, Bruce C, Hewson R. Rapid molecular detection of Lujo virus RNA. *J Virol Methods.* 2014;195:170–173.

46. Ishii A, Thomas Y, Moonga L, et al. Molecular surveillance and phylogenetic analysis of Old World arenaviruses in Zambia. *J Gen Virol.* 2012;93(Pt 10):2247–2251.

47. Cajimat MN, Milazzo ML, Rollin PE, et al. Genetic diversity among Bolivian arenaviruses. *Virus Res.* 2009;140(1–2):24–31.

48. Delgado S, Erickson BR, Agudo R, et al. Chapare virus, a newly discovered arenavirus isolated from a fatal hemorrhagic fever case in Bolivia. *PLoS Pathog.* 2008;4(4):e1000047.

49. Pritt BS, Sloan LM, Johnson DK, et al. Emergence of a new pathogenic *Ehrlichia* species, Wisconsin and Minnesota, 2009. *N Engl J Med.* 2011;365(5):422–429.

50. Thirumalapura NR, Qin X, Kuriakose JA, Walker DH. Complete genome sequence of *Ehrlichia muris* strain AS145T, a model monocytotropic *Ehrlichia* strain. *Genome Announc.* 2014;2(1).

51. Bermudez de Leon M, Penuelas-Urquides K, Aguado-Barrera ME, et al. In vitro transcribed RNA molecules for the diagnosis of pandemic 2009 influenza A(H1N1) virus by real-time RT-PCR. *J Virol Methods.* 2013;192(2):487–491.

52. Tellez-Sosa J, Rodriguez MH, Gomez-Barreto RE, et al. Using high-throughput sequencing to leverage surveillance of genetic diversity and oseltamivir resistance: a pilot study during the 2009 influenza A (H1N1) pandemic. *PLoS One.* 2013;8(7):e67010.

53. Pekova S, Vydra J, Kabickova H, et al. *Candidatus* Neoehrlichia mikurensis infection identified in 2 hematoooncologic patients: benefit of molecular techniques for rare pathogen detection. *Diagn Microbiol Infect Dis.* 2011;69(3):266–270.

54. Welinder-Olsson C, Kjellin E, Vaht K, Jacobsson S, Wenneras C. First case of human "*Candidatus* Neoehrlichia mikurensis" infection in a febrile patient with chronic lymphocytic leukemia. *J Clin Microbiol.* 2010;48(5):1956–1959.

55. Wen HL, Zhao L, Zhai S, et al. Severe fever with thrombocytopenia syndrome, Shandong Province, China, 2011. *Emerg Infect Dis.* 2014;20(1):1–5.

56. Yu XJ, Liang MF, Zhang SY, et al. Fever with thrombocytopenia associated with a novel bunyavirus in China. *N Engl J Med.* 2011;364(16):1523–1532.

57. Lu X, Whitaker B, Sakthivel SK, et al. Real-time reverse transcription-PCR assay panel for Middle East respiratory syndrome coronavirus. *J Clin Microbiol.* 2014;52(1):67–75.

58. Memish ZA, Al-Tawfiq JA, Makhdoom HQ, et al. Respiratory tract samples, viral load, and genome fraction yield in patients with middle East respiratory syndrome. *J Infect Dis.* 2014;210(10):1590–1594.

59. Lam TT, Wang J, Shen Y, et al. The genesis and source of the H7N9 influenza viruses causing human infections in China. *Nature.* 2013;502(7470):241–244.

agents of human infection is that of hepatitis C virus (HCV). After the discoveries of hepatitis A and B viruses, it was clear that the majority of cases of post-transfusion hepatitis were due to a condition designated non-A non-B hepatitis. The disease was transmissible to chimpanzees. In 1989, plasma from an infected chimpanzee was pelleted by ultracentrifugation and nucleic acids extracted from the pellet. cDNA was synthesized from both RNA and DNA with random primers and reverse transcriptase. Screening identified an RNA-encoded clone that expressed an antigen that reacted with antibodies of infected subjects. Eventually the complete genomes of all of the genotypes of HCV were determined, and a novel species most closely related to flaviviruses was established [12,13].

Another dramatic emergence of a viral disease occurred in 1993 in the Four Corners region of the southwestern United States. A mysterious highly lethal respiratory illness was investigated by a team from the Centers for Disease Control and Prevention (CDC). Extensive serologic screening of numerous antigens revealed unexpected reactivity with antigen of hantaviruses from other parts of the world that caused renal disease and hemorrhagic fever, and immunohistochemistry detected hantaviral antigen in pulmonary endothelium. Regions within the M segment of the RNA hantaviral genomes encoding G2 protein that are highly conserved were targeted by primers for nested PCR after reverse transcriptase generation of cDNA. Tissues from infected patients were analyzed, and the PCR products sequenced revealing a novel hantavirus subsequently named Sin Nombre virus. Viral sequences were identified in other patients and in *Peromyscus maniculatus* rodents, the reservoir. The story of hantaviral pulmonary syndrome unfolded to reveal related agents in many locations in North, Central, and South America [14,15].

A novel coronavirus in association with cases of severe acute respiratory syndrome (SARS-CoV) emerged in southern China in late 2002 and spread to 37 countries in five continents with 8273 confirmed cases and 775 deaths. No further cases have been reported since July 2003 [16]. RT-PCR, cloning, and sequencing contributed to identification of the SARS-CoV within weeks of the first cases reported in 2003 [17–20] and enabled rapid development of effective molecular diagnostic assays for routine clinical use [21,22]. SARS-CoV is associated with high mortality. Thus, timely and accurate diagnosis is needed to prevent the spread of this contagious disease. SARS-CoV spreads by respiratory secretions and airborne transmission. Early in the illness, SARS cannot be distinguished from common respiratory infections based on clinical symptoms [16]. During the SARS epidemic,

PCR-based molecular testing was helpful because of its ability to rapidly screen for many viruses. After the identification of SARS-CoV, specific RT-PCR and serological assays were developed, and RT-PCR detected infection before the appearance of antibodies when the risk of transmission is greatest [16–22].

The bacterial *rrs* gene encoding 16S rRNA was recognized as a valuable phylogenetic tool for discrimination and identification of bacterial species. David Relman crafted this tool into an approach to identify an unknown etiologic agent by PCR of the *rrs* gene with primers that corresponded to genomic regions that were conserved among eubacteria. Using this approach, he amplified and determined bacterial DNA sequences from bacillary angiomatosis lesions of patients with AIDS. Comparison with a bacterial gene database revealed that the DNA sequences matched bacteria that are currently named *Bartonella henselae* and *B. quintana*. Serendipitous testing of a patient who also had been diagnosed with cat scratch disease led to the recognition that *B. henselae* was also the long sought-after etiology of this well-characterized disease [23,24].

The same approach to discovery using *rrs* gene amplification and DNA sequencing led to the identification of what is currently classified as *Anaplasma phagocytophilum* as the etiologic agent of tick-transmitted human granulocytotropic anaplasmosis [25,26]. Subsequently *Ehrlichia ewingii* was recognized as another human tick-borne pathogen among patients evaluated in a molecular diagnostics laboratory who tested negative for *E. chaffeensis* infection [27–29]. More recently, Bobbi Pritt at Mayo Clinic noted that the melting curve of the DNA amplicons in a real-time PCR assay for Anaplasmataceae differed from the expected curves of known pathogens for a group of patients in Wisconsin and Minnesota. Sequence analysis identified another novel tick-borne pathogen tentatively designated *Ehrlichia muris*-like agent [8,9].

The discovery of a novel bunyavirus that has caused thousands of human infections with a case fatality rate of 12% in 15 provinces in China relied upon a molecular approach to identify the viral agent. Xue-Jie Yu investigated an outbreak in China that was thought to be due to severe infection with *A. phagocytophilum*. He noted that some of the clinical manifestations differed from those of anaplasmosis. He observed cytopathic effect in DH-82 cells inoculated with clinical samples rather than the typical morulae formed by *Anaplasma* species in infected cells. Ultrastructural analysis suggested that the pathogen causing the outbreak was a virus that belongs to the family of bunyaviruses. Based on the known sequence of bunyaviruses, PCR primers were designed, which yielded no amplicons. Subsequently, he began sequencing the RNA of heavily infected cells and discarded the sequences of the culture host species, *Canis familiaris*. This approach enabled him to determine that he had recovered a novel *Phlebovirus* of the family Bunyaviridae. He accomplished this feat without the use of next-generation sequencing (NGS) [10]. The application of NGS now allows us to obtain an abundance of viral gene sequences from infected host cells and the discovery of further novel viral and bacterial agents.

MOLECULAR EPIDEMIOLOGICAL STUDIES OF EMERGING INFECTIOUS PATHOGENS

Molecular technologies have been critical in the initial discovery of agents of emerging infectious diseases. These methods have also been routinely used for further characterization of pathogen strains and sequence variations. Molecular data are now widely used in molecular epidemiological studies and phylogenetic analyses, and sequence comparisons have been performed to facilitate the specific detection of genetically diverse strains/sequences and investigate the origin, transmission, distribution, biology, and diversity of these pathogens [12,13,21,30−33], which are fundamentally important in the prevention and tracking of disease outbreaks. Knowledge of sequence variations is used in the development of accurate diagnostic assays and for the design of effective treatment strategies of diseases caused by these agents. Molecular epidemiological studies are critical for public health surveillance [14,15,34−65]. We provide here examples of how molecular tests contributed to public health surveillance and patient care.

Influenza A

Seasonal and pandemic influenza A represents one of the greatest threats to global health [66−68]. Continuing challenges in influenza include the sporadic human cases of highly pathogenic avian H5N1 influenza, emergence of pandemic H1N1 influenza in 2009 [62,69], and human infections with avian H7N9 influenza in 2013 [11]. Influenza A virus undergoes continuous antigenic drift and sporadic antigenic shifts in the viral surface glycoproteins, hemagglutinin (H) and neuraminidase (N). Influenza A has 15 H and 9 N subtypes. Antigenic H and N subtypes to which humans lack immunity are introduced by reassortment of virus genes and cause pandemics, whereas H and N antigenic variants determined by point mutations cause seasonal influenza epidemics [66,67].

Molecular assays are the preferred method for identification and surveillance of new strains of influenza A infections [11,62,67]. Influenza A has no pathognomonic symptoms, and diagnosis based on clinical signs is correct in only two-thirds of patients [68,70]. Therefore, sensitive and rapid laboratory tests are required to diagnose and guide antiviral treatment. Recently, multiplex molecular assays for respiratory viruses including influenza viruses have been developed, and several have received approval/clearance by the US Food and Drug Administration (FDA) for routine clinical use (Table 15.2, listed under pandemic H1N1 influenza virus). These assays provide rapid and sensitive tests for respiratory viral infections.

Human Immunodeficiency Virus 1

Human immunodeficiency virus 1 (HIV-1) was discovered in 1983 (Table 15.1). It is a single-stranded, positive-sense, enveloped RNA retrovirus (http://www.hiv.lanl.gov/). HIV-1 can cause AIDS, a chronic disease leading to immunodeficiency and susceptibility to opportunistic infections (http://www.who.int/hiv/en/). Three groups of HIV-1 have been identified based on sequence similarity, including M (main), O (outlier), and N (non-M/non-O) (http://www.hiv.lanl.gov/). Of the three groups of HIV-1, group M dominates the global epidemic and is further classified into subtypes A, B, C, D, F, G, H, J, and K. In addition, circulating recombinant forms (CRFs), mosaic viruses formed between subtypes during co- or superinfection, have also been recognized (http://www.hiv.lanl.gov/). Although subtype B is predominant in North America and Europe, non-B variants represent more than 90% of HIV-1 circulating globally [71]. In recent years, the prevalence of non-B subtypes and CRFs in the United States is steadily increasing due to increased international travel and immigration [72−74]. Sequencing data of HIV-1 genomes have been used for tracking HIV epidemics and for the design of accurate viral detection, viral load, and HIV-1 drug-resistance genotype assays to guide clinical use of antiretroviral treatment [38,75]. The recent availability of the NGS approach has greatly facilitated generation of HIV-1 sequences and detection of quasispecies, which can improve understanding of HV-1 infection, pathogenesis, and epidemics [38,76,77].

Hepatitis C Virus

It is believed that 150 million people worldwide are infected with HCV (http://www.who.int/mediacentre/factsheets/fs164/en/). Between 70% and 80% of people infected with HCV will develop chronic infection. Chronic hepatitis C is closely associated with the development of cirrhosis and hepatocellular carcinoma and is the most common cause of adult liver transplantation in the United States and the world (http://www.cdc.gov/hepatitis/hcv/). A comparison of HCV genomic sequences from around the world revealed substantial heterogeneity of nucleotide sequences. Phylogenetic analyses have shown that HCV strains can be classified into six genotypes (numbered 1–6) and a large number of subtypes within each genotype [78]. HCV genotypes 1, 2, and 3 appear to have a worldwide distribution, but their relative prevalence varies from one geographic area to another. HCV genotype 1 is reported to be the most common in the United States [79–81]. HCV virus genome sequencing has been used to study HCV genotypes, subtypes, quasispecies, and mutations. The information is important for epidemiological studies, to trace the source of infection, for development of direct acting antiviral (DAA) therapy, and for understanding of susceptibility and resistance to antiviral treatment [82–85].

MOLECULAR DIAGNOSTICS OF EMERGING INFECTIOUS PATHOGENS

Many methods have been employed for the clinical diagnostics of emerging pathogens including microscopy, bacterial culture, cell culture, and serologic tests. However, each of these methods has its own limitations that must be considered by the clinical laboratory. For example, even though cell culture could be considered as the gold standard in diagnosis of infection with emerging obligate intracellular bacteria such as *Rickettsia* or *Ehrlichia*, the requirement for biological safety laboratory level 3 (BSL-3) (for *Rickettsia*) or BSL-2 (for *Ehrlichia*) makes this test difficult to implement in many conventional clinical microbiology laboratories. Further, the prolonged turnaround time (TAT) (eg, detection by culture at 7–10 days after sample processing) makes this approach impractical. Results from such a test are not clinically useful due to failure to guide therapy during the early stages of infection when appropriate antibiotic treatment is highly effective. Similar to culture, serologic tests such as indirect immunofluorescence assays, which rely on detection of antigen-specific antibodies, have several limitations such as low sensitivity during the early stages of infection when there is a low level of specific antibodies and false-positive results due to cross-reaction of antibodies to antigens from closely-related bacterial species. In addition, diagnosis of acute infection by IgG serology using single or paired (acute and convalescent) serum samples has the limitation of lack of a standardized cutoff titer among laboratories if a single sample is obtained, or the frequent inability to obtain convalescent serum when paired samples are required. In the latter case, while IgG serology could be useful for epidemiologic surveillance, paired sera are not optimal for timely diagnosis and treatment of acute infection. Thus, the emergence of molecular methods including probe hybridization, target or signal amplification, and sequencing provides better diagnostic advantages compared to microscopy, culture, and serology such as rapid TAT, higher sensitivity and higher specificity in different patient populations, and using different specimen types (eg, blood, plasma, cerebrospinal fluid, tissues, fluids). These molecular tests have become the gold standards due to their high negative and positive predictive values and their ability to detect and characterize newly emerging pathogens for clinical purposes [1–11].

Molecular assays are routinely used in clinics for the diagnosis, prognosis, and treatment decisions of various emerging infectious diseases [12,13,38,75, 86–91] (Tables 15.1 and 15.2). As listed in Tables 15.1 and 15.2, there are US FDA-approved/cleared tests for some of these pathogens, and CDC has tests for all these agents. There are also laboratory-developed tests brought to clinical use after significant research and development and validation studies by individual laboratories [12,13,38,75,86–91]. As in other infectious diseases, clinical molecular tests for emerging infectious diseases include (1) nucleic acid detection assays with defined limit of detection cutoffs, (2) quantitative methods with broad dynamic ranges, lower and higher limit of quantification values, (3) genotyping and subtyping assays, and drug resistance mutation assays at even single base-pair resolution are used for disease prognosis and guiding treatment strategies [71,81,92]. General quality management protocols that cover preanalytic, analytic, and postanalytic phases also apply to molecular tests of emerging pathogens.

Following the discovery of HIV-1 in 1983 and HCV in 1989, molecular tests were developed and implemented for routine clinical use to detect viral infection, monitor viral load, and examine specific HIV-1 drug-resistant mutations and HCV genotypes to guide patient management. Several practice guidelines have incorporated HIV-1 and HCV molecular tests (eg, http://www.who.int/hiv/pub/guidelines/en/; http://www.hcvguidelines.org/full-report-view). For example, because detection of HCV RNA, not IgG antibody, is diagnostic of current HCV infection, and HCV genotype 1 is more difficult to treat than genotype 2 or 3, testing for HCV genotype is recommended to guide selection of the most appropriate treatment regimen. HCV RNA detection and genotyping assays are routinely performed in clinical diagnostics laboratories (http://www.hcvguidelines.org/full-report-view).

Over the years, with advances in molecular technology, HIV-1 and HCV clinical molecular tests have improved significantly with respect to performance characteristics including sensitivity, specificity, and dynamic range. Currently, there are several FDA-approved/cleared molecular tests for HIV-1 and HCV (Table 15.1), and new methods are continuously developed and evaluated for better care of patients with HIV-1 and HCV infection [38,75,89,93].

LIMITATIONS OF CURRENT TESTING AND FUTURE PROSPECTS

A high portion of emerging infectious diseases are vector-borne zoonoses that have emerged from natural cycles. The underlying causes of their emergence are a combination of environmental changes, such as increased populations and geographic distribution of their reservoir hosts and vectors, and development of new scientific tools that contribute to their detection and identification. For example, PCR-based molecular methods have enabled the discovery of a large number of bacterial and viral organisms in ticks, which preceded the identification of these organisms as etiologic agents of emerging infectious diseases.

Among these emerging infectious diseases are two contrasting tick-borne infections, Lyme borreliosis [37,47], and human monocytotropic ehrlichiosis (HME) [27–29]. Lyme disease is well known, feared, at times inappropriately diagnosed, and very rarely fatal. HME is largely unknown, frequently misdiagnosed as another tick-borne disease such as Rocky Mountain spotted fever or a viral infection, and is often life-threatening. Lyme borreliosis occurs particularly in suburban populations in the northeastern United States and has been investigated extensively in prominent academic medical institutions in this region. HME occurs particularly in the rural southeastern United States and has not been the focus of in-depth clinical studies in academic medical centers in this region. Both Lyme borreliosis and HME have high incidence although that of HME is not well recognized.

The effects of these conditions on the development and application of diagnostic tests including molecular diagnostics are far from satisfying. Diagnosis of Lyme borreliosis depends heavily on serological assays. Patients with Lyme disease frequently have developed antibodies to Borrelia burgdorferi by the time in their course of illness when they present for medical attention. These patients and those with a classic bulls-eye appearing rash are diagnosed, treated effectively with appropriate antibiotics and recover. As with other infections antibodies take time to be stimulated and produced. Thus, some patients' diagnoses may be delayed. Molecular methods seldom provide a diagnosis owing to the paucity of organisms in the blood and other readily obtained clinical samples [94].

A tremendous problem is the large number of persons with atypical symptoms of a wide range that includes those similar to chronic fatigue syndrome or fibromyalgia who are convinced that they are suffering from chronic Lyme disease but whose results of validated tests do not support the diagnosis. Many of them are convinced that the tests are inadequate and that better tests are needed [94]. In contrast, patients with HME often have not developed antibodies to the etiologic agent, E. chaffeensis, at the time when they present for medical attention. The bacteria can infect mononuclear phagocytes and are present in circulating monocytes providing an often effective target for molecular diagnostics at a time when appropriate antibiotic treatment results in rapid recovery from an otherwise life-threatening infection [29]. Yet HME, which likely has an incidence similar to Lyme disease, lacks a readily available point-of-care diagnostic test. Effective molecular target genes have been identified, and in-house assays provide proof-of-concept that molecular diagnostics offer an effective approach [28,95]. Moreover, low-cost instrument-free devices for nucleic acid amplification and specific identification have been developed that would be appropriate for point-of-care diagnosis.

Why have no more effective efforts been made to devise, develop, and commercialize molecular approaches to these two important emerging infections? For Lyme borreliosis, molecular diagnostics may not possess the solution when too few or no Borrelia are present. For HME, the issues lie in the realms of clinical practice, public health, and business. Physicians who are unaware of HME and note that febrile illnesses during the tick season often respond to doxycycline therapy are not inclined to order send out tests that would cost the patient. Serology that is based on comparing IgG antibody levels in paired sera often fails to provide a diagnosis of acute infection as it relies on the seldom-obtained convalescent serum. Public health agencies are powerless to address effectively a disease that is not diagnosed, and if diagnosed, is not reported. The epidemiologic reports depend on the data obtained by passive surveillance. In fact, active, prospective, population-based surveillance in endemic regions such as Missouri suggested that HME is a highly prevalent disease [96]. The combination of nonspecific clinical manifestations of HME, test under-utilization, lack of a gold standard test that is effective when therapeutic decisions are made, and problems in interpretation of diagnostic tests such as serology, and misleading epidemiologic data have accounted for reported low incidence of HME. This situation has failed to stimulate interest in commercial development and marketing of a useful point-of-care assay, although there could be an adequate pull from the potential users of the test.

The advances of sequencing technology, nanotechnology, and bioinformatics have driven molecular tests including assays for emerging pathogens to be more comprehensive and precise. For example, the availability of various sequence databases permits quick identification of sequence identity and variations. For example, the HIV database http://www.hiv.lanl.gov/ contains data on HIV genetic sequences and drug resistance associated mutations. It is valuable for HIV epidemiological studies, research, development, and clinical validation studies of HIV clinical assays [38,71]. It is well known that there are significant variations of clinical phenotypes in the presence of emerging infections ranging from asymptomatic carrier to lethal infection. Recently, assays to examine multiple pathogen panels have been developed [97–102], which should increase the diagnostic yield for many pathogens. A critical need for emerging pathogen analysis is quicker, easier, cost-effect assays that can be used in a point-of-care setting. New assays that are performed on platforms with a small footprint and detect pathogens quickly (in minutes instead of hours or days) have entered clinical use. For example, the FilmArray (BioFire Diagnostics, Inc.) and Simplexa (Focus Diagnostics, Inc.) molecular assays can generate results in approximately 60 min. The user-friendly Alere i (Alere Inc.) and Cobas Liat (Roche Molecular Systems) platforms are compact and portable, generate rapid molecular results in 15–20 min, can use electricity or rechargeable battery, and therefore are completely mobile and suited for point-of-care testing. It is obvious that the current rapid development of new technologies will further enhance the utility of molecular diagnostics in various emerging infectious diseases.

The advancement of molecular methods for emerging infections comes hand-in-hand with other areas including general infectious diseases, genetics and genomics, and oncology. There are needs to develop unified sequence databases for the input and search of emerging pathogens and other sequences, to understand pathogen/genotype/sequence correlation with phenotypes (eg, lethality or carrier with an emerging infection), to develop panels to more effectively diagnose patients based on shared clinical signs and symptoms, and to develop point-of-care molecular platforms and assays for emerging infectious diseases.

Over the last two decades, sequencing technology has evolved from labor-intensive and time-consuming methodologies to automatic and real-time sequence detection. Recent development and use of NGS has revolutionized the landscape of microbiology and infectious disease. The availability of sequencing data has speeded up pathogen discovery, and also helped improve diagnosis, typing of pathogens, detection of virulence and drug resistance, and development of new vaccines and targeted treatment [103–106].

With the ever-extending use of NGS on a variety of clinical samples, rapid progress on determining the composition of the human microbiome and its impact upon human health are to be expected in the coming years. This deluge of sequencing data requires a consolidated and curated database to input and search sequences, sequence variations, associated symptoms and diseases, available tests, and treatment options. A unified reporting guideline for molecular epidemiology has been proposed recently [107]. Adoption of this guideline by the research and clinical communities should help to integrate the effort for the comprehension of genomics and metagenomics relevant to the field of medical microbiology, and to improve management of infectious diseases.

Traditional pathogen detection methods in infectious diseases rely upon the identification of agents associated with a particular clinical syndrome. The availability of a significant amount of sequence information and the emerging field of metagenomics using NGS have the potential to revolutionize pathogen detection by allowing the simultaneous detection of all microorganisms in a clinical sample, without a priori knowledge of their identities. This can identify new sequences and organisms that may be initially considered nonpathogenic and may cause infections in different human populations and health conditions. They may cause diseases not previously thought to have a microbial component, and the methods may determine previously unknown etiology of infections. For example, infection with certain emerging pathogens may only cause disease symptoms in patients with AIDS or immune suppression after organ transplantation, or in travelers not previously exposed to the agents. Further biological and clinical studies are necessary to categorize sequence information and interpret clinical relevance when a pathogen sequence is detected, which is critical for diagnosis, treatment, and public health surveillance of emerging infectious diseases.

Assays to examine multiple pathogen panels have been developed [97–102]. These assays are designed either to detect many infections that can cause similar symptoms (eg, FDA-approved/cleared respiratory viral panels as listed in Table 15.2, multiple viruses that can trigger gastrointestinal symptoms) [100,101], pathogens that share homologous sequences, for example, 16S rRNA sequencing [98,99,102,108] or are expected to occur under the circumstances of biothreat [97]. The availability of more pathogen sequences and further understanding of their correlation with clinical symptoms are necessary for the rational design of panels that can fit various needs.

New technological developments including microfluidics, nanotechnology, and lab-on-a-chip technologies have enabled development of user-friendly, easy, and quick point-of-care molecular tests including Alere i (Alere Inc.) and Cobas Liat (Roche Molecular Systems). In the setting of emerging infectious diseases, rapid and accurate identification of the causative agent is critical to facilitate effective patient management and enable prompt initiation of infection controls. Point-of-care assays are especially needed in resource-limited settings and in situations with lack of access to centralized medical facilities. Further development of point-of-care molecular tests for emerging pathogens is critical to timely diagnosis, treatment, and subsequent control of emerging infectious disease.

References

[1] Delgado S, Erickson BR, Agudo R, Blair PJ, Vallejo E, Albarino CG, et al. Chapare virus, a newly discovered arenavirus isolated from a fatal hemorrhagic fever case in Bolivia. PLoS Pathog 2008;4:e1000047.

[2] Djavani M, Lukashevich IS, Sanchez A, Nichol ST, Salvato MS. Completion of the Lassa fever virus sequence and identification of a RING finger open reading frame at the L RNA 5′ End. Virology 1997;235:414–18.

[3] Harcourt BH, Tamin A, Ksiazek TG, Rollin PE, Anderson LJ, Bellini WJ, et al. Molecular characterization of Nipah virus, a newly emergent paramyxovirus. Virology 2000;271:334–49.

[4] Lee E, Stocks C, Lobigs P, Hislop A, Straub J, Marshall I, et al. Nucleotide sequence of the Barmah Forest virus genome. Virology 1997;227:509–14.

[5] Madani TA, Azhar EI, Abuelzein el TM, Kao M, Al-Bar HM, Farraj SA, et al. Complete genome sequencing and genetic characterization of Alkhumra hemorrhagic fever virus isolated from Najran, Saudi Arabia. Intervirology 2014;57(5):300–10.

[6] Matsutani M, Ogawa M, Takaoka N, Hanaoka N, Toh H, Yamashita A, et al. Complete genomic DNA sequence of the East Asian spotted fever disease agent *Rickettsia japonica*. PLoS One 2013;8:e71861.

[7] Pekova S, Vydra J, Kabickova H, Frankova S, Haugvicova R, Mazal O, et al. Candidatus Neoehrlichia mikurensis infection identified in 2 hematooncologic patients: benefit of molecular techniques for rare pathogen detection. Diagn Microbiol Infect Dis 2011;69:266–70.

[8] Pritt BS, Sloan LM, Johnson DK, Munderloh UG, Paskewitz SM, McElroy KM, et al. Emergence of a new pathogenic *Ehrlichia* species, Wisconsin and Minnesota, 2009. N Engl J Med 2011;365:422–9.

[9] Thirumalapura NR, Qin X, Kuriakose JA, Walker DH. Complete genome sequence of *Ehrlichia muris* strain AS145T, a model monocytotropic *Ehrlichia* strain. Genome Announc 2014;2:e01234–13.

[10] Yu XJ, Liang MF, Zhang SY, Liu Y, Li JD, Sun YL, et al. Fever with thrombocytopenia associated with a novel bunyavirus in China. N Engl J Med 2011;364:1523–32.

[11] Lam TT, Wang J, Shen Y, Zhou B, Duan L, Cheung CL, et al. The genesis and source of the H7N9 influenza viruses causing human infections in China. Nature 2013;502:241–4.

[12] Houghton M. The long and winding road leading to the identification of the hepatitis C virus. J Hepatol 2009;51:939–48.

[13] Houghton M. Discovery of the hepatitis C virus. Liver Int 2009;29:82–8.

[14] Black WC, Doty JB, Hughes MT, Beaty BJ, Calisher CH. Temporal and geographic evidence for evolution of Sin Nombre virus using molecular analyses of viral RNA from Colorado, New Mexico and Montana. Virol J 2009;6:102.

[15] Henderson WW, Monroe MC, St Jeor SC, Thayer WP, Rowe JE, Peters CJ, et al. Naturally occurring Sin Nombre virus genetic reassortants. Virology 1995;214:602–10.

[16] Payne B, Bellamy R. Novel respiratory viruses: what should the clinician be alert for? Clin Med 2014;14:s12–16.

[17] Drosten C, Gunther S, Preiser W, van der Werf S, Brodt HR, Becker S, et al. Identification of a novel coronavirus in patients with severe acute respiratory syndrome. N Engl J Med 2003;348:1967–76.

[18] Ksiazek TG, Erdman D, Goldsmith CS, Zaki SR, Peret T, Emery S, et al. A novel coronavirus associated with severe acute respiratory syndrome. N Engl J Med 2003;348:1953–66.

[19] Peiris JS, Lai ST, Poon LL, Guan Y, Yam LY, Lim W, et al. Coronavirus as a possible cause of severe acute respiratory syndrome. Lancet 2003;361:1319–25.

[20] Rota PA, Oberste MS, Monroe SS, Nix WA, Campagnoli R, Icenogle JP, et al. Characterization of a novel coronavirus associated with severe acute respiratory syndrome. Science 2003;300:1394–9.

[21] Adachi D, Johnson G, Draker R, Ayers M, Mazzulli T, Talbot PJ, et al. Comprehensive detection and identification of human coronaviruses, including the SARS-associated coronavirus, with a single RT-PCR assay. J Virol Methods 2004;122:29–36.

[22] Huang JL, Lin HT, Wang YM, Yeh YC, Peck K, Lin BL, et al. Rapid and sensitive detection of multiple genes from the SARS-coronavirus using quantitative RT-PCR with dual systems. J Med Virol 2005;77:151–8.

[23] Lantos PM, Maggi RG, Ferguson B, Varkey J, Park LP, Breitschwerdt EB, et al. Detection of *Bartonella* species in the blood of veterinarians and veterinary technicians: a newly recognized occupational hazard? Vector Borne Zoonotic Dis 2014;14:563–70.

[24] Psarros G, Riddell IV J, Gandhi T, Kauffman CA, Cinti SK. *Bartonella henselae* infections in solid organ transplant recipients: report of 5 cases and review of the literature. Medicine 2012;91:111–21.

[25] Chan K, Marras SA, Parveen N. Sensitive multiplex PCR assay to differentiate Lyme spirochetes and emerging pathogens *Anaplasma phagocytophilum* and *Babesia microti*. BMC Microbiol 2013;13:295.

[26] Stuen S, Granquist EG, Silaghi C. *Anaplasma phagocytophilum*—a widespread multi-host pathogen with highly adaptive strategies. Front Cell Infect Microbiol 2013;3:31.

[27] Breitschwerdt EB, Hegarty BC, Qurollo BA, Saito TB, Maggi RG, Blanton LS, et al. Intravascular persistence of *Anaplasma platys*, *Ehrlichia chaffeensis*, and *Ehrlichia ewingii* DNA in the blood of a dog and two family members. Parasit Vectors 2014;7:298.

[28] Doyle CK, Labruna MB, Breitschwerdt EB, Tang YW, Corstvet RE, Hegarty BC, et al. Detection of medically important *Ehrlichia* by quantitative multicolor TaqMan real-time polymerase chain reaction of the *dsb* gene. J Mol Diagn 2005;7:504–10.

[29] Walker DH, Paddock CD, Dumler JS. Emerging and re-emerging tick-transmitted rickettsial and ehrlichial infections. Med Clin North Am 2008;92:1345–61.

[30] Hadjinicolaou AV, Farcas GA, Demetriou VL, Mazzulli T, Poutanen SM, Willey BM, et al. Development of a molecular-

beacon-based multi-allelic real-time RT-PCR assay for the detection of human coronavirus causing severe acute respiratory syndrome (SARS-CoV): a general methodology for detecting rapidly mutating viruses. Arch Virol 2011;156:671–80.

[31] Lan YC, Liu TT, Yang JY, Lee CM, Chen YJ, Chan YJ, et al. Molecular epidemiology of severe acute respiratory syndrome-associated coronavirus infections in Taiwan. J Infect Dis 2005;191:1478–89.

[32] Tang JW, Cheung JL, Chu IM, Ip M, Hui M, Peiris M, et al. Characterizing 56 complete SARS-CoV S-gene sequences from Hong Kong. J Clin Virol 2007;38:19–26.

[33] Tang JW, Cheung JL, Chu IM, Sung JJ, Peiris M, Chan PK. The large 386-nt deletion in SARS-associated coronavirus: evidence for quasispecies? J Infect Dis 2006;194:808–13.

[34] Bonvicini F, Manaresi E, Bua G, Venturoli S, Gallinella G. Keeping pace with parvovirus B19 genetic variability: a multiplex genotype-specific quantitative PCR assay. J Clin Microbiol 2013;51:3753–9.

[35] Cajimat MN, Milazzo ML, Rollin PE, Nichol ST, Bowen MD, Ksiazek TG, et al. Genetic diversity among Bolivian arenaviruses. Virus Res 2009;140:24–31.

[36] Christensen A, Dollner H, Skanke LH, Krokstad S, Moe N, Nordbo SA. Detection of spliced mRNA from human bocavirus 1 in clinical samples from children with respiratory tract infections. Emerg Infect Dis 2013;19:574–80.

[37] Clark KL, Leydet BF, Threlkeld C. Geographical and genospecies distribution of *Borrelia burgdorferi* sensu lato DNA detected in humans in the USA. J Med Microbiol 2014;63:674–84.

[38] Di Giallonardo F, Zagordi O, Duport Y, Leemann C, Joos B, Kunzli-Gontarczyk M, et al. Next-generation sequencing of HIV-1 RNA genomes: determination of error rates and minimizing artificial recombination. PLoS One 2013;8:e74249.

[39] Dingle KE, Elliott B, Robinson E, Griffiths D, Eyre DW, Stoesser N, et al. Evolutionary history of the *Clostridium difficile* pathogenicity locus. Genome Biol Evol 2014;6:36–52.

[40] Donaldson CD, Clark DA, Kidd IM, Breuer J, Depledge DD. Genome sequence of human herpesvirus 7 strain UCL-1. Genome Announc 2013;1 e00830–13.

[41] Eyre DW, Cule ML, Wilson DJ, Griffiths D, Vaughan A, O'Connor L, et al. Diverse sources of *C. difficile* infection identified on whole-genome sequencing. N Engl J Med 2013;369:1195–205.

[42] Fulhorst CF, Cajimat MN, Milazzo ML, Paredes H, de Manzione NM, Salas RA, et al. Genetic diversity between and within the arenavirus species indigenous to western Venezuela. Virology 2008;378:205–13.

[43] Galvan A, Magnet A, Izquierdo F, Fenoy S, Henriques-Gil N, del Aguila C. Variability in minimal genomes: analysis of tandem repeats in the microsporidia *Encephalitozoon intestinalis*. Infect Genet Evol 2013;20:26–33.

[44] Gomez-Valero L, Rusniok C, Rolando M, Neou M, Dervins-Ravault D, Demirtas J, et al. Comparative analyses of *Legionella* species identifies genetic features of strains causing Legionnaires inverted question mark disease. Genome Biol 2014;15:505.

[45] Gonzalez JP, Bowen MD, Nichol ST, Rico-Hesse R. Genetic characterization and phylogeny of Sabia virus, an emergent pathogen in Brazil. Virology 1996;221:318–24.

[46] Ishii A, Thomas Y, Moonga L, Nakamura I, Ohnuma A, Hang'ombe BM, et al. Molecular surveillance and phylogenetic analysis of Old World arenaviruses in Zambia. J Gen Virol 2012;93:2247–51.

[47] Jacquot M, Gonnet M, Ferquel E, Abrial D, Claude A, Gasqui P, et al. Comparative population genomics of the *Borrelia burgdorferi* species complex reveals high degree of genetic isolation among species and underscores benefits and constraints to studying intra-specific epidemiological processes. PLoS One 2014;9:e94384.

[48] Lares-Jimenez LF, Booton GC, Lares-Villa F, Velazquez-Contreras CA, Fuerst PA. Genetic analysis among environmental strains of *Balamuthia mandrillaris* recovered from an artificial lagoon and from soil in Sonora, Mexico. Exp Parasitol 2014;145: S57–61.

[49] Le Gal F, Gault E, Ripault MP, Serpaggi J, Trinchet JC, Gordien E, et al. Eighth major clade for hepatitis delta virus. Emerg Infect Dis 2006;12:1447–50.

[50] Mazurie AJ, Alves JM, Ozaki LS, Zhou S, Schwartz DC, Buck GA. Comparative genomics of cryptosporidium. Int J Genomics 2013;2013:832756.

[51] Muhlberger E, Trommer S, Funke C, Volchkov V, Klenk HD, Becker S. Termini of all mRNA species of Marburg virus: sequence and secondary structure. Virology 1996;223:376–80.

[52] Pang B, Zheng X, Diao B, Cui Z, Zhou H, Gao S, et al. Whole genome PCR scanning reveals the syntenic genome structure of toxigenic *Vibrio cholerae* strains in the O1/O139 population. PLoS One 2011;6:e24267.

[53] Pessoa R, Watanabe JT, Nukui Y, Pereira J, Kasseb J, de Oliveira AC, et al. Molecular characterization of human T-cell lymphotropic virus type 1 full and partial genomes by Illumina massively parallel sequencing technology. PLoS One 2014;9: e93374.

[54] Poidinger M, Roy S, Hall RA, Turley PJ, Scherret JH, Lindsay MD, et al. Genetic stability among temporally and geographically diverse isolates of Barmah Forest virus. Am J Trop Med Hyg 1997;57:230–4.

[55] Pombert JF, Selman M, Burki F, Bardell FT, Farinelli L, Solter LF, et al. Gain and loss of multiple functionally related, horizontally transferred genes in the reduced genomes of two microsporidian parasites. Proc Natl Acad Sci USA 2012; 109:12638–43.

[56] Ratner L, Philpott T, Trowbridge DB. Nucleotide sequence analysis of isolates of human T-lymphotropic virus type 1 of diverse geographical origins. AIDS Res Hum Retroviruses 1991; 7:923–41.

[57] Riner DK, Nichols T, Lucas SY, Mullin AS, Cross JH, Lindquist HD. Intragenomic sequence variation of the ITS-1 region within a single flow-cytometry-counted *Cyclospora cayetanensis* oocysts. J Parasitol 2010;96:914–19.

[58] Rump LV, Gonzalez-Escalona N, Ju W, Wang F, Cao G, Meng S, et al. Genomic diversity and virulence characterization of historical *Escherichia coli* O157 strains isolated from clinical and environmental sources. Appl Environ Microbiol 2014;81: 569–77.

[59] Sanchez-Buso L, Comas I, Jorques G, Gonzalez-Candelas F. Recombination drives genome evolution in outbreak-related *Legionella pneumophila* isolates. Nature Genet 2014;46:1205–11.

[60] Sekeyova Z, Roux V, Raoult D. Phylogeny of *Rickettsia* spp. inferred by comparing sequences of 'gene D', which encodes an intracytoplasmic protein. Int J Syst Evol Microbiol 2001;51: 1353–60.

[61] Strouts FR, Power P, Croucher NJ, Corton N, van Tonder A, Quail MA, et al. Lineage-specific virulence determinants of *Haemophilus influenzae* biogroup *aegyptius*. Emerg Infect Dis 2012;18:449–57.

[62] Tellez-Sosa J, Rodriguez MH, Gomez-Barreto RE, Valdovinos-Torres H, Hidalgo AC, Cruz-Hervert P, et al. Using high-throughput sequencing to leverage surveillance of genetic diversity and oseltamivir resistance: a pilot study during the 2009 influenza A(H1N1) pandemic. PLoS One 2013;8:e67010.

[63] Towner JS, Khristova ML, Sealy TK, Vincent MJ, Erickson BR, Bawiec DA, et al. Marburgvirus genomics and association with

a large hemorrhagic fever outbreak in Angola. J Virol 2006;80:6497−516.

[64] Yu M, Hansson E, Shiell B, Michalski W, Eaton BT, Wang LF. Sequence analysis of the Hendra virus nucleoprotein gene: comparison with other members of the subfamily Paramyxovirinae. J Gen Virol 1998;79:1775−80.

[65] Zhou Y, Lv B, Wang Q, Wang R, Jian F, Zhang L, et al. Prevalence and molecular characterization of *Cyclospora cayetanensis*, Henan, China. Emerg Infect Dis 2011;17:1887−90.

[66] Nicholson KG, Wood JM, Zambon M. Influenza. Lancet 2003;362:1733−45.

[67] Webster RG, Govorkova EA. Continuing challenges in influenza. Ann NY Acad Sci 2014;1323:115−39.

[68] Ortiz JR, Neuzil KM, Shay DK, Rue TC, Neradilek MB, Zhou H, et al. The burden of influenza-associated critical illness hospitalizations. Crit Care Med 2014;42(11):2325−32.

[69] Bermudez de Leon M, Penuelas-Urquides K, Aguado-Barrera ME, Curras-Tuala MJ, Escobedo-Guajardo BL, Gonzalez-Rios RN, et al. In vitro transcribed RNA molecules for the diagnosis of pandemic 2009 influenza A(H1N1) virus by real-time RT-PCR. J Virol Methods 2013;193:487−91.

[70] Ebell MH, Afonso AM, Gonzales R, Stein J, Genton B, Senn N. Development and validation of a clinical decision rule for the diagnosis of influenza. J Am Board Fam Med 2012;25:55−62.

[71] Xu F, Schwab C, Liang X, Weaver S, Li A, Sanborn MR, et al. Low Prevalence of non-subtype B HIV-1 strains in the Texas prisoner population. J Mol Genet 2010;2:41−4.

[72] Brennan CA, Stramer SL, Holzmayer V, Yamaguchi J, Foster GA, Notari EP, et al. Identification of human immunodeficiency virus type 1 non-B subtypes and antiretroviral drug-resistant strains in United States blood donors. Transfusion 2009;49:125−33.

[73] Lin HH, Gaschen BK, Collie M, El-Fishaway M, Chen Z, Korber BT, et al. Genetic characterization of diverse HIV-1 strains in an immigrant population living in New York City. J Acquir Immune Defic Syndr 2006;41:399−404.

[74] Peeters M, Aghokeng AF, Delaporte E. Genetic diversity among human immunodeficiency virus-1 non-B subtypes in viral load and drug resistance assays. Clin Microbiol Infect 2010;16:1525−31.

[75] Casabianca A, Orlandi C, Canovari B, Scotti M, Acetoso M, Valentini M, et al. A real time PCR platform for the simultaneous quantification of total and extrachromosomal HIV DNA forms in blood of HIV-1 infected patients. PLoS One 2014;9:e111919.

[76] de Goede AL, Vulto AG, Osterhaus AD, Gruters RA. Understanding HIV infection for the design of a therapeutic vaccine. Part I: Epidemiology and pathogenesis of HIV infection. Ann Pharm Fr 2014;73:87−99.

[77] Young SD. A "big data" approach to HIV epidemiology and prevention. Prev Med 2014;70C:17−18.

[78] Simmonds P, Bukh J, Combet C, Deleage G, Enomoto N, Feinstone S, et al. Consensus proposals for a unified system of nomenclature of hepatitis C virus genotypes. Hepatology 2005;42:962−73.

[79] Nainan OV, Alter MJ, Kruszon-Moran D, Gao FX, Xia G, McQuillan G, et al. Hepatitis C virus genotypes and viral concentrations in participants of a general population survey in the United States. Gastroenterology 2006;131:478−84.

[80] Rustgi VK. The epidemiology of hepatitis C infection in the United States. J Gastroenterol 2007;42:513−21.

[81] Clement CG, Yang Z, Mayne JC, Dong J. HCV genotype and subtype distribution of patient samples tested at University of Texas Medical Branch in Galveston, Texas. J Mol Genet 2010;2:36−40.

[82] Aherfi S, Solas C, Motte A, Moreau J, Borentain P, Mokhtari S, et al. Hepatitis C virus NS3 protease genotyping and drug concentration determination during triple therapy with telaprevir or boceprevir for chronic infection with genotype 1 viruses, southeastern France. J Med Virol 2014;86:1868−76.

[83] Campo DS, Skums P, Dimitrova Z, Vaughan G, Forbi JC, Teo CG, et al. Drug resistance of a viral population and its individual intrahost variants during the first 48 hours of therapy. Clin Pharmacol Ther 2014;95:627−35.

[84] Irving WL, Rupp D, McClure CP, Than LM, Titman A, Ball JK, et al. Development of a high-throughput pyrosequencing assay for monitoring temporal evolution and resistance associated variant emergence in the hepatitis C virus protease coding-region. Antivir Res 2014;110:52−9.

[85] Svarovskaia ES, Dvory-Sobol H, Parkin N, Hebner C, Gontcharova V, Martin R, et al. Infrequent development of resistance in genotype 1-6 hepatitis C virus-infected subjects treated with sofosbuvir in phase 2 and 3 clinical trials. Clin Infect Dis 2014;59:1666−74.

[86] Benitez AJ, Thurman KA, Diaz MH, Conklin L, Kendig NE, Winchell JM. Comparison of real-time PCR and a microimmunofluorescence serological assay for detection of *Chlamydophila pneumoniae* infection in an outbreak investigation. J Clin Microbiol 2012;50:151−3.

[87] Ravindranath BS, Krishnamurthy V, Krishna V, Sunil Kumar C. In silico synteny based comparative genomics approach for identification and characterization of novel therapeutic targets in *Chlamydophila pneumoniae*. Bioinformation 2013;9:506−10.

[88] Fevery B, Susser S, Lenz O, Cloherty G, Perner D, Picchio G, et al. HCV RNA quantification with different assays: implications for protease-inhibitor-based response-guided therapy. Antivir Ther 2014;19:559−67.

[89] Quer J, Gregori J, Rodriguez-Frias F, Buti M, Madejon A, Perez-Del-Pulgar S, et al. High-resolution hepatitis C virus (HCV) subtyping, using NS5B deep sequencing and phylogeny, an alternative to current methods. J Clin Microbiol 2014;53:219−26.

[90] Klemenc J, Asad Ali S, Johnson M, Tollefson SJ, Talbot HK, Hartert TV, et al. Real-time reverse transcriptase PCR assay for improved detection of human metapneumovirus. J Clin Virol 2012;54:371−5.

[91] Roussy JF, Carbonneau J, Ouakki M, Papenburg J, Hamelin ME, De Serres G, et al. Human metapneumovirus viral load is an important risk factor for disease severity in young children. J Clin Virol 2014;60:133−40.

[92] Yang ZM, Morrison R, Oates C, Sarria J, Patel J, Habibi A, et al. HIV-1 genotypic resistance testing on low viral load specimens using the Abbott ViroSeq HIV-1 Genotyping System. LabMedicine 2008;39:671−3.

[93] Vollmer T, Knabbe C, Dreier J. Comparison of real-time PCR and antigen assays for detection of hepatitis E virus in blood donors. J Clin Microbiol 2014;52:2150−6.

[94] Aguero-Rosenfeld ME, Wormser GP. Lyme disease: diagnostic issues and controversies. Exp Rev Mol Diagn 2015;15:1−4.

[95] Killmaster LF, Loftis AD, Zemtsova GE, Levin ML. Detection of bacterial agents in *Amblyomma americanum* (Acari: Ixodidae) from Georgia, USA, and the use of a multiplex assay to differentiate *Ehrlichia chaffeensis* and *Ehrlichia ewingii*. J Med Entomol 2014;51:868−72.

[96] Olano J, Masters E, Hogrefe W, Walker DH. Human monocytotropic ehrlichiosis, Missouri. Emerg Infect Dis 2003;9:1579−86.

[97] Euler M, Wang Y, Heidenreich D, Patel P, Strohmeier O, Hakenberg S, et al. Development of a panel of recombinase polymerase amplification assays for detection of biothreat agents. J Clin Microbiol 2013;51:1110−17.

[98] Koehler JW, Hall AT, Rolfe PA, Honko AN, Palacios GF, Fair JN, et al. Development and evaluation of a panel of filovirus sequence capture probes for pathogen detection by next-generation sequencing. PLoS One 2014;9:e107007.

[99] Lindsay B, Pop M, Antonio M, Walker AW, Mai V, Ahmed D, et al. Survey of culture, goldengate assay, universal biosensor assay, and 16S rRNA gene sequencing as alternative methods of bacterial pathogen detection. J Clin Microbiol 2013;51:3263–9.

[100] Martinez MA, Soto-Del Rio MD, Gutierrez RM, Chiu CY, Greninger AL, Contreras JF, et al. DNA microarray for detection of gastrointestinal viruses. J Clin Microbiol 2014;53:136–45.

[101] Moore NE, Wang J, Hewitt J, Croucher D, Williamson DA, Paine S, et al. Metagenomic analysis of viruses in feces from unsolved outbreaks of gastroenteritis in humans. J Clin Microbiol 2014;53:15–21.

[102] Wei S, Zhao H, Xian Y, Hussain MA, Wu X. Multiplex PCR assays for the detection of *Vibrio alginolyticus*, *Vibrio parahaemolyticus*, *Vibrio vulnificus*, and *Vibrio cholerae* with an internal amplification control. Diagn Microbiol Infect Dis 2014;79:115–18.

[103] Cox MJ, Cookson WO, Moffatt MF. Sequencing the human microbiome in health and disease. Hum Mol Genet 2013;22: R88–94.

[104] Lecuit M, Eloit M. The human virome: new tools and concepts. Trends Microbiol 2013;21:510–15.

[105] Miller RR, Montoya V, Gardy JL, Patrick DM, Tang P. Metagenomics for pathogen detection in public health. Genome Med 2013;5:81.

[106] Padmanabhan R, Mishra AK, Raoult D, Fournier PE. Genomics and metagenomics in medical microbiology. J Microbiol Methods 2013;95:415–24.

[107] Field N, Cohen T, Struelens MJ, Palm D, Cookson B, Glynn JR, et al. Strengthening the Reporting of Molecular Epidemiology for Infectious Diseases (STROME-ID): an extension of the STROBE statement. Lancet Infect Dis 2014;14:341–52.

[108] Zhao J, Kang L, Hu R, Gao S, Xin W, Chen W, et al. Rapid oligonucleotide suspension array-based multiplex detection of bacterial pathogens. Foodborne Pathog Dis 2013;10:896–903.

SECTION III

MOLECULAR TESTING
IN GENETIC DISEASE

16

Noninvasive Prenatal Screening (NIPS) for Fetal Aneuploidies

W.E. Highsmith[1,2], M.A. Allyse[3], K.S. Borowski[2,4] and M.J. Wick[2,4]

[1]Department of Laboratory Medicine and Pathology, Mayo Clinic School of Medicine, Rochester, MN, United States
[2]Department of Medical Genetics, Mayo Clinic School of Medicine, Rochester, MN, United States [3]Department of Health Sciences Research, Mayo Clinic School of Medicine, Rochester, MN, United States [4]Department of Obstetrics and Gynecology, Mayo Clinic School of Medicine, Rochester, MN, United States

INTRODUCTION

Prenatal screening for fetal anomalies began in the mid-1970s with the observation that maternal serum alpha-fetal protein (AFP) levels were elevated in cases with fetal neural tube defects [1]. In the 1980s, researchers found a link between decreased serum AFP and Down syndrome, but the sensitivity of the test was low, approximately 21% at a 5% screen positive (or false-positive) rate [2]. However, the attributable risk due to maternal age and the risk defined by second trimester serum AFP were shown to be independent and could thus be combined to calculate a net risk for a pregnancy affected with Down syndrome. The combination of age and serum marker resulted in a higher detection rate, at a constant cutoff, than either parameter alone [3].

Over the next decade, a variety of biomarkers were proposed and evaluated for second trimester screening. At the time of writing, the standard second trimester serum screen in use is the so-called quad screen—testing for (1) serum AFP, (2) free beta hCG, (3) unconjugated estriol, and (4) inhibin A. The FASTER trial, a multi-institutional trial of Down syndrome screening that tested over 35,000 singleton pregnancies, found a sensitivity of 83% for the quad screen at a 5% false-positive rate [4]. For a recent review, see Ref. [5].

In the original description of Down syndrome (the eponym proposed by the WHO in the 1960s over the ethnically-biased term Mongoloid) by Langston Down in 1866, Dr. Down noted that, among other features, the skin of individuals with Down syndrome "…appeared to be too large for the body…" [6]. In the 1990s, it was proposed that this could be due, at least in part, to the accumulation of fluid and edema of the back of the fetal neck, observed by ultrasound imaging as an increase in nuchal translucency (NT) [7]. Increased NT has since proven to be a sensitive indicator of the presence of a fetus with trisomy 21 (T21, Down syndrome). However, it is not completely specific. NT is also increased in a number of other fetal anomalies and genetic conditions, including Turner syndrome, cardiac defects, and hydrops. Since the optimal gestational age for the detection of Down syndrome by NT measurement is between 11 and 14 weeks, there was incentive to identify biochemical markers, such as those used in the quad test, that can be used in the first trimester and be used in conjunction with the ultrasound NT measurements. Of the four quad screen markers, only free beta-hCG correlates with Down syndrome risk in the first trimester. However, an additional biomarker, pregnancy-associated plasma protein A (PAPP-A) was shown to also be associated with Down syndrome in the first trimester. At the time of writing, the standard first trimester screen consists of a first trimester serum screen (free β-hCG and PAPP-A) plus NT measurement (in locations that staff experienced sonographers) [7].

The multi-institutional FASTER trial was designed to test the first and second trimester screens, in various

© 2017 Elsevier Inc. All rights reserved.

combinations, and provide data on each screening approach from a large cohort of prospectively followed women. The study concluded that detection of serum biomarkers, when performed in both the first and second trimester and combined with ultrasound, could achieve sensitivities for the detection of Down syndrome of 90–95% with a specificity of 95–97% (3–5% false positive). These markers therefore provided very good screening approaches to Down syndrome, but there was room for improvement [4].

It should be noted that the sensitivity of all of the screening modalities could be increased by decreasing the thresholds for positive results However, the great majority of these high-risk cases would test negative on definitive testing by amniocentesis or chorionic villus sampling (CVS). The 5% cutoff is a generally agreed upon compromise between the sensitivity of the screen and the number of women who could reasonably receive good genetic counseling and elect for an invasive procedure.

THE MOLECULAR TARGET: APPROACHES USING PLASMA NUCLEIC ACIDS

In 1997, Lo et al. demonstrated the existence of placental DNA in maternal circulation [8]. They immediately realized the potential of this discovery for prenatal diagnostics. The Lo laboratory, at the Chinese University of Hong Kong, and many others began intensive efforts to utilize this resource.

The application of circulating cell-free fetal DNA (cffDNA) to the prenatal detection of Down syndrome was presented in two ground-breaking publications that appeared 2 months apart in the journal *Proceedings of the National Academy of Sciences*: Fan et al., from the Quake laboratory at Stanford University, appeared in the October 21, 2008 issue [9] and Chui et al., from the Lo laboratory and Sequenom, Inc., was published in the December 23, 2008 issue [10]. Both studies utilized similar methodology—shotgun massively parallel sequencing (MPS; also commonly termed next-generation sequencing (NGS)) of DNA isolated from the plasma of pregnant women. Both studies sequenced unselected DNA fragments from plasma and mapped the reads to specific loci in the human genome. After mapping, the reads per chromosome were simply counted and ratios of the counts per chromosome versus total counts for all the chromosomes were calculated. If the woman was carrying a euploid fetus, then the ratios of reads on each chromosome to each other and to the total would be the same as either DNA from

nonpregnant plasma, or even genomic DNA. However, if the fetus had an additional copy of chromosome 21, then the relative number of hits on chromosome 21 was recorded as higher than normal. Both groups calculated Z-scores to quantify the relative ratios of hits on chromosome 21 to the total. Both groups found that the variance of the method was a function of the GC content of each chromosome, which was near a minimum for chromosome 21 [9,10]. Both groups turned to private companies to commercialize their findings. The Stanford group licensed their intellectual property to a new start-up company, Verinata (later acquired by Illumina, Inc.). The Illumina test is branded as the Verify test. The Hong Kong group licensed their intellectual property to a publically traded company, Sequenom, Inc. The Sequenom test is known as the MaterniT21 test.

Two other companies developed technology to utilize MPS and cffDNA from maternal circulation. Ariosa (originally Aria, Inc.) launched a test, branded as the Harmony Prenatal test, which differs from the Verify and MaterniT21 tests in that Ariosa applies a capture step and sequences DNA that is greatly enriched for DNA sequences from chromosomes 13, 18, and 21. Natera, a Silicon Valley company that had previously focused on preimplantation diagnosis, added a prenatal aneuploidy test, Panorama, to its test menu. The Panorama test differs in that it interrogates several thousand targeted single-nucleotide polymorphisms (SNPs) on chromosomes 13, 18, 21, X, and Y and uses a proprietary algorithm to identify aneuploidy and copy number variation.

As one might expect in a new, potentially lucrative, field featuring young companies with similar methodologies, there were immediate claims of intellectual property infringement from all four companies. A discussion of intellectual property and the current landscape of patent litigation is beyond the scope of this chapter and well beyond the expertise of its authors. At the time of writing, Illumina and Sequenom had settled their outstanding lawsuits and pooled their intellectual property investments. Litigation against and between Ariosa and Natera continues. In addition, the four primary companies have begun aggressive licensing campaigns to allow other testing companies to rebrand the primary tests for distribution. LabCorp, Inc., for instance, now offers an in-house test called InformaSeq, and Counsyl (a preconception screening company) offers a version of Verify. Although there are now additional companies offering MPS-based prenatal screening for fetal anueploidy internationally, the following discussion will focus on the four largest companies based in the United States.

MOLECULAR TECHNOLOGIES AND CLINICAL UTILITY: FOUR APPROACHES BY FOUR COMMERCIAL LABORATORIES

Sequenom

Sequenom, Inc. (San Diego, CA) was the first entity to launch a commercial MPS test for Down syndrome screening in the United States. Sequenom (as well as the other three US companies) has presented a substantial validation of their test's performances in a series of publications in the peer-reviewed literature. The initial trial consisted of 480 high-risk women (high risk is typically defined as: positive serum screening test, maternal age >35 years, ultrasound findings, or history of pregnancy affected with Down syndrome). After eliminating samples that had low plasma volume, processing errors, or quality control issues, Sequenom analyzed 449 samples. All cases had a diagnostic amniocentesis with karyotyping. Forty samples were called positive for T21 by the MPS assay, 39 of which were confirmed by karyotyping. One sample was shown to be euploid (false positive). Four hundred and nine samples were called normal, none of which were false negatives. The calculated sensitivity of 100% and specificity of 99.7% demonstrated the large improvement of plasma DNA-based testing over serum and ultrasound-based screening methods [11]. Shortly after this study appeared, a larger, international case-control study with 212 Down syndrome cases and 1484 euploid births was published. The sensitivity and specificity reported in this trial were 98.6% and 99.8% [12]. The same group mined the sequencing data to show that the MPS method reliably detected fetal trisomies for chromosomes 13 (T13, Patau syndrome) and 18 (T18, Edwards syndrome). The sensitivity and specificity for T13 ($n = 12$) were 91.7% and 99%, respectively, and 100% and 99.7% for T18 ($n = 59$) [13]. These results confirmed the applicability of MPS of cell-free plasma DNA to T13 and T18 previously reported by Fan et al. [9] and the Lo laboratory [14]. The commercial test, MaterniT21, was expanded to include testing for all three trisomies. Shortly thereafter, Sequenom demonstrated that trisomies could be detected in pregnancies with multiple gestations [15]. Using an improved methodology, which featured the use of robotics for sample preparation, higher level multiplexing in the MPS, and improved bioinformatics in cohorts of 1587 (blinded) and 1269 (unblinded) samples, Sequenom reported 100% sensitivity for the detection of all three trisomies, with false-positive rates of less than 0.1% [16]. Finally, in 2013, Sequenom added testing for sex chromosome aneuploidies (SCAs) [17]. A second international collaborative trial of 137 fetuses with T21, 39 with T18, 16 with T13, and 15 with SCAs reported 100% sensitivity for T21 and a specificity of 99.9%. Sensitivity and specificity were 92.3% and 100% for T18, and 87.5% and 100% for T13. All 15 SCA cases were detected. However, there were 11 false-positive results for 45, X. All of the pregnancies reported in this paper were singleton and 54 cases with complex karyotypes on amniocentesis were excluded [18].

Illumina (Formerly Verinata Health, Inc.)

Investigators at Verinata Health extended the work of Fan et al. [9] and published a proof-of-concept paper featuring an improved data analysis strategy in which the counts from each chromosome of interest (13, 18, 21, X, and Y) were normalized, not to the total number of counts in the genome, but to specific denominator chromosomes which were matched in terms of sequencing efficiency to the chromosome of interest. This maneuver decreased the variance of the method and allowed detection of aneuploidies for five chromosomes (13, 18, 21, X, and Y). After analyzing a pilot set of 71 samples (26 aneuploid), the group tested their method on a set of 48 samples (27 aneuploid). Trisomies of 21 ($n = 13$) and 18 ($n = 8$) were all correctly identified. The sole T13 case could not be interpreted and yielded a no-call result. Two of the three 45, X cases were identified, and the third was a no-call [19].

To further characterize the MPS methodology, Verinata funded a prospective, multicenter observational study with a blinded, nested, case-control analysis. Between June 2010 and August 2011, the MELISSA trial (MatErnal bLood IS Source to Accurately diagnose fetal aneuploidy) recruited women who were undergoing an invasive procedure for fetal karyotyping and had at least one of the following high-risk criteria: (1) age 38 years or older, (2) positive serum and/or ultrasound screening for Down syndrome, (3) ultrasound finding suggestive of an aneuploidy, or (4) a prior aneuploid pregnancy. Blood samples were collected from 2882 women. Two thousand six hundred and twenty-five samples were eligible for analysis and 221 had abnormal karyotypes. In keeping with the predefined statistical plan, 534 samples were selected for sequencing (two were excluded due to sample tracking errors). After sequencing, each sample was analyzed for six categories: aneuploidy for chromosomes 13, 18, and 21; fetal sex, male or female; and monosomy X. The bioinformatics analyses were as described by Sehnert et al. [20], but with new normalizing chromosomes (due to the different instrumentation used in the larger study) and more samples multiplexed together [20]. Of 89 nonmosaic T21 karyotypes, the MPS method detected 100% and found no false positives in the 404 samples without T21. Interestingly, three of three mosaic T21 cases were also identified as T21. Seven samples did not return a result and were unclassified. With regard to T18, all 37 cases were detected, five samples were unclassified, and there was one false positive in 461 non-T18 samples. Eleven of 14 T13 cases were detected and three false positives were called in 488 non-T13 samples (two samples were unclassified). Of 433 samples, all were correctly called male or not male, two were miscalled female or not female. Fifteen of 20 cases of Turner syndrome were correctly identified, there was one false-positive 45XO of 417 cases with normal sex chromosomes. Forty-nine cases were unclassified for the monosomy X analysis [21].

Bianchi et al. [21] noted the very high, but not perfect, sensitivity and specificity of this method for aneuploidy of chromosomes 13, 18, 21, fetal sex, and monosomy X identification and emphasized that theirs is a screening, rather than diagnostic, test. However, they observed that this technology may well reduce the number of invasive procedures performed due to the poor specificity of the currently available serum and ultrasound markers. They also urged caution in extrapolating these results to low-risk pregnancies and recommended detailed genetic counseling prior to testing. It is useful to note that this study excluded pregnancies with multiples, as well as some complex karyotypes that were identified prior to MPS testing. Some of these karyotypes included balanced and unbalanced translocations, trisomies of other chromosome (16 and 22), and triploidy. These observations highlight the fact that, while extremely powerful and useful, plasma DNA screening will not substitute for, nor eliminate, invasive testing [21].

In a follow-up study, Bianchi and colleagues reported separately on a group of 113 cases from the MELISSA study that had cystic hygromas observed on ultrasound. The authors confirmed that the MPS method had high sensitivity and specificity in this group, 29 of 30 karyotype T21 cases were confirmed, as were 20 of 21 Turner syndrome (45, X) cases. Of the 44 cases that did not have abnormal karyotypes after invasive procedures, there were no false positives. Overall, the fraction of cases with cystic hygroma that had chromosome abnormalities was 61%, which is in accord with previous literature [22].

Based on the results of the MELISSA study, Verinata launched its version of an MPS cell-free DNA aneuploidy test in February of 2012. In 2013, the laboratory presented a summary of its clinical testing of almost 6000 samples. Aneuploidy was reported in 284 cases, for which confirmation by karyotype was available in 77. The fraction of submitted samples that were positive for aneuploidy was 4.8%, which was very close to

that seen in the MELISSA study, indicating that the test was being used appropriately in high-risk populations. As in the MELISSA study, the positive rate (30%) was much greater than that seen in the general population. However, it was approximately half the rate seen in the trial. Verinata includes an unclassified zone between the chromosome count ratios that define high risk or low risk for trisomy. In the follow-up study, there were 170 cases with an individual chromosome that were unclassified. Interestingly, there were unusual histories or negative outcomes, including co-twin demise, false-negative trisomies, pregnancy loss, and severe ultrasound abnormalities, in 51% of the unclassified cases, suggesting that there may be useful information in the unclassified condition [23].

In response to calls for evidence of MPS test performance in the general obstetrical practice, as opposed to high-risk women only, and several publications on this population in non-US cohorts, Illumina funded a second, prospective, blinded, multiinstitutional, study—the Comparison of Aneuploidy Risk Evaluations (CARE) study. The purpose of the CARE study was to compare the performance characteristics of the MPS test to the standard first and second trimester serum and ultrasound methods in the general, or low-risk, obstetrical population. The study design allowed for cffDNA specimen collection in the third trimester if the patient had completed either first or second trimester standard screening. Five hundred and forty-four samples of the total 1914 eligible samples were collected in the third trimester. Both the standard and MPS methods detected all cases of T21 ($n = 5$), T18 ($n = 2$), and T13 ($n = 1$), for a sensitivity and negative predictive value (NPV) of 100%. However, the MPS test performed with better specificity: there were 6 false positives in the MPS group for T21 and 69 for the standard testing (0.3% vs 3.6%). Thus, the positive predictive value (PPV) was 45% with MPS and only 4% with standard screening. Performance was similar for aneuploidy of chromosomes 13 and 18. The authors concluded that the performance of the MPS assay was equivalent in a low-risk population to that seen by multiple studies in high-risk ones. They note that the increase in PPV test by 10-fold would reduce the theoretical invasive procedure volume by almost 90%. The study concluded that plasma DNA testing "...merits serious consideration as a primary screening method for fetal aneuploidy..." [24].

Ariosa Diagnostics (Previously Aria Diagnostics)

The MPS assays offered by Sequenom and Illumina have some differences, primarily in their bioinformatic analyses, but they take a similar overall approach—total cell-free plasma DNA is isolated, made into a library, and sequenced without manipulation or attempts to target specific regions of the genome. By contrast, investigators at Aria Diagnostics specifically targeted chromosomes 18 and 21. In a proof-of-concept paper, they outlined their method, which they termed digital analysis of selected regions (DANSR). In this method, 384 regions on each chromosome are designated a set of three oligonucleotides, two outside and one middle probe, that are designed to hybridize to the region in a head-to-tail configuration. After hybridization, a DNA ligase is added and the probes that specifically hybridize to their complementary genomic sequences are ligated together. Because the two outside oligonucleotides are tagged with universal PCR primer sequences (and one of them additionally tagged with a multiplexing index tag), the ligation mix can be amplified by PCR, while simultaneously adding the Illumina sequencing tags. The mixture is then sequenced on an Illumina sequencer, such as the HiSeq 2000. In this approach, if the mother carries a fetus with a trisomy of either chromosome 18 or 21, the ratio of counts would be disturbed, and the magnitude of the shift can be measured with a Z-score statistic. In the proof-of-principle experiment, the Aria investigators evaluated samples from 289 pregnancies. All 39 T21 cases and all 7 T18 cases were identified. Although accuracy increased with increasing read depth (ie, decreased number of samples multiplexed in a sequencing lane), the effect was very small after a certain depth (420,000 total counts). The authors noted that it would be possible to multiplex as many as 96 samples in one lane [25]. In a follow-up paper, Aria scientists included polymorphic loci on chromosomes 18 and 21 in order to calculate the percentage of the total cell-free DNA that was fetal in origin (fetal fraction). Fetal fraction is an important parameter because the accurate call rate for all MPS methods varies with fetal fraction. The authors introduced a novel analysis algorithm, named fetal-fraction optimized risk of trisomy evaluation (FORTE), which includes fetal fraction and outputs the result in terms of a probability of having either T18 or T21. A cohort consisting of 250 normal, 72 T21, and 16 T18 cases, all confirmed by invasive procedure and karyotyping, were split into training and validation sets. After optimization of the FORTE algorithm in the training set, all of the trisomies and all of the normal samples in the validation set were correctly identified. The authors noted that this approach allows the combination of multiple parameters, such as fetal fraction and maternal age, into a final probability of the pregnancy being trisomic or euploid. They also emphasized that the ability to highly multiplex the sequencing leads to markedly lower costs than the shotgun sequencing methods [26].

To further characterize their assay, Ariosa funded a prospective, multiinstitutional, international, blinded, cohort study—the noninvasive chromosomal evaluation (NICE) study. Samples were collected from 4002 pregnant women who were planning to have invasive diagnostic for any reason. After setting aside approximately 400 samples for method development, and excluding ineligible samples, 3080 paired DANSR-FORTE analyses/karyotype pairs were analyzed. All of the 81 T21 cases were identified, and there was one false positive in the 2888 karyotype normal samples. Thirty-seven of 38 T18 cases were identified, and there were 2 false positives in 2888 normal samples. For the majority of samples, there was a greater than 100,000-fold magnitude separation in the probabilities of trisomy versus euploidy. In 17 cases (0.5% of the total), intermediate risk scores were obtained. However, these were typically large enough to make a high-risk/low-risk call. However, in the end the authors of the NICE study returned to the theme raised by their counterparts in other validation studies, noting that almost 40% of the abnormal karyotypes seen in the study had abnormalities other than T18 or T21. They reiterated that cffDNA analysis for aneuploidy is a screening test, and not a replacement for invasive testing and karyotyping or chromosome microarray [27].

Reports extending this methodology to chromosome 13, and the sex chromosomes X and Y, soon followed [28,29]. However, these authors raised important biological difficulties in calling fetal SCAs.

Natera

Natera uses a different approach for the detection of aneuploidy. All the methods discussed so far involve counting MPS reads that map to chromosomes of interest and utilize a ratio of those counts to the number of counts obtained from one or more different chromosomes. As described in the proof-of-concept paper by Zimmermann et al. [30], Natera utilized MPS to sequence and obtain plasma DNA, maternal, and paternal genotypes from a set of 11,000 multiplexed PCR products, each of which span well-characterized SNP loci. After sequencing, alignment, and SNP calling, Natera's method employs a proprietary algorithm, termed Parental Support. This algorithm uses parental genotypes and the information is then used to calculate theoretical mixtures of maternal and fetal genotypes at various fetal fractions if each chromosome of interest was monosomic, disomic, or trisomic. Each of these genotypes, or hypotheses, is compared to what is observed from the cffDNA. The hypothesis with the maximum likelihood is selected as the copy number and fetal fraction of the sample. Importantly, the Parental Support algorithm calculates the accuracy of its prediction by comparing the observed distribution of alleles and determining the likelihood of the predicted distribution. In this study, 166 samples were tested, including 11 T21, 3 T18, 2 T13, one 45, X, two 47, XXY, 57 normal female fetuses, and 69 normal male fetuses. Twenty-one samples did not pass the rigorous quality control metrics for inclusion. The remaining 145 samples were tested for five different chromosomes: 13, 18, 21, X, and Y, generating 725 total chromosome results. All results were correct, yielding a sensitivity and specificity of 100% [30]. The authors concluded that the Parental Support method was a promising method for prenatal detection of fetal aneuploidies. However, they noted that confined placental mosaicism is a fundamental biological issue that can confound any method based on cffDNA. In addition, the authors commented that their 12.6% no-call rate was high. However, these no-calls were on samples that did not pass QC metrics such as DNA quality or concentration. Apparently, there were no samples with sufficient DNA and fetal fraction that did not return a result. Studies evaluating other methods screened out samples of insufficient quality as well.

An improvement to the Parental Support algorithm termed NATUS (next-generation sequencing aneuploidy test using SNPs) was published in a paper focused on the detection of the rarest of the common trisomies, T13. The new algorithm performed flawlessly, identifying all 17 T13 cases and 51 age-matched controls [31]. In 2013, Nicolaides et al. [32] employed the NATUS algorithm in additional validation studies. Samples were collected from 242 pregnant women who were planning invasive diagnostic procedures and the number of multiplexed SNP assays was increased from 11,000 to 19,488. The NATUS algorithm identified all cases of T13, T18, T21, and 45, X with no false positives. Fetal sex was correctly called in all cases. One case of triploidy was included in this series and multiple alleles were detected, consistent with either twins or triploidy. After ultrasound confirmed a single fetus, the diagnosis of 69, XXX was correctly made. The SNP-based approach is the only current method that reliably detects triploidy in cffDNA. Interestingly, while the test did not identify complex karyotypes ($n = 5$), the copy number enumeration for the chromosomes of interest was all correct, with no interference [32].

A further study by Natera included samples from women who were not at high risk for having a baby with an aneuploidy. In this study, 1064 samples were collected, roughly half from high-risk and half from low-risk women. As in previous studies, the sensitivity and specificity for the detection of T21 ($n = 58$) and T13 ($n = 12$) were 100%. However, the NATUS method had one false positive and one false negative each for T18 and monosomy X. The false-negative T18 was shown to be due to mosaicism in the placenta. Although the number of aneuploidy pregnancies was significantly less that in the high-risk group, all five (1 T21, 2 T13, 2 45, X) were correctly identified, leading the investigators to conclude that the sensitivity and specificity of the method was the same in both groups [33].

Natera has published its clinical experience after receiving 31,030 samples for noninvasive prenatal testing (NIPT) for aneuploidy. Of these, 1966 failed sample QC, and 28,739 received a result. Approximately 38% of the samples were referred from outside partner laboratories, so follow-up clinical or karyotype information was not available. Almost 18,000 samples were available for clinical follow-up. Of these, 356 cases with an aneuploidy were identified and data regarding invasive testing results was available in 222 (62.4%; data was not available for cases where invasive testing was declined, or for the spontaneous fetal demise cases, or elective terminations). Of the 222 cases with confirmed outcomes, 184 (83%) were true positives and 38 (18%) were false positives. The PPV test was very good for T21 (91%) and T18 (93%), but lower for T13 (38%) and monosomy X (50%). The report on live clinical samples therefore demonstrated some degradation in performance compared to earlier validation studies. Nevertheless, the performance of the NATUS methodology is much better than current serum and ultrasound testing and is in-line with the performance of tests using counting methods [34].

There were a number of interesting observations made in this study. For instance, approximately 30% of cases included a paternal sample. Inclusion of the paternal sample was associated with a decreased sample QC failure rate. As noted by others, the authors observed that the percent fetal fraction was inversely proportional to maternal body mass index and directly proportional to gestational age.

LIMITATIONS OF TESTING

The principle limitation of noninvasive cffDNA-based prenatal screening is the fact that this is a screen for specific aneuploidies and not a definitive or diagnostic test. Thus, the term noninvasive prenatal screening (NIPS) is preferred over the original term NIPT.

Additional limitations are associated with the fact that circulating fetal DNA in maternal circulation is actually a misnomer. The pregnancy-derived DNA in maternal circulation is predominantly placental DNA [8].

Thus, there is a theoretical limit on clinical sensitivity given by the population frequency of confined placental mosaicism (CPM), in which the fetus and placenta have differing karyotypes. Because CVS, an invasive diagnostic test, directly samples the placenta, a review of the CVS literature reveals the levels of CPM of single trisomic chromosomes expected in the population. A review of over 90,000 CVS cases in Europe between 1986 and 1994 showed mosaicism or results discordant with the fetus in 1415 cases (1.5%). Of these, approximately half (0.75%) had mosaic or discordant results for one trisomic chromosome. Of 192 cases that were extensively evaluated, including karyotyping at least three different cells types (fetal and placental), 42 (22%) involved chromosomes 13, 18, or 21 [35]. Thus, the minimal level of CPM expected for cffDNA screening, potentially resulting in false-positive or false-negative results, is approximately 0.16% (0.75% $\times$ 22%). Because all NIPS methods described here interrogate circulating placental DNA, this figure is likely to represent a hard limit on the sensitivity and specificity of the assays. Hence, claims of greater than 99.9% sensitivity and specificity are therefore unlikely to be correct.

Clearly, a test with sensitivity and specificity of 99% is extremely good. Nevertheless, the truly useful measures, the PPV or NPV test, are functions of both the specificity and sensitivity, respectively, and the incidence of the condition being tested. In the context of trisomies, the incidence of trisomies for chromosomes 13, 18, and 21 is very different. For example, in the CARE study, the sensitivity of the NGS method for the detection of T21 and T18 was 100%, giving an NPV test of 100%. However, there were false positives for both T21 and T18, resulting in PPVs of 45.5% and 40.0%, respectively [24]. The Natera study found a better PPV for T18 than the CARE study, but a very similar one for T13 [34]. These results emphasize the fact that NIPS is a screening test, and that follow-up diagnostic testing is essential. Even more essential is appropriate pretest and posttest genetic counseling.

False-positive cffDNA screening results have been reported in the peer-reviewed literature. Typically, these are case reports [36] or small series from one or two collaborating institutions [37]. At the time of writing, the largest series of cffDNA screening results discordant with follow-up karyotyping or chromosome microarray is a series of 109 cases reported from the commercial reference laboratory, Quest Diagnostics. Platform presentations at national genetics conferences [38,39] prompted Quest investigators to review 109 consecutive cases submitted to their laboratory for confirmation of a positive NIPS test results. One advantage of this report is inclusion of cases from all four commercial laboratories providing NIPS. This paper reported that the positive rate across all providers was 93% for T21 (38/41 NIPS positive for T21), 64% for T18 (16/25), 44% for T13 (7/16), and only 38% for SCAs. Results discordant with NIPS findings were observed for monosomy 21, trisomy 16, triploidy, and 22q11.2 microdeletion syndrome. Interestingly, only one discordant case was due to a low-level mosaic fetus, and confined placental mosaicism was confirmed in only two cases. The cause of the majority of false-positive results was unexplained. The authors note that the observed true-positive rate was similar to that predicted for tests with 99.9% sensitivity and specificity in a population of 38-year-old women based on the incidence of the trisomies at birth The observed true positive rate of 93% for T21 versus the predicted 84% (for an incidence of Down syndrome of 1 in 185 births) may be due to additional selection via serum screening. The authors emphasized that it is critical to educate providers regarding the difference between sensitivity/specificity, which are analytical properties of the assays, and predictive value of test results, which is strongly influenced by incidence. They noted that many physicians expect false-positive rates of less than 1% for tests with sensitivities greater than 99%. However, this is not the case for rare disorders, such as T13 and T18 [40].

CONCLUSIONS AND FUTURE DIRECTIONS

The field of prenatal screening utilizing cffDNA in maternal circulation is rapidly evolving. Although it is difficult to estimate the number of tests performed in the United States at this time, conservative estimates are at greater than 200,000 per year. In addition, the number and type of disorders included in cffDNA-based screening is growing. Screening currently includes not only the common trisomies for chromosome 13, 18, and 21, but also rare trisomies such as T16 and T22. In addition, copy number variation for sex chromosome (Klienfelter and Turner syndromes, and related conditions) is now routinely included in NIPS results, as well as for subchromosomal deletions, such as those responsible for DiGeorge, Wolf-Hirschhorn, and Cri-du-chat syndromes. How providers and patients react to these increased offerings, and whether they can be place into proper context of the predictive values for increasingly rare disorders remains to be determined. Also remaining to be seen is whether there will be a paradoxical increase in the number of invasive procedures (CVS and amniocentesis) to confirm the expected increase number of positive results for the additions conditions.

This is one of the few, if not the only, widespread clinical tests that remain (at this writing) solely in the providence of commercial laboratories that are focused (exclusively or almost exclusively) on this single test. The influence of corporate advertising on the analytical sensitivity and specificity versus the more relevant predictive values is an open question, as is the rapid expansion of the testing to include conditions other than Down syndrome or the common trisomies. This rapid expansion is reminiscent of the arms race for ever-larger numbers of mutations in cystic fibrosis carrier screening panels [41].

What is unequivocal at this point is that the NIPS approach is superior to the serum and ultrasound-based screening for detection of Down syndrome. Although this, in and of itself, raises crucial ethical considerations regarding the increasing importance of pretest genetic counseling and how to deliver such in increasingly busy clinical practices, there is little doubt that, if the price of the testing is competitive with serum-based testing, the plasma nucleic acid approach is very likely to replace current prenatal screening algorithms for Down syndrome.

References

[1] Macri JN, Weiss RR, Tillitt R, Balsam D, Elligers KW. Prenatal diagnosis of neural tube defects. JAMA 1976;236:1251−4.

[2] Cuckle HS, Wald NJ, Lindenbaum RH. Maternal serum alpha-fetoprotein measurement: a screening test for Down syndrome. Lancet 1984;1:926−9.

[3] Hershey DW, Crandall BF, Perdue S. Combining maternal age and serum alpha-fetoprotein to predict the risk of Down syndrome. Obstet Gynecol 1986;68:177−80.

[4] Malone FD, Canick JA, Ball RH, Nyberg DA, Comstock CH, Bukowski R, et al. First-trimester or second-trimester screening, or both, for Down's syndrome. N Engl J Med 2005;353:2001−11.

[5] Chasen ST. Maternal serum analyte screening for fetal aneuploidy. Clin Obstet Gynecol 2014;57:182−8.

[6] Dunn PM. Dr Langdon Down (1828−1896) and 'mongolism'. Arch Dis Child 1991;66:827−8.

[7] Nicolaides KH. Screening for fetal aneuploidies at 11 to 13 weeks. Prenat Diagn 2011;31:7−15.

[8] Lo YM, Corbetta N, Chamberlain PF, Rai V, Sargent IL, Redman CW, et al. Presence of fetal DNA in maternal plasma and serum. Lancet 1997;350:485−7.

[9] Fan HC, Blumenfeld YJ, Chitkara U, Hudgins L, Quake SR. Noninvasive diagnosis of fetal aneuploidy by shotgun sequencing DNA from maternal blood. Proc Natl Acad Sci USA 2008;105:16266−71.

[10] Chiu RW, Chan KC, Gao Y, Lau VY, Zheng W, Leung TY, et al. Noninvasive prenatal diagnosis of fetal chromosomal aneuploidy by massively parallel genomic sequencing of DNA in maternal plasma. Proc Natl Acad Sci USA 2008;105:20458−63.

[11] Ehrich M, Deciu C, Zwiefelhofer T, Tynan JA, Cagasan L, Tim R, et al. Noninvasive detection of fetal trisomy 21 by sequencing of DNA in maternal blood: a study in a clinical setting. Am J Obstet Gynecol 2011;204:e1−e11, 205.

[12] Palomaki GE, Kloza EM, Lambert-Messerlian GM, Haddow JE, Neveux LM, Ehrich M, et al. DNA sequencing of maternal plasma to detect Down syndrome: an international clinical validation study. Genet Med 2011;13:913−20.

[13] Palomaki GE, Deciu C, Kloza EM, Lambert-Messerlian GM, Haddow JE, Neveux LM, et al. DNA sequencing of maternal plasma reliably identifies trisomy 18 and trisomy 13 as well as Down syndrome: an international collaborative study. Genet Med 2012;14:296−305.

[14] Chen EZ, Chiu RW, Sun H, Akolekar R, Chan KC, Leung TY, et al. Noninvasive prenatal diagnosis of fetal trisomy 18 and trisomy 13 by maternal plasma DNA sequencing. PLoS One 2011;6:e21791.

[15] Canick JA, Kloza EM, Lambert-Messerlian GM, Haddow JE, Ehrich M, van den Boom D, et al. DNA sequencing of maternal plasma to identify Down syndrome and other trisomies in multiple gestations. Prenat Diagn 2012;32:730−4.

[16] Jensen TJ, Zwiefelhofer T, Tim RC, Dzakula Z, Kim SK, Mazloom AR, et al. High-throughput massively parallel sequencing for fetal aneuploidy detection from maternal plasma. PLoS One 2013;8:e57381.

[17] Mazloom AR, Dzakula Z, Oeth P, Wang H, Jensen T, Tynan J, et al. Noninvasive prenatal detection of sex chromosomal aneuploidies by sequencing circulating cell-free DNA from maternal plasma. Prenat Diagn 2013;33:591−7.

[18] Porreco RP, Garite TJ, Maurel K, Marusiak B, Ehrich M, van den Boom D, et al. Noninvasive prenatal screening for fetal trisomies 21, 18, 13 and the common sex chromosome aneuploidies from maternal blood using massively parallel genomic sequencing of DNA. Am J Obstet Gynecol 2014;211:e1−e12 365.

[19] Rosenberg S, Elashoff MR, Beineke P, Daniels SE, Wingrove JA, Tingley WG, et al. Multicenter validation of the diagnostic accuracy of a blood-based gene expression test for assessing obstructive coronary artery disease in nondiabetic patients. Ann Intern Med 2010;153:425−34.

[20] Sehnert AJ, Rhees B, Comstock D, de Feo E, Heilek G, Burke J, et al. Optimal detection of fetal chromosomal abnormalities by massively parallel DNA sequencing of cell-free fetal DNA from maternal blood. Clin Chem 2011;57:1042−9.

[21] Bianchi DW, Platt LD, Goldberg JD, Abuhamad AZ, Sehnert AJ, Rava RP. Genome-wide fetal aneuploidy detection by maternal plasma DNA sequencing. Obstet Gynecol 2012;119:890−901.

[22] Bianchi DW, Prosen T, Platt LD, Goldberg JD, Abuhamad AZ, Rava RP, et al. Massively parallel sequencing of maternal plasma DNA in 113 cases of fetal nuchal cystic hygroma. Obstet Gynecol 2013;121:1057−62.

[23] Futch T, Spinosa J, Bhatt S, de Feo E, Rava RP, Sehnert AJ. Initial clinical laboratory experience in noninvasive prenatal testing for fetal aneuploidy from maternal plasma DNA samples. Prenat Diagn 2013;33:569−74.

[24] Bianchi DW, Parker RL, Wentworth J, Madankumar R, Saffer C, Das AF, et al. DNA sequencing versus standard prenatal aneuploidy screening. N Engl J Med 2014;370:799−808.

[25] Sparks AB, Wang ET, Struble CA, Barrett W, Stokowski R, McBride C, et al. Selective analysis of cell-free DNA in maternal blood for evaluation of fetal trisomy. Prenat Diagn 2012;32:3−9.

[26] Sparks AB, Struble CA, Wang ET, Song K, Oliphant A. Noninvasive prenatal detection and selective analysis of cell-free DNA obtained from maternal blood: evaluation for trisomy 21 and trisomy 18. Am J Obstet Gynecol 2012;206:e1−9, 319.

[27] Norton ME, Brar H, Weiss J, Karimi A, Laurent LC, Caughey AB, et al. Non-Invasive Chromosomal Evaluation (NICE) Study: results of a multicenter prospective cohort study for detection of fetal trisomy 21 and trisomy 18. Am J Obstet Gynecol 2012;207:e1−8, 137.

[28] Ashoor G, Syngelaki A, Wang E, Struble C, Oliphant A, Song K, et al. Trisomy 13 detection in the first trimester of pregnancy using a chromosome-selective cell-free DNA analysis method. Ultrasound Obstet Gynecol 2013;41:21−5.

[29] Hooks J, Wolfberg AJ, Wang ET, Struble CA, Zahn J, Juneau K, et al. Non-invasive risk assessment of fetal sex chromosome aneuploidy through directed analysis and incorporation of fetal fraction. Prenat Diagn 2014;34:496−9.

[30] Zimmermann B, Hill M, Gemelos G, Demko Z, Banjevic M, Baner J, et al. Noninvasive prenatal aneuploidy testing of chromosomes 13, 18, 21, X, and Y, using targeted sequencing of polymorphic loci. Prenat Diagn 2012;32:1233−41.

[31] Hall MP, Hill M, Zimmermann B, Sigurjonsson S, Westemeyer M, Saucier J, et al. Non-invasive prenatal detection of trisomy 13 using a single nucleotide polymorphism- and informatics-based approach. PLoS One 2014;9:e96677.

[32] Nicolaides KH, Syngelaki A, Gil M, Atanasova V, Markova D. Validation of targeted sequencing of single-nucleotide polymorphisms for non-invasive prenatal detection of aneuploidy of chromosomes 13, 18, 21, X, and Y. Prenat Diagn 2013;33:575−9.

[33] Pergament E, Cuckle H, Zimmermann B, Banjevic M, Sigurjonsson S, Ryan A, et al. Single-nucleotide polymorphism-based noninvasive prenatal screening in a high-risk and low-risk cohort. Obstet Gynecol 2014;124:210−18.

[34] Dar P, Curnow KJ, Gross SJ, Hall MP, Stosic M, Demko Z, et al. Clinical experience and follow-up with large scale single-nucleotide polymorphism-based noninvasive prenatal aneuploidy testing. Am J Obstet Gynecol 2014;211:e1−e17, 527.

[35] Hahnemann JM, Vejerslev LO. European collaborative research on mosaicism in CVS (EUCROMIC)—fetal and extrafetal cell lineages in 192 gestations with CVS mosaicism involving single autosomal trisomy. Am J Med Genet 1997;70:179−87.

[36] Clark-Ganheart CA, Iqbal SN, Brown DL, Black S, Fries MH. Understanding the limitations of circulating cell free fetal DNA: an example of two unique cases. J Clin Gynecol Obstet 2014; 3:38−70.

[37] Mennuti MT, Cherry AM, Morrissette JJ, Dugoff L. Is it time to sound an alarm about false-positive cell-free DNA testing for fetal aneuploidy? Am J Obstet Gynecol 2013;209:415−19.

[38] Choy KW, Kwok KY, Lau ET, Tang MH, Pursley A, Smith J, et al. Discordant karyotype results among non-invasive prenatal screening positive cases. Presented at American Society for Human Genetics 2013 national meeting. Boston, MA; 2013. <http://www.ashg.org/2013meeting/abstracts/fulltext/f130121408.htm/>.

[39] Meck JM, Dugan EK, Aviram A, Trunca C, Riethmaier D, Pineda-Alvarez D, et al. Non-invasive prenatal screening: a cytogenetic perspective. Presented at oral platform presentations: cytogenetics. 2014. <http://ww2.aievolution.com/acm1401/index.cfm?do=abs.viewAbs&abs=2009/>.

[40] Wang JC, Sahoo T, Schonberg S, Kopita KA, Ross L, Patek K, et al. Discordant noninvasive prenatal testing and cytogenetic results: a study of 109 consecutive cases. Genet Med 2015; 17:234−6.

[41] Grody WW, Cutting GR, Watson MS. The cystic fibrosis mutation "arms race": when less is more. Genet Med 2007;9:739−44.

17

Molecular Testing in Inherited Cardiomyopathies

H.M. McLaughlin[1,2,3] and B.H. Funke[1,2,3]

[1]Laboratory for Molecular Medicine, Partners Personalized Medicine, Boston, MA, United States [2]Department of Pathology, Harvard Medical School, Boston, MA, United States [3]Department of Pathology, Massachusetts General Hospital, Boston, MA, United States

INTRODUCTION

Inherited cardiomyopathies are a clinically and genetically heterogeneous group of disorders characterized by the disruption of normal structure and/or function of the myocardium [1]. Cardiomyopathies can lead to serious clinical complications including arrhythmia, thromboembolism, heart failure, and even sudden cardiac death (SCD). Indeed, undiagnosed cardiomyopathy accounts for a significant portion of SCD in young adults and athletes [2,3]. Cardiomyopathies primarily characterized by structural abnormalities of the myocardium include hypertrophic cardiomyopathy (HCM), restrictive cardiomyopathy (RCM), dilated cardiomyopathy (DCM), arrhythmogenic right ventricular cardiomyopathy (ARVC), and left ventricular noncompaction cardiopathy (LVNC). Together, structural cardiomyopathies are estimated to effect approximately 1/390 individuals [4]. However, this may be an underestimate because mildly affected individuals often remain undiagnosed [5]. This chapter will review the clinical features of the major inherited cardiomyopathies, outline the cellular and molecular mechanisms of disease associated with each disorder, and explore the clinical utility and limitations of the current genetic testing modalities being utilized.

MAJOR FORMS OF CARDIOMYOPATHY

Hypertrophic Cardiomyopathy

HCM is the most common inherited cardiomyopathy and is estimated to affect approximately 1/500 individuals [1]. HCM is characterized by maximal left ventricular wall thickness greater than or equal to 15 mm, myocyte disarray, and fibrosis [1,6,7] in the absence of other known causes of hypertrophy (eg, chronic hypertension). Clinically, HCM typically manifests between 20 and 40 years of age, but has also been identified in infants and in the elderly [8–10]. HCM is most commonly inherited in an autosomal dominant manner, although autosomal recessive forms have been identified (Table 17.1). Some storage disorders, such as Fabry disease or Danon disease, can present with apparently isolated left ventricular hypertrophy [13,14]. However, these disorders are typically accompanied by additional clinical features including acroparesthesia, angiokeratomas, sweating abnormalities, and renal disease in individuals affected with Fabry disease [15], and skeletal myopathy and intellectual disability in individuals affected with Danon disease [16]. In severe or end-stage cases of HCM, the phenotype may mimic DCM [17].

Restrictive Cardiomyopathy

RCM is a rare myocardial disorder characterized by myocardial stiffness, impaired left ventricular diastolic filling, and reduced diastolic volume in the presence of normal systolic function [18]. RCM is commonly caused by myocardial infiltration or fibrosis. However, RCM has been documented in the absence of these features and is usually referred to as idiopathic cardiomyopathy [19]. Idiopathic RCM is

Diagnostic Molecular Pathology
DOI: http://dx.doi.org/10.1016/B978-0-12-800886-7.00017-0

© 2017 Elsevier Inc. All rights reserved.

TABLE 17.1   Genes Implicated in Inherited Cardiomyopathies

| Gene | Protein | MOI | Cardiomyopathy | | | | | Protein function/ cellular location | Syndromic disorders** |
			HCM	RCM	DCM	ARVC	LVNC		
ABCC9	ATP-binding cassette, subfamily c, member 9	AD			X			Potassium channel	
ACTC1	Actin, alpha, cardiac muscle	AD	X	X	X		X	Sarcomere	
ACTN2	Actinin, alpha-2	AD	X		X			Z-disc	
ANKRD1	Ankyrin repeat domain-containing protein 1	Unknown	X		X			Z-disc	
BAG3	BCL2-associated athanogene 3	AD	X	X	X			Z-disc	Myofibrillar myopathy
					**				
CASQ2	Calsequestrin 2	AD					X	Sarcoplasmatic reticulum	
CAV3	Caveolin 3	AD	X		X			Plasma membrane	
CRYAB	Crystallin, alpha-B	Unknown			X			Chaperone	Myofibrillar myopathy
		AD/AR			**				
CSRP3	Cysteine- and glycine-rich protein 3	AD	X		X			Z-disc	
DES	Desmin	AD		X	X	X		Intermediate filament	Myofibrillar myopathy
		AD/AR			**				
DMD	Dystrophin	XL			**			Dystrophin-associated protein complex	Duchenne muscular dystrophy
DSC2	Desmocollin 2	AD			X	X		Desmosome	
DSG2	Desmoglein 2	AD			X	X		Desmosome	
DSP	Desmoplakin	AD			X	X		Desmosome	Carjaval syndrome
		AR			**	**			
DTNA	Dystrobrevin, alpha	AD					X	Dystrophin-associated protein complex	
EMD	Emerin	XL			**			Nuclear membrane	Emery–Dreifuss muscular dystrophy
FHL2	Four-and-a-half LIM domains 2	Unknown			X			Z-disc	
GATAD1	GATA zinc finger domain-containing protein 1	AR			X			Gene expression regulation	
GLA	Galactosidase, alpha	XL	**					Lysosome	Fabry disease
JUP	Junction plakoglobin	AD				X		Desmosome	
LAMA4	Laminin, alpha-4	Unknown			X			Basement membrane	
LAMP2	Lysosome-associated membrane protein 2	XL	**		**			Lysosome	Danon disease
LDB3	LIM domain-binding 3	AD	X		X		X	Z-disc	
LMNA	Lamin A/C	AD			X		X	Nuclear membrane	Myopathy, muscular dystrophy, lipodystrophy
		AD/AR			**				
MYBPC3	Myosin-binding protein C, cardiac	AD	X	X	X		X	Sarcomere	

(Continued)

TABLE 17.1 (Continued)

Gene	Protein	MOI	Cardiomyopathy					Protein function/cellular location	Syndromic disorders**
			HCM	RCM	DCM	ARVC	LVNC		
MYH6	Myosin, heavy chain 6, cardiac muscle, alpha	AD	X		X			Sarcomere	
MYH7	Myosin, heavy chain 7, cardiac muscle, beta	AD	X	X	X		X	Sarcomere	
MYL2	Myosin, light chain 2, regulatory, cardiac, slow	AD	X					Sarcomere	
MYL3	Myosin, light chain 3, alkali, ventricular, skeletal, slow	AD	X	X				Sarcomere	
MYLK2	Myosin light chain kinase 2	Unknown	X					Kinase	
MYOZ2	Myozenin 2	AD	X					Z-disc	
NEBL	Nebulette	Unknown			X			Z-disc	
NEXN	Nexilin (F actin binding protein)	Unknown	X		X			Z-disc	
PKP2	Plakophilin 2	AD			X	X		Desmosome	
PLN	Phospholamban	AD	X		X	X		Sarcoplasmatic reticulum	
PRKAG2	Protein kinase, AMP-activated, noncatalytic, gamma-2	AD	**					Kinase	Wolff-Parkinson-White syndrome, glycogen storage disease
RBM20	RNA-binding motif protein 20	AD			X			RNA-binding motif protein	
RYR2	Ryanodine receptor 2 (cardiac)	AD	X			X		Ryanodine receptor	
SCN5A	Sodium channel, voltage-gated, type V, alpha subunit	AD			X	X		Sodium channel	
SGCD	Sarcoglycan, delta	AD			X			Dystrophin-associated protein complex	Limb-girdle muscular dystrophy
		AR			**				
TAZ	Tafazzin	XL			**		**	Mitochondrium	Barth syndrome
TCAP	Titin-cap	Unknown	X					Z-disc	Limb-girdle muscular dystrophy
		AR			**				
TMEM43	Transmembrane protein 43	AD				X		Transmembrane protein	
TNNC1	Troponin C type 1 (slow)	AD	X		X			Sarcomere	
TNNI3	Troponin I type 3 (cardiac)	AD	X	X	X			Sarcomere	
TNNT2	Troponin T type 2 (cardiac)	AD	X	X	X		X	Sarcomere	
TPM1	Tropomyosin 1 (alpha)	AD	X	X	X			Sarcomere	
TTN	Titin	AD	X		X	X		Sarcomere	
TTR	Transthyretin	AD	**					Transport protein	Amyloidosis
VCL	Vinculin	AD	X		X		X	Z-disc	

MOI, mode of inheritance; AD, autosomal dominant; AR, autosomal recessive; XL, X-linked; HCM, hypertrophic cardiomyopathy; DCM, dilated cardiomyopathy; ARVC, arrhythmogenic right ventricular cardiomyopathy; LVNC, left ventricular noncompaction; **, cardiomyopathy seen as part of other disease/syndrome.
All genes are reviewed in Refs. [1] Ackerman MJ, Priori SG, Willems S, et al. HRS/EHRA expert consensus statement on the state of genetic testing for the channelopathies and cardiomyopathies this document was developed as a partnership between the Heart Rhythm Society (HRS) and the European Heart Rhythm Association (EHRA). Heart Rhythm 2011;8:1308−39; [11] Oechslin E, Jenni R. Left ventricular non-compaction revisited: a distinct phenotype with genetic heterogeneity? Eur Heart J 2011;32:1446−56; [12] Wilde AA, Behr ER. Genetic testing for inherited cardiac disease. Nat Rev Cardiol 2013;10:571−83.

III. MOLECULAR TESTING IN GENETIC DISEASE

most commonly sporadic, but autosomal dominant inheritance has been reported in individuals with a family history of disease [20,21].

Dilated Cardiomyopathy

DCM is characterized by left ventricular enlargement or dilatation and systolic dysfunction [1,7,22,23]. The prevalence of DCM was originally estimated to be 1/2500 individuals. However, more recent appraisals indicate that the true prevalence is closer to 1/250 [23]. DCM often occurs as a result of external factors such as hypertension, ischemia, inflammation, and drug or alcohol abuse [23]. If none of these external causes can be identified, idiopathic DCM is diagnosed and a genetic cause can be considered. It is estimated that 20–35% of individuals with idiopathic DCM have a family history of the disease [24,25]. Familial DCM is typically inherited in an autosomal dominant manner with isolated cardiomyopathy. However, DCM can also be inherited in an autosomal recessive or X-linked manner and some affected individuals may also present with arrhythmia and/or myopathic involvement. Several syndromic disorders include DCM as a cardinal feature including Barth syndrome, Duchenne muscular dystrophy, Emery–Dreifuss muscular dystrophy, and myofibrillar myopathy [26–29]. There is phenotypic overlap between DCM and severe or end-stage HCM [17] and there is also an increasing appreciation for phenotypic similarities and genetic overlap between ARVC and DCM—leading to the increasing adoption of the term arrhythmogenic cardiomyopathy [30–34].

Arrhythmogenic Right Ventricular Cardiomyopathy

ARVC is characterized by gradual replacement of myocytes with fibro-fatty tissue [35,36]. This infiltration occurs primarily in the right ventricle, though the left ventricle is also involved in a proportion of cases [37–39]. Left ventricular involvement and dilation often confounds an accurate diagnosis because fibro-fatty filtration can only be diagnosed after cardiac biopsy. Moreover, there is an increasing appreciation for phenotypic similarities and genetic overlap between ARVC and DCM [30–34]. ARVC may lead to secondary complications including arrhythmia, syncope, and an increased risk for SCD, especially in the young [40,41]. ARVC, which was originally referred to as arrhythmogenic right ventricular dysplasia, is estimated to affect 1/2000–1/5000 individuals and is typically inherited in an autosomal dominant manner. Up to 50% of individuals with ARVC have a family

history of disease, though penetrance is age-dependent and often incomplete [42,43].

Left Ventricular Noncompaction Cardiomyopathy

LVNC is characterized by left ventricular hypertrophy with deep ventricular trabeculations. However, right ventricular involvement has been described in approximately 50% of cases [11,44]. Clinical complications of LVNC include heart failure, arrhythmia, and an increased risk for thromboembolic events [45]. LVNC is thought to occur due to failure of the myocardium to properly compact into mature musculature during embryonic development, though this hypothesis is still controversial because adult-onset LVNC has been identified in individuals without trabeculations at birth [44,46]. Approximately 45% of patients have a family history of the disease and though the prevalence of the disorder has not been well studied, it is estimated to affect 7/50,000–13/1000 individuals [11,47].

MOLECULAR TARGET

Hypertrophic Cardiomyopathy

Most disease-causing variants in HCM occur in genes encoding components of the sarcomere, a structure that serves as the contractile unit of the myocyte (Fig. 17.1) [1,49,50]. While disease-causing variants have been identified in more than 20 sarcomere-encoding genes (Table 17.1), pathogenic variants in the *MYH7* gene, encoding a cardiac beta-myosin heavy chain, and the *MYBPC3* gene, encoding a cardiac myosin binding protein, account for the majority (~80%) of disease-causing variants in HCM [6,49,51,52]. Most HCM-associated variants, including those in *MYH7*, are missense variants that elicit a dominant-negative effect on sarcomere function [53,54]. Disease-associated *MYBPC3* variants are the clear exception, with loss-of-function variants leading to haploinsufficiency serving as the predominant mechanism of disease [55].

Restrictive Cardiomyopathy

RCM is a rare disorder and the genetic etiology of this cardiomyopathy is not well defined. Variants in the *ACTC1*, *MYPBC3*, *MYH7*, *MYL3*, *TNNI3*, *TNNT2*, and *TPM1* genes have been identified in individuals with RCM (Table 17.1), suggesting that some cases are related to sarcomere dysfunction [1,46]. In addition, variants in the *DES* and *BAG3* genes have been identified in several families with myopathy +/− RCM

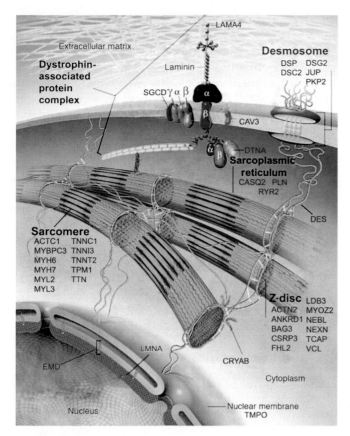

FIGURE 17.1 Cellular locations of proteins involved in inherited cardiomyopathies. Structural components of the myocyte that are impacted by cardiomyopathy include the sarcomere, Z-disc, nuclear lamina proteins, intermediate filaments, the desmosome, and the dystrophin-associated glycoprotein complex. Genes linked to cardiomyopathy and the locations of their protein products are indicated. *Source: Adapted with permission from Elsevier and the American Society for Clinical Investigation ([46] Teekakirikul P, Kelly MA, Rehm HL, Lakdawala NK, Funke BH. Inherited cardiomyopathies: molecular genetics and clinical genetic testing in the postgenomic era. J Mol Diagn 2013;15:158−70; [48] Morita H, Seidman J, Seidman CE. Genetic causes of human heart failure. J Clin Invest 2005;115:518−26).*

(Table 17.1) [56,57]. Clinically significant variants in these genes are identified in approximately 35% of individuals with RCM [46].

Dilated Cardiomyopathy

DCM has much greater locus heterogeneity compared to the other cardiomyopathies discussed in this chapter. More than 30 genes associated with DCM encode proteins with a wide variety of cellular locations including the sarcomere, Z-disc, desmosome, intermediate filaments, and various cellular membranes (Fig. 17.1; Table 17.1) [58,59]. Truncating variants in the *TTN* gene, which encodes the largest human protein, are estimated to account for roughly 12−25% of inherited DCM [59,60]. Disease-associated variants in the *MYH7*, *LMNA*, *TNNT2*, and *RBM20* genes make up for another 10−20% of familial DCM cases [58,59].

Arrythmogenic Right Ventricular Cardiomyopathy

ARVC is primarily caused by pathogenic variation in genes associated with the desmosome, a cytoskeleton-interacting cell−cell adhesion complex that is important for myocardial mechanical activity (Fig. 17.1). Loss-of-function variants in the *PKP2* gene, encoding the plakophilin 2 protein, account for the majority of ARVC cases [36], although disease-associated variants in several other desmosomal genes including *DSC2*, *DCG2*, and *DSP* have also been identified (Table 17.1).

Left Ventricular Noncompaction Cardiomyopathy

Because LVNC is so rare, the genetic etiology of this cardiomyopathy is less well understood. However, disease-associated variants have been identified in a number of genes encoding sarcomeric and Z-disc proteins including *ACTC1*, *DTNA*, *LDB3*, *MYBPC3*, *MYH7*, *TNNT2*, and *VCL* (Table 17.1).

MOLECULAR TECHNOLOGIES

Over the last decade, molecular genetic discoveries have uncovered the molecular basis for many cardiomyopathies and over 55 disease-associated genes have been identified to date (Table 17.1). These discoveries have resulted in a dramatic increase in the utilization of clinical genetic testing by physicians caring for patients with these disorders [12,61,62]. Genetic testing not only provides molecular diagnoses for affected individuals, it can also refine clinical diagnoses for individuals with inconclusive diagnoses and can provide presymptomatic testing for individuals with a family history of disease.

Diagnostic Testing

Due to widespread allelic heterogeneity and the fact that very few recurrent pathogenic variants exist for inherited cardiomyopathy, clinical laboratories typically sequence all coding exons and consensus splice sites (+/− 1,2) for each disease-associated gene tested. The extreme locus heterogeneity that exists for these disorders also necessitates that multiple genes be sequenced for each cardiomyopathy.

Laboratories may use Sanger sequencing to interrogate multiple genes, but large next-generation sequencing (NGS) panels that allow for simultaneous, cost-effective interrogation of many disease-associated genes are becoming the norm [46,61]. NGS cardiomyopathy panels may be offered as large tests that span many cardiomyopathies with the option to order disease-specific subpanels focusing on genes implicated in a specific cardiomyopathy. Disease-specific subpanels often use the same NGS exon capture kit as the full panel, allowing for reflex testing to the remaining genes if the initial result is negative. Copy number variation (CNV) testing may also be offered as part of the NGS test or as a separate assay.

Presymptomatic Testing

Genetic testing for individuals with a family history of disease is typically offered via Sanger sequencing of pathogenic or suspected pathogenic variants that have been previously identified in an affected relative.

Emerging Technologies

As the number of genes in NGS cardiomyopathy assays continues to grow and new disease-associated genes are discovered, whole exome sequencing (WES) and whole genome sequencing (WGS) technologies are beginning to gain momentum in the field of inherited cardiomyopathy. While the cost of these assays currently inhibits widespread adoption, it is likely that utilization of WES and WGS testing will continue to grow as costs decline because they allow great flexibility and circumvent multiple rounds of stepwise testing that often occurs in patients with equivocal phenotypes.

CLINICAL UTILITY

One of the most important clinical utilities of genetic testing in the field of cardiomyopathy is the ability to provide at-risk family members with presymptomatic testing when the disease-causing variant is definitively established within a family. Identification of individuals with pathogenic variants allows for timely clinical evaluation and management in presymptomatic individuals. Genetic testing in cardiomyopathies also allows for more informed reproductive decision-making in individuals with known pathogenic variants, allowing for preimplantation genetic diagnosis and proper genetic counseling.

Clinical genetic testing can confirm a suspected clinical diagnosis in cases where the phenotype is clear and may define or refine clinical diagnoses in situations where a patient's clinical features are equivocal [46]. Achieving an accurate diagnosis is important for physicians treating patients with cardiomyopathy because treatment options can vary widely depending on the diagnosis. Genetic diagnoses may provide some patients with therapeutic opportunities (eg, enzyme replacement therapy in patients with Fabry disease or implantable cardioverter defibrillator placement in patients with cardiomyopathy) and certain diagnoses associated with syndromic disorders may alert the physician to screen for features of the disease that may benefit from medical management (eg, renal disease in Fabry disease) [13,63]. Unfortunately, the prognostic value of genetic testing for inherited cardiomyopathy is currently poor and there is insufficient evidence supporting genotype–phenotype correlations.

LIMITATIONS OF TESTING

Technical Limitations

There are several inherent technical limitations to genetic testing in cardiomyopathies. First, not all disease-associated genes have been identified, and many NGS tests focus solely on gene-coding regions. Therefore, pathogenic variation may be missed in genes not included on the assay and variants in functionally important nonexonic regions (eg, regulatory regions) may be missed. Secondly, NGS may not provide complete coverage of disease-associated genes and false positives may occur. Sanger sequencing is therefore often required to sequence genes to completion and confirm variants detected via NGS. Finally, CNVs are not always readily detected using sequencing technologies and some disease-associated variants, especially those in genes where loss-of-function is an established disease mechanism, may be missed.

Interpretive Limitations

While our ability to provide genetic diagnoses for inherited cardiomyopathies is continuously improving, there are several factors that make interpretation of these genetic tests difficult. First, the majority of cardiomyopathies exhibit a high degree of both locus and allelic heterogeneity with many genes and variants implicated with disease. Second, multiple modes of inheritance for these disorders exist and autosomal dominant, autosomal recessive, and even X-linked forms have been described. Finally, the penetrance of these disorders is age-dependent and incomplete and some individuals with pathogenic variants may not exhibit clinical features of the disease. Variable expressivity has also been described and affected family members carrying an identical pathogenic variant may have differing clinical features or ages of onset. All of these issues require genetic test results to be interpreted within the context of a patient's clinical features and family history.

References

[1] Ackerman MJ, Priori SG, Willems S, et al. HRS/EHRA expert consensus statement on the state of genetic testing for the channelopathies and cardiomyopathies this document was developed as a partnership between the Heart Rhythm Society (HRS) and the European Heart Rhythm Association (EHRA). Heart Rhythm 2011;8:1308–39.

[2] Maron BJ, Doerer JJ, Haas TS, Tierney DM, Mueller FO. Sudden deaths in young competitive athletes: analysis of 1866 deaths in the United States, 1980–2006. Circulation 2009;119:1085–92.

[3] Maron BJ, Haas TS, Murphy CJ, Ahluwalia A, Rutten-Ramos S. Incidence and causes of sudden death in U.S. college athletes. J Am Coll Cardiol 2014;63:1636–43.

[4] Raju H, Alberg C, Sagoo GS, Burton H, Behr ER. Inherited cardiomyopathies. BMJ 2011;343:d6966.

[5] Mahon NG, Murphy RT, MacRae CA, Caforio AL, Elliott PM, McKenna WJ. Echocardiographic evaluation in asymptomatic relatives of patients with dilated cardiomyopathy reveals preclinical disease. Ann Intern Med 2005;143:108–15.

[6] Gersh BJ, Maron BJ, Bonow RO, et al. ACCF/AHA guideline for the diagnosis and treatment of hypertrophic cardiomyopathy: a report of the American College of Cardiology Foundation/American Heart Association Task Force on Practice Guidelines. J Thorac Cardiovasc Surg 2011 2011;142: e153–203.

[7] Hershberger RE, Lindenfeld J, Mestroni L, Seidman CE, Taylor MR, Towbin JA. Genetic evaluation of cardiomyopathy—a Heart Failure Society of America practice guideline. J Card Fail 2009;15:83–97.

[8] Maron BJ, Tajik AJ, Ruttenberg HD, et al. Hypertrophic cardiomyopathy in infants: clinical features and natural history. Circulation 1982;65:7–17.

[9] Kubo T, Kitaoka H, Okawa M, Nishinaga M, Doi YL. Hypertrophic cardiomyopathy in the elderly. Geriatr Gerontol Int 2010;10:9–16.

[10] Maron BJ. Hypertrophic cardiomyopathy: a systematic review. JAMA 2002;287:1308–20.

[11] Oechslin E, Jenni R. Left ventricular non-compaction revisited: a distinct phenotype with genetic heterogeneity? Eur Heart J 2011;32:1446–56.

[12] Wilde AA, Behr ER. Genetic testing for inherited cardiac disease. Nat Rev Cardiol 2013;10:571–83.

[13] Brito D, Miltenberger-Miltenyi G, Moldovan O, Navarro C, Madeira HC. Cardiac Anderson-Fabry disease: lessons from a 25-year-follow up. Rev Port Cardiol 2014;33(247):e1–7.

[14] Cheng Z, Fang Q. Danon disease: focusing on heart. J Hum Genet 2012;57:407–10.

[15] Thomas AS, Hughes DA. Fabry disease. Pediatr Endocrinol Rev 2014;12:88–101.

[16] D'Souza RS, Levandowski C, Slavov D, et al. Danon disease: clinical features, evaluation, and management. Circ Heart Fail 2014;7:843–9.

[17] Olivotto I, Cecchi F, Poggesi C, Yacoub MH. Patterns of disease progression in hypertrophic cardiomyopathy: an individualized approach to clinical staging. Circ Heart Fail 2012;5:535–46.

[18] Richardson P, McKenna W, Bristow M, et al. Report of the 1995 World Health Organization/International Society and Federation of Cardiology Task Force on the Definition and Classification of cardiomyopathies. Circulation 1996;93:841–2.

[19] Kushwaha SS, Fallon JT, Fuster V. Restrictive cardiomyopathy. N Engl J Med 1997;336:267–76.

[20] Aroney C, Bett N, Radford D. Familial restrictive cardiomyopathy. Aust N Z J Med 1988;18:877–8.

[21] Zachara E, Bertini E, Lioy E, Boldrini R, Prati PL, Bosman C. Restrictive cardiomyopathy due to desmin accumulation in a family with evidence of autosomal dominant inheritance. G Ital Cardiol 1997;27:436–42.

[22] Hershberger RE, Morales A, Siegfried JD. Clinical and genetic issues in dilated cardiomyopathy: a review for genetics professionals. Genet Med 2010;12:655–67.

[23] Hershberger RE, Hedges DJ, Morales A. Dilated cardiomyopathy: the complexity of a diverse genetic architecture. Nat Rev Cardiol 2013;10:531–47.

[24] Grunig E, Tasman JA, Kucherer H, Franz W, Kubler W, Katus HA. Frequency and phenotypes of familial dilated cardiomyopathy. J Am Coll Cardiol 1998;31:186–94.

[25] Keeling PJ, Gang Y, Smith G, et al. Familial dilated cardiomyopathy in the United Kingdom. Br Heart J 1995;73:417–21.

[26] Decostre V, Ben Yaou R, Bonne G. Laminopathies affecting skeletal and cardiac muscles: clinical and pathophysiological aspects. Acta Myologica 2005;24:104–9.

[27] Kaspar RW, Allen HD, Montanaro F. Current understanding and management of dilated cardiomyopathy in Duchenne and Becker muscular dystrophy. J Am Acad Nurse Pract 2009; 21:241–9.

[28] Clarke SL, Bowron A, Gonzalez IL, et al. Barth syndrome. Orphanet J Rare Dis 2013;8:23.

[29] Sanbe A. Dilated cardiomyopathy: a disease of the myocardium. BiolPharm Bull 2013;36:18–22.

[30] Marcus FI, McKenna WJ, Sherrill D, et al. Diagnosis of arrhythmogenic right ventricular cardiomyopathy/dysplasia: proposed modification of the Task Force Criteria. Eur Heart J 2010;31:806–14.

[31] Elmaghawry M, Migliore F, Mohammed N, Sanoudou D, Alhashemi M. Science and practice of arrhythmogenic cardiomyopathy: a paradigm shift. Glob Cardiol Sci Pract 2013;2013:63–79.

[32] Rizzo S, Pilichou K, Thiene G, Basso C. The changing spectrum of arrhythmogenic (right ventricular) cardiomyopathy. Cell Tissue Res 2012;348:319–23.

[33] Saffitz JE. Arrhythmogenic cardiomyopathy: advances in diagnosis and disease pathogenesis. Circulation 2012;124:e390–2.

[34] Sen-Chowdhry S, Morgan RD, Chambers JC, McKenna WJ. Arrhythmogenic cardiomyopathy: etiology, diagnosis, and treatment. Annu Rev Med 2012;61:233–53.

[35] Marcus FI, Fontaine GH, Guiraudon G, et al. Right ventricular dysplasia: a report of 24 adult cases. Circulation 1982;65:384–98.

[36] Marcus FI, Edson S, Towbin JA. Genetics of arrhythmogenic right ventricular cardiomyopathy: a practical guide for physicians. J Am Coll Cardiol 2013;61:1945–8.

[37] Gallo P, d'Amati G, Pelliccia F. Pathologic evidence of extensive left ventricular involvement in arrhythmogenic right ventricular cardiomyopathy. Hum Pathol 1992;23:948–52.

[38] Michalodimitrakis M, Papadomanolakis A, Stiakakis J, Kanaki K. Left side right ventricular cardiomyopathy. Med Sci Law 2002;42:313–17.

[39] Lindstrom L, Nylander E, Larsson H, Wranne B. Left ventricular involvement in arrhythmogenic right ventricular cardiomyopathy—a scintigraphic and echocardiographic study. Clin Physiol Funct Imaging 2005;25:171–7.

[40] Corrado D, Thiene G, Nava A, Rossi L, Pennelli N. Sudden death in young competitive athletes: clinicopathologic correlations in 22 cases. Am J Med 1990;89:588−96.

[41] Nava A, Bauce B, Basso C, et al. Clinical profile and long-term follow-up of 37 families with arrhythmogenic right ventricular cardiomyopathy. J Am Coll Cardiol 2000;36:2226−33.

[42] Corrado D, Thiene G. Arrhythmogenic right ventricular cardiomyopathy/dysplasia: clinical impact of molecular genetic studies. Circulation 2006;113:1634−7.

[43] Quarta G, Muir A, Pantazis A, et al. Familial evaluation in arrhythmogenic right ventricular cardiomyopathy: impact of genetics and revised task force criteria. Circulation 2011;123:2701−9.

[44] Udeoji DU, Philip KJ, Morrissey RP, Phan A, Schwarz ER. Left ventricular noncompaction cardiomyopathy: updated review. Ther Adv Cardiovasc Dis 2013;7:260−73.

[45] Shemisa K, Li J, Tam M, Barcena J. Left ventricular noncompaction cardiomyopathy. Cardiovasc Diagn Ther 2013;3:170−5.

[46] Teekakirikul P, Kelly MA, Rehm HL, Lakdawala NK, Funke BH. Inherited cardiomyopathies: molecular genetics and clinical genetic testing in the postgenomic era. J Mol Diagn 2013;15:158−70.

[47] Ichida F, Hamamichi Y, Miyawaki T, et al. Clinical features of isolated noncompaction of the ventricular myocardium: long-term clinical course, hemodynamic properties, and genetic background. J Am Coll Cardiol 1999;34:233−40.

[48] Morita H, Seidman J, Seidman CE. Genetic causes of human heart failure. J Clin Invest 2005;115:518−26.

[49] Richard P, Charron P, Carrier L, et al. Hypertrophic cardiomyopathy: distribution of disease genes, spectrum of mutations, and implications for a molecular diagnosis strategy. Circulation 2003;107:2227−32.

[50] Maron BJ, Maron MS, Semsarian C. Genetics of hypertrophic cardiomyopathy after 20 years: clinical perspectives. J Am Coll Cardiol 2012;60:705−15.

[51] Van Driest SL, Ackerman MJ, Ommen SR, et al. Prevalence and severity of "benign" mutations in the beta-myosin heavy chain, cardiac troponin T, and alpha-tropomyosin genes in hypertrophic cardiomyopathy. Circulation 2002;106:3085−90.

[52] Alfares A, Kelly MA, McDermott G, et al. Results of clinical genetic testing of 2912 probands with hypertrophic cardiomyopathy: expanded panels offer limited additional sensitivity. Genet Med 2015;17:880−8.

[53] Tardiff JC, Hewett TE, Factor SM, Vikstrom KL, Robbins J, Leinwand LA. Expression of the beta (slow)-isoform of MHC in the adult mouse heart causes dominant-negative functional effects. Am J Physiol Heart Circ Physiol 2000;278: H412−19.

[54] Oberst L, Zhao G, Park JT, et al. Dominant-negative effect of a mutant cardiac troponin T on cardiac structure and function in transgenic mice. J Clin Invest 1998;102:1498−505.

[55] Marston S, Copeland O, Gehmlich K, Schlossarek S, Carrier L. How do MYBPC3 mutations cause hypertrophic cardiomyopathy? J Muscle Res Cell Motil 2012;33:75−80.

[56] Dalakas MC, Park KY, Semino-Mora C, Lee HS, Sivakumar K, Goldfarb LG. Desmin myopathy, a skeletal myopathy with cardiomyopathy caused by mutations in the desmin gene. N Engl J Med 2000;342:770−80.

[57] Goldfarb LG, Park KY, Cervenakova L, et al. Missense mutations in desmin associated with familial cardiac and skeletal myopathy. Nat Genet 1998;19:402−3.

[58] Lakdawala NK, Funke BH, Baxter S, et al. Genetic testing for dilated cardiomyopathy in clinical practice. J Card Fail 2012;18:296−303.

[59] Pugh TJ, Kelly MA, Gowrisankar S, et al. The landscape of genetic variation in dilated cardiomyopathy as surveyed by clinical DNA sequencing. Genet Med 2014;16:601−8.

[60] Herman DS, Lam L, Taylor MR, et al. Truncations of titin causing dilated cardiomyopathy. N Engl J Med 2012;366:619−28.

[61] Lebo MS, Baxter SM. New molecular genetic tests in the diagnosis of heart disease. Clin Lab Med 2014;34:137−56.

[62] Tester DJ, Ackerman MJ. Genetic testing for potentially lethal, highly treatable inherited cardiomyopathies/channelopathies in clinical practice. Circulation 2011;123:1021−37.

[63] Anselme F, Moubarak G, Savoure A, et al. Implantable cardioverter-defibrillators in lamin A/C mutation carriers with cardiac conduction disorders. Heart Rhythm 2013;10:1492−8.

18

Molecular Diagnostics for Coagulopathies

M.B. Smolkin and P.L. Perrotta

Department of Pathology, West Virginia University, Morgantown, WV, United States

INTRODUCTION

The most important elements of the hemostatic system include endothelial cells, platelets, and the clotting proteins that comprise the classical coagulation cascade (Fig. 18.1). Disruption of any of these elements can alter the hemostatic balance, resulting in bleeding or thrombotic tendencies. Genes that code for coagulation proteins have been studied using DNA technologies to detect specific mutations that result in the formation of smaller amounts of protein or a molecule with less functional activity. Several of these technologies are used clinically to assess individuals for their risk for either bleeding or thrombosis.

The major clotting proteins that have been characterized at the molecular level include the procoagulant proteins factor VII (FVII), factor IX (FIX), and factor XI (FXI), and the anticoagulant proteins protein C (PC), protein S (PS), and antithrombin (AT) (Table 18.1). Mutations that result in PC, PS, and AT deficiency are found in a small minority of patients with thrombotic disease, and these mutations are spread throughout the corresponding genes. Both PC and PS deficiency are transmitted in an autosomal dominant fashion. Heterozygotes have mild protein deficiency and are at risk for thrombosis, whereas homozygotes have severe deficiency and a life-threatening clotting disorder called neonatal purpura fulminans [1]. With the exception of specific mutations that lead to reduced affinity for heparin, homozygous deletions for AT appear to be lethal.

This chapter highlights the use of molecular technologies in evaluating individuals with disorders of hemostasis. This will include disorders that increase the risk of thrombosis such as the factor V Leiden (FVL) and prothrombin (PT) G20210A gene mutations and genetic diseases that cause significant bleeding including hemophilia A (HA), hemophilia B (HB), and von Willebrand disease (vWD). Other less common coagulation disorders in which molecular mechanisms have been elucidated will be discussed including inherited disorders of platelet function.

MOLECULAR TARGETS

The pathogenesis of venous thromboembolism, a condition that affects a large number of individuals, is often associated with acquired and/or genetic risk factors [2]. These risk factors are generally associated with either the overexpression of procoagulant proteins or a reduction in proteins with anticoagulant properties. The most common acquired conditions associated with thrombosis include oral contraceptive use, estrogen therapy, pregnancy including the postpartum state, the antiphospholipid syndrome, chemotherapy, and malignancy. The latter includes Janus kinase 2 (*JAK2*) mutations found in certain cancers. Inherited germline mutations have also been linked to thrombophilia and these disorders will be discussed in detail.

Factor 5 (F5) Gene Mutations

Factor V (FV) is a procoagulant molecule that interacts with other clotting proteins including activated factor X and PT to increase the production of thrombin, the key hemostatic enzyme that converts soluble fibrinogen to a fibrin clot [3]. Mutations in the gene that encodes FV, *F5*, have been extensively studied as risk factors for thrombosis. Activated protein C (APC) plays an important role as an anticoagulant because it inactivates FV through a specific cleavage site. A single nucleotide mutation in the *F5* gene (FVL mutation)

© 2017 Elsevier Inc. All rights reserved.

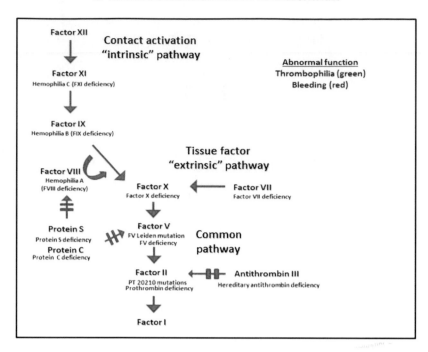

FIGURE 18.1 Classical clotting cascade and associated defects.

TABLE 18.1 Major Coagulation Proteins and Genes

Gene name	Gene symbol	Gene location	Genetic disorders (phenotype)
Coagulation factor VIII (antihemophilic factor, procoagulant component)	F8	Xq28	Hemophilia A (FVIII deficiency, bleeding)
Coagulation factor IX (Christmas factor)	F9	Xq27.1−q27.2	Hemophilia B (FIX deficiency, bleeding)
Coagulation factor V (proaccelerin, labile factor)	F5	1q23	FVL mutation (thrombophilia)
			FV deficiency (bleeding)
Coagulation factor XI	F11	4q35.2	Hemophilia C (FXI deficiency, bleeding)
Coagulation factor II (prothrombin, PT)	F2	11p11	PT 20210 mutations (thrombophilia)
			Prothrombin deficiency (bleeding)
Coagulation factor XII (Hageman factor)	F12	5q35.3	Hereditary angioedema type III (swelling, angioedema)
Coagulation factor X	F10	13q34	Factor X deficiency (bleeding)

(Continued)

TABLE 18.1 (Continued)

Gene name	Gene symbol	Gene location	Genetic disorders (phenotype)
Coagulation factor VII	F7	13q34	Factor VII deficiency (bleeding)
Kallikrein B, plasma (Fletcher factor) 1	KLKB1	4q35	KLKB1 mutations (generally asymptomatic)
Protein S (alpha)	PROS1	3q11.2	Protein S deficiency (thrombophilia)
Protein C (inactivator of coagulation factors Va and VIIIa)	PROC	2q13-q14	Protein C deficiency (thrombophilia)
Serpin peptidase inhibitor, clade C (antithrombin)	SERPINC1	1q25.1	Hereditary antithrombin deficiency (thrombophilia)

results in an arginine (R) to glutamine (Q) substitution at position 506 of the FV protein (R506Q). The mutant FV protein is relatively resistant to degradation by APC. The lower rate of FV inactivation increases the risk of thromboembolism by increasing thrombin production and subsequent clot formation [4,5]. Over 5% of FVL heterozygotes in the general population will experience venous thrombosis during their lifetime [6]. General recommendations for FVL mutation testing have been proposed (Table 18.2) [7,8]. Other rare F5

TABLE 18.2 Recommendations for FVL Mutation Testing

GENERAL RECOMMENDATIONS FOR FVL MUTATION TESTING

- Venous thrombosis in patient <50 years old
- Recurrent venous thrombosis
- Relatives of patients who had venous thrombosis at <50 years of age
- Venous thrombosis and strong family history of thrombosis
- Venous thrombosis during pregnancy or while taking oral contraceptives
- Myocardial infarction in women <50 years old who smoke
- Thrombosis in unusual locations (eg, mesenteric, cerebral, hepatic veins).

MUTATION TESTING MAY BE CONSIDERED IN THE FOLLOWING SITUATIONS

- Venous thrombosis, age >50, except when active malignancy is present
- Relatives of individuals known to have FVL. Knowledge that they have FVL may influence management of pregnancy and may be a factor in decision-making regarding oral contraceptive use
- Women with recurrent pregnancy loss or unexplained severe preeclampsia, placental abruption, intrauterine fetal growth retardation, or stillbirth. Knowledge of FVL carrier status may influence management of future pregnancies.

variants, such as the heterozygous G1689A [9] and C1690T [10] mutations, are not typically detected by some commercially available in vitro diagnostic assays and if present, may produce an invalid or error result. In these circumstances, bidirectional sequencing may be necessary for confirmation. However, these mutations may also result in FV proteins that are resistant to APC. For this reason, some laboratories may screen patients with thrombophilia for APC resistance using coagulation-based assays as a first-line testing instead molecular assays that detect only the FVL mutation.

Factor 2 (F2) Gene Mutations

PT (coagulation factor II) is the precursor of thrombin, the end-product of the coagulation cascade that converts soluble fibrinogen to a fibrin clot. PT is encoded by the *F2* gene, and a specific mutation in the *F2* gene, PT G20210A, is associated with an increased risk of deep venous and cerebral venous thromboembolism attributed to enhanced plasma PT activity [11]. The point mutation, lying within the 3′ untranslated region, results in mRNA accumulation and increased synthesis of PT and subsequently, thrombin [12,13]. Approximately 2% of the US population are carriers of this mutation, which is most frequently observed in Caucasians [14]. PT protein activity can be measured directly in plasma but these tests cannot be used to detect PT G20210A gene mutation carriers because their PT activity overlaps those seen in individuals

lacking the mutation; molecular analysis is required [15]. It is recommended that testing for PT G20210A mutations be considered whenever testing for FVL mutations is indicated (Table 18.2) [7]. In addition to the G20210A mutation, other rare mutations in the *F2* gene have been reported (eg, PT C20209) that are of uncertain significance [16]. It is important that clinical assays be designed to clearly distinguish clinically significant G20210A abnormalities from nonsignificant polymorphisms. As with *F5* mutations, these other mutations can interfere with the performance of some assays used by clinical laboratories, producing invalid or error results. Again, resolving these cases may require repeating the test using different primers or an alternate technology such as bidirectional sequencing.

Methylenetetrahydrofolate Reductase Mutations

The conversion of 5,10-methylenetetrahydrofolate to folate, a cosubstrate for homocysteine remethylation to methionine, is catalyzed by the enzyme 5,10-methylenetetrahydrofolate reductase (MTHFR). Methionine is then converted to *S*-adenosylmethionine, which functions as an essential methyl donor. A thermolabile variant C665T (p.Ala222Val, more commonly referred to as C677T) and the A1286C (p.Glu429Ala) variant are two common polymorphic genetic variants that encode for forms of this enzyme with decreased enzymatic activity [17,18]. Approximately 25% of Hispanics and 10–15% of North American Caucasians are homozygous for the thermolabile variant [19]. C665T and C1286A are in linkage disequilibrium with each other. Therefore, a combination of both variants is usually seen only in individuals who are compound heterozygotes in the *trans* position. Combined homozygosity for one variant and heterozygosity for the other variant is not uncommon [20]. It was originally thought that reduced MTHFR activity led to hyperhomocysteinemia, which may lead to an increased risk for coronary heart disease, venous thromboembolism, and recurrent pregnancy loss [21–23]. A recent meta-analysis challenged the hypothesis that long-term moderately elevated homocysteine levels have any effect on cardiovascular disease [24]. As a result of these and other studies, the American College of Medical Genetics (ACMG) has issued a practice guideline that does not recommend *MTHFR* testing as part of the routine evaluation of patients with thrombophilia testing because of the lack of clinical utility [25].

AT Gene (AT Deficiency)

AT is a serine protease inhibitor that inactivates thrombin and factor Xa. AT deficiency is an uncommon disorder that may be congenital or acquired and has a prevalence of approximately of 1:500 to 1:5000 [26−28]. Congenital AT deficiency is typically inherited as an autosomal dominant disorder and has variable clinical penetrance [29]. The gene for AT, *SERPINC1*, is located on chromosome 1q23.1−23 and is 13.5 kb in length containing seven exons [30]. Over 120 mutations associated with congenital AT deficiency have been identified. Type I AT deficiencies are caused by heterozygous mutations that cause reduced synthesis of a functionally active AT protein [31]. Immunologic (antigenic) and functional AT levels are 50% or less of normal. Homozygous type I AT deficiency is thought to be incompatible with life [32]. Type II deficiencies are associated with a qualitative defect in which AT is dysfunctional. These mutations are heterozygous and patients typically have normal AT antigen levels but decreased functional AT activity.

There are three subtypes of type II deficiency (IIa, IIb, IIc), which are categorized based on the mutation-binding sites. Type IIb, the most common type II deficiency, is due to a defect in the heparin-binding region of AT. Although less common, the IIa type caused by mutations in the thrombin-binding site is more thrombogenic. Type IIc deficiencies include a pleotropic group of mutations located close to the reactive loop site in the AT gene [33]. Genetic testing is not usually performed in routine clinical practice. However, genetic testing may be considered to confirm the presence of a congenital defect in patients with quantitative or qualitative deficiencies as determined by antigenic and/or functional assays. Since type II deficiencies have mutations located in specific areas of the *SEPINC1* gene, targeted mutation testing may be appropriate. Type I deficiency is associated with mutations throughout the AT gene and may require full gene sequencing for detection [34].

Cytochrome P4502C9 and Vitamin K Epoxide Reductase

Warfarin has a narrow therapeutic range and untoward effects including serious bleeding are common. At least 30 genes may be involved in warfarin metabolism and differences in expression of these genes may explain the wide variation in warfarin doses needed to achieve a therapeutic effect [35]. By identifying genetic polymorphisms that affect the metabolism or action of warfarin, it may be possible to tailor dosing to more rapidly achieve a therapeutic drug level as measured by the international normalized ratio (INR). Maintaining a therapeutic INR may minimize both bleeding risk from over-anticoagulation and thrombotic risk from under-coagulation. The two most

important genes implicated in warfarin sensitivity are cytochrome P4502C9 (*CYP2C9*) and vitamin K epoxide reductase complex 1 (*VKORC1*) [36].

The *VKORC1* gene located on chromosome 16 produces an enzyme, VKOR, that reduces vitamin K 2,3-epoxide to an enzymatically activated form required for the posttranslational modification of vitamin K dependent coagulation factors. Warfarin decreases VKOR activity, which lowers levels of these coagulation factors. Thus, polymorphisms of the *VKORC1* gene, including a promoter polymorphism ($-1639G > A$) and an exonic polymorphism (C1173T), result in warfarin sensitivity in which lower drug doses are needed to produce an anticoagulant effect. The contribution of the *VKORC1* genotype to the variability in warfarin response has been estimated between 15% and 30% [37]. Variation is also observed between ethnic groups, with very few African-Americans but approximately 80% of Chinese being homozygous carriers of the warfarin-sensitive *VKORC1* mutant AA genotype [38].

The highly polymorphic *CYP2C9* gene located on chromosome 10 encodes an enzyme that metabolizes the more potent S-isomer of warfarin. Polymorphisms of the *CYP2C9* gene are associated with reduced enzymatic activity, resulting in significant reductions in drug metabolism [39,40]. With reduced warfarin clearance, lower warfarin doses are required to achieve a therapeutic INR. The most common allelic variants in Caucasians are *CYP2C9*2* (C430T) and *CYP2C9*3* (A1075C), which result in about 70% and 20% of the wild-type enzyme activity, respectively. These polymorphisms can be detected using various methodologies.

Pharmacogenomic (PGx) algorithms incorporating *VKORC1*, *CYP2C9*2*, and *CYP2C9*3* polymorphism testing have been developed to help guide warfarin dosing based on phenotypic differences in drug metabolism [41−45]. Adding genetic data appears to partially explain variation in warfarin response not predicted based on clinical parameters alone, even across racially diverse groups [46,47]. A retrospective international trial used mathematical models to dose warfarin that were based on either clinical factors alone or a combination of clinical and genetic factors. This study found that stable warfarin doses were better predicted by the PGx algorithm—which included both clinical and genetic factors—than by clinical factors alone [48]. Several studies suggest that warfarin can be dosed more precisely when genotypic information is available, including a small number of prospective trials that have associated genotype-specific dosing with decreased time to achieve a target INR with fewer dose changes. However, the utility of PGx-based warfarin dosing remains controversial. A large randomized controlled trial did not observe a reduction in nontherapeutic INRs when comparing standard and PGx-guided warfarin dosing, although PGx-guided dosing did provide smaller and fewer dosing changes [41]. Two larger studies comparing PGx-guided dosing to standard-of-care dosing reached opposite conclusions [49,50]. The conflicting studies regarding PGx-based drug dosing guided by *CYP2C9* and *VORC1* polymorphism status could be due to yet unidentified factors that alter warfarin efficacy [51]. Most PGx trials exclude individuals with coexisting morbidity or who are receiving other medications, factors that clearly affect drug metabolism. Algorithms used to dose warfarin may become more complex when patients are further stratified based on comorbid conditions and other genetic predispositions for clotting such as FVL and PT gene mutations [52].

An additional obstacle for the widespread adoption of PGx testing for dosing warfarin is the difficulty proving improvements in health outcomes or cost. Relative to the targeted therapies used in oncology, warfarin is a relatively inexpensive drug. Many clinicians who prescribe warfarin are not convinced that PGx testing will improve their ability to achieve warfarin therapeutic levels, despite the inclusion of this information in warfarin labeling by the FDA. Several professional societies including the ACMG and the American College of Chest Physicians do not currently recommend routine screening of patients prior to warfarin treatment [53,54]. The increasing use of platelet inhibitors and alternative anticoagulants (eg, direct oral anticoagulants) that do not require monitoring has also hindered adoption of testing for warfarin sensitivity [55].

Factor 8 Gene (HA)

HA is an X-linked recessive bleeding disorder most commonly caused by mutations in the *F8* gene that result in reduced production of FVIII. The severity of the clinical phenotype correlates with residual FVIII activity, ranging from less than 1% for severe disease, 1–5% for moderate bleeding, and 5–40% for milder forms [56]. In general, carriers of HA mutations have FVIII levels of over 35% which may not be detected by routine screening tests like the activated partial thromboplastin time (aPTT) that has varying sensitivity to FVIII levels. As an X-linked recessive disorder, HA is much more common in men than women. However, approximately 10% of female carriers of HA mutations have low FVIII clotting activity that causes unexpected bleeding. This includes women with Turner syndrome (45X) or abnormal X-inactivation. Other very rare causes of reduced FVIII activity in women include homozygosity, a translocation between the X chromosome and an autosome involving a breakpoint within the *F8* gene, and uniparental disomy [57].

The *F8* gene, located on the short arm of the X chromosome, contains 26 relatively short exons and encompasses about 186 kb. More than 1200 HA mutations of various types have been identified in the *F8* gene. The intron 22 inversion (int22) is the most common *F8* gene defect encountered in severe HA, accounting for approximately 40% of all observed mutations [58]. Within this intron are two genes, the 2 kb *F8-associated gene A*, which is transcribed in the opposite direction of *F8*, and the 2.5 kb *F8-associated gene B*, which is transcribed in the same direction as *F8*. The functions of these genes are not known. Additionally, the *F8-associated gene A* is replicated in two other locations external to the *F8* gene (*INT22H2* and *INT22H3*) [59,60]. The net effect of this abnormality is to confer the *F8-associated gene A* sequences the potential to interact through homologous recombination to create a gene rearrangement. With this rearrangement, the int22 inversion places exons 1–22 of the *F8* gene about 500 kb upstream of exon 23–26 and oriented in the opposite direction [61,62]. Due to significant gene disruption, patients with int22 inversions typically demonstrate a severe form of HA. The recombinatorial events leading to int22 inversions generally occur during spermatogenesis [63].

Another common mutation includes an inversion involving exon 1 that is present in about 5% of patients with severe HA [64]. Many other mutations in the *F8* gene have been reported in HA, which include reading frame shifts leading to nonfunctional gene products [65,66]. Large deletions in the *F8* gene occur in approximately 15% of HA cases. The severity of bleeding in patients with large *F8* deletions depends on the location of the deletion and the resulting impact on exon splicing. HA patients with large *F8* deletions or nonsense mutations also have a significant (40–60%) risk of developing FVIII inhibitory antibodies [67,68]. In these patients with large deletions, immune tolerance therapy used to eliminate or weaken FVIII inhibitors is less successful than for other mutations [69]. Thus, in some cases sequencing of the *F8* gene to determine the specific molecular alteration responsible for HA may help to predict clinical outcomes.

Diagnosing HA is typically straightforward and is based on clinical and family bleeding history, and measurements of plasma FVIII clotting activity. Molecular testing is not usually indicated but may be helpful in some circumstances. For example, difficulties may arise in patients with mild FVIII deficiencies and when other conditions coexist that can transiently increase FVIII levels such as inflammatory processes, pregnancy, oral contraceptive use, and exercise. Patients with vWD may have low FVIII levels because FVIII circulates bound to von Willebrand factor (vWF) such that low vWF levels result in a shorter FVIII half-life. Rarely, a mutation in the *ERGIC-53* gene that encodes for a transport protein utilized by both FVIII and FV may lead to a combined mild deficiency of these two factors [70,71]. FVIII activity assays alone cannot identify carriers of *F8* mutations because reference ranges for FVIII testing are wide and some carriers may have apparently normal FVIII levels. Thus, molecular testing is required when clinicians wish to determine carrier status. Molecular testing may also be helpful identifying patients at higher risk for developing FVIII inhibitors or characterizing the specific *F8* mutation in family members of a female carrier who may become pregnant [72]. The molecular diagnosis of HA in cases not caused by the common inversions may be more challenging because of the size of the *F8* gene and the heterogeneity of the mutations spread across the gene. A database listing previously characterized mutations and polymorphisms in the *F8* gene is available to use in conjunction with sequence data to diagnose HA [73].

In rare cases when extensive molecular testing cannot identify a mutation, linkage analysis may be used to trace the mutation through a pedigree. This method is also employed in developing countries with limited access to advanced molecular testing [74]. Linkage analysis is a relatively cumbersome technique that relies on access to accurate pedigree data and to samples from the proband and both parents. Furthermore, linkage analysis confers the risk of revealing nonpaternity and genetic counseling should be offered before the analysis is undertaken. Approximately 30% of patients have spontaneous mutations and no family history of disease. These cases presumably result from de novo *F8* mutations or from gene transmission through several generations of asymptomatic female carriers. Other diagnostic difficulties arise when a mother is germline mosaic for an *F8* mutation. This occurs when the mutation is present in only a subpopulation of her germ cells. In this situation, the mutation cannot be easily detected in blood samples and the risk for her offspring cannot be precisely quantified, although the risk appears to be less than 50% [75]. If a family-specific mutation is detected, at-risk male relatives can be assessed using an FVIII activity assay or a molecular assay specific for the familial mutation. Testing of female relatives is performed using molecular techniques.

Factor 9 Gene (HB)

HB is an X-linked recessive bleeding disorder caused by a deficiency of coagulation factor 9 (FIX) that is clinically identical to HA. The *F9* gene is located on Xq27.1−q27.2 and contains eight exons. Various *F9* mutations have been described including several promoter region mutations (eg, HB Leyden 1, 2, 3) that are associated with a phenotype characterized by severe disease at birth [76,77]. Several mutations that affect the alanine−10 loci confer increased sensitivity to warfarin treatment and are associated with a disproportionately prolonged aPTT [78].

The diagnosis of HB is established by clinical history and by measuring plasma FIX activity, in the same way history and FVIII levels are used to diagnose HA. Vitamin K deficiency should be excluded because deficiency of this vitamin results in lower levels of FIX and other vitamin-K-dependent proteins. As in HA, molecular testing is required to identify carriers of FIX mutations with normal or near-normal FIX activity. Knowledge of mutation status will not directly affect clinical management. However, it may help predict the risk of developing FIX inhibitors or anaphylactic reactions during FIX replacement therapy [79]. About 25% of HB patients have one of three founder mutations (Gly60Ser, Ile397Thr, or Thr296Met), and other mutations have been reported [80]. *F9* mutation testing may incorporate an initial test for the three most common mutations. If these mutations are not found in a patient with severe HB, the functional regions of the *F9* gene can be screened. Sequencing of the *F9* gene is clinically available.

Factor 11 Gene (Hemophilia C)

Factor 11 (FXI) is involved in normal hemostasis and upon activation by factor 12a, thrombin, or by self-activation, FXIa cleaves FIX in the intrinsic coagulation pathway [81]. FXI deficiency can be congenital or acquired. Congenital FXI deficiency is a rare bleeding disorder with an overall estimated prevalence of one in a million [82], with a higher frequency (1:450) among Ashkenazim [83]. Generally, FXI deficiency is inherited as an autosomal recessive trait. However, due to the dimeric structure of circulating FXI, an autosomal dominant form has been described presumably due to a dominant negative effect [84]. Severe FXI deficiency is associated with a plasma FXI activity less than 20% of normal and is often discovered during presurgical testing when the patient's aPTT is elevated [85]. Patients with FXI deficiency may be asymptomatic and those with abnormal bleeding usually have symptoms milder than those seen in HA or HB. Bleeding typically occurs following injury and the extent of bleeding correlates better with the site of injury rather than the genotype itself [86]. Interestingly, FXI deficiency may have a protective effect against ischemic stroke [87]. The *F11* gene is 23.6 kb, contains 15 exons, and is located at the end of the long arm of chromosome 4 (4q35.2) [88]. More than 220 mutations have been described within the *F11* gene. In Ashkenazi Jewish populations, there are two common mutations accounting for 95% of cases, Glu117-stop (type II) and Phe283Leu (type III). However, in the non-Ashkenazim population, there is a relatively even distribution of mutations throughout the gene [82]. Testing for the underlying *F11* mutations may be warranted in patients with low FXI activity levels (~1%) as these patients may be prone to develop inhibitors against FXI following factor replacement therapy. Genetic testing for *F11* mutations typically focuses on the founder mutations that are common in Ashkenazi Jewish patients—expanded gene sequencing may be required in other populations.

vWF Gene (vWD)

vWD is the most commonly recognized inherited bleeding disorder caused by either quantitative (types 1 and 3) or qualitative (type 2) abnormalities of vWF, a multimeric glycoprotein that plays a critical role in primary hemostasis. Following binding to the platelet glycoprotein Ib receptor, vWF forms a bridge between the platelet surface and the subendothelial collagen exposed by vascular injury. vWF consists of low, intermediate, and high-molecular-weight multimers. The higher weight forms are more efficient adhesive molecules [89]. Acquired forms of vWD caused by structural or functional defects in vWF that are secondary to autoimmune and malignant diseases will not be discussed. The initial diagnosis of inherited vWD is made using laboratory tests for vWF activity (eg, ristocetin cofactor activity), vWF antigen levels, FVIII clotting activity, vWF multimer analysis, and platelet function tests [90]. The *vWF* gene, located on the short arm of chromosome 12 (12p13.3), is approximately 180 kb in size and consists of 52 exons.

Type I vWD is the most common form of vWD and is associated with a quantitative defect in structurally normal vWF. The disorder is usually inherited in an autosomal dominant manner with variable penetrance and clinical heterogeneity that can complicate pedigree analysis. Symptoms of type I vWD are often mild and patients may not be diagnosed until after they bleed excessively following trauma or surgery. Laboratory testing for type I vWD can be complicated by the fact the vWF levels can increase in response to inflammation, stress, infection, hormone therapy, pregnancy, exercise, surgical procedures, and liver disease. Therefore, low normal levels of vWF may not exclude mild vWD. Over 100 different mutations have been associated with type 1 vWD. However, the mutations leading to type 1 vWD are not well

characterized in terms of genotype–phenotype correlation. Locus heterogeneity for type 1 vWD involving genes other than the vWF gene further complicates the interpretation of genetic testing [91,92]. For these reasons, genetic testing for type 1 vWD is usually of little clinical value.

The type 2 vWD subtypes (2A, 2B, 2M, 2N) result from a primary qualitative abnormality of vWF. The most common type 2 subtypes, 2A and 2B, are inherited in an autosomal dominant fashion. Type 2A cases typically manifest normal or mildly reduced plasma levels of vWF antigen and FVIII, but have reduced vWF activity levels that are discordant with the vWF antigen levels. In addition, type 2A cases demonstrate significant reductions in high- and intermediate-weight multimer complexes. Numerous mutations have been described in type 2A vWD, the majority of which are missense mutations. Approximately 80% of the identified mutations are found in exon 28. Common mutations in exon 28 include 4517C > T, 4789C > T, and 4790G > A [93]. These mutations effect the A2 domain of vWF. Large deletions have not been reported in association with type 2A vWD.

Type 2B patients have a qualitatively abnormal vWF with an increased affinity for the platelet glycoprotein Ib. As the platelets bind to vWF multimers they are rapidly cleared from the circulation, resulting in thrombocytopenia. Like type 2A, there is significant reduction in high- and intermediate-weight multimer complexes and a disproportional decrease in vWF activity compared to vWF antigen levels. As platelet counts can vary over time in individuals with type 2B vWD, platelet aggregation studies are necessary to distinguish these patients from those with type 2A vWD. There are approximately 25 mutations currently associated with type 2B vWD and all are located in exon 28. The most commonly reported mutations include C3916T, C3922T, G3946A, and G4022A. These gain-of-function mutations are located in the Gp Ib-binding site in the A1 domain of the protein. It is thought that these mutations may inactivate specific ligand-binding sites or disrupt the regulation of vWF binding to platelets [90]. Sequencing exon 28 of the vWD gene should identify the majority of mutations in type 2A and 2B vWD. Distinguishing type 2B from 2A vWD is important due to the differences in clinical management. Specifically, patients with type 2B vWD may experience worsening thrombocytopenia when administered desamino-8-D-arginine vasopressin [94,95]. Since most mutations in both type 2A and 2B vWD are clustered in exon 28 of the *vWF* gene, sequencing studies may be specifically targeted to this exon [96]. However, genotype–phenotype correlation is not always possible since many loci have not yet been characterized at a functional level.

Type 2M vWD patients have decreased vWF activity levels with normal vWF antigen levels and FVIII activity, and multimeric studies are usually normal [97]. Over 20 mutations are associated with type 2M vWD, with more than 80% of these located in exon 28. Type 2N (ie, Normandy variant) vWD is an autosomal recessive disorder characterized by a decreased half-life of FVIII. Approximately 30 different mutations have been identified with type 2N vWD, and 40% of these are found in exon 18. Common mutations reported include 2372C > T in exon 18, 2446C > T in exon 19, and 2561G > A in exon 20. These mutations correspond to deficiencies in the D'-D3 region of vWF which involves the FVIII-binding domain [98]. This results in decreased or absent binding of FVIII to vWF, causing rapid degeneration of FVIII. Patients typically present with low FVIII activity levels of approximately 5–40% of normal. In these cases, it is often important to differentiate this disorder from HA through use of molecular techniques [99].

Type 3 vWD is a severe autosomal recessive bleeding disorder. Most affected individuals are compound heterozygotes for mutations in the vWF gene. However, homozygous individuals have been identified in consanguineous pedigrees. Patients with type 3 vWD demonstrate an absence of vWF with markedly decreased FVIII levels. Mutations in type 3 vWD are well characterized, with over 90 types reported. Diagnosis and clinical management of type 3 vWD does not require molecular testing, although knowledge of specific mutations may be useful in genetic counseling and prenatal diagnosis [100].

Genes Associated with Polycythemia and Coagulopathy

Polycythemia is defined as an increase in peripheral blood red blood cells (RBCs) as demonstrated by increased hemoglobin content, hematocrit, and RBC counts after adjusting for altitude, gender, and/or race [101]. Polycythemia can be primary/congenital (germline mutations) or secondary/acquired (somatic mutations). Molecular abnormalities have been characterized in specific entities of both types. Primary polycythemias associated with coagulopathies include a broad group of hereditary disorders including primary familial polycythemia (benign erythrocytosis) and Chuvash polycythemia. Acquired polycythemia includes the myeloproliferative neoplasms associated with JAK2 mutations.

Acquired polycythemia is seen in patients with myeloproliferative neoplasms (MPNs) such as polycythemia vera (PV) and essential thrombocytosis (ET). Thrombosis is a significant cause of morbidity and mortality in these patients and the initial thrombotic event may occur at the time of diagnosis or during the treatment period [102]. A recent study from the Italian Group for Haematological Diseases described a recurrence rate of 5.6% per patient-year and a cumulative 10-year probability of approximately 50% for a thrombotic event in patients with PV and ET [103]. Cardiovascular events appear to significantly affect patients with MPN and evidence is building for supporting a role of JAK2V617F mutation as a risk factor for thrombosis irrespective of the patient's specific type of MPN [104–106]. Additionally, thrombosis has been associated with both JAK2 exon 12 mutations [107] and calreticulin gene mutations, although the thrombotic risk appears smaller in the latter [108]. Testing for JAK2 mutations is routinely performed as part of the evaluation of patients with an MPN. Although there are no FDA-approved assays for JAK2 mutations currently available, testing is performed by many reference laboratories using laboratory-developed tests.

Congenital polycythemia is suspected in patients with either an early onset of or family history of polycythemia. Congenital erythrocytosis includes patients with germline mutations that result in: (1) enhanced responsiveness of the erythropoietin (EPO) receptor, (2) disrupted intracellular oxygen sensing as occurs in mutations involving the von Hippel-Lindau (VHL) tumor suppressor gene, or (3) increased affinity of hemoglobin for oxygen associated with some hemoglobinopathies [101]. Of these disorders, primary familial polycythemia and Chuvash-type polycythemia (CTP) can be associated with thrombosis.

Primary familial polycythemia (benign erythrocytosis) is an autosomal dominant disorder which may result from mutations in the gene coding for the EPO receptor. It is characterized by an increased absolute RBC mass due to uncontrolled RBC production in the background of low EPO levels [109]. Over 10 different mutations have been described in single families [110]. Most identified EPO receptor mutations are linked to a truncated C-terminal cytoplasmic domain of the receptor that results in heightened sensitivity to circulating EPO because of a lack of negative feedback regulation [111,112]. These patients are at an increased risk of thrombotic events [113].

CTP, or autosomal recessive benign congenital polycythemia, is a rare disorder associated with mutations in the VHL gene present on chromosome 3p25. It occurs worldwide and is endemic in the Chuvash region of central Russia [114]. This disorder is associated with a C→T missense mutation at amino acid residue 200 resulting in an arginine to tryptophan substitution. VHL protein is involved in modulating ubiquitination and the resulting destruction of hypoxia-inducible factor

1 subunit alpha (HIF1α). This mutation is thought to decrease interactions between VHL protein and HIF1α, thereby decreasing the rate of HIF1α degradation. This can lead to increased expression of downstream gene targets including *EPO*, solute carrier family 2 (GLUT1), transferrin, transferrin receptor (p90, CD71), and vascular endothelial growth factor [115]. Other rare mutations in the *VHL* gene associated with CTP have been identified including 235C > T, 562C > G, and 598C > T mutations [110]. CTP is clinically associated with arterial and venous thrombosis and also with major bleeding episodes [114]. Resources are available describing the clinical utility of mutation testing for familial polycythemias including CTP [116].

Genes Affecting Platelet Function

Platelets play a key role in primary hemostasis and variations in platelet genes that result in abnormal platelet function and/or thrombocytopenia can lead to a bleeding phenotype. The loci responsible for many of these platelet function defects have been identified (Table 18.3) [117–121]. Of these, the most studied gene variants are those that encode platelet receptors, such as the integrin αIIb gene (*ITGA2B*) or the integrin β3 gene (*ITGB3*). Abnormalities of these genes result in Glanzmann thrombasthenia, a disorder of platelet aggregation in which platelets cannot bind fibrinogen normally. In Bernard–Soulier syndrome, mutations in genes encoding platelet receptors (*GP1BA*, *GP1BB*, *GP9*) that interact with vWF lead to pathologic bleeding. Mutations in several genes involved in the formation of lysosomal-related organelles are found in patients with Hermansky–Pudlak syndrome (HPS), a bleeding disorder associated with oculocutaneous albinism. Next-generation sequencing (NGS) has been used to identify a pathogenic single-nucleotide variation in one of the HPS genes (*HPS4*) in a patient with this disorder. However, this technology is not routinely used when evaluating platelet function disorders [122]. RNA and exome sequencing were used to identify *NBEAL2* as the causative gene in gray platelet syndrome [123], a disease in which platelets are deficient in granules that contain proteins (eg, platelet factor 4, vWF) critical for normal platelet responses to injury. Wiskott-Aldrich syndrome (WAS) results from mutations in the *WAS* gene located on the X chromosome, which results in disruption of actin cytoskeletal organization and signaling. Gene sequencing has been used to identify 62 unique *WAS* mutations including 17 novel sequence variants in 87 affected males and 48 female carriers [124]. Overall, characterization of inherited platelet disorders has been challenging because of the rarity of these diseases and often lack of correspondence between a genetic polymorphism and the platelet phenotype. For these and other reasons, testing for common sequence variants is currently of little clinical utility.

CLINICAL UTILITY

The clinical utility of molecular-based assays used to test patients with hemostatic disorders is most clearly established for thrombotic disorders such as FVL and factor II PT mutations. The role of molecular testing for bleeding disorders is less clear as many of these disorders are often screened for and diagnosed using immunogenic and functional assays. Although molecular testing has become more available for some

TABLE 18.3 Inherited Platelet Disorders

Platelet disorder	Clinical findings	Inheritance	Etiology	Molecular findings
Glanzmann thrombasthenia	Spontaneous mucocutaneous bleeding, excessive trauma related bleeding	Autosomal recessive	Quantitative or qualitative deficiencies of integrins αIIb and β3	Mutations in *ITGA2B* and *ITGB3*
Bernard–Soulier syndrome	Low platelet count, abnormally large platelets, often severe bleeding	Autosomal recessive (biallelic) Autosomal dominant (monoallelic)	Defects in glycoproteins Ibα, Ibβ, IX	Mutations in *GP1BA*, *GP1BB*, *GP9*
Hermansky–Pudlak syndrome	Oculocutaneous albinism, bleeding	Autosomal recessive	Platelet storage pool defect and lysosomal accumulation of ceroid lipofuscin	Mutations in multiple genes (*AP3B1*, *BLOC*, *HPS* family)
Gray platelet syndrome	Mild thrombocytopenia, enlarged platelets, mild/moderate bleeding	Autosomal recessive	Reduced platelet α-granules	Mutations in *NBEAL2*
Wiskott-Aldrich syndrome	Thrombocytopenia, small platelets, eczema, immune disorders, malignancies	X-linked	Abnormal actin cytoskeletal organization and signaling	Mutations in the *WAS* gene

hemostatic disorders, these techniques are still reserved for a small number of cases. Molecular testing must be coordinated with clinical colleagues who will need to interpret the results within the clinical context and offer counseling as appropriate. With the advent of targeted therapies based on specific genetic findings, testing may become more commonplace for some of the clotting disorders.

LIMITATIONS OF TESTING

In general, the basic technology utilized in molecular testing for coagulopathies (eg, FVL, factor II) has not changed significantly in the past few years. However, some testing has been discontinued due to concerns regarding clinical utility. The greatest technological changes have arisen with the commercial adoption of NGS, allowing for expanded menu offerings utilizing this newer technology. As NGS technologies become more widely accepted and affordable in the clinical laboratory, targeted mutation assays or even whole exome sequencing could be applied to hemostasis testing [125]. These technologies could detect many of the mutations described in this chapter in a single assay. One important aspect that requires further clarification in hemostatic and other disorders is the challenge determining the difference between mutations that confer a disease phenotype and those genetic polymorphisms that produce no adverse effects [126]. These technologies also have the potential to identify new genetic variants of unclear significance [127]. Careful use of informatics approaches will be needed to analyze novel sequence data related to coagulopathies and to determine their clinical significance.

References

[1] Marlar RA, Neumann A. Neonatal purpura fulminans due to homozygous protein C or protein S deficiencies. Semin Thromb Hemost 1990;16:299–309.

[2] Dahlback B. Advances in understanding pathogenic mechanisms of thrombophilic disorders. Blood 2008;112:19–27.

[3] Kalafatis M, Mann KG. Factor V Leiden and thrombophilia. Arterioscler Thromb Vasc Biol 1997;17:620–7.

[4] Bertina RM, Koeleman BP, Koster T, et al. Mutation in blood coagulation factor V associated with resistance to activated protein C. Nature 1994;369:64–7.

[5] Kujovich JL. Factor V Leiden thrombophilia. Genet Med 2011;13:1–16.

[6] Heit JA, Sobell JL, Li H, Sommer SS. The incidence of venous thromboembolism among factor V Leiden carriers: a community-based cohort study. J Thromb Haemost 2005;3:305–11.

[7] Grody WW, Griffin JH, Taylor AK, Korf BR, Heit JA, Group AFVLW. American College of Medical Genetics consensus statement on factor V Leiden mutation testing. Genet Med 2001;3:139–48.

[8] Spector EB, Grody WW, Matteson CJ, et al. Technical standards and guidelines: venous thromboembolism (factor V Leiden and prothrombin 20210G >A testing): a disease-specific supplement to the standards and guidelines for clinical genetics laboratories. Genet Med 2005;7:444–53.

[9] Lyondagger E, Millsondagger A, Phan T, Wittwer CT. Detection and identification of base alterations within the region of factor V Leiden by fluorescent melting curves. Mol Diagn 1998;3:203–9.

[10] Mihalatos M, Apessos A, Dauwerse H, et al. Rare mutations predisposing to familial adenomatous polyposis in Greek FAP patients. BMC Cancer 2005;5:40.

[11] Almawi WY, Tamim H, Kreidy R, et al. A case control study on the contribution of factor V-Leiden, prothrombin G20210A, and MTHFR C677T mutations to the genetic susceptibility of deep venous thrombosis. J Thromb Thrombolysis 2005;19:189–96.

[12] Gehring NH, Frede U, Neu-Yilik G, et al. Increased efficiency of mRNA 3' end formation: a new genetic mechanism contributing to hereditary thrombophilia. Nat Genet 2001;28:389–92.

[13] Danckwardt S, Hartmann K, Gehring NH, Hentze MW, Kulozik AE. 3' end processing of the prothrombin mRNA in thrombophilia. Acta Haematol 2006;115:192–7.

[14] Rosendaal FR, Doggen CJ, Zivelin A, et al. Geographic distribution of the 20210 G to A prothrombin variant. Thromb Haemost 1998;79:706–8.

[15] Kujovich JL, Factor V. Leiden thrombophilia. In: Pagon RA, Adam MP, Ardinger HH, et al., editors. GeneReviews(R). Seattle, WA: University of Washington; 1993.

[16] Ozelo MC, Annichino-Bizzacchi JM, Pollak ES, Russell JE. Rapid detection of the prothrombin C20209T variant by differential sensitivity to restriction endonuclease digestion. J Thromb Haemost 2003;1:2683–5.

[17] Weisberg I, Tran P, Christensen B, Sibani S, Rozen R. A second genetic polymorphism in methylenetetrahydrofolate reductase (MTHFR) associated with decreased enzyme activity. Mol Genet Metab 1998;64:169–72.

[18] Frosst P, Blom HJ, Milos R, et al. A candidate genetic risk factor for vascular disease: a common mutation in methylenetetrahydrofolate reductase. Nat Genet 1995;10:111–13.

[19] Den Heijer M, Lewington S, Clarke R. Homocysteine, MTHFR and risk of venous thrombosis: a meta-analysis of published epidemiological studies. J Thromb Haemost 2005;3:292–9.

[20] Brown NM, Pratt VM, Buller A, et al. Detection of 677CT/1298AC "double variant" chromosomes: implications for interpretation of MTHFR genotyping results. Genet Med 2005;7:278–82.

[21] den Heijer M, Koster T, Blom HJ, et al. Hyperhomocysteinemia as a risk factor for deep-vein thrombosis. N Engl J Med 1996;334:759–62.

[22] Wald DS, Morris JK, Wald NJ. Reconciling the evidence on serum homocysteine and ischaemic heart disease: a meta-analysis. PLoS One 2011;6:e16473.

[23] Homocysteine Studies C. Homocysteine and risk of ischemic heart disease and stroke: a meta-analysis. JAMA 2002;288:2015–22.

[24] Clarke R, Bennett DA, Parish S, et al. Homocysteine and coronary heart disease: meta-analysis of MTHFR case-control studies, avoiding publication bias. PLoS Med 2012;9:e1001177.

[25] Hickey SE, Curry CJ, Toriello HV. ACMG Practice Guideline: lack of evidence for MTHFR polymorphism testing. Genet Med 2013;15:153–6.

[26] Khor B, Van Cott EM. Laboratory tests for antithrombin deficiency. Am J Hematol 2010;85:947–50.

[27] Tait RC, Walker ID, Perry DJ, et al. Prevalence of antithrombin deficiency in the healthy population. Br J Haematol 1994;87:106–12.

[28] Wells PS, Blajchman MA, Henderson P, et al. Prevalence of antithrombin deficiency in healthy blood donors: a cross-sectional study. Am J Hematol 1994;45:321–4.

[29] Patnaik MM, Moll S. Inherited antithrombin deficiency: a review. Haemophilia 2008;14:1229–39.

[30] Bock SC, Harris JF, Balazs I, Trent JM. Assignment of the human antithrombin III structural gene to chromosome 1q23-25. Cytogenet Cell Genet 1985;39:67–9.

[31] Fitches AC, Appleby R, Lane DA, De Stefano V, Leone G, Olds RJ. Impaired cotranslational processing as a mechanism for type I antithrombin deficiency. Blood 1998;92:4671–6.

[32] Ishiguro K, Kojima T, Kadomatsu K, et al. Complete antithrombin deficiency in mice results in embryonic lethality. J Clin Invest 2000;106:873–8.

[33] Picard V, Nowak-Gottl U, Biron-Andreani C, et al. Molecular bases of antithrombin deficiency: twenty-two novel mutations in the antithrombin gene. Hum Mutat 2006;27:600.

[34] Cooper PC, Coath F, Daly ME, Makris M. The phenotypic and genetic assessment of antithrombin deficiency. Int J Lab Hematol 2011;33:227–37.

[35] Wadelius M, Pirmohamed M. Pharmacogenetics of warfarin: current status and future challenges. Pharmacogenomics J 2007;7:99–111.

[36] Wadelius M, Chen LY, Lindh JD, et al. The largest prospective warfarin-treated cohort supports genetic forecasting. Blood 2009;113:784–92.

[37] Hynicka LM, Cahoon Jr. WD, Bukaveckas BL. Genetic testing for warfarin therapy initiation. Ann Pharmacother 2008;42:1298–303.

[38] Cavallari LH, Perera MA. The future of warfarin pharmacogenetics in under-represented minority groups. Future Cardiol 2012;8:563–76.

[39] Geisen C, Watzka M, Sittinger K, et al. VKORC1 haplotypes and their impact on the inter-individual and inter-ethnical variability of oral anticoagulation. Thromb Haemost 2005;94:773–9.

[40] Rieder MJ, Reiner AP, Gage BF, et al. Effect of VKORC1 haplotypes on transcriptional regulation and warfarin dose. N Engl J Med 2005;352:2285–93.

[41] Anderson JL, Horne BD, Stevens SM, et al. Randomized trial of genotype-guided versus standard warfarin dosing in patients initiating oral anticoagulation. Circulation 2007;116:2563–70.

[42] Sconce EA, Khan TI, Wynne HA, et al. The impact of CYP2C9 and VKORC1 genetic polymorphism and patient characteristics upon warfarin dose requirements: proposal for a new dosing regimen. Blood 2005;106:2329–33.

[43] Voora D, Eby C, Linder MW, et al. Prospective dosing of warfarin based on cytochrome P-450 2C9 genotype. Thromb Haemost 2005;93:700–5.

[44] Hillman MA, Wilke RA, Yale SH, et al. A prospective, randomized pilot trial of model-based warfarin dose initiation using CYP2C9 genotype and clinical data. Clin Med Res 2005;3:137–45.

[45] Gage BF, Eby C, Johnson JA, et al. Use of pharmacogenetic and clinical factors to predict the therapeutic dose of warfarin. Clin Pharmacol Ther 2008;84:326–31.

[46] Wu AH, Wang P, Smith A, et al. Dosing algorithm for warfarin using CYP2C9 and VKORC1 genotyping from a multi-ethnic population: comparison with other equations. Pharmacogenomics 2008;9:169–78.

[47] Limdi NA, Beasley TM, Crowley MR, et al. VKORC1 polymorphisms, haplotypes and haplotype groups on warfarin dose among African-Americans and European-Americans. Pharmacogenomics 2008;9:1445–58.

[48] International Warfarin Pharmacogenetics Consortium, Klein TE, Altman RB, et al. Estimation of the warfarin dose with clinical and pharmacogenetic data. N Engl J Med 2009;360:753–64.

[49] Kimmel SE, French B, Kasner SE, et al. A pharmacogenetic versus a clinical algorithm for warfarin dosing. N Engl J Med 2013;369:2283–93.

[50] Pirmohamed M, Burnside G, Eriksson N, et al. A randomized trial of genotype-guided dosing of warfarin. N Engl J Med 2013;369:2294–303.

[51] Sconce EA, Daly AK, Khan TI, Wynne HA, Kamali F. APOE genotype makes a small contribution to warfarin dose requirements. Pharmacogenet Genomics 2006;16:609–11.

[52] Leung A, Huang CK, Muto R, Liu Y, Pan Q. CYP2C9 and VKORC1 genetic polymorphism analysis might be necessary in patients with factor V Leiden and prothrombin gene G2021A mutation(s). Diagn Mol Pathol 2007;16:184–6.

[53] Flockhart DA, O'Kane D, Williams MS, et al. Pharmacogenetic testing of CYP2C9 and VKORC1 alleles for warfarin. Genet Med 2008;10:139–50.

[54] Hirsh J, Guyatt G, Albers GW, Harrington R, Schunemann HJ. American College of Chest Physicians. Executive summary: American College of Chest Physicians evidence-based clinical practice guidelines (8th Edition). Chest 2008;133:71S–109S.

[55] Baker WL, Chamberlin KW. New oral anticoagulants vs. warfarin treatment: no need for pharmacogenomics? Clin Pharmacol Ther 2014;96:17–19.

[56] White 2nd GC, Rosendaal F, Aledort LM, et al. Definitions in hemophilia. Recommendation of the scientific subcommittee on factor VIII and factor IX of the scientific and standardization committee of the International Society on Thrombosis and Haemostasis. Thromb Haemost 2001;85:560.

[57] Nussbaum RL, McInnes RR, Willard HF, Thompson MW, Hamosh A. Thompson & Thompson genetics in medicine. 7th ed. Philadelphia, PA: Saunders/Elsevier; 2007.

[58] Naylor JA, Green PM, Rizza CR, Giannelli F. Factor VIII gene explains all cases of haemophilia A. Lancet 1992;340:1066–7.

[59] Levinson B, Kenwrick S, Lakich D, Hammonds Jr. G, Gitschier J. A transcribed gene in an intron of the human factor VIII gene. Genomics 1990;7:1–11.

[60] Naylor JA, Buck D, Green P, Williamson H, Bentley D, Giannelli F. Investigation of the factor VIII intron 22 repeated region (int22h) and the associated inversion junctions. Hum Mol Genet 1995;4:1217–24.

[61] Lakich D, Kazazian Jr. HH, Antonarakis SE, Gitschier J. Inversions disrupting the factor VIII gene are a common cause of severe haemophilia A. Nat Genet 1993;5:236–41.

[62] Naylor J, Brinke A, Hassock S, Green PM, Giannelli F. Characteristic mRNA abnormality found in half the patients with severe haemophilia A is due to large DNA inversions. Hum Mol Genet 1993;2:1773–8.

[63] Van de Water NS, Williams R, Nelson J, Browett PJ. Factor VIII gene inversions in severe hemophilia A patients. Pathology 1995;27:83–5.

[64] Bagnall RD, Waseem N, Green PM, Giannelli F. Recurrent inversion breaking intron 1 of the factor VIII gene is a frequent cause of severe hemophilia A. Blood 2002;99:168–74.

[65] Repesse Y, Slaoui M, Ferrandiz D, et al. Factor VIII (FVIII) gene mutations in 120 patients with hemophilia A: detection of 26 novel mutations and correlation with FVIII inhibitor development. J Thromb Haemost 2007;5:1469–76.

[66] Santacroce R, Acquila M, Belvini D, et al. Identification of 217 unreported mutations in the F8 gene in a group of 1,410 unselected Italian patients with hemophilia A. J Hum Genet 2008;53:275–84.

[67] Schwaab R, Brackmann HH, Meyer C, et al. Haemophilia A: mutation type determines risk of inhibitor formation. Thromb Haemost 1995;74:1402–6.

III. MOLECULAR TESTING IN GENETIC DISEASE

[68] Gilles JG, Peerlinck K, Arnout J, Vermylen J, Saint-Remy JM. Restricted epitope specificity of anti-FVIII antibodies that appeared during a recent outbreak of inhibitors. Thromb Haemost 1997;77:938−43.

[69] Salviato R, Belvini D, Radossi P, et al. F8 gene mutation profile and ITT response in a cohort of Italian haemophilia A patients with inhibitors. Haemophilia 2007;13:361−72.

[70] Neerman-Arbez M, Johnson KM, Morris MA, et al. Molecular analysis of the ERGIC-53 gene in 35 families with combined factor V-factor VIII deficiency. Blood 1999;93:2253−60.

[71] Nichols WC, Seligsohn U, Zivelin A, et al. Mutations in the ER-golgi intermediate compartment protein ERGIC-53 cause combined deficiency of coagulation factors V and VIII. Cell 1998;93:61−70.

[72] Kessler L, Adams R, Mighion L, Walther S, Ganguly A. Prenatal diagnosis in haemophilia A: experience of the genetic diagnostic laboratory. Haemophilia 2014;20:e384−91.

[73] CDC Hemophilia A Mutation Project (CHAMP). <http://www.cdc.gov/ncbddd/hemophilia/champs.html/> [accessed 02.02.15].

[74] Peyvandi F, Jayandharan G, Chandy M, et al. Genetic diagnosis of haemophilia and other inherited bleeding disorders. Haemophilia 2006;12:82−9.

[75] Leuer M, Oldenburg J, Lavergne JM, et al. Somatic mosaicism in hemophilia A: a fairly common event. Am J Hum Genet 2001;69:75−87.

[76] Veltkamp JJ, Meilof J, Remmelts HG, van der Vlerk D, Loeliger EA. Another genetic variant of haemophilia B: haemophilia B Leyden. Scand J Haematol 1970;7:82−90.

[77] Reitsma PH, Mandalaki T, Kasper CK, Bertina RM, Briet E. Two novel point mutations correlate with an altered developmental expression of blood coagulation factor IX (hemophilia B Leyden phenotype). Blood 1989;73:743−6.

[78] Oldenburg J, Quenzel EM, Harbrecht U, et al. Missense mutations at ALA-10 in the factor IX propeptide: an insignificant variant in normal life but a decisive cause of bleeding during oral anticoagulant therapy. Br J Haematol 1997;98:240−4.

[79] Thorland EC, Drost JB, Lusher JM, et al. Anaphylactic response to factor IX replacement therapy in haemophilia B patients: complete gene deletions confer the highest risk. Haemophilia 1999;5:101−5.

[80] Ketterling RP, Bottema CD, Phillips 3rd JA, Sommer SS. Evidence that descendants of three founders constitute about 25% of hemophilia B in the United States. Genomics 1991;10:1093−6.

[81] Emsley J, McEwan PA, Gailani D. Structure and function of factor XI. Blood 2010;115:2569−77.

[82] Mitchell M, Mountford R, Butler R, et al. Spectrum of factor XI (F11) mutations in the UK population—116 index cases and 140 mutations. Hum Mutat 2006;27:829.

[83] Peretz H, Mulai A, Usher S, et al. The two common mutations causing factor XI deficiency in Jews stem from distinct founders: one of ancient Middle Eastern origin and another of more recent European origin. Blood 1997;90:2654−9.

[84] Dai L, Rangarajan S, Mitchell M. Three dominant-negative mutations in factor XI-deficient patients. Haemophilia 2011;17:e919−22.

[85] Gomez K, Bolton-Maggs P. Factor XI deficiency. Haemophilia 2008;14:1183−9.

[86] Salomon O, Steinberg DM, Seligshon U. Variable bleeding manifestations characterize different types of surgery in patients with severe factor XI deficiency enabling parsimonious use of replacement therapy. Haemophilia 2006;12:490−3.

[87] Salomon O, Steinberg DM, Koren-Morag N, Tanne D, Seligsohn U. Reduced incidence of ischemic stroke in patients with severe factor XI deficiency. Blood 2008;111:4113−17.

[88] Kato A, Asakai R, Davie EW, Aoki N. Factor XI gene (F11) is located on the distal end of the long arm of human chromosome 4. Cytogenet Cell Genet 1989;52:77−8.

[89] James PD, Goodeve AC von. Willebrand disease. Genet Med 2011;13:365−76.

[90] Sadler JE, Budde U, Eikenboom JC, et al. Update on the pathophysiology and classification of von Willebrand disease: a report of the Subcommittee on von Willebrand Factor. J Thromb Haemost 2006;4:2103−14.

[91] James PD, Paterson AD, Notley C, et al. Genetic linkage and association analysis in type 1 von Willebrand disease: results from the Canadian type 1 VWD study. J Thromb Haemost 2006;4:783−92.

[92] Eikenboom J, Van Marion V, Putter H, et al. Linkage analysis in families diagnosed with type 1 von Willebrand disease in the European study, molecular and clinical markers for the diagnosis and management of type 1 VWD. J Thromb Haemost 2006;4:774−82.

[93] von Willebrand factor Variant Database (VWFdb). <http://www.vwf.group.shef.ac.uk/> [accessed 02.02.15].

[94] Casonato A, Pontara E, Dannhaeuser D, et al. Re-evaluation of the therapeutic efficacy of DDAVP in type IIB von Willebrand's disease. Blood Coagul Fibrinolysis 1994;5:959−64.

[95] Mauz-Korholz C, Budde U, Kruck H, Korholz D, Gobel U. Management of severe chronic thrombocytopenia in von Willebrand's disease type 2B. Arch Dis Child 1998;78:257−60.

[96] Ahmad F, Jan R, Kannan M, et al. Characterisation of mutations and molecular studies of type 2 von Willebrand disease. Thromb Haemost 2013;109:39−46.

[97] Rayes J, Hommais A, Legendre P, et al. Effect of von Willebrand disease type 2B and type 2M mutations on the susceptibility of von Willebrand factor to ADAMTS-13. J Thromb Haemost 2007;5:321−8.

[98] Fressinaud E, Mazurier C, Meyer D. Molecular genetics of type 2 von Willebrand disease. Int J Hematol 2002;75:9−18.

[99] Schneppenheim R, Budde U, Krey S, et al. Results of a screening for von Willebrand disease type 2N in patients with suspected haemophilia A or von Willebrand disease type 1. Thromb Haemost 1996;76:598−602.

[100] Lillicrap D von. Willebrand disease: advances in pathogenetic understanding, diagnosis, and therapy. Blood 2013;122 3735−40.

[101] Patnaik MM, Tefferi A. The complete evaluation of erythrocytosis: congenital and acquired. Leukemia 2009;23:834−44.

[102] Falanga A, Marchetti M. Thrombosis in myeloproliferative neoplasms. Semin Thromb Hemost 2014;40:348−58.

[103] De Stefano V, Za T, Rossi E, et al. Recurrent thrombosis in patients with polycythemia vera and essential thrombocythemia: incidence, risk factors, and effect of treatments. Haematologica 2008;93:372−80.

[104] Vannucchi AM. Insights into the pathogenesis and management of thrombosis in polycythemia vera and essential thrombocythemia. Intern Emerg Med 2010;5:177−84.

[105] Barbui T, Finazzi G, Falanga A. Myeloproliferative neoplasms and thrombosis. Blood 2013;122:2176−84.

[106] Vannucchi AM, Guglielmelli P. JAK2 mutation-related disease and thrombosis. Semin Thromb Hemost 2013;39:496−506.

[107] Passamonti F, Elena C, Schnittger S, et al. Molecular and clinical features of the myeloproliferative neoplasm associated with JAK2 exon 12 mutations. Blood 2011;117:2813−16.

[108] Andrikovics H, Krahling T, Balassa K, et al. Distinct clinical characteristics of myeloproliferative neoplasms with calreticulin mutations. Haematologica 2014;99:1184—90.

[109] Huang LJ, Shen YM, Bulut GB. Advances in understanding the pathogenesis of primary familial and congenital polycythaemia. Br J Haematol 2010;148:844—52.

[110] Bento C, Percy MJ, Gardie B, et al. Genetic basis of congenital erythrocytosis: mutation update and online databases. Hum Mutat 2014;35:15—26.

[111] Cario H. Childhood polycythemias/erythrocytoses: classification, diagnosis, clinical presentation, and treatment. Ann Hematol 2005;84:137—45.

[112] Perrotta S, Cucciolla V, Ferraro M, et al. EPO receptor gain-of-function causes hereditary polycythemia, alters CD34 cell differentiation and increases circulating endothelial precursors. PLoS One 2010;5:e12015.

[113] Kiladjian JJ, Gardin C, Renoux M, Bruno F, Bernard JF. Long-term outcomes of polycythemia vera patients treated with pipobroman as initial therapy. Hematol J 2003;4:198—207.

[114] Gordeuk VR, Prchal JT. Vascular complications in Chuvash polycythemia. Semin Thromb Hemost 2006;32:289—94.

[115] Ang SO, Chen H, Gordeuk VR, et al. Endemic polycythemia in Russia: mutation in the VHL gene. Blood Cells Mol Dis 2002;28:57—62.

[116] Hussein K, Granot G, Shpilberg O, Kreipe H. Clinical utility gene card for: familial polycythaemia vera. Eur J Hum Genet 2013;21. Available from: http://dx.doi.org/10.1038/ejhg.2012.216.

[117] Andrews RK, Berndt MC. Bernard-Soulier syndrome: an update. Semin Thromb Hemost 2013;39:656—62.

[118] Hurford MT, Sebastiano C. Hermansky-pudlak syndrome: report of a case and review of the literature. Int J Clin Exp Pathol 2008;1:550—4.

[119] Nurden AT, Nurden P. The gray platelet syndrome: clinical spectrum of the disease. Blood Rev 2007;21:21—36.

[120] Nurden AT, Pillois X, Wilcox DA. Glanzmann thrombasthenia: state of the art and future directions. Semin Thromb Hemost 2013;39:642—55.

[121] Zhu Q, Watanabe C, Liu T, et al. Wiskott-Aldrich syndrome/X-linked thrombocytopenia: WASP gene mutations, protein expression, and phenotype. Blood 1997;90:2680—9.

[122] Jones ML, Murden SL, Bem D, et al. Rapid genetic diagnosis of heritable platelet function disorders with next-generation sequencing: proof-of-principle with Hermansky-Pudlak syndrome. J Thromb Haemost 2012;10:306—9.

[123] Albers CA, Cvejic A, Favier R, et al. Exome sequencing identifies NBEAL2 as the causative gene for gray platelet syndrome. Nat Genet 2011;43:735—7.

[124] Gulacsy V, Freiberger T, Shcherbina A, et al. Genetic characteristics of eighty-seven patients with the Wiskott-Aldrich syndrome. Mol Immunol 2011;48:788—92.

[125] Iglesias AI, Springelkamp H, van der Linde H, et al. Exome sequencing and functional analyses suggest that SIX6 is a gene involved in an altered proliferation-differentiation balance early in life and optic nerve degeneration at old age. Hum Mol Genet 2014;23:1320—32.

[126] Cooper DN, Krawczak M, Polychronakos C, Tyler-Smith C, Kehrer-Sawatzki H. Where genotype is not predictive of phenotype: towards an understanding of the molecular basis of reduced penetrance in human inherited disease. Hum Genet 2013;132:1077—130.

[127] MacArthur DG, Manolio TA, Dimmock DP, et al. Guidelines for investigating causality of sequence variants in human disease. Nature 2014;508:469—76.

III. MOLECULAR TESTING IN GENETIC DISEASE

19

Molecular Diagnosis of Cystic Fibrosis

Y. Si[1] and D.H. Best[1,2]

[1]Department of Pathology, University of Utah School of Medicine, Salt Lake City, UT, United States
[2]Molecular Genetics and Genomics, ARUP Laboratories, University of Utah School of Medicine, Salt Lake City, UT, United States

INTRODUCTION

Cystic fibrosis (CF) is one of the most common life-threatening autosomal recessive disorders affecting approximately 1 in 2500 people of European Caucasian descent [1]. Worldwide there are approximately 70,000 individuals affected with CF [2,3]. In the United States alone, 30,000 individuals are affected with CF and an additional 12 million people are CF carriers [4]. Typically, individuals affected with classic CF disease survive into their 40s [5]. However, studies have shown that early diagnosis and treatment of CF can improve both survival and quality of life [6–8].

CF is caused by loss-of-function mutations in the CF transmembrane conductance regulator (*CFTR*) gene [9,10]. The *CFTR* gene encodes an ATP-binding cassette (ABC) transporter protein called CFTR that is expressed on the apical side of epithelia [11,12]. CFTR functions as a low conductance chloride-selective channel and mediates electrolyte transport across epithelial membranes. Therefore, the loss of CFTR function can affect any organ system with epithelia. As a result, the clinical manifestations of CF are widespread and can include findings in the lungs, pancreas, intestine, hepatobiliary systems, exocrine sweat glands, and male genital tract [6,13]. Due to the broad variability of clinical findings and disease severity, a clinical designation of *CFTR*-related disorders has gained acceptance in the general medical community to refer to the myriad effects of the loss of CFTR function [14,15]. The wide spectrum of clinical findings and disease severity in CF and related disorders is not only due to multisystem involvement, but also by the complexity of the genetic variations in *CFTR* gene [13].

Clinical Criteria for Diagnosis of CF

Clinical diagnosis of CF requires that the patient demonstrate one or more of the characteristic phenotypic features of CF (Table 19.1) plus evidence of abnormality in CFTR protein function based on one of the three laboratory criteria: [16,17] (1) presence of two disease-causing mutations in the *CFTR* gene, (2) two abnormal quantitative sweat chloride values (>60 mEq/L) [18], and/or (3) transepithelial nasal potential difference measurements characteristic of CF [19]. Positive test results for any of these tests in a symptomatic patient confirm a clinical diagnosis. It is important to note that in prenatal samples and newborns, a diagnosis of CF can be made based solely on the identification of two disease-causing mutations in the *CFTR* gene or (in the case of newborns) multiple abnormal sweat chloride values. Sweat chloride testing is considered the gold standard for confirming a CF diagnosis and for the majority of cases molecular testing is not strictly required to support clinical diagnosis [18]. However, in such situations molecular confirmation still carries clinical utility in terms of confirming diagnosis and enabling carrier testing and prenatal diagnosis within the extended family. In some situations, individuals may demonstrate atypical (monosymptomatic) disease symptoms and have sweat chloride values that are inconclusive. These cases require genetic analysis to identify the disease-causing mutations and confirm a diagnosis of atypical CF. In addition, the recent development of FDA-approved targeted drug therapy necessitates accurate *CFTR* genetic testing [20–22].

© 2017 Elsevier Inc. All rights reserved.

MOLECULAR TARGET

CFTR Gene and Mutation Spectrum

The *CFTR* gene was first identified in 1989 and was met with great excitement that this discovery would provide the tools necessary to do population-based screening for CF [23]. However, due to gene complexity and the wide spectrum of *CFTR* gene mutations, CF screening did not become a reality for a decade after the gene was described. *CFTR* (7q31.2) is a large gene that spans 250 kb and contains a total of 27 exons [24]. The most common (and first identified) *CFTR* gene mutation, p.Phe508del (historically called ΔF508), accounts for 70% of CF mutations among the non-Jewish white population [25–27]. However, the distribution of known *CFTR* mutations encompasses all exons, introns, and regulatory regions of the gene. Mutations in the *CFTR* gene are not limited to any specific class and can include missense, nonsense, frameshift, in-frame or out-of-frame insertions/deletions, splicing, and promoter mutations.

Over 1900 *CFTR* gene variants have been cataloged in the Sickkids *CFTR* mutation database [5]. However,

only a subset of the described variants are known to be disease causing [28]. *CFTR* mutations have traditionally been classified into one of five categories. Recently, class VI was added to the classification system (Table 19.2). These classifications are solely based on the effect that the mutation has on the CFTR protein (eg, premature termination). Briefly, the six classes of mutations are as follows:

Class I—Mutations that cause a premature stop codon and result in defective CFTR protein synthesis. Nonsense mutations, frameshift mutations, large deletions, and splice junction mutations are included in this category. Collectively, they account approximately 10% of *CFTR* mutations [29].
Class II—Mutations that lead to defective protein trafficking or protein processing. The most common *CFTR* mutation, p.Phe508del, some missense mutations, and all in-frame deletion mutations fall into this category.
Class III—Mutations that result in a defect in channel regulation affecting chloride transport.
Class IV—Mutations that result in defective channel conduction due to channel narrowing.
Class V—Mutations caused by splicing defects resulting in the improper processing of mRNA. This leads to a reduced amount of protein capable of reaching the cell surface.
Class VI—Mutations that cause reduced stability of the CFTR protein at the cell surface.

It is worth noting that some mutations may fall into multiple categories. The most common *CFTR* gene mutation, p.Phe508del, mainly results in misfolded protein and is typically categorized as a class II mutation. However, in some cases the CFTR protein with a p.Phe508del mutation makes it to the apical surface, but exhibits gating and conductance defects and would classify as a class III and VI mutation [30,31].

TABLE 19.1 Common Clinical Symptoms of Cystic Fibrosis Patients

- Pulmonary findings
  - Chronic cough
  - Recurrent lung infection
  - Exertional dyspnea
  - Bronchiectasis
- Gastrointestinal findings
  - Meconium ileus
  - Rectal prolapse
  - Pancreatic insufficiency/recurrent pancreatitis
  - Diabetes mellitus
- Infertility
  - Azoospermia
  - Congenital absence of the vas deferens

TABLE 19.2 Classification of *CFTR* Gene Mutations

	Class I	Class II	Class III	Class IV	Class V	Class VI
Functional consequence	No protein synthesis	Defective trafficking or protein processing	Defect in channel regulating or "gating"	Altered channel conductance	Reduced protein synthesis	Reduced protein stability
Molecular defect	Nonsense, frameshift, deletions, splicing	Missense, in-frame deletions	Missense	Missense	Splicing, missense	Varies
Examples	p.Trp1282* p.Arg553* p.Gly542* c.1717-1G > A	p.Phe508del p.Asn1303Lys	p.Gly551Asp p.Gly551Ser p.Gly1349Asp	p.Arg117His p.Arg334Trp p.Arg347Pro	c.2789 + 5G > A p.Ala455Glu	p.Asn287Tyr c.4278insA

Mutations in classes I, II, III, and VI are commonly associated with severe impairment of channel function and a severe disease phenotype. In contrast, CFTR protein function is usually preserved at some level in class IV and V mutations. Individuals with mutations in these two classes typically exhibit a milder phenotype.

Genotype–Phenotype Correlations

Genotype–phenotype correlations for CF are established in the context of pancreatic function. The most common CF-causing mutations can be categorized as either pancreatic sufficient (PS) or pancreatic insufficient (PI). Typically, mutations in classes I, II, III, and VI predict a PI phenotype, while classes IV and V predict a PS phenotype [32,33]. Individuals with at least one mild mutation are usually PS, indicating that the milder of the two mutations is *dominant* with respect to pancreatic function. In contrast to the fairly accurate genotype–pancreatic phenotype correlation, genotype–phenotype correlation for lung function in CF patients is not an adequate method to predict disease progression. Multiple studies have demonstrated that individuals carrying identical CF mutations, even those in the same family, can have different pulmonary manifestations [4,34].

The mildest form of CF is a form of infertility known as congenital bilateral absence of the vas deferens (CBAVD). CBAVD is most commonly observed as an isolated phenotype, but has also been described in individuals with respiratory and/or pancreatic issues [35,36]. Usually, individuals with CBAVD carry both a severe and a mild (or very mild) *CFTR* gene mutation.

It is important to note that clinical manifestations of CF are highly variable, complex, and influenced by environmental factors and additional genetic factors (eg, modifier genes) [37]. Caution should always be applied when trying to utilize genotype information (or mutation classification) to predict the clinical course of the disease.

ACMG Guideline for CFTR Molecular Testing

As one of the most common severe autosomal recessive disorders (see Table 19.3 for the carrier frequencies in various ethnic groups), CF carrier screening has been recommended by a number of professional organizations [38–41]. In 1997, the NIH published recommendations suggesting that all couples seeking reproductive counseling should be offered genetic testing for CF [42]. The NIH recommendations also suggested that any individual with a family history of CF or reproductive partners of those with CF should be offered genetic testing [43]. Subsequently, both the American College of

TABLE 19.3 *CFTR* Mutation Carrier Risk by Ethnicity [44]

Ethnic group	Detection rate	Risk before testing	Risk after negative result
Ashkenazi Jewish	95%	1 in 25	~1 in 930
European Caucasian	89%	1 in 25	~1 in 220
African American	65%	1 in 65	~1 in 207
Hispanic American	73%	1 in 46	~1 in 105
Asian American	~30%	1 in 90	***

***Based on p.Phe508del only. No additional data available.

TABLE 19.4 Mutations Recommended for CF Screening by the ACMG

Missense	Deletions	Nonsense	Frameshift	Splicing
p.Gly85Glu	p.Ile507del	p.Gly542*	c.2184delA	c.621 + 1G > T
p.Arg117His	p.Phe508del	p.Arg553*	c.3659delC	c.711 + 1G > T
p.Arg334Trp		p.Arg1162*		c.1717 − 1G > A
p.Arg347Pro		p.Trp1282*		c.1898 + 1G > A
p.Ala455Glu				c.2789 + 5G > A
p.Gly551Asp				c.3120 + 1G > A
p.Arg560Thr				c.3849 + 10 kb
p.Asn1303Lys				

Medical Genetics (ACMG) and the American College of Obstetricians and Gynecologists (ACOG) published statements that recommended a universal, pan-ethnic *CFTR* mutation screening panel for all individuals planning pregnancy or to pregnant females [38]. Based on phenotypic severity and a population frequency more than or equal to 0.1% in the general US population affected with CF, a group of 25 mutations was selected for the core mutation panel [44]. In 2004, the core panel was updated and reduced to 23 *CFTR* gene mutations (Table 19.4) [28].

Expanded Mutation Panels

Although expanded *CFTR* mutation panels are not recommended by the ACMG, these panels have been available since the earliest days of CF molecular testing [45]. Currently, most clinical molecular diagnostics laboratories offer a CF screening panel containing more than the 23 ACMG-recommended mutations. In many cases, these expanded panels are targeted to cover the mutations most commonly seen in the local ethnic groups (eg, Hispanic) resulting in an increase in clinical sensitivity [46,47]. However, some reference laboratories offer expanded CF mutation screening panels that test for variants that are not known with certainty to be pathogenic. As a result, the ACMG has published statements expressing concern about expanded *CFTR* mutation panels [48,49]. The primary concerns conveyed by the ACMG were that the lack of available information on some variants may increase the complexity of variant reporting and that a negative test result may give patients a false sense of security as no assay provides 100% clinical sensitivity [48].

MOLECULAR TECHNOLOGIES

Platforms for Targeted CF Mutation Testing

There are multiple FDA-approved molecular testing platforms that can be used for clinical molecular genetic testing for CF. The decision of which testing platform to implement in a clinical laboratory must take many factors into account. Some factors that should be taken into consideration include (but are not limited to):

Required instrumentation—Is it already present in the laboratory or will new equipment purchases be necessary?
Flexibility of panel design—Is it a set mutation panel or can it be tailored to include mutations relevant to your patient population?
Analytical sensitivity/specificity
Dedicated technologist time—Is the process largely automated or will it require more hands-on time from a technologist?
Assay repeat rate

Most CF testing platforms are either multiplex PCR-based or use a form of liquid array. Some of the commercially available platforms available for CF testing include (but are not limited to) the following. Most of these platforms have comparable analytical sensitivity/specificity and no-call rates [50].

eSensor: This is a multiplex PCR-based method in which patient DNA is amplified, converted to single-stranded DNA, and loaded onto a microfluidics-based cartridge. The cartridge is loaded onto a proprietary detection platform where the single-stranded DNA is hybridized to a signal probe and patient genotype is generated through electrochemical detection [50,51].

InPlex CF: This is a multiplex method that utilizes targeted oligos that anneal to target DNA sequences. The oligos are specifically designed such that a one-base overlapping structure is formed any time the oligo is bound to template DNA. A proprietary enzyme is used to cleave the overlapping structure which then binds to

a fluorescence resonance energy transfer probe to produce a signal that is used for genotyping [52].

Oligonucleotide ligation assay: This is a multiplex method in which a set of probes target each single-nucleotide polymorphism (SNP) region. One probe is specific for the SNP site such that the terminal 3' base sits directly on the targeted mutation while a second probe is designed to an upstream wild-type sequence. When both probes hybridize to the target region ligation can occur (Fig. 19.1). Ligated products are then

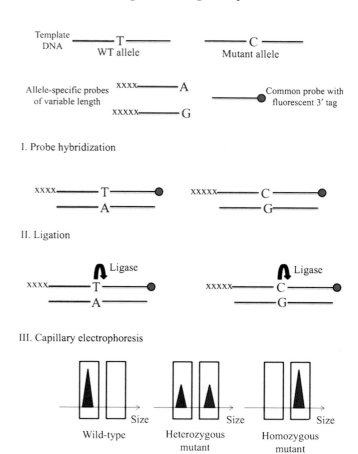

FIGURE 19.1 The oligonucleotide ligation assay (OLA). OLA requires adjacent hybridization of both the allele-specific oligonucleotide (ASO) probe and a common probe. The specificity of the ASO probe is determined by the terminal nucleotide at its 3'end which is complementary to either the mutant nucleotide or the wild-type allele. The common probe contains a fluorescent tag at the 3' end and this probe hybridizes to the nucleotide position adjacent to the area of interest. In this example, (I) we show template DNA with both a wild-type allele (T) and mutant allele (C). An ASO is designed to hybridize to either the wild type or mutant allele. The 5' end of ASO can be attached with a nongenomic sequence to modify the size of the generated product for separation allowing for the interrogation of multiple targets in a single reaction. (II) The ASO hybridize to their target allowing ligation with the adjacent common probe. (III) The ligated allele-specific, fluorescent-labeled fragments can be separated by capillary electrophoresis. Based on the size of the fragments, the mutant and normal alleles can be determined and genotype can be assigned accordingly. *Source: Adapted from Abbott Molecular: http://www.abbottmolecular.com/us/products/genetics/sequencing/cystic-fibrosis.html.*

separated electrophoretically and visualized. By using different fluorescent labels and different product lengths, it is possible to interrogate a number of targets in a single reaction [44,53].

xTAG liquid bead array: A multiplex method that combines PCR, primer extension, and flow cytometry. The liquid bead array starts with a multiplex PCR targeting all gene regions of interest. The PCR reaction undergoes allele-specific primer extension where products are tagged with an xTAG universal sequence. The universal xTAG sequence hybridizes to beads and is sorted using a flow cell [44,52].

MassARRAY: A multiplex method that combines PCR with matrix-assisted laser desorption/ionization—time-of-flight mass spectrometry (MALDI-TOF). Briefly, a PCR reaction is performed for the SNP-containing region (typically designed to generate an amplicon of approximately 100 bp). The amplicon is subjected to a single-base extension or fragmentation reaction to generate DNA fragments of different mass and analyzed by MALDI-TOF. This method quantitatively measures genetic material and is able to detect nucleic acid variation through the measurement of sample mass [54].

Full Gene Sanger Sequencing and Multiplex Ligation-Dependent Probe Amplification

CFTR full gene Sanger sequencing is considered to be the gold standard for CF diagnostic testing. An earlier study showed that more than 98% of *CFTR* mutations can be identified by Sanger sequencing all exons, intron exon junctions, regulatory, and promoter regions, as well as specific intronic regions [55]. However, full gene sequencing is typically considered to be a second-tier test as the majority of classically affected individuals (depending on ethnicity) are detected using a panel containing the 23 ACMG-recommended mutations. It should also be noted that individuals with atypical (mild or monosymptomatic) CF may require sequencing for the detection of milder mutations not included on screening panels [16].

CFTR gene deletion and duplication analysis is usually performed as the next tier of diagnostic testing following the negative findings of full gene Sanger sequencing. Multiplex ligation-dependent probe amplification (MLPA) can detect large exonic level deletions and duplications and is the most common method used in clinical laboratories for this type of testing [16,56—58]. As the name suggests, MLPA is a multiplex method that utilizes allele-specific probes that are attached to a universal primer sequence and a stuffer sequence of varying lengths (Fig. 19.2). The allele-specific probes hybridize to adjacent areas of the target

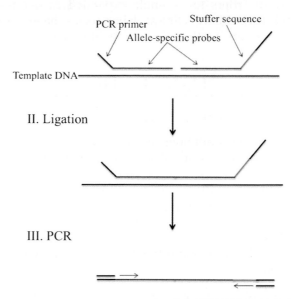

FIGURE 19.2 Multiplex ligation-dependent probe amplification (MLPA). For each target sequence/region, a pair of target-specific probes (red) are designed for both the 5′ and 3′ sequences in the region of interest such that they will hybridize adjacent to each other. Attached to each of the probes is a universal primer sequence (black) that allows for a multiplex PCR of all ligated probes in a single reaction. Additionally, the 3′ probe also contains a stuffer sequence that varies in length for every area of interest. This facilitates the physical separation of each ligated product for analysis. Each MLPA reaction occurs in a series of steps: (I) The target probes hybridize to denatured (single-stranded) template DNA. (II) Ligation occurs only when both of the targeted probes hybridize to target sequence. As a result the left and right probes are ligated to yield a continuous sequence flanked by the PCR primer sequences. (III) A multiplex PCR is performed using the universal primer set attached to each ligated probe set. The amplified products can then be separated by size using capillary electrophoresis. *Source: A detailed summary of the MLPA assay can be found at https://www.mlpa.com/WebForms/WebFormMain.aspx.*

DNA and subjected to a ligation reaction. Ligation only occurs when both allele-specific probes are hybridized to the target DNA. Following ligation, a PCR reaction is performed and amplicons are analyzed by capillary electrophoresis for visualization of results [59].

Future Platforms

In 2013 the FDA approved the first high-throughput next-generation sequencer [60]. Next-generation sequencing (NGS) is rapidly evolving to the point where it is possible to sequence a patient's entire genome at a relatively low expense and in a time frame fitting to a clinical laboratory. Therefore, it is likely that NGS-based gene panels may become the preferred platform for diagnosis in many clinical scenarios, including CF. The adaptation of an NGS-based panel allows for grouping together genes that cause many disorders for a single expanded carrier testing panel [61]. Many clinical laboratories have already embraced this methodology to great success.

CLINICAL UTILITY

Molecular Genetic Testing

Testing for mutations in the *CFTR* gene is one of the most common forms of genetic testing performed in the United States. Genetic testing for CF is typically divided into carrier screening, diagnostic testing (in symptomatic individuals, prenatal samples in at-risk pregnancies, and presymptomatic individuals), and newborn screening [31]. In this section, we will discuss these types of tests and the different testing modalities utilized.

Carrier Screening for CF

CF carrier screening is one of the most commonly ordered molecular genetic tests and has been considered a standard of care for prenatal patients in the United States for many years. Given that the selection criteria for the *CFTR* mutation screening panel established by the ACMG and ACOG [38,40], it is crucial to know a patient's ethnicity and family history of CF. For each individual the pretest carrier risk and test sensitivity vary based on ethnicity (Table 19.3). It is also important to note that given the complex ethnic makeup in the United States, often the self-identified ethnicity of the tested individual can be misleading [62,63]. As such, it is good practice to provide general ethnic carrier frequencies and residual carrier risks on most negative CF carrier screening reports.

Carrier testing is preferably carried out before conception. According to the ACOG recommendation [38,41], CF carrier testing using a panel that contains the 23 ACMG-recommended mutations should be offered to all women of reproductive age, regardless of ancestry. It is also recommended that a pretest evaluation be carried out to identify the couples' ethnicity and family history of CF. Based on the ethnicity, a negative test result can significantly reduce the risk of having a child affected with CF, but does not completely eliminate the possibility. It should be noted that regardless of the testing platform utilized, there is always residual risk that the patient is a *CFTR* mutation carrier [48,64].

Family History of CF

When an individual has a family history of CF, it is important to determine and document the *CFTR* gene mutations associated with disease in the family.

The risk of being a *CFTR* mutation carrier depends on the degree of relationship between an affected individual and the patient. For example, the unaffected sibling of an affected individual has a 2/3 chance of being a *CFTR* mutation carrier. Likewise, a second-degree relatives has 1/4 chance of being carrier. Individuals with a family history of CF should be strongly encouraged to seek genetic counseling so that an accurate assessment of their *a priori* risk can be obtained.

Diagnostic Testing of Symptomatic Individuals

Individuals with symptoms of classic CF or atypical CF (eg, CBAVD, pancreatitis) should be offered diagnostic testing for mutations in the *CFTR* gene. Comprehensive *CFTR* gene testing in these individuals can confirm or rule out a suspected diagnosis of CF. In most clinical scenarios, diagnostic testing for *CFTR* gene mutations uses a tiered approach. The first step in this approach is to offer a targeted *CFTR* mutation panel including the 23 ACMG-recommended mutations. If this panel is negative or only a single mutation identified then the patient undergoes full gene sequencing of the *CFTR* gene which detects more than 98% of causative *CFTR* mutations [55]. However, deep intronic mutations and large gene deletions cannot be detected by DNA sequencing. Hence, the final step in comprehensive CF testing is typically deletion/duplication analysis that is able to identify an additional 1–3% of *CFTR* mutations [65]. This three-tiered testing method has a high clinical sensitivity (>98%) across all ethnicities [16,55] and is typically the preferred approach for CF molecular diagnosis. It is important to note that this tiered testing strategy is typically not indicated for carrier screening because it is costly and time-consuming.

Prenatal Diagnosis

When reproductive partners are both carriers of *CFTR* gene mutation, they have 25% risk of having a child affected with CF. In this clinical scenario, the couple is often offered CF prenatal genetic testing. Prenatal diagnosis of CF allows the medical team to be prepared for any possible complications at the time of birth and for disease management after birth. Additionally, prenatal diagnosis allows the couple to make termination decisions early in the pregnancy. It is important to note that *CFTR* genotype alone cannot predict the clinical course of disease. It is not uncommon for a child with a molecular diagnosis of CF to be born without clinical symptoms. Previous studies have demonstrated that prenatal diagnosis identifying two pathogenic mutations in the *CFTR* gene establishes approximately 4.0% of newly diagnosed individuals [5].

Newborn Screening

Studies indicate that the early diagnosis of CF and the subsequent monitoring of affected individuals improve the overall health of these patients and result in lower medical costs [8]. As a result, in 2009 newborn screening for CF was added to the list of mandatory diseases tested for in the United States [66]. Testing methodologies used for newborn screening are typically capable of identifying both affected individuals, as well as asymptomatic carriers. However, it is often difficult to distinguish CF carriers from affected individuals using newborn screening methods alone. Therefore, it is necessary for individuals suspected to be either carriers or affected to move on from newborn screening to a second-tier testing protocol that typically involves a molecular panel that includes the 23 ACMG-recommended mutations. An earlier publication demonstrated that through newborn screening, 12.8% of newly diagnosed individuals were identified in 2002 [5].

Targeted Therapy

Over the past several years, new advances have been made in the development of novel therapeutic agents that target the pathophysiological process at the CFTR chloride channel. In the wake of these findings, CF may no longer be considered a disease with only supportive therapy. Three main categories of new drugs (potentiators, read-through agents, and correctors) have been developed to target the different classes of CFTR gene mutations and are already in phase 2 and phase 3 clinical trials [67,68]. For example, the potentiator category of drug targets class III mutations and is intended to interact with mutant CFTR protein at the apical membrane and enhance the ability of the protein to transport chloride. Likewise, read-through agents target class I mutations by promoting polymerase read-through of nonsense mutations. Finally, correctors act like a pharmacological chaperone and promote trafficking rescue of the mutated CFTR protein.

In 2014, the first mutation targeted drug, Ivacaftor, was approved by the FDA for the treatment of CF in patients 6 years of age or older [21]. Ivacaftor belongs to the potentiator category of CF drugs and works to enhance the ability of the mutant CFTR channel to transport chloride in those patients with class III–V mutations [69]. Specifically, this drug has been shown to be effective in the treatment of patients carrying at least one copy of the p.Gly551Asp (a class III) mutation [70]. Ivacaftor has also been approved for the use in patients who may carry any one of eight additional mutations: p.Gly178Arg, p.Ser549Asn, p.Ser549Arg, p.Gly551Ser, p.Gly1244Glu, p.Ser1251Asn, p.Ser1255Pro, p.Arg117His, and p.Gly1349Asp. It is recommended that CFTR genetic testing be performed to determine a patient's genotype before therapy is initiated. Recently, the clinical pharmacogenetics implementation consortium published guidelines regarding Ivacaftor therapy and CFTR genotyping [71,72].

Molecular Testing Result Interpretation

CF is inherited in an autosomal recessive manner. Therefore, the detection of two pathogenic mutations in a symptomatic patient (assuming they are on opposite alleles) confirms a diagnosis of CF. It is important to note that CF result reporting relies heavily on the clinical scenario for which the testing is being performed. For example, if a single CFTR gene mutation is identified in a healthy individual undergoing routine pre- or postnatal testing then they are reported to be a carrier of CF. In contrast, if a symptomatic patient is tested by a screening panel and only a single mutation is identified then it is typically recommended that the patient then undergo full gene Sanger sequencing to identify the second mutation [73]. Sequence variants detected by Sanger analysis should be interpreted according to the ACMG standards and guidelines [74].

In the majority of cases, the results of CF screening are negative and straightforward to report. However, it should be noted that no CFTR mutation screening panel detects 100% of mutations and consequently there is always a residual risk that a screened patient is still a carrier of CF. Hence, it is important that a report on a patient who has undergone screening includes information on residual carrier risk so that proper counseling can be performed by the patient's physician. Table 19.3 illustrates the residual risk present after screening performed using only the ACMG-recommended 23 mutation panel and can be used as a model for information that should be included on carrier screening reports.

LIMITATIONS OF TESTING

Diagnostic errors can occur due to rare sequence variations in primer/probe sites in almost all of the technologies mentioned. In addition, a major limitation of any targeted CF-mutation panel is that only the listed CFTR mutations, such as the ACMG-recommended 23 mutations, will be detected. There are no commercially available CF screening panels that detect 100% of the CF-causing mutations. Even with comprehensive CFTR full gene Sanger sequencing it is possible to miss CF disease-causing mutations as regulatory regions (promoters/enhancers) and introns are typically not sequenced so rare mutations in these areas will not be detected. It should also be noted that rare sequence variants located in allele-specific probe hybridizing regions can result in interference with the probe hybridization process and lead to a false-positive result. As such, MLPA results should always be correlated with *CFTR* sequencing data for proper interpretation.

References

[1] Ratjen F, Doring G. Cystic fibrosis. Lancet 2003;361:681–9.

[2] Farrell PM. The prevalence of cystic fibrosis in the European Union. J Cyst Fibros 2008;7:450–3.

[3] Gibson RL, Burns JL, Ramsey BW. Pathophysiology and management of pulmonary infections in cystic fibrosis. Am J Respir Crit Care Med 2003;168:918–51.

[4] Cutting GR. Modifier genetics: cystic fibrosis. An Rev Genomics Hum Genet 2005;6:237–60.

[5] Registry CFFP. Cystic Fibrosis Mutation Database. 2014 [cited 12.10.14]. Available from: http://www.genet.sickkids.on.ca/app.

[6] O'Sullivan BP, Freedman SD. Cystic fibrosis. Lancet 2009;373:1891–904.

[7] Koscik RL, Douglas JA, Zaremba K, Rock MJ, Splaingard ML, Laxova A, et al. Quality of life of children with cystic fibrosis. J Pediatr 2005;147:S64–8.

[8] Farrell PM, Kosorok MR, Laxova A, Shen G, Koscik RE, Bruns WT, et al. Nutritional benefits of neonatal screening for cystic fibrosis. Wisconsin Cystic Fibrosis Neonatal Screening Study Group. N Engl J Med 1997;337:963–9.

[9] Lubamba B, Dhooghe B, Noel S, Leal T. Cystic fibrosis: insight into CFTR pathophysiology and pharmacotherapy. Clin Biochem 2012;45:1132–44.

[10] Gregory RJ, Cheng SH, Rich DP, Marshall J, Paul S, Hehir K, et al. Expression and characterization of the cystic fibrosis transmembrane conductance regulator. Nature 1990;347:382–6.

[11] Frizzell RA, Hanrahan JW. Physiology of epithelial chloride and fluid secretion. Cold Spring Harb Perspect Med 2012;2: a009563.

[12] Trezise AE, Buchwald M. In vivo cell-specific expression of the cystic fibrosis transmembrane conductance regulator. Nature 1991;353:434–7.

[13] Tsui LC, Dorfman R. The cystic fibrosis gene: a molecular genetic perspective. Cold Spring Harb Perspect Med 2013;3: a009472.

[14] Groman JD, Karczeski B, Sheridan M, Robinson TE, Fallin MD, Cutting GR. Phenotypic and genetic characterization of patients with features of "nonclassic" forms of cystic fibrosis. J Pediatr 2005;146:675–80.

[15] Bombieri C, Claustres M, De Boeck K, Derichs N, Dodge J, Girodon E, et al. Recommendations for the classification of diseases as CFTR-related disorders. J Cyst Fibros 2011;10:S86–102.

[16] Moskowitz M. GeneReviews: CFTR-related disorders includes: congenital absence of the vas deferens, cystic fibrosis. [cited 12.10.2014]. Available from: http://www.ncbi.nlm.nih.gov/books/NBK1250/.

[17] Farrell PM, Rosenstein BJ, White TB, Accurso FJ, Castellani C, Cutting GR, et al. Guidelines for diagnosis of cystic fibrosis in newborns through older adults: Cystic Fibrosis Foundation consensus report. J Pediatr 2008;153:S4–14.

[18] Taylor CJ, Hardcastle J, Southern KW. Physiological measurements confirming the diagnosis of cystic fibrosis: the sweat test and measurements of transepithelial potential difference. Paediatr Respir Rev 2009;10:220–6.

[19] Rosenstein BJ, Cutting GR. The diagnosis of cystic fibrosis: a consensus statement. Cystic Fibrosis Foundation Consensus Panel. J Pediatr 1998;132:589–95.

[20] Bell SC, De Boeck K, Amaral MD. New pharmacological approaches for cystic fibrosis: promises, progress, pitfalls. Pharmacol Ther 2015;145:19–34.

[21] Boyle MP, De Boeck K. A new era in the treatment of cystic fibrosis: correction of the underlying CFTR defect. Lancet Respir Med 2013;1:158–63.

[22] Accurso FJ, Rowe SM, Clancy JP, Boyle MP, Dunitz JM, Durie PR, et al. Effect of VX-770 in persons with cystic fibrosis and the G551D-CFTR mutation. N Engl J Med 2010;363:1991–2003.

[23] Rommens JM, Iannuzzi MC, Kerem B, Drumm ML, Melmer G, Dean M, et al. Identification of the cystic fibrosis gene: chromosome walking and jumping. Science 1989;245:1059–65.

[24] Drumm ML, Collins FS. Molecular biology of cystic fibrosis. Mol Genet Med 1993;3:33–68.

[25] Kerem B, Rommens JM, Buchanan JA, Markiewicz D, Cox TK, Chakravarti A, et al. Identification of the cystic fibrosis gene: genetic analysis. Science 1989;245:1073–80.

[26] Morral N, Bertranpetit J, Estivill X, Nunes V, Casals T, Gimenez J, et al. The origin of the major cystic fibrosis mutation (delta F508) in European populations. Nat Genet 1994;7:169–75.

[27] Bobadilla JL, Macek Jr. M, Fine JP, Farrell PM. Cystic fibrosis: a worldwide analysis of CFTR mutations—correlation with incidence data and application to screening. Hum Mutat 2002;19:575–606.

[28] Watson MS, Cutting GR, Desnick RJ, Driscoll DA, Klinger K, Mennuti M, et al. Cystic fibrosis population carrier screening: 2004 revision of American College of Medical Genetics mutation panel. Genet Med 2004;6:387–91.

[29] Rogan MP, Stoltz DA, Hornick DB. Cystic fibrosis transmembrane conductance regulator intracellular processing, trafficking, and opportunities for mutation-specific treatment. Chest 2011;139:1480–90.

[30] Dalemans W, Barbry P, Champigny G, Jallat S, Dott K, Dreyer D, et al. Altered chloride ion channel kinetics associated with the delta F508 cystic fibrosis mutation. Nature 1991;354:526–8.

[31] Cutting GR. Cystic fibrosis genetics: from molecular understanding to clinical application. Nat Rev Genet 2015;16:45–56.

[32] Ahmed N, Corey M, Forstner G, Zielenski J, Tsui LC, Ellis L, et al. Molecular consequences of cystic fibrosis transmembrane regulator (CFTR) gene mutations in the exocrine pancreas. Gut 2003;52:1159–64.

[33] Kristidis P, Bozon D, Corey M, Markiewicz D, Rommens J, Tsui LC, et al. Genetic determination of exocrine pancreatic function in cystic fibrosis. Am J Hum Genet 1992;50:1178–84.

[34] Braun AT, Farrell PM, Ferec C, Audrezet MP, Laxova A, Li Z, et al. Cystic fibrosis mutations and genotype-pulmonary phenotype analysis. J Cyst Fibros 2006;5:33–41.

[35] Dork T, Dworniczak B, Aulehla-Scholz C, Wieczorek D, Bohm I, Mayerova A, et al. Distinct spectrum of CFTR gene mutations in congenital absence of vas deferens. Hum Genet 1997;100:365–77.

[36] Gilljam M, Moltyaner Y, Downey GP, Devlin R, Durie P, Cantin AM, et al. Airway inflammation and infection in congenital bilateral absence of the vas deferens. Am J Respir Crit Care Med 2004;169:174–9.

[37] Rowntree RK, Harris A. The phenotypic consequences of CFTR mutations. Ann Hum Genet 2003;67:471–85.

[38] Grody WW, Cutting GR, Klinger KW, Richards CS, Watson MS, Desnick RJ, et al. Laboratory standards and guidelines for population-based cystic fibrosis carrier screening. Genet Med 2001;3:149–54.

[39] Castellani C, Macek Jr. M, Cassiman JJ, Duff A, Massie J, ten Kate LP, et al. Benchmarks for cystic fibrosis carrier screening: a European consensus document. J Cyst Fibros 2010;9:165–78.

[40] American College of Obstetricians and Gynecologists Committee on Genetics. ACOG Committee Opinion No. 486: update on carrier screening for cystic fibrosis. Obstet Gynecol 2011;117:1028–31.

[41] Committee on Genetics, American College of Obstetricians and Gynecologists. ACOG Committee Opinion. Number 325, December 2005. Update on carrier screening for cystic fibrosis. Obstet Gynecol 2005;106:1465–8.

[42] Genetic testing for cystic fibrosis. National Institutes of Health Consensus Development Conference Statement on genetic testing for cystic fibrosis. Arch Intern Med 1999;159:1529–39.

[43] Statement NIoHCDC. Genetic testing for cystic fibrosis. Apr 14–16, 1997.

[44] Richards CS, Bradley LA, Amos J, Allitto B, Grody WW, Maddalena A, et al. Standards and guidelines for CFTR mutation testing. Genet Med 2002;4:379–91.

[45] Heim RA, Sugarman EA, Allitto BA. Improved detection of cystic fibrosis mutations in the heterogeneous U.S. population using an expanded, pan-ethnic mutation panel. Genet Med 2001;3:168–76.

[46] Dequeker E, Stuhrmann M, Morris MA, Casals T, Castellani C, Claustres M, et al. Best practice guidelines for molecular genetic diagnosis of cystic fibrosis and CFTR-related disorders—updated European recommendations. Eur J Hum Genet 2009;17:51–65.

[47] Rohlfs EM, Zhou Z, Heim RA, Nagan N, Rosenblum LS, Flynn K, et al. Cystic fibrosis carrier testing in an ethnically diverse US population. Clin Chem 2011;57:841–8.

[48] Grody WW, Cutting GR, Watson MS. The cystic fibrosis mutation "arms race": when less is more. Genet Med 2007;9:739–44.

[49] Grody WW. Expanded carrier screening and the law of unintended consequences: from cystic fibrosis to fragile X. Genet Med 2011;13:996–7.

[50] Johnson MA, Yoshitomi MJ, Richards CS. A comparative study of five technologically diverse CFTR testing platforms. J Mol Diagn 2007;9:401–7.

[51] Bernacki SH, Farkas DH, Shi W, Chan V, Liu Y, Beck JC, et al. Bioelectronic sensor technology for detection of cystic fibrosis and hereditary hemochromatosis mutations. Arch Pathol Lab Med 2003;127:1565–72.

[52] ACMG. Technical standards and guidelines for CFTR mutation testing. 2006 edition. 2006.

[53] Brinson EC, Adriano T, Bloch W, Brown CL, Chang CC, Chen J, et al. Introduction to PCR/OLA/SCS, a multiplex DNA test, and its application to cystic fibrosis. Genet Test 1997;1:61–8.

[54] Jurinke C, van den Boom D, Cantor CR, Koster H. The use of MassARRAY technology for high throughput genotyping. Adv Biochem Eng Biotechnol 2002;77:57–74.

[55] Strom CM, Huang D, Chen C, Buller A, Peng M, Quan F, et al. Extensive sequencing of the cystic fibrosis transmembrane regulator gene: assay validation and unexpected benefits of developing a comprehensive test. Genet Med 2003;5:9–14.

[56] Schrijver I, Rappahahn K, Pique L, Kharrazi M, Wong LJ. Multiplex ligation-dependent probe amplification identification of whole exon and single nucleotide deletions in the CFTR gene of Hispanic individuals with cystic fibrosis. J Mol Diagn 2008;10:368–75.

[57] Sellner LN, Taylor GR. MLPA and MAPH: new techniques for detection of gene deletions. Hum Mutat 2004;23:413–19.

[58] Schouten JP, McElgunn CJ, Waaijer R, Zwijnenburg D, Diepvens F, Pals G. Relative quantification of 40 nucleic acid sequences by multiplex ligation-dependent probe amplification. Nucleic Acids Res 2002;30:e57.

[59] Stuppia L, Antonucci I, Palka G, Gatta V. Use of the MLPA assay in the molecular diagnosis of gene copy number alterations in human genetic diseases. Int J Mol Sci 2012;13:3245–76.

[60] Collins FS, Hamburg MA. First FDA authorization for next-generation sequencer. N Engl J Med 2013;369:2369–71.

[61] Aziz N, Zhao Q, Bry L, Driscoll DK, Funke B, Gibson JS, et al. College of American pathologists' laboratory standards for next-generation sequencing clinical tests. Arch Pathol Lab Med 2015;139:481–93.

[62] Palomaki GE, FitzSimmons SC, Haddow JE. Clinical sensitivity of prenatal screening for cystic fibrosis via CFTR carrier testing in a United States panethnic population. Genet Med 2004;6:405–14.

[63] Palomaki GE, Knight GJ, Roberson MM, Cunningham GC, Lee JE, Strom CM, et al. Invasive trophoblast antigen (hyperglycosylated human chorionic gonadotropin) in second-trimester maternal urine as a marker for down syndrome: preliminary results of an observational study on fresh samples. Clin Chem 2004;50:182–9.

[64] Lebo RV, Grody WW. Testing and reporting ACMG cystic fibrosis mutation panel results. Genet Test 2007;11:11–31.

[65] Ferec C, Casals T, Chuzhanova N, Macek Jr. M, Bienvenu T, Holubova A, et al. Gross genomic rearrangements involving deletions in the CFTR gene: characterization of six new events from a large cohort of hitherto unidentified cystic fibrosis chromosomes and meta-analysis of the underlying mechanisms. Eur J Hum Genet 2006;14:567–76.

[66] Wagener JS, Zemanick ET, Sontag MK. Newborn screening for cystic fibrosis. Curr Opin Pediatr 2012;24:329–35.

[67] Clancy JP, Jain M. Personalized medicine in cystic fibrosis: dawning of a new era. Am J Respir Crit Care Med 2012;186:593–7.

[68] Okiyoneda T, Veit G, Dekkers JF, Bagdany M, Soya N, Xu H, et al. Mechanism-based corrector combination restores DeltaF508-CFTR folding and function. Nat Chem Biol 2013;9:444–54.

[69] Condren ME, Bradshaw MD. Ivacaftor: a novel gene-based therapeutic approach for cystic fibrosis. J Pediatr Pharmacol Ther 2013;18:8–13.

[70] Kalydeco [package insert]. Cambridge, MA: Vertex Pharmaceuticals Incorporated; 2012.

[71] Clancy JP, Johnson SG, Yee SW, McDonagh EM, Caudle KE, Klein TE, et al. Clinical Pharmacogenetics Implementation Consortium (CPIC) guidelines for Ivacaftor therapy in the context of CFTR genotype. Clin Pharmacol Ther 2014;95:592–7.

[72] CPIC Ivacaftor Supplement. Update for: Clinical Pharmacogenetics Implementation Consortium (CPIC). Guidelines for Ivacaftor therapy in the context of CFTR genotype. 2014.

[73] Lyon E, Schrijver I, Weck KE, Ferreira-Gonzalez A, Richards CS, Palomaki GE. Molecular genetic testing for cystic fibrosis: laboratory performance on the College of American Pathologists external proficiency surveys. Genet Med 2015;17:219–25.

[74] Richards S, Aziz N, Bale S, Bick D, Das S, Gastier-Foster J, et al. Standards and guidelines for the interpretation of sequence variants: a joint consensus recommendation of the American College of Medical Genetics and Genomics and the Association for Molecular Pathology. Genet Med 2015;17:405–23.

III. MOLECULAR TESTING IN GENETIC DISEASE

20

Molecular Testing in Hemochromatosis

P. Brissot[1,2], O. Loréal[1,2] and A.-M. Jouanolle[1,3]

[1]National Center of Reference for Rare Genetic Iron Overload Diseases, Pontchaillou University Hospital, Rennes, France [2]Inserm-UMR 991, University of Rennes 1, Rennes, France [3]Laboratory of Molecular Genetics and Genomics, Pontchaillou University Hospital, Rennes, France

INTRODUCTION

For a long time, this disease corresponding to systemic iron overload of genetic origin has been confined to a single entity called hereditary hemochromatosis (HC), primary HC, or idiopathic HC. The discovery of the *HFE* gene in 1996 by Feder et al. [1], and later on of multiple molecular actors of iron metabolism, opened the field to the identification of a variety of non-*HFE*-related genetic iron overload disorders [2–4]. Therefore, since this is a collection of diseases, it should be referred to as HCs rather than by the singular HC. Molecular testing has become the critical step, not only for diagnostic purposes in a given individual, but also for prevention in the context of the family study. Molecular diagnostics must be precisely guided by the phenotype and is part of an overall diagnostic approach, which has become essentially noninvasive.

BACKGROUND ON DISEASE MECHANISMS

Mechanistically, two main types of HC can be considered (Fig. 20.1): (1) HC due to increased iron entry into targeted cells and (2) HC due to decreased iron egress from the cells.

HCs Due to Increased Iron Entry into Targeted Cells

Causes and mechanisms of hepcidin deprivation. HC due to increased iron entry into target cells encompasses most HC and the common causal feature is cellular hepcidin deficiency. Most forms involve *quantitative* hepcidin deficiency. In these conditions, there is decreased production, by the hepatocytes, of hepcidin which is the iron hormone regulating systemic iron homeostasis [5–9]. Four genes are of greatest significance: (1) most importantly, the *HFE* gene. Mutations of the *HFE* gene may impact the ERK/MAPK and BMP-SMAD signaling pathways, leading to decreased hepcidin mRNA transcription [10–12]. The main mutation profile is C282Y homozygosity (*C282Y/C282Y* or, according to the recommended nomenclature, *p.Cys282Tyr/p.Cys282Tyr*). However, some other rare genetic *HFE* profiles can cause HC, especially compound heterozygosity where one allele carries the mutation C282Y and the other exhibits a rare mutation [13–15] or deletion [16–18]. In contrast, the frequently occurring compound heterozygote *C282Y/H63D* (*p.His63Asp*) should not been considered as responsible for significant iron overload and clinical HC [19]. The same holds true for *H63D* homozygosity [20,21]. Likewise, the variant *S65C* (*p.Ser65Cys*) is no longer considered clinically significant [22]. (2) The hemojuvelin (*HJV* or *HFE2*) gene [23]. Mutations in *HJV* (homozygous or compound heterozygous) affect the BMP-SMAD signaling pathway, leading to hepcidin decrease. (3) The transferrin receptor 2 (*TFR2*) gene. Mutations in *TFR2* may exert their effect on hepcidin via the ERK/MAPK pathway (like *HFE* mutations) [24–26]. (4) Mutations of *HAMP* which encodes hepcidin can directly result in hepcidin deficiency and cause HC [27,28].

A peculiar and rare form of HC is related to *qualitative* hepcidin deficiency. Hepcidin production is not

Diagnostic Molecular Pathology
DOI: http://dx.doi.org/10.1016/B978-0-12-800886-7.00020-0

© 2017 Elsevier Inc. All rights reserved.

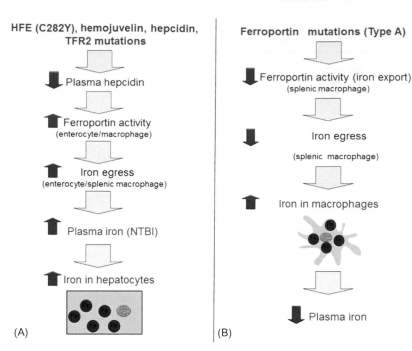

FIGURE 20.1 Hemochromatosis pathophysiological cascade. The left panel shows hemochromatosis with hepcidin deficiency. The right panel shows hemochromatosis with ferroportin deficiency. *NTBI*, non-transferrin bound iron.

TABLE 20.1   Characteristics of the Major Genetic Iron Disorders

Type	Chromosome	Gene	Mechanism (iron excess)	Phenotype severity	Mode of inheritance
HC 1	6	*HFE*	HD	+ +	R
HC 2A	1	*HJV*	HD	+ + + +	R
HC 2B	19	*HAMP*	HD	+ + + +	R
HC 3	7	*TFR2*	HD	+ +	R
HC 4A	2	*SLC40A1*	FD	±	D
HC 4B	2	*SLC40A1*[a]	HD	+ +	D
HA	3	*CP*	FD + HD(?)	+ + +	R

[a]*Hepcidin-resistant ferroportin.*

HC, hemochromatosis; *HJV*, hemojuvelin; *HAMP*, hepcidin; *TFR2*, transferrin receptor2; HA, hereditary aceruloplasminemia; *CP*, ceruloplasmin; HD, hepcidin-deficient phenotype; FD, ferroportin-deficient phenotype; IO, iron overload; R, recessive mode of transmission; D, dominant mode of transmission.

affected, but its effect on iron metabolism is hampered by a state of resistance to this hormone due to mutations of the ferroportin gene (*SLC40A1*) which affect the receptor function of ferroportin to circulating hepcidin [29–31].

The following names have been assigned to those different forms of HC: type 1 for *HFE*-related HC, type 2A for *HJV* (*HFE2*)-related HC, type 2B for *HAMP*-related HC, type 3 for *TFR2*-related HC, and type 4B for *FPN* (*SLC40A1*)-related HC (Table 20.1).

Consequences of hepcidin deprivation. All these forms of HC related to hepcidin deficiency deprivation share a common phenotype resulting from the pathophysiology underlying the development of systemic iron overload (Fig. 20.1). Due to low plasma concentration of hepcidin (HC 1, 2A, 2B, 3) or inefficient plasma hepcidin (HC 4B) [32], the iron exportation property of ferroportin is increased [8]. Ferroportin has two functions: (1) to act as the hepcidin receptor and (2) cellular iron exporter. Increasing the iron exportation properties of ferroportin leads to increased entry of iron into the plasma at both the enterocyte level (corresponding to enhanced intestinal absorption of iron at the duodenal level) and the macrophage level, especially in the spleen. Chronic hypersideremia leads to increased transferrin saturation (TS) and to the subsequent appearance in the blood of abnormal iron forms [33,34]. One form, which may appear when TS is over

45%, corresponds to non-transferrin bound iron (NTBI). NTBI has the special kinetic properties to be very avidly taken up by parenchymal cells, especially in the liver, explaining iron deposition within the hepatocytes. In contrast, most transferrin iron targets the bone marrow. The other abnormal iron form, present when TS is over 75%, is termed labile plasma iron (LPI) or reactive plasma iron (RPI). RPI corresponds to a potentially toxic species of plasma iron due to its high propensity to produce reactive oxygen species [35–37]. RPI-related cellular damage is likely responsible for the main syndromes observed in hepcidin-related HC, such as liver, pancreas, and heart diseases. Therefore, the common phenotype to hepcidin-related HC is characterized by high plasma iron, high plasma TS, major hepatocytic iron deposition, and lack of hepatic and splenic macrophagic iron. However, differences exist in terms of phenotype severity, HC type 2 and HC type 3 (to a certain extent) correspond to more severe forms affecting children or adolescents (also referred to as juvenile HC) with dominant endocrine (pituitary deficiency) and cardiac lesions [2].

HC Due to Decreased Iron Egress from Cells

HC due to ferroportin deficiency [38–40]. In HC due to ferroportin deficiency, the iron export property of ferroportin is altered, leading to iron trapping inside cells, especially in macrophages which express a high level of ferroportin (Fig. 20.1). Therefore, the phenotype is associated with low plasma iron, low plasma TS, preferential iron deposition in the macrophages and within the spleen. Since TS is low, no abnormal forms of plasma iron, dominant NTBI-related hepatocytic iron excess, or RPI-related cellular damage are expected. The corresponding HC form is HC type 4A, also termed ferroportin disease, which is a relatively mild disease despite pronounced systemic iron overload [41].

HC due to ceruloplasmin deficiency. Hereditary aceruloplasminemia (HA), due to mutations of the ceruloplasmin gene (*CP*) [42,43], should also be classified as HC since it corresponds to systemic iron overload of genetic origin. Ceruloplasmin-related ferroxidase activity is critical for cellular iron egress, probably also at the brain level, and may account for both systemic and cerebral iron excess. The phenotype is characterized by very low levels of plasma iron and TS, with frequent anemia, and organ iron excess (mainly the liver, pancreas, and the brain).

In summary, *HFE*-related HC is the most frequent HC form, exclusively observed in Caucasian populations. The non-*HFE*-related HCs are rare diseases but with a broader geographical distribution [44].

CLINICAL STRATEGY LEADING TO DIAGNOSIS OF HC

The clinical strategy to diagnose HC is based on a three-step process: (1) to identify iron overload, (2) to confirm iron overload, and (3) to exclude nongenetic iron overload.

To Identify Iron Overload

Clinical context. A number of clinical symptoms are suggestive of iron overload, including chronic unexplained fatigue, erectile dysfunction, joint pains, melanodermia, osteoporosis, diabetes, liver disease (hepatomegaly, mild transaminase increase), cardiac signs (rhythm disturbances, heart failure). Rarely, anemia and/or neurological symptoms are associated with and reflect iron overload (HA).

Biological context. In clinical practice, the most frequent biochemical parameter suggesting iron excess is hyperferritinemia (usually >300 µg/L in men and >200 µg/L in women). It is of utmost importance to interpret this increase correctly [45]. While low plasma ferritin levels always indicate iron deficiency, high levels are not always associated with iron excess. The following situations can produce hyperferritinemia independently of significant iron excess: (1) inflammatory syndrome, (2) alcoholism, (3) dysmetabolic syndrome, and (4) various other causes. Ferritin is an acute phase protein. Hence, in inflammatory syndrome it is important to systematically evaluate plasma levels of C-reactive protein. In alcoholism, the apparently spontaneous fluctuations of ferritin levels typically correspond to the fluctuations of alcohol consumption. Besides the clinical data which will support this cause, checking for macrocytosis and increased plasma gamma-glutamyltranspeptidase levels is biologically very useful. Dysmetabolic syndrome is by far the most frequent cause of hyperferritinemia. Plasma ferritin concentrations are usually less than 1000 µg/L, TS is normal (<45%), and the metabolic context is highly suggestive (overweight, increased blood pressure, noninsulin dependent diabetes, hyperlipidemia, hyperuricemia, hepatic steatosis). A mild hepatic iron increase is sometimes noticed, corresponding to the syndrome of dysmetabolic hepatosiderosis [46], but always contrasts with the sharpness of the plasma ferritin rise. Among the other causes of hyperferritinemia are marked cytolysis (requiring evaluation of plasma transaminase activities), macrophage activation syndrome, Still disease, ferritin-cataract syndrome [47,48], and mutations of the L-ferritin gene (*FTL*) [48]. A peculiar form of genetic iron overload in which ferritin levels are unexpectedly low compared to iron excess is represented by the exceptional divalent metal transporter (DMT1)-related iron overload, due to *SLC11A2* gene mutations, and usually expressed as microcytic anemia present from birth and which is refractory to oral iron supplementation [49,50].

To Confirm Iron Overload

Regardless of the clinical and/or biochemical context leading to suspicion of iron overload, confirmation of body iron excess requires direct investigation. For many years, liver biopsy was required to show excessive iron deposition, to provide a means for semiquantification, and to determine cellular iron distribution (hepatocytic vs macrophagic) [51–54]. The cellular iron distribution is important to orientate the pathophysiology underlying the iron excess. Today, liver biopsy can be replaced by magnetic resonance imaging (MRI) [52–54]. MRI has several advantages. It is a noninvasive method that enables detection and quantification of iron excess, not only in the liver but also in the pancreas, spleen, heart, and pituitary gland. Moreover, through determination of the ratio between hepatic and splenic iron excess, valuable pathophysiological information is obtained: hepatic iron overload is mainly due to parenchymal iron deposition affecting the hepatocytes, whereas splenic iron indicates reticuloendothelial iron excess affecting the macrophages.

To Exclude Nongenetic Iron Overload

When considering nongenetic iron overload, the differential diagnosis is represented by chronic parenteral iron administration, either due to excessive iron injections for supplementation purpose or to repeated transfusions, knowing that each blood unit provides 200–250 mg of iron. Those transfusions are performed in the context of chronic anemias, such as observed in the myelodysplastic syndrome [55], in hemolytic anemias (thalassemia [56] and sickle cell disease [57]), or in aplasia related to bone marrow transplantation procedures [58]. The differential diagnosis with HC is usually easy by considering the context of chronic anemia. However, it should be recalled that HA is a form of genetic iron excess which can give chronic anemia, but in HA TS is low, in contrast with the situations where iron overload is due to excessive iron input.

Dyserythropoiesis, as seen in myelodysplastic syndrome and in chronic hemolytic anemias (especially non-transfusion-dependent thalassemia [59]), is another mechanism accounting for iron overload in these diseases, before any transfusion. Iron excess is explained by the bone marrow production of the hormone erythroferrone, which exerts an inhibiting effect on the hepatic synthesis of hepcidin [60]. Therefore, the iron overload phenotype mimicks that of hepcidin deprivation related HC (high plasma iron, high plasma TS, and hepatocytic iron deposition).

CLINICAL STRATEGY FOR MOLECULAR TESTING IN INDIVIDUALS

The clinical strategy for molecular testing differs depending upon whether one considers an individual diagnosis or family screening (Fig. 20.2) [22,61]. In a given individual, the phenotype must guide the genotyping approach. Four main bioclinical data strongly orientate the diagnosis: ethnicity, age, neurologic symptoms, and TS.

Ethnicity

HFE-related HC is expected only in Caucasian individuals. In contrast, non-*HFE* HC can be observed in non-Caucasian populations [44].

Age

Juvenile HC, corresponding to fully developed HC under the age of 30, is not expected in the typical *HFE*-related HC. In contrast, several forms of non-*HFE* HC can manifest in younger individuals. Juvenile onset is the rule for *HJV* and *HAMP*-related HC (HC types 2A and 2B) and may also occur in *TFR2*-related HC (HC type 3).

Neurological Symptoms

The only HC condition with neurological damage is represented by HA.

Transferrin Saturation

Whenever marked organ iron overload has been established (by MRI and/or liver biopsy), the plasma TS value is pivotal for the genetic classification. When TS is normal or low (<45%), the two main likely diagnoses are the ferroportin disease (HC type 4A) in which iron excess predominates in the spleen and HA. It should be noted that, in the latter situation, a simple biochemical test is essential to support the diagnosis, namely, the determination of plasma ceruloplasmin concentration which is dramatically decreased. When TS is elevated (>45%, but often >60% or close to 100%), the most frequent diagnosis (in Caucasian individuals and over 30 years of age) is *HFE* (*C282Y/C282Y*)-related HC. If the *C282Y* mutation is absent,

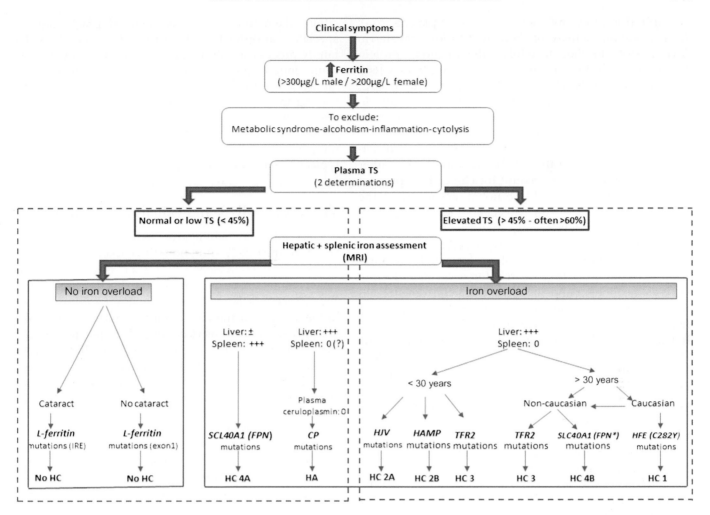

FIGURE 20.2 Diagnostic strategy for hemochromatosis. *TS*, transferrin saturation; *MRI*, magnetic resonance imaging; *HC*, hemochromatosis; *HA*, hereditary aceruloplasminemia; *FPN*, ferroportin gene; *CP*, ceruloplasmin gene; *HJV*, hemojuvelin gene; *HAMP*, hepcidin gene; *TFR2*, transferrin receptor 2 gene.

the genetic search should concern non-*HFE* mutations (preferentially *HJV* and *HAMP* mutations in younger people, and *TFR2* mutations in adults). In all these situations, iron excess spares the spleen and affects the liver, and to a lesser degree the pancreas and heart. If the C282Y mutation is heterozygous, it cannot be held responsible for organ iron overload [62], and an associated rare *HFE* mutation [13–15,63] should be sought, before searching for associated non-*HFE* mutations.

Limitations of Diagnostic Clues

Ethnicity. Due to increasing racial mixing, it may be difficult to ascertain that a given person is purely of Caucasian or non-Caucasian origin.

Age. If the threshold of 30 years old is practically useful to differentiate between adult and juvenile HC, this threshold is closely dependent on the definition that is assigned to HC. This notion joins those of

phenotypic variability and partial penetrance. For instance, in the case of C282Y homozygosity, five stages of expression have been defined [64], and if stage 0 (no TS or ferritin elevation, no clinical symptoms) rules out HC, all the other stages (stage 1—increased TS, no ferritin increase, no clinical symptoms; stage 2—increased TS, increased ferritin, no clinical symptoms; stage 3—increased TS and ferritin, mild clinical symptoms; stage 4—increased TS and ferritin, severe clinical symptoms) can be assimilated to biochemically and/or clinically expressed HC. Therefore, a milder phenotype may be observed in younger HC patients. A further limitation concerning the age threshold is that some cases of highly expressed C282Y/C282Y homozygotes may be related to associated rare mutations [65].

Transferrin saturation. Before using this parameter as a key indicator of genetic etiology, it is essential to (1) obtain coherent results from at least two dosages, considering the spontaneous fluctuations of TS values (notably due to the nycthemeral cycle of plasma iron); (2) exclude an increase of TS related to low transferrin levels, due to hepatocellular dysfunction, protein losses, or rarely to transferrin (*TF*) gene mutations [66]; (3) rule out factors which could, independently of HC, interfere with TS results (false-positive values due to concomitant cytolysis or hemolysis, false-negative values due to coexisting fortuitous inflammation).

HFE mutations. Some rare profiles of compound heterozygosity involving the *C282Y* mutation and another rare *HFE* mutation can account for clinically expressed *HFE*-related HC. It should be noted, according to the recent EMQN (European Molecular Genetics Quality Network) recommendations, that the most frequent *HFE* compound heterozygosity, *C282/H63D*, cannot explain the classical form of *HFE*-related HC and should rather lead to a search for associated factors such as alcoholism and the metabolic syndrome [67,68] or for non-*HFE* mutations. At most, this compound heterozygosity can by itself give some increase of TS but not a significant increase of plasma ferritin (in the absence of those associated cofactors). It is therefore advised either to not perform the search for *H63D* mutation whenever *C282Y* is present in the heterozygous state (as is the case in France), or, if performed, to be very cautious in the interpretation of results. With regard to the *S65C* variant, it is recommended not to test for this variation, and if this variant is detected not to report the result.

Non-HFE mutations. Non-*HFE* mutations must be explored by expert laboratories, ideally in the frame of reference centers for rare iron overload diseases of genetic origin.

CLINICAL STRATEGY FOR MOLECULAR TESTING IN FAMILY SCREENING

In the context of family screening, the genotyping approach is most often used, but investigators should also consider phenotypic iron markers. Once the diagnosis of HC has been made in a given individual, the involved genotype will serve as a flag to evaluate at-risk family members. The screening strategy must obviously consider the dominant or recessive mode of transmission of the disease, knowing that the only HC form with a dominant transmission is ferroportin disease (HC type 4). In the frequent and typical situation of recessive transmission as observed in *C282Y/C282Y*-related HC, the following procedure can be adopted. The proband must be informed that he (she) is responsible for getting in touch with his (her) family members to convey the importance of genetic screening (the general practitioner or medical specialist is not supposed to contact family members directly) and to advise them to contact their general practitioner (or specialist) in order to test for the *C282Y* mutation. The family process should ideally be conducted by genetics counselors, in the frame of specialized centers for HC family screening, who will gather the results (often coming from various geographical areas), and once all the family data have been collected will synthetize the results and propose to the general practitioner a plan for personalized management. The overall procedure typically takes several months to complete.

The Most At-Risk Individuals Are Siblings

Once all major siblings of an index case are genotyped for the *C282Y* mutation, three main situations can be observed: (1) absence of *C282Y* mutation, (2) the *C282Y* mutation is present in the heterozygous state, and (3) there is homozygosity for the *C282Y* mutation. In the absence of the *C282Y* mutation, the individual is not at-risk for iron overload and does not require any surveillance related to iron metabolism. If the *C282Y* mutation is present in the heterozygous state, there is no risk of developing clinically significant iron excess and no special follow-up is necessary. As to the first-degree relatives, systematic genetic testing is not currently recommended [22]. However, it should be remembered that a heterozygote has a 50% risk of transmitting the mutation to his (her) offspring, and that, if the spouse is also heterozygote, there exists a risk of homozygosity in their children. Therefore, heterozygote subjects should be informed that their offspring would benefit from checking TS and ferritin when they become over 18 years old and, in case of abnormal results, the *C282Y* mutation will need to be evaluated. If there is homozygosity for the C282Y mutation, the individual is at-risk of iron overload. Phenotypic evaluation is required both clinically and biologically (initially involving plasma TS and plasma ferritin). If phenotypic parameters are normal, regular follow-up of the plasma TS and ferritin is required. However, follow-up should be conducted knowing that *C282Y* homozygosity only expresses partial penetrance. In fact, it has been reported that less than 30% of men and 1% of women will develop clinically diagnosed iron overload. The proposed frequency of follow-up is every 3 years for stage 0 individuals and every year for stage 1 patients.

Testing Procedures for Offspring and Parents

Although HC type 1 is a recessive disorder, it is advised to test offspring given the high prevalence of the *C282Y* mutation in Caucasians (at least 1 in 10 subjects). Therefore, there is significant risk (1/10) that a given homozygote will have a heterozygote spouse and half of their children will be homozygous and the other half heterozygous (pseudo-dominant transmission). Again, according to the guidelines from the American College of Medical Genetics and the European Society of Human Genetics, no genetic testing is advised in minor subjects. This is often poorly accepted by parents who are anxious to know whether or not their children are homozygous. It should be conveyed to the parents that identifying homozygosity in a minor may not be socially harmless (subsequent risk of societal discrimination) and is not medically justified (given the low penetrance of the genotype). There are two solutions to this issue: (1) advise the parents to check for plasma TS and ferritin after the child reaches puberty, or (2) genotype the spouse while remaining aware of the risk of biological paternity.

For parents, a simple phenotypic study (plasma TS and ferritin) is usually recommended. Genotyping studies are performed only in case of phenotypic abnormalities.

MOLECULAR BIOLOGY METHODS AND STRATEGY

HFE Testing

Real-time PCR is the most common testing method for *C282Y* and *H63D*. Other currently performed methodological approaches are PCR and restriction fragment length polymorphism, PCR and reverse hybridization, and direct sequencing. Testing for the *C282Y* and *H63D* variants can be simultaneous or sequential. Adopting a sequential strategy, testing begins with *C282Y*. This approach enables avoidance of the controversial *H63D* homozygosity (in terms of iron overload risk), but reflects a longer and more expensive procedure. Compound heterozygosity *C282Y/H63D* must be reported as excluding the most common form of *HFE*-related HC, and as being compatible only with mild iron overload in the setting of associated excessive alcohol consumption, fatty liver disease, and/or metabolic syndrome. Testing laboratories are recommended to be accredited to international standards (ISO 15189 or equivalent).

Non-HFE Testing

Non-*HFE* testing should be performed by specialty centers in close connection with expert clinicians, constituting, at best, national reference centers, whose missions will also be to disseminate recommendations for good medical practice and to stimulate clinical and basic research in this area of rare genetic iron overload diseases. These laboratories usually offer a panel of iron-related genes (*HFE, HAMP, HJV, TFR2, SLC40A1, CP, TMPRSS6, SLC40A12, FTL*, and others), increasingly utilizing next-generation sequencing (NGS) technology.

PERSPECTIVES

The traditional limited-throughput sequencing based on the Sanger method is now giving way to the arrival of NGS. These technologies offer the possibility to perform billions of sequencing reactions, opening the way not only for quick simultaneous testing of numerous target genes, but also for performing whole exome or genome sequencing. The advantage of this approach is the huge amount of data produced. However, information technology management is required to handle the data and its analysis. Likewise, the identification of many variants of uncertain significance through NGS raises medical and ethical issues. Hence, newly identified variants require functional studies before they can be used in clinical decision making.

CONCLUSIONS

Molecular testing for the *HFE* gene is one of the most frequently prescribed genetic tests. In the context of documented tissue iron excess with high plasma TS, the presence of the *p.Cys282Tyr* mutation in the homozygous state confirms the diagnosis of *HFE*-related HC. If there is only heterozygosity for *p.Cys282Tyr*, *HFE* sequencing should be performed by a specializing center to search for rare compound heterozygosity. If no *p.Cys282Tyr* mutation is present, rare non-*HFE* mutations should be tested. NGS technologies will facilitate the study of known genes, but also open the field for discovering new iron-related genes.

References

[1] Feder JN, Gnirke A, Thomas W, et al. A novel MHC class I-like gene is mutated in patients with hereditary haemochromatosis. Nat Genet 1996;13:399–408.

[2] Brissot P, Bardou-Jacquet E, Jouanolle AM, Loreal O. Iron disorders of genetic origin: a changing world. Trends Mol Med 2011;17:707–13.

[3] Bardou-Jacquet E, Ben Ali Z, Beaumont-Epinette MP, Loreal O, Jouanolle AM, Brissot P. Non-HFE hemochromatosis: pathophysiological and diagnostic aspects. Clin Res Hepatol Gastroenterol 2014;38:143–54.

[4] Pietrangelo A, Caleffi A, Corradini E. Non-HFE hepatic iron overload. Semin Liver Dis 2011;31:302–18.

[5] Pigeon C, Ilyin G, Courselaud B, et al. A new mouse liver-specific gene, encoding a protein homologous to human antimicrobial peptide hepcidin, is overexpressed during iron overload. J Biol Chem 2001;276:7811–19.

[6] Nicolas G, Bennoun M, Devaux I, et al. Lack of hepcidin gene expression and severe tissue iron overload in upstream stimulatory factor 2 (USF2) knockout mice. Proc Natl Acad Sci USA 2001;98:8780–5.

[7] Ganz T. Systemic iron homeostasis. Physiol Rev 2013;93:1721–41.

[8] Nemeth E, Tuttle MS, Powelson J, et al. Hepcidin regulates cellular iron efflux by binding to ferroportin and inducing its internalization. Science 2004;306:2090–3.

[9] Bardou-Jacquet E, Philip J, Lorho R, et al. Liver transplantation normalizes serum hepcidin level and cures iron metabolism alterations in HFE hemochromatosis. Hepatology 2014;59:839–47.

[10] Hentze MW, Muckenthaler MU, Galy B, Camaschella C. Two to tango: regulation of mammalian iron metabolism. Cell 2010;142:24–38.

[11] Ramey G, Deschemin JC, Vaulont S. Cross-talk between the mitogen activated protein kinase and bone morphogenetic protein/hemojuvelin pathways is required for the induction of hepcidin by holotransferrin in primary mouse hepatocytes. Haematologica 2009;94:765–72.

[12] Wu XG, Wang Y, Wu Q, et al. HFE interacts with the BMP type I receptor ALK3 to regulate hepcidin expression. Blood 2014;124:1335–43.

[13] Aguilar-Martinez P, Grandchamp B, Cunat S, et al. Iron overload in HFE C282Y heterozygotes at first genetic testing: a strategy for identifying rare HFE variants. Haematologica 2011;96:507–14.

[14] Merryweather-Clarke AT, Cadet E, Bomford A, et al. Digenic inheritance of mutations in HAMP and HFE results in different types of haemochromatosis. Hum Mol Genet 2003;12:2241–7.

[15] Cezard C, Rabbind Singh A, Le Gac G, Gourlaouen I, Ferec C, Rochette J. Phenotypic expression of a novel C282Y/R226G compound heterozygous state in HFE hemochromatosis: molecular dynamics and biochemical studies. Blood Cells Mol Dis 2014;52:27–34.

[16] Pelucchi S, Mariani R, Bertola F, Arosio C, Piperno A. Homozygous deletion of HFE: the Sardinian hemochromatosis? Blood 2009;113:3886.

[17] Le Gac G, Congiu R, Gourlaouen I, Cau M, Ferec C, Melis MA. Homozygous deletion of HFE is the common cause of hemochromatosis in Sardinia. Haematologica 2010;95:685–7.

[18] Cukjati M, Koren S, Curin Serbec V, Vidan-Jeras B, Rupreht R. A novel homozygous frameshift deletion c.471del of HFE associated with hemochromatosis. Clin Genet 2007;71:350–3.

[19] Gurrin LC, Bertalli NA, Dalton GW, et al. HFE C282Y/H63D compound heterozygotes are at low risk of hemochromatosis-related morbidity. Hepatology 2009;50:94–101.

[20] Kelley M, Joshi N, Xie Y, Borgaonkar M. Iron overload is rare in patients homozygous for the H63D mutation. Can J Gastroenterol Hepatol 2014;28:198–202.

[21] Neghina AM, Anghel A. Hemochromatosis genotypes and risk of iron overload—a meta-analysis. Ann Epidemiol 2011;21:1–14.

[22] Porto G, Brissot P, Swinkels DW, Zoller H, Alonso I, Morris MA, et al. EMQN best practive guidelines for the molecular genetic diagnosis of hereditary hemochromatosis (HH). Eur J Hum Genet 2016;24:479–95.

[23] Papanikolaou G, Samuels ME, Ludwig EH, et al. Mutations in HFE2 cause iron overload in chromosome 1q-linked juvenile hemochromatosis. Nature Genet 2004;36:77–82.

[24] Poli M, Luscieti S, Gandini V, et al. Transferrin receptor 2 and HFE regulate furin expression via mitogen-activated protein kinase/extracellular signal-regulated kinase (MAPK/Erk) signaling. Implications for transferrin-dependent hepcidin regulation. Haematologica 2010;95:1832–40.

[25] Camaschella C, Roetto A, Cali A, et al. The gene TFR2 is mutated in a new type of haemochromatosis mapping to 7q22. Nat Genet 2000;25:14–15.

[26] Radio FC, Majore S, Binni F, et al. TFR2-related hereditary hemochromatosis as a frequent cause of primary iron overload in patients from Central-Southern Italy. Blood Cells Mol Dis 2014;52:83–7.

[27] Roetto A, Papanikolaou G, Politou M, et al. Mutant antimicrobial peptide hepcidin is associated with severe juvenile hemochromatosis. Nat Genet 2003;33:21–2.

[28] Hattori A, Tomosugi N, Tatsumi Y, et al. Identification of a novel mutation in the HAMP gene that causes non-detectable hepcidin molecules in a Japanese male patient with juvenile hemochromatosis. Blood Cells Mol Dis 2012;48:179–82.

[29] Sham RL, Phatak PD, West C, Lee P, Andrews C, Beutler E. Autosomal dominant hereditary hemochromatosis associated with a novel ferroportin mutation and unique clinical features. Blood Cells Mol Dis 2005;34:157–61.

[30] Sham RL, Phatak PD, Nemeth E, Ganz T. Hereditary hemochromatosis due to resistance to hepcidin: high hepcidin concentrations in a family with C326S ferroportin mutation. Blood 2009;114:493–4.

[31] Fernandes A, Preza GC, Phung Y, et al. The molecular basis of hepcidin-resistant hereditary hemochromatosis. Blood 2009;114:437–43.

[32] Drakesmith H, Schimanski LM, Ormerod E, et al. Resistance to hepcidin is conferred by hemochromatosis-associated mutations of ferroportin. Blood 2005;106:1092–7.

[33] Hershko C, Graham G, Bates GW, Rachmilewitz EA. Non-specific serum iron in thalassaemia: an abnormal serum iron fraction of potential toxicity. Br J Haematol 1978;40:255–63.

[34] Brissot P, Ropert M, Le Lan C, Loreal O. Non-transferrin bound iron: a key role in iron overload and iron toxicity. Biochim Biophys Acta 2012;1820:403–10.

[35] Cabantchik ZI, Breuer W, Zanninelli G, Cianciulli P. LPI-labile plasma iron in iron overload. Best Pract Res Clin Haematol 2005;18:277–87.

[36] Esposito BP, Breuer W, Sirankapracha P, Pootrakul P, Hershko C, Cabantchik ZI. Labile plasma iron in iron overload: redox activity and susceptibility to chelation. Blood 2003;102:2670–7.

[37] Le Lan C, Loreal O, Cohen T, et al. Redox active plasma iron in C282Y/C282Y hemochromatosis. Blood 2005;105:4527–31.

[38] Pietrangelo A. The ferroportin disease. Blood Cells Mol Dis 2004;32:131–8.

[39] Njajou OT, Vaessen N, Joosse M, et al. A mutation in SLC11A3 is associated with autosomal dominant hemochromatosis. Nat Genet 2001;28:213–14.

[40] Montosi G, Donovan A, Totaro A, et al. Autosomal-dominant hemochromatosis is associated with a mutation in the ferroportin (SLC11A3) gene. J Clin Invest 2001;108:619–23.

[41] Le Lan C, Mosser A, Ropert M, et al. Sex and acquired cofactors determine phenotypes of ferroportin disease. Gastroenterology 2011;140:1199–207.

[42] Miyajima H. Aceruloplasminemia. Neuropathology 2014;35:83–90.

[43] Kono S. Aceruloplasminemia: an update. Int Rev Neurobiol 2013;110:125–51.

[44] McDonald CJ, Wallace DF, Crawford DH, Subramaniam VN. Iron storage disease in Asia-Pacific populations: the importance of non-HFE mutations. J Gastroenterol Hepatol 2013;28:1087–94.

[45] Aguilar-Martinez P, Schved JF, Brissot P. The evaluation of hyperferritinemia: an updated strategy based on advances in detecting genetic abnormalities. Am J Gastroenterol 2005;100:1185–94.

[46] Mendler MH, Turlin B, Moirand R, et al. Insulin resistance-associated hepatic iron overload. Gastroenterology 1999;117:1155−63.

[47] Yin D, Kulhalli V, Walker AP. Raised serum ferritin concentration in hereditary hyperferritinemia cataract syndrome is not a marker for iron overload. Hepatology 2014;59:1204−6.

[48] Kannengiesser C, Jouanolle AM, Hetet G, et al. A new missense mutation in the L ferritin coding sequence associated with elevated levels of glycosylated ferritin in serum and absence of iron overload. Haematologica 2009;94:335−9.

[49] Iolascon A, De Falco L. Mutations in the gene encoding DMT1: clinical presentation and treatment. Semin Hematol 2009;46:358−70.

[50] Bardou-Jacquet E, Island ML, Jouanolle AM, et al. A novel N491S mutation in the human SLC11A2 gene impairs protein trafficking and in association with the G212V mutation leads to microcytic anemia and liver iron overload. Blood Cells Mol Dis 2011;47:243−8.

[51] Deugnier Y, Turlin B. Pathology of hepatic iron overload. Semin Liver Dis 2011;31:260−71.

[52] Gandon Y, Olivie D, Guyader D, et al. Non-invasive assessment of hepatic iron stores by MRI. Lancet 2004;363:357−62.

[53] St Pierre TG, Clark PR, Chua-anusorn W, et al. Noninvasive measurement and imaging of liver iron concentrations using proton magnetic resonance. Blood 2005;105:855−61.

[54] Wood JC. Impact of iron assessment by MRI. Hematology Am Soc Hematol Educ Program 2011;2011:443−50.

[55] Steensma DP, Gattermann N. When is iron overload deleterious, and when and how should iron chelation therapy be administered in myelodysplastic syndromes? Best Pract Res Clin Haematol 2013;26:431−44.

[56] Porter JB, Shah FT. Iron overload in thalassemia and related conditions: therapeutic goals and assessment of response to chelation therapies. Hematol Oncol Clin North Am 2010;24:1109−30.

[57] Porter J, Garbowski M. Consequences and management of iron overload in sickle cell disease. Hematology Am Soc Hematol Educ Program 2013;2013:447−56.

[58] Sivgin S, Eser B. The management of iron overload in allogeneic hematopoietic stem cell transplant (alloHSCT) recipients: where do we stand? Ann Hematol 2013;92:577−86.

[59] Musallam KM, Cappellini MD, Taher AT. Iron overload in beta-thalassemia intermedia: an emerging concern. Curr Opin Hematol 2013;20:187−92.

[60] Kautz L, Jung G, Valore EV, Rivella S, Nemeth E, Ganz T. Identification of erythroferrone as an erythroid regulator of iron metabolism. Nature Genet 2014;46:678−84.

[61] Jouanolle AM, Gerolami V, Ged C, et al. Molecular diagnosis of HFE mutations in routine laboratories. Results of a survey from reference laboratories in France. Ann Biol Clin (Paris) 2012;70:305−13.

[62] Zaloumis SG, Allen KJ, Bertalli NA, et al. The natural history of HFE simple heterozygosity for C282Y and H63D: a prospective twelve year study. J Gastroenterol Hepatol 2015;30:719−25.

[63] Barton JC, West C, Lee PL, Beutler E. A previously undescribed frameshift deletion mutation of HFE (c.del277; G93fs) associated with hemochromatosis and iron overload in a C282Y heterozygote. Clin Genet 2004;66:214−16.

[64] HAS. French recommendations for management of HFE hemochromatosis. Haute Autorité de Santé 2005; <www.has-sante.fr>.

[65] Island ML, Jouanolle AM, Mosser A, et al. A new mutation in the hepcidin promoter impairs its BMP response and contributes to a severe phenotype in HFE related hemochromatosis. Haematologica 2009;94:720−4.

[66] Beaumont-Epinette MP, Delobel JB, Ropert M, et al. Hereditary hypotransferrinemia can lead to elevated transferrin saturation and, when associated to HFE or HAMP mutations, to iron overload. Blood Cells Mol Dis 2015;54:151−4.

[67] Cheng R, Barton JC, Morrison ED, et al. Differences in hepatic phenotype between hemochromatosis patients with HFE C282Y homozygosity and other HFE genotypes. J Clin Gastroenterol 2009;43:569−73.

[68] Walsh A, Dixon JL, Ramm GA, et al. The clinical relevance of compound heterozygosity for the C282Y and H63D substitutions in hemochromatosis. Clin Gastroenterol Hepatol 2006;4:1403−10.

III. MOLECULAR TESTING IN GENETIC DISEASE

MOLECULAR TESTING IN ONCOLOGY

21

Molecular Testing in Breast Cancer

K.H. Allison

Department of Pathology, Stanford University School of Medicine, Stanford, CA, United States

INTRODUCTION

Molecular techniques have both changed our understanding of the basic biology of breast cancer and provided the foundation for new methods of personalized diagnostics and treatment strategies. Traditionally, clinicians treating breast cancer have relied on prognostic information such as patient age, cancer grade, and stage at presentation to make clinical decisions about how aggressively to treat individual patients. The field has advanced significantly by using diagnostic tests to determine not just prognosis—but to predict benefit from a particular therapeutic modality. This has placed the pathologist in the role of a diagnostic oncologist, with the interpretation of ancillary tissue-based test results now critical to determining what specific therapies will be used to maximize survival benefit [1].

The first of these predictive tests detected the level of estrogen receptor (ER) and progesterone receptor (PR) in breast cancers to predict benefit from hormone-targeting therapies. Hormone receptor testing was originally performed by ligand-binding assays on fresh tissue samples. Cases that were ER or PR positive were associated with both a significantly better prognosis and showed benefit from hormone-targeted therapies like tamoxifen [2–4]. While the ligand-binding technique offered a quantitative result, disadvantages included having to use fresh tissue, the inability to separate out the contribution of noninvasive tissue to results, and the use of radioactivity in the assay. It was replaced by immunohistochemistry (IHC) techniques because of the ability to score only the invasive cancer and improvement in the ease of testing. IHC has remained the standard for hormone receptor testing today, with guidelines for testing and interpretation established by laboratory-accrediting agencies like the College of American Pathologists (CAP) [5]. However, some of the panel-based RT-PCR molecular assays also report quantitative hormone receptor results.

Non-immunohistochemistry-based standardized molecular testing in breast cancer first began with fluorescence in situ hybridization (FISH) methods to test for gene amplification of the human epidermal growth factor receptor 2 (HER2) gene. This molecular method is used to identify patients with cancers with an aggressive biology that may respond to HER2-targeted biologic therapies. HER2 testing has become standard practice for all newly diagnosed or newly metastatic breast cancers [6–8].

In addition to single-marker testing to identify the most common drug targets in breast cancer (hormone receptors and HER2), there has been an increase in the use of panel-based molecular tests for both prognostic and predictive testing [9–12]. These tests were developed after molecular signatures were described that could separate breast cancers into groups associated with unique signatures and associated outcomes. The clinical utility of these molecular signatures will be discussed in this chapter.

Additional molecular genetic techniques, such as testing for germline mutations in breast cancer related genes like BRCA1 and BRCA2 and genetic testing for resistance to endocrine therapy, have contributed to risk management and treatment strategies in breast cancer. However, this chapter will focus on tissue-based molecular testing in breast cancer.

HER2 TESTING BY FLUORESCENCE IN SITU HYBRIDIZATION

Molecular Target and Clinical Utility

HER2 is a transmembrane receptor tyrosine kinase that is over-expressed in 10–15% of breast cancers and is believed to be responsible for driving an aggressive clinical course in this subset of breast cancers [13–20].

© 2017 Elsevier Inc. All rights reserved.

When over-expressed, there can be a 40-fold to 100-fold increase in protein expression [20,21]. The primary mechanism of protein over-expression is gene amplification of the encoding region on chromosome 17.

HER2 testing of breast cancers was traditionally performed for the purposes of identifying cancers with more aggressive biology and worse outcomes. Because of the association with poor survival rates, a positive HER2 test often leads treating oncologists to use more aggressive chemotherapy regimens. However, when antibody-based therapies specifically targeting HER2-positive cancer cells showed dramatic increases in survival rates for patients with HER2-positive cancers (used in conjunction with chemotherapy), HER2 testing became essential to predict if patients would benefit from these targeted drugs [22–25].

The first HER2-targeted drug to be approved in the adjuvant setting in 2006 was the humanized monoclonal antibody, trastuzumab, which is believed to bind HER2 and block downstream signaling pathways [26]. Other HER2-targeted drugs have also been developed that use different mechanisms of action. Some of these drugs have been approved for use in combination with each other in the setting of traditional neoadjuvant chemotherapy [27,28]. In addition, agents combining HER2-targeted biologic drugs with highly toxic chemotherapy, such as ado-trastuzumab emtansine (T-DM1), are undergoing trials to determine if more targeted chemotherapy delivery can offer a reduced side effect profile and additional survival advantages [29]. Accurate HER2 testing to predict response to these drugs on the initial core biopsy sample has become critical in determining which agents a patient may receive.

In order to guide appropriate treatment decisions, testing for HER2 gene amplification or HER2 protein over-expression is required for all primary and metastatic breast cancers [6]. Because of this, CAP and The American Society of Clinical Oncology (ASCO) published HER2 testing guidelines in 2007, with a more recent update in 2013, in an attempt to set standards for HER2 testing in breast cancer [6,7]. According to these guidelines, HER2 testing can be performed by looking for protein over-expression by IHC, but if there is an equivocal result, the test needs to be confirmed by molecular methods such as FISH or other in situ hybridization methods to look for HER2 gene amplification.

There is concern for the variability in HER2 IHC test performance and interpretation, particularly with false-positive results from over-staining or over-interpretation [30,31]. The 2007 CAP/ASCO HER2 testing guidelines attempted to address these issues by creating standards for tissue fixation, test interpretation, test validation, and requiring proficiency testing for HER2 test interpreters. Greater than 95%

concordance was required between HER2 IHC and FISH-negative and -positive results (or another previously validated test). There is evidence that these guidelines may have reduced false-positive rates and increased concordance rates between laboratories, but some issues persist [32].

Molecular Testing Technique

HER2 FISH testing can be performed using a dual- or single-probe technique on formalin-fixed paraffin-embedded tissue specimens. Both dual- and single-probe assays use a fluorescent-labeled DNA probe to detect the HER2 gene, with the dual-probe assays using a second probe to centromere of chromosome 17 (CEP17) (Fig. 21.1). Intended as form of internal control, the CEP17 probe is then compared to the number of HER2 signals per cell and the results reported as an HER2:CEP17 ratio as well as the absolute HER2 and CEP17 per cell counts. Single-probe assays use only an HER2 gene probe, and only the HER2 copy number is reported.

After probe hybridization, the FISH assay is interpreted using a fluorescence microscope. It is critical to have a pathologist identify which area of the slide to score such that only the invasive carcinoma is scored (Fig. 21.2). The presence of large areas of in situ carcinoma or other findings can confound results if this is not done carefully and only invasive carcinoma should be scored. Next, the areas with invasion on the slide are scanned at low power under the fluorescence microscope to get an overview of the findings and to identify if there is more than one population of cells with variable numbers of signals per cell. According to the CAP/ASCO HER2 testing guidelines, the pathologist must perform this review prior to counting of the FISH slide or IHC can be used as the pathologist's screening method to identify any areas of protein over-expression that should be counted separately [6]. If present in more than 10% of the overall cell population, these clustered foci of cells with amplified HER2 are counted separately from the rest of the invasive cancer cell population. Clustered heterogeneity is rare, estimated to occur in less than 5% to 10% of cases. However, it is important to rule out [33,34]. Most cases have more uniform findings.

After scanning the slide, at least two areas are counted at high power (oil immersion) for a total of a minimum of 20 cells (at least 10 cells from each area counted). The individual cell counts are recorded, but in contrast to some FISH assays in which any detectable abnormality counts as a positive result, HER2 FISH results are reported as a cell population average. The criteria for positive, negative,

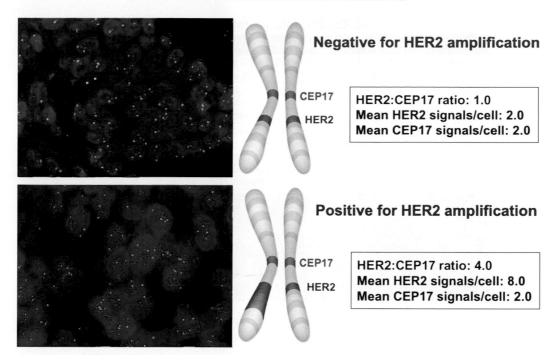

FIGURE 21.1   Example of *HER2* FISH testing. The red fluorescent probe is hybridized to the *HER2* gene, located on chromosome 17. The green fluorescent probe is hybridized with DNA in the centromeric region of chromosome 17 (*CEP17*) and is intended to serve as an internal control. In the *HER2*-negative example (upper panels), there are two copies of both *HER2* and the *CEP17* genes per cell (normal). In the *HER2*-amplified example (lower panels), there are many more copies of the *HER2* gene present, causing both high mean *HER2* signals per cell and an *HER2:CEP17* ratio greater than 2.0.

Initial scoring of HER2 FISH test

1 Pathologist review of H&E and/or IHC-stained slide to localize invasive cancer to evaluate (exclude DCIS).

2 Review controls (repeat if not as expected).

3 Review entire slide to examine for heterogeneity or use IHC stain to guide where to count FISH. If more than one distinctly clustered population has different levels of protein expression or gene amplification, they should be scored separately.

4 Count a minimum of 20 nonoverlapping cells in at least two separate areas (at least 10 cells/area).

5 If close to the threshold for positive (ratio 1.8–2.2 or between 4 and 6 HER2 signals/cell) have an additional observer count of at least an additional 20 cells.

6 Pathologist review of cell counts and confirmation that the appropriate area was scored. Correlation with histology and additional findings before case interpreted and reported.

FIGURE 21.2   Steps involved in the initial scoring of an *HER2* FISH test are outlined. The pathologist is involved in selecting the appropriate area to score (only invasive carcinoma), reviewing controls, evaluating for *HER2* heterogeneity, requesting additional scoring of cases close to the positive thresholds or with other issues, and review of the final scoring of the case with correlation of any additional case findings.

and equivocal results by FISH testing have changed slightly between the 2007 and the 2013 update to the HER2 ASCO/CAP guidelines [6,35]. The 2013 *HER2* FISH interpretation criteria are shown in Fig. 21.3. In the case of the dual-probe assays, these reported results include the overall *HER2:CEP17* ratio, the mean *HER2* signals per cell and the mean *CEP17* signals per cell (Fig. 21.4).

HER2 FISH Interpretation Criteria

FISH result	Criteria	Notes
Positive for HER2 gene amplification	Ratio ≥2.0 OR ≥6.0 gene copies	≥6.0 gene copies can be present with a ratio <2.0
Equivocal for HER2 gene amplification	4.0–5.9 gene copies and ratio <2.0	The guideline suggests counting additional cells for FISH, retesting on another block, or performing IHC
Negative for HER2 gene amplification	Ratio <2.0 and <4.0 gene copies	Cases with <4.0 gene copies and ratio ≥2.0 are considered eligible for HER2-targeted therapies but other features of the case should be considered when making treatment decisions

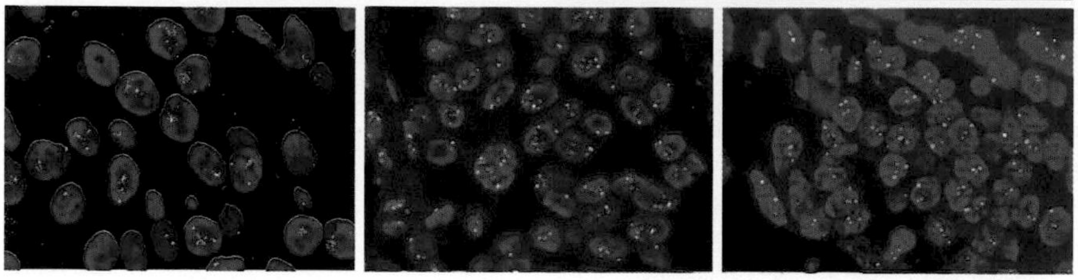

FIGURE 21.3   HER2 FISH interpretation criteria according to the 2013 CAP/ASCO guideline recommendations [36].

HER2 FISH reporting

Cell	HER2	CEP17
1	15	2
2	9	2
3	7	1
4	12	2
5	10	2
6	10	1
7	8	3
8	2	2
9	2	2
10	8	2
11	15	1
12	12	3
13	8	2
14	2	2
15	7	2
16	9	2
17	12	1
18	12	2
19	15	2
20	10	3
Mean	**9.25**	**1.95**
Ratio	**4.74**	

INTERPRETATION:

Positive for HER2 gene amplification

HER2:CEP17 ratio: 4.7
Number of cells counted: 20
Mean HER2 signals/cell: 9.3
Mean CEP17 signals/cell: 2.0

FIGURE 21.4   Example of the individual cell counting sheet results, used to calculate the mean *HER2* signals per cell, mean *CEP17* signals per cell, and the overall *HER2:CEP17* ratio results. The pathologist is required to include these results in their report with an interpretation as positive, negative, equivocal, or indeterminate.

While classic *HER2* amplification (ratio ≥ 2.0 and ≥ 6 mean *HER2* signals/cell) and classic nonamplified cases (ratio <2.0 and <4 *HER2* signals/cell) are typically straightforward, there are cases with less clear results that warrant additional consideration by pathologists and treating clinicians. The 2013 update to the CAP/ASCO *HER2* Testing Guidelines emphasized the critical role of the pathologist in correlating morphologic and clinical findings with *HER2* test results. These updates emphasized that the pathologist should recognize *HER2* testing results that would be considered discordant that might require additional testing or explanation [6,37,38]. In addition, the guidelines went further to clarify cases with nonclassical results that could be considered *HER2* positive and eligible for treatment with HER2-targeted agents. These cases are clarified more extensively in the supplemental material of the guidelines and they include the following: (1) "polysomy" or coamplified cases (≥ 6 *HER2* signals/cell with concurrent increases in *CEP17*), (2) cases amplified only by ratio but with less than 4 mean *HER2* signals per cell ("monosomy"), (3) "low-amplified" cases with ratio more than 2.0 but between 4.0 and 5.9 *HER2* signals per cell, (4) "heterogeneous" cases with a clustered subpopulation of cells with *HER2* amplification representing at least 10% of the total population. All of these categories are considered *HER2*-positive results by the 2013 guidelines update and eligible for HER2-targeted treatment. However, correlation with other features of the case and clinical setting should be emphasized in these nonclassical amplified cases.

Cases that are equivocal by both IHC and FISH testing results remain in a gray zone if repeat testing does not resolve the equivocal result. Some of these cases are "low amplified" with low-level increases in both *HER2* and *CEP17* signals, which may have amplified ratios if an alternative probe for *CEP17* is attempted; however, the use of alternative probes is currently not standardized [39]. FISH equivocal cases also frequently have intermixed, nonclustered cells with *HER2* amplification that do not meet criteria for a heterogeneously amplified case [40,41]. Discussion of "nonclassical" *HER2* FISH positive cases, as well as FISH-equivocal cases, in a multidisciplinary setting can be helpful when treatment decisions are challenging.

The guidelines also note that *HER2* testing can be performed on either the initial core biopsy sample or the surgical specimen as long as preanalytical variables are properly controlled. The core biopsy sample offers the advantage of better control of the ischemic time and adequate penetration of the formalin for appropriate tissue fixation in addition to the benefit of knowing the HER2 status at initial diagnosis if neoadjuvant therapy is being considered. However, to ensure appropriate sampling, consideration for repeat testing on the surgical specimen is recommended when the core biopsy sample was very limited, equivocal, or had high-risk features such as a grade 3 cancer. While repeat testing after a core biopsy with a negative HER2 result is not mandated, its purpose is to ensure that in high-risk patients, heterogeneity for HER2 is not missed [37,38].

Test Limitations

While HER2 testing in breast cancers is straightforward in the majority of cases, there can be issues with HER2 concordance by different testing methods and borderline or equivocal results. Discussion of "nonclassical" *HER2* FISH positive cases, as well as FISH-equivocal cases, in a multidisciplinary setting can be helpful when treatment decisions are challenging. HER2 testing guidelines will also continue to evolve as additional evidence helps clarify which patients may benefit most from treatment.

BREAST CANCER MOLECULAR SUBTYPE

Molecular Target and Clinical Utility

Studies evaluating the gene expression profiles of breast cancers have segregated them into intrinsic or molecular subtypes based on the relatedness of their gene expression patterns [42–44]. These techniques confirmed that the major drivers of breast cancer biology are hormone receptor-related genes and *HER2*-related genes and highlighted the importance of proliferation-related genes [45–50]. They support classification of breast cancers into four main molecular/intrinsic breast cancer subtypes that have prognostic relevance to survival. These subtypes have been termed luminal A, luminal B, HER2-related, and basal-like (a fifth normal breast-like category has not been reproducibly identified). Originally defined by gene expression profiling, similar groupings using different platforms, including genomic DNA copy number arrays, DNA methylation, exome sequencing, microRNA sequencing, and reverse-phase protein assays are also seen [51]. Although significant heterogeneity still exists within these four main groupings of breast cancer (especially in the basal-like category), and additional categories have also been described, these four main groupings provide a major classification framework for further exploration of the biology of breast cancers, especially in the clinical trial setting [10]. Fig. 21.5 shows examples of the four main molecular subtypes and their additional clinicopathological features.

There are several studies that suggest breast cancer subtyping with multigene tests can outperform standard clinical and pathologic prognostic predictors [52–55]. However, their value in guiding treatment decisions is still unclear. In clinical practice, the typical breast panel markers and grade are often used as a surrogate for molecular subtypes, with ER-positive breast cancers correlating with the luminal subtypes, HER2-positive cancers with the HER2 subtype, and

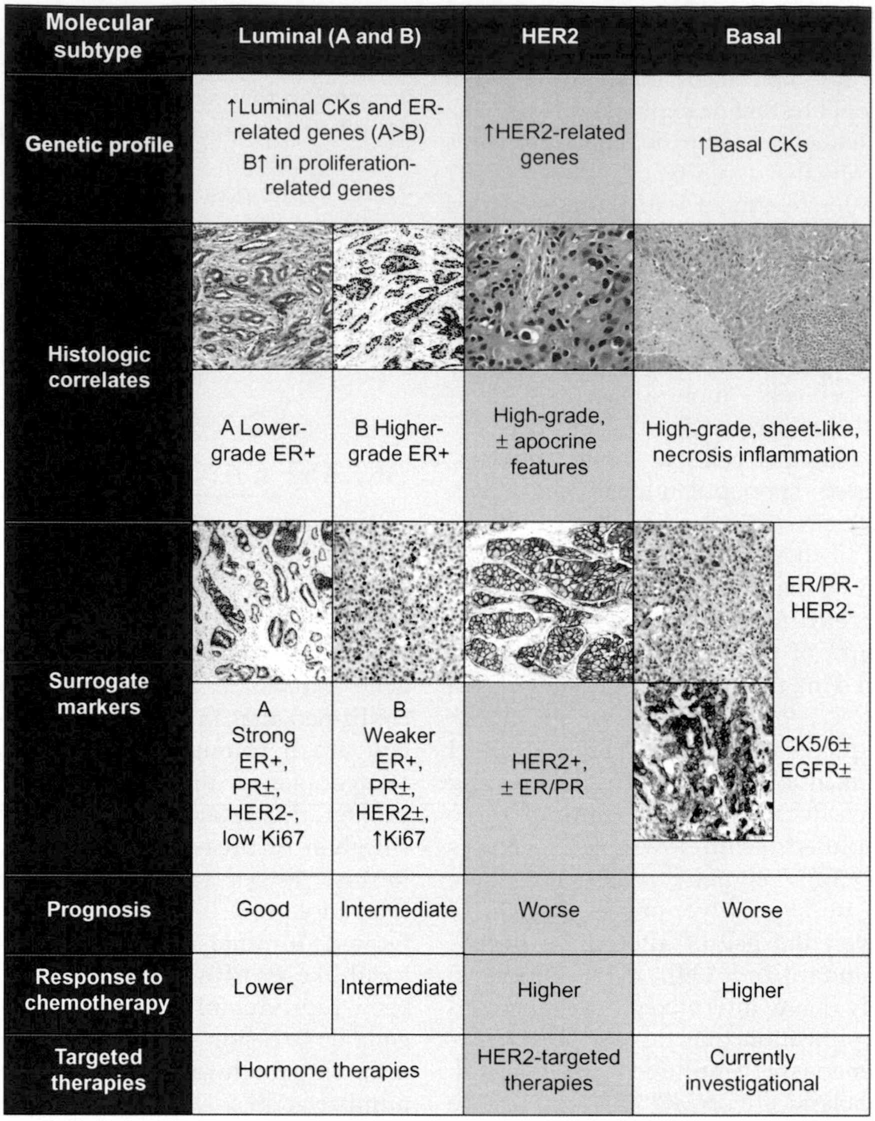

Molecular subtype	Luminal (A and B)		HER2	Basal
Genetic profile	↑Luminal CKs and ER-related genes (A>B) B↑ in proliferation-related genes		↑HER2-related genes	↑Basal CKs
Histologic correlates				
	A Lower-grade ER+	B Higher-grade ER+	High-grade, ± apocrine features	High-grade, sheet-like, necrosis inflammation
Surrogate markers				ER/PR- HER2-
	A Strong ER+, PR±, HER2-, low Ki67	B Weaker ER+, PR±, HER2±, ↑Ki67	HER2+, ± ER/PR	CK5/6± EGFR±
Prognosis	Good	Intermediate	Worse	Worse
Response to chemotherapy	Lower	Intermediate	Higher	Higher
Targeted therapies	Hormone therapies		HER2-targeted therapies	Currently investigational

FIGURE 21.5 The four main molecular subtypes of breast cancer and their additional clinicpathological features. The luminal category is unified by its ER expression and is divided into the lower-risk, lower-grade, lower-proliferation luminal A cancers and the higher-risk, higher-grade, higher-proliferation luminal B cancers. The HER2 subtype is unified by increased expression of *HER2*-related genes, higher grade, more aggressive behavior, but increased rates of response to HER2-targeted therapies and chemotherapies. The basal-like cancers are also characterized by high grade, aggressive behavior, and better response to chemotherapy, but no additional targeted therapy since they are typically hormone receptor-negative and *HER2*-negative.

triple-negative cancers with the basal-like subtype [56,57]. However, these surrogate markers are imperfect, classification schemes vary, and they do not always correlate with molecular subtype results. Therefore, in certain settings, molecular testing to determine subtype may add additional data to the clinical picture.

The most common of these scenarios is distinguishing between the two main luminal subtypes because of the worse outcomes and potential benefit from chemotherapy associated with luminal B cancers when compared to luminal A cancers (both of which are ER positive) [58]. The differences between these two groups

are largely based on differences in proliferation-related genes, with the luminal B cancers having higher proliferation rates and often lower levels of hormone receptors [45,59,60]. Distinguishing which ER-positive luminal breast cancers may benefit from chemotherapy as an adjunct to hormone-targeted therapy is an area of great interest, and several clinical assays have been developed with this specific subset of breast cancers patients in mind.

Outside of the setting of a clinical trial, determining if a triple-negative breast cancer by standard markers also has a basal-like molecular profile has not proven useful since there are currently no differences in standard treatment protocols. Since a subset of basal-like cancers are actually weak–moderately ER-positive, some oncologists have an interest in using molecular profiling to distinguish between a luminal B and basal-like subtype in these cancers for prognostic purposes [61]. However, there is limited data on how to use this information in management decisions. Basal-like cancers are also noted to occur more often in younger patients and African-American women and are associated with a worse prognosis than the luminal subtypes [62,63]. Interestingly, the vast majority of BRCA1-associated breast cancers appear to have a basal-like molecular profile, suggesting a common pathway of carcinogenesis in these patients [64–67]. However, a basal-like or triple-negative profile by itself does not necessarily predict BRCA1 mutation status (ie, many of these cancers are in BRCA1-negative patients), so it is currently not used clinically as a screen for this genetic test [68,69].

Similarly, molecular subtyping of HER2-positive breast cancer (by standard assays) has not provided clear clinical utility beyond standard HER2 testing. Although studies looking at molecular subtype find that many HER2-positive carcinomas fall into the luminal B subtype by molecular profiling, and therefore may identify patients with a lower likelihood of responding to targeted therapy, HER2-targeted treatments should not be withheld from these patients on this basis [70].

Molecular Technique

The original studies that classified breast cancers into molecular subtypes used cDNA microarrays and unsupervised hierarchical clustering analysis [42,43]. Commercially-available assays have also been developed with more selected panels to classify breast cancers on an individual case basis into molecular subtypes. The two most widely used of these assays are the Prosigna PAM-50 assay (Nanostring Technology, Seattle, WA) and Agendia's BluePrint and MammaPrint assays (Agendia, Inc., Irvine, CA), both of which have now been validated for use in formalin-fixed paraffin-embedded tissue.

The Prosigna PAM-50 assay is a commercially-available quantitative reverse transcriptase PCR assay that uses a 50-gene set to classify cancers into the four molecular types (luminal A, luminal B, HER2-enriched, basal-like) [71]. NanoString's nCounter technology uses a digital barcode that allows for direct multiplexed measurement of gene expression using color-coded probe pairs, without requiring an amplification step [72]. After an overnight hybridization step using a single tube, samples are processed and purified in the nCounter and then analyzed in the digital analyzer. Results are then classified into molecular subtype based on their gene expression patterns. It also has been validated and FDA approved to generate and report a risk of recurrence (ROR) score that uses the expression profile in combination with a proliferation score, gross tumor size, and nodal status to estimate risk of distant recurrence within 10 years for postmenopausal women with hormone receptor positive, early-stage breast cancers (with 0–3 positive lymph nodes) [70]. In the United States, only the ROR is reported, while in non-US countries NanoString has approval to also report the intrinsic molecular subtype. There are laboratories with nCounter technology in both the United States and Europe that offer testing and institutions can purchase the technology to set up testing in-house.

BluePrint is another recently developed microarray-based assay that examines the mRNA levels of 80 genes to classify cancers into luminal type, HER2 type, or basal type [73]. Using further analysis with MammaPrint's 70-gene classifier, the luminal subtype cases can be further divided into the low-risk luminal A and high-risk luminal B subtypes [52,70]. MammaPrint/BluePrint testing is only performed at Agendia laboratories and unlike PAM-50 testing, cannot be brought in-house.

When compared directly to each other, the PAM-50 assay classifies more patients in the low-risk luminal A category than BluePrint/MammaPrint [74]. It is argued that when used clinically, these assays may reclassify a large percentage of breast cancer cases with prognostic and treatment implications [75].

Test Limitations

Because molecular subtypes were developed on the basis of hierarchical clustering rather than as a predictive test on an individual sample, these tests have been criticized as single-sample predictors [76,77]. When multiple platforms are compared with the same samples, there is only moderate agreement among them [77–81]. PAM-50 and BluePrint correlation is reported to be as low as 59% [74]. While the classification of the basal-like subtype appears to be the most reliable, the critical distinction between luminal A and B subtypes produces variable results. The impact of intermixed normal tissue has also been suggested as a source of interference with gene expression profiling used as a predictive single-sample test [82,83].

Although molecular/intrinsic subtypes have emphasized the importance of the biology driving different breast cancers, it remains to be seen whether molecular assays for subtyping will prove to be reproducible, clinically useful, and practical as a part of standard clinical practice. However, results from these tests may help add data to complex clinical pictures and inform treatment decisions and clinical trials.

PROGNOSTIC SIGNATURES AND PREDICTION OF BENEFIT FROM CHEMOTHERAPY

Molecular Target and Clinical Utility

It has become clear that not all breast cancers presenting at the same stage have the same underlying biology or clinical behavior [84,85]. While hormone receptor and *HER2* testing can also help define which cancers may respond to targeted therapies, predicting which will respond to chemotherapy is more challenging. Traditionally, breast cancer subtypes known to have worse outcomes are treated with chemotherapy. However, worse prognosis and benefit from chemotherapy are not exactly the same outcome measure. In the ER-positive group of cancers, this distinction is a particular challenge since most of these are good prognosis luminal A type cancers would receive little benefit if given chemotherapy in addition to hormone therapies. There are a variety of prognostic assays using gene expression signatures to help determine which breast cancers are at higher ROR. However, assays that predict patient benefit from chemotherapy (rather than just prognosis) have become the most

widely adopted by oncologists treating breast cancer because of their predictive value.

OncotypeDX and MammaPrint are the assays most widely used, both of which are now available in formalin-fixed paraffin-embedded tissue. These assays have their primary utility in the hormone receptor positive cancers and are often used by oncologists to determine which ER-positive, lymph node-negative breast cancer patients may *not* benefit from the addition chemotherapy to hormone therapy treatment regimens. Thus, these tests are most frequently used when chemotherapy is being considered in ER-positive breast cancer, with a low recurrence score (RS) supporting a decision to opt-out of chemotherapy.

Although both OncotypeDX and MammaPrint assays have many genes in their panels (21 for OncotypeDX and 70 for MammaPrint), their recurrence risk scores are heavily weighted by the proliferation-related genes in their panels. This makes sense since chemotherapy targets more rapidly dividing cells. Multigene prognostic assays are now endorsed by the ASCO, St. Gallen, and National Comprehensive Cancer Network guidelines as information that could assist therapeutic decision-making in ER-positive cancers. A summary of the most commonly used panel-based assays is shown in Table 21.1.

Molecular Testing Technique

OncotypeDX is a quantitative reverse transcriptase PCR assay that quantifies the expression of 16

TABLE 21.1 Common Panel-Based Molecular Tests Used to Predict Outcomes in Breast Cancer

Test (company)	Centralized testing?	Number of genes tested	Patient population validated for	Results reported	Clinical utility
OncotypeDX (Genomic Health, Redwood City, CA)	Yes	Expression of 16 cancer-related genes (5 reference genes)	ER + ; 0–3 positive lymph nodes	1. RS and low-, intermediate-, or high-risk category 2. Quantitative ER, PR, and HER2 levels	Estimation of recurrence risk (three categories) and benefit of chemotherapy (in addition to hormonal therapy)
MammaPrint BluePrint TargetPrint (Agendia, Amsterdam, the Netherlands)	Yes	Expression of 70 genes	ER + or ER − ; stage 1–2, 0–3 positive lymph nodes	1. Low or high recurrence risk categories 2. Molecular subtype 3. Quantitative ER, PR, and HER2 levels	Estimation of recurrence risk (high vs low)—most useful in ER + patients; and subtype used as outcome indicator
PAM50/Prosigna (NanoString Technologies, Seattle, WA)	No (laboratories can purchase testing system)	Expression of 50 cancer-related genes	ER + , stage 1–2, lymph node negative or 1–3 positive lymph nodes	1. Molecular subtype (*only reported in non-US countries*) 2. ROR score	Estimation of recurrence risk (three categories) and subtype used as outcome indicator

cancer-related genes and 5 control genes [86–88]. The genes quantified include ER, PR, HER2, and Ki67, in addition to several other proliferation-related genes. The quantitative results are used in an equation that is then used to calculate an RS. This calculation gives the greatest weight to the panel of five proliferation-related genes. The RS is divided into low (RS < 18), intermediate (RS = 18–30), and high (RS > 30) recurrence risk categories. In validation studies, cancers with a low-risk RS received no benefit from adding chemotherapy to hormone therapy, while cancers in the high-risk category received significant benefit from chemotherapy treatment [86]. However, a significant percent of cancers fall into the intermediate-risk category, which has unclear clinical implications for treatment [89]. Despite these limitations, the RS has been reported to change treatment decisions in 16–49% of patients undergoing testing [89]. The net result of OncotypeDX testing appears to be a reduction in chemotherapy use, a cost-saving that some argue justifies the high cost of testing [90].

Agendia's MammaPrint was one of the first prognostic gene expression assays available for clinical use [91]. One advantage over OncotypeDX is its ability to stratify patients into low-risk and high-risk categories without an intermediate-risk category. Originally, this assay used microarray technology that required fresh tissue, which limited its clinical utility. But the test is now available and FDA approved for use in formalin-fixed paraffin-embedded tissues.

MammaPrint's 70-gene profile was developed by examining over 25,000 genes in lymph node-negative patients (<55 years old). Supervised classification identified 70 genes that correlated with poor prognosis [91,92]. The prognostic capabilities of the 70-gene classifier was validated in a series of 295 patients (some of which were lymph node positive) to show that those with the low-risk signatures had a 10-year recurrence risk less than 15% while those with the high-risk signature had a 50% 10-year risk of distant metastasis [91,93]. More recent meta-analysis data supports that MammaPrint is also predictive of chemotherapy benefit [94]. However, since almost all ER-negative cancers have high-risk signatures, the test is most valuable in discriminating which ER-positive cancers are high risk and low risk. While most studies to date have been retrospective analyses, there are now some limited prospective 5-year survival data showing the ability of MammaPrint to restratify 20% of clinically high-risk patients into the low-risk category, which was associated with a 97% distance recurrence-free interval at 5 years [95].

In addition to the RS result and risk category, a quantitative result for the traditional markers ER, PR, and HER2 can be reported by both the OncotypeDX assay and MammaPrint's TargetPrint assay [96,97]. While these quantitative results were validated by traditional testing methods (IHC and ISH testing), they have not been validated as predictive of response to targeted hormonal and HER2-directed therapies and need to be used with caution rather than as replacements for traditional testing methods [98,99].

Test Limitations

To date, OncotypeDX and MammaPrint have been validated primarily on retrospective cohorts. Similar to the molecular subtype assays, they have been criticized for not being as thoroughly examined as prospective predictors of outcome on an individual patient basis. High discordance rates between OncotypeDX and MammaPrint risk categories when tested on the same samples have been reported as well, with as many as 30% of MammaPrint high-risk cases reclassified as low risk by OncotypeDX [100]. Studies looking at the concordance of multiple gene signatures including PAM50, MammaPrint, and Oncotype DX have also found that while the each had significant prognostic value, the individual risk assessments were often discordant with each other (Cohen's kappa values ranging from 0.24 to 0.70) [100–102]. Results of large multi-institutional studies including the TAILORx, MINDACT, and ISPY-2 trials are anticipated to better evaluate the prospective value of these assays.

One of the inherent limitations of an assay that is not interpreted in situ (such as IHC and ISH testing) is the possibility of the contribution of noncancer tissue to confound results. Intermixed inflammation, in situ carcinoma, or desmoplastic stroma may influence results on an individual case basis [103,104]. Therefore, the results need to be correlated with morphologic findings and taken in the context of other data points about the behavior of the cancer rather than taking any panel-based test results as a new gold standard.

It is of interest that no molecular markers have emerged from these panels that are associated with tumor size or nodal status, suggesting that prognostic information captured by these histopathologic variables is not captured in current prognostic gene signatures. These first-generation prognostic/predictive assays have also been critiqued as being poor predictors of late recurrences, which account for most deaths in the ER-positive group of cancers. Other assays are being developed that aim to determine which ER-positive patients remain at high risk after treatment and may benefit from more prolonged endocrine, chemotherapeutic regimens, or clinical trials.

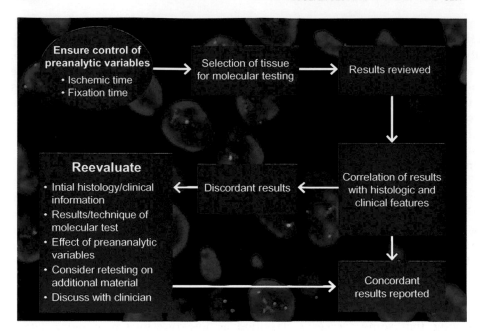

FIGURE 21.6 The general role of the pathologist in evaluation of molecular testing in breast cancer. The pathologist is involved not only in selection of the best tissue for additional molecular testing, but also in controlling for preanalytical variables, correlation of test results with other clinical and histologic features, and recognition and troubleshooting of discordant results.

CONCLUSIONS

Molecular testing has transformed the way we think about breast cancer biology and contributed to the way we diagnose and treat this disease. However, as with any test, molecular tests need to provide clear clinical utility in a cost-effective manner and standards need to be in place to ensure testing is performed accurately and reproducibly. Pathologists have a critical role in this process whether or not they are from the laboratory performing the molecular test. Fig. 21.6 outlines the role of the pathologist in ensuring accurate molecular testing in cancer, including ensuring appropriate tissue handling, selection of the most appropriate tissue sample to test, and being able to shed light on molecular test results within their clinical context such that discordant results can be recognized and dealt with appropriately. Molecular testing in breast cancer will continue to evolve and bring new advances to the field.

References

[1] Allison KH. Molecular pathology of breast cancer: what a pathologist needs to know. Am J Clin Pathol 2012;138:770−80.

[2] Harvey JM, Clark GM, Osborne CK, Allred DC. Estrogen receptor status by immunohistochemistry is superior to the ligand-binding assay for predicting response to adjuvant endocrine therapy in breast cancer. J Clin Oncol 1999;17:1474−81.

[3] Mohsin SK, Weiss H, Havighurst T, et al. Progesterone receptor by immunohistochemistry and clinical outcome in breast cancer: a validation study. Mod Pathol 2004;17:1545−54.

[4] Elledge RM, Green S, Pugh R, et al. Estrogen receptor (ER) and progesterone receptor (PgR), by ligand-binding assay compared with ER, PgR and pS2, by immuno-histochemistry in predicting response to tamoxifen in metastatic breast cancer: a Southwest Oncology Group Study. Int J Cancer 2000;89:111−17.

[5] Hammond ME, Hayes DF, Dowsett M, et al. American Society of Clinical Oncology/College of American Pathologists guideline recommendations for immunohistochemical testing of estrogen and progesterone receptors in breast cancer (unabridged version). Arch Pathol Lab Med 2010;134:e48−72.

[6] Wolff AC, Hammond ME, Hicks DG, et al. Recommendations for human epidermal growth factor receptor 2 testing in breast cancer: American Society of Clinical Oncology/College of American Pathologists clinical practice guideline update. J Clin Oncol 2013;31:3997−4013.

[7] Wolff AC, Hammond ME, Schwartz JN, et al. American Society of Clinical Oncology/College of American Pathologists guideline recommendations for human epidermal growth factor receptor 2 testing in breast cancer. J Clin Oncol 2007;25:118−45.

[8] Ballinger TJ, Sanders ME, Abramson VG. Current HER2 testing recommendations and clinical relevance as a predictor of response to targeted therapy. Clin Breast Cancer 2015;15:171−80.

[9] Ross JS, Hatzis C, Symmans WF, Pusztai L, Hortobagyi GN. Commercialized multigene predictors of clinical outcome for breast cancer. Oncologist 2008;13:477−93.

[10] Hicks DG, Turner B. Pathologic diagnosis, immunohistochemistry, multigene assays and breast cancer treatment: progress toward "precision" cancer therapy. Biotech Histochem 2015;90:81−92.

[11] De Abreu FB, Schwartz GN, Wells WA, Tsongalis GJ. Personalized therapy for breast cancer. Clin Genet 2014;86:62−7.

[12] Paoletti C, Hayes DF. Molecular testing in breast cancer. Annu Rev Med 2014;65:95−110.

[13] Coussens L, Yang-Feng TL, Liao YC, et al. Tyrosine kinase receptor with extensive homology to EGF receptor shares chromosomal location with neu oncogene. Science 1985;230:1132−9.

[14] King CR, Kraus MH, Aaronson SA. Amplification of a novel v-erbB-related gene in a human mammary carcinoma. Science 1985;229:974−6.

[15] Yarden Y. Biology of HER2 and its importance in breast cancer. Oncology 2001;61:1−13.

[16] Barron JJ, Cziraky MJ, Weisman T, Hicks DG. HER2 testing and subsequent trastuzumab treatment for breast cancer in a managed care environment. Oncologist 2009;14:760−8.

[17] Slamon DJ, Clark GM, Wong SG, Levin WJ, Ullrich A, McGuire WL. Human breast cancer: correlation of relapse and survival with amplification of the HER-2/neu oncogene. Science 1987;235:177–82.

[18] Yaziji H, Goldstein LC, Barry TS, et al. HER-2 testing in breast cancer using parallel tissue-based methods. JAMA 2004;291:1972–7.

[19] Press MF, Bernstein L, Thomas PA, et al. HER-2/neu gene amplification characterized by fluorescence in situ hybridization: poor prognosis in node-negative breast carcinomas. J Clin Oncol 1997;15:2894–904.

[20] Pauletti G, Godolphin W, Press MF, Slamon DJ. Detection and quantitation of HER-2/neu gene amplification in human breast cancer archival material using fluorescence in situ hybridization. Oncogene 1996;13:63–72.

[21] Akiyama T, Sudo C, Ogawara H, Toyoshima K, Yamamoto T. The product of the human c-erbB-2 gene: a 185-kilodalton glycoprotein with tyrosine kinase activity. Science 1986;232:1644–6.

[22] Slamon DJ, Leyland-Jones B, Shak S, et al. Use of chemotherapy plus a monoclonal antibody against HER2 for metastatic breast cancer that overexpresses HER2. N Engl J Med 2001;344:783–92.

[23] Goldhirsch A, Gelber RD, Piccart-Gebhart MJ, et al. 2 years versus 1 year of adjuvant trastuzumab for HER2-positive breast cancer (HERA): an open-label, randomised controlled trial. Lancet 2013;382:1021–8.

[24] Pogue-Geile KL, Kim C, Jeong JH, et al. Predicting degree of benefit from adjuvant trastuzumab in NSABP trial B-31. J Natl Cancer Inst 2013;105:1782–8.

[25] Perez EA, Romond EH, Suman VJ, et al. Four-year follow-up of trastuzumab plus adjuvant chemotherapy for operable human epidermal growth factor receptor 2-positive breast cancer: joint analysis of data from NCCTG N9831 and NSABP B-31. J Clin Oncol 2011;29:3366–73.

[26] Molina MA, Codony-Servat J, Albanell J, Rojo F, Arribas J, Baselga J. Trastuzumab (herceptin), a humanized anti-Her2 receptor monoclonal antibody, inhibits basal and activated Her2 ectodomain cleavage in breast cancer cells. Cancer Res 2001;61:4744–9.

[27] Swain SM, Kim SB, Cortes J, et al. Pertuzumab, trastuzumab, and docetaxel for HER2-positive metastatic breast cancer (CLEOPATRA study): overall survival results from a randomised, double-blind, placebo-controlled, phase 3 study. Lancet Oncol 2013;14:461–71.

[28] Baselga J, Bradbury I, Eidtmann H, et al. Lapatinib with trastuzumab for HER2-positive early breast cancer (NeoALTTO): a randomised, open-label, multicentre, phase 3 trial. Lancet 2012;379:633–40.

[29] Verma S, Miles D, Gianni L, et al. Trastuzumab emtansine for HER2-positive advanced breast cancer. N Engl J Med 2012;367:1783–91.

[30] Grimm EE, Schmidt RA, Swanson PE, Dintzis SM, Allison KH. Achieving 95% cross-methodological concordance in HER2 testing: causes and implications of discordant cases. Am J Clin Pathol 2010;134:284–92.

[31] Perez EA, Suman VJ, Davidson NE, et al. HER2 testing by local, central, and reference laboratories in specimens from the North Central Cancer Treatment Group N9831 intergroup adjuvant trial. J Clin Oncol 2006;24:3032–8.

[32] McCullough AE, Dell'orto P, Reinholz MM, et al. Central pathology laboratory review of HER2 and ER in early breast cancer: an ALTTO trial [BIG 2-06/NCCTG N063D (Alliance)] ring study. Breast Cancer Res Treat 2014;143:485–92.

[33] Starczynski J, Atkey N, Connelly Y, et al. HER2 gene amplification in breast cancer: a rogues' gallery of challenging diagnostic cases: UKNEQAS interpretation guidelines and research recommendations. Am J Clin Pathol 2012;137:595–605.

[34] Lee HJ, Seo AN, Kim EJ, et al. HER2 heterogeneity affects trastuzumab responses and survival in patients with HER2-positive metastatic breast cancer. Am J Clin Pathol 2014; 142:755–66.

[35] Wolff AC, Hammond ME, Schwartz JN, et al. American Society of Clinical Oncology/College of American Pathologists guideline recommendations for human epidermal growth factor receptor 2 testing in breast cancer. Arch Pathol Lab Med 2007;131:18–43.

[36] Wolff AC, Hammond ME, Hicks DG, et al. Recommendations for human epidermal growth factor receptor 2 testing in breast cancer: American Society of Clinical Oncology/College of American Pathologists clinical practice guideline update. Arch Pathol Lab Med 2014;138:241–56.

[37] Wolff AC, Hammond ME, Hicks DG, et al. Reply to E.A. Rakha et al. J Clin Oncol 2015;33:1302–4.

[38] Rakha EA, Pigera M, Shaaban A, et al. National guidelines and level of evidence: comments on some of the new recommendations in the American Society of Clinical Oncology and the College of American Pathologists human epidermal growth factor receptor 2 guidelines for breast cancer. J Clin Oncol 2015;33:1301–2.

[39] Hanna WM, Ruschoff J, Bilous M, et al. HER2 in situ hybridization in breast cancer: clinical implications of polysomy 17 and genetic heterogeneity. Mod Pathol 2014;27:4–18.

[40] Allison KH, Dintzis SM, Schmidt RA. Frequency of HER2 heterogeneity by fluorescence in situ hybridization according to CAP expert panel recommendations: time for a new look at how to report heterogeneity. Am J Clin Pathol 2011;136:864–8671.

[41] Bartlett AI, Starczynski J, Robson T, et al. Heterogeneous HER2 gene amplification: impact on patient outcome and a clinically relevant definition. Am J Clin Pathol 2011;136:266–74.

[42] Perou CM, Sorlie T, Eisen MB, et al. Molecular portraits of human breast tumours. Nature 2000;406:747–52.

[43] Sorlie T, Perou CM, Tibshirani R, et al. Gene expression patterns of breast carcinomas distinguish tumor subclasses with clinical implications. Proc Natl Acad Sci USA 2001;98:10869–74.

[44] Sorlie T, Tibshirani R, Parker J, et al. Repeated observation of breast tumor subtypes in independent gene expression data sets. Proc Natl Acad Sci USA 2003;100:8418–23.

[45] Reis-Filho JS, Pusztai L. Gene expression profiling in breast cancer: classification, prognostication, and prediction. Lancet 2011;378:1812–23.

[46] Hu Z, Fan C, Oh DS, et al. The molecular portraits of breast tumors are conserved across microarray platforms. BMC Genomics 2006;7:96.

[47] Sotiriou C, Neo SY, McShane LM, et al. Breast cancer classification and prognosis based on gene expression profiles from a population-based study. Proc Natl Acad Sci USA 2003;100:10393–8.

[48] Sotiriou C, Wirapati P, Loi S, et al. Gene expression profiling in breast cancer: understanding the molecular basis of histologic grade to improve prognosis. J Natl Cancer Inst 2006;98:262–72.

[49] Weigelt B, Reis-Filho JS. Molecular profiling currently offers no more than tumour morphology and basic immunohistochemistry. Breast Cancer Res 2010;12:S5.

[50] Weigelt B, Reis-Filho JS. Histological and molecular types of breast cancer: is there a unifying taxonomy? Nat Rev Clin Oncol 2009;6:718–30.

[51] Cancer Genome Atlas Network. Comprehensive molecular portraits of human breast tumours. Nature 2012;490:61–70.

[52] Nguyen B, Cusumano PG, Deck K, et al. Comparison of molecular subtyping with BluePrint, MammaPrint, and TargetPrint to local clinical subtyping in breast cancer patients. Ann Surg Oncol 2012;19:3257–63.

[53] Buyse M, Loi S, van't Veer, et al. Validation and clinical utility of a 70-gene prognostic signature for women with node-negative breast cancer. J Natl Cancer Inst 2006;98:1183–92.

[54] Mook S, Schmidt MK, Weigelt B, et al. The 70-gene prognosis signature predicts early metastasis in breast cancer patients between 55 and 70 years of age. Ann Oncol 2010;21:717–22.

[55] Nielsen TO, Parker JS, Leung S, et al. A comparison of PAM50 intrinsic subtyping with immunohistochemistry and clinical prognostic factors in tamoxifen-treated estrogen receptor-positive breast cancer. Clin Cancer Res 2010;16:5222–32.

[56] Nielsen TO, Hsu FD, Jensen K, et al. Immunohistochemical and clinical characterization of the basal-like subtype of invasive breast carcinoma. Clin Cancer Res 2004;10:5367–74.

[57] Tang P, Skinner KA, Hicks DG. Molecular classification of breast carcinomas by immunohistochemical analysis: are we ready? Diagn Mol Pathol 2009;18:125–32.

[58] Desmedt C, Haibe-Kains B, Wirapati P, et al. Biological processes associated with breast cancer clinical outcome depend on the molecular subtypes. Clin Cancer Res 2008;14:5158–65.

[59] Reis-Filho JS, Weigelt B, Fumagalli D, Sotiriou C. Molecular profiling: moving away from tumor philately. Sci Transl Med 2010;2(47):47ps3.

[60] Wirapati P, Sotiriou C, Kunkel S, et al. Meta-analysis of gene expression profiles in breast cancer: toward a unified understanding of breast cancer subtyping and prognosis signatures. Breast Cancer Res 2008;10:R65.

[61] Prabhu JS, Korlimarla A, Desai K, et al. A majority of low (1–10%) ER positive breast cancers have like hormone receptor negative tumors. J Cancer 2014;5:156–65.

[62] Carey LA, Perou CM, Livasy CA, et al. Race, breast cancer subtypes, and survival in the Carolina Breast Cancer Study. JAMA 2006;295:2492–502.

[63] Morris GJ, Naidu S, Topham AK, et al. Differences in breast carcinoma characteristics in newly diagnosed African-American and Caucasian patients: a single-institution compilation compared with the National Cancer Institute's Surveillance, Epidemiology, and End Results database. Cancer 2007;110:876–84.

[64] Foulkes WD, Stefansson IM, Chappuis PO, et al. Germline BRCA1 mutations and a basal epithelial phenotype in breast cancer. J Natl Cancer Inst 2003;95:1482–5.

[65] Joosse SA, Brandwijk KI, Mulder L, Wesseling J, Hannemann J, Nederlof PM. Genomic signature of BRCA1 deficiency in sporadic basal-like breast tumors. Genes Chromosomes Cancer 2011;50:71–81.

[66] Turner NC, Reis-Filho JS. Tackling the diversity of triple-negative breast cancer. Clin Cancer Res 2013;19:6380–8.

[67] Turner NC, Reis-Filho JS, Russell AM, et al. BRCA1 dysfunction in sporadic basal-like breast cancer. Oncogene 2007;26:2126–32.

[68] Collins LC, Martyniak A, Kandel MJ, et al. Basal cytokeratin and epidermal growth factor receptor expression are not predictive of BRCA1 mutation status in women with triple-negative breast cancers. Am J Surg Pathol 2009;33:1093–7.

[69] Lakhani SR, Reis-Filho JS, Fulford L, et al. Prediction of BRCA1 status in patients with breast cancer using estrogen receptor and basal phenotype. Clin Cancer Res 2005;11:5175–80.

[70] Gluck S, de Snoo F, Peeters J, Stork-Sloots L, Somlo G. Molecular subtyping of early-stage breast cancer identifies a group of patients who do not benefit from neoadjuvant chemotherapy. Breast Cancer Res Treat 2013;139:759–67.

[71] Parker JS, Mullins M, Cheang MC, et al. Supervised risk predictor of breast cancer based on intrinsic subtypes. J Clin Oncol 2009;27:1160–7.

[72] Geiss GK, Bumgarner RE, Birditt B, et al. Direct multiplexed measurement of gene expression with color-coded probe pairs. Nat Biotechnol 2008;26:317–25.

[73] Krijgsman O, Roepman P, Zwart W, et al. A diagnostic gene profile for molecular subtyping of breast cancer associated with treatment response. Breast Cancer Res Treat 2012;133:37–47.

[74] Bayraktar S, Royce M, Stork-Sloots L, de Snoo F, Gluck S. Molecular subtyping predicts pathologic tumor response in early-stage breast cancer treated with neoadjuvant docetaxel plus capecitabine with or without trastuzumab chemotherapy. Med Oncol 2014;31:163.

[75] Whitworth P, Stork-Sloots L, de Snoo FA, et al. Chemosensivity predicted by BluePrint 80-gene functional subtype and MammaPrint in the Prospective Neoadjuvant Breast Registry Symphony Trial (NBRST). Ann Surg Oncol 2014;21:3261–7.

[76] Pusztai L, Mazouni C, Anderson K, Wu Y, Symmans WF. Molecular classification of breast cancer: limitations and potential. Oncologist 2006;11:868–77.

[77] Weigelt B, Mackay A, A'Hern R, et al. Breast cancer molecular profiling with single sample predictors: a retrospective analysis. Lancet Oncol 2010;11:339–49.

[78] Kapp AV, Tibshirani R. Are clusters found in one dataset present in another dataset? Biostatistics 2007;8:9–31.

[79] Lusa L, McShane LM, Reid JF, et al. Challenges in projecting clustering results across gene expression-profiling datasets. J Natl Cancer Inst 2007;99:1715–23.

[80] Mackay A, Weigelt B, Grigoriadis A, et al. Microarray-based class discovery for molecular classification of breast cancer: analysis of interobserver agreement. J Natl Cancer Inst 2011; 103:662–73.

[81] Fan C, Oh DS, Wessels L, et al. Concordance among gene-expression-based predictors for breast cancer. N Engl J Med 2006;355:560–9.

[82] Elloumi F, Hu Z, Li Y, et al. Systematic bias in genomic classification due to contaminating non-neoplastic tissue in breast tumor samples. BMC Med Genomics 2011;4:54.

[83] Cleator SJ, Powles TJ, Dexter T, et al. The effect of the stromal component of breast tumours on prediction of clinical outcome using gene expression microarray analysis. Breast Cancer Res 2006;8:R32.

[84] Foulkes WD, Reis-Filho JS, Narod SA. Tumor size and survival in breast cancer—a reappraisal. Nat Rev Clin Oncol 2010; 7:348–53.

[85] Foulkes WD, Grainge MJ, Rakha EA, Green AR, Ellis IO. Tumor size is an unreliable predictor of prognosis in basal-like breast cancers and does not correlate closely with lymph node status. Breast Cancer Res Treat 2009;117:199–204.

[86] Paik S, Tang G, Shak S, et al. Gene expression and benefit of chemotherapy in women with node-negative, estrogen receptor-positive breast cancer. J Clin Oncol 2006;24:3726–34.

[87] Albain KS, Barlow WE, Shak S, et al. Prognostic and predictive value of the 21-gene recurrence score assay in postmenopausal women with node-positive, oestrogen-receptor-positive breast cancer on chemotherapy: a retrospective analysis of a randomised trial. Lancet Oncol 2010;11:55–65.

[88] Cronin M, Sangli C, Liu ML, et al. Analytical validation of the Oncotype DX genomic diagnostic test for recurrence prognosis and therapeutic response prediction in node-negative, estrogen receptor-positive breast cancer. Clin Chem 2007;53:1084–91.

[89] Markopoulos C. Overview of the use of Oncotype DX® as an additional treatment decision tool in early breast cancer. Expert Rev Anticancer Ther 2013;13:179–94.

[90] Hornberger J, Chien R, Krebs K, Hochheiser L. US insurance program's experience with a multigene assay for early-stage breast cancer. J Oncol Pract 2011;7 e38s–e345

[91] van't Veer LJ, Dai H, van de Vijver MJ, et al. Gene expression profiling predicts clinical outcome of breast cancer. Nature 2002;415:530–6.

[92] Tian S, Roepman P, Van't Veer LJ, Bernards R, de Snoo F, Glas AM. Biological functions of the genes in the mammaprint breast cancer profile reflect the hallmarks of cancer. Biomark Insights 2010;5:129—38.

[93] van de Vijver MJ, He YD, van't Veer LJ, et al. A gene-expression signature as a predictor of survival in breast cancer. N Engl J Med 2002;347:1999—2009.

[94] Knauer M, Mook S, Rutgers EJ, et al. The predictive value of the 70-gene signature for adjuvant chemotherapy in early breast cancer. Breast Cancer Res Treat 2010;120:655—61.

[95] Drukker CA, Bueno-de-Mesquita JM, Retel VP, et al. A prospective evaluation of a breast cancer prognosis signature in the observational RASTER study. Int J Cancer 2013;133:929—36.

[96] Roepman P, Horlings HM, Krijgsman O, et al. Microarray-based determination of estrogen receptor, progesterone receptor, and HER2 receptor status in breast cancer. Clin Cancer Res 2009;15:7003—11.

[97] Badve SS, Baehner FL, Gray RP, et al. Estrogen- and progesterone-receptor status in ECOG 2197: comparison of immunohistochemistry by local and central laboratories and quantitative reverse transcription polymerase chain reaction by central laboratory. J Clin Oncol 2008;26:2473—81.

[98] Kraus JA, Dabbs DJ, Beriwal S, Bhargava R. Semi-quantitative immunohistochemical assay versus oncotype DX® qRT-PCR assay for estrogen and progesterone receptors: an independent quality assurance study. Mod Pathol 2012;25:869—76.

[99] Dabbs DJ, Klein ME, Mohsin SK, Tubbs RR, Shuai Y, Bhargava R. High false-negative rate of HER2 quantitative reverse transcription polymerase chain reaction of the Oncotype DX test: an independent quality assurance study. J Clin Oncol 2011;29:4279—85.

[100] Iwamoto T, Lee JS, Bianchini G, et al. First generation prognostic gene signatures for breast cancer predict both survival and chemotherapy sensitivity and identify overlapping patient populations. Breast Cancer Res Treat 2011;130:155—64.

[101] Prat A, Parker JS, Fan C, et al. Concordance among gene expression-based predictors for ER-positive breast cancer treated with adjuvant tamoxifen. Ann Oncol 2012;23:2866—73.

[102] Kelly CM, Bernard PS, Krishnamurthy S, et al. Agreement in risk prediction between the 21-gene recurrence score assay (Oncotype DX®) and the PAM50 breast cancer intrinsic Classifier in early-stage estrogen receptor-positive breast cancer. Oncologist 2012;17:492—8.

[103] Allison KH, Kandalaft PL, Sitlani CM, Dintzis SM, Gown AM. Routine pathologic parameters can predict Oncotype DX™ recurrence scores in subsets of ER positive patients: who does not always need testing? Breast Cancer Res Treat 2012; 131:413—24.

[104] Massink MP, Kooi IE, van Mil SE, et al. Proper genomic profiling of (BRCA1-mutated) basal-like breast carcinomas requires prior removal of tumor infiltrating lymphocytes. Mol Oncol 2015; 9:877—88.

22

Molecular Pathology of Prostate Cancer

A.M. Udager[1], S.C. Smith[2] and S.A. Tomlins[1,3,4,5]

[1]Department of Pathology, University of Michigan Medical School, Ann Arbor, MI, United States [2]Department of Pathology, Virginia Commonwealth University School of Medicine, Richmond, VA, United States [3]Michigan Center for Translational Pathology, University of Michigan Medical School, Ann Arbor, MI, United States [4]Department of Urology, University of Michigan Medical School, Ann Arbor, MI, United States [5]Comprehensive Cancer Center, University of Michigan Medical School, Ann Arbor, MI, United States

INTRODUCTION

In the United States, prostate cancer is the most common noncutaneous malignancy in men [1]. In 2015, an estimated 220,800 men will be diagnosed with prostate cancer, and approximately 27,540 men will die from the disease. Despite the relatively high number of cancer-related deaths each year, the majority of prostate cancers are clinically localized, indolent, and can be treated effectively with surgery (ie, radical prostatectomy), with or without adjuvant radiation or hormone therapy. However, a subset of patients will present with nonlocalized disease or recur and progress after primary treatment. For these patients, the development of bone or visceral organ metastases carries a very poor prognosis, given that nearly all metastatic prostate cancers will eventually become resistant to anti-androgen therapies and progress despite castrate levels of serum testosterone (ie, metastatic castration-resistant prostate cancer). Finally, a small proportion of patients will be diagnosed with high-grade neuroendocrine prostate cancer (NEPC), an aggressive form of prostate cancer with unique clinical and molecular features, either de novo (ie, small cell carcinoma of the prostate) or secondary to treatment of conventional prostatic adenocarcinoma [2].

Prior to widespread introduction of serum prostate-specific antigen (PSA) as a protein-based early detection biomarker, prostate cancer patients commonly presented with symptomatic and/or nonlocalized disease. However, over the past several decades, the high incidence in older men, coupled with the effectiveness of

definitive therapy for localized disease, has spurred development of early detection programs. Currently, serum PSA is the most widely utilized prostate cancer early detection biomarker, although there is significant controversy regarding its ability to prevent prostate cancer-specific death (the goal of early detection) and whether it should be used as a screening test [3]. Once a patient has been identified for evaluation of possible prostate cancer, the gold standard for diagnosis is pathological examination of core biopsy tissue from a well-sampled prostate gland. Prostatic biopsy is usually performed with ultrasound guidance using a transrectal approach, although advances in multiparametric MRI will likely dramatically impact prostate biopsy in the near future. At present, prostate cancer is graded according to the Gleason grading system yielding an overall score of 2–10, although nearly all cancers diagnosed currently fall between Gleason score 6 and 10 (higher scores indicate more aggressive disease). Localized prostate cancer may be definitively treated using surgery (radical prostatectomy) or radiation therapy. If the patient chooses radical prostatectomy, standard pathological examination of the resected prostate gland yields essential information regarding the need for additional adjuvant therapy and risk of future recurrence and/or progression. Alternatively, in response to acknowledged overtreatment of lower risk prostate cancer, a growing number of patients with indolent tumors are being offered active surveillance, a monitoring program which defers definitive therapy unless and until the tumor progresses to a more aggressive form [4].

Diagnostic Molecular Pathology
DOI: http://dx.doi.org/10.1016/B978-0-12-800886-7.00022-4

© 2017 Elsevier Inc. All rights reserved.

Advances in our understanding of the molecular underpinnings of prostate cancer have created new opportunities for early detection, diagnosis, prognostication, and targeted therapeutics. In addition, new technologies are rapidly changing the landscape of clinical molecular diagnostics and generating new paradigms for the treatment of a wide variety of oncologic diseases, including prostate cancer [5]. Despite these exciting developments, few molecular assays are currently utilized in routine clinical care of prostate cancer patients—although this will surely change in the era of personalized medicine.

MOLECULAR TARGETS

Prostate-Specific Kallikreins

Kallikreins are a large family of related serine proteases with diverse roles in a variety of human tissues, including prostate and breast. These proteins are predominantly localized to the cytoplasm of glandular epithelial cells and may be present in excreted fluids of these tissues [3,6,7]. Two kallikreins are regulated by androgen signaling and expressed at high levels specifically in the prostate gland: (1) PSA (which is encoded by *KLK3*) and (2) human kallikrein 2 (or hK2, which is encoded by *KLK2*). Similar to other serine proteases, PSA and hK2 are produced as preproenzymes and require posttranslational modification to become catalytically active—specifically, proteolytic cleavage of the preproenzyme signal sequence and release of the N-terminal activation domain. While the mechanisms of hK2 processing are not fully understood, for PSA, the preproenzyme (preproPSA) is processed by a signal peptidase into the proenzyme (proPSA), which is then enzymatically cleaved by trypsin-like proteases (including hK2) to yield active PSA. However, for a subset of molecules, proPSA is truncated into a stable, catalytically inactive form (ie, [-2]proPSA) that is not further processed into active PSA, and in seminal fluid, a proportion of active PSA molecules are proteolytically inactivated by different proteases.

When present in the peripheral blood, the vast majority of active PSA molecules are bound to protease inhibitors (eg, alpha-1 antitrypsin). The remainder of the PSA molecules, including inactive PSA and proPSA forms, circulate unbound as free PSA. Detailed analysis of these free PSA forms has revealed a number of unique circulating molecules, including multichain and nicked PSA forms. To distinguish among these different forms of free PSA, all inactive, single-chain, free PSA molecules are collectively referred to as intact PSA. Thus, for a given patient, multiple serum values can be measured, including total PSA, free PSA, intact PSA, and [-2]proPSA (among others). In normal patients, total PSA is usually low, serumfree PSA is often relatively high ($\sim 10-30\%$), and the proportion of intact PSA is typically low. However, in patients with prostate cancer, total PSA is often high, serumfree PSA is usually relatively decreased, and the proportion of intact PSA, including proPSA forms such as [-2] proPSA, is frequently high. For a complete review of the biology and biochemistry of prostate-specific kallikreins, we recommend recent comprehensive reviews [3,6,7].

ETS Gene Rearrangements

Over the past decade, significant progress has been made in defining the molecular oncogenesis of prostate cancer [8,9]. High-throughput gene expression profiling studies bore early fruit in the study of prostate cancer, through the seminal discovery 10 years ago of highly prevalent fusions of androgen-responsive genes and ETS family proto-oncogenic transcription factors [10]. Bioinformatic analyses of a large group of microarray-based gene expression datasets identified outlier expression of ETS family transcription factors (including *ERG*) in prostate cancer, and subsequent experiments demonstrated recurrent ETS gene fusions in nearly 50% of human prostate cancers [11]. The most prevalent ETS gene fusion in prostate cancer is *TMPRSS2:ERG* [12], which results from an intrachromosomal rearrangement of chromosome 21. This gene fusion event joins the androgen signaling responsive elements in the *TMPRSS2* 5′-untranslated region to the protein-coding exons of the *ERG* transcript, essentially bringing the full-length ERG oncoprotein under androgenic regulation. Subsequent studies have demonstrated that ETS gene rearrangements are early, clonal events in the pathogenesis of prostate cancer [13], which activate distinct oncogenic transcriptional programs [14].

PTEN Deletion

Another frequent genomic alteration in prostate cancer is *PTEN* deletion. In fact, apart from the *TMPRSS2: ERG* gene rearrangement, *PTEN* deletion is the most common recurrent genetic aberration in prostate cancer [15]. *PTEN*, a tumor-suppressor gene, encodes a protein and lipid phosphatase that negatively regulates the PI3K signaling pathway. Thus, *PTEN* deletion in prostate cancer leads to dysregulated PI3K signaling, resulting in increased proliferation and decreased apoptosis [16]. Interestingly, while *PTEN* deletion occurs in both ETS-positive and ETS-negative prostate cancers, it is much more common in ETS-positive tumors; and

PTEN deletion and ETS gene rearrangement may synergize to promote prostate cancer development and tumor progression [17–20].

Long Noncoding RNA

Long noncoding RNA (lncRNA) is an emerging area of focus in prostate cancer biology, and recent transcriptomic profiling studies have identified sets of lncRNAs associated with prostate cancer [21,22]. Some lncRNAs, such as *PCAT-1* and *PCAT29*, have novel oncogenic or tumor-suppressor roles in prostate cancer progression [21,23–25]. *SChLAP1*, for example, is an lncRNA with outlier expression in a subset of prostate cancers. It antagonizes the SWI/SNF chromatin-modifying complex, promotes tumor cell invasion and metastasis, and is strongly associated with aggressive, lethal disease [26,27]. In contrast, while its precise function in prostate cancer oncogenesis is not known, the lncRNA *PCA3* is a sensitive and specific biomarker for prostate cancer detection (as assessed in the urine) [28–30].

Germline Mutations

Recent genomic sequencing has identified germline mutations in several genes, including *HOXB13*, *BRCA1*, and *BRCA2*, that predispose patients to prostate cancer [31,32]. *HOXB13* encodes a homeobox transcription factor with central roles in prostate gland development [33]. A recurrent *HOXB13* G84E germline mutation is associated with a significantly increased risk of early-onset prostate cancer [31]; and interestingly, these tumors have a low frequency of *ERG* gene rearrangement and correspondingly high rate of *SPINK1* overexpression [34]. *BRCA1* and *BRCA2* are tumor-suppressor genes with roles in multiple intracellular processes, including DNA damage repair, transcriptional regulation, and chromatin remodeling, and germline mutations in these genes are associated with an increased risk of prostate cancer and may predispose to aggressive disease [32].

Rare Potentially Targetable Alterations

High-throughput characterization of the prostate cancer genome and transcriptome has not identified frequent alteration of easily druggable targets (ie, tyrosine kinases or G-protein-coupled receptors). However, characterization of large prostate cancer cohorts has identified rare potentially targetable recurrent alterations. For example, Palanisamy et al. used RNAseq to characterize a cohort of ETS fusion negative prostate cancers and identified gene rearrangements involving RAF family members (*BRAF* or *RAF1*), which were shown to drive tumor development in model systems [35]. Subsequent studies identified additional RAF family fusions and activating mutations (eg, *BRAF* T599_V600insHT) in 2–4% of prostate cancers, although *BRAF* V600E mutations are very infrequent, at least in Caucasian cohorts [36–38]. Likewise, using comprehensive exome and transcriptome sequencing as part of the MI-ONCOSEQ program, Wu et al. identified an androgen-driven *SLC45A3:FGFR2* fusion in a patient with ETS fusion negative CRPC [39]. Modeling of *FGFR2* fusions (observed in prostate and other cancers) supports these fusions as drivers targetable by FGFR inhibitors, and in silico analysis of publically available gene expression data suggested a frequency of *FGFR2* fusions in prostate cancer of approximately 0.1%. Lastly, recurrent *IDH1* R132H mutations have been identified in 1–2% of prostate cancers (exclusively ETS fusion negative) across multiple independent cohorts [40–43]. These results suggest that rare potentially targetable alterations may occasionally drive prostate cancer development, particularly in tumors lacking ETS fusions.

MOLECULAR TECHNOLOGIES

Prostate-Specific Kallikreins

Prostate-specific kallikreins can be detected in samples of peripheral blood or urine by a variety of immunoassays (ie, enzyme-linked immunosorbent assay), although urine samples are not routinely analyzed in clinical practice (Fig. 22.1A). Monoclonal antibodies provide specificity for hK2 and the various PSA isoforms.

ETS Gene Rearrangements

As shown in Fig. 22.1B, multiple technologies have been used to detect *TMPRSS2:ERG* transcripts in the urine of patients with prostate cancer, including reverse transcription qPCR after whole transcriptome amplification and transcription-mediated amplification (TMA) [44–47]. TMA is a specialized nucleic acid based amplification method that utilizes RNA polymerase and reverse transcriptase to rapidly and specifically produce RNA amplicons, which can subsequently be quantitated by a number of detection methods. In standard tissue sections, ETS gene rearrangements, including the dominant *TMPRSS2:ERG* gene fusion, can be detected with high sensitivity and specificity by interphase fluorescent in situ hybridization (FISH) [48,49]. In addition, in clinically localized prostate cancer, immunohistochemistry with a monoclonal ERG antibody has a high sensitivity and specificity for *ERG* gene rearrangements [50], again using standard tissue

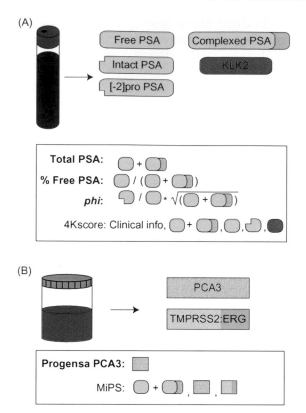

FIGURE 22.1 Clinically available prostate cancer early detection biomarkers. (A) Multiple serum- and (B) urine-based prostate cancer early detection biomarkers are available (FDA-approved biomarkers are in bold). (A) Multiple forms of prostate-specific antigen (PSA; KLK3) can be quantified from serum, including free PSA, intact PSA (a form of free PSA), [-2]pro PSA (another form of free PSA), and complexed PSA. KLK2 (hK2) is a related enzyme that can also be quantified from serum. Multiple combinations are used clinically, including: total PSA (free + complexed PSA); % free PSA (free/total PSA); *phi* ([-2]pro PSA/free PSA * $\sqrt{\text{total PSA}}$); and 4Kscore (logistic regression model including clinical information, total PSA, free PSA, intact PSA, and KLK2). (B) In urine, prostate cancer specific transcripts from *PCA3* (a long noncoding RNA) and the *TMPRSS2:ERG* gene fusion are used as biomarkers. The Progensa *PCA3* assay quantifies urine *PCA3* transcripts (normalized to urine *PSA* transcripts) through a transcription-mediated amplification approach to generate a PCA3 score, while the Mi-Prostate Score (MiPS) test uses a logistic regression model incorporating serum PSA and urine *PCA3* and *TMPRSS2:ERG* scores (generated by the Progensa assay and an analogous assay for *TMPRSS2:ERG*).

sections (Figs. 22.2 and 22.3). Given the lack of validated antibodies against non-*ERG* ETS genes involved in recurrent prostate cancer fusions (*ETV1*, *ETV4*, *ETV5*, and *FLI1*), RNA-based assays, including emerging chromogenic in situ hybridization (ISH), are being developed for the detection of ETS gene rearrangements in tissue [51]. These technologies employ small target sequence-specific probes and specialized amplification methods to allow visualization of chromogenic dyes in tissue by routine light microscopy. Lastly, a variety of high-throughput approaches, including qRT-PCR, multiplexed PCR based RNAseq, RNAseq, and

whole genome sequencing, have been used to characterize ETS gene fusions in fresh frozen and formalin-fixed paraffin-embedded (FFPE) tissue samples, in both the translational and clinical settings [38,40,41,52–54].

PTEN Deletion

In prostate cancer, *PTEN* deletion results in either heterozygous or homozygous *PTEN* loss, and in tissue, these genomic aberrances can be detected with high sensitivity and specificity by interphase FISH [55–57], as shown in Fig. 22.4. In addition, *PTEN* expression can be assessed by immunohistochemistry with a monoclonal PTEN antibody. Although this protein-based assay cannot discriminate between heterozygous and homozygous genomic loss, lack of PTEN expression by IHC has been analytically validated as sensitive and specific for *PTEN* loss [57,58].

Long Noncoding RNA

Similar to detection of *TMPRSS2:ERG* transcripts (Fig. 22.1B), *PCA3* transcript levels can be assessed in urine specimens with a TMA assay [59,60]. In addition, in standard tissue sections, both *PCA3* and *SChLAP1* expression (Fig. 22.5) can be examined using an RNA-based ISH assay [27,60–62].

Germline Mutations

Although not currently utilized in routine clinical practice (unless as part of a workup for hereditary cancer), testing for germline mutations in *HOXB13*, *BRCA1*, and/or *BRCA2* could be performed using a variety of standard DNA-based methods, including Sanger sequencing, allele-specific PCR, single-strand conformation polymorphism, melting point analysis, and/or targeted next-generation sequencing, depending on the mutation type and frequency.

Rare Targetable Oncogenic Alterations

A variety of techniques can be used to identify targetable alterations (depending on alteration type) in prostate cancer tissue specimens, including FISH, Sanger sequencing, and next-generation sequencing, as shown in Fig. 22.6.

CLINICAL UTILITY

Early Detection

For prostate cancer, early detection efforts have essentially two main goals: (1) recognizing patients with clinically significant disease and (2) identifying

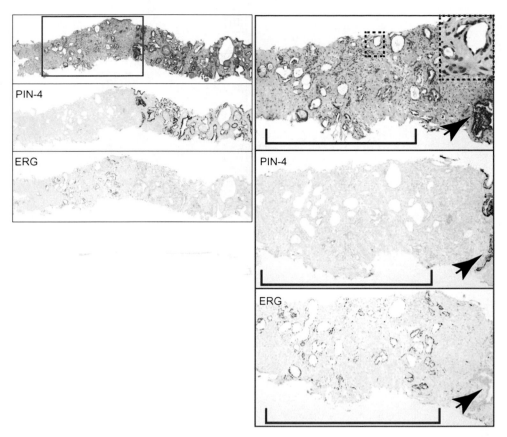

FIGURE 22.2 Diagnostic utility of immunohistochemistry (IHC) for ERG protein expression as a surrogate for *ERG* gene fusions. Multiple monoclonal antibodies directed against ERG have been validated with *ERG* rearrangement status by FISH and shown to have more than 99.99% specificity for prostate cancer at the tissue level. Low power views (left panels) of hematoxylin and eosin (H&E), PIN-4 IHC (basal markers and AMACR), and ERG IHC for a core containing a suspicious but atrophic focus as indicated in the orange box. Higher power view (right panels) demonstrates some glands with nuclear atypia (inset shows blue dashed box) while others are more benign appearing (including areas consistent with complete atrophy). IHC demonstrates lack of basal cell markers and AMACR (PIN-4), while ERG is diffusely positive, consistent with a focus of atrophic carcinoma (green bracket). Note benign glands (black arrows) showing retained basal cells and lack of AMACR and ERG expression.

patients at increased risk for developing disease. While current prostate cancer early detection programs are focused on recognizing patients with clinically significant disease (goal #1), in the future, these programs will likely be expanded and refined in order to identify patients at increased risk for developing disease (goal #2). Serum PSA—more specifically, total PSA—is the current standard for the early detection of prostate cancer. However, while an elevated PSA is sensitive for prostate cancer, it is certainly not specific. Certain benign conditions, such as prostatitis and benign prostatic hyperplasia, can lead to elevated PSA levels (ie, a false-positive test result) and subsequent unnecessary prostate biopsies. In addition, an elevated PSA is not specific (or even necessarily sensitive) for high-grade cancers. Therefore, standard PSA-based early detection programs may identify many low-grade, indolent cancers, which may be unnecessarily biopsied and treated, and may fail to detect some high-grade, aggressive cancers.

With these limitations in mind, additional protein-based assays have been developed to improve the early detection of clinically significant prostate cancer. Such assays include—but are not limited to—assessment of serum-free PSA, intact PSA, [-2]proPSA, hK2 levels, and various combinations (Fig. 22.1A). For example, the Prostate Health Index (*phi*) test, which is FDA approved for prostate cancer risk estimation in men with a serum total PSA between 4 and 10 ng/mL, combines total PSA, free PSA, and [-2]proPSA levels into a single score, and in a multiinstitutional prospective clinical trial, *phi* outperformed any of its individual elements for the detection of clinically significant prostate cancer [63]. Another assay, the four kallikrein (or 4K) panel, combines total PSA, free PSA, intact PSA, and human kallikrein 2 into a single score. Compared to total PSA alone, the 4K assay has shown clinical improvement for the detection of high-grade cancer in multiple independent large European and American cohorts [64–66]. Importantly, in another

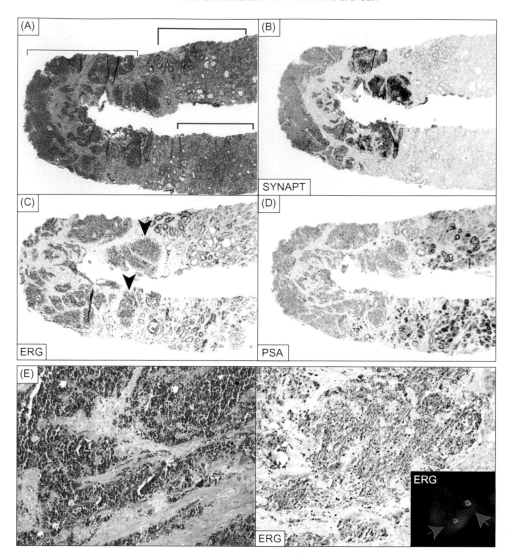

FIGURE 22.3   Utility of *ERG* assessment by fluorescence in situ hybridization (FISH) and immunohistochemistry (IHC) to identify androgen receptor signaling positive (AR +) and negative (AR −) prostate cancer. Expression of androgen-driven ETS gene fusions (most commonly *TMPRSS2:ERG*) in cases harboring these alterations at the genomic level informs on AR signaling status. (A) Areas of conventional acinar adenocarcinoma (orange brackets) and neuroendocrine small cell carcinoma (SCC, black bracket) are present in a prostate biopsy core. (B) Neuroendocrine differentiation is supported by synaptophysin (SYNAPT) expression exclusively in the SCC component. (C) ERG protein expression is present diffusely in both components, consistent with clonality; however, ERG expression is weaker in the SCC component (note black arrows). (D) Reduced ERG expression is consistent with decreased AR signaling in the SCC component, supported by lack of PSA expression in this component. (E) Presence of *ERG* gene rearrangement supports prostatic origin of poorly differentiated carcinoma and neuroendocrine/SCC. A neuroendocrine SCC was identified on a transurethral resection specimen from the urinary bladder in a man with a history of prostatic adenocarcinoma. Although ERG staining was negative (right panel), an *ERG* rearrangement was present by FISH using split signal probes (right panel inset). For *ERG* FISH, 5′ (red) and 3′ (green) probes flanking *ERG* were used. A single normal signal (colocalized 5′/3′ signals, yellow arrow) and loss of a 5′ signal (unpaired 3′ signal, green arrow) are consistent with the presence of a *TMPRSS2:ERG* rearrangement through deletion and support prostatic origin.

recent large European prostate biopsy cohort, the 4K and *phi* assays demonstrated similar overall improved detection of high-grade cancer (GS ≥ 7), compared to total PSA alone [67].

In addition to serum protein biomarkers, risk calculators that incorporate demographic and clinical information have been developed to improve the early detection of clinically significant prostate cancer. For example, the Prostate Cancer Prevention Trial Risk Calculator 2.0 (PCPTRC 2.0) demonstrated excellent discrimination between no disease, low-grade disease (GS < 7), and high-grade disease (GS ≥ 7) in a large American cohort with total PSA alone [68]. This calculator was subsequently validated in 10 large independent American and European cohorts, and improved performance of PCPTRC 2.0 after inclusion of free PSA

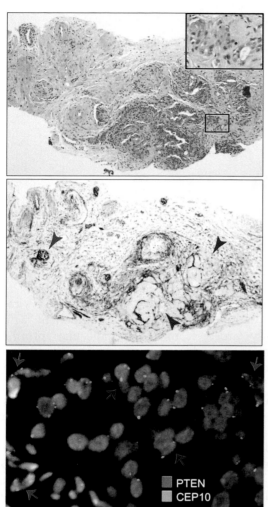

FIGURE 22.4 Evaluation of *PTEN* in prostate cancer tissue specimens by immunohistochemistry (IHC) and fluorescence in situ hybridization (FISH). *PTEN* is one of the most frequently deleted genes in prostate cancer, with numerous studies demonstrating an association between *PTEN* deletion and aggressiveness. Here, consecutive sections from a routine biopsy specimen with high-grade prostate cancer were evaluated by histology (top panel), IHC for PTEN (middle panel) expression and quantum dot FISH for *PTEN* copy number (bottom panel). By IHC, note retained PTEN expression in stromal cells and blood vessels (green arrowheads) with complete loss of PTEN expression in cancer (orange arrowheads). For quantum dot fish, probes for *PTEN* and the centromeric region of chromosome 10 (*CEP10*) were labeled in magenta and cyan, respectively, with nuclei stained by DAPI in gray. Note equal numbers of *PTEN* and *CEP10* signals in benign stromal cells (green arrows) and complete loss of *PTEN* signals in cancerous cells (orange arrows).

was demonstrated in two large American cohorts [68]. A similar integrated, protein-based risk calculator (4Kscore), which incorporates the 4K assay and routine clinical and demographic information, has been validated for the detection of high-grade prostate cancer in large prospective American cohort [66]. When compared directly to a modified PCPTRC 2.0 (without

family history), 4Kscore showed superior discrimination for detecting high-grade cancer (GS ≥ 7) and had a higher net benefit at all thresholds by decision curve analysis [66].

Several new RNA-based biomarkers have been proposed for the early detection of prostate cancer, including urine assessment of *PCA3* and *TMPRSS2:ERG* transcripts (Fig. 22.1B). Importantly, in contrast to existing protein-based assays, these RNA molecules are highly specific for prostate cancer and are likely to improve specificity of early detection efforts [66,69]. TMA assays are currently clinically available for the quantification of *PCA3* and *TMPRSS2:ERG* assays in postattentive digital rectal exam urine [46,47]. When incorporated into prostate biopsy nomograms that include total PSA and standard clinical and demographic information, urine PCA3 significantly improved the predictive accuracy for prostate cancer in multiple large international cohorts [70,71]. Although the Progensa *PCA3* assay is FDA approved for prostate cancer risk assessment in men with a prior negative biopsy, a recent validation study demonstrated clinical utility in men presenting for initial biopsy [72]. Similarly, inclusion of urine *PCA3* results in PCPTRC 1.0 yielded better diagnostic accuracy for prostate cancer than either total PSA, PCPTRC 1.0, or urine *PCA3* [73]. Given sensitivity concerns, urine *TMPRSS2:ERG* quantification/detection has most commonly been combined with more sensitive markers of prostate cancer (ie, total PSA and/or urine *PCA3*) and standard clinical and demographic information. For example, urine *TMPRSS2:ERG* assessed by qRT-PCR after whole transcriptome amplification significantly improved detection of prostate cancer in patients with total PSA less than 10 ng/mL [74]. In addition, in multiple independent large prostate biopsy cohorts, urine *TMPRSS2:ERG* and *PCA3* quantification by TMA assays improved performance of clinical risk assessment tools, including PCTPRC 1.0 and European Randomized Study of Screening for Prostate Cancer risk calculator, and increased detection of clinically significant and high-grade prostate cancer [45,47]. At the University of Michigan Health System, this combination of assays (total serum PSA, and urine *PCA3* and *TMPRSS2:ERG* by TMA) has been developed into a clinically available test (Mi-Prostate Score) for the early detection of prostate cancer (Fig. 22.1B).

Finally, while no molecular tests are currently used in routine clinical practice to identify patients at increased risk for developing prostate cancer, in the future, targeted or whole genome sequencing approaches may be utilized to screen specific cohorts of clinically identifiable high-risk patients (ie, strong family history) for germline mutations in prostate cancer associated genes, including *HOXB13*, *BRCA1*, and

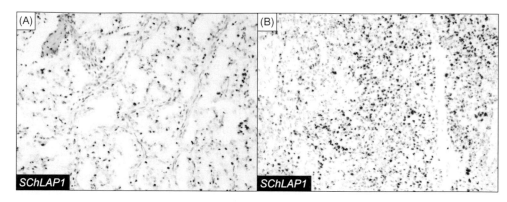

FIGURE 22.5 Detection of the long noncoding RNA (lncRNA) *SChLAP1* in prostate cancer. Next-generation sequencing and DNA microarray based approaches have identified numerous prostate and/or prostate cancer specific lncRNAs. *SChLAP1* has been identified as a prostate cancer specific lncRNA that is strongly associated with aggressive prostate cancer. *SChLAP1* expression by RNA in situ hybridization (brown signal) is shown in (A) localized and (B) metastatic castration resistant prostate cancer.

BRCA2. Our understanding of hereditary prostate cancer is as-of-yet incomplete, which limits the potential utility of this molecular data in current clinical practice. However, emerging research suggests that hereditary prostate cancer may have unique clinicopathological characteristics with important prognostic and therapeutic implications [32,34]. Likewise, numerous single-nucleotide polymorphisms (SNPs), which are associated with relatively modest increases in prostate cancer risk, have been identified through genome-wide association studies [75]. Assessment of such markers may be useful in stratifying men for early detection efforts in the near future, although typically only a small subset of patients have combinations of SNPs with moderate effect size (eg, >2 odds ratios compared to men at the median risk) [76,77].

Diagnosis

To date, molecular advances have not provided a disruptive innovation for the diagnosis of prostate cancer, where traditional histopathologic review remains the gold standard. This usually entails surgical pathological review of needle core biopsies sampled from an anatomic template of six or more areas of the gland, although increasingly, sampling is performed in conjunction with the guidance of advanced imaging technologies (including multiparametric MRI), which enable correlation of specific samples to imaged lesions. Nonetheless, there are a few specific scenarios for which molecular techniques may be helpful diagnostically. The diagnosis of prostate cancer in core biopsy specimens has become generally straightforward, especially with the support of reliable, widely available ancillary immunohistochemistry (ie, PIN-4 cocktail). It bears mention that the now quotidian

PIN-4 cocktail utilizes alpha methyl acyl-coA racemase (AMACR) as a cancer-specific component, a use that actually arose from early molecular profiling (ie, microarray-based gene expression profiling) of normal and cancerous prostate tissue [78]. Even greater precision is provided by contemporary prostate cancer immunohistochemistry for the *ERG* gene product, the expression of which can be detected in approximately half of prostate cancers. For example, in some cases, small foci of atypical prostate glands—generally termed atypical small acinar proliferations (ASAP)—may be suspicious for but difficult to definitively classify as cancer, due to quantitative (ie, amount sampled) or qualitative (ie, degree of atypia) limitations of the focus. Since benign prostate epithelium, including benign mimickers that may express AMACR, essentially never expresses ERG, the immunohistochemical detection of ERG protein expression in atypical prostate glands that are morphologically suspicious for cancer (ie, ASAP) confirms the diagnosis of prostate cancer [79–81]. Fig. 22.2 shows a focus of atrophic prostate cancer that is negative for AMACR but diffusely overexpresses ERG. However, it is important to note that because only approximately half of prostate cancers have *ERG* gene rearrangements, in this scenario, the lack of ERG protein expression in ASAP neither supports nor excludes a diagnosis of prostate cancer. Hence it remains a marker that is exquisitely cancer specific but only 40–50% sensitive.

Another situation in which molecular techniques may be helpful in prostate cancer diagnostics is distinguishing de novo high-grade NEPC (ie, small cell carcinoma of the prostate) from small cell carcinomas of other organs. While in many cases it is relatively straightforward to diagnose NEPC based on anatomic location, small cell carcinomas involving the prostatic base/bladder neck may be of either prostatic or

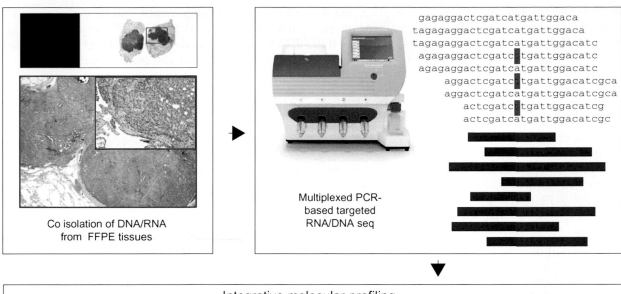

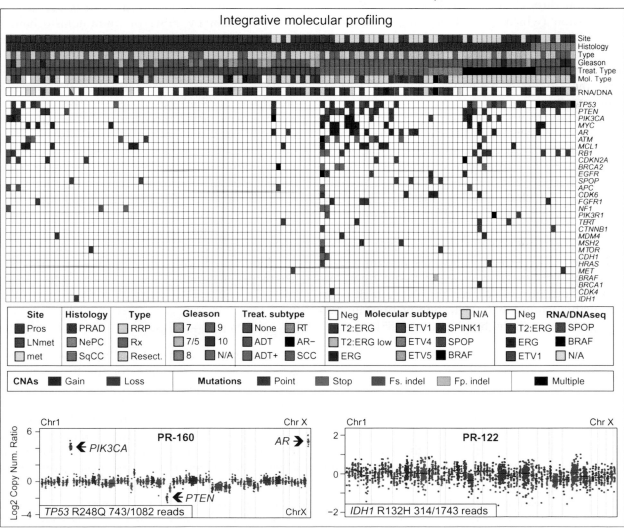

FIGURE 22.6 Targeted next-generation sequencing (NGS) to enable precision medicine for prostate cancer. NGS can be used to identify germline and somatic mutations, copy number alterations (CNAs) and gene fusions that may targetable by approved or investigational therapies. A schematic of an integrative NGS-based system to identify potentially actionable somatic alterations in prostate cancer using the Oncomine Comprehensive Panel (OCP) [41] is shown. DNA/RNA is co-isolated from macrodissected, routine FFPE tissue specimens

bladder origin. In these cases, as in Fig. 22.3, documenting the presence of an *ERG* gene rearrangement, either by FISH or immunohistochemistry, supports the diagnoses of NEPC, since *ERG* gene rearrangements are found in greater than 50% of NEPC but not in small cell carcinoma of the bladder or other primary sites relevant to this differential [82]. Similarly, evidence of an *ERG* gene rearrangement by FISH or immunohistochemistry would strongly support the diagnosis of metastatic prostatic cancer in patients for whom a prostatic primary is known or suspected [83]. Of course, in either case, the caveat applies that negative *ERG* FISH or ERG immunohistochemistry does not exclude a diagnosis of prostate cancer. Furthermore, it is important to note that because androgen signaling may be dysregulated in NEPC or metastatic prostate cancer (indeed, often in an iatrogenic fashion through androgen blockade), ERG protein expression (which is androgen regulated in the *TMPRSS2:ERG* gene fusion) in such cancers may be falsely negative by immunohistochemistry, as shown in Fig. 22.3 [82,84]. In these cases, *ERG* FISH may be the preferred method to detect *ERG* gene rearrangements.

Finally, although not used routinely in current clinical practice, there is some utility for the ConfirmMDx assay in patients with a negative biopsy for whom prostate cancer is strongly suspected on clinical grounds [85]. This assay detects methylation at multiple gene loci, including *APC* and *GSTP1*, using methylation-specific PCR [86]. Hypermethylation of *APC* and *GSTP1* is significantly more frequent in prostate cancer than benign prostatic tissue [86], and the absence of *APC* hypermethylation (as determined by the ConfirmMDx assay) in prostatic tissue from core biopsies without prostate cancer has a very high negative predictive value for the detection of prostate cancer on repeat biopsy in high-risk patients [85].

Prognostication

Biomarkers for prostate cancer prognostication are an emerging area of focus in genitourinary pathology, and both single-gene and multigene assays have been developed. Genomic loss of *PTEN*, the second most common genetic aberration in prostate cancer after

ERG gene rearrangements, can be detected by FISH or immunohistochemistry [55–58], as shown in Fig. 22.4. *PTEN* deletion, as detected by FISH, is associated with advanced disease at the time of radical prostatectomy (ie, extraprostatic extension, seminal vesicle invasion), as well as a decreased time to biochemical recurrence [87]. In a large European cohort of patients treated with radical prostatectomy, *PTEN* deletion was significantly associated with decreased time to biochemical recurrence in univariate and multivariate analyses, and this effect was independent of *ERG* gene rearrangement status [88]. Similarly, loss of *PTEN* by immunohistochemistry, which shows high concordance with *PTEN* FISH, is associated with high Gleason score, advanced pathological stage, and decreased time to metastasis in high-risk patients after radical prostatectomy [58]. Finally, for conservatively managed, clinically localized disease, the detection of *PTEN* deletion by FISH or immunohistochemistry is highly predictive of cancer-related death in low-risk but not high-risk patients [89].

Increasing awareness of oncogenic roles for lncRNA in prostate cancer raises the intriguing possibility of prognostic lncRNA-based assays. For example, the lncRNA *SChLAP1* demonstrates outlier expression in a subset of prostate cancers, promotes cancer cell invasion and metastasis, and is strongly associated metastatic progression in a large multi-institutional analysis [26,27]. *SChLAP1* expression can be detected by ISH in FFPE tissue [22,26,27] (Fig. 22.5), and high *SChLAP1* expression by ISH in radical prostatectomy specimens is associated with poor outcome, after univariate and multivariate analyses [62]. Future studies will explore the potential clinical utility for *SChLAP1* ISH in prostate biopsy material.

Targeted Therapeutics

In current clinical practice, there is very limited targeted therapeutic selection for the treatment of prostate cancer, and only recently have clinical trials been developed to assess possible targeted therapeutic strategies [83]. Given that nearly half of all prostate cancers harbor ETS gene rearrangements, the development of therapeutics targeting the molecular mechanisms of ETS-related oncogenesis has been area of active

◀ (upper left panel); a representative specimen, demonstrating a castration-resistant prostate cancer (CRPC) lung metastasis is shown. Samples were assessed by NGS using multiplexed PCR-based DNA and RNA sequencing (top right). Data was analyzed using a highly automated pipeline to generate an integrative molecular profile for each sample consisting of predefined potentially actionable somatic point mutations, CNAs, and gene fusions (bottom panel). Sample information is shown in the header according to the legend at the bottom. The heatmap shows individual alterations according to the bottom legend. Integrative profiles for two samples (PR-160 and PR-122) are shown, with copy number plots visualized (colored points represent individual amplicons for a given gene; black bars represent gene level copy number ratios) and prioritized somatic mutations indicated. Of note, PR-160, a CRPC metastasis, shows an acquired androgen receptor (AR) amplification, a known adaptive response to anti-androgen therapy. PR-122 harbors an *IDH1* R132H mutation and no other prioritized alterations.

research [90]. Poly (ADP-ribose) polymerase 1 (PARP1) is a chromatin-associated enzyme with roles in multiple intracellular processes. In ETS-positive prostate cancer, PARP1 physically interacts with ETS proteins and is required for ETS-mediated transcription, and in preclinical models, pharmacologic inhibition of PARP1 reduces ETS-dependent cellular proliferation [91]. PARP1 is also involved in the DNA damage response pathway, and inhibition of PARP1 induces DNA damage mediated cellular apoptosis, which sensitizes tumor cells to radiotherapy and platinum-based alkylating agents [92,93]. Mutations in *BRCA1* and/or *BRCA2* also impact DNA damage response via inhibition of homologous recombination, which leads to an accumulation of DNA double-stranded breaks and sensitivity to PARP1 inhibitors [90,93–95]. Therefore, while sporadic *BRCA1* and/or *BRCA2* mutations are uncommon in prostate cancer, patients with *BRCA*-associated hereditary prostate cancer could benefit from PARP1 inhibition—with or without radiotherapy and platinum-based chemotherapy. Interestingly, *PCAT-1*, an lncRNA with outlier overexpression in a subset of prostate cancers, represses *BRCA2* expression in prostate cancer cell lines, resulting in a functional BRCA-deficient phenotype (BRCAness). Similar to tumors with *BRCA1* and/or *BRCA2* mutations, this BRCAness bestows sensitivity to PARP1 inhibitors and suggests that *PCAT-1* may be a biomarker for predicting response to PARP1 inhibitors in prostate cancer [25]. Aside from ETS gene rearrangements, several of the recurrent genomic aberrations in prostate cancer are likely amenable to targeted therapy in the future. For example, *PTEN* deletion is the second most common molecular abnormality in prostate cancer and leads to dysregulation of the PI3K signaling pathway [15]. Drugs that inhibit the PI3K pathway are currently in clinical trials for a range of malignancies and may be useful for prostate cancers harboring *PTEN* deletions.

Prostate cancer harbors rare but potentially targetable alterations, including mutations or fusions involving RAF family members (including *BRAF* and *RAF1*), *FGFR2*, and *IDH1* [35–39,41]. Importantly, molecularly informed clinical trials enrolling patients with prostate or any cancer (so-called basket trials) are increasingly available. In the near future, we anticipate that both targeted and comprehensive molecular profiling efforts may be used to identify potential treatment strategies for patients with aggressive or advanced prostate cancer. For example, we recently utilized a pan-cancer, multiplexed PCR-based targeted DNA/RNA sequencing based approach to identify potentially targetable alterations in a prostate cancer cohort [41]. Examples of patients harboring *AR/PIK3CA* amplifications and *IDH1* R132H mutations are shown in Fig. 22.6.

LIMITATIONS OF TESTING

Despite tremendous recent progress in our understanding of the molecular mechanisms of prostate cancer, as well as emerging molecular-based clinical assays for the early detection, diagnosis, prognostication, and treatment of prostate cancer, the role for molecular pathology in routine clinical practice is still limited, due to many factors. These include (1) the fragmentation of care for men at risk for or with prostate cancer (eg, early detection by general practitioners, biopsy by urologists, definitive treatment usually by urologists or radiation oncologists, and medical therapy by oncologists), (2) the lack of predictive biomarkers, (3) the relatively modest additional benefit that prognostic biomarkers provide when added to standard clinical, radiological, and pathological evaluation, (4) the challenges of working with small diagnostic FFPE biopsies, and (5) the lack of routine biopsy of metastatic tissues. In particular, perhaps with the exception of *PTEN* deletion, currently available single-gene assays lack the ability to add to optimized clinico-pathological models, either at the time of biopsy or after radical prostatectomy. Critically, this is the key question when evaluating a new prognostic biomarker: does its inclusion improve the performance of an optimized model (ie, a nomogram) that incorporates all relevant and easily assessable clinicopathological parameters (eg, grade, stage, serum PSA, margin status and node status in the prostatectomy setting) [96]?

Multigene Assays

Several multigene assays for prostate cancer prognostication have been introduced over the past several years, including Oncotype DX Prostate, GenomeDx Decipher, and Prolaris [97,98]. Oncotype DX Prostate, designed for use with prostate biopsy material, is an RT-PCR-based assay that measures expression of 12 prostate cancer related genes and 5 references genes to generate the Genomic Prostate Score (GPS) [97]. In a large retrospective American cohort of patients who met clinical criteria for active surveillance, the biopsy GPS was associated with high grade and stage cancers at radical prostatectomy, even after multivariate analysis, and by decision curve analysis, the addition of biopsy GPS to the Cancer of the Prostate Risk Assessment (CAPRA) score provided increased net benefit over clinical information alone [99]. In an independent American cohort, the biopsy GPS was similarly associated with high grade and stage cancers at radical prostatectomy (after multivariate analysis), as well as increased risk of biochemical recurrence (after univariate analysis) [100].

Similar to Oncotype DX Prostate, Prolaris is an RT-PCR-based gene expression assay. The Polaris test integrates expression of 31 genes involved in cell cycle progression (CCP) to generate a CCP score [101]. The CCP score was useful for predicting biochemical recurrence after radical prostatectomy, as well as cancer-related death after transurethral resection of the prostate in a conservatively managed prostate cancer cohort [102]. Similarly, in a large independent American cohort, addition of the CCP score increased the predictive capability of the Cancer of the Prostate Risk Assessment post-Surgical (CAPRA-S) score for recurrence after radical prostatectomy, and the combined CCP and CAPRA-S score outperformed either score alone in decision curve analysis [103]. In two large independent prostate biopsy cohorts, the CCP score is significantly associated with adverse outcomes, including biochemical recurrence after radical prostatectomy and prostate cancer related death after conservative management [102,104]. An independent meta-analysis confirmed that the CCP score demonstrates robust prognostic value for predicting biochemical recurrence of prostate cancer, after univariate and multivariate analyses [105].

In contrast to Oncotype DX Prostate and Prolaris, GenomeDx Decipher is an RNA microarray based test that integrates expression of 22 RNA molecules to generate a genomic classifier (GC) score [106]. The GC score was developed from a radical prostatectomy cohort enriched for patients with metastatic disease and was validated for the prediction of metastatic progression after radical prostatectomy in a cohort of high-risk prostate cancer patients [106,107]. While GC and CAPRA-S scores were significant independent predictors of cancer-related death after radical prostatectomy in a cohort of high-risk prostate cancer, integration of GC and CAPRA-S scores identified a subset of patients with very high prostate cancer related mortality [108].

Most recently, an eight biomarker multiplexed immunofluorescence (IF) based assay on FFPE prostate biopsy tissues has recently been validated for predicting prostatectomy pathology [109]. This panel, which uses quantitative multiplex proteomics imaging, integrates morphological object recognition and molecular biomarker measurements from tumor epithelium at the individual slide level. In a blinded study of 276 cases using trained logistic regression models, the eight biomarker multiplexed IF assay improved the AUC for predicting favorable disease at prostatectomy from 0.69 (NCCN guideline classification) to 0.75. Net reclassification and decision curve analysis demonstrated benefit from the combined eight biomarker IF assay and NCCN classification, compared to NCCN classification alone [109].

Future Directions

In summary, over the past several years, a number of promising new molecular assays have been developed for early detection, diagnosis, and prognostication of prostate cancer; and in the near future, these tests will find increasing utility in routine clinical practice. The challenge moving forward will be to understand what test needs to be used for which patients at a given stage in the clinical presentation of prostate cancer and, furthermore, to demonstrate that the overall net benefit for these assays justifies their routine use and cost to patients and payers. In particular, it will be important to determine the true added value of these tests over the current clinical risk stratification tools, which are effective, widely available, and essentially free.

Based on currently available molecular assays, one can imagine the future state of prostate cancer early detection, diagnosis, and prognostication: (1) for patients with a strong family history of prostate cancer, germline HOXB13, BRCA1, and/or BRCA2 sequencing identifies those for close clinical monitoring; (2) for otherwise low-risk patients in the general population, advanced serum-based protein and/or urine-based RNA assays select those for subsequent prostate biopsy; (3) for patients with a high clinical suspicion of prostate cancer but negative prostate biopsy, a tissue-based DNA methylation assay detects those for repeat prostate biopsy; (4) for patients with relatively low-grade prostate cancer on prostate biopsy, tissue-based RNA assays distinguish those who would benefit from radical prostatectomy over active surveillance; and (5) for patients with high grade and stage prostate cancer at radical prostatectomy, a tissue-based RNA assay identifies those at high risk for metastatic progression.

Given our rapidly evolving understanding of the molecular underpinnings of prostate cancer, we are poised to enter an era of targeted therapeutics for prostate cancer treatment. Emerging therapies for prostate cancer with ETS gene rearrangements and PTEN deletions are promising advances. Currently, molecular tests that might be useful to detect molecular aberrations for the selection of targeted therapeutics are overwhelming single-gene assays (ie, ERG FISH, PTEN FISH, or immunohistochemistry). However, significant technical advances over the past several years promise to fundamentally change the field of clinical molecular diagnostics, through targeted next-generation and/or integrative comprehensive sequencing [41,110,111]. In contrast to current single-gene assays, these advanced molecular approaches will provide mutational information for hundreds to thousands of genes simultaneously, a development that will support the dual clinical goals of personalized medicine and targeted therapeutics.

Acknowledgments

S.A.T. is supported in part by the Evans Foundation/Prostate Cancer Foundation, the A. Alfred Taubman Medical Research Institute, and the National Institutes of Health (R01 CA183857).

DISCLOSURES

The University of Michigan has been issued a patent on the detection of ETS gene fusions in prostate cancer, on which S.A.T. is listed as a coinventor. The University of Michigan licensed the diagnostic field of use to Hologic/Gen-Probe, Inc, which sublicensed some rights to Ventana Medical Systems, Inc. The University of Michigan has filed a patent on SPINK1 in prostate cancer, on which S.A.T. is listed as a coinventor. S.A.T. has received honoraria from, and served as a consultant to, Ventana Medical Systems. S.A.T. had a sponsored research agreement with Thermo Fisher Scientific/Life Technologies that supported next-generation sequencing of prostate cancer samples referenced herein.

References

[1] Siegel RL, Miller KD, Jemal A. Cancer statistics, 2015. CA Cancer J Clin 2015;65:5−29.

[2] Beltran H, Tomlins S, Aparicio A, Arora V, Rickman D, Ayala G, et al. Aggressive variants of castration-resistant prostate cancer. Clin Cancer Res 2014;20:2846−50.

[3] Lilja H, Ulmert D, Vickers AJ. Prostate-specific antigen and prostate cancer: prediction, detection and monitoring. Nat Rev 2008;8:268−78.

[4] Amin MB, Lin DW, Gore JL, Srigley JR, Samaratunga H, Egevad L, et al. The critical role of the pathologist in determining eligibility for active surveillance as a management option in patients with prostate cancer: consensus statement with recommendations supported by the College of American Pathologists, International Society of Urological Pathology, Association of Directors of Anatomic and Surgical Pathology, the New Zealand Society of Pathologists, and the Prostate Cancer Foundation. Arch Pathol Lab Med 2014;138:1387−405.

[5] Beltran H, Rubin MA. New strategies in prostate cancer: translating genomics into the clinic. Clin Cancer Res 2013;19:517−23.

[6] Balk SP, Ko YJ, Bubley GJ. Biology of prostate-specific antigen. J Clin Oncol 2003;21:383−91.

[7] Thorek DL, Evans MJ, Carlsson SV, Ulmert D, Lilja H. Prostate-specific kallikrein-related peptidases and their relation to prostate cancer biology and detection. Established relevance and emerging roles. Thromb Haemost 2013;110:484−92.

[8] Roychowdhury S, Chinnaiyan AM. Advancing precision medicine for prostate cancer through genomics. J Clin Oncol 2013;31:1866−73.

[9] Barbieri CE, Tomlins SA. The prostate cancer genome: perspectives and potential. Urol Oncol 2014;32 53 e15−e22

[10] Kumar-Sinha C, Tomlins SA, Chinnaiyan AM. Recurrent gene fusions in prostate cancer. Nat Rev 2008;8:497−511.

[11] Tomlins SA, Rhodes DR, Perner S, Dhanasekaran SM, Mehra R, Sun XW, et al. Recurrent fusion of TMPRSS2 and ETS transcription factor genes in prostate cancer. Science 2005;310:644−8.

[12] Pettersson A, Graff RE, Bauer SR, Pitt MJ, Lis RT, Stack EC, et al. The TMPRSS2:ERG rearrangement, ERG expression, and prostate cancer outcomes: a cohort study and meta-analysis. Cancer Epidemiol Biomarkers Prev 2012;21:1497−509.

[13] Mehra R, Han B, Tomlins SA, Wang L, Menon A, Wasco MJ, et al. Heterogeneity of TMPRSS2 gene rearrangements in multifocal prostate adenocarcinoma: molecular evidence for an independent group of diseases. Cancer Res 2007;67:7991−5.

[14] Tomlins SA, Laxman B, Varambally S, Cao X, Yu J, Helgeson BE, et al. Role of the TMPRSS2-ERG gene fusion in prostate cancer. Neoplasia 2008;10:177−88.

[15] Phin S, Moore MW, Cotter PD. Genomic rearrangements of PTEN in prostate cancer. Front Oncol 2013;3:240.

[16] Sarker D, Reid AH, Yap TA, de Bono JS. Targeting the PI3K/AKT pathway for the treatment of prostate cancer. Clin Cancer Res 2009;15:4799−805.

[17] Krohn A, Freudenthaler F, Harasimowicz S, Kluth M, Fuchs S, Burkhardt L, et al. Heterogeneity and chronology of PTEN deletion and ERG fusion in prostate cancer. Mod Pathol 2014;27:1612−20.

[18] Gumuskaya B, Gurel B, Fedor H, Tan HL, Weier CA, Hicks JL, et al. Assessing the order of critical alterations in prostate cancer development and progression by IHC: further evidence that PTEN loss occurs subsequent to ERG gene fusion. Prostate Cancer Prostatic Dis 2013;16:209−15.

[19] Carver BS, Tran J, Gopalan A, Chen Z, Shaikh S, Carracedo A, et al. Aberrant ERG expression cooperates with loss of PTEN to promote cancer progression in the prostate. Nat Genet 2009;41:619−24.

[20] Han B, Mehra R, Lonigro RJ, Wang L, Suleman K, Menon A, et al. Fluorescence in situ hybridization study shows association of PTEN deletion with ERG rearrangement during prostate cancer progression. Mod Pathol 2009;22:1083−93.

[21] Prensner JR, Iyer MK, Balbin OA, Dhanasekaran SM, Cao Q, Brenner JC, et al. Transcriptome sequencing across a prostate cancer cohort identifies PCAT-1, an unannotated lncRNA implicated in disease progression. Nat Biotechnol 2011;29:742−9.

[22] Bottcher R, Hoogland AM, Dits N, Verhoef EI, Kweldam C, Waranecki P, et al. Novel long non-coding RNAs are specific diagnostic and prognostic markers for prostate cancer. Oncotarget 2015;6:4036−50.

[23] Malik R, Patel L, Prensner JR, Shi Y, Iyer MK, Subramaniyan S, et al. The lncRNA PCAT29 inhibits oncogenic phenotypes in prostate cancer. Mol Cancer Res 2014;12:1081−7.

[24] Prensner JR, Chen W, Han S, Iyer MK, Cao Q, Kothari V, et al. The long non-coding RNA PCAT-1 promotes prostate cancer cell proliferation through cMyc. Neoplasia 2014;16:900−8.

[25] Prensner JR, Chen W, Iyer MK, Cao Q, Ma T, Han S, et al. PCAT-1, a long noncoding RNA, regulates BRCA2 and controls homologous recombination in cancer. Cancer Res 2014;74:1651−60.

[26] Prensner JR, Iyer MK, Sahu A, Asangani IA, Cao Q, Patel L, et al. The long noncoding RNA SChLAP1 promotes aggressive prostate cancer and antagonizes the SWI/SNF complex. Nat Genet 2013;45:1392−8.

[27] Prensner JR, Zhao S, Erho N, Schipper M, Iyer MK, Dhanasekaran SM, et al. RNA biomarkers associated with metastatic progression in prostate cancer: a multi-institutional high-throughput analysis of SChLAP1. Lancet Oncol 2014;15:1469−80.

[28] de Kok JB, Verhaegh GW, Roelofs RW, Hessels D, Kiemeney LA, Aalders TW, et al. DD3(PCA3), a very sensitive and specific marker to detect prostate tumors. Cancer Res 2002;62:2695−8.

[29] Bussemakers MJ, van Bokhoven A, Verhaegh GW, Smit FP, Karthaus HF, Schalken JA, et al. DD3: a new prostate-specific

gene, highly overexpressed in prostate cancer. Cancer Res 1999;59:5975—9.

[30] Deras IL, Aubin SM, Blase A, Day JR, Koo S, Partin AW, et al. PCA3: a molecular urine assay for predicting prostate biopsy outcome. J Urol 2008;179:1587—92.

[31] Ewing CM, Ray AM, Lange EM, Zuhlke KA, Robbins CM, Tembe WD, et al. Germline mutations in HOXB13 and prostate-cancer risk. N Engl J Med 2012;366:141—9.

[32] Castro E, Eeles R. The role of BRCA1 and BRCA2 in prostate cancer. Asian J Androl 2012;14:409—14.

[33] Decker B, Ostrander EA. Dysregulation of the homeobox transcription factor gene HOXB13: role in prostate cancer. Pharmgenomics Pers Med 2014;7:193—201.

[34] Smith SC, Palanisamy N, Zuhlke KA, Johnson AM, Siddiqui J, Chinnaiyan AM, et al. HOXB13 G84E-related familial prostate cancers: a clinical, histologic, and molecular Survey. Am J Surg Pathol 2014;38:615—26.

[35] Palanisamy N, Ateeq B, Kalyana-Sundaram S, Pflueger D, Ramnarayanan K, Shankar S, et al. Rearrangements of the RAF kinase pathway in prostate cancer, gastric cancer and melanoma. Nat Med 2010;16:793—8.

[36] Barbieri CE, Baca SC, Lawrence MS, Demichelis F, Blattner M, Theurillat JP, et al. Exome sequencing identifies recurrent SPOP, FOXA1 and MED12 mutations in prostate cancer. Nat Genet 2012;44:685—9.

[37] Beltran H, Yelensky R, Frampton GM, Park K, Downing SR, Macdonald TY, et al. Targeted next-generation sequencing of advanced prostate cancer identifies potential therapeutic targets and disease heterogeneity. Eur Urol 2013;63:920—6.

[38] Grasso CS, Cani AK, Hovelson DH, Quist MJ, Douville NJ, Yadati V, et al. Integrative molecular profiling of routine clinical prostate cancer specimens. Ann Oncol 2015;26:1110—18.

[39] Wu YM, Su FY, Kalyana-Sundaram S, Khazanov N, Ateeq B, Cao XH, et al. Identification of targetable FGFR gene fusions in diverse cancers. Cancer Discov 2013;3:636—47.

[40] Baca SC, Prandi D, Lawrence MS, Mosquera JM, Romanel A, Drier Y, et al. Punctuated evolution of prostate cancer genomes. Cell. 2013;153:666—77.

[41] Hovelson DH, McDaniel AS, Cani AK, Johnson B, Rhodes K, Williams PD, et al. Development and validation of a scalable next-generation sequencing system for assessing relevant somatic variants in solid tumors. Neoplasia 2015;17:385—99.

[42] Mauzo SH, Lee M, Petros J, Hunter S, Chang CM, Shu HK, et al. Immunohistochemical demonstration of isocitrate dehydrogenase 1 (IDH1) mutation in a small subset of prostatic carcinomas. Appl Immunohistochem Mol Morphol 2014;22:284—7.

[43] Ghiam AF, Cairns RA, Thoms J, Dal Pra A, Ahmed O, Meng A, et al. IDH mutation status in prostate cancer. Oncogene 2012;31:3826.

[44] Laxman B, Morris DS, Yu J, Siddiqui J, Cao J, Mehra R, et al. A first-generation multiplex biomarker analysis of urine for the early detection of prostate cancer. Cancer Res 2008;68:645—9.

[45] Leyten GH, Hessels D, Jannink SA, Smit FP, de Jong H, Cornel EB, et al. Prospective multicentre evaluation of PCA3 and TMPRSS2-ERG gene fusions as diagnostic and prognostic urinary biomarkers for prostate cancer. Eur Urol 2014;65:534—42.

[46] Salagierski M, Schalken JA. Molecular diagnosis of prostate cancer: PCA3 and TMPRSS2:ERG gene fusion. J Urol 2012;187:795—801.

[47] Tomlins SA, Aubin SM, Siddiqui J, Lonigro RJ, Sefton-Miller L, Miick S, et al. Urine TMPRSS2:ERG fusion transcript stratifies prostate cancer risk in men with elevated serum PSA. Sci Transl Med 2011;3:94ra72.

[48] Mehra R, Tomlins SA, Shen R, Nadeem O, Wang L, Wei JT, et al. Comprehensive assessment of TMPRSS2 and ETS family gene aberrations in clinically localized prostate cancer. Mod Pathol 2007;20:538—44.

[49] Yoshimoto M, Joshua AM, Chilton-Macneill S, Bayani J, Selvarajah S, Evans AJ, et al. Three-color FISH analysis of TMPRSS2/ERG fusions in prostate cancer indicates that genomic microdeletion of chromosome 21 is associated with rearrangement. Neoplasia 2006;8:465—9.

[50] Park K, Tomlins SA, Mudaliar KM, Chiu YL, Esgueva R, Mehra R, et al. Antibody-based detection of ERG rearrangement-positive prostate cancer. Neoplasia 2010;12:590—8.

[51] Kunju LP, Carskadon S, Siddiqui J, Tomlins SA, Chinnaiyan AM, Palanisamy N. Novel RNA hybridization method for the in situ detection of ETV1, ETV4, and ETV5 gene fusions in prostate cancer. Appl Immunohistochem Mol Morphol 2014;22: e32—40.

[52] Berger MF, Lawrence MS, Demichelis F, Drier Y, Cibulskis K, Sivachenko AY, et al. The genomic complexity of primary human prostate cancer. Nature 2011;470:214—20.

[53] Maher CA, Kumar-Sinha C, Cao X, Kalyana-Sundaram S, Han B, Jing X, et al. Transcriptome sequencing to detect gene fusions in cancer. Nature 2009;458:97—101.

[54] Maher CA, Palanisamy N, Brenner JC, Cao X, Kalyana-Sundaram S, Luo S, et al. Chimeric transcript discovery by paired-end transcriptome sequencing. Proc Natl Acad Sci USA 2009;106:12353—8.

[55] Verhagen PC, van Duijn PW, Hermans KG, Looijenga LH, van Gurp RJ, Stoop H, et al. The PTEN gene in locally progressive prostate cancer is preferentially inactivated by bi-allelic gene deletion. J Pathol 2006;208:699—707.

[56] Yoshimoto M, Cutz JC, Nuin PA, Joshua AM, Bayani J, Evans AJ, et al. Interphase FISH analysis of PTEN in histologic sections shows genomic deletions in 68% of primary prostate cancer and 23% of high-grade prostatic intra-epithelial neoplasias. Cancer Genet Cytogenet 2006;169:128—37.

[57] Sathyanarayana UG, Birch C, Nagle RB, Tomlins SA, Palanisamy N, Zhang W, et al. Determination of optimum formalin fixation duration for prostate needle biopsies for immunohistochemistry and quantum dot FISH analysis. Appl Immunohistochem Mol Morphol 2015;23:364—73.

[58] Lotan TL, Gurel B, Sutcliffe S, Esopi D, Liu W, Xu J, et al. PTEN protein loss by immunostaining: analytic validation and prognostic indicator for a high risk surgical cohort of prostate cancer patients. Clin Cancer Res 2011;17:6563—73.

[59] Hessels D, Klein Gunnewiek JM, van Oort I, Karthaus HF, van Leenders GJ, van Balken B, et al. DD3(PCA3)-based molecular urine analysis for the diagnosis of prostate cancer. Eur Urol 2003;44:8—15.

[60] Warrick JI, Tomlins SA, Carskadon SL, Young AM, Siddiqui J, Wei JT, et al. Evaluation of tissue PCA3 expression in prostate cancer by RNA in situ hybridization—a correlative study with urine PCA3 and TMPRSS2-ERG. Mod Pathol 2014;27:609—20.

[61] Popa I, Fradet Y, Beaudry G, Hovington H, Tetu B. Identification of PCA3 (DD3) in prostatic carcinoma by in situ hybridization. Mod Pathol 2007;20:1121—7.

[62] Mehra R, Shi Y, Udager AM, Prensner JR, Sahu A, Iyer MK, et al. A novel RNA in situ hybridization assay for the long noncoding RNA SChLAP1 predicts poor clinical outcome after radical prostatectomy in clinically localized prostate cancer. Neoplasia 2014;16:1121—7.

[63] Loeb S, Sanda MG, Broyles DL, Shin SS, Bangma CH, Wei JT, et al. The prostate health index selectively identifies clinically significant prostate cancer. J Urol 2015;193:1163—9.

[64] Vickers AJ, Cronin AM, Aus G, Pihl CG, Becker C, Pettersson K, et al. A panel of kallikrein markers can reduce unnecessary biopsy for prostate cancer: data from the European

Randomized Study of Prostate Cancer Screening in Goteborg, Sweden. BMC Med 2008;6:19.

[65] Vickers A, Cronin A, Roobol M, Savage C, Peltola M, Pettersson K, et al. Reducing unnecessary biopsy during prostate cancer screening using a four-kallikrein panel: an independent replication. J Clin Oncol 2010;28:2493–8.

[66] Parekh DJ, Punnen S, Sjoberg DD, Asroff SW, Bailen JL, Cochran JS, et al. A multi-institutional prospective trial in the USA confirms that the 4Kscore accurately identifies men with high-grade prostate cancer. Eur Urol 2015;68:464–70.

[67] Nordstrom T, Vickers A, Assel M, Lilja H, Gronberg H, Eklund M. Comparison between the four-kallikrein panel and prostate health index for predicting prostate cancer. Eur Urol 2015;68:139–46.

[68] Ankerst DP, Hoefler J, Bock S, Goodman PJ, Vickers A, Hernandez J, et al. Prostate Cancer Prevention Trial risk calculator 2.0 for the prediction of low- vs high-grade prostate cancer. Urology 2014;83:1362–7.

[69] Hessels D, Smit FP, Verhaegh GW, Witjes JA, Cornel EB, Schalken JA. Detection of TMPRSS2-ERG fusion transcripts and prostate cancer antigen 3 in urinary sediments may improve diagnosis of prostate cancer. Clin Cancer Res 2007;13:5103–8.

[70] Chun FK, de la Taille A, van Poppel H, Marberger M, Stenzl A, Mulders PF, et al. Prostate cancer gene 3 (PCA3): development and internal validation of a novel biopsy nomogram. Eur Urol 2009;56:659–67.

[71] Auprich M, Haese A, Walz J, Pummer K, de la Taille A, Graefen M, et al. External validation of urinary PCA3-based nomograms to individually predict prostate biopsy outcome. Eur Urol 2010;58:727–32.

[72] Wei JT, Feng Z, Partin AW, Brown E, Thompson I, Sokoll L, et al. Can urinary PCA3 supplement PSA in the early detection of prostate cancer? J Clin Oncol 2014;32:4066–72.

[73] Ankerst DP, Groskopf J, Day JR, Blase A, Rittenhouse H, Pollock BH, et al. Predicting prostate cancer risk through incorporation of prostate cancer gene 3. J Urol 2008;180:1303–8.

[74] Salami SS, Schmidt F, Laxman B, Regan MM, Rickman DS, Scherr D, et al. Combining urinary detection of TMPRSS2:ERG and PCA3 with serum PSA to predict diagnosis of prostate cancer. Urol Oncol 2013;31:566–71.

[75] Al Olama AA, Kote-Jarai Z, Berndt SI, Conti DV, Schumacher F, Han Y, et al. A meta-analysis of 87,040 individuals identifies 23 new susceptibility loci for prostate cancer. Nat Genet 2014;46:1103–9.

[76] Pashayan N, Duffy SW, Neal DE, Hamdy FC, Donovan JL, Martin RM, et al. Implications of polygenic risk-stratified screening for prostate cancer on overdiagnosis. Genet Med 2015;17:789–95.

[77] Amin Al Olama A, Benlloch S, Antoniou AC, Giles GG, Severi G, Neal D, et al. Risk analysis of prostate cancer in PRACTICAL, a multinational consortium, using 25 known prostate cancer susceptibility loci. Cancer Epidemiol Biomarkers Prev 2015;24:1121–9.

[78] Rubin MA, Zhou M, Dhanasekaran SM, Varambally S, Barrette TR, Sanda MG, et al. Alpha-methylacyl coenzyme A racemase as a tissue biomarker for prostate cancer. JAMA 2002;287:1662–70.

[79] Tomlins SA, Palanisamy N, Siddiqui J, Chinnaiyan AM, Kunju LP. Antibody-based detection of ERG rearrangements in prostate core biopsies, including diagnostically challenging cases: ERG staining in prostate core biopsies. Arch Pathol Lab Med 2012;136:935–46.

[80] Shah RB, Tadros Y, Brummell B, Zhou M. The diagnostic use of ERG in resolving an "atypical glands suspicious for cancer" diagnosis in prostate biopsies beyond that provided by basal cell and alpha-methylacyl-CoA-racemase markers. Hum Pathol 2013;44:786–94.

[81] Shah RB. Clinical applications of novel ERG immunohistochemistry in prostate cancer diagnosis and management. Adv Anat Pathol 2013;20:117–24.

[82] Lotan TL, Gupta NS, Wang W, Toubaji A, Haffner MC, Chaux A, et al. ERG gene rearrangements are common in prostatic small cell carcinomas. Mod Pathol 2011;24:820–8.

[83] Udager AM, Alva A, Mehra R. Current and proposed molecular diagnostics in a genitourinary service line laboratory at a tertiary clinical institution. Cancer J 2014;20:29–42.

[84] Udager AM, Shi Y, Tomlins SA, Alva A, Siddiqui J, Cao X, et al. Frequent discordance between ERG gene rearrangement and ERG protein expression in a rapid autopsy cohort of patients with lethal, metastatic, castration-resistant prostate cancer. Prostate 2014;74:1199–208.

[85] Trock BJ, Brotzman MJ, Mangold LA, Bigley JW, Epstein JI, McLeod D, et al. Evaluation of GSTP1 and APC methylation as indicators for repeat biopsy in a high-risk cohort of men with negative initial prostate biopsies. BJU Int 2012;110:56–62.

[86] Enokida H, Shiina H, Urakami S, Igawa M, Ogishima T, Li LC, et al. Multigene methylation analysis for detection and staging of prostate cancer. Clin Cancer Res 2005;11:6582–8.

[87] Yoshimoto M, Cunha IW, Coudry RA, Fonseca FP, Torres CH, Soares FA, et al. FISH analysis of 107 prostate cancers shows that PTEN genomic deletion is associated with poor clinical outcome. Br J Cancer 2007;97:678–85.

[88] Krohn A, Diedler T, Burkhardt L, Mayer PS, De Silva C, Meyer-Kornblum M, et al. Genomic deletion of PTEN is associated with tumor progression and early PSA recurrence in ERG fusion-positive and fusion-negative prostate cancer. Am J Pathol 2012;181:401–12.

[89] Cuzick J, Yang ZH, Fisher G, Tikishvili E, Stone S, Lanchbury JS, et al. Prognostic value of PTEN loss in men with conservatively managed localised prostate cancer. Br J Cancer 2013;108:2582–9.

[90] Feng FY, Brenner JC, Hussain M, Chinnaiyan AM. Molecular pathways: targeting ETS gene fusions in cancer. Clin Cancer Res 2014;20:4442–8.

[91] Brenner JC, Ateeq B, Li Y, Yocum AK, Cao Q, Asangani IA, et al. Mechanistic rationale for inhibition of poly(ADP-ribose) polymerase in ETS gene fusion-positive prostate cancer. Cancer Cell 2011;19:664–78.

[92] Han S, Brenner JC, Sabolch A, Jackson W, Speers C, Wilder-Romans K, et al. Targeted radiosensitization of ETS fusion-positive prostate cancer through PARP1 inhibition. Neoplasia 2013;15:1207–17.

[93] Do K, Chen AP. Molecular pathways: targeting PARP in cancer treatment. Clin Cancer Res 2013;19:977–84.

[94] Lee JM, Ledermann JA, Kohn EC. PARP Inhibitors for BRCA1/2 mutation-associated and BRCA-like malignancies. Ann Oncol 2014;25:32–40.

[95] Zhang J. Poly (ADP-ribose) polymerase inhibitor: an evolving paradigm in the treatment of prostate cancer. Asian J Androl 2014;16:401–6.

[96] Kattan MW. Judging new markers by their ability to improve predictive accuracy. J Natl Cancer Inst 2003;95:634–5.

[97] Knezevic D, Goddard AD, Natraj N, Cherbavaz DB, Clark-Langone KM, Snable J, et al. Analytical validation of the Oncotype DX prostate cancer assay—a clinical RT-PCR assay optimized for prostate needle biopsies. BMC Genomics 2013;14:690.

[98] Nguyen HG, Welty CJ, Cooperberg MR. Diagnostic associations of gene expression signatures in prostate cancer tissue. Curr Opin Urol 2015;25:65–70.

[99] Klein EA, Cooperberg MR, Magi-Galluzzi C, Simko JP, Falzarano SM, Maddala T, et al. A 17-gene assay to predict prostate cancer aggressiveness in the context of Gleason grade heterogeneity, tumor multifocality, and biopsy undersampling. Eur Urol 2014;66:550—60.

[100] Cullen J, Rosner IL, Brand TC, Zhang N, Tsiatis AC, Moncur J, et al. A biopsy-based 17-gene genomic prostate score predicts recurrence after radical prostatectomy and adverse surgical pathology in a racially diverse population of men with clinically low- and intermediate-risk prostate cancer. Eur Urol 2015;68:123—31.

[101] Cuzick J, Swanson GP, Fisher G, Brothman AR, Berney DM, Reid JE, et al. Prognostic value of an RNA expression signature derived from cell cycle proliferation genes in patients with prostate cancer: a retrospective study. Lancet Oncol 2011;12:245—55.

[102] Cuzick J, Berney DM, Fisher G, Mesher D, Moller H, Reid JE, et al. Prognostic value of a cell cycle progression signature for prostate cancer death in a conservatively managed needle biopsy cohort. Br J Cancer 2012;106:1095—9.

[103] Cooperberg MR, Simko JP, Cowan JE, Reid JE, Djalilvand A, Bhatnagar S, et al. Validation of a cell-cycle progression gene panel to improve risk stratification in a contemporary prostatectomy cohort. J Clin Oncol 2013;31:1428—34.

[104] Bishoff JT, Freedland SJ, Gerber L, Tennstedt P, Reid J, Welbourn W, et al. Prognostic utility of the cell cycle progression score generated from biopsy in men treated with prostatectomy. J Urol 2014;192:409—14.

[105] Sommariva S, Tarricone R, Lazzeri M, Ricciardi W, Montorsi F. Prognostic value of the cell cycle progression score in patients with prostate cancer: a systematic review and meta-analysis. Eur Urol 2016;69:107—15.

[106] Erho N, Crisan A, Vergara IA, Mitra AP, Ghadessi M, Buerki C, et al. Discovery and validation of a prostate cancer genomic classifier that predicts early metastasis following radical prostatectomy. PLoS One 2013;8:e66855.

[107] Karnes RJ, Bergstralh EJ, Davicioni E, Ghadessi M, Buerki C, Mitra AP, et al. Validation of a genomic classifier that predicts metastasis following radical prostatectomy in an at risk patient population. J Urol 2013;190:2047—53.

[108] Cooperberg MR, Davicioni E, Crisan A, Jenkins RB, Ghadessi M, Karnes RJ. Combined value of validated clinical and genomic risk stratification tools for predicting prostate cancer mortality in a high-risk prostatectomy cohort. Eur Urol 2015;67:326—33.

[109] Blume-Jensen P, Berman D, Rimm DL, Shipitsin M, Putzi M, Nifong TP, et al. Development and clinical validation of an in situ biopsy based multi-marker assay for risk stratification in prostate cancer. Clin Cancer Res 2015;21:2591—600.

[110] Roychowdhury S, Iyer MK, Robinson DR, Lonigro RJ, Wu YM, Cao X, et al. Personalized oncology through integrative high-throughput sequencing: a pilot study. Sci Transl Med 2011;3:111ra21.

[111] Chang F, Li MM. Clinical application of amplicon-based next-generation sequencing in cancer. Cancer Genet 2013; 206:413—19.

23

Molecular Testing in Lung Cancer

C.J. Shiau and M.-S. Tsao

Department of Pathology, University Health Network, Toronto, ON, Canada

INTRODUCTION

Lung cancer continues to be the most common cause of cancer incidence and mortality worldwide with an estimated 1.8 million new cases reported in 2012 and 1.59 million deaths (19.4% of all cancer-related deaths) [1]. Similar to other cancers, lung cancer patients who present with advanced clinical stage show significantly poorer 5-year survival rates (7–9% for stage IIIB and 2–13% for stage IV) as compared to early-stage presentations (50–73% for stage IA and 43–58% for stage IB) [2]. Although the current 5-year overall survival for all lung cancer patients remains low at 16.8%, recent results from the randomized National Lung Screening Trial show reduction in lung cancer related mortality with low-dose CT screening for early detection [3].

The current WHO classification of lung cancer includes over 30 histologic subtypes, reflecting the impressive heterogeneity of cancers of pulmonary origin [4]. The most clinically relevant division of cancers is between small cell lung cancer (SCLC, 15%) and non-small cell lung cancer (NSCLC, 85%) given the availability of chemotherapy agents for SCLC and the historically similar prognosis and treatment options available for NSCLC. However, the NSCLC category is exceptionally heterogeneous and can be further broadly divided to include adenocarcinoma (ADC, 40%), squamous cell carcinoma (SQCC, 25%), large cell carcinoma (10%), adenosquamous carcinoma, and other less common subtypes [5,6]. Recent advances in our understanding of the molecular biology in NSCLC and the ever-evolving availability of targeted agents have now increased the demand on pathologists to distinguish between subtypes of NSCLC. In 2011, a collaborative multidisciplinary reclassification of lung ADC was undertaken by the International Association for the Study of Lung Cancer (IASLC), American Thoracic Society, and European Respiratory Society in an effort to standardize terminology and match histologic subtype morphology with cancer genetics [7]. The usage of the term "NSCLC—not otherwise specified (NOS)" by pathologists in the setting of a poorly differentiated cancer has been strongly discouraged. The usage of just three markers—thyroid transcription factor-1 (TTF-1), a mucin stain, and either p63 or p40—can reduce the number of NSCLC-NOS diagnoses to less than 5–10% [8]. For well-differentiated lesions, cancers previously diagnosed as bronchioloalveolar carcinoma are now reported as ADC with lepidic predominant pattern as part of an effort to standardize reporting of the predominant pattern in ADC (lepidic, acinar, papillary, micropapillary, solid, mucinous), even within small biopsy samples (Table 23.1).

Advancements in systemic treatment for lung cancer have been revolutionary within the last few decades with numerous potential oncogenic targets identified in NSCLC (Fig. 23.1) [9]. The discovery of genetic aberrations involving epidermal growth factor receptor (*EGFR*) and anaplastic lymphoma kinase (*ALK*) [7,10,11] as drivers of tumorigenesis in NSCLC has established the basis for personalized medicine in lung cancer patients. Together with the development of targeted tyrosine kinase inhibitors (TKIs), there has been an added pressure on laboratories to be able to test available cancer tissue for a wide array of genetic mutations in order to direct clinical treatment.

The overexpression of *EGFR* (7p11.2) in NSCLC has been well established within the literature as early as the 1980s [12]. In 2004, oncogenic somatic mutations in the kinase domain of *EGFR* gene were identified in association with cancer response to EGFR-TKIs

© 2017 Elsevier Inc. All rights reserved.

TABLE 23.1   Classification of NSCLC

Large resection specimens[a]	Small biopsy or cytology specimens
Adenocarcinoma	
Preinvasive lesion	ADC with lepidic pattern (invasive component cannot be excluded)
Atypical adenomatous hyperplasia	
ADC in situ (≤3 cm, formerly BAC)	Formerly BAC
Nonmucinous, mucinous, mixed	Nonmucinous, mucinous, mixed
Minimally invasive ADC (≤3 cm, lepidic predominant, ≤5 mm of invasion)	
Nonmucinous, mucinous, mixed	
Invasive ADC	ADC (no predominant pattern identified)
Lepidic predominant (formerly BAC)	
Acinar predominant	ADC with acinar pattern
Papillary predominant	ADC with papillary pattern
Micropapillary predominant	ADC with micropapillary pattern
Solid predominant with mucin production	ADC with solid pattern
Variants of invasive ADC	
Invasive mucinous ADC (formerly mucinous BAC)	Mucinous ADC
Colloid	ADC with colloid pattern
Fetal (low and high grade)	ADC with fetal pattern
Enteric	ADC with enteric pattern
Squamous cell carcinoma	SQCC
Papillary, clear cell, small cell, basaloid types	
Small cell carcinoma	SCC
Large cell carcinoma	NSCLC-NOS
Large cell neuroendocrine (NE) carcinoma	NSCLC with NE morphology
Large cell carcinoma with NE features	
Adenosquamous carcinoma	NSCLC with SCC and ADC patterns
Sarcomatoid carcinoma	Poorly differentiated NSCLC with spindle or giant cell features

[a]Based on 2004 WHO classification of lung tumors [4].
BAC, bronchioloalveolar carcinoma.
Source: Adapted from Travis WD, Brambilla E, Noguchi M, et al. International Association for the Study of Lung Cancer/American Thoracic Society/European Respiratory Society international multidisciplinary classification of lung adenocarcinoma. J Thorac Oncol 2011;6:244–85.

[13–15]. The Iressa Pan-ASia Study (IPASS) was the first randomized trial demonstrating improvement in progression-free survival (PFS) when comparing EGFR-TKI to chemotherapy (carboplatin/paclitaxel) in an Asian patient population selected for high probability of their cancer harboring the EGFR tyrosine kinase (TK) domain mutation [16]. Retrospective correlative studies of IPASS identified that the *EGFR* mutation status of a patient's cancer was a strong predictor of response to EGFR-TKI [16]. Several randomized trials

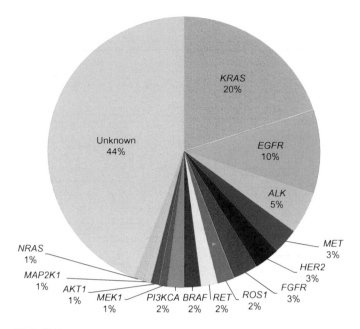

FIGURE 23.1   Many somatic oncogenic drivers have been identified in NSCLC with varying percentages. The majority of these mutations are mutually exclusive. *Source: Adapted from Faugeroux V, Pailler E, Auger N, et al. Clinical utility of circulating tumor cells in ALK-positive non-small-cell lung cancer. Front Oncol 2014;4:281.*

of gefitinib (WJTOG3405 [17], NEJ002 [18]) and erlotinib (OPTIMAL [19], EURTAC [20]) in chemotherapy-naïve (first-line) advanced lung ADC patients with *EGFR* mutation showed significantly improved PFS when compared to chemotherapy, despite the lack of demonstrated improvement in overall survival. Additional studies demonstrated that with rare exception, these sensitizing somatic mutations are identified predominantly in lung ADC or cancers with features of ADC (including adenosquamous carcinoma and NSCLC-NOS). In order to prospectively identify patients who would most benefit from targeted therapy, the College of American Pathologists (CAP), IASLC, and Association of Molecular Pathology (AMP) released clinical guidelines in 2013 recommending that *EGFR* and *ALK* testing be performed on lung ADC or cancer biopsy samples in which ADC component cannot be excluded [21].

The fusion of *ALK* (2p23) with echinoderm microtubule-associated protein-like 4 gene—*EML4* (2p21) was identified in a small subset (5–7%) of lung ADC as of 2007 [22,23]. Further investigation showed the fusion of *ALK* with other genes as well as other variants of the *ALK–EML4* fusion product. For this reason, ALK-TKIs have received greater interest as a targeted therapy for cancers showing *ALK* fusion products [24–26]. Commercially available fluorescence in situ hybridization (FISH) probes exist to facilitate the identification of patients who would benefit from this targeted therapy (Vysis ALK Break-Apart FISH Probe Kit, Abbott Molecular, Des Plaines, IL).

MOLECULAR TARGETS

Epidermal Growth Factor Receptor

EGFR or ErbB-1 is a transmembrane receptor tyrosine kinase (RTK) and is commonly overexpressed in a wide range of cancers from lung, brain, breast, head and neck, colorectal, pancreas, and bladder [27]. EGFR forms homodimers and heterodimers with other members of the ErbB family of receptors, including HER2/c-neu (ErbB-2), HER3 (ErbB-3), and HER4 (ErbB-4). Upon binding to one of its many ligands (epidermal growth factor—EGF, transforming growth factor alpha—TGFα, amphiregulin, betacellulin, heparin-binding like EGF factor, or epiregulin—EREG) [28], EGFR forms a homodimer or heterodimer resulting in subsequent activation of intracellular protein TK, leading to autophosphorylation of tyrosine residues in the C-terminal domain of EGFR. Downstream activation of multiple signaling pathways, including RAS/RAF/MAPK, PI3K/AKT, JNK, and JAK/STAT, lead to cell proliferation and survival [29,30]. Thus, mutations in the extracellular portion of EGFR leading to constitutive activation of receptor (independent of growth factor ligand) or leading to overexpression and over-activity of EGFR have been associated with carcinogenesis. In particular, mutations that alter the kinase domain (ATP-binding cleft) are of particular interest as these may confer sensitivity to targeted EGFR-TKIs.

Mutations in EGFR

The TK domain of EGFR is coded for by exons 18–21 of the *EGFR* gene (Fig. 23.2) [31]. Mutations in this region can be classified as: (1) in-frame deletions of exon 19, (2) missense (point) mutations in exons 18–21, and (3) insertion mutations in exon 20 [32]. Approximately 90% of the sensitizing mutations in *EGFR* are comprised of exon 19 deletions (15 bp and 18 bp deletions most commonly) as well as a single point mutation L858R in exon 21. Additional mutations with a frequency of at least 1% of all mutation-positive ADC include exon 18 point mutations (E709 and G719, 5%), exon 20 point mutations (S768 and T790M), exon 20 insertions, and exon 21 point mutations (T858R and L861Q, 3%) [10]. Among the various mutations, there are some point mutations in exon 20 that confer primary (insertions involving D770, P772, and V774) or secondary (T790M) resistance to TKIs [33,34]. While most rapid screening high-sensitivity methods identify exon 19 deletions and the specific exon 21 L858R mutation, detection of the multitude of other mutations requires an alternate testing platform (Fig. 23.2) [31,35].

EGFR mutations are more commonly associated with certain clinical characteristics (female, East Asian demographics, and nonsmokers or light smokers of ≤10 pack-years) and histologic characteristics (ADC subtype, positive TTF-1 immunohistochemistry

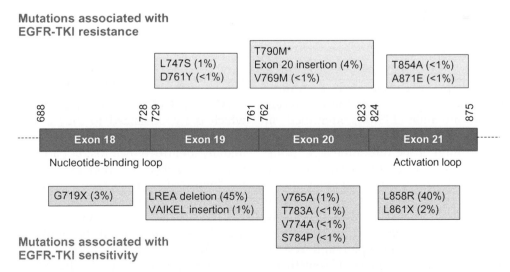

FIGURE 23.2 Mutations identified in the intracellular TK domain of *EGFR* gene by exon. The most common sensitizing mutations include deletions in exon 19 involving a string of amino acids (LREA—leucine, arginine, glutamic acid, alanine; VAIKEL—valine, alanine, isoleucine, lysine, glutamic acid, leucine) and the point mutation in exon 21 at position 858 resulting in an amino acid substitution from leucine (L) to arginine (R). Other mutations cause primary or acquired resistance to EGFR-TKIs. The most common acquired mutation following treatment by EGFR-TKIs is identified as a point mutation in exon 20 at position 790, resulting in an amino acid substitution from threonine (T) to methionine (M)*. Current guidelines recommend identification of all *EGFR* mutations with prevalence of 1% among *EGFR*-mutated lung ADC [21]. *Source: Adapted from Sharma SV, Bell DW, Settleman J, et al. Epidermal growth factor receptor mutations in lung cancer. Nat Rev Cancer 2007;7:169–81.*

staining, and nonmucinous morphology) [10,14,21,36]. The prevalence of *EGFR* mutation in East Asian population studies varies between 30% and 50% compared to Caucasian populations showing a prevalence of 10–20% [37]. The association with nonsmokers has also raised the possibility of *EGFR* mutation being a carcinogenic pathway that is less likely related to cigarette smoke exposure [38]. Although these mutations are also more commonly identified in cancers with lepidic growth pattern (formerly known as bronchioloalveolar carcinoma pattern) and papillary or acinar patterns, it occurs in poorly differentiated ADCs (often classified as NSCLC), adenosquamous carcinomas, and rarely in cases of SCC especially with an admixed ADC component [39,40]. Earlier studies suggesting the prevalence of *EGFR* mutation in squamous carcinomas as approximately 5% have been questioned as a potential result of incomplete sampling in cancers of adenosquamous histology [41]. However, further studies of advanced squamous cell lung cancer in Asian populations have demonstrated the presence of EGFR sensitizing mutations in 6–10% of patients, although EGFR-TKIs are less effective in this setting [42,43]. In the case of TTF-1 immunohistochemistry and nonmucinous morphology, both have been proposed as a potential surrogate marker for streamlining *EGFR* mutation testing [44,45]. However, approximately 7% of TTF-1-negative cancers and 9% of mucinous cancers show *EGFR* mutation [36,46]. As a significant number of cancers are found to be mutation positive despite the proposed clinicopathological characteristics, these are considered insufficient for excluding a cancer from molecular testing [21].

TKI-Resistant Mutations in EGFR

Patients who show an initial response to first-generation TKI therapy eventually show cancer progression. Approximately 50% of these cancers show the missense mutation T790M, identified in cancers with a preexisting sensitizing *EGFR* mutation [35,47,48]. The substitution of the threonine side group for the larger methionine side group results in some steric hindrance of the kinase receptor domain and is postulated to cause difficulty for TKI binding while maintaining the ability to bind ATP. This effectively reduces the potency of the TKI and returns the receptor back to the effectiveness of wild-type EGFR despite the presence of a molecularly detected sensitizing mutation [49]. The ability to test for the presence of the T790M mutation at the time of initial identification of a sensitizing mutation can help identify patients who may show poorer response to traditional EGFR-TKIs or who may benefit from novel agents designed to target *EGFR*-sensitizing mutation in the setting of T790M mutation (AZD9291 [50], HM61713 [51], CO-1686 [52]).

In approximately 5–10% of patients who develop resistance to TKIs, there is focal amplification of c-MET, allowing reactivation of the signaling pathways for cell proliferation despite the presence of EGFR inhibition [53]. This has raised the possibility of considering inhibition of multiple kinases in order to produce a more durable treatment response [54].

EGFR Copy Number and EGFR Protein Expression

There are multiple methods available to detect *EGFR* gene copy number (GCN) including the most commonly used FISH, as well as silver in situ hybridization, and real-time quantitative PCR. Multiple studies have identified that cancers with high *EGFR* GCN are often associated with the presence of an *EGFR* mutation (80%) [55]. In theory, a high GCN may result in increased translation of the mutant protein product or increase in the dependence of the cancer cell on the EGFR signaling pathway. Although initial studies proposed high GCN as a marker of higher response rates with TKI treatment, follow-up clinical trials have not reliably demonstrated this relationship. Thus, assessment of *EGFR* GCN is not routinely recommended for the selection of patients for targeted therapy.

Expression of EGFR protein may be demonstrated using immunohistochemistry showing membranous staining of the cancer cells and has been available for many years. However, the expression of this stain does not correlate with the presence of sensitizing *EGFR* mutations nor with *EGFR* GCN or treatment response. Thus, immunohistochemistry studies for EGFR protein expression are not recommended, especially in the setting of biopsy samples with limited cancer DNA.

Anaplastic Lymphoma Kinase

ALK or CD246 is a transmembrane TK receptor which is hypothesized to play a role in the development of the peripheral nervous system [56]. *ALK* was initially identified as a fusion partner with nucleophosmin in a subset of anaplastic large-cell lymphoma, resulting in its nomenclature. ALK has an intracellular domain that is involved in the RAS/RAF/MAPK, PI3K/AKT, and JAK-STAT pathways. It is coded by the *ALK* gene located on the short arm of chromosome 2 (2p23).

While mutations in *ALK* have been identified in a variety of cancers, it is the fusion of *ALK* with a variety of other genes that is identified in a small subset of NSCLC (5–7%) [22,23]. The most common fusion partner is *EML4* resulting from a small inversion on the short arm of chromosome 2 (Fig. 23.3).

ALK rearrangements are associated with certain clinical features (nonsmoking or light-smoking history

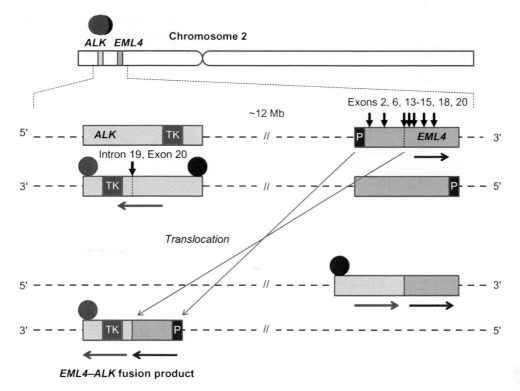

FIGURE 23.3 *EML4—ALK* fusion products result from a translocation on the short arm of chromosome 2. The break point of *ALK* is well conserved within exon 20 (most commonly) or intron 19 with the fusion partner *EML4* demonstrating multiple break points [22,57]. The translocation results in fusion of the 5′ portion of *EML4* including the promoter (P) region of the gene to the 3′ portion of *ALK* including the intracellular TK portion of the gene. The commercially available FISH break-apart probes are used to detect the fusion product when the SpectrumOrange signal (3′ telomeric side of *ALK*) is identified separate from the SpectrumGreen signal (5′ centromeric side of *ALK*) rather than the fused yellow signal seen in the nonrearranged *ALK* [58]. Other identified fusion partners include KIF5B, TFG, and KLC1 [24,57,59,60]. In all cases, the genomic breakpoint of *ALK* is well conserved, resulting in fusion of the partner protein to the intracellular domain of ALK. The fusion promotes aberrant dimerization of ALK with other receptors and leads to constitutive activation of the ALK kinase activity, leading to uncontrolled cellular proliferation.

of ≤10 pack-years, and younger age) and histologic characteristics (ADC histology, mucinous morphology, or signet-ring cell morphology) [61,62]. Similar to EGFR, clinicopathological features should not be used to exclude a cancer from *ALK* molecular testing. *ALK* rearrangements are mutually exclusive of other oncogenic driver mutations. Cancers with *ALK* rearrangement show dependency on the continued signaling of the fusion protein and are thus highly susceptible to targeted therapy [63].

TKI-Resistant Mutations in ALK

ALK-positive NSCLC shows variable PFS as cancers eventually develop resistance to first-generation ALK inhibitor crizotinib within 1—2 years. Like *EGFR*, there are resistance mutations within the TK domain of ALK, most commonly L1196M [64,65]. This mutation causes steric hindrance at the binding site resulting in decreased potency of response to crizotinib. However, unlike in EGFR-TKI resistance, there are multiple kinase domain mutations (G1269A, G1202R, S1206Y, F1174C/L, D1203N) as well as mutations away from

the binding site (threonine insertion at 1151, C1156Y, L1152R) that have been identified in the setting of ALK-TKI resistance (Fig. 23.4) [56,66—68]. In vitro studies indicate that the different resistance mutations confer different levels of resistance to structurally different TKIs, highlighting the need to identify the secondary resistance mutation through a repeat cancer sample at the time of acquired resistance and a more detailed sequencing method rather than FISH analysis [69,70]. In addition, there are other cases of ALK-TKI resistance that show only amplification of the fusion product by FISH testing, some which show one of the identified resistance mutations and others that show only amplification.

It has been noted in a small series of patients that wild-type EGFR may be activated in the setting of ALK inhibition [66,71]. Similar bypass tracks have been identified through amplification of cKIT in a smaller proportion of cases [66]. Both tracks effectively activate signaling pathways for cell proliferation in the presence of ALK inhibition. Initial in vitro studies demonstrated that inhibition of EGFR in these cases

(A)

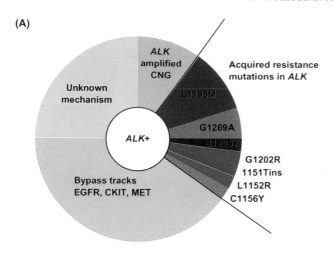

(B)

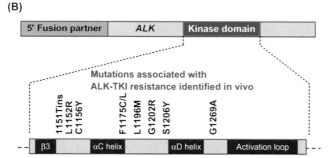

FIGURE 23.4 (A) Mechanisms of resistance to crizotinib therapy include amplification or copy number gains (CNG) of the ALK fusion product, bypass of the inhibited ALK protein by activating an alternate pathway in approximately 47% of cases (EGFR, CKIT, MET), or the evolution of a resistance mutation. In approximately 25% of cases, the mechanism of resistance is unknown. (B) Schematic of the kinase domain of ALK showing the distribution of resistance mutations identified in patients previously treated with ALK-TKIs. Many additional resistance mutations have been identified in experimental studies on cell lines. The mutations shown here have been identified in vivo. *Source: Adapted (A) from Shaw AT, Engelman JA. ALK in lung cancer: past, present, and future. J Clin Oncol 2013;31:1105—11; Kataama R, Shaw AT, Khan TM et al. Mechanisms of acquired crizotinib resistance in ALK-rearranged lung cancers. Science Transl Med 2012;4:120ra17; (B) from Camidge DR, Doebele RC. Treating ALK-positive lung cancer—early successes and future challenges. Nat Rev Clin Oncol 2012;9:268—77.*

resensitizes the cancer cells to crizotinib, strengthening the case for multiple kinase inhibition as a means of addressing secondary acquired resistance [68].

Kirsten Rat Sarcoma

Kirsten rat sarcoma (KRAS) viral oncogene homolog is an intracellular GTPase that is tethered to the cell membrane. KRAS acts as an early player in the signaling pathways of PI3K/AKT, RAF/MEK/ERK, and RLF/RAL, leading to cell proliferation and survival when activated [72]. Normally, KRAS has intrinsic enzymatic activity that allows it to cleave GTP to GDP, effectively stopping the downstream signaling pathways. The

KRAS gene is located on the short arm of chromosome 12 (12p12.1). Mutations in codons 12 and 13 of exon 2 of KRAS result in an inability to hydrolyze GTP to GDP, leaving KRAS constitutively activated.

Approximately 25—35% of lung ADC show the presence of mutation in KRAS [73]. Interestingly, since KRAS is considered downstream from EGFR, mutations in both genes are generally considered mutually exclusive. Cancers with KRAS mutation are often associated with a history of smoking as well as mucinous morphology. Recent studies with 5-year follow-up demonstrate no difference in prognosis of KRAS-mutated NSCLC versus cancers with wild-type KRAS. Thus, KRAS mutation status is not recommended for usage in selecting patients for adjuvant chemotherapy or predicting response to EGFR-TKI treatment [74,75].

ROS1 Gene

The ROS1 gene is located on the long arm of chromosome 6 (6q22) and codes a type-1 integral membrane protein with TK activity. Activation of the protein product results in growth and differentiation of a cell through the MAPK signaling pathway and phosphorylation of RAS.

Rearrangements in ROS1 have been identified in up to 2.5% of lung ADC [76] with the most common fusion partner identified as CD74 [t(5;6)(q32:q22)]. Patients with ROS1-rearranged NSCLC are often younger age with nonsmoking or light-smoking history, Asian ethnicity, and ADC histology [77]. Rearrangements can be detected utilizing FISH with dual break-apart probes. In general, ROS1-rearranged and ALK-rearranged cancers are mutually exclusive, despite similarities in clinicopathological characteristics.

Although ROS1-rearranged NSCLC represents a small subset of cancers, their sensitivity to crizotinib has led to growing interest as both a first-line treatment as well as an alternate target for multiple kinase inhibition [78,79]. Acquired resistance to crizotinib treatment has been identified in one patient with a resistance mutation (G2032R) in a cancer with known CD75—ROS1 fusion product [80].

BRAF Gene

The BRAF gene is located on the long arm of chromosome 7 (7q34) and codes for the serine/threonine protein kinase, B-Raf. B-Raf is a member of the Raf kinase family and is a downstream target of RAS, playing a pivotal role in the MAPK/ERK signaling pathway.

Activating mutations in BRAF have been well described in melanoma with the specific V600E mutation being the most common. BRAF mutations have been identified in approximately 1—4% of NSCLC with V600E and non-V600E mutations showing relatively equal frequency [81—83]. V600E mutations are more often associated with younger, nonsmoking female patients and show micropapillary ADC pattern. In contrast, non-V600E mutations are associated with smoking history and Caucasian ethnicity. However, currently available case numbers remain small and these clinicopathological characteristics may not hold true in larger population based studies.

Multiple BRAF inhibitors are available with specific targeted affinity for the V600E mutation of *BRAF*, and NSCLC with this specific mutation has been reported to show good response [84]. Patients with non-V600E mutations may show resistance to specific BRAF inhibition, but may potentially respond to MEK inhibition as a downstream target [85]. Thus, a new highly selective MEK inhibitor (selumetinib—AZD6244) is currently being tested in phase 2 clinical trials for patient with *BRAF* mutation.

MOLECULAR TECHNOLOGIES

Methodology for Molecular Testing of Multiple Targets

Sanger sequencing is the most widely used testing platform for mutation detection in various cancer settings, as it provides a comprehensive examination of all genetic aberrations in the sample material. However, the disadvantage of this testing platform is the requirement for 40–50% tumor cellularity in the test sample (20–25% mutated allele assuming heterozygosity at the targeted chromosomal site), with tested samples of lower tumor cellularity showing a higher number of false-negative results [86–88]. Furthermore, while FFPE samples generally provide sufficient cancer DNA quality for most molecular analysis methods, artifacts of deamination at cytosine and adenine bases (with transition to uracil and hypoxanthine residues) can occur resulting in small numbers of artifactual mutations [89]. Multiple rounds of PCR on a sample containing an artifactual mutation may result in amplification of this aberration and a subsequent false-positive result if the original sample is of limited quantity (low DNA copy number) [90–92]. To avoid this artifact, the sample can be pretreated with uracil-N-glycosylase and a minimum amount of template DNA (at least 1 μg) should be used in PCR reactions [92]. Any novel mutations identified using Sanger sequencing on a sample of low cancer DNA content should be cross-referenced with known artifactual mutations or considered for an alternate molecular analysis method. Given that the majority of lung cancer patients are diagnosed based on a small-volume cancer sample, achieving the minimum specimen requirements for this methodology to avoid both false-negative and false-positive results would exclude a large proportion of available test material, necessitating a second procedure to obtain additional tumor tissue, and yet with no guarantee of meeting the tumor cellularity requirements with the new sample [36].

Next-generation sequencing (NGS) methods (also known as massive parallel sequencing) show great promise in replacing Sanger sequencing [93,94]. Recent head-to-head comparisons between Sanger sequencing and NGS show improved sensitivity of mutation detection utilizing NGS in cases with tumor cellularity less than 40% [95–101]. In addition, NGS has demonstrated a 100-fold improvement in throughput over Sanger sequencing and has the capability to detect multiple forms of genetic aberrations (single base-pair substitution, copy number alterations, rearrangements) [102]. The main disadvantage of NGS is the prohibitive start-up cost of this methodology for most small laboratories to obtain both appropriate hardware for this test platform as well as the bioinformatics support for analysis of the generated data. However, use of a centralized laboratory and batching of samples to reduce costs may help facilitate future utilization of this testing methodology [103].

Multiple additional ultrasensitive testing platforms are available that offer the advantages of high caseload throughput and rapid turnaround time. These include amplification refractory mutation system, length analysis, restriction length polymorphism, real-time PCR, high-resolution melting curve analysis, single-base extension genotyping, mass spectrometry, and denaturing high-performance liquid chromatography [104–111]. The main advantage of these methods is the ability to detect mutations in test samples with very low cancer cellularity (<10%) as well as the rapid turnaround time and cost-effective application of some of these techniques, allowing smaller centers to offer molecular testing without the delay of transporting a sample to an off-site laboratory. However, many of these methods allow detection of only a limited scope of very specific mutations and may not be able to detect all sensitizing mutations as recommended by the CAP/IASLC/AMP guidelines [21].

It has been proposed that having a two-tiered testing strategy may be helpful, with an ultrasensitive method offered first as a means of expediting treatment in cases that show the most common detectable mutations, and a second step using a sequencing methodology that would allow for a more comprehensive genetic analysis of the test sample [21,100]. However, the challenge of limited tumor material in lung cancer diagnoses as well as the increase in turnaround time remains the limiting step as we attempt to glean the most genetic information from small-volume tumor samples.

Methodology for *EGFR* Testing

CAP/IASLC/AMP guidelines recommend that all cancers with ADC histology or for which the presence of an ADC component cannot be excluded should be tested for the presence of all individual *EGFR* mutations with a frequency of 1% of *EGFR*-mutated lung ADC [21]. In addition, given the frequency of *EGFR* mutations compared to all other currently identified molecular targets, prioritization of cancer DNA in limited samples for the identification of *EGFR* mutations is recommended.

There are mutation-specific immunohistochemical stains available that may be utilized to detect specifically the exon 21 L858R mutation and the exon 19 15 bp deletion [112–115]. While the mutation-specific antibodies show high specificity (97.8% for exon 21 L858R and 95.5% for exon 19 deletion), the sensitivity remains insufficient to use as the sole method of testing (75.6% and 42.2%, respectively). However, the availability of these immunohistochemical tests may allow rapid identification of a TKI-sensitizing mutation to facilitate early initiation of treatment, especially in the setting of a small volume tumor sample that may contain insufficient tumor DNA for molecular testing.

Methodology for ALK Testing

CAP/IASLC/AMP guidelines recommend that all cancers with ADC histology or ADC component be tested for the presence of *ALK* rearrangements [21]. In small biopsy samples that are identified as wild-type *EGFR*, prioritization of the remaining cancer DNA for the identification of *ALK* rearrangements is suggested. The recommended testing platform is a dual-labeled break-apart FISH assay [58] for which there is a Food and Drug Administration (FDA) approved diagnostic commercial assay (Abbott Molecular, Des Plaines, IL). Detection of a fusion product is identified when the SpectrumOrange-labeled 5′ telomeric end of *ALK* is identified separate from the SpectrumGreen labeled 3′ centromeric end (Figure 23.3).

Given the success of FISH testing for *ALK*-positive NSCLC, the samples amenable to testing are not as limited by low cancer cellularity compared to *EGFR* testing. It is more important to ensure that an area of tumor cells can be distinguished from surrounding nontumor cells and that the tumor DNA is well preserved. As such, FISH testing should be performed and supervised by pathologists and technologists with dedicated training in solid-tumor FISH testing. Cases are considered positive if 15% or more of 50 tumor nuclei are identified to show a split signal [116]. Formalin-fixed specimens that are preferred as alcohol fixatives may interfere with FISH probes.

More recent publications have reported higher sensitivity and specificity for ALK immunohistochemistry to detect *ALK*-rearranged lung cancers [116–119]. This requires optimization of the staining protocol that includes signal amplification with linked-polymer methods, as the routinely used protocol for detection of anaplastic large cell lymphoma (CD246 clone ALK1) demonstrates poor sensitivity, likely due to the low expression of protein product in NSCLC compared to the lymphomas [117,120]. Newer antibodies (ALK mouse monoclonal clone 5A4, rabbit monoclonal clones D5F3 and D9E4) have been developed that show increased sensitivity and specificity for fusion products [120]. A multicenter Canadian study examining *ALK* testing has shown 100% sensitivity and specificity for immunohistochemistry utilizing the mouse monoclonal 5A4 antibody compared to FISH in 373 routine clinically tested cases [118]. Given the rapid turnaround time and cost-effectiveness of an immunohistochemical stain compared to cytogenetics testing, mutation-specific immunohistochemistry as an initial screening test may serve to streamline *ALK* testing, allowing preservation of valuable cancer tissue for additional tests that may be required.

LIMITATIONS OF TESTING

Sample Limitations

Approximately 70% of lung cancer patients are present with surgically unresectable and locally advanced (stage IIIB) or metastatic disease (stage IV) [21,121,122]. Establishing a tissue diagnosis for this patient population involves predominantly small-volume tumor samples, including transbronchial biopsies, core needle biopsies, fine needle aspiration biopsies, and pleural effusion samples [36,123]. This introduces an added layer of complexity in pursuing molecular testing when the amount of available tumor tissue may be very limited. In addition, the vast majority of tissue samples undergo traditional processing, resulting in formalin-fixed paraffin-embedded (FFPE) specimens from which to procure tumor DNA for analysis.

In current pathology practice, a nonspecific diagnosis of NSCLC has become insufficient to guide treatment, as the specific histologic subtype may help prioritize molecular testing in the setting of a small biopsy sample. To preserve as much tumor DNA as possible for mutation testing, judicious usage of immunohistochemical stains for diagnostic purposes has been recommended [7,21]. When a diagnosis cannot be established based on morphologic features alone, a limited panel of TTF-1 or mucin stain (to establish ADC lineage) and p63/p40 or CK5/6 (to establish SCC lineage) is recommended [8]. If strong TTF-1 staining is present, the lesion should be classified as ADC (or favor ADC), even when squamous markers are positive in the same cancer cells. Once a tissue diagnosis has been established, the remaining cancer tissue should be made available for further molecular testing. Currently, molecular testing is recommended for patients presenting with advanced-stage disease (stage IV) who are suitable for targeted therapy or at the time of recurrence or progression in patients who presented at an earlier stage [21]. In the latter case, an archived resection specimen of the primary cancer may be available for testing [124].

The genetic material present within the sample may be a mixture of tumor DNA and nontumor DNA, leading to potential dilution of the desired testing target. Thus, review of the corresponding slides is important to estimate the relative abundance of tumor DNA present within the material submitted for mutation testing. This is most commonly done through an estimation of tumor cellularity, defined as the percentage of epithelial tumor cells to all cells in the submitted test material [36,125]. Efforts to enrich the tumor DNA content may be made through manual macrodissection of a designated area of a histologic section (scraping of the glass slides), core sampling of a tissue block, or laser capture microdissection (for cytologic specimens). However, often the tumor cells are admixed with an abundance of stromal fibroblasts or inflammatory cells, resulting in limited tumor DNA enrichment despite best efforts.

Although mutations have been detected in as little as 1–3% tumor DNA content, there are increased chances of false-negative results in these cases. A minimum number of tumor cells has not been well established in the literature or within the CAP/IASLC/AMP guidelines. However, some studies have indicated that mutations can be detected with as few as 100–400 tumor cells total [126,127].

Each testing laboratory must identify the sensitivity limits of their mutation testing platform and communicate this with the clinician or pathologist requesting the molecular test, noting that a negative result in such cases may still warrant a repeat test on an alternate sample [21,36]. In addition, it is important for clinicians to be aware of the testing platform being utilized for molecular testing as different platforms require variable specimen characteristics in order to produce reliable results.

Selecting the Best Sample to Test

Ideally, molecular testing should be carried out on a preexisting tissue sample with good preservation of

tumor DNA, high tumor cellularity, and limited necrosis and mucin [21]. Multiple studies have shown that cytology specimens (fine needle aspiration biopsies, pleural effusion samples) perform equivalently to histology biopsy samples in mutation testing [36,128–132]. The majority of studies have been performed utilizing FFPE samples although fresh-frozen tissue and samples in alcohol-based fixative (including cytology preparations) are also considered appropriate for testing. Fixatives containing tumor heavy metals (Zenker, B5, B-plus, acid-zinc formalin) may interfere with molecular testing and are not routinely used in everyday surgical pathology practice [133]. Acidic decalcifying solutions that may be used in postfixation processing may result in extensive DNA fragmentation and are also not recommended for molecular testing [134].

For patients with multiple available cancer samples, the most recent adequate sample is the best choice, especially in the case of a patient with recurrence following lower-stage disease. There are exceptionally rare occurrences of differences in mutation status between primary cancer and metastatic foci [135,136], as well as synchronous primary lung lesions [137]. Thus, if previous mutation tests have been negative, a patient with a new primary lesion or metastatic focus not responding to treatment may be considered for repeat molecular testing.

Currently, there is no requirement to test different areas of the same tumor as the issue of intra-tumor heterogeneity remains controversial [125,138]. Studies indicating *EGFR* testing on biopsies compared to final resection specimens show good concordance of results [36,139], while other studies have shown an association of intra-tumor heterogeneity with decreased tumor response to EGFR-TKIs [140,141]. Subpopulations of a tumor may also show variation in the type of mutation detected by high-sensitivity sequencing methodology, although only 4% of tumors showed a significant subclone (>2% of tumor DNA) with a different mutation [98]. This may partially contribute to the variable response to targeted therapy and the emergence of resistance over time.

Circulating Tumor Cells and Circulating Tumor DNA

There has been growing interest in the use of circulating tumor cells (CTCs) to monitor the presence and genetic evolution of solid tumors as CTCs can be detected in up to 70% of patients with metastatic disease [142,143]. While this testing methodology offers the main advantage of being a blood test rather than a tissue biopsy procedure, the limited amount of tumor DNA and the verification of the detected transcripts as having arisen from true CTCs remain a challenge. Most methods utilized to detect the presence of NSCLC in blood exploit the presence of specific nucleic acid sequences (mRNA released from CTCs), epithelial proteins (immunocytometric strategy, examining for TTF-1 or CK19), or other distinctive characteristics (tumor cell size) [144,145]. Currently, studies examining the presence of specific mutations in NSCLC CTCs have shown good specificity, but poor sensitivity, likely due to the small volume of tumor DNA available through this sampling method [146,147]. However, CTCs may still play a role in real-time monitoring of treatment response and evolution of acquired resistance to TKIs as the technology continues to evolve [148,149].

Circulating DNA fragments are 140−170 bp in length and present within the plasma or cell-free fraction of blood [150,151]. There may be a few-thousand amplifiable copies of DNA per milliliter of blood, of which a small fraction (0.01−0.1%) may represent DNA from a solid tumor (circulating tumor DNA (ctDNA)) [152,153]. This also represents a potential noninvasive source of tumor DNA that can be assessed by highly sensitive sequencing techniques to identify potential oncogenic driver mutations. ctDNA can be identified in the plasma of patients who have only localized disease and do not show CTCs, and thus may represent a separate underlying biological process rather than cancer cell metastasis [154]. Unlike analysis of CTCs, deep sequencing or NGS analysis of ctDNA from NSCLC represents both a specific and highly sensitive biomarker that can be used to detect cancer burden, oncogenic mutations, and TKI-resistance mutation subclones before clinically or radiologically apparent [155−157]. This has the potential to identify patients on TKI treatment who may require early salvage therapy or could be considered for an alternate targeted agent.

CLINICAL UTILITY

The last two decades have shown incredible advances in our understanding of oncogenic driver mutations in lung cancer, specifically in ADC or tumors with an ADC component. Together with targeted designer agents with high selectivity for the aberrant protein products, there has been renewed focus and hope for advanced-stage NSCLC. Although great improvements in 5-year overall survival remain a challenge, the improvements in PFS and management of drug resistance are continuously being generated by researchers worldwide. Effective testing platforms with high sensitivity and specificity and efficient turnaround times between tissue diagnosis and molecular analysis are necessary to facilitate the early initiation of targeted therapy for patients with tumors harboring specific genetic aberrations. While there are many agents that exist in particular for EGFR- and ALK-mutated cancers, our ability to identify targets in both treatment-naïve cancers and previously treated cancers can help to facilitate development of new agents. Thus, in the current clinical practice, the ability to test and identify these potential targets—especially for EGFR mutations and ALK rearrangements—has become the standard of care that will continue to direct clinical treatment options for patients.

EGFR-Targeted Therapy

First-generation 4-anilinoquinazoline small-molecule reversible TKIs (gefitinib and erlotinib) competitively bind at the kinase domain of EGFR [158]. Detection of one of the sensitizing mutations in a timely manner is important in order to initiate therapy with these drugs. Approximately 60−80% of EGFR mutation-positive chemotherapy-naïve patients show response to TKIs [16−20,159]. In general, the treatment is well tolerated with the main adverse events being a mild to moderate skin rash and mild diarrhea, often occurring within the first month of treatment. Regardless of the initial effectiveness of the treatment, all patients eventually progress due to acquired drug resistance.

Second-generation quinazoline-based small-molecule irreversible TKIs (afatinib, dacomitinib) have been studied as potential agents to overcome the drug resistance identified in first-generation TKIs [160−162]. Afatinib (BIBW-2992) has demonstrated increased PFS in patients with acquired resistance to first-generation EGFR-TKI therapy [163,164] and may show improved response when used in combination with paclitaxel (LUX-Lung 5 trial) [165]. The small benefit in overall survival identified in phase 3 trials of afatinib was ascribed mainly to the cohort of patient with exon 19 deletion in EGFR (compared to the exon 21 L858R mutation) [166]. In addition, the combination of afatinib and cetuximab has demonstrated a synergistic effect where afatinib targets the phosphorylated EGFR and cetuximab affects total EGFR protein expression [167]. With these promising results from phase 3 studies [165,168,169], afatinib has also been approved for first-line therapy for advanced EGFR-mutated lung cancer patients. The most common adverse events associated with afatinib include skin rash or acne, severe diarrhea, and paronychia. Dacomitinib (PF-00299804) showed a modest improvement in PFS compared to erlotinib in EGFR-TKI-naïve patient population in phase 2 trial, particularly in patients with EGFR mutation [170]. Unfortunately, subsequent phase

3 trials (ARCHER-1009, NCIC CTG BR.26) did not show increased benefit of dacomitinib compared to erlotinib in second- or third-line treatment [171,172]. Currently, a phase 3 trial (ARCHER-1050) is underway comparing dacomitinb to gefitinb as a first-line treatment for *EGFR*-mutated lung cancer [173,174].

Third-generation non-quinazoline-based TKIs (AZD9291, rociletinib [CO-1686], HM61713) have been designed in response to the challenges of second-generation agents which irreversibly bind both mutant and wild-type EGFR [50–52,175]. These new agents show preferential activity against T790M mutant receptors over wild-type EGFR. As such, the adverse gastrointestinal and dermatologic effects commonly noted in first-generation and second-generation TKIs are much less common or milder in severity. Response rates for these agents in phase 1 trials have been promising (64% for AZD9291, 58% for CO-1686, 29.2% for HM61713). AZD9291 is currently being studied in phase 2 and phase 3 trials (AURA 2—NCT02094261 and AURA 3—NCT0215198) comparing AZD9291 to pemetrexed plus platinum chemotherapy in patients with acquired resistance to EGFR-TKI therapy due to the T790M mutation. Additional phase 1 studies utilizing AZD9291 in combination with MEDI4736 (an anti-PD-L1 antibody), selumetinib (MEK inhibitor), or AZD6094 (MET-TKI) are also ongoing. Rociletinib (CO-1686) is also being studied in phase 2 trials (TIGER 1—NCT 02186301 and TIGER 2—NCT02147990) for patients with progression on first-generation EGFR-TKI therapy.

A second class of targeted agents include monoclonal antibodies that bind to the extracellular domain of EGFR (including cetuximab and panitumumab). These antibodies function as receptor antagonists to inhibit binding of growth factor ligands and may lead to internalization and breakdown of EGFR [176]. Unfortunately, the clinical studies (FLEX) combining cetuximab with conventional therapy have only shown modest improvement in overall survival and were correlated with increased protein expression of EGFR determined by immunohistochemistry rather than molecular mutation analysis [157]. The difficulties in reliably identifying patients who will benefit from this treatment and the minimal improvement in survival have led to the worldwide failure of regulatory approval for this class of anti-EGFR agents as a monotherapy agent in the setting of advanced NSCLC. However, the possibility of using cetuximab in combination with afatinib is still under evaluation [167].

ALK-Targeted Therapy

Crizotinib is a small-molecule first-generation TKI that was initially designed as an inhibitor of c-MET but shows activity against other TKs including ALK and ROS1 [177]. Identification of the *EML4–ALK* fusion product in NSCLC was identified during the initial phase 1 multicenter trial [23] and the first two patients with *ALK*-positive lung cancer were enrolled, showing significant improvement in symptoms. Further studies have shown a response rate of approximately 50–60% to patients with *ALK*-positive NSCLC with PFS of 8–9 months (compared to 2–3 months with single agent chemotherapy) [178–180]. As a result of the response rates demonstrated and in light of the historical significance of EGFR-TKIs, crizotinib received accelerated approval from the FDA for first-line use in *ALK*-positive NSCLC in August 2011. The medication is generally well tolerated with common adverse effects including nausea, vomiting, vision problems, and dizziness. Unfortunately, patients ultimately develop acquired resistance to crizotinib therapy within 1 year [181].

Second-generation ALK-TKIs (ceritinib, alectinib, AP26113) with greater selectivity or potency for the ALK-TK domain have been developed to address the issue of acquired resistance [69,70]. Ceritinib shows a 20-fold greater potency than crizotinib and demonstrated a response rate of 55.4–56% in crizotinib-resistant patients during phase 1 and 2 trials, and prolonged PFS in ALK-TKI-naïve patients [182]. Thus, ceritinib has been granted FDA approval for treatment of ALK-rearranged tumors following failure or intolerance to crizotinib as of August 2014. Phase 3 trials comparing ceritinib to standard chemotherapy are ongoing (NCT01828112 patients previously treated with chemotherapy, NCT01828099 chemotherapy-naïve patients). Alectinib is a potent and selective ALK-TKI with activity against the L1196M and G1269A resistance mutations in *ALK*. Phase 2 trials have demonstrated response rate of 93.5% in a crizotinib-naïve patient population and 55% in crizotinib-resistant patients [183,184]. Thus, alectinib has been granted breakthrough therapy designation by FDA for ALK-rearranged cancers following progression on crizotinib. Ongoing phase 3 trials including comparison of ceritinib and alectinib are underway (Alex study—NCT02075840). AP26113 is a dual TKI with activity against native *ALK*, L1196M *ALK*, mutated *EGFR*, and T790M *EGFR*. Preliminary results from phase 2 studies have demonstrated a response rate of 72.2% in crizotinib-resistant patients [185].

As an alternate strategy to combat crizotinib acquired resistance, there is ongoing investigation of using crizotinib in combination with a second agent. Pemetrexed (antifolate) may augment the effect of crizotinib as *ALK*-positive cancers have been identified as being more sensitive to pemetrexed [186]. A targeted Hsp90 inhibitor (ganetespib, AUY922) is also an ideal candidate as Hsp90 is a chaperone protein involved in the degradation of several oncogenic proteins including the *ALK* fusion product [187].

In order to optimize clinical therapy, it is rapidly becoming insufficient to simply identify the presence of an *ALK* fusion product. At the time of acquired resistance to crizotinib, the precise resistance mutation that has evolved may help direct the clinical choice of a second-line agent. Further evolution of clinical testing assays is required to monitor the genetic changes in cancers throughout the treatment process.

KRAS/MEK-Targeted Therapy

Although no specific targeted agents are available for *KRAS*-mutated NSCLC, new agents targeting the downstream MEK1/MEK2 have shown promise in phase 2 trials. Selumetinib, a MEK inhibitor, has demonstrated modest improvement in PFS when used in combination with docetaxel (162 days vs 63 days in placebo group) although no improvement in overall survival has been identified [188]. The adverse events included diarrhea, vomiting, stomatitis, and dry skin with increased neutropenic effects above what would be expected for docetaxel alone. Further study with larger cohorts will be needed to determine if specific *KRAS* mutations can further predict for response to MEK inhibition.

Acknowledgment

Dr. Tsao is the M. Qasim Choksi Chair in Lung Cancer Translational Research.

References

[1] International Agency for Research on Cancer. Globocan 2012: estimated cancer incidence, mortality and prevalence worldwide in 2012—lung cancer fact sheet, http://globocan.iarc.fr/Pages/fact_sheets_cancer.aspx.

[2] Goldstraw P, Crowley J, Chansky K, et al. The IASLC Lung Cancer Staging Project: proposals for the revision of the TNM stage groupings in the forthcoming (seventh) edition of the TNM classification of malignant tumours. J Thorac Oncol. 2007;2:706–14.

[3] McWilliams A, Tammemagi MC, Mayo JR, et al. Probability of cancer in pulmonary nodules detected on first screening CT. N Engl J Med 2013;369:910–19.

[4] Travis WD, Brambilla E, Müller-Hermelink HK, et al. Pathology and genetics of tumours of the lung, pleura, thymus, and heart. Chapter 1: Tumours of the lung. World Health Organization Classification of Tumours. Lyon, France: IARC Scientific Publications; 2004.

[5] Travis WD, Colby TV, Corrin B, et al. Histological typing of lung and pleural tumors. 3rd ed. Berlin: Springer-Verlag; 1999.

[6] Curado MP, Edwards B, Shin HR, et al. Cancer incidence in five continents, vol. 9. Lyon, France: IARC Scientific Publications; 2007.

[7] Travis WD, Brambilla E, Noguchi M, et al. International Association for the Study of Lung Cancer/American Thoracic Society/European Respiratory Society international multidisciplinary classification of lung adenocarcinoma. J Thorac Oncol 2011;6:244–85.

[8] Loo PS, Thomas SC, Nicolson MC. Subtyping of undifferentiated non-small cell carcinomas in bronchial biopsy specimens. J Thorac Oncol 2010;5:442–7.

[9] Faugeroux V, Pailler E, Auger N, et al. Clinical utility of circulating tumor cells in ALK-positive non-small-cell lung cancer. Front Oncol 2014;4:281.

[10] Cheng L, Alexander RE, MacLennan GT, et al. Molecular pathology of lung cancer: key to personalized medicine. Mod Pathol. 2012;25:347–69.

[11] West L, Vidwans SJ, Campbell NP, et al. A novel classification of lung cancer into molecular subtypes. PLoS One. 2012;7: e31906.

[12] Sobol RE, Astarita RW, Hofeditz C, et al. Epidermal growth factor receptor expression in human lung carcinomas defined by a monoclonal antibody. J Natl Cancer Inst. 1987;79:403–7.

[13] Lynch TJ, Bell DW, Sordella R, et al. Activating mutations in the epidermal growth factor receptor underlying responsiveness of nonsmall-cell lung cancer to gefitinib. N Engl J Med. 2004;350:2129–39.

[14] Paez JG, Janne PA, Lee JC, et al. EGFR mutations in lung cancer: correlation with clinical response to gefitinib therapy. Science 2004;304:1497–500.

[15] Pao W, Miller V, Zakowski M, et al. EGF receptor gene mutations are common in lung cancers from "never smokers" and are associated with sensitivity of tumors to gefitinib and erlotinib. Proc Natl Acad Sci USA 2004;101:13306–11.

[16] Mok TS, Wu YL, Thongprasert S, et al. Gefitinib or carboplatin-paclitaxel in pulmonary adenocarcinoma. N Engl J Med 2009;361:947–57.

[17] Mitsudomi T, Morita S, Yatabe Y, et al. Gefitinib versus cisplatin plus docetaxel in patients with non-small-cell lung cancer harbouring mutations of the epidermal growth factor receptor (WJTOG3405): an open label, randomized phase 3 trial. Lancet Oncol 2010;11:121–8.

[18] Maemondo M, Inoue A, Kobayashi K, et al. Gefitinib or chemotherapy for non-small-cell lung cancer with mutated EGFR. N Engl J Med 2010;362:2380–8.

[19] Zhou C, Wu YL, Chen G, et al. Erlotinib versus chemotherapy as first-line treatment for patients with advanced EGFR mutation-positive non-small-cell lung cancer (OPTIMAL, CTONG-0802): a multicentre, open-label, randomised, phase 3 study. Lancet Oncol 2011;12:735–42.

[20] Rosell R, Carcereny E, Gervais R, et al. Erlotinib versus standard chemotherapy as first-line treatment for European patients with advanced EGFR mutation-positive non-small-cell lung cancer (EURTAC): a multicentre, open-label, randomised phase 3 trial. Lancet Oncol 2012;13:239–46.

[21] Lindeman NI, Cagle PT, Beasley MB, et al. Molecular testing guideline for selection of lung cancer patients for EGFR and ALK tyrosine kinase inhibitors. Arch Lab Med Pathol 2013;137:828–60.

[22] Soda M, Choi YL, Enomoto M, et al. Identification of the transforming EML4-ALK fusion gene in non-small-cell lung cancer. Nature. 2007;448:561–6.

[23] Kwak EL, Bang YJ, Camidge DR, et al. Anaplastic lymphoma kinase inhibition in non-small cell lung cancer. N Engl J Med. 2010;363:1693–703.

[24] Togashi Y, Soda M, Sakata S, et al. KLC1-ALK: a novel fusion in lung cancer identified using a formalin-fixed paraffin-embedded tissue only. PLoS One. 2012;7:e31323.

[25] Shaw AT, Yeap BY, Mino-Kenudson M, et al. Clinical features and outcome of patients with non-small cell lung cancer who harbor EML4-ALK. J Clin Oncol 2009;27:4247−53.

[26] Shaw AT, Yeap BY, Solomon BJ, et al. Effect of crizotinib on overall survival in patients with advanced non-small-cell lung cancer harbouring ALK gene rearrangement: a retrospective analysis. Lancet Oncol. 2011;12:1004−12.

[27] Hembrough T, Thyparambil S, Liao WL, et al. Selected reaction monitoring (SRM) analysis of epidermal growth factor receptor (EGFR in formalin fixed tumor tissue. Clin Proteomics 2012;9:5.

[28] Hobor S, Van Emburgh BO, Crowley E, et al. TGFα and amphiregulin paracrine network promotes resistance to EGFR blockade in colorectal cancer cells. Clin Cancer Res 2014;20:6429−38.

[29] Oda K, Matsuoka Y, Funahashi A, et al. A comprehensive pathway map of epidermal growth factor receptor signaling. Mol Syst Biol 2005;1:2005.0010.

[30] Normanno N, De Luca A, Bianco C, et al. Epidermal growth factor receptor (EGFR) signaling in cancer. Gene 2006;366:2−16.

[31] Sharma SV, Bell DW, Settleman J, et al. Epidermal growth factor receptor mutations in lung cancer. Nat Rev Cancer 2007;7:169−81.

[32] Pao W, Girard N. New driver mutations in non-small-cell lung cancer. Lancet Oncol 2011;12:175−80.

[33] Wu JY, Wu SG, Yang CH, et al. Lung cancer with epidermal growth factor receptor exon 20 mutations is associated with poor gefitinib treatment response. Clin Cancer Res 2008;14:4877−82.

[34] Balak MN, Gong Y, Riely GJ, et al. Novel D761Y and common secondary T790M mutations in epidermal growth factor receptor-mutant lung adenocarcinomas with acquired resistance to kinase inhibitors. Clin Cancer Res 2006;12:6494−501.

[35] Stewart EL, Tan SZ, Liu G, et al. Known and putative mechanisms of resistance to EGFR targeted therapies in NSCLC patients with EGFR mutations—a review. Transl Lung Cancer Res 2015;4:67−81.

[36] Shiau CJ, Babwah JP, da Cunha Santos G, et al. Sample features associated with success rates in population-based EGFR mutation testing. J Thorac Oncol 2014;9:947−56.

[37] Tsao MS, Sakurada A, Cutz JC, et al. Erlotinib in lung cancer—molecular and clinical predictors of outcome. N Engl J Med 2005;353:133−44.

[38] Yatabe Y, Kosaka T, Takahasthi T, et al. EGFR mutation is specific for terminal respiratory unit type adenocarcinoma. Am J Surg Pathol 2005;29:633−9.

[39] Shiao TH, Chang YL, Yu CJ, et al. Epidermal growth factor receptor mutations in small cell lung cancer: a brief report. J Thorac Oncol 2011;6:195−8.

[40] Tatematsu A, Shimizu J, Murakami Y, et al. Epidermal growth factor receptor mutations in small cell lung cancer. Clin Cancer Res 2008;14:6092−6.

[41] Marchetti A, Martella C, Felicioni L, et al. EGFR mutations in non-small cell lung cancer: analysis of a large series of cases and development of a rapid and sensitive method for diagnostic screening with potential implications on pharmacologic treatment. J Clin Oncol 2005;23:857−65.

[42] Qiong Z, Na WY, Bo W, et al. Alterations of a spectrum of driver genes in female Chinese patients with advanced or metastatic squamous cell carcinoma of the lung. Lung Cancer 2015;87:117−21.

[43] Kenmotsu H, Serizawa M, Koh Y, et al. Prospective genetic profiling of squamous cell lung cancer and adenosquamous carcinoma in Japanese patients by multitarget assays. BMC Cancer 2014;14:786.

[44] Vincenten J, Smit EF, Vos W, et al. Negative NKX2-1 (TTF-1) as temporary surrogate marker for treatment selection during EGFR-mutation analysis in patients with non-small-cell lung cancer. J Thorac Oncol 2012;7:1522−7.

[45] Chung KP, Huang YT, Chang YL, et al. Clinical significance of thyroid transcription factor-1 in advanced lung adenocarcinoma under epidermal growth factor receptor tyrosine kinase inhibitor treatment. Chest 2012;141:420−8.

[46] Thunnissen E, Boers E, Heideman DA, et al. Correlation of immunohistochemical staining p63 and TTF-1 with EGFR and K-ras mutational spectrum and diagnostic reproducibility in non small cell lung carcinoma. Virchows Arch 2012; 461:629−38.

[47] Pao W, Miller VA, Politi KA, et al. Acquired resistance of lung adenocarcinoma to gefitinib or erlotinib is associated with a second mutation in the EGFR kinase domain. PLoS Med 2005;2: e73.

[48] Oxnard GR, Arcila ME, Sima CS, et al. Acquired resistance to EGFR tyrosine kinase inhibitors in EGFR-mutant lung cancer: distinct natural history of patients with tumors harboring the T790M mutation. Clin Cancer Res 2011;17: 1616−22.

[49] Yoshikawa S, Kukimoto-Niino M, Parker L, et al. Structural basis for the altered drug sensitivities of the non-small cell lung cancer-associated mutants of human epidermal growth factor receptor. Oncogene 2013;32:27−38.

[50] Jänne PA, Yang JC, Kim DW, et al. AZD9291 in EGFR inhibitor-resistant non-small-cell lung cancer. N Engl J Med 2015;372:1689−99.

[51] Kim DW, Lee DH, Kang JH, et al. Clinical activity and safety of HM61713, an EGFR-mutant selective inhibitor, in advanced non-small cell lung cancer (NSCLC) patients (pts) with EGFR mutations who had received EGFR tyrosine kinase inhibitors (TKIs). J Clin Oncol 2014;32 Abstract 8009

[52] Sequist LV, Soria JC, Goldman JW, et al. Rociletinib in EGFR-mutated non-small-cell lung cancer. N Engl J Med. 2015;372:1700−9.

[53] Engelman JA, Zejnullahu K, Mitsudomi T, et al. MET amplification leads to gefitinib resistance in lung cancer by activating ERBB3 signaling. Science 2007;316:1039−43.

[54] Brugger W, Thomas M. EGFR-TKI resistant non-small cell lung cancer (NSCLC): new developments and implications for future treatment. Lung Cancer 2012;77:2−8.

[55] Soh J, Okumura N, Lockwood WW, et al. Oncogene mutations, copy number gains, and mutant allele specific imbalance (MASI) frequently occur together in tumor cells. PLoS One 2009;4:e7464.

[56] Shaw AT, Engelman JA. ALK in lung cancer: past, present, and future. J Clin Oncol 2013;31:1105−11.

[57] Choi YL, Takeuchi K, Soda M, et al. Identification of novel isoforms of the EML4-ALK transforming gene in non-small cell lung cancer. Cancer Res 2008;68:4971−6.

[58] Perner S, Wagner PL, Demichelis F, et al. EML4-ALK fusion lung cancer: a rare acquired event. Neoplasia 2008;10: 298−302.

[59] Takeuchi K, Choi YL, Togashi Y, et al. KIF5B-ALK, a novel fusion oncokinase identified by an immunohistochemistry-based diagnostic system for ALK-positive lung cancer. Clin Cancer Res 2009;15:3143−9.

[60] Rikova K, Guo A, Zeng Q, et al. Global survey of phophotyrosine signaling identifies oncogenic kinases in lung cancer. Cell 2007;131:1190−203.

[61] Shaw AT, Yeap BY, Mino-Kenudson M, et al. Clinical features and outcome of patients with non-small-cell lung cancer who harbor EML4-ALK. J Clin Oncol 2009;27:4247−53.

[62] Wong DW, Leung EL, So KK, et al. The EML4-ALK fusion gene is involved in various histologic types of lung cancers from non-smokers with wild-type EGFR and KRAS. Cancer 2009;115:1723–33.

[63] McDermott U, Iafrate AJ, Gray NS, et al. Genomic alterations of anaplastic lymphoma kinase may sensitize tumors to anaplastic lymphoma kinase inhibitors. Cancer Res 2008;68:3389–95.

[64] Choi YL, Soda M, Yamashita Y, et al. EML4-ALK mutations in lung cancer that confer resistance to ALK inhibitors. N Engl J Med 2010;363:1734–9.

[65] Katayama R, Khan TM, Benes C, et al. Therapeutic strategies to overcome crizotinib resistance in non-small cell lung cancers harboring the fusion oncogene EML4-ALK. Proc Natl Acad Sci USA 2011;108:7535–40.

[66] Kataama R, Shaw AT, Khan TM, et al. Mechanisms of acquired crizotinib resistance in ALK-rearranged lung cancers. Sci Transl Med 2012;4:120ra17.

[67] Camidge DR, Doebele RC. Treating ALK-positive lung cancer—early successes and future challenges. Nat Rev Clin Oncol 2012;9:268–77.

[68] Doebele RC, Pilling AB, Aisner DL, et al. Mechanisms of resistance to crizotinib in patients with ALK gene rearranged non-small cell lung cancer. Clin Cancer Res 2012;18:1472–82.

[69] Heuckmann JM, Hölzel M, Sos ML, et al. ALK mutations conferring differential resistance to structurally diverse ALK inhibitors. Clin Cancer Res 2011;17:7394–401.

[70] Friboulet L, Li N, Katayama R, et al. The ALK inhibitor ceritinib overcomes crizotinib resistance in non-small cell lung cancer. Cancer Discov 2014;4:662–73.

[71] Sasaki T, Koivunen J, Ogino A, et al. A novel ALK secondary mutation and EGFR signaling cause resistance to ALK kinase inhibitors. Cancer Res 2011;71:6051–60.

[72] Vincent MD, Kuruvilla MS, Leighl NB, et al. Biomarkers that currently affect clinical practice: EGFR, ALK, MET, KRAS. Curr Oncol 2012;19:S33–44.

[73] Raparia K, Villa C, DeCamp MM, et al. Molecular profiling in non-small cell lung cancer—a step toward personalized medicine. Arch Pathol Lab Med 2013;137:481–91.

[74] Shepherd FA, Domerg C, Hainaut P, et al. Pooled analysis of the prognostic and predictive effects of KRAS mutation status and KRAS mutation subtype in early-staged resected non-small-cell lung cancer in four trials of adjuvant chemotherapy. J Clin Oncol 2013;31:2173–81.

[75] Martin P, Leighl NB, Tsao M-S, et al. KRAS mutations as prognostic and predictive markers in non-small cell lung cancer. J Thorac Oncol 2013;8:530–42.

[76] Yoshida A, Kohno T, Tsuta K, et al. ROS1-rearranged lung cancer: a clinicopathologic and molecular study of 15 surgical cases. Am J Surg Pathol 2013;37:554–62.

[77] Bergethon K, Shaw AT, Ou SH, et al. ROS1 rearrangements define a unique molecular class of lung cancers. J Clin Oncol 2012;30:863–70.

[78] Davies KD, Le AT, Theodoro MF, et al. Identifying and targeting ROS1 gene fusions in non-small cell lung cancer. Clin Cancer Res 2012;18:4570–9.

[79] Shaw AT, Ou SH, Bang YJ, et al. Crizotinib in ROS1-rearranged non-small cell lung cancer. N Eng J Med 2014;371:1963–71.

[80] Awad MM, Katayama R, McTigue M, et al. Acquired resistance to crizotinib from a mutation in CD74-ROS1. N Engl J Med 2013;368:2395–401.

[81] Marchetti A, Felicioni L, Malatesta S, et al. Clinical features and outcome of patients with non–small-cell lung cancer harboring BRAF mutations. J Clin Oncol 2011;29:3574–9.

[82] Paik PK, Arcila ME, Fara M, et al. Clinical characteristics of patients with lung adenocarcinomas harboring BRAF mutations. J Clin Oncol 2011;29:2046–51.

[83] Sasaki H, Kawano O, Endo K, et al. Uncommon V599E BRAF mutations in Japanese patients with lung cancer. J Surg Res 2006;133:203–6.

[84] Peters S, Michielin O, Zimmermann S. Dramatic response induced by vemurafenib in a BRAF V600E-mutated lung adenocarcinoma. J Clin Oncol 2013;31:e341–4.

[85] Trejo CL, Juan J, Vicent S, et al. MEK1/2 inhibition elicits regression of autochthonous lung tumors induced by KRASG12D or BRAFV600E. Cancer Res 2012;72:3048–59.

[86] Zhu CQ, da Cunha Santos G, Ding K, et al. Role of KRAS and EGFR as biomarkers of response to erlotinib in National Cancer Institute of Canada Clinical Trials Group Study BR.21. J Clin Oncol 2008;26:4268–75.

[87] Jänne PA, Borras AM, Kuang Y, et al. A rapid and sensitive enzymatic method for epidermal growth factor receptor mutation screening. Clin Cancer Res 2006;12:751–8.

[88] Liu X, Lu Y, Zhu G, et al. The diagnostic accuracy of pleural effusion and plasma samples versus tumor tissue for detection of EGFR mutation in patients with advanced non-small cell lung cancer: comparison of methodologies. J Clin Pathol 2013;66:1065–9.

[89] Srinivasan M, Sedmak D, Jewell S. Effect of fixatives and tissue processing on the content and integrity of nucleic acids. Am J Pathol 2002;161:1961–71.

[90] Akbari M, Hansen MD, Halgunset J, et al. Low copy number DNA template can render polymerase chain reaction error prone in a sequence-dependent manner. J Mol Diagn 2005;7:36–9.

[91] Williams C, Ponten F, Moberg C, et al. A high frequency of sequence alterations is due to formalin fixation of archival specimens. Am J Pathol 1999;155:1467–71.

[92] Marchetti A, Felicioni L, Buttitta F. Assessing EGFR mutations. N Engl J Med 2006;354:526–7.

[93] Metzker ML. Sequencing technologies—the next generation. Nat Rev Genet 2010;11:31–46.

[94] Altimari A, de Biase D, De Maglio G, et al. 454 next generation-sequencing outperforms allele-specific PCR, Sanger sequencing, and pyrosequencing for routine KRAS mutation analysis of formalin-fixed, paraffin-embedded samples. Onco Targets Ther 2013;6:1057–64.

[95] Tuononen K, Mäki-Nevala S, Sarhadi VK, et al. Comparison of targeted next-generation sequencing (NGS) and real-time PCR in the detection of EGFR, KRAS, and BRAF mutations on formalin-fixed, paraffin-embedded tumor material of non-small cell lung carcinoma-superiority of NGS. Genes Chromosomes Cancer 2013;52:503–11.

[96] Querings S, Altmüller J, Ansén S, et al. Benchmarking of mutation diagnostics in clinical lung cancer specimens. PLoS One 2011;6:e19601.

[97] Moskalev EA, Stöhr R, Rieker R, et al. Increased detection rates of EGFR and KRAS mutations in NSCLC specimens with low tumour cell content by 454 deep sequencing. Virchows Arch 2013;462:409–19.

[98] Marchetti A, Del Grammastro M, Filice G, et al. Complex mutations & subpopulations of deletions at exon 19 of EGFR in NSCLC revealed by next generation sequencing: potential clinical implications. PLoS One 2012;7:e42164.

[99] Buttitta F, Felicioni L, Del Grammastro M, et al. Effective assessment of EGFR mutation status in bronchoalveolar lavage and pleural fluids by next-generation sequencing. Clin Cancer Res 2013;19:691–8.

[100] de Biase D, Visani M, Malapelle U, et al. Next-generation sequencing of lung cancer EGFR exons 18-21 allows effective molecular diagnosis of small routine samples (cytology and biopsy). PLoS One 2013;8(12):e83607.

[101] Warth A, Penzel R, Brandt R, et al. Optimized algorithm for Sanger sequencing-based EGFR mutation analyses in NSCLC biopsies. Virchows Arch 2012;460:407—14.

[102] Margulies M, Egholm M, Altman WE, et al. Genome sequencing in microfabricated high-density picolitre reactors. Nature 2005;437:376—80.

[103] MacConaill LE. Existing and emerging technologies for tumor genomic profiling. J Clin Oncol 2013;31:1815—24.

[104] Pan Q, Pao W, Ladanyi M. Rapid polymerase chain reaction-based detection of epidermal growth factor receptor gene mutations in lung adenocarcinomas. J Mol Diagn 2005;7:396—403.

[105] Kamel-Reid S, Chong G, Ionescu DN, et al. EGFR tyrosine kinase mutation testing in the treatment of non-small-cell lung cancer. Curr Oncol 2012;19:e67—74.

[106] Molina-Vila MA, Bertran-Alamillo J, Reguart N, et al. A sensitive method for detecting EGFR mutations in non-small cell lung cancer samples with few tumor cells. J Thorac Oncol 2008;3:1224—35.

[107] Yatabe Y, Hida T, Horio Y, et al. A rapid sensitive assay to detect EGFR mutation in small biopsy specimens from lung cancer. J Mol Diagn 2006;8:335—41.

[108] Kawada I, Soejima K, Watanabe H, et al. An alternative method for screening EGFR mutation using RFLP in non-small cell lung cancer patients. J Thorac Oncol 2008;3:1096—103.

[109] Do H, Krypuy M, Mitchell PL, et al. High resolution melting analysis for rapid and sensitive EGFR and KRAS mutation detection in formalin fixed paraffin embedded biopsies. BMC Cancer 2008;8:142.

[110] Sueoka N, Sato A, Eguchi H, et al. Mutation profile of EGFR gene detected by denaturing high-performance liquid chromatography in Japanese lung cancer patients. J Cancer Res Clin Oncol 2007;133:93—102.

[111] Fukui T, Ohe Y, Tsuta K, et al. Prospective study of the accuracy of EGFR mutational analysis by high-resolution melting analysis in small samples obtained from patients with non-small cell lung cancer. Clin Cancer Res 2008;14:4751—7.

[112] Yu J, Kane S, Wu J, et al. Mutation-specific antibodies for the detection of EGFR mutations in non-small-cell lung cancer. Clin Cancer Res 2009;15:3023—8.

[113] Kato Y, Peled N, Wynes MW, et al. Novel epidermal growth factor receptor mutation-specific antibodies for non-small cell lung cancer: immunohistochemistry as a possible screening method for epidermal growth factor receptor mutations. J Thorac Oncol 2010;5:1551—8.

[114] Kozu Y, Tsuta K, Kohno T, et al. The usefulness of mutation-specific antibodies in detecting epidermal growth factor receptor mutations and in predicting response to tyrosine kinase inhibitor therapy in lung adenocarcinoma. Lung Cancer 2011;73:45—50.

[115] Allo G, Bandarchi B, Yanagawa N, et al. Epidermal growth factor receptor mutation-specific immunohistochemical antibodies in lung adenocarcinoma. Histopathology 2014;64:826—39.

[116] Camidge DR, Kono SA, Flacco A, et al. Optimizing the detection of lung cancer patients harboring anaplastic lymphoma kinase (ALK) gene rearrangements potentially suitable for ALK inhibitor treatment. Clin Cancer Res 2010;16:5581—90.

[117] Tsao MS, Hirsch FR, Yatabe Y, editors. IASLC ATLAS of ALK testing in lung cancer. International Association for the Study of Lung Cancer (IASLC) Press; 2013.

[118] Cutz JC, Craddock KJ, Torlakovic E, et al. Canadian anaplastic lymphoma kinase study: a model for multicenter standardization and optimization of ALK testing in lung cancer. J Thorac Oncol 2014;9:1255—63.

[119] Wynes MW, Sholl LM, Dietel M, et al. An international interpretation study using the ALK IHC antibody D5F3 and a sensitive detection kit demonstrates high concordance between ALK IHC and ALK FISH and between evaluators. J Thorac Oncol 2014;9:631—8.

[120] Mino-Kenudson M, Chirieac LR, Law K, et al. A novel, highlight sensitive antibody allows for the courtine detection of ALK-rearranged lung adenocarcinomas by standard immuno-histochemistry. Clin Cancer Res 2010;16:1561—71.

[121] Schrump DS, Altorki NK, Henscheke CL, et al. Non-small cell lung cancer. In: DeVita VT, Hellman S, Rosenberg SA, editors. Cancer: principles and practices. Philadelphia, PA: Lippincott Williams & Wilkins; 2005. p. 753—810.

[122] Rivera MP, Mehta AC. Initial diagnosis of lung cancer: ACCP evidence-based clinical practice guidelines (ed 2). Chest. 2007;132:S131—48.

[123] da Cunha Santos G, Shepherd FA, Tsao MS. EGFR mutations and lung cancer. Annu Rev Pathol. 2011;6:49—69.

[124] Ellis PM, Verma S, Sehdev S, et al. Challenges to implementation of an epidermal growth factor receptor testing strategy for non-small-cell lung cancer in a publicly funded health system. J Thorac Oncol 2013;8:1136—41.

[125] Kim L, Tsao MS. Tumor tissue sampling for lung cancer management in the era of personalised therapy: what is good enough for molecular testing? Eur Resp J 2014;44:1011—22.

[126] Pirker R, Herth FJ, Kerr KM, et al. Consensus for EGFR mutation testing in non-small cell lung cancer: results from a European workshop. J Thorac Oncol 2010;5:1706—13.

[127] Savic S, Tapia C, Grilli B, et al. Comprehensive epidermal growth factor receptor gene analysis from cytological specimens of non-small-cell lung cancers. Br J Cancer 2008;98:154—60.

[128] Salto-Tellez M, Tsao MS, Shih JY, et al. Clinical and testing protocols for the analysis of epidermal growth factor receptor mutations in East Asian patients with non-small cell lung cancer: a combined clinical-molecular pathological approach. J Thorac Oncol 2011;6:1663—9.

[129] Hagiwara K, Kobayashi K. Importance of the cytological samples for the epidermal growth factor receptor gene mutation test for non-small cell lung cancer. Cancer Sci 2013;104:291—7.

[130] Hlinkova K, Babal P, Berzinec P, et al. Evaluation of 2-year experience with EGFR mutation analysis of small diagnostic samples. Diagn Mol Pathol 2013;22:70—5.

[131] Billah S, Stewart J, Staerkel G, et al. EGFR and KRAS mutations in lung carcinoma: molecular testing by using cytology specimens. Cancer Cytopathol 2011;119:111—17.

[132] Chowdhuri SR, Xi L, Pham TH-T, et al. EGFR and KRAS mutation analysis in cytologic samples of lung adenocarcinoma enabled by laser capture microdissection. Mod Pathol 2012;25:548—55.

[133] Gillespie JW, Best CJ, Bichsel VE, et al. Evaluation of non-formalin tissue fixation for molecular profiling studies. Am J Pathol 2002;160:449—57.

[134] Wilson IG. Inhibition and facilitation of nucleic acid amplification. Appl Environ Microbiol 1997;63:3741—51.

[135] Munfus-McCray D, Harada S, Adams C, et al. EGFR and KRAS mutations in metastatic lung adenocarcinoma. Hum Pathol 2011;42:1447–53.

[136] Han H-S, Eom D-W, Kim JH, et al. EGFR mutation status in primary lung adenocarcinomas and corresponding metastatic lesions: discordance in pleural metastases. Clin Lung Cancer 2011;12:380–6.

[137] Arai J, Tsuchiya T, Oikawa M, et al. Clinical and molecular analysis of synchronous double lung cancers. Lung Cancer 2012;77:281–7.

[138] Sakurada A, Lara-Guerra H, Liu N, et al. Tissue heterogeneity of EGFR mutation in lung adenocarcinoma. J Thorac Oncol 2008;3:527–9.

[139] Han HS, Lim SN, An JY, et al. Detection of EGFR mutation status in lung adenocarcinoma specimens with different proportions of tumor cells using two methods of differential sensitivity. J Thorac Oncol 2012;7:355–64.

[140] Taniguchi K, Okami J, Kodama K, et al. Intratumor heterogeneity of epidermal growth factor receptor mutations in lung cancer and its correlation to the response to gefitinib. Cancer Sci 2008;99:929–35.

[141] Bai H, Wang Z, Wang Y, et al. Detection and clinical significance of intratumoral EGFR mutational heterogeneity in Chinese patients with advanced non-small cell lung cancer. PLoS One 2013;8:e54170.

[142] Cristofanilli M, Budd GT, Ellis MJ, et al. Circulating tumour cells, disease progression, and survival in metastatic breast cancer. N Engl J Med 2004;351:781–91.

[143] Pachmann K, Camara O, Kavallaris A, et al. Monitoring the response of circulating epithelial tumour cells to adjuvant chemotherapy in breast cancer allows detection of patients at risk of early relapse. J Clin Oncol. 2008;26:1208–15.

[144] Hosokawa M, Kenmotsu H, Koh Y, et al. Size-based isolation of circulating tumor cells in lung cancer patients using a microcavity array system. PLoS One 2013;8(6):e67466.

[145] Krebs MG, Hou JM, Sloane R, et al. Analysis of circulating tumor cells in patients with non-small cell lung cancer using epithelial marker-dependent and-independent approaches. J Thorac Oncol 2012;7:306–15.

[146] Punnoose EA, Atwal S, Liu W, et al. Evaluation of circulating tumour cells and circulating tumour DNA in non–small cell lung cancer: association with clinical endpoints in a phase II clinical trial of pertuzumab and erlotinib. Clin Cancer Res 2012;18:2391–401.

[147] Costa DB. Identification of somatic genomic alteration in circulating tumors cells: Another step forward in non-small-cell lung cancer? J Clin Oncol 2013;31:2236–9.

[148] Maheswaran S, Sequist LV, Nagrath S, et al. Detection of mutations in EGFR in circulating lung cancer cells. N Engl J Med 2008;359:366–77.

[149] Pailler E, Adam J, Barthélémy A, et al. Detection of circulating tumour cells harboring a unique ALK rearrangement in ALK-positive non-small-cell lung cancer. J Clin Oncol 2013;31:2273–81.

[150] Diehl F, Li M, Dressman D, et al. Detection and quantification of mutations in the plasma of patients with colorectal tumors. Proc Natl Acad Sci USA. 2005;102:16368–73.

[151] Diehl F, Schmidt K, Choti MA, et al. Circulating mutant DNA to assess tumor dynamics. Nat Med 2008;14:985–90.

[152] Gormally E, Caboux E, Vineis P, et al. Circulating free DNA in plasma or serum as biomarker of carcinogenesis: practical aspects and biological significance. Mutat Res 2007;635:105–17.

[153] Schwarzenbach H, Hoon DS, Pantel K. Cell-free nucleic acids as biomarkers in cancer patients. Nat Rev Cancer 2011;11:426–37.

[154] Bettegowda C, Sausen M, Leary RJ, et al. Detection of circulating tumor DNA in early- and late-stage human malignancies. Sci Transl Med 2014;6:224ra24.

[155] Forshew T, Murtaza M, Parkinson C, et al. Noninvasive identification and monitoring cancer mutations by targeted deep sequencing of plasma DNA. Sci Transl Med 2012;4:136ra68.

[156] Newman AM, Bratman SV, To J, et al. An ultrasensitive method for quantitating circulating tumor DNA with broad patient coverage. Nat Med 2014;20:548–54.

[157] Oxnard GR, Paweletz CP, Kuang Y, et al. Noninvasive detection of response and resistance in EGFR-mutant lung cancer using quantitative next-generation genotyping of cell-free plasma DNA. Clin Cancer Res 2014;20:1698–705.

[158] Sordella R, Bell DW, Haber DA, et al. Gefitinib-sensitizing EGFR mutation in lung cancer activated anti-apoptotic pathways. Science 2004;305:1163–7.

[159] Pirker R, Pereira JR, von Pawel J, et al. EGFR expression as a predictor of survival for first-line chemotherapy plus cetuximab in patients with advanced non-small-cell lung cancer: analysis of data from the phase 3 FLEX study. Lancet Oncol 2012;13:33–42.

[160] Kwak EL, Sordella R, Bell DW, et al. Irreverisble inhibitors of the EGF receptor may circumvent acquired resistance to gefitinib. Proc Natl Acad Sci USA 2005;102:7665–70.

[161] Godin-Heymann N, Ulkus L, Brannigan BW, et al. The T790M "gatekeeper" mutation in EGFR mediates resistance to low concentrations of an irreversible EGFR inhibitor. Mol Cancer Ther 2008;7:874–9.

[162] Sequist LV. Second-generation epidermal growth factor receptor tyrosine kinase inhibitors in non-small cell lung cancer. Oncologist. 2007;12:325–30.

[163] Miller VA, Hirsh V, Cadranel J, et al. Afatinib versus placebo for patients with advanced, metastatic non-small-cell lung cancer after failure of erlotinib, gefitinib, or both, and one or two lines of chemotherapy (LUXLung 1): a phase 2b/3 randomised trial. Lancet Oncol. 2012;13:528–38.

[164] Katakami N, Atagi S, Goto K, et al. LUX-Lung 4: a phase II trial of afatinib in patients with advanced non-small-cell lung cancer who progressed during prior treatment with erlotinib, gefitinib, or both. J Clin Oncol. 2013;31:3335–41.

[165] Schuler MH, Yang CH, Park K, et al. Continuation of afatinib beyond progression: results of a randomized, open-label, phase III trial of afatinib plus paclitaxel (P) versus investigator's choice chemotherapy (CT) in patients (pts) with metastatic non-small cell lung cancer (NSCLC) progressed on erlotinib/gefitinib (E/G) and afatinib—LUX-Lung 5 (LL5). J Clin Oncol 2014;32 Abstract 8019

[166] Yang JC, Wu Y-L, Schuler M, et al. Afatinib versus cisplatin-based chemotherapy for EGFR mutation-positive lung adenocarcinoma (LUX-Lung 3 and LUX-Lung 6): analysis of overall survival data from two randomised, phase 3 trials. Lancet Oncol 2015;16:141–51.

[167] Janjigian YY, Smit EF, Groen HJ, et al. Dual inhibition of EGFR with afatinib and cetuximab in kinase inhibitor-resistant EGFR-mutant lung cancer with and without T790M mutations. Cancer Discov 2014;4:1036–45.

[168] Wu YL, Zhou C, Hu CP, et al. Afatinib versus cisplatin plus gemcitabine for first-line treatment of Asian patients with advanced non-small-cell lung cancer harbouring EGFR mutations (LUX-Lung 6): an open-label, randomised phase 3 trial. Lancet Oncol 2014;15:213–22.

[169] Yang JC, Shih JY, Su WC, et al. Afatinib for patients with lung adenocarcinoma and epidermal growth factor receptor mutations (LUX-Lung 2): a phase 2 trial. Lancet Oncol 2012;13:539–48.

[170] Reckamp KL, Giaccone G, Camidge DR, et al. A phase 2 trial of dacomitinib (PF-00299804), an oral, irreversible pan-HER (human epidermal growth factor receptor) inhibitor, in patients with advanced non-small cell lung cancer after failure of prior chemotherapy and erlotinib. Cancer 2014;120:1145−54.

[171] Ellis PM, Shepherd FA, Millward M, et al. Dacomitinib compared with placebo in pretreated patients with advanced or metastatic non-small-cell lung cancer (NCIC CTG BR.26): a double-blind, randomised, phase 3 trial. Lancet Oncol 2014;15:1379−88.

[172] Ramalingam SS, Jänne PA, Mok T, et al. Dacomitinib versus erlotinib in patients with advanced-stage, previously treated non-small-cell lung cancer (ARCHER 1009): a randomised, double-blind, phase 3 trial. Lancet Oncol 2014;15:1369−78.

[173] Jänne PA, Ou SH, Kim DW, et al. Dacomitinib as first-line treatment in patients with clinically or molecularly selected advanced non-small-cell lung cancer: a multicentre, open-label, phase 2 trial. Lancet Oncol 2014;15:1433−14141.

[174] Mok T, Nakagawa K, Rosell R, et al. Phase III randomized open label study (ARCHER 1050) of first-line dacomitinib versus gefitinib for advanced non-small cell lung cancer (NSCLC) in patients with epidermal growth factor receptor activating mutation(s). J Clin Oncol 2013;31:TPS8123.

[175] Cross DA, Ashton SE, Ghiorghiu S, et al. AZD9291, an irreversible EGFR TKI, overcomes T790M-mediated resistance to EGFR inhibitors in lung cancer. Cancer Discov 2014;4:1046−61.

[176] Wheeler DL, Dunn EF, Harari PM. Understanding resistance to EGFR inhibitors—impact on future treatment strategies. Nat Rev Clin Oncol. 2010;7:493−501.

[177] Cui JJ, Tran-Dubé M, Shen H, et al. Structure based drug design of crizotinib (PF-02341066), a potent and selective dual inhibitor of mesenchymal epithelial transition factor (c-MET) kinase and anaplastic lymphoma kinase (ALK). J Med Chem 2011;54:6342−63.

[178] Camidge DR, Bang YJ, Kwak EL, et al. Activity and safety of crizotinib in patients with ALK-positive non-small-cell lung cancer: updated results from phase 1 study. Lancet Oncol 2012;13:1011−19.

[179] Kim D-W, Ahn M-J, Shi Y, et al. Results of a global phase II study with crizotinib in advanced ALK-positive non-small cell lung cancer. J Clin Oncol 2012;30:488s.

[180] Shaw AT, Kim DW, Nakagawa K, et al. Crizotinib versus chemotherapy in advanced ALK-positive lung cancer. N Engl J Med 2013;368:2385−94.

[181] Solomon BJ, Mok T, Kim DW, et al. First-line crizotinib versus chemotherapy in ALK-positive lung cancer: results of a phase III study (PROFILE 1014). N Engl J Med 2014;371:2167−77.

[182] Shaw AT, Kim DW, Mehra R, et al. Ceritinib in ALK-rearranged non-small-cell lung cancer. N Engl J Med 2014;370:1189−97.

[183] Seto T, Kiura K, Nishio M, et al. CH5424802 (RO5424802) for patients with ALK-rearranged advanced non-small-cell lung cancer (AF-001JP study): a single-arm, open-label, phase 1−2 study. Lancet Oncol 2013;14:590−8.

[184] Gadgeel SM, Gandhi L, Riely GJ, et al. Safety and activity of alectinib against systemic disease and brain metastases in patients with crizotinib-resistant ALK-rearranged non-small-cell lung cancer (AF-002JG): results from the dose-finding portion of a phase 1/2 study. Lancet Oncol 2014;15:1119−28.

[185] Gettinger SN, Bazhenova L, Salgia R, et al. ALK inhibitor AP26113 in patients with advanced malignancies, including ALK + non-small cell lung cancer (NSCLC): updated efficacy and safety data. Ann Oncol 2014;25:1292.

[186] Lee JO, Kim TM, Lee SH, et al. Anaplastic lymphoma kinase translocation: a predictive biomarker of pemetrexed in patients with non-small cell lung cancer. J Thorac Oncol 2011;6:1474−80.

[187] Felip E, Carcereny E, Barlesi F, et al. Phase II activity of the Hsp90 inhibitor AUY922 in patients with ALK-rearranged (ALK +) or EGFR-mutated advanced non-small cell lung cancer. Ann Oncol 2012;23:438.

[188] Jänne PA, Shaw AT, Pereira JR, et al. Selumetinib plus docetaxel for KRAS-mutant advanced non-small-cell lung cancer: a randomised, multicentre, placebo-controlled, phase 2 study. Lancet Oncol 2013;14:38−47.

24

Molecular Testing in Colorectal Cancer

J.S. Thomas and Chanjuan Shi

Department of Pathology, Microbiology, and Immunology, Vanderbilt University Medical Center,
Nashville, TN, United States

INTRODUCTION

Colorectal cancer (CRC) is one of the most commonly diagnosed malignancies worldwide and continues to be a major global public health problem. Despite continued advances in detection and treatment, CRC ranks third in both incidence and cause of cancer deaths in the United States [1,2]. CRC may present as sporadic or inherited/familial disease and has long been considered a single disease process with shared causality, clinical characteristics, and prognosis. The prototypical genetic model for the tumorigenesis of CRC, termed the adenoma-carcinoma sequence [3], describes a multistep process resulting from the accumulation of mutations in oncogenes and tumor suppressor genes in the cells of the colonic mucosa. However, through recent advances in, and applications of, molecular technologies, coupled with extensive analysis of precursor lesions and hereditary forms of the disease, it is now clear that CRC is a heterogeneous and complex disorder that develops as a consequence of accumulation of both genetic and epigenetic genomic alterations [4]. Three distinct molecular pathways have now been described that all lead to the development of CRC, based on different global cellular events that occur during the development of CRC: (1) the conventional suppressor pathway or chromosomal instability (CIN) pathway, (2) the serrated pathway or CpG island methylator phenotype (CIMP), and (3) the microsatellite instability (MSI) pathway [3,5–7]. Different genes may be mutated or altered in CRCs arising from the same genetic pathway. Comprehensive exome sequencing studies have demonstrated that individual CRCs harbor an average of 76 gene mutations [8], which adds to the complexity of the disease.

The role for molecular diagnostics in the diagnosis and management of CRC is increasing. The risk of recurrence and subsequent death in patients with CRC is known to be closely related to the stage of the disease at the time of first diagnosis [9]. Therefore, considerable effort has been directed toward identifying specific molecular alterations and biomarkers that support early diagnosis and selection of effective therapeutic strategies. Current indications for standard-of-care molecular testing in CRCs include identification of hereditary colon cancer syndromes, examination of molecular biomarkers to predict prognosis and response to antiepidermal growth factor receptor (EGFR) therapies, and testing of MSI status. As growth continues in the area of personalized cancer medicine, additional molecular testing will likely become increasingly recommended in clinical decision making for patients with CRC.

GENETIC PATHWAYS TO CRC

Conventional Suppressor Pathway or CIN Pathway

CIN is the most common type of genomic instability identified in CRC. CIN demonstrates accelerated gains or losses of large portions or whole chromosomes and commonly results in gross chromosomal or karyotypic abnormalities [10,11]. The consequence of CIN is a high frequency of aneuploidy (an imbalance in chromosome number), genomic amplifications, and loss of heterozygosity [10]. Approximately 60% of all CRCs, including those arising from Familial Adenomatous Polyposis (FAP), follow this conventional suppressor

© 2017 Elsevier Inc. All rights reserved.

pathway [12]. CRC caused by CIN has a poor prognosis, regardless of tumor stage or therapy [13,14].

The accumulation of mutations in specific oncogenes and tumor suppressor genes, coupled with the gross chromosomal abnormalities in CIN cancers, leads to the activation of pathways that cause the initiation and progression of CRC over a period of years to decades. The conventional suppressor or CIN pathway is initiated by inactivation of the adenomatous polyposis coli (APC)/β-catenin/Wnt signaling pathway, typically by mutation of one copy of the APC tumor suppressor gene. This is followed by a second event that leads to inactivation of the second APC allele through allelic deletion or additional mutations. These alterations of the APC gene lead to the development of dysplasia in aberrant crypt foci and early adenomas. Sequential accumulation of additional genetic events, including mutations in the oncogene KRAS, as well as in the tumor suppressor genes DCC, SMAD4, and TP53, drives tumor progression and the development of CRC.

MSI Pathway

MSI is present in approximately 15% of all CRCs [15]. Microsatellites are stretches of short-tandem DNA sequence repeats, approximately 1−6 bases in length, and are found throughout the human genome [16]. Microsatellites are prone to the accumulation of mutations, largely due to inefficient binding of DNA polymerases to these sequence motifs [17]. The mismatch repair (MMR) system is mainly composed of five genes (MSH2, MLH1, MSH3, MSH6, and PSM2) that encode proteins which are critical to the proper repair of DNA sequence mismatch errors missed by DNA polymerases and the preservation of genomic integrity [16]. As microsatellites are present in the coding regions of key genes for regulation of cell growth and apoptosis, loss of MMR function may result in frameshift mutations due to expansion or contraction of these regions and create an environment of uncontrolled cell survival and carcinogenesis [18]. One example is frameshift mutations in the TGF-βRII gene, which have been reported in 90% of CRCs with MSI [19]. Inactivation of the MMR system may be caused by a germline mutation in one of four MMR genes (MSH2, MLH1, MSH6, and PSM2), as seen in Lynch Syndrome (hereditary nonpolyposis colorectal cancer/HNPCC), or by aberrant epigenetic methylation of MLH1, as seen in sporadic CRCs with MSI [9,13]. Two of the MMR genes, MSH3 and MSH6, contain coding region microsatellites themselves, which may be mutated in MSI-high (MSI-H) CRCs [12]. When compared to microsatellite-stable (MSS) CRCs, MSI-

associated CRCs demonstrate a better prognosis at all stages, despite their known resistance to some chemotherapy regimens [4,16].

Serrated Pathway or CIMP Pathway

DNA methylation, an epigenetic modification that regulates gene expression, is essential for normal embryonic development and functions in X-chromosome inactivation and genomic imprinting [20,21]. The normal mammalian genome contains methylated CpG islands in nonpromoter regions. However, approximately half of CpG islands are located in promoter regions around transcription start sites and are unmethylated in normal cells [21,22]. Genes that contain these unmethylated CpG islands will undergo normal transcription in the presence of transcriptional activators. The CIMP pathway refers to widespread CpG island methylation within promoter regions of tumor suppressor genes [13]. In cancer cells, hypermethylation of CpG islands within these promoter regions leads to transcriptional silencing of tumor suppressor genes and loss of gene function, contributing to the tumorigenic process [16]. The CpG island methylator phenotype (CIMP+) accounts for approximately 35% of all CRCs.

The most common carcinomas arising through CIMP pathway begin with sessile serrated adenomas (SSAs), which frequently harbor an activating mutation in the BRAF gene [23,24] (Fig. 24.1). SSAs are prone to hypermethylation of a number of genes rich in CpG islands within their promoter regions. Depending on which genes are silenced by hypermethylation, the arising carcinoma may be microsatellite-stable (60% of CIMP+ CRCs) or MSI-H (40% of CIMP+ CRCs). Most sporadic MSI-H CRCs result from epigenetic silencing of hMLH1 due to hypermethylation of CpG islands in the promoter region. It has been proposed that the loss of hMLH1 protein function in SSAs leads to rapid accumulation of additional mutations in other genes, such as transforming growth factor-β (TGFβ) and BAX, which then drive cancer progression [12]. Morphologically, SSAs with hMLH1 hypermethylation are characterized by cytologic dysplasia, which is followed by the rapid development of malignant transformation. CpG island hypermethylation may also occur in tumor suppressor genes other than hMLH1, resulting in CIMP+ MSS CRCs.

HEREDITARY COLON CANCER SYNDROMES

Several hereditary colon cancer and polyposis syndromes have been characterized, which are all

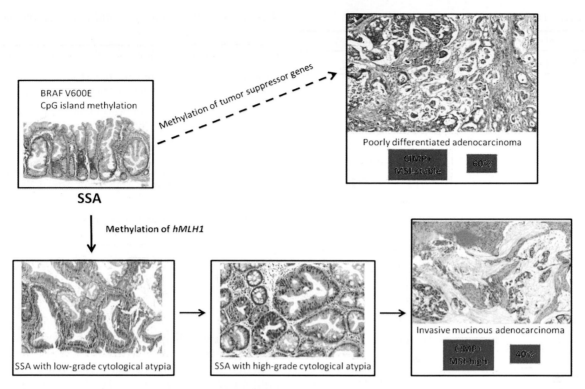

FIGURE 24.1 The serrated or CIMP genetic pathway of tumorigenesis in CRC with associated progression of a sessile serrated adenoma (SSA) bearing an activating mutation in BRAF V600E and demonstrating aberrant CpG island hypermethylation. Depending upon which genes are silenced by hypermethylation, the arising CIMP + CRC may be MSS or MSI-H. Loss of tumor suppressor genes leads to a MSS CRC (60% of CIMP + CRCs), represented morphologically by progression of an SSA to a poorly differentiated adenocarcinoma. Loss of *hMLH1* by hypermethylation leads to an MSI-H CRC (40% of CIMP + CRCs), which morphologically display increasing cytologic dysplasia (Images: SSA with low-grade cytological atypia, SSA with high-grade cytological atypia) followed by rapid development of frank malignant transformation (Image: invasive mucinous adenocarcinoma).

associated with a high risk of developing CRC. Together, these account for <10% of all CRCs. The hereditary colon cancer syndromes are classified based upon the clinical presence or absence of colonic polyps as a major disease manifestation and the presence of known causative genetic mutation, which are summarized in Table 24.1 [25,26].

Lynch Syndrome

Lynch Syndrome is the most common hereditary colon cancer syndrome, accounting for approximately 2–7% of all CRCs [7,27]. It is an autosomal dominant disorder that carries an increased risk for the development of CRC, endometrial cancer, and other cancers (Table 24.1) [28]. Lynch Syndrome is typically caused by a germline mutation in one of the MMR genes, including *MSH2*, *MLH1*, *MSH6*, and *PSM2*. Loss of MMR function leads to development of CRC through the MSI pathway. In Lynch Syndrome, 90% of MMR mutations involve either *MLH1* or *MSH2* genes [18,21]. The two-hit hypothesis of tumorigenesis applies, where germline mutation in one copy of

one MMR gene represents the first hit, and somatic inactivation of the remaining wild-type allele represents the second hit. Additionally, a novel mechanism has recently been identified in a subset of Lynch Syndrome families, which includes hypermethylation of the *MSH2* promoter without MMR gene mutations and germline deletions in the 3′ region of the epithelial cell adhesion molecule (EPCAM) gene [21,29,30].

Individuals at risk for Lynch Syndrome are identified in clinical practice using the Amsterdam criteria and the revised Bethesda guidelines, which recommend MSI testing of CRCs in individuals as outlined in Table 24.2 [31]. Diagnosis of Lynch Syndrome requires assessment of patient tissue samples for defective MMR proteins by molecular testing for MSI or immunohistochemical (IHC) methods, and further testing is always necessary to differentiate sporadic from hereditary MSI-H CRCs. Unfortunately, application of the revised Bethesda criteria as a screening tool for Lynch Syndrome is not sufficient for identification of all affected patients [12]. Current National Comprehensive Cancer Network (NCCN) guidelines now recommend evaluation of MSI status by

TABLE 24.1   Hereditary CRC Syndromes

Syndrome	Inheritance	Gene(s)	Associated cancers
NONPOLYPOSIS			
Lynch Syndrome	Autosomal Dominant	MLH1	Colon
		MLH2	Endometrium
		MSH6	Stomach
		PMS2	Ovary
		EpCAM	Hepatobiliary tract
			Upper urinary tract
			Pancreatic
			Small bowel
			CNS (glioblastoma)
ADENOMATOUS POLYPOSIS			
FAP	Autosomal Dominant	APC	Colon
			Duodenum
			Stomach
			Pancreas
			Thyroid
			Liver (hepatoblastoma)
			CNS (medulloblastoma)
AFAP	Autosomal Dominant	APC	Colon
			Duodenum
MAP	Autosomal Recessive	MUTYH	Colon
			Duodenum
PPAP	Autosomal Dominant	POLE	Colon
		POLD1	Endometrium (with POLD1 mutation)
HMPS	Autosomal Dominant	GREM1	Colon
HAMARTOMATOUS POLYPOSIS			
PJS	Autosomal Dominant	STK11	Breast
			Colon
			Pancreas
			Stomach
			Ovary
			Lung
			Small bowel
			Uterine/Cervix
			Testicle
JPS	Autosomal Dominant	SMAD4	Colon
		BMPR1A	Stomach
		ENG	Pancreas
			Small bowel

FAP, familial adenomatous polyposis; AFAP, attenuated familial adenomatous polyposis; MAP, MUTYH-associated polyposis; PPAP, polymerase proofreading associated polyposis; HMPS, hereditary mixed polyposis syndrome; PJS, Peutz-Jeghers syndrome; JPS, juvenile polyposis syndrome; CNS, central nervous system.

TABLE 24.2   Revised Bethesda Guidelines for MSI Testing in CRC

CRC diagnosed in a patient younger than 50 years of age

Presence of multiple, synchronous or metasynchronous CRC, or other Lynch-related tumors[a] in a patient of any age

CRC with MSI-H histology (presence of tumor-infiltrating lymphocytes, Crohn's-like lymphocytic response, mucinous/signet ring differentiation, or medullary growth pattern) in a patient younger than 60 years of age

CRC diagnosed in a patient with one or more first-degree relatives with an Lynch-related tumor,[a] with one of the cancers diagnosed at younger than 50 years of age

CRC diagnosed in a patient with two or more first- or second-degree relatives with Lynch-related tumors[a] diagnosed at any age

[a]Lynch-related tumors include endometrial, small bowel, gastric, ovarian, pancreatic, biliary, ureteral, or renal pelvis carcinomas, brain tumors, sebaceous gland adenomas, and keratoacanthomas.
MSI, microsatellite instability; CRC, colorectal cancer.

molecular or IHC methods on all resected CRCs, or CRCs diagnosed in patients <70 years of age and in patients ≥70 years of age who meet the Bethesda guidelines. Gold-standard germline testing is then typically offered if one of the screening tests (MSI or IHC) is positive and further testing supports a hereditary MSI-H CRC.

Familial Adenomatous Polyposis

Familial Adenomatous Polyposis (FAP) is the most common polyposis syndrome and is associated with approximately 0.5% of all CRCs [13,32]. FAP displays autosomal dominant inheritance, though up to 25% of CRCs classified as FAP are caused by de novo germline mutations [17]. FAP is characterized by the presence of hundreds to thousands of adenomas and carries a 100% lifetime risk of CRC with the median age for CRC development being 36 years old. Classic FAP typically shows the presence of >100 colonic polyps, autosomal dominant inheritance, other cancers (Table 24.1), and extracolonic findings such as congenital hypertrophy of the retinal pigment epithelium, osteomas, supernumerary teeth, desmoids, and small bowel adenomas. Attenuated FAP (AFAP) is a less severe form of the disease which presents with less than 100 adenomas that tend to demonstrate a flat morphology. AFAP is characterized by a lifetime risk for the development of CRC of up to 69% [33].

FAP and AFAP are caused by germline mutations of the APC gene on chromosome 5q21, which encodes a tumor suppressor gene. The development of CRC in FAP/AFAP follows the conventional suppressor (CIN) pathway. Distinctive phenotypic correlations have been described for specific mutations of the

APC gene [32,33]. To date, more than 3000 unique disease-causing mutations of *APC* have been reported, not all of which result in FAP/AFAP (COSMIC—http://cancer.sanger.ac.uk/cosmic) [34].

Other Hereditary Gastrointestinal Polyposis Syndromes

In addition to FAP and AFAP, there are three other inherited adenomatous polyposis syndromes that have been described (Table 24.1): (1) *MUTYH*-associated polyposis (MAP), (2) polymerase proofreading associated polyposis (PPAP), and (3) hereditary mixed polyposis syndrome (HMPS). MAP is inherited in an autosomal recessive manner. While the true incidence is not yet known, MAP may account for 0.5–1% of all CRCs [13,21]. MAP has a cumulative risk for CRC development of 80% by 70 years of age [35]. MAP is caused by a biallelic germline mutation in *MUTYH*, a base-excision repair gene for oxidative DNA damage [28,31]. Testing for MAP is recommended in patients with more than 10 adenomatous polyps, particularly those with a family history of CRC consistent with recessive inheritance and test negative for *APC* mutations [31,32]. PPAP is inherited in an autosomal dominant manner and is characterized by the presence of multiple colorectal adenomas and early onset CRC [36]. PPAP is associated with germline mutations in the proofreading domains of *POLE* and *POLD1*, two DNA polymerases with exonuclease activity [37,38]. HMPS is inherited in an autosomal dominant manner and presents with polyps that display multiple and mixed morphologies, including serrated polyps, Peutz-Jeghers polyps, juvenile polyps, and conventional adenomas, that may progress to CRC without extracolonic features [39]. HMPS is associated with a 40 kb duplication present at the 3′ end of the *SCG5* gene and upstream of the *GREM1* locus, leading to increased *GREM1* expression [39].

Two major inherited hamartomatous polyposis syndromes have also been described: Peutz-Jeghers syndrome (PJS) and juvenile polyposis syndrome (JPS). Both PJS and JPS are inherited in an autosomal dominant manner and carry an increased risk for development of CRC, as well as pancreatic and other gastrointestinal cancers (Table 24.1) [40]. Lifetime risk for development of CRC is 39% in PJS patients and 10–38% in JPS patients [21,41]. PJS is associated with germline mutations or deletions in *STK11*, which is a serine-threonine kinase that regulates *TP53*-mediated apoptosis [32]. JPS is associated with mutations in *SMADH4*, *BMPR1A*, and *ENG*, all related to the transforming growth factor-β/SMAD pathway [21,32].

BIOMARKERS FOR TARGETED THERAPIES IN CRC

The EGFR signaling pathway plays an integral role in carcinogenesis and progression of cancer, including CRC, which makes it an important target for therapeutic drugs. EGFR activation leads to autophosphorylation of its c-terminal tyrosine residues, which serve as docking sites that bind to several intracellular proteins and activate a number of downstream signaling pathways, including the RAS-RAF-MAP kinase signaling pathway and the PI3K-AKT-mammalian target of rapamycin (mTOR) signaling pathway (Fig. 24.2) [12]. The PI3K-AKT-mTOR signaling pathway may also be activated by KRAS. These signaling pathways are involved in cancer cell proliferation, invasion, migration, and inhibition of apoptosis.

EGFR

EGFR is upregulated in up to 85% of CRCs. Increased copy numbers of the *EGFR* gene are present in some CRCs and overexpression of EGFR may be demonstrated by IHC methods [42,43]. However, there is no apparent correlation between EGFR expression and response to therapy with anti-EGFR monoclonal antibodies (MoAbs), such as cetuximab and panitumumab [44]. *EGFR* gene mutations are rare in CRCs, unlike lung cancers, and have not been shown to

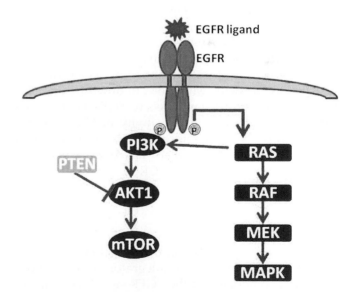

FIGURE 24.2 Epidermal growth factor receptor (EGFR) signaling pathway. Activation of EFGR by its ligands leads to activation of the RAS-RAF-MAP kinase signaling pathway and the PI3K-AKT1-mTOR signaling pathway, which are ultimately involved in tumor cell proliferation, invasion, migration, and inhibition of apoptosis. Mutation of genes within these signaling pathways in CRCs may predict disease prognosis and resistance to anti-EGFR monoclonal antibody therapy.

TABLE 24.3 Gene Mutations in the EGFR Signaling Pathway in CRCs

Gene	Mutation frequency (%)	Location of mutations	Marker of poor prognosis	Resistance to anti-EGFR therapy
EGFR	0.3	Exon 18	Unknown	Unknown
KRAS	36–40	Codon 12, 13[a]	No, remains controversial	Yes
		Codon 61		
		Codon 117[b]		
		Codon 146		
NRAS	1–6	Codon 12	Unclear, needs validation	Yes, reported by most studies
		Codon 61		
BRAF	5–22	V600E[a]	Yes, in MSS tumors	Unclear, reported in some studies
		Others[b]		
PIK3CA	10–30	Exon 9[a]	Unclear, suggested by some studies	Unclear, reported with Exon 20
		Exon 20		
		Exon 1, 2[b]		
PTEN	5–14 (higher in MSI-H CRCs)	Exon 6	Unclear, needs validation	Unclear, needs validation
		Exon 7		
AKT1	1–6	Exon 4	Unclear, needs validation	Unclear, needs validation

[a]Most common mutation.
[b]Rare mutation.
CRC, colorectal cancer; EGFR, epidermal growth factor receptor; MSI-H, microsatellite instability-high; MSS, microsatellite-stable.
Source: Modified from Walther A, Houlston R, Tomlinson I. Association between instability and prognosis in colorectal cancer: a meta-analysis. Gut 2008;57:941−50 and Forbes, S.A. et al, COSMIC: exploring the world's knowledge of somatic mutations in human cancer Nucl. Acids Res. 2015;43 (D1): D805−D811 (COSMIC).

predict response to therapy [45]. Nonetheless, numerous clinical trials have demonstrated that anti-EGFR MoAbs are effective in treating some metastatic CRCs [46]. Anti-EGFR MoAbs play a role in inhibition of downstream signaling pathways and may also have therapeutic effects through antibody-dependent cell-mediated cytotoxicity. Mutations in downstream genes like KRAS and expression levels of EGFR ligands have been shown to affect the sensitivity of CRCs to anti-EGFR therapy [47]. Therefore, a focus on downstream signaling pathway biomarkers has been explored to predict response to therapy in CRCs.

KRAS and NRAS

KRAS is a small G protein that functions as a signal transducer and downstream integrator of EGFR and is an integral component of the EGFR signaling pathway. Activating KRAS mutations result in constitutively active RAS proteins that stimulate the RAS/mitogen-activated protein (MAP) kinase signaling pathway independent of EGFR signaling. Activating KRAS mutations involving codons 12, 13, or 61 have been detected in 40–50% of sporadic CRCs, with approximately 90% of mutations occurring in codons 12 and

13 [48] (Table 24.3). Activating mutations in codon 146 of exon 4 in the KRAS gene have also been described in 1–6% of CRCs [49,50]. NRAS, a member of the RAS family, has also been shown to have activating mutations in 1–6% of CRCs [51]. These mutations serve to constitutively activate the downstream signaling pathway. KRAS mutations have been demonstrated in MSS, sporadic MSI-H, and HNPCC CRCs [12,52]. However, sporadic MSI-H cancers have a lower frequency of KRAS mutations and a higher frequency of BRAF mutations, which are mutually exclusive mutations in a single tumor.

KRAS mutations, in general, have not been shown to have a significant prognostic value in CRC [13,53]. However, KRAS mutations have been associated with poor response to anti-EGFR MoAbs in randomized clinical trials. Only cancers with wild-type KRAS show a significant response to these agents. Review of randomized and nonrandomized clinical studies suggests that patients with CRCs that contain KRAS mutations should not receive anti-EGFR MoAb therapy [48,54−56]. Similarly, NRAS mutations have also been associated with poor response to anti-EGFR MoAb therapy [57]. Therefore, KRAS and NRAS mutational testing is increasingly recommended for selection of appropriate therapeutic treatment in CRC. Current

NCCN guidelines recommend *KRAS* and *NRAS* genotyping of cancer tissue for all patients with metastatic CRC—if *KRAS* or *NRAS* mutations are present, anti-EGFR MoAb therapy is not recommended.

BRAF

BRAF is a serine/threonine protein kinase and an immediate downstream effector of KRAS in the MAP kinase signaling pathway. *BRAF* mutations are present in 5–22% of CRCs (Table 24.3), with the most common mutation being V600E [51,53,58]. Mutated *BRAF* is detected at a higher frequency in sporadic MSI-H cancers than in MSS cancers [59,60] and is never detected in HNPCC [60]. The current clinical importance of *BRAF* mutational testing in CRC is in differentiating sporadic MSI-H cancers from HNPCC, though its role as a prognostic and predictive biomarker for anti-EGFR MoAb therapy continues to evolve.

Some studies have indicated that *BRAF* mutations are associated with a more aggressive phenotype and overall shorter survival in CRCs [58,59,61,62]. However, others have reported that the prognostic effect of *BRAF* mutation is actually related to the MSI status of the cancer, with MSI-low (MSI-L) or MSS cancers demonstrating a poor prognosis and MSI-H cancers demonstrating no difference in prognosis compared to *BRAF* wild-type cancers [59,60]. *BRAF* status may also predict response to anti-EGFR therapy. In a retrospective study, wild-type *BRAF* was required for response to anti-EGFR MoAbs in metastatic CRC patients [61]. Current NCCN guidelines state that *BRAF* mutational testing should be performed when *KRAS* testing indicates a wild-type *KRAS* gene; however, *BRAF* mutational analysis is not currently required for treatment decisions, though it is useful for determining prognosis.

Small molecular *BRAF* inhibitors are currently being investigated as potential therapeutics in CRC, as many have shown efficacy in the treatment of *BRAF* mutant melanoma. Preclinical models using a *BRAF* inhibitor, vemurafenib, with standard-of-care or novel targeted therapies in *BRAF* mutated CRCs have shown enhanced antitumor efficacy [63]. However, CRC patients with V600E mutations showed only a limited response to the inhibitor, likely due to rapid feedback activation of EGFR [64]. As additional clinical studies are performed using these targeted BRAF inhibitors, new recommendations regarding mutational analysis of *BRAF* in CRC may be forthcoming.

PIK3A

Class 1A PI3Ks are heterodimeric proteins composed of two subunits: the p85 regulatory subunit and the p110 catalytic subunit. Three catalytic isoforms exist, which are the products of three genes—*PIK3CA*, *PIK3CB*, and *PIK3CD*. Class I PI3Ks are activated by G-protein coupled receptors, receptor tyrosine kinases such as KRAS and EGFR, and certain oncoproteins such as RAS. Mutations in *PIK3CA* are the most common genetic alterations in the PI3K signaling pathway seen in CRCs and are detected at a frequency of 10–30%, mostly in exon 9 (60–65%) and exon 20 (20–25%) [65,66] (Table 24.3). Multiple studies have suggested that *PIK3CA* mutation is associated with a poor prognosis in CRCs, though some studies have not supported this relationship to prognosis [67,68].

The clinical effect of *PIK3CA* mutations on anti-EGFR MoAb therapy is an area of current investigation. Most early clinical studies demonstrated contradictory results and did not evaluate the individual *PIK3CA* mutations separately. It is now known that the exon 9 gain of function mutation is RAS-dependent while the exon 20 mutation is not [13]. Recent studies have shown that CRCs with *PIK3CA* exon 20 mutations may be associated with a lack of response to anti-EGFR therapy, while the response to therapy is retained in *KRAS* wild-type CRCs with *PIK3CA* exon 9 mutations [51]. *PIK3CA* mutated CRCs may also have either *KRAS* or *BRAF* mutations, making the use of *PIK3CA* as a single marker for predicting response to anti-EGFR therapy of little current clinical value in patients with CRC.

PTEN

PTEN is a tumor suppressor gene in the PI3K/AKT pathway that negatively regulates the PI3K-AKT-mTOR signaling pathway [69]. Loss of PTEN via mutations, deletions, or promoter methylation of the *PTEN* gene is seen in 20–40% of CRCs and allows for constitutive activation of the signaling pathway [12] (Table 24.3). CRCs with inactivated PTEN may also have mutations of *KRAS*, *BRAF*, or *PIK3CA*. In addition to downstream involvement in the EGFR pathway, loss of PTEN has been demonstrated to confer resistance to cetuximab in vitro [70]. Additionally, loss of PTEN expression may predict a lack of response and shorter survival in patients with wild-type *KRAS* metastatic CRCs treated with anti-EGFR therapy [71–73].

MOLECULAR TECHNOLOGIES, CLINICAL UTILITY, AND LIMITATIONS OF TESTING

Molecular diagnostic evaluation of sporadic CRCs now commonly includes eight genes to identify MSI status and to guide evaluation of prognosis and therapeutic interventions, including those for the family of MMR proteins: MLH1, MSH2, MSH6, and PMS2, as well as *KRAS*, *BRAF*, *PIK3CA*, and *PTEN*. Molecular evaluation of the known hereditary colon cancer syndromes involves testing for germline alterations in the MMR genes, as well as five additional genes: *APC, MUTYH, STK11, BMPR1A,* and *SMAD4*. Mutational analyses of many additional molecular biomarkers in CRCs will likely become increasingly recommended as new targeted therapies make their way into clinical practice. As the various molecular technologies and use of distinct validated intratumor molecular assays for single genetic mutations in CRC have increased, many companies and academic medical centers now offer clinical molecular testing on a variety of different platforms. Many multiplexed assays for more comprehensive and simultaneous assessment of multiple genes involved in CRCs on a single patient specimen are currently available. However, it is important to recognize that a number of different methodologies are employed at individual institutions to meet current clinical testing recommendations and guidelines for CRC, based on what technologies are available.

Molecular Testing in Hereditary CRC Syndromes

The genes associated with each of the inherited colon cancer syndromes have been identified, and molecular testing is available for the diagnosis of these conditions [40,74,75]. Inherited conditions typically arise from mutation of a single gene, though different families with the same disorder may have separate and distinct mutations of the implicated gene. Genetic testing for each of the hereditary colon cancer syndromes follows a similar approach, with the exception of Lynch Syndrome. Testing is typically performed on DNA obtained from a peripheral blood sample. DNA sequencing (Sanger sequencing) is the standard method utilized for initial identification of mutations, though additional methods may be utilized prior to sequencing to detect and localize the mutation [76]. Once the disease-causing mutation is identified in the index patient, family members may be tested for the presence of that specific mutation (mutation-specific testing) using various methodologies. Depending on the syndrome, the likelihood of finding the disease-causing mutation in the index patient may range from 50% to greater than 90% [76].

Semiautomated Sanger sequencing of DNA has been utilized for many years in clinical genetic testing and is still considered the gold standard, though there are limitations to this methodology, including failure to detect large deletions, low throughput, and cost. DNA sequencing methodologies have evolved quickly, and whole exome sequencing of all DNA-coding regions is now often utilized clinically [77]. With

Sanger sequencing, initial DNA screening methods for localization of the area within the implicated gene to be sequenced may be used to increase the efficiency and likelihood of mutation detection. These screening methods include conformation-specific gel electrophoresis, single-strand conformation polymorphism, and denaturing gradient gel electrophoresis [76]. Large deletions or rearrangements within the gene, which may not be detected by sequencing, often require southern blot hybridization analysis or multiplex ligation-dependent probe amplification (MLPA) for detection. If MLPA is used as the original method of detection, then Southern blot or quantitative polymerase chain reaction (PCR) are performed for confirmation [78]. Chromosomal karyotyping and fluorescent in situ hybridization are used to detect gross chromosomal alterations caused by large deletions and gene rearrangements. Protein truncation testing may also be used to detect the presence of a truncated protein by gel electrophoresis, generally caused by nonsense or frameshift mutations within the gene.

Genetic testing should be considered in individuals with more than 10 adenomas to identify the familial disease-causing mutation in the *APC* gene, due to the clinical concern for FAP, AFAP, or MAP in these patients [79,80]. The majority of disease-causing mutations that have been identified in the *APC* gene are nonsense or frameshift mutations resulting in a termination codon and a truncated protein, 95% of which may be identified by sequencing [81]. Mutations contributing to classic FAP typically occur between exon 5 and the 5′ portion of exon 15 (codons 169–1393), and mutations associated with AFAP typically occur at the 5′ or 3′ end of the gene beyond codon 1595 [21,78]. The gold standard for detection of *APC* mutations is direct DNA sequencing. Protein truncation testing may be performed prior to sequencing in suspected cases of FAP/AFAP. Linkage testing of *APC* may be performed if other molecular approaches are unsuccessful. This method utilizes multiple DNA markers near or within the gene among multiple affected members of a family to correlate with disease phenotype. The currently available DNA markers for linkage analysis may be used in 90–95% of FAP families with >98% accuracy [74].

In patients where FAP or AFAP is suspected and an *APC* mutation cannot be determined, MAP testing should be performed as *MUTYH* mutations are found in approximately 10–20% of these patients [80,82]. Mutation specific testing of *MUTYH* is performed prior to sequencing for suspected cases of MAP, as 80% of affected individuals have one of two specific mutations: Y165C and G382D [80,83]. If one of these *MUTYH* mutations is present, sequencing is performed to identify an inactivating mutation in the second

MUTYH allele. Both alleles must be mutated to inactivate the gene and cause disease [76]. Sequencing of the *MUTYH* gene for less common mutations is performed if neither of the two common mutations is identified by mutation specific testing and MAP is still clinically suspected.

MSI Testing and Diagnosis of Lynch Syndrome

MSI testing is conducted using a PCR-based assay and/or IHC staining for MMR protein expression. Examination of the expression of MLH1, MSH2, MSH6, and PMS2 in tissues with commercially available antibodies is the most common IHC method used for identification of MSI-H CRCs (Fig. 24.3), with a sensitivity of 90–95% and specificity of 100% [7]. In addition, a two-antibody panel (MSH6 and PMS2), has been shown to exhibit similar sensitivity to the four-protein antibody panel [84], as loss of MLH1 and MSH2 causes concurrent loss of MSH6 and PMS2, respectively. Intact expression of all four proteins indicates that the MMR proteins are present, but does not entirely exclude Lynch Syndrome. Missense mutations, particularly in *MLH1*, are present in 5–8% of Lynch Syndrome families and may lead to nonfunctional proteins with retained antigenicity [12]. Defects in lesser-known MMR proteins may also show a similar IHC result. Additionally, IHC labeling for the MMR proteins may be heterogeneous, which may compromise the sensitivity of the test. Further, caution should be taken in examination of MMR protein expression following therapy, as neoadjuvant treatment may reduce protein expression and lead to a false-positive IHC result [85]. Loss of MLH1 expression may be the result of promoter methylation (as in sporadic MSI-H CRC) or Lynch Syndrome. Loss of MSH2/MSH6 expression almost always indicates Lynch Syndrome [84].

Molecular testing for MSI relies on the evaluation of loci within the human genome known to harbor microsatellites. As microsatellites vary in size between individuals, DNA is typically extracted from both normal and cancer tissue for evaluation, obtained from either

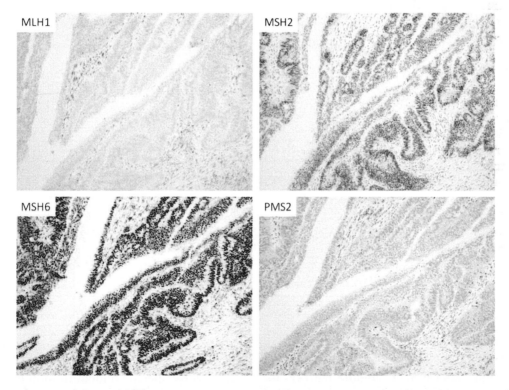

FIGURE 24.3 MSI testing by tissue immunohistochemistry. IHC staining for expression of MLH1, MSH2, MSH6, and PMS2 proteins in tissue using commercially available antibodies may be used to identify MSI in CRCs. MMR protein immunohistochemistry images shown, original magnification 200×. Intact MMR protein expression or a positive result is indicated by the dark brown staining nuclei of the colonic epithelial cells, as shown in the image panels MSH2 and MSH6. Loss of MMR protein expression or a negative result is indicated by no staining of the nuclei within the colonic epithelial cells, as shown in the image panels MLH1 and PMS2. This pattern of MMR protein expression shown in tissue immunohistochemistry from a CRC (with intact expression of MSH2 and MSH6 proteins, and loss of expression of MLH1 and PMS2 protein expression) would be reported as abnormal expression of MLH1 and PMS2 proteins. Depending upon the patient history, morphologic features of the tumor, and results of any other testing performed, the pattern of MMR protein expression shown would be concerning for an MSI-H CRC or Lynch Syndrome.

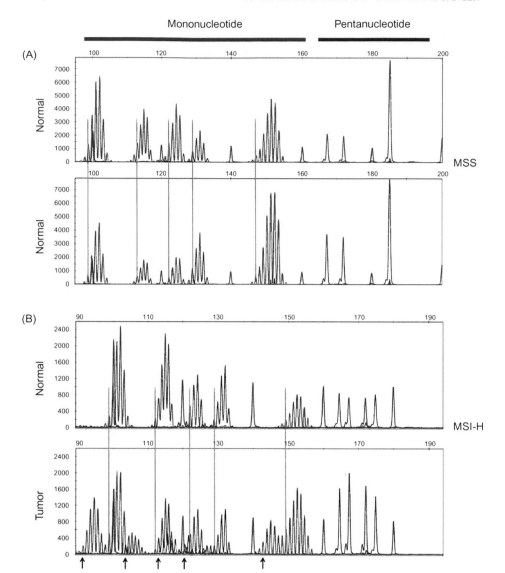

FIGURE 24.4 Fluorescent multiplex PCR-based MSI testing. (A) An example of a MSS tumor sample with chromatography results that are identical to those of a matched normal tissue sample. (B) An example of an MSI-H tumor sample with chromatography results showing deletions (black arrows) in five of five mononucleotide microsatellite loci compared to a matched normal tissue sample. The pentanucleotide loci are utilized as specimen identification markers.

fresh tissue or formalin-fixed paraffin-embedded (FFPE) tissue blocks. Following PCR amplification of the selected microsatellite, the sizes of PCR products obtained from normal and cancer tissue are compared. MSI is considered to be present when there is a change of any length within the PCR products obtained from cancer tissue, as compared to the normal tissue. The original Bethesda guidelines for identification of individuals with HNPCC proposed a panel of five mononucleotide and dinucleotide microsatellite markers for PCR-based detection of MSI [86]. Cancers are classified as MSI-H if instability is present at two or more loci, MSI-L if instability is present at one locus (with additional testing of other loci for definitive classification), and MSS if no instability is present. The revised Bethesda guidelines recommended a secondary panel of mononucleotide markers be used in MSI-L cases in which only a dinucleotide marker is mutated, as mononucleotide markers are more sensitive than dinucleotide or trinucleotide markers [87]. Currently, a fluorescent multiplex PCR-based method is typically used for detecting MSI, with amplification of five to seven microsatellite markers (Fig. 24.4). Many molecular diagnostic laboratories utilize a commercially available kit with five mononucleotide markers (BAT-25, BAT-26, MON0-27, NR-21, and NR-24) for detection of MSI and two pentanucleotide markers (Penta C and Penta D) for specimen identification to ensure that cancer and normal DNA samples are derived from a single patient. Fluorescently-labeled PCR products are sized by capillary electrophoresis. The patterns of normal and cancer genotypes are compared for each marker and scored as MSI-H, MSI-L, or MSS (Fig. 24.4). The sensitivity of PCR-based testing is similar to that of the IHC method. Low cancer cellularity may contribute some false-negative PCR-based testing results. Cases with a high index of suspicion that show a negative MSI result by PCR should be tested with the IHC method, and vice versa.

Additional methods of MSI detection using diagnostic microarrays are in development and may have potential for clinical use, but are not yet commonly offered. One such assay was developed based on a 64-gene expression signature using genes that correlated with MSI status by full-genome expression data [88]. This assay has shown high accuracy in identification of MSI in patients with CRC (sensitivity of 90.3—94.3% and overall accuracy of 84.8—90.6%) and may have utility in identification of MSI in patients not recognized by traditional PCR or IHC methods [88]. The use of miRNA expression profiles for determination of MSI status by spotted locked nucleic acid-based oligonucleotide microarrays has also been investigated [89].

CRCs identified as MSI-H require further testing to differentiate sporadic from hereditary MSI-H CRCs. Analysis for somatic mutations in the V600E hot spot of *BRAF* may be indicated for CRCs that show MSI-H or loss of MLH1 expression, as this mutation is present in sporadic MSI-H cancers but not in HNPCC-associated cancers [90]. The mutational analysis of *BRAF* prior to germline genetic testing for HNPCC in patients with MSI-H CRCs is a cost-effective method of identifying patients with sporadic CRC, in whom further testing is not indicated [12]. The presence of *BRAF* mutations may be assessed by many methods, including IHC staining of tissue with a recently developed mutation-specific antibody for *BRAF* V600E [91]. If *BRAF* is wild-type, methylation analysis of the *hMLH1* promoter may be performed using methylation-specific MLPA or methylation-specific PCR testing, as the *hMLH1* promoter is rarely methylated in HNPCCs.

Germline mutation analysis is required in MSI-H CRCs that are *BRAF* wild-type and lack *hMLH1* promoter methylation due to the high probability of Lynch Syndrome in these cases. MMR genes may contain pathologic mutations, including nonsense or frameshift mutations that cause protein truncation, missense mutations that lead to a dysfunctional protein, and large deletions. Sequence analysis of exons and intron—exon boundaries of the implicated gene is performed to detect small insertions or deletions and missense mutations. Large deletions may be detected using MLPA or other methods.

Molecular Technologies for Detection of Specific Gene Mutations in CRCs

The vast majority of gene mutations that occur in CRCs are point mutations (single base substitutions),

including those identified in sporadic CRCs, inherited CRC syndromes, and molecular biomarkers utilized for evaluation of prognosis and therapeutic responses in CRCs. Multiple technologies are available for molecular detection of gene mutations, including Sanger DNA sequencing, allele-specific PCR, melting-curve real-time PCR (RT-PCR), pyrosequencing, single base extension, and mass spectrometry. These methods offer varying degrees of sensitivity and have disadvantages associated with their use. For example, these methods may be time-consuming (Sanger), labor-intensive with high costs, have short read length limits (pyrosequencing), and lack multiplex and high-throughput capabilities. Therefore, the field of molecular diagnostics has seen continuous development of new technologies and implementation of multigene and multiplexed assays for improved diagnosis and management of patients.

Standard DNA sequencing (Sanger sequencing) may be used to detect mutations in the gene of interest, but sensitivity is low and generally requires greater than 25% cancer cells in a specimen for detection [12]. More sensitive technologies have been developed and are widely utilized to detect point mutations, including allele-specific PCR, quantitative PCR with melting-curve analysis, and pyrosequencing. These techniques are most useful clinically to specifically determine the presence of a single mutation, such as allele-specific PCR for reflex *BRAF* mutation analysis in MSI-H CRCs, as they have limited multiplex and high throughput capabilities. More recently, multigene assays have been developed using various technologies for the simultaneous detection of multiple mutations. These include single nucleotide extension assays, such as mass spectrometry—based assays and the SNaPSHOT platform (Life Technologies). These assays are performed using multiplex PCR amplification of specific gene targets such that different alleles will result in PCR products of differing and predictable sizes, followed by multiplex single base extension of oligonucleotide primers using fluorescently-labeled ddNTPs for detection of single nucleotide polymorphisms (SNPs). In the mass spectrometry—based assay, genotype is determined by single base differences of the extension products distinguished by their mass-charge ratio. On the SNaPSHOT platform, capillary electrophoresis allows size determination and genotype is determined by the color of the fluorescently labeled nucleotide added during single base extension (Fig. 24.5).

Multiplex platforms for gene express profiling stratification have been developed based on known biomarkers of prognosis and treatment responses in CRC to identify patients with Stage II CRC who are more likely to develop recurrent disease and may be candidates for adjuvant chemotherapy [12]. While not yet

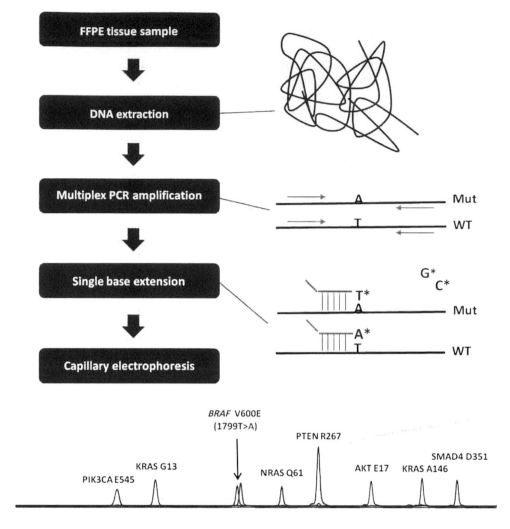

FIGURE 24.5 Detection of point mutations or SNPs using a multiplexed single base extension assay (SNaPshot assay platform, Life Technologies). DNA is extracted from FFPE tumor tissues. A multiplexed PCR is performed to amplify multiple loci in which specific mutations or SNPs may have diagnostic, therapeutic, or prognostic significance in cancer, often performed in tumor-specific panels (ie, CRC). At each of the loci of interest, mutant (Mut) or wild-type (WT) alleles may be amplified, which serve as templates for single-base extension using an oligonucleotide primer that binds immediately upstream of a known mutational hot spot or SNP. This is followed by multiplexed single-base primer extension using fluorescently labeled dideoxynucleotides (ddNTPs). Labeled primer extension products are analyzed by capillary electrophoresis; the color and relative size of the fluorescent peak determines which nucleotide was added during single-base extension and the resultant genotype at each locus. An example shows detection of a *BRAF* V600E mutation (1799T > A, black arrow).

standard-of-care or widely used, these tests are urgently needed, as clinicopathologic features alone have not been shown to effectively identify high-risk Stage II patients who may show survival benefit from adjuvant chemotherapy, similar to that demonstrated in patients with Stage III metastatic CRC [12,92,93]. These include the ColoPrint assay (Agendia, Irvine, CA) which is an oligonucleotide microarray gene expression profile composed of 18 genes and developed using gene expression data from whole genome oligonucleotide arrays [94,95], and the Oncotype DX colon cancer test (Genomic Health, Redwood City, CA) which is a quantitative multigene RT-PCR-based gene expression assay developed to assess recurrence risk and treatment benefits based on expression levels of seven genes associated with recurrence and six genes associated with treatment benefit in Stage II CRC patients [96].

The use of next-generation sequencing technologies (NGS), which enable high-throughput massively parallel sequencing of nucleic acids at lower cost, faster speeds, higher sensitivity, and with reduced error rates compared to traditional Sanger sequencing, are increasingly utilized for molecular testing in clinical laboratories [97,98]. Three levels of analysis can be performed by NGS, including targeted-gene panels, exome sequencing, and whole-genome sequencing. Multiple NGS platforms have been developed with the capacity to massively sequence millions of DNA fragments in parallel. These platforms differ in sequence read length, total sequence capacity, run-times, and quality and accuracy of the data produced [98]. NGS assays require genomic DNA extracted from a patient sample and enriched for a subset of targets by PCR amplicon-based or hybridization capture approaches [99]. As such, it is now possible to sequence all genes implicated in CRC at a lower cost than that for performing Sanger sequencing. However, it is important to note that the data analysis is complex for NGS, requires significant bioinformatics input, and a major effort is required for annotation and variant classification. Recently, standard and professional practice guidelines were established for NGS clinical applications to assist clinical laboratories with the validation of NGS methods and platforms, monitoring of NGS testing, and interpretation of variants found using these technologies [77].

Multiple multigene panels for CRC have been developed for clinical use with various NGS technologies at select companies and academic institutions and include genes for which the FDA has approved single-gene companion diagnostic assays. These include hereditary colon cancer panels for CRC, such as ColoNext (Ambry Genetics) and ColoSeq (Washington University) [100,101]. NGS panels for personalized CRC diagnosis and therapy have also been developed using genetic biomarkers associated with response to anti-EGFR therapy, chemotherapy, or other targeted therapies. These are currently being evaluated for use in metastatic CRCs [98], including the Molecular Intelligence for CRC (CARIS Life Sciences) and FoundationOne pan-cancer test (Foundation Medicine). A recent study by Cragun et al. [101] reported the prevalence of clinically significant mutations and variants of uncertain significance among 586 patients who underwent ColoNext panel testing for hereditary colon cancer. Sixty-one patients (10.4%) had genetic alterations consistent with pathogenic mutations or likely pathogenic variants and 42 patients (7.2%) were considered to have actionable mutations. In addition, 118 patients (20.1%) had at least one variant of uncertain significance, including 14 patients who had at least one variant of uncertain significance in addition to a pathologic mutation. Of the 42 patients with an actionable pathologic mutation, most (30 patients, 71%) clearly met NCCN guidelines for syndrome-based testing, screening, or diagnosis, based on the available clinical and family history. Therefore, at this time, the true clinical utility of NGS multigene panels for CRC may be in building a comprehensive knowledge base of genes and mutations which may direct patients to future targeted therapies.

There has been an increasing role for molecular testing in the diagnosis and management of CRC and a rapid expansion in the molecular technologies available for clinical use. Current standard-of-care recommendations for molecular testing in CRC will likely rapidly evolve with the vast amount of information now available through the molecular technologies currently in use, advances in our understanding of molecular mechanisms of disease, and the rapid development of new targeted therapies for CRC.

References

[1] Siegel R, Desantis C, Jemal A. Colorectal cancer statistics. CA Cancer J Clin 2014;64:104–17.

[2] Siegel R, Ma J, Zou Z, Jemal A. Cancer statistics. CA Cancer J Clin 2014;64:9–29 214.

[3] Vogelstein B, Fearon ER, Hamilton SR, et al. Genetic alterations during colorectal-tumor development. N Engl J Med 1988;319:525–32.

[4] Coppede F, Lopomo A, Spisni R, Migliore L. Genetic and epigenetic biomarkers for diagnosis, prognosis and treatment of colorectal cancer. World J Gastroenterol 2014;20:943–56.

[5] Ogino S, Goel A. Molecular classification and correlates in colorectal cancer. J Mol Diagn 2008;10:13–27.

[6] Ahnen DJ. The American College of Gastroenterology Emily Couric Lecture—the adenoma-carcinoma sequence revisited: has the era of genetic tailoring finally arrived? Am J Gastroenterol 2011;106:190–8.

[7] Boland CR, Goel A. Microsatellite instability in colorectal cancer. Gastroenterology 2010;138:2073–87.

[8] Wood LD, Parsons DW, Jones S, et al. The genomic landscapes of human breast and colorectal cancers. Science 2007;318:1108–13.

[9] Tsang AH, Cheng KH, Wong AS, et al. Current and future molecular diagnostics in colorectal cancer and colorectal adenoma. World J Gastroenterol 2014;20:3847–57.

[10] Pino MS, Chung DC. The chromosomal instability pathway in colon cancer. Gastroenterology 2010;138:2059–72.

[11] Rajagopalan H, Nowak MA, Vogelstein B, Lengauer C. The significance of unstable chromosomes in colorectal cancer. Nature Rev Cancer 2003;3:695–701.

[12] Shi C, Washington K. Molecular testing in colorectal cancer: Diagnosis of Lynch syndrome and personalized cancer medicine. Am J Clin Pathol 2012;137:847–59.

[13] Legolvan MP, Taliano RJ, Resnick MB. Application of molecular techniques in the diagnosis, prognosis and management of patients with colorectal cancer: a practical approach. Hum Pathol 2012;43:1157–68.

[14] Walther A, Houlston R, Tomlinson I. Association between instability and prognosis in colorectal cancer: a meta-analysis. Gut 2008;57:941–50.

[15] Pritchard CC, Grady WM. Colorectal cancer molecular biology moves into clinical practice. Gut 2010;60:116–29.

[16] Armaghany T, Wilson JD, Chu Q, Mills G. Genetic alterations in colorectal cancer. Gastrointest Cancer Res 2012;5:19–27.

[17] Fearon ER. Molecular genetics of colorectal cancer. Annu Rev Pathol 2010;6:479–507.

[18] Umar A, Risinger JI, Hawk ET, Barrett JC. Testing guidelines for hereditary non-polyposis colorectal cancer. Nature Rev Cancer 2004;4:153–8.

[19] Narayan S, Roy D. Role of APC and DNA mismatch repair genes in the development of colorectal cancers. Mol Cancer 2003;2:41.

[20] Daniel FI, Cherubini K, Yurgel LS, de Figueiredo MA, Salum FG. The role of epigenetic transcription repression and DNA methyltransferases in cancer. Cancer 2010;117:677–87.

[21] Kim ER, Kim YH. Clinical application of genetics in management of colorectal cancer. Intest Res 2014;12:184–93.

[22] Jones PA, Baylin SB. The fundamental role of epigenetic events in cancer. Nature Rev Genet 2002;3:415–28.

[23] East JE, Saunders BP, Jass JR. Sporadic and syndromic hyperplastic polyps and serrated adenomas of the colon: classification, molecular genetics, natural history, and clinical management. Gastroenterol Clin North Am 2008;37:25–46.

[24] Grady WM, Carethers JM. Genomic and epigenetic instability in colorectal cancer pathogenesis. Gastroenterology 2008;135:1079–99.

[25] Rustgi AK. The genetics of hereditary colon cancer. Genes Dev 2007;21:2525–38.

[26] Lipton L, Tomlinson I. The genetics of FAP and FAP-like syndromes. Fam Cancer 2006;5:221–6.

[27] Heinen CD. Genotype to phenotype: analyzing the effects of inherited mutations in colorectal cancer families. Mutat Res 2009;693:32–45.

[28] Bogaert J, Prenen H. Molecular genetics of colorectal cancer. Ann Gastroenterol 2014;27:9–14.

[29] Kovacs ME, Papp J, Szentirmay Z, Otto S, Olah E. Deletions removing the last exon of TACSTD1 constitute a distinct class of mutations predisposing to Lynch syndrome. Hum Mutat 2009;30:197–203.

[30] Ligtenberg MJ, Kuiper RP, Chan TL, et al. Heritable somatic methylation and inactivation of MSH2 in families with Lynch syndrome due to deletion of the 3′ exons of TACSTD1. Nature Genet 2009;41:112–17.

[31] Jasperson KW, Tuohy TM, Neklason DW, Burt RW. Hereditary and familial colon cancer. Gastroenterology 2010;138:2044–58.

[32] Gala M, Chung DC. Hereditary colon cancer syndromes. Semin Oncol 2011;38:490–9.

[33] Galiatsatos P, Foulkes WD. Familial adenomatous polyposis. Am J Gastroenterol 2006;101:385–98.

[34] Forbes SA, Beare D, Gunasekaran P, et al. COSMIC: exploring the 'world's knowledge of somatic mutations in human cancer. Nucleic Acids Res 2015;43:D805–11.

[35] Goodenberger M, Lindor NM. Lynch syndrome and MYH-associated polyposis: review and testing strategy. J Clin Gastroenterol 2011;45:488–500.

[36] Briggs S, Tomlinson I. Germline and somatic polymerase epsilon and delta mutations define a new class of hypermutated colorectal and endometrial cancers. J Pathol 2013;230:148–53.

[37] Seshagiri S. The burden of faulty proofreading in colon cancer. Nature Genet 2013;45:121–2.

[38] Palles C, Cazier JB, Howarth KM, et al. Germline mutations affecting the proofreading domains of POLE and POLD1 predispose to colorectal adenomas and carcinomas. Nature Genet 2012;45:136–44.

[39] Jaeger E, Leedham S, Lewis A, et al. Hereditary mixed polyposis syndrome is caused by a 40-kb upstream duplication that leads to increased and ectopic expression of the BMP antagonist GREM1. Nature Genet 2012;44:699–703.

[40] Schreibman IR, Baker M, Amos C, McGarrity TJ. The hamartomatous polyposis syndromes: a clinical and molecular review. Am J Gastroenterol 2005;100:476–90.

[41] Johns LE, Houlston RS. A systematic review and meta-analysis of familial colorectal cancer risk. Am J Gastroenterol 2001;96:2992–3003.

[42] Moroni M, Veronese S, Benvenuti S, et al. Gene copy number for epidermal growth factor receptor (EGFR) and clinical response to antiEGFR treatment in colorectal cancer: a cohort study. Lancet Oncol 2005;6:279–86.

[43] Shia J, Klimstra DS, Li AR, et al. Epidermal growth factor receptor expression and gene amplification in colorectal carcinoma: an immunohistochemical and chromogenic in situ hybridization study. Mod Pathol 2005;18:1350–6.

[44] Dienstmann R, Vilar E, Tabernero J. Molecular predictors of response to chemotherapy in colorectal cancer. Cancer J 2011;17:114–26.

[45] Tsuchihashi Z, Khambata-Ford S, Hanna N, Janne PA. Responsiveness to cetuximab without mutations in EGFR. N Engl J Med 2005;353:208–9.

[46] Cunningham D, Humblet Y, Siena S, et al. Cetuximab monotherapy and cetuximab plus irinotecan in irinotecan-refractory metastatic colorectal cancer. N Engl J Med 2004;351:337–45.

[47] Khambata-Ford S, Garrett CR, Meropol NJ, et al. Expression of epiregulin and amphiregulin and K-ras mutation status predict disease control in metastatic colorectal cancer patients treated with cetuximab. J Clin Oncol 2007;25:3230–7.

[48] Karapetis CS, Khambata-Ford S, Jonker DJ, et al. K-ras mutations and benefit from cetuximab in advanced colorectal cancer. N Engl J Med 2008;359:1757–65.

[49] Loupakis F, Ruzzo A, Cremolini C, et al. KRAS codon 61, 146 and BRAF mutations predict resistance to cetuximab plus irinotecan in KRAS codon 12 and 13 wild-type metastatic colorectal cancer. Br J Cancer 2009;101:715–21.

[50] Smith G, Bounds R, Wolf H, Steele RJ, Carey FA, Wolf CR. Activating K-Ras mutations out with 'hotspot' codons in sporadic colorectal tumours - implications for personalised cancer medicine. Br J Cancer 2010;102:693–703.

[51] De Roock W, Claes B, Bernasconi D, et al. Effects of KRAS, BRAF, NRAS, and PIK3CA mutations on the efficacy of cetuximab plus chemotherapy in chemotherapy-refractory metastatic colorectal cancer: a retrospective consortium analysis. Lancet Oncol 2010;11:753–62.

[52] Oliveira C, Velho S, Moutinho C, et al. KRAS and BRAF oncogenic mutations in MSS colorectal carcinoma progression. Oncogene 2007;26:158–63.

[53] Roth AD, Tejpar S, Delorenzi M, et al. Prognostic role of KRAS and BRAF in stage II and III resected colon cancer: results of the translational study on the PETACC-3, EORTC 40993, SAKK 60-00 trial. J Clin Oncol 2009;28:466–74.

[54] Jimeno A, Messersmith WA, Hirsch FR, Franklin WA, Eckhardt SG. KRAS mutations and sensitivity to epidermal growth factor receptor inhibitors in colorectal cancer: practical application of patient selection. J Clin Oncol 2009;27:1130–6.

[55] Amado RG, Wolf M, Peeters M, et al. Wild-type KRAS is required for panitumumab efficacy in patients with metastatic colorectal cancer. J Clin Oncol 2008;26:1626–34.

[56] Lievre A, Bachet JB, Le Corre D, et al. KRAS mutation status is predictive of response to cetuximab therapy in colorectal cancer. Cancer Res 2006;66:3992–5.

[57] Douillard JY, Oliner KS, Siena S, et al. Panitumumab-FOLFOX4 treatment and RAS mutations in colorectal cancer. N Engl J Med 2013;369:1023–34.

[58] Richman SD, Seymour MT, Chambers P, et al. KRAS and BRAF mutations in advanced colorectal cancer are associated with poor prognosis but do not preclude benefit from oxaliplatin or irinotecan: results from the MRC FOCUS trial. J Clin Oncol 2009;27:5931–7.

[59] Samowitz WS, Albertsen H, Herrick J, et al. Evaluation of a large, population-based sample supports a CpG island methylator phenotype in colon cancer. Gastroenterology 2005;129:837–45.

[60] Deng G, Bell I, Crawley S, et al. BRAF mutation is frequently present in sporadic colorectal cancer with methylated hMLH1, but not in hereditary nonpolyposis colorectal cancer. Clin Cancer Res 2004;10:191–5.

[61] Di Nicolantonio F, Martini M, Molinari F, et al. Wild-type BRAF is required for response to panitumumab or cetuximab in metastatic colorectal cancer. J Clin Oncol 2008;26:5705–12.

[62] Yokota T, Ura T, Shibata N, et al. BRAF mutation is a powerful prognostic factor in advanced and recurrent colorectal cancer. Br J Cancer 2011;104:856–62.

[63] Yang H, Higgins B, Kolinsky K, et al. Antitumor activity of BRAF inhibitor vemurafenib in preclinical models of BRAF-mutant colorectal cancer. Cancer Res 2011;72:779–89.

[64] Prahallad A, Sun C, Huang S, et al. Unresponsiveness of colon cancer to BRAF(V600E) inhibition through feedback activation of EGFR. Nature 2012;483:100–3.

[65] Zhao L, Vogt PK. Hot-spot mutations in p110alpha of phosphatidylinositol 3-kinase (pI3K): Differential interactions with the regulatory subunit p85 and with RAS. Cell Cycle 2009;9:596–600.

[66] Velho S, Oliveira C, Ferreira A, et al. The prevalence of PIK3CA mutations in gastric and colon cancer. Eur J Cancer 2005;41:1649–54.

[67] Kato S, Iida S, Higuchi T, et al. PIK3CA mutation is predictive of poor survival in patients with colorectal cancer. Int J Cancer 2007;121:1771–8.

[68] Barault L, Veyrie N, Jooste V, et al. Mutations in the RAS-MAPK, PI(3)K (phosphatidylinositol-3-OH kinase) signaling network correlate with poor survival in a population-based series of colon cancers. Int J Cancer 2008;122:2255–9.

[69] McCubrey JA, Steelman LS, Kempf CR, et al. Therapeutic resistance resulting from mutations in Raf/MEK/ERK and PI3K/PTEN/Akt/mTOR signaling pathways. J Cell Physiol 2011;226:2762–81.

[70] Jhawer M, Goel S, Wilson AJ, et al. PIK3CA mutation/PTEN expression status predicts response of colon cancer cells to the epidermal growth factor receptor inhibitor cetuximab. Cancer Res 2008;68:1953–61.

[71] Frattini M, Saletti P, Romagnani E, et al. PTEN loss of expression predicts cetuximab efficacy in metastatic colorectal cancer patients. Br J Cancer 2007;97:1139–45.

[72] Bardelli A, Siena S. Molecular mechanisms of resistance to cetuximab and panitumumab in colorectal cancer. J Clin Oncol 2010;28:1254–61.

[73] Negri FV, Bozzetti C, Lagrasta CA, et al. PTEN status in advanced colorectal cancer treated with cetuximab. Br J Cancer 2009;102:162–4.

[74] Grady WM. Genetic testing for high-risk colon cancer patients. Gastroenterology 2003;124:1574–94.

[75] Lindor NM. Recognition of genetic syndromes in families with suspected hereditary colon cancer syndromes. Clin Gastroenterol Hepatol 2004;2:366–75.

[76] Burt R, Neklason DW. Genetic testing for inherited colon cancer. Gastroenterology 2005;128:1696–716.

[77] Rehm HL, Bale SJ, Bayrak-Toydemir P, et al. ACMG clinical laboratory standards for next-generation sequencing. Genet Med 2013;15:733–47.

[78] Hegde M, Ferber M, Mao R, Samowitz W, Ganguly A. ACMG technical standards and guidelines for genetic testing for inherited colorectal cancer (Lynch syndrome, familial adenomatous polyposis, and MYH-associated polyposis). Genet Med 2013;16:101–16.

[79] Burt RW, Leppert MF, Slattery ML, et al. Genetic testing and phenotype in a large kindred with attenuated familial adenomatous polyposis. Gastroenterology 2004;127:444–51.

[80] Wang L, Baudhuin LM, Boardman LA, et al. MYH mutations in patients with attenuated and classic polyposis and with young-onset colorectal cancer without polyps. Gastroenterology 2004;127:9–16.

[81] Giardiello FM, Brensinger JD, Petersen GM. AGA technical review on hereditary colorectal cancer and genetic testing. Gastroenterology 2001;121:198–213.

[82] Venesio T, Molatore S, Cattaneo F, Arrigoni A, Risio M, Ranzani GN. High frequency of MYH gene mutations in a subset of patients with familial adenomatous polyposis. Gastroenterology 2004;126:1681–5.

[83] Gismondi V, Meta M, Bonelli L, et al. Prevalence of the Y165C, G382D and 1395delGGA germline mutations of the MYH gene in Italian patients with adenomatous polyposis coli and colorectal adenomas. Int J Cancer 2004;109:680–4.

[84] Shia J, Tang LH, Vakiani E, et al. Immunohistochemistry as first-line screening for detecting colorectal cancer patients at risk for hereditary nonpolyposis colorectal cancer syndrome: a 2-antibody panel may be as predictive as a 4-antibody panel. Am J Surg Pathol 2009;33:1639–45.

[85] Vilkin A, Halpern M, Morgenstern S, et al. How reliable is immunohistochemical staining for DNA mismatch repair proteins performed after neoadjuvant chemoradiation? Hum Pathol 2014;45:2029–36.

[86] Boland CR, Thibodeau SN, Hamilton SR, et al. A National Cancer Institute Workshop on Microsatellite Instability for cancer detection and familial predisposition: development of international criteria for the determination of microsatellite instability in colorectal cancer. Cancer Res 1998;58:5248–57.

[87] Umar A, Boland CR, Terdiman JP, et al. Revised Bethesda Guidelines for hereditary nonpolyposis colorectal cancer (Lynch syndrome) and microsatellite instability. J Natl Cancer Inst 2004;96:261–8.

[88] Tian S, Roepman P, Popovici V, et al. A robust genomic signature for the detection of colorectal cancer patients with microsatellite instability phenotype and high mutation frequency. J Pathol 2012;228:586–95.

[89] Schepeler T, Reinert JT, Ostenfeld MS, et al. Diagnostic and prognostic microRNAs in stage II colon cancer. Cancer Res 2008;68:6416–24.

[90] Domingo E, Niessen RC, Oliveira C, et al. BRAF-V600E is not involved in the colorectal tumorigenesis of HNPCC in patients with functional MLH1 and MSH2 genes. Oncogene 2005;24:3995–8.

[91] Day F, Muranyi A, Singh S, et al. A mutant BRAF V600E-specific immunohistochemical assay: correlation with molecular mutation status and clinical outcome in colorectal cancer. Target Oncol 2015;10:99–109.

[92] Benson 3rd AB, Schrag D, Somerfield MR, et al. American Society of Clinical Oncology recommendations on adjuvant chemotherapy for stage II colon cancer. J Clin Oncol 2004;22:3408–19.

[93] O'Connor ES, Greenblatt DY, LoConte NK, et al. Adjuvant chemotherapy for stage II colon cancer with poor prognostic features. J Clin Oncol 2011;29:3381–8.

[94] Salazar R, Roepman P, Capella G, et al. Gene expression signature to improve prognosis prediction of stage II and III colorectal cancer. J Clin Oncol 2007;29:17–24.

[95] Maak M, Simon I, Nitsche U, et al. Independent validation of a prognostic genomic signature (ColoPrint) for patients with stage II colon cancer. Ann Surg 2013;257:1053–8.

[96] Gray RG, Quirke P, Handley K, et al. Validation study of a quantitative multigene reverse transcriptase-polymerase chain reaction assay for assessment of recurrence risk in patients with stage II colon cancer. J Clin Oncol 2011;29:4611–19.

[97] Metzker ML. Sequencing technologies – the next generation. Nature Rev Genet 2009;11:31–46.

[98] Deeb KK, Sram JP, Gao H, Fakih MG. Multigene assays in metastatic colorectal cancer. J Natl Compr Canc Netw 2013;11: S9–17.

[99] Mamanova L, Coffey AJ, Scott CE, et al. Target-enrichment strategies for next-generation sequencing. Nat Methods 2010;7:111–18.

[100] LaDuca H, Stuenkel AJ, Dolinsky JS, et al. Utilization of multigene panels in hereditary cancer predisposition testing: analysis of more than 2,000 patients. Genet Med 2014;16:830–7.

[101] Cragun D, Radford C, Dolinsky JS, Caldwell M, Chao E, Pal T. Panel-based testing for inherited colorectal cancer: a descriptive study of clinical testing performed by a US laboratory. Clin Genet 2014;86:510–20.

25

Molecular Pathology and Testing in Melanocytic Tumors

A.A. Hedayat[1], S. Yan[1] and G.J. Tsongalis[2]

[1]Department of Pathology, Dartmouth-Hitchcock Medical Center, Lebanon, NH, United States [2]Laboratory for Clinical Genomics and Advanced Technology (CGAT), Department of Pathology and Laboratory Medicine, Dartmouth-Hitchcock Medical Center and Norris Cotton Cancer Center, Geisel School of Medicine at Dartmouth, Hanover, NH, United States

INTRODUCTION

Cutaneous melanoma is thought to arise from melanocytes, which are neural crest-derived cells located in the epidermis and skin appendages. The major role of melanocytes is synthesizing melanin pigments that are scattered among keratinocytes, and protect the skin from ultraviolet (UV) light and solar radiation [1]. Many melanin-containing tumors that demonstrate nuclear pleomorphism and cherry-red macronucleoli may be composed of various cancers, but are currently considered under the broad category of melanoma. Therefore, melanoma is not a single entity, but rather a genetically heterogeneous group of neoplasms that differ in cells of origin, age of onset, clinical and histologic presentation, pattern of metastasis, ethnic distribution, causative role of UV radiation, predisposing germline alterations, mutational processes, and patterns of somatic mutations [2]. Bastian and colleagues evaluated 37 melanoma cases and demonstrated distinct evolutionary trajectories for different melanoma subtypes [2]. Additionally, UV light has been shown to be a major factor in nevus formation and progression to melanoma [3].

Incidence rates for melanoma are gradually increasing. In 2014, approximately 76,100 Americans (43,890 men and 32,210 women) were diagnosed with melanoma. With an estimated 9710 deaths, melanoma accounts for 1.7% of all cancer-related deaths in the United States [4]. Identification of at-risk patients, implementation of preventive actions, early screening and diagnosis, and prompt treatment remain the mainstays of care for melanoma. Implementation of genome-wide approaches, systems biology and technologies such as array-based platforms, immunohistochemistry (IHC), and fluorescence in situ hybridization (FISH) has enabled evaluation of cancers for various mutations and identification of key signaling molecules and pathways. These methods have contributed to the current concepts of melanoma pathogenesis.

This chapter is devoted to the recent diagnostic methods in molecular diagnostics of melanoma and some of the proliferative disturbances of the melanocyte—nevus cell system. It is not intended to be an all-encompassing treatise on melanocytic lesions as those are well developed in other texts and publications.

MELANOMA-ASSOCIATED CONDITIONS AND SYNDROMES

Melanoma has been associated with many other conditions and syndromes, including squamous and basal carcinoma [5,6], xeroderma pigmentosum [6], ocular melanoma [7], albinism [8—12], retinoblastoma [13—16], Li—Fraumeni syndrome [17], pancreatic carcinoma [18—20], chronic inflammatory demyelinating polyneuropathy [20,21], mesothelioma [22], meningioma [23], renal cell carcinoma [24,25], LEOPARD syndrome [26], Birt—Hogg—Dubé syndrome [27—29], Hailey—Hailey disease [30], Gorlin syndrome [31], Cowden disease [32], neurofibromatosis [33,34], Beckwith—Wiedemann syndrome [35], Lynch syndrome, pulmonary carcinoid [36], soft tissue sarcomas [37—39], gastrointestinal stromal

© 2017 Elsevier Inc. All rights reserved.

tumors [40], thyroid carcinoma [41,42], Merkel cell carcinoma [43], mycosis fungoides [44,45], other leukemias and lymphomas [46–54], Castleman disease [55], systemic mastocytosis [56], bronchogenic cysts [57], cystic teratomas [58], epidermal inclusion cysts [59], phacomatosis pigmentovascularis [60], Turner syndrome [61,62], Down syndrome, Parkinson disease, Charcot–Marie–Tooth disease, retinoblastoma, dermatomyositis, erythema dyschromicum perstans, sarcoidosis, stasis dermatitis, chronic wounds/ulcers and lymphedema, burn scars, cesarean skin scars [63], tattoos [64–67], trauma (subungual melanomas) [68,69], lichen sclerosus of the vulva [70,71], epidermolysis bullosa simplex [72], Marjolin ulcers [41], human papillomavirus infections linked to epidermodysplasia verruciformis (HPV16) [73,74], and infections with the human immunodeficiency virus [75,76]. Genes associated with these syndromes may either represent a chance occurrence with melanomas or may be mutually exclusive. Therefore, it is unclear whether these genes have a causal effect for melanomagenesis.

It is generally accepted that about 10% of melanomas are hereditary with a strong family history for melanoma-related germline mutation. The familial type is typically inherited in an autosomal dominant manner affecting every generation. A positive family history increases the risk of melanoma twofold [42,77]. Over the last two decades much progress has been made in discovering and expanding panels of somatic driver mutations in melanocytic neoplasms. These genes are known to activate oncogenes or inactivate tumor-suppressor genes leading to the pathogenesis of melanoma.

MOLECULAR ANALYSIS OF MELANOMA

Introduction of molecular testing in the context of melanoma has been a breakthrough to better understand the histogenesis. Genomic aberrations are a hallmark of cancer cells including malignant melanoma (MM). The following mutations have been associated with melanoma and are divided into two groups: (1) genes associated with initiating events in melanocytic lesions and (2) genes associated with progression events in melanocytic lesions.

Initiating Events in Melanoma: Oncogenes

The following oncogenic events are shared by almost all classes of melanocytic lesions and are known as the initiating oncogenes (Fig. 25.1).

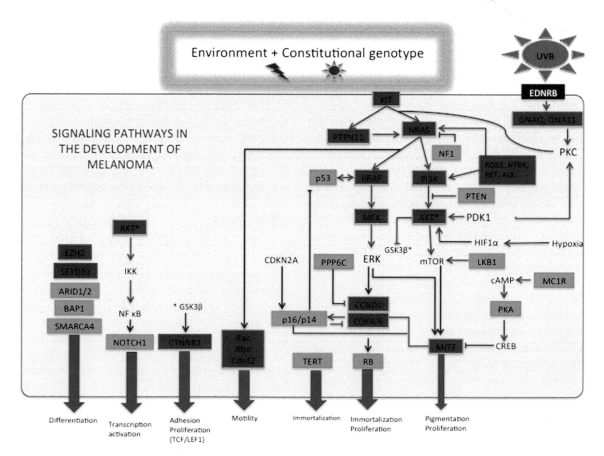

FIGURE 25.1　Major pathways disrupted and involved in development of melanoma. →: Activating signal; ⊥: Inhibiting signal; blue boxes: proteins affected by loss of function; orange boxes: proteins affected by gain of function.

ROS proto-oncogene 1, receptor tyrosine kinase, (ROS1). ROS1 is a kinase fusion mutation that affects the mitogen kinase-activated protein kinase (MAPK), phosphoinositides 3-kinase (PI3K), and signal transducer and activator of transcription 3 (STAT3). Wiesner and colleagues reported *ROS1* rearrangements in 19 of 73 (26%) Spitz nevi, 3 of 34 (8%) atypical Spitz tumors, and 3 of 33 (9%) Spitzoid melanoma [78].

RET proto-oncogene (RET). RET is a kinase fusion mutation that affects the MAPK, PI3K, and STAT3 pathways. Wiesner and colleagues reported *RET* mutations in 2 of 75 (3%) Spitz nevi, 1 of 32 (3%) atypical Spitz tumors, and 1 of 33 (3%) Spitzoid melanoma [78].

Neurotrophic tyrosine kinase, receptor, type 1 (NTRK1). NTRK1 is a kinase fusion mutation that affects MAPK, PI3K, and STAT3 pathways. Wiesner and colleagues reported *NTRK1* rearrangements in 8 of 75 (10.7%) Spitz nevi, 8 of 32 (25%) atypical Spitz tumors, and 7 of 33 (21%) Spitzoid melanoma [78,79].

Anaplastic lymphoma receptor tyrosine kinase (ALK). ALK is a kinase fusion mutation that affects MAPK, PI3K, and STAT3 pathways. Wiesner and colleagues reported *ALK* fusions in 8 of 75 (10.7%) Spitz nevi, 5 of 32 (15.6%) atypical Spitz tumors, and 1 of 33 (3%) Spitzoid melanoma [78].

V-kit Hardy-Zuckerman 4 feline sarcoma viral oncogene homolog (KIT). KIT is a cell surface receptor tyrosine kinase. *KIT* is affected through an amplification mutation that can affect the PI3K, MAPK, and STAT3 pathways. Curtin and colleagues examined 102 primary melanomas and found mutations and/or copy number increases of *KIT* in 39% of mucosal, 36% of acral, and 28% of melanomas on chronically sun-damaged skin, but not in any (0%) melanomas on skin without chronic sun damage. Seventy-nine percent of tumors with *KIT* mutations and 53% of tumors with multiple copies of *KIT* demonstrated increased KIT protein levels [79]. Traditional Sanger sequencing is typically used for the purpose of evaluating multiple exon regions of the *KIT* gene for somatic point *KIT* mutations. This is primarily due to discovery of the mutations in exons 9, 11, 13, and 17 without a predominant mutation [80].

Guanine nucleotide binding protein, q polypeptide (GNAQ) and guanine nucleotide binding protein alpha 11 (GNA11). GNAQ and GNA11 are subject to point mutations that affect protein kinase C and MAPK pathways by encoding α subunits of G-protein-coupled receptors [80]. Frequent somatic mutations of *GNAQ* and *GNA11* are reported in blue nevi, blue nevus-like melanomas, and uveal melanomas [81,82]. Van Raamsdonk and colleagues evaluated 186 uveal melanomas, of which 83% had somatic mutations in *GNAQ* or *GNA11* [81,82]. Constitutive activation of the pathway involving these two genes appears to be a major contributor to the development of uveal melanoma [83].

B-Raf proto-oncogene, serine/threonine kinase (BRAF). The *BRAF* gene encodes a serine/threonine protein kinase downstream of the epidermal growth factor receptor and RAS family of small G-proteins. *BRAF* is mutated in 40–60% melanomas [84,85]. *BRAF* mutation is a point mutation and kinase fusion that affects MAPK pathway. *BRAF* is a member of the RAF kinase family, which acts in the ERK/MAPK pathway, a signaling cascade that regulates cellular proliferation, differentiation, and survival. It is typically involved in acquired nevi, Spitz tumors, acral melanomas, non-cumulative sun-damaged skin (non-CSD), and cumulative sun-damaged skin (CSD) [85–87]. The RAS/RAF/MEK/ERK signaling pathway has been linked to induction and maintenance of melanoma, particularly *BRAF*, which is mutated in approximately 44% of melanoma cases [86]. Oncogenic *BRAF* signaling leads to progression of melanoma through the activation of tumor progression-related genes that are downstream of *BRAF*. *BRAF* can also induce autocrine vascular endothelial growth factor secretion leading to angiogenesis and promoting tumor survival and growth [86–88].

The most common *BRAF* mutation, p.V600E (c.1799 T > A) mutation, causes upregulation of the kinase activity [89]. In 2003, Pollock and colleagues examined *BRAF* mutations on microdissected melanoma and nevus samples [90]. This study showed that a mutation in *BRAF* plays an important role in melanocytic lesion development but do not necessarily contributes to progression to melanoma [90]. *BRAF V600E* mutations are associated with lesions arising in the background of benign nevus and are more common in younger patients on intermittently sun-exposed areas [3]. In addition, BRAF V600E are more common in older patients on chronically sun-exposed areas associated with melanomas arising in a background of intermediate lesions [3]. In 2011, the Food and Drug Administration of the United States approved vemurafenib for treatment of unresectable or metastatic melanomas with a *BRAF p.V600E* mutation. Therefore, clinical detection of the *BRAF* p.V600E mutation has become the standard of care for patients with advanced melanoma in order to predict response to vemurafenib, dabrafenib, and trametinib [89,91–93].

Neuroblastoma RAS viral (v-ras) oncogene homolog (NRAS). NRAS is subject to point mutation that affects the MAPK and PI3K pathways and are involved in cell growth and differentiation [94]. *NRAS* mutations, upstream of *BRAF*, are usually mutually exclusive of *BRAF* mutations. *BRAF* may be activated by the common V600E mutations, as well as by upstream *NRAS* mutations. *NRAS* mutations are the predominant mutation in large congenital nevi [95]. *NRAS* mutations are also reported to occur in nodular melanomas and melanomas arising in CSD [88].

Harvey rat sarcoma viral oncogene homolog (HRAS). HRAS gene mutation also affects MAPK and PI3K pathways. Involvement of Q61 of exon 3, with replacement of glutamine by lysine is the most common mutation [96—98]. van Dijk and colleagues reported HRAS mutations in 29% of Spitz nevi, in 14% of atypical Spitz nevi, 7% Spitzoid tumors suspected for melanoma [97]. However, these mutations rarely occur in melanomas [97]. Morphologically, the bulky Spitz nevi with sclerosis of the deep dermal component usually have been reported to carry the HRAS mutation. However, only about 20% of Spitz nevi carry this mutation [96].

Considering that Spitz melanomas rarely carry this mutation, HRAS mutation analysis may be a useful diagnostic tool in differentiating between Spitz nevus and Spitzoid melanoma, and to help predict the biological behavior of Spitz tumor of unknown malignant potential [97,98]. However, mutational analysis of BRAF, NRAS, and HRAS may not be useful in differentiating between Spitzoid melanoma and Spitz nevus in children [96,97,99].

Genes Associated with Progression Events in Melanocytic Lesions: Loss-of-Function Mutations

Neurofibromin 1 (NF1). NF1 deletion affects the MAPK pathway. According to Nissan et al., loss of NF1 is common in cutaneous melanoma and is associated with RAS activation, MEK-dependence, and resistance to RAF inhibition [100]. Additionally, loss of the NF1 gene was reported in melanomas lacking BRAF or NRAS mutations as well as a subset with RAS/BRAF mutation. Wiesner and colleagues performed sequencing on 15 desmoplastic melanomas (DMs) and 20 non-DMs and reported NF1 mutation in 93% of DMs and 20% non-DMs [101]. Therapeutically, melanomas with NF1 gene mutations were more resistant to treatments that target BRAF [100].

BRCA1-associated protein 1 (BAP1). BAP1 is a tumor-suppressor gene that affects chromatin modulation and transcriptional regulation. BAP1 is a nuclear protein encoded by the tumor-suppressor gene located on chromosome 3p21.1 [102]. Somatic BAP1 mutations have been reported to increase susceptibility for the development of cutaneous melanocytic tumors (including epithelioid atypical Spitz tumors and melanoma), uveal melanoma, mesothelioma, clear-cell renal cell carcinoma, and other tumors [102—104].

BAP1 hereditary cancer predisposition syndrome was first described in 2011 [104]. It is an autosomal dominant tumor syndrome caused by inactivating germline mutations of the BAP1 gene. Affected individuals have increased risk of developing mesothelioma and uveal melanoma [103]. The combination of BRAF mutation and loss of BAP1 nuclear expression have been shown to characterize a subset of atypical Spitz tumors with distinct features, although their clinical prognostic significance is not yet clear [22]. Evaluation of melanocytic lesions for BAP1 protein expression using IHC or chromosomal microarray tools such as single nucleotide polymophism array or comparative genomic hybridization array may serve as a rapid and cost-effective means of identifying BAP1-deficient melanocytic lesions [104].

Cyclin-dependent kinase inhibitor 2 (CDKN2A). CDKN2A is the first familial melanoma gene identified and accounts for the majority of high-density melanoma-prone families [105—107]. CDKN2A is a gene located at chromosome 9p21 and it encodes for two tumor-suppressor proteins—p14^{CDKN2A} and p16^{CDKN2A} [108]. p16^{CDKN2A} inhibits CDK4, which in turn phosphorylates pRB leading to immortalization and proliferation of cancer cells [2,109]. p14^{CDKN2A} inhibits the oncogenic actions of the downstream MDM2 protein, whose direct interaction with p53 blocks any p53-mediated activity and targets the p53 protein for rapid degradation [109]. Fargnoli and colleagues performed a meta-analysis study and showed that multiple variants of the melanocortin-1 receptors (MC1R) gene increase the risk of melanoma in CDKN2A mutation carriers [110]. Loss of CDKN2A is reported to be almost exclusively associated with invasive melanoma [3].

Genes with Nonspecific Melanocytic Association

SWI/SNF-related, matrix-associated, actin-dependent regulator of chromatin, subfamily a, member 4 (SMARCA4). SMARCA4 is also known as BRG1. BRG1 can be affected by deletions affecting the chromatin-remodeling pathway leading to differentiation of melanoma cells. Lin and colleagues evaluated BRG1 immunostaining in 48 dysplastic nevi, 90 primary melanomas, and 47 metastatic melanomas [111]. This study showed that BRG1 is significantly increased in primary cutaneous melanoma and metastatic melanoma compared to dysplastic nevi [111]. BRG1 is essential for melanoma cell proliferation and for normal melanocyte development [112]. In 2012, Zhang and colleagues evaluated multiple markers to distinguish melanoma from dysplastic nevi [113]. This study showed BRG1 is one of the markers that can optimally aid in the clinical diagnosis of melanoma from dysplastic nevi [113]. Furthermore, SWI/SNF is shown to be predominantly associated with invasive melanomas [3].

Phosphatase and tensin homologue deleted from chromosome 10 (PTEN). PTEN is reported to be an important tumor-suppressor gene in melanoma. PTEN functions as a lipid phosphatase opposing the action of PI3K [114]. PTEN has a key role in cellular signal transduction by decreasing intracellular phosphatidylinositol that is produced by the activation of PI3K. *PI3K* activation leads to conformational change of AKT. Activated AKT, then phosphorylates serine/threonine kinase mTOR, leading to increased synthesis of target proteins that regulate cell division and apoptosis, pigmentation, and proliferation [115,116]. In melanoma, activation of AKT and/or MAPK pathways has been reported to increase degradation of inhibitor-of-κB kinase effector, with subsequent release of NF-κB, which may thus move to the nucleus and activate transcription. Additionally, AKT has the ability to suppress apoptosis through phosphorylation and inactivation of many pro-apoptotic proteins, such as Bcl-2 antagonist of cell death and MDM2, as well as activation of NF-κB [117]. Inactivation of *PTEN* gene is mainly due to epigenetic mechanisms involving DNA hypermethylation, with less than 10% of inactivation events involving somatic mutation [118]. Alteration of the BRAF–MAPK pathway is frequently associated with PTEN–AKT impairment [119]. Shain and colleagues reported loss of PTEN and TP53 in thicker invasive melanoma cases [3].

Tumor protein p53 (TP53). TP53 is inactivated through a loss-of-function mutation that affects the p53 pathway (see CDKN2A for more information).

AT-rich interaction domain (ARID1A, ARID1B, and ARID2) genes. The *ARID1A, ARID1B, ARID2* genes contain deletion mutations that affect chromatin remodeling and lead to cellular differentiation.

Serine/threonine-protein phosphatase 6 catalytic (PPP6C). The *PPP6C* gene encodes subunit the enzyme *PPP6C*. Mutation in this gene can affect the cell cycle regulation and inhibit *CCND1* (see CCND1 and Fig. 25.1 for more information).

Genes Associated with Progression Events in Melanocytic Lesions: Gain-of-Function Mutations

Cyclin-dependent kinase 4 (CDK4). CDK4 activates the G1-S transition that phosphorylates RB. (see CDKN2A for more information).

Microphthalmia-associated transcription factor (MITF). MITF is a helix-loop-helix leucine zipper protein important for melanocyte development and differentiation and is considered to be a master regulator of melanocyte biology [2]. *MITF* is activated by cAMP,

MAPK (KIT/NRAS), and CDKN2A pathways. Garraway and colleagues reported that MITF amplification is correlated with resistance to chemotherapy and decreased overall survival [120]. Conversely, reduction of MITF activity sensitizes melanoma cells to chemotherapeutic agents [120].

MITF can act as both an inducer and a repressor in cellular proliferation. High levels of MITF leads to G1 cell cycle arrest and differentiation, through induction of p16^{CDKN2A} and p21 [121,122]. Low MITF expression has been associated with apoptosis predisposition, while intermediate MITF expression levels promote cell proliferation [120–122]. Therefore, melanoma cells are thought to have acquired strategies to maintain MITF levels in the intermediate range compatible with tumorigenesis. Additionally, constitutive activation of *MEK* with *BRAF^{V600E}* in melanoma cells is associated with MITF ubiquitin-dependent degradation [123].

MEK1 and MEK2. Mutations in *MEK1* and *MEK2* result in gain of function and involve the MAPK pathway [2]. Fernandes and colleagues assessed the status of the MAPK pathways in the pathogenesis of acral melanomas by examining the components of the RAS–RAF–MEK–ERK cascades in a series of 16 primary acral melanomas by tissue microarray [124]. They demonstrated absence of RAS and presence of MEK2, ERK1, and ERK2 in every invasive case with high thickness (Fig. 25.1) [124].

Catenin beta 1 (CTNNB1). The *CTNNB1* gene encodes a protein called β-catenin. Mutation in this gene primarily affects the WNT signaling pathway [2]. Glycogen synthase kinase-3 (GSK-3) is a widely expressed serine/threonine protein kinase. *CTNNB1* mutation leads to stabilization of the β-catenin protein and an increased transcription [125]. Wnt/β-catenin pathway is crucial in both embryonic development and adult homeostasis. Aberrant Wnt/β-catenin pathway may lead to developmental malformations and is associated with many various malignancies including melanoma. For this reason much effort has been made to specifically target the Wnt/β-catenin pathway with anticancer drugs [126].

Cyclin D1 (CCND1). Gain-of-function mutation affects the RB pathway. In 1996, Maelandsmo and colleagues reported an over-expression of *CCND1*, suggesting that functional inactivation of *pRB* through this pathway is involved in the development or progression of sporadic human melanomas [127].

Protein phosphatase 6, catalytic subunit (PPP6C). See CDKN2A and Fig. 25.1 for more information.

Enhancer of zeste 2 polycomb repressive complex 2 subunit (EZH2). EZH2 is affected by a gain-of-function mutation that is involved in chromatin remodeling leading to cellular differentiation (Fig. 25.1).

Ras-related C3 botulinum toxin substrate 1 (RAC1). RAC1 is subject to point mutation. RAC1 is involved in cell adhesion, migration, invasion, and motility. It is related to and activated by NRAS (Fig. 25.1).

Telomerase reverse transcriptase (TERT). TERT promoter mutations and amplification have been shown to affect telomerase elongation [128] and are also associated with fast growing melanoma [129]. TERT promoter mutations are reported to be the earliest secondary molecular alterations during melanomagenesis, emerging in intermediate lesions and melanoma in situ [3]. Griewank and colleagues studied 410 melanoma cases and reported that UV-induced TERT promoter mutations are one of the most frequent genetic alterations [128]. Furthermore this study demonstrated that in non-acral cutaneous melanomas, the presence of TERT promoter mutations is independently associated with poor prognosis [128]. Shain and colleagues demonstrated that melanoma precursors that acquire TERT mutations also acquire subsequent mutations, with progression toward melanoma [3].

Cyclin-dependent kinase 4 (CDK4). CDK4 is subject to gain-of-function amplification mutation leading to influence *RB* (see CDKN2A for more information).

Microenvironment Changes and Melanoma

Hypoxia has been shown to be prerogative of advanced neoplasia and a cradle for melanoma development and progression [130]. Notch1 has been demonstrated to be a key effector of both AKT and hypoxia in melanoma [131]. Furthermore, physiologic tissue hypoxia together with activity of HIF1α and stimulation of KIT can function as a promoting factor in melanogenesis (Fig. 25.1) [132]. Notch signaling pathway has been shown to be a potential therapeutic target in melanoma treatment [131].

MOLECULAR TESTING IN MELANOMA

The current gold standard for melanoma diagnosis is histopathology. The majority of melanocytic lesions can be diagnosed based on Hematoxylin and Eosin (H&E). However, when faced with a challenging lesion or when there is discrepancy in a small percentage of lesions, there are ancillary studies available that may help in making a definitive diagnosis.

Immunohistochemistry

Melanomas mimic the histologic features of other malignancies such as lymphomas, poorly differentiated carcinomas, and sarcomas [133]. IHC is a useful tool to differentiate melanomas from other tumors that they mimic. For a more comprehensive detailed review of IHC markers utilized in the diagnosis of melanoma, see Ref. [134].

S100. In cutaneous melanomas (with the exception of the S100-negative melanoma), S100 is present in both nucleus and cytoplasm (Fig. 25.2) with a sensitivity of 97−100% [135]. However, the specificity of S100 is low (75−87%) as it can be expressed in other cells such as nerve sheath cells, myoepithelial cells, Langerhans cells, and dermal dendrocytes [134,136]. Also considering that S100 can be expressed in spindle cells in dermal scars, it could be a diagnostic pitfall, especially when evaluating desmoplastic melanomas [134]. Therefore, other specific markers should be used in addition to S100 to differentiate melanomas from other malignancies.

HMB45. HMB45 is a cytoplasmic marker (Fig. 25.2) and compared to S100, it is more specific than sensitive. The reported sensitivity for HMB45 for cutaneous melanoma is 69−93% [134]. Immunopositivity is the highest in primary melanoma. However, it has reduced expression in metastatic melanomas. This decreased sensitivity in cases of metastatic melanoma necessitates utilization of other markers when there is doubt in the diagnosis [134]. Uguen and colleagues demonstrated that HMB45 is helpful in distinguishing melanomas versus benign nevi [137]. Benign nevi show positive HMB45 staining in the superficial portion, but negative in the deep portion. The HMB45 gradient is considered to be positive when the most superficial cells are HMB45 stained. It is considered negative when the staining involves equally the superficial and deep parts of the tumor. Absence of HMB45 gradient has been described to be more frequent in melanomas [137].

MART-1 and Melan-A. Melanoma antigen recognized by T cell-1 (MART-1) and Melan-A (Fig. 25.2) is cytoplasmic protein of melanosomal differentiation recognized by T cells [135]. Two clones of the antibody to this cytoplasmic protein are M2-7C10 (also referred to as MART-1) and A103 (also referred to as Melan-A) [138]. The sensitivity and specificity of MART-1/MelanA in cutaneous melanomas are 75−92% and 95−100%, respectively [134]. Similar to HMB45, there is decreased sensitivity in metastatic melanoma compared to the primary lesions [139]. Conversely, compared to HMB45, it has a more diffuse and more intense staining, making it a better marker in distinguishing metastatic melanomas [134,138]. MART1/Melan-A is also helpful in distinguishing in situ and invasive melanoma and measuring thickness [140]. Another application of MART1/Melan-A is differentiating neurotized melanocytic nevi, which are immunopositive, from neurofibroma (immunonegative) [141].

SOX-10. SRY-related HMG-box 10 (SOX10) is a crucial transcription factor in the specification of the

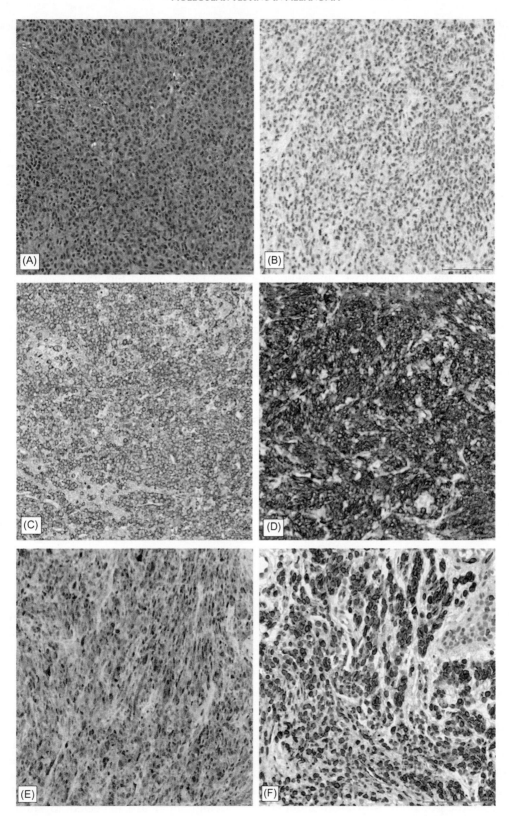

FIGURE 25.2 Immunostaining with various IHC stains in one melanoma case. (A) H&E of metastatic melanoma, (B) SOX10, (C) HMB45, (D) S100, (E) MITF, and (F) Melan-A.

neural crest and maintenance of Schwann cells and melanocytes [142]. It is expressed in the nuclei of melanocytes (Fig. 25.2) and breast myoepithelial cells. SOX10 has been shown to be a sensitive and specific marker of MM of multiple histologic types. SOX10 shows an increased specificity for soft tissue tumors of neural crest origin compared with S100 [143]. It is proposed in the literature that it could supplement or potentially replace the traditionally utilized immunohistochemical stains such as S100.

Tyrosinase. Tyrosinase (also known as T311) is a protein that hydroxylates tyrosine as the first step in melanin synthesis [134,135] and is important for melanosome formation. This antibody shows strong and diffuse cytoplasmic stain. Furthermore, in cutaneous melanomas, there is diffuse reactivity compared to diminish expression toward the base in nevi or melanomas with paradoxical maturation. Additionally, it may be useful in evaluating melanomas with extensive necrosis [144]. The specificity of tyrosinase for melanoma is 97−100% [134,145]. Its sensitivity decreases in metastatic lesions or with increased clinical stage (79−93%) [134,139,146].

MITF. MITF (also known as *MI, WS2, CMM8, WS2A, bHLHe32*) encodes a transcription factor on chromosome 3p14.1, which is important for development and survival of melanocytes [147]. This nuclear marker is used in diagnosis of melanoma and other melanogenic tumors (Fig. 25.2). The sensitivity and specificity of this immunostain have been reported as 81−100% and about 88%, respectively [134,146,147]. The specificity of MITF is even lower in spindle cell neoplasms [134]. Additionally, this stain has been reported to be reliably positive in S100-negative melanomas [148]. One of the important limitations of MART-1/Melan-A is that it can overestimate the number of melanocytes in actinically damaged skin. In such cases, MITF, showing nuclear staining pattern, is more accurate for evaluation of intraepidermal melanocytes. Therefore, MITF is useful in assisting diagnosis in such cases [149].

Ki67. Ki67 has utility in determination of proliferation index and has been reported to be useful in distinguishing benign nevi from MM. Ki67 index is reported to be less than 5% in benign nevi versus 13−30% in melanomas [134,150,151]. Additionally, Ki67 immunopositivity is increased in Spitz neoplasms and atypical nevi [152−156]. Ki67 staining index in melanoma is correlated with tumor mitotic figures, depth of invasion, tumorigenic vertical growth phase, vascular invasion, and metastatic potential. Ki67 as an independent prognostic indicator in overall survival has shown variable significance [157−160].

BAP1. Inactivating somatic mutations in *BAP1*, the gene encoding BRCA-associated protein 1 in the predominantly metastasizing (class 2) uveal melanoma

was reported by Koopmans and colleagues [161,162]. BAP1 is a nuclear immunostain and should be interpreted in the context of the H&E features. For example, Spitzoid neoplasms and a subset of combined nevi with a population of large epithelioid cells have been reported to demonstrate loss of BAP1 immunostaining [163]. Additionally, when BAP1 mutation is confirmed in a tumor, the possibility of a BAP1 germline mutation should be raised and addressed with potential genetic counseling and further testing of the family [105]. This is especially important as loss of BAP1 may be involved in the progression of uveal melanoma to an aggressive, metastatic phenotype [162].

p16. p16 is a tumor-suppressor gene that is located in 9p21 that leads to inhibiting of cyclin-dependent kinases in phosphorylating the retinoblastoma protein (see CDKN2A and Fig. 25.1). Gerami and colleagues in 2013 showed that homozygous loss of 9p21 in atypical Spitzoid melanocytic neoplasms is highly associated with clinically aggressive behavior and death [164]. Loss of nuclear expression of p16 by IHC method has been shown to be associated with poor outcome in melanoma patients [165−167].

pHH3. Phosphohistone H3 (pHH3) is a histone protein that is closely related to the mitotic process and IHC with antibody to PHH3 is used for labeling mitotic figures in all phases of mitosis, including early prophase [168−170]. The mitotic rate is an important prognostic criterion in patients with thin melanoma not more than 1 mm and is part of the American Joint Committee on Cancer guidelines. IHC with antibody to pHH3 is a useful tool for pathologists and dermatopathologists to confirm mitotic figures [171].

Array Comparative Genomic Hybridization

Comparative genomic hybridization (CGH) is a molecular testing method for detecting copy number changes throughout the genome [172−174]. DNA is isolated from tissue samples and is labeled with fluorochromes before being hybridized to a microarray of mapped genomic DNA clones or probes. Hybridization signals are then digitized and analyzed using software that generates a virtual karyotype highlighting regions of copy number changes. One reported limitation of this technology is that if the copy number is present only at low quantities in a tumor, they can potentially escape detection. This can limit the capability of detecting tumor cell heterogeneity in a subset of tumors [175].

Benign nevi are expected to show minimal to no copy number variations, whereas melanomas are expected to have numerous copy number variations [3,172,176−178]. For example, melanomas have been shown to have chromosomal gains (1q, 6p, 7, 8q, 17q, and 20q) and losses (6q, 8p, 9p, and 10q). When dealing with a challenging melanocytic lesion histologically or where there is ambiguity in diagnosis, application of array CGH (aCGH) may be helpful in supporting the diagnosis of melanoma (when there is copy number changes) versus a benign lesion (when there is a lack of copy number aberrations).

While it is expected to see copy number changes in melanoma, occasionally no abnormality is detected by aCGH in a histologically obvious MM. This finding may be related to test sensitivity or intrinsic reasons [80]. Additional limitations of aCGH include the quantity of DNA needed, a lower limit of resolution of 400 kb throughout the genome [179]. Therefore, if the quality or the amount of extracted DNA is inadequate or if the melanocytes harboring aberrations are not present in the sampled tissue, CGH results may be falsely negative. Additionally, if a very large sample is dissected that includes other cells such as lymphocytes, stromal cells, and nevomelanocytes, the isolated DNA can be too dilute to reliably and accurately show copy number aberration in the abnormal melanocytes. Rarely a uniform CGH-negative MM has been observed [80].

Another limitation of aCGH is differentiating heterozygous versus homozygous deletions. For example, a single copy loss of 9p21 can be seen in both melanomas and atypical Spitz nevi. However, a homozygous loss of 9p21, which encodes the *CDKN2A* gene, is a specific finding in melanoma and is associated with aggressive behavior in melanocytic lesions [164,180]. The presence of a single copy number aberration does not prove malignancy. The homozygous versus heterozygous loss of 9p21 is not distinguishable using the aCGH technology. Therefore, homozygosity versus heterozygosity of 9p21 should be further studied with FISH ancillary study, when clinically indicated.

Spitz nevi have been shown to harbor a variety of chromosome losses [78,177,181,182]. Therefore, each case should be carefully examined taking into account the age, clinical setting, specific copy number aberration, light microscopic features and morphology, and immunohistochemical staining patterns [80].

Single Nucleotide Polymorphism Array

Genome positions at which there are two distinct nucleotide residues that each appears in a significant portion of the human population are termed single nucleotide polymorphisms (SNPs). They comprise a major part of the DNA variants. There are approximately 10 million SNPs in the human genome [183]. Manufacturers typically arbitrarily assign the two alleles of an SNP as A and B. As each individual inherits one copy of each SNP position from each parent, the individual's genotype at an SNP site is

therefore *AA*, *AB*, or *BB*. The Oncoscan SNP array utilizes over 220,000 probes across all chromosomes, compared to the FISH technology with only a few specific probes. Additionally, it provides information of copy number changes and losses of heterozygosity (LOH). LOH in melanocytic lesions occurs when one allele is deleted and the normally functioning allele is lost. Using SNP-array technology can help in identifying patterns of allelic imbalance in melanocytic lesions and potentially provide prognostic and diagnostic value.

Literature regarding the affected genes in melanoma is developing at a rapid rate. SNP array is especially useful when the melanocytic lesions appear questionable and ambiguous. To further delineate the behavior of these lesions, SNP array can show if there are copy number changes or copy-neutral LOH. In benign nevi, minimal to no changes in copy number are expected. Conversely in melanoma, it is expected to see copy number changes and or copy-neutral aberrations [3,172,176−178]. Fig. 25.3 shows a benign melanocytic nevus with features of congenital onset, with a corresponding SNP array schematic showing no chromosomal aberrations. Conversely, Fig. 25.4 shows an MM with multiple chromosomal copy number changes and LOH.

Fluorescence In Situ Hybridization

Gerami and colleagues introduced the FISH technology in the context of diagnostic melanocytic lesions [184]. In their study they determined cut-off values which permitted a test sensitivity of 85% and specificity of 95% for unambiguous benign versus malignant lesions using various FISH probes. FISH targets a few individual chromosomes and specific regions of the chromosome in contrast with SNP array that utilizes over 220,000 probes across all chromosomes. Specific fluorescence-labeled oligonucleotide probes bind to their complementary DNA sequence resulting in labeling of that region which can be visualized under a fluorescence microscope.

There are six probes available that are designed specifically for melanocytic lesions. FISH testing in the context of melanoma uses the following probes [178,185,186]:

1. Ras-responsive element-binding protein 1 (RREB1)
2. Myeloblastosis (MYB)
3. Cyclin D1 or chromosome 11q13 (CCND1) [203].
4. Centromeric enumeration probe control for chromosome 6 (CEP6)
5. 9p21 (CDKN2A) (useful in diagnosing conventional and Spitzoid melanomas [181,187])
6. 8q24 (useful in diagnosing nodular amelanotic and nevoid melanomas [188,203])

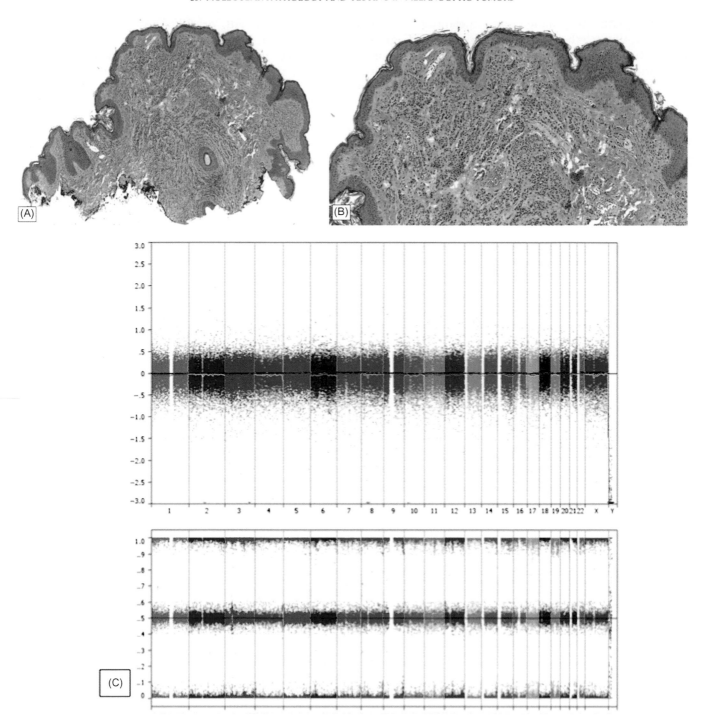

FIGURE 25.3 Benign nevus with some congenital features. (A) H&E, 20×; (B) H&E, 100×; and (C) SNP array. SNP array shows a normal diploid SNP array with no copy number changes or copy-neutral LOH.

FISH interpretation requires analysis of 30 lesional melanocytes for each case. Gerami and colleagues in 2009 proposed that a positive lesion needs one of these criteria [178]:

1. Gain in 6p25 (RREB1) relative to CEP6 greater than 55%
2. Gain in 6p25 (RREB1) greater than 29%

3. Loss in 6q23 (MYB) relative to CEP6 greater than 42%
4. Gain in 11q13 (CCND1) greater than 38%

A negative FISH test result does not automatically exclude the diagnosis of melanoma. Melanoma can have other copy number changes that are not specifically targeted by the few FISH probes [187]. This makes the sensitivity of this testing method as a

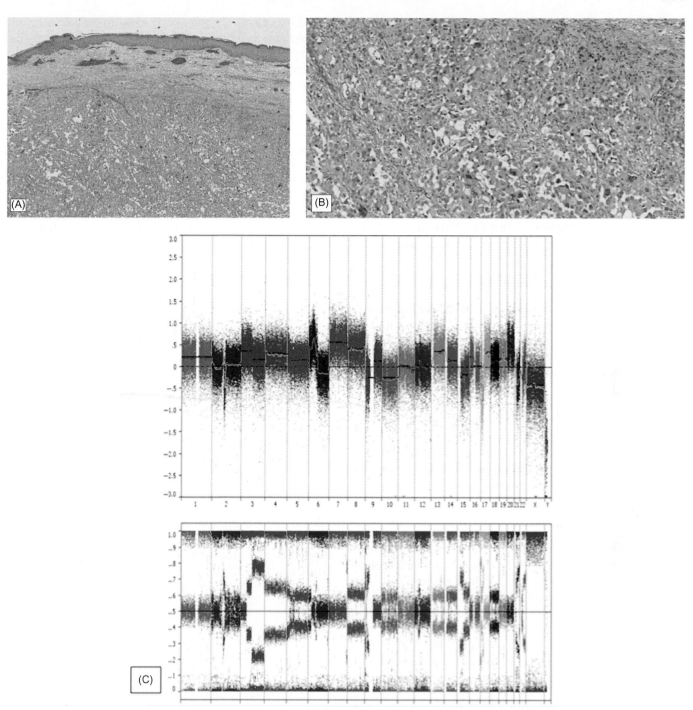

FIGURE 25.4 Metastatic melanoma. (A) H&E, 40×; (B) H&E, 200×; and (C) SNP array. SNP array shows multiple copy number gains, losses, and multiple copy-neutral LOH.

first-line diagnostic tool inferior to other molecular testing methods such as SNP array and aCGH. Furthermore, not all melanomas harbor copy number changes. Hence, this could be another reason for a FISH-negative melanoma and should be pondered in the context of morphology. One of the pitfalls of FISH-positive lesions is polyploidy. For example, some Spitz nevi have been reported to be tetraploid [188,189].

Hence, FISH results should be interpreted accordingly in order to correct for this. Another possible interpretive error of false-positive FISH is choosing only the large nuclei at different foci in a lesion instead of counting all nuclei in a given area [80].

Clear cell sarcoma (CCS), formerly known as MM of soft parts, is a rare malignant soft tissue tumor that resembles cutaneous MM morphologically [190]. CSS cells can produce melanin and have been reported to extend into the subcutis and dermis [191]. However, a reciprocal translocation t(12;22)(q13;q12) is observed in more than 90% of CCS cases [192]. This translocation results in fusion of activating transcription factor gene (*ATF1*) on the 12q13 and the Ewing's sarcoma oncogene R1 (*EWSR1*) on 22q12 [192,193]. It is important to distinguish these two entities, as the treatment options are different. MM and CCS can be distinguished via FISH to detect *EWS* gene rearrangement [194]. In addition, diagnosis of MM can be made in the presence of *BRAF* mutation [195]. Therefore, molecular diagnosis such as chromosome analysis, reverse transcriptase polymerase chain reaction, and FISH are key to differentiate it from cutaneous or mucosal melanoma.

Targeted Next-Generation Sequencing

Next-generation sequencing (NGS) or massively parallel sequencing has become a critical tool in assessing numerous genes for somatic mutations and for the potential selection of targeted therapy [196,197]. Recent discoveries of novel mutations in melanoma and their impact on response to therapy have made somatic mutation analysis a significant part of the routine workup for melanoma. NGS can sequence thousands to millions of DNA fragments in parallel, in contrast to the traditional Sanger-based sequencing. This results in substantially more throughputs and the capability to sequence numerous genes simultaneously from multiple patient samples. Details for performing NGS can be found from our validation study of a 50-gene cancer hotspot panel [198].

Molecular testing is routinely performed for patients with advanced melanoma. In 2015, Siroy and colleagues reported results of clinical testing of 699 advanced melanoma patients using a pan-cancer NGS panel of hotspot regions in 46 genes [199]. The most common mutations were reported to be $BRAF^{V600E}$ (36%), *NRAS* (21%), *TP53* (16%), $BRAF^{Non-V600}$ (6%), and *KIT* (4%). Additionally this study showed that $BRAF^{V600E}$ and *KIT* mutations were significantly associated with melanoma subtypes, whereas $BRAF^{V600E}$ and *TP53* mutations were significantly associated with cutaneous primary tumor location [199].

NGS technology has helped with identifying key genetic mutations for targeted therapy. Historically, the goal of the treatment has been to reduce tumor burden together with palliative care, with little hope for prolonged survival. Systemic treatment of patients with advanced melanoma has been focused on cytostatic chemotherapy, such as dacarbazine or other alkylating agents such as temozolomide, fotemustine, or taxanes [200]. Chemotherapy has remained the standard of care for advanced melanoma, but it has not altered survival [200–202]. Therefore, a major consideration for physicians is to integrate chemotherapy agents into clinical practice. Immune-based therapies, such as IFN-α and IL-2 have not yielded significant impact on the overall survival [201].

New research has led to the development of new treatments with different mechanisms of action such as ipilimumab, nivolumab, and pembrolizumab (immunotherapies) and vemurafenib, dabrafenib, and trametinib (targeted therapies). Details on treatment options and drug method of action is beyond the scope of this text and more information regarding the most recent therapeutic agents can be found from the Michielin and Hoeller update on new options and opportunities for the treatment of advanced melanoma or other texts focused on melanoma treatment [202].

CONCLUSIONS

Pathogenesis of melanoma and histogenesis of melanoma has been a hot topic in molecular pathology. Accumulation of genetic and acquired alterations may lead to unrestricted cell proliferation and melanomagenesis. Several key signaling molecules and pathways have been associated with complex disease process such as PTEN/AKT, CDKN2A, and MAPK. Other signaling pathways and melanomas have been proposed that are triggered by environmental factors and UV radiation. In other words, the number of mutations and amount of UV exposure is related to the progression of melanomas. Giant leaps in molecular pathology have led to better diagnosis and treatment of melanocytic lesions. Much work is ahead in discovering and bridging the gaps in knowledge.

References

[1] Noonan FP, Zaidi MR, Wolnicka-Glubisz A, et al. Melanoma induction by ultraviolet A but not ultraviolet B radiation requires melanin pigment. Nat Commun 2012;3:884.

[2] Bastian BC. The molecular pathology of melanoma: an integrated taxonomy of melanocytic neoplasia. Annu Rev Pathol Mech Dis 2014;9:239–71.

[3] Shain AH, Yeh I, Kovalyshyn I, et al. The genetic evolution of melanoma from precursor lesions. N Engl J Med 2015;373: 1926–36.

[4] Gandhi SA, Kampp J. Skin cancer epidemiology, detection, and management. Med Clin North Am 2015;99:1323–35.

[5] Cornejo KM, Deng AC. Malignant melanoma within squamous cell carcinoma and basal cell carcinoma. Am J Dermatopathol 2013;35:226–34.

[6] Kraemer KH. The role of sunlight and DNA repair in melanoma and nonmelanoma skin cancer. Arch Dermatol 1994;130:1018.

[7] Hurst EA. Ocular melanoma. Arch Dermatol 2003;139:1067.

[8] Stern JB, Peck GL, Haupt HM, Hollingsworth HC, Beckerman T. Malignant melanoma in xeroderma pigmentosum: search for a precursor lesion. J Am Acad Dermatol 1993;28:591—4.

[9] Binesh F, Akhavan A, Navabii H. Nevoid malignant melanoma in an albino woman. Case Rep 2010;2010 bcr0820103262.

[10] George AO, Ogunbiyi AO, Daramola OO, Campbell OB. Albinism among Nigerians with malignant melanoma. Trop Doct 2005;35:55—6.

[11] Young TE. Malignant melanoma in an albino; report of a case. Am Med Assoc Arch Pathol 1957;64:186—91.

[12] Kennedy BJ, Zelickson AS. Melanoma in an albino. JAMA 1963;186:839—41.

[13] Bataille V, Hiles R, Bishop JAN. Retinoblastoma, melanoma and the atypical mole syndrome. Br J Dermatol 1995;132:134—8.

[14] Albert LS, Sober AJ, Rhodes AR. Cutaneous melanoma and bilateral retinoblastoma. J Am Acad Dermatol 1990;23:1001—4.

[15] Eng C, Li FP, Abramson DH, et al. Mortality from second tumors among long-term survivors of retinoblastoma. J Natl Cancer Inst 1993;85:1121—8.

[16] Kleinerman RA. Risk of new cancers after radiotherapy in long-term survivors of retinoblastoma: an extended follow-up. J Clin Oncol 2005;23:2272—9.

[17] Curiel-Lewandrowski C, Speetzen LS, Cranmer L, Warneke JA, Loescher LJ. Multiple primary cutaneous melanomas in Li—Fraumeni Syndrome. Arch Dermatol 2011;147:248.

[18] Schenk M, Severson RK, Pawlish KS. The risk of subsequent primary carcinoma of the pancreas in patients with cutaneous malignant melanoma. Cancer 1998;82:1672—6.

[19] Parker JF. Pancreatic carcinoma surveillance in patients with familial melanoma. Arch Dermatol 2003;139:1019.

[20] Rutter JL, Bromley CM, Goldstein AM, et al. Heterogeneity of risk for melanoma and pancreatic and digestive malignancies. Cancer 2004;101:2809—16.

[21] Dbouk MB, Nafissi S, Ghorbani A. Chronic inflammatory demyelinating polyneuropathy following malignant melanoma. Neurosciences 2012;17:167—70.

[22] Cheung M, Talarchek J, Schindeler K, et al. Further evidence for germline BAP1 mutations predisposing to melanoma and malignant mesothelioma. Cancer Genet 2013;206:206—10.

[23] Nielsen K, Ingvar C, Masback A, et al. Melanoma and nonmelanoma skin cancer in patients with multiple tumours-evidence for new syndromes in a population-based study. Br J Dermatol 2004;150:531—6.

[24] Matin RN, Szlosarek P, McGregor JM, Cerio R, Harwood CA. Synchronous melanoma and renal carcinoma: a clinicopathological study of five cases. Clin Exp Dermatol 2012;38:47—9.

[25] Dhandha M, Chu MB, Richart JM. Coexistent metastatic melanoma of the kidney with unknown primary and renal cell carcinoma. Case Rep 2012;2012 bcr2012007286

[26] Cheng Y-P, Chiu H-Y, Hsiao T-L, et al. Scalp melanoma in a woman with LEOPARD syndrome: possible implication of PTPN11 signaling in melanoma pathogenesis. J Am Acad Dermatol 2013;69:e186—7.

[27] Mota-Burgos A, Acosta EH, Márquez FV, Mendiola M, Herrera-Ceballos E. Birt-Hogg-Dubé syndrome in a patient with melanoma and a novel mutation in the FCLN gene. Int J Dermatol 2013;52:323—6.

[28] Fontcuberta IC, Salomão DR, Quiram PA, Pulido JS. Choroidal melanoma and lid fibrofoliculomas in Birt-Hogg-Dubé Syndrome. Ophthalmic Genet 2011;32:143—6.

[29] Cocciolone RA, Crotty KA, Andrews L, Haass NK, Moloney FJ. Multiple desmoplastic melanomas in Birt-Hogg-Dubé Syndrome and a proposed signaling link between folliculin, the mTOR pathway, and melanoma susceptibility. Arch Dermatol 2010;146.

[30] Mohr MR, Erdag G, Shada AL, et al. Two patients with Hailey-Hailey Disease, multiple primary melanomas, and other cancers. Arch Dermatol 2011;147:211.

[31] Gregoriou S, Kazakos C, Belyaeva H, et al. Hypomelanotic nail melanoma in a patient with Gorlin Syndrome. J Cutan Med Surg 2012;16:143—4.

[32] Greene SL, Thomas 3rd JR, Doyle JA. Cowden's disease with associated malignant melanoma. Int J Dermatol 1984;23:466—7.

[33] Duve S, Rakoski J. Cutaneous melanoma in a patient with neurofibromatosis: a case report and review of the literature. Br J Dermatol 1994;131:290—4.

[34] Stokkel MPM, Kroon BBR, Van Der Sande JJ, Neering H. Malignant cutaneous melanoma associated with neurofibromatosis in two sisters from a family with familial atypical multiple mole melanoma syndrome: case reports and review of the literature. Cancer 1993;72:2370—5.

[35] Buckley C, Thomas V, Crow J, et al. Cancer family syndrome associated with multiple malignant melanomas and a malignant fibrous histiocytoma. Br J Dermatol 1992;126:83—5.

[36] Rajaratnam R, Marsden JR, Marzouk J, Hero I. Pulmonary carcinoid associated with melanoma: two cases and a review of the literature. Br J Dermatol 2007;156:738—41.

[37] Berking C, Brady MS. Cutaneous melanoma in patients with sarcoma. Cancer 1997;79:843—8.

[38] Gupta R, Kathiah R, Prakash G, Venkatasubramaniam B. Metachronous pleomorphic liposarcoma and melanoma: a rare case report. Indian J Pathol Microbiol 2011;54:196.

[39] De Giorgi V, Santi R, Grazzini M, et al. Synchronous angiosarcoma, melanoma and morphea of the breast skin 14 years after radiotherapy for mammary carcinoma. Acta Derm Venereol 2010;90:283—6.

[40] Matin RN, Gonzalez D, Thompson L, et al. KIT and BRAF mutational status in a patient with a synchronous lentigo maligna melanoma and a gastrointestinal stromal tumor. Am J Clin Dermatol 2012;13:64—5.

[41] Gan BS, Colcleugh RG, Scilley CG, Craig ID. Melanoma arising in a chronic (Marjolin's) ulcer. J Am Acad Dermatol 1995;32:1058—9.

[42] Ford D, Bliss JM, Swerdlow AJ, et al. Risk of cutaneous melanoma associated with a family history of the disease. Int J Cancer 1995;62:377—81.

[43] Forman SB, Vidmar DA, Ferringer TC. Collision tumor composed of Merkel cell carcinoma and lentigo maligna melanoma. J Cutan Pathol 2008;35:203—6.

[44] Evans AV, Scarisbrick JJ, Child FJ, et al. Cutaneous malignant melanoma in association with mycosis fungoides. J Am Acad Dermatol 2004;50:701—5.

[45] Flindt-Hansen H, Brandrup F. Malignant melanoma associated with mycosis fungoides. Dermatology 1984;169:167—8.

[46] Koeppel MC, Grego F, Andrac L, Berbis P. Primary cutaneous large B-cell lymphoma of the legs and malignant melanoma: coincidence or association? Br J Dermatol 1998;139:751—2.

[47] McKenna DB, Doherty VR, McLaren KM, Hunter JAA. Malignant melanoma and lymphoproliferative malignancy: is there a shared aetiology? Br J Dermatol 2000;143:171—3.

[48] Goggins WB, Finkelstein DM, Tsao H. Evidence for an association between cutaneous melanoma and non-Hodgkin lymphoma. Cancer 2001;91:874—80.

[49] Tsao H, Kwitkiwski K, Sober AJ. A single-institution case series of patients with cutaneous melanoma and non-Hodgkin's lymphoma. J Am Acad Dermatol 2002;46:55—61.

[50] Wu Y-H, Kim GH, Wagner JD, Hood AF, Chuang T-Y. The association between malignant melanoma and noncutaneous malignancies. Int J Dermatol 2006;45:529—34.

[51] Dueber JC, Coffin CM. Collision of chronic lymphocytic leukemia/small lymphocytic lymphoma and melanoma. Blood 2013;121:4819.

[52] Vlaskamp M, de Wolff-Rouendaal D, Jansen PM, Luyten GPM. Concomitant choroidal melanoma and non-Hodgkin lymphoma in two adult patients: case report. Case Rep Ophthalmol 2012;3:209–13.

[53] Cantor AS, Moschos S, Jukic DM. A principal case of multiple lymphoid collision tumors involving both B-cell chronic lymphocytic leukemia and metastatic malignant melanoma. Dermatol Online J 2010;16:6.

[54] Addada J, Anoop P, Swansbury JG, et al. Synchronous mantle cell lymphoma, chronic lymphocytic leukaemia and melanoma in a single lymph node. Acta Haematol 2010;123:194–6.

[55] Shahani L. Castleman's disease in a patient with melanoma: the role of VEGF. Case Rep 2012;2012 bcr0720114519

[56] Todd P, Garioch J, Seywright M, Rademaker M, Thomson J. Malignant melanoma and systemic mastocytosis—a possible association? Clin Exp Dermatol 1991;16:455–7.

[57] Tanita M. Malignant melanoma arising from cutaneous bronchogenic cyst of the scapular area*1. J Am Acad Dermatol 2002;46:S19–21.

[58] Hyun HS, Mun ST. Primary malignant melanoma arising in a cystic teratoma. Obstet Gynecol Sci 2013;56:201.

[59] Bajoghli A, Agarwal S, Goldberg L, Mirzabeigi M. Melanoma arising from an epidermal inclusion cyst. J Am Acad Dermatol 2013;68:e6–7.

[60] Shields CL. Phacomatosis pigmentovascularis of cesioflammea type in 7 patients. Arch Ophthalmol 2011;129:746.

[61] Zvulunov A, Wyatt DT, Laud PW, Esterly NB. Influence of genetic and environmental factors on melanocytic nevi: a lesson from Turner's syndrome. Br J Dermatol 1998;138:993–7.

[62] Gibbs P, Brady BM, Gonzalez R, Robinson WA. Nevi and melanoma: lessons from Turner's Syndrome. Dermatology 2001;202:1–3.

[63] Brandt JS, Fishman S, Magro CM. Cutaneous melanoma arising from a cesarean delivery skin scar. J Perinatol 2012;32:807–9.

[64] Kircik L, Armus S, Broek H. Malignant melanoma in a tattoo. Int J Dermatol 1993;32:297–8.

[65] Singh RS, Hafeez Diwan A, Prieto VG. Potential diagnostic pitfalls in melanoma arising in a cutaneous tattoo. Histopathology 2007;51:283–5.

[66] Nolan KA, Kling M, Birge M, et al. Melanoma arising in a tattoo: case report and review of the literature. Cutis 2013;92:227–30.

[67] Varga E, Korom I, Varga J, et al. Melanoma and melanocytic nevi in decorative tattoos: three case reports. J Cutan Pathol 2011;38:994–8.

[68] Mohrle M, Hafner HM. Is subungual melanoma related to trauma? Dermatology 2002;204:259–61.

[69] Rangwala S, Hunt C, Modi G, Krishnan B, Orengo I. Amelanotic subungual melanoma after trauma: an unusual clinical presentation. Dermatol Online J 2011;17:8.

[70] Hassanein AM, Mrstik ME, Hardt NS, Morgan LA, Wilkinson EJ. Malignant melanoma associated with lichen sclerosus in the vulva of a 10-year-old. Pediatr Dermatol 2004;21:473–6.

[71] Rosamilia LL, Schwartz JL, Lowe L, et al. Vulvar melanoma in a 10-year-old girl in association with lichen sclerosus. J Am Acad Dermatol 2006;54:S52–3.

[72] Hocker TL, Fox MC, Kozlow JH, et al. Malignant melanoma arising in the setting of epidermolysis bullosa simplex. JAMA Dermatol 2013;149:1195.

[73] Takamiyagi A, Asato T, Nakashima Y, Nonaka S. Association of human papillomavirus type 16 with malignant melanoma. Am J Dermatopathol 1998;20:69–73.

[74] Rohwedder A, Philips B, Malfetano J, Kredentser D, Carlson JA. Vulvar malignant melanoma associated with human papillomavirus DNA. Am J Dermatopathol 2002;24:230–40.

[75] van Ginkel CJW, Lim Sang RT, Blaauwgeers JLG, et al. Multiple primary malignant melanomas in an HIV-positive man. J Am Acad Dermatol 1991;24:284–5.

[76] Tindall B, Finlayson R, Mutimer K, et al. Malignant melanoma associated with human immunodeficiency virus infection in three homosexual men. J Am Acad Dermatol 1989; 20:587–91.

[77] Gandini S, Sera F, Cattaruzza MS, et al. Meta-analysis of risk factors for cutaneous melanoma: III. Family history, actinic damage and phenotypic factors. Eur J Cancer 2005;41:2040–59.

[78] Wiesner T, He J, Yelensky R, et al. Kinase fusions are frequent in Spitz tumours and spitzoid melanomas. Nat Commun 2014;5:3116.

[79] Curtin JA, Busam K, Pinkel D, Bastian BC. Somatic activation of KIT in distinct subtypes of melanoma. J Clin Oncol 2006;24:4340–6.

[80] Gerami P, Busam KJ. Cytogenetic and mutational analyses of melanocytic tumors. Dermatol Clin 2012;30:555–66.

[81] Van Raamsdonk CD, Bezrookove V, Green G, et al. Frequent somatic mutations of GNAQ in uveal melanoma and blue naevi. Nature 2008;457:599–602.

[82] Van Raamsdonk CD, Griewank KG, Crosby MB, et al. Mutations in GNA11 in uveal melanoma. N Engl J Med 2010;363:2191–9.

[83] Banerji U, Affolter A, Judson I, Marais R, Workman P. BRAF and NRAS mutations in melanoma: potential relationships to clinical response to HSP90 inhibitors. Mol Cancer Ther 2008;7:737–9.

[84] Millington GW. Mutations of the BRAF gene in human cancer, by Davies et al. (Nature 2002; 417: 949-54). Clin Exp Dermatol 2013;38:222–3.

[85] Chapman PB, Hauschild A, Robert C, et al. Improved survival with vemurafenib in melanoma with BRAF V600E mutation. N Engl J Med 2011;364:2507–16.

[86] Dhomen N, Marais R. BRAF signaling and targeted therapies in melanoma. Hematol Oncol Clin North Am 2009;23:529–45.

[87] Wellbrock C, Hurlstone A. BRAF as therapeutic target in melanoma. Biochem Pharmacol 2010;80:561–7.

[88] Lee JH, Choi JW, Kim YS. Frequencies of BRAF and NRAS mutations are different in histological types and sites of origin of cutaneous melanoma: a meta-analysis. Br J Dermatol 2011;164:776–84.

[89] Wan PT, Garnett MJ, Roe SM, et al. Mechanism of activation of the RAF-ERK signaling pathway by oncogenic mutations of B-RAF. Cell 2004;116:855–67.

[90] Pollock PM, Harper UL, Hansen KS, et al. High frequency of BRAF mutations in nevi. Nat Genet 2003;33:19–20.

[91] Sosman JA, Kim KB, Schuchter L, et al. Survival in BRAF V600-mutant advanced melanoma treated with vemurafenib. N Engl J Med 2012;366:707–14.

[92] Hauschild A, Grob JJ, Demidov LV, et al. Dabrafenib in BRAF-mutated metastatic melanoma: a multicentre, open-label, phase 3 randomised controlled trial. Lancet 2012;380:358–65.

[93] Flaherty KT, Robert C, Hersey P, et al. Improved survival with MEK inhibition in BRAF-mutated melanoma. N Engl J Med 2012;367:107–14.

[94] Jakob JA, Bassett Jr. RL, Ng CS, et al. NRAS mutation status is an independent prognostic factor in metastatic melanoma. Cancer 2012;118:4014–23.

[95] Bauer J, Curtin JA, Pinkel D, Bastian BC. Congenital melanocytic nevi frequently harbor NRAS mutations but no BRAF mutations. J Invest Dermatol 2007;127:179–82.

[96] van Engen-van Grunsven AC, van Dijk MC, Ruiter DJ, et al. HRAS-mutated Spitz tumors: a subtype of Spitz tumors with distinct features. Am J Surg Pathol 2010;34:1436–41.

[97] van Dijk MC, Bernsen MR, Ruiter DJ. Analysis of mutations in B-RAF, N-RAS, and H-RAS genes in the differential diagnosis of Spitz nevus and spitzoid melanoma. Am J Surg Pathol 2005;29:1145–51.

[98] Busam KJ. Molecular pathology of melanocytic tumors. Semin Diagn Pathol 2013;30:362–74.

[99] Gill M, Cohen J, Renwick N, et al. Genetic similarities between Spitz nevus and Spitzoid melanoma in children. Cancer 2004;101:2636–40.

[100] Nissan MH, Pratilas CA, Jones AM, et al. Loss of NF1 in cutaneous melanoma is associated with RAS activation and MEK dependence. Cancer Res 2014;74:2340–50.

[101] Wiesner T, Kiuru M, Scott SN, et al. NF1 mutations are common in desmoplastic melanoma. Am J Surg Pathol 2015;39:1357–62.

[102] Bott M, Brevet M, Taylor BS, et al. The nuclear deubiquitinase BAP1 is commonly inactivated by somatic mutations and 3p21.1 losses in malignant pleural mesothelioma. Nat Genet 2011;43:668–72.

[103] Testa JR, Cheung M, Pei J, et al. Germline BAP1 mutations predispose to malignant mesothelioma. Nat Genet 2011;43:1022–5.

[104] Shah AA, Bourne TD, Murali R. BAP1 protein loss by immunohistochemistry: a potentially useful tool for prognostic prediction in patients with uveal melanoma. Pathology 2013;45:651–6.

[105] Murali R, Wiesner T, Scolyer RA. Tumours associated with BAP1 mutations. Pathology 2013;45:116–26.

[106] Hussussian CJ, Struewing JP, Goldstein AM, et al. Germline p16 mutations in familial melanoma. Nat Genet 1994;8:15–21.

[107] Kamb A, Shattuck-Eidens D, Eeles R, et al. Analysis of the p16 gene (CDKN2) as a candidate for the chromosome 9p melanoma susceptibility locus. Nat Genet 1994;8:22–6.

[108] Pomerantz J, Schreiber-Agus N, Liegeois NJ, et al. The Ink4a tumor suppressor gene product, p19Arf, interacts with MDM2 and neutralizes MDM2's inhibition of p53. Cell 1998;92:713–23.

[109] Zhang Y, Xiong Y, Yarbrough WG. ARF promotes MDM2 degradation and stabilizes p53: ARF-INK4a locus deletion impairs both the Rb and p53 tumor suppression pathways. Cell 1998;92:725–34.

[110] Fargnoli MC, Gandini S, Peris K, Maisonneuve P, Raimondi S. MC1R variants increase melanoma risk in families with CDKN2A mutations: a meta-analysis. Eur J Cancer 2010;46:1413–20.

[111] Lin H, Wong RP, Martinka M, Li G. BRG1 expression is increased in human cutaneous melanoma. Br J Dermatol 2010;163:502–10.

[112] Laurette P, Strub T, Koludrovic D, et al. Transcription factor MITF and remodeller BRG1 define chromatin organisation at regulatory elements in melanoma cells. eLife 2015;4:e06857.

[113] Zhang G, Li G. Novel multiple markers to distinguish melanoma from dysplastic nevi. PLoS One 2012;7:e45037.

[114] Maehama T, Dixon JE. The tumor suppressor, PTEN/MMAC1, dephosphorylates the lipid second messenger, phosphatidylinositol 3,4,5-trisphosphate. J Biol Chem 1998;273:13375–8.

[115] Wu H, Goel V, Haluska FG. PTEN signaling pathways in melanoma. Oncogene 2003;22:3113–22.

[116] Li J, Yen C, Liaw D, et al. PTEN, a putative protein tyrosine phosphatase gene mutated in human brain, breast, and prostate cancer. Science 1997;275:1943–7.

[117] Plas DR, Thompson CB. Akt-dependent transformation: there is more to growth than just surviving. Oncogene 2005;24:7435–42.

[118] Mirmohammadsadegh A, Marini A, Nambiar S, et al. Epigenetic silencing of the PTEN gene in melanoma. Cancer Res 2006;66:6546–52.

[119] Tsao H, Goel V, Wu H, Yang G, Haluska FG. Genetic interaction between NRAS and BRAF mutations and PTEN/MMAC1 inactivation in melanoma. J Invest Dermatol 2004;122:337–41.

[120] Garraway LA, Widlund HR, Rubin MA, et al. Integrative genomic analyses identify MITF as a lineage survival oncogene amplified in malignant melanoma. Nature 2005;436:117–22.

[121] Carreira S, Goodall J, Aksan I, et al. Mitf cooperates with Rb1 and activates p21Cip1 expression to regulate cell cycle progression. Nature 2005;433:764–9.

[122] Loercher AE, Tank EM, Delston RB, Harbour JW. MITF links differentiation with cell cycle arrest in melanocytes by transcriptional activation of INK4A. J Cell Biol 2005;168:35–40.

[123] Wellbrock C, Marais R. Elevated expression of MITF counteracts B-RAF-stimulated melanocyte and melanoma cell proliferation. J Cell Biol 2005;170:703–8.

[124] Fernandes JD, Hsieh R, de Freitas LA, et al. MAP kinase pathways: molecular roads to primary acral lentiginous melanoma. Am J Dermatopathol 2015;37:892–7.

[125] Worm J, Christensen C, Gronbaek K, Tulchinsky E, Guldberg P. Genetic and epigenetic alterations of the APC gene in malignant melanoma. Oncogene 2004;23:5215–26.

[126] Zhang X, Hao J. Development of anticancer agents targeting the Wnt/beta-catenin signaling. Am J Cancer Res 2015;5:2344–60.

[127] Maelandsmo GM, Florenes VA, Hovig E, et al. Involvement of the pRb/p16/cdk4/cyclin D1 pathway in the tumorigenesis of sporadic malignant melanomas. Br J Cancer 1996;73:909–16.

[128] Griewank KG, Murali R, Puig-Butille JA, et al. TERT promoter mutation status as an independent prognostic factor in cutaneous melanoma. J Natl Cancer Inst 2014;106.

[129] Nagore E, Heidenreich B, Requena C, et al. TERT promoter mutations associate with fast growing melanoma. Pigment Cell Melanoma Res 2015;29:236–8.

[130] Michaylira CZ, Nakagawa H. Hypoxic microenvironment as a cradle for melanoma development and progression. Cancer Biol Ther 2006;5:476–9.

[131] Bedogni B, Warneke JA, Nickoloff BJ, Giaccia AJ, Powell MB. Notch1 is an effector of Akt and hypoxia in melanoma development. J Clin Invest 2008;118:3660–70.

[132] Bedogni B, Powell MB. Skin hypoxia: a promoting environmental factor in melanomagenesis. Cell Cycle 2006;5:1258–61.

[133] Banerjee SS, Harris M. Morphological and immunophenotypic variations in malignant melanoma. Histopathology 2000;36:387–402.

[134] Ohsie SJ, Sarantopoulos GP, Cochran AJ, Binder SW. Immunohistochemical characteristics of melanoma. J Cutan Pathol 2008;35:433–44.

[135] Dabbs DJ. Diagnostic immunohistochemistry. Philadelphia: Churchill Livingstone; 2002.

[136] McKee PH, Calonje E, Granter SR. Pathology of the skin with clinical correlations. 3rd ed. Philadelphia: Elsevier Mosby; 2005.

[137] Uguen A, Talagas M, Costa S, et al. A p16-Ki-67-HMB45 immunohistochemistry scoring system as an ancillary diagnostic tool in the diagnosis of melanoma. Diagn Pathol 2015;10:195.

[138] Fetsch PA, Marincola FM, Abati A. The new melanoma markers: MART-1 and Melan-A (the NIH experience). Am J Surg Pathol 1999;23:607–10.

[139] Orchard GE. Comparison of immunohistochemical labelling of melanocyte differentiation antibodies Melan-A, tyrosinase and HMB 45 with NKIC3 and S100 protein in the evaluation of benign nevi and malignant melanoma. Histochem J 2000;32:475–81.

[140] Drabeni M, Lopez-Vilaro L, Barranco C, et al. Differences in tumor thickness between hematoxylin and eosin and Melan-A immunohistochemically stained primary cutaneous melanomas. Am J Dermatopathol 2013;35:56–63.

[141] Chen Y, Klonowski PW, Lind AC, Lu D. Differentiating neurotized melanocytic nevi from neurofibromas using Melan-A (MART-1) immunohistochemical stain. Arch Pathol Lab Med 2012;136:810–15.

[142] Nonaka D, Chiriboga L, Rubin BP. Sox10: a pan-schwannian and melanocytic marker. Am J Surg Pathol 2008;32:1291–8.

[143] Karamchandani JR, Nielsen TO, van de Rijn M, West RB. Sox10 and S100 in the diagnosis of soft-tissue neoplasms. Appl Immunohistochem Mol Morphol 2012;20:445–50.

[144] Nonaka D, Laser J, Tucker R, Melamed J. Immunohistochemical evaluation of necrotic malignant melanomas. Am J Clin Pathol 2007;127:787–91.

[145] Busam KJ, Kucukgol D, Sato E, et al. Immunohistochemical analysis of novel monoclonal antibody PNL2 and comparison with other melanocyte differentiation markers. Am J Surg Pathol 2005;29:400–6.

[146] Busam KJ, Iversen K, Coplan KC, Jungbluth AA. Analysis of microphthalmia transcription factor expression in normal tissues and tumors, and comparison of its expression with S-100 protein, gp100, and tyrosinase in desmoplastic malignant melanoma. Am J Surg Pathol 2001;25:197–204.

[147] King R, Googe PB, Weilbaecher KN, Mihm Jr. MC, Fisher DE. Microphthalmia transcription factor expression in cutaneous benign, malignant melanocytic, and nonmelanocytic tumors. Am J Surg Pathol 2001;25:51–7.

[148] Miettinen M, Fernandez M, Franssila K, et al. Microphthalmia transcription factor in the immunohistochemical diagnosis of metastatic melanoma: comparison with four other melanoma markers. Am J Surg Pathol 2001;25:205–11.

[149] Nybakken GE, Sargen M, Abraham R, et al. MITF accurately highlights epidermal melanocytes in atypical intraepidermal melanocytic proliferations. Am J Dermatopathol 2013;35:25–9.

[150] Tran TA, Ross JS, Carlson JA, Mihm Jr. MC. Mitotic cyclins and cyclin-dependent kinases in melanocytic lesions. Hum Pathol 1998;29:1085–90.

[151] Chorny JA, Barr RJ, Kyshtoobayeva A, Jakowatz J, Reed RJ. Ki-67 and p53 expression in minimal deviation melanomas as compared with other nevomelanocytic lesions. Mod Pathol 2003;16:525–9.

[152] Vogt T, Zipperer KH, Vogt A, et al. p53-protein and Ki-67-antigen expression are both reliable biomarkers of prognosis in thick stage I nodular melanomas of the skin. Histopathology 1997;30:57–63.

[153] Hazan C, Melzer K, Panageas KS, et al. Evaluation of the proliferation marker MIB-1 in the prognosis of cutaneous malignant melanoma. Cancer 2002;95:634–40.

[154] Henrique R, Azevedo R, Bento MJ, et al. Prognostic value of Ki-67 expression in localized cutaneous malignant melanoma. J Am Acad Dermatol 2000;43:991–1000.

[155] Moretti S, Spallanzani A, Chiarugi A, Fabiani M, Pinzi C. Correlation of Ki-67 expression in cutaneous primary melanoma with prognosis in a prospective study: different correlation according to thickness. J Am Acad Dermatol 2001;44:188–92.

[156] Kanter-Lewensohn L, Hedblad MA, Wejde J, Larsson O. Immunohistochemical markers for distinguishing Spitz nevi from malignant melanomas. Mod Pathol 1997;10:917–20.

[157] Gimotty PA, Van Belle P, Elder DE, et al. Biologic and prognostic significance of dermal Ki67 expression, mitoses, and tumorigenicity in thin invasive cutaneous melanoma. J Clin Oncol 2005;23:8048–56.

[158] Boni R, Doguoglu A, Burg G, Muller B, Dummer R. MIB-1 immunoreactivity correlates with metastatic dissemination in primary thick cutaneous melanoma. J Am Acad Dermatol 1996;35:416–18.

[159] Ladstein RG, Bachmann IM, Straume O, Akslen LA. Ki-67 expression is superior to mitotic count and novel proliferation markers PHH3, MCM4 and mitosin as a prognostic factor in thick cutaneous melanoma. BMC Cancer 2010;10:140.

[160] Niezabitowski A, Czajecki K, Rys J, et al. Prognostic evaluation of cutaneous malignant melanoma: a clinicopathologic and immunohistochemical study. J Surg Oncol 1999;70: 150–60.

[161] Harbour JW, Onken MD, Roberson ED, et al. Frequent mutation of BAP1 in metastasizing uveal melanomas. Science 2010;330:1410–13.

[162] Koopmans AE, Verdijk RM, Brouwer RW, et al. Clinical significance of immunohistochemistry for detection of BAP1 mutations in uveal melanoma. Mod Pathol 2014;27:1321–30.

[163] Busam KJ, Wanna M, Wiesner T. Multiple epithelioid Spitz nevi or tumors with loss of BAP1 expression: a clue to a hereditary tumor syndrome. JAMA Dermatol 2013; 149:335–9.

[164] Gerami P, Scolyer RA, Xu X, et al. Risk assessment for atypical spitzoid melanocytic neoplasms using FISH to identify chromosomal copy number aberrations. Am J Surg Pathol 2013; 37:676–84.

[165] Straume O, Sviland L, Akslen LA. Loss of nuclear p16 protein expression correlates with increased tumor cell proliferation (Ki-67) and poor prognosis in patients with vertical growth phase melanoma. Clin Cancer Res 2000;6:1845–53.

[166] Piras F, Perra MT, Murtas D, et al. Combinations of apoptosis and cell-cycle control biomarkers predict the outcome of human melanoma. Oncol Rep 2008;20:271–7.

[167] Straume O, Akslen LA. Alterations and prognostic significance of p16 and p53 protein expression in subgroups of cutaneous melanoma. Int J Cancer 1997;74:535–9.

[168] Shibata K, Ajiro K. Cell cycle-dependent suppressive effect of histone H1 on mitosis-specific H3 phosphorylation. J Biol Chem 1993;268:18431–4.

[169] Hendzel MJ, Wei Y, Mancini MA, et al. Mitosis-specific phosphorylation of histone H3 initiates primarily within pericentromeric heterochromatin during G2 and spreads in an ordered fashion coincident with mitotic chromosome condensation. Chromosoma 1997;106:348–60.

[170] Schimming TT, Grabellus F, Roner M, et al. pHH3 immunostaining improves interobserver agreement of mitotic index in thin melanomas. Am J Dermatopathol 2012;34:266–9.

[171] Ottmann K, Tronnier M, Mitteldorf C. Detection of mitotic figures in thin melanomas—immunohistochemistry does not replace the careful search for mitotic figures in hematoxylin-eosin stain. J Am Acad Dermatol 2015;73:637–44.

[172] Bastian BC, LeBoit PE, Pinkel D. Mutations and copy number increase of HRAS in Spitz nevi with distinctive histopathological features. Am J Pathol 2000;157:967–72.

[173] Bastian BC, Olshen AB, LeBoit PE, Pinkel D. Classifying melanocytic tumors based on DNA copy number changes. Am J Pathol 2003;163:1765—70.

[174] Kallioniemi A, Kallioniemi OP, Sudar D, et al. Comparative genomic hybridization for molecular cytogenetic analysis of solid tumors. Science 1992;258:818—21.

[175] Gerami P, Zembowicz A. Update on fluorescence in situ hybridization in melanoma: state of the art. Arch Pathol Lab Med 2011;135:830—7.

[176] Bauer J, Bastian BC. Distinguishing melanocytic nevi from melanoma by DNA copy number changes: comparative genomic hybridization as a research and diagnostic tool. Dermatol Ther 2006;19:40—9.

[177] Bastian BC, Wesselmann U, Pinkel D, Leboit PE. Molecular cytogenetic analysis of Spitz nevi shows clear differences to melanoma. J Invest Dermatol 1999;113:1065—9.

[178] Gerami P, Wass A, Mafee M, et al. Fluorescence in situ hybridization for distinguishing nevoid melanomas from mitotically active nevi. Am J Surg Pathol 2009;33:1783—8.

[179] Miller DT, Adam MP, Aradhya S, et al. Consensus statement: chromosomal microarray is a first-tier clinical diagnostic test for individuals with developmental disabilities or congenital anomalies. Am J Hum Genet 2010;86:749—64.

[180] Gammon B, Beilfuss B, Guitart J, Gerami P. Enhanced detection of spitzoid melanomas using fluorescence in situ hybridization with 9p21 as an adjunctive probe. Am J Surg Pathol 2012;36:81—8.

[181] Gerami P, Yelamos O, Lee CY, et al. Multiple cutaneous melanomas and clinically atypical moles in a patient with a novel germline BAP1 mutation. JAMA Dermatol 2015;151:1235—9.

[182] Wiesner T, Obenauf AC, Murali R, et al. Germline mutations in BAP1 predispose to melanocytic tumors. Nat Genet 2011;43:1018—21.

[183] Kruglyak L, Nickerson DA. Variation is the spice of life. Nat Genet 2001;27:234—6.

[184] Gerami P, Jewell SS, Morrison LE, et al. Fluorescence in situ hybridization (FISH) as an ancillary diagnostic tool in the diagnosis of melanoma. Am J Surg Pathol 2009;33:1146—56.

[185] North JP, Garrido MC, Kolaitis NA, et al. Fluorescence in situ hybridization as an ancillary tool in the diagnosis of ambiguous melanocytic neoplasms: a review of 804 cases. Am J Surg Pathol 2014;38:824—31.

[186] Gerami P, Li G, Pouryazdanparast P, et al. A highly specific and discriminatory FISH assay for distinguishing between benign and malignant melanocytic neoplasms. Am J Surg Pathol 2012;36:808—17.

[187] Fang Y, Dusza S, Jhanwar S, Busam KJ. Fluorescence in situ hybridization (FISH) analysis of melanocytic nevi and melanomas: sensitivity, specificity, and lack of association with sentinel node status. Int J Surg Pathol 2012;20:434—40.

[188] Isaac AK, Lertsburapa T, Pathria Mundi J, et al. Polyploidy in spitz nevi: a not uncommon karyotypic abnormality identifiable by fluorescence in situ hybridization. Am J Dermatopathol 2010;32:144—8.

[189] Boone SL, Busam KJ, Marghoob AA, et al. Two cases of multiple spitz nevi: correlating clinical, histologic, and fluorescence in situ hybridization findings. Arch Dermatol 2011;147:227—31.

[190] Enzinger FM. Clear-cell sarcoma of tendons and aponeuroses. An analysis of 21 cases. Cancer 1965;18:1163—74.

[191] Chung EB, Enzinger FM. Malignant melanoma of soft parts. A reassessment of clear cell sarcoma. Am J Surg Pathol 1983;7:405—13.

[192] Langezaal SM, Graadt van Roggen JF, Cleton-Jansen AM, Baelde JJ, Hogendoorn PC. Malignant melanoma is genetically distinct from clear cell sarcoma of tendons and aponeurosis (malignant melanoma of soft parts). Br J Cancer 2001;84:535—8.

[193] Wang WL, Mayordomo E, Zhang W, et al. Detection and characterization of EWSR1/ATF1 and EWSR1/CREB1 chimeric transcripts in clear cell sarcoma (melanoma of soft parts). Mod Pathol 2009;22:1201—9.

[194] Patel RM, Downs-Kelly E, Weiss SW, et al. Dual-color, break-apart fluorescence in situ hybridization for EWS gene rearrangement distinguishes clear cell sarcoma of soft tissue from malignant melanoma. Mod Pathol 2005;18:1585—90.

[195] Hocar O, Le Cesne A, Berissi S, et al. Clear cell sarcoma (malignant melanoma) of soft parts: a clinicopathologic study of 52 cases. Dermatol Res Pract 2012;2012:984096.

[196] Jeck WR, Parker J, Carson CC, et al. Targeted next generation sequencing identifies clinically actionable mutations in patients with melanoma. Pigment Cell Melanoma Res 2014;27:653—63.

[197] Meyerson M, Gabriel S, Getz G. Advances in understanding cancer genomes through second-generation sequencing. Nat Rev Genet 2010;11:685—96.

[198] Tsongalis GJ, Peterson JD, de Abreu FB, et al. Routine use of the Ion Torrent AmpliSeq Cancer Hotspot Panel for identification of clinically actionable somatic mutations. Clin Chem Lab Med 2014;52:707—14.

[199] Siroy AE, Boland GM, Milton DR, et al. Beyond BRAF(V600): clinical mutation panel testing by next-generation sequencing in advanced melanoma. J Invest Dermatol 2015;135:508—15.

[200] Eggermont AM, Robert C. New drugs in melanoma: it's a whole new world. Eur J Cancer 2011;47:2150—7.

[201] Garbe C, Terheyden P, Keilholz U, Kolbl O, Hauschild A. Treatment of melanoma. Deutsches Arzteblatt Int 2008;105:845—51.

[202] Michielin O, Hoeller C. Gaining momentum: new options and opportunities for the treatment of advanced melanoma. Cancer Treat Rev 2015;41:660—70.

[203] Gerami P, Jewell SS, Pouryazdanparast P, et al. Copy number gains in 11q13 and 8q24 are highly linked to prognosis in cutaneous malignant melanoma. J Mol Diagn 2011;13:352—8.

26

Molecular Testing for Glioblastoma

D.G. Trembath

Division of Neuropathology, Department of Pathology and Laboratory Medicine, The University of North Carolina at Chapel Hill, Chapel Hill, NC, United States

MOLECULAR TARGET

Epidermal Growth Factor Receptor

The *epidermal growth factor receptor* (EGFR) gene is located at 7p12 and encodes a 170 kDa transmembrane receptor protein which transfers signals from extracellular molecules such as epidermal growth factor (EGF) and transforming growth factor α (TGF-α) to activate downstream signaling molecules such a phosphatidylinositide 3-kinase (PI3K) and AKT (protein kinase B) (Fig. 26.1) [2]. *EGFR* is the most frequently amplified gene in glioblastoma (GBM) and is often associated with structural abnormalities, including the mutant protein EGFRvIII, characterized by deletion of the extracellular domain. 30−40% of GBMs show amplification of *EGFR* and approximately half of these will show the *EGFRvIII* mutation [3−5]. Patients with *EGFR* amplification have decreased overall survival (OS) compared to other types of GBM [6].

O^6-Methylguanine Methyltransferase

O^6-Methylguanine methyltransferase (MGMT) is a dealkylating enzyme that removes methyl groups from the O6 position of guanine, inducing resistance to alkylating chemotherapeutic agents [7]. Initial studies demonstrated that inactivation of MGMT was associated with tumor regression and prolonged overall and disease-free survival in GBM [8]. Later studies showed that methylation of specific promoter sites in the *MGMT* promoter decreased protein expression and were associated with increased progression-free and OS [7]. It now appears that the presence of a methylated *MGMT* promoter (Fig. 26.2) confers survival advantage, regardless of therapy [10].

Isocitrate Dehydrogenase

Isocitrate dehydrogenase (IDH) enzymes catalyze the oxidative decarboxylation of isocitrate to form α-ketoglutarate, during which NADPH is produced (Fig. 26.3) [12]. NADPH is critical for the synthesis of fatty acids and cholesterol, the oxidative metabolism of drugs by cytochrome P450 enzymes, and the generation of nitric oxide and reactive oxygen species by neutrophils.

Three isocitrate dehydrogenases are present within the human genome with *IDH1* and *IDH2* share approximately 70% sequence similarity. *IDH1* plays a crucial role in lipid metabolism, promoting glycogenesis during hypoxia by catalyzing the reductive carboxylation of α-ketoglutarate dehydrogenase to acetyl-CoA for lipid biosynthesis. *IDH1* is the main source of NADPH in the human brain where it is protective against oxidative damage [12].

In 2008, sequencing of gliomas as part of the TCGA project revealed recurrent missense mutations in *IDH1* in GBM [13]. Additional studies revealed that *IDH1* and *IDH2* mutations occurred in a high percentage of WHO grade II/III astrocytomas, oligodendrogliomas, and oligoastrocytomas [14]. *IDH* mutations occurr in a number of other malignancies including acute myeloid leukemia (AML), intrahepatic cholangiocarcinoma, cartilaginous tumors, and melanoma. In malignant gliomas, mutant IDH proteins are invariably expressed within tumor cells and the mutation appears to be an early event in the genesis of brain tumors [14]. Mutations in IDH are generally heterozygous missense mutations, and mutations in *IDH1* and *IDH2* occur at a specific arginine residues in the enzyme's active site—the most common alteration is R132H (approximately 85% of mutations in *IDH1*; Table 26.1).

Diagnostic Molecular Pathology
DOI: http://dx.doi.org/10.1016/B978-0-12-800886-7.00026-1

© 2017 Elsevier Inc. All rights reserved.

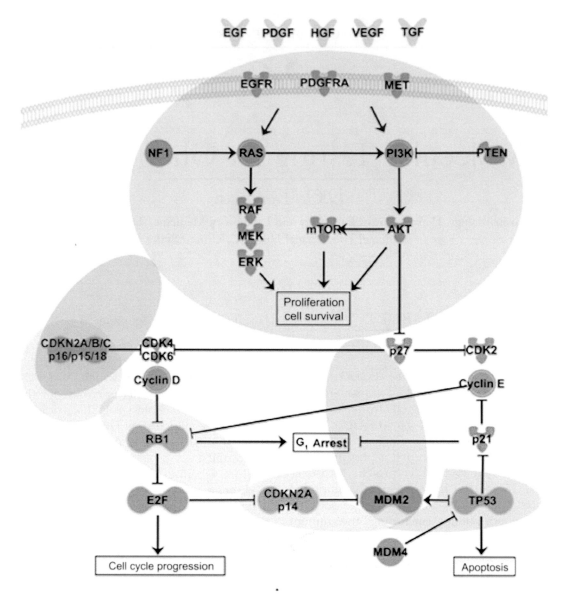

FIGURE 26.1 The *Epidermal growth factor receptor* **(EGFR) pathway.** EGFR binds to signaling molecules such as epidermal growth factor (EGF) or transforming growth factor (TGF). This binding induces autophosphorylation that activates downstream signaling transduction pathways that control cell proliferation, differentiation, and survival [1].

1p/19q Loss

Oligodendrogliomas are diffusely infiltrating well-differentiated low-grade gliomas that typically occur in adults and have a predilection for the cerebral hemispheres. The classic histologic features of an oligodendroglioma include the *fried egg* appearance of the tumor cells with the nuclei surrounded by clear cytoplasm, secondary to formalin fixation. Oligodendrogliomas were one of the first gliomas characterized by a distinct genetic alteration—the loss of chromosomal arms 1p and 19q. This characteristic co-deletion is observed in 50−90% of cases, particularly those with the classic histology described above [15−17]. In addition to having diagnostic value, the loss of 1p/19q is associated with prolonged survival time and favorable response to procarbazine, CCNU (1-(2-chloroethyl)-3-cyclohexyl-1-nitrosourea), vincristine (PCV), and temozolomide chemotherapy and radiation therapy [18,19].

In GBM, the most common loss of heterozygosity (LOH) occurs due to loss of chromosome 10 [20,21]. While the loss of chromosomal arms 1p and 19q is not common in GBM, there is a variant of this tumor where oligodendroglial-like features predominate. These lesions are often assessed for the loss of chromosomal arms 1p and 19q in the hopes that they will be more responsive to chemotherapy compared to the

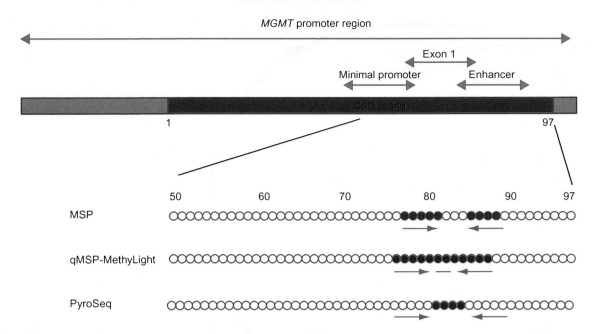

FIGURE 26.2 *O^6-Methylguanine methyltransferase* **(MGMT) promoter region.** Diagram of the *MGMT* promoter region, demonstrating sites interrogated by methylation-specific PCR (MSP), MethylLight qMSP, and pyrosequencing [9].

standard GBM. LOH of chromosome 1p is seen in both primary and secondary GBM at similar frequencies (12—15%) [2]. LOH of 19q is seen in 20—25% of GBM and is more frequently present in secondary rather than primary GBM [22—25].

MOLECULAR TECHNOLOGIES

Epidermal Growth Factor Receptor

Determining the amplification of *EGFR* has classically relied upon cytogenetics, specifically fluorescent in situ hybridization (FISH) studies. In situ hybridization uses the localization and detection of specific DNA or RNA sequences in cells, preserved tissue sections, or, occasionally, entire specimens [26]. The sequence of interest is located by hybridizing the complimentary strand of a nucleotide probe to the sequence [26]. These tests can be performed on formalin-fixed paraffin embedded (FFPE) tissue as well as cultured cells. Testing in FFPE tissue allows for testing to be performed after standard microscopic analysis has determined the identity of the tumor.

FISH requires pretreatment of the target tissue—agents such as isothiocyanate permeabilize the tissue, followed by acid hydrolysis, and treatment with proteases [27]. DNA probes can be generated via cloning and amplification while single-stranded cDNA probes can be created by reverse transcriptase PCR (RT-PCR). The probes are conjugated to fluorescent labels,

generally rhodamine or fluorescein. Today, most probes are prepared either by selecting clones for the gene/region of interest from appropriate yeast, bacterial, or P1 artificial chromosomes, or by amplifying the target sequence using PCR. Hybridization stringency is controlled by adjusting factors such as pH, salt and formamide concentrations, and temperature, in order to minimize cross-hybridization of probes to nonspecific targets. Interpreting FISH signals requires experienced personnel. FISH signals can vary in terms of size and shape, extensive denaturation can lead to loss of signal, and samples must not be over-layered or inadequately counterstained which interferes with accurate signal enumeration. In addition to the standard cytogenetic materials, equipment requirements include a fluorescent microscopic with an oil immersion lens and the necessary excitation and emission filters for the probe in question.

Determining the presence or absence of EGFRvIII is frequently performed by immunohistochemistry using specific EGFRvIII antibodies or by RT-PCR to detect rearranged RNA transcripts [4,28,29]. Immunohistochemistry is generally performed on FFPE tissue with antibodies either produced in-house or purchased from commercial vendors. Antibodies can be either polyclonal or monoclonal [30]. Generally two antibodies are applied to the fixed tissue; the primary antibody which recognizes the antigen of interest (eg, EGFRvIII) and the secondary antibody which is conjugated to a fluorescent dye, biotin, or to an enzyme that can produce a visible result after reaction with a chromogenic substrate [30]. Antigen

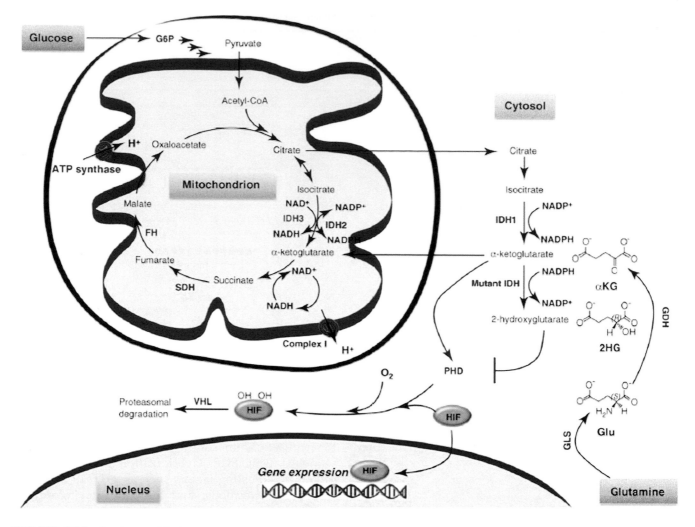

FIGURE 26.3 *Isocitrate dehydrogenase* **(IDH) and** *IDH* **mutations in metabolism.** Normally, *IDH1* catalyzes the production of α-ketoglutarate from isocitrate, creating NADPH in the process. When *IDH* is mutated, the initial reaction still occurs, but α-ketoglutarate is converted into 2-hydroxyglutarate, which consumes NADPH (and leaves the cell vulnerable to oxidative stress). It is hypothesized that the accumulation of 2-hydroxyglutarate may then lead to the accumulation of HIF-1α and predispose the cell to oncogenesis [11].

TABLE 26.1   IDH1 and IDH2 Mutations in Gliomas [12]

Gene	Mutation	Amino acid change	Frequency (%)
IDH1	c. 395G > A	R132H	83.5−88.9
	c. 394C > T	R132C	3.9−4.1
	c. 394C > A	R132S	1.5−2.4
	c. 394C > G	R132G	0.6−1.3
	c. 395G > T	R132L	0.3−4.1
IDH2	c. 515G > A	R172K	2.4−2.7
	c. 515G > T	R172M	0.8−1.8
	c. 514A > T	R172W	0.0−0.7
	c. 514A > G	R172G	0.0−1.2

retrieval is frequently performed using a technique known as heat-induced epitope retrieval. Briefly, antigens are uncovered by exposing the tissue to different temperatures in different buffers, with different heating methods used depending upon the antigen in question [30]. In today's laboratories, labeled antigen detection methods, using avidin−biotin, streptavidin−biotin, and other complexes are generally used, followed by signal amplification using agents such as tyramide [30].

O^6-Methylguanine Methyltransferase

Various methodologies exist to test *MGMT* promoter methylation and there has yet to be an agreed standardized testing methodology [31]. However, regardless of the methodology used nearly all share an initial critical step—the bisulfite modification of DNA which converts unmethylated cytosine to uracil, leaving methylated cytosine unchanged [9]. This modified DNA is then the basis for all subsequent testing and it is critical to have appropriate controls present to determine if complete conversion has occurred.

The most common methodology is methylation-specfic PCR (MSP). In this technique, bisulfite-treated DNA is amplified using methylation-specific primers. The primers are designed to bind only to the bisulfite-modified products of methylated or unmethylated DNA (Fig. 26.2). The product of the PCR generally covers several possible CpG sites in the *MGMT* promoter [31]. PCR products can be qualitatively analyzed via gel-based techniques or quantitatively via real-time PCR. An example of the latter is the MethyLight assay which is based on the TaqMan PCR principle [9]. Forward and reverse primers are designed to the methylated *MGMT* sequence as well as an oligomeric probe that emits fluorescence following degradation by the 5'-3' exonuclease activity of Taq polymerase [9].

Sequencing is another approach for analyzing the methylation of the *MGMT* promoter. Pyrosequencing following bisulfite conversion provides a semiquantitative analysis of multiple sites in the promoter (Fig. 26.4). Pyrosequencing uses the real-time detection of inorganic pyrophosphate that is released by successful incorporation of nucleotides during DNA synthesis [32]. Inorganic pyrophosphate is converted to ATP by ATP sulfurylase and then the amount of ATP generated is detected by the luciferase-producing photons. The light produced from this procedure is observed as sequence signal peaks in pyrograms. ATP and nonincorporated nucleotides are degraded by the enzyme apyrase, then the sequencing reaction continues. The sequence signals are proportional to the number of bases incorporated [32]. Pyrosequencing has the advantage of being semiquantitative and can be standardized, but there are debates as to what level should be used as cutoffs to determine if the *MGMT* promoter is methylated.

Combined bisulfite restriction analysis (COBRA) relies on bisulfite conversion of DNA followed by PCR and then digestion of the purified PCR product with the endonucleases *Taq*α1 and *Bst*UI. Methylation at a limited number of CpG sites can be determined following enucleation digestion and comparison of fragment percentages [33].

Isocitrate Dehydrogenase

IDH mutations are typically initially detected using immunohistochemistry with sequencing as an accompanying methodology. Monoclonal antibodies have been developed that are specific for the *IDH1* R132H

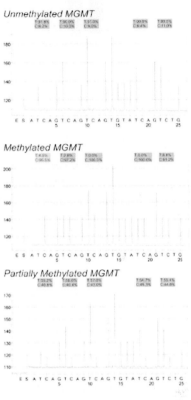

FIGURE 26.4 Pyrosequencing for detection of *O⁶-Methylguanine methyltransferase* (**MGMT**) promoter methylation. The first pyrogram shows an unmethylated *MGMT* promoter (with the cut off set at <20%), the middle pyrogram a methylated promoter (>80% methylation) and the bottom pyrogram a partially methylated promoter [9].

mutation [34,35]. The application of IDH1 antibodies is similar to that of other antibody techniques. Unstained slides are cut, dried, pretreated, and then incubated with the antibody of choice. This incubation is then followed by standard signal amplification, washing, and counter staining with appropriate agents such as 3, 3-diaminobenzidine (DAB) and hematoxylin.

Detection of *IDH* mutations can be accomplished via using either typical Sanger sequencing, pyrosequencing, or multiplex allele-specific PCR. An initial step for each of these approaches is to macro-dissect the tissue to enhance the amount of tumor present. Following DNA extraction, PCR is performed with appropriate primers that encompass codon 132 in

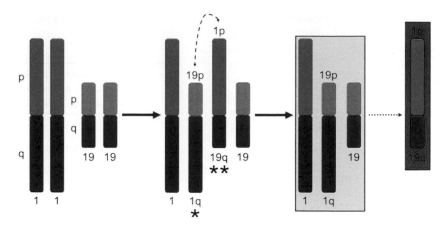

FIGURE 26.5 Whole arm translocation leading to 1p/19q deletion. Two derivative chromosomes are believed to be created following the translocation between chromosomal arms 1p and 19q, creating the chromosome designated with an asterisk (*) corresponding to der(1;19) (p10;q10) and the chromosome designated by the double asterisk (**) corresponding to der(1;19)(q10;p10). Following the loss of der(1;19)(p10; q10) (orange box), an abnormal chromosome configuration (shown in the gray box) remains. *Source: This figure was adapted from Pinkham MB, Telford N, Whitfield GA, Colaco RJ, O'Neill F, McBain CA. FISHing tips: what every clinician should know about 1p19q analysis in gliomas using fluorescence in situ hybridisation. Clin Oncol 2015; 27:445–453 [36].*

IDH1 and if necessary codon 172 in *IDH2*. Commercial sequencing kits can then be used and the PCR products analyzed on a commercial sequencing instrument followed by sequence analysis for the presence/absence of the mutation.

1p/19q

To assess for the loss of chromosome arms 1p and 19q (Fig. 26.5), the typical approaches involve either FISH or LOH studies. Both of these methodologies have inherent advantages and disadvantages.

In FISH, fluorescent-labeled DNA probes bind to regions of interest in tumor nuclei. For oligodendrogliomas and other tumors where there is concern for 1p/19q loss, there are commercially-available probe pairs that target 1p36/1p25 and 19q13/19p13 [37]. Technically, FISH is performed using standard methodology. Slides from FFPE tissue are deparaffinized, typically using xylene and ethanol. Following water rinses, molecular targets are retrieved using a citrate buffer and heat, enzymatic digestion performed, the probe mixture applied, and specimens denatured overnight. The next day the slides are washed in appropriate buffers and counterstained, generally with DAB, and visualized under a fluorescent microscope [38].

The most common alternative to FISH for determining 1p/19q loss is PCR-based analysis of LOH studies, using microsatellite DNA repeats. The number of repeats in each microsatellite varies; when both alleles (maternal and paternal) of a microsatellite have different numbers of repeats, this is considered an informative microsatellite as the PCR products from each allele will migrate differently on an agarose or acrylamide gel. When a tumor shows only a single PCR product size in comparison to the results for the informative microsatellite from the patient's normal tissue, one allele is assumed to be lost [37].

To perform LOH studies, one needs paired tumor and normal tissue (or a constitutional DNA sample), the latter can be DNA from a tube of the patient's blood. Microsatellite markers are selected that are localized to 1p and 19q. Frequently used microsatellite markers include D1S1184, D1S1592, D19S4311, and D19S718. Primers are designed to each microsatellite and PCR performed using standard methodology. Classically, PCR products were analyzed on a polyacrylamide gel and band intensity determined by ethidium bromide staining. Today, there are numerous commercially-available single nucleotide polymorphism (SNP) chips to which the PCR products can be hybridized, and the loss of microsatellite sequence determined in that manner.

CLINICAL UTILITY

While *EGFR* is an extensively studied molecule in GBM, its clinical utility at this time is more questionable. A meta-analysis of studies that examined either the presence/absence of *EGFR* amplification or the presence/absence of the *EGFRvIII* mutation did not find conclusive evidence that either change carried prognostic value in GBM patients [39]. However, there are numerous therapeutic approaches being explored that target the *EGFR* molecule. Tyrosine kinase inhibitors (TKIs) have been efficacious in other tumors, such as non-small cell lung cancer (NSLC) [40]. However, although these TKIs have been shown in preclinical results to inhibit tumor cell growth, angiogenesis, survival, and proliferation in EFGR transfected cell lines, these results have not been replicated in clinical trials [29,41].

Similarly, antibody-targeting of EGFR that has been successful in other tumors has not been efficacious in treatment of GBM. *EGFR*-amplified GBM cells in mouse xenografts that are treated with cetuximab demonstrate decreased proliferation and increased OS. However, increased survival benefit has not been demonstrated in Phase II studies, stratifying patients with GBM by their *EGFR* gene amplification status [42,43]. Other approaches have used antibodies conjugated to toxins or radioisotopes as adjuvant therapy. ^{125}I-Mab425 is a radiolabeled antibody which has shown survival benefit in Phase II trials when combined with temozolamide (TMZ), increasing median survival by approximately 6 months [44].

One promising area in EGFR-based therapies appears to be in using vaccines to target cells expressing EGFRvIII. Vaccines based upon the unique amino acid sequence created by the in-frame deletion of EGFRvIII, chemically conjugated to keyhole limpet hemocyanin (Rindopepimut) have been tested, with the goal to create a specific immune response against EGFRvIII + tumor cells [45]. Phase II trials indicated that rindopepimut is well-tolerated in patients with resected EGFRvIII + GBM who demonstrated improvement in both progression-free survival (PFS) and OS [46,47]. The follow-up study confirmed these initial findings with increase in PFS and OS with a double-blind phase III trial underway [45].

O[6]-Methylguanine Methyltransferase

The clinical utility of *MGMT* lies in the fact that it carries both prognostic and predictive value. The predictive value of *MGMT* was the first to be realized when it was demonstrated that patients with low levels of *MGMT* expression showed greater OS and PFS when treated with temozolomide than those with high levels of *MGMT* expression [7]. The difference in *MGMT* expression levels correlates with levels of methylation in the *MGMT* promoter [7]. More recently, analysis of long-term survival demonstrated that patients with methylated *MGMT* showed improved survival compared to patients with a unmethylated promoter [10].

Isocitrate Dehydrogenase

IDH mutations have proven to have both diagnostic and prognostic value in gliomas. Characterization and

classification of gliomas is classically by histologic criteria, using guidelines created by the World Health Organization. Unfortunately, this creates diagnostic dilemmas due to the inherent subjectivity in the assessment of histologic features. The presence of *IDH* mutations in a large subset of gliomas, particularly WHO grade II/III gliomas, is of diagnostic utility as the presence of the mutation is strongly suggestive that, regardless of what histologic features are present, the changes are that of a glioma. The presence of an *IDH* mutation also carries prognostic value. *IDH1/2* mutations are associated with a younger age of diagnosis, which is a favorable prognostic factor, in both low-grade and high-grade gliomas, compared to tumors with wild-type *IDH* [48,49]. *IDH* mutations are generally associated with a favorable prognosis, particularly in patients with GBM and anaplastic astrocytoma when compared to tumors of the same histologic type that lack the *IDH* mutation [50–53].

1p/19q

The clinical utility of 1p/19q is unquestioned in terms of oligodendrogliomas and testing for this alteration is frequently requested by clinicians for GBM that show any oligodendroglial features. It has been demonstrated that the loss of 1p/19q is associated with better OS and response to chemotherapy and radiation therapy [17,18]. While the same results have not been demonstrated consistently in epidemiologic studies of GBM, given the dismal outcome associated with this tumor, oncologists frequently request 1p/19q analysis in the hopes of finding more efficacious therapy for their patients.

LIMITATIONS OF TESTING

The amount of tissue obtained and the type of tissue, in terms of how it is prepared in the laboratory, is frequently the predominant limitation to molecular testing in gliomas and other tumors. For successful sequencing to assess the presence of promoter methylation in *MGMT* and the presence of the *IDH1* or *IDH2* mutations, at least 20–30% of the specimen must consist of tumor cells. This limitation can be surmounted in many cases by macro-dissecting the tumor away from the surrounding brain in order to keep contamination by normal tissue to a minimum.

The type of tissue available also limits the types of studies that can be performed. FFPE tissue can be used for DNA extraction, but is less robust in producing usable RNA. Some cytogenetic studies (eg, FISH) can be performed on FFPE, but other studies cannot. Thus it is necessary for the clinician and pathologist to work together to determine what type of tissue will be needed for which tests before performing a biopsy.

For determining promoter methylation in *MGMT*, the critical step is successful bisulfite conversion of the tumor DNA. Appropriate controls must be performed in all cases to ensure that the conversion is as complete as possible. For *IDH1* and *IDH2* mutations, the limitations are created predominantly by the type of tissue available and what particular techniques are chosen to detect the presence or absence of the mutation. The *IDH1* R132H mutation can be detected by immunohistochemistry, but also by sequencing. The former obviously only detects the presence or absence of the R132H mutation, while the latter will demonstrate the presence of the most common mutation but also that of the less frequent variants (see Table 26.1 for percentages).

The advantage to using FISH for determining 1p/19q status is that a small number (60–100) of tumor nuclei are required and thus FISH can be performed on very small tissue biopsies. The drawbacks to FISH are that there can be loss of tumor cells in the cutting of unstained slides. One also needs nonneoplastic brain tissue as a negative validation control. There is internal lab variation in terms of the cutoff criteria that should be used to determine whether true deletion of 1p/19q is present. FISH probes are limited by size and cannot be used to analyze all of the chromosomal arm of 1p and 19q; only smaller representative areas can be assayed [37].

The limitations to LOH studies are predominantly secondary to possible contamination of the tumor tissue with normal tissue that may skew the results of the PCR. PCR-based LOH studies also tend to be more expensive than FISH due to the more sophisticated equipment required.

References

[1] Crespo I, Vital AL, Gonzalez-Tablas M, et al. Molecular and genomic alterations in glioblastoma multiforme. Am J Pathol 2015;185:1820–33.

[2] Louis DN, Ohgaki H, Wiestler OD, Cavenne WK, editors. WHO classification of tumors of the central nervous system. Lyon: IARC; 2007.

[3] Low SY, Ho YK, Too HP, Yap CT, Ng WH. MicroRNA as potential modulators in chemoresistant high-grade gliomas. J Clin Neurosci 2014;21:395–400.

[4] Ekstrand AJ, Sugawa N, James CD, Collins VP. Amplified and rearranged epidermal growth factor receptor genes in human glioblastomas reveal deletions of sequences encoding portions of the N- and/or C-terminal tails. Proc Natl Acad Sci USA 1992;89:4309–13.

[5] Appin CL, Brat DJ. Molecular pathways in gliomagenesis and their relevance to neuropathologic diagnosis. Adv Anat Pathol 2015;22:50–8.

[6] Taylor TE, Furnari FB, Cavenee WK. Targeting EGFR for treatment of glioblastoma: molecular basis to overcome resistance. Curr Cancer Drug Targets 2012;12:197–209.

[7] Hegi ME, Diserens AC, Gorlia T, et al. MGMT gene silencing and benefit from temozolomide in glioblastoma. N Engl J Med 2005;352:997–1003.

[8] Esteller M, Garcia-Foncillas J, Andion E, et al. Inactivation of the DNA-repair gene MGMT and the clinical response of gliomas to alkylating agents. N Engl J Med 2000;343:1350–4.

[9] Cankovic M, Nikiforova MN, Snuderl M, et al. The role of MGMT testing in clinical practice: a report of the association for molecular pathology. J Mol Diagn 2013;15:539–55.

[10] Reifenberger G, Weber RG, Riehmer V, et al. Molecular characterization of long-term survivors of glioblastoma using genome- and transcriptome-wide profiling. Int J Cancer 2014;135:1822–31.

[11] Dang L, Jin S, Su SM. IDH mutations in glioma and acute myeloid leukemia. Trends Mol Med 2010;16:387–97.

[12] Waitkus MS, Diplas BH, Yan H. Isocitrate dehydrogenase mutations in gliomas. Neuro Oncol 2016;18:16–26.

[13] Parsons DW, Jones S, Zhang X, et al. An integrated genomic analysis of human glioblastoma multiforme. Science 2008;321:1807–12.

[14] Yan H, Parsons DW, Jin G, et al. IDH1 and IDH2 mutations in gliomas. N Engl J Med 2009;360:765–73.

[15] Perry A, Fuller CE, Banerjee R, Brat DJ, Scheithauer BW. Ancillary FISH analysis for 1p and 19q status: preliminary observations in 287 gliomas and oligodendroglioma mimics. Front Biosci 2003;8:a1–9.

[16] Burger PC, Minn AY, Smith JS, et al. Losses of chromosomal arms 1p and 19q in the diagnosis of oligodendroglioma. A study of paraffin-embedded sections. Mod Pathol 2001;14:842–53.

[17] Aldape K, Burger PC, Perry A. Clinicopathologic aspects of 1p/19q loss and the diagnosis of oligodendroglioma. Arch Pathol Lab Med 2007;131:242–51.

[18] Smith JS, Perry A, Borell TJ, et al. Alterations of chromosome arms 1p and 19q as predictors of survival in oligodendrogliomas, astrocytomas, and mixed oligoastrocytomas. J Clin Oncol 2000;18:636–45.

[19] Cairncross JG, Ueki K, Zlatescu MC, et al. Specific genetic predictors of chemotherapeutic response and survival in patients with anaplastic oligodendrogliomas. J Natl Cancer Inst 1998;90:1473–9.

[20] Fults D, Pedone C. Deletion mapping of the long arm of chromosome 10 in glioblastoma multiforme. Genes Chromosomes Cancer 1993;7:173–7.

[21] Fults D, Pedone CA, Thompson GE, et al. Microsatellite deletion mapping on chromosome 10q and mutation analysis of MMAC1, FAS, and MXI1 in human glioblastoma multiforme. Int J Oncol 1998;12:905–10.

[22] von Deimling A, Bender B, Jahnke R, et al. Loci associated with malignant progression in astrocytomas: a candidate on chromosome 19q. Cancer Res 1994;54:1397–401.

[23] von Deimling A, Louis DN, von Ammon K, Petersen I, Wiestler OD, Seizinger BR. Evidence for a tumor suppressor gene on chromosome 19q associated with human astrocytomas, oligodendrogliomas, and mixed gliomas. Cancer Res 1992;52:4277–9.

[24] von Deimling A, Nagel J, Bender B, et al. Deletion mapping of chromosome 19 in human gliomas. Int J Cancer 1994;57:676–80.

[25] Nakamura M, Yang F, Fujisawa H, Yonekawa Y, Kleihues P, Ohgaki H. Loss of heterozygosity on chromosome 19 in secondary glioblastomas. J Neuropathol Exp Neurol 2000;59:539–43.

[26] Meyer M, Reimand J, Lan X, et al. Single cell-derived clonal analysis of human glioblastoma links functional and genomic heterogeneity. Proc Natl Acad Sci USA 2015;112:851−6.

[27] McNicol AM, Farquharson MA. In situ hybridization and its diagnostic applications in pathology. J Pathol 1997;182:250−61.

[28] Aldape KD, Ballman K, Furth A, et al. Immunohistochemical detection of EGFRvIII in high malignancy grade astrocytomas and evaluation of prognostic significance. J Neuropathol Exp Neurol 2004;63:700−7.

[29] Gan HK, Kaye AH, Luwor RB. The EGFRvIII variant in glioblastoma multiforme. J Clin Neurosci 2009;16:748−54.

[30] Coleman WB, Tsongalis GJ, editors. Molecular diagnostics for the clinical laboratorian. 2nd ed. Totowa, N.J: Humana Press; 2006.

[31] Weller M, Stupp R, Reifenberger G, et al. MGMT promoter methylation in malignant gliomas: ready for personalized medicine? Nat Rev Neurol 2010;6:39−51.

[32] Ronaghi M, Shokralla S, Gharizadeh B. Pyrosequencing for discovery and analysis of DNA sequence variations. Pharmacogenomics 2007;8:1437−41.

[33] Mikeska T, Bock C, El-Maarri O, et al. Optimization of quantitative MGMT promoter methylation analysis using pyrosequencing and combined bisulfite restriction analysis. J Mol Diagn 2007;9:368−81.

[34] Capper D, Zentgraf H, Balss J, Hartmann C, von Deimling A. Monoclonal antibody specific for IDH1 R132H mutation. Acta Neuropathol 2009;118:599−601.

[35] Capper D, Weissert S, Balss J, et al. Characterization of R132H mutation-specific IDH1 antibody binding in brain tumors. Brain Pathol 2010;20:245−54.

[36] Pinkham MB, Telford N, Whitfield GA, Colaco RJ, O'Neill F, McBain CA. FISHing tips: what every clinician should know about 1p19q analysis in gliomas using fluorescence in situ hybridisation. Clin Oncol 2015;27:445−53.

[37] Horbinski C. Something old and something new about molecular diagnostics in gliomas. Surg Pathol Clin 2012;5:919−39.

[38] Jha P, Sarkar C, Pathak P, et al. Detection of allelic status of 1p and 19q by microsatellite-based PCR versus FISH: limitations and advantages in application to patient management. Diagn Mol Pathol 2011;20:40−7.

[39] Chen JR, Xu HZ, Yao Y, Qin ZY. Prognostic value of epidermal growth factor receptor amplification and EGFRvIII in glioblastoma: meta-analysis. Acta Neurol Scand 2015;132:310−22.

[40] Liang W, Wu X, Fang W, et al. Network meta-analysis of erlotinib, gefitinib, afatinib and icotinib in patients with advanced non-small-cell lung cancer harboring EGFR mutations. PLoS One 2014;9:e85245.

[41] Padfield E, Ellis HP, Kurian KM. Current therapeutic advances targeting EGFR and EGFRvIII in glioblastoma. Front Oncol 2015;5:5.

[42] Eller JL, Longo SL, Hicklin DJ, Canute GW. Activity of anti-epidermal growth factor receptor monoclonal antibody C225 against glioblastoma multiforme. Neurosurgery 2002;51: 1005−13.

[43] Neyns B, Sadones J, Joosens E, et al. Stratified phase II trial of cetuximab in patients with recurrent high-grade glioma. Ann Oncol 2009;20:1596−603.

[44] Li L, Quang TS, Gracely EJ, et al. A Phase II study of anti-epidermal growth factor receptor radioimmunotherapy in the treatment of glioblastoma multiforme. J Neurosurg 2010;113:192−8.

[45] Schuster J, Lai RK, Recht LD, et al. A phase II, multicenter trial of rindopepimut (CDX-110) in newly diagnosed glioblastoma: the ACT III study. Neuro Oncol 2015;17:854−61.

[46] Sampson JH, Heimberger AB, Archer GE, et al. Immunologic escape after prolonged progression-free survival with epidermal growth factor receptor variant III peptide vaccination in patients with newly diagnosed glioblastoma. J Clin Oncol 2010; 28:4722−9.

[47] Sampson JH, Aldape KD, Archer GE, et al. Greater chemotherapy-induced lymphopenia enhances tumor-specific immune responses that eliminate EGFRvIII-expressing tumor cells in patients with glioblastoma. Neuro Oncol 2011;13: 324−33.

[48] Balss J, Meyer J, Mueller W, Korshunov A, Hartmann C, von Deimling A. Analysis of the IDH1 codon 132 mutation in brain tumors. Acta Neuropathol 2008;116:597−602.

[49] Lai A, Kharbanda S, Pope WB, et al. Evidence for sequenced molecular evolution of IDH1 mutant glioblastoma from a distinct cell of origin. J Clin Oncol 2011;29:4482−90.

[50] Killela PJ, Pirozzi CJ, Healy P, et al. Mutations in IDH1, IDH2, and in the TERT promoter define clinically distinct subgroups of adult malignant gliomas. Oncotarget 2014;5:1515−25.

[51] Sanson M, Marie Y, Paris S, et al. Isocitrate dehydrogenase 1 codon 132 mutation is an important prognostic biomarker in gliomas. J Clin Oncol 2009;27:4150−4.

[52] Jiao Y, Killela PJ, Reitman ZJ, et al. Frequent ATRX, CIC, FUBP1 and IDH1 mutations refine the classification of malignant gliomas. Oncotarget 2012;3:709−22.

[53] Juratli TA, Kirsch M, Geiger K, et al. The prognostic value of IDH mutations and MGMT promoter status in secondary high-grade gliomas. J Neurooncol 2012;110:325−33.

27

Molecular Testing in Pancreatic Cancer

M.J. Bartel[1], S. Chakraborty[2] and M. Raimondo[1]

[1]Division of Gastroenterology & Hepatology, Mayo Clinic, Jacksonville, FL, United States [2]Division of Gastroenterology & Hepatology, Mayo Clinic, Rochester, MN, United States

INTRODUCTION

Pancreatic ductal adenocarcinoma (PDAC) is a devastating disease with high mortality. Current therapeutic modalities offer patients with PDAC only a chance for cure as long the disease is limited to the pancreas. The chances of cure decrease with cancer involvement of adjacent vessel structures and lymph nodes, and eventually progresses to a terminal disease in the setting of unresectable locally advanced cancer and the presence of distant metastasis. Therefore the current goal is to detect PDAC in its earliest stages. Diagnostic biomarkers for pancreatic cancer must not only fulfill this criterion with a high sensitivity, specificity, and accuracy, but also be able to distinguish PDAC from other pancreatic conditions which have increased risk for developing PDAC but do not have invasive disease at the time of testing, in order to avoid unnecessary invasive testing and surgery. These pancreatic conditions include chronic pancreatitis, mucinous cystic neoplasms (MCN), and intraductal papillary mucinous neoplasms (IPMN). Although genetic alterations are well described in chronic pancreatitis, MCN, and IPMN, the natural progression of these lesions to PDAC is not well understood.

Most PDAC arise from premalignant lesions termed pancreatic intraepithelial neoplasia (PanIN), which are further divided into PanIN-1a, PanIN-1b, PanIN-2, and PanIN-3, based on their morphologic dysplasia and accumulation of genetic mutations. In this context, Plectin-1 has been described as marker for PanIN-3 lesions, which are believed to be the earliest stage of PDAC [1].

Over the last three decades, the search for adequate diagnostic biomarkers for PDAC followed closely the establishment of new biotechnological methods. Early methods detected protein biomarkers utilizing ELISA and Western blots. Subsequently, the PCR was developed and found utility for detection of genetic alterations. The latest chromatographic methods, gene-chips, and protein-chips enable a very broad search for diagnostic biomarkers including genome-wide screening. Most recently, epigenetic changes which encompass DNA methylation status, histone posttranslational modifications, and microRNA expression levels were found to be significantly altered in PDAC drawing significant attention to this field.

Several studies have shown that panels of biomarkers increase sensitivity, specificity, and diagnostic accuracy in the diagnosis of PDAC and its discrimination from chronic pancreatitis and IPMN [2–4]. This observation probably reflects the fact that PDAC exhibits significant heterogeneity. Despite similar anatomic location, cancers can have different biologic backgrounds reflecting the molecular pathways to tumorigenesis. Cancers that arise through distinct molecular pathways require different diagnostic biomarkers [5].

PDAC BIOMARKERS IN SERUM

Carbohydrate antigen 19-9 (CA19-9) is the most accurate and well-studied serum biomarker to distinguish PDAC from other diseases. A systematic review of literature that included over 2000 patients concluded that CA19-9 has a median sensitivity of 79% and specificity of 82% for the diagnosis of PDAC in symptomatic individuals [6]. Numerous other potential markers have been tested and compared against CA19-9, but none has proven more accurate. Obstructive jaundice (due to any cause) is commonly associated with an increase in levels of most investigated biomarkers

© 2017 Elsevier Inc. All rights reserved.

(including CA19-9). It is increasingly evident that no single marker performs better than CA19-9 for the diagnosis of PDAC. This has led to studies where a novel biomarker(s) was combined with CA19-9 in an attempt to improve its performance. One such combination tested consisted of alpha-1 chymotrypsin (AACT), thrombospondin-1 (THBS1), and haptoglobin (HPT) with CA19-9. This combination of biomarkers demonstrated improved ability to distinguish PDAC from healthy controls (AUC 0.99 versus 0.89 for CA19-9 alone), diabetics without PDAC (AUC 0.90 versus 0.85 for CA19-9), pancreatic cysts (AUC 0.90 versus 0.81), CP (AUC 0.90 versus 0.79), obstructive jaundice from other causes (AUC 0.74 versus 0.68), and other conditions (AUC 0.92 versus 0.81) [7]. In another study, CA19-9 alone distinguished pretreatment PDAC from benign conditions with an AUC of 0.80 while in combination with CA125 and LAMC2 the AUC increased to 0.87. The combination also performed better than CA19-9 to distinguish early stage PDAC from benign conditions (AUC 0.76 versus 0.69) and from chronic pancreatitis (AUC 0.74 versus 0.59) [8]. Similar improvements have been reported by combining CA19-9 with other serum biomarkers [9], including microRNAs [10,11].

A key shortcoming of most studies investigating novel biomarkers for PDAC is that these studies were conducted using samples from patients who already had the malignancy at the time of serum collection. Few studies have examined whether these same markers retain their accuracy when tested in patients in the precancer stage. In one study ICAM-1 and TIMP-1 were tested in patients between 0 and 12 months before development of PDAC and in the same patients after they developed PDAC. While both proteins were significantly elevated in serum of PDAC cancer patients, no difference was observed in the precancer samples from the same patients [12]. Prospective studies (ie, on patients who did not have cancer at the time of collection of serum, but developed it during follow-up) would be ideal to understand the natural history of PDAC and of potentially useful biomarkers. In a study employing serum collected from participants in the Prostate Lung Colorectal and Ovarian Cancer Screening Trial (PLCO), the investigators tested 67 potential biomarkers in serum collected either 1–12 months, >12 months, or >24 months prior to diagnosis of PDAC. There was significant overlap between healthy controls and those who developed PDAC. No biomarker was better than CA19-9 in distinguishing PDAC from non-PDAC cases. However, a combination of CA19-9, CEA, and Cyfra 21-1 with cut-off set to specificity of 95% was about 32% and 30% sensitive in identifying PDAC in samples collected <1 and >1 year prior to diagnosis of PDAC (compared to sensitivity of 26% and 17% for CA19-9 alone) [13].

It is estimated that for a rare disease like PDAC, with an estimated prevalence of 40 per 100,000 (http://seer.cancer.gov/statfacts/html/pancreas.html) a test with a specificity of 90% used for screening would identify nearly 10 million false positives. By increasing the specificity of the test to 99.9%, we can reduce the number of false positives to an acceptable 100,000 while achieving a sensitivity of 99.99%. Given the heterogeneity of the disease, finding a single marker that is applicable in all patients and identifies cancer in all stages (preclinical, clinical but treatable, and precancerous) is unlikely. Another problem is that while mean levels of the serum marker may be higher in PDAC than in controls, there is significant overlap between individual PDAC samples and controls in nearly 100% of studies, thereby weakening the power of the test. Mathematical modeling has revealed that if we take 40 biomarkers whose levels do not correlate (ie, correlation coefficient zero) with one another and delineate a cut-off such that each marker has a sensitivity of at least 32% and forced specificity of at least 95%, then if at least 7 biomarkers are above the threshold, the panel will distinguish PDAC from non-PDAC cases with a sensitivity and specificity of 99%. If the correlation coefficient is 0.05 or 0.15 the sensitivity of the panel decreased to 94% and 85%, respectively. One could still achieve nearly 100% sensitivity (for the combination), but this requires larger number of biomarkers. The study suggests that the identity of the biomarker does not matter, and importantly a group of uncorrelated weak classifiers can be combined to generate a strong classifier. Thus a panel comprising of a mix of different biomarkers such as proteins, miRNAs, mRNAs, genetic mutations, and metabolic products could be envisioned that can then be employed to screen asymptomatic populations [14].

PDAC BIOMARKERS IN PANCREATIC JUICE

Similar to the search for diagnostic biomarkers of PDAC in serum, numerous investigators have explored diagnostic biomarkers of PDAC in pancreatic juice (PJ). The rational for this is that approximately 1.5 L PJ are excreted daily into the small bowel. Multiple investigators hypothesized that secreted pancreatic proteins and shed cells, which represent potential diagnostic biomarkers for PDAC, should be detected in higher concentrations in PJ due to its proximity to the pancreas and the pancreatic ductal system [15,16]. However, PJ as a source of biomarkers has similar limitations compared to other common specimens reflecting the need to distinguish between pancreatic malignancies and benign pancreatic conditions. An early study addressing K-ras mutations in PJ detected its occurrence in 61% of PDAC-derived specimens and in 10% of chronic pancreatitis cases [17].

PJ is collected following intravenous secretin injection. A common limitation of previous studies is that the PJ collection required cannulation of the pancreatic duct, which is usually performed in the setting of an Endoscopic Retrograde Cholangio-Pancreatography (ERCP). This method is invasive and associated with significant risks and adverse outcomes, which limits this approach for screening purposes. Therefore, the results of studies analyzing PJ following pancreatic duct cannulation must be interpreted with caution. The goal is to identify a reliable general diagnostic biomarker for PDAC that can also be detected in specimens that are collected using less invasive methods. Of note, the investigative group headed by Raimondo overcame the previously mentioned limitations by collecting PJ following intravenous secretin injection without cannulation of the pancreatic duct, utilizing PJ collection techniques employed for the measurement of pancreatic exocrine function [18].

Two recent studies evaluated the utility of bile collected during ERCP as a biomarker carrier for PDAC without the utilization of Secretin. Interestingly, in this context nasoduodenal tubes have not been evaluated as a tool to obtain PJ or bile specimens. Farina et al. measured potential protein biomarkers in bile collected during the workup for biliary strictures. The authors utilized liquid chromatography—tandem mass spectrometry with subsequent confirmation by immunoblotting and ELISA using specimens from 41 patients of which 23 had PDAC. Ten patients had a benign stricture, including eight patients with chronic pancreatitis. In this study, biliary carcinoembryonic cell adhesion molecule 6 (CEAM6) levels produced an AUC of 0.92 reflecting a sensitivity and specificity of 93% and 83%, respectively, for distinguishing benign and malignant biliary strictures. However, combining biliary CEAM6 and serum Ca19-9 level showed only a minimal improvement in diagnostic efficiency [19]. In another study, Zabron et al. measured neutrophil gelatinase-associated lipocalin (NGAL) level in bile from 16 patients with pancreatobiliary malignancies (including 8 PDAC) and 22 patients with benign disease (choledocholithiasis and chronic pancreatitis). NGAL levels were significantly higher in bile from patients with pancreatobiliary malignances. The calculated AUC of 0.76, reflected a sensitivity, specificity, positive predictive value, and negative predictive value of 94%, 55%, 60%, and 94%, respectively for distinguishing malignant from benign pancreatobiliary conditions. Combining biliary NGAL level and serum CA 19-9 improved the sensitivity, specificity, positive predictive value, and negative predictive value to 85%, 82%, 79%, and 87%, respectively [20].

Historically, cytology of shed pancreatic ductal cells in PJ obtained by ERCP in most cases was studied for its utility as diagnostic biomarker of PDAC. Unfortunately, its test performance was insufficient. Despite the drawbacks, several investigators improved this technique by increasing the PJ volume, by adding brushing cytology, or by combining cytology with known genetic alterations in PDAC. Both Iiboshi and Mikata placed a nasopancreatic drainage tube during ERCP and collected PJ 5 times (mean) or for a maximal of 6 times, respectively. In Iiboshi's study all 14 patients with PDAC yielded a positive cytology for malignancy. Sensitivity, specificity, and accuracy to detect PDAC were 100%, 83%, and 95%. However, the stage of PDAC was not specified [21]. Mikata's study included 40 patients with PDAC and 20 patients with benign pancreatic diseases and had sensitivity, specificity, positive predictive value, negative predictive value, and accuracy for PDAC of 80%, 100%, 100%, 71%, and 87% respectively [22]. The addition of brush cytology to PJ analysis in 127 patients with PDAC and 74 patients with benign pancreatic strictures increased the sensitivity for PDAC diagnosis compared with PJ alone from 21% to 62% [23]. Combining PJ cytology and endoscopic ultrasound fine needle aspiration (EUS-FNA) biopsy in 90 patients with pancreatic mass increased the sensitivity, specificity, positive predictive value, negative predictive value, and accuracy for PDAC from 86%, 100%, 100%, 70%, and 89% for EUS-FNA alone to 92%, 100%, 100%, 92%, and 96%, respectively. Of note, 29 of 90 patients had stage 1 and 2 PDAC [24]. Nakashima et al. studied human telomerase reverse transcriptase (hTERT) in PJ samples of 97 patients who underwent pancreatic resection, including 48 PDAC, 43 IPMN, and 6 patients with chronic pancreatitis and compared the test performance with cytology. hTERT is a catalytic subunit of human telomerase which is known to be activated in pancreatic cancer. Again, the sensitivity, specificity, positive predictive value, negative predictive value, and accuracy for PDAC increased from 47%, 89%, 93%, 35%, and 57% for cytology alone to 92%, 75%, 92%, 75%, and 88%, respectively for combined cytology and hTERT immunohistochemistry of the cytology specimen [25,26].

DNA mutations are well-documented in the carcinogenesis of PDAC, most commonly involving K-*ras*, *CDKN2A*, *SMAD4*, and *BRCA*. Multiple investigators addressed the utility of these mutations as biomarkers for PDAC using various specimens including PJ. The K-*ras* gene is the most extensively studied, and K-*ras* mutations are detected in up to 90% of PDAC tissue specimens. Of interest, a metaanalysis by Liu et al. included seven studies which measured K-*ras* mutations in PJ. The calculated pooled sensitivity, specificity, positive, and negative likelihood ratio, and diagnostic odds ratio to diagnose PDAC based on presence of K-*ras* mutations were 57%, 84%, 2.87, 0.55, and 6.05 respectively, which emphasize that the presence of K-*ras* mutation alone in PJ is insufficient to serve as a diagnostic biomarker for PDAC [27]. Subsequently, several investigators addressed whether the number of detected K-*ras* mutations in PJ proves to be of utility to distinguish PDAC from nonmalignant pancreatic diseases. Shi et al. found more than 0.5% mutant K-*ras* in relation to wild-type K-*ras* in 16 of 17 (94%) patients with PDAC compared to one of 9 (11%) patients with chronic pancreatitis, producing a sensitivity and specificity of 94% and 89%, respectively. Of note, 10 patients with PDAC, who did not have this particular K-*ras* mutation, were not included in the final calculation [28]. Eshleman et al. examined K-*ras* mutations in PJ of 194 patients with family history or genetic predisposition for PDAC, 30 patients with PDAC, and 30 controls with pancreatic cysts, pancreatitis, or normal pancreata. Three or more K-*ras* mutations were detected in 47% of patients with PDAC, 21% patients with risk for PDAC, and in 6% of the control cohort. The investigators suspected that the measured K-*ras* mutations likely arise from PanIN lesions. However, these lesions are not incorporated in any guidelines as of now, and clinical consequences for the general population are not understood, which again emphasizes that K-*ras* alone has no utility as diagnostic biomarker for PDAC [29].

Beyond known genetic mutations, multiple investigators detected differences in gene expression patterns in PDAC. However, none of these gene expression signatures serves as a reliable diagnostic biomarker for PDAC in PJ. The search for biomarkers in PJ is further complicated by the fact that mRNA expression levels frequently do not correlate between expression in PDAC tissue and in PJ, as shown by Oliveira-Cunha et al. In this study, only ANXA1 mRNA expression (of 30 genes examined) correlated between PDAC tissue and PJ. Of note, all of the genes analyzed displayed similar gene expression levels between PDAC and chronic pancreatitis [30]. S100A6 mRNA was found to be overexpressed in PDAC tissue, including PanIN lesions [31]. S100A6 mRNA expression was also measured in PJ and was found to be expressed at higher levels in patients with PDAC (mainly stage 4) and IPMN than in patients with chronic pancreatitis. A ROC analysis measured an AUC of 0.864 distinguishing PDAC from chronic pancreatitis [32].

Methylated DNA levels have been investigated as potential biomarkers for PDAC. The initial landmark studies by Tan et al. and Omura et al. evaluated different methylation profiles in pancreatic cancer cell lines, PDAC tissue, and normal pancreata. Of a total of 807 genes, Tan found 23 hypermethylated and hypomethylated genes in

PDAC [33]. Omura et al. analyzed the methylation status of promoters and CpG islands of 606 genes and identified hypermethylation most significantly in *MDFI*, hsa-miR-9-1, *ZNF415*, *CNTNAP2*, and *ELOVL4* [34].

PJ was evaluated in multiple studies for aberrantly methylated DNA levels as a potential diagnostic biomarker for PDAC. Matsubayashi et al. profiled aberrantly methylated DNA for 17 genes in PJ from patients with PDAC, IPMN, chronic pancreatitis, and controls with positive family history for PADC. Six genes, namely *Cyclin D2*, *FOXE1*, *NPTX2*, *ppENK*, *p16*, and *TFPI2* were further quantified using a cutoff of >1% methylated DNA. 82% (9 of 11) of patients with cancer had >1% methylation for two or more genes, whereas none of 64 individuals without neoplasia demonstrated this level of DNA methylation. Calculated sensitivity and specificity to predict PDAC was 82% and 100%, respectively. A limitation to the findings of this study was that the investigators noted a higher prevalence of methylated DNA in PJ of patients with chronic pancreatitis [35].

Watanabe et al. evaluated the methylation status of secreted apoptosis-related protein-2 (SARP2) gene in PJ after previous studies demonstrated aberrantly methylated SARP2 frequently in PDAC tissue, but not in normal pancreatic tissue. PJ was collected from 33 patients with PDAC, 20 IPMN, 19 chronic pancreatitis, and 10 control patients. Methylated *SARP2* was detected in 26 of 33 (79%) PDAC-derived PJ samples and 17 of 20 (85%) from IPMN. However, only 1 of 19 (5%) PJ samples from chronic pancreatitis patients exhibited methylated *SARP2*, and none of the 10 controls exhibited methylated *SARP2*. Although significant differences were also detected in terms of the concentration of methylated *SARP2*, testing performance was not calculated [36].

Most recently, Yokoyama et al. analyzed the methylation status of Mucins (*MUC*) genes in 45 patients with PDAC and IPMN. MUCs are known to play crucial role in carcinogenesis. Utilizing methylation-specific electrophoresis, the DNA methylation status of *MUC1*, *MUC2*, and *MUC4* in PJ differentiated PDAC with a specificity and sensitivity of 87% and 80%, discriminated intestinal-type IPMN with a specificity and sensitivity of 100% and 88%, and distinguished gastric-type IPMN with a specificity and sensitivity of 88% and 77% [37]. Additional studies found aberrant DNA methylation of several genes in IPMN, MCN, and PDAC. However, their test performances as diagnostic biomarkers were less promising [38–43].

Protein-based biomarkers have also been analyzed in PJ. Chromatography techniques are ideal for this type of discovery research. The following studies are of particular interest. The proteome of PJ was initially profiled by Gronborg et al. who performed a liquid chromatography—tandem mass spectrometry on PJ of three patients with PDAC. One hundred and seventy proteins were identified including CEA, MUC1, HIP/PAP, and PAP-2 of which 23 were present in all 3 patients with PDAC. However, a comparison with a control group was not performed [44]. Chen et al. used an isotope-code affinity tag (ICAT) technology with mass spectrometry for PJ from patients with PDAC and normal controls, and found 30 proteins to be at least two-fold upregulated in PDAC. Of these proteins, insulin-like growth factor binding protein 2 (IGFBP-2) was confirmed by immunoblotting utilizing PJ and pancreatic tissue [15]. A subsequent analysis by Chen et al. utilizing similar methods identified 27 proteins to be abundant in PJ from 1 patient with chronic pancreatitis compared with PJ of 10 normal controls. When analyzing both studies, Chen et al. identified 21 proteins with altered levels only in PJ from PDAC [15,45].

Utilizing a two-dimensional electrophoresis with subsequent MALDI-TOF mass spectrometry of PJ, Park et al. identified 26 upregulated proteins in PDAC compared with chronic pancreatitis and controls. Three of these proteins, namely lithostathine-1-alpha (REG1alpha), brefeldin A-inhibited guanine nucleotide-exchange protein 2 (BIG2), and peroxiredoxin 6 (PRDX6), were confirmed with immunohistochemistry in PDAC tissue. REG1alpha was further evaluated as a biomarker in serum, and yielded a sensitivity and specificity of 83% and 81%, respectively, for distinguishing PDAC from normal controls. Lower performance was noticed for chronic pancreatitis [46]. Using a comparable two-dimensional electrophoresis for PJ, Gao et al. found serine proteinase 2 (PRSS2) preproprotein and pancreatic lipase-related protein-1 (PLRP1) to be upregulated, and chymotrypsinogen B (CTRB) precursor and elastase 3B (ELA3B) preproprotein to be downregulated in PDAC compared with chronic pancreatitis and controls [16].

Distinguishing premalignant lesions from PDAC is crucial for a reliable diagnostic biomarker. In this context, two studies evaluated biomarkers for PanIN and IPMN. Chen et al. identified 20 proteins that were 2-fold to 10-fold increase in PJ of 3 patients with PanIN-3 compared with 5 controls. Of these proteins, the anterior gradient-2 (AGR2) protein was further confirmed to be elevated by ELISA in PJ in a cohort of 25 patients with PanIN-2, PanIN-3, and IPMN, as well as 8 patients with PDAC (stage 2—4). AGR2 reached a sensitivity and specificity of 67% and 90%, respectively, differentiating PanIN-3 from nonmalignant conditions [47]. Focusing on premalignant IPMN lesions, Shirai utilized surface-enhanced laser desorption and ionization time-of-flight (SELDI-TOF) mass spectrometry for PJ from 33 patients with IPMN, 54 with PDAC, and 31 with chronic pancreatitis. Spectrometry identified a significantly higher peak at 6240-Da in

IPMN derived PJ than in PDAC and chronic pancreatitis. Further analysis targeted pancreatic secretory trypsin inhibitor (PSTI). Utilizing a diagnostic cutoff value of 25,000 ng/mL for PSTI, sensitivity, specificity, positive and negative predictive value were 48%, 98%, 89%, and 83% respectively, to diagnose IPMN [48].

Addressing specific protein biomarkers in PJ, Rosty et al. identified hepatocarcinoma-intestine-pancreas/pancreatitis-associated protein I (HIP/PAP-I) in 10 of 15 (67%) PJ samples from PDAC patients with mostly advanced disease stage utilizing SELDI mass spectrometry and Protein Chip technology. Subsequently, ELISA of PJ from 28 patients with PDAC and 15 controls was utilized and HIP/PAP-I levels were found to be significantly higher in PJ-derived from PDAC. The investigators calculated the sensitivity and specificity of 75% and 87% respectively, for differentiating PDAC from controls [49]. Tian et al. identified 14 upregulated proteins inducing matrix metalloproteinase-9 (MMP-9), oncogene DJ1 (DJ-1), and alpha-1B-glycoprotein precursor (A1BG) based on difference gel electrophoresis (DIGE) and tandem mass spectrometry in PJ from 9 patients with PDAC compared with 9 cancer-free controls. The protein was also confirmed with Western blot, but test performance based on these biomarkers was not calculated [50]. Kaur et al. measured NGAL, macrophage inhibitory cytokine 1 (MIC-1), and CA19-9 in PJ from 58 patients with PDAC, 24 with chronic pancreatitis, and 23 with no pancreatic disease. NGAL reached the highest sensitivity and specificity of 79% and 83% differentiating PDAC from controls, whereas only MIC-1 level was significantly different between PDAC and chronic pancreatitis [51].

Of note, biliary obstruction was shown to affect the protein composition of PJ significantly. Zhou et al. performed a two-dimensional electrophoresis on PJ and identified proteins by MALDI-TOF mass spectrometry in five patients with PDAC, six patients with benign pancreatic disease, and three patients with cholelithiasis. In this study, biliary obstruction affected the protein composition of PJ significantly [52]. Similarly, Yan et al. found that only transthyretin in PJ but not apolipoprotein A1 or apolipoprotein E were associated with PDAC. Moreover, differences of apolipoprotein A1 and apolipoprotein E between PJ of PDAC and control cohorts originated from biliary obstruction [53].

MicroRNAs are small RNAs which regulate gene expression. Alterations in their levels have been recently found in multiple cancers, including PDAC [54,55]. Sadakari et al. analyzed miR-21 and miR-155 levels (which are known to be overexpressed in PDAC) in PJ of 16 patients with PDAC (mostly stage 3 and 4) and 5 patients with chronic pancreatitis. Real-time reverse transcription-PCR found significant elevation of both microRNAs in PDAC-derived PJ compared with the control cohort [56]. Another study by Wang et al. was of particular interest. First, the investigators identified alterations of microRNAs in PJ from PDAC patients by microarray analysis. Then, the results were validated using a panel of microRNAs from 50 patients with PDAC, 19 with chronic pancreatitis, and 19 controls. Combining miR-205, miR-210, miR-492, miR-1427, and serum 19-9 levels predicted PDAC with a sensitivity and specificity of 91% and 100% [57]. Habbe et al. evaluated microRNA levels in PJ from patients with IPMN which can be, second to chronic pancreatitis, particularly challenging to distinguishing from PDAC. Again miR-21 (mean 11.6-fold) and miR-155 (mean 12.1-fold) were found in significantly higher levels in PJ of 10 patients with IPMN compared to controls. Although the authors concluded that aberrant miRNA expression is an early event in pancreatic cancer, it became clear that microRNA cannot distinguish IPMN from PDAC [58].

Beyond protein, genetic, and epigenetic biomarkers, several investigators have focused on PJ-derived biomarkers which are seldom recognized as a diagnostic biomarker for malignancies. Most interestingly, Noh et al. measured cytokine levels of interleukin-8 (IL8), interleukin-6 (IL6), transforming growth factor-beta1 (TGF-β1), and intercellular adhesion molecule 1 (ICAM-1) in PJ of 38 patients with PDAC, 39 with chronic pancreatitis, and 41 normal controls. A multivariate analysis detected only IL8 as a potential diagnostic biomarker, distinguishing PDAC from normal pancreas and from chronic pancreatitis with an AUC of 0.9 and 0.67, respectively [18]. Another study of interest measured the concentration of heavy metals in PJ. The investigators found elevated chromium, selenium, and molybdenum concentrations in 35 patients with PDAC compared with 30 patients with chronic pancreatitis, and 35 controls, but test accuracy was not calculated [59]. The most significant differentially expressed proteins, miRNA, and methylated DNA in PJ and bile derived from patients with PDAC are summarized in Table 27.1. For comparison, the most significant differentially expressed proteins, miRNA, and methylated DNA in serum and plasma derived from patients with PDAC are summarized in Table 27.2.

PDAC BIOMARKERS IN STOOL

Similar to the rational search for pancreatic biomarkers in PJ, several investigators hypothesized that pancreatic biomarkers should also be detectable in feces since 1.5 L of PJ are secreted into the intestine daily of which a fraction undergoes fecal excretion [60–63]. Tobi et al. analyzed the utility of Adnab-9 monoclonal antibodies as biomarker in feces for PDAC. This antibody was previously shown to be a potential diagnostic biomarker for colorectal neoplasia. The authors analyzed 1132 stool samples from the United States and another 249 stool samples from China which were initially utilized for colorectal cancer screening purposes. 15 patients in the Chinese patient cohort eventually developed pancreatic cancer, at a median of 2.3 years following stool submission. Interestingly, 12 of those patients (80%) had positive stool binding for Adnab-9, suggesting Adnab-9 as a potential screening tool. However, calculated sensitive and specificity ranged only from 67% to 80% and 87% to 91%, respectively [64].

An early study by Lu et al. addressed the utility of K-*ras* and *p53* mutations as biomarkers for PDAC. Both mutations were previously detected in pancreatic tissue specimens of PDAC. The investigators collected feces of patients undergoing resection for PDAC and 60 controls (chronic pancreatitis, pancreatic adenoma, and pancreatic endocrine neoplasm). K-*ras* mutation was detected in 66 of 75 (88%) patients with PDAC and in 24 of 47 patients (52%) of the control cohort. Similarly, a significant overlap between PDAC and controls was observed for *p53* mutations. *p53* mutations were detected in 23 of 62 (37%) patients with PDAC, and 4 of 21 (19%) patients with chronic pancreatitis. Of note, the authors also measured *p53* mutations in PJ, which were detected in 47% of PDAC and 13% of control PJ specimens [63]. Focusing on genetic markers, Kisiel et al. analyzed epigenetic alterations in fecal specimens as a potential biomarker for PDAC. The investigators measured the concentration of four methylated target genes which were previously detected in elevated levels in PDAC tissue specimens, namely *EYA4*, *MDFI*, *UCHL1*, and *BMP3*. The investigators included 58 stool samples of patients with PDAC (stage 1: 5%, stage 2: 35%, stage 3: 26%, stage 4: 33%) and 65 age-matched and sex-matched healthy controls. *BMP3* yielded the best performance, with a sensitivity of 51% and specificity of 90%. Utilizing a panel consisting of two K-*ras* mutations, age, and methylated *BMP3* as biomarkers, sensitivities of 52–79% were reached. At a set specificity of 90%, the sensitivity was 67% [61].

Link et al. focused on microRNA detection in fecal samples. The authors measured seven microRNAs which are known to be frequently deregulated in PDAC (miR-21, miR-143, miR-155, miR-196a, miR-210, miR-216a, and miR-375) in stool samples derived from 15 patients with PDAC (11 of 15 with stage 4), 15 patients with chronic pancreatitis, and 15 healthy controls. miR-143, miR-155, miR-196a, and miR-216a were found in lower concentrations in the stool of patients with PDAC compared with controls and chronic pancreatitis. Combined microRNA expression of miR-143,

TABLE 27.1 Summary of the Most Significantly Differentially Expressed (A) proteins, (B) miRNA, and (C) Methylated DNA in Pancreatic Juice (PJ) and Bile Derived from Patients with PDAC, in Comparison with patients with Benign Pancreatic Conditions or Normal Controls. Most Markers Were Identified from Proteomics and Microarray Analysis

A—PROTEINS IN PJ OR BILE DIFFERENTIALLY EXPRESSED IN PDAC

14-3-3 Protein sigma

Alpha-1B-glycoprotein precursor (A1BG)

Annexin A4 (ANXA4)

Anterior gradient homolog 2(AGR2)

Anterior gradient-2 (AGR2) protein

Beta 2 microglobulin

Brefeldin A-inhibited guanine nucleotide-exchange protein 2 (BIG2)

Carbohydrate antigen (CA19-9)

Carcinoembryonic antigen (CEA)

Carcinoembryonic cell adhesion molecule 6 (CEAM6)

Chymotrypsinogen B (CTRB) precursor

Collagen alpha-1 (IV) chain (COL6A1)

Elastase 3B (ELA3B)

Hepatocarcinoma-intestine-pancreas/pancreatitis-associated protein I (HIP/PAP-I)

Insulin-like growth factor binding protein 2 (IGFBP-2)

Lithostathine-1-beta precursor

Lithostathine-1-alpha (REG1alpha)

Macrophage inhibitory cytokine 1 (MIC-1)

Matrix metalloproteinase-9 (MMP-9)

Mucin 1 (MUC1)

Neutrophil gelatinase-associated lipocalin (NGAL)

Olfactomedin 4 (OLFM4)

Oncogene DJ1 (DJ-1)

Pancreatic elastase 3B (CEL3B)

Pancreatic lipase-related protein-1 (PLRP1)

Pancreatic secretory trypsin inhibitor (PSTI)

Pancreatitis-associated protein 2 (PAP-2)

Peroxiredoxin 6 (PRDX6)

S100A10

S100A8

S100A9

Serine proteinase 2 (PRSS2)

Syncollin (SYNC)

Transthyretin (TTR)

(Continued)

TABLE 27.1 (Continued)

B—MIRNA IN PJ OR BILE DIFFERENTIALLY EXPRESSED IN PDAC

miR-18a

miR-21

miR-31

miR-93

miR-155

miR-196a

miR-205

miR-210

miR-216 (downregulated)

miR-217 (downregulated)

miR-221

miR-224

miR-492

miR-1427

C—METHYLATED DNA IN PJ OR BILE DIFFERENTIALLY EXPRESSED IN PDAC

Apoptosis-related protein-2 gene (SARP2)

Contactin associated protein-like 2 (CNTNAP2)

Cyclin D2

Cyclin-dependent kinase inhibitor 2A (p16)

FOXE1 (Forkhead box E1)

Mucin 1 (MUC1)

Mucin 2 (MUC2)

Mucin 4 (MUC4)

MyoD family inhibitor (MDFI)

Neuronal pentraxin II (NPTX2)

Proenkephalin (ppENK)

Tissue factor pathway inhibitor 2 (TFPI2)

Zinc finger protein 415 (ZNF415)

miR-155, miR-196a, and miR-216a resulted in differences between the three cohorts with greater significance, although the sensitivity and specificity were not calculated [62].

In summary, fecal material has gained wide acceptance as screening specimen since its introduction as a screening method for colorectal neoplasia. Preliminary reports of fecal material as a biomarker source in PDAC are encouraging, and test performances resemble the early studies of serum and PJ as PDAC biomarker sources.

TABLE 27.2 Summary of the Most Significantly Differentially Expressed (A) Proteins, (B) miRNA, and (C) Methylated DNA in Serum and Plasma Derived from Patients with PDAC, in Comparison with Patients with Benign Pancreatic Conditions or Normal Controls

A—PROTEINS IN SERUM OR PLASMA DIFFERENTIALLY EXPRESSED IN PDAC

Activated leukocyte cell adhesion molecule (ALCAM)

Adaptor-related protein complex 2, alpha 1 (AP2A1)

Adrenomedullin (ASM)

Alpha-1 chymotrypsin (AACT)

Annexin A1 (ANXA1)

Annexin A2 (ANXA2)

Anterior gradient-2 (AGR2)

Basigin (BSG)

Caldesmon 1 (CALD1)

Carbohydrate antigen (CA19-9)

Carcinoembryonic antigen (CEA)

Carcinoembryonic antigen-related cell adhesion molecule 1 (CEACAM1)

Carcinoembryonic antigen-related cell adhesion molecule 5 (CEACAM5)

Collagen alpha-1(VI) (COL6A1)

CYFRA 21-1

Cytokeratin-18 (CK 18)

Fascin actin-bundling protein 1 (FSCN1)

Heat shock protein 27 (Hsp 27)

Heat shock protein 70 (HSP 70)

Hematopoietic cell kinase (Hck)

Insulin-like growth factor binding protein 2 (IGFBP-2)

Insulin-like growth factor binding protein 4 (IGFBP4)

Intercellular adhesion molecule 1 (ICAM-1)

L1 cell adhesion molecule (L1CAM)

Laminin, gamma 2 (LAMC2)

Lipocalin 2 (LCN2)

Macrophage inhibitory cytokine 1 (MIC-1)

Matrix metallopeptidase 7 (MMP7)

Matrix metallopeptidase 9 (MMP-9)

Matrix metallopeptidase 11 (MMP11)

Menkes protein (ATP7A, MNK)

Mesothelin (MSLN)

Mucin 4 (MUC4)

Mucin 5AC (MUC5AC)

(Continued)

TABLE 27.2   (Continued)

Neuropilin 1 (NRP1)

Olfactomedin 4 (OLFM4)

Osteoprotegerin (OPG)

Plasminogen activator urokinase

Plectin-1

Polymeric immunoglobulin receptor (pIgR)

Proliferation-inducing ligand (APRIL)

Regenerating islet-derived 1alpha (REG1alpha)

Regenerating islet-derived 3alpha (REG3A)

S100A6

S100P

Secreted phosphoprotein 1 (SPP1)

Serine protease inhibitor (SPINK1)

SNAIL

Spark-like 1 (SPARCL1)

Syncollin (SYNC)

Tenascin-C (TNC)

Thrombospondin-1 (THBS1)

Thrombospondin 2 (THBS2)

Tissue inhibitor of metalloproteinase 1 (TIMP-1)

Trefoil factor 1 (TFF1)

UL16 binding protein 2 (ULBP2)

B—MIRNA IN SERUM OR PLASMA DIFFERENTIALLY EXPRESSED IN PDAC

miR-10

miR-18a

miR-20a

miR-21

miR-24

miR-25

miR-99a

miR-100a

miR-155

miR-191

miR-196a

miR-200a

miR-200b

miR-210

miR-216

miR-217

(Continued)

TABLE 27.2 (Continued)

C—METHYLATED DNA IN SERUM OR PLASMA DIFFERENTIALLY EXPRESSED IN PDAC

Cyclin-Dependent Kinase Inhibitor 1C (P57)

Glypican 3 (GPC3)

HIN

Hippel-Lindau disease (VHL)

HSHLTF1

HUM

Human MyoD1 (MYF3)

Mucin 2 (MUC2)

Serum deprivation response factor (SRBC)

TMS

Most of these markers were identified from proteomics and microarray analysis.

PDAC BIOMARKERS IN URINE

Urine is a commonly used specimen for a multitude of testing given its noninvasive nature. Despite the advantages of urine-based testing, significantly fewer studies have evaluated urine as a potential source for diagnostic biomarkers for PDAC compared with serum or PJ. Weeks et al. performed two-dimensional differential gel electrophoresis on urine samples from patients with PDAC, chronic pancreatitis, and healthy controls. one hundred and twenty seven differentially expressed protein spots were detected, of which 101 were further identified as annexin A2, gelsolin, and CD59 using MALDI-TOF mass spectrometry. Due to technical challenges, the investigators did not validate the markers with immunoblotting, which prevented a calculation of their test performance [65].

PDAC BIOMARKERS IN SALIVA

Another promising approach is the detection of cancer biomarkers in saliva given its simple specimen collection. Zhang et al. performed a gene chip analysis on saliva derived from PDAC patients. Upregulated mRNAs were identified. Four of them were used as a panel (K-*ras*, *MBD3L2*, *ACRV1*, *DPM*) to distinguish 30 patients with PDAC from 30 patients with chronic pancreatitis and 30 healthy controls. ROC analysis calculated a sensitivity and specificity of 90% and 95%, respectively. This study provided proof of principle and demonstrated the feasibility of detecting PDAC biomarkers in saliva [66].

Roy et al. analyzed urinary levels of matrix metalloproteases (uMMPs) and their endogenous inhibitors, tissue inhibitor of metalloproteases (uTIMPs), by ELISA. Both protein groups have been previously described to be elevated in PDAC. The investigators included urine samples of 51 patients with PDAC, 28 with pancreatic neuroendocrine tumors, and 60 healthy controls. Using a multivariable logistic regression analyses controlling for age, sex, uMMP-2, and uTIMP-1, significant differences were identified between PDAC patients and healthy controls. ROC analysis revealed a sensitivity and specificity of 91% and 75%, respectively, for the panel of uMMP-2 and uTIMP-1 [67]. Most recently, Davis et al. performed a metabolomic study on urine of 32 patients with PDAC (mostly stage 2), 25 patients with chronic pancreatic and cystic neoplasms, and 32 healthy controls. The investigators achieved a particularly high AUC distinguishing PDAC from controls. These very promising results are yet to be confirmed in large cohorts [68].

SUMMARY

Tremendous progress has been made over the last two decades to detect reliable diagnostic biomarkers for PDAC. Common obstacles encountered include: (1) the heterogeneity of PDAC which makes a single diagnostic marker unlikely to be successful, and (2) the close molecular biological relationship to known premalignant conditions, namely chronic pancreatitis, MCN, and IPMN. Studies utilizing a panel of biomarkers achieved better test performance than single biomarker studies. An additional challenge for interpretation of these studies reflects the common inclusion of patients with advanced stage PDAC which are not curable. The future search for diagnostic biomarkers for PDAC will require that studies are limited to the earliest invasive stage of PDAC, that all premalignant PDAC conditions are included as controls, and that utilization of a panel of biomarkers is required.

References

[1] Bausch D, Thomas S, Mino-Kenudson M, Fernandez-del CC, Bauer TW, Williams M, et al. Plectin-1 as a novel biomarker for pancreatic cancer. Clin Cancer Res 2011;17:302–9.

[2] Brand RE, Nolen BM, Zeh HJ, Allen PJ, Eloubeidi MA, Goldberg M, et al. Serum biomarker panels for the detection of pancreatic cancer. Clin Cancer Res 2011;17:805–16.

[3] Firpo MA, Gay DZ, Granger SR, Scaife CL, DiSario JA, Boucher KM, et al. Improved diagnosis of pancreatic adenocarcinoma using haptoglobin and serum amyloid A in a panel screen. World J Surg 2009;33:716–22.

[4] Zhou W, Sokoll LJ, Bruzek DJ, Zhang L, Velculescu VE, Goldin SB, et al. Identifying markers for pancreatic cancer by gene expression analysis. Cancer Epidemiol Biomarkers Prev 1998;7:109–12.

[5] Jones S, Zhang X, Parsons DW, Lin JC, Leary RJ, Angenendt P, et al. Core signaling pathways in human pancreatic cancers revealed by global genomic analyses. Science 2008;321:1801–6.

[6] Goonetilleke KS, Siriwardena AK. Systematic review of carbohydrate antigen (CA 19-9) as a biochemical marker in the diagnosis of pancreatic cancer. Eur J Surg Oncol 2007;33:266–70.

[7] Nie S, Lo A, Wu J, Zhu J, Tan Z, Simeone DM, et al. Glycoprotein biomarker panel for pancreatic cancer discovered by quantitative proteomics analysis. J Proteome Res 2014;13:1873–84.

[8] Chan A, Prassas I, Dimitromanolakis A, Brand RE, Serra S, Diamandis EP, et al. Validation of biomarkers that complement CA19.9 in detecting early pancreatic cancer. Clin Cancer Res 2014;20:5787–95.

[9] Makawita S, Dimitromanolakis A, Soosaipillai A, Soleas I, Chan A, Gallinger S, et al. Validation of four candidate pancreatic cancer serological biomarkers that improve the performance of CA19.9. BMC Cancer 2013;13:404.

[10] Schultz NA, Dehlendorff C, Jensen BV, Bjerregaard JK, Nielsen KR, Bojesen SE, et al. MicroRNA biomarkers in whole blood for detection of pancreatic cancer. JAMA 2014;311:392–404.

[11] Wang WS, Liu LX, Li GP, Chen Y, Li CY, Jin DY, et al. Combined serum CA19-9 and miR-27a-3p in peripheral blood mononuclear cells to diagnose pancreatic cancer. Cancer Prev Res 2013;6:331–8.

[12] Jenkinson C, Elliott V, Menon U, Apostolidou S, Fourkala OE, Gentry-Maharaj A, et al. Evaluation in pre-diagnosis samples discounts ICAM-1 and TIMP-1 as biomarkers for earlier diagnosis of pancreatic cancer. J Proteomics 2014;113C:400–2.

[13] Nolen BM, Brand RE, Prosser D, Velikokhatnaya L, Allen PJ, Zeh HJ, et al. Prediagnostic serum biomarkers as early detection tools for pancreatic cancer in a large prospective cohort study. PLoS One 2014;9:e94928.

[14] Firpo MA, Boucher KM, Mulvihill SJ. Prospects for developing an accurate diagnostic biomarker panel for low prevalence cancers. Theor Biol Med Model 2014;11:34.

[15] Chen R, Pan S, Yi EC, Donohoe S, Bronner MP, Potter JD, et al. Quantitative proteomic profiling of pancreatic cancer juice. Proteomics 2006;6:3871–9.

[16] Gao J, Zhu F, Lv S, Li Z, Ling Z, Gong Y, et al. Identification of pancreatic juice proteins as biomarkers of pancreatic cancer. Oncol Rep 2010;23:1683–92.

[17] Costentin L, Pages P, Bouisson M, Berthelemy P, Buscail L, Escourrou J, et al. Frequent deletions of tumor suppressor genes in pure pancreatic juice from patients with tumoral or nontumoral pancreatic diseases. Pancreatology 2002;2:17–25.

[18] Noh KW, Pungpapong S, Wallace MB, Woodward TA, Raimondo M. Do cytokine concentrations in pancreatic juice predict the presence of pancreatic diseases? Clin Gastroenterol Hepatol 2006;4:782–9.

[19] Farina A, Dumonceau JM, Antinori P, Annessi-Ramseyer I, Frossard JL, Hochstrasser DF, et al. Bile carcinoembryonic cell adhesion molecule 6 (CEAM6) as a biomarker of malignant biliary stenoses. Biochim Biophys Acta 2014;1844:1018–25.

[20] Zabron AA, Horneffer-van der Sluis VM, Wadsworth CA, Laird F, Gierula M, Thillainayagam AV, et al. Elevated levels of neutrophil gelatinase-associated lipocalin in bile from patients with malignant pancreatobiliary disease. Am J Gastroenterol 2011;106:1711–17.

[21] Iiboshi T, Hanada K, Fukuda T, Yonehara S, Sasaki T, Chayama K. Value of cytodiagnosis using endoscopic nasopancreatic drainage for early diagnosis of pancreatic cancer: establishing a new method for the early detection of pancreatic carcinoma in situ. Pancreas 2012;41:523–9.

[22] Mikata R, Ishihara T, Tada M, Tawada K, Saito M, Kurosawa J, et al. Clinical usefulness of repeated pancreatic juice cytology via endoscopic naso-pancreatic drainage tube in patients with pancreatic cancer. J Gastroenterol 2013;48:866–73.

[23] Yamaguchi T, Shirai Y, Nakamura N, Sudo K, Nakamura K, Hironaka S, et al. Usefulness of brush cytology combined with pancreatic juice cytology in the diagnosis of pancreatic cancer: significance of pancreatic juice cytology after brushing. Pancreas 2012;41:1225–9.

[24] Matsumoto K, Takeda Y, Harada K, Horie Y, Yashima K, Murawaki Y. Effect of pancreatic juice cytology and/or endoscopic ultrasound-guided fine-needle aspiration biopsy for pancreatic tumor. J Gastroenterol Hepatol 2014;29:223–7.

[25] Nakashima A, Murakami Y, Uemura K, Hayashidani Y, Sudo T, Hashimoto Y, et al. Usefulness of human telomerase reverse transcriptase in pancreatic juice as a biomarker of pancreatic malignancy. Pancreas 2009;38:527–33.

[26] Suehara N, Mizumoto K, Tanaka M, Niiyama H, Yokohata K, Tominaga Y, et al. Telomerase activity in pancreatic juice differentiates ductal carcinoma from adenoma and pancreatitis. Clin Cancer Res 1997;3:2479–83.

[27] Liu SL, Chen G, Zhao YP, Wu WM, Zhang TP. Diagnostic accuracy of K-ras mutation for pancreatic carcinoma: a meta-analysis. Hepatobiliary Pancreat Dis Int 2013;12:458–64.

[28] Shi C, Fukushima N, Abe T, Bian Y, Hua L, Wendelburg BJ, et al. Sensitive and quantitative detection of KRAS2 gene mutations in pancreatic duct juice differentiates patients with pancreatic cancer from chronic pancreatitis, potential for early detection. Cancer Biol Ther 2008;7:353–60.

[29] Eshleman JR, Norris AL, Sadakari Y, Debeljak M, Borges M, Harrington C, et al. KRAS and guanine nucleotide-binding protein mutations in pancreatic juice collected from the duodenum of patients at high risk for neoplasia undergoing endoscopic ultrasound. Clin Gastroenterol Hepatol 2015;13:963–9.

[30] Oliveira-Cunha M, Byers RJ, Siriwardena AK. Poly(A) RT-PCR measurement of diagnostic genes in pancreatic juice in pancreatic cancer. Br J Cancer 2011;104:514–19.

[31] Ohuchida K, Mizumoto K, Ishikawa N, Fujii K, Konomi H, Nagai E, et al. The role of S100A6 in pancreatic cancer development and its clinical implication as a diagnostic marker and therapeutic target. Clin Cancer Res 2005;11:7785–93.

[32] Ohuchida K, Mizumoto K, Yu J, Yamaguchi H, Konomi H, Nagai E, et al. S100A6 is increased in a stepwise manner during pancreatic carcinogenesis: clinical value of expression analysis in 98 pancreatic juice samples. Cancer Epidemiol Biomarkers Prev 2007;16:649–54.

[33] Tan AC, Jimeno A, Lin SH, Wheelhouse J, Chan F, Solomon A, et al. Characterizing DNA methylation patterns in pancreatic cancer genome. Mol Oncol 2009;3:425–38.

[34] Omura N, Li CP, Li A, Hong SM, Walter K, Jimeno A, et al. Genome-wide profiling of methylated promoters in pancreatic adenocarcinoma. Cancer Biol Ther 2008;7:1146–56.

[35] Matsubayashi H, Canto M, Sato N, Klein A, Abe T, Yamashita K, et al. DNA methylation alterations in the pancreatic juice of patients with suspected pancreatic disease. Cancer Res 2006;66:1208–17.

[36] Watanabe H, Okada G, Ohtsubo K, Yao F, Jiang PH, Mouri H, et al. Aberrant methylation of secreted apoptosis-related protein 2 (SARP2) in pure pancreatic juice in diagnosis of pancreatic neoplasms. Pancreas 2006;32:382–9.

[37] Yokoyama S, Kitamoto S, Higashi M, Goto Y, Hara T, Ikebe D, et al. Diagnosis of pancreatic neoplasms using a novel method of DNA methylation analysis of mucin expression in pancreatic juice. PloS One 2014;9:e93760.

[38] Sato N, Ueki T, Fukushima N, Iacobuzio-Donahue CA, Yeo CJ, Cameron JL, et al. Aberrant methylation of CpG islands in intraductal papillary mucinous neoplasms of the pancreas. Gastroenterology 2002;123:365–72.

[39] Hong SM, Kelly D, Griffith M, Omura N, Li A, Li CP, et al. Multiple genes are hypermethylated in intraductal papillary mucinous neoplasms of the pancreas. Modern Pathology 2008;21:1499–507.

[40] Kim SG, Wu TT, Lee JH, Yun YK, Issa JP, Hamilton SR, et al. Comparison of epigenetic and genetic alterations in mucinous cystic neoplasm and serous microcystic adenoma of pancreas. Mod Pathol 2003;16:1086–94.

[41] Fukushima N, Sato N, Ueki T, Rosty C, Walter KM, Wilentz RE, et al. Aberrant methylation of preproenkephalin and p16 genes in pancreatic intraepithelial neoplasia and pancreatic ductal adenocarcinoma. Am J Pathol 2002;160:1573–81.

[42] Sato N, Fukushima N, Hruban RH, Goggins M. CpG island methylation profile of pancreatic intraepithelial neoplasia. Mod Pathol 2008;21:238–44.

[43] Sato N, Fukushima N, Matsubayashi H, Iacobuzio-Donahue CA, Yeo CJ, Goggins M. Aberrant methylation of Reprimo correlates with genetic instability and predicts poor prognosis in pancreatic ductal adenocarcinoma. Cancer 2006;107:251–7.

[44] Gronborg M, Bunkenborg J, Kristiansen TZ, Jensen ON, Yeo CJ, Hruban RH, et al. Comprehensive proteomic analysis of human pancreatic juice. J Proteome Res 2004;3:1042–55.

[45] Chen R, Pan S, Cooke K, Moyes KW, Bronner MP, Goodlett DR, et al. Comparison of pancreas juice proteins from cancer versus pancreatitis using quantitative proteomic analysis. Pancreas 2007;34:70–9.

[46] Park JY, Kim SA, Chung JW, Bang S, Park SW, Paik YK, et al. Proteomic analysis of pancreatic juice for the identification of biomarkers of pancreatic cancer. J Cancer Res Clin Oncol 2011;137:1229–38.

[47] Chen R, Pan S, Duan X, Nelson BH, Sahota RA, de Rham S, et al. Elevated level of anterior gradient-2 in pancreatic juice from patients with pre-malignant pancreatic neoplasia. Mol Cancer 2010;9:149.

[48] Shirai Y, Sogawa K, Yamaguchi T, Sudo K, Nakagawa A, Sakai Y, et al. Protein profiling in pancreatic juice for detection of intraductal papillary mucinous neoplasm of the pancreas. Hepatogastroenterology 2008;55:1824–9.

[49] Rosty C, Christa L, Kuzdzal S, Baldwin WM, Zahurak ML, Carnot F, et al. Identification of hepatocarcinoma-intestine-pancreas/pancreatitis-associated protein I as a biomarker for pancreatic ductal adenocarcinoma by protein biochip technology. Cancer Res 2002;62:1868–75.

[50] Tian M, Cui YZ, Song GH, Zong MJ, Zhou XY, Chen Y, et al. Proteomic analysis identifies MMP-9, DJ-1 and A1BG as overexpressed proteins in pancreatic juice from pancreatic ductal adenocarcinoma patients. BMC Cancer 2008;8:241.

[51] Kaur S, Baine MJ, Guha S, Ochi N, Chakraborty S, Mallya K, et al. Neutrophil gelatinase-associated lipocalin, macrophage inhibitory cytokine 1, and carbohydrate antigen 19-9 in pancreatic juice: pathobiologic implications in diagnosing benign and malignant disease of the pancreas. Pancreas 2013;42:494–501.

[52] Zhou L, Lu Z, Yang A, Deng R, Mai C, Sang X, et al. Comparative proteomic analysis of human pancreatic juice: methodological study. Proteomics 2007;7:1345–55.

[53] Yan L, Tonack S, Smith R, Dodd S, Jenkins RE, Kitteringham N, et al. Confounding effect of obstructive jaundice in the interpretation of proteomic plasma profiling data for pancreatic cancer. J Proteome Res 2009;8:142–8.

[54] Szafranska AE, Davison TS, John J, Cannon T, Sipos B, Maghnouj A, et al. MicroRNA expression alterations are linked to tumorigenesis and non-neoplastic processes in pancreatic ductal adenocarcinoma. Oncogene 2007;26:4442–52.

[55] Wang J, Chen J, Chang P, LeBlanc A, Li D, Abbruzzesse JL, et al. MicroRNAs in plasma of pancreatic ductal adenocarcinoma patients as novel blood-based biomarkers of disease. Cancer Prev Res 2009;2:807–13.

[56] Sadakari Y, Ohtsuka T, Ohuchida K, Tsutsumi K, Takahata S, Nakamura M, et al. MicroRNA expression analyses in preoperative pancreatic juice samples of pancreatic ductal adenocarcinoma. J OP 2010;11:587–92.

[57] Wang J, Raimondo M, Guha S, Chen J, Diao L, Dong X, et al. Circulating microRNAs in pancreatic juice as candidate biomarkers of pancreatic cancer. J Cancer 2014;5:696–705.

[58] Habbe N, Koorstra JB, Mendell JT, Offerhaus GJ, Ryu JK, Feldmann G, et al. MicroRNA miR-155 is a biomarker of early pancreatic neoplasia. Cancer Biol Ther 2009;8:340–6.

[59] Carrigan PE, Hentz JG, Gordon G, Morgan JL, Raimondo M, Anbar AD, et al. Distinctive heavy metal composition of pancreatic juice in patients with pancreatic carcinoma. Cancer Epidemiol Biomarkers Prev 2007;16:2656–63.

[60] Haug U, Wente MN, Seiler CM, Jesenofsky R, Brenner H. Stool testing for the early detection of pancreatic cancer: rationale and current evidence. Expert Rev Mol Diagn 2008;8:753–9.

[61] Kisiel JB, Yab TC, Taylor WR, Chari ST, Petersen GM, Mahoney DW, et al. Stool DNA testing for the detection of pancreatic cancer: assessment of methylation marker candidates. Cancer 2012;118:2623–31.

[62] Link A, Becker V, Goel A, Wex T, Malfertheiner P. Feasibility of fecal microRNAs as novel biomarkers for pancreatic cancer. PloS One 2012;7:e42933.

[63] Lu X, Xu T, Qian J, Wen X, Wu D. Detecting K-ras and p53 gene mutation from stool and pancreatic juice for diagnosis of early pancreatic cancer. Chin Med J 2002;115:1632–6.

[64] Tobi M, Kim M, Weinstein DH, Rambus MA, Hatfield J, Adsay NV, et al. Prospective markers for early diagnosis and prognosis of sporadic pancreatic ductal adenocarcinoma. Dig Dis Sci 2013;58:744–50.

[65] Weeks ME, Hariharan D, Petronijevic L, Radon TP, Whiteman HJ, Kocher HM, et al. Analysis of the urine proteome in patients with pancreatic ductal adenocarcinoma. Proteomics Clin Appl 2008;2:1047–57.

[66] Zhang L, Farrell JJ, Zhou H, Elashoff D, Akin D, Park NH, et al. Salivary transcriptomic biomarkers for detection of resectable pancreatic cancer. Gastroenterology 2010;138:949–57.

[67] Roy R, Zurakowski D, Wischhusen J, Frauenhoffer C, Hooshmand S, Kulke M, et al. Urinary TIMP-1 and MMP-2 levels detect the presence of pancreatic malignancies. Br J Cancer 2014;111:1772–9.

[68] Davis VW, Schiller DE, Eurich D, Bathe OF, Sawyer MB. Pancreatic ductal adenocarcinoma is associated with a distinct urinary metabolomic signature. Ann Surg Oncol 2013;20:S415–23.

Molecular Testing in Gynecologic Cancer

S.E. Kerr

Department of Laboratory Medicine and Pathology, College of Medicine, Mayo Clinic, Rochester, MN, United States

BACKGROUND

The diagnosis of cancers specific to women has been traditionally within the anatomic pathologist's realm. However, gynecologic pathologists have made increasing use of molecular tests to refine microscopic diagnosis as the genetic understanding of female cancer improves. Additionally, the prevention and early detection of some gynecologic cancers have been enhanced by molecular techniques. The model of molecular screening has been human papilloma virus (HPV) testing in uterine cervical cancer screening [1–3], but molecular tests may eventually aid in the detection of endometrial and ovarian cancer. Finally, cancers of women have been traditionally treated with a one-size-fits-all chemotherapeutic approach of platinum-based regimens. However, molecular data is increasingly being recognized as a potential way to personalize therapy for cancers of women by histologic and molecular subtype [4,5]. This chapter will explore molecular techniques currently useful in the diagnosis of gynecologic cancers, but will present only a glimpse into the rapidly evolving future of gynecologic cancer screening, prognostics, and theranostics.

HPV AND UTERINE CERVICAL CANCER

Background

HPV is now recognized as the almost exclusive primary cause of uterine cervical cancer. HPV-related cervical cancer can have morphologic and clinical overlap with other cancers such as endometrial cancer [6,7], and uterine cervical cancer may not have distinguishing morphologic features when presenting at a metastatic site. Additionally, benign entities such as seborrheic keratosis, genital warts, squamous atrophy, immature squamous metaplasia, and tuboendometrioid metaplasia can show morphologic overlap with precursors to cervical cancer (high-grade squamous intraepithelial lesion (HSIL) and adenocarcinoma in situ) [8–11]. In these situations, biomarkers of high-risk HPV viral integration and/or direct detection of HPV DNA and/or RNA can be useful in making a more accurate diagnosis. This section describes the use of molecular testing in the context of surgical pathology.

Overview of Molecular Mechanisms and Traditional testing

Overall, uterine cervical cancer is the second most common cancer in women worldwide [12]. In the third world, cervical cancer is the most common cancer of women, whereas in developed countries cervical cancer incidence has reduced to 10th in rank [12,13] due to the successes of cervical cancer screening for precursor lesions. The diagnosis of cervical cancer and its precursors has traditionally relied upon the viral cytopathic effects of HPV, including disorganized and decreased squamous maturity, nuclear hyperchromasia, enlargement, and membrane irregularities, increased/abnormal mitotic activity, and abnormal cytoplasmic clearing (koilocytosis) [9]. The diagnosis of metastatic lesions has required morphologic comparison to the primary uterine cervical tumor.

Molecular Target

The two most common molecular targets used to aid in cervical cancer diagnosis are p16 (CDKN2A) protein and HPV DNA and/or RNA. Briefly, E6 and

© 2017 Elsevier Inc. All rights reserved.

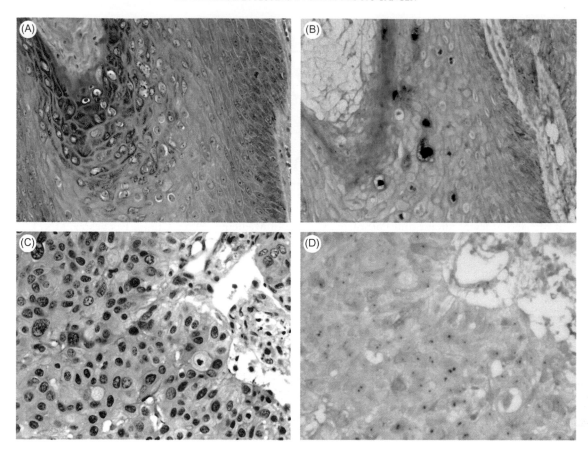

FIGURE 28.1 HPV DNA chromogenic in situ hybridization. (A) Vulvar condyloma (H&E, 400×) and (B) HPV DNA family 6, 11 CISH (400×). Note the homogeneous nuclear pattern of staining with the HPV CISH typical of an episomal HPV copies in a low-risk HPV infection. (C) Sacral recurrence of uterine cervical squamous cell carcinoma (H&E, 600×) and (D) HPV DNA family 16, 18 CISH (600×). Note the discrete nuclear probe signals typical of HPV that has been integrated into the host genome.

E7 HPV DNA, mRNA, and protein are conserved in relevant human HPV types [14]. These proteins have the ability to take control of host cell machinery. E6 causes p53 (TP53) to be targeted for degradation via the ubiquitin pathway, preventing normal DNA repair and/or apoptosis. E7 binds to retinoblastoma family proteins, causing the release of Rb-bound E2F transcription factor, driving forward the cell cycle. p16 is a nonspecific biomarker of HPV, because CDKN2A expression is upregulated by the HPV-infected cell in an attempt to (unsuccessfully) overcome the drive to cell cycle progression [15,16]. In most early research studies of the relationship of HPV with cervical cancer, the conserved L1 DNA region (encoding a component of the viral capsid) was used as a target to detect a broad range of HPV types [12].

the clinical surgical pathology laboratory, HPV DNA and/or E6/E7 RNA are usually detected by chromogenic in situ hybridization (CISH) directed to FFPE tissues (Fig. 28.1) that can be made specific for HPV types. Assays are usually divided into low-risk HPV (family 6 and 11, found in benign genital warts) and high-risk HPV (16, 18, 31, 33, 35, 39, 45, 51, 52, 56, 58, 59, 67, 68, found in cervical cancer) types. PCR amplification of the conserved L1 DNA region with MY09/11 and GP5/6 primers has been the gold standard assay for detecting HPV in epidemiologic studies due to high sensitivity (~99% of cervical cancer is positive for HPV by L1 PCR) [12], but this approach has not been widely adopted in clinical laboratories due to the ubiquitous nature of the HPV carrier state in sexually active populations [17,18].

Molecular Technologies

CDKN2A upregulation can be detected by immunohistochemistry (IHC) directed to p16 protein in formalin-fixed paraffin-embedded (FFPE) tissues. In

Clinical Utility

In the Lower Anogenital Squamous Terminology (LAST) standardization project guidelines [9], p16 was recommended for use in lower anogenital tract biopsies in the following situations: (1) The pathologist is considering a diagnosis of intraepithelial neoplasia grade 2 (moderate dysplasia or "IN2"). In these situations, morphologic IN2 lesions represent a mixture of low-risk and high-risk HPV lesions. A positive p16 stain can be used to confirm a diagnosis of HSIL rather than low-grade squamous intraepithelial lesion (LSIL). (2) The morphologic differential diagnosis is between a benign lesion (such as immature squamous metaplasia or atrophy) and HSIL. A positive p16 stain should be used as evidence of HSIL. (3) p16 can be used to adjudicate professional disagreement regarding morphologic interpretation in small biopsies. To qualify for a positive stain, p16 must show a diffuse block positive reaction pattern (Fig. 28.2) within the lesional tissue. The stain should be interpreted as negative if there is patchy staining.

The LAST guidelines did not make any recommendations for the use of HPV ISH or PCR in the context of making a diagnosis of HSIL. Several situations arise in routine surgical pathology practice that make HPV ISH useful. First, warty lesions such as seborrheic keratosis, traumatized skin tags (acrochordon), and rarely epidermal nevus can show significant morphologic overlap with anogenital condylomas [10,11]. The diagnosis of condyloma in some situations could have profound social consequences (eg, in a child where sexual abuse may or may not be suspected [19,20] or in certain religious groups). In these situations, it may be preferable to document the presence of an HPV-associated lesion by morphology and a second molecular method such as HPV ISH. Second, it can be difficult to morphologically distinguish endocervical and endometrial adenocarcinoma when the cancer is well differentiated and there is not a precursor lesion (HSIL/AIS or atypical complex endometrial hyperplasia) for reference. IHC can be useful [7], but results can be misleading in some cancers. Documenting the presence of HPV by ISH is specific for endocervical cancer (Fig. 28.2). Finally, metastatic HPV-associated cancers can present a diagnostic challenge when other primary sites are included in the differential diagnosis.

Limitations of Testing

Carrier status for anogenital HPV is approximately 50% in college-aged women [17], while the prevalence of high-risk disease in this age group is low [21]. Highly sensitive techniques for detecting HPV are not very specific for high-risk disease. p16 IHC may be more specific for high-risk disease in the context of anogenital tract biopsy specimens, but approximately 30% of expert adjudicated LSIL is p16 positive [9], demonstrating the need to interpret HPV and biomarker testing in context.

Future Directions

Despite the successes of cervical cancer screening, the high sensitivity and low specificity of current screening methods result in overtreatment of clinically insignificant lesions in some women. The costs and morbidity associated with colposcopy and cervical excision procedures are not trivial [22–25]. This situation leaves room for the development of more specific biomarkers of cervical lesions at higher risk of developing into invasive carcinoma if left untreated.

GESTATIONAL TROPHOBLASTIC DISEASE

Background

Gestational trophoblastic disease (GTD) is defined by tumor-like growths and frank malignancies that arise from abnormal placental tissue. GTD, if defined genetically, affects at least 1 in 100 pregnancies, but many less pregnancies result in clinically significant disease [26–28]. GTD can be subdivided into complete hydatidiform molar pregnancy, partial mole, and malignant neoplasms (choriocarcinoma, epithelioid trophoblastic tumor (ETT), and placenta site trophoblastic tumor (PSTT)). Partial mole is estimated to affect 1 in 100 pregnancies [27], while complete mole affects approximately 1 in 1000 pregnancies [28]. PSTT and ETT are rare [29,30]. Choriocarcinoma, the most rapidly progressive form of malignant GTD, usually arises after molar pregnancy, but occasionally these cancers arise in the mature placenta [31]. The incidence of choriocarcinoma after molar pregnancy appears to vary by population, but is higher after complete mole (~3–4%) than after partial mole (<1%) [27,28].

With rare exceptions, GTD seems to be driven by paternal genetic (androgenetic) material present in excess, or to the exclusion of, genetic material from the woman [26]. Fig. 28.3 outlines the most common types of conceptions leading to GTD. Complete hydatidiform mole (Fig. 28.4) is an overgrowth of abnormal placental tissue that occurs after fertilization of an empty ovum by one (or rarely more) sperm that reduplicates to form a diandrogenetic diploid conception. A fetus does not form, but the placenta grows rapidly, potentially leading to death of the woman from local growth or metastasis, or progression to choriocarcinoma (Fig. 28.5), if undetected. Partial mole (Fig. 28.6), like complete mole, is a growth of abnormal placental tissue that occurs after fertilization of an ovum with a normal haploid maternal chromosome complement by two (or rarely more) sperm, resulting in a diandrogenetic triploid conception. An abnormal fetus usually

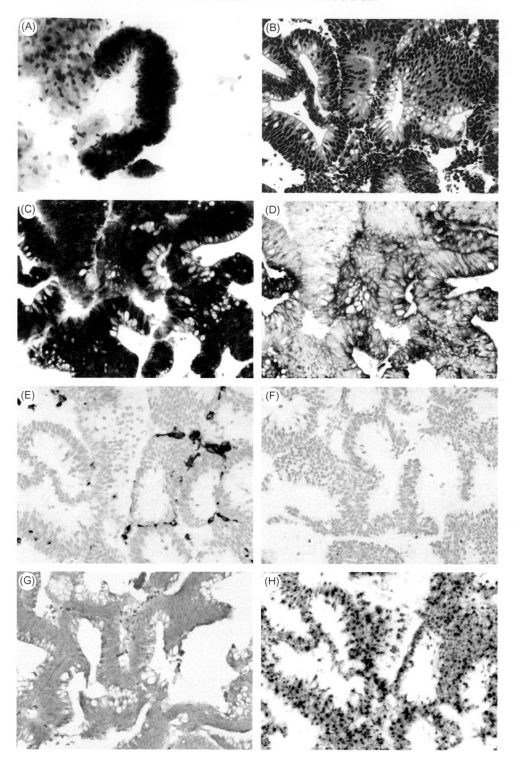

FIGURE 28.2   IHC and HPV testing in the differential diagnosis of endocervical versus endometrial adenocarcinoma. (A) A cervical/endocervical cytology screening (Pap) test showed atypical glandular and squamous cells in a 51-year-old woman (Papanicolau stained liquid-based preparation, 600×). Concurrent HPV testing for the Pap test was negative. (B) The patient underwent fractional endometrial and endocervical curettage, both of which showed well-differentiated adenocarcinoma (H&E, 400×). (C) IHC was positive with antibodies directed to p16 (400×) and (D) polyclonal carcinoembryonic antigen (400×), and (E) was negative for vimentin (400×) and (F) estrogen receptor (400×). (G) HPV DNA CISH (400×) was negative. (H) HPV E6/E7 RNA CISH was positive (400×).

Mechanisms of molar pregnancy

Partial mole

Diandrogenetic triploid (dispermic)

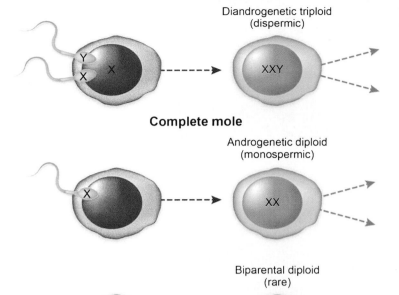

Complete mole

Androgenetic diploid (monospermic)

Biparental diploid (rare)

FIGURE 28.3 Mechanisms of molar pregnancy. Partial moles are usually conceived as a result of two different sperm fertilizing a normal ovum, but occasionally partial moles arise through other mechanisms (eg, triandrogenetic tetraploidy) that produce paternal chromosome complements in excess of maternal chromosome complements. Complete moles are usually conceived as a result of one sperm fertilizing an empty ovum and then reduplicating its chromosome complement, but complete moles also can be dispermic diploid, tetraandrogenetic tetraploid, or biparental. Biparental complete moles should raise concern for familial molar pregnancy syndrome, caused by biallelic maternal mutations in the *NLRP7* gene (autosomal recessive inheritance).

forms in addition to the abnormal placental tissue of partial mole. In contrast to the androgenetic mechanism of most molar pregnancies, some complete moles are biparental, arising from a seemingly normal diploid conception. Some of these women have been found to have mutations in the *NLRP7* gene with an autosomal recessive pattern of inheritance [32]. Termed familial molar pregnancy, women with two deleterious *NLRP7* mutations do not have normal pregnancies and usually report only miscarriages and molar pregnancies. The mechanism of GTD in these women is not yet clear. Some data suggest that carriers of *NLRP7* mutations also have more frequent miscarriages than the general population, but these women can have normal children [33].

Earlier prenatal care has led to the earlier detection and cure of most GTD. Even choriocarcinoma, one of the most rapidly progressive and aggressive human cancers, is usually curable if detected before advanced complications occur [33]. It is usually recommended that a woman be cured of GTD prior to becoming pregnant again, because serum beta-HCG levels are the mainstay of monitoring patients during and after treatment [34]. Unfortunately, the clinical and pathologic diagnosis of molar pregnancy is imperfect due to overlapping clinical and morphologic features with a normal missed abortion (Fig. 28.7), and so the traditional approach of caution in monitoring women with only suspected molar pregnancy as a cause of miscarriage has often become unacceptable, especially as women are more frequently attempting pregnancy within smaller reproductive windows later in life. As such, molecular techniques have become part of the standard of care for appropriately diagnosing GTD and triaging women into the appropriate management stratified by risk.

Molecular Target

The molecular diagnosis of GTD capitalizes on two important principles: paternity and ploidy. As outlined in Fig. 28.1, the molecular distinction between nonmolar, partial molar, and complete molar pregnancy can be usually deduced using one or both of these principles. For example, approximately 85% of all triploid conceptions are diandric triploid partial moles (the remaining are digynic triploids and do not carry the risk of subsequent recurrence and GTD of partial mole) [26]. Detection of triploidy alone in a suspicious morphologic context is virtually diagnostic of partial mole. However, ploidy does not separate complete mole from normal and this is a diagnostic problem that commonly arises in early pregnancy when the

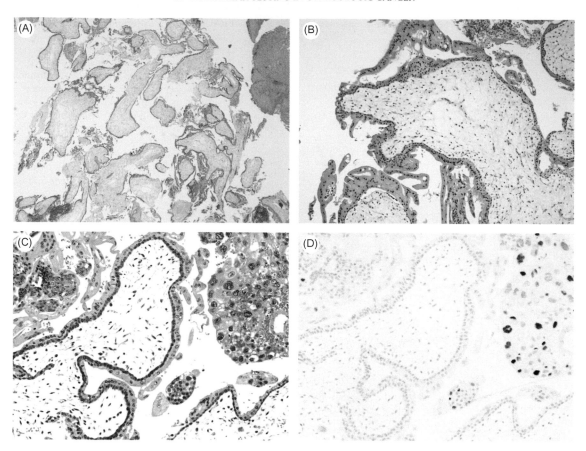

FIGURE 28.4 Complete hydatidiform mole. (A) Complete moles are recognized by abnormal enlargement and irregularity in shape of chorionic villi at low power magnification (H&E, 20 ×). (B and C) Circumferential atypical trophoblastic proliferation, peripheral fibroplasia of villous stroma, and abortive or no fetal vessels are typical (H&E, 100 × and 200 ×, respectively). (D) IHC for p57 protein (p57 IHC, 200 ×), expressed maternally from the paternally imprinted CDKN1C gene, is lost in the villous stroma and cytotrophoblast, but expression is maintained in the intermediate trophoblast.

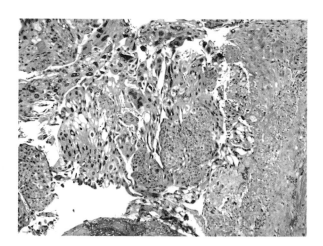

FIGURE 28.5 Choriocarcinoma (H&E, 100 ×). Choriocarcinoma shows similar features to the atypical trophoblast of complete mole, characterized by a biphasic proliferation of mononuclear trophoblast and syncytiotrophoblast. To make a diagnosis of choriocarcinoma, immature chorionic villi should be absent. Occasionally, choriocarcinoma arises in the mature placenta.

morphologic differences are subtle [35]. Techniques that determine either parental imprinting or genotype can be used to distinguish diandric diploid complete moles from normal biparental diploid conceptions [36].

Molecular Technologies

p57 IHC. Because the highest risk of recurrent GTD and choriocarcinoma is after complete molar pregnancy, it follows that accurate and timely diagnosis of complete mole is of primary importance. p57 IHC is a rapid and accurate technique [36] that utilizes the concept of imprinting. It is important to note that p57 protein, encoded by the *CDKN1C* gene locus, does not play a specific role in the mechanism of molar pregnancy and is only used as a convenient marker of maternal contribution of DNA. Because *CDKN1C* is normally paternally imprinted, the lack of maternal p57 protein expression can be used as a

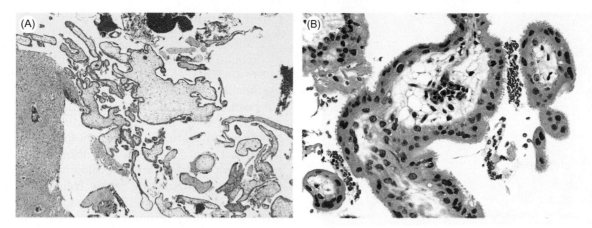

FIGURE 28.6 Partial mole. (A) Partial mole shows some features in common with complete mole, including abnormal villous enlargement and irregularity in villous shape (H&E, 20×). The villous outlines in partial mole may be more dramatically scalloped, and the degree of trophoblastic hyperplasia is usually less. (B) A fetus or other evidence of fetal development may be present (H&E, 400×, demonstrating erythroblasts in fetal vessels).

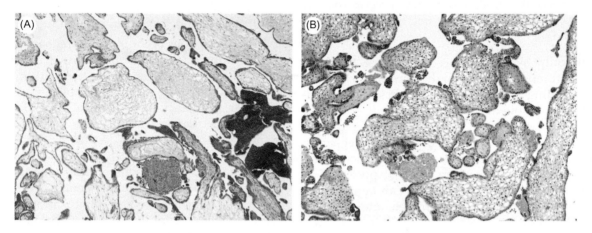

FIGURE 28.7 Hydropic nonmolar pregnancy. Hydropic change occurs in all immature placentas after fetal death and becomes more dramatic as the postmortem interval increases. (A) This can result in villous enlargement (H&E, 40×). (B) Additionally, single chromosome anomalies can be associated with villous hydrops and architectural abnormalities (H&E, 100×, first trimester placenta with trisomy 21).

surrogate for androgenetic complete molar pregnancy. In developing placental tissue, paternal imprinting of *CDKN1C* occurs in the villous stroma and cytotrophoblast, but not in the intermediate trophoblast. Therefore, the intermediate trophoblast can be used as a good internal control of antigenicity (Fig. 28.4). One can imagine a variety of pitfalls using this approach (abnormal imprinting of *CDKN1C*, monosomy or uniparental disomy of chromosome 11, biparental complete mole, etc.) that could cause false-negative or positive results [36].

Ploidy analysis by flow cytometry, digital image analysis, and FISH. Ploidy is useful for detecting partial moles, which are usually triploid. Flow cytometry and digital image analysis (DIA) have largely been replaced by fluorescence in situ hybridization (FISH) for determining ploidy on FFPE tissue [37] (Fig. 28.8).

Flow cytometry can estimate the ploidy of a conception by measuring the amount of incorporated fluorescent dye binding to double-stranded DNA per cell as compared to normal diploid cells. Likewise, DIA uses a Fuelgen DNA stain to bind FFPE sections of placental tissue. Light is shined through the stained slide, and the amount of light detected after passing through the slide is normalized to the amount of light passing through control diploid cells to determine ploidy. Virtually any FISH assay used to detect copy number can be used to deduce ploidy of placental tissue. For example, a simple HER2 FISH assay used to detect *ERBB2* amplification in breast cancer can also be used to detect triploid conceptions [38]. A single-locus probe could give false-positive results in a case of single chromosome trisomy, and so using probes on multiple chromosomes increases the specificity for triploidy.

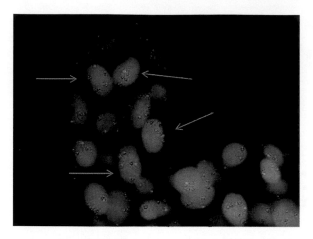

FIGURE 28.8 Partial mole demonstrating triploid chromosome complement by multiplex FISH. Centromeric probes are directed to chromosomes X (green), Y (red), and 18 (aqua). The arrows indicate cells with an XXY, +18 signal pattern. All other tested FISH probes (13, 15, 16, 21, 22) also showed three copies, indicating triploidy. Correlation with morphology and/or genotyping is required to distinguish diandrogenetic triploid partial moles from digynic nonmolar triploid conceptions.

Some clinically available FISH assays test the most common chromosome gains and losses in early pregnancy, providing information about these abnormalities in addition to the overall ploidy [37].

Genotyping. Genotyping has become the gold standard in molecular diagnosis of molar pregnancy, because it can be used to both determine ploidy and parent of origin of the genetic material present [36]. The combination of these two factors can be used in a single test that distinguishes complete mole, partial mole, and nonmolar pregnancy in most settings. Genotyping is most simply accomplished by short tandem repeat (STR, also known as microsatellite repeat) testing. This technology is widely available for other purposes such as paternity testing, bone marrow engraftment, forensic and clinical specimen identity testing, zygosity determination in twins, and maternal cell contamination in prenatal specimens. Each STR is PCR amplified and the product measured for fragment size, usually by capillary electrophoresis. Measuring multiple highly polymorphic loci can be used to determine the number of chromosome complements that are likely to be present and the parent of origin for each haploid set. A complete mole would be expected to show two doses of paternal marker contribution (usually identical) with no maternal contribution. A partial mole would show two (usually nonidentical) paternal contributions and one maternal contribution at each marker. A nonmolar pregnancy would typically show biparental inheritance at each marker. Genotyping can also be used to determine whether a malignant tumor is GTD [39] (Fig. 28.9).

Clinical Utility

Testing for molar pregnancy is optimally done using a combination of clinical, morphologic, and molecular test result data. Guidelines for when and when not to use ancillary testing in addition to morphology have not been established. In general, an experienced surgical pathologist can distinguish between products of conception that are clearly morphologically normal and those that are not classically normal. Because even experts in placental and gynecologic pathology are not perfect in distinguishing molar from nonmolar pregnancy in difficult cases [35], consideration should be given to ancillary techniques when morphologic clues are equivocal or there is clinical suspicion for GTD. Most referral laboratory practices use morphology and p57 IHC as a rapid triage to diagnose complete moles, with reflex to genotyping or ploidy testing when the morphology is abnormal but p57 is normal.

Limitations of Testing

Many exceptions occur in early pregnancy losses that may contribute to false assumptions if molecular tests are not used correctly in conjunction with morphologic and clinical data. For example, complete moles in familial molar pregnancy are biparental diploid, producing normal molecular test results. Likewise, ploidy testing cannot be used blindly as a test for partial molar pregnancy, as approximately 15% of triploid conceptions and most tetraploid conceptions represent nonmolar pregnancies without excess paternal contribution. Genotype testing comes with its own technical problems. Ideally one would have pure maternal, paternal, and placental DNA to compare by STR testing, but gathering parental blood samples is inconvenient. Such testing usually compares placental tissue and maternal decidual tissue dissected from the FFPE blocks, a process that requires skilled laboratory staff for sample preparation and interpretation of specimens that are invariably contaminated due to the natural admixture of maternal and placental tissue in endometrial curettage specimens.

Future Directions

Array comparative genomic hybridization and noninvasive prenatal screening have the potential to determine ploidy and genotype in the prenatal setting, sooner than currently available test methods for diagnosing molar pregnancy. As these tests develop, it is likely that many molar pregnancies will be detected even before a surgical pathologist examines the products of conception, and so the pathologist may move into an equally important confirmatory and backup role to the prenatal tests.

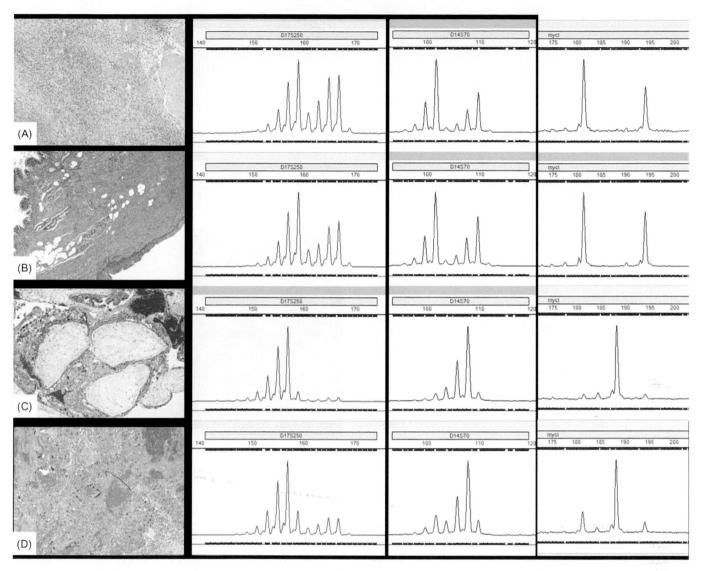

FIGURE 28.9 Genotyping in diagnosis of gestational trophoblastic disease. A young woman presented with choriocarcinoma in an ovarian teratoma, initially diagnosed as mixed germ cell tumor. It came to light that she had a molar pregnancy 4 years prior. Genotyping analysis was performed using DNA extracted from (A) normal ovary, (B) teratoma, (C) prior molar pregnancy, and (D) choriocarcinoma. Both the molar pregnancy and choriocarcinoma showed allelotypes that matched and were different from the woman's ovary and teratoma, proving gestational choriocarcinoma occurring in the ovary arising from the prior molar pregnancy. The molar pregnancy and choriocarcinoma were homozygous at all alleles, suggesting a monospermic mechanism of conception typical of complete hydatidiform mole. Contaminating inflammatory cells in the choriocarcinoma are responsible for the minor contaminating maternal peaks in the electropherograms for the choriocarcioma (even smaller contaminating peaks from maternal decidua are present in the electropherograms from the molar pregnancy).

ENDOMETRIAL STROMAL SARCOMA

Background

A limited number of mesenchymal tumors of the gynecologic tract have reproducible chromosomal translocations that can be used to make a diagnosis when morphology or context is challenging. Endometrial stromal sarcoma (ESS) [40], inflammatory myofibroblastic tumor (IMT) [41], aggressive angiomyxoma [42], and rare perivascular epithelioid cell (PEC) tumors [43] are examples. ESS is a rare, malignant neoplasm arising in the endometrial stroma, accounting for less than 1% of all uterine malignancies [44]. In the usual case with classic morphology mimicking normal endometrial stroma, a diagnosis can be confidently rendered by morphology alone. IHC (desmin, caldesmon, smooth muscle actin, CD10, ALK, HMB45) can be helpful in distinguishing ESS from smooth muscle tumors, PEComa, and IMT, but there can be immunophenotypic overlap, in particular

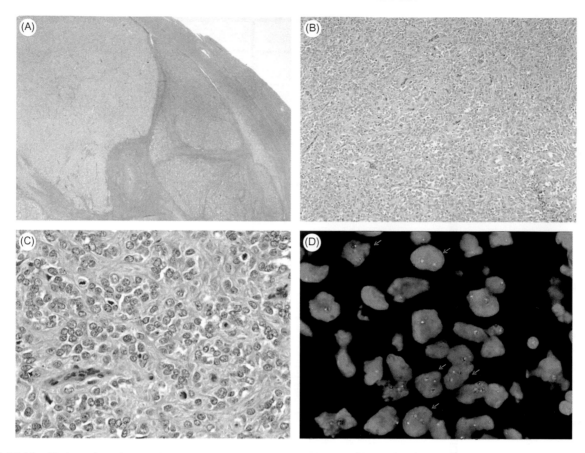

FIGURE 28.10 High-grade endometrial stromal sarcoma. (A–C) High-grade endometrial stromal sarcoma (H&E, 20×, 100×, and 400×, respectively). (D) FISH break-apart strategy demonstrated rearrangement of the 17p13.3 YWHAE locus (3′ probe red, 5′probe green, fusion yellow).

between pure smooth muscle tumors and ESS with smooth muscle differentiation [45,46]. CD10 has been the most useful positive marker for ESS, but CD10 can show variable expression in other cancers, and some ESS are CD10 negative. Even more challenging are scenarios where the diagnosis of ESS was initially missed in a hysterectomy specimen (usually diagnosed as leiomyoma) or documentation has been lost, and a woman presents with metastatic disease in an unexpected location such as peritoneum, liver, or lung. Recurrences can notoriously happen years to even decades after the primary cancer is removed [47–49]. Molecular testing can be useful in confirming a suspected diagnosis of ESS in these scenarios. Additionally, some rearrangements have relative prognostic value [50].

Molecular Target

The most common rearrangement in ESS occurs between the short arm of chromosome 7 and the long arm of chromosome 17, t(7;17)(p15;q21), resulting in a fusion gene between *JAZF1* and *JJAZ1* [51]. Other reported translocations involve a variety of genes, including *YWHAE*,

PHF1, *EPC1*, *NUTM2A/B*, *SUZ12*, *MEAF6*, *ZC3H7B*, *BCOR*, *MBTD1*, *CXorf67*. Approximately 60% of ESS harbor a *JAZF1*, *YWHAE*, or *PHF1* rearrangement [40,48].

Molecular Technologies

A variety of technologies are now available to detect chromosomal translocations. Break-apart FISH strategies can be used to detect rearrangements in the most common involved genes, *JAZF1*, *YWHAE*, and *PHF1* (Fig. 28.10). Reverse transcriptase PCR can also be used to amplify the transcripts of gene fusions, but a multiplex approach would be needed to detect all common rearrangements.

Clinical Utility

Most ESS can be diagnosed by a surgical pathologist using gross morphology and microscopy. In morphologically difficult cases, molecular testing can be useful to confirm the suspected diagnosis (excellent specificity), but current approaches lack sensitivity for use of molecular testing in isolation. Molecular testing can be particularly useful in confirming the diagnosis of ESS when suspected based on morphology at an extrauterine site, especially when a clinical history of ESS has not been previously documented [46–48]. Additionally, some data suggest that the *YWHAE* translocation is confined to high-grade ESSs having worse prognosis, suggesting that testing may even have prognostic value at diagnosis, especially if tumor grade is in question [40,50].

Limitations of Testing

Unfortunately, a large subset of ESS do not harbor one of the common recurrent translocations, making currently available tests less useful in the primary diagnostic setting. Additionally, benign endometrial stromal nodules can also harbor *JAZF1* rearrangements [45,51]. This is particularly important in the setting of endometrial biopsy or morcellated hysterectomy showing an endometrial stromal tumor [49,52]. A *JAZF1* translocation cannot be used as evidence of malignancy in these settings where determination of invasion is also challenging (if not impossible).

Future Directions

Next-generation sequencing (NGS) technologies (both DNA- and RNA- based) have the potential to increase the sensitivity of molecular testing for ESS by widening the net of variant translocations that can be detected in a single assay. As costs for NGS go down, it is likely that these technologies will increasingly be used by surgical pathologists to make more accurate and prognostically useful diagnoses for unusual mesenchymal tumors in the gynecologic tract.

ADULT GRANULOSA CELL TUMOR

Background

Adult granulosa cell tumor (GCT) is a neoplasm that is thought to arise from the granulosa cell tissue in developing ovarian follicles, corpus lutea, or persistent follicle cysts. GCT cells usually have a similar appearance to their benign counterparts, but undergo neoplastic transformation and tumorigenesis to produce a cancer with malignant potential through unknown mechanisms. GCT is the most common of sex-cord stromal tumors of the ovary, affecting a broad age range of adult women [53,54]. Classic morphology tumors do not present a diagnostic challenge, but there is morphologic overlap with other entities, including

benign ovarian fibroma. Similar to ESS, GCT often recurs years after the initial presentation [53,54] and can occasionally provide significant diagnostic challenge in the metastatic setting.

Molecular Target

Whole transcriptome RNA sequencing of GCT has shown a single hotspot mutation in *FOXL2*, c.402C > G (p.C134W) in nearly all GCT with classic morphology [55]. Rarely, juvenile GCTs, fibromas, and thecomas have been reported to have the *FOXL2* mutation, but no other neoplasms have been shown to be mutated at this locus. The presence of *FOXL2* mutations in cancers other than GCT is controversial due to the morphologic overlap with juvenile GCT and fibroma, calling into question the diagnostic gold standard [56,57]. *FOXL2* mutations have been shown to rarely occur in granulosa cell proliferations in epithelial ovarian tumors, suggesting the possibility of hybrid epithelial and sex-cord stromal neoplasia [58].

Molecular Technologies

Any molecular test designed to detect DNA point mutations in FFPE tissue could be used to detect the hotspot *FOXL2* mutation. Fig. 28.11 provides an example of a pyrosequencing approach.

Clinical Utility

The c.402C > G *FOXL2* mutation is present in nearly all classic morphology GCT, but such tumors are not a diagnostic challenge for the experienced surgical pathologist. Variant morphology GCT (diffuse GCT, luteinized GCT) can show overlap with fibrothecoma and other sex-cord stromal tumors, making the test possibly helpful in such cases. When GCT presents as recurrence in an extraovarian location, *FOXL2* testing can be very useful for confirming the diagnosis when other available information (morphology, IHC) are relatively nonspecific.

Limitations of Testing

It should be noted that variant morphology GCT, as defined by expert pathologists has a *FOXL2* mutation rate that is much less than classic GCT, begging the question of gold standard for these tumors [55–57,59–61]. Until a better understanding of these variant morphology neoplasms is obtained, a negative FOXL2 mutation test cannot be used as evidence to exclude the diagnosis of GCT, but a positive test may aid in expert diagnostic consensus and may have

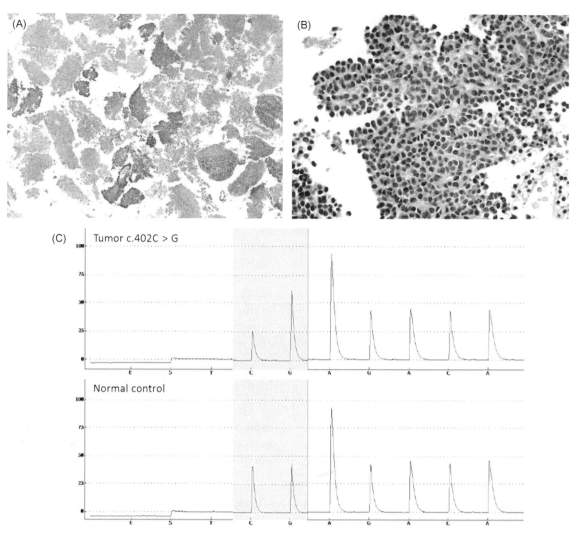

FIGURE 28.11 Adult granulosa cell tumor. (A and B) Retroperitoneal mass, core biopsy (H&E, 40× and 400×, respectively). This post-menopausal woman presented with retroperitoneal mass after hysterectomy and bilateral salpingo-oophorectomy 5 years prior to remove an infarcted ovarian mass. (C) A *FOXL2* c.402C > G mutation was detected by pyrosequencing in DNA extracted from the retroperitoneal tumor tissue, confirming a diagnosis of metastatic adult granulosa cell tumor. The prior infarcted ovarian mass was likely the primary site.

prognostic significance [60,61]. IHC for FOXL2 protein is available, but expression is seen in GCT and a variety of other sex-cord tumors [62].

Future Directions

Discovery of *FOXL2* as an early driver of adult granulosa cell tumorigenesis may be important to further refining morphologic cancer classification in this category. Other, as yet undiscovered, drivers may be prognostically important in malignancies currently classified as adult GCTs that are *FOXL2* negative. Additionally, discovery of these drivers could result in targeted therapy for these cancers that currently do not have specialized therapy available beyond what has been the standard for epithelial ovarian cancer.

LYNCH SYNDROME SCREENING IN ENDOMETRIAL CANCER

Background

Approximately 50,000 women in the United States are diagnosed with endometrial cancer each year, and 10,000 will die as a result of their disease. The lifetime incidence is approximately 1 in 37 women [63]. Although major risk factors revolve around estrogen excess (unopposed estrogen therapy, obesity), approximately 3–7% of cases are thought to be attributable to hereditary predisposition [64–68]. The most common hereditary predisposition is germline mismatch repair (MMR) deficiency, also known as Lynch syndrome. Though the prototypic Lynch syndrome cancer is colorectal adenocarcinoma, over half of women with

Lynch syndrome will present first with endometrial cancer [64,66,68]. Ovarian cancer is less common as an incident Lynch syndrome cancer. Because over half of women with Lynch syndrome and incident endometrial cancer do not have a personal or family history suggestive of Lynch syndrome [64,66,68,69], universal screening of endometrial adenocarcinoma for evidence of MMR defects has become increasingly accepted. Detection of affected women can reduce morbidity and mortality of the proband and affected family members through appropriate cancer screening and preventative measures [70,71].

Overview of Molecular Mechanisms and Traditional Testing

Microsatellite instability (MSI), shortening or lengthening of small repetitive elements in DNA, is a result of the inability of MMR enzymes to repair random mutations that occur during DNA synthesis [72,73]. This increased mutation rate is thought to be a major driver of tumorigenesis in endometrial cancer, as approximately 25–30% of endometrial cancers show MSI [64,66,74,75]. MMR proteins can be thought of as tumor suppressors, and deficiency can be caused by damaging point mutations, insertions and deletions, promoter methylation, and rearrangement. These alterations can happen in the germline, such as in Lynch syndrome, or sporadically. Though a large proportion of endometrial cancers show MSI, germline mutations in MMR genes are only detected in about 2–4% of patients with endometrial cancer [64,65,68,69]. Most of these cancers are of endometrioid type, but other histologies have been reported. Factors that increase the probability of Lynch syndrome in a woman with incident endometrial cancer include location in the lower uterine segment, dedifferentiated or mixed endometrioid and clear-cell histology tumors, synchronous ovarian cancer (especially clear cell carcinoma), and detailed cancer characteristics, including tumor infiltrating lymphocytes and mucinous differentiation [76,77].

Molecular Target

The first tests that were used to screen for MSI in cancer targeted a variety of types of microsatellites, including dinucleotide and pentanucleotide DNA repeats [72], but current methods target mononucleotide repeats that are usually nonpolymorphic in the population [78–80]. MMR proteins most commonly altered in incident endometrial cancer are MLH1, MSH6, MSH2, and PMS2. *MSH2* and *MLH1* germline mutations, as in colorectal cancer, are usually associated with a much younger age of endometrial cancer onset. Finally, most microsatellite instability high (MSI-H) endometrial cancers are sporadic and driven by the methylation pathway, showing methylation of the *MLH1* gene promoter as the cause MLH1 expression loss [75].

Molecular Technologies

MSI can be detected by PCR through amplification of microsatellites that are scattered throughout the genome. Mononucleotide repeats are preferred due to the sensitivity of these regions to MMR deficiency, relatively non-polymorphic nature of the loci, and ease of interpretation of mononucleotide repeats in microsatellite unstable cancers due to the tendency for these regions to uniformly shorten (rather than lengthen) in the face of MMR deficiency [78–80]. A normal comparison is used to prevent false-positive results that could otherwise occur if a patient shows microsatellite length variation polymorphically in the germline. Variation in microsatellite length can be detected by electrophoresis of amplification products. An MSI-H cancer is defined as a cancer with greater than 20% of tested markers showing instability (eg, two of five markers) (Fig. 28.12). Alternatively, IHC can be used to detect loss of expression of MMR proteins (Fig. 28.13). *MLH1* promoter hypermethylation is usually detected by using bisulfite treatment of cancer DNA followed by PCR using primers directed to wild-type DNA and the sequence change caused by bisulfite treatment of methylated DNA in the *MLH1* promoter. A difference in amplicon length and fluorescent color can be created in the primer design to detect *MLH1* promoter methylation.

Clinical Utility

Guidelines are rapidly evolving to suggest that all women with endometrial cancer should receive Lynch syndrome screening in some form. Clinical screening (taking a detailed family and personal history) should be done for all women with endometrial or ovarian cancer. Some screening proponents have suggested that all women with endometrial cancer should have their cancers screened for evidence of Lynch syndrome, but in a resource-constrained environment, an upper age cutoff of 60 or 70 years may be reasonable. Screening with both MMR IHC and MSI PCR is the most sensitive due to uncommon false-negative results with both methods, but either method is considered to be acceptable. IHC offers the advantage of identifying the most likely affected gene so that women with sporadic *MLH1* promoter hypermethylation can be identified quickly and triaged to methylation testing. Women with MMR IHC suggestive of Lynch syndrome can be referred for genetic counseling and germline mutation testing (Fig. 28.14).

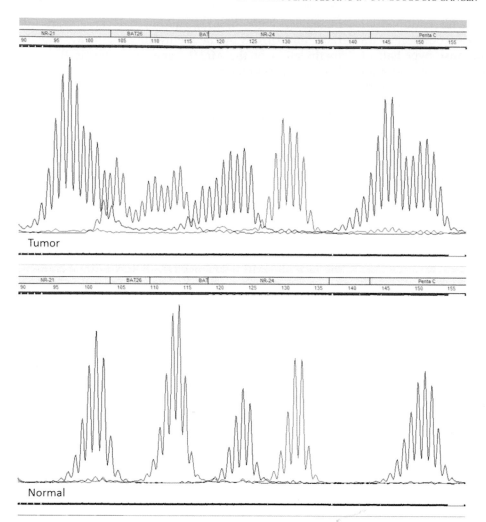

FIGURE 28.12 PCR detection of MSI in cancer. The electropherogram shows five mononucleotide repeat markers in the tumor (top) that demonstrate shortening in the tumor versus the matched normal tissue (bottom). When 20% or greater of tested markers show instability, the cancer is considered to be MSI-H.

A significant proportion of women identified with probable Lynch syndrome by universal screening do not have a detectable germline mutation, suggesting that other somatic alterations may be at play in some cancers [64,66,68,81]. It is recommended that women diagnosed with Lynch syndrome undergo enhanced colon cancer screening. Annual transvaginal ultrasound and/or endometrial biopsy, urine cytology, and prophylactic hysterectomy and oophorectomy can be considered [70,82,83]. Potentially affected family members can also be tested for the familial mutation and undergo preventative screening and surgeries if confirmed to be affected. Increased frequency of colonoscopy has been shown to decrease the incidence of colon cancer in patients with Lynch syndrome, justifying the cost of inclusive universal tumor screening to identify these patients [84–86].

Limitations of Testing

Microsatellite fragment length shifts may be more subtle in endometrial cancer patients than in colon cancer patients [87], especially in MSH6 carriers, potentially resulting in false-negative MSI PCR results. Conversely, MMR IHC is normal in the cancers of some patients with Lynch syndrome [64], usually due to a germline missense mutation that is associated with normal expression of a dysfunctional protein. It is estimated that over half of women having an MMR IHC pattern suggestive of Lynch syndrome do not have a detectable germline mutation by current methods [64,66,68]. Some of these women have uncommon mutations (eg, gene inversions or deep intronic mutations) [81,88,89], but more of these women probably have uncommon somatic mutations as the cause of their disease [81].

Future Directions

Recently, biallelic MMR gene mutations have been shown to explain some endometrial and colon cancers with abnormal MMR IHC patterns that would otherwise be suggestive of Lynch syndrome [81]. Some of these patients have somatic or germline mutations in *POLE*, a DNA polymerase [74,81,90]. When *POLE* is mutated in its exonuclease domain, the result is an

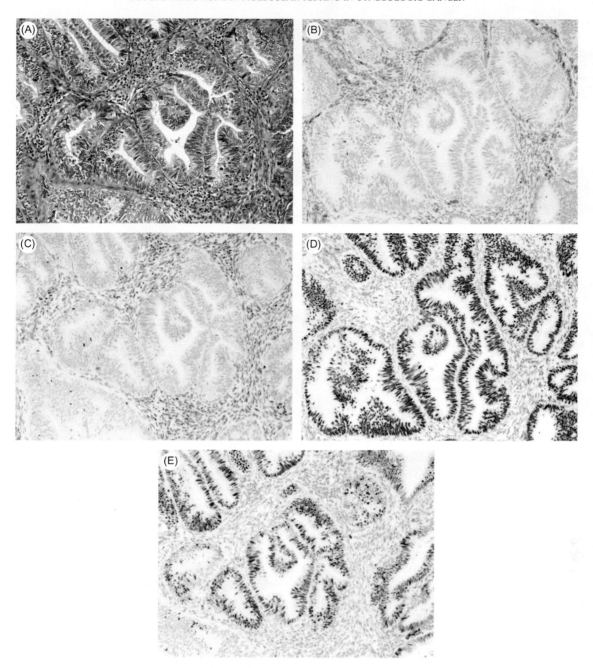

FIGURE 28.13 Mismatch repair IHC in an endometrial cancer. (A) Endometrial cancer (H&E, 200×) demonstrating loss of expression of (B) MLH1 and (C) PMS2 with normal expression of (D) MSH2 and (E) MSH6. This pattern of MMR protein expression is most commonly the result of somatic hypermethylation of the *MLH1* gene promoter in the cancer, but can also be seen in women with Lynch syndrome that have an *MLH1* germline mutation.

ultramutated cancer. If secondary biallelic MMR gene mutations occur, this results in cancer MSI. NGS has now made possible the detection of MSI, *POLE*, and MMR gene defects all in one test. It is not yet clear if this type of testing will be advantageous at the initial cancer screening step or should be used when a positive screen by MSI and/or MMR IHC is not explained by *MLH1* promoter hypermethylation and/or germline MMR gene testing.

FUTURE DIRECTIONS IN MOLECULAR TESTING IN GYNECOLOGIC CANCER

The Cancer Genome Atlas has analyzed extensive genome level sequencing, expression, and copy number variation analysis in endometrial carcinoma [74], carcinosarcoma, and high-grade serous ovarian carcinoma [91]. These studies have shown new ways of thinking about cancer classification (other than histologic

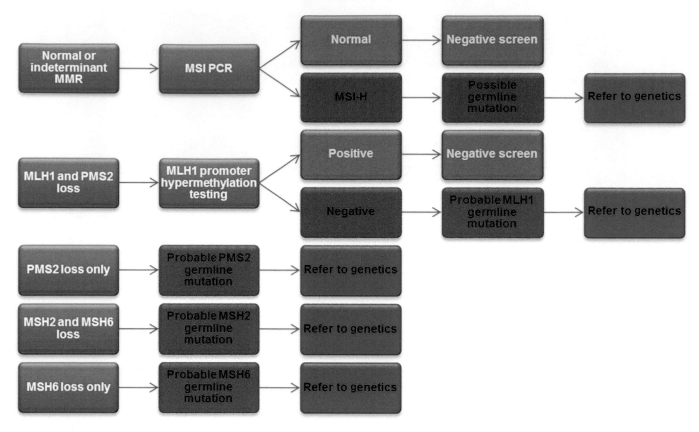

FIGURE 28.14 Example endometrial cancer screening algorithm for Lynch syndrome. This example starts with MMR IHC for MLH1, PMS2, MSH2, and MSH6 on all incident endometrial cancers. Reflex testing, such as MSI PCR or *MLH1* promoter hypermethylation testing, is performed depending upon MMR results. Referral to genetic counseling is made as appropriate for consideration of germline mutation testing.

differences) that are probably prognostically and theranostically useful. Though molecular classifications often largely overlap with traditional histopathology classification of tumors, these studies have shown distinct molecular subtypes that explain cancers that behave in an unexpectedly good or poor manner based on histology alone. It is likely that all common gynecologic cancers in the future will undergo some combination of pathology review and molecular testing to better refine cancer classification, prognosis, and management.

References

[1] Castle PE, Stoler MH, Wright Jr. TC, Sharma A, Wright TL, Behrens CM. Performance of carcinogenic human papillomavirus (HPV) testing and HPV16 or HPV18 genotyping for cervical cancer screening of women aged 25 years and older: a subanalysis of the ATHENA study. Lancet Oncol 2011;12:880−90.

[2] Stoler MH, Wright Jr. TC, Cuzick J, Dockter J, Reid JL, Getman D, et al. APTIMA HPV assay performance in women with atypical squamous cells of undetermined significance cytology results. Am J Obstet Gynecol 2013;208 144.e1−8.

[3] Wright Jr. TC, Stoler MH, Sharma A, Zhang G, Behrens C, Wright TL. Evaluation of HPV-16 and HPV-18 genotyping for the triage of women with high-risk HPV + cytology-negative results. Am J Clin Pathol 2011;136:578−86.

[4] Kurman RJ, Shih Ie M. Molecular pathogenesis and extraovarian origin of epithelial ovarian cancer—shifting the paradigm. Hum Pathol 2011;42:918−31.

[5] Lalwani N, Prasad SR, Vikram R, Shanbhogue AK, Huettner PC, Fasih N. Histologic, molecular, and cytogenetic features of ovarian cancers: implications for diagnosis and treatment. Radiographics 2011;31:625−46.

[6] Kong CS, Beck AH, Longacre TA. A panel of 3 markers including p16, ProExC, or HPV ISH is optimal for distinguishing between primary endometrial and endocervical adenocarcinomas. Am J Surg Pathol 2010;34:915−26.

[7] Jones MW, Onisko A, Dabbs DJ, Elishaev E, Chiosea S, Bhargava R. Immunohistochemistry and HPV in situ hybridization in pathologic distinction between endocervical and endometrial adenocarcinoma: a comparative tissue microarray study of 76 tumors. Int J Gynecol Cancer 2013;23:380−4.

[8] Cameron RI, Maxwell P, Jenkins D, McCluggage WG. Immunohistochemical staining with MIB1, bcl2 and p16 assists in the distinction of cervical glandular intraepithelial neoplasia from tubo-endometrial metaplasia, endometriosis and microglandular hyperplasia. Histopathology 2002;41: 313−21.

[9] Darragh TM, Colgan TJ, Thomas Cox J, Heller DS, Henry MR, Luff RD, et al. The lower anogenital squamous terminology standardization project for HPV-associated lesions: background and consensus recommendations from the College of American Pathologists and the American Society for Colposcopy and Cervical Pathology. Int J Gynecol Pathol 2013;32:76–115.

[10] Shimizu A, Kato M, Ishikawa O. Pigmented condyloma acuminatum. J Dermatol 2014;41:337–9.

[11] Reutter JC, Geisinger KR, Laudadio J. Vulvar seborrheic keratosis: is there a relationship to human papillomavirus? J Low Genit Tract Dis 2014;18:190–4.

[12] Bosch FX, Manos MM, Munoz N, Sherman M, Jansen AM, Peto J, et al. Prevalence of human papillomavirus in cervical cancer: a worldwide perspective. International biological study on cervical cancer (IBSCC) Study Group. J Natl Cancer Inst 1995;87:796–802.

[13] Peirson L, Fitzpatrick-Lewis D, Ciliska D, Warren R. Screening for cervical cancer: a systematic review and meta-analysis. Syst Rev 2013;2:35.

[14] Stoler MH. Human papillomaviruses and cervical neoplasia: a model for carcinogenesis. Int J Gynecol Pathol 2000;19:16–28.

[15] Doorbar J. The papillomavirus life cycle. J Clin Virol 2005;32: S7–15.

[16] Doorbar J. Papillomavirus life cycle organization and biomarker selection. Dis Markers 2007;23:297–313.

[17] Giuliano AR, Harris R, Sedjo RL, Baldwin S, Roe D, Papenfuss MR, et al. Incidence, prevalence, and clearance of type-specific human papillomavirus infections: The Young Women's Health Study. J Infect Dis. 2002;186:462–9.

[18] Syrjanen K. Mechanisms and predictors of high-risk human papillomavirus (HPV) clearance in the uterine cervix. Eur J Gynaecol Oncol 2007;28:337–51.

[19] Percinoto AC, Danelon M, Crivelini MM, Cunha RF, Percinoto C. Condyloma acuminata in the tongue and palate of a sexually abused child: a case report. BMC Res Notes 2014;7:467.

[20] Varma S, Lathrop E, Haddad LB. Pediatric condyloma acuminata. J Pediatr Adolesc Gynecol 2013;26:e121–2.

[21] Wright Jr. TC, Stoler MH, Behrens CM, Apple R, Derion T, Wright TL. The ATHENA human papillomavirus study: design, methods, and baseline results. Am J Obstet Gynecol 2012;206 46.e1–e11.

[22] Acharya G, Kjeldberg I, Hansen SM, Sorheim N, Jacobsen BK, Maltau JM. Pregnancy outcome after loop electrosurgical excision procedure for the management of cervical intraepithelial neoplasia. Arch Gynecol Obstet 2005;272:109–12.

[23] Kietpeerakool C, Srisomboon J, Khobjai A, Chandacham A, Tucksinsook U. Complications of loop electrosurgical excision procedure for cervical neoplasia: a prospective study. J Med Assoc Thai 2006;89:583–7.

[24] Crane JM. Pregnancy outcome after loop electrosurgical excision procedure: a systematic review. Obstet Gynecol 2003;102:1058–62.

[25] Liu Y, Qiu HF, Tang Y, Chen J, Lv J. Pregnancy outcome after the treatment of loop electrosurgical excision procedure or cold-knife conization for cervical intraepithelial neoplasia. Gynecol Obstet Invest 2014;77:240–4.

[26] Lindor NM, Ney JA, Gaffey TA, Jenkins RB, Thibodeau SN, Dewald GW. A genetic review of complete and partial hydatidiform moles and nonmolar triploidy. Mayo Clin Proc 1992;67:791–9.

[27] Matsui H, Iizuka Y, Sekiya S. Incidence of invasive mole and choriocarcinoma following partial hydatidiform mole. Int J Gynaecol Obstet 1996;53:63–4.

[28] Palmer JR. Advances in the epidemiology of gestational trophoblastic disease. J Reprod Med 1994;39:155–62.

[29] Hyman DM, Bakios L, Gualtiere G, Carr C, Grisham RN, Makker V, et al. Placental site trophoblastic tumor: analysis of presentation, treatment, and outcome. Gynecol Oncol 2013;129:58–62.

[30] Davis MR, Howitt BE, Quade BJ, Crum CP, Horowitz NS, Goldstein DP, et al. Epithelioid trophoblastic tumor: a single institution case series at the New England Trophoblastic Disease Center. Gynecol Oncol 2015;137:456–61.

[31] Kanehira K, Starostik P, Kasznica J, Khoury T. Primary intraplacental gestational choriocarcinoma: histologic and genetic analyses. Int J Gynecol Pathol 2013;32:71–5.

[32] Nguyen NM, Slim R. Genetics and epigenetics of recurrent hydatidiform moles: basic science and genetic counselling. Curr Obstet Gynecol Rep 2014;3:55–64.

[33] Ulker V, Gurkan H, Tozkir H, Karaman V, Ozgur H, Numanoglu C, et al. Novel NLRP7 mutations in familial recurrent hydatidiform mole: are NLRP7 mutations a risk for recurrent reproductive wastage? Eur J Obstet Gynecol Reprod Biol 2013;170:188–92.

[34] Seckl MJ, Sebire NJ, Fisher RA, Golfier F, Massuger L, Sessa C. Gestational trophoblastic disease: ESMO Clinical Practice Guidelines for diagnosis, treatment and follow-up. Ann Oncol 2013;24:vi39–50.

[35] Gupta M, Vang R, Yemelyanova AV, Kurman RJ, Li FR, Maambo EC, et al. Diagnostic reproducibility of hydatidiform moles: ancillary techniques (p57 immunohistochemistry and molecular genotyping) improve morphologic diagnosis for both recently trained and experienced gynecologic pathologists. Am J Surg Pathol 2012;36:1747–60.

[36] Banet N, DeScipio C, Murphy KM, Beierl K, Adams E, Vang R, et al. Characteristics of hydatidiform moles: analysis of a prospective series with p57 immunohistochemistry and molecular genotyping. Mod Pathol 2014;27:238–54.

[37] Kipp BR, Ketterling RP, Oberg TN, Cousin MA, Plagge AM, Wiktor AE, et al. Comparison of fluorescence in situ hybridization, p57 immunostaining, flow cytometry, and digital image analysis for diagnosing molar and nonmolar products of conception. Am J Clin Pathol 2010;133:196–204.

[38] LeGallo RD, Stelow EB, Ramirez NC, Atkins KA. Diagnosis of hydatidiform moles using p57 immunohistochemistry and HER2 fluorescent in situ hybridization. Am J Clin Pathol 2008;129:749–55.

[39] Zhao J, Xiang Y, Wan XR, Feng FZ, Cui QC, Yang XY. Molecular genetic analyses of choriocarcinoma. Placenta 2009;30:816–20.

[40] Lee CH, Nucci MR. Endometrial stromal sarcoma—the new genetic paradigm. Histopathology 2014;29:12594.

[41] Parra-Herran C, Quick CM, Howitt BE, Dal Cin P, Quade BJ, Nucci MR. Inflammatory myofibroblastic tumor of the uterus: clinical and pathologic review of 10 cases including a subset with aggressive clinical course. Am J Surg Pathol 2015;39:157–68.

[42] Medeiros F, Erickson-Johnson MR, Keeney GL, Clayton AC, Nascimento AG, Wang X, et al. Frequency and characterization of HMGA2 and HMGA1 rearrangements in mesenchymal tumors of the lower genital tract. Genes Chromosomes Cancer 2007;46:981–90.

[43] Agaram NP, Sung YS, Zhang L, Chen CL, Chen HW, Singer S, et al. Dichotomy of Genetic abnormalities in PEComas with therapeutic implications. Am J Surg Pathol 2015;25:25.

[44] D'Angelo E, Prat J. Uterine sarcomas: a review. Gynecol Oncol 2010;116:131–9.

[45] Oliva E, de Leval L, Soslow RA, Herens C. High frequency of JAZF1-JJAZ1 gene fusion in endometrial stromal tumors with smooth muscle differentiation by interphase FISH detection. Am J Surg Pathol 2007;31:1277–84.

[46] Stewart CJ, Leung YC, Murch A, Peverall J. Evaluation of fluorescence in-situ hybridization in monomorphic endometrial stromal neoplasms and their histological mimics: a review of 49 cases. Histopathology 2014;65:473—82.

[47] Aubry MC, Myers JL, Colby TV, Leslie KO, Tazelaar HD. Endometrial stromal sarcoma metastatic to the lung: a detailed analysis of 16 patients. Am J Surg Pathol 2002;26:440—9.

[48] Oliva E, Clement PB, Young RH. Endometrial stromal tumors: an update on a group of tumors with a protean phenotype. Adv Anat Pathol 2000;7:257—81.

[49] Kho KA, Lin KY, Hechanova ML, Richardson DL. Risk of occult uterine sarcoma in women undergoing hysterectomy for benign indications. Obstet Gynecol 2015;125:4S.

[50] Lee CH, Marino-Enriquez A, Ou W, Zhu M, Ali RH, Chiang S, et al. The clinicopathologic features of YWHAE-FAM22 endometrial stromal sarcomas: a histologically high-grade and clinically aggressive tumor. Am J Surg Pathol 2012;36:641—53.

[51] Koontz JI, Soreng AL, Nucci M, Kuo FC, Pauwels P, van Den Berghe H, et al. Frequent fusion of the JAZF1 and JJAZ1 genes in endometrial stromal tumors. Proc Natl Acad Sci USA 2001;98:6348—53.

[52] Singh SS, Scott S, Bougie O, Leyland N, Wolfman W, Allaire C, et al. Technical update on tissue morcellation during gynaecologic surgery: its uses, complications, and risks of unsuspected malignancy. J Obstet Gynaecol Can 2015;37:68—81.

[53] Ud Din N, Kayani N. Recurrence of adult granulosa cell tumor of the ovary: experience at a tertiary care center. Ann Diagn Pathol 2014;18:125—58.

[54] Bryk S, Farkkila A, Butzow R, Leminen A, Heikinheimo M, Anttonen M, et al. Clinical characteristics and survival of patients with an adult-type ovarian granulosa cell tumor: a 56-year single-center experience. Int J Gynecol Cancer 2015;25:33—41.

[55] Shah SP, Kobel M, Senz J, Morin RD, Clarke BA, Wiegand KC, et al. Mutation of FOXL2 in granulosa-cell tumors of the ovary. N Engl J Med 2009;360:2719—29.

[56] McCluggage WG, Singh N, Kommoss S, Huntsman DG, Gilks CB. Ovarian cellular fibromas lack FOXL2 mutations: a useful diagnostic adjunct in the distinction from diffuse adult granulosa cell tumor. Am J Surg Pathol 2013;37:1450—5.

[57] Kommoss S, Gilks CB, Penzel R, Herpel E, Mackenzie R, Huntsman D, et al. A current perspective on the pathological assessment of FOXL2 in adult-type granulosa cell tumours of the ovary. Histopathology 2014;64:380—8.

[58] Singh N, Gilks CB, Huntsman DG, Smith JH, Coutts M, Ganesan R, et al. Adult granulosa cell tumour-like areas occurring in ovarian epithelial neoplasms: report of a case series with investigation of FOXL2 mutation status. Histopathology 2014;64:626—32.

[59] Rosario R, Wilson M, Cheng WT, Payne K, Cohen PA, Fong P, et al. Adult granulosa cell tumours (GCT): clinicopathological outcomes including FOXL2 mutational status and expression. Gynecol Oncol 2013;131:325—9.

[60] Maillet D, Goulvent T, Rimokh R, Vacher-Lavenu MC, Pautier P, Alexandre J, et al. Impact of a second opinion using expression and molecular analysis of FOXL2 for sex cord-stromal tumors. A study of the GINECO group & the TMRO network. Gynecol Oncol 2014;132:181—7.

[61] D'Angelo E, Mozos A, Nakayama D, Espinosa I, Catasus L, Munoz J, et al. Prognostic significance of FOXL2 mutation and mRNA expression in adult and juvenile granulosa cell tumors of the ovary. Mod Pathol 2011;24:1360—7.

[62] Al-Agha OM, Huwait HF, Chow C, Yang W, Senz J, Kalloger SE, et al. FOXL2 is a sensitive and specific marker for sex cord-stromal tumors of the ovary. Am J Surg Pathol 2011;35:484—94.

[63] Calin GA, Ferracin M, Cimmino A, Di Leva G, Shimizu M, Wojcik SE, et al. A microRNA signature associated with prognosis and progression in chronic lymphocytic leukemia. N Engl J Med 2005;353:1793—801.

[64] Hampel H, Frankel W, Panescu J, Lockman J, Sotamaa K, Fix D, et al. Screening for Lynch syndrome (hereditary nonpolyposis colorectal cancer) among endometrial cancer patients. Cancer Res 2006;66:7810—17.

[65] Svampane L, Strumfa I, Berzina D, Svampans M, Miklasevics E, Gardovskis J. Epidemiological analysis of hereditary endometrial cancer in a large study population. Arch Gynecol Obstet 2014;289:1093—9.

[66] Mills AM, Liou S, Ford JM, Berek JS, Pai RK, Longacre TA. Lynch syndrome screening should be considered for all patients with newly diagnosed endometrial cancer. Am J Surg Pathol 2014;38:1501—9.

[67] Batte BA, Bruegl AS, Daniels MS, Ring KL, Dempsey KM, Djordjevic B, et al. Consequences of universal MSI/IHC in screening ENDOMETRIAL cancer patients for Lynch syndrome. Gynecol Oncol 2014;134:319—25.

[68] Bruegl AS, Djordjevic B, Batte B, Daniels M, Fellman B, Urbauer D, et al. Evaluation of clinical criteria for the identification of Lynch syndrome among unselected patients with endometrial cancer. Cancer Prev Res 2014;7:686—97.

[69] Ferguson SE, Aronson M, Pollett A, Eiriksson LR, Oza AM, Gallinger S, et al. Performance characteristics of screening strategies for Lynch syndrome in unselected women with newly diagnosed endometrial cancer who have undergone universal germline mutation testing. Cancer 2014;120:3932—9.

[70] McCann GA, Eisenhauer EL. Hereditary cancer syndromes with high risk of endometrial and ovarian cancer: surgical options for personalized care. J Surg Oncol 2015;111:118—24.

[71] Ketabi Z, Gerdes AM, Mosgaard B, Ladelund S, Bernstein I. The results of gynecologic surveillance in families with hereditary nonpolyposis colorectal cancer. Gynecol Oncol 2014;133: 526—30.

[72] Thibodeau SN, Bren G, Schaid D. Microsatellite instability in cancer of the proximal colon. Science 1993;260:816—19.

[73] Kim H, Jen J, Vogelstein B, Hamilton SR. Clinical and pathological characteristics of sporadic colorectal carcinomas with DNA replication errors in microsatellite sequences. Am J Pathol 1994;145:148—56.

[74] Kandoth C, Schultz N, Cherniack AD, Akbani R, Liu Y, Shen H, et al. Integrated genomic characterization of endometrial carcinoma. Nature 2013;497:67—73.

[75] Bruegl AS, Djordjevic B, Urbauer DL, Westin SN, Soliman PT, Lu KH, et al. Utility of MLH1 methylation analysis in the clinical evaluation of Lynch syndrome in women with endometrial cancer. Curr Pharm Des 2014;20:1655—63.

[76] Ryan P, Mulligan AM, Aronson M, Ferguson SE, Bapat B, Semotiuk K, et al. Comparison of clinical schemas and morphologic features in predicting Lynch syndrome in mutation-positive patients with endometrial cancer encountered in the context of familial gastrointestinal cancer registries. Cancer 2012;118:681—8.

[77] Rabban JT, Calkins SM, Karnezis AN, Grenert JP, Blanco A, Crawford B, et al. Association of tumor morphology with mismatch-repair protein status in older endometrial cancer patients: implications for universal versus selective screening strategies for Lynch syndrome. Am J Surg Pathol 2014;38: 793—800.

[78] Ferreira AM, Westers H, Sousa S, Wu Y, Niessen RC, Olderode-Berends M, et al. Mononucleotide precedes dinucleotide repeat instability during colorectal tumour development in Lynch syndrome patients. J Pathol 2009;219:96—102.

[79] Sinn DH, Chang DK, Kim YH, Rhee PL, Kim JJ, Kim DS, et al. Effectiveness of each Bethesda marker in defining microsatellite instability when screening for Lynch syndrome. Hepatogastroenterology 2009;56:672–6.

[80] Pagin A, Zerimech F, Leclerc J, Wacrenier A, Lejeune S, Descarpentries C, et al. Evaluation of a new panel of six mononucleotide repeat markers for the detection of DNA mismatch repair-deficient tumours. Br J Cancer 2013;108:2079–87.

[81] Haraldsdottir S, Hampel H, Tomsic J, Frankel WL, Pearlman R, de la Chapelle A, et al. Colon and endometrial cancers with mismatch repair deficiency can arise from somatic, rather than germline, mutations. Gastroenterology 2014;147:1308–16.

[82] Helder-Woolderink JM, De Bock GH, Sijmons RH, Hollema H, Mourits MJ. The additional value of endometrial sampling in the early detection of endometrial cancer in women with Lynch syndrome. Gynecol Oncol 2013;131:304–8.

[83] Win AK, Lindor NM, Winship I, Tucker KM, Buchanan DD, Young JP, et al. Risks of colorectal and other cancers after endometrial cancer for women with Lynch syndrome. J Natl Cancer Inst 2013;105:274–9.

[84] Nebgen DR, Lu KH, Rimes S, Keeler E, Broaddus R, Munsell MF, et al. Combined colonoscopy and endometrial biopsy cancer screening results in women with Lynch syndrome. Gynecol Oncol 2014;135:85–9.

[85] Haanstra JF, Vasen HF, Sanduleanu S, van der Wouden EJ, Koornstra JJ, Kleibeuker JH, et al. Quality colonoscopy and risk of interval cancer in Lynch syndrome. Int J Colorectal Dis 2013;28:1643–9.

[86] de Vos tot Nederveen Cappel WH, Jarvinen HJ, Lynch PM, Engel C, Mecklin JP, Vasen HF. Colorectal surveillance in Lynch syndrome families. Fam Cancer 2013;12:261–5.

[87] Kuismanen SA, Moisio AL, Schweizer P, Truninger K, Salovaara R, Arola J, et al. Endometrial and colorectal tumors from patients with hereditary nonpolyposis colon cancer display different patterns of microsatellite instability. Am J Pathol 2002;160:1953–8.

[88] van der Klift H, Wijnen J, Wagner A, Verkuilen P, Tops C, Otway R, et al. Molecular characterization of the spectrum of genomic deletions in the mismatch repair genes MSH2, MLH1, MSH6, and PMS2 responsible for hereditary nonpolyposis colorectal cancer (HNPCC). Genes Chromosomes Cancer 2005;44:123–38.

[89] Wagner A, van der Klift H, Franken P, Wijnen J, Breukel C, Bezrookove V, et al. A 10-Mb paracentric inversion of chromosome arm 2p inactivates MSH2 and is responsible for hereditary nonpolyposis colorectal cancer in a North-American kindred. Genes Chromosomes Cancer 2002;35:49–57.

[90] Konstantinopoulos PA, Matulonis UA. POLE mutations as an alternative pathway for microsatellite instability in endometrial cancer: implications for Lynch syndrome testing. Cancer 2015;121:331–4.

[91] Cancer Genome Atlas Research Network. Integrated genomic analyses of ovarian carcinoma. Nature 2011;474:609–15.

29

The Emerging Genetic Landscape in Renal Cell Carcinoma

K. Willoughby and H.A. Drabkin

Department of Medicine, Division of Hematology/Oncology, Medical University of South Carolina, Charleston, SC, United States

INTRODUCTION

Renal cell carcinoma (RCC) represents a group of diseases arising from renal tubule epithelial cells, which historically has been resistant to available therapies. The major socioeconomic impact of this disease continues to grow with its rising incidence worldwide. More than 63,000 new diagnoses of malignant tumors of the kidney are projected in the United States in 2014, and an estimated 13,860 patient deaths will be attributed to this disease [1]. RCC consists of three main histopathologic subgroups with clear-cell RCC (ccRCC) being the predominant histology, representing approximately 70–75% of reported cases. In contrast, papillary (pRCC) and chromophobe RCCs account for approximately 10–15% and only 5% of cases, respectively [2]. The remaining histologic subtypes are rarer entities including: multiloculated clear cell, collecting duct, mucinous tubular, spindle cell, medullary, Xp11 translocation RCC, carcinomas associated with neuroblastoma, RCC unclassified, and newer varieties being introduced in recent years [3–5].

Historically, histology has been the primary means of risk-stratifying patients and it continues to play an important role in clinical management. ccRCC is known to carry an unfavorable prognosis compared with other, less common, histologic subtypes [6]. Like other malignant diseases, the multitude of separate histologic entities and their associated variable prognosis has led to the hypothesis that a variety of molecular derangements underlie RCC oncogenesis and that elucidation of these mechanisms will lead to the development of more effective targeted therapies.

The study of hereditary RCC models has revealed the importance of genetics and has been applied to develop a deeper understanding of somatic mutations and the interplay within the landscape of cancer development and propagation. Hereditary RCC accounts for 3–5% of all cases of kidney cancer [7]. There are 10 described cancer susceptibility syndromes with 12 associated germline mutations [7–11] (Table 29.1). In recent years, mutated genes identified through the study of familial kidney cancers have been of paramount importance in understanding carcinogenesis of sporadic kidney cancers. Von Hipple-Lindau (VHL) disease was the first of these conditions to be recognized and the genomic discoveries related to this entity serve as a foundation on which the molecular understanding of RCC is based.

MOLECULAR TARGETS AND TECHNOLOGIES

VHL Disease

VHL disease is an autosomal-dominant cancer syndrome that predisposes involved patients to hemangioblastomas of the cranial neural axis and retina, as well as pancreatic and kidney cysts, pancreatic neuroendocrine tumors, endolymphatic sac tumors, pheochromocytomas, and ccRCC [15]. Early cases of this syndrome were described in 1860 in France. However, the hereditary nature of the syndrome was not recognized until 30 years later by von Hipple with later contributions by Lindau. There is significant variation in the phenotype of affected patients, which we now know is

Diagnostic Molecular Pathology
DOI: http://dx.doi.org/10.1016/B978-0-12-800886-7.00029-7

© 2017 Elsevier Inc. All rights reserved.

TABLE 29.1 Kidney Cancer Syndrome Gene Targets [7–14]

Gene	Protein	Disease/syndrome	Histology
VHL	pVHL	VHL disease	Clear cell
MET	c-MET	Hereditary papillary renal cell carcinoma	Papillary type 1
BAP-1	BRCA-associated protein	BAP1-mutant disease	Clear cell
FLCN	Folliculin	Birt–Hogg–Dube syndrome	Oncocytic, chromophobe
TSC1	Hamartin	Tuberous sclerosis complex	Anigomyolipoma
TSC2	Tuberin		
FH	Fumarate hydratase	Hereditary leiomyomatosis renal cell carcinoma	Papillary type 2
t(3;8)(p14.2)(q24.1)	TRC8	Familial clear-cell kidney cancer with chromosome 3 translocation	Clear cell
t(3;6)(q12;q15)			
t(1;3)(q32-q41;q13q21)			
t(2;3)(q35;q21)	DIRC2		
Other chromosome 3 translocations			
PTEN	PTEN	PTEN hamartoma syndrome	Clear cell
SDHB	Succinate dehydrogenase subunits B, C, D	SDH-associated kidney cancer	Clear cell, chromophobe, oncoytoma
SDHC			
SDHD			

attributable to the heterogeneity of mutations in the target gene. The mapping and identification of the mutated gene to 3p25–26 occurred in the early 1990s and has vastly broadened the understanding of both familial and sporadic kidney cancer [15].

The *VHL* gene is a classic tumor suppressor in that mutations of both copies are required for cancer development. It encodes the substrate recognition component of an E3 ligase, the major targets of which are hypoxia-inducible factor-1α (HIF1α) and hypoxia-inducible factor-2α (HIF2α) [16]. HIF1α and HIF2α are transcription factors that regulate hypoxia-responsive genes, including, among others, vascular endothelial growth factor (VEGF), platelet-derived growth factor (PDGF), TGFα (a ligand of the epidermal growth factor receptor), the glucose transporter, GLUT1, and carbonic anhydrase IX [17]. In normoxia, the VHL complex, which is formed by the interaction between VHL, elongin B/C, and Cul2, targets HIF1α and HIF2α for polyubiquitylation and subsequent proteasome-mediated degradation [16]. However, in the presence of hypoxia and absence of HIF modification by prolyl hydroxylases, the complex does not bind HIF, allowing for robust transcription of downstream HIF-dependent

genes [18]. Mutations that inactivate the VHL protein, which occur typically in the alpha domain that binds elongin C/B and Cul2, or in the beta domain that targets HIF, result in constitutive production of HIF target genes leading to cell proliferation and angiogenesis [19]. In addition, there are non-HIF targets of VHL involving the regulation of microtubules, integrin maturation [20–22], NF-κB activity [23], and p53 stability [24] that likely contribute to RCC pathogenesis.

VHL alterations represent the classic paradigm of a hereditary cancer gene that is often somatically mutated in sporadic forms of kidney cancer. Nickerson et al. detected somatic mutations of *VHL* in 82.4% of tumors from patients with ccRCC [25]. This study also revealed that 8.3% of tumors had *VHL* promoter sites that were hypermethylated and silenced. In total, 91% of ccRCCs studied expressed alterations of the gene through genetic or epigenetic mechanisms. These findings have been corroborated in larger scale genetic sequencing studies [26,27].

Other inherited cancer susceptibility syndromes associated with an increased risk of kidney cancer have played a role in identifying molecular targets for therapeutic intervention. Hereditary pRCC (type I papillary) is characterized by activating mutations of the *MET* gene on 7q31 with a reported rate of mutation in sporadic cases between 4% and 10% [28]. Alterations of the tuberous sclerosis complex, comprised of *TSC1* and *TSC2*, have recently been found to occur in approximately 5% of ccRCCs and are expected to predict sensitivity to mTOR inhibitor treatment in patients [29]. Conversely, somatic mutations of other noted germline targets have been rarely identified, including *FLCN*, *FH*, and *SDHB* [30]. Likewise, commonly altered genes in other cancers, for example, *RAS*, *BRAF*, *TP53*, *RB*, *CDKN2A*, *PIK3CA*, *PTEN*, *EGFR*, and *ERBB2*, are rarely directly genetically affected in the oncogenesis of kidney cancer.

Chromatin Remodeling/Histone Modification: SETD2, JARID1C (KDM5C), UTX, MLL2

With the advancement of genomic technologies, such as next-generation sequencing (NGS) and high-density single-nucleotide polymorphism (SNP) arrays, further identification of somatic gene mutations has become possible. These previously unidentified biologic mechanisms and genetic pathways are, in essence, restructuring the way we look at kidney cancer and creating ways for newer classification systems and novel treatments. In one of the first large-scale sequencing studies in RCC, Dalgliesh et al. examined the coding exons of 3544 genes in 101 kidney cancers [31]. In addition to *NF2* alterations, inactivating mutations were identified in genes involved in chromatin remodeling and histone methylation, that is, *SETD2*, *JARID1C* (also known as *KDM5C*), *UTX* (*KMD6A*), and *MLL2*. *SETD2* is a histone H3 lysine 36 methyltransferase, *JARID1C* is a histone H3 lysine 4 demethylase, *UTX* is a histone H3 lysine 27 demethylase, and *MLL2* is a histone H3 lysine 4 methyltransferase. Methylation of histone H3 lysine residues regulates chromatin structure and is implicated in transcriptional control. The clustering of these mutations around genes that control epigenetic modifications of histone H3 has opened the door to a new target pathway in the pathogenesis or progression of kidney cancer. However, these mutations encompassed less than 15% of samples in the reported cohort, which suggested that unidentified genes still existed.

Chromatin Remodeling/Histone Modification: PBRM1

Varela et al., using protein-coding exome sequencing of ccRCCs and matched normal tissue, identified yet another target, *PBRM1* [32]. The *PBRM1* gene maps to 3p21 and encodes the BAF180 protein [33], which is the chromatin-targeting subunit of the Polybromo BRG1-

associated factor complex (PBAF, SWI/SNF-B). The PBAF complex is active in nucleosome remodeling. In general, DNA tightly wrapped around a nucleosome is inaccessible for transcription. This is due, in part, to strong interactions between unmodified lysine residues on histones and the sugar phosphate backbone of DNA. However, posttranslational modifications of lysine residues such as acetylation and methylation can loosen these interactions and yield a more open configuration to provide access for transcription factors. Nucleosome remodeling complexes function to restructure the DNA and histone interaction and regulate transcription. The SWI/SNF complexes are comprised of multiple subunits. BAF180 contains multiple bromodomains that bind acetylated lysine residues on histone tails. In this cancer cohort, *PBRM1* truncating mutations were identified in a substantial 41% of samples [32]. Most mutations occurred together with *VHL* alterations, and nearly all *PBRM1*-mutant cancers examined (36/38) exhibited a hypoxia gene expression signature. The SWI/SNF complex is involved in the normal cellular response to hypoxia and a dysfunctional complex may leave cells resistant to cell-cycle arrest. Initial reports indicated that *PBRM1* mutations correlate with phenotypic advanced disease stage, high Fuhrman grade, and poor overall survival [34]. However, more recent reports suggest that these mutations are seen at similar rates regardless of stage and may not have an adverse impact on survival [35,36].

Chromatin Remodeling/Histone Modification: BAP1

Carrying this methodological approach forward, Peña-Llopis and colleagues performed whole-genome and exome sequencing followed by tumor graft analyses and identified several putative tumor-suppressor genes. Mutations of the *BRCA1-associated protein-1* (*BAP1*) gene were identified at a rate of 14% [35]. *BAP1* encodes a nuclear protein containing a ubiquitin carboxy-terminal hydrolase domain reported to target histone H2A. Approximately 53% of cancers in this study exhibited *PBRM1* mutations. However, only 4 of 21 *BAP1*-mutant cancers were also deficient in PBRM1. These results suggested that *PBRM1* and *BAP1* are mutually exclusive driver mutations. Since both genes are located on the short arm of chromosome 3 (3p21) and represent classical tumor-suppressor genes, the investigators proposed that the loss of 3p subsequent to *VHL* mutation represents the initial event in tumorigenesis. This would leave cells vulnerable to loss of the remaining *PBRM1* or *BAP1* allele, and acquisition of further mutations in these genes would determine the course of ccRCC.

BAP1-mutated and *PBRM1*-mutated cancers are phenotypically distinct [36]. *BAP1*-mutated carcinomas demonstrate higher tumor grade and mTOR activation compared to *PBRM1*-mutated counterparts [36]. In further support of the important role of *BAP1*, mice with combined cre-mediated loss of *VHL* and *BAP1* develop progressive atypical cysts and tumor formation resembling patients with VHL disease [37]. The investigators propose that the lack of synteny, with *BAP1* and *VHL* residing on different chromosomes in the mouse, is responsible for the lack of RCC in the mouse model of VHL.

Other Mutation Studies Involving Whole-Genome and Exome Sequencing

More recently, in an effort to further standardize the approach to this disease, the Cancer Genome Atlas Research Network (TCGA) published results on more than 400 histologically confirmed ccRCCs and analyzed cases for clinical and pathologic features, genomic alterations, DNA methylation profiles, and RNA and proteomic signatures [26]. The results showed a lack of focal somatic copy number alterations (SCNAs). However, SCNAs were detected involving entire whole chromosomes and chromosomal arms, most notably, frequent loss of chromosome 3p (91% of samples). This was expected, as *PBRM1*, *BAP1*, *SETD2*, and *VHL* all map to this region. Other significantly identified chromosomal alterations included: loss of chromosome 14q (45% of samples) and gains of 5q (67% of samples). Whole-exome sequencing validated 19 genes with high mutation frequencies. Eight of these were identified as significantly mutated genes (SMGs): *VHL*, *PBRM1*, *SETD2*, *KDM5D*, *PTEN*, *BAP1*, *mTOR*, and *TP53*. Among the genes classified as SMGs, *BAP1* correlated with poor survival outcomes. By examining DNA methylation profiles, the authors were able to elucidate yet another novel target: *UQCRH* (ubiquinol-cytochrome c reductase hinge protein). This gene was hypermethylated in 36% of cancers and increased promoter methylation correlated with higher stage and grade disease. However, at this time, little more is known about the function of this gene.

Sato et al. reported on a Japanese cohort, involving 100 ccRCC cases, analyzed by whole-genome or exome sequencing, RNA sequencing with microarray-based gene expression, DNA methylation, genomic copy number analyses, and immunohistochemistry, published around the same time as the Cancer Genome Atlas Research Network [27]. They identified 28 genes that were significantly mutated compared to background mutations rates. Of the top mutated genes, *VHL*, *PBRM1*, *BAP1*, and *SETD2* were all located within the common site of LOH at 3p between 3p25

and 3p21. Other important identified targets included: *TCEB1*, which is the gene-encoding Elongin C, known to be an essential part of the VHL complex; *TET2*, which encodes an α-ketoglutarate-dependent oxygenase catalyzing a critical step in DNA demethylation; Kelch-like ECH-associated protein 1 (*Keap1*), which is a key component of a cullin–RING ubiquitin ligase complex that targets nuclear factor erythroid 2 related factor (NRF2) in oxidative stress responses; and finally *mTOR*, *PTEN*, *PIK3CA*, *MTORC1*, *PIK3CG*, *RPS6KA2*, *TSC1*, *TSC2*, and others, which together comprised 26% of reported cases involving the PI3K/AKT/mTORC1 signaling pathway.

PI3K/AKT/mTORC1 Signaling Pathway

mTOR pathway is subject to mutational activation in RCC (Fig. 29.1). Although *mTOR* itself is mutated in only 5–6% of ccRCCs [26,27], cumulative mutations within the pathway are significantly more frequent [27]. The *mTOR* gene encodes a serine/threonine protein kinase that belongs to the phosphatidylinositol 3-kinase family and is the catalytic subunit in both mTORC1 and mTORC2 complexes. *mTOR* mutations evaluated in vitro have been found to increase mTORC1 activity without doing the same to mTORC2 activity [38]. Mutations within the protein typically converge on two domains, the kinase and FAT domains, and often sensitize cells to sirolimus analogs [39]. The two FAT domains flank the kinase domain and bind the inhibitor, Deptor. Mutations proximal to mTOR within the mTORC1 pathway have also been identified. The *PIK3CA* gene, which encodes the catalytic subunit p110α of PI3K, is mutated in 2–5% of ccRCC [26,27]. PI3K catalyzes the formation of phosphatidylinositol-3,4,5-triphosphate (PIP3), which is a lipid secondary messenger that is down-regulated by the tumor suppressor, PTEN. *PTEN* mutations have also been identified in ccRCC, at a frequency of 1–5% [26,27]. Both activating mutations of *PI3K* and inactivating mutations of *PTEN* lead to increased levels of PIP3 at the plasma membrane, which result in increased binding and translocation of AKT to the membrane. AKT is activated through a dual phosphorylation mechanism and subsequently phosphorylates multiple substrates, one of which is TSC2. TSC2 forms a complex with TSC1 and together this complex (TSC1/TSC2) functions as a tumor suppressor [29]. The TSC1/TSC2 complex is inactivated in the tuberous sclerosis hereditary cancer syndrome, which leads to the development of renal angiomyolipomas in a large proportion of patients afflicted, while a TSC2 mutation in the Eker rat model leads to ccRCC [40]. The TSC1/TSC2 complex acts as a GTPase-activating protein, which hydrolyzes GTP from

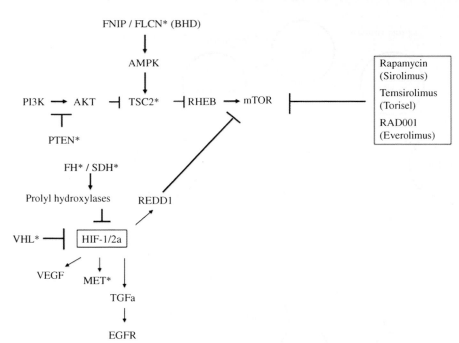

FIGURE 29.1 The mTOR pathway schema highlighting commonly mutated genes in kidney cancer.

the protein, Ras homologue enriched in brain (Rheb). When Rheb is in its GTP-bound state, it binds and activates mTORC1. Mutations in *TSC1* have been reported in 4% of ccRCCs [29] with *Rheb* mutations representing only four cases as reported by the Cancer Genome Atlas [26]. Thus, mutations in the *TSC1/TSC2* complex, *mTOR*, and *Rheb* represent mechanisms to activate the mTORC1 pathway. Collectively, the identification of cancers with mTOR activation is clinically important since those tumors may be exquisitely sensitive to mTORC1 inhibitors [29].

In addition to the above mechanisms of mTORC1 control, another protein, regulated in development and DNA damage response 1 (REDD1), is involved in regulation as a response to hypoxia. *REDD1* is directly induced by HIF1α and HIF2α in ccRCC, and REDD1 induction is sufficient to inhibit mTORC1 [41,42]. As approximately 90% of ccRCC tumors harbor a VHL alteration [25], HIF1α and HIF2α are often upregulated and this explains the consistently increased REDD1 activity seen in most ccRCC [29]. However, based on these findings, mTORC1 activity in ccRCC would be low, unless harboring an inactivating escape mutation within the *TSC1/TSC2* complex or *PTEN*. We know that these mutations occur, but at low frequency [26,27] and as such, other mechanisms of mTORC1 activation must be present that remain yet unknown.

Intratumoral Heterogeneity

Large-scale sequencing endeavors over the last few years have substantially increased the knowledge of mutations in ccRCC and have implicated new potential therapeutic targets. However, intratumoral heterogeneity, as described by Gerlinger et al. in 2012, adds yet another layer to the genomic landscape [43]. Based on multiple samples from primary and metastatic sites, considerable variability in gene expression signatures, allelic imbalances, and somatic mutations was identified. This study classified somatic mutations as ubiquitous, shared, or private. Ubiquitous mutations were seen in all subclones analyzed, whereas shared and private mutations were seen in more distant subclones—purposing a branching phylogenetic tree of tumor regions by clonal ordering [43]. In short, a single biopsy is unlikely to adequately represent the mutational spectrum of these heterogeneous cancers, implying that therapeutic approaches should take these findings into consideration.

Metabolism in RCC

Returning to hereditary cancer syndromes as a roadmap for uncovering pathways to carcinogenesis, a close link between metabolism and RCC was identified. SDH-associated kidney cancer and hereditary leiomyomatosis renal cell carcinoma (HLRCC) are both genetic cancer syndromes that involve germline mutations in genes-encoding enzymes of the Krebs cycle [44] (Fig. 29.2). Succinate dehydrogenase (SDH) is a complex of four polypeptides (SDH A–D) that catalyzes the conversion of succinate to fumarate. The genes encoding the subunits of SDH are mutated in the SDH-associated kidney cancer syndrome. Fumarate hydratase (FH) catalyzes

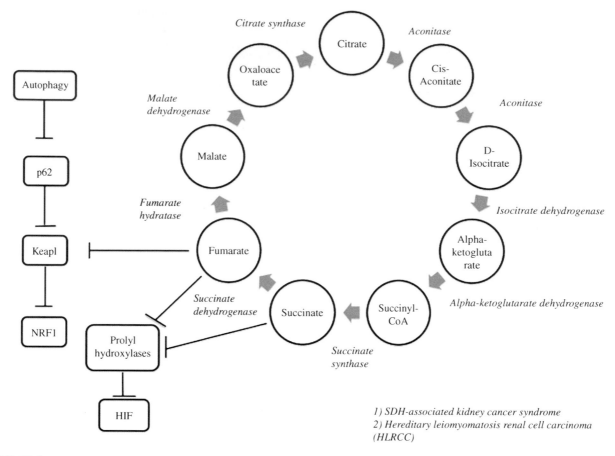

FIGURE 29.2 This diagram demonstrates the interaction between Krebs cycle intermediates, the Keap1/NRF1 pathway, and prolyl hydroxylases.

the conversion of fumarate to malate and germline mutations in the gene encoding this enzyme cause HLRCC, which predisposes those affected to the development of pRCC type 2. The *FH* and *SDH* genes function as tumor suppressors [45] and loss-of-function mutations lead to accumulation of fumarate and succinate, respectively [44]. Increased levels of succinate and fumarate inhibit HIFα prolyl hydroxylase domain (PHD) containing enzymes, which are responsible for HIF hydroxylation and binding by VHL [44,46–48]. Alpha-ketoglutarate, an upstream intermediate, is a known substrate for PHD and increased levels competitively inhibit succinate and fumarate and restore PHD and HIF1α to normal level activity [47]. However, fumarate and succinate at excess levels also interact with other proteins, including Keap1 [49]. Keap1 is a component of an E3 ubiquitin ligase that targets NRF2, a master regulator of antioxidant response. NRF2, under normal conditions, is anchored in the cytoplasm by binding to Keap1. Upon succination of Keap1, the complex dissociates, NRF2 accumulates in the nucleus and leads to increased expression of antioxidant and antiinflammatory responses [50].

Autophagy: SQSTM1 and NRF2

Autophagy is a mechanism of protein degradation responsible for the disposal of damaged organelles and clearance of aggregated proteins. During this process, nuclear membranes engulf cytoplasmic substrates forming autophagosomes that then fuse with lysosomes and lead to protein degradation. Of note, autophagy-inducing drugs were reported to have selective toxicity for VHL-deficient cells [51]. Sequestosome 1 (p62) is a scaffold protein and both an essential component and a target of autophagy. When autophagy is impaired, p62 accumulates and inactivates Keap1, allowing NRF2 to translocate to the nucleus and induce the transcription of genes involved in redox response [52]. Increased levels of p62 are toxic to normal cells [52] and also promote tumorigenesis [53]. The *SQSTM1* gene, which maps to chromosome 5q35, has been reported to be the target of copy number gains occurring in approximately 70% of RCCs [26,27,54,55]. The p62 protein interacts with a variety of proteins involved in the NRF2, NF-κB, and mTOR pathways [55,56]. Thus, identifying RCC cases with p62 overexpression is likely to have future therapeutic implications as more details emerge [57].

CLINICAL UTILITY

Gene expression analyses have provided important insight into the heterogeneity among kidney cancer and within the clear-cell subtype. Some groups have proposed a new classification of clear-cell kidney cancer using gene expression analysis [54,58]. Brannon et al. demonstrated that two or more molecular sub-classifications of ccRCC exist [58]. A meta-analysis performed in 2011 by the same group identified two distinct subsets, ccA and ccB, as well as a third divergent group characterized by wild-type VHL and a clear-cell papillary histology [59]. Subtypes ccA and ccB were associated with a significant difference in outcome, with ccA patients having a median overall survival of 103 months compared to 24 months for ccB [60]. RNA expression profiles reported from the Cancer Genome Atlas showed four stable subsets of mRNA and miRNA expression data sets, which were compatible with the new subtypes proposed by Brannon et al. [26]. A similar survival advantage was again observed in the mRNA profile m1 comparable to ccA subtype and these cases were characterized by a higher frequency of mutations within the chromatin remodeling system. The mRNA profile m3, comparable with the previously described ccB subtype, was characterized by CDKN2A and PTEN mutations, while the m4 profile included cancers with a higher rate of BAP1 and mTOR mutations.

Gene expression profiling by microarray analysis provides important information for prognostication and treatment. However, gene expression analysis is costly and may be difficult to implement on a large scale. Therefore, we would advocate for mutation testing of ccRCCs for driver mutations to at least include VHL, PBMR1, BAP1, SETD2, mTORC, and NF2. With the increased availability of platforms for the simultaneous exon sequencing of many mutated cancer genes, and with the development of new therapeutic agents that can target specific epigenetic alterations, an RCC chip that includes most recurrently mutated genes in this disease should provide the most rational approach to improving outcome. Upfront identification of these mutations will provide valuable prognostic information and may predict response to available therapies. For example, the identification of cancers with mTOR pathway activation is rationale for the optimal use of mTOR inhibitors [61,62]. Conversely, NF2 mutations, which lack a hypoxic gene expression signature, may not respond to anti-VEGF-targeted therapies, currently the standard first-line treatment for patients with metastatic disease [63]. In acute lymphoblastic leukemia, mutations in the histone H3 lysine 27 demethylase gene, UTX, confer sensitivity to the histone demethylase inhibitor, GSKJ4 [64], although responses in RCCs with this mutation have not been reported. Other therapeutic agents, such as EZH2 inhibitors, potentially may have selective toxicity in cancers with high EZH2 expression [33,65] or in cancers with loss-of-function mutations involving the SWI/SNF complex [33], which demonstrates epigenetic antagonism with the Polycomb PRC2 complex [66]. These and other therapeutic strategies will require careful testing given the complexity of epigenetic modifications. As we learn more about the genetic landscape of renal cell cancer and develop more targeted therapies, determining the appropriate targeted therapy for each individual patient should become achievable.

LIMITATIONS OF TESTING

Universal guidelines for molecular testing in RCC have not yet been established. Current techniques broadly used include histologic review and immunohistochemical staining for hypoxia-responsive proteins such as carbonic anhydrase IX, GLUT, and HIF1α. This analysis alone is limited and quickly becoming arcane given the recent discoveries of a number of molecular targets. One method for further stratifying ccRCCs is use of whole-genome RNA profiling. Brannon et al. [58,59] and TCGA investigators [26] have published subtyping algorithms that are associated with prognostication [67]. However, the process of determining these molecular subtypes will require validation. Therefore, at this time we do not recommend routine use of this technique outside of the setting of a clinical trial. Molecular testing for single biomarkers, such as BAP1 and PBRM1, has been shown to confer prognostic information [68] and is a cost-conscious approach for the moment. Lastly, we would be remiss to not again emphasize the importance of the work of Gerlinger et al. [43] and its implication in the future treatment of RCC. This work gives us a greater sense of the complexity and intratumoral heterogeneity that exists within this disease and reminds us that molecular analysis of any one biologic sample at any one time represents only a piece of a much larger puzzle.

References

[1] Siegel R, Ma J, Zou Z, Jemal A. Cancer statistics, 2014. CA Cancer J Clin 2014;64:9–29.

[2] Soung Sullivan P, Rao J, Cheng L, Cote RJ. Classical pathology versus molecular pathology in renal cell carcinoma. Curr Urol Rep 2007;8:5–11.

[3] Srigley JR, Delahunt B, Eble JN, Egevad L, Epstein JI, Grignon D, et al. The International Society of Urological Pathology (ISUP) Vancouver classification of renal neoplasia. Am J Surg Pathol 2013;37:1469–89.

[4] Lopez-Beltran A, Scarpelli M, Montironi R, Kirkali Z. WHO classification of the renal tumors of the adults. Eur Urol 2004;49:798–805 2006.

[5] Delahunt B. Urologic pathology: SY25-1 prognostic factors in renal cancer. Pathology 2014;46:S44.

[6] Lohse CM, Cheville JC. A review of prognostic pathologic features and algorithms for patients treated surgically for renal cell carcinoma. Clin Lab Med 2005;25:433–64.

[7] Haas NB, Nathanson KL. Hereditary kidney cancer syndromes. Adv Chronic Kidney Dis 2014;21:81–90.

[8] Bodmer D, Eleveld M, Kater-Baats E, Janssen I, Janssen B, Weterman M, et al. Disruption of a novel MFS transporter gene, DIRC2, by a familial renal cell carcinoma-associated t(2;3)(q35;q21). Hum Mol Genet 2002;11:641–9.

[9] Bodmer D, Janssen I, Jonkers Y, van den Berg E, Dijkhuizen T, Debiec-Rychter M, et al. Molecular cytogenetic analysis of clustered sporadic and familial renal cell carcinoma-associated 3q13 approximately q22 breakpoints. Cancer Genet Cytogenet 2002;136:95–100.

[10] Bonne AC, Bodmer D, Schoenmakers EF, van Ravenswaaij CM, Hoogerbrugge N, van Kessel AG. Chromosome 3 translocations and familial renal cell cancer. Curr Mol Med 2004;4:849–54.

[11] Eleveld MJ, Bodmer D, Merkx G, Siepman A, Sprenger SH, Weterman MA, et al. Molecular analysis of a familial case of renal cell cancer and a t(3;6)(q12;q15). Genes, Chromosomes Cancer 2001;31:23–32.

[12] Foster RE, Abdulrahman M, Morris MR, Prigmore E, Gribble S, Ng B, et al. Characterization of a 3;6 translocation associated with renal cell carcinoma. Genes Chromosomes Cancer 2007;46:311–17.

[13] Kanayama H, Lui WO, Takahashi M, Naroda T, Kedra D, Wong FK, et al. Association of a novel constitutional translocation t(1q;3q) with familial renal cell carcinoma. J Med Genet 2001;38:165–70.

[14] Gemmill RM, West JD, Boldog F, Tanaka N, Robinson LJ, Smith DI, et al. The hereditary renal cell carcinoma 3;8 translocation fuses FHIT to a patched-related gene, TRC8. Proc Natl Acad Sci USA 1998;95:9572–7.

[15] Bausch B, Jilg C, Glasker S, Vortmeyer A, Lutzen N, Anton A, et al. Renal cancer in von Hippel-Lindau disease and related syndromes. Nat Rev Nephrol 2013;9:529–38.

[16] Shen C, Kaelin Jr. WG. The VHL/HIF axis in clear cell renal carcinoma. Semin Cancer Biol 2013;23:18–25.

[17] Lawrentschuk N, Lee FT, Jones G, Rigopoulos A, Mountain A, O'Keefe G, et al. Investigation of hypoxia and carbonic anhydrase IX expression in a renal cell carcinoma xenograft model with oxygen tension measurements and (1)(2)(4)I-cG250 PET/CT. Urol Oncol 2011;29:411–20.

[18] Brugarolas J. Molecular genetics of clear-cell renal cell carcinoma. J Clin Oncol 2014;32:1968–76.

[19] Linehan WM, Bratslavsky G, Pinto PA, Schmidt LS, Neckers L, Bottaro DP, et al. Molecular diagnosis and therapy of kidney cancer. Annu Rev Med 2010;61:329–43.

[20] Hergovich A, Lisztwan J, Barry R, Ballschmieter P, Krek W. Regulation of microtubule stability by the von Hippel-Lindau tumour suppressor protein pVHL. Nat Cell Biol 2003;5:64–70.

[21] Esteban-Barragan MA, Avila P, Alvarez-Tejado M, Gutierrez MD, Garcia-Pardo A, Sanchez-Madrid F, et al. Role of the von Hippel-Lindau tumor suppressor gene in the formation of beta1-integrin fibrillar adhesions. Cancer Res 2002;62:2929–36.

[22] Ji Q, Burk RD. Downregulation of integrins by von Hippel-Lindau (VHL) tumor suppressor protein is independent of VHL-directed hypoxia-inducible factor alpha degradation. Biochem Cell Biol 2008;86:227–34.

[23] Yang H, Minamishima YA, Yan Q, Schlisio S, Ebert BL, Zhang X, et al. pVHL acts as an adaptor to promote the inhibitory phosphorylation of the NF-kappaB agonist Card9 by CK2. Mol Cell 2007;28:15–27.

[24] Roe JS, Kim H, Lee SM, Kim ST, Cho EJ, Youn HD. p53 stabilization and transactivation by a von Hippel-Lindau protein. Mol Cell 2006;22:395–405.

[25] Nickerson ML, Jaeger E, Shi Y, Durocher JA, Mahurkar S, Zaridze D, et al. Improved identification of von Hippel-Lindau gene alterations in clear cell renal tumors. Clin Cancer Res 2008;14:4726–34.

[26] Cancer Genome Atlas Research Network. Comprehensive molecular characterization of clear cell renal cell carcinoma. Nature 2013;499:43–9.

[27] Sato Y, Yoshizato T, Shiraishi Y, Maekawa S, Okuno Y, Kamura T, et al. Integrated molecular analysis of clear-cell renal cell carcinoma. Nat Genet 2013;45:860–7.

[28] Schmidt L, Duh FM, Chen F, Kishida T, Glenn G, Choyke P, et al. Germline and somatic mutations in the tyrosine kinase domain of the MET proto-oncogene in papillary renal carcinomas. Nat Genet 1997;16:68–73.

[29] Kucejova B, Pena-Llopis S, Yamasaki T, Sivanand S, Tran TA, Alexander S, et al. Interplay between pVHL and mTORC1 pathways in clear-cell renal cell carcinoma. Mol Cancer Res 2011;9:1255–65.

[30] Kiuru M, Lehtonen R, Arola J, Salovaara R, Jarvinen H, Aittomaki K, et al. Few FH mutations in sporadic counterparts of tumor types observed in hereditary leiomyomatosis and renal cell cancer families. Cancer Res 2002;62:4554–7.

[31] Dalgliesh GL, Furge K, Greenman C, Chen L, Bignell G, Butler A, et al. Systematic sequencing of renal carcinoma reveals inactivation of histone modifying genes. Nature 2010;463:360–3.

[32] Varela I, Tarpey P, Raine K, Huang D, Ong CK, Stephens P, et al. Exome sequencing identifies frequent mutation of the SWI/SNF complex gene PBRM1 in renal carcinoma. Nature 2011;469:539–42.

[33] Brugarolas J. PBRM1 and BAP1 as novel targets for renal cell carcinoma. Cancer J 2013;19:324–32.

[34] Pawlowski R, Muhl SM, Sulser T, Krek W, Moch H, Schraml P. Loss of PBRM1 expression is associated with renal cell carcinoma progression. Int J Cancer 2013;132:E11–17.

[35] Peña-Llopis S, Vega-Rubin-de-Celis S, Liao A, Leng N, Pavia-Jimenez A, Wang S, et al. BAP1 loss defines a new class of renal cell carcinoma. Nat Genet 2012;44:751–9.

[36] Kapur P, Pena-Llopis S, Christie A, Zhrebker L, Pavia-Jimenez A, Rathmell WK, et al. Effects on survival of BAP1 and PBRM1 mutations in sporadic clear-cell renal-cell carcinoma: a retrospective analysis with independent validation. Lancet Oncol 2013;14:159–67.

[37] Wang SS, Gu YF, Wolff N, Stefanius K, Christie A, Dey A, et al. Bap1 is essential for kidney function and cooperates with Vhl in renal tumorigenesis. Proc Natl Acad Sci USA 2014;111:16538–43.

[38] Vogelstein B, Papadopoulos N, Velculescu VE, Zhou S, Diaz Jr. LA, Kinzler KW. Cancer genome landscapes. Science 2013;339: 1546–58.

[39] Voss MH, Hakimi AA, Pham CG, Brannon AR, Chen YB, Cunha LF, et al. Tumor genetic analyses of patients with metastatic renal cell carcinoma and extended benefit from mTOR inhibitor therapy. Clin Cancer Res 2014;20:1955–64.

[40] Yeung RS, Xiao GH, Jin F, Lee WC, Testa JR, Knudson AG. Predisposition to renal carcinoma in the Eker rat is determined by germ-line mutation of the tuberous sclerosis 2 (TSC2) gene. Proc Natl Acad Sci USA 1994;91:11413–16.

[41] Brugarolas J, Lei K, Hurley RL, Manning BD, Reiling JH, Hafen E, et al. Regulation of mTOR function in response to hypoxia by REDD1 and the TSC1/TSC2 tumor suppressor complex. Genes Dev 2004;18:2893–904.

[42] Vadysirisack DD, Ellisen LW. mTOR activity under hypoxia. Methods Mol Biol 2012;821:45–58.

[43] Gerlinger M, Rowan AJ, Horswell S, Larkin J, Endesfelder D, Gronroos E, et al. Intratumor heterogeneity and branched evolution revealed by multiregion sequencing. N Engl J Med 2012; 366:883–92.

[44] Pollard PJ, Briere JJ, Alam NA, Barwell J, Barclay E, Wortham NC, et al. Accumulation of Krebs cycle intermediates and overexpression of HIF1alpha in tumours which result from germline FH and SDH mutations. Hum Mol Genet 2005;14:2231–9.

[45] Gottlieb E, Tomlinson IP. Mitochondrial tumour suppressors: a genetic and biochemical update. Nat Rev Cancer 2005;5:857–66.

[46] Selak MA, Armour SM, MacKenzie ED, Boulahbel H, Watson DG, Mansfield KD, et al. Succinate links TCA cycle dysfunction to oncogenesis by inhibiting HIF-alpha prolyl hydroxylase. Cancer Cell 2005;7:77–85.

[47] MacKenzie ED, Selak MA, Tennant DA, Payne LJ, Crosby S, Frederiksen CM, et al. Cell-permeating alpha-ketoglutarate derivatives alleviate pseudohypoxia in succinate dehydrogenase-deficient cells. Mol Cell Biol 2007;27:3282–9.

[48] Isaacs JS, Jung YJ, Mole DR, Lee S, Torres-Cabala C, Chung YL, et al. HIF overexpression correlates with biallelic loss of fumarate hydratase in renal cancer: novel role of fumarate in regulation of HIF stability. Cancer Cell 2005;8:143–53.

[49] Adam J, Hatipoglu E, O'Flaherty L, Ternette N, Sahgal N, Lockstone H, et al. Renal cyst formation in Fh1-deficient mice is independent of the Hif/Phd pathway: roles for fumarate in KEAP1 succination and Nrf2 signaling. Cancer Cell 2011;20:524–37.

[50] Kinch L, Grishin NV, Brugarolas J. Succination of Keap1 and activation of Nrf2-dependent antioxidant pathways in FH-deficient papillary renal cell carcinoma type 2. Cancer Cell 2011;20:418–20.

[51] Turcotte S, Chan DA, Sutphin PD, Hay MP, Denny WA, Giaccia AJ. A molecule targeting VHL-deficient renal cell carcinoma that induces autophagy. Cancer Cell 2008;14:90–102.

[52] Komatsu M, Kurokawa H, Waguri S, Taguchi K, Kobayashi A, Ichimura Y, et al. The selective autophagy substrate p62 activates the stress responsive transcription factor Nrf2 through inactivation of Keap1. Nat Cell Biol 2010;12:213–23.

[53] Mathew R, Karp CM, Beaudoin B, Vuong N, Chen G, Chen HY, et al. Autophagy suppresses tumorigenesis through elimination of p62. Cell 2009;137:1062–75.

[54] Beroukhim R, Brunet JP, Di Napoli A, Mertz KD, Seeley A, Pires MM, et al. Patterns of gene expression and copy-number alterations in von-Hippel Lindau disease-associated and sporadic clear cell carcinoma of the kidney. Cancer Res 2009;69: 4674–81.

[55] Li L, Shen C, Nakamura E, Ando K, Signoretti S, Beroukhim R, et al. SQSTM1 is a pathogenic target of 5q copy number gains in kidney cancer. Cancer Cell 2013;24:738–50.

[56] Duran A, Linares JF, Galvez AS, Wikenheiser K, Flores JM, Diaz-Meco MT, et al. The signaling adaptor p62 is an important NF-kappaB mediator in tumorigenesis. Cancer Cell 2008;13: 343–54.

[57] Bray K, Mathew R, Lau A, Kamphorst JJ, Fan J, Chen J, et al. Autophagy suppresses RIP kinase-dependent necrosis enabling survival to mTOR inhibition. PLoS One 2012;7:e41831.

[58] Brannon AR, Reddy A, Seiler M, Arreola A, Moore DT, Pruthi RS, et al. Molecular stratification of clear cell renal cell carcinoma by consensus clustering reveals distinct subtypes and survival patterns. Genes Cancer 2010;1:152–63.

[59] Brannon AR, Haake SM, Hacker KE, Pruthi RS, Wallen EM, Nielsen ME, et al. Meta-analysis of clear cell renal cell carcinoma gene expression defines a variant subgroup and identifies gender influences on tumor biology. Eur Urol 2012;61:258–68.

[60] Banks RE, Tirukonda P, Taylor C, Hornigold N, Astuti D, Cohen D, et al. Genetic and epigenetic analysis of von Hippel-Lindau (VHL) gene alterations and relationship with clinical variables in sporadic renal cancer. Cancer Res 2006;66:2000–11.

[61] Conti A, Santoni M, Amantini C, Burattini L, Berardi R, Santoni G, et al. Progress of molecular targeted therapies for advanced renal cell carcinoma. BioMed Res Int 2013;2013:419176.

[62] Calvo E, Grunwald V, Bellmunt J. Controversies in renal cell carcinoma: treatment choice after progression on vascular endothelial growth factor-targeted therapy. Eur J Cancer 2014; 50:1321–9.

[63] Hwang C, Heath EI. The Judgment of Paris: treatment dilemmas in advanced renal cell carcinoma. J Clin Oncol 2014;32: 729–34.

[64] Ntziachristos P, Tsirigos A, Welstead GG, Trimarchi T, Bakogianni S, Xu L, et al. Contrasting roles of histone 3 lysine 27 demethylases in acute lymphoblastic leukaemia. Nature 2014;514:513–17.

[65] Wagener N, Macher-Goeppinger S, Pritsch M, Husing J, Hoppe-Seyler K, Schirmacher P, et al. Enhancer of zeste homolog 2 (EZH2) expression is an independent prognostic factor in renal cell carcinoma. BMC Cancer 2010;10:524.

[66] Wilson BG, Wang X, Shen X, McKenna ES, Lemieux ME, Cho YJ, et al. Epigenetic antagonism between polycomb and SWI/SNF complexes during oncogenic transformation. Cancer Cell 2010;18:316–28.

[67] Eckel-Passow JE, Igel DA, Serie DJ, Joseph RW, Ho TH, Cheville JC, et al. Assessing the clinical use of clear cell renal cell carcinoma molecular subtypes identified by RNA expression analysis. Urol Oncol 2015;33:e17–23.

[68] Joseph RW, Kapur P, Serie DJ, Eckel-Passow JE, Parasramka M, Ho T, et al. Loss of BAP1 protein expression is an independent marker of poor prognosis in patients with low-risk clear cell renal cell carcinoma. Cancer 2014;120:1059–67.

30

Molecular Testing in Thyroid Cancer

S.J. Hsiao[1] and Y.E. Nikiforov[2]

[1]Department of Pathology & Cell Biology, Columbia University Medical Center, New York, NY, United States
[2]Division of Molecular & Genomic Pathology, Department of Pathology, University of Pittsburgh School of Medicine, Pittsburgh, PA, United States

INTRODUCTION

Thyroid cancer is one of the most common endocrine tumors, with an estimated annual incidence of 12.2 cases per 100,000 people in the United States [1]. Women are approximately three times more likely to develop thyroid cancer than men [1]. Risk factors for thyroid cancer development include environmental factors such as exposure to radiation and diets low in iodine, as well as genetic factors such as hereditary syndromes that predispose to thyroid cancer (eg, MEN2A/2B or Cowden syndrome), or positive family history of thyroid cancer. Indeed, although the genes underlying thyroid cancer predisposition remain yet to be fully characterized, thyroid cancer in a first-degree relative results in an increased risk of 4- to 10-fold [2,3].

Thyroid cancer typically presents as a nodule and rarely may be accompanied by symptoms such as difficulty swallowing, hoarseness, or pain. Most thyroid nodules (present in up to 50% of patients older than 60 years of age) are discovered incidentally, and only approximately 5% are malignant [4–7]. Most thyroid cancers are well-differentiated tumors and are associated with a low mortality rate, particularly in patients with stage I or II disease (survival rate >98%) [8]. However, although the mortality rate is low and has remained relatively stable, the incidence of thyroid cancer has been increasing over the last four decades [9–11]. The increased incidence of thyroid cancer, in particular tumors of small (subcentimeter) size, may be primarily due to improved detection of thyroid nodules by thyroid ultrasound, but improved detection alone may not fully explain the increased incidence of thyroid cancer as the incidence of tumors of all sizes has been reported to be increased [9,12]. The increase in thyroid cancer has been

mostly attributed to papillary thyroid carcinoma, and specifically, RAS mutation positive, follicular variant of papillary thyroid carcinomas [12].

The increased incidence of thyroid cancer has increased the challenge in clinical management. Identifying patients with low-risk disease who may be appropriately managed with active surveillance, and those patients with risk factors for more aggressive disease and higher mortality rates can guide clinical treatment and management. Molecular diagnostics are increasingly being incorporated into routine clinical management of thyroid cancer patients.

MOLECULAR TARGETS IN THYROID CANCER DIAGNOSTICS

The molecular mechanisms underlying the majority of thyroid cancers have been well characterized through work from many laboratories over the last few decades. Indeed, for papillary thyroid carcinomas, recent data from 496 papillary thyroid carcinomas sequenced by The Cancer Genome Atlas (TCGA) initiative identified confirmed or likely oncogenic driver mutations in 96.5% of cases [13].

Papillary thyroid carcinomas have very low overall mutation frequency as compared to most other solid cancers [13]. In most well-differentiated follicular thyroid tumors, a single early oncogenic driver mutation is likely to be sufficient to initiate tumorigenesis, while aggressive or poorly differentiated tumors often harbor multiple driver mutations. Most genes described to play a role in thyroid cancer development involve the mitogen-activated protein kinase (MAPK)

© 2017 Elsevier Inc. All rights reserved.

and phosphatidylinositol-3 kinase (PI3K) pathways. Dysregulation of the MAPK pathway in thyroid cancer frequently occurs through mutation of *BRAF* or *RAS* genes, or through rearrangements involving *RET/PTC* and *TRK* [14–17]. PI3K pathway dysregulation may occur through activating mutations of *PIK3CA* and *AKT1* or through inactivating mutations of *PTEN*. Certain molecular alterations, on the other hand, are typically present in benign nodules and may be helpful in differentiating tumor from benign disease (Fig. 30.1 and Table 30.1).

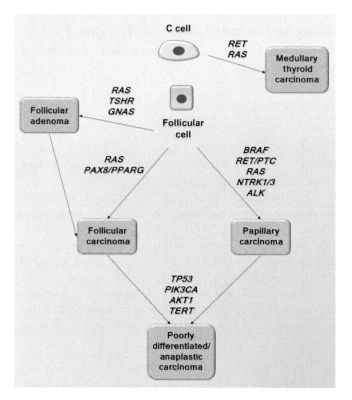

FIGURE 30.1 Scheme of major molecular events in thyroid tumorigenesis and putative tumor progression.

Mutation of *BRAF*, a serine–threonine kinase that functions in the MAPK pathway, occurs in 40–45% of papillary thyroid cancers [16,18]. In more than 95% of cases, *BRAF* activation results from the V600E mutation, whereas other activating *BRAF* mutations (such as K601E mutation and small in-frame insertions or deletions) are seen in the remaining cases [17,19–21]. Typically, the *BRAF* V600E mutation is seen in classical papillary thyroid cancer and tall-cell variant of papillary thyroid cancer [14,22,23]. The *BRAF* K601E mutation, however, is frequently seen in the follicular variant of papillary thyroid cancer [24]. *BRAF* activation may also occur through chromosomal rearrangement. The *AKAP9–BRAF* fusion has been detected in papillary thyroid cancers associated with radiation exposure and, rarely, in sporadic papillary cancers [25]. Several other *BRAF* fusions (eg, *SND1–BRAF* and *MKRN1–BRAF*) have been recently reported [13].

Other genes frequently found mutated in thyroid lesions are the *RAS* genes (*NRAS*, *HRAS*, and *KRAS*). The RAS proteins signal to both the MAPK and PI3K pathways. Activating mutations of the *RAS* genes typically occur at codons 12, 13, and 61. *NRAS* mutations are the most frequent reported change, followed by *HRAS* mutations, and then by *KRAS* mutation. *RAS* mutations have been reported in both benign follicular adenomas as well as follicular carcinomas [26–28]. Thyroid nodules with *KRAS* codon 12 or 13 mutation may have a lower risk of carcinoma than thyroid nodules with *NRAS* codon 61 mutation [29]. *RAS* mutations are also seen in papillary thyroid carcinomas, usually the follicular variant of papillary thyroid cancer, as well as some sporadic medullary thyroid cancers [14,30,31].

In both familial and sporadic medullary thyroid carcinomas, *RET* mutation is frequently seen. RET is a receptor tyrosine kinase expressed in thyroid C cells. RET is typically activated by mutation in the tyrosine

TABLE 30.1 Prevalence of Mutations in Thyroid Tumors

Papillary thyroid carcinoma	Follicular adenoma	Follicular carcinoma	Poorly differentiated carcinoma	Anaplastic carcinoma	Medullary carcinoma
BRAF V600E (40–45%)	*RAS* (20–30%)	*RAS* (40%)	*RAS* (20–30%)	*TP53* (70–80%)	*RET* (60–70%)
RET/PTC (10%)	*PAX8/PPARG* (5–10%)	*PAX8/PPARG* (30–35%)	*TP53* (20–30%)	*RAS* (30–40%)	*RAS* (5–10%)
RAS (20%)	*PTEN* (40%)	*PIK3CA* (<10%)	*BRAF* (10–15%)	*BRAF* (20–30%)	
NTRK1/3 (5%)	*TSHR* (50–80%)[a]	*PTEN* (<10%)	*TERT* (30–40%)	*PIK3CA* (10–20%)	
ALK (1–2%)	*GNAS* (3–6%)[a]	*TERT* (15–30%)	*CTNNB1* (<10%)	*PTEN* (10–20%)	
TERT (10–20%)				*AKT1* (10–20%)	
				TERT (30–40%)	
				CTNNB1 (10%)	

[a]*Incidence in hyperfunctioning adenomas.*

kinase domain (most commonly M918T mutation) or by mutation of cysteine residues in the extracellular domain. The M918T mutation is associated with sporadic medullary thyroid carcinomas or with medullary carcinomas arising in MEN2B syndrome [32–36]. Cysteine residue mutations, which confer the mutant RET protein the ability to undergo ligand-independent dimerization with another mutant RET protein, are seen in familial medullary thyroid carcinoma and MEN2A syndrome [37,38].

Poorly differentiated and anaplastic thyroid cancers are thought to typically arise from dedifferentiation of a well-differentiated cancer (Fig. 30.1). In many cases, both differentiated and undifferentiated components have been observed in these tumors, and these components share mutations in RAS or BRAF. However, poorly differentiated and anaplastic tumors often acquire additional mutations, most commonly in TP53, PIK3CA, and AKT1. TP53 is a tumor suppressor with important roles in cell cycle regulation and apoptosis. The most commonly seen mutations in TP53 are point mutations within the DNA-binding domain. TP53 mutations have been reported in 20–30% of poorly differentiated carcinomas and 70–80% of anaplastic carcinomas [39–43]. Activating mutations of PIK3CA typically occur in poorly differentiated and anaplastic thyroid carcinomas, and AKT1 mutations can be found more often in advanced, metastatic, and dedifferentiating thyroid cancer [44–46].

PTEN, a negative regulator of the PI3K/AKT pathway, may be mutated in both follicular thyroid carcinomas and follicular adenomas [45,47–50]. In addition to PTEN, other genes that have been reported to be altered in benign lesions include TSHR and GNAS. Activating mutations of TSHR, a membrane receptor whose function is mediated by G proteins, may be seen in 50–80% of hyperfunctioning nodules [51,52]. GNAS, an alpha subunit of heterotrimeric G protein complexes, is mutated in 3–6% of hyperfunctioning nodules [53–55]. TSHR and GNAS mutation are found predominantly in benign hyperfunctioning nodules, and very rarely in follicular carcinomas that may also present as hot nodules [50].

Finally, other novel mutations have been recently described in thyroid tumors. Telomerase (TERT) promoter mutations, c.1-124C > T (C228T) and c.1-146C > T (C250T), have been described in several tumors including thyroid cancer and are thought to increase promoter activity [56,57]. These TERT promoter mutations are found in follicular cell derived thyroid cancers (but not in medullary carcinoma) and tend to occur at the highest frequency in poorly differentiated and anaplastic carcinomas [58–61]. The Cancer Genome Atlas study identified three novel significantly mutated genes, EIF1AX, PPM1D, and CHEK2 [13]. EIF1AX mutations were found to be mutually exclusive with other known driver mutations, while PPM1D and CHEK2 mutations were found to co-occur with driver mutations [13]. The role of these genes in thyroid tumorigenesis is yet to be fully characterized.

In addition to gene mutations, chromosomal rearrangements are also important in thyroid cancer. In papillary thyroid carcinomas, RET/PTC1 (fusion of RET with CCDC6) and RET/PTC3 (fusion of RET with NCOA4) are the most common rearrangements and are currently seen in approximately 10% of cases [62,63], down from 20% to 30% frequency observed two decades earlier [12]. The PAX8/PPARG rearrangement is a common event in follicular carcinoma, being found in 30–40% of these tumors [64–66]. This rearrangement may also be seen in the follicular variant of papillary thyroid carcinoma and in follicular adenomas [64–68]. Rearrangements involving the NTRK genes (NTRK1 and NTRK3) are seen in up to 5% of papillary thyroid cancers. Several fusion partners (TPM3, TPR, and TFG) have been identified for NTRK1 and one fusion partner, ETV6, has been identified for NTRK3 [69–73]. Other fusions, such as ALK fusions are found in approximately 2% of papillary carcinomas and with higher frequency in anaplastic and particularly poorly differentiated thyroid carcinomas [74].

MOLECULAR TECHNOLOGIES

Several molecular approaches are available and used in molecular diagnostics of thyroid cancer to detect point mutations, small insertions/deletions, and chromosomal rearrangements. As with all assays, high sensitivity and specificity are desirable in choosing a molecular diagnostic test. However, specific to testing of thyroid specimens is the typically limited amount of material available in thyroid fine-needle aspiration (FNA) biopsy. As such, molecular technologies that require small amounts of DNA or RNA are best suited. Other sample types commonly encountered in thyroid testing include formalin-fixed paraffin-embedded (FFPE) tissue and fresh/frozen tissue.

Testing for recurrent mutations in hotspots in oncogenes important in thyroid cancer such as codons 12/13 or 61 in the RAS genes, or codon 600 or 601 in BRAF can be accomplished by a variety of assays including real-time PCR, sequencing (Sanger and next generation), or single-base (primer) extension assays. Real-time PCR typically involves hybridization of fluorescently labeled probes (TaqMan or FRET probes) to detect and quantify PCR products. The TaqMan (Applied Biosystems) assay utilizes the 5′ nuclease activity of Taq polymerase and allele-specific TaqMan probes to detect mutations. In this assay, the region of interest is amplified, and an allele-specific probe (specific to wild-type or mutant sequence) with a fluorophore at the 5′ end and a fluorescence quencher at the 3′ end is hybridized to the DNA. As Taq polymerase extends DNA and reaches the probe, the hybridized allele-specific probe functions as a substrate for the 5′ nuclease activity of Taq polymerase, and the 5′ fluorophore is released. Separation of the 5′ fluorophore from the 3′ fluorescence quencher allows fluorescence to be emitted and measured. The fluorescence emitted is proportional to the amount of amplified product. Other real-time PCR assays such as LightCycler real-time PCR (Roche) utilize FRET probes. In this assay, two probes are required: one probe is labeled at the 3′ end with a donor fluorophore and the other probe is labeled at the 5′ end with an acceptor fluorophore. The region of interest is amplified, and when both probes are bound to the target, they are brought into close proximity, allowing emission through fluorescence resonance energy transfer. A post-PCR melting curve can then be generated to determine whether any mismatches are present between the probe and the target sequence.

Single-base (primer) extension assays are another type of assay commonly used to detect point mutations. Following PCR amplification of the region of interest, these assays utilize a probe in which the 3′ end of the probe is a single base upstream of the nucleotide to be interrogated. The probe is extended by a single dideoxynucleotide base, and the incorporated base determined. In the SNaPshot (Life Technologies) assay, the incorporated base is fluorescently labeled and in the MassARRAY system (Sequenom), the incorporated base is unlabeled but the identity is determined by mass spectrometry. These assays are very sensitive and may be multiplexed, with multiple hotspots interrogated in a single reaction.

Finally, sequencing analysis is useful both in detection of mutations in hotspots and in detection of nonrecurrent mutations, for example, inactivating mutations in tumor suppressors which can occur at many different locations across a gene. Both Sanger sequencing and next-generation sequencing (NGS) technologies are widely used. NGS allows high-throughput, massively parallel sequencing, which allows many genes to be analyzed simultaneously in a cost-effective way. Commonly used NGS technologies include the Ion Torrent (Life Technologies) and Illumina platforms. Both technologies are based on the sequencing by synthesis methodology. With the Ion Torrent platform, incorporation of a nucleotide results in the release of a hydrogen ion that is detected by an ion sensor, and with the Illumina platform, the incorporated nucleotide is fluorescently labeled. NGS technology is particularly well suited for use in thyroid cancer, as the genes involved in the majority of tumors have been described,

and these genes can be sequenced simultaneously, using only small amounts of DNA.

Chromosomal rearrangements such as *RET/PTC1*, *RET/PTC3*, and *PAX8/PPARG* can be detected using techniques such as reverse transcription PCR (RT-PCR), fluorescence in situ hybridization (FISH), or RNA sequencing. In RT-PCR, RNA is transcribed into cDNA by reverse transcriptase. The cDNA is then amplified using primers specific to the fusion gene product. With real-time PCR amplification, use of internal probes further increases specificity. RT-PCR works best on fresh or frozen thyroid tissue. In FFPE, RNA is often degraded, which may limit assay sensitivity. FISH analysis may be performed on fresh, frozen, or FFPE tissue. In FISH, a fluorescently labeled probe is hybridized to DNA and visualized by microscopy. Probes can be designed to each fusion partner, each labeled with a different color, and fusion detected by the overlap of colors. A break-apart probe strategy can also be utilized to detect translocations for all possible fusion partners. The break-apart probe is designed to a known partner, for example, *RET*, which if undergoes translocation, will break apart and split the signal. Finally, NGS-based analyses of RNA are being increasingly used to detect gene fusions. RNA is reverse transcribed to cDNA, and adapters are ligated. Short sequence reads are obtained and used to search for fusion genes and measure gene expression levels. Targeted RNA NGS panels can be used to detect known fusions and measure gene expression levels and are ideal for the limited amounts of nucleic acids typically present in FNA specimens. Alternatively, RNA sequencing of total RNA can be performed which in addition to detection of known fusions, allows for discovery of new fusions. However, this approach usually requires a larger amount of input RNA.

CLINICAL UTILITY

The majority of thyroid nodules are classified as benign or malignant through ultrasound examination and FNA biopsy. However, 20–30% of nodules are cytologically indeterminate [75,76]. These indeterminate nodules, which fall into the categories of Atypia of Undetermined Significance or Follicular Lesion of Undetermined Significance (AUS/FLUS) (Bethesda category III), Follicular Neoplasm or Suspicious for a Follicular Neoplasm (FN/SFN) (Bethesda category IV), and Suspicious for Malignancy (SUSP) (Bethesda category V), carry a risk of malignancy ranging from 5–15% to 50–75% [75,77]. Molecular testing of indeterminate thyroid nodules can help to rule in or rule out cancer and can guide appropriate clinical management (active surveillance vs surgery). Prognostic information to predict the aggressiveness of a tumor can also be obtained, as well as potential targets for therapy.

Molecular testing of thyroid nodules has been shown to have utility in the clinical setting as a diagnostic tool. Testing for the *BRAF* V600E mutation, a mutation highly specific for malignancy in thyroid nodules, has been shown to increase the sensitivity of the FNA biopsy [78–80]. Further increases in sensitivity and specificity have been achieved through the use of multigene panels. In three prospective studies, a multigene panel consisting of the most commonly mutated genes (*BRAF, KRAS, HRAS, NRAS, PAX8/PPARG, RET/PTC1, RET/PTC3* (and *TRK* rearrangement)) was shown to have high specificity for thyroid cancer [81–83]. Recently, a multigene NGS thyroid panel was reported [84]. This panel includes mutational analysis of 13 genes for point mutations and small indels (*BRAF, KRAS, HRAS, NRAS, RET, GNAS, TSHR, CTNNB1, TP53, AKT1, PTEN, PIK3CA,* and *TERT*), and detection of 42 types of rearrangements involving *RET, BRAF, PPARG, NTRK1, NTRK3, ALK,* and *THADA* genes. Cancer risk prediction based on the results of testing is based on the mutational hotspot, allelic frequency, and consideration of somatic versus germline nature of the particular variant. In indeterminate thyroid nodules with FN/SFN cytology, this panel demonstrated high sensitivity (90%), specificity (93%), PPV (83%), and NPV (96%) for cancer detection [84].

In addition to aiding in establishing a diagnosis, molecular findings have been shown to have utility in guiding clinical management of patients with thyroid nodules. Patients with indeterminate cytology nodules may undergo repeat FNA (AUS/FLUS nodules), surgical lobectomy (FN/SFN nodules), or near-total thyroidectomy, or lobectomy (SUSP nodules). Most nodules on diagnostic lobectomies are found to be benign, but in 10–40% of cases, the nodule is found to be malignant [85–87]. If the cancer is found to be greater than 1 cm in size, patients typically undergo a completion thyroidectomy. Multigene panels have been shown to have high negative predictive value and thus could be used to rule out malignancy [84]. Patients who are negative for mutations could potentially avoid unnecessary surgery. Multigene panels also have been demonstrated to have high specificity and positive predictive value [81–84]. Detection of mutation or gene fusion (with the possible exception of genes that may be found mutated in both benign and malignant lesions (such as *RAS* or *TSHR*)) could be used as an indication to recommend a total thyroidectomy prospectively, rather than a diagnostic lobectomy that would have to be followed by a completion lobectomy. Use of a seven-gene thyroid mutation panel has been shown to reduce the likelihood of requiring a

two-step surgery by 2.5-fold [88]. Similar results were seen in a small series of cases in pediatric patients (who tend to have higher frequencies of indeterminate nodules) [89].

Prognostic information is also obtainable through molecular testing of thyroid nodules. Prognostic information could be used preoperatively to plan a more extensive initial resection that includes central compartment lymph node dissection and postoperatively for close clinical follow-up postsurgery. *BRAF* V600E mutation is associated with cancer recurrence or persistent disease (25% of *BRAF* V600E positive tumors vs 13% of *BRAF* mutation negative tumors) and with a small, but significant, increased risk of mortality (5% of *BRAF* V600E mutation positive tumors vs 1% of *BRAF* mutation negative tumors), but *BRAF* V600E alone lacks sufficient specificity to guide more aggressive management of patients with thyroid cancer [90,91]. While detection of *BRAF* V600E mutation in thyroid FNA biopsy has been used as a highly accurate diagnostic marker of cancer, it needs to be used in combination with other mutations for predicting aggressive tumor behavior with high specificity [87].

TP53 mutation and *TERT* promoter mutation are two other promising prognostic molecular markers. *TERT* promoter mutations and to lesser extent *TP53* mutation occur in some well-differentiated carcinomas, but can be found at much higher frequencies in poorly differentiated and anaplastic thyroid carcinomas [39,40,58–61]. *TP53* mutation, typically a late event in tumorigenesis, is important in tumor dedifferentiation, and thus detection of a *TP53* mutation may signal an aggressive tumor. *TERT* promoter mutation has been found to be an independent predictive marker of disease recurrence, distant metastases, and disease-related mortality in well-differentiated thyroid cancer [61]. Cancers carrying both *TERT* and *BRAF* V600E mutation may represent the most aggressive subset of well-differentiated papillary carcinomas [58,60].

An additional marker of aggressive tumors is the presence of multiple mutations. The vast majority of thyroid cancers will have a single mutation, but observation of mutation in an early driver gene such as *BRAF* or *RAS* along with mutation in genes such as *PIK3CA, AKT1,* or *TP53,* thought to be acquired as a late event, has been reported in radioiodine refractory, poorly differentiated and anaplastic carcinomas [44,45,92]. NGS panels which are able to examine multiple genes simultaneously are ideal for this type of analysis.

Finally, molecular diagnostics may be helpful in guiding therapy, especially in locally advanced and inoperable tumors or those with distant metastases that are refractory to radioactive iodine treatment.

Current FDA-approved therapies include sorafenib and lenvatinib (multi-tyrosine kinase inhibitors) for the treatment of locally recurrent or metastatic, progressive differentiated thyroid carcinoma and vandetanib and cabozantinib, which are tyrosine kinase inhibitors with activity against *RET*, for medullary thyroid cancer. Many therapies are under investigation and target the MAPK and PI3K/AKT pathways. Specific targeted therapies that could be considered for utilization include vemurafenib and dabrafenib, which are *BRAF* inhibitors, and crizotinib or other ALK inhibitors in patients with advanced thyroid cancer positive for *STRN−ALK* or other *ALK* fusions [74,93,94]. In addition, other therapies currently being studied include PPARG agonists in thyroid cancers that are positive for *PAX8−PPARG* fusion and NTRK inhibitors in thyroid tumors with *NTRK1* or *NTRK3* fusions.

LIMITATIONS OF TESTING

The limitations to molecular testing of thyroid specimens relate both to sample limitations as well as limitations of technology. Testing of thyroid FNA biopsies may be challenging because of limited specimen available, but also because of difficulties in assessing sample adequacy. Sufficient material from thyroid FNA biopsy for molecular testing can be obtained from the residual material and needle washes from the first two FNA passes [82,83]. Although this material is generally representative of the material used in cytological examination, determination of thyroid cell percentage (vs other cells such as lymphocytes) is useful to ensure an adequate sample and reduce the risk of false negatives. This can be done by measuring the expression of genes expressed in thyroid epithelial cells (such as *KRT7*, *KRT19*, *TPO*, *TG*, or *TTF1*) and comparing the expression level with that of a housekeeping gene (such as *GAPDH* or *PGK1*).

Other limitations of testing arise from the sensitivity of the testing method used. Many techniques may be used to detect point mutations and chromosomal rearrangements, such as Sanger sequencing, real-time PCR, RT-PCR, and NGS. The analytical sensitivity of these assays range from ~5% to 20%. Although in general NGS-based assays have very high sensitivity, and are more cost-effective than assaying each gene separately, NGS requires specialized equipment and expertise. NGS generates complex information and specialized requirements for analysis and reporting. However, this information can be managed with bioinformatics tools (such as SeqReporter) [95]. Use of bioinformatics tools can reduce the difficulty in analysis and reporting and turnaround times comparable to those for conventional testing can be achieved.

Another limitation of testing thyroid nodules is that it is intended to address somatic mutations, but identification of germline mutations can have important implications. For example, germline *RET* mutations have implications for intraoperative management of parathyroid glands, for surveillance and management of other tumors, and for screening and prophylactic thyroidectomy of family members [32]. Clinical characteristics or family history generally guide recommendations for genetic counseling and germline testing. However, in some cases, specific mutations are present that are characteristic of patients with familial forms of carcinoma. In MEN2A, 90% of mutations occur at codon 634 of *RET* [35−37]. Detection of such a mutation could be an additional indication for germline testing.

References

[1] Howlader N, Noone AM, Krapcho M, et al. SEER cancer statistics review, 1975−2010. Bethesda, MD: National Cancer Institute; 2013<http://seer.cancer.gov/csr/1975_2010/>, based on November 2012 SEER data submission, posted to the SEER web site, April 2013.

[2] Frich L, Glattre E, Akslen LA. Familial occurrence of nonmedullary thyroid cancer: a population-based study of 5673 first-degree relatives of thyroid cancer patients from Norway. Cancer Epidemiol Biomarkers Prev 2001;10:113−17.

[3] Hemminki K, Eng C, Chen B. Familial risks for nonmedullary thyroid cancer. J Clin Endocrinol Metab 2005;90:5747−53.

[4] Guth S, Theune U, Aberle J, Galach A, Bamberger CM. Very high prevalence of thyroid nodules detected by high frequency (13 MHz) ultrasound examination. Eur J Clin Invest 2009;39: 699−706.

[5] Mazzaferri EL. Thyroid cancer in thyroid nodules: finding a needle in the haystack. Am J Med 1992;93:359−62.

[6] Mazzaferri EL. Management of a solitary thyroid nodule. N Engl J Med 1993;328:553−9.

[7] Brito JP, Yarur AJ, Prokop LJ, McIver B, Murad MH, Montori V. Prevalence of thyroid cancer in multinodular goiter versus single nodule: a systematic review and meta-analysis. Thyroid 2013;23:449−55.

[8] Motomura T, Nikiforov YE, Namba H, et al. Ret rearrangements in Japanese pediatric and adult papillary thyroid cancers. Thyroid 1998;8:485−9.

[9] Albores-Saavedra J, Henson DE, Glazer E, Schwartz AM. Changing patterns in the incidence and survival of thyroid cancer with follicular phenotype—papillary, follicular, and anaplastic: a morphological and epidemiological study. Endocr Pathol 2007;18:1−7.

[10] Burgess JR, Tucker P. Incidence trends for papillary thyroid carcinoma and their correlation with thyroid surgery and thyroid fine-needle aspirate cytology. Thyroid 2006;16:47−53.

[11] Davies L, Welch HG. Increasing incidence of thyroid cancer in the United States, 1973−2002. JAMA 2006;295:2164−7.

[12] Jung CK, Little MP, Lubin JH, et al. The increase in thyroid cancer incidence during the last four decades is accompanied by a high frequency of BRAF mutations and a sharp increase in RAS mutations. J Clin Endocrinol Metab 2014;99:E276−85.

[13] Cancer Genome Atlas Research Network. Integrated genomic characterization of papillary thyroid carcinoma. Cell 2014;159: 676−90.

[14] Adeniran AJ, Zhu Z, Gandhi M, et al. Correlation between genetic alterations and microscopic features, clinical manifestations, and prognostic characteristics of thyroid papillary carcinomas. Am J Surg Pathol 2006;30:216–22.

[15] Frattini M, Ferrario C, Bressan P, et al. Alternative mutations of BRAF, RET and NTRK1 are associated with similar but distinct gene expression patterns in papillary thyroid cancer. Oncogene 2004;23:7436–40.

[16] Kimura ET, Nikiforova MN, Zhu Z, Knauf JA, Nikiforov YE, Fagin JA. High prevalence of BRAF mutations in thyroid cancer: genetic evidence for constitutive activation of the RET/PTC-RAS-BRAF signaling pathway in papillary thyroid carcinoma. Cancer Res 2003;63:1454–7.

[17] Soares P, Trovisco V, Rocha AS, et al. BRAF mutations and RET/PTC rearrangements are alternative events in the etiopathogenesis of PTC. Oncogene 2003;22:4578–80.

[18] Cohen Y, Xing M, Mambo E, et al. BRAF mutation in papillary thyroid carcinoma. J Natl Cancer Inst 2003;95:625–7.

[19] Chiosea S, Nikiforova M, Zuo H, et al. A novel complex BRAF mutation detected in a solid variant of papillary thyroid carcinoma. Endocr Pathol 2009;20:122–6.

[20] Ciampi R, Nikiforov YE. Alterations of the BRAF gene in thyroid tumors. Endocr Pathol 2005;16:163–72.

[21] Hou P, Liu D, Xing M. Functional characterization of the T1799-1801del and A1799-1816ins BRAF mutations in papillary thyroid cancer. Cell Cycle 2007;6:377–9.

[22] Nikiforova MN, Kimura ET, Gandhi M, et al. BRAF mutations in thyroid tumors are restricted to papillary carcinomas and anaplastic or poorly differentiated carcinomas arising from papillary carcinomas. J Clin Endocrinol Metab 2003;88:5399–404.

[23] Xing M. BRAF mutation in thyroid cancer. Endocr Relat Cancer 2005;12:245–62.

[24] Trovisco V, Soares P, Preto A, et al. Type and prevalence of BRAF mutations are closely associated with papillary thyroid carcinoma histotype and patients' age but not with tumour aggressiveness. Virchows Arch 2005;446:589–95.

[25] Ciampi R, Knauf JA, Kerler R, et al. Oncogenic AKAP9-BRAF fusion is a novel mechanism of MAPK pathway activation in thyroid cancer. J Clin Invest 2005;115:94–101.

[26] Lemoine NR, Mayall ES, Wyllie FS, et al. High frequency of ras oncogene activation in all stages of human thyroid tumorigenesis. Oncogene 1989;4:159–64.

[27] Namba H, Rubin SA, Fagin JA. Point mutations of ras oncogenes are an early event in thyroid tumorigenesis. Mol Endocrinol 1990;4:1474–9.

[28] Suarez HG, du Villard JA, Severino M, et al. Presence of mutations in all three ras genes in human thyroid tumors. Oncogene 1990;5:565–70.

[29] Radkay LA, Chiosea SI, Seethala RR, et al. Thyroid nodules with KRAS mutations are different from nodules with NRAS and HRAS mutations with regard to cytopathologic and histopathologic outcome characteristics. Cancer Cytopathol 2014;122:873–82.

[30] Zhu Z, Gandhi M, Nikiforova MN, Fischer AH, Nikiforov YE. Molecular profile and clinical-pathologic features of the follicular variant of papillary thyroid carcinoma. An unusually high prevalence of ras mutations. Am J Clin Pathol 2003;120:71–7.

[31] Agrawal N, Jiao Y, Sausen M, et al. Exomic sequencing of medullary thyroid cancer reveals dominant and mutually exclusive oncogenic mutations in RET and RAS. J Clin Endocrinol Metab 2013;98:E364–9.

[32] Kloos RT, Eng C, Evans DB, et al. Medullary thyroid cancer: management guidelines of the American Thyroid Association. Thyroid 2009;19:565–612.

[33] Eng C, Smith DP, Mulligan LM, et al. Point mutation within the tyrosine kinase domain of the RET proto-oncogene in multiple endocrine neoplasia type 2B and related sporadic tumours. Hum Mol Genet 1994;3:237–41.

[34] Hofstra RM, Landsvater RM, Ceccherini I, et al. A mutation in the RET proto-oncogene associated with multiple endocrine neoplasia type 2B and sporadic medullary thyroid carcinoma. Nature 1994;367:375–6.

[35] Elisei R, Romei C, Cosci B, et al. RET genetic screening in patients with medullary thyroid cancer and their relatives: experience with 807 individuals at one center. J Clin Endocrinol Metab 2007;92:4725–9.

[36] de Groot JW, Links TP, Plukker JT, Lips CJ, Hofstra RM. RET as a diagnostic and therapeutic target in sporadic and hereditary endocrine tumors. Endocr Rev 2006;27:535–60.

[37] Hansford JR, Mulligan LM. Multiple endocrine neoplasia type 2 and RET: from neoplasia to neurogenesis. J Med Genet 2000;37:817–27.

[38] Mulligan LM, Marsh DJ, Robinson BG, et al. Genotype–phenotype correlation in multiple endocrine neoplasia type 2: report of the International RET Mutation Consortium. J Intern Med 1995;238:343–6.

[39] Fagin JA, Matsuo K, Karmakar A, Chen DL, Tang SH, Koeffler HP. High prevalence of mutations of the p53 gene in poorly differentiated human thyroid carcinomas. J Clin Invest 1993;91:179–84.

[40] Donghi R, Longoni A, Pilotti S, Michieli P, Della Porta G, Pierotti MA. Gene p53 mutations are restricted to poorly differentiated and undifferentiated carcinomas of the thyroid gland. J Clin Invest 1993;91:1753–60.

[41] Dobashi Y, Sugimura H, Sakamoto A, et al. Stepwise participation of p53 gene mutation during dedifferentiation of human thyroid carcinomas. Diagn Mol Pathol 1994;3:9–14.

[42] Ho YS, Tseng SC, Chin TY, Hsieh LL, Lin JD. p53 gene mutation in thyroid carcinoma. Cancer Lett 1996;103:57–63.

[43] Takeuchi Y, Daa T, Kashima K, Yokoyama S, Nakayama I, Noguchi S. Mutations of p53 in thyroid carcinoma with an insular component. Thyroid 1999;9:377–81.

[44] Garcia-Rostan G, Costa AM, Pereira-Castro I, et al. Mutation of the PIK3CA gene in anaplastic thyroid cancer. Cancer Res 2005;65:10199–207.

[45] Hou P, Liu D, Shan Y, et al. Genetic alterations and their relationship in the phosphatidylinositol 3-kinase/Akt pathway in thyroid cancer. Clin Cancer Res 2007;13:1161–70.

[46] Ricarte-Filho JC, Ryder M, Chitale DA, et al. Mutational profile of advanced primary and metastatic radioactive iodine-refractory thyroid cancers reveals distinct pathogenetic roles for BRAF, PIK3CA, and AKT1. Cancer Res 2009;69:4885–93.

[47] Dahia PL, Marsh DJ, Zheng Z, et al. Somatic deletions and mutations in the Cowden disease gene, PTEN, in sporadic thyroid tumors. Cancer Res 1997;57:4710–13.

[48] Gustafson S, Zbuk KM, Scacheri C, Eng C. Cowden syndrome. Semin Oncol 2007;34:428–34.

[49] Wang Y, Hou P, Yu H, et al. High prevalence and mutual exclusivity of genetic alterations in the phosphatidylinositol-3-kinase/akt pathway in thyroid tumors. J Clin Endocrinol Metab 2007;92:2387–90.

[50] Nikiforova MN, Wald AI, Roy S, Durso MB, Nikiforov YE. Targeted next-generation sequencing panel (ThyroSeq) for detection of mutations in thyroid cancer. J Clin Endocrinol Metab 2013;98:E1852–60.

[51] Garcia-Jimenez C, Santisteban P. TSH signalling and cancer. Arq Bras Endocrinol Metabol 2007;51:654–71.

[52] Nishihara E, Amino N, Maekawa K, et al. Prevalence of TSH receptor and Gsalpha mutations in 45 autonomously functioning thyroid nodules in Japan. Endocr J 2009;56:791−8.

[53] Fuhrer D, Holzapfel HP, Wonerow P, Scherbaum WA, Paschke R. Somatic mutations in the thyrotropin receptor gene and not in the Gs alpha protein gene in 31 toxic thyroid nodules. J Clin Endocrinol Metab 1997;82:3885−91.

[54] Trulzsch B, Krohn K, Wonerow P, et al. Detection of thyroid-stimulating hormone receptor and Gsalpha mutations: in 75 toxic thyroid nodules by denaturing gradient gel electrophoresis. J Mol Med 2001;78:684−91.

[55] Parma J, Duprez L, Van Sande J, et al. Diversity and prevalence of somatic mutations in the thyrotropin receptor and Gs alpha genes as a cause of toxic thyroid adenomas. J Clin Endocrinol Metab 1997;82:2695−701.

[56] Horn S, Figl A, Rachakonda PS, et al. TERT promoter mutations in familial and sporadic melanoma. Science 2013;339:959−61.

[57] Huang FW, Hodis E, Xu MJ, Kryukov GV, Chin L, Garraway LA. Highly recurrent TERT promoter mutations in human melanoma. Science 2013;339:957−9.

[58] Landa I, Ganly I, Chan TA, et al. Frequent somatic TERT promoter mutations in thyroid cancer: higher prevalence in advanced forms of the disease. J Clin Endocrinol Metab 2013;98:E1562−6.

[59] Liu T, Wang N, Cao J, et al. The age- and shorter telomere-dependent TERT promoter mutation in follicular thyroid cell-derived carcinomas. Oncogene 2014;33:4978−84.

[60] Liu X, Bishop J, Shan Y, et al. Highly prevalent TERT promoter mutations in aggressive thyroid cancers. Endocr Relat Cancer 2013;20:603−10.

[61] Melo M, Rocha AG, Vinagre J, et al. TERT promoter mutations are a major indicator of poor outcome in differentiated thyroid carcinomas. J Clin Endocrinol Metab 2014;99:E754−65.

[62] Nikiforov YE. RET/PTC rearrangement—a link between Hashimoto's thyroiditis and thyroid cancer...or not. J Clin Endocrinol Metab 2006;91:2040−2.

[63] Zhu Z, Ciampi R, Nikiforova MN, Gandhi M, Nikiforov YE. Prevalence of RET/PTC rearrangements in thyroid papillary carcinomas: effects of the detection methods and genetic heterogeneity. J Clin Endocrinol Metab 2006;91:3603−10.

[64] Dwight T, Thoppe SR, Foukakis T, et al. Involvement of the PAX8/peroxisome proliferator-activated receptor gamma rearrangement in follicular thyroid tumors. J Clin Endocrinol Metab 2003;88:4440−5.

[65] French CA, Alexander EK, Cibas ES, et al. Genetic and biological subgroups of low-stage follicular thyroid cancer. Am J Pathol 2003;162:1053−60.

[66] Nikiforova MN, Lynch RA, Biddinger PW, et al. RAS point mutations and PAX8-PPAR gamma rearrangement in thyroid tumors: evidence for distinct molecular pathways in thyroid follicular carcinoma. J Clin Endocrinol Metab 2003;88:2318−26.

[67] Marques AR, Espadinha C, Catarino AL, et al. Expression of PAX8-PPAR gamma 1 rearrangements in both follicular thyroid carcinomas and adenomas. J Clin Endocrinol Metab 2002;87:3947−52.

[68] Nikiforova MN, Biddinger PW, Caudill CM, Kroll TG, Nikiforov YE. PAX8-PPARgamma rearrangement in thyroid tumors: RT-PCR and immunohistochemical analyses. Am J Surg Pathol 2002;26:1016−23.

[69] Greco A, Pierotti MA, Bongarzone I, Pagliardini S, Lanzi C, Della Porta G. TRK-T1 is a novel oncogene formed by the fusion of TPR and TRK genes in human papillary thyroid carcinomas. Oncogene 1992;7:237−42.

[70] Greco A, Mariani C, Miranda C, et al. The DNA rearrangement that generates the TRK-T3 oncogene involves a novel gene on chromosome 3 whose product has a potential coiled-coil domain. Mol Cell Biol 1995;15:6118−27.

[71] Martin-Zanca D, Hughes SH, Barbacid M. A human oncogene formed by the fusion of truncated tropomyosin and protein tyrosine kinase sequences. Nature 1986;319:743−8.

[72] Radice P, Sozzi G, Miozzo M, et al. The human tropomyosin gene involved in the generation of the TRK oncogene maps to chromosome 1q31. Oncogene 1991;6:2145−8.

[73] Leeman-Neill RJ, Kelly LM, Liu P, et al. ETV6-NTRK3 is a common chromosomal rearrangement in radiation-associated thyroid cancer. Cancer 2014;120:799−807.

[74] Kelly LM, Barila G, Liu P, et al. Identification of the transforming STRN-ALK fusion as a potential therapeutic target in the aggressive forms of thyroid cancer. Proc Natl Acad Sci USA 2014;111:4233−8.

[75] Baloch ZW, LiVolsi VA, Asa SL, et al. Diagnostic terminology and morphologic criteria for cytologic diagnosis of thyroid lesions: a synopsis of the National Cancer Institute Thyroid Fine-Needle Aspiration State of the Science Conference. Diagn Cytopathol 2008;36:425−37.

[76] Ohori NP, Schoedel KE. Variability in the atypia of undetermined significance/follicular lesion of undetermined significance diagnosis in the Bethesda system for reporting thyroid cytopathology: sources and recommendations. Acta Cytol 2011;55:492−8.

[77] Ali SZ, Cibas ES. The Bethesda system for reporting thyroid cytopathology. New York, NY: Springer; 2010.

[78] Kim SW, Lee JI, Kim JW, et al. BRAFV600E mutation analysis in fine-needle aspiration cytology specimens for evaluation of thyroid nodule: a large series in a BRAFV600E-prevalent population. J Clin Endocrinol Metab 2010;95:3693−700.

[79] Zatelli MC, Trasforini G, Leoni S, et al. BRAF V600E mutation analysis increases diagnostic accuracy for papillary thyroid carcinoma in fine-needle aspiration biopsies. Eur J Endocrinol 2009;161:467−73.

[80] Marchetti I, Lessi F, Mazzanti CM, et al. A morpho-molecular diagnosis of papillary thyroid carcinoma: BRAF V600E detection as an important tool in preoperative evaluation of fine-needle aspirates. Thyroid 2009;19:837−42.

[81] Cantara S, Capezzone M, Marchisotta S, et al. Impact of proto-oncogene mutation detection in cytological specimens from thyroid nodules improves the diagnostic accuracy of cytology. J Clin Endocrinol Metab 2010;95:1365−9.

[82] Nikiforov YE, Ohori NP, Hodak SP, et al. Impact of mutational testing on the diagnosis and management of patients with cytologically indeterminate thyroid nodules: a prospective analysis of 1056 FNA samples. J Clin Endocrinol Metab 2011;96:3390−7.

[83] Nikiforov YE, Steward DL, Robinson-Smith TM, et al. Molecular testing for mutations in improving the fine-needle aspiration diagnosis of thyroid nodules. J Clin Endocrinol Metab 2009;94:2092−8.

[84] Nikiforov YE, Carty SE, Chiosea SI, et al. Highly accurate diagnosis of cancer in thyroid nodules with follicular neoplasm/suspicious for a follicular neoplasm cytology by ThyroSeq v2 next-generation sequencing assay. Cancer 2014;120:3627−34.

[85] Alexander EK, Kennedy GC, Baloch ZW, et al. Preoperative diagnosis of benign thyroid nodules with indeterminate cytology. N Engl J Med 2012;367:705−15.

[86] Buryk MA, Monaco SE, Witchel SF, et al. Preoperative cytology with molecular analysis to help guide surgery for pediatric thyroid nodules. Int J Pediatr Otorhinolaryngol 2013;77:1697−700.

[87] Xing M, Clark D, Guan H, et al. BRAF mutation testing of thyroid fine-needle aspiration biopsy specimens for preoperative risk stratification in papillary thyroid cancer. J Clin Oncol 2009;27(18):2977—82.

[88] Yip L, Wharry L, Armstrong M, et al. A clinical algorithm for fine-needle aspiration molecular testing effectively guides the appropriate extent of initial thyroidectomy. Ann Surg 2014;260:163—8.

[89] Monaco SE, Pantanowitz L, Khalbuss WE, et al. Cytomorphological and molecular genetic findings in pediatric thyroid fine-needle aspiration. Cancer Cytopathol 2012;120:342—50.

[90] Tufano RP, Teixeira GV, Bishop J, Carson KA, Xing M. BRAF mutation in papillary thyroid cancer and its value in tailoring initial treatment: a systematic review and meta-analysis. Medicine 2012;91:274—86.

[91] Xing M, Alzahrani AS, Carson KA, et al. Association between BRAF V600E mutation and mortality in patients with papillary thyroid cancer. JAMA 2013;309:1493—501.

[92] Liu Z, Hou P, Ji M, et al. Highly prevalent genetic alterations in receptor tyrosine kinases and phosphatidylinositol 3-kinase/akt and mitogen-activated protein kinase pathways in anaplastic and follicular thyroid cancers. J Clin Endocrinol Metab 2008;93: 3106—16.

[93] Perot G, Soubeyran I, Ribeiro A, et al. Identification of a recurrent STRN/ALK fusion in thyroid carcinomas. PLoS One 2014;9:e87170.

[94] Demeure MJ, Aziz M, Rosenberg R, Gurley SD, Bussey KJ, Carpten JD. Whole-genome sequencing of an aggressive BRAF wild-type papillary thyroid cancer identified EML4-ALK translocation as a therapeutic target. World J Surg 2014;38: 1296—305.

[95] Roy S, Durso MB, Wald A, Nikiforov YE, Nikiforova MN. SeqReporter: automating next-generation sequencing result interpretation and reporting workflow in a clinical laboratory. J Mol Diagn 2014;16:11—22.

31

Molecular Testing in Pediatric Cancers

K.R. Crooks[1] and K.W. Rao[2,3,†]

[1]Department of Pathology, University of Colorado, Anschutz Medical Campus,
Aurora, CO, United States [2]Departments of Pediatrics, Pathology and Laboratory Medicine, and Genetics,
University of North Carolina School of Medicine, Chapel Hill, NC, United States [3]Cytogenetics Laboratory, McLendon
Clinical Laboratories, UNC Hospitals, Chapel Hill, NC, United States

INTRODUCTION

Pediatric cancer, while rare, remains the leading cause of childhood death past infancy [1]. It has long been appreciated that childhood cancer differs remarkably from that in adulthood in incidence, treatment, and outcome. Growth and development present opportunities for malignant transformation in the populations of dividing and differentiating cells that are unique to childhood. In this chapter, we present three of the most common malignancies that are largely restricted to the pediatric population—retinoblastoma (RB), Wilms tumor (WT, nephroblastoma), and neuroblastoma (NB)—and discuss the role of molecular genetic testing in each.

RETINOBLASTOMA

Background

RB is an intraocular tumor that arises from immature cells of the retina, the thin membrane at the back of the eye. Although it is the most common childhood ocular tumor and represents approximately 3% of all childhood cancer diagnoses, it remains a rare malignancy, with a stable worldwide incidence of 1/15,000 births [2]. Because the tumor begins in developing cells, nearly all cases occur during the first 5 years of life [2,3]. Early detection and treatment (including laser ablation and enucleation) are important, because RB is an aggressive tumor that can spread rapidly along the optic nerve to the brain. Once extraocular cancer is present, survival probability declines significantly.

RB can present unilaterally or bilaterally. Among the unilateral cases, most are sporadic, but approximately 10% have been found to be heritable [4]. All cases of bilateral RB are assumed to be heritable. In the majority of the heritable cases, the germline mutation occurs de novo, and the remainder are parentally inherited in an autosomal-dominant manner [5]. Penetrance in heritable disease varies but is generally high, estimated in some instances to approach 95%, and those with a germline mutation are at increased risk of developing additional malignancies throughout life [5,6]. Trilateral RB is the co-occurrence of bilateral RB with pineoblastoma and represents the most common extraocular tumor in childhood, while adults are at risk for a number of other cancers, particularly lung and bladder cancer, osteosarcoma, and soft tissue sarcomas [7,8]. Since those with sporadic RB have not been found to be at increased risk for these or other malignancies, determining whether a case is heritable or sporadic has important implications for medical management and counseling.

Molecular Target

The primary genetic locus for RB is the retinoblastoma 1 gene (RB1) on chromosome 13 [9]. Genetic variation at this locus accounts for approximately 98% of RB [10]. The protein product, pRB, is a classic tumor suppressor involved in cell-cycle regulation. As predicted by Knudson's two-hit hypothesis, RB arises following biallelic loss-of-function mutation of RB1 [5]. In sporadic cases, the two mutations are acquired somatically, while in heritable cases, there is one germline

†Deceased

© 2017 Elsevier Inc. All rights reserved.

mutation and one somatic mutation. The molecular insults that cause loss or inactivation of pRB are varied and include cytogenetically-visible deletions, smaller deletions, and many types of sequence variants (eg, nonsense, frameshift, missense, splicing, and promoter) [11]. Importantly, penetrance in germline cases correlates with the nature of the genetic insult. Nearly complete penetrance is observed in cases arising from premature termination mutations or large deletions, while incomplete penetrance and variable expressivity have been observed in families transmitting certain missense, splicing, or promoter mutations [4,12].

Comprehensive *RB1* testing is negative in approximately 2% of RB cases. Amplification of the *MYCN* oncogene has been identified in tumorigenesis in approximately half of such cases, but the clinical implications of *MYCN* amplification (MYCNA) in RB remain to be identified [10].

Molecular Technologies

The appropriate genetic testing approach for newly-diagnosed RB is determined by the likelihood that a germline *RB1* mutation is present. In familial and bilateral cases, the priority is to identify the putative germline mutation, typically by testing a peripheral blood sample. In the unilateral sporadic cases, testing may be first performed on tumor tissue, with the goal of identifying both *RB1* mutations. Peripheral blood is tested next, in order to determine whether either mutation is present in the germline. Tumor tissue may only be available following enucleation. RB biopsy is rarely performed, largely because the diagnosis is usually certain by clinical exam, but also due to the concern of causing tumor dispersal during the procedure.

Genetic testing is typically undertaken in a reflexive manner, beginning with sequence analysis to detect the point mutations and small deletions and insertions that comprise approximately 75% of *RB1* mutations. To date, over 1500 such mutations have been reported and are distributed throughout all 27 exons over the 178 kb gene [4]. Clinical sequence analysis by both Sanger sequencing and next-generation sequencing is commercially available. Targeted analysis for the most common mutations is offered, but sensitivity is low, often approximately 25%.

If sequence analysis is negative, then deletion/duplication analysis is usually performed, most commonly by multiplex ligation dependent probe amplification (MLPA), to identify the single-exon or multi-exon deletions, and whole-gene deletions that account for approximately 15% of causative mutations. Approximately half of these are large cytogenetically-visible rearrangements involving the *RB1* locus, and most are associated with 13q14 deletion syndrome (OMIM #613884), characterized by dysmorphic features,

mild to moderate intellectual disability, and growth retardation. Larger deletions that extend distally into 13q32 are associated with major malformations of the brain, genitourinary organs, and gastrointestinal tract. In some instances, karyotype or chromosomal microarray analysis (CMA) for congenital abnormalities has identified the deletion and RB susceptibility before the tumor was present [13]. The smaller, submicroscopic whole-gene deletions can be detected by fluorescence in situ hybridization (FISH) analysis, but this technique has been largely replaced by the higher resolution MLPA. FISH analysis remains useful for the rare instances (<5%) in which *RB1* is interrupted by a submicroscopic balanced translocation or insertion. If both sequence analysis and deletion/duplication testing are negative, methylation analysis will identify an additional approximately 10% with hypermethylation of the *RB1* promoter region resulting in gene silencing. Together, these tests have been demonstrated to identify more than 95% of causative *RB1* mutations [14].

Clinical Utility

The primary goal of *RB1* genetic testing is to identify patients with a hereditary predisposition for RB in order to properly screen them for RB and other malignancies. An intensive protocol of surveillance is recommended in young children at risk for RB (e.g., those known to carry a hereditary *RB1* mutation, and unilateral RB survivors of unknown mutation status) [15]. Ophthalmologic examination, typically under general anesthesia, is performed every 3–4 weeks until the child is 3 years of age or until there has been no tumor activity for 8–12 months. Once old enough to cooperate, children are subjected to dilated eye exam without anesthesia in decreasing intervals as the risk of new tumors decreases, typically every 6 months until age 9.

In patients with unilateral disease, identifying the two *RB1* mutations in the tumor and demonstrating their absence in a peripheral blood sample reduce the risk of germline mutation from approximately 15% to less than 1% and justify omitting them from the extensive RB screening protocol, although it is recommended that they undergo repeated dilated eye exam. Since most RB patients do not carry a germline mutation, the cost of genetic *RB1* testing is generally outweighed by the substantial health care savings of avoiding intensive tumor surveillance [16].

Limitations of Testing

Although *RB1* genetic testing is clinically useful in most cases, there are a number of circumstances in which it is less effective. First, the reported 95% sensitivity of comprehensive *RB1* testing is high but incomplete, so a negative result for an *RB1* mutation does not eliminate the possibility of hereditary RB. Second, the necessity of identifying both *RB1* mutations in the tumor of patients with unilateral RB can be problematic. Current treatments aim to spare the eye, so enucleation is not performed in all cases, and tumor may not be available for testing. Similarly, in a subset of patients with sporadic, unilateral RB, testing will fail to identify both *RB1* mutations in the tumor, and germline testing will be uninformative. These patients will need to remain on the intensive screening protocol although the *a priori* risk for hereditary RB is low. Last, in cases of hereditary RB, the recurrence risk for the patient and for his or her relatives is dependent to some extent on the nature of the germline mutation. Evidence that certain alleles are associated with reduced penetrance and/or variable expressivity is emerging, but there are as yet insufficient data to support modifying clinical care or risk estimates based on the type of mutation identified.

WILMS TUMOR

Background

Renal tumors represent approximately 7% of all childhood cancer diagnoses, and WT, or nephroblastoma, is by far the most common, accounting for 95% of cases among children younger than 15 years [17]. WT is thought to develop from nephrogenic rests, collections of abnormally persistent fetal blastema cells. These cells, which usually do not persist beyond 36 weeks' gestation, represent precursor lesions for which there are multiple potential outcomes. The vast majority spontaneously regress by early childhood, some remain stable, and only a very few will proliferate and undergo tumorigenesis. Nephrogenic rests have been identified in approximately 1% of newborns, 40% of unilateral WT patients, and 90% of bilateral WT patients [18,19].

In most cases, WT occurs as an isolated finding in a previously healthy child. However, approximately 5% of cases occur among children with a congenital syndrome associated with an increased risk of WT [20,21]. Although WT has been reported in association with a wide range of Mendelian diseases and constitutional chromosome abnormalities, the number of conditions for which multiple studies have convincingly demonstrated an increased risk of WT is limited. Those known to have the highest risk of WT include Wilms tumor—aniridia—genitourinary abnormalities—mental retardation (WAGR) syndrome, Denys—Drash syndrome (DDS), Beckwith—Wiedemann syndrome (BWS), Perlman syndrome, and Fanconi anemia (FA) subtypes D2 and N. An additional 1—2% is familial, occurring among children who have one or more relatives with WT but no evidence of a WT-associated syndrome [22].

Molecular Target

The molecular targets in WT testing are varied and include germline and mosaic mutations leading to tumor predisposition as well as somatic mutations acquired within the cancer cells.

WT1. The Wilms tumor 1 gene, *WT1*, encodes a transcription factor that can act as either an enhancer or a repressor and is required for proper kidney development [23,24]. Mutation or deletion of *WT1* has been identified in approximately 20% of WT patients [25]. The majority of these are somatic mutations in nonsyndromic sporadic cases and include deletions, insertions, missense, and splice site mutations. Germline mutation of *WT1* is found in approximately 5% of cases, including both congenital syndromic cases and cases of heritable WT without other features [20,25].

The *WT1*-associated predisposition syndromes include WAGR (OMIM #194072) and DDS (OMIM #194080). Both are characterized by a spectrum of genitourinary abnormalities and WT predisposition. There is a strong genotype—phenotype correlation among these cases. WAGR is clinically the more severe, and as its name suggests, causes WT predisposition, aniridia, genitourinary abnormalities such as ambiguous genitalia, undescended testicles or hypospadias in males, and internal genital or urinary anomalies in females, and varying degrees of intellectual disability. It is caused by a deletion at 11p13 that comprises both the *WT1* and *PAX6* genes. The estimated risk of WT in these patients is consistently high, varying between approximately 45% and 60% [26,27]. In contrast, DDS is associated with undermasculinized external genitalia in 46,XY individuals that can range from ambiguous to a normal female appearance, while affected 46,XX individuals have normal female genitalia. Patients with DDS have an exceptionally high risk of WT (>90%) and of early-onset renal failure [28,29]. DDS is most often associated with missense mutation in exon 8 or 9 of *WT1* [28]. Frasier syndrome (OMIM #136680) is an allelic disorder which is similarly associated with undermasculinization of male external genitalia but has a much lower risk of WT and renal failure. Frasier syndrome is caused by splicing mutations of IVS9 [30]. In addition to the syndromic cases, germline *WT1* mutations have been identified in patients with WT and no other features [31]. These mutations include missense, frameshift, nonsense, and splicing and have been found to be distributed throughout the gene [32].

11p15 locus. Converging lines of evidence indicate that one or more genes at 11p15 influence WT risk. Somatic abnormalities of 11p15 have been demonstrated in a subset of sporadic WT [33] and constitutional chromosomal abnormalities of 11p15 are associated with both syndromic and nonsyndromic WT [31,33].

The syndrome associated with 11p15 abnormalities, BWS, is characterized by overgrowth and elevated incidence of multiple types of malignancies, of which WT is the most common with an approximately 9% risk [34]. The genetic abnormalities that underlie BWS are complex, and their thorough consideration is outside the scope of this chapter. Put very simply, the BWS critical region at 11p15 includes two imprinting centers, IC1 and IC2, which are differentially methylated on the maternal and paternal alleles. These imprinting centers regulate the parental-specific expression of five genes: at IC1, *IGF2* is paternally expressed while *H19* is maternally expressed, and at IC2, *KCNQ10T1* is paternally expressed, while both *CDKN1C* and *KCNQ1* are maternally expressed. Taken together, maternal hypermethylation of IC1, maternal hypomethylation of IC2, mutation of the maternal CDKN1C allele, paternal uniparental disomy of the BWS critical region, and cytogenetic abnormalities disrupting the BWS critical region account for approximately 80% of BWS [34]. Interestingly, constitutional 11p15 defects have been identified in approximately 3% of children with WT and no evidence of overgrowth [33].

Other WT predisposition syndromes. A number of additional Mendelian diseases are clearly associated with an increased risk of WT. Those with the highest risk of WT include Perlman syndrome, a severe autosomal recessive overgrowth syndrome that is often fatal in infancy, and two FA complementation groups, FANCD1 and FANCN. Those with lower WT risk include the rare overgrowth conditions, Simpson—Golabi—Behmel syndrome and Bohring—Opitz syndrome, and the cancer predisposition conditions, Bloom syndrome and Li-Fraumeni syndrome. The genes associated with these syndromes and the WT risk for each are detailed in Table 31.1.

Acquired cytogenetic abnormalities. Tumor-specific recurrent cytogenetic abnormalities have been observed, and some of these abnormalities have prognostic implications. Among the most common are gain of 1q and loss of heterozygosity (LOH) for chromosomes 1p and 16q, which are poor prognostic indicators [36,37].

Molecular Technologies

Genetic testing in WT is determined in part by the likelihood that a patient has an unrecognized WT predisposition syndrome. Any patient presenting with apparently isolated WT should be evaluated for subtle manifestations of one of these syndromes. Some, such as WAGR, are of course highly unlikely in an otherwise healthy and typically developing child, but others, particularly DDS in females, can be difficult to discern. In one study, approximately 17% of patients presenting with WT were clinically diagnosed with a WT-associated syndrome [38]. If clinical suspicion is high, genetic testing of a peripheral blood sample is indicated.

For suspected cases of WAGR, FISH analysis for the associated deletion at 11p15.5 is the most common method of diagnosis, but chromosomal microarray is also available. Although the reported deletion sizes vary greatly, most are too small to be detected by routine cytogenetic analysis. For DDS, sequence analysis of the relevant regions of *WT1* is clinically available by Sanger sequencing, massively parallel sequencing, and genotyping assays for the more common mutations. For sporadic WT cases, full sequence analysis of the coding region of WT1 is available by Sanger sequencing and massively parallel sequencing.

Genetic evaluation of BWS is most often performed in a tiered fashion. Because methylation defects are responsible for the majority of cases, analysis of the

TABLE 31.1 Congenital Syndromes Associated With an Increased Risk of WT

Syndrome	Locus	Estimated WT risk	Inheritance[a]
DDS	*WT1*	90% [28]	AD
WAGR	11p13; *WT1* and *PAX6*	45−60% [27]	AD
FA, N	*PALB2*	38% [35]	AR
FA, D1	*BRCA2*	30% [35]	AR
Perlman syndrome	*DIS3L2*	30% [35]	AR
BWS	11p15.5	9% [34]	N/A[b]
Simpson—Golabi—Behmel syndrome	*GPC3*	6% [35]	XLR
Bloom syndrome	*BLM*	3% [35]	AR
Frasier syndrome	*WT1*	<1%	AD

[a]*AD, autosomal dominant; AR, autosomal recessive; XLR, X-linked recessive.*
[b]*Only a few percent of BWS cases, such as those with mutations in CDKN1C or with cytogenetic abnormalities involving 11p15.5 are heritable, and these are typically transmitted in an AD manner.*

two imprinting centers at 11p15.5 by methylation-sensitive MLPA is usually performed first. This test will identify hypomethylation of IC2 in approximately 50% of cases and hypermethylation of IC1 in approximately 5%. Paternal uniparental disomy, which accounts for an additional approximately 20%, is suspected if both hypomethylation of IC2 and hypermethylation of IC1 are detected, and confirmatory testing can be performed using highly informative short tandem repeat (STR) markers. Among patients with normal methylation results, sequence analysis of the coding region of *CDKN1C* may be performed. Causative mutations within this gene are identified in approximately 5% of sporadic BWS, but in up to 40% of cases with a positive family history [39]. Routine karyotype analysis has been shown to detect relevant cytogenetic rearrangements involving 11p15.5 in a small fraction ($\sim$1–2%) of patients [40,41].

Perlman syndrome is caused by biallelic mutation in *DIS3L2*, and both sequence analysis and deletion/duplication testing are clinically available [42].

For the FAs, FANCD2, and FANCN, the initial genetic diagnosis is by chromosome breakage assay. In this test, cultured lymphocytes are exposed to a DNA cross-linking or alkylating agent, such as diepoxybutane or mitomycin C, and the number of DNA abnormalities (e.g., chromosome and chromatid breaks, radial formation, fragmentation) is quantified and compared to that of normal controls. Cells from FA patients are deficient in DNA repair, and an increase in the number of abnormalities is diagnostic for FA. Once a diagnosis of FA has been established, the particular subgroup can be identified. This is commonly done by massively parallel sequencing, which permits simultaneous analysis of all FA-associated genes and has largely replaced the historical complementation group analysis.

Genetic testing of the tumor sample, whether a biopsy or an excision, is most often undertaken by karyotype analysis of cultured cells to identify numerical and structural chromosome abnormalities. CMA is a useful tool for identifying smaller unbalanced structural abnormalities and regions of LOH.

Clinical Utility

Genetic testing in the setting of WT is used in the estimation of recurrence risk for the patient and for his or her relatives, to determine whether additional tumor surveillance is appropriate, and to inform prognosis.

Most cases of WT are unilateral ($\sim$90%) and sporadic ($\sim$98%), and after treatment the patient's risk of recurrence or development of additional malignancies is low. Distinguishing these patients from the few with WT predisposition syndromes permits the former to avoid unnecessarily intensive tumor surveillance. In certain cases, genetic testing may be the best way to differentiate between clinically similar syndromes with different health risks, as in DDS and Frasier syndrome. Even among patients with the same diagnosis, the exact nature of the causative mutation can have profound clinical implications. For example, the overall risk of WT in BWS is routinely cited as approximately 5%, but to date no cases of WT have been reported in association with the most common cause of BWS, isolated hypomethylation of IC2, or with mutation in *CDKN1C*.

Regardless of whether a patient has a genetic predisposition syndrome, karyotype analysis of tumor cells to identify the presence or absence of important recurrent cytogenetic abnormalities can aid accurate prognosis. WT karyotypes are often complex, with multiple numerical and structural abnormalities. Fig. 31.1 shows a karyotype from a WT sample exhibiting a number of common recurrent abnormalities, including isochromosomes 1q and 16p, and gain of an additional copy each of chromosomes 1, 7, and 8. As gain of 1q and loss of 16q are associated with inferior outcome, this patient may be expected to require aggressive treatment. In addition, CMA for the detection of tumor-specific LOH at 1p or 16q is used to stratify prognosis.

Limitations of Testing

Among the primary limitations of genetic testing for WT is our incomplete understanding of the genetic mechanisms that underlie WT risk and tumorigenesis. In one study, 52 WT patients were observed to have associated malformations indicative of a congenital syndrome, but genetic testing identified a causative mutation among the known syndromic predisposition genes in only 14 ($\sim$27%) [20]. A related concern is the difficulty in calculating risk when many of the associated syndromes are characterized by variable expressivity and reduced penetrance. Interpretation is further complicated by the large number of private variants observed in certain genes, particularly of *WT1* variants in nonsyndromic familial cases. Although these are active areas of research, the low incidence of the individual predisposition syndromes makes it difficult to obtain datasets of sufficient power for prospective study.

The interpretation of tumor-specific acquired genetic abnormalities has similar limitations. Although a handful of poor prognostic indicators have been discovered, there are a large number of recurrent abnormalities with unknown clinical implication. In addition, karyotype analysis is generally limited to imbalances of greater than approximately 5 Mb in good preparations, and WT metaphase spreads, like those of most solid

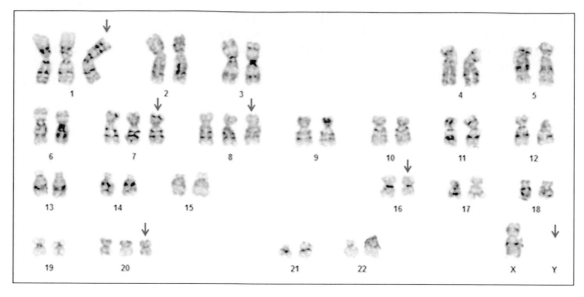

FIGURE 31.1 49,X,-Y,+1,i(1)(q10),+7,+8,i(16)(p10),+20 karyotype of a tumor cell from a young male WT patient. Blue arrows indicate cytogenetic abnormalities. Structural abnormalities involving 1q and 16p, as well as trisomy 8, are recurrent abnormalities [43]. Gain of 1q and loss of 16q are poor prognostic indicators [36].

tumors, often result in short chromosomes that further increase the limit of detection. Thus, many abnormalities may be too subtle to be observed.

NEUROBLASTOMA

Background

NB is a solid tumor arising from embryonic neural crest cells. It accounts for approximately 8–10% of all childhood cancers and 15% of all deaths related to cancer in children. NB is the most common cancer diagnosis in infants with the majority of patients diagnosed between birth and age 4. Although NB can be seen in older children and young adults, fewer than 5% of NB diagnoses are made after age 10. NB is characterized as heterogeneous in terms of both presentation and outcome. Tumors can arise in a variety of locations including the adrenal gland (most frequent site), mediastinum, connective, subcutaneous, and soft tissues, retroperitonium, central and autonomic nervous systems, and other locations. Outcomes can range from spontaneous regression of the tumor to very aggressive metastatic disease and death [2,44].

The diagnosis of NB is generally based either on tissue biopsy and histopathology or the finding of NB cells in the bone marrow and the elevated presence of specific serum or urine analytes [45]. Staging of the tumor at diagnosis takes into consideration whether the tumor is localized, involves vital structures, crosses the midline, and the presence of involved lymph nodes

and distant metastases [46]. Age at diagnosis has long been recognized as an important prognostic indicator with infants having a superior outcome compared to patients older than 1–2 years at ascertainment. Current treatment trials are examining the efficacy of reducing the toxicity of treatment regimens for patients diagnosed before 18 months of age [47].

Treatment algorithms are generally based on risk classifications developed as a result of large clinical trials performed by collaborative groups like the Children's Oncology Group (COG) and data analysis projects that combine data from multiple trials. The International Neuroblastoma Risk Group (INRG) Project risk classification combines staging, age, histology, and genetic characteristics of the tumor to classify patients, pretreatment, into very low, low, intermediate, and high-risk categories [48]. In one COG study (ANBL00B1), infants less than 6 months of age at diagnosis with small localized adrenal tumors demonstrated spontaneous regression in 81% of cases without surgery or other treatment. Patients with intermediate-risk classification were treated with chemotherapy alone, and those at high risk had surgery, chemotherapy, stem cell transplant, immunotherapy, and/or treatment with other biological agents [49–51].

Molecular Targets and Technologies

Genomic testing for NB falls into two categories: (1) testing for germline mutations in predisposition genes and (2) testing for somatic alterations in tumor tissue that may have prognostic significance or suggest treatment options.

Approximately 1–2% of NB cases are thought to be familial with an autosomal dominant pattern of inheritance and unknown penetrance. The two genes most often identified as predisposition genes for NB are *Phox2b* and *ALK*. Patients with familial NB are often present with multifocal tumors and may also have, or have a family history of, other disorders of the sympathetic nervous system, such as Hirschsprung disease (HSCR), congenital central hypoventilation syndrome (CCHS), or neurofibromatosis type 1 (NF1). Germline mutations in *Phox2b* (at 4p12–13) are seen in patients with NB and HSCR or CCHS. Pedigree analysis in familial cases of NB led to the identification of germline mutations in the tyrosine kinase domain of the *ALK* oncogene in some patients. *ALK* is involved in the development of the nervous system. Some clinical laboratories are offering germline testing for *Phox2b* and *ALK* mutations using next-generation sequencing or other sequencing approaches. Approximately 12% of high-risk NB tumors also contain somatic *ALK* mutations. In their 2013 paper describing the genetic landscape of high-risk NB tumors using a combination of whole-exome and whole-genome sequencing, Pugh et al. identified five genes with germline mutations that could potentially predispose to NB: *ALK*, *CHEK2*, *PINK1*, *TP53*, and *BARD1*. They concluded that germline mutations may play a larger role in NB pathogenesis than previously reported. In the same paper, the authors also identified five genes with a relatively high number of somatic mutations in these tumors and biologic evidence for involvement in the development of NB: *ALK*, *PTPN11*, *ATRX*, *MYCN*, and *NRAS* [52–54].

Somatic genomic testing in NB tumors has traditionally addressed cytogenetic markers. MYCNA status has been studied since the mid-1980s and is highly correlated with high-risk disease [55]. In the INRG cohort of 8800 NB patients, 16% of NB tumors exhibited MYCNA, but the majority of metastatic tumors did not have MYCNA [48]. Ploidy status is also an independent indicator of outcome and many aggressive NB tumors without MYCNA have diploid (2N) or tetraploid (4N) karyotypes. Hyperdiploid or near triploid NB tumor karyotypes are generally associated with a good outcome [56]. More recently, segmental chromosome abnormalities (SCAs) have been identified as indicators of poor outcome in patients without MYCNA. SCAs are losses, gains, or rearrangements of chromosomal regions. Specific SCAs have been identified as having prognostic significance in tumors without MYCNA. Gain of 17q, loss of 1p, and loss of 11q are the most common SCAs in tumors without MYCNA and the presence of any one of these abnormalities is predictive of a higher risk of relapse and poorer outcome. Less frequent, but also recurring SCAs include deletions in 3p and 4p, and gains of 1q and 2p [57].

In nonresearch settings, karyotype and FISH assays are most commonly used to distinguish among tumors with diploid, triploid, MYCNA, and SCA genomes. MYCNA is usually demonstrated with an FISH assay that combines a probe for the MYCN gene in one color and a different color probe for a similar-sized gene on the long arm or for the centromere region of chromosome 2 (Fig. 31.2). A positive MYCNA result is defined as more than four times as many *MYCN* signals as control probe signals in the same cell, while a lower discrepancy between *MYCN* and control signals is termed *MYCN* gain. Although *MYCN* gain has been shown to be associated with a poor outcome in other studies, the INRG study did not have sufficient data on *MYCN* status to confirm this observation [48]. Clinical laboratories often use either karyotypic evidence of 11q23 deletion or a loss of an FISH probe in the 11q23 region to demonstrate the 11q SCA. Similarly, 1p loss or 17q gain can be shown by the use of FISH probes that map to the relevant region. The prognostic significance of

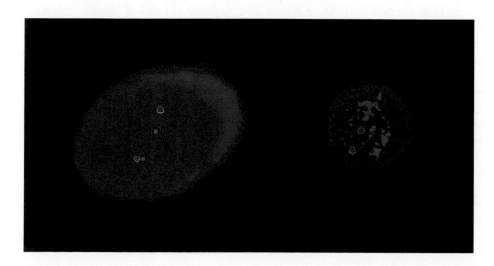

FIGURE 31.2 *MYCN* amplification. Two interphase nuclei from an NB tumor probed with the Abbott/Vysis LSI probe set for *MYCN* (green) and the centromere region (orange) of chromosome 2. The nucleus on the left is normal with two copies each of *MYCN* (2p24) and the chromosome 2 specific centromere alpha satellite sequences. The nucleus on the right has over 100 copies of *MYCN* and two copies of the centromere signal. This degree of *MYCN* amplification is seen in some high-risk NB tumors.

SCAs reported by Schleiermacher et al. [57] from the INRG cohort and the inclusion of LOH for these regions in their definition of SCA suggest that pangenomic assays such as SNP microarray may be a more efficient and thorough approach for this analysis, and this technology is also in use in clinical laboratories for NB tumor classification [49].

The observation that *ALK* is one of the most commonly mutated genes in NB has led to clinical trials with the small-molecule tyrosine kinase inhibitor crizotinib. Approximately 7–10% of NB tumors have activating mutations in the tyrosine kinase domain of *ALK*, and mutations in three amino acid domains account of approximately 86% of these mutations. Clinical trials are in progress using varying dosage levels of crizotinib depending on the degree of sensitivity of the different mutations to the treatment. Sequencing of tumor tissue for the presence of these mutations is taking place in the clinical research setting. Initial trials have focused on patients with refractory or relapsed disease [58,59].

Clinical Utility

The clinical utility of germline testing is somewhat limited by the small number of NB tumors that are thought to arise in carriers of NB-predisposing germline mutations. Overall it has been estimated that about 2% of NB tumors fall into this category, and the majority of these mutations are in the *ALK* gene, with 6% of the 2% estimated to occur in *PHOX2B* [60]. The question as to whether this is an underestimate of the prevalence of hereditary NB has been raised by at least one sequencing study that identified five candidate genes that may be involved in predisposition to NB tumors [52]. Because NB is one of the most common tumors of childhood and infancy, population screening for NB in infants has been attempted, but with mixed results. The presence of elevated urinary catecholamines is a significant marker for NB. A Japanese program of universal screening of asymptomatic 6-month-old infants resulted in a significant increase in the incidence of NB in that population, but the tumors identified had biological markers associated with good outcomes. Similar results were seen in two other studies in Germany and Quebec, and overall, because high-risk tumors were not identified, population screening has not impacted mortality rates. The use of a similar screening approach for monitoring patients at increased risk for NB because of inherited mutations may prove to be more effective [61–63].

Low-risk NB patients enrolled in the COG ANBL00B1 biology study which opened in 2001, had a 3-year event-free survival (EFS) of over 95%. Intermediate-risk patients had a 3-year EFS of 80–95%. Treatment strategies were in part based on the biology of the tumors, including some genetic characteristics. However, patients with high-risk tumors had only a 40–50% 3-year EFS, although even that disappointing statistic was a marked improvement over similar data on patients treated before the development of biology-based treatment algorithms. Current efforts to improve treatment for low- and intermediate-risk patients are focused on reducing the treatment toxicity while maintaining excellent survival [49]. Patients who have received the most toxic levels of treatment, usually the high-risk patients who received high levels of chemotherapy and/or radiation, have been observed to have some late effects which can include hearing loss, renal impairment, and other organ system dysfunction. Some NB survivors have also developed secondary malignancies including myelodysplastic syndrome, acute myeloid leukemia, and some solid tumors [64,65]. In terms of clinical utility, treatment algorithms that are based on tumor biology have worked well for low- and intermediate-risk patients, but high-risk patients with higher intensity treatments have seen relatively smaller benefits and an increased risk of treatment-related morbidity.

Limitations of Testing

The primary limitation of testing to date seems to reside with the heterogeneity of this tumor, the paucity of known driver mutations, and the relatively small number of recurring somatic mutations identified in high-risk NB tumors. While excellent outcomes have been achieved for low- and intermediate-risk patients, the goal of finding targeted therapies for the approximately 40–50% of patients in the high-risk group has not made much headway.

References

[1] Ward E, DeSantis C, Robbins A, Kohler B, Jemal A. Childhood and adolescent cancer statistics, 2014. CA Cancer J Clin 2014; 64:83–103.

[2] Ries LAG, Smith MA, Gurney JG, Linet M, Tamra T, Young JL, Bunin GR, editors. Cancer incidence and survival among children and adolescents: United States SEER program 1975–1995. Bethesda, MD: National Cancer Institute, SEER Program. NIH Pub. No. 99–4649; 1999.

[3] Wong JR, Tucker MA, Kleinerman RA, Devesa SS. Retinoblastoma incidence patterns in the US Surveillance, Epidemiology, and End Results program. JAMA Ophthalmol 2014;132:478–83.

[4] Dommering CJ, Mol BM, Moll AC, et al. RB1 mutation spectrum in a comprehensive nationwide cohort of retinoblastoma patients. J Med Genet 2014;51:366–74.

[5] Knudson Jr. AG. Mutation and cancer: statistical study of retinoblastoma. Proc Natl Acad Sci USA 1971;68:820–3.

[6] Matsunaga E. Hereditary retinoblastoma: penetrance, expressivity and age of onset. Hum Genet 1976;33:1–15.

[7] Marees T, Moll AC, Imhof SM, de Boer MR, Ringens PJ, van Leeuwen FE. Risk of second malignancies in survivors of retinoblastoma: more than 40 years of follow-up. J Natl Cancer Inst 2008;100:1771−9.

[8] Blach LE, McCormick B, Abramson DH, Ellsworth RM. Trilateral retinoblastoma-incidence and outcome: a decade of experience. Int J Radiat Oncol Biol Phys 1994;29:729−33.

[9] Dunn JM, Phillips RA, Becker AJ, Gallie BL. Identification of germline and somatic mutations affecting the retinoblastoma gene. Science 1988;241:1797−800.

[10] Rushlow DE, Mol BM, Kennett JY, et al. Characterisation of retinoblastomas without RB1 mutations: genomic, gene expression, and clinical studies. Lancet Oncol 2013;14:327−34.

[11] Richter S, Vandezande K, Chen N, et al. Sensitive and efficient detection of RB1 gene mutations enhances care for families with retinoblastoma. Am J Hum Genet 2003;72:253−69.

[12] Otterson GA, Modi S, Nguyen K, Coxon AB, Kaye FJ. Temperature-sensitive RB mutations linked to incomplete penetrance of familial retinoblastoma in 12 families. Am J Hum Genet 1999;65:1040−6.

[13] Jones K, Minassian BA. Genetic testing in infantile spasms identifies a chromosome 13q deletion and retinoblastoma. Pediatr Neurol 2014;50:522−4.

[14] Price EA, Price K, Kolkiewicz K, et al. Spectrum of RB1 mutations identified in 403 retinoblastoma patients. J Med Genet 2014;51:208−14.

[15] Valenzuela A, Chan HS, Héon E, Gallie BL. A language for retinoblastoma: guidelines and standard operating procedures. In: Reynolds J, Olistsky S, editors. Pediatric retina. Springer; 2011. p. 205−34.

[16] Dhar SU, Chintagumpala M, Noll C, Chévez-Barrios P, Paysse EA, Plon SE. Outcomes of integrating genetics in management of patients with retinoblastoma. Arch Ophthalmol 2011;129:1428−34.

[17] Breslow N, Olshan A, Beckwith JB, Green DM. Epidemiology of Wilms tumor. Med Pediatr Oncol 1993;21:172−81.

[18] Beckwith JB, Kiviat NB, Bonadio JF. Nephrogenic rests, nephroblastomatosis, and the pathogenesis of Wilms' tumor. Pediatr Pathol 1990;10:1−36.

[19] Breslow NE, Beckwith JB, Perlman EJ, Reeve AE. Age distributions, birth weights, nephrogenic rests, and heterogeneity in the pathogenesis of Wilms tumor. Pediatr Blood Cancer 2006; 47:260−7.

[20] Dumoucel S, Gauthier-Villars M, Stoppa-Lyonnet D, et al. Malformations, genetic abnormalities, and Wilms tumor. Pediatr Blood Cancer 2014;61:140−4.

[21] Szychot E, Brodkiewicz A, Pritchard-Jones K. Review of current approaches to the management of Wilms' tumor. Int J Clin Rev 2012;10:07.

[22] Breslow N, Olson J, Moksness J, Beckwith JB, Grundy P. Familial Wilms' tumor: a descriptive study. Med Pediatr Oncol 1996;27:398−403.

[23] Ellisen LW. Regulation of gene expression by WT1 in development and tumorigenesis. Int J Hematol 2002;76:110−16.

[24] Kreidberg JA, Sariola H, Loring JM, et al. WT-1 is required for early kidney development. Cell 1993;74:679−91.

[25] Huff V. Wilms tumor genetics. Am J Med Genet 1998;79:260−7.

[26] Muto R, Yamamori S, Ohashi H, Osawa M. Prediction by FISH analysis of the occurrence of Wilms tumor in aniridia patients. Am J Med Genet 2002;108:285−9.

[27] Fischbach BV, Trout KL, Lewis J, Luis CA, Sika M. WAGR syndrome: a clinical review of 54 cases. Pediatrics 2005; 116:984−8.

[28] Pelletier J, Bruening W, Kashtan CE, et al. Germline mutations in the Wilms' tumor suppressor gene are associated with abnormal urogenital development in Denys-Drash syndrome. Cell 1991;67:437−47.

[29] Breslow NE, Collins AJ, Ritchey ML, Grigoriev YA, Peterson SM, Green DM. End stage renal disease in patients with Wilms tumor: results from the National Wilms Tumor Study Group and the United States Renal Data System. J Urol 2005;174:1972−5.

[30] Barbaux S, Niaudet P, Gubler M, et al. Donor splice-site mutations in WT1 are responsible for Frasier syndrome. Nat Genet 1997;17:467−70.

[31] Little SE, Hanks SP, King-Underwood L, et al. Frequency and heritability of WT1 mutations in nonsyndromic Wilms' tumor patients: a UK Children's Cancer Study Group Study. J Clin Oncol 2004;22:4140−6.

[32] Lehnhardt A, Karnatz C, Ahlenstiel-Grunow T, et al. Clinical and molecular characterization of patients with heterozygous mutations in Wilms tumor suppressor gene 1. Clin J Am Soc Nephrol 2015;10:825−31.

[33] Scott RH, Douglas J, Baskcomb L, et al. Constitutional 11p15 abnormalities, including heritable imprinting center mutations, cause nonsyndromic Wilms tumor. Nat Genet 2008;40:1329−34.

[34] Tan TY, Amor DJ. Tumour surveillance in Beckwith-Wiedemann syndrome and hemihyperplasia: a critical review of the evidence and suggested guidelines for local practice. J Paediatr Child Health 2006;42:486−90.

[35] Scott RH, Rahman N. Genetic predisposition to Wilms tumour. In: Pritchard-Jones K, Dome J, editors. Renal tumors of childhood: biology and therapy. Springer; 2014. p. 19−38.

[36] Gratias EJ, Jennings LJ, Anderson JR, Dome JS, Grundy P, Perlman EJ. Gain of 1q is associated with inferior event-free and overall survival in patients with favorable histology Wilms tumor: a report from the Children's Oncology Group. Cancer 2013;119:3887−94.

[37] Grundy PE, Telzerow PE, Breslow N, Moksness J, Huff V, Paterson MC. Loss of heterozygosity for chromosomes 16q and 1p in Wilms' tumors predicts an adverse outcome. Cancer Res 1994;54:2331−3.

[38] Merks JH, Caron HN, Hennekam R. High incidence of malformation syndromes in a series of 1,073 children with cancer. Am J Med Genet A 2005;134:132−43.

[39] Lam WW, Hatada I, Ohishi S, et al. Analysis of germline CDKN1C (p57KIP2) mutations in familial and sporadic Beckwith-Wiedemann syndrome (BWS) provides a novel genotype-phenotype correlation. J Med Genet 1999;36:518−23.

[40] Cooper WN, Luharia A, Evans GA, et al. Molecular subtypes and phenotypic expression of Beckwith−Wiedemann syndrome. Eur J Hum Genet 2005;13:1025−32.

[41] Hoovers JM, Kalikin LM, Johnson LA, et al. Multiple genetic loci within 11p15 defined by Beckwith-Wiedemann syndrome rearrangement breakpoints and subchromosomal transferable fragments. Proc Natl Acad Sci USA 1995;92:12456−60.

[42] Astuti D, Morris MR, Cooper WN, et al. Germline mutations in DIS3L2 cause the Perlman syndrome of overgrowth and Wilms tumor susceptibility. Nat Genet 2012;44:277−84.

[43] Sheng WW, Soukup S, Bove K, Gotwals B, Lampkin B. Chromosome analysis of 31 Wilms' tumors. Cancer Res 1990;50:2786−93.

[44] London WB, Castleberry RP, Matthay KK, et al. Evidence for an age cutoff greater than 365 days for neuroblastoma risk group stratification in the Children's Oncology Group. J Clin Oncol 2005;23:6459−65.

[45] Irwin MS, Park JR. Neuroblastoma: paradigm for precision medicine. Pediatr Clin N Am 2015;62:225−56.

[46] Brodeur GM, Pritchard J, Berthold F, et al. Revisions of the international criteria for neuroblastoma diagnosis, staging, and response to treatment. J Clin Oncol 1993;11:1466−77.

[47] Park JR, Eggert A, Caron H. Neuroblastoma: biology, prognosis, and treatment. Hematol Oncol Clin North Am 2010;24:65—86.

[48] Ambros PF, Ambros IM, Brodeur GM, et al. International consensus for neuroblastoma molecular diagnostics: report from the International Neuroblastoma Risk Group (INRG) Biology Committee. Br J Cancer 2009;100:1471—82.

[49] Park JR, Bagatell R, London WB, et al. Children's Oncology Group's 2013 blueprint for research: neuroblastoma. Pediatr Blood Cancer 2013;60:985—93.

[50] Alvarado CS, London WB, Look AT, et al. Natural history and biology of stage A neuroblastoma: a Pediatric Oncology Group study. J Pediatr Hematol Oncol 2000;22:197—205.

[51] Simon T, Spitz R, Faldum A, et al. New definition of low-risk neuroblastoma using stage, age, and 1p and MYCN status. J Pediatr Hematol Oncol 2004;26:791—6.

[52] Pugh TJ, Morozova O, Attiyeh EF, et al. The genetic landscape of high-risk neuroblastoma. Nat Genet 2013;45:279—84.

[53] Mosse YP, Laudenslager M, Khazi D, et al. Germline PHOX2B mutation in hereditary neuroblastoma. Am J Hum Genet 2004;75:727—30.

[54] Mosse YP, Laudenslager M, Longo L, et al. Identification of ALK as a major familial neuroblastoma predisposition gene. Nature 2008;455:930—5.

[55] Brodeur GM, Seeger RC, Schwab M, et al. Amplification of N-myc in untreated human neuroblastomas correlates with advanced disease stage. Science 1984;224:1121—4.

[56] Look AT, Hayes FA, Shuster JJ, et al. Clinical relevance of tumor cell ploidy and N-myc gene amplification in childhood neuroblastoma: a Pediatric Oncology Group study. J Clin Oncol 1991;9:581—91.

[57] Schleiermacher G, Mosseri V, London WB, et al. Segmental chromosomal alterations have prognostic impact in neuroblastoma: a report from the INRG project. Br J Cancer 2012;107:1418—22.

[58] Carpenter EL, Mosse YP. Targeting ALK in neuroblastoma—preclinical and clinical advancements. Nat Rev Clin Oncol 2012;9:391—9.

[59] Bresler SC, Wood AC, Haglund EA, et al. Differential inhibitor sensitivity of anaplastic lymphoma kinase variants found in neuroblastoma. Sci Transl Med 2011;3:108ra114.

[60] Barone G, Anderson J, Pearson AD, et al. New strategies in neuroblastoma: therapeutic targeting of MYCN and ALK. Clin Cancer Res 2013;19:5814—21.

[61] Yamamoto K, Ohta S, Ito E, et al. Marginal decrease in mortality and marked increase in incidence as a result of neuroblastoma screening at 6 months of age: cohort study in seven prefectures in Japan. J Clin Oncol 2002;20:1209—14.

[62] Schilling FH, Spix C, Berthold F, et al. Neuroblastoma screening at one year of age. N Engl J Med 2002;346:1047—53.

[63] Woods WG, Gao RN, Shuster JJ, et al. Screening of infants and mortality due to neuroblastoma. N Engl J Med 2002;346:1041—6.

[64] Federico SM, Allewelt HB, Spunt SL, et al. Subsequent malignant neoplasms in pediatric patients initially diagnosed with neuroblastoma. J Pediatr Hematol Oncol 2015;37:e6—e12.

[65] Cohen LE, Gordon JH, Popovsky EY, et al. Late effects in children treated with intensive multimodal therapy for high-risk neuroblastoma: high incidence of endocrine and growth problems. Bone Marrow Transplant 2014;49:502—8.

MOLECULAR TESTING IN HEMATOPATHOLOGY

32

Molecular Testing in Chronic Myelogenous Leukemia

N.A. Brown and B.L. Betz

Department of Pathology, University of Michigan, Ann Arbor, MI, United States

INTRODUCTION

Chronic myelogenous leukemia (CML) is a myeloproliferative neoplasm that has become a paradigm of molecular diagnosis, targeted therapy, and disease monitoring. The disease is defined by the *BCR−ABL1* fusion gene, resulting from the translocation of chromosomes 9 and 22 [1]. This fusion leads to constitutive activation of the ABL1 tyrosine kinase, resulting in growth factor independent myelopoiesis [2].

CML has an incidence of 1−2 cases per 100,000 population and is diagnosed at a median age of approximately 50 years [3,4]. The natural course of CML can be divided into three phases. A chronic phase lasting approximately 4−5 years is characterized by leukocytosis with marked proliferation of granulocytes and their precursors, particularly, myelocytes and segmented neutrophils. Basophilia, eosinophilia, and thrombocytosis are also common, but blast counts are typically less than 2%. Progression to accelerated phase has been defined by (1) leukocytosis, splenomegaly, or thrombocytosis uncontrolled by therapy; (2) thrombocytopenia unrelated to therapy; (3) clonal cytogenetic evolution; (4) basophilia comprising 20% or more of cells in the peripheral blood; or (5) 10−19% myeloblasts in the blood or bone marrow [1]. Blast phase is the most advanced stage and is marked by an acute leukemia (>20% blasts) of either myeloid or lymphoid phenotype.

The development of imatinib (STI571) and other tyrosine kinase inhibitors (TKIs) with activity against *BCR−ABL1* has revolutionized the treatment of patients with CML [5,6]. Following publication of findings from the International Randomized Study of Interferon versus STI571 (IRIS) trial in 2003, imatinib has quickly replaced interferon-α as the standard of care [7]. Since 2003, additional TKIs such as dasatinib and nilotinib have also gained FDA approval for treatment of CML. Molecular testing has evolved in parallel with these treatments and has become standard of care in the diagnosis and monitoring of CML, as well as the selection of specific therapeutic agents.

MOLECULAR BIOLOGY OF CML

At the time of diagnosis, approximately 95% of CML cases demonstrate the characteristic t(9;22)(q34;q11.2) reciprocal translocation resulting in the Philadelphia chromosome [der(22q)] and the fusion of the 5′ portion of the *BCR* gene to the 3′ portion of *ABL1* [8] (Fig. 32.1). The remaining cases have either cytogenetically cryptic *BCR−ABL1* fusions or variant *BCR−ABL1* translocations involving other chromosomes in addition to 9 and 22. While *BCR−ABL1* is present in all cases of CML, this fusion is not unique to CML and can also be found in many cases of de novo B lymphoblastic leukemia/lymphoma (B-ALL) [9].

The specific chromosomal breakpoints giving rise to the *BCR−ABL1* fusion transcript can vary [10,11] (Fig. 32.2). Breakpoints in *ABL1* are almost always 5′ (upstream) of the second exon, leading to juxtaposition of *ABL1* exon 2 (previously known as a2) with one of several possible *BCR* exons. Variant *ABL1* breakpoints that are 5′ to exon 3 (a3) have also been described but these are rare. *BCR* breakpoints most commonly occur in the major breakpoint region (M-bcr) which spans exons 12−16 (previously known as b1−b5). Within the M-bcr, the vast majority of translocations involve either exon 13 (e13 previously known as b2) or exon 14 (e14 previously known as b3). The juxtaposition of these

Diagnostic Molecular Pathology
DOI: http://dx.doi.org/10.1016/B978-0-12-800886-7.00032-7

413

© 2017 Elsevier Inc. All rights reserved.

BCR b2 or b3 breakpoints with *ABL1* a2 creates e13a2 (b2a2) or e14a2 (b3a2) fusion transcripts and a BCR−ABL1 protein of 210 kDa (p210). This p210 protein is present in the vast majority of CML cases

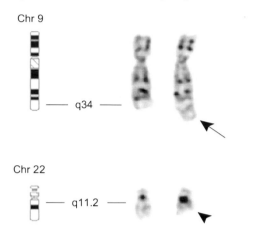

FIGURE 32.1 Metaphase cytogenetics. CML is characterized by a reciprocal translocation involving 9q34 and 22q11.2, resulting in two derivative chromosomes. The derivative chromosome 22 [der(22q)] is known as the Philadelphia chromosome (shown with arrowhead) and contains a fusion of the 5′ portion of the *BCR* gene to the 3′ portion of *ABL1*. The derivative chromosome 9 is also shown (arrow). *Source: Image courtesy of Diane Roulston.*

(∼98%) as well as approximately 50% of adult and 20% of pediatric t(9;22)-positive B-ALL [9,10]. A shorter fusion protein (p190) results from breaks in the minor breakpoint region (m-bcr) and the juxtaposition of *BCR* exon 1 (e1) with *ABL1* exon 2 (a2) leading to e1a2. This fusion is rare in CML (∼1%) but occurs in approximately 50% of adult and 80% of pediatric patients with t(9;22)-positive B-ALL [9]. Rarely, fusions involve an extreme 3′ region of *BCR* designated the micro breakpoint region (μ-bcr). Breaks in this region lead to the fusion of *BCR* exon 19 (e19) with *ABL1* exon 2 (a2) and a larger protein product (p230). This event is commonly associated with a variant of CML in which mature granulocytes predominate [13]. Other extremely rare variants have been described but are infrequently countered in clinical practice [14].

MOLECULAR METHODS

Traditional laboratory methods such as morphology, hematologic assays, and conventional cytogenetics remain important tools in the diagnosis and monitoring of CML. However, several molecular techniques have improved the sensitivity and precision of detecting and

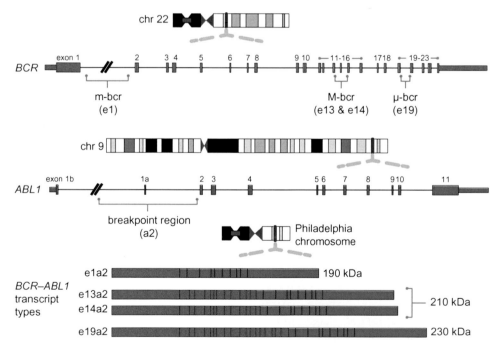

FIGURE 32.2 The t(9;22)(q34;q11) translocation and associated BCR−ABL1 fusion products. The translocation of chromosomes 9 and 22 leads to the juxtaposition of the *BCR* and *ABL1* genes and is cytogenetically recognizable by the presence of the Philadelphia (Ph) chromosome. A large breakpoint region upstream of *ABL1* exon 2 is joined with one of several *BCR* breakpoint regions. In CML, the M-bcr (major) breakpoint region is most common (∼98% cases) and leads to juxtaposition of either *BCR* exon 13 or 14 to *ABL1* exon 2, producing e13a2 or e14a2 transcripts and a 210 kDa BCR−ABL1 fusion protein (p210). This transcript is also found in B-ALL/LBL. Fusion of the *BCR* minor breakpoint region (m-bcr) with *ABL1* exon 2 leads to e1a2 transcripts and a 190 kDa protein (p190). This transcript is associated with B-ALL/LBL and only rarely occurs in CML. The micro breakpoint region (μ-bcr) is rare and juxtaposes *BCR* exon 19 with *ABL1* exon 2, resulting in e19a2 transcript and a 230 kDa (p230) protein. *Source: Reprinted with kind permission of [12] Behdad A, Betz BL, Lim MS, Bailey NG. Molecular testing in hematologic malignancies. In: Yousef GM, Jothy, editors. Molecular testing in cancer. New York, NY: Springer Science + Business Media; 2014. pp.135−68.*

monitoring CML and offer additional information that informs targeted treatment decisions.

Fluorescence In Situ Hybridization

Fluorescence in situ hybridization (FISH) assays employ fluorescently labeled probes to interrogate the BCR and ABL1 loci. A dual-color, dual-fusion approach is commonly used and can detect all BCR−ABL1 fusions with excellent sensitivity and specificity including cytogenetically cryptic translocations (Fig. 32.3). The dual-fusion design also significantly reduces the risk of false positivity due to random colocalization of probe signals [15]. The limit of detection using this approach is approximately 1−2% CML cells in a background of normal cells when 200 cells are scored and can improve to approximately 0.5% when 500 cells are scored. Despite the improved sensitivity over conventional karyotyping, FISH is considered a complimentary method for monitoring treatment response in CML due to lack of standardized response criteria. Further, technologies with higher sensitivity are required to effectively monitor minimal residual disease for signs of early relapse in patients undergoing kinase inhibitor therapies. Therefore, FISH is more useful in the initial diagnosis of CML rather than disease monitoring.

Reverse Transcription Polymerase Chain Reaction

In reverse transcription polymerase chain reaction (RT-PCR), RNA is isolated from cells, reverse transcribed into cDNA, and subjected to PCR amplification. Because intronic sequence is spliced out of RNA transcripts, this approach enables the detection of gene fusions regardless of the specific breakpoint within a given intron. RT-PCR assays are generally designed to detect both M-bcr and m-bcr rearrangements accounting for the large

majority of BCR−ABL1 rearrangements and can be designed to be either qualitative or quantitative.

Qualitative assays offer a sensitive means of detecting BCR−ABL1 with a rapid turnaround time and are therefore well suited for initial diagnosis. These assays should always include a control RT-PCR reaction for another mRNA such as GAPDH or ABL1 to ensure the integrity of the extracted RNA and to assess for the presence of RT-PCR inhibitors. Qualitative RT-PCR can be performed with a simple, nested, or multiplex approach. While multiplexing may reduce the sensitivity of the individual reactions, this effect is not likely to be relevant in the diagnostic setting. Qualitative assays can also be designed to identify rare variant breakpoints such as those involving BCR μ-bcr or ABL1 exon 3 (a3) that may not be amenable to monitoring by quantitative PCR assays.

Quantitative RT-PCR using real-time RT-PCR has emerged as the preferred method for posttherapeutic monitoring of CML [16]. This approach offers a sensitive, reproducible, and rapid means of quantifying BCR−ABL1 fusion transcripts. BCR−ABL1 can be quantitated over a broad range of values spanning many orders of magnitude and the lower limit of detection real-time RT-PCR has been reported at less than one CML cell in a background of 100,000 [17]. Common chemistries employed include Taqman and FRET probe methods to enable florescent detection of accumulated amplification product after each PCR cycle [18,19]. Quantitation relies on the fact that the PCR cycle at which the reaction product crosses a specified fluorescent threshold (C_t) is inversely proportional to log of the number of template copies. Accurate quantitation requires the generation of a standard curve using standards of known concentration (Fig. 32.4A). Based on this curve, the number of copies in an unknown patient sample can be determined. The number of BCR−ABL transcripts is normalized against an internal reference transcript—most commonly ABL1—and reported out as a BCR−ABL1/ABL1 ratio. In

(A)   (B)

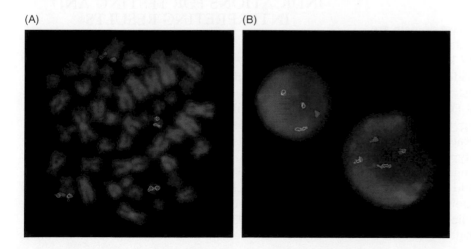

FIGURE 32.3 FISH for detection of BCR−ABL1. (A) Metaphase FISH using dual-fusion probes. The green fluorescent probe is designed to span the breakpoint region of the BCR gene on chromosome 22q11.2 and the red probe spans the ABL1 breakpoint region on chromosome 9q34. The reciprocal t(9;22)(q34; q11.2) results in colocalized green/red fusion signals on both derivative chromosomes: a BCR−ABL1 fusion signal on der(22q) and an ABL1−BCR fusion signal on der(9q). (B) Interphase FISH using the same dual-fusion probes. Both cells harbor the translocation as indicated by two fusion signals. *Source: Images courtesy of Diane Roulston.*

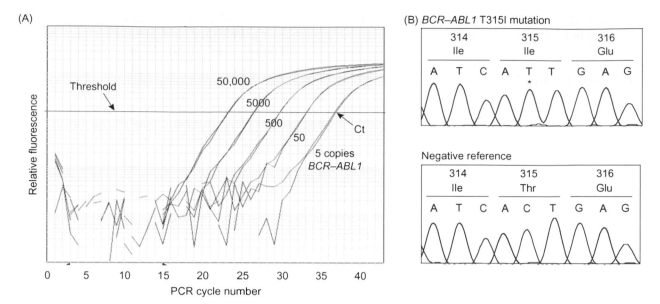

FIGURE 32.4 PCR tests for BCR−ABL1. (A) Quantitative *BCR−ABL1* testing. Real-time RT-PCR is a sensitive means to detect and quantify *BCR−ABL1* transcripts across a 4−6 log range of *BCR−ABL1* levels. Amplification products can be detected during each PCR cycle using a fluorescent probe specific to the PCR product. The accumulated fluorescence in log(10) value is plotted against the number of PCR cycles. For a given specimen the PCR cycle number is measured when the increase in fluorescence is exponential and exceeds a threshold. This point is called the quantification or threshold cycle (C_t), which is inversely proportional to the amount of PCR target in the specimen (ie, lower C_t values indicate a greater amount of target). Calibration standards of known quantity are used in standard curves to calculate the amount of target in a tested specimen. Shown are real-time RT-PCR plots of calibration standards for *BCR−ABL1* quantitation. Note that PCR increases the amount of amplification product by a factor of 2 with each PCR cycle. Therefore, specimens that produce a C_t value that is 1 cycle lower are expected to have a two-fold higher concentration of target. Specimens that differ in target concentration by a factor of 10 (as shown) are expected to be 3.3 cycles apart ($2^{3.3} = 10$). Note the calibration samples with 500 and 50 copies of *BCR−ABL1* produced C_t values of 29.7 and 33.0, respectively. (B) *ABL1* kinase mutation testing. A variety of substitution mutations within the ABL1 kinase domain of BCR−ABL1 can lead to differential resistance to TKI therapies. Sanger sequencing of this region within the *BCR−ABL1* transcript is a preferred method to detect the variety of mutations. Shown is a sequencing trace of a C to T nucleotide transition leading to a threonine (Thr) to isoleucine (Ile) substitution at amino acid 315 (T315I). A wild-type trace is included for reference. *Source: Reprinted with kind permission of Behdad A, Betz BL, Lim MS, Bailey NG. Molecular testing in hematologic malignancies. In: Yousef GM, Jothy, editors. Molecular testing in cancer. New York, NY: Springer Science + Business Media; 2014. pp.135−68.*

addition to calibration standards, quantitative controls (typically both a high positive and a low positive control) are required with each run and should be closely monitored to ensure adequate performance of the assay and reliability of the quantitative results.

ABL1 Mutational Analysis

Substitution mutations in the *ABL1* kinase domain are a common mechanism for acquired resistance to kinase inhibitor therapy. Sequencing of the *ABL1* kinase domain (Fig. 32.4B) is required to identify the wide array of different mutations that confer variable resistance to various kinase inhibitor therapies [16]. A nested PCR strategy is typically employed prior to sequencing to specifically amplify the kinase domain from translocated *ABL1*, thus avoiding sequencing of the native wild-type *ABL1* allele. The first PCR reaction is performed using forward *BCR* primers and reverse *ABL1* primers to yield a *BCR−ABL1* product containing the *ABL1* kinase domain. One or more additional PCR reactions are utilized to further amplify the kinase domain of translocated *ABL1* to a

sufficient copy number. This amplicon is then sequenced using standard Sanger sequencing to identify the specific resistance mutation if present.

INDICATIONS FOR TESTING AND INTERPRETING RESULTS

Initial Diagnosis

CML is frequently suspected based on characteristic clinical and pathologic findings. However, many conditions such as a leukemoid reaction or another myeloproliferative neoplasm can mimic CML. Therefore, a definitive diagnosis requires the identification of the *BCR−ABL1* fusion [1]. In addition, the diagnosis of many non-CML myeloproliferative neoplasms requires the exclusion of *BCR−ABL1*. Conventional cytogenetics can be used to detect the t(9;22) or variant translocation in approximately 95% of cases. Cytogenetic evaluation also enables the detection of additional chromosomal abnormalities, providing a baseline that can later be used to determine if a patient has undergone cytogenetic evolution and disease progression. However, approximately 5% of BCR−ABL1 fusions are cytogenetically cryptic and require detection using molecular methods. Both FISH and RT-PCR can be used to detect BCR−ABL1 fusions with greater sensitivity and typically have a shorter test turnaround time compared with metaphase cytogenetics. While FISH identifies essentially all BCR−ABL1 fusions, RT-PCR will only identify specific fusions for which the primers were designed and may fail to detect BCR−ABL1 with rare breakpoints. RT-PCR may afford a shorter turnaround time and lower limit of detection, though high sensitivity is not typically needed at the time of diagnosis. Quantitative RT-PCR testing should be performed at the time of diagnosis, prior to initiating therapy, in order to ensure the presence of quantifiable BCR−ABL1 transcripts and to obtain a baseline BCR−ABL1/ABL1 ratio [19]. Qualitative RT-PCR with breakpoint analysis may be useful to identify rare BCR−ABL1 breakpoints that are not amenable to monitoring by quantitative RT-PCR tests.

Monitoring of Disease

Following diagnosis, CML patients are initiated on TKI therapy such as imatinib mesylate or newer agents. The degree of remission achieved following such therapy can be graded according to the sensitivity of the parameter used to monitor disease response. The earliest form of remission occurs when blood counts and spleen size normalize signifying a complete hematologic remission. Cytogenetic response is graded based on the percentage of residual t(9;22)-positive cells with major complete response defined as 0% t(9;22)-positive cells, major partial response 1−34%, minor response 35−94%, and no response 95% or greater [15]. However, the use TKI therapy has made this schema less relevant. FISH is a more sensitive approach than metaphase cytogenetics, but FISH is seldom used in monitoring disease due to its limited analytic sensitivity.

Quantitative RT-PCR has emerged as the preferred method for posttherapeutic monitoring of CML [13]. Once the diagnosis of CML has been established, TKI therapy is initiated and BCR−ABL1 transcripts are monitored by quantitative RT-PCR every 3 months [20]. Historically, the wide variation in RT-PCR testing platforms, control genes, and result reporting made interlaboratory comparisons of quantitative RT-PCR results difficult. However, recent guidelines and the widespread implementation of the International Scale (IS) have significantly improved the consistency, quality, and comparability of RT-PCR testing across different laboratories [21]. The IS was originally derived from a pool of 30 patient samples with untreated, newly-diagnosed chronic phase CML in the IRIS trial [22]. The median baseline BCR−ABL1 level from this patient cohort was taken to represent a value of 100% on the IS. The same IRIS study was the first to establish a more favorable outcome in patients who achieved a 3 log reduction in BCR−ABL1 values, which was defined as a major molecular response (MMR). Consequently, a 3 log reduction in BCR−ABL1 transcripts and achieving MMR is equal to a value of 0.1% on the IS. Complete molecular response is defined as a 4.5 log reduction. Recent data also point to the favorable prognostic significance of achieving a 1 log reduction (10% IS) in BCR−ABL1 levels by 3 months and 6 months after initiation of TKI, and this is now included as a response milestone at these time points [23]. To translate their results into an IS value, laboratories must establish a conversion factor for their individual quantitative BCR−ABL1 assay by testing commercially available reference materials or exchanging specimens with an IS-calibrated reference laboratory.

Therapy Refractoriness

TKI resistance can be suspected if an appropriate initial response to TKI therapy is not obtained at 3 months, there is a 1 log increase in BCR−ABL1 transcripts later in therapy, or there is other evidence of disease progression such or hematologic or cytogenetic relapse [24]. While TKI resistance is multifactorial, approximately half to three-fourths of patients have point mutations within the ABL1 kinase domain that contribute to resistance [25]. Numerous mutations have been described that span the entire kinase domain. ABL1 kinase domain sequencing offers an unbiased approach to detect all mutations and may even detect mutations prior to relapse [26] (Fig. 32.4). Individual mutations display a range of effects on TKI refractoriness and prognosis with some mutations conferring resistance against some TKIs but preserved susceptibility to others. Other mutations confer moderate TKI resistance which can be overcome by higher doses. In addition, more than one resistant clone may exist with each clonal component bearing a different mutation. One of the most common and notorious mutations involves a substitution of isoleucine for the threonine at position 315 (T315I), which imparts resistance to multiple TKIs including imatinib, dasatinib, and nilotinib [20]. Recently, novel TKIs have been developed that have shown promise in targeting the T315I mutation [27]. Identification of ABL1 mutations allows appropriate, patient-specific selection of TKIs with activity against specific resistance mutations. Patients who are determined to be resistant to all available TKI may be referred for hematopoietic stem cell transplantation or clinical trial.

CONCLUSIONS

CML is a myeloproliferative neoplasm that has become a paradigm of targeted molecular therapy as well as molecular diagnosis and disease monitoring. The disease is defined by the BCR−ABL1 fusion which can be detected by a variety of techniques. While conventional morphology and hematology as well as cytogenetics remain important diagnostic tools, both FISH and RT-PCR for the BCR−ABL1 fusion offer increased sensitivity. TKI therapy has revolutionized the treatment of CML and quantitative RT-PCR has become the standard of care for monitoring patients after the initiation of therapy. Finally, the identification of ABL1 kinase domain mutations using sequencing allows appropriate, patient-specific selection of TKIs in patients who are refractory to first-line treatment.

References

[1] Swerdlow SH, International Agency for Research on Cancer, World Health Organization. WHO classification of tumours of haematopoietic and lymphoid tissues. 4th ed. Lyon, France: International Agency for Research on Cancer; 2008.

[2] Ren R. Mechanisms of BCR-ABL in the pathogenesis of chronic myelogenous leukaemia. Nat Rev Cancer 2005;5:172−83.

[3] Jemal A, Tiwari RC, Murray T, Ghafoor A, Samuels A, Ward E, et al. Cancer statistics, 2004. CA Cancer J Clin 2004;54:8−29.

[4] Redaelli A, Bell C, Casagrande J, Stephens J, Botteman M, Laskin B, et al. Clinical and epidemiologic burden of chronic myelogenous leukemia. Expert Rev Anticancer Ther 2004; 4:85−96.

[5] Druker BJ, Tamura S, Buchdunger E, Ohno S, Segal GM, Fanning S, et al. Effects of a selective inhibitor of the Abl tyrosine kinase on the growth of Bcr-Abl positive cells. Nat Med 1996; 2:561−6.

[6] Deininger M, Buchdunger E, Druker BJ. The development of imatinib as a therapeutic agent for chronic myeloid leukemia. Blood 2005;105:2640−53.

[7] O'Brien SG, Guilhot F, Larson RA, Gathmann I, Baccarani M, Cervantes F, et al. Imatinib compared with interferon and low-dose cytarabine for newly diagnosed chronic-phase chronic myeloid leukemia. N Engl J Med 2003;348:994−1004.

[8] Rowley JD. Letter: a new consistent chromosomal abnormality in chronic myelogenous leukaemia identified by quinacrine fluorescence and Giemsa staining. Nature 1973;243: 290−3.

[9] Faderl S, Kantarjian HM, Talpaz M, Estrov Z. Clinical significance of cytogenetic abnormalities in adult acute lymphoblastic leukemia. Blood 1998;91:3995−4019.

[10] Deininger MWN, Goldman JM, Melo JV. The molecular biology of chronic myeloid leukemia. Blood 2000;96:3343−56.

[11] Kurzrock R, Gutterman JU, Talpaz M. The molecular genetics of Philadelphia chromosome−positive leukemias. N Engl J Med 1988;319:990−8.

[12] Behdad A, Betz BL, Lim MS, Bailey NG. Molecular testing in hematologic malignancies. In: Yousef GM, Jothy S, editors. Molecular testing in cancer. New York, NY: Springer Science + Business Media; 2014. p. 135−68.

[13] Pane F, Frigeri F, Sindona M, Luciano L, Ferrara F, Cimino R, et al. Neutrophilic-chronic myeloid leukemia: a distinct disease with a specific molecular marker (BCR/ABL with C3/A2 junction). Blood 1996;88:2410−14.

[14] Barnes DJ, Melo JV. Cytogenetic and molecular genetic aspects of chronic myeloid leukaemia. Acta Haematol 2002;108:180−202.

[15] Kaeda J, Chase A, Goldman JM. Cytogenetic and molecular monitoring of residual disease in chronic myeloid leukaemia. Acta Haematol 2002;107:64−75.

[16] Hughes T, Deininger M, Hochhaus A, Branford S, Radich J, Kaeda J, et al. Monitoring CML patients responding to treatment with tyrosine kinase inhibitors: review and recommendations for harmonizing current methodology for detecting BCR-ABL transcripts and kinase domain mutations and for expressing results. Blood 2006;108:28−37.

[17] Luu MH, Press RD. BCR-ABL PCR testing in chronic myelogenous leukemia: molecular diagnosis for targeted cancer therapy and monitoring. Expert Rev Mol Diagn 2013;13:749−62.

[18] Eder M, Battmer K, Kafert S, Stucki A, Ganser A, Hertenstein B. Monitoring of BCR-ABL expression using real-time RT-PCR in CML after bone marrow or peripheral blood stem cell transplantation. Leukemia 1999;13:1383−9.

[19] Emig M, Saussele S, Wittor H, Weisser A, Reiter A, Willer A, et al. Accurate and rapid analysis of residual disease in patients with CML using specific fluorescent hybridization probes for real time quantitative RT-PCR. Leukemia 1999;13:1825−32.

[20] Baccarani M, Pileri S, Steegmann J-L, Muller M, Soverini S, Dreyling M, et al. Chronic myeloid leukemia: ESMO Clinical Practice Guidelines for diagnosis, treatment and follow-up. Ann Oncol 2012;23:vii72−7.

[21] Zhen C, Wang YL. Molecular monitoring of chronic myeloid leukemia: international standardization of BCR-ABL1 quantitation. J Mol Diagn 2013;15:556−64.

[22] Hughes TP, Kaeda J, Branford S, Rudzki Z, Hochhaus A, Hensley ML, et al. Frequency of major molecular responses to imatinib or interferon alfa plus cytarabine in newly diagnosed chronic myeloid leukemia. N Engl J Med 2003;349:1423−32.

[23] Marin D, Ibrahim AR, Lucas C, Gerrard G, Wang L, Szydlo RM, et al. Assessment of BCR-ABL1 transcript levels at 3 months is the only requirement for predicting outcome for patients with chronic myeloid leukemia treated with tyrosine kinase inhibitors. J Clin Oncol 2012 Jan 20;30(3):232−8.

[24] Saglio G, Fava C. Practical monitoring of chronic myelogenous leukemia: when to change treatment. J Natl Compr Canc Netw 2012;10:121−9.

[25] Milojkovic D, Apperley J. Mechanisms of resistance to imatinib and second-generation tyrosine inhibitors in chronic myeloid leukemia. Clin Cancer Res 2009;15:7519−27.

[26] Soverini S, Hochhaus A, Nicolini FE, Gruber F, Lange T, Saglio G, et al. BCR-ABL kinase domain mutation analysis in chronic myeloid leukemia patients treated with tyrosine kinase inhibitors: recommendations from an expert panel on behalf of European Leukemia Net. Blood 2011;118:1208−15.

[27] Cortes JE, Kantarjian H, Shah NP, Bixby D, Mauro MJ, Flinn I, et al. Ponatinib in refractory Philadelphia chromosome−positive leukemias. N Engl J Med 2012;367:2075−88.

33

Molecular Testing in Acute Myeloid Leukemia

A. Behdad[1] and B.L. Betz[2]

[1]Division of Hematopathology, Northwestern University, Feinberg School of Medicine, Northwestern Memorial Hospital, Chicago, IL, United States [2]Department of Pathology, University of Michigan, Ann Arbor, MI, United States

INTRODUCTION

Acute myeloid leukemia (AML) is the most common type of leukemia in adults. The American Cancer Society estimates 18,860 new cases of AML in adults with an estimated 10,460 deaths from the disease in 2014 [1]. AML is a clinically heterogeneous disease, and prognosis and response to therapy is variable among patients. Early attempts to classify this cancer were based on the morphologic and immunochemical characteristics of the leukemic cells. The most widely accepted classification prior to the integration of genetic information, was the French–American–British (FAB) system, which divided AML based on the type of cell from which the leukemia cells developed and the degree of maturation. Due to this, with few exceptions, the majority of groups in the FAB classification do not show a significant prognostic difference [2].

Recent advances in our understing of the genetic basis of cancer have revolutionized the diagnosis and treatment of this disease. In AML, perhaps more than any other neoplasm, genetic findings have been incorporated into the diagnostic classification scheme. Genetic-based classification has facilitated the prognostic stratification of various AML subtypes to favorable, intermediate, and poor risk categories which is a key factor in determining the mode and intensity of therapy. This information also provides the opportunity to utilize therapy that is targeted against the underlying genetic aberration. A classic example is acute promyelocytic leukemia (APL) in which the mainstay of treatment, all-*trans* retinoic acid (ATRA), specifically and effectively targets the underlying pathogenic aberration, the *PML–RARA* fusion gene product.

AML is the result of somatic genetic alterations in hematopoietic progenitor cells that affect normal mechanisms of proliferation, self-renewal, and differentiation [3]. In rare instances, AML can arise in the background of an inherited mutation. Examples of inherited genes that have been linked to AML include: *CEBPA, SRP72, DDX41, RUNX1,* and *GATA2.* The process of leukemogenesis is believed to be multistep as evidenced by the fact that introduction of single-gene mutations is insufficient to induce AML in animal models. Through study of families with inherited germline mutations [4,5], Gilliland and Griffin proposed a two-hit model for AML pathogenesis that divides the genetic aberrations involved in AML into two classes [6]. Class 1 mutations comprise those that activate signaling pathways to promote proliferation and survival of hematopoietic progenitors. In contrast, class 2 mutations affect transcription factors that impair hematopoietic differentiation. Examples of class 1 mutated genes include receptor tyrosine kinases (*FLT3, KIT*) and downstream signaling genes (*NRAS, KRAS*), while class 2 mutations include gene fusions such as t(8;21) *RUNX1/RUNX1T1* and inv(16) *CBFB/MYH11* and gene mutations in *NPM1, CEBPA,* and *RUNX1* [7]. Whereas class 1 mutations are typically later events, the class 2 mutations occur early in leukemogenesis, are stable during the disease course, and consequently are believed to be initiating (founder) mutations. Furthermore, class 2 mutations usually do not coexist and the presence of each is frequently associated with specific clinicopathological features, suggesting that each defines a distinct entity [8]. These observations provided rationale for AML classification under the World Health Organization (WHO) scheme (Table 33.1).

© 2017 Elsevier Inc. All rights reserved.

TABLE 33.1 AML Classification Based on Recurrent Cytogenetic and Molecular Abnormalities [9]

Abnormality	Affected genes
t(8;21)(q22;q22)	*RUNX1−RUNX1T1*
inv(16)(p13.1q22) or t(16;16)(p13.1;q22)	*CBFB−MYH11*
t(15;17)(q22;q12)	*PML−RARA*
t(9;11)(p22;q23)	*MLLT3-KMT2A(MLL)*
t(6;9)(p23;q34)	*DEK−NUP214*
inv(3)(q21q26.2) or t(3;3)(q21;q26.2)	*GATA2, MECOM(EVI1)*
t(1;22)(p13;q13)	*RBM15−MKL1*
AML with *NPM1* mutation	*NPM1*
AML with biallelic *CEBPA* mutations	*CEBPA*

DIAGNOSTIC WORKUP OF AML

The first step in diagnosis of AML is morphologic evaluation of the neoplasm, which establishes the presence of leukemia. Diagnosis in most cases requires the presence of at least 20% myeloid blasts and monocytic progenitors. The exception to this rule is AML with cytogenetic abnormalities t(8;21), inv(16)/t(16;16), and t(15;17), which can be rendered based solely on the presence of the cytogenetic abnormality. Certain morphologic characteristics of the leukemic cells (such as presence of Auer rods) can be diagnostic of a myeloid lineage. However, immunophenotypic analysis of the leukemic cells is generally needed to distinguish AML from lymphoblastic leukemia. This is often achieved using flow cytometry, but other ancillary techniques such as cytochemistry and/or immunohistochemistry may also be utilized. After the diagnosis of AML is established, further subclassification requires cytogenetic and/or molecular diagnostic information. From a technical standpoint, the presence of at least 20% neoplastic cells in a diagnostic specimen is beneficial since it facilitates the use of common cytogenetic and molecular diagnostic assays that have a relatively low analytical sensitivity (such as karyotyping, fluorescence in situ hybridization (FISH), and Sanger sequencing).

Suitable diagnostic specimens include blood and bone marrow aspirate. When the diagnosis of AML is suspected and particularly if a bone marrow biopsy is performed, various dedicated specimens are sent to the hematology, flow cytometry, cytogenetics, and molecular diagnostic laboratories. The sequence in which various tests are performed can vary among laboratories. Flow cytometry, immunophenotyping, and cytogenetic analysis are typically performed on all cases. In current practice, molecular diagnostic tests for gene mutations at the time of diagnosis are typically indicated only in

AML cases without recurrent cytogenetic abnormalities. An exception is *KIT* mutation testing in core-binding factor AML (CBF-AML). While the significance of cytogenetic and many molecular assays (eg, mutation testing for *FLT3, NPM1, CEBPA, KIT*) is primarily relevant for postremission therapeutic decisions, testing is typically performed on diagnostic samples when the burden of disease is usually sufficiently high to allow reliable detection of the relevant genetic aberrations.

CYTOGENETIC ABNORMALITIES IN AML

Cytogenetic karyotyping remains the most powerful prognostic factor in AML and provides the framework for disease classification and approach to treatment (Table 33.2). The WHO classification recognizes several diagnostic entities that are based on cytogenetic abnormalities [9]. Here we discuss these chromosomal abnormalities as they are currently relevant to the clinical management of patients with AML.

Core-Binding Factor AML

CBF is a transcription factor complex that plays a key role in hematopoiesis [14]. The CBF complex includes

TABLE 33.2 AML Risk Categories Based on Established Cytogenetic and Molecular Abnormalities [10−13]

Risk category	Cytogenetics	Gene mutations
Favorable	t(15;17)	Normal cytogenetics without *FLT3*-ITD and with either *NPM1* or biallelic *CEBPA* mutations
	t(8;21)	
	inv(16)/t(16;16)	
Intermediate	t(9;11)	t(8;21) or inv(16)/t(16;16) with *KIT* mutation
	Normal cytogenetics	
	+8 alone	
	Other karyotype	
Poor	Complex cytogenetics (≥3 abnormalities)	Normal cytogenetics with *FLT3*-ITD mutation
	inv(3)/t(3;3)	
	t(6;9)	
	11q23 abnormalities other than t(9;11)	
	t(9;22)	
	−5, del(5q)	
	−7, del(7q)	

two subunits: RUNX1 (AML1, CBFA2) and CBFB. Disruption of CBF can lead to various hematopoietic neoplasms such as myelodysplastic syndrome (MDS), acute lymphoblastic leukemia (ALL), and AML [15]. CBF-AML usually results from translocations that disrupt CBF function and accounts for the most common cytogenetic subtype of AML. Two chromosomal abnormalities are associated with CBF-AML: (1) translocation (8;21) that creates a RUNX1−RUNX1T1 (AML1−ETO) gene fusion and (2) inversion or translocation within chromosome 16 [inv(16) or t(16;16)] that results in the CBFB−MYH11 fusion gene. AML with either t(8;21) or inv(16)/t(16;16) is associated with excellent prognosis and better response to chemotherapy [16].

Conventional cytogenetic analysis may not detect a small subset of AML cases with t(8;21) or inv(16)/t(16;16) [17]. In these instances, FISH or reverse transcriptase-polymerase chain reaction (RT-PCR) can be utilized to evaluate specimens for these abnormalities when CBF-AML is suspected by morphology. This is particularly important in AML with inv(16)/t(16;16) as the karyotypic finding may be subtle.

CBF-AMLs can be accompanied by other chromosomal abnormalities or gene mutations. Several animal studies have demonstrated that although essential, RUNX1−RUNX1T1 and CBFB−MYH11 are not adequate for leukemogenesis and require cooperative genetic/epigenetic events [18]. Secondary genetic mutations affecting KIT, FLT3, or RAS can be frequently seen in CBF-AML.

Acute Promyelocytic Leukemia

APL is a morphologically and genetically unique subtype that accounts for approximately 12% of AML. Before the utilization of cytogenetic information for classification, APL was clinically suspected by bleeding diathesis and diagnosed by its unique cytomorphologic features (Fig. 33.1A). The characteristic t(15;17) chromosomal rearrangement was first observed in the mid-1970s, followed later by the identification of the specific chromosomal breakpoints and involved genes in the 1990s (Fig. 33.1B) [19−21].

Translocation (15;17) creates a PML−RARA fusion gene that halts myeloid progenitor differentiation and leads to expansion of neoplastic promyelocytes [22]. The RARA (retinoic acid receptor alpha) and PML (promyelocytic leukemia) gene products both play roles in normal hematopoiesis [23,24]. The chromosome 17 breakpoints associated with t(15;17) always occur within intron 2 of the RARA gene. By contrast, three distinct chromosome 15 breakpoints are involved, all occurring within the PML gene: intron 6 (bcr1; 55% of cases), exon 6 (bcr2; 5%), and intron 3

(bcr3; 40%). As a consequence, there are three possible PML−RARA isoforms, referred to as long (L, or bcr1), variant (V, or bcr2), and short (S, or bcr3). The variant (bcr2) isoform is so-called because of the variable breakpoint positions in exon 6 leading to variably sized RT-PCR amplification products. A minor subset of APL cases harbor variant translocations that fuse RARA to alternative partners including NPM1, NUMA, FIP1L1, BCOR, ZBTB16 (PLZF), PRKAR1A, and STAT5B [10,25]. Cases with these variant translocations are designated as AML with a variant RARA translocation in the WHO classification [9].

There are several clinical features of APL that set it apart from other AMLs and highlight the importance of rapid and accurate diagnosis. Although a highly curable disease with treatment, most APL mortalities occur during the first few days following initial diagnosis, predominantly due to coagulopathy. In order to reduce early mortalities, APL-specific treatment including ATRA must be initiated as early as possible. ATRA, often in combination with a second drug, is able to induce complete remission in almost all APL patients. The discovery and utilization of ATRA for APL was a historic advance in cancer treatment as it was one of the first examples of a highly effective targeted therapy [26,27]. Although APL is frequently suspected based on the clinical presentation, as well as morphologic and flow cytometric findings, definitive diagnosis requires detection of the PML−RARA fusion. Traditional karyotyping cannot offer a sufficiently short turnaround time for optimal management of these patients. In addition, karyotyping may fail to detect the classic reciprocal t(15;17) at diagnosis in a sizable percentage of APL cases with PML−RARA [25]. Many of these harbor PML−RARA fusion through other mechanisms including cryptic microinsertions or complex translocations. However, FISH and RT-PCR will detect the fusion in most of these instances (Fig. 33.1C and D). Rare cases with microinsertions may test negative by FISH using standard probes, but these are usually detectable by RT-PCR [28,29]. These findings highlight the importance of testing via multiple methods in suspicious cases. Molecular assays such as FISH or RT-PCR can be performed on an urgent basis to help confirm or rule out this diagnosis when needed. Notably, some AMLs with RARA variant translocations (PLZF−RARA and STAT5B−RARA) have been associated with poor outcomes and resistance to ATRA [30,31].

Residual disease monitoring for early relapse is important in the posttherapy setting for APL. RT-PCR represents the most sensitive means to detect PML−RARA, and achieving molecular remission with a negative RT-PCR test is a treatment milestone

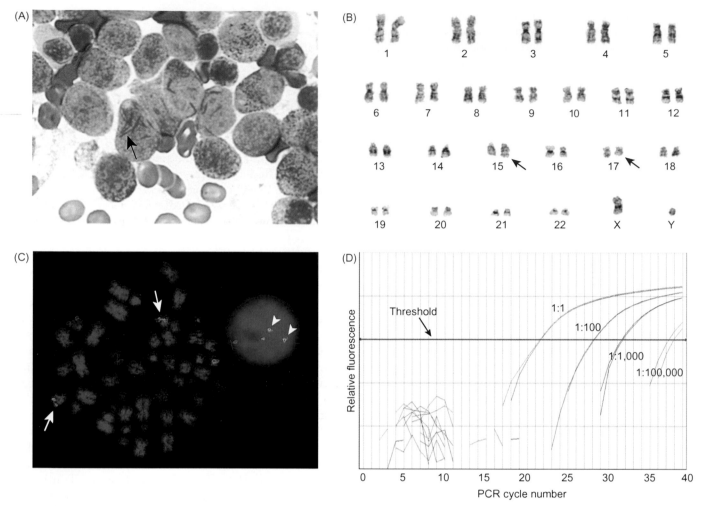

FIGURE 33.1 Diagnostic testing in AML. (A) The presence of leukemic blasts in blood or bone marrow aspirate is typically the first step in establishing a diagnosis of AML. The presence of certain morphologic features is characteristic of some AML subtypes, such as Auer rods in APL (shown with arrow). However, further subclassification of the AML usually requires other ancillary techniques. (B) Cytogenetic analysis provides a framework for both classification and disease prognosis. Recurrent genetic alterations are detected by karyotyping in approximately 45% of AML cases. APL is characterized by fusion of the *PML* and *RARA* genes that is usually the result of a reciprocal translocation between chromosomes 15 and 17 (arrows). Some gene fusions occur as cryptic submicroscopic insertions that are undetectable by karyotype. Identifying these requires FISH or RT-PCR. (C) Dual-color dual-fusion FISH is useful to detect gene fusions when specific translocation partners are consistently involved, such as *PML−RARA*. This assay utilizes a red FISH probe specific to the *PML* gene locus and a green probe specific to the *RARA* gene, each spanning the respective breakpoint regions. Colocalization of the red and green probes is observed when a *PML−RARA* gene fusion is present. FISH can be performed on metaphase chromosomes or interphase nuclei as shown (arrows on the metaphase spread and arrowheads on the interphase nucleus indicate the fused probes). The two sets of colocalized probes within a metaphase spread or interphase nucleus indicate both rearranged chromosomes [der(15) and der(17)] resulting from the reciprocal translocation; the functional *PML−RARA* fusion is on der(15). The isolated red and green signals indicate the remaining normal *PML* and *RARA* alleles, respectively. (D) Real-time RT-PCR achieves a high analytic sensitivity and therefore is useful for posttreatment disease monitoring for detection of early relapse. There are three alternate breakpoint regions within the *PML* gene (intron 6, exon 6, and intron 3) that fuse to a conserved breakpoint region within *RARA* intron 2. Depending on which *PML* breakpoint is used, the resulting *PML−RARA* fusion transcript is referred to as the long (bcr1), variant (bcr2), or short (bcr3) subtype. Real-time PCR assays are designed to detect each fusion transcript type such that amplification products are detected during each PCR cycle using a fluorescent probe specific to the PCR product. The accumulated fluorescence is plotted against the number of PCR cycles. For a given specimen, the PCR cycle number is measured when the increase in fluorescence is exponential and exceeds a threshold. This point is called the quantification or threshold cycle (C_t) and is inversely proportional to the quantity of fusion transcript (a high C_t corresponds to a low level of fusion transcript). Shown is a real-time RT-PCR plot for the *PML−RARA* bcr1 transcript, demonstrating a limit of detection down to 1 leukemic cell in a background of 100,000 normal cells.

following consolidation therapy [11]. Patients with detectable *PML–RARA* in two consecutive tests, with one being in the bone marrow, will require therapeutic intervention for relapse. In contrast, consecutive negative RT-PCR results are associated with remission, long-term survival, and possible cure.

Other Cytogenetic Abnormalities

Several other recurrent cytogenetic abnormalities are included in the WHO classification [9]. These include *KMT2A(MLL)* rearrangements, t(6;9)(p23;q34), inv(3)(q21q26.2), t(3;3)(q21;q26.2), and t(1;22)(p13;q13). Over 60 *KMT2A(MLL)* translocation partner genes have been identified which are in general associated with aggressive leukemia and poor prognosis. The exception is the most common *KMT2A(MLL)* rearrangement t(9;11)/ MLLT3-*KMT2A(MLL)*, which is associated with intermediate risk disease [9,32–34]. Not all *KMT2A(MLL)* translocations are detected by conventional karyotyping, in which cases FISH analysis is useful [35]. AML with t(6;9) and inv(3)/t(3;3) are both associated with poor prognosis. AML with inv(3)/t(3;3) may occur de novo, or secondary to prior MDS. AML with t(1;22) is very rare, occurring predominantly in children less than 3 years old and is associated with acute megakaryoblastic leukemia.

GENE MUTATIONS IN AML

Approximately half of AML cases lack chromosomal abnormalities; these are grouped under cytogenetically normal AML (CN-AML) [36]. This is a heterogeneous category, which can be further stratified based on the presence of nucleotide-level gene mutations. Three genes, *NPM1*, *FLT3*-ITD, and *CEBPA*, have established clinical utility and testing is now standard of care in patients with CN-AML. *KIT* mutation testing is also relevant in CBF-AML. The presence or absence of mutations in each of these four genes has established prognostic significance and plays an essential role in guiding postremission therapy (Table 33.2). The latest WHO classification now includes AML with *NPM1* or *CEBPA* as provisional entities, highlighting the pathogenic and clinical importance of these mutations (Table 33.1) [9].

The list of mutations in AML is rapidly expanding, particularly with the increasing utilization of massively parallel sequencing to evaluate large gene panels. Many of these additional mutations are believed to play a fundamental role in AML pathogenesis, including recurrent mutations in several genes involved in epigenetic regulation of transcription. To translate this plethora of new data into clinically actionable information remains a challenge, a feat even more difficult given the heterogeneity of this disease. However, this is an active area of study, and the expectation is that the list of clinically relevant mutations will increase with further refinement of prognostic stratification and approaches to therapy.

To facilitate reader's review of this section we divide gene mutations in AML into two categories: (1) those that are well established and have known clinical utility and (2) those that are under investigation and may gain clinical relevance in the near future (Table 33.3).

Gene Mutations with Well-Established Clinical Utility

NPM1 mutations. Mutations in the nucleophosmin (*NPM1*) gene are the most common genetic abnormality in AML, occurring in approximately 30% of all AMLs and in 50% of CN-AML [37,38]. NPM1 is a nucleolar phosphoprotein that serves as a shuttle between the nucleus and the cytoplasm and regulates the transport of preribosomal particles through the nuclear membrane [39,40]. Mutations in *NPM1* were first discovered in AML following the observation of abnormal cytoplasmic localization of the mutant NPM1 protein, rather than that its normal nuclear localization [41]. Genetic evaluation of leukemic blasts with cytoplasmic NPM1 led to the discovery of a variety of frameshift mutations clustering within exon 12 of *NPM1* gene. As depicted in Fig. 33.2A, virtually all of these mutations result in a net insertion of four nucleotides that lead to a consequent shift in the translational reading frame. The functional result is loss of a nucleolar localization signal and generation of a novel nuclear export signal, both of which contribute to the aberrant cytoplasmic localization of the protein.

Clinically, *NPM1* mutation in the absence of *FLT3*-ITD mutation is associated with better overall survival, event-free survival, and response to treatment [37,38,43]. Patients with this genotype are classified as favorable risk and are therefore not typically candidates for allogeneic stem cell transplantation [11]. Of note, concurrent testing for both *NPM1* and *FLT3*-ITD is indicated in cases with CN-AML since *NPM1* is prognostically favorable only in the absence of *FLT3*-ITD.

Sensitive and reliable detection of *NPM1* mutations can be achieved with a simple PCR fragment-sizing assay that detects the larger amplification products that result from the 4 bp insertion [44]. The affected region within exon 12 is amplified by PCR and the amplification products are resolved and analyzed by capillary electrophoresis (Fig. 33.2B and C). This approach is advantageous, as it will detect all reported

TABLE 33.3   Recurrent Gene Mutations in AML

Name	Physiologic function	Frequency in CN-AML	Prognosis
MUTATIONS IN CURRENT CLINICAL PRACTICE			
NPM1	Nuclear-cytoplasmic shuttling phosphoprotein	50%	Favorable
FLT3-ITD	Receptor tyrosine kinase	25–30%	Unfavorable
CEBPA	Transcription factor	10–15%	Favorable
KIT	Receptor tyrosine kinase	30% of CBF-AML	Unfavorable
MUTATIONS UNDER INVESTIGATION			
FLT3-TK	Receptor tyrosine kinase	5–10%	Inconclusive
RUNX1	Transcription factor	10–15%	Unfavorable
IDH1/ IDH2	Epigenetic modifier	15–30%	Inconclusive
DNMT3A	Epigenetic modifier	20–30%	Inconclusive
TET2	Epigenetic modifier	10%	Inconclusive
ASXL1	Chromatin modifier	5–10%	Unfavorable
KMT2A (MLL)-PTD	Epigenetic modifier	5–10%	Inconclusive
WT1	Transcription factor	10–15%	Inconclusive
TP53	Cell cycle regulator	2–5%	Unfavorable
RAS	Membrane-associated signaling	10%	Neutral
PHF6	Chromatin modifier	<5%	Unfavorable

ITD, internal tandem duplication; TK, tyrosine kinase; PTD, partial tandem duplication.

NPM1 mutations, of which greater than 50 variants have been reported. Used in this way, capillary fragment sizing can achieve an analytic sensitivity down to approximately 2% mutant allele, which is easily sufficient for diagnostic specimens that will typically contain greater than 20% leukemic blasts. Exquisite sensitivity down to 10^{-5} can be achieved with real-time PCR assays using mutation-specific primers [45,46]. However, this technique will only detect the specific *NPM1* mutations targeted by the assay and thus is better suited for posttherapy minimal residual disease monitoring.

FLT3 mutations. Mutations in FMS-like tyrosine kinase 3 (*FLT3*) are the second most common mutation in AML, occurring in approximately 25–30% of all patients [47]. FLT3 is a receptor tyrosine kinase that is involved in regulating proliferation of hematopoietic progenitor cells. Two classes of activating *FLT3* mutations occur in AML: (1) internal tandem duplication (*FLT3*-ITD) which occur in 20–25% of patients and

(2) tyrosine kinase domain mutations (*FLT3*-TKD) which are seen in 5–10% of patients [48]. Both classes of mutations lead to ligand-independent constitutive activation of the FLT3 receptor. ITD mutations are the result of duplicating insertions that occur within exon 14 or 15 of the *FLT3* gene [49]. These are always in-frame and vary in length from 3 to several hundred base pairs [50]. *FLT3*-TKD mutations are typically missense mutations that affect codon 835 aspartic acid (D835) within exon 20, and less frequently deletions of amino acid I836 [51].

The prognostic significance of *FLT3*-ITD mutations is well established. The presence of this mutation is consistently associated with inferior outcomes in CN-AML [47,52,53]. *FLT3*-ITD also appears to have a negative prognostic impact when occurring concurrently with *NPM1* or *CEBPA* mutations, which in isolation are each associated with favorable prognosis [12]. From a therapeutic standpoint, testing for *FLT3*-ITD mutations is important for two reasons. Firstly, AML patients with *FLT3*-ITD may benefit from hematopoietic stem cell transplantation [12]. Additionally, there are several clinical trials currently exploring the utility of nonspecific tyrosine kinase inhibitors (TKIs) as well as FLT3-specific TKIs in *FLT3*-mutant AMLs [54]. In contrast to *FLT3*-ITD, the prognostic impact of *FLT3*-TKD mutation is less clear due to conflicting clinical data [51,55,56]. Consequently, *FLT3*-TKD testing is not included in current National Comprehensive Cancer Network (NCCN) guidelines [11].

As with *NPM1* testing, detection of the variably sized *FLT3*-ITD mutations is efficiently performed with PCR fragment-sizing assays that detect larger PCR products indicative of duplication (Fig. 33.3). This approach can reliably detect ITDs of all sizes. Other techniques such as next-generation sequencing (NGS) based assays may have difficulty in detecting larger duplications. Concurrent detection of *FLT3*-TKD mutations can be accomplished in multiplex with the addition of a PCR amplicon within exon 20 that is evaluated for resistance to digestion by EcoRV at the D835 mutation site [57]. The wild-type D835 sequence contains an EcoRV restriction site that is eliminated in the presence of a mutation, leading to a larger amplicon that can be detected on the capillary electropherogram.

Despite being present early in leukemogenesis, *FLT3*-ITD mutations may be lost at relapse or vice versa [58,59]. These observations suggest that ITD mutations may be a secondary event, either present in a subclone of leukemic cells at diagnosis or acquired during disease progression. For this reason, and due to the limited analytic sensitivity of available tests, *FLT3*-ITD mutation is not an ideal minimal residual disease marker in follow-up testing. A caveat in test interpretation in diagnostic specimens relates to the relative level

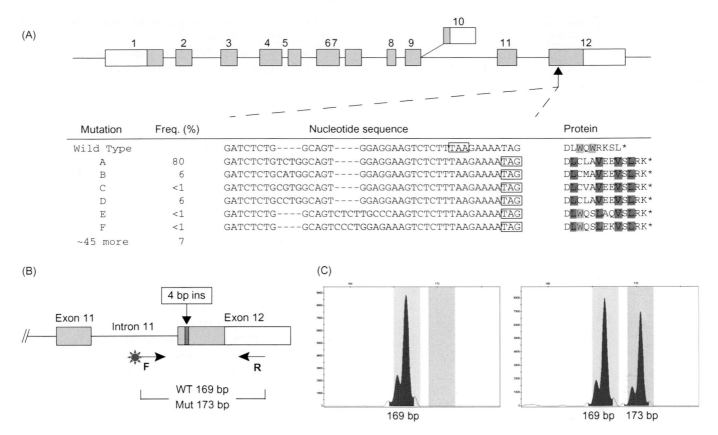

FIGURE 33.2 Molecular testing for *NPM1* mutation. (A) The *NPM1* gene consists of 12 exons which encode three alternatively spliced transcripts, the most prevalent of which excludes exon 10. Greater than 50 different mutations have been described in AML, however virtually all lead to a net insertion of 4 bp within exon 12. The three most common mutations (types A, B, and D) constitute approximately 90–95% of mutations. Wild-type NPM1 contains two tryptophan residues (W, shaded pink) that are important for its normal nucleolar localization. All of the mutations lead to a frameshift that disrupt one or both W residues while also generating a novel leucine-rich nuclear export signal (shaded green). Together, these mutations disrupt wild-type NPM1 function through aberrant localization of mutant NPM1 to the cytoplasm. (B) Detection of *NPM1* mutations can be accomplished with a simple PCR fragment-sizing assay to detect the larger amplification products resulting from the insertion mutation. Forward and reverse primers are designed to flank the mutation region, generating a PCR amplicon of known size (169 bp in this example). One primer is fluorescently labeled to permit detection and sizing via capillary electrophoresis. (C) Capillary electropherograms demonstrating mutation negative (left panel) and positive (right panel) case examples. A positive result is indicated by the presence of a mutant amplicon peak that is usually 4 bp larger than the wild-type peak. Of note, a track of repetitive T nucleotides within intron 11 causes the PCR polymerase to stutter during amplification which is observed as shoulder peaks on the electropherogram. Furthermore, the poly-T tract is also polymorphic within the normal population, ranging from 12 to 14 nucleotides in length. This variation does not affect the sensitivity or specificity for detection of the 4 bp insertion mutation but must be taken into account when interpreting the assay. *Panel A is adapted from published Refs. [41] Falini B, Mecucci C, Tiacci E, Alcalay M, Rosati R, Pasqualucci L, et al. Cytoplasmic nucleophosmin in acute myelogenous leukemia with a normal karyotype. N Engl J Med 2005;352:254–66; [42] Thiede C, Koch S, Creutzig E, Steudel C, Illmer T, Schaich M, et al. Prevalence and prognostic impact of NPM1 mutations in 1485 adult patients with acute myeloid leukemia (AML). Blood 2006;107:4011–20.*

of *FLT3*-ITD mutation. Several studies have suggested that very high-level (biallelic) *FLT3*-ITD mutations are associated with even worse outcomes when compared to lower level mutations [47,52]. Although fragment analysis assays are not typically utilized in a quantitative manner, diagnosis of a high-level *FLT3*-ITD mutation can be rendered if the amplitude of the mutant peak is higher than the wild-type peak (Fig. 33.3D).

CEBPA mutations. CCAAT/enhancer-binding protein alpha (CEBPA) is a member of the basic region leucine zipper (bZIP) family of transcription factors and plays an important role in the differentiation of myeloid progenitors [60]. Mutations of *CEBPA* occur

in 8–15% of AML and are most commonly seen CN-AML [12,61,62]. AML can carry one or two *CEBPA* mutations. The majority of cases with two mutations have a combination of an N-terminal frameshift mutation on one allele and an in-frame C-terminal mutation on the other (Fig. 33.4A). This pattern of mutations results in loss of function of the differentiation promoting p42 isoform while allowing expression of a smaller pro-proliferative p30 isoform [61]. Numerous studies have demonstrated that only double-mutated *CEBPA* is associated with favorable prognosis in CN-AML [63–65]. From a practical standpoint, testing can be limited to CN-AML that lacks *NPM1* and *FLT3*-ITD

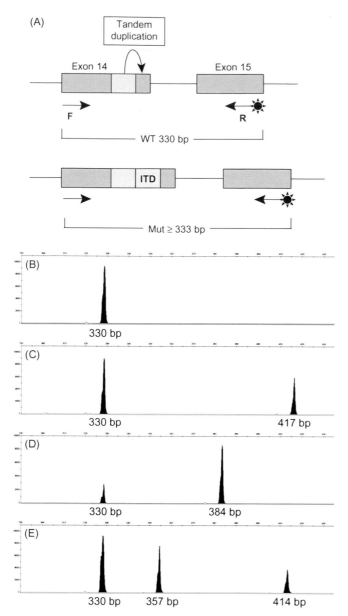

FIGURE 33.3 Molecular testing for *FLT3*-ITD mutation. (A) ITD mutations result from the duplication and tandem insertion of a variably sized fragment of the *FLT3* gene. The ITDs range in size from 3 to several hundred base pairs and most commonly occur within exon 14, but also affect exon 15. Despite the variety, *FLT3*-ITDs are always in-frame since they are gain-of-function mutations that lead to constitutive activation of the FLT3 kinase. Detection of *FLT3*-ITDs is frequently accomplished with a PCR fragment-sizing assay that utilizes primers that flank the mutation region to detect the larger mutant fragments. Fluorescent labeling of one primer permits detection and sizing via capillary electrophoresis. (B) Capillary electropherogram demonstrating a negative result as indicated by the presence of only the 330 bp wild-type amplicon. (C) *FLT3*-ITD positive result showing an additional amplicon of 417 bp in size, consistent with the presence of an ITD mutation of 87 bp. (D) High-level (biallelic) ITD mutation is associated with even worse prognosis and is indicated by a mutant peak of higher amplitude than the WT peak. (E) Multiple ITD mutations are detected in approximately 15% of ITD positive cases. The presence of multiple ITD peaks at different allelic levels exemplifies the clonal heterogeneity that can be observed with these mutations.

mutation since *CEBPA* is prognostically relevant only in this subgroup of cases.

Unlike *FLT3* and *NPM1* in which similar types of mutations cluster within specific gene regions, *CEBPA* mutations are highly variable and span the entire coding region. Detection of these mutations therefore requires a technology capable of interrogating the entire coding region for a wide range of nucleotide alterations. As most *CEBPA* mutations are length-affecting insertions and deletions, fragment-sizing assays have been utilized [67]. But as the specific type of nucleotide alteration has clinical relevance, cases with a positive PCR result require further characterization by a sequencing-based assay. Further, the presence of benign length-affecting polymorphisms and cases with only substitution mutations present confounding issues for fragment analysis based approaches. Although more laborious, the utilization of Sanger sequencing assays with guidance on mutation interpretation currently represents the preferred method for *CEBPA* testing (Fig. 33.4B) [66]. Sequencing of *CEBPA* as part of a large NGS-based panel can be difficult due to the high GC content of the gene. Additional reagents such as DMSO are frequently employed in Sanger-based assays to overcome this challenge. Sanger sequencing inherently suffers from a lower analytic sensitivity ($\sim$10% mutation), but this limitation is diminished since diagnostic specimens typically contain at least 20% of leukemic blasts.

KIT mutations. Activating mutations in the *KIT* receptor tyrosine kinase are common in CBF-AML (AML with t(8;21) or inv(16)/t(16;16)) occurring with a frequency of approximately 30% in these subtypes [68–70]. *KIT* mutations predominantly occur as small length-affecting mutations in exon 8 that affect the extracellular domain, and point mutations in exon 17 at D816 and N822 that affect the tyrosine kinase domain. Sanger sequencing of *KIT* exons 8 and 17 remains the most widely used method for mutation testing given the variety of mutations (Fig. 33.5).

CBF-AML is generally classified as favorable risk with a lower rate of relapse. However, several studies have shown that *KIT* mutations in CBF-AML, particularly D816 mutations in adult AML with t(8;21), are associated with a higher risk of relapse [68,71–73]. Current NCCN guidelines have defined both t(8;21) and inv(16)/t(16;16) AMLs with *KIT* mutations as intermediate risk [11]. However, it should be noted that the prognostic impact of *KIT* mutations in AML with inv(16) or t(16;16) and in pediatric CBF leukemia is less clear than in adult AML with t(8;21). In addition, it is not clear if less frequent *KIT* mutations have the same prognostic significance as D816 mutations.

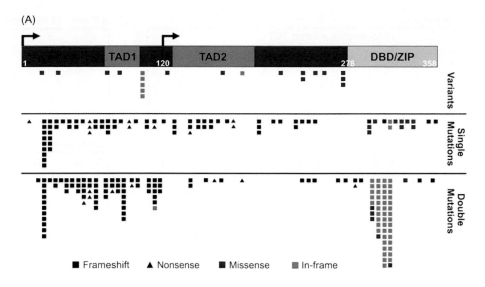

(A)

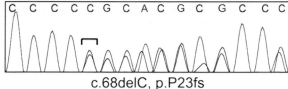

(B) N-Terminal Frameshift Mutation

c.68delC, p.P23fs

C-Terminal In-frame Mutation

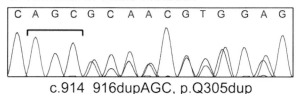

c.914_916dupAGC, p.Q305dup

FIGURE 33.4 Molecular testing for *CEBPA* mutations. (A) Distribution of *CEBPA* mutations. The locations of mutations and variants are shown with respect to the CEBPA protein and functional regions. Arrows depict the two translation initiation sites at amino acids (aa) 1 and 120. For interpretative purposes, the coding region of *CEBPA* is divided into three regions: N-terminal (aa 1–120), mid-region (aa 121–277), and C-terminal (aa 278–358). The majority of *CEBPA* double-mutation positive cases harbor a combination of a truncating frameshift or nonsense mutation in the N-terminal region and an in-frame insertion/deletion or missense mutation in the C-terminal region (lower panel). Mutations in *CEBPA* single-mutant cases are distributed in the entire coding region with a greater proportion in the mid-region (mid panel). Missense and in-frame mutations of the N-terminal and mid-regions are typically classified as variants of unknown significance (upper panel). TAD1, transactivation domain 1 (aa 70–97); TAD2, transactivation domain 2 (aa 127–200); DBD/ZIP, DNA binding and dimerization domain (aa 278–358). (B) Sequencing is the preferred method to detect *CEBPA* mutations since they are highly variable and occur throughout the coding region. Depicted in the figure are Sanger sequence examples of an N-terminal frameshift mutation and a C-terminal in-frame mutation. Brackets in each panel indicate the deleted and duplicated nucleotide sequences, respectively. *Panels A and B are reprinted from Ref. [66] Behdad A, Weigelin HC, Elenitoba-Johnson KS, Betz BL. A clinical grade sequencing-based assay for CEBPA mutation testing: report of a large series of myeloid neoplasms. J Mol Diagn 2015;17:76–84.*

Genes Under Investigation in AML

Recent progress in genomic sequencing has led to the recognition of an increasing number of gene mutations in AML. Several of these are involved in epigenetic regulation of transcription, which may constitute a third class of AML mutations [74]. In this section, we discuss gene mutations that are not incorporated in current testing guidelines, but are increasingly gaining clinical relevance and are often incorporated into multigene testing panels.

RUNX1. The Runt-related transcription factor 1 (*RUNX1*) plays a critical role in hematopoietic differentiation as it is required for hematopoiesis [75].

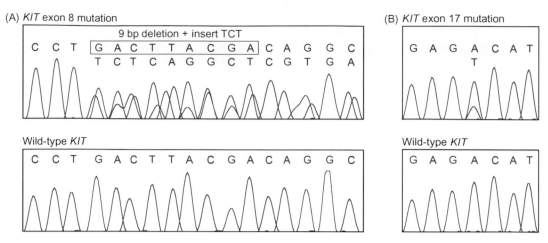

FIGURE 33.5 Molecular testing for *KIT* mutations. Activating *KIT* gene mutations occur in CBF-AML with t(8;21) or inv(16)/ t(16;16). *KIT* mutations mostly occur in exon 8 or 17 and are commonly detected by Sanger sequencing. (A) Exon 8 mutations are predominately small in-frame insertions/deletions that affect codon 419. Depicted is a Sanger sequencing example of a c.1248_1256delinsTCT (p.T417_D419delinsL) mutation within *KIT* exon 8. This is a 9 bp deletion (GACTTACGA) with insertion of three nucleotides (TCT) that results in the in-frame dele-tion of amino acids T417-D419, substituting them with an L (Leu) amino acid. Overlapping peaks in the sequence chromatogram indicate the mutation. A wild-type trace is shown for reference. (B) Exon 17 mutations are typically substitution mutations at codon 816 or 822. Shown is a case with a c.2447A > T (p.D816V) substitution mutation. The presence of a *KIT* mutation, particularly at codon 816 in AML with t(8; 21), is associated with a higher risk of relapse in this otherwise favorable risk disease.

In addition to being involved in CBF-AML through the t(8;21) *RUNX1−RUNX1T1* chromosomal translocation, *RUNX1* is implicated in CN-AML by virtue of recur-rent gene mutations in 10−15% of cases [76−78]. *RUNX1* mutations can also be seen in MDS and AML with myelodysplasia-related changes [79,80]. In AML, *RUNX1* mutations have been proposed to be initiating events since they are generally found in the absence of recurrent gene fusions and *NPM1* and double *CEBPA* mutations [78,81]. Further, *RUNX1* mutations are asso-ciated with an adverse prognosis [78,81,82]. Given the emerging role of *RUNX1* mutation in AML pathogene-sis and prognosis, it is possible that AML with *RUNX1* mutation may be recognized as a distinct clinicopatho-logical entity in the near future. Testing for *RUNX1* mutation requires sequencing of a significant portion of the gene since mutations are highly variable (substi-tutions, insertions, deletions) and are distributed throughout most of the coding region [81].

IDH1 and *IDH2*. Mutations in isocitrate dehydroge-nase genes *IDH1* and *IDH2* were first discovered in glioma and subsequently described in AML [83,84]. *IDH1* and *IDH2* mutations are cumulatively detected in up to 15−30% of AML patients, occurring most commonly in CN-AML [85]. All *IDH* mutations described in AML are substitution mutations affecting three codons—R132 of *IDH1*, and R140 or R172 of *IDH2*. These mutations affect the active site of the IDH enzyme leading to high levels of 2-hydroxyglutarate (2HG) [83,86]. *IDH1* and *IDH2* mutations occur in a mutually exclusive fashion suggesting functional overlap.

Most studies suggest that *IDH1* and *IDH2* mutations in AML are associated with adverse outcome, unlike the favorable outcome in glioma [85,87−89]. Notably, the presence of an *IDH1* mutation in CN-AML that is *NPM1* mutant, *FLT3* wild type is associated with a worse prognosis in this otherwise favorable risk group [88]. However, a recent study of prognostic relevance of integrated genetic profiling demonstrated improved overall survival for AML with *IDH2* mutations, as well as *NPM1*-mutated AMLs with *IDH1* or *IDH2* mutation [90]. Additionally, mutant IDH is a target for selective inhibitors and clinical trials are now emerging. Many clinical laboratories currently offer targeted *IDH1* and *IDH2* testing, which can be utilized in both AML and glioma. Sequencing-based assays are typically the pre-ferred method given the diversity of different substitu-tion mutations across the three affected codons.

DNMT3A. Somatic mutations in the DNA (cytosine-5-)-methyltransferase 3 alpha (*DNMT3A*) gene in AML were first identified utilizing massively parallel DNA sequencing [91,92]. These mutations are encountered in approximately 20% of AML patients and are most com-mon in CN-AML. DNA methylation regulated by DNMT3A is an important epigenetic modification that is critical in regulating gene expression. Most studies have demonstrated that *DNMT3A* mutations are associ-ated with adverse outcome [91−94]. However, this was not confirmed in a recent large study [90]. The most common *DNMT3A* mutation is a substitution mutation at arginine residue 882 (R882), which accounts for approximately 60% of all mutations. The remaining mutations are scattered throughout much of the

DNMT3A-coding region and include both in-frame and frameshift mutations [92]. Testing requires coverage of the majority of the coding exons given the wide distribution of mutations in this large gene. This is frequently achieved as part of NGS testing panels. Targeted mutation testing is also available at some centers using high-resolution melting and/or Sanger sequencing assays that interrogate the R882 site within exon 23.

KMT2A(MLL). Chromosomal translocations involving the *KMT2A(MLL)* gene were discussed earlier in this chapter. Another type of recurrent genetic alteration involving this gene is the partial tandem duplication (PTD) which results from intragenic duplication of a genomic region between exons 5–11 or 5–12. *KMT2A(MLL)*-PTDs are found in 5–10% of patients with CN-AML [95,96]. While earlier studies demonstrated an inconsistent prognostic role for *KMT2A (MLL)*-PTD [95–97], growing evidence suggests this aberration confers an adverse prognosis, irrespective of the presence of *FLT3*-ITD mutation [90].

ASXL1. Mutations of the additional sex combs like-1 (*ASXL1*) gene have been described in various myeloid neoplasms including AML and MDS. The *ASXL1* gene belongs to the enhancer of trithorax and polycomb (ETP) family and functions as a transcriptional regulator through its chromatin-binding activity. *ASXL1* mutations are more common in older patients and are seen in 5–10% of AML cases [98,99]. Almost all reported mutations are frameshift or nonsense truncating mutations in exon 12 and have been associated with inferior outcomes in most studies [90,98,99].

WT1. Wilms tumor 1 (*WT1*) gene mutations occur in approximately 10–15% of CN-AML [100,101]. The *WT1* gene is located at chromosomal band 11p13 and encodes a potent transcription factor that can serve as either a tumor suppressor or an oncogene. Germline deletion of *WT1* was first identified in patients with WAGR syndrome characterized by Wilms tumor, aniridia, genitourinary malformations, and mental retardation. A connection between *WT1* mutation and AML was made in patients with WAGR syndrome who developed secondary AML [102]. Various studies have found a contradictive prognostic role for somatic *WT1* mutations in CN-AML [100,101,103]. The majority of *WT1* mutations are frameshift insertions and deletions in exon 7, but various mutations occur throughout other exons [103].

TET2. Mutations in the ten-eleven-translocation 2 (TET2) gene have been identified in MDS and AML. TET2 is an enzyme that converts 5-methylcytosine to 5-hydroxymethylcytosine and thus is implicated in regulating the demethylation of DNA. The *TET2* mutations identified in AML are inactivating and have been associated with increased hematopoietic stem cell renewal and myeloproliferation [104,105]. The prognostic role of *TET2* mutation remains controversial.

TP53. Mutations of the tumor-suppressor gene *TP53* have been described in a wide variety of cancers due to its important role in cell cycle regulation. The prevalence of *TP53* mutation is low in AML as a whole (2–5%) [106]. However, *TP53* mutation is strongly associated with complex karyotype AML, occurring in approximately 75% of these cases where it is associated with poor outcome [107–109].

RAS. The RAS family of oncogenes contains several well-known membrane-associated signaling proteins with key roles in regulating proliferation, differentiation, and apoptosis. Oncogenic activating *KRAS* and *NRAS* mutations are found in approximately 20% of human cancers, including approximately 10% of CN-AML. Their presence in AML has no known prognostic value [12,110,111].

Gene expression profiling. The relative expression of specific genes has also been utilized to predict prognosis in AML. Overexpression of *BAALC* (brain and acute leukemia, cytoplasmic), *MN1* (meningioma 1), *ERG* (v-ets erythroblastosis virus E26 oncogene homolog, avian), and *EVI1* (ecotropic viral integration site 1) has been associated with worse outcome in most studies [112–115]. Among these, *BAALC* is the most extensively studied and has demonstrated to be an independent negative prognostic indicator in CN-AML [113,116]. On a broader scale, genome-wide expression profiling has been investigated for AML classification for over a decade [117–119]. Despite the longstanding history of use as investigational tools, gene expression analyses have not been incorporated into clinical practice for classification, prognostication, or to guide treatment in AML.

MINIMAL RESIDUAL DISEASE TESTING

Detection and monitoring of low-level residual disease requires quantitative technology with high analytical sensitivity. Compared to flow cytometry, which is widely used for MRD detection in AML, molecular techniques have the potential to provide higher sensitivity. In particular, the stability of the class 2 mutations (recurrent gene fusions, *NPM1*, *CEBPA*) throughout the disease course makes them excellent markers for MRD testing. This has been exemplified in APL (*PML–RARA*) and CBF-AML (*RUNX1–RUNX1T1* and *CBFB–MYH11*) where serial monitoring of fusion transcripts by RT-PCR has been extensively studied as a marker of early relapse [120–122]. Molecular testing in these instances is facilitated by recurrent exon to exon fusions that allow RT-PCR assays to be designed for each fusion transcript. However, in CN-AML molecular testing for MRD monitoring has proven more difficult and molecular testing is not widely utilized for this purpose at present. A major limiting factor is the challenge associated with developing sensitive PCR assays for gene mutations that are highly variable. This is particularly true in the case of *CEBPA*, where hundreds of different mutations prevent the practical implementation of mutation-specific PCR assays. In the case of *NPM1*, several mutation-specific PCR assays can be designed to target the most common mutations, a strategy which has shown promise in clinical trials [46]. The high frequency of *FLT3* mutations in CN-AML and other subtypes make this an attractive candidate for MRD monitoring. However, clonal evolution and lack of sufficiently sensitive testing platforms prevent the use of *FLT3* mutations as reliable MRD markers.

NEW HORIZONS

Gene mutations do not occur in isolation, and a patient with de novo AML, harbors an average of 13 mutations, 5 of which are in recurrently mutated genes [123]. Defining the prognostic and predictive value of each may frequently depend on the existence of other mutations. A prime example of this paradigm is the nullified prognostic value of *NPM1* and *CEBPA* mutations in the presence of *FLT3*-ITD mutation. This emphasizes the need for comprehensive mutational analysis, which becomes cumbersome and expensive using single-gene assays given the growing list of clinically relevant genes. High-throughput sequencing (NGS), which was previously utilized only as a research tool in mutation discovery, is now being vigorously tested for clinical application. By combining comprehensive mutational analysis and cytogenetic data, a recent study by Patel et al., devised a new risk stratification scheme for AML [90]. The integrated classification in this study enabled better stratification of AML by subclassifying more patients from the heterogeneous intermediate risk group to favorable and unfavorable groups. This study provides a compelling example how the power of integrated genetic information may enable better patient care.

In addition to providing a broad platform for comprehensive genetic testing in AML, NGS has improved analytic sensitivity and holds potential for quantitative trending of variant allele frequency. The study of genetic heterogeneity and clonal evolution in AML is currently an active area of investigation, and it has become clear that most AMLs acquire additional mutations during the disease course [124,125]. Detecting and trending therapy-resistant subclones may gain importance in improving long-term outcomes [126].

The role of microRNAs in AML pathogenesis is another active area. MicroRNAs are small noncoding RNAs that regulate posttranscriptional control of gene expression and play a central role in many cellular processes including normal hematopoiesis [127]. Aberrant expression of multiple microRNAs has recently been reported in AML, and microRNA-expression signatures have been shown to distinguish AML subtypes and provide prognostic value [128,129].

AML diagnosis is increasingly based on the underlying genetic characteristics. While cytogenetic analysis still comprises the backbone of AML classification, evaluation of several gene mutations is now standard of care for diagnosis, prognostic stratification, and differentially tailored treatment strategies. Genomic technologies have rapidly increased our understanding of the molecular pathogenesis of AML, and this new information is being actively evaluated in comprehensive mutation panels with promise to further improve patient outcomes. Expanded targeted sequencing tests are already implemented in some centers and this will soon become commonplace, as interpretative algorithms, sequencing infrastructure, and bioinformatics pipelines are further refined. Looking toward the near future, molecular laboratories will be relied upon to play an even more expanded role in AML care with the increasing application of molecular testing in disease monitoring, emerging technologies involving microRNAs, and the adoption of new targeted therapies.

References

[1] American Cancer Society. Leukemia: acute myeloid (myelogenous) overview. 2014 [cited 3/31/2015]. Available from: www.cancer.org

[2] Haferlach T, Schoch C, Loffler H, Gassmann W, Kern W, Schnittger S, et al. Morphologic dysplasia in de novo acute myeloid leukemia (AML) is related to unfavorable cytogenetics but has no independent prognostic relevance under the conditions of intensive induction therapy: results of a multiparameter analysis from the German AML Cooperative Group studies. J Clin Oncol 2003;21:256–65.

[3] Frohling S, Scholl C, Gilliland DG, Levine RL. Genetics of myeloid malignancies: pathogenetic and clinical implications. J Clin Oncol 2005;23(26):6285–95.

[4] Pabst T, Eyholzer M, Haefliger S, Schardt J, Mueller BU. Somatic CEBPA mutations are a frequent second event in families with germline CEBPA mutations and familial acute myeloid leukemia. J Clin Oncol 2008;26:5088–93.

[5] Preudhomme C, Renneville A, Bourdon V, Philippe N, Roche-Lestienne C, Boissel N, et al. High frequency of RUNX1 biallelic alteration in acute myeloid leukemia secondary to familial platelet disorder. Blood 2009;113:5583–7.

[6] Gilliland DG, Griffin JD. The roles of FLT3 in hematopoiesis and leukemia. Blood 2002;100:1532–42.

[7] Takahashi S. Current findings for recurring mutations in acute myeloid leukemia. J Hematol Oncol 2011;4:36.

[8] Betz BL, Hess JL. Acute myeloid leukemia diagnosis in the 21st century. Arch Pathol Lab Med 2010;134:1427–33.

[9] Swerdlow SHN, Jaffe E, Pileri S, Stein H, Thiele J, Vardiman J, editors. WHO classification of tumours of the haematopoietic and lymphoid tissues. 4 ed. Lyon, France: IARC; 2008.

[10] Grimwade D, Mrozek K. Diagnostic and prognostic value of cytogenetics in acute myeloid leukemia. Hematol Oncol Clin North Am 2011;25:1135–61.

[11] Acute myeloid leukemia. NCCN clinical practice guidelines in oncology (NCCN Guidelines®). Version 1.2015 ed. National Comprehensive Cancer Network; 2015.

[12] Schlenk RF, Dohner K, Krauter J, Frohling S, Corbacioglu A, Bullinger L, et al. Mutations and treatment outcome in cytogenetically normal acute myeloid leukemia. N Engl J Med 2008;358:1909–18.

[13] Mrozek K, Marcucci G, Nicolet D, Maharry KS, Becker H, Whitman SP, et al. Prognostic significance of the European LeukemiaNet standardized system for reporting cytogenetic and molecular alterations in adults with acute myeloid leukemia. J Clin Oncol 2012;30:4515–23.

[14] Goyama S, Mulloy JC. Molecular pathogenesis of core binding factor leukemia: current knowledge and future prospects. Int J Hematol 2011;94:126–33.

[15] Speck NA, Gilliland DG. Core-binding factors in haematopoiesis and leukaemia. Nat Rev Cancer 2002;2:502–13.

[16] Byrd JC, Mrozek K, Dodge RK, Carroll AJ, Edwards CG, Arthur DC, et al. Pretreatment cytogenetic abnormalities are predictive of induction success, cumulative incidence of relapse, and overall survival in adult patients with de novo acute myeloid leukemia: results from Cancer and Leukemia Group B (CALGB 8461). Blood 2002;100:4325–36.

[17] Mrozek K, Prior TW, Edwards C, Marcucci G, Carroll AJ, Snyder PJ, et al. Comparison of cytogenetic and molecular genetic detection of t(8;21) and inv(16) in a prospective series of adults with de novo acute myeloid leukemia: a Cancer and Leukemia Group B Study. J Clin Oncol 2001;19:2482–92.

[18] Mrozek K, Marcucci G, Paschka P, Bloomfield CD. Advances in molecular genetics and treatment of core-binding factor acute myeloid leukemia. Curr Opin Oncol 2008;20:711–18.

[19] Rowley JD, Golomb HM, Dougherty C. 15/17 translocation, a consistent chromosomal change in acute promyelocytic leukaemia. Lancet 1977;1:549–50.

[20] de The H, Chomienne C, Lanotte M, Degos L, Dejean A. The t(15;17) translocation of acute promyelocytic leukaemia fuses the retinoic acid receptor alpha gene to a novel transcribed locus. Nature 1990;347:558–61.

[21] Borrow J, Goddard AD, Sheer D, Solomon E. Molecular analysis of acute promyelocytic leukemia breakpoint cluster region on chromosome 17. Science 1990;249:1577–80.

[22] Brown D, Kogan S, Lagasse E, Weissman I, Alcalay M, Pelicci PG, et al. A PMLRARalpha transgene initiates murine acute promyelocytic leukemia. Proc Natl Acad Sci USA 1997;94:2551–6.

[23] Tsai S, Collins SJ. A dominant negative retinoic acid receptor blocks neutrophil differentiation at the promyelocyte stage. Proc Natl Acad Sci USA 1993;90(15):7153–7.

[24] Wang ZG, Delva L, Gaboli M, Rivi R, Giorgio M, Cordon-Cardo C, et al. Role of PML in cell growth and the retinoic acid pathway. Science 1998;279:1547–51.

[25] Grimwade D, Biondi A, Mozziconacci MJ, Hagemeijer A, Berger R, Neat M, et al. Characterization of acute promyelocytic leukemia cases lacking the classic t(15;17): results of the European Working Party. Groupe Francais de Cytogenetique Hematologique, Groupe de Francais d'Hematologie Cellulaire, UK Cancer Cytogenetics Group and BIOMED 1 European Community-Concerted Action "Molecular Cytogenetic Diagnosis in Haematological Malignancies". Blood 2000;96:1297–308.

[26] Warrell Jr. RP, Frankel SR, Miller Jr. WH, Scheinberg DA, Itri LM, Hittelman WN, et al. Differentiation therapy of acute promyelocytic leukemia with tretinoin (all-trans-retinoic acid). N Engl J Med 1991;324:1385–93.

[27] Degos L, Wang ZY. All trans retinoic acid in acute promyelocytic leukemia. Oncogene 2001;20:7140–5.

[28] Campbell LJ, Oei P, Brookwell R, Shortt J, Eaddy N, Ng A, et al. FISH detection of PML-RARA fusion in ins(15;17) acute promyelocytic leukaemia depends on probe size. BioMed Res Int 2013;2013:164501.

[29] Kim MJ, Cho SY, Kim MH, Lee JJ, Kang SY, Cho EH, et al. FISH-negative cryptic PML-RARA rearrangement detected by long-distance polymerase chain reaction and sequencing analyses: a case study and review of the literature. Cancer Genet Cytogenet 2010;203:278–83.

[30] Dong S, Tweardy DJ. Interactions of STAT5b-RARalpha, a novel acute promyelocytic leukemia fusion protein, with retinoic acid receptor and STAT3 signaling pathways. Blood 2002;99:2637–46.

[31] Licht JD, Chomienne C, Goy A, Chen A, Scott AA, Head DR, et al. Clinical and molecular characterization of a rare syndrome of acute promyelocytic leukemia associated with translocation (11;17). Blood 1995;85(4):1083–94.

[32] De Braekeleer M, Morel F, Le Bris MJ, Herry A, Douet-Guilbert N. The MLL gene and translocations involving chromosomal band 11q23 in acute leukemia. Anticancer Res 2005;25:1931–44.

[33] Meyer C, Hofmann J, Burmeister T, Groger D, Park TS, Emerenciano M, et al. The MLL recombinome of acute leukemias in 2013. Leukemia 2013;27:2165–76.

[34] Hess JL. MLL: a histone methyltransferase disrupted in leukemia. Trends Mol Med 2004;10:500–7.

[35] Shih LY, Liang DC, Fu JF, Wu JH, Wang PN, Lin TL, et al. Characterization of fusion partner genes in 114 patients with de novo acute myeloid leukemia and MLL rearrangement. Leukemia 2006;20:218–23.

[36] Mrozek K, Heerema NA, Bloomfield CD. Cytogenetics in acute leukemia. Blood Rev 2004;18:115–36.

[37] Verhaak RG, Goudswaard CS, van Putten W, Bijl MA, Sanders MA, Hugens W, et al. Mutations in nucleophosmin (NPM1) in acute myeloid leukemia (AML): association with other gene abnormalities and previously established gene expression signatures and their favorable prognostic significance. Blood 2005;106:3747–54.

[38] Schnittger S, Schoch C, Kern W, Mecucci C, Tschulik C, Martelli MF, et al. Nucleophosmin gene mutations are predictors of favorable prognosis in acute myelogenous leukemia with a normal karyotype. Blood 2005;106:3733–9.

[39] Falini B, Nicoletti I, Bolli N, Martelli MP, Liso A, Gorello P, et al. Translocations and mutations involving the nucleophosmin (NPM1) gene in lymphomas and leukemias. Haematologica 2007;92:519–32.

[40] Borer RA, Lehner CF, Eppenberger HM, Nigg EA. Major nucleolar proteins shuttle between nucleus and cytoplasm. Cell 1989;56:379–90.

[41] Falini B, Mecucci C, Tiacci E, Alcalay M, Rosati R, Pasqualucci L, et al. Cytoplasmic nucleophosmin in acute myelogenous leukemia with a normal karyotype. N Engl J Med 2005;352:254–66.

[42] Thiede C, Koch S, Creutzig E, Steudel C, Illmer T, Schaich M, et al. Prevalence and prognostic impact of NPM1 mutations in 1485 adult patients with acute myeloid leukemia (AML). Blood 2006;107:4011–20.

[43] Haferlach C, Mecucci C, Schnittger S, Kohlmann A, Mancini M, Cuneo A, et al. AML with mutated NPM1 carrying a normal or aberrant karyotype show overlapping biologic, pathologic, immunophenotypic, and prognostic features. Blood 2009;114: 3024–32.

[44] Szankasi P, Jama M, Bahler DW. A new DNA-based test for detection of nucleophosmin exon 12 mutations by capillary electrophoresis. J Mol Diagn 2008;10(3):236–41.

[45] Schnittger S, Kern W, Tschulik C, Weiss T, Dicker F, Falini B, et al. Minimal residual disease levels assessed by NPM1 mutation-specific RQ-PCR provide important prognostic information in AML. Blood 2009;114:2220–31.

[46] Kronke J, Schlenk RF, Jensen KO, Tschurtz F, Corbacioglu A, Gaidzik VI, et al. Monitoring of minimal residual disease in NPM1-mutated acute myeloid leukemia: a study from the German-Austrian acute myeloid leukemia study group. J Clin Oncol 2011;29:2709–16.

[47] Thiede C, Steudel C, Mohr B, Schaich M, Schakel U, Platzbecker U, et al. Analysis of FLT3-activating mutations in 979 patients with acute myelogenous leukemia: association with FAB subtypes and identification of subgroups with poor prognosis. Blood 2002;99:4326–35.

[48] Stirewalt DL, Radich JP. The role of FLT3 in haematopoietic malignancies. Nat Rev Cancer 2003;3:650–65.

[49] Schnittger S, Schoch C, Dugas M, Kern W, Staib P, Wuchter C, et al. Analysis of FLT3 length mutations in 1003 patients with acute myeloid leukemia: correlation to cytogenetics, FAB subtype, and prognosis in the AMLCG study and usefulness as a marker for the detection of minimal residual disease. Blood 2002;100:59–66.

[50] Gale RE, Green C, Allen C, Mead AJ, Burnett AK, Hills RK, et al. The impact of FLT3 internal tandem duplication mutant level, number, size, and interaction with NPM1 mutations in a large cohort of young adult patients with acute myeloid leukemia. Blood 2008;111:2776–84.

[51] Bacher U, Haferlach C, Kern W, Haferlach T, Schnittger S. Prognostic relevance of FLT3-TKD mutations in AML: the combination matters—an analysis of 3082 patients. Blood 2008;111:2527–37.

[52] Whitman SP, Archer KJ, Feng L, Baldus C, Becknell B, Carlson BD, et al. Absence of the wild-type allele predicts poor prognosis in adult de novo acute myeloid leukemia with normal cytogenetics and the internal tandem duplication of FLT3: a cancer and leukemia group B study. Cancer Res 2001;61:7233–9.

[53] Kottaridis PD, Gale RE, Frew ME, Harrison G, Langabeer SE, Belton AA, et al. The presence of a FLT3 internal tandem duplication in patients with acute myeloid leukemia (AML) adds important prognostic information to cytogenetic risk group and response to the first cycle of chemotherapy: analysis of 854 patients from the United Kingdom Medical Research Council AML 10 and 12 trials. Blood 2001;98:1752–9.

[54] Leung AY, Man CH, Kwong YL. FLT3 inhibition: a moving and evolving target in acute myeloid leukaemia. Leukemia 2013;27:260–8.

[55] Frohling S, Schlenk RF, Breitruck J, Benner A, Kreitmeier S, Tobis K, et al. Prognostic significance of activating FLT3 mutations in younger adults (16 to 60 years) with acute myeloid leukemia and normal cytogenetics: a study of the AML Study Group Ulm. Blood 2002;100:4372–80.

[56] Mead AJ, Linch DC, Hills RK, Wheatley K, Burnett AK, Gale RE. FLT3 tyrosine kinase domain mutations are biologically distinct from and have a significantly more favorable prognosis than FLT3 internal tandem duplications in patients with acute myeloid leukemia. Blood 2007;110:1262–70.

[57] Murphy KM, Levis M, Hafez MJ, Geiger T, Cooper LC, Smith BD, et al. Detection of FLT3 internal tandem duplication and D835 mutations by a multiplex polymerase chain reaction and capillary electrophoresis assay. J Mol Diagn 2003;5:96–102.

[58] Shih LY, Huang CF, Wu JH, Lin TL, Dunn P, Wang PN, et al. Internal tandem duplication of FLT3 in relapsed acute myeloid leukemia: a comparative analysis of bone marrow samples from 108 adult patients at diagnosis and relapse. Blood 2002;100:2387–92.

[59] Levis M, Murphy KM, Pham R, Kim KT, Stine A, Li L, et al. Internal tandem duplications of the FLT3 gene are present in leukemia stem cells. Blood 2005;106:673–80.

[60] Nerlov C. CEBPalpha mutations in acute myeloid leukaemias. Nat Rev Cancer 2004;4:394–400.

[61] Pabst T, Mueller BU, Zhang P, Radomska HS, Narravula S, Schnittger S, et al. Dominant-negative mutations of CEBPA, encoding CCAAT/enhancer binding protein-alpha (C/EBPalpha), in acute myeloid leukemia. Nat Genet 2001;27:263–70.

[62] Frohling S, Schlenk RF, Stolze I, Bihlmayr J, Benner A, Kreitmeier S, et al. CEBPA mutations in younger adults with acute myeloid leukemia and normal cytogenetics: prognostic relevance and analysis of cooperating mutations. J Clin Oncol 2004;22:624–33.

[63] Green CL, Koo KK, Hills RK, Burnett AK, Linch DC, Gale RE. Prognostic significance of CEBPA mutations in a large cohort of younger adult patients with acute myeloid leukemia: impact of double CEBPA mutations and the interaction with FLT3 and NPM1 mutations. J Clin Oncol 2010;28:2739–47.

[64] Dufour A, Schneider F, Metzeler KH, Hoster E, Schneider S, Zellmeier E, et al. Acute myeloid leukemia with biallelic CEBPA gene mutations and normal karyotype represents a distinct genetic entity associated with a favorable clinical outcome. J Clin Oncol 2010;28:570–7.

[65] Fasan A, Haferlach C, Alpermann T, Jeromin S, Grossmann V, Eder C, et al. The role of different genetic subtypes of CEBPA mutated AML. Leukemia 2014;28:794–803.

[66] Behdad A, Weigelin HC, Elenitoba-Johnson KS, Betz BL. A clinical grade sequencing-based assay for CEBPA mutation testing: report of a large series of myeloid neoplasms. J Mol Diagn 2015;17:76–84.

[67] Fuster O, Barragan E, Bolufer P, Such E, Valencia A, Ibanez M, et al. Fragment length analysis screening for detection of CEBPA mutations in intermediate-risk karyotype acute myeloid leukemia. Ann Hematol 2012;91:1–7.

[68] Boissel N, Leroy H, Brethon B, Philippe N, de Botton S, Auvrignon A, et al. Incidence and prognostic impact of c-Kit, FLT3, and Ras gene mutations in core binding factor acute myeloid leukemia (CBF-AML). Leukemia 2006;20:965–70.

[69] Goemans BF, Zwaan CM, Miller M, Zimmermann M, Harlow A, Meshinchi S, et al. Mutations in KIT and RAS are frequent events in pediatric core-binding factor acute myeloid leukemia. Leukemia 2005;19:1536–42.

[70] Paschka P, Du J, Schlenk RF, Gaidzik VI, Bullinger L, Corbacioglu A, et al. Secondary genetic lesions in acute myeloid leukemia with inv(16) or t(16;16): a study of the German-Austrian AML Study Group (AMLSG). Blood 2013;121:170–7.

[71] Paschka P, Marcucci G, Ruppert AS, Mrozek K, Chen H, Kittles RA, et al. Adverse prognostic significance of KIT mutations in adult acute myeloid leukemia with inv(16) and t(8;21): a Cancer and Leukemia Group B study. J Clin Oncol 2006;24: 3904–11.

[72] Park SH, Chi HS, Min SK, Park BG, Jang S, Park CJ. Prognostic impact of c-KIT mutations in core binding factor acute myeloid leukemia. Leuk Res 2011;35:1376–83.

[73] Qin YZ, Zhu HH, Jiang Q, Jiang H, Zhang LP, Xu LP, et al. Prevalence and prognostic significance of c-KIT mutations in core binding factor acute myeloid leukemia: a comprehensive large-scale study from a single Chinese center. Leuk Res 2014;38:1435−40.

[74] Meyer SC, Levine RL. Translational implications of somatic genomics in acute myeloid leukaemia. Lancet Oncol 2014;15: e382−94.

[75] Ito Y. Oncogenic potential of the RUNX gene family: 'overview'. Oncogene 2004;23:4198−208.

[76] Preudhomme C, Warot-Loze D, Roumier C, Grardel-Duflos N, Garand R, Lai JL, et al. High incidence of biallelic point mutations in the Runt domain of the AML1/PEBP2 alpha B gene in Mo acute myeloid leukemia and in myeloid malignancies with acquired trisomy 21. Blood 2000;96:2862−9.

[77] Langabeer SE, Gale RE, Rollinson SJ, Morgan GJ, Linch DC. Mutations of the AML1 gene in acute myeloid leukemia of FAB types M0 and M7. Genes Chromosomes Cancer 2002;34:24−32.

[78] Tang JL, Hou HA, Chen CY, Liu CY, Chou WC, Tseng MH, et al. AML1/RUNX1 mutations in 470 adult patients with de novo acute myeloid leukemia: prognostic implication and interaction with other gene alterations. Blood 2009;114:5352−61.

[79] Christiansen DH, Andersen MK, Pedersen-Bjergaard J. Mutations of AML1 are common in therapy-related myelodysplasia following therapy with alkylating agents and are significantly associated with deletion or loss of chromosome arm 7q and with subsequent leukemic transformation. Blood 2004;104:1474−81.

[80] Harada H, Harada Y, Niimi H, Kyo T, Kimura A, Inaba T. High incidence of somatic mutations in the AML1/RUNX1 gene in myelodysplastic syndrome and low blast percentage myeloid leukemia with myelodysplasia. Blood 2004;103:2316−24.

[81] Schnittger S, Dicker F, Kern W, Wendland N, Sundermann J, Alpermann T, et al. RUNX1 mutations are frequent in de novo AML with noncomplex karyotype and confer an unfavorable prognosis. Blood 2011;117:2348−57.

[82] Mendler JH, Maharry K, Radmacher MD, Mrozek K, Becker H, Metzeler KH, et al. RUNX1 mutations are associated with poor outcome in younger and older patients with cytogenetically normal acute myeloid leukemia and with distinct gene and microRNA expression signatures. J Clin Oncol 2012;30:3109−18.

[83] Mardis ER, Ding L, Dooling DJ, Larson DE, McLellan MD, Chen K, et al. Recurring mutations found by sequencing an acute myeloid leukemia genome. N Engl J Med 2009;361:1058−66.

[84] Yan H, Parsons DW, Jin G, McLendon R, Rasheed BA, Yuan W, et al. IDH1 and IDH2 mutations in gliomas. N Engl J Med 2009;360:765−73.

[85] Marcucci G, Maharry K, Wu YZ, Radmacher MD, Mrozek K, Margeson D, et al. IDH1 and IDH2 gene mutations identify novel molecular subsets within de novo cytogenetically normal acute myeloid leukemia: a Cancer and Leukemia Group B study. J Clin Oncol 2010;28:2348−55.

[86] Ward PS, Patel J, Wise DR, Abdel-Wahab O, Bennett BD, Coller HA, et al. The common feature of leukemia-associated IDH1 and IDH2 mutations is a neomorphic enzyme activity converting alpha-ketoglutarate to 2-hydroxyglutarate. Cancer Cell 2010;17(3):225−34.

[87] Paschka P, Schlenk RF, Gaidzik VI, Habdank M, Kronke J, Bullinger L, et al. IDH1 and IDH2 mutations are frequent genetic alterations in acute myeloid leukemia and confer adverse prognosis in cytogenetically normal acute myeloid leukemia with NPM1 mutation without FLT3 internal tandem duplication. J Clin Oncol 2010;28:3636−43.

[88] Boissel N, Nibourel O, Renneville A, Gardin C, Reman O, Contentin N, et al. Prognostic impact of isocitrate dehydrogenase enzyme isoforms 1 and 2 mutations in acute myeloid leukemia: a study by the Acute Leukemia French Association group. J Clin Oncol 2010;28:3717−23.

[89] Wagner K, Damm F, Gohring G, Gorlich K, Heuser M, Schafer I, et al. Impact of IDH1 R132 mutations and an IDH1 single nucleotide polymorphism in cytogenetically normal acute myeloid leukemia: SNP rs11554137 is an adverse prognostic factor. J Clin Oncol 2010;28:2356−64.

[90] Patel JP, Gonen M, Figueroa ME, Fernandez H, Sun Z, Racevskis J, et al. Prognostic relevance of integrated genetic profiling in acute myeloid leukemia. N Engl J Med 2012;366:1079−89.

[91] Yan XJ, Xu J, Gu ZH, Pan CM, Lu G, Shen Y, et al. Exome sequencing identifies somatic mutations of DNA methyltransferase gene DNMT3A in acute monocytic leukemia. Nature Genet 2011;43:309−15.

[92] Ley TJ, Ding L, Walter MJ, McLellan MD, Lamprecht T, Larson DE, et al. DNMT3A mutations in acute myeloid leukemia. N Engl J Med 2010;363:2424−33.

[93] Renneville A, Boissel N, Nibourel O, Berthon C, Helevaut N, Gardin C, et al. Prognostic significance of DNA methyltransferase 3A mutations in cytogenetically normal acute myeloid leukemia: a study by the Acute Leukemia French Association. Leukemia 2012;26:1247−54.

[94] Thol F, Damm F, Ludeking A, Winschel C, Wagner K, Morgan M, et al. Incidence and prognostic influence of DNMT3A mutations in acute myeloid leukemia. J Clin Oncol 2011;29:2889−96.

[95] Dohner K, Tobis K, Ulrich R, Frohling S, Benner A, Schlenk RF, et al. Prognostic significance of partial tandem duplications of the MLL gene in adult patients 16 to 60 years old with acute myeloid leukemia and normal cytogenetics: a study of the Acute Myeloid Leukemia Study Group Ulm. J Clin Oncol 2002;20:3254−61.

[96] Steudel C, Wermke M, Schaich M, Schakel U, Illmer T, Ehninger G, et al. Comparative analysis of MLL partial tandem duplication and FLT3 internal tandem duplication mutations in 956 adult patients with acute myeloid leukemia. Genes Chromosomes Cancer 2003;37:237−51.

[97] Caligiuri MA, Strout MP, Lawrence D, Arthur DC, Baer MR, Yu F, et al. Rearrangement of ALL1 (MLL) in acute myeloid leukemia with normal cytogenetics. Cancer Res 1998;58 (1):55−9.

[98] Metzeler KH, Becker H, Maharry K, Radmacher MD, Kohlschmidt J, Mrozek K, et al. ASXL1 mutations identify a high-risk subgroup of older patients with primary cytogenetically normal AML within the ELN favorable genetic category. Blood 2011;118:6920−9.

[99] Pratcorona M, Abbas S, Sanders MA, Koenders JE, Kavelaars FG, Erpelinck-Verschueren CA, et al. Acquired mutations in ASXL1 in acute myeloid leukemia: prevalence and prognostic value. Haematologica 2012;97:388−92.

[100] Paschka P, Marcucci G, Ruppert AS, Whitman SP, Mrozek K, Maharry K, et al. Wilms' tumor 1 gene mutations independently predict poor outcome in adults with cytogenetically normal acute myeloid leukemia: a cancer and leukemia group B study. J Clin Oncol 2008;26:4595−602.

[101] Virappane P, Gale R, Hills R, Kakkas I, Summers K, Stevens J, et al. Mutation of the Wilms' tumor 1 gene is a poor prognostic factor associated with chemotherapy resistance in normal

karyotype acute myeloid leukemia: the United Kingdom Medical Research Council Adult Leukaemia Working Party. J Clin Oncol 2008;26:5429–35.

[102] King-Underwood L, Renshaw J, Pritchard-Jones K. Mutations in the Wilms' tumor gene WT1 in leukemias. Blood 1996;87:2171–9.

[103] Gaidzik VI, Schlenk RF, Moschny S, Becker A, Bullinger L, Corbacioglu A, et al. Prognostic impact of WT1 mutations in cytogenetically normal acute myeloid leukemia: a study of the German-Austrian AML Study Group. Blood 2009;113:4505–11.

[104] Weissmann S, Alpermann T, Grossmann V, Kowarsch A, Nadarajah N, Eder C, et al. Landscape of TET2 mutations in acute myeloid leukemia. Leukemia 2012;26:934–42.

[105] Moran-Crusio K, Reavie L, Shih A, Abdel-Wahab O, Ndiaye-Lobry D, Lobry C, et al. Tet2 loss leads to increased hematopoietic stem cell self-renewal and myeloid transformation. Cancer Cell 2011;20:11–24.

[106] Fenaux P, Preudhomme C, Quiquandon I, Jonveaux P, Lai JL, Vanrumbeke M, et al. Mutations of the P53 gene in acute myeloid leukaemia. Br J Haematol 1992;80:178–83.

[107] Bowen D, Groves MJ, Burnett AK, Patel Y, Allen C, Green C, et al. TP53 gene mutation is frequent in patients with acute myeloid leukemia and complex karyotype, and is associated with very poor prognosis. Leukemia 2009;23:203–6.

[108] Rucker FG, Schlenk RF, Bullinger L, Kayser S, Teleanu V, Kett H, et al. TP53 alterations in acute myeloid leukemia with complex karyotype correlate with specific copy number alterations, monosomal karyotype, and dismal outcome. Blood 2012;119:2114–21.

[109] Haferlach C, Dicker F, Herholz H, Schnittger S, Kern W, Haferlach T. Mutations of the TP53 gene in acute myeloid leukemia are strongly associated with a complex aberrant karyotype. Leukemia 2008;22:1539–41.

[110] Bowen DT, Frew ME, Hills R, Gale RE, Wheatley K, Groves MJ, et al. RAS mutation in acute myeloid leukemia is associated with distinct cytogenetic subgroups but does not influence outcome in patients younger than 60 years. Blood 2005;106:2113–19.

[111] Bacher U, Haferlach T, Schoch C, Kern W, Schnittger S. Implications of NRAS mutations in AML: a study of 2502 patients. Blood 2006;107:3847–53.

[112] Heuser M, Argiropoulos B, Kuchenbauer F, Yung E, Piper J, Fung S, et al. MN1 overexpression induces acute myeloid leukemia in mice and predicts ATRA resistance in patients with AML. Blood 2007;110:1639–47.

[113] Langer C, Radmacher MD, Ruppert AS, Whitman SP, Paschka P, Mrozek K, et al. High BAALC expression associates with other molecular prognostic markers, poor outcome, and a distinct gene-expression signature in cytogenetically normal patients younger than 60 years with acute myeloid leukemia: a Cancer and Leukemia Group B (CALGB) study. Blood 2008;111:5371–9.

[114] Lugthart S, van Drunen E, van Norden Y, van Hoven A, Erpelinck CA, Valk PJ, et al. High EVI1 levels predict adverse outcome in acute myeloid leukemia: prevalence of EVI1 overexpression and chromosome 3q26 abnormalities underestimated. Blood 2008;111:4329–37.

[115] Metzeler KH, Dufour A, Benthaus T, Hummel M, Sauerland MC, Heinecke A, et al. ERG expression is an independent prognostic factor and allows refined risk stratification in cytogenetically normal acute myeloid leukemia: a comprehensive analysis of ERG, MN1, and BAALC transcript levels using oligonucleotide microarrays. J Clin Oncol 2009;27:5031–8.

[116] Baldus CD, Tanner SM, Ruppert AS, Whitman SP, Archer KJ, Marcucci G, et al. BAALC expression predicts clinical outcome of de novo acute myeloid leukemia patients with normal cytogenetics: a Cancer and Leukemia Group B Study. Blood 2003;102:1613–18.

[117] Golub TR, Slonim DK, Tamayo P, Huard C, Gaasenbeek M, Mesirov JP, et al. Molecular classification of cancer: class discovery and class prediction by gene expression monitoring. Science 1999;286:531–7.

[118] Haferlach T, Kohlmann A, Schnittger S, Dugas M, Hiddemann W, Kern W, et al. Global approach to the diagnosis of leukemia using gene expression profiling. Blood 2005;106:1189–98.

[119] Haferlach T, Kohlmann A, Wieczorek L, Basso G, Kronnie GT, Bene MC, et al. Clinical utility of microarray-based gene expression profiling in the diagnosis and subclassification of leukemia: report from the International Microarray Innovations in Leukemia Study Group. J Clin Oncol 2010;28:2529–37.

[120] Duployez N, Willekens C, Marceau-Renaut A, Boudry-Labis E, Preudhomme C. Prognosis and monitoring of core-binding factor acute myeloid leukemia: current and emerging factors. Expert Rev Hematol 2015;8:43–56.

[121] Grimwade D, Jovanovic JV, Hills RK, Nugent EA, Patel Y, Flora R, et al. Prospective minimal residual disease monitoring to predict relapse of acute promyelocytic leukemia and to direct pre-emptive arsenic trioxide therapy. J Clin Oncol 2009;27:3650–8.

[122] Yin JA, O'Brien MA, Hills RK, Daly SB, Wheatley K, Burnett AK. Minimal residual disease monitoring by quantitative RT-PCR in core binding factor AML allows risk stratification and predicts relapse: results of the United Kingdom MRC AML-15 trial. Blood 2012;120:2826–35.

[123] Cancer Genome Atlas Research Network. Genomic and epigenomic landscapes of adult de novo acute myeloid leukemia. N Engl J Med 2013;368:2059–74.

[124] Ding L, Ley TJ, Larson DE, Miller CA, Koboldt DC, Welch JS, et al. Clonal evolution in relapsed acute myeloid leukaemia revealed by whole-genome sequencing. Nature 2012;481:506–10.

[125] Welch JS, Ley TJ, Link DC, Miller CA, Larson DE, Koboldt DC, et al. The origin and evolution of mutations in acute myeloid leukemia. Cell 2012;150:264–78.

[126] Jan M, Majeti R. Clonal evolution of acute leukemia genomes. Oncogene 2013;32:135–40.

[127] Vasilatou D, Papageorgiou S, Pappa V, Papageorgiou E, Dervenoulas J. The role of microRNAs in normal and malignant hematopoiesis. Eur J Haematol 2010;84:1–16.

[128] Marcucci G, Haferlach T, Dohner H. Molecular genetics of adult acute myeloid leukemia: prognostic and therapeutic implications. J Clin Oncol 2011;29:475–86.

[129] Marcucci G, Mrozek K, Radmacher MD, Garzon R, Bloomfield CD. The prognostic and functional role of microRNAs in acute myeloid leukemia. Blood 2011;117:1121–9.

34

Molecular Testing in Myeloproliferative Neoplasms

L.V. Furtado[1] and B.L. Betz[2]

[1]Department of Pathology, University of Chicago, Chicago, IL, United States [2]Department of Pathology, University of Michigan, Ann Arbor, MI, United States

INTRODUCTION

Myeloproliferative neoplasms (MPNs), previously known as myeloproliferative disorders, are acquired clonal hematopoietic malignancies, characterized by the abnormal and excessive proliferation of one or more of myeloid cell lines (granulocytes, erythrocytes, and/or platelets) with no marked alterations in cellular maturation [1]. This heterogeneous group of disorders is classified by the World Health Organization (WHO) 2008 as Philadelphia chromosome (BCR−ABL1)-positive MPNs which include chronic myelogenous leukemia, and Philadelphia chromosome (BCR−ABL1)-negative MPNs that among other entities include polycythemia vera (PV), essential thrombocythemia (ET), and primary myelofibrosis (PMF).

The primary characteristics of PV and ET are the increased production of red blood cells and platelets, which predispose these patients to thrombosis or hemorrhage. Patients with PV and ET can also present with constitutional symptoms, such as night sweats, fever, pruritus, and splenomegaly [2]. These diseases tend to take a protracted chronic course. As long-term sequelae, some patients progress to an accelerated myelofibrosis phase, marked by cytopenias and insufficient extramedullary hematopoiesis that is often clinically and morphologically indistinguishable from PMF [3]. PMF is characterized by progressive marrow fibrosis and variable degree of megakaryocyte and granulocyte proliferation [4]. Compared to PV and ET, PMF patients tend to exhibit more severe disease-associated symptoms and more rapid disease progression, which markedly impact their quality of life [5]. MPNs may also progress to myelodysplastic syndrome (MDS) or transform to acute myeloid leukemia (AML) as long-term sequelae of their chronic phase or secondary to cytoreductive therapies such as alkylating agents or radioactive phosphorus [6−9]. The reported risk of leukemic transformation is approximately 20% for PMF, 4.5% for PV, and less than 1% for strictly WHO-classified ET cases [10−12].

The vast majority of MPN cases are sporadic and the disease has an overall incidence of 5 in 100,000 individuals, with approximately 15,000 new cases in the United States each year [13,14]. MPNs are typically diagnosed in the fifth to sixth decade of life, although it can be diagnosed in younger individuals, especially when there is a familial predisposition [15]. The life expectancy of all MPN subtypes is reduced when compared with the general population [16] with the shortest survival rates being observed among PMF patients [17].

MOLECULAR PATHOGENESIS OF MPNs

JAK2 V617F Mutation

In 2005, several independent groups identified a single somatic activating mutation in the Janus Kinase 2 (*JAK2*) gene on chromosome 9p24 that had a high incidence in PV, ET, and PMF [18−21]. JAK2 is a cytoplasmic tyrosine kinase protein, which mediates signal transducing downstream of various cytokine receptors implicated in erythropoietin receptor signaling and hematopoiesis [22,23]. The protein is composed of 1132 amino acids and has four functional domains: (1) FERM domain, (2) SH2 domain, (3) pseudokinase (JH2)

© 2017 Elsevier Inc. All rights reserved.

domain, and (4) tyrosine kinase (JH1) domain [24]. The JH2 domain negatively regulates the JAK function [25].

The G to T substitution mutation in *JAK2* exon 14 (nucleotide 1849) replaces a phenylalanine with a valine at amino acid 617 in the JH2 pseudokinase domain of the JAK2 protein. As a result, the *JAK2* V617F mutation leads to constitutive activation of JAK–STAT, PI3K, and AKT pathways, as well as mitogen-activated protein kinase and extracellular signal regulated kinase [18–20,26]. Consequently, hematopoietic cells harboring the *JAK2* V617F mutation show cytokine hypersensitivity and cytokine-independent growth [27]. The role of the *JAK2* V617F mutation in the pathogenesis of MPN has been validated in vitro and in vivo by means of murine models of myeloproliferative phenotype driven by this mutation [28,29].

The *JAK2* V617F mutation is found in approximately 95% of PV, 55% of ET, and 65% of PMF cases [18,20,21]. A significant number of PV and PMF patients have biallelic *JAK2* V617F mutation as a result of mitotic recombination involving chromosome 9p that leads to uniparental disomy (UPD) [18–21], but this event is uncommon in ET patients [30]. Some evidence suggests that both an advanced disease stage and some MPN complications, such as marrow fibrosis, thrombotic propensity, and overall survival, correlate with the overall proportion of the mutant allele in circulating clonal granulocytes [31–35]. Other hematological neoplasms (eg, hypereosinophilic syndrome, chronic myelomonocytic leukemia, chronic neutrophilic leukemia, myelodysplasia, acute lymphoblastic leukemia, or AML) may also harbor *JAK2* V617F mutation in a low proportion of cases [36–41].

JAK2 Exon 12 Mutations

The *JAK2* V617F mutation is identified in approximately 95% of patients with PV, but the molecular basis of cases lacking this mutation was unclear until 2007, when mutations affecting *JAK2* exon 12 were reported in most of the remaining patients [42–45]. Exon 12 mutations cluster in a distinct region of JAK2 that is adjacent to the JH2 pseudokinase domain where V617F is located. Despite this difference, exon 12 mutations lead to a similar constitutive activation of erythropoietin signaling that results in a myeloproliferative phenotype, as demonstrated by in vitro analysis and murine model experiments [42].

Unlike the V617F mutation that is found in several MPNs, *JAK2* exon 12 mutations are restricted only to cases of PV [42]. At the nucleotide level, at least 37 different *JAK2* exon 12 mutations have been reported to date [46]. Despite this variety, all exon 12 mutations cluster within a 36 nucleotide stretch of the exon that

spans codons 536–547 [42,45,47]. The majority are small in-frame deletions of three to nine nucleotides (6 bp deletions are most common). Less frequent are substitutions leading to K539L, and in-frame duplications usually 33 bp in length. With respect to the protein, most *JAK2* exon 12 mutations fall within three main types: (1) deletions that include E543, (2) amino acid substitution or deletion mutations that involve K539, and (3) duplications of 10–12 amino acids that occur in the region of V536 to F547 [45].

Exon 12 mutations can be detected in variable percentages of peripheral blood granulocytes, monocytes, and platelets, but rarely lymphocytes [45,48]. Most cases exhibit heterozygous exon 12 mutation with only rare reports of biallelic mutation [42,45,48–50]. This is in contrast to V617F-positive PV in which biallelic mutations are frequent and mutation burdens are generally higher [48,50]. Patients with exon 12 mutations frequently present with erythrocytosis as the predominant feature, but without concurrent elevations in the megakaryocytic or granulocytic lineages as seen in V617F-positive PV [42,49,51]. Consequently some may have previously received a diagnosis of idiopathic erythrocytosis. *JAK2* exon 12 and V617F mutations are mutually exclusive. Additional *JAK2* variants have been reported in exons 13, 14 (other than V617F), and 15, but their biological and clinical significance is unclear at present [52].

MPL Mutations

In 2006, a search for genetic alterations in *JAK2* V617F negative patients with ET and PMF revealed mutations in the myeloproliferative leukemia virus oncogene (*MPL*) [53,54]. The *MPL* gene, located on chromosome 1p34, has 12 exons and encodes the thrombopoietin receptor. *MPL* mutations associated with ET and PMF are gain-of-function and lead to receptor activation in the absence of thrombopoietin binding with constitutional activation of the JAK–STAT signaling pathway [42,53]. Five recurrent *MPL* mutations have been reported in ET and PMF patients, all clustering in exon 10 (juxtamembrane domain) and affecting two amino acids (W515 and S505). Of those, W515L and W515K represent the vast majority of reported *MPL* mutations, whereas the W515A, W515R, and S505N mutations are less commonly reported [53,55–59]. Notably, the S505N mutation has been reported as both a germline and an acquired (sporadic) mutation in ET and PMF [55,59,60]. Several reports of patients harboring two concurrent *MPL* mutations have been described, including W515L + W515K, or W515L + S505N [53,57,61,62], but the pathogenic implications of these findings remain to be elucidated. A few *MPL* mutations outside exon 10 have also been reported, but their biological and clinical significance is unclear [63,64].

MPL mutations can be detected in progenitors of both myeloid and lymphoid lineages [65]. The mutant allele burden in specimens harboring *MPL* mutations is frequently greater than 50%, suggesting that biallelic mutation (or loss of heterozygosity) is somewhat common [55,56,58]. However, mutations in patients with MPNs have also been reported at lower levels (5% and less) [55,56,58,66]. *MPL* mutations occur in ET and PMF with an approximate frequency of 3% and 10%, respectively [67,68], but have not been reported in PV [53]. Significantly, *MPL* and *JAK2* mutations are not mutually exclusive and can occasionally occur in conjunction [53,55].

Clinically, patients with *MPL* mutations tend to be more anemic, present at older age, have higher platelet counts, and have a higher risk of developing arterial thrombosis than those with a *JAK2* V617F mutation [55,69,70]. *MPL* mutations do not seem to have a significant influence on hemorrhagic or venous thrombotic events or progression to myelofibrosis, nor do they seem to impact survival. However, in PMF patients, *MPL* mutations are associated with a more severe phenotype, older age, female gender, and lower hemoglobin levels [55,69].

CALR Mutations

In 2013, two independent research groups identified *CALR* somatic mutations in *JAK2/MPL* wild-type patients with ET and PMF [71,72]. *CALR* is located on chromosome 19p13.2 and has nine exons, which encode the endoplasmic reticulum associated, calcium-binding protein calreticulin. The CALR protein is composed of three domains: N-domain (residues 1−180), P-domain (residues 181−290), and C-domain (residues 291−400). Thus far, more than 36 different *CALR* frameshift insertions or deletions have been reported, all clustering within exon 9 [71,72]. Of those, type 1 mutations (52 bp deletion, L367fs*46) and type 2 mutations (5 bp TTGTC insertion, K385fs*47) account for greater than 80% of all *CALR* mutations. *CALR* mutations are detected at a frequency of 20−30% in ET and PMF. In the subset of cases negative for *JAK2* V617F, *CALR* mutation incidence is 49−71% in ET and 56−88% in PMF [71−75]. Although *CALR* mutations have been regarded as mutually exclusive with mutations in both *JAK2* and *MPL*, a few ET and PMF patients with concurrent *JAK2* V617F and *CALR* mutations have been reported [74,76]. Most patients carry heterozygous mutations with an allele burden of 40−50% [77,78], although a low allele burden (2%) of the *CALR* type 2 mutation has been reported in a PMF patient [79]. Homozygous *CALR* mutations have also

been reported in rare instances, almost exclusively in type 2 mutations. Due to clonal evolution, more than one type of *CALR* mutation can be detected. *CALR* mutations have been reported primarily in MPN cases and have only infrequently been detected in MDS and atypical chronic myeloid leukemia [72,80]. Notably, germline polymorphisms (in-frame deletions) have been reported in healthy individuals [71,72].

Clinically, ET and PMF patients with *CALR* mutations tend to present with lower hemoglobin levels, lower leukocyte count, higher platelet count, and lower risk of thrombosis than *JAK2*-mutated patients [71−73,77]. *CALR* mutations also correlate with male gender and younger age at presentation [73,77], although a better overall survival has been observed only in PMF patients [73,74,77]. Overall, the clinical course of sporadic *CALR*-mutated patients tends to be more indolent than that of *JAK2*-mutated patients [71,81]. However, emerging evidence suggests that the prognostic advantage of calreticulin mutations in PMF might be confined to type 1 *CALR* variants [75,82]. Further, in *CALR*-mutated PMF, the concomitant presence of *ASXL1* mutations has been associated with an unfavorable prognosis [83].

Other Genetic Alterations

In addition to the common *JAK2*, *MPL*, and *CALR* mutations, several other recurrent mutations of *TET2*, *ASXLI*, *IDH1/2*, *CBL*, *LNK*, *NRAS*, *SF3B1*, *DNMT3A*, and *EZH2* genes have been reported in MPN [84]. The relevance of these mutations in the pathogenesis of MPN is currently an active area of investigation.

A small number of MPN patients have karyotypic abnormalities, the most frequent being gain of chromosome 9 in PV, which is correlated with a copy number gain of *JAK2* V617F [18−21]. Gain of chromosome 8, partial trisomy for 1q, and interstitial deletions of 13q and 20q have been noted in all MPN subtypes. Acquired UPD 1p, 4q 7q, 9p, and 11q is commonly associated with homozygosity for *MPL*, *TET2*, *EZH2*, *JAK2*, and *CBL* mutations, respectively [85].

Deregulated miRNA profiles have also been reported in MPN patient samples [86,87] and in various MPN cell lines [88,89]. Differential miRNA expression has been observed not only between MPN patients and healthy donors but also among the three MPN entities (PV, ET, and PMF).

MOLECULAR TESTING IN MPNs

Indications for Testing

Molecular testing has become standard of care for the diagnostic workup of any suspected MPN. The primary indication for molecular analysis in this setting is to support the diagnosis of a clonal MPN over a secondary or reactive condition in the context of unexplained polycythemia, thrombocytosis, or neutrophilia. However, molecular testing cannot subclassify MPNs because *JAK2* V617F, *CALR*, and *MPL* mutations are not MPN specific. The exception is *JAK2* exon 12 mutations, which are exclusively detected in PV. Further, other hematological neoplasms (eg, MDS, chronic myelomonocytic leukemia, AML, and acute lymphoblastic leukemia) may harbor *JAK2* V617F, *CALR*, and *MPL* mutations in low frequencies [37–41]. For this reason, the detection of mutations in any of these genes in isolation does not warrant a diagnosis of MPN. Likewise, the absence of mutations in *JAK2*, *CALR*, or *MPL* does not rule out a diagnosis of ET and PMF, because a triple-negative genotype accounts for approximately 10–15% of these patients [74,90]. Consequently, molecular testing is an adjunct, not a replacement to bone marrow morphology and other clinical and laboratory data for the diagnostic workup of a suspected MPN.

The usefulness of follow-up testing for *JAK2*, *CALR*, and *MPL* mutations in routine clinical practice has not yet been established. Quantitative monitoring of *JAK2* V617F allele burden has been reserved to patients enrolled in some clinical trials [91]. Even though a decrease in allele burden has been demonstrated after therapy with some of the available JAK inhibitors, there is no consistent correlation of allele burden with clinical response.

Testing Algorithm

Due to its high prevalence among all MPN subtypes, it is recommended that *JAK2* V617F mutation screening is carried out first when a diagnosis of MPN is suspected. In patients with suspected PV due to abnormal hemoglobin (hematocrit) levels of greater than 18.5 g/dL (>52%) in men and greater than 16.5 g/dL (>48%) in women and subnormal serum erythropoietin level, *JAK2* exon 12 mutation analysis is recommended following a negative *JAK2* V617F result [47,92]. The vast majority (>98%) of PV cases harbor a *JAK2* mutation [44]. Consequently, both *JAK2* V617F and exon 12 mutation testing is an important diagnostic adjunct to rule out suspected PV. In contrast, *CALR* and *MPL* mutation screening is not routinely indicated in the workup of a case suspicious of PV. Likewise, bone marrow examination is not essential for the diagnosis of PV, and in the appropriate clinical context, the absence of *JAK2* mutations virtually rules out this diagnosis. Of note, *CALR* type 1 deletions have been reported in two *JAK2* V617F negative PV patients, raising the possibility that *CALR* mutations can be associated with a PV phenotype [93].

Molecular workup of ET and PMF should also start with the assessment of *JAK2* V617F, followed by *CALR* mutation analysis in patients negative for *JAK2* V617F. Screening for *CALR* mutation is expected to be formally incorporated in the diagnostic workup of MPN in future revisions of the WHO classification system [94]. *MPL* mutation screening should be reserved in patients who are negative for both *JAK2* V617F and *CALR* mutations [94]. *JAK2* exon 12 mutation testing is not indicated as a part of the molecular testing algorithm for patients with a suspicion for ET or PMF since these mutations are restricted to cases of PV [42]. Significantly, the diagnosis of ET and PMF is based on combined assessment of bone marrow morphology, laboratory values, and clinical history. The presence of *JAK2* V617F, *CALR*, or *MPL* mutation is supportive but not essential or specific for the diagnosis of ET or PMF. The absence of these mutations does not rule out a diagnosis of ET or PMF, because approximately 10% of PMF and 13% of ET patients are triple negative [74,90].

With the advent of next-generation sequencing (NGS)-based multigene panels, it is becoming feasible to screen for pathogenic mutations in these and many other genes simultaneously. The rapid technological advances, falling cost, and increased accessibility of these platforms may supplant the need for a stepwise approach for MPN molecular analysis in the near future.

Preanalytical Considerations

Both peripheral blood and bone marrow aspirate specimens are suitable for *JAK2*, *CALR*, and *MPL* mutation analysis. The qualitative and quantitative assessments of *JAK2* V617F mutant allele burden in peripheral blood have been shown to be equivalent to that in bone marrow aspirate specimens when sensitive molecular assays are employed [95,96]. This is because granulocytes (which carry the mutations) constitute the predominant cell population in both of these specimen types. Blood and bone marrow samples should be drawn into anticoagulated tubes. The preferred anticoagulant for these molecular assays is ethylenediaminetetraacetic acid (EDTA, lavender top tube). Heparinized tubes should be avoided because heparin inhibits the polymerase enzyme utilized in PCR, which may lead to assay failure. Blood and bone marrow samples can be transported at ambient temperature. Blood samples should not be frozen prior to separation of cellular elements because this causes hemolysis, which interferes with DNA amplification. Most clinical laboratories perform DNA-based analyses for MPN-related genes, because with sensitive assays there is not expected to be substantial different in the clinical sensitivity of mutation testing in DNA compared to RNA.

For DNA extraction, the preferred age for blood and bone marrow samples is less than 5 days. Bone marrow biopsy specimens that are subjected to acid decalcification during processing are usually unsuitable for PCR-based testing due to extensive DNA degradation.

Molecular analysis for MPN-related mutations are adequately performed in total white blood cells from peripheral blood, because this cell population is largely composed of granulocytes. Granulocyte isolation increases time, labor, and cost on a routine basis, and the increase in analytical sensitivity by this approach is less remarkable than the one achieved with the use of sensitive PCR-based methods [97].

MOLECULAR METHODS FOR MPN TESTING

Several methods have been used to detect *JAK2*, *CALR*, and *MPL* mutations, each with different analytical and diagnostic sensitivities. Currently, there are no Food and Drug Administration (FDA)-cleared tests and no standardized test platforms for these analyses. The choice of which assay to implement should take into consideration the type, spectrum, distribution, frequency, and allelic burden of mutations for each disease-associated gene, as well as financial and practical aspects, such as laboratory infrastructure, workflow, and expertise.

JAK2 V617F Mutation Analysis

The V617F mutant allele burden varies widely among MPN patients (between 0.1% and 100%) and often reaches levels below 10−25% [35,98]. The analytical sensitivity of a clinical *JAK2* assay should be at least 1% to ensure that more than 90% of *JAK2* V617F mutations are detected [99−101]. Detection methods currently utilized for the *JAK2* V617F mutation include: Sanger sequencing, pyrosequencing, restriction fragment length polymorphism analysis, amplification refractory mutation system, allele-competitive blocker PCR, and high-resolution melting [26,36,39,102−104]. Most of the scanning technologies that have been reported for V617F detection, such as high-resolution melting, pyrosequencing, and Sanger sequencing, typically do not achieve sensitivities of less than 5−15% of mutant allele frequency and have the potential to generate false-negative results. For instance, a study using Sanger sequencing detected *JAK2* V617F mutation in 65% of PV, 23% of ET, and 30% of PMF cases [21]. Reassessment using more sensitive methods showed higher *JAK2* V617F mutation frequencies: 95−97% in PV, approximately 65% in PMF, and approximately 55% in ET, corroborating that these analytically insensitive techniques do not appear suitable for clinical diagnostic testing.

Currently, allele-specific quantitative PCR (qPCR) is the most commonly used technique for V617F testing (Fig. 34.1). This approach is advantageous for several reasons: It has high specificity and analytical sensitivity (down to 0.1% mutant allele), allows for straightforward interpretation, provides a rapid turnaround time in a closed tube format, can be easily deployed in most clinical laboratories, and is adaptable to a quantitative assay. However, false-positive results may be generated with assays that detect less than 0.1% of the mutant allele, because such low levels of *JAK2* V617F mutation have been reported in the peripheral blood of healthy individuals [101]. In addition, the risk of false-positive results due to amplicon contamination within the laboratory is another important issue when using extremely sensitive techniques. Further, allele-specific PCR may not provide detection of rare V617F mutations other than the canonical c.1849G > T mutation for which the specific primers are designed, and variants present at oligonucleotide-binding sites may negatively interfere with primer hybridization potentially resulting in false-negative results.

The clinical utility of *JAK2* V617F quantitative assays have not yet been validated in routine clinical practice. Quantitative assays have been employed to guide adoptive immunotherapy, such as donor lymphocyte infusion [105] in patients who have received allogeneic stem cell transplant. Currently, there are several JAK2 kinase inhibitors at various stages of clinical development for therapeutic management of MPN. For instance, Ruxolitinib is an oral JAK1 and JAK2 inhibitor that has been approved by the US FDA for treatment of patients with myelofibrosis. While serial monitoring of posttreatment levels of *JAK2* V617F is controversial at present, quantification of *JAK2* V617F is being considered as an end point in several clinical trials with novel JAK inhibitors. As new *JAK2*-specific inhibitors and other new drugs become available for therapeutic use, the relevance of monitoring disease response using quantitative assays may increase. Notably, there are currently no reference standards available for quantitative *JAK2* testing. As a result, *JAK2* quantitative assays are not standardized and test results may not be comparable across different laboratories.

JAK2 Exon 12 Mutation Analysis

Strategies to detect *JAK2* exon 12 mutations must have sufficient analytic sensitivity to detect mutations that can occur at low allelic levels (<15%) in affected individuals, as well as the ability to detect the wide variety of mutations that occur in this exon.

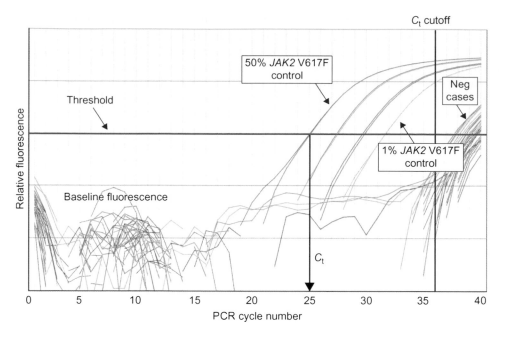

FIGURE 34.1 *JAK2* V617F mutation testing. Allele-specific real-time PCR is an effective and widely used technique to detect the *JAK2* c.1849G > T (V617F) substitution mutation. Allele-specific PCR refers to the selective amplification of targets that contain a specific allele. In the case of *JAK2* V617F, detection of the G > T nucleotide mutation is achieved by designing a PCR primer that matches the mutant base (T) at the most 3′ end of the primer. An intentional mismatch is also introduced at the -2 or -3 position from the 3′ end of the primer to maximize specificity by decreasing the efficiency of mismatched amplification products. Sensitivity of the assay is limited to the number of cycles of amplification before known negatives give rise to detectable mismatched PCR products. Detection of amplification products in real--time PCR occurs during each PCR cycle using fluorescent reporter probes or dyes. The accumulated fluorescence in log(10) value (y axis) is plotted against the number of PCR cycles (x axis). For a given specimen, the PCR cycle number at which the exponential increase in fluorescence exceeds a threshold line above the baseline signal is called cycle threshold (C_t). The C_t value is inversely proportional to the amount of PCR target in the specimen (ie, higher C_t values for a given specimen indicate lesser amount of target). Shown are real-time PCR plots of an allele-specific assay for the *JAK2* c.1849G > T (V617F) mutation. In this example, specimens with C_t values less than 36 are considered positive for the mutation. The C_t cutoff of 36 was determined during the preclinical validation of the assay to accurately distinguish mutant from wild-type cases while maintaining a clinically relevant limit of detection, which in this test approaches 0.1% mutant allele. Note that PCR increases the amount of amplification product by a factor of 2 with each PCR cycle. Therefore, specimens with a twofold lower concentration of target will exceed the fluorescent threshold 1 cycle later to produce a C_t value that is 1 cycle higher. This inherent feature of real-time PCR permits these assays to be utilized for quantitative testing.

Sanger sequencing has been widely used for *JAK2* exon 12 mutation analysis. However, this approach is relatively expensive, labor intensive, and has poor analytical sensitivity (~15%). Other methods, such as allele-specific PCR [42], quantitative real-time PCR [106], quantitative bead-based assay [107], and clamp-based methods [108], can achieve low analytical sensitivity (~2–0.1%), but they do not query the wide variety of possible exon 12 mutations, leading to potential false-negative results. Some *JAK2* exon 12 testing approaches have included a combination of mutation screening methods, such as high-resolution melting curve analysis which provides detection down to 5–10% mutant allele (Fig. 34.2A), followed by a Sanger sequencing confirmatory assay. A limitation of this approach is that equivocal melt profiles can occur, particularly in cases with a low mutant allelic burden that may not be confirmable by direct sequencing [109]. Therefore, it is important to use a sensitive screening method for exon 12 mutations that provides a

straightforward interpretation and can be reliably used as a standalone test, especially for cases with mutation levels that approach the detection limit of the assay. To this end, our group has recently reported a multiplex fragment analysis based assay that combines a length mutation assay to detect deletion and duplication mutations with an allele-specific PCR assay to detect K539L substitution mutations (Fig. 34.2B) [109]. This assay detects nearly all *JAK2* exon 12 mutations associated with PV with a high analytic sensitivity (1–2%) that exceeds direct sequencing and high-resolution melting. Notably, an exquisite analytic sensitivity may not be critical for exon 12 mutation analysis, given that diagnostic certainty of *JAK2* mutations is enhanced in the presence of greater than 1% mutant allele burden [101].

MPL Mutation Analysis

Similar to *JAK2* exon 12 mutation screening, strategies to detect *MPL* mutations must have sufficient

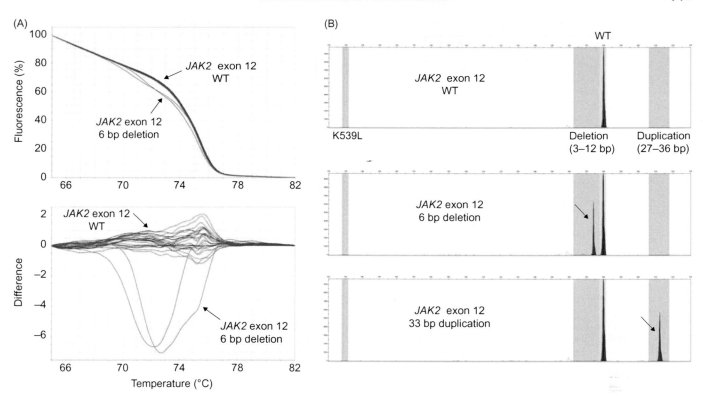

FIGURE 34.2 *JAK2* exon 12 mutation testing. (A) High-resolution melting (HRM) curve analysis. HRM can be an efficient method to detect a variety of different mutations that cluster in a specific region, such as occurs with *JAK2* exon 12 mutations in polycythemia vera. PCR is utilized to amplify the target region in the presence of a fluorescent reporter molecule. Following amplification, the double-stranded PCR products are melted (ie, denatured) with increasing temperature, during which time the decay of emitted fluorescence is measured. Plotting fluorescence versus temperature generates a characteristic melt curve (top plot). The presence of a mutation alters the melt profile due to mismatched double-stranded heteroduplexes of mutant and wild-type (WT) fragments. A difference plot in which sample curves are subtracted from a WT control can accentuate the melt profile differences (bottom plot). (B) Multiplex fragment length analysis. Most *JAK2* exon 12 mutations are small length-affecting deletions of 3–12 bp, duplications of 27–36 bp, or substitutions leading to K539L. Sensitive detection of these mutations can be accomplished by combining a PCR fragment sizing assay to detect deletions and duplications with an allele-specific PCR assay for detection of the K539L mutation. PCR products are resolved by capillary electrophoresis to distinguish the size of the amplicons. Shown is a negative case with only WT *JAK2* exon 12 amplicon (top) and cases with a 6 bp deletion mutation (middle), and a 33 bp duplication mutation (bottom). Pink analysis bins represent the locations of expected mutant fragment sizes. Arrows indicate the mutant fragments.

analytical sensitivity, as well as the ability to detect the variety of clinically relevant nucleotide alterations that occur within *MPL* exon 10.

Sanger sequencing appears insufficiently sensitive for robust *MPL* mutation testing, given the frequency of MPN cases that harbor mutations at levels below 15% mutant allele [55,56,58]. Because the detection of an *MPL* mutation at any level is significant in the setting of possible ET or PMF, and allows for appropriate medical decision making, sensitive assays that are capable of capturing cases with low mutant allele frequency should be employed for *MPL* mutation analysis. Several methods described for the detection of *MPL* mutations, such as high-resolution melting [53,57], qPCR [61,110], bead-based assay with locked nucleic acid modified probes [111], amplification refractory mutation system qPCR [112], pyrosequencing [113], and singleplex allele-specific PCR [55], can achieve low

analytical sensitivity (~3–0.1%). However, most of these previously reported assays do not include the S505N mutation [53,61,110–112], which can been identified in a significant proportion (10.3%) of *MPL* mutation positive cases [62]. We have described an allele-specific PCR assay capable of detecting nearly all *MPL* exon 10 mutations associated with PMF and ET (W515L, W515K, W515A, and S505N) at high analytic sensitivity (~2.5%) that can be easily deployed in most clinical laboratories (Fig. 34.3) [62].

CALR Mutation Analysis

Several methods have been reported for *CALR* mutation analysis, including PCR followed by Sanger sequencing [71,72,73,77], fragment length analysis assay [71], high-resolution melting curve analysis [114], and NGS.

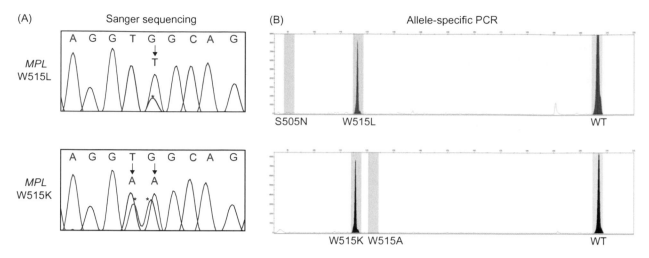

FIGURE 34.3 *MPL* mutation testing. (A) Sanger sequencing can detect the various nucleotide substitution mutations in *MPL*, albeit with limited sensitivity down to 10–15% mutant allele. Overlapping peaks in the DNA sequence chromatogram indicate the presence of a mutation. Cases with c.1544 G > T (W515L) and c.1543_1544delinsAA (W515K) mutations are shown. Mutant peaks are highlighted with asterisks. (B) Allele-specific PCR. Primers specific to various *MPL* substitution mutations can be utilized in a multiplexed PCR assay using capillary electrophoresis for detection. Primers designed to amplify a wild-type *MPL* product are included as a control. This technique is capable of detecting mutations down to 2% mutant allele burden and lower. Corresponding allele-specific PCR results are shown for the two cases that were Sanger sequenced. The size location of the expected mutation-specific PCR products is indicated with pink analysis bins.

Sanger sequencing is a gold standard technique for *CALR* mutation analysis. This method has the ability to detect the wide variety of nucleotide alterations, can determine the exact change in DNA sequence, and can distinguish in-frame length variants (germline polymorphisms) from pathogenic frameshift mutations. However, direct sequencing may be insufficiently sensitive in cases with low *CALR* mutant rate (<15%) [79]. Fragment analysis (Fig. 34.4A) can achieve higher analytical sensitivity (2–5%), but a validation of the different amplicon fragments sizes that correspond to specific *CALR* insertions and deletions based on direct sequencing results is required for accurate differentiation of length-affecting polymorphisms from mutations. In specimens with low mutant allelic burden, mutations identified by fragment analysis may not be confirmable by an alternate method, such as Sanger sequencing. NGS represents a viable approach for *CALR* mutation screening (Fig. 34.4B). It has higher analytical sensitivity than Sanger sequencing (~5–10% mutant allelic frequency) and can simultaneously interrogate multiple MPN-related genes during diagnostic workup.

Test Interpretation and Reporting

As with other molecular pathology tests, reports for MPN molecular testing should include information about preanalytic (eg, specimen type, indication for testing), analytic (eg, test methodology, analytic result, and test limitations), and postanalytic (eg, interpretative comments) components of the assay [115].

Mutations identified by sequencing-based methods should be reported using standard Human Genome Variation Society (HGVS) nucleotide and amino acid nomenclature (HGVS, http://www.hgvs.org/mutnomen). The mutant allelic frequency should be reported for quantitative assays, but the terms heterozygous and homozygous should not be used to describe cases with allele burdens of less than 50% and greater than or equal to 50%, respectively, because the mutation phase cannot be ascertained by traditional qPCR methods [116]. *CALR* in-frame length-affecting polymorphisms should not be reported to avoid confusion and potential misinterpretation by the clinical team. However, information about the type of *CALR* mutation detected (type 1 vs type 2) should be incorporated into the clinical report due to growing evidence suggesting differential impact on disease prognosis between these two variant types [75].

FUTURE DIRECTIONS

Advances in defining the mutational landscape of MPNs over the past decade have revolutionized the molecular diagnosis of these entities. Looking to the future, there is vast potential for expanded application of NGS technologies toward discovering novel MPN-related genes and defining new clinically relevant genetic subtypes of this heterogeneous group of neoplasms. With NGS making its way into clinical laboratories, testing will progressively move from mutational

(A)

(B)

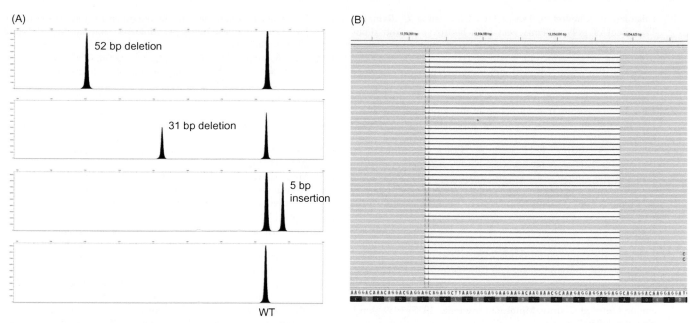

FIGURE 34.4 *CALR* mutation testing. (A) Capillary fragment length analysis. Greater than 36 different *CALR* mutations have been identified in ET and PMF. Despite the variety, all are length-affecting insertions or deletions that cluster in exon 9 of the gene. Detection of these mutations can be accomplished by a PCR fragment sizing assay utilizing PCR primers that flank the mutation region. The amplification products are sized and detected using capillary electrophoresis. Cases with 52 bp deletion, 31 bp deletion, and 5 bp insertion mutations are shown. A negative control with wild-type sized *CALR* is shown for reference. (B) Next-generation sequencing (NGS). Hundreds of thousands of sequence reads are simultaneously generated, then mapped and horizontally aligned to specific targeted regions in the reference genome (sequence shown on bottom). Software-assisted analysis assists in the detection of mutations. Wild-type sequence within each read is displayed in gray color while missing bases indicative of a deletion are indicated as contiguous gaps. Mutation frequency correlates to the number of reads demonstrating the mutant sequence compared to the total number of reads at that nucleotide position. Shown are NGS results from a case harboring a 52 bp deletion in *CALR*.

analysis of single genes toward a multigene panel analysis. The simultaneous evaluation of multiple MPN-associated genes has the potential to better refine disease diagnosis and to allow for the development of more personalized approaches for disease prognostication, risk stratification, management, and minimal residual disease monitoring.

CONCLUSIONS

Mutation testing is standard of care for diagnostic workup of any suspected MPN, with *JAK2* V617F being the most useful first test for PV, ET, and PMF. Well-established second-order tests include: *JAK2* exon 12 mutation analysis (suspected PV) and *CALR* and *MPL* mutation screening (suspected PMF or ET). The presence of a mutation establishes a clonal (neoplastic) proliferation and rules out a secondary/reactive condition. While the absence of *JAK2* mutations virtually rules out PV, the absence of *JAK2* V617F, *CALR*, and *MPL* does not rule out ET or PMF, given that approximately 13% ET and 10% PMF cases are triple negative [74,90]. Clinicopathological correlation is required for a final diagnosis of ET and PMF.

References

[1] Spivak JL. The chronic myeloproliferative disorders: clonality and clinical heterogeneity. Semin Hematol 2004;41:1−5.

[2] Tibes R, Bogenberger JM, Benson KL, Mesa RA. Current outlook on molecular pathogenesis and treatment of myeloproliferative neoplasms. Mol Diagn Ther 2012;16:269−83.

[3] Cervantes F, Dupriez B, Pereira A, Passamonti F, Reilly JT, Morra E, et al. New prognostic scoring system for primary myelofibrosis based on a study of the International Working Group for Myelofibrosis Research and Treatment. Blood 2009;113:2895−901.

[4] Vardiman JW, Thiele J, Arber DA, Brunning RD, Borowitz MJ, Porwit A, et al. The 2008 revision of the World Health Organization (WHO) classification of myeloid neoplasms and acute leukemia: rationale and important changes. Blood 2009; 114:937−51.

[5] Mesa RA, Schwager S, Radia D, Cheville A, Hussein K, Niblack J, et al. The Myelofibrosis Symptom Assessment Form (MFSAF): an evidence-based brief inventory to measure quality of life and symptomatic response to treatment in myelofibrosis. Leuk Res 2009;33:1199−203.

[6] Passamonti F, Rumi E, Arcaini L, Boveri E, Elena C, Pietra D, et al. Prognostic factors for thrombosis, myelofibrosis, and leukemia in essential thrombocythemia: a study of 605 patients. Haematologica 2008;93:1645−51.

[7] Bjorkholm M, Derolf AR, Hultcrantz M, Kristinsson SY, Ekstrand C, Goldin LR, et al. Treatment-related risk factors for transformation to acute myeloid leukemia and myelodysplastic syndromes in myeloproliferative neoplasms. J Clin Oncol 2011; 29:2410−15.

[8] Kiladjian JJ, Chevret S, Dosquet C, Chomienne C, Rain JD. Treatment of polycythemia vera with hydroxyurea and pipobroman: final results of a randomized trial initiated in 1980. J Clin Oncol 2011;29:3907–13.

[9] Campbell PJ, Green AR. The myeloproliferative disorders. N Engl J Med 2006;355:2452–66.

[10] Mesa RA, Li CY, Ketterling RP, Schroeder GS, Knudson RA, Tefferi A. Leukemic transformation in myelofibrosis with myeloid metaplasia: a single-institution experience with 91 cases. Blood 2005;105:973–7.

[11] Crisa E, Venturino E, Passera R, Prina M, Schinco P, Borchiellini A, et al. A retrospective study on 226 polycythemia vera patients: impact of median hematocrit value on clinical outcomes and survival improvement with anti-thrombotic prophylaxis and non-alkylating drugs. Ann Hematol 2010; 89:691–9.

[12] Barbui T, Thiele J, Passamonti F, Rumi E, Boveri E, Ruggeri M, et al. Survival and disease progression in essential thrombocythemia are significantly influenced by accurate morphologic diagnosis: an international study. J Clin Oncol 2011;29:3179–84.

[13] Ania BJ, Suman VJ, Sobell JL, Codd MB, Silverstein MN, Melton III LJ. Trends in the incidence of polycythemia vera among Olmsted County, Minnesota residents, 1935–1989. Am J Hematol 1994;47:89–93.

[14] Mesa RA, Silverstein MN, Jacobsen SJ, Wollan PC, Tefferi A. Population-based incidence and survival figures in essential thrombocythemia and agnogenic myeloid metaplasia: an Olmsted County Study, 1976–1995. Am J Hematol 1999;61:10–15.

[15] Bellanne-Chantelot C, Chaumarel I, Labopin M, Bellanger F, Barbu V, De Toma C, et al. Genetic and clinical implications of the Val617Phe JAK2 mutation in 72 families with myeloproliferative disorders. Blood 2006;108:346–52.

[16] Hultcrantz M, Kristinsson SY, Andersson TM, Landgren O, Eloranta S, Derolf AR, et al. Patterns of survival among patients with myeloproliferative neoplasms diagnosed in Sweden from 1973 to 2008: a population-based study. J Clin Oncol 2012;30: 2995–3001.

[17] Hoffman R, Rondelli D. Biology and treatment of primary myelofibrosis. Hematology Am Soc Hematol Educ Program 2007;346–54.

[18] Levine RL, Wadleigh M, Cools J, Ebert BL, Wernig G, Huntly BJ, et al. Activating mutation in the tyrosine kinase JAK2 in polycythemia vera, essential thrombocythemia, and myeloid metaplasia with myelofibrosis. Cancer Cell 2005;7:387–97.

[19] Kralovics R, Passamonti F, Buser AS, Teo SS, Tiedt R, Passweg JR, et al. A gain-of-function mutation of JAK2 in myeloproliferative disorders. N Engl J Med 2005;352:1779–90.

[20] James C, Ugo V, Le Couedic JP, Staerk J, Delhommeau F, Lacout C, et al. A unique clonal JAK2 mutation leading to constitutive signalling causes polycythaemia vera. Nature 2005;434:1144–8.

[21] Baxter EJ, Scott LM, Campbell PJ, East C, Fourouclas N, Swanton S, et al. Acquired mutation of the tyrosine kinase JAK2 in human myeloproliferative disorders. Lancet 2005;365: 1054–61.

[22] Witthuhn BA, Quelle FW, Silvennoinen O, Yi T, Tang B, Miura O, et al. JAK2 associates with the erythropoietin receptor and is tyrosine phosphorylated and activated following stimulation with erythropoietin. Cell 1993;74:227–36.

[23] Ugo V, Marzac C, Teyssandier I, Larbret F, Lecluse Y, Debili N, et al. Multiple signaling pathways are involved in erythropoietin-independent differentiation of erythroid progenitors in polycythemia vera. Exp Hematol 2004;32:179–87.

[24] Yamaoka K, Saharinen P, Pesu M, Holt III VE, Silvennoinen O, O'Shea JJ. The Janus kinases (Jaks). Genome Biol 2004;5:253.

[25] Saharinen P, Takaluoma K, Silvennoinen O. Regulation of the Jak2 tyrosine kinase by its pseudokinase domain. Mol Cell Biol 2000;20:3387–95.

[26] Zhao R, Xing S, Li Z, Fu X, Li Q, Krantz SB, et al. Identification of an acquired JAK2 mutation in polycythemia vera. J Biol Chem 2005;280:22788–92.

[27] Morgan KJ, Gilliland DG. A role for JAK2 mutations in myeloproliferative diseases. Ann Rev Med 2008;59:213–22.

[28] Tiedt R, Hao-Shen H, Sobas MA, Looser R, Dirnhofer S, Schwaller J, et al. Ratio of mutant JAK2-V617F to wild-type Jak2 determines the MPD phenotypes in transgenic mice. Blood 2008;111:3931–40.

[29] Xing S, Wanting TH, Zhao W, Ma J, Wang S, Xu X, et al. Transgenic expression of JAK2V617F causes myeloproliferative disorders in mice. Blood 2008;111:5109–17.

[30] Scott LM, Scott MA, Campbell PJ, Green AR. Progenitors homozygous for the V617F mutation occur in most patients with polycythemia vera, but not essential thrombocythemia. Blood 2006;108:2435–7.

[31] Passamonti F, Cervantes F, Vannucchi AM, Morra E, Rumi E, Pereira A, et al. A dynamic prognostic model to predict survival in primary myelofibrosis: a study by the IWG-MRT (International Working Group for Myeloproliferative Neoplasms Research and Treatment). Blood 2010;115:1703–8.

[32] Campbell PJ, Griesshammer M, Dohner K, Dohner H, Kusec R, Hasselbalch HC, et al. V617F mutation in JAK2 is associated with poorer survival in idiopathic myelofibrosis. Blood 2006; 107:2098–100.

[33] Guglielmelli P, Barosi G, Specchia G, Rambaldi A, Lo Coco F, Antonioli E, et al. Identification of patients with poorer survival in primary myelofibrosis based on the burden of JAK2V617F mutated allele. Blood 2009;114:1477–83.

[34] Kittur J, Knudson RA, Lasho TL, Finke CM, Gangat N, Wolanskyj AP, et al. Clinical correlates of JAK2V617F allele burden in essential thrombocythemia. Cancer 2007; 109:2279–84.

[35] Vannucchi AM, Antonioli E, Guglielmelli P, Longo G, Pancrazzi A, Ponziani V, et al. Prospective identification of high-risk polycythemia vera patients based on JAK2(V617F) allele burden. Leukemia 2007;21:1952–9.

[36] Jones AV, Kreil S, Zoi K, Waghorn K, Curtis C, Zhang L, et al. Widespread occurrence of the JAK2 V617F mutation in chronic myeloproliferative disorders. Blood 2005;106:2162–8.

[37] Steensma DP, Dewald GW, Lasho TL, Powell HL, McClure RF, Levine RL, et al. The JAK2 V617F activating tyrosine kinase mutation is an infrequent event in both "atypical" myeloproliferative disorders and myelodysplastic syndromes. Blood 2005; 106:1207–9.

[38] Scott LM, Campbell PJ, Baxter EJ, Todd T, Stephens P, Edkins S, et al. The V617F JAK2 mutation is uncommon in cancers and in myeloid malignancies other than the classic myeloproliferative disorders. Blood 2005;106:2920–1.

[39] Jelinek J, Oki Y, Gharibyan V, Bueso-Ramos C, Prchal JT, Verstovsek S, et al. JAK2 mutation 1849G > T is rare in acute leukemias but can be found in CMML, Philadelphia chromosome-negative CML, and megakaryocytic leukemia. Blood 2005;106:3370–3.

[40] Lee JW, Kim YG, Soung YH, Han KJ, Kim SY, Rhim HS, et al. The JAK2 V617F mutation in de novo acute myelogenous leukemias. Oncogene 2006;25:1434–6.

[41] Mulligan CG, Zhang J, Harvey RC, Collins-Underwood JR, Schulman BA, Phillips LA, et al. JAK mutations in high-risk childhood acute lymphoblastic leukemia. Proc Natl Acad Sci USA 2009;106:9414–18.

[42] Scott LM, Tong W, Levine RL, Scott MA, Beer PA, Stratton MR, et al. JAK2 exon 12 mutations in polycythemia vera and idiopathic erythrocytosis. N Engl J Med 2007;356:459–68.

[43] Scott LM, Beer PA, Bench AJ, Erber WN, Green AR. Prevalence of JAK2 V617F and exon 12 mutations in polycythaemia vera. Br J Haematol 2007;139(3):511–12.

[44] Pardanani A, Lasho TL, Finke C, Hanson CA, Tefferi A. Prevalence and clinicopathologic correlates of JAK2 exon 12 mutations in JAK2V617F-negative polycythemia vera. Leukemia 2007;21:1960–3.

[45] Pietra D, Li S, Brisci A, Passamonti F, Rumi E, Theocharides A, et al. Somatic mutations of JAK2 exon 12 in patients with JAK2 (V617F)-negative myeloproliferative disorders. Blood 2008;111: 1686–9.

[46] Scott LM. The JAK2 exon 12 mutations: a comprehensive review. Am J Hematol 2011;86:668–76.

[47] Tefferi A, Thiele J, Vardiman JW. The 2008 World Health Organization classification system for myeloproliferative neoplasms: order out of chaos. Cancer 2009;115:3842–7.

[48] Li S, Kralovics R, De Libero G, Theocharides A, Gisslinger H, Skoda RC. Clonal heterogeneity in polycythemia vera patients with JAK2 exon12 and JAK2-V617F mutations. Blood 2008;111:3863–6.

[49] Passamonti F, Elena C, Schnittger S, Skoda RC, Green AR, Girodon F, et al. Molecular and clinical features of the myeloproliferative neoplasm associated with JAK2 exon 12 mutations. Blood 2011;117:2813–16.

[50] Schnittger S, Bacher U, Haferlach C, Geer T, Muller P, Mittermuller J, et al. Detection of JAK2 exon 12 mutations in 15 patients with JAK2V617F negative polycythemia vera. Haematologica 2009;94:414–18.

[51] Levine RL. Mechanisms of mutations in myeloproliferative neoplasms. Best Pract Res Clin Haematol 2009;22:489–94.

[52] Ma W, Kantarjian H, Zhang X, Yeh CH, Zhang ZJ, Verstovsek S, et al. Mutation profile of JAK2 transcripts in patients with chronic myeloproliferative neoplasias. J Mol Diagn 2009; 11:49–53.

[53] Pardanani AD, Levine RL, Lasho T, Pikman Y, Mesa RA, Wadleigh M, et al. MPL515 mutations in myeloproliferative and other myeloid disorders: a study of 1182 patients. Blood 2006;108:3472–6.

[54] Pikman Y, Lee BH, Mercher T, McDowell E, Ebert BL, Gozo M, et al. MPLW515L is a novel somatic activating mutation in myelofibrosis with myeloid metaplasia. PLoS Med 2006;3:e270.

[55] Beer PA, Campbell PJ, Scott LM, Bench AJ, Erber WN, Bareford D, et al. MPL mutations in myeloproliferative disorders: analysis of the PT-1 cohort. Blood 2008;112:141–9.

[56] Schnittger S, Bacher U, Haferlach C, Beelen D, Bojko P, Burkle D, et al. Characterization of 35 new cases with four different MPLW515 mutations and essential thrombocytosis or primary myelofibrosis. Haematologica 2009;94:141–4.

[57] Boyd EM, Bench AJ, Goday-Fernandez A, Anand S, Vaghela KJ, Beer P, et al. Clinical utility of routine MPL exon 10 analysis in the diagnosis of essential thrombocythaemia and primary myelofibrosis. Br J Haematol 2010;149:250–7.

[58] Millecker L, Lennon PA, Verstovsek S, Barkoh B, Galbincea J, Hu P, et al. Distinct patterns of cytogenetic and clinical progression in chronic myeloproliferative neoplasms with or without JAK2 or MPL mutations. Cancer Genet Cytogenet 2010;197:1–7.

[59] Ding J, Komatsu H, Wakita A, Kato-Uranishi M, Ito M, Satoh A, et al. Familial essential thrombocythemia associated with a dominant-positive activating mutation of the c-MPL gene, which encodes for the receptor for thrombopoietin. Blood 2004;103:4198–200.

[60] Liu K, Martini M, Rocca B, Amos CI, Teofili L, Giona F, et al. Evidence for a founder effect of the MPL-S505N mutation in eight Italian pedigrees with hereditary thrombocythemia. Haematologica 2009;94:1368–74.

[61] Pancrazzi A, Guglielmelli P, Ponziani V, Bergamaschi G, Bosi A, Barosi G, et al. A sensitive detection method for MPLW515L or MPLW515K mutation in chronic myeloproliferative disorders with locked nucleic acid-modified probes and real-time polymerase chain reaction. J Mol Diagn 2008;10:435–41.

[62] Furtado LV, Weigelin HC, Elenitoba-Johnson KS, Betz BL. Detection of MPL mutations by a novel allele-specific PCR-based strategy. J Mol Diagn 2013;15:810–18.

[63] Williams DM, Kim AH, Rogers O, Spivak JL, Moliterno AR. Phenotypic variations and new mutations in JAK2 V617F-negative polycythemia vera, erythrocytosis, and idiopathic myelofibrosis. Exp Hematol 2007;35:1641–6.

[64] Chaligne R, James C, Tonetti C, Besancenot R, Le Couedic JP, Fava F, et al. Evidence for MPL W515L/K mutations in hematopoietic stem cells in primitive myelofibrosis. Blood 2007;110: 3735–43.

[65] Pardanani A, Lasho TL, Finke C, Mesa RA, Hogan WJ, Ketterling RP, et al. Extending Jak2V617F and MplW515 mutation analysis to single hematopoietic colonies and B and T lymphocytes. Stem Cells 2007;25:2358–62.

[66] Brisci A, Damin F, Pietra D, Galbiati S, Boggi S, Casetti I, et al. COLD-PCR and innovative microarray substrates for detecting and genotyping MPL exon 10 W515 substitutions. Clin Chem 2012;58:1692–702.

[67] Tefferi A. Novel mutations and their functional and clinical relevance in myeloproliferative neoplasms: JAK2, MPL, TET2, ASXL1, CBL, IDH and IKZF1. Leukemia 2010;24:1128–38.

[68] Vainchenker W, Delhommeau F, Constantinescu SN, Bernard OA. New mutations and pathogenesis of myeloproliferative neoplasms. Blood 2011;118:1723–35.

[69] Guglielmelli P, Pancrazzi A, Bergamaschi G, Rosti V, Villani L, Antonioli E, et al. Anaemia characterises patients with myelofibrosis harbouring Mpl mutation. Br J Haematol 2007;137:244–7.

[70] Vannucchi AM, Antonioli E, Guglielmelli P, Pancrazzi A, Guerini V, Barosi G, et al. Characteristics and clinical correlates of MPL 515W > L/K mutation in essential thrombocythemia. Blood 2008;112:844–7.

[71] Klampfl T, Gisslinger H, Harutyunyan AS, Nivarthi H, Rumi E, Milosevic JD, et al. Somatic mutations of calreticulin in myeloproliferative neoplasms. N Engl J Med 2013;369:2379–90.

[72] Nangalia J, Massie CE, Baxter EJ, Nice FL, Gundem G, Wedge DC, et al. Somatic CALR mutations in myeloproliferative neoplasms with nonmutated JAK2. N Engl J Med 2013;369: 2391–405.

[73] Rotunno G, Mannarelli C, Guglielmelli P, Pacilli A, Pancrazzi A, Pieri L, et al. Impact of calreticulin mutations on clinical and hematological phenotype and outcome in essential thrombocythemia. Blood 2014;123:1552–5.

[74] Tefferi A, Lasho TL, Finke CM, Knudson RA, Ketterling R, Hanson CH, et al. CALR vs JAK2 vs MPL-mutated or triple-negative myelofibrosis: clinical, cytogenetic and molecular comparisons. Leukemia 2014;28:1472–7.

[75] Tefferi A, Lasho TL, Finke C, Belachew AA, Wassie EA, Ketterling RP, et al. Type 1 vs type 2 calreticulin mutations in primary myelofibrosis: differences in phenotype and prognostic impact. Leukemia 2014;28:1568–70.

[76] McGaffin G, Harper K, Stirling D, McLintock L. JAK2 V617F and CALR mutations are not mutually exclusive; findings from retrospective analysis of a small patient cohort. Br J Haematol 2014;167:276–8.

[77] Rumi E, Pietra D, Ferretti V, Klampfl T, Harutyunyan AS, Milosevic JD, et al. JAK2 or CALR mutation status defines subtypes of essential thrombocythemia with substantially different clinical course and outcomes. Blood 2014;123:1544—51.

[78] Chi J, Nicolaou KA, Nicolaidou V, Koumas L, Mitsidou A, Pierides C, et al. Calreticulin gene exon 9 frameshift mutations in patients with thrombocytosis. Leukemia 2014;28:1152—4.

[79] Wojtaszewska M, Iwola M, Lewandowski K. Frequency and molecular characteristics of calreticulin gene (CALR) mutations in patients with JAK2-negative myeloproliferative neoplasms. Acta Haematol 2014;133:193—8.

[80] Broseus J, Lippert E, Harutyunyan AS, Jeromin S, Zipperer E, Florensa L, et al. Low rate of calreticulin mutations in refractory anaemia with ring sideroblasts and marked thrombocytosis. Leukemia 2014;28:1374—6.

[81] Tefferi A, Guglielmelli P, Larson DR, Finke C, Wassie EA, Pieri L, et al. Long-term survival and blast transformation in molecularly annotated essential thrombocythemia, polycythemia vera, and myelofibrosis. Blood 2014;124:2507—13.

[82] Tefferi A, Lasho TL, Tischer A, Wassie EA, Finke CM, Belachew AA, et al. The prognostic advantage of calreticulin mutations in myelofibrosis might be confined to type 1 or type 1-like CALR variants. Blood 2014;124:2465—6.

[83] Tefferi A, Guglielmelli P, Lasho TL, Rotunno G, Finke C, Mannarelli C, et al. CALR and ASXL1 mutations-based molecular prognostication in primary myelofibrosis: an international study of 570 patients. Leukemia 2014;28:1494—500.

[84] Milosevic JD, Kralovics R. Genetic and epigenetic alterations of myeloproliferative disorders. Int J Hematol 2013;97:183—97.

[85] Score J, Cross NC. Acquired uniparental disomy in myeloproliferative neoplasms. Hematol Oncol Clin North Am 2012; 26:981—91.

[86] Bruchova H, Yoon D, Agarwal AM, Mendell J, Prchal JT. Regulated expression of microRNAs in normal and polycythemia vera erythropoiesis. Exp Hematol 2007;35:1657—67.

[87] Guglielmelli P, Tozzi L, Pancrazzi A, Bogani C, Antonioli E, Ponziani V, et al. MicroRNA expression profile in granulocytes from primary myelofibrosis patients. Exp Hematol 2007;35(11):1708—18.

[88] Bruchova-Votavova H, Yoon D, Prchal JT. miR-451 enhances erythroid differentiation in K562 cells. Leuk Lymphoma 2010;51(4):686—93.

[89] Bortoluzzi S, Bisognin A, Biasiolo M, Guglielmelli P, Biamonte F, Norfo R, et al. Characterization and discovery of novel miRNAs and moRNAs in JAK2V617F-mutated SET2 cells. Blood 2012;119(13):e120—30.

[90] Tefferi A, Wassie EA, Lasho TL, Finke C, Belachew AA, Ketterling RP, et al. Calreticulin mutations and long-term survival in essential thrombocythemia. Leukemia 2014;28:2300—3.

[91] Pardanani A, Gotlib JR, Jamieson C, Cortes JE, Talpaz M, Stone RM, et al. Safety and efficacy of TG101348, a selective JAK2 inhibitor, in myelofibrosis. J Clin Oncol 2011;29:789—96.

[92] Tefferi A, Vardiman JW. Classification and diagnosis of myeloproliferative neoplasms: the 2008 World Health Organization criteria and point-of-care diagnostic algorithms. Leukemia 2008;22:14—22.

[93] Broseus J, Park JH, Carillo S, Hermouet S, Girodon F. Presence of calreticulin mutations in JAK2-negative polycythemia vera. Blood 2014;124:3964—6.

[94] Tefferi A, Thiele J, Vannucchi AM, Barbui T. An overview on CALR and CSF3R mutations and a proposal for revision of WHO diagnostic criteria for myeloproliferative neoplasms. Leukemia 2014;28:1407—13.

[95] Takahashi K, Patel KP, Kantarjian H, Luthra R, Pierce S, Cortes J, et al. JAK2 p.V617F detection and allele burden

[96] measurement in peripheral blood and bone marrow aspirates in patients with myeloproliferative neoplasms. Blood 2013; 122:3784—6.

[96] Larsen TS, Pallisgaard N, Moller MB, Hasselbalch HC. Quantitative assessment of the JAK2 V617F allele burden: equivalent levels in peripheral blood and bone marrow. Leukemia 2008;22:194—5.

[97] Hermouet S, Dobo I, Lippert E, Boursier MC, Ergand L, Perrault-Hu F, et al. Comparison of whole blood vs purified blood granulocytes for the detection and quantitation of JAK2 (V617F). Leukemia 2007;21:1128—30.

[98] Lippert E, Boissinot M, Kralovics R, Girodon F, Dobo I, Praloran V, et al. The JAK2-V617F mutation is frequently present at diagnosis in patients with essential thrombocythemia and polycythemia vera. Blood 2006;108:1865—7.

[99] Cankovic M, Whiteley L, Hawley RC, Zarbo RJ, Chitale D. Clinical performance of JAK2 V617F mutation detection assays in a molecular diagnostics laboratory: evaluation of screening and quantitation methods. Am J Clin Pathol 2009;132:713—21.

[100] Antonioli E, Guglielmelli P, Poli G, Bogani C, Pancrazzi A, Longo G, et al. Influence of JAK2V617F allele burden on phenotype in essential thrombocythemia. Haematologica 2008;93:41—8.

[101] Martinaud C, Brisou P, Mozziconacci MJ. Is the JAK2(V617F) mutation detectable in healthy volunteers? Am J Hematol 2010;85:287—8.

[102] Er TK, Lin SF, Chang JG, Hsieh LL, Lin SK, Wang LH, et al. Detection of the JAK2 V617F missense mutation by high resolution melting analysis and its validation. Clin Chim Acta 2009;408:39—44.

[103] Kannim S, Thongnoppakhun W, Auewarakul CU. Two-round allele specific-polymerase chain reaction: a simple and highly sensitive method for JAK2V617F mutation detection. Clin Chim Acta 2009;401:148—51.

[104] Tan AY, Westerman DA, Dobrovic A. A simple, rapid, and sensitive method for the detection of the JAK2 V617F mutation. Am J Clin Pathol 2007;127:977—81.

[105] Kroger N, Badbaran A, Holler E, Hahn J, Kobbe G, Bornhauser M, et al. Monitoring of the JAK2-V617F mutation by highly sensitive quantitative real-time PCR after allogeneic stem cell transplantation in patients with myelofibrosis. Blood 2007;109:1316—21.

[106] Kjaer L, Westman M, Hasselbalch Riley C, Hogdall E, Weis Bjerrum O, Hasselbalch H. A highly sensitive quantitative real-time PCR assay for determination of mutant JAK2 exon 12 allele burden. PLoS One 2012;7:e33100.

[107] Shivarov V, Ivanova M, Yaneva S, Petkova N, Hadjiev E, Naumova E. Quantitative bead-based assay for detection of JAK2 exon 12 mutations. Leuk Lymphoma 2013;54:1343—4.

[108] Laughlin TS, Moliterno AR, Stein BL, Rothberg PG. Detection of exon 12 mutations in the JAK2 gene: enhanced analytical sensitivity using clamped PCR and nucleotide sequencing. J Mol Diagn 2010;12:278—82.

[109] Furtado LV, Weigelin HC, Elenitoba-Johnson KS, Betz BL. A multiplexed fragment analysis-based assay for detection of JAK2 exon 12 mutations. J Mol Diagn 2013;15:592—9.

[110] Alchalby H, Badbaran A, Bock O, Fehse B, Bacher U, Zander AR, et al. Screening and monitoring of MPL W515L mutation with real-time PCR in patients with myelofibrosis undergoing allogeneic-SCT. Bone Marrow Transplant 2010;45:1404—7.

[111] Ivanova MI, Shivarov VS, Hadjiev EA, Naumova EJ. Novel multiplex bead-based assay with LNA-modified probes for detection of MPL exon 10 mutations. Leuk Res 2011;35:1120—3.

[112] Zhuge J, Zhang W, Zhang W, Xu M, Hoffman R. Sensitive detection of MPLW515L/K mutations by amplification

refractory mutation system (ARMS)-PCR. Clin Chim Acta 2010;411:122−3.

[113] Hussein K, Bock O, Theophile K, Schulz-Bischof K, Porwit A, Schlue J, et al. MPLW515L mutation in acute megakaryoblastic leukaemia. Leukemia 2009;23:852−5.

[114] Bilbao-Sieyro C, Santana G, Moreno M, Torres L, Santana-Lopez G, Rodriguez-Medina C, et al. High resolution melting analysis: a rapid and accurate method to detect CALR mutations. PLoS One 2014;9:e103511.

[115] Gulley ML, Braziel RM, Halling KC, Hsi ED, Kant JA, Nikiforova MN, et al. Clinical laboratory reports in molecular pathology. Arch Pathol Lab Med 2007;131: 852−63.

[116] Gong JZ, Cook JR, Greiner TC, Hedvat C, Hill CE, Lim MS, et al. Laboratory practice guidelines for detecting and reporting JAK2 and MPL mutations in myeloproliferative neoplasms: a report of the Association for Molecular Pathology. J Mol Diagn 2013;15: 733−44.

35

Molecular Testing in the Assessment of Bone Marrow Transplant Engraftment

J.K. Booker

Department of Pathology and Laboratory Medicine; Department of Genetics,
University of North Carolina at Chapel Hill, Chapel Hill, NC, United States

INTRODUCTION

Allogeneic hematopoietic stem cell transplantation (HSCT) has been used for decades to treat a wide variety of inherited and acquired disorders. Inherited disorders treated with HSCT include hemoglobinopathies, bone marrow failure syndromes, immunodeficiencies, and inborn metabolic diseases [1,2]. HSCT is also used in the treatment of acquired hematological malignancies and solid tumors. The application for solid tumors is for the benefit from graft-versus-tumor activity [3] and as an adjunct to high-dose chemotherapy [4].

The source of hematopoietic stem cells for HSCT can be from peripheral blood (obtained by apheresis), bone marrow, or cord blood. The best source of stem cells is somewhat controversial in the literature and is likely dependent on the mobilization regimen, preconditioning regimen, and reason for transplant [1,5]. While earlier HSCT used bone marrow as a donor source, the most recent literature suggests that HSCT utilizing stem cells from peripheral blood may actually be advantageous [6,7]. T-cell depletion prior to transplant can minimize the risk of graft-versus-host disease (GVHD), but increases the risk of graft rejection and also diminishes graft-versus-leukemia (GVL) activity [8,9].

Engraftment kinetics depends upon the type of pretransplant regimen the patient receives. HSCT was originally developed for patients who had received myeloablative doses of chemotherapy and radiotherapy [10]. Myeloablative conditioning was considered optimal for patients receiving HSCT for hematopoietic malignancies until the benefit of graft-versus-tumor activity was recognized [11]. Myeloablative conditioning regimens are expected to ablate marrow hematopoiesis and do not allow for autologous hematologic recovery. Nonmyeloablative conditioning regimens use lower doses of chemotherapeutic agents and total body irradiation which results in minimal cytopenias, without the requirement for stem cell support. Reduced intensity conditioning (RIC) falls between myeloablative and nonmyeloablative, resulting in prolonged cytopenias that do require hematopoietic stem cell support, although the cytopenia may not be irreversible [12]. While myeloablative regimens are still used, RIC and nonmyeloablative regimens continue to evolve as they are optimized for types and stages of disease. The less intensive regimens, which essentially represent a continuum, make HSCT available to patients who would not be expected to tolerate myeloablative conditioning.

Regardless of the pretransplant conditioning regimen, when engraftment is successful the recipient becomes a chimera (an organism comprised of more than one individual). Longitudinal monitoring of chimerism is used for the early detection of impending graft rejection, information relevant to GVHD, and when the transplant was performed as a treatment for a malignant hematological disorder, monitoring chimerism can be used for early detection of relapse [13].

Monitoring engraftment of HSCT is routinely accomplished through identity testing, also known as DNA fingerprinting. It is generally the same methodology employed for paternity testing, identification of maternal cell contamination in fetal samples, and most significantly, in the field of forensics [14]. Forensic

© 2017 Elsevier Inc. All rights reserved.

analysis requires a highly robust and dependable method that can absolutely discriminate between two individuals, even when samples are minute or compromised.

MOLECULAR TARGET

The predominant method currently used for monitoring of HSCT engraftment utilizes short tandem repeats (STRs), also known as microsatellites. STRs are repetitive DNA sequences between 1 and 6 bp in length that are repeated in tandem [15]. The number of times any given STR is repeated is highly polymorphic between individuals, making them an ideal target for identity testing. These sequences account for about 3% of the human genome and are scattered throughout the genome [16]. The majority occur in noncoding regions of the genome, with 8% located within coding regions [17].

Alternative methods utilizing single-nucleotide polymorphic (SNP) markers include pyrosequencing [18], allele-specific real-time polymerase chain reaction (PCR) [19,20], and TaqMan real-time PCR [21,22]. Additionally, allele-specific PCR of insertion/deletion markers can be used [23–28]. SNP and indel markers require analysis of a greater number of specific markers compared to STRs due to their relatively low discriminating power. However, their greater sensitivity is advantageous for monitoring minimal residual disease (MMR) [29].

MOLECULAR TECHNOLOGIES

Monitoring HSCT engraftment is typically performed using DNA extracted from leukocytes present in peripheral blood and bone marrow. Analysis can be accomplished with unfractionated samples, as well as from specific cellular subsets purified by flow cytometry or immunomagnetic methods. There are commercially-available kits for the multiplex amplification of STRs as well as laboratory-developed assays for single marker and multiplex analysis. PCR primers are designed to flank the DNA region containing the tandem repeats, positioned in regions that do not contain SNPs that might interfere with annealing and amplification. For any given marker, the size of the PCR product is determined by the number of repeats present on the two alleles of an individual. Typically, each forward primer is labeled with a fluorescent dye so that the resultant amplification product is fluorescently labeled. During assay development, the overall size range that includes all repeat number known for any given marker can be controlled by the placement of the primers. Primers that are positioned close to the repeat region will generate products that are overall smaller than those generated by primers that are positioned further away from the repeat region.

Multiplex assays are designed such that the range in size of the alleles of each marker is distinguishable from the range of alleles of other markers by the size range of the amplified products as well as the fluorescent dyes that are used. Resultant fluorescently-labeled PCR products are separated by capillary gel electrophoresis with a size standard included with each sample and an allelic ladder included on each run. An allelic ladder for each STR (consisting of all available alleles for that marker) enables software to identify the alleles present in each sample. Software is used to measure the height and area of each resultant peak. Calculations to determine the relative contribution of cells from donor and recipient can be accomplished based upon either the peak area or peak height.

Monitoring of engraftment begins with the analysis of donor and pretransplant recipient samples to identify informative STR markers. An informative marker is one that provides clear differentiation between the donor and recipient. Because STR markers have been selected for their high degree of polymorphism, most individuals inherit two different alleles for each marker (one from each parent). As long as the donor and recipient differ by at least one allele of a specific marker, that marker is informative for monitoring engraftment in that individual.

A technical issue that impacts selection of informative markers for monitoring engraftment is the phenomenon of stutter which occurs during PCR amplification [30]. Due to the repetitive nature of the STRs, there is some slippage during amplification which results in a product one repeat smaller or larger than the true allele and is referred to as stutter. The stutter product can be as great as 10% of the amplified product of the true allele and can occur in either direction (one repeat smaller or larger than the true allele), but is more commonly one repeat smaller. If a stutter product is large enough, it can generate an additional stutter product which would be two repeats smaller than the true allele. The consequence of stutter is that it limits the selection of informative markers. If for any given marker the donor and recipient share an allele that differs by only a single repeat, it can impact usage. For example, consider the THO1 marker and a donor whose two alleles have 8 and 10 repeats and a recipient whose two alleles have 7 and 10 repeats. If a small amount of seven repeats is present in a posttransplant sample, it cannot be determined whether this represents cells of recipient origin, or if it is stutter from the eight-repeat allele in the donor. Therefore, THO1 would not be an informative marker for this particular donor and recipient.

In theory, amplification from an individual with two alleles for any given marker should be roughly equal. What is seen in practice is variability, sometimes with preferential amplification of the smaller allele, which might be expected, and sometimes with preferential amplification seen for the larger allele. Because of the variability, which can result from technical artifact or a true biological process, multiple informative markers are analyzed for each posttransplant sample and averaged, for the most accurate estimate of chimerism. Results are reported as percent cells of donor origin and percent cells of recipient origin. Sensitivity of chimerism testing by STR is 1—5% [31], meaning that for a patient who is mostly engrafted with donor cells, STR analysis has the ability to detect as little as 1—5% of cells of recipient origin in a posttransplant sample from that patient.

It is not uncommon to have transplant patients transfer care over time, and when that happens, the new laboratory does not always receive a pretransplant recipient or donor sample. While not ideal, buccal or saliva samples can be used as a substitute for a pretransplant recipient sample as the cells obtained from these samples are primarily epithelial and will not have been replaced with donor cells. In reality, there are often lymphocytes which, if the patient is well engrafted, would be from the donor [32]. Comparison of the recipient's buccal or saliva sample with bone marrow or peripheral blood generally makes it possible to identify informative alleles. For any marker in the buccal or saliva sample that has four alleles, two can be matched with the donor present in peripheral blood or bone marrow, and the other two can be attributed to the recipient. For any marker that has two alleles that are not present in the peripheral blood or bone marrow, they can also be attributed to the recipient. This is possible if the patient is well engrafted, but becomes less reliable if there is a significant recipient contribution in the posttransplant peripheral blood or bone marrow sample.

CLINICAL UTILITY

Quantitative assessment of the state of chimerism in a transplanted individual's blood or bone marrow can provide critical information regarding transplant rejection, GVHD, and in some cases, relapse. While monitoring patients at critical points such as a change in clinical status or altered therapeutics can provide valuable information, it is more informative if the patient is monitored longitudinally so the results can be interpreted in the context of the patient's history and previous levels of chimerism. Additionally, results from longitudinal testing can be compared to engraftment kinetics of other HSCT patients with the same treatment and prognosis [33].

Following HSCT, there can be a delicate balance between the risks of graft failure or relapse and GVHD [34]. Immunosuppressive therapy can be increased to prevent graft rejection when the transplant was performed for nonmalignant disease [35]. With malignant diseases, decreasing immunosuppressive therapy can optimize graft-versus-tumor activity, along with donor lymphocyte infusions and the use of immunomodulatory cytokines [36].

The kinetics of donor engraftment are dependent upon the pretransplant conditioning regimen. With less intense pretransplant conditioning regimens, a higher degree of mixed chimerism is expected, at least shortly after the transplant. In patients who received RIC and have been transplanted to treat inherited or acquired nonmalignant disease, it is not necessary to completely replace the recipient's hematopoietic system and mixed chimerism may exist indefinitely. Studies have shown that mixed chimerism persists in 50% of such patients and 10—20% of donor cells are necessary to produce a significant clinical effect [37]. When HSCT is utilized in the treatment of malignant disease, persistent or increasing recipient cells can either indicate relapse or survival of host hematopoietic cells.

Longitudinal monitoring of engraftment from peripheral blood or bone marrow can provide valuable information and guide clinical decisions. Chimerism studies on different cell populations within the myeloid and lymphoid lineages can provide additional information about graft rejection, GVHD, and relapse. Isolation of desired subsets of cells can be performed by flow cytometry or immunomagnetic bead-based techniques. Once the targeted population has been obtained, the procedure and analysis are identical to that for unfractionated samples—DNA is extracted from the isolated cell population, STR markers are amplified by PCR, and the resultant products are visualized by capillary gel electrophoresis. The relative contribution of donor and recipient cells is calculated and reported, along with results from an unfractionated sample. Interpretation depends on the status of the patient, previous results, reason for the transplant, and the pretransplant conditioning regimen that was employed.

Chimerism studies for specific cell populations are most frequently performed for CD3+ cells as increasing or consistently high levels of recipient CD3+ T cells and CD56+ NK cells are associated with graft rejection [38,39]. This is in contrast to complete donor engraftment of cells from the myeloid lineage which can be seen at the same time. Monitoring CD3+ T-cell and CD56+ NK-cell populations along with unfractionated samples provides an early indication of impending graft rejection, which is often treated with a donor lymphocyte infusion. There are many other opportunities for performing lineage-specific chimerism studies. For example, chimerism studies on the CD34+ subset in patients with acute myeloid leukemia and B-cell acute lymphoblastic leukemia are useful in detecting imminent relapse earlier than studies on unfractionated cells from the same patients [40,41].

Utilizing engraftment studies for early indications of relapse can be challenging as the sensitivity of chimerism testing by STR analysis is generally in the range of 1−5% which is several orders of magnitude higher than what is optimal for monitoring of MMR, which aims to be in the range of 10^{-3} to 10^{-6}. In cases where there is a molecular target for a malignant population, the tumor-specific molecular marker will likely provide a more sensitive method for MMR detection. However, in many malignancies there are no known tumor-specific molecular markers and chimerism studies may be the only means for monitoring of MMR. In such cases, sensitivity can be significantly increased by lineage-specific testing. Depending on the contribution of a specific lineage to the overall leukocyte population, sensitivity within subsets can be in the range of 0.1−0.01% [42].

LIMITATIONS OF TESTING

The greatest limitation of chimerism testing is sensitivity. STR markers are distributed throughout the genome and can be selected to cumulatively represent multiple chromosomes. Care should be taken when HSCT is used in the case of a hematological malignancy characterized by or susceptible to chromosomal aneuploidy [43]. If chromosomal aneuploidy is present in the malignancy, and relapse occurs, it is possible to either miss it or have the calculated chimerism values be incorrect if the only informative markers are on the impacted chromosome(s). When multiple markers are analyzed, and a single marker yields a value that is an outlier, the presence of chromosomal aneuploidy should be considered.

A challenge presented by the literature on monitoring engraftment in HSCT is that there are so many variables present both for the donor and recipient prior to transplant as well as follow-up treatments after the transplant, it is difficult to interpret any given single chimerism result [44]. With longitudinal testing, results are interpreted in the context of a specific patient, their disease, type of pretransplant conditioning, preparation of transplanted cells, posttransplant therapy, and previous chimerism results. It is in this context that chimerism testing offers the greatest potential for impacting patient care.

Technologies are rapidly changing and there will likely be more efficient, automated, and sensitive methods for monitoring engraftment in the future. It is important to keep in mind that results from different methods, or even the same method performed in different laboratories, are not necessarily interchangeable, which becomes relevant when a patient transfers to a different laboratory. A final consideration for HSCT recipients who achieve complete donor engraftment is that any future genetic testing that is performed on peripheral leukocytes will yield results on the donor, rather than the recipient.

References

[1] Karakukcu M, Unal E. Stem cell mobilization and collection from pediatric patients and healthy children. Transfus Apher Sci 2015;53:17−22.

[2] Peffault de Latour R, Peters C, Gibson B, et al. Recommendations on hematopoietic stem cell transplantation for inherited bone marrow failure syndromes. Bone Marrow Transplant 2015;50:1168−72.

[3] Marraco SAF, Verdeil TMG, Speiser DE. From T cell "exhaustion" to anti-cancer immunity. Biochim Biophys Acta 2016;1865:49−57.

[4] Pedrazzoli P, Martino M, Delfanti S, et al. High-dose chemotherapy with autologous hematopoietic stem cell transplantation for high-risk primary breast cancer. J Natl Cancer Inst Monogr 2015;2015:70−5.

[5] Champlin RE, Schmitz N, Horowitz MM, et al. Blood stem cells compared with bone marrow as a source of hematopoietic cells for allogeneic transplantation. IBMTR Histocompatibility and Stem Cell Sources Working Committee and the European Group for Blood and Marrow Transplantation (EBMT). Blood 2000;95:3702−9.

[6] Wu S, Zhang C, Zhang X, Xu Y, Deng T. Is peripheral blood or bone marrow a better source of stem cells for transplantation in cases of HLA-matched unrelated donors? A meta-analysis. Crit Rev Oncol 2015;96:20−33.

[7] Holtick U, Albrecht M, Chemnitz JM, et al. Comparison of bone marrow versus peripheral blood allogeneic hematopoietic stem cell transplantation for hematological malignancies in adults—a systematic review and meta-analysis. Crit Rev Oncol 2015;94: 179−88.

[8] Aversa F. T-cell depletion: from positive selection to negative depletion in adult patients. Bone Marrow Transplant 2015;50: S11−13.

[9] Triplett B, Shook D, Eldridge P, et al. Rapid memory T-cell reconstitution recapitulating CD45RA-depleted haploidentical transplant graft content in patients with hematologic malignancies. Bone Marrow Transplant 2015;50:1012.

[10] Pingali S, Champlin R. Pushing the envelope—nonmyeloablative and reduced intensity preparative regimens for allogeneic hematopoietic transplantation. Bone Marrow Transplant 2015; 50:1157−67.

[11] Gyurkocza B, Sandmaier BM. Conditioning regimens for hematopoietic cell transplantation: one size does not fit all. Blood 2014;124:344−53.

[12] Bacigalupo A, Ballen K, Rizzo D, et al. Defining the intensity of conditioning regimens: working definitions. Biol Blood Marrow Transplant 2009;15:1628−33.

[13] Zielińska P, Markiewicz M, Dzierżak-Mietła M, et al. Assessment of lineage-specific chimerism after allogeneic stem cell transplantation. Acta Haematol Pol 2014;45(4):360–9.

[14] Buttler JM. Commonly used short tandem repeat markers. Forensic DNA typing. San Diego, CA: Academic Press; 2001. p. 53–79.

[15] Tautz D. Notes on the definition and nomenclature of tandemly repetitive DNA sequences. DNA fingerprinting: state of the science. Springer; 1993. p. 21–8.

[16] Fan H, Chu J. A brief review of short tandem repeat mutation. Genomics Proteomics Bioinformatics 2007;5:7–14.

[17] Ellegren H. Heterogeneous mutation processes in human microsatellite DNA sequences. Nat Genet 2000;24:400–2.

[18] Hochberg EP, Miklos DB, Neuberg D, et al. A novel rapid single nucleotide polymorphism (SNP)-based method for assessment of hematopoietic chimerism after allogeneic stem cell transplantation. Blood 2003;101:363–9.

[19] Eshel R, Vainas O, Shpringer M, Naparstek E. Highly sensitive patient-specific real-time PCR SNP assay for chimerism monitoring after allogeneic stem cell transplantation. Lab Hematol 2006;12:39–46.

[20] Maas F, Schaap N, Kolen S, et al. Quantification of donor and recipient hemopoietic cells by real-time PCR of single nucleotide polymorphisms. Leukemia 2003;17:621–9.

[21] Oliver DH, Thompson RE, Griffin CA, Eshleman JR. Use of single nucleotide polymorphisms (SNP) and real-time polymerase chain reaction for bone marrow engraftment analysis. J Mol Diagn 2000;2:202–8.

[22] Harries L, Wickham C, Evans J, Rule S, Joyner M, Ellard S. Analysis of haematopoietic chimerism by quantitative real-time polymerase chain reaction. Bone Marrow Transplant 2005;35:283–90.

[23] Alizadeh M, Bernard M, Danic B, et al. Quantitative assessment of hematopoietic chimerism after bone marrow transplantation by real-time quantitative polymerase chain reaction. Blood 2002;99:4618–25.

[24] Masmas TN, Madsen HO, Petersen SL, et al. Evaluation and automation of hematopoietic chimerism analysis based on real-time quantitative polymerase chain reaction. Biol Blood Marrow Transplant 2005;11:558–66.

[25] Jimenez-Velasco A, Barrios M, Roman-Gomez J, et al. Reliable quantification of hematopoietic chimerism after allogeneic transplantation for acute leukemia using amplification by real-time PCR of null alleles and insertion/deletion polymorphisms. Leukemia 2005;19:336–43.

[26] Koldehoff M, Steckel NK, Hlinka M, Beelen DW, Elmaagacli AH. Quantitative analysis of chimerism after allogeneic stem cell transplantation by real-time polymerase chain reaction with single nucleotide polymorphisms, standard tandem repeats, and Y-chromosome-specific sequences. Am J Hematol 2006;81:735–46.

[27] Bach C, Tomova E, Goldmann K, et al. Monitoring of hematopoietic chimerism by real-time quantitative PCR of micro insertions/deletions in samples with low DNA quantities. Transfus Med Hemother 2015;42:38–45.

[28] Frankfurt O, Zitzner JR, Tambur AR. Real-time qPCR for chimerism assessment in allogeneic hematopoietic stem cell transplants from unrelated adult and double umbilical cord blood. Hum Immunol 2015;76:155–60.

[29] Murphy KM. Chimerism analysis following hematopoietic stem cell transplantation. Hematological malignancies. Springer; 2013. p. 137–49.

[30] Brookes C, Bright J, Harbison S, Buckleton J. Characterising stutter in forensic STR multiplexes. Forensic Sci Int Genet 2012;6:58–63.

[31] Lion T. Summary: reports on quantitative analysis of chimerism after allogeneic stem cell transplantation by PCR amplification of microsatellite markers and capillary electrophoresis with fluorescence detection. Leukemia 2003;17:252–4.

[32] Berger B, Parson R, Clausen J, Berger C, Nachbaur D, Parson W. Chimerism in DNA of buccal swabs from recipients after allogeneic hematopoietic stem cell transplantations: implications for forensic DNA testing. Int J Legal Med 2013;127:49–54.

[33] Kristt D, Stein J, Yaniv I, Klein T. Assessing quantitative chimerism longitudinally: technical considerations, clinical applications and routine feasibility. Bone Marrow Transplant 2007;39:255–68.

[34] Clark JR, Scott SD, Jack AL, et al. Monitoring of chimerism following allogeneic haematopoietic stem cell transplantation (HSCT): technical recommendations for the use of short tandem repeat (STR) based techniques, on behalf of the United Kingdom National External Quality Assessment Service for Leucocyte Immunophenotyping Chimerism Working Group. Br J Haematol 2015;168:26–37.

[35] Lawler M, McCann SR, Marsh JC, et al. Serial chimerism analyses indicate that mixed haemopoietic chimerism influences the probability of graft rejection and disease recurrence following allogeneic stem cell transplantation (SCT) for severe aplastic anaemia (SAA): indication for routine assessment of chimerism post SCT for SAA. Br J Haematol 2009;144:933–45.

[36] Kolb HJ. Graft-versus-leukemia effects of transplantation and donor lymphocytes. Blood 2008;112:4371–83.

[37] Bader P, Niethammer D, Willasch A, Kreyenberg H, Klingebiel T. How and when should we monitor chimerism after allogeneic stem cell transplantation? Bone Marrow Transplant 2005;35:107–19.

[38] Martínez-Laperche C, Noriega V, Kwon M, et al. Achievement of early complete donor chimerism in CD25 -activated leukocytes is a strong predictor of the development of graft-versus-host-disease after stem cell transplantation. Exp Hematol 2015;43:4–13.

[39] Preuner S, Lion T. Post-transplant monitoring of chimerism by lineage-specific analysis. Bone marrow and stem cell transplantation. Springer; 2014. p. 271–91.

[40] Hoffmann JC, Stabla K, Burchert A, et al. Monitoring of acute myeloid leukemia patients after allogeneic stem cell transplantation employing semi-automated CD34 donor cell chimerism analysis. Ann Hematol 2014;93:279–85.

[41] Yang Y, Wang X, Qin Y, Wan L, Jiang Y, Wang C. Is there a role for B lymphocyte chimerism in the monitoring of B-acute lymphoblastic leukemia patients receiving allogeneic stem cell transplantation? Chronic Dis Transl Med 2015;1:48–54.

[42] Lion T, Daxberger H, Dubovsky J, et al. Analysis of chimerism within specific leukocyte subsets for detection of residual or recurrent leukemia in pediatric patients after allogeneic stem cell transplantation. Leukemia 2001;15:307–10.

[43] Zhou M, Sheldon S, Akel N, Killeen AA. Chromosomal aneuploidy in leukemic blast crisis: a potential source of error in interpretation of bone marrow engraftment analysis by VNTR amplification. Mol Diagn 1999;4:153–7.

[44] Ofran Y, Lazarus H, Rapoport A, Rowe J. Interpreting outcome data in hematopoietic cell transplantation for leukemia: tackling common biases. Bone Marrow Transplant 2015;50:324–33.

V. MOLECULAR TESTING IN HEMATOPATHOLOGY

MOLECULAR TESTING IN PERSONALIZED MEDICINE

36

Personalized Medicine in Cardiovascular Disease

R.M. Turner, M. Bula and M. Pirmohamed

The Wolfson Centre for Personalised Medicine, Institute of Translational Medicine,
University of Liverpool, Liverpool, United Kingdom

INTRODUCTION

Cardiovascular disease is a leading cause of morbidity and mortality worldwide. Conservative, interventional (eg, cardioversion), and surgical approaches are implemented to ameliorate the burden of cardiovascular disease, although the mainstay physician-directed strategy remains pharmacological therapy. However, notable interindividual variation in cardiovascular drug response exists, which manifests clinically as reduced effectiveness and/or adverse drug reactions (ADRs). Variable drug response is a large healthcare problem. It has been previously estimated that the proportion of patients who respond beneficially to the first drug offered in the treatment of a wide range of diseases, including cardiac arrhythmias, is typically just 50–75% [1]. In the hospital setting, 6.5% of admissions are related to ADRs [2] and 14.7% of inpatients experience an ADR [3]. For any given drug, the etiology underlying variable response can be parsed into three categories: (1) drug-specific (eg, drug regimen differences, including variable adherence), (2) human body (the system), and (3) environmental (eg, smoking, drug–drug, and food–drug interactions). The human body has a compartmentalized and hierarchical design composed of different levels of biological organization including genomic, epigenomic, transcriptomic, proteomic, metabolomic, tissue, and organ. The wide range of intrinsic factors present in each level, the complex interplay between factors within (*horizontal* interaction) and between (*vertical* interaction) biological levels, and the capacity for drugs and environmental factors to provoke network-based system-wide perturbations plausibly underpin the broad spectrum of emergent, complex phenotypes seen in clinical practice, including variable health/disease status and drug response.

Pharmacogenomics is the study of genetic determinants of interindividual variation in response to a given drug and aims to optimize drug efficacy and/or minimize ADRs through the stratification of medicines to prospectively-identified genetically-distinct patient subgroups by genotype(s)-informed drug and/or dose selection strategies. Pharmacogenomics research has successfully decreased the incidence of abacavir hypersensitivity syndrome in clinical practice through prospective genotyping for *HLA-B*57:01* [4]. Furthermore, oncology therapies targeted to patients with specific cancer biomarkers are becoming increasingly common. For example, the tyrosine kinase inhibitor, crizotinib, is licensed for treatment of non-small-cell lung cancers carrying oncogenic anaplastic lymphoma kinase (*ALK*) gene rearrangements [5].

Pharmacogenomic information is included within the product labels of 10 US Food and Drug Administration (FDA)-approved cardiovascular drugs and 128 other drugs [6]. However, there is no cardiovascular pharmacogenomic test currently used in widespread clinical practice. Clinical translation of pharmacogenomic research is hampered by evidential, logistical, financial, and healthcare provider knowledge barriers [7]. In particular, inconsistent study findings, a lack of prospective evidence of patient benefit, and the focus on approved

© 2017 Elsevier Inc. All rights reserved.

cardiovascular drugs all hamper cardiovascular trans- lational pharmacogenomics. The latter is because commonly prescribed cardiovascular drugs are often off-patent, inexpensive, and prescribed by a broad range of community and hospital physicians in contrast to, for example, new expensive targeted antineoplastic agents that are managed exclusively by oncologists. Therefore, even if robust and consistent randomized controlled trial (RCT) evidence supporting a cardiovas- cular pharmacogenomic test emerges, cardiovascular medicine faces the extra challenges of changing accepted clinical practice and existing physician behav- ior. Perhaps the most disconcerting observation is that although cardiovascular drug—gene associations typi- cally have larger effect sizes than the expanding number of cardiovascular disease susceptibility loci [7], pharmacogenomic predictive values remain predomi- nantly inadequate for clinical utility. This indicates that they do not explain enough interindividual variation and so pharmacogenomics alone, while intrinsically necessary, is unlikely to be sufficient to adequately parse interindividual response variability for most drugs. As a consequence, investigation into and incor- poration of novel factors identified using other omics technologies, alongside pharmacogenomic associ- ations, are being perceived as increasingly necessary. Nevertheless, pharmacogenomics is integral to interindividual drug response variability, multiple car- diovascular pharmacogenomics associations have been determined and the field is continuing to advance through new technologies (eg, next-generation sequenc- ing—NGS) (Fig. 36.1).

This chapter focuses on contemporary cardiovascu- lar pharmacogenomics, but is supplemented where appropriate with novel insight from alternate omics- levels. Table 36.1 provides a summary of established and novel pharmacogenomic associations for cardio- vascular drugs including warfarin, clopidogrel, bucin- dolol, statins, and antiarrhythmics [8—23].

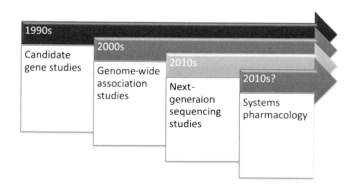

FIGURE 36.1 The evolution of pharmacogenomics studies from candidate gene studies to the implementation of NGS. Systems phar- macology holds promise and may become the norm for future stud- ies but poses many challenges.

WARFARIN

The anticoagulant warfarin is a widely prescribed coumarin-derived racemic mixture, first approved for use in humans in 1954 [24], and is indicated in throm- boembolism prophylaxis (eg, in atrial fibrillation (AF) and following mechanical heart valve implantation) and venous thromboembolism treatment. The degree of anticoagulation is determined by the international normalized ratio (INR): the target INR range for most warfarinized patients is 2.0—3.0. There is widespread interindividual variation in warfarin stable dose (WSD) requirements, ranging from 0.6 to 15.5 mg/day [25] and attributable to incompletely characterized genetic, clini- cal, and environmental components. Patients conven- tionally spend approximately 45—63% of the time within the therapeutic range [26,27] and elevated INRs are asso- ciated with an increased risk of warfarin-associated hemorrhage [28]. Therefore, the high prevalence of use, narrow therapeutic index, and multifactorial etiology of warfarin exposure make warfarin the leading cause of preventable ADRs [29] and a priority for pharmacoge- nomic studies [30].

VKORC1

VKORC1 encodes vitamin K epoxide reductase com- plex subunit 1, the warfarin target that catalyzes the rate-limiting step of the vitamin K cycle [25], and facili- tates posttranslational γ-carboxylation to produce functional clotting factors II, VII, IX, X, and proteins C and S (Fig. 36.2) [31]. The single-nucleotide polymor- phism (SNP) rs9923231 (−1639G > A; G3673A) alters a transcription factor binding site in the *VKORC1* pro- moter region and −1639A is associated with decreased gene expression [32]. The allele frequency of −1639A in African-American, Asian, and Caucasian populations is approximately 0.13, 0.92, and 0.40, respectively, indicating minor allele reversal in Asian populations. Carrying −1639A has been associated with decreased WSD requirements in several popula- tions [33] and over-coagulation [14], although not with hemorrhage [14]. While rs9923231 alone explains approximately 20—25% of WSD variation in Asian and Caucasian populations, this decreases to approximately 6% in African-Americans [34], perhaps due to the lower frequency of −1639A and/or a collective larger influence of additional factors in Africa-Americans. Interestingly, rare nonsynonymous *VKORC1* variants, such as rs61742245 (D36Y), have been identified which are associated with warfarin resistance and higher WSD requirements [35].

As well as genetic factors, *VKORC1* expression appears to be under epigenetic regulation. In vitro

TABLE 36.1 Select Examples of Pharmacogenomics Variants Significantly Associated with Cardiovascular Drug Response

Clinical outcome	Study	Locus/Gene	Variant(s)	Effect size[a]	Reference
WARFARIN					
(a) Dose requirement	MA	VKORC1	−1639G > A	GA vs GG: ~1.5 mg/day increase	[8]
				AA vs GG: ~2–3 mg/day reduction	[8]
	MA	CYP2C9	*2	*1/*2 vs *1/*1: ~1 mg/day reduction	[8]
				*2/*2 vs *1/*1: ~1.5 mg/day reduction	[8]
	MA		*3	*1/*3 vs *1/*1: ~1.5 mg/day reduction	[8]
				*3/*3 vs *1/*1: ~2.5 mg/day reduction	[8]
	CG		*5, *6, *8, *11	*5, *6, *8, or *11 carriers vs *1/*1: ~1 mg/day reduction	[9]
	GWAS	CYP4F2	1297G > A (V433M)	A carriers vs GG: ~0.2 mg/day increase	[10]
	GWAS	CYP2C cluster	rs12777823 (G > A)	AG vs GG: ~1 mg/day reduction	[11]
				AA vs GG: ~1.5 mg/day reduction	[11]
	ES	FPGS	rs7856096 (A > G)	AG: ~1 mg/day reduction	[12]
				GG: ~1.5 mg/day reduction	[12]
	CG	GATA4	rs867858 (G > T) + rs10090884 (A > C)	GG/AA vs all other genotype combinations: ~1 mg/day reduction	[13]
			rs2645400 (G > T) + rs4841588 (G > T)	GG/GT,TT vs all other genotype combinations: ~2 mg/day reduction	[13]
(b) Hemorrhage	MA	CYP2C9	*3	*1/*3 vs *1/*1: HR 2.05 (95% CI 1.36, 3.10)	[14]
				*3/*3 vs *1/*1: HR 4.87 (95% CI 1.38, 17.14)	[14]
(c) Over-anticoagulation (INR > 4)	MA	CYP2C9	*2	*2 vs *1: HR 1.52 (95% CI 1.11, 2.09)	[14]
			*3	*3 vs *1: HR 2.37 (95% CI 1.46, 3.83)	[14]
	MA	VKORC1	−1639G > A	GA vs GG: HR 1.49 (95% CI 1.15, 1.92)	[14]
CLOPIDOGREL					
(a) Stent thrombosis	MA	CYP2C19	*2, *3, *4−*8	ROF alleles present vs noncarriers: HR 2.81 (95% CI 1.81, 4.37)	[15]
				1 ROF allele vs noncarriers: HR 2.67 (95% CI 1.69, 4.22)	[15]
				2 ROF alleles vs noncarriers: HR 3.97 (95% CI 1.75, 9.02)	[15]
(b) MACE in patients at high risk of MACE (eg, requiring PCI)	MA	CYP2C19	*2, *3, *4−*8	ROF alleles present vs noncarriers: HR 1.57 (95% CI 1.13, 2.16)	[15]
				1 ROF allele vs noncarriers: HR 1.55 (95% CI 1.11, 2.27)	[15]
				2 ROF alleles vs noncarriers: HR 1.76 (95% CI 1.24, 2.50)	[15]
(c) MACE	MA	CYP2C19	*17	HR 0.82 (95% CI 0.72, 0.94)	[16]
SIMVASTATIN					
Myopathy	GWAS	SLCO1B1	rs4149056, T > C	Per copy of C allele: OR 4.5 (95% CI 2.6, 7.7)	[17]
				CC vs TT: OR 16.9 (95% CI 4.7, 61.1)	[17]

(Continued)

TABLE 36.1 (Continued)

Clinical outcome	Study	Locus/Gene	Variant(s)	Effect size[a]	Reference
BUCINDOLOL					
(a) All-cause mortality	RCT			B vs P overall in RCT[b]: HR 0.90 (95% CI 0.78, 1.02), NS	[18]
	CGS of RCT	*ADRB1*	Arg389Gly	B vs P if Arg389Arg: HR 0.62 (95% CI 0.39, 0.99)	[19]
	CGS of RCT	*ADRA2C*	Ins322−325Del	B vs P if Ins322−325Ins: HR 0.70 (95% CI 0.51, 0.96)	[20]
(b) VT/VF	RCT genetic substudy			B vs P overall[b]: HR 0.42 (95% CI 0.27, 0.64)	[21]
	CGS of RCT	*ADRB1*	Arg389Gly	B vs P if Arg389Arg: HR 0.26 (95% CI 0.14, 0.50)	[21]
(c) New onset atrial fibrillation	RCT			B vs P overall in RCT[b]: HR 0.59 (95% CI 0.44, 0.79)	[22]
	CGS of RCT	*ADRB1*	Arg389Gly	B vs P if Arg389Arg: HR 0.26 (95% CI 0.12, 0.57)	[22]
AMIODARONE					
Torsades de Pointes	CG	*NOS1AP*	rs10919035 (C > T)	T carriers vs CC: OR 2.81 (95% CI 1.62, 4.89)	[23]

[a]*All effect sizes are statistically significant unless otherwise stated.*

[b]*Overall risk estimate of drug versus placebo in RCT independent of genotype, provided when available to aid interpretation of corresponding statistically significant* within *genotype analyses.*

B, bucindolol; βB, beta-blocker; CG, candidate gene study; CGS, candidate gene substudy; CI, confidence interval; ES, exome sequencing study; GWAS, genome-wide association study; HR, hazard ratio; MA, meta-analysis; MACE, major adverse cardiovascular events; NS, not statistically significant; OR, odds ratio; P, placebo; PCI, percutaneous coronary intervention; RCT, randomized controlled trial; ROF, reduction-of-function; VT/VF, ventricular tachycardia/ventricular fibrillation.

Adapted from Turner RM, Pirmohamed M. Cardiovascular pharmacogenomics: expectations and practical benefits. Clin Pharmacol Ther 2014;95:281−93.

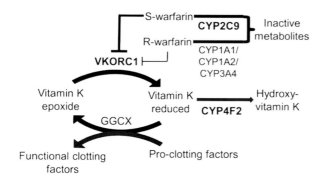

FIGURE 36.2 Warfarin is a racemic mixture of two enantiomers that perturb the vitamin K cycle through inhibition of vitamin K epoxide reductase complex subunit 1 (VKORC1). This decreases the regeneration of reduced vitamin K, an essential cofactor for γ-glutamyl carboxylase (GGCX), decreasing the posttranslational activating γ-carboxylation of glutamate residues in clotting factors II, VII, IX, and X. The more potent S-warfarin is metabolized by cytochrome P450 (CYP) 2C9, while R-warfarin is metabolized by CYP1A1, CYP1A2, and CYP3A4. CYP4F2 depletes the vitamin K cycle of reduced vitamin K [31]. The therapeutic action of warfarin is modulated by established pharmacokinetic (*CYP2C9*) and pharmacodynamic (*VKORC1, CYP4F2*) pharmacogenes.

assays have reported that the hepatic microRNA, miR-133a, interacts with the 3′UTR of *VKORC1* mRNA, decreases *VKORC1* mRNA dose-dependently, and in ex vivo liver samples from healthy subjects, mirR-133a levels are inversely correlated to *VKORC1* mRNA levels [36]. However, the influence of miR-133a on WSD requirements is yet to be determined.

CYP2C9

CYP2C9 metabolizes the S-warfarin enantiomer, which is three- to five-fold more potent than R-warfarin. The main *CYP2C9* variants, *CYP2C9*2, *3, *5, *6, *8*, and *11*, are all nonsynonymous reduction-of-function (ROF) SNPs resulting in variously attenuated S-warfarin metabolism except *CYP2C9*6*, which is an exonic single nucleotide deletion that shifts the reading frame causing loss of function [37]. *CYP2C9*2* and *3* are the most common Caucasian variants (minor allele frequencies (MAFs) 0.13 and 0.07, respectively). *CYP2C9*2* is very rare and *CYP2C9*3* has a low frequency (0.04) in Asians, while in African populations they are both absent or rare (0–0.036 and 0.003–0.02, respectively) [38]. CYP2C9*2 and *3 impair S-warfarin metabolism by approximately 30–40% and 80–90%, respectively [39], decrease WSD requirements [40], and a meta-analysis has reported a gene–dose trend for increased risk of hemorrhage with CYP2C9*3 in heterozygotes (*1/*3) and homozygotes (*3/*3) compared to wild-type homozygous (*1/*1) patients. The hazard ratios for bleeding were 2.05 (95% confidence interval (CI) 1.36, 3.10) and 4.87 (95% CI 1.38, 17.14), respectively [14].

CYP2C9*5, *6, *8, and *11 occur predominantly in African populations [34] and approximately 20% of African-Americans carry at least one [37]. Although there is insufficient evidence currently for CYP2C9*6, the others (CYP2C9*5, *8, and *11) have all been associated with decreased WSD requirements in African-American patients [37]. A recent genome-wide association study (GWAS) of African-American patients identified and replicated a novel noncoding variant, rs12777823, near CYP2C18 on chromosome 10. The minor A allele was associated with reduced S-warfarin clearance and decreased WSD requirements [11]. Interestingly, although rs12777823 is present in other ethnicities it has never before been associated with WSD, suggesting it is in linkage disequilibrium with an underlying African-specific causal variant [11]. However, further research into the underlying mechanism is required.

GATA4 encodes a liver-specific transcription factor involved in the regulation of CYP2C9 expression [41] and combinations of noncoding GATA4 SNPs have recently been associated with WSD requirements in a cohort of Korean patients with prosthetic cardiac valves [13]. Of note, patients homozygous wild type for both rs867858 and rs10090884 had lower WSD requirements compared to patients with other allele combinations of these SNPs, independent of age, CYP2C9*3, VKORC1, and CYP4F2 genotypes [13]. However, these novel associations require independent replication.

CYP4F2

CYP4F2 metabolizes reduced vitamin K to hydroxyvitamin K, removing reduced (active) vitamin K from the vitamin K cycle. GWAS investigations have identified the variant allele of rs2108622 (1297G > A, V433M) to be independently associated with higher WSD requirements in Caucasian [42] and Asian [10] individuals, but not African-American patients [11]. rs2108622 is correlated with lower CYP4F2 hepatic concentrations [31] and greater vitamin K availability, although it only explains an additional 1–2% of observed WSD variability [10,42].

Clinical Utility

There is unequivocal evidence that genetic variants influence warfarin dose requirements [37] and multivariable pharmacogenomics warfarin dosing algorithms, incorporating clinical and pharmacogenomic determinants (mainly CYP2C9*2, *3, and VKORC1 −1639G > A) have been constructed [43,44]. However, they are predominantly derived from retrospective studies insufficiently robust for cardiovascular clinical translation. Importantly, two multicenter prospective large warfarin pharmacogenomic RCTs, COAG [27], and EU-PACT [45], have recently been published. Briefly, COAG was US-based (n = 1015), compared a pharmacogenomic to clinical algorithm and found no difference in the mean time in the therapeutic INR range (TTR) between days 4 and 5 after warfarin initiation through to 4 weeks (45.2% and 45.4%, respectively, p = 0.91) [27]. EU-PACT was based in the United Kingdom and Sweden (n = 455), compared a pharmacogenomic algorithm to standard dosing and found that the TTR during the first 12 weeks after warfarin initiation was significantly higher in the pharmacogenomic (67.4%) compared to standard dosing arm (60.3%, p < 0.001) [45].

There are many reason(s) for these divergent results, which have been recently reviewed [46]. An important reason is ethnicity-specific pharmacogenomics. The pharmacogenomic algorithms used best explain warfarin variability in Caucasians. EU-PACT recruited predominantly Caucasians, but approximately 30% of COAG participants were African-American and approximately 6% Hispanic [27]. Importantly, African-American COAG participants had a lower TTR and higher incidence of INR values greater than 3 with pharmacogenomic dosing compared to clinical algorithmic dosing [27], emphasizing the necessity to utilize ethnicity-specific algorithms. Therefore, there is a need to identify additional pharmacogenomic variants in non-Caucasian populations. Interestingly, exome sequencing in 103 African-American patients with extreme WSD requirements (≤35 and ≥49 mg/week) recently associated the population-specific regulatory SNP, rs7856096, with lower WSD requirements [12]. rs7856096 is located within the folate homeostasis gene, folylpolyglutamate synthase (FPGS), and correlates with FPGS gene expression, although the reason for its impact on WSD requirements remains unresolved [12].

Five overlapping aggregate data meta-analyses of RCTs [47–50], or RCTs with prospective cohort studies [51], have been published since COAG/EU-PACT were reported, comparing pharmacogenomics to clinical dosing (Table 36.2) [47–51]. For deaths and thromboembolic events, when analyzed individually, no

TABLE 36.2   A Table of Recently Published Meta-Analyses That Report Outcomes for Warfarin Pharmacogenomics Dosing Compared to Clinical Dosing

Meta-analysis	Included study design	Summary estimate for difference in time in therapeutic INR range (95% CI) with pharmacogenomics compared to clinical dosing	Risk estimate for clinical adverse events with pharmacogenomics compared to clinical dosing
Franchini et al. (2014)[a] [47]	RCT	WMD: 4.25 (−1.95, 10.45), $I^2 = 89.4\%$, $n = 2812$	Major bleeding: 0.47 (0.23, 0.96), $I^2 = 0\%$
			Thromboembolism: 0.98 (0.45, 2.11), $I^2 = 0\%$
			Death: 0.71 (0.19, 2.60), $I^2 = 0\%$
Goulding et al. (2014) [48]	RCT	MD: 6.67 (1.34, 12.0), $I^2 = 80\%$, $n = 1952$	Bleeding or thromboembolism: 0.57 (0.33, 0.99), $I^2 = 60\%$, $n = 2211$
Liao et al. (2014) [49]	RCT	Overall SMD: 0.08 (−0.02, 0.17), $I^2 = 65\%$, $n = 1729$	Composite of adverse events[b]: 0.94 (0.84, 1.04), $I^2 = 0\%$, $n = 1763$
		SMD if initial dose fixed in clinical dosing arm: 0.24 (0.09, 0.40), $I^2 = 47.8\%$	Death: 1.36 (0.46, 4.05), $I^2 = 10.4\%$, $n = 1571$
		SMD if initial dose nonfixed in clinical arm: −0.02 (−0.14, 0.10), $I^2 = 0\%$	
Stergiopoulos and Brown (2014)[a] [50]	RCT	SDM: 0.14 (−0.10, 0.39), $I^2 = 88\%$, $n = 2812$	Major bleeding: 0.60 (0.29, 1.22), $I^2 = 0\%$, $n = 2586$
			Thromboembolism: 0.97 (0.46, 2.05), $I^2 = 0\%$, $n = 2586$
Tang et al. (2014) [51]	RCT and prospective cohort	Overall MD: 5.72 (1.84, 9.59), $I^2 = 84\%$, $n = 5148$	Major bleeding: 0.47 (0.24, 0.91), $I^2 = 0\%$, $n = 2614$
		MD during first 1−4 weeks: 4.64 (−0.31, 9.60), $I^2 = 88\%$	Thromboembolism: 0.79 (0.38, 1.63), $I^2 = 4\%$, $n = 2423$
		MD during 5−8 weeks: 7.99 (1.35, 14.63), $I^2 = 69\%$	

[a]Of the same nine RCTs included in each of these two meta-analyses [47,50], one included RCT investigated warfarin analogues (acenocoumarol, phenprocoumon). Exclusion of this trial was reported by Stergiopoulos and Brown to not change the results for any endpoint in their study [50].

[b]Composite of adverse events includes major bleeding, thromboembolism, myocardial infarction, death, clinically relevant nonmajor bleeding, or other conditions needed for emergency medical management (including an elevated INR) [49].

CI, confidence interval; INR, international normalized ratio; MD, mean difference; SDM, standardized difference in means; WMD, weighted mean difference. I^2 statistic = a measure of heterogeneity. It determines the proportion of total variation observed between trials attributable to differences between trials rather than sampling error [48].

meta-analysis has demonstrated a significant difference between dosing strategies. Beyond this, conclusions are elusive: for TTR and major bleeding, different meta-analyses both support pharmacogenomic [48,51] and show no overall benefit [47,49,50]. Heterogeneity is high for TTR, there are low absolute numbers of major bleeding events and individual studies are reused between meta-analyses, further limiting conclusions. One meta-analysis stratified by comparator and reported superiority of pharmacogenomics to fixed initial-dose standard practice but not when pharmacogenomics-guided dosing was compared to nonfixed initial dosing (eg, clinical algorithms) [49]. However, given that the standard dosing EU-PACT

arm had a higher TTR than the clinical algorithmic COAG arm (60.3−45.4%, respectively) an individual patient data meta-analysis might be illuminating. There are other trials ongoing such as GIFT [52], but whether they will shed light on the use of warfarin pharmacogenomics in a global sense seems unlikely. A major confounder is that the way in which warfarin is used globally varies widely: these include differences in initial dosing (ie, use of loading doses), the dosing used for early maintenance phases, the frequency of INR monitoring, whether computerized dosing software is used, how anticoagulation services are delivered, and who delivers them. Given these practice differences, no generalized conclusions can be made about the implementation of genotype-guided warfarin dosing, and it will be important that local initiatives are developed which allow a full evaluation of the utility of genotype-guided warfarin dosing when this is embedded in (local) clinical care pathways.

CLOPIDOGREL

The second-generation thienopyridine prodrug, clopidogrel, is used following ischemic stroke, acute coronary syndrome (ACS), and percutaneous coronary intervention (PCI). Although newer antiplatelet agents (eg, prasugrel, ticagrelor) have been developed, it is still anticipated that clopidogrel prescribing will remain high in cardiovascular medicine for the foreseeable future [53]. Approximately 85% of absorbed clopidogrel is hydrolyzed by hepatic carboxylesterase (CES1) to an inactive metabolite [54], and approximately 15% is oxidized by the CYP system in a two-step process involving CYP1A2, CYP3A4/5, CYP2B6, CYP2C9, and CYP2C19 to produce the active 5-thiol metabolite (R-130964) [54]. This active metabolite irreversibly inhibits the platelet purinergic P2Y$_{12}$ receptor, ameliorating adenosine diphosphate (ADP)-induced platelet aggregation. In patients with myocardial infarction undergoing PCI, up to 25% may exhibit clopidogrel resistance as determined by ex vivo platelet function testing [55], although definitions of clopidogrel resistance vary between studies. Importantly, clopidogrel nonresponse can be associated with an increased risk of myocardial infarction, stent thrombosis, and death [56]. Clinical factors associated with ex vivo high on-treatment platelet reactivity (HTPR) include older age (>65 years), increased body mass index, diabetes mellitus, reduced left ventricular function, and renal failure [57].

CYP2C19

The two-step process of clopidogrel biotransformation is largely catalyzed by CYP2C19 [54], which is encoded by polymorphic CYP2C19 and enzymatic activity is inherited in an autosomal codominant manner. Over 25 CYP2C19 variants have been identified. The most common ROF CYP2C19 variant is CYP2C19*2 (rs4244285, c.681G >A), which has an allelic frequency of approximately 0.15 in Africans and Caucasians, and less than or equal to 0.35 in Asians [53]. In Asians CYP2C19*3 (rs4986893, c.636G >A) is also common with an appreciable MAF of 2—9% [53]. However, the majority of other variants with decreased enzymatic activity (eg, *4—*8) are rare (MAF < 1%) [57]. The defining SNPs of CYP2C19*2 and CYP2C19*3 lead to a cryptic splice variant and a premature stop codon, respectively,

reducing conversion of clopidogrel to its active metabolite [57]. Several studies have demonstrated that CYP2C19 ROF alleles are associated with increased HTPR compared to *1/*1 wild-type homozygotes [53]. In fact CYP2C19*2 explains more variability in ex vivo platelet function response than any established clinical predictor, although CYP2C19*2 still only accounts for approximately 5—12% of ADP-induced aggregation variability [57]. Another common variant is CYP2C19*17 (rs12248560, c.-806C > T) that has estimated allelic frequencies of 0.16, 0.027, and 0.18 in Africans, Asians, and Caucasians, respectively [7]. CYP2C19*17 is associated with increased CYP2C19 transcription, resulting in a modest gain of function [57]. Furthermore, CYP2C19*17 has been associated with a decreased prevalence of HTPR, although the magnitude of effect is lower than for the ROF alleles [34]. CYP2C19 metabolizer phenotype is predicted from CYP2C19 genotype: *1/*1 (predicts the normal or extensive metabolizer phenotype (EM)), ROF heterozygotes (intermediate metabolizer—IM), ROF/ROF (poor metabolizer—PM), and both *1/*17 and *17/*17 (ultrarapid metabolizer—UM) [53].

Multiple meta-analyses investigating the impact of CYP2C19 variation on clinical response to clopidogrel have been undertaken [7]. These meta-analyses uniformly and robustly demonstrate that CYP2C19 ROF alleles (principally CYP2C19*2) significantly increase the risk of stent thrombosis [7] and furthermore, a gene—dose trend is evident [15]. Second, the interaction between clopidogrel and CYP2C19 genotype on the risk of major adverse cardiovascular events (MACE—predominantly cardiovascular death and nonfatal myocardial infarction or ischemic stroke) is more evident in patients undergoing PCI, compared to indications of lower baseline MACE risk [34,58]. Stratification by PCI indication appears to not further influence the effect of CYP2C19 ROF alleles on clopidogrel clinical response [58]. Interestingly, one meta-analysis has reported that, even when high-dose clopidogrel therapy is used in patients receiving PCI, CYP2C19 ROF alleles are significantly associated with increased risks of stent thrombosis and MACE, compared to wild-type homozygotes [59]. Finally, meta-analyses that have stratified by ethnicity have reported that the risk of MACE [58,60] and stent thrombosis [58] in clopidogrel-treated patients carrying CYP2C19 ROF alleles compared to wild-type homozygotes appears higher in Asian compared to Caucasian populations. However, it is currently unclear whether these observed ethnicity differences are due to the higher prevalence of CYP2C19 ROF alleles in Asian populations or are attributable to other genetic and/or clinical differences (eg, Asian studies have a trend for increased use of drug-eluting stents) [58].

On balance, CYP2C19*17 may confer a marginally decreased risk of MACE [7]. No meta-analysis has demonstrated that it reduces stent thrombosis and it has been inconsistently linked with bleeding [7]. However, the gain-of-function CYP2C19*17 allele is in linkage disequilibrium with the wild-type allele at the CYP2C19*2 locus and therefore its clinically observed effects may, in part, be due to the absence of CYP2C19*2 [53].

Carboxylesterase 1

Besides CYP2C19, other genes including adenosine triphosphate binding cassette subfamily B member 1 (ABCB1) and paraoxonase-1 (PON-1) have been inconsistently associated with clopidogrel pharmacogenomics [7]. There is growing interest in CES1, which is important in the metabolism of approximately 85% of absorbed parent clopidogrel, but it also metabolizes the intermediate product (2-oxo-clopidogrel) and 5-thiol active metabolite produced during the CYP-mediated bioactivation steps, limiting clopidogrel activity [54]. The nonsynonymous CES1 SNP, G143E (rs71647871), is significantly associated with decreased in vitro clopidogrel and 2-oxo-clopidogrel hydrolysis [54], higher plasma levels of the clopidogrel active metabolite in healthy volunteers ($n = 506$), reduced ADP-induced platelet aggregation in healthy volunteers ($n = 566$), and clopidogrel-treated coronary heart disease patients ($n = 350$) carrying 143E [61]. A nonsignificant trend for reduced 1-year cardiovascular events associated with 143E carriage was observed, although this comparison was relatively underpowered [61]. Recently, a study of 162 clopidogrel-treated patients reported that carrying CES1A2 A(-816)C, a SNP in the promoter region associated with increased carboxylesterase transcription efficiency, was associated with significantly higher ex vivo platelet reactivity [62], implying increased carboxylesterase-mediated clopidogrel metabolism and so reduced circulating clopidogrel active metabolite. Interestingly, several angiotensin-converting enzyme inhibitors (ACEIs) are metabolized by CES. In vitro, enalapril inhibits CES1-mediated hydrolysis and enalapril and trandolapril are associated with increased formation of 2-oxo-clopidogrel and clopidogrel active metabolite in human liver s9 fractions [63]. A large pharmacoepidemiological study ($n = 70,934$) has suggested that cotreatment of selected ACEIs and clopidogrel may increase bleeding risk ($p = 0.002$) [63], although the role of CES1 variants on this potential adverse drug–drug interaction requires further study.

miR-223

Platelets harbor a diverse and abundant miRNA repertoire [64] and miR-223 has been shown in a reporter gene activity assay to have the potential to repress platelet P2Y$_{12}$ mRNA expression [64]. Interestingly, low circulating plasma miR-223 levels in unstable angina clopidogrel-treated patients of Chinese ancestry ($n = 62$) have been significantly correlated with increased ex vivo platelet reactivity [65]. A smaller study ($n = 21$) has reported that elevated plasma miR-223 is significantly associated with decreased platelet reactivity in patients on P2Y$_{12}$ inhibitors [66]. Both these studies, although focusing on opposing miR-223 levels, are consistent with the hypothesis that miR-223 and P2Y$_{12}$ expression are inversely related and may influence antiplatelet response.

Clinical Utility

The 2012 American College of Cardiology Foundation/American Heart Association (ACCF/AHA) guidelines for patients with unstable angina/non-ST-elevation myocardial infarction do not recommend routine CYP2C19 genotyping in clinical practice largely due to insufficient supportive data from adequately powered prospective randomized genotype-directed clopidogrel trials [67]. However, the ACCF/AHA guidelines permit CYP2C19 genotyping on a case-by-case basis, such as for patients who experience recurrent ACS events despite clopidogrel therapy [67]. Overall, the current evidence base suggests that patients undergoing PCI (especially those of Asian ancestry) and carrying CYP2C19 ROF alleles, compared to wild-type homozygotes, are at an increased risk of stent thrombosis and MACE. For the increasing number of patients in whom genetic data is already available, the 2013 Clinical Pharmacogenetics Implementation Consortium (CPIC) guidelines recommend consideration of CYP2C19 genotype in ACS patients undergoing PCI and the prescribing of an alternative antiplatelet agent (prasugrel, ticagrelor) in place of clopidogrel in patients predicted to be IMs or PMs [53]. Ultimately the goal of antiplatelet therapy is thrombotic event prevention without hemorrhagic complications and it remains to be established whether pre-clopidogel genotyping with antiplatelet stratification, compared to routine contemporary prasugrel/ticagrelor therapy, is clinically and cost-effective. These questions are being addressed in the large prospective ongoing POPular Genetics trial in patients with ST elevation myocardial infarction undergoing primary PCI [68].

STATINS

Statins are hypolipidemic drugs efficacious in the prevention of cardiovascular disease and are the most commonly prescribed class of medication worldwide [69]. Over 40 candidate genes have been associated with differential statin effects on lipid-lowering efficacy and/or cardiovascular endpoints [70]. However, the association between simvastatin-induced myopathy and *SLCO1B1* rs4149056 (521T > C, V174A) [17] has the largest effect size of any known statin—gene association and is highlighted here.

Statins are associated with a spectrum of clinical skeletal muscle ADRs ranging from myalgias (~1–5%) [71] to increasingly severe myopathies with elevated plasma creatine kinase (CK) levels (eg, ~0.11% of patients with CK >10-fold but <50-fold the upper limit of normal—ULN) [72], to potentially fatal rhabdomyolysis (~0.1–8.4/100,000 patient-years) [72], and to the rare autoimmune-mediated necrotizing myositis (~2/million/year) [72]. Clinical correlates of statin myotoxicity include female gender, older age, low body mass index, untreated hypothyroidism, and concomitant drug therapies such as gemfibrozil [71]. As well as directly affecting patients, statin-associated (mild) muscle side effects confer an increased risk of statin discontinuation and nonadherence [73], and variable statin adherence is associated with an increased risk of cardiovascular events [74].

SLCO1B1

A case-control genetic substudy of the SEARCH RCT compared 85 cases of myopathy (CK > 3 × ULN) to 90 controls. All cases/controls were prescribed simvastatin 80 mg daily [17]. This seminal GWAS identified a single strong association between the intronic SNP rs4363657 and myopathy. The regional analysis found rs4363657 to be in almost complete linkage disequilibrium with the nonsynonymous SNP rs4149056. The risk of myopathy conferred by the rs4149056 C allele (MAF 0.15) showed a gene—dose trend: odds ratio (OR) 4.5 (95% CI 2.6, 7.7) per copy of the C allele and 16.9 (95% CI 4.7, 61.1) in CC homozygotes, compared to TT wild-type patients [17]. This pharmacogenomic association has been replicated [17] and confirmed by meta-analysis [75], although the effect size is lower in patients administered simvastatin 40 mg daily [17]. Furthermore, rs4149056 has been associated with milder statin (predominantly simvastatin)-associated events suggestive of intolerance including a composite of discontinuation, dose reduction, switching lipid-lowering therapy associated with biochemical testing, and mild biochemical abnormalities (CK 1–3 × ULN and/or an elevated alanine aminotransferase level) [76].

SLCO1B1 encodes the organic anion-transporting polypeptide 1B1 (OATP1B1), a hepatocyte-specific sinusoidal xenobiotic influx transporter. Mechanistically, rs4149056 does not cause OATP1B1 mis-localization and so plausibly decreases OATP1B1 intrinsic transport function [77]. In healthy volunteers, rs4149056 had no

significant impact on the pharmacokinetics of the lipophilic parent compound, simvastatin lactone, but CC homozygotes experienced a mean increase of 221% in the area under the plasma concentration—time curve (AUC) of the active metabolite, simvastatin acid, compared to TT wild-type homozygotes [78]. Therefore, decreased OATP1B1-mediated hepatic uptake of simvastatin acid is hypothesized to increase muscle exposure, predisposing to myotoxicity through ill-defined mechanisms. This elevated muscle exposure hypothesis accounts for the increased effect size of rs4149056 with higher simvastatin doses [17]. However, rs4149056 has not been significantly associated with pravastatin [79] and rosuvastatin [80] muscle ADRs, while the effect with atorvastatin [75] seems to be less marked, suggesting that the *SLCO1B1* variation is either simvastatin specific, or the effect size of rs4149056 for muscle ADRs with other statins is smaller. The latter hypothesis is consistent with: (1) rs4149056 increases the AUC of several statins (except fluvastatin) but to a lesser degree than simvastatin acid [81], (2) simvastatin is intrinsically more myotoxic than other approved statins [82], and (3) smaller effect sizes are more challenging to demonstrate empirically.

Other Genes Associated with Statin Myopathy

Although rs4149056 is the most well-validated statin—gene association, other genes have been implicated in exploratory analyses [83]. Interestingly, the incidence of patients with drug (predominantly statin)-induced myopathy carrying aberrant metabolic myopathy genes, including *CPT2* (encoding carnitine palmitoyltransferase 2), appears higher compared to a control population [84]. Furthermore, in vitro transcriptomics revealed *CPT2* to be in the top 1% of genes whose mRNA levels are perturbed by 75 rhabdomyolysis-inducing drugs [85]. Deleterious mutations in *RYR1* (encoding ryanodine receptor 1) predispose to anesthesia-induced malignant hyperthermia [86] and *RYR1* candidate mutations are more frequent in statin myopathy patients than controls [87]. Although these findings remain at the discovery stage, they suggest that targeted NGS of influential skeletal muscle genes (and genes whose proteins interact with the protein products of these skeletal muscle genes in the protein—protein interaction network) in large sample sizes with rare variant gene enrichment analyses may help identify novel pharmacogenomic genes associated with statin (likely severe) myotoxicity.

Clinical Utility

The CPIC has proposed recommendations that incorporate rs4149056 when considering simvastatin initiation in patients whose rs4149056 genotype is already known [71]. However, clinical uptake of *SLCO1B1* genotyping has not entered widespread clinical practice. Although highly validated, no definitive prospective study has been undertaken and on its own it has a low positive predictive value [88]. However, incorporation of rs4149056 into the Qstatin risk score for myopathy, which is a model derived from electronic medical records, based on clinical predictors [89] and of borderline clinical utility, could be potentially beneficial [88].

BUCINDOLOL

Beta-adrenoreceptor (β-AR) antagonists (β-blockers) are indicated in the management of several cardiovascular pathologies including heart failure, hypertension, ACS, angina pectoris, and arrhythmias. Here, we highlight the contemporary pharmacogenomics of bucindolol.

Bucindolol is a nonselective β-AR inhibitor with sympatholytic activity (ie, reduces circulating noradrenaline levels) and weak alpha-1 (α_1)-AR antagonism [90], which is currently unlicensed. The BEST RCT evaluated the clinical effectiveness of bucindolol compared to placebo in patients with New York Heart Association functional class III and IV heart failure due to primary or secondary dilated cardiomyopathy with impaired left ventricular ejection fractions (≤ 0.35) [18]. However, BEST was ceased early due to the accrual of contemporary data establishing the usefulness of β-blocker therapy in chronic heart failure and the equipoise of continuing the trial. At the time of termination, a mean of 2.0 years follow-up had been undertaken and bucindolol conferred no overall survival benefit ($p = 0.13$), but marginally reduced cardiovascular deaths ($p = 0.04$) [18].

Subsequent investigations using the 1040 patient BEST genetic substudy have revealed that bucindolol efficacy is modulated by Arg389Gly (rs1801253) in *ADRB1* (encoding β₁-AR) and to a lesser extent by the Ins322−325Del polymorphism of *ADRA2C* (encoding α₂C-AR) [19]. Patients can be stratified into three clusters based on genotype that delineate the effectiveness of bucindolol in reducing six clinical endpoints (including death) into those with enhanced (Arg389Arg homozygotes + any Ins322−325Del allele), intermediate (Gly389 carriers + Ins322−325Ins wild-type homozygotes), and no bucindolol response (Gly389 carriers + Del322−325 carriers), compared to placebo [19]. Further work using the BEST genetic substudy demonstrated that this three-genotype construct can differentiate the risks of new-onset AF (interaction test; $p = 0.016$) [22] and ventricular arrhythmias

(interaction test; $p = 0.028$) [21]. In particular, the subgroup of homozygous Arg389Arg heart failure patients on bucindolol experienced a clear reduction in the incidence of both arrhythmia types in contrast to placebo [21,22]. New-onset AF in patients with heart failure is associated with increased mortality and increased hospitalization days [91].

The β₁-AR is the prevalent cardiomyocyte β-AR subtype and functionally Arg389Arg human nonfailing left ventricular membranes have higher affinities for noradrenaline than membranes expressing the Gly389 β₁-AR form [19]. The Arg389 β₁-AR is also associated with enhanced downstream signaling. It is therefore conceptually plausible that β-blocker therapy might be more beneficial in the presence of the Arg389 β₁-AR. Indeed, a meta-analysis has reported that in heart failure patients on β-blocker therapy, the Arg389Arg genotype is associated with a significantly improved left-ventricular ejection fraction compared to Gly389 carriers, although there was no difference in clinical endpoints [92].

α₂C-AR is a presynaptic receptor that mediates negative feedback of noradrenaline release. The ROF Del322−325 minor allele is associated with adrenergic dysfunction and an exaggerated sympatholytic response (ie, increased reduction in circulating noradrenaline) to bucindolol [20]. It is hypothesized that carrying both *ADRB1* Gly389 and *ADRA2C* Del322−325 cancels all bucindolol efficacy because the marked sympatholysis leads to insufficient noradrenaline for the hypofunctional Gly389 β₁-AR to adequately support the failing myocardium [19].

These genetic substudy findings have led to the GENETIC-AF phase IIB/III RCT [93], which has recently begun recruiting patients. This RCT aims to determine whether bucindolol is superior to metoprolol in Arg389Arg heart failure patients who have persistent symptomatic AF requiring electrical cardioversion to stable sinus rhythm for reducing the time to event of recurrent symptomatic AF/atrial flutter or all-cause mortality during 24 weeks follow-up [93]. If bucindolol superiority is demonstrated, bucindolol may become the first β-blocker licensed for a genotype-specific patient subgroup.

Antiarrhythmics

Prolongation of the QT-interval can be congenital or acquired, lead to the long QT syndrome (LQTS), and is associated with ventricular arrhythmias [94] and mortality [95]. The principle mechanism of drug-induced QT prolongation is blockade of the rapidly activating delayed rectifier potassium current (I_{Kr}), perturbing cardiac repolarization and predisposing to drug-induced Torsades de Pointes (DITdP) [96]. DITdP is a rare, unpredictable, potentially fatal drug-induced ventricular arrhythmia and a leading cause of drug withdrawal [97]. Most DITdP is related to antiarrhythmic medications (eg, amiodarone, flecainide, sotalol). However, multiple classes of noncardiac drugs are also associated with QT prolongation, although only a proportion have been implicated in DITdP (eg, erythromycin, chlorpromazine, domperidone) [97]. Besides drugs, other clinical factors prolong the QT-interval including electrolyte disturbances (especially hypokalemia), bradycardia, and heart failure [97], and in many DITdP cases, one or more of these secondary risk factors is also present.

A GWAS meta-analysis of greater than 100,000 individuals of European ancestry has identified 35 common variant loci associated with QT-interval variation in the general population, collectively explaining approximately 8–10% of QT-interval variation [98]. However, two recent pharmacogenomics GWAS that were designed to identify genetic variants modifying the effect of drugs on the QT-interval [99] or increasing DITdP risk [100] uncovered no variants of genome-wide significance. The latter GWAS compared 216 North-western European cases of DITdP due to any culprit drug to 771 ancestry-matched controls. No SNP reached genome-wide significance, despite an 80% power to detect a variant at genome-wide significance with an MAF of 0.1 conferring an OR of greater than or equal to 2.7. Subgroup analyses restricting DITdP cases to specific drugs (sotalol, amiodarone, or quinidine) were similarly insignificant [100]. These pharmacogenomics GWAS argue that in Caucasian patients, common pharmacogenomic variants do not predispose to drug-induced QT prolongation or DITdP, unlike the pharmacogenomic associations of other cardiovascular drugs. Nevertheless, it is still possible that they could be underpowered for individual drugs. Interestingly, a recent candidate gene study identified common noncoding SNPs in NOS1AP (encoding nitric oxide synthase 1 adapter protein) to be significantly associated with amiodarone-related TdP in Caucasians, after correcting for multiple testing [23]. The most significant SNP, rs10919035, was present in approximately 13% of controls and conferred an OR of 2.81 (95% CI 1.62, 4.89) [23]. NOS1AP is known to modulate baseline QT-interval [98]. Furthermore, the significance of variants in other ethnicities remains undefined. rs7626962 (S1103Y) in SCN5A (encoding sodium voltage-gated channel, type V α-subunit) has been implicated in arrhythmias including DITdP in African-American patients [101]. The MAF of S1103Y is approximately 0.05 in African-American patients, but is rare in other ethnicities. Rare SCN5A mutations cause approximately 10% of congenital LQTS [102].

Candidate gene studies have also associated low frequency and rare variants with DITdP in Caucasians. A large candidate gene study of Caucasian individuals found rs1805128 (D85N) in KCNE1 (encoding potassium voltage-gated channel subfamily E member 1) to be significantly associated with DITdP. D85N was present in 8.6% of cases, 2.9% of drug-exposed controls, 1.8% of population controls, and conferred an OR of 9.0 (95% CI 3.5, 22.9) [103]. However, only a nonsignificant trend was present when D85N was genotyped in the validation cohort [103]. Rare KCNE1 mutations cause approximately 1% of congenital LQTS [102]. Finally, NGS found that 23.1% of Caucasian individuals (6 of 26) with DITdP carried a highly conserved nonsynonymous variant within 22 congenital arrhythmia genes (including the 13 congenital LQTS genes) compared to 1.7% in 60 control subjects from the 1000 Genomes CEU data [104].

Clinical Utility

DITdP is challenging to study given its rare and capricious phenotype. Recent pharmacogenomic GWAS have reported no genome-wide significant results, arguing against common genomic variation predisposing to DITdP in Caucasians. On the other hand, candidate gene studies have suggested rare and ethnically-restricted DITdP associations, although they remain in the discovery phase. Meanwhile, a clinical decision support algorithm has been implemented at the Mayo Clinic (USA) which alerts a physician if they attempt to prescribe a medication that can predispose to DITdP for a patient with a history of QT prolongation (QTc > 500 ms) [105]. This system has significantly reduced exposure to QT-prolonging medications in patients at high DITdP risk [105], although its impact on clinical endpoints will require further observation.

CONCLUSION AND FUTURE PERSPECTIVE

Large-scale initiatives, including the 1000 Genomes projects [106], have led to a prolific increase in our knowledge of common and infrequent human genetic variation. The Exome Sequencing Project [107] and recently commenced Rare Diseases Genomes Project, which will sequence the genomes of 10,000 rare disease patients in the next 3 years [108], are cataloging rare and potentially deleterious genetic variation. Concurrent to and facilitated by such initiatives, intensive pharmacogenomics research is ongoing and several cardiovascular drug–gene associations have been established. However, as yet no cardiovascular pharmacogenomic biomarker has been translated into routine clinical practice. Undoubtedly logistical, financial, and knowledge barriers exist [7], but the major current obstacle remains a lack of robust evidence demonstrating patient benefit. Therefore, the near future results of ongoing warfarin, clopidogrel, and bucindolol pharmacogenomics trials, and the forthcoming real-world observational data from pharmacogenomics early adopter sites [109], are anticipated with cautious optimism.

Another reason for optimism is the emerging, novel, interdisciplinary field of systems pharmacology, which includes but goes beyond pharmacogenomics (Fig. 36.1). The wide range of drug-specific, environmental, and multilevel human factors and their complicated interrelationships indicate the complexity underlying interindividual drug response variation and the potential benefits of a multifaceted approach. Systems pharmacology represents a paradigm shift toward multilevel systems biology and its integration with quantitative pharmacological modeling (pharmacometrics) [110], aiming to identify novel drug targets, enhance drug development, and facilitate precision medicine [111]. Systems pharmacology recognizes that drug response is an emergent phenotype resulting from drug-induced perturbations occurring at different biological levels (eg, proteomic, metabolomic, and organ) on different spatial and temporal scales, which are further shaped by the drug-specific (eg, dose regimen) and environmental factors (eg, smoking). Within and between levels, molecules are interlinked to form biological networks (eg, macromolecular protein–protein structure and gene regulatory networks) that coalesce to ultimately form the overall human system. Biological network properties include redundancy and robustness and drug-induced network perturbations can be additive, synergistic, or opposing. Therefore, rather than just understanding the effect of a drug on a single biological component (eg, molecule) in isolation, systems pharmacology aspires to adequately understand drug action on the interlinked system as a whole to increase predictive utility and clinical application [111]. One approach is construction of multiscale models, whose development is envisaged through an iterative cycle of new empirical data acquisition leading to model refinement and subsequent model predictions (and prioritization of unknown parameters) driving the next round of empirical investigations [111]. For example, a quantitative multiscale model of calcium homeostasis and bone remodeling has been constructed that can predict nonlinear longitudinal changes to the clinical surrogate endpoint of lumbar spine bone mineral density during and following

discontinuation and reinstitution of the antiresorptive drug, denosumab [112]. However, many technical hurdles exist including the difficulties of multiscale modeling and producing a user-friendly, updateable software platform capable of integrating data from the growing array of publically-available biological databases to maximize data usefulness. The key to systems pharmacology will be ever greater intradisciplinary and interdisciplinary collaboration to pool financial and distinct skillsets, increase sample sizes, standardize phenotypes, and facilitate multi-omics analyses, in silico network-based investigations and pharmacological modeling.

References

[1] Spear BB, Heath-Chiozzi M, Huff J. Clinical application of pharmacogenetics. Trends Mol Med 2001;7:201–4.
[2] Pirmohamed M, James S, Meakin S, et al. Adverse drug reactions as cause of admission to hospital: prospective analysis of 18 820 patients. BMJ 2004;329:15–19.
[3] Davies EC, Green CF, Taylor S, Williamson PR, Mottram DR, Pirmohamed M. Adverse drug reactions in hospital in-patients: a prospective analysis of 3695 patient-episodes. PLoS One 2009;4:e4439.
[4] Martin MA, Kroetz DL. Abacavir pharmacogenetics—from initial reports to standard of care. Pharmacotherapy 2013;33:765–75.
[5] Landi L, Cappuzzo F. Management of NSCLC: focus on crizotinib. Expert Opin Pharmacother 2014;15:2587–97.
[6] US Food and Drug Administration (FDA). Table of pharmacogenomic biomarkers in drug labeling, <http://www.fda.gov/drugs/scienceresearch/researchareas/pharmacogenetics/ucm083378.htm>; 2014 [accessed 26.10.14].
[7] Turner RM, Pirmohamed M. Cardiovascular pharmacogenomics: expectations and practical benefits. Clin Pharmacol Ther 2014;95:281–93.
[8] Jorgensen AL, FitzGerald RJ, Oyee J, Pirmohamed M, Williamson PR. Influence of CYP2C9 and VKORC1 on patient response to warfarin: a systematic review and meta-analysis. PLoS One 2012;7:e44064.
[9] Cavallari LH, Langaee TY, Momary KM, et al. Genetic and clinical predictors of warfarin dose requirements in African Americans. Clin Pharmacol Ther 2010;87:459–64.
[10] Cha PC, Mushiroda T, Takahashi A, et al. Genome-wide association study identifies genetic determinants of warfarin responsiveness for Japanese. Hum Mol Genet 2010;19:4735–44.
[11] Perera MA, Cavallari LH, Limdi NA, et al. Genetic variants associated with warfarin dose in African-American individuals: a genome-wide association study. Lancet 2013;382:790–6.
[12] Daneshjou R, Gamazon ER, Burkley B, et al. Genetic variant in folate homeostasis is associated with lower warfarin dose in African Americans. Blood 2014;124:2298–305.
[13] Jeong E, Lee KE, Jeong H, Chang BC, Gwak HS. Impact of GATA4 variants on stable warfarin doses in patients with prosthetic heart valves. Pharmacogenomics J 2015;15:33–7.
[14] Yang J, Chen Y, Li X, et al. Influence of CYP2C9 and VKORC1 genotypes on the risk of hemorrhagic complications in warfarin-treated patients: a systematic review and meta-analysis. Int J Cardiol 2013;168:4234–43.

[15] Mega JL, Simon T, Collet JP, et al. Reduced-function CYP2C19 genotype and risk of adverse clinical outcomes among patients treated with clopidogrel predominantly for PCI: a meta-analysis. JAMA 2010;304:1821–30.

[16] Li Y, Tang HL, Hu YF, Xie HG. The gain-of-function variant allele CYP2C19*17: a double-edged sword between thrombosis and bleeding in clopidogrel-treated patients. J Thromb Haemost 2012;10:199–206.

[17] Link E, Parish S, Armitage J, et al. SLCO1B1 variants and statin-induced myopathy—a genomewide study. N Engl J Med 2008;359:789–99.

[18] Beta-Blocker Evaluation of Survival Trial Investigators. A trial of the beta-blocker bucindolol in patients with advanced chronic heart failure. N Engl J Med 2001;344:1659–67.

[19] O'Connor CM, Fiuzat M, Carson PE, et al. Combinatorial pharmacogenetic interactions of bucindolol and beta1, alpha2C adrenergic receptor polymorphisms. PLoS One 2012;7:e44324.

[20] Bristow MR, Murphy GA, Krause-Steinrauf H, et al. An alpha2C-adrenergic receptor polymorphism alters the norepinephrine-lowering effects and therapeutic response of the beta-blocker bucindolol in chronic heart failure. Circ Heart Fail 2010;3:21–8.

[21] Aleong RG, Sauer WH, Robertson AD, Liggett SB, Bristow MR. Adrenergic receptor polymorphisms and prevention of ventricular arrhythmias with bucindolol in patients with chronic heart failure. Circ Arrhythm Electrophysiol 2013;6:137–43.

[22] Aleong RG, Sauer WH, Sauer WH, et al. Prevention of atrial fibrillation by bucindolol is dependent on the beta(1)389 Arg/Gly adrenergic receptor polymorphism. JACC Heart Fail 2013;1:338–44.

[23] Jamshidi Y, Nolte IM, Dalageorgou C, et al. Common variation in the NOS1AP gene is associated with drug-induced QT prolongation and ventricular arrhythmia. J Am Coll Cardiol 2012;60:841–50.

[24] Pirmohamed M. Warfarin: almost 60 years old and still causing problems. Br J Clin Pharmacol 2006;62:509–11.

[25] Owen RP, Gong L, Sagreiya H, Klein TE, Altman RB. VKORC1 pharmacogenomics summary. Pharmacogenet Genomics 2010;20:642–4.

[26] Caraco Y, Blotnick S, Muszkat M. CYP2C9 genotype-guided warfarin prescribing enhances the efficacy and safety of anticoagulation: a prospective randomized controlled study. Clin Pharmacol Ther 2008;83:460–70.

[27] Kimmel SE, French B, Kasner SE, et al. A pharmacogenetic versus a clinical algorithm for warfarin dosing. N Engl J Med 2013;369:2283–93.

[28] Marie I, Leprince P, Menard JF, Tharasse C, Levesque H. Risk factors of vitamin K antagonist overcoagulation. QJM 2012;105:53–62.

[29] Lovborg H, Eriksson LR, Jonsson AK, Bradley T, Hagg S. A prospective analysis of the preventability of adverse drug reactions reported in Sweden. Eur J Clin Pharmacol 2012;68:1183–9.

[30] Shaw K, Amstutz U, Castro-Pastrana L, et al. Pharmacogenomic investigation of adverse drug reactions (ADRs): the ADR prioritization tool, APT. J Popul Ther Clin Pharmacol 2013;20:e110–27.

[31] McDonald MG, Rieder MJ, Nakano M, Hsia CK, Rettie AE. CYP4F2 is a vitamin K1 oxidase: an explanation for altered warfarin dose in carriers of the V433M variant. Mol Pharmacol 2009;75:1337–46.

[32] Yuan HY, Chen JJ, Lee MT, et al. A novel functional VKORC1 promoter polymorphism is associated with inter-individual and inter-ethnic differences in warfarin sensitivity. Hum Mol Genet 2005;14:1745–51.

[33] Lee MT, Klein TE. Pharmacogenetics of warfarin: challenges and opportunities. J Hum Genet 2013;58:334–8.

[34] Johnson JA, Cavallari LH. Pharmacogenetics and cardiovascular disease—implications for personalized medicine. Pharmacol Rev 2013;65:987–1009.

[35] Loebstein R, Dvoskin I, Halkin H, et al. A coding VKORC1 Asp36Tyr polymorphism predisposes to warfarin resistance. Blood 2007;109:2477–80.

[36] Perez-Andreu V, Teruel R, Corral J, et al. miR-133a regulates vitamin K 2,3-epoxide reductase complex subunit 1 (VKORC1), a key protein in the vitamin K cycle. Mol Med 2012;18:1466–72.

[37] Johnson JA, Cavallari LH. Warfarin pharmacogenetics. Trends Cardiovasc Med 2015;25:33–41.

[38] Suarez-Kurtz G, Botton MR. Pharmacogenomics of warfarin in populations of African descent. Br J Clin Pharmacol 2013;75:334–46.

[39] Lee CR, Goldstein JA, Pieper JA. Cytochrome P450 2C9 polymorphisms: a comprehensive review of the in-vitro and human data. Pharmacogenetics 2002;12:251–63.

[40] Johnson JA, Gong L, Whirl-Carrillo M, et al. Clinical Pharmacogenetics Implementation Consortium Guidelines for CYP2C9 and VKORC1 genotypes and warfarin dosing. Clin Pharmacol Ther 2011;90:625–9.

[41] Mwinyi J, Nekvindova J, Cavaco I, et al. New insights into the regulation of CYP2C9 gene expression: the role of the transcription factor GATA-4. Drug Metab Dispos 2010;38:415–21.

[42] Takeuchi F, McGinnis R, Bourgeois S, et al. A genome-wide association study confirms VKORC1, CYP2C9, and CYP4F2 as principal genetic determinants of warfarin dose. PLoS Genet 2009;5:e1000433.

[43] Gage BF, Eby C, Johnson JA, et al. Use of pharmacogenetic and clinical factors to predict the therapeutic dose of warfarin. Clin Pharmacol Ther 2008;84:326–31.

[44] Klein TE, Altman RB, Eriksson N, et al. Estimation of the warfarin dose with clinical and pharmacogenetic data. N Engl J Med 2009;360:753–64.

[45] Pirmohamed M, Burnside G, Eriksson N, et al. A randomized trial of genotype-guided dosing of warfarin. N Engl J Med 2013;369:2294–303.

[46] Pirmohamed M, Kamali F, Daly A, Wadelius M. Oral anticoagulation: a critique of recent advances and controversies. Trends Pharmacol Sci 2015;36:153–63.

[47] Franchini M, Mengoli C, Cruciani M, Bonfanti C, Mannucci PM. Effects on bleeding complications of pharmacogenetic testing for initial dosing of vitamin K antagonists: a systematic review and meta-analysis. J Thromb Haemost 2014;12:1480–7.

[48] Goulding R, Dawes D, Price M, Wilkie S, Dawes M. Genotype-guided drug prescribing: a systematic review and meta-analysis of randomized control trials. Br J Clin Pharmacol 2015;80:868–77.

[49] Liao Z, Feng S, Ling P, Zhang G. Meta-analysis of randomized controlled trials reveals an improved clinical outcome of using genotype plus clinical algorithm for warfarin dosing. J Thromb Thrombolysis 2015;39:228–34.

[50] Stergiopoulos K, Brown DL. Genotype-guided vs clinical dosing of warfarin and its analogues: meta-analysis of randomized clinical trials. JAMA Intern Med 2014;174:1330–8.

[51] Tang Q, Zou H, Guo C, Liu Z. Outcomes of pharmacogenetics-guided dosing of warfarin: a systematic review and meta-analysis. Int J Cardiol 2014;175:587–91.

[52] Do EJ, Lenzini P, Eby CS, et al. Genetics informatics trial (GIFT) of warfarin to prevent deep vein thrombosis (DVT): rationale and study design. Pharmacogenomics J 2012;12(5):417–24.

[53] Scott SA, Sangkuhl K, Stein CM, et al. Clinical pharmacogenetics implementation consortium guidelines for CYP2C19 genotype and clopidogrel therapy: 2013 update. Clin Pharmacol Ther 2013;94:317–23.

[54] Zhu HJ, Wang X, Gawronski BE, Brinda BJ, Angiolillo DJ, Markowitz JS. Carboxylesterase 1 as a determinant of clopidogrel metabolism and activation. J Pharmacol Exp Ther 2013; 344:665–72.

[55] Matetzky S, Shenkman B, Guetta V, et al. Clopidogrel resistance is associated with increased risk of recurrent atherothrombotic events in patients with acute myocardial infarction. Circulation 2004;109:3171–5.

[56] Sharma RK, Reddy HK, Singh VN, Sharma R, Voelker DJ, Bhatt G. Aspirin and clopidogrel hyporesponsiveness and nonresponsiveness in patients with coronary artery stenting. Vasc Health Risk Manag 2009;5:965–72.

[57] Trenk D, Hochholzer W. Genetics of platelet inhibitor treatment. Br J Clin Pharmacol 2014;77:642–53.

[58] Sorich MJ, Rowland A, McKinnon RA, Wiese MD. CYP2C19 genotype has a greater effect on adverse cardiovascular outcomes following percutaneous coronary intervention and in Asian populations treated with clopidogrel: a meta-analysis. Circ Cardiovasc Genet 2014;7:895–902.

[59] Zhang L, Yang J, Zhu X, et al. Effect of high-dose clopidogrel according to CYP2C19*2 genotype in patients undergoing percutaneous coronary intervention—a systematic review and meta-analysis. Thromb Res 2015;135:449–58.

[60] Jang JS, Cho KI, Jin HY, et al. Meta-analysis of cytochrome P450 2C19 polymorphism and risk of adverse clinical outcomes among coronary artery disease patients of different ethnic groups treated with clopidogrel. Am J Cardiol 2012;110:502–8.

[61] Lewis JP, Horenstein RB, Ryan K, et al. The functional G143E variant of carboxylesterase 1 is associated with increased clopidogrel active metabolite levels and greater clopidogrel response. Pharmacogenet Genomics 2013;23:1–8.

[62] Xie C, Ding X, Gao J, et al. The effects of CES1A2 A(-816)C and CYP2C19 loss-of-function polymorphisms on clopidogrel response variability among Chinese patients with coronary heart disease. Pharmacogenet Genomics 2014;24:204–10.

[63] Kristensen KE, Zhu HJ, Wang X, et al. Clopidogrel bioactivation and risk of bleeding in patients cotreated with angiotensin-converting enzyme inhibitors after myocardial infarction: a proof-of-concept study. Clin Pharmacol Ther 2014;96:713–22.

[64] Landry P, Plante I, Ouellet DL, Perron MP, Rousseau G, Provost P. Existence of a microRNA pathway in anucleate platelets. Nat Struct Mol Biol 2009;16:961–6.

[65] Zhang YY, Zhou X, Ji WJ, et al. Decreased circulating microRNA-223 level predicts high on-treatment platelet reactivity in patients with troponin-negative non-ST elevation acute coronary syndrome. J Thromb Thrombolysis 2014;38:65–72.

[66] Chyrchel B, Toton-Zuranska J, Kruszelnicka O, et al. Association of plasma miR-223 and platelet reactivity in patients with coronary artery disease on dual antiplatelet therapy: a preliminary report. Platelets 2014;28:1–5.

[67] Jneid H, Anderson JL, Wright RS, et al. ACCF/AHA focused update of the guideline for the management of patients with unstable angina/non-ST-elevation myocardial infarction (updating the 2007 guideline and replacing the 2011 focused update): a report of the American College of Cardiology Foundation/American Heart Association Task Force on Practice Guidelines. J Am Coll Cardiol 2012;60:645–81.

[68] Bergmeijer TO, Janssen PW, Schipper JC, et al. CYP2C19 genotype-guided antiplatelet therapy in ST-segment elevation myocardial infarction patients-Rationale and design of the Patient Outcome after primary PCI (POPular) Genetics study. Am Heart J 2014;168 16-22.e1.

[69] Postmus I, Verschuren JJ, de Craen AJ, et al. Pharmacogenetics of statins: achievements, whole-genome analyses and future perspectives. Pharmacogenomics 2012;13:831–40.

[70] Verschuren JJ, Trompet S, Wessels JA, et al. A systematic review on pharmacogenetics in cardiovascular disease: is it ready for clinical application? Eur Heart J 2012;33:165–75.

[71] Ramsey LB, Johnson SG, Caudle KE, et al. The clinical pharmacogenetics implementation consortium guideline for SLCO1B1 and simvastatin-induced myopathy: 2014 update. Clin Pharmacol Ther 2014;96:423–8.

[72] Alfirevic A, Neely D, Armitage J, et al. Phenotype standardization for statin-induced myotoxicity. Clin Pharmacol Ther 2014; 96:470–6.

[73] Wei MY, Ito MK, Cohen JD, Brinton EA, Jacobson TA. Predictors of statin adherence, switching, and discontinuation in the USAGE survey: understanding the use of statins in America and gaps in patient education. J Clin Lipidol 2013;7:472–83.

[74] Phan K, Gomez YH, Elbaz L, Daskalopoulou SS. Statin treatment non-adherence and discontinuation: clinical implications and potential solutions. Curr Pharm Des 2014;20:6314–24.

[75] Carr DF, O'Meara H, Jorgensen AL, et al. SLCO1B1 genetic variant associated with statin-induced myopathy: a proof-of-concept study using the clinical practice research datalink. Clin Pharmacol Ther 2013;94:695–701.

[76] Donnelly LA, Doney AS, Tavendale R, et al. Common nonsynonymous substitutions in SLCO1B1 predispose to statin intolerance in routinely treated individuals with type 2 diabetes: a go-DARTS study. Clin Pharmacol Ther 2011;89:210–16.

[77] Nies AT, Niemi M, Burk O, et al. Genetics is a major determinant of expression of the human hepatic uptake transporter OATP1B1, but not of OATP1B3 and OATP2B1. Genome Med 2013;5:1.

[78] Pasanen MK, Neuvonen M, Neuvonen PJ, Niemi M. SLCO1B1 polymorphism markedly affects the pharmacokinetics of simvastatin acid. Pharmacogenet Genomics 2006;16:873–9.

[79] Voora D, Shah SH, Spasojevic I, et al. The SLCO1B1*5 genetic variant is associated with statin-induced side effects. J Am Coll Cardiol 2009;54:1609–16.

[80] Danik JS, Chasman DI, MacFadyen JG, Nyberg F, Barratt BJ, Ridker PM. Lack of association between SLCO1B1 polymorphisms and clinical myalgia following rosuvastatin therapy. Am Heart J 2013;165:1008–14.

[81] Elsby R, Hilgendorf C, Fenner K. Understanding the critical disposition pathways of statins to assess drug-drug interaction risk during drug development: it's not just about OATP1B1. Clin Pharmacol Ther 2012;92:584–98.

[82] Skottheim IB, Gedde-Dahl A, Hejazifar S, Hoel K, Asberg A. Statin induced myotoxicity: the lactone forms are more potent than the acid forms in human skeletal muscle cells in vitro. Eur J Pharm Sci 2008;33:317–25.

[83] Needham M, Mastaglia FL. Statin myotoxicity: a review of genetic susceptibility factors. Neuromuscul Disord 2014;24: 4–15.

[84] Vladutiu GD, Simmons Z, Isackson PJ, et al. Genetic risk factors associated with lipid-lowering drug-induced myopathies. Muscle Nerve 2006;34:153–62.

[85] Hur J, Liu Z, Tong W, Laaksonen R, Bai JP. Drug-induced rhabdomyolysis: from systems pharmacology analysis to biochemical flux. Chem Res Toxicol 2014;27:421–32.

[86] Robinson R, Carpenter D, Shaw MA, Halsall J, Hopkins P. Mutations in RYR1 in malignant hyperthermia and central core disease. Hum Mutat 2006;27:977—89.

[87] Vladutiu GD, Isackson PJ, Kaufman K, et al. Genetic risk for malignant hyperthermia in non-anesthesia-induced myopathies. Mol Genet Metab 2011;104:167—73.

[88] Stewart A. SLCO1B1 polymorphisms and statin-induced myopathy. PLoS Curr 2013;5. Available from: http://dx.doi.org/10.1371/currents.eogt.d21e7f0c58463571bb0d9d3a19b82203.

[89] Collins GS, Altman DG. Predicting the adverse risk of statin treatment: an independent and external validation of Qstatin risk scores in the UK. Heart 2012;98:1091—7.

[90] Smart NA, Kwok N, Holland DJ, Jayasighe R, Giallauria F. Bucindolol: a pharmacogenomic perspective on its use in chronic heart failure. Clin Med Insights Cardiol 2011;5: 55—66.

[91] Aleong RG, Sauer WH, Davis G, Bristow MR. New-onset atrial fibrillation predicts heart failure progression. Am J Med 2014; 127:963—71.

[92] Liu WN, Fu KL, Gao HY, et al. beta1 adrenergic receptor polymorphisms and heart failure: a meta-analysis on susceptibility, response to beta-blocker therapy and prognosis. PLoS One 2012;7:e37659.

[93] ClinicalTrials.gov. Genetically targeted therapy for the prevention of symptomatic atrial fibrillation in patients with heart failure (GENETIC-AF). ARCA Biopharma, Inc; 2014<https://www.clinicaltrials.gov/ct2/show/study/NCT01970501?term=bucindolol&rank=1>; [accessed 27.01.15].

[94] Moss AJ. The QT interval and torsade de pointes. Drug Saf 1999;21:5—10.

[95] Zhang Y, Post WS, Blasco-Colmenares E, Dalal D, Tomaselli GF, Guallar E. Electrocardiographic QT interval and mortality: a meta-analysis. Epidemiology 2011;22(5):660—70.

[96] Roden DM, Viswanathan PC. Genetics of acquired long QT syndrome. J Clin Invest 2005;115:2025—32.

[97] Behr ER, Roden D. Drug-induced arrhythmia: pharmacogenomic prescribing? Eur Heart J 2013;34:89—95.

[98] Arking DE, Pulit SL, Crotti L, et al. Genetic association study of QT interval highlights role for calcium signaling pathways in myocardial repolarization. Nat Genet 2014;46: 826—36.

[99] Avery CL, Sitlani CM, Arking DE, et al. Drug-gene interactions and the search for missing heritability: a cross-sectional pharmacogenomics study of the QT interval. Pharmacogenomics J 2014;14:6—13.

[100] Behr ER, Ritchie MD, Tanaka T, et al. Genome wide analysis of drug-induced torsades de pointes: lack of common variants with large effect sizes. PLoS One 2013;8:e78511.

[101] Splawski I, Timothy KW, Tateyama M, et al. Variant of SCN5A sodium channel implicated in risk of cardiac arrhythmia. Science 2002;297:1333—6.

[102] Abbott GW. KCNE genetics and pharmacogenomics in cardiac arrhythmias: much ado about nothing? Expert Rev Clin Pharmacol 2013;6:49—60.

[103] Kaab S, Crawford DC, Sinner MF, et al. A large candidate gene survey identifies the KCNE1 D85N polymorphism as a possible modulator of drug-induced torsades de pointes. Circ Cardiovasc Genet 2012;5:91—9.

[104] Ramirez AH, Shaffer CM, Delaney JT, et al. Novel rare variants in congenital cardiac arrhythmia genes are frequent in drug-induced torsades de pointes. Pharmacogenomics J 2013; 13:325—9.

[105] Sorita A, Bos JM, Morlan BW, Tarrell RF, Ackerman MJ, Caraballo PJ. Impact of clinical decision support preventing the use of QT-prolonging medications for patients at risk for torsade de pointes. J Am Med Inform Assoc 2015;22:e21—7.

[106] Abecasis GR, Auton A, Brooks LD, et al. An integrated map of genetic variation from 1,092 human genomes. Nature 2012; 491:56—65.

[107] The National Heart Lung and Blood Institute (NHLBI) Exome Sequencing Project (ESP). <http://hmg.oxfordjournals.org/content/early/2014/09/12/hmg.ddu450.long>; [accessed 19.12.14].

[108] Moran N. 10,000 rare-disease genomes sequenced. Nat Biotech 2014;32:114—17.

[109] Van Driest SL, Shi Y, Bowton EA, et al. Clinically actionable genotypes among 10,000 patients with preemptive pharmacogenomic testing. Clin Pharmacol Ther 2014;95:423—31.

[110] van der Graaf PH, Benson N. Systems pharmacology: bridging systems biology and pharmacokinetics-pharmacodynamics (PKPD) in drug discovery and development. Pharm Res 2011;28:1460—4.

[111] Sorger KS, Allerheiligen SRB. Quantitative and systems pharmacology in the post-genomic era: new approaches to discovering drugs and understanding therapeutic mechanisms. <http://www.nigms.nih.gov/training/documents/systemspharmawpsorger2011.pdf>; 2011 [accessed 26.09.14].

[112] Peterson MC, Riggs MM. Predicting nonlinear changes in bone mineral density over time using a multiscale systems pharmacology model. Pharmacometrics Syst Pharmacol 2012;1:e14.

37

Personalized Medicine for Coagulopathies

J. Fareed and O. Iqbal

Department of Pathology, Loyola University Health System, Maywood, IL, United States

INTRODUCTION

The concept of personalized medicine evolved several centuries ago and Hippocrates (c. 460–c. 370 BC) once stated "...It's far more important to know what person the disease has than what disease the person has..." [1]. The present working definition of personalized medicine based on The Personalized Medicine Coalition document on *The Case for Personalized Medicine* states "...The molecular methods that make personalized medicine possible include testing for variation in genes, gene expression, proteins, and metabolites, as well as new treatments that target molecular mechanisms. Test results are correlated with clinical factors – such as disease state, prediction of future disease states, drug response, and treatment prognosis – to help physicians individualize treatment for each patient..." [2]. A more comprehensive definition states "...Personalized Medicine is the concept that managing a patient's health should be based on the individual patient's specific characteristics, including age, gender, height/weight, diet, and environment, and so on. Recent developments in genetic testing allow the development of 'Genomic Personalized Medicine' and Predictive Medicine, which is the combination of comprehensive genetic testing with proactive, personalized preventive medicine. Personalized medicine is not solely about genomics, however, as personalized medicine is about you, the health consumer. Personalized medicine also allows your healthcare provider, such as your physician, to focus their attention on what makes you, instead of abiding by generalities..." [3]. The European Union defined personalized medicine as "...Providing the right treatment to the right patient, at the right dose at the right time...." President's Council of Advisors on Science and Technology defined it as "...the tailoring of the medical treatment to the individual characteristics of each patient...." The American Medical Association defined it as "...Health Care that is informed by each person's unique clinical and genetic and environmental information...." The National Cancer Institute (NCI) of National Institutes of Health (NIH) defined personalized medicine as "...A form of medicine that uses information about a person's genes, proteins and environment to prevent diagnose and treat disease...."

The discipline of mapping/sequencing (including analysis of the information), born from a marriage of molecular and cell biology with classical genetics and fostered by computational science, was termed *genomics* by T.H. Roderick of the Jackson Laboratory (Bar Harbor, ME) [4]. In the interesting course of History of Genomics (Table 37.1), the year 2003 is significantly marked by the announcement of the completion of the Human Genome Project. The practice of personalized medicine entails the application of genomics, the concept which is defined in Webster's Dictionary as "...a branch of biotechnology concerned with applying the techniques of genetics and molecular biology to the genetic mapping and DNA sequencing of sets of genes or the complete genomes of selected organisms, with organizing the results in databases and with applications of the data (as in medicine or biology)...." According to the World Health Organization definitions, genetics is the study of heredity [4] and genomics is the study of genes and their functions and related techniques [5,6]. While genetics scrutinizes the functioning and composition of the single gene, genomics addresses all genes and their interrelationships in order to identify their combined influence on the growth and development of the organism [5,6].

Hemostasis is physiologic homeostasis resulting from a dynamic equilibrium between coagulation and fibrinolysis. An intact endothelium is by far the largest

© 2017 Elsevier Inc. All rights reserved.

TABLE 37.1   Milestones of Genomic Development

Years	Name	Event
1745	Maupertuis	Adaptationist account of organic design
1859	Darwin	The origin of species
1865	Mendel	Combinatorial rules of inherited traits
1869	Miescher	Discovered "Nuclein" (DNA) in pus cells
1874	Miescher	Separated nucleic acids into a protein and an acid molecule
1918	Muller	Formulation of the chief principles of spontaneous gene mutations
1920s		Nucleic acid found to be a major component of the chromosome
1930s		Tetranucleotide—adenylic, guanylic, thymidylic, cytidylic acids
1940s		Molecular weight of nucleic acid much higher than tetranucleotide hypothesis
1944	Avery	Identified nucleic acids as active principle in bacterial transformation
1950	Chargaff	Nucleotide composition differs according to its biologic source
1951		First protein sequence (insulin)
1952	Hershey and Chase	Bacteriophage—80% of viral DNA entry in cell and 80% protein outside
1953	Watson and Crick	Discovery of double helix structure of DNA
1960s		Elucidation of the genetic code
1975	King and Wilson	Discovery of regulatory genes
1976		First cloning of human genes and output of structural genes
1977		Advent of DNA sequencing
1980s	McClintock	Discovered transposable strands of genes in maize
1984	McGinnis	Discovered homeotic (HOX) regulatory genes—basic body plan of animals
1986		Fully automated DNA sequencing
1995		Identification of first whole genome (*Haemophilus influenzae*)
1999		Discovery of first human chromosome (Chr#22)
2000		*Drosophila/Arabidopsis* genomes
2000		The human genome project presents preliminary results
2001		Human and mouse genome
2003		US White House announces completion of the human genome project

(Continued)

TABLE 37.1   (Continued)

Years	Name	Event
2004		NHGRI and DOE publish scientific description of human genome sequence
2005		International Hap Map Consortium—catalog of human genetic variation
2006		NCI and NHGRI—study of first three cancers first phase of Cancer Genome Atlas
2007		NIH announces the official launch of the human microbiome project
2008		President Bush signed into law the Genetic Information Nondiscrimination Act
2009		Therapeutics for Rare & Neglected Diseases Program or TRND
2010		NIH awards to support the genotype tissue expression (GTEx) project
2011		NHGRI's new strategic plan for the future of human genome research
2012		The Encyclopedia of DNA Elements project—working of human genome
2013		10th anniversary of human genome project (HGP)
2013		Smithsonian Institute (Washington, DC) presents "Unlocking the life's code"
2014		NIH issues final genomic data saving (GDS) policy
2015		NIHGRI's Workshop—research directions in genetically-mediated SJS/TEN
2015		Stanford Researchers suss out cancer mutations in genome's dark spots

endocrine, paracrine, and autocrine gland ever known to man. Alteration of this delicate hemostatic balance may lead to bleeding disorders. Thanks to Vogel who for the first time introduced the term *pharmacogenomics* in 1959 [7], pharmacogenomics-guided drug development has enabled personalized or individualized therapy for clotting or bleeding disorders, given to the right patient, at the right dosage, and at the right time. Completion of the Human Genome Project has been a significant achievement. Gene expression profiling and identification of single-nucleotide polymorphisms (SNPs) will facilitate the diagnoses of various hemostatic and other disorders. Pharmacogenetics involves the entire library of genes that determine drug efficacy and safety. According to the SNP Map Working Group [8], there are 1.42 million SNPs: one SNP per 1900 bases; 60,000 SNPs within exons; two exonic SNPs per gene (1/1080 bases); and 93% of genetic loci contain two SNPs. Because each person is different at one in 1000–2000 bases, SNPs are responsible for human individuality. A list of genes involved in hemostatic disorders is given in Table 37.2.

TABLE 37.2 Genes Related to Coagulation

Clone ID	Name	Gene title
22040	MMP9	Matrix metalloproteinase 9 (gelatinase B)
26418	EDG1	Endothelial differentiation, sphingolipid G-protein-coupled receptor
32609	LAMA4	Laminin, alpha 4
34778	VEGF	Vascular endothelial growth factor
40463	PDGFRB	Platelet-derived growth factor (PDGF) receptor, betapolypeptide
41898	PTGDS	Prostaglandin D2 synthase (21 kD, brain)
44477	VCAM1	Vascular cell adhesion molecule 1
45138	VEGFC	Vascular endothelial growth factor C
49164	VCAM1	Vascular cell adhesion molecule 1
49509	EPOR	Erythropoietin receptor
49665	EDNRB	Endothelin receptor type B
49920	PTDSS1	Phosphatidylserine synthase 1
51447	FCGR3B	Fc fragment of IgG, low affinity III B, receptor for Z (CD16)
66982	PLGL	Plasminogen like
67654	PDGFB	PDGF betapolypeptide (Simian sarcoma viral (v-sim) oncogene homolog)
71101	PROCR	Protein C receptor, endothelial (EPCR)
71626	ZNF268	Zinc finger protein 268
768246	G6PD	Glucose-6-phosphate dehydrogenase
85678	F2	Coagulation factor 2
85979	PLG	Plasminogen
120189	PSG4	Pregnancy-specific beta 1-glycoprotein 4
121218	PF4	Platelet factor 4
127928	HBP1	HMG-box containing protein 1
130541	PECAM1	Platelet/endothelial-cell adhesion molecule (CD31 antigen)
131839	FOLR1	Folate receptor 1 (adult)
135221	S100P	S100 calcium-binding protein P
136821	TGFB1	Transforming growth factor, beta 1
137836	PDCD10	Programmed cell death 10
138991	COL6A3	Collagen, type VI, alpha 3
139009	PN1	Fibronectin 1
142556	PSG2	Pregnancy-specific beta-glycoprotein 2

(Continued)

TABLE 37.2 (Continued)

Clone ID	Name	Gene title
143287	PSG11	Pregnancy-specific beta-1 glycoprotein 11
143443	TBXASa	Thromboxane A synthase 1
149910	SELL	Selectin E (endothelial adhesion molecule 1)
151662	P11	Protease, serine, 22
155287	HSPA1A	Heatshock 70 Kd protein 1A
160723	LAMC1	Laminin, gamma 19—formerly, LAMB2
179276	FASN	Fatty acid synthase
180864	ICAM5	Intercellular adhesion molecule 5, telencephalin
184038	SPTBN2	Spectrin beta, nonerythrocytic 2
191664	THBS2	Thrombospondin 2
194804	PTTPN	Phosphatidylinositol transfer protein
196612	MMP12	Matrix metalloproteinase 1 (interstitial collagenase)
199945	TGM2	Transglutaminase 2
205185	THBD	Thrombomodulin
210687	AGTR1	Angiotensin receptor 1
212429	TF	Transferrin
212649	GRG	Histidine rich glycoprotein
234736	GATA6	GATA-binding protein 6
240249	APLP2	Amyloid beta (a4) precursor-like protein 2
241788	FGB	Fibrinogen, B, betapolypeptide
243816	CD36	CD36 antigen (collagen type 1 receptor, thrombospondin receptor)
245242	CPB2	Carboxypeptidase B2 (plasma carboxypeptidase U)
260325	ALB	Albumin
261519	TNFRSF5	TNF receptor (superfamily, member 5)
292306	LPC	Lipase, hepatic
296198	CHS1	Chediak–Higashi syndrome 1
310519	F10	Coagulation Factor X
360644	ITGB8	Integrin, beta 8
343072	ITGB1	Integrin, beta 1
345430	PIK3CA	Phosphoinositide 3 kinase, catalytic, alpha polypeptide
589115	MMP1	Matrix metalloproteinase 1 (interstitial collagenase)
666218	TGFB2	Transforming growth factor, beta 2
712641	PRG4	Proteoglycan 4 (megakaryocyte stimulating factor)
714106	PLAU	Plasminogen activator, urokinase

(Continued)

TABLE 37.2 (Continued)

Clone ID	Name	Gene title
726086	TFP12	Tissue factor pathway inhibitor (TFPI) 2
727551	IRF2	Interferon regulatory factor 2
753211	PTGER3	Prostaglandin E receptor 3 (subtype EP30)
753418	VASP	Vasodilator stimulated phosphoprotein
753430	ATRX	Alpha thalassemia
754080	ICAM3	Intercellular adhesion molecule 3
755054	IL18R1	Interleukin 18 receptor 1
758266	THBS4	Thrombospondin 4
770462	CPZ	Carboxypeptidase Z
770670	TNFAIP3	Tumor necrosis factor (TNF), alpha-induced protein 3
770859	ITGB5	Integrin, beta 5
776636	BHMT	Betaine-homocysteine methyltransferase
782789	AVPR1A	Arginine vasopressor receptor 1A
785975	F13A1	Coagulation factor XIII. A1 polypeptide
788285	EDNR A	Endothelial receptor type A
809938	TACSTD2	Matrix metalloproteinase 7 (matrilysin, uterine)
810010	PDGFRL	PDGF-receptor-like
810017	PLAUR	Plasminogen activator, urokinase receptor
810117	ANXA11	Annexin A11
810124	PAFAH1B3	Platelet-activating factor acetylhydrolase, isoform 1b, gamma subunit (29 kD)
810242	C3AR1	Complement component 3a receptor 1
810512	THBS1	Thrombospondin 1
810891	LAMA5	Laminin, alpha 5
811096	ITGB4	Integrin, beta 4
811792	GSS	Glutathione synthetase
812276	SNCA	Synuclein, alpha (non-A4 component of amyloid precursor)
813757	FOLR2	Folate receptor 2 (fetal)
813841	PLAT	Plasminogen activator, tissue serine (or cysteine) proteinase inhibitor
814378	SPINT2	Serine proteinase inhibitor, Kunitz type 2
814615	MTHFD	Methylene tetrahydrofolate dehydrogenase (NAD dependent)
825295	LDLR	Low-density lipoprotein receptor (familial hypercholesterolemia)
840486	vWF	Von Willebrand factor
842846	TIMP2	Tissue inhibitor of metalloproteinase-2
1813254	F2R	Coagulation factor II (thrombin) receptor

There are approximately three billion base pairs of DNA in the human genome that code for at least 30,000 genes. Although the sequence of the majority of base pairs is identical from individual to individual, natural variation occurs and 0.1% of DNA base pairs contribute to individual differences. Three consecutive base pairs form a codon that specifies an amino acid constitute of an encoded protein. Genes represent a series of codons that specify the amino acid sequence of a particular protein. At each gene locus, an individual carries two alleles, one from each parent. If there are two identical alleles, it is referred to as a homozygous genotype, and if the alleles are different, it is referred to as a heterozygous genotype. Genetic variations usually occur as SNPs and occur on average at least once every 1000 base pairs, reflecting approximately three million base pairs distributed throughout the entire genome. Genetic variations that occur at a frequency of at least 1% in the human population are referred to as polymorphisms. Genetic polymorphisms are inherited and monogenic—they involve one locus and have interethnic differences in frequency. Rare mutations occur at a frequency of less than 1% in the human population. Other examples of genetic variations include insertion–deletion polymorphisms, tandem repeats, defective splicing, aberrant splice site, and premature stop codon polymorphisms. Through the discovery of new genetic targets, pharmacogenomics is expected to improve the quality of life and control healthcare costs by treating specific genetic subgroups, avoiding adverse drug reactions, and by decreasing the number of treatment failures.

GENOMICS OF COAGULATION DISORDERS

Arterial and venous thromboembolism are major causes of significant morbidity and mortality especially in developed countries. With respect to arterial thrombosis, factor XIII 34Leu was reported in several studies to have protective effect on the development of myocardial infarction [9]. No other single polymorphism is considered as a significant risk factor for arterial thrombosis, although factor V Arg506Gln, factor VII Arg353Gln, and vWF Thr789Ala may be involved in patient subgroups [9]. Factor V Arg506Gln and prothrombin 20210 mutations are known to play a role in venous thrombosis.

Factor V Leiden R506Q

The factor V Leiden R506Q mutation (G1691A) occurs in 8% of the population with a specific G → A substitution at nucleotide 1691 in the gene for factor V. The defective protein product is cleaved less efficiently (10%) by activated protein C, resulting in deep vein thrombosis (DVT), recurrent miscarriages, portal vein thrombosis in cirrhotic patients, early kidney transplant loss, and other forms of venous thromboembolism (VTE) [10–13]. A dramatic increase in the incidence of thrombosis is seen in women who are taking oral contraceptives. Both prothrombin G20210 and factor V Leiden mutation in the presence of major risk factors may contribute to atherothrombosis (a thrombus that forms due to rupture of an atherosclerotic plaque). Antithrombin drugs play a crucial role in the management of these thrombotic disorders. The factor V Leiden allele is common in Europe, with a population frequency of 4.4%. The mutation is very rare outside Europe, with a frequency of 0.6% in Asia Minor [14].

Factor VII

Polymorphism in the factor VII gene, especially the Arg-353Gln mutation in exon 8 located in the catalytic domain of factor VII, influences plasma FVII levels. The Gln-353 allele is associated with strong protective effect against the occurrence of myocardial infarction [15]. Since FVIIa/tissue factor (TF) is involved in the initial coagulation cascade, much attention has been given to blocking this pathway by developing FVIIa inhibitors and tissue factor pathway inhibitor (TFPI) [16].

Prothrombin 20210A Mutation

The coagulation factor II (prothrombin) G20210 mutation occurs in 2% of the population and is located in the 3′ UTR of the coagulation factor II polypeptide near a putative polyadenylation site [17]. This mutation is associated with increased levels of prothrombin resulting in DVT, recurrent miscarriages, and portal vein thrombosis in these conditions. The interactive role of hormone replacement therapy and prothrombotic mutations has been reported to cause the risk of nonfatal myocardial infarction in postmenopausal women [18].

Laki Lorand Factor (FXIII)

Factor XIII SNP G→T in exon 2 causes a Val/Leu change at position 34. The Val/Leu polymorphism increases the rate of thrombin activation by factor XIII and causes increased and faster clot stabilization [19,20]. The Leu34 allele has been shown to play a protective role against arterial and venous thrombosis [21,22].

Natural Anticoagulant System

Genetic defects in the antithrombin, protein C, and protein S systems are very rare, but result in increased risk of venous thrombosis. The role of inherited deficiencies of antithrombin, protein C, and protein S in arterial disease is not completely understood and may not contribute to the risk of arterial thrombosis.

Thrombomodulin

Thrombomodulin mutations are more important in arterial diseases than in venous diseases. The thrombomodulin polymorphism, G→A substitution at nucleotide position 127 in the gene has been studied regarding its relation with the arterial disease. The 25Thr allele was reported to be more prevalent in male patients with myocardial infarction than the control population [23]. Polymorphisms in the thrombomodulin gene promoter (−33G/A) influence the soluble thrombomodulin levels in plasma and cause increased risk of coronary heart disease [24]. Carriers of the −33A allele were reported to exhibit increased occurrence of carotid atherosclerosis in patients less than 60 years old [25].

Tissue Factor Pathway Inhibitor

TFPI or lipoprotein-associated coagulation inhibitor is one of the coagulation protease inhibitors which combines and inhibits FVIIa–TF complex and FXa (Figs. 37.1 and 37.2). Sequence variation in the TFPI gene has been reported. The four different polymorphisms reported include: (1) Pro-151Leu, (2) Val-264Met, (3) T384C exon 4, and (4) C33T intron 7 [26,27] The Val264Met mutation causes decreased TFPI levels [27]. It is reported that the Pro-151Leu replacement is a risk factor for venous thrombosis [28]. A polymorphism in the 3′ UTR of the TFPI gene (−287T/C) did not alter the TFPI levels and did not influence the risk of coronary atherothrombosis [29]. It has been recently reported that the −33T→C polymorphism in the intron 7 of the TFPI gene influences the risk of VTE, independently of the factor V Leiden and prothrombin mutations and its effect is mediated by increased total TFPI levels [30].

People who have a genetic predisposition to thrombophilia may harbor DNA mutations that result in deficiency of endogenous anticoagulants such as protein C, protein S, and antithrombin [31,32]. There could be at least 100 different types of mutations including point mutations, deletions, or insertions with each of the endogenous anticoagulants which would make genetic testing for diagnosis not feasible. Therefore, functional assays are employed for diagnostic purposes.

Anticoagulant drugs

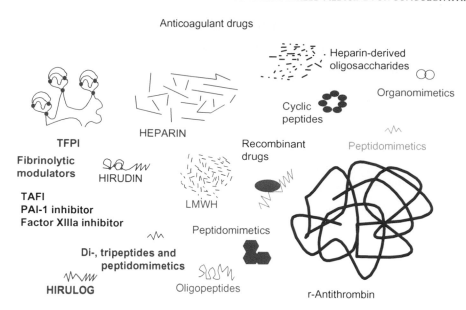

FIGURE 37.1 Structural representation of different anticoagulant drugs. This schematic shows a structural representation of heparin comprising of a heterogeneous mixture of low-, medium-, and high-molecular-weight fractions. Structural representations of other anticoagulants (such as direct thrombin inhibitors like hirudin and hirulog) can be seen. Other endogenous anticoagulants such as TFPI and antithrombin are also shown. *TFPI*, tissue factor pathway inhibitor; *TAFI*, thrombin-activatable fibrinolysis inhibitor-1; *PAI-1*, plasminogen activator inhibitor-1.

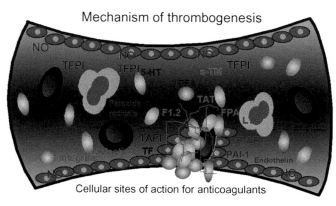

FIGURE 37.2 Diagrammatic representation of the interior of a blood vessel showing the mechanism of thrombogenesis. This schematic shows the intact endothelium lining the interior of the blood vessel on one side and the platelet aggregation forming a primary hemostatic plug on the other side where the structural integrity of endothelium has been breached.

Polymorphism Associated with VTE

VTE is a complex thrombotic disorder with environmental and genetic determinants, an annual incidence of one per 1000, and is the third leading cause of cardiovascular morbidity. Rudolph Virchow initially identified three factors contributing to risk of VTE, commonly referred to as Virchow's triad: (1) endothelial injury or activation, (2) reduced blood flow, and (3) hypercoagulability of blood [33]. While genome-wide association strategy is an effective way to identify common SNPs associated with VTE, previous genome-wide association studies (GWAS) have not included more than 1961 subjects [34,35]. The well-established genetic risk factors for VTE include heterozygous deficiencies of the endogenous anticoagulants such as antithrombin, protein C, and protein S (relatively rare affecting <1% of the general population), as well as deficiencies of factor V (FV) (MIM 612309) Leiden, prothrombin (MIM 176930) G20210A, fibrinogen γ (FGG) (MIM 134850) rs2066865, and blood group non-O (which are more frequent). A recent meta-analysis of 65,734 individuals reported identification of SNPs affecting TSPAN15 (MIM 613140), SLC44A2 (MIM 606106), and ZFPM (MIM 603140) as risk factors for VTE [36]. SNPs associated with each of these loci were selected for validation in three independent case-control studies totaling 3009 VTE-affected individuals and 2586 controls subjects, leading to identification of TSPAN15 and SLC44A2 as susceptibility loci for VTE [36]. Furthermore, the lead SNP at the TSPAN15 locus was reported to be the intronic rs78707713 and that of SLC44A2 as the nonsynonymous rs2288904 which was previously reported to be associated with transfusion-related acute lung injury. This study also identified six other susceptibility loci which were already known to be associated with VTE, namely, ABO (MIM 110300), FII, FV, FXI (MIM 264900), FGG, and endothelial protein C receptor (PROCR) (MIM600646) [36]. Investigators associated with the Leiden Thrombophilia study and the Multiple Environmental and Genetic Assessment of Risk Factors for Venous Thrombosis study examined two SNPs (rs2289252 and rs2036914) in factor XI and concluded that these SNPs are associated with increased plasma factor XI levels and are independent risk factors for DVT [37].

Hyperhomocysteinemia and Thrombosis

Polymorphism in methylenetetrahydrofolate (MTHFR C677T) has been shown to be associated with arterial and venous thromboses [38,39]. However, but recent studies reported uncertainties of such an association [40,41].

Polymorphism in Fibrinogen and Thrombosis

SNPs in the β-chain of fibrinogen (SNP 455G/A) were found to be associated with increased plasma fibrinogen levels and increased risk for stroke [42]. The association between fibrinogen gene mutation and arterial thrombosis is controversial because of different findings in different studies. The α-chain Thr-312Ala polymorphism is reported to increase the stability of clots [43].

PHARMACOGENOMICS OF ANTIPLATELET AND ANTICOAGULANT DRUGS

Anticoagulant, antiplatelet, and thrombolytic drugs are very commonly used agents for VTE (Figs. 37.1–37.3). Warfarin, aspirin, and clopidogrel are some of the very commonly used drugs. There are well-known interindividual patient responses to these agents which may pose a challenge to medical practice [44].

Warfarin

Warfarin is the most commonly used oral anticoagulant drug. Several limitations of its use include: (1) frequent anticoagulant monitoring using prothrombin time and international normalized ratio, (2) drug−food interactions and drug−drug interactions, (3) allergic manifestations, (4) Warfarin-induced skin necrosis, (5) bleeding complications, and (6) wide interindividual differences in anticoagulant response due to VKORC polymorphism.

Warfarin has a relatively narrow therapeutic index and as a result underdosing may cause thrombosis and overdosing may result in bleeding complications. Newer oral anticoagulant drugs such as dabigatran, rivaroxaban, apixaban, and edoxaban have been approved as alternatives to warfarin and several anticoagulant reversal agents are in advanced phases of clinical development. However, the use of warfarin will continue. Warfarin is derived from dicoumarol, a natural product isolated from sweet clover and its synthetic form consists of R and a more active S enantiomer. The S and R enantiomers of warfarin have different mechanisms of metabolism. S-warfarin is primarily metabolized by cytochrome P450 2C9 (CYP2C9), and R-warfarin by CYP3A4 [45]. Warfarin inhibits vitamin K epoxide reductase complex by binding to the VKORC1 subunit, thereby preventing reduced vitamin K dependent gamma-carboxylation of coagulation factors II, VII, IX, and X, as well as protein C and protein S (Fig. 37.3) [46]. Variation in three genes account for 40−54% of the observed interindividual response to warfarin dosing. GWAS have provided further insights into warfarin pharmacogenomics. CYP2C9*2 (C430T, rs1799853) and CYP2C9*3 (A1075C, rs1057910) are the most common alleles associated with decreased function and patients who harbor them require smaller doses of warfarin due to their greater sensitivity to the drug [47,48].

Aspirin

Aspirin is one of the most commonly used drugs for cardioprotection. It has been reported that 0.4−70.1%

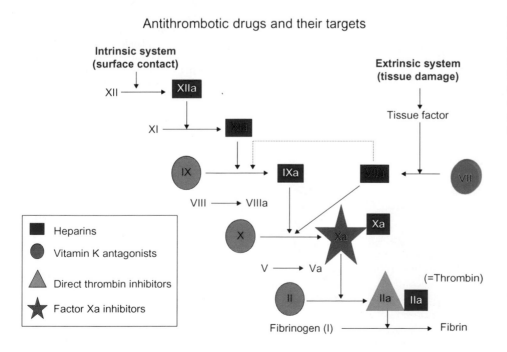

Antithrombotic drugs and their targets

FIGURE 37.3 Coagulation cascade showing sites of actions of different anticoagulant drugs. The schematic shows the coagulation cascade comprising the intrinsic and extrinsic pathways, and sites of actions for various anticoagulant drugs including heparin, vitamin K antagonists, direct thrombin inhibitors, and factor Xa inhibitors.

of patients respond poorly to aspirin or exhibit aspirin resistance [49]. It has been reported that genetic susceptibility may be the key to aspirin sensitivity and that genetic polymorphisms of HO-1 and COX-1 are associated with aspirin resistance defined by light transmittance aggregation in Chinese Han patients [50].

Clopidogrel

Clopidogrel is a thienopyridine that irreversibly inhibits the P2Y12 ADP receptors on the surface of platelets (Fig. 37.4) [51]. Following oral absorption and introduction into the blood circulation, 85% of the drug is hydrolyzed into inactive metabolites by carboxylesterases, mainly carboxyesterase 1 in the liver [52]. The remainder of the drug is biotransformed into an active drug by two steps that involve CYP2C19, CYP2B6, and others (including CYP1A2, CYP2C9, CYP3A4/5, and PON1) [53−55]. The active metabolite oxidizes cysteine residues and irreversibly blocks platelet P2Y12 ADP receptors. Patients treated with clopidogrel show interindividual variability in response [56−61]. Common loss-of-function variants in CYP2C19 represent the genetic determinants of clopidogrel responsiveness. CYP2C19*2 (rs4244285) is the most common loss-of-function variant with allele frequencies of 29% in Asians, and 15% in Caucasians and Africans. Other alleles (such as CYP2C19*2 and CYP2C19*3) are rare. Patients harboring one CYP2C19 loss-of-function allele are considered intermediate metabolizers and patients with two loss-of-function alleles are considered poor metabolizers. Patients with loss-of-function variants of CYP2C19 have lower clopidogrel active metabolite concentrations [62−64], greater on-treatment residual platelet function [65−68], and poorer cardiovascular outcomes in percutaneous coronary intervention patients on clopidogrel therapy [62,68−72].

GENOMICS OF BLEEDING DISORDERS AND PERSONALIZED MEDICINE IN HEMOPHILIA CARE

The inherited bleeding disorders include coagulation factor and platelet bleeding disorders and genetic analysis for hemophilia A, hemophilia B, and von Willebrand disease is routinely performed in most laboratories [73−77]. Next-generation sequencing that enables parallel sequencing of many genes regions at once is becoming available in diagnostic laboratories [78]. Replacement therapy of FVIII/FIX in hemophiliac patients either having absence or functionally inactive factors causes the development of allo-antibodies (inhibitors) as a part of the immune response in the individual patient. Understanding of why the inhibitors develop in only 25−30% of patients and not in all is not completely understood [79]. It is generally thought that the immune response triggered is T-helper cell mediated and that it involves the processing of

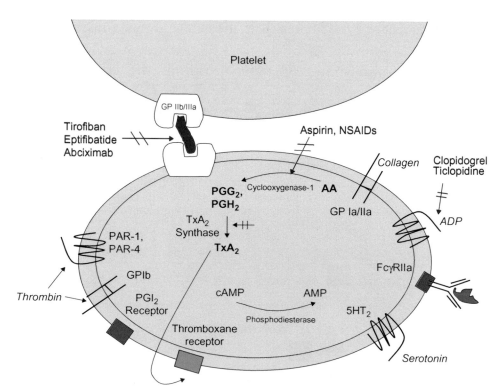

FIGURE 37.4 Diagrammatic representation of the platelet and the different platelet receptors on its surface and the sites of actions of different antiplatelet drugs. This schematic shows the platelet and various receptors on its surface with the sites of actions of different antiplatelet drugs such as aspirin, NSAIDS, clopidogrel, ticlopidine, and other GPIIb/IIIa receptor antagonists such as tirofiban, eptifibatide, and abciximab. *Source: Adapted from HL Messmore, et al. Antiplatelet agents: current drugs and future trends. Heme Onc Clin N Am 2005;19:87−118.*

proteins by antigen-presenting cells and subsequent association of these peptides to HLA molecules [80,81], especially to HLA class II alleles such as HLA-DRB1*14, DRB1*15, HLA-DQB1*06:02, and DQB1*06:03. Genetic polymorphism immune-response associated genes such as *IL1b, IL4, IL10, TNFα*, and *CTLA4* have been analyzed [81]. In the Hemophilia Inhibitor Genetic study, new genes, such as *CD44, CSF1R, DOCK2, MAPK9*, and *IQGAP2*, responsible for inhibitor development have been identified [82]. However, DRB1*16 and DQB1*05:02 alleles had lower inhibitor development risk [83—86].

Genotype-based individualized patient care for the treatment of thrombotic and bleeding disorders is expected to reduce the number of adverse drug—drug interactions, and drug failure rates, besides providing the right dosage to the right person thereby improving the quality of life and reducing healthcare expenditure. Next-generation DNA sequencing will enable parallel sequencing of many genes at once for a defined panel of coagulation and bleeding disorders.

References

[1] Abrahams E, Silver M. The history of personalized medicine. In: Gordon E, Koslow S, editors. Integrative neuroscience and personalized medicine. New York, NY: Oxford University Press; 2010.

[2] Personalized Medicine Coalition. www.personalizedmedicine-coalition.org/sciencepolicy/personalmed.

[3] Wikipedia. Personalized medicine, http://en.wikipedia.org/wiki/Personalized_medicine.

[4] McKusick VK, Ruddle FH. A new discipline, a new name, a new journal (editorial). Genomics 1987;1:1—2.

[5] Genomics and World Health. Report of the advisory committee on health research. Geneva: World Health Organization; 2002.

[6] WHA 57.13: Genomics and World Health. Fifty seventh world health assembly resolution. 22 May 2004.

[7] Vogel F. Moderne problem der humangenetik. Ergebn Inn Med Klinderheik 1959;12:52—125.

[8] Sachidanandam R, Weissman D, Schmidt SC, et al. A map of human genome sequence variation containing 1.42 million single nucleotide polymorphisms. Nature 2001;409:928—33.

[9] Endler G, Mannhalter C. Polymorphisms in coagulation factor genes and their impact on arterial and venous thrombosis. Clin Chim Acta 2003;330:31—55.

[10] Manucci PM. The molecular basis of inherited thrombophilia. Vo-Sang 2000;78:39—45.

[11] Foka ZJ, Lambropoulos AF, Saravelos H, et al. Factor V Leiden and prothrombin G20210A mutation but no methylenetetrahydrofolate reductase C677T are associated with recurrent miscarriages. Hum Reprod 2000;15:458—62.

[12] Amirrano L, Brancaccio V, Guardascione MA, et al. Inherited coagulation disorders in cirrhotic patients with portal vein thrombosis. Hepatology 2000;31:345—8.

[13] Ekberg H, Svensson PJ, Simanaitis M, Dahlback B. Factor V R506Q mutation (activated protein C resistance) is additional risk factor for early renal graft loss associated with acute vascular rejection. Transplantation 2000;69:1577—81.

[14] Rees DC, Cox M, Clegg JB. World distribution of factor V Leiden. Lancet 1995;346:1133—4.

[15] Iacovelli L, Di Castelnuovo A, de Knijiff P, et al. Alu-repeat polymorphism in the tissue-type plasminogen activator (tPA) gene, tPA levels and risk of familial myocardial infarction (MI). Fibrinolysis 1996;10:13—16.

[16] Furie B, Burie BC. Molecular and cellular biology of blood coagulation. N Engl J Med 1992;326:800—6.

[17] Poort SW, Rosendaal FR, Reitsma PH, Bertina RM. A common genetic variation in the 3′-untranslated region of the prothrombin gene is associated with elevated plasma prothrombin levels and an increase in venous thrombosis. Blood 1996;88:3698—703.

[18] Psaty BM, Smith NL, Lemairre RN, et al. Hormone replacement therapy, prothrombotic mutations and the risk of incident nonfatal myocardial infarction in postmenopausal women. JAMA 2001;285:906—13.

[19] Wartiovaarta U, Mikkola H, Szoke G, et al. Effect of Val34Leu polymorphism on the activation of coagulation factor XIII-A. Thromb Haemost 2000;8:595—600.

[20] Ariens RAS, Philippou H, Nagaswami C, et al. The factor XIII V34L polymorphism accelerates thrombin activation of factor XIII and affects crosslinked fibrin structure. Blood 2000;96:988—95.

[21] Kohler HP, Stickland MH, Ossei-Gernig N, et al. Association of a common polymorphism in the factor XIII gene with myocardial infarction. Thromb Haemost 1998;79:8—13.

[22] Wartiovaara U, Perola M, Mikkola H, et al. Association of factor XIII Va34Leu with decrease risk of myocardial infarction in Finnish males. Atherosclerosis 1999;142:295—300.

[23] Doggen CJM, Kunz G, Rosebdaal FR, et al. A mutation in the thrombomodulin gene, 127G to A coding for Ala25Thr and the risk of myocardial infarction in men. Thromb Haemost 1998;80:743—8.

[24] Li YH, Chen JH, Wu HL, et al. G-33A mutation in the promoter region of thrombomodulin gene and its association with coronary artery disease and plasma soluble thrombomodulin levels. Am J Cardiol 2000;85:8—12.

[25] Li YH, Chen CH, Yeh PS, et al. Functional mutation in the promoter region of thrombomodulin gene in relation to carotid atherosclerosis. Atherosclerosis 2001;154:713—19.

[26] Kleesiek K, Schmidt M, Gotting C, et al. A first mutation in the human tissue factor pathway inhibitor gene encoding[P151L] TFPI. Blood 1998;92:3976—7.

[27] Moati D, Seknaddji P, Galand C, et al. Polymorphisms of the tissue factor pathway inhibitor (TFPI) gene in patients with acute coronary syndromes and in healthy subjects: impact of the V264M substitution on plasma levels of TFPI. Arterioscler Thromb Haemost 1999;19:862—9.

[28] Kleesiek K, Schmidt M, Gotting C, et al. The 536C→T transition in the human tissue factor pathway inhibitor (TFPI) gene is statistically associated a higher risk for venous thrombosis. Thromb Haemost 1999;82:1—5.

[29] Moati D, Haider B, Fumeron F, et al. A new T-287C polymorphism in the 5′ regulatory region of the tissue factor pathway inhibitor gene. Association study of the T-287C and 3-399T polymorphisms with coronary artery disease and plasma TFPI levels. Thromb Haemost 2000;84:244—9.

[30] Arneziane N, Seguin C, Borgel D, et al. The -33T→C polymorphism in intron 7 of the TFPI gene influences the risk of venous thromboembolism independently of factor V Leiden and prothrombin mutations. Thromb Haemost 2002;88:195—9.

[31] Aiach M, Borgel D, Gaussem P, et al. Protein C and protein S deficiencies. Semin Hematol 1997;34:205—16.

[32] Vinazzer H. Hereditary and acquired antithrombin deficiency. Semin Thromb Hemost 1999;25:257—63.

[33] Virchow R. Gessamelte Abhandlungen zur Wissenschaftlichen Medzin. Frankfurt: Meidinger; 1856.

[34] Tang W, Teichert M, Chasman DI, et al. A genome-wide association study for venous thromboembolism: the extended cohorts for heart and aging research in genomic epidemiology (CHARGE) Consortium. Genet Epidemiol 2013;37:512−21.

[35] Germain M, Saut N, Oudot-Mellakh T, et al. Caution in interpreting results from imputation analysis when linkage disequilibrium extends over a large distance: a case study on venous thrombosis. PLoS One 2012;7:e38538.

[36] Germain M, Chasman DI, Haan HD, et al. Meta-analysis of 65,734 individuals identifies TSPAN15 and SLC44A2 as two susceptibility loci for venous thromboembolism. Am J Hum Genet 2015;96:532−42.

[37] Li Y, Bezemer I, Rowland CM, et al. Genetic variants associated with deep vein thrombosis: the f11 locus. J Thromb Haemost 2009;7:1802−8.

[38] Almavwwi WY, Tamin H, Kreidy R, et al. A case-control study on the contribution of factor V Leiden prothrombin G20210A, and MTHFR C677T mutations to the genetic susceptibility of deep venous thrombosis. J Thromb Thrombolysis 2005;19:189−96.

[39] Miccoll MD, Chalmers EA, Thomas A, et al. Factor V Leiden, prothrombin 20210G→A and the MTHFR C677T mutations in childhood stroke. Thromb Haemost 1999;8:690−4.

[40] Domagala TB, Adamek L, Nizankowska E, et al. Mutations C677T and A1298C of the 10-methylenetetrahydrofolate reductase gene and fasting plasma homocysteine levels are not associated with the increased risk of venous thromboembolic disease. Blood Coagul Fibrinolysis 2002;13:423−31.

[41] Klujimans LA, Den Heijer M, Reitsma PH, et al. Thermolabile methylenetetrahydrofolate reductase and factor V Leiden in the risk of deep vein thrombosis. Thromb Haemost 1998;79:254−8.

[42] Siegerink B, Rosendaal FR, Algra A. Genetic variation in fibrinogen; its relationship to fibrinogen levels and the risk of myocardial infarction and ischemic stroke. J Thromb Haemost 2009;7:385−91.

[43] Muszbeck L, Adany R, Mikkola H. Novel aspects of blood coagulation factor XIII. Structure, distribution, activation and function. Crit Rev Clin Lab Sci 1996;33:357−421.

[44] Shin J. Clinical pharmacogenomics of warfarin and clopidogrel. J Pharm Pract 2012;25:428−38.

[45] Kaminsky LS, Zhang ZY. Human P450 metabolism of warfarin. Pharmacol Ther 1997;73:67−74.

[46] Ansell J, Hirsh J, Poller L, Bussey H, et al. The pharmacology and management of the vitamin K antagonists: the Seventh ACCP Conference on Antithrombotic and Thrombolytic Therapy. Chest 2004;126:204S−233SS.

[47] Higashi MK, Veenstra DL, Kondo LM, et al. Association between CYP2C9 genetic variants and anticoagulation-related outcomes during warfarin therapy. JAMA 2002;287:1690−8.

[48] Gage BF, Eby C, Milligan PE, et al. Use of pharmacogenetics and clinical factors to predict the maintenance dose of warfarin. Thromb Haemost 2004;91:87−94.

[49] Shuldiner AR, O'Connell JR, Bliden KP, et al. Association of cytochrome P450 2C19 genotype with the antiplatelet effect and clinical efficacy of clopidogrel therapy. JAMA 2009;302:849−57.

[50] Rafferty M, Walters MR, Dawson J. Antiplatelet therapy and aspirin resistance clinically and chemically relevant? Curr Med Chem 2010;17:4578−86.

[51] Wang L, McLeod HL, Weinshilboum RM. Genomics and drug response. N Engl J Med 2011;364:1144−53.

[52] Ancrenaz V, Daali Y, Fontana P, et al. Impact of genetic polymorphisms and drug−drug interactions on clopidogrel and prasugrel response variability. Curr Drug Metab 2010;11:667−77.

[53] Richter T, Murdter TE, Heinkele G, et al. Potent-mechanism-based inhibition of human CYP2B6 by clopidogrel and ticlopidine. J Pharmacol Exp Ther 2004;308:189−97.

[54] Savi P, Combalbert J, Gaich C, et al. The antiaggregating activity of clopidogrel is due to a metabolic activation by the hepatic cytochrome P450-1A. Thromb Haemost 1994;72:313−17.

[55] Turpeinen M, Tolonen A, Uusitalo J, et al. Effect of clopidogrel and ticlopidine on cytochrome P450 2B6 activity as measured by bupropion hydroxylation. Clin Pharmacol Ther 2005;77:553−9.

[56] Gum PA, Kotkeke-Merchant K, Welsh PA, et al. A prospective blinded determination of the natural history of aspirin resistance among stable patients with cardiovascular disease. J Am Coll Cardiol 2003;41:961−5.

[57] Gurbel PA, Bliden KP, Guyer K, et al. Platelet reactivity in patients and recurrent events post-stenting results of the PREPARE POST-STENTING Study. J Am Coll Cardiol 2005;46:1820−6.

[58] Gurbel PA, Becker RC, Mann KG, et al. Platelet function monitoring in patients with coronary artery disease. J Am Coll Cardiol 2007;50:1822−34.

[59] Angiolillo DJ, Alfonso F. Platelet function testing and cardiovascular outcomes: steps forward in identifying the best predictive measure. Thromb Haemost 2007;98(4):707−9.

[60] O'Donoghue M, Wiviott SD. Clopidogrel response variability and future therapies: clopidogrel: does one size fit all? Circulation 2006;114:e600−6.

[61] Wang TH, Bhatt DL, Topol EJ. Aspirin and clopidogrel resistance: an emerging clinical entity. Eur Heart J 2006;27:647−54.

[62] Mega JL, Close SL, Wiviot SD, et al. Cytochrome P-450 polymorphisms and response to clopidogrel. N Engl J Med 2009;360:354−62.

[63] Kim KA, Park PW, Hong SJ, Park JY. The effect of CYP2C19 polymorphism on the pharmacokinetic and pharmacodynamics of clopidogrel: a possible mechanism for clopidogrel resistance. Clin Pharmacol Ther 2008;84:236−42.

[64] Umemura K, Furuta T, Kondo K. The common gene variants of CYP2C19 affect pharmacokinetics and pharmacodynamics in an active metabolite of clopidogrel in healthy subjects. J Thromb Haemost 2008;6:1439−41.

[65] Fontana P, Hulot JS, De Moerloose P, et al. Influence of CYP2C19 and CYP3A4 gene polymorphisms on clopidogrel responsiveness in healthy subjects. J Thromb Haemost 2007;5:2153−5.

[66] Brandt JT, Close SL, Iturria SJ, et al. Common polymorphisms of CYP2C19 and CYP2C9 after the pharmacokinetic and pharmacodynamics response to clopidogrel but not prasugrel. J Thromb Haemost 2007;5:2429−36.

[67] Trenk D, Hochholzer W, Fromm MF, et al. Cytochrome P450 2C19 681G > a polymorphism and high on-clopidogrel platelet reactivity associated with adverse 1-year clinical outcome of elective percutaneous coronary intervention with drug-eluting or bare-metal stents. J Am Coll Cardiol 2008;5:1925−34.

[68] Coller JP, Hulot JS, Pena A, et al. Cytochrome P450 2C19 polymorphism in young patients treated with clopidogrel after myocardial infarction: a cohort study. Lancet 2009;373:309−17.

[69] Simon T, Vertuyft C, Mary-Krause M, et al. Genetic determinants of response to clopidogrel and cardiovascular events. N Engl J Med 2009;360:363−75.

[70] Sibbing D, Stegherr J, Latz W, et al. Cytochrome P450 2C19 loss-of-function polymorphism and stent thrombosis following percutaneous coronary intervention. Eur Heart J 2009;30:916−22.

[71] Price MJ, Murray SS, Angiulillo DJ, et al. Influence of genetic polymorphisms on the effects of high and standard dose clopidogrel after percutaneous coronary intervention: the GIFT (Genotype Information and Functional Testing) study. J Am Coll Cardiol 2012;59:1928−37.

[72] Li X, Cao J, Fan L, et al. Genetic polymorphisms of HO-1 and COX-1 are associated with aspirin resistance defined by light transmittance aggregation in Chinese Han patients. Clin Appl Thromb Hemost 2013;19:513−21.

[73] Nichols WC, Seligsohn U, Zivelin A, et al. Mutations in the ER Golgi intermediate compartment protein ERGIC-53 cause combined deficiency of coagulation factors V and VIII. Cell 1998;93:61−70.

[74] Zhang B, McGee B, Yamaoka JS, et al. Combined deficiency of factor V and factor VIII is due to mutations in either LMAN1 or MCFD2. Blood 2006;107:1903−7.

[75] Rost S, Fregin A, Ivaskevicius V, et al. Mutations in VKORC1 cause warfarin resistance and multiple coagulation factor deficiency type 2. Nature 2004;427:537−41.

[76] Santagostino E, Mancuso ME, Tripodi A, et al. Several hemophilia with mild bleeding phenotype: molecular characterization and global coagulation profile. J Thromb Haemost 2010;8:737−43.

[77] Carcao MD, van den Berg HM, Ljung R, et al. PedNet and the Rodin Study Group. Correlation between phenotype and genotype in a large unselected cohort and children with severe hemophilia A. Blood 2013;121:3946−52.

[78] Watson SP, Lowe GC, Lordkipanidze M, et al. Genotyping and phenotyping of platelet function disorders. J Thromb Haemost 2013;11:351−63.

[79] Goodeve AC, Pavlova A, Oldenburg J. Genomics of bleeding disorders. Haemophilia 2014;20:50−3.

[80] Reiper BM, Allacher P, Hausl C, et al. Modulation of factor VIII-specific memory B cells. Haemophilia 2010;16:25−30.

[81] Van Helden PM, Kaijen PH, Fijnvandraat K, et al. Factor VIII specific B cells in patients with hemophilia A. J Thromb Haemost 2007;5:2306−8.

[82] Hay CR, Ollier W, Pepper L, et al. HLA class II profile: a weak determinant of factor VIII inhibitor development in severe hemophilia A. UKHCDO Inhibitor Working Party. Thromb Haemost 1997;77:234−7.

[83] Oldenburg J, Picard JK, Schwaab R, et al. HLA genotype of patients with severe haemophilia A due to intron 22 inversion with and without inhibitors of factor VIII. Thromb Haemost 1997;77:238−42.

[84] Hay CR. The epidemiology of factor VIII inhibitors. Haemophilia 2006;12:23−8.

[85] Pavlova A, Delev D, Lacroix-Desmazes S, et al. Impact of the polymorphisms of the major histocompatibility complex class II, interleukin 10, tumor necrosis factor alpha and cytotoxic T-lymphocyte antigen-4 genes on inhibitor development in severe hemophilia A. J Thromb Haemost 2009;7:2006−15.

[86] Astermark J, Donfield SM, Gomperts ED, et al. Haemophilia Inhibitor Genetics Study (HIGS) Combined Cohort. The polygenic nature of inhibitors in haemophilia A: results from the Haemophilia Inhibitor Genetics Study (HIGS) Combined Cohort. Blood 2013;121:1446−54.

38

Personalized Medicine for Hepatitis C Virus

A. Ferreira-Gonzalez

Division of Molecular Diagnostics, Department of Pathology, Virginia Commonwealth University,
Richmond, VA, United States

INTRODUCTION

Hepatitis C is a viral infection of the liver that ultimately causes the liver to become swollen and inflammation occurs. Hepatitis C virus (HCV) is an infectious particle that causes cirrhosis and hepatocellular carcinoma around the globe [1]. HCV infects an estimated 185 million individuals worldwide and up to 4.4 million in the United States. HCV prevalence is approximately 5% in the general population and 57% in the injecting drug use (IDU) population. Approximately 3% of the population of world is affected by HCV and it is estimated that end-stage liver disease develops in 30% of these patients. The incidence of HCV infection is elevated among IDU people. In 2010, an estimated 10 million IDU individuals were positive for HCV antibodies and the global prevalence of HCV was 67% among that population, especially for people who inject drugs [2]. In several parts of the world, IDU has become the most widespread risk factor for current cases, particularly in the United States. Sexual transmission of HCV does not appear to be so common, as the studies proved its spread in less than 1% of couples yearly, among monogamous heterosexual partners. Reinfection has been observed in 26% of patients who had previously eliminated the initial infection. Many researchers have demonstrated reinfection and superinfection. Today, HCV mostly spreads through sharing of syringes and needles. Transmission of infection from mother-to-child is more common among patients with hepatitis B or HIV than those with hepatitis C, but it can occur in HCV-infected mothers nonetheless. This type of viral infection transference occurs mostly in HCV viremic women. The most common risk factor that becomes the cause of HCV transference from mother-to-child is coinfection of HIV in mother with HCV viremia detectable during pregnancy. This transmission has not been found to be related to breast feeding or type of delivery (whether Cesarean or vaginal). HCV infection is a serious problem in centers where hemodialysis is done. Hence, many steps are taken by these centers to prevent HCV infection. These preventative steps include: (1) HCV-infected patients are grouped or isolated in separate rooms of dialysis center, (2) adherence to infection-control rules has been increased (eg, the screening for HCV at regular intervals), (3) not to reuse the shared vials or syringes, and (4) regularly vaccinating HCV-infected patients for hepatitis A virus (HAV) and hepatitis B virus (HBV) [3,4].

In the United States, HCV infection is the most common blood-borne infection. The best estimates of HCV prevalence derive from analysis of serum specimens taken from participants in the National Health and Nutrition Examination Survey (NHANES). The first estimate of HCV prevalence in the United States was generated from NHANES conducted between 1988 and 1994 and it estimated 2.7 million persons had chronic HCV. In a similar subsequent NHANES analysis, which involved a survey conducted between 1999 and 2002, investigators estimated 3.2 million persons had chronic hepatitis C, which corresponded to approximately 1.3% of the population [4]. A recent follow-up NHANES study that involved survey data from 2003 to 2010 estimated 2.7 million persons are chronically infected with HCV, corresponding to a population prevalence of chronic hepatitis C of 1%. However, these NHANES surveys did not sample certain populations, including the incarcerated, homeless, nursing home residents, persons on active military duty, and immigrants. HCV prevalence is highest among persons born from 1945 to 1965. Indeed, the

© 2017 Elsevier Inc. All rights reserved.

CDC estimates that approximately 75% of all persons living with HCV infection in the United States were born between 1945 and 1965. The relatively high prevalence of HCV infection among persons born from 1945 to 1965 reflects the high HCV incidence (new infections) that occurred among young adults in the 1970s and 1980s [4,5].

An estimated 40–85% of persons infected with HCV are unaware of their HCV infection status. Persons with HCV infection have all-cause mortality greater than twice that of HCV-negative persons. In the United States, hepatitis C is the cause of death or contributing cause of death in approximately 15,000 people per year. From 1997 to 2007, the number of annual deaths related to hepatitis C increased substantially, and in 2007 the number of deaths related to hepatitis C exceeded those related to HIV [6,7]. The number of hepatitis C related deaths is at least eightfold greater than those related to hepatitis B. Investigators have identified factors associated with an increased risk of death in persons with chronic hepatitis C infection: chronic liver disease, coinfection with HBV, alcohol-related conditions, minority status, and coinfection with HIV. Among the HCV-related deaths in recent years, more than 70% have involved persons 45–64 years of age. Overall, approximately 20% of persons infected with HCV will develop cirrhosis after 20 years of infection if not treated, and this number increases as the duration of infection increases. Hepatitis C associated liver disease is the number one indication for liver transplantation and approximately one-third of all persons on liver transplantation waiting lists have hepatitis C associated liver disease. In addition, hepatitis C associated liver disease is the number one cause of hepatocellular carcinoma, accounting for approximately 50% of cases of hepatocellular carcinoma [7]. Modeling studies have projected a dismal future in the next 40–50 years related to HCV-related disease burden. In general, these models make forecasts based on current conditions of low rates of screening and treatment and do not include a widespread program of identifying and treating the large proportion of undiagnosed HCV-infected individuals [8]. Factoring in treatment gives lower estimates of death. Investigators predict that 1.76 million persons with chronic HCV infection (if not treated) will develop cirrhosis during the next 40–50 years, with a peak prevalence of about 1 million in the mid-2020s. Several research studies are being conducted for developing new therapeutic and prophylactic vaccines for HCV [9,10].

HCV is a small, enveloped, positive-sense single-stranded RNA virus that belongs to the genus *Hepacivirus* and it is a member of the family *Flaviviridae* [1]. The HCV particle consists of a core of genetic material (RNA), surrounded by an icosahedral protective shell of protein, and further encased in a lipid bilayer envelope of cellular origin. Two viral envelope glycoproteins, E1 and E2, are embedded in the lipid envelope. HCV core genetic material has a positive-sense single-stranded RNA genome (Fig. 38.1) [1,11,12]. The genome consists of a single open reading frame that is 9600 nucleotide bases in length. This single open reading frame is translated to produce a single protein product, which is then further processed to produce smaller active proteins. At the 5′ and 3′ ends of the RNA are the untranslated regions (or UTRs), which are not translated into proteins but are important to translation and replication of the viral RNA. The 5′UTR has a ribosome-binding site that starts the translation of a very long protein containing about 3000 amino acids. The large pre-protein is later cut by cellular and viral proteases into the 10 smaller proteins that allow viral replication within the host cell or assemble into the mature viral particles. Structural proteins made by the HCV include core protein, E1, and E2, and nonstructural proteins include NS2, NS3, NS4A, NS4B, NS5A, and NS5B. The proteins of this virus are arranged along the genome in the following order: N-terminal-core-envelope (E1)–E2–p7-nonstructural protein 2 (NS2)–NS3–NS4A–NS4B–NS5A–NS5B–C-terminal. The generation of mature nonstructural proteins (NS2 to NS5B) relies on the activity of viral proteinases. The NS2-NS3 junction is cleaved by a metal-dependent autocatalytic proteinase encoded within NS2 and the N-terminus of NS3. The remaining cleavages downstream from this site are catalyzed by a serine proteinase also contained within the N-terminal region of NS3. NS2 protein is a transmembrane protein with protease activity. NS3 is a protein whose N-terminal has serine protease activity and whose C-terminal has NTPase/helicase activity. It is located within the endoplasmic reticulum and forms a heterodimeric complex with NS4A—a membrane protein that acts as a cofactor of the proteinase. NS4B is a small hydrophobic integral membrane protein with four transmembrane domains. NS5A is a hydrophilic phosphoprotein which plays an important role in viral replication, modulation of cell signaling pathways, and the interferon response. The NS5B protein is the viral RNA dependent RNA polymerase. NS5B has the key function of replicating the HCV's viral RNA by using the viral-positive RNA strand as its template and catalyzes the polymerization of ribonucleoside triphosphates (rNTP) during RNA replication synthesis. HCV encodes two proteases, the NS2 cysteine auto protease and the NS3–4A serine protease. The NS proteins recruit the viral genome into an RNA replication complex, which is associated with rearranged cytoplasmic membranes. RNA replication takes places via the viral RNA dependent RNA polymerase NS5B, which produces a negative-strand RNA intermediate. The negative-strand RNA then serves as a template for

Hepatitis C virus

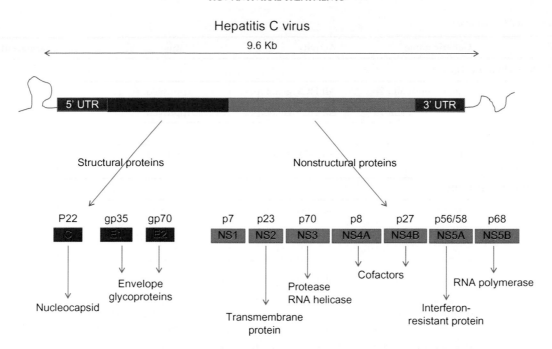

FIGURE 38.1 Translation and processing of the HCV polyprotein. The HCV polyprotein is processed cotranslationally and posttranslationally by host and viral proteases into at least 10 different proteins, which are arranged in the order of NH2–C–E1–E2–p7–NS2–NS3–NS4A–NS4B–NS5A–NS5B-COOH. Host signal peptidase is required for the cleavages at C-E1, E1–E2, E2–p7, and p7–NS2 junctions. NS2 cleaves the site between NS2 and NS3. NS3-4A serine protease cleaves the sites at NS3–NS4A, NS4A–NS4B, NS4B–NS5A, and NS5A–NS5B junctions. The wavy lines mark the UTR of HCV genomic RNA while the rectangle represents the polyprotein derived from the long open reading frame.

the production of new positive-strand viral genomes. HCV replicates mainly in the hepatocytes of the liver, where it is estimated that daily each infected cell produces approximately 50 virions with a calculated total of 1 trillion virions generated [13,14]. The virus may also replicate in peripheral blood mononuclear cells, potentially accounting for the high levels of immunological disorders found in patients chronically infected with HCV. Once inside the hepatocyte, HCV takes over portions of the intracellular machinery to replicate. Nascent genomes can then be translated, further replicated, or packaged within new virus particles. New virus particles are thought to bud into the secretory pathway and are released at the cell surface [14–19].

Based on genetic differences between HCV isolates, the HCV species is classified into seven genotypes (numbered 1–7) with several subtypes within each genotype (represented by lower-case letters) [18,20]. Subtypes are further broken down into quasispecies based on their genetic diversity. Genotypes differ at 30–35% of the nucleotide sites over the complete genome. The difference in genomic composition of subtypes of a genotype is usually 20–25%. HCV has a wide variety of genotypes and mutates rapidly due to the high error rate on the part of the virus' RNA-dependent RNA polymerase. The mutation rate produces so many variants of the virus it is considered a

quasispecies rather than a conventional virus species. HCV 1a and 1b are the most prevalent HCV genotypes in Western Europe and the United States and these genotypes account for 60% of all HCV cases. HCV genotypes 2 and 3 are less prevalent. HCV genotype 4 is widespread in Egypt, genotype 5 is common in South Africa, and genotype 6 is found in Southeast Asia. In patients from Canada and Belgium, another seventh genotype has also been identified. In the United States, approximately 70% of chronic HCV infections are caused by genotype 1, 15–20% by genotype 2, 10–12% by genotype 3, 1% by genotype 4, and less than 1% by genotypes 5 and 6. Among the HCV infections associated with genotype 1, approximately 55% correspond to genotype 1a and 35% to genotype 1b [18–20].

HCV ANTIVIRAL TREATMENTS

Treatment of HCV Using Indirect-Acting Antivirals

Until 2011, the standard of care treatment for chronic HCV was the combination of weekly pegylated interferon alpha (INF) and daily doses of Ribavirin (RBV) in a 24- or 48-week course (Table 38.1) [20,21]. INF and RBV dual therapy is associated with several important side effects, including anemia, depression,

TABLE 38.1   HCV Therapies

Brand name	Generic name	Activity	Status	Pharmaceutical company
PEGYLATED INTERFERON ALPHA				
PegIntron	Peginterferon alfa-2b	All HCV genotypes	Approved	Merck
Pagasys	Peginterferon alfa-2a	All HCV genotypes	Approved	Genentech
NUCLEOSIDE ANALOG				
Copegus, Rebetol, Ribasphere, and Virazole	Ribavirin	All HCV genotypes	Approved	Genentech, Merck, Kadmon
NS3/4A PROTEASE INHIBITORS				
Incivek	Telaprevir	HCV genotypes 1a and 1b	Approved (to be discontinued as of 10/16/2014)	Vertex
Victrelis	Boceprevir	HCV genotypes 1a and 1b	Approved	Merck
Olysio	Simeprevir (TMC435)	HCV genotypes 1a and 1b	Approved	Janssen and Medivir AB
Sunvepra	Asunaprevir (BMS-650032)	HCV genotypes 1a and 1b	Phase III (not pursuing approval in the United States)	Bristol-Myers Squibb
n/a	Vaniprevir (MK-7009)	HCV genotypes 1a and 1b	Phase III	Merck
n/a	ABT-450	HCV genotypes 1a and 1b	Phase III	AbbVie
n/a	MK-5172	HCV genotypes 1a and 1b	Phase III	Merck
NUCLEOSIDE AND NUCLEOTIDE NS5B POLYMERASE INHIBITORS				
Sovaldi	Sofosbuvir (GS-7977)	All HCV genotypes	Approved	Gilead Sciences
n/a	Mericitabine (RG-7128)	All HCV genotypes	Phase III	Roche
NS5A INHIBITORS				
Daklinza	Daclatasvir (BM-790052)	All HCV genotypes	Phase III (submitted to FDA 4/7/2014)	Bristol-Myers Squibb
n/a	Ledipasvir (GS-5885)	All HCV genotypes	Phase III	Gilead Sciences
n/a	Ombitasvir (ABT-267)	All HCV genotypes	Phase III	AbbVie
n/a	GS-5816	All HCV genotypes	Phase III	Gilead Sciences
n/a	Elbasvir (MK-8742)	All HCV genotypes	Phase III	Merck
NON-NUCLEOSIDE NS5B POLYMERASE INHIBITORS				
Exviera	Dasabuvir (ABT-333)	HCV genotypes 1a and 1b	Phase III (submitted to FDA 4/2/2014)	AbbVie
n/a	Beclabuvir (BMS-791325)	HCV genotypes 1a and 1b	Phase III	Bristol-Myers Squibb
n/a	ABT-072	HCV genotypes 1a and 1b	Phase II	AbbVie
MULTICLASS COMBINATION DRUGS				
Viekirax	Ombitasvir (ABT-267) + Paritavir (ABT-450) + Ritonavir		Phase III (submitted to FDA 4/21/2014)	AbbVie
n/a	Asunaprevir + Daclatasvir + BMS-791325		Phase III	Bristol-Myers Squib
Harvoni	Sofosbuvir + Ledipasvir (GS-7977 + GS-5885)		Approved	Gilead Sciences
n/a	Grazoprevir (ML-8742) + Elbasvir (MK-5172)		Phase III	Merck

and nausea, which can lead to discontinuation of therapy. Cure of chronic HCV infection is tantamount to the sustained virologic response (SVR), which is defined as undetectable HCV RNA in the blood at the end of treatment and again 6 months later [22]. HCV RNA viral load measurement at baseline, weeks 4, 12, and 24, at the end of treatment, and 24 weeks after treatment withdrawal are used to characterize the virological response: (1) the rapid virologic response (RVR) is defined as undetectable HCV viral load at week 4, (2) early virologic response (EVR) is defined as HCV viral load detectable at 4 weeks, but undetectable at week 12, and (3) the slow or delayed virologic response (DVR) is defined as detectable at week 12, but undetectable after week 24 of therapy. Patients who show a RVR and a low baseline HCV RNA viral load need 24 weeks of therapy, patients who achieved an EVR require 48 weeks of treatment, whereas patients with DVR appear to benefit from extending treatment to 72 weeks. Patients with less than a 2 log decline in HCV viral load level at week 12 are unlikely to experience an SVR and could be taken off therapy. SVR rates vary among individuals infected with different HCV genotypes. SVR rates of 70–90% are observed in patients infected with HCV genotypes 2, 3, 5, and 6, but with SVR rates of less than 50% for genotypes 1 and 4 [21,22].

Treatment of HCV Using Direct-Acting Antivirals

Major advances in the understanding of the molecular virology of HCV came with the development of genotype 1 subgenomic and genomic replicon system and the identification of the genotype 2a JFH1 clone that leads to productive infection in cell culture after transfection. With these new model systems and the resolution of the three-dimensional structure of key HCV enzymes, the steps of the HCV life cycle were unraveled, identifying multiple targets for drug development. The development of HCV direct antiviral agents (DAAs) has occurred extremely rapidly, moving from approval of the first-wave, first-generation protease inhibitors that were combined with INF and RBV, to oral single tablet DAA combinations in less than 4 years (Table 38.1) [23–25]. Despite the dramatic success of DAA development, there were challenges along the way. The two first-wave, first-generation HCV NS3–4A protease inhibitors, Telaprevir (TVR) and Boceprevir (BOC), were approved in combination with INF and RBV for the treatment of chronic HCV genotype 1 in 2011 [26,27]. These two drugs are linear ketoamine inhibitors, which form a reversible but covalent complex with the HCV NS3-4A serine

protease catalytic site and block posttranslational processing of the viral polyprotein. Adding one or two NS3-4A inhibitors to dual INF plus RBV therapy increased SVR to 75% in treatment naïve and to 64% for previous nonresponders to INF–RBV dual therapy. Initial treatment regimens were complicated and proved less effective and more toxic in the real world than in clinical trials. Multiple promising agents were abandoned for toxicity, including fatal complications in a small number of patients. Due to these issues TVR was discontinued in October 2014 and BOC was discontinued in December 2015 [26–29].

Viral polymerases are prime targets for the development of antiviral drugs since their enzymatic sites are highly conserved between different genotypes. In addition, mutations in the active site of the viral polymerases are rarely well tolerated, because they are often associated with reduced viral replication. In 2014, the marketing and approval of the first nucleotide NS5B polymerase inhibitor Sofosbuvir represented a major milestone in the treatment of chronic hepatitis C [30]. Considered safe and well tolerated with pan-genotypic activity and a high barrier for resistance, the once daily in combination with INF and RBV dual therapy for 12 weeks improved SVR rates to 82–100% in treatment-naïve patients infected with HCV genotypes 1, 4, 5, and 6. The overall SVR was 92% in patients without cirrhosis versus 80% SVR in those with cirrhosis. Furthermore, phase II clinical trials in treatment-experienced patients showed SVR of 96% and 83% in patients infected with HCV genotype 2 or 3, respectively. The triple combination INF, RBV, and the second wave, first-generation NS3-4A protease inhibitor Simeprevir was also approved in 2014 for patients infected with HCV genotype 1 or 4 [28,29].

A further step forward toward the next generation of HCV treatment represented the first DAA-only regimens (Table 38.1). Daclastasvir and Lepdipasvir are inhibitors of the HCV NS5A protein, which play an important role in the viral replication and assembly [31–34]. These NS5A inhibitors bind the domain 1 of the NS5A protein and block its ability to regulate replication within the replication complex. In addition, NS5A inhibitors inhibit assembly and release of viral particles. Some of these NS5A inhibitors have pan-genotypic activity whereas others are poorly active against genotype 3. Second-generation NS5A inhibitors have pan-genotypic activity [30,31].

Combinations of agents targeting different stages of the life cycle have proven highly effective. INF-sparing regimens will probably be completely replaced in 2015 by INF-free regimens with improved efficacy and tolerability for both previously untreated patients and previously INF nonresponder patients [34,35]. These regimens comprise the following: (1) a protease inhibitor or an NS5A inhibitor plus a nucleoside NS5B inhibitor, with or without RBV; (2) a protease inhibitor, an NS5A inhibitor, and an NS5B non-nucleoside inhibitor, with or without RBV; or (3) a protease inhibitor and NS5A inhibitor, with or without RBV. Two new INF-free DAA-based combinations were approved in late 2014/early 2015 in the United States and Europe. The combination pill of Sofosbuvir and Ledipasvir administered daily, with or without RBV according to severity of liver disease, is approved for HCV genotypes 1 in the United States, and HCV genotypes 1, 3, and 4 in Europe. The triple combination of Ritonavir-boosted Paritaprevir and Ombitasvir in a single pill plus Dasabuvir in another pill with or without RBV according to HCV subtype and the presence of cirrhosis were approved for HCV genotype 1 early this year [32–37].

WHEN AND IN WHOM TO INITIATE HCV TREATMENT

The goal of treatment of HCV-infected individuals is to reduce all-cause mortality and liver-related health adverse consequences, including end-stage liver disease and hepatocellular carcinoma, by achieving an SVR to treatment [22]. Successful HCV treatment results in SVR, which is tantamount to virologic cure, and as such, is expected to benefit nearly all chronically infected persons [29,32,36,37]. The proximate goal of HCV therapy is SVR, defined as the continued absence of detectable HCV RNA at least 12 weeks after completion of therapy. SVR is a marker for cure of HCV infection and has been shown to be durable in large prospective studies in more than 99% of patients followed up for 5 years or more. In other words, SVR is achieved in patients who have HCV antibodies, but no longer detectable HCV RNA in circulation, liver tissue, or mononuclear cells, and achieve a substantial improvement in liver histology. Assessment of viral response, including documented SVR, requires specific nucleic acid based test (NAT) such as those approved by the Food and Drug Administration (FDA) or FDA qualitative or quantitative NAT with detection levels of 25 IU/mL or lower. Evidence clearly supports treatment in all HCV-infected individuals, except those with limited life expectancy (<12 months) due to non-liver related comorbidity. Urgent initiation of treatment is only recommended for those patients with advanced fibrosis or compensated fibrosis [22,33,35,37,38].

Immediate treatment is recommended for patients with chronic HCV infection with advanced fibrosis (Metavir F3), those with compensated cirrhosis (Metavir F4), liver transplants recipients, and patients with severe extrahepatic HCV, such as type 2 and type 3 essential mixed cryoglobulinemia with end-organ manifestations, proteinuria, nephritic syndrome, or membranoproliferative glomerulonephritis [38]. Treatment should be considered for patients with fibrosis (Metavir F2), HIV-1 infection, HBV coinfection, other coexisting liver disease (eg, NASH), debilitating fatigue, type 2 diabetes mellitus, and porphyria cutanea tarda [38,39]. A third group that should be considered for treatment includes individuals who pose elevated risk of HCV transmission and in whom treatment may yield transmission reduction benefits. This population includes men who have sex with men with high-risk sexual practices, active injection users, incarcerated individuals, persons on long-term hemodialysis, HCV-infected women of child-bearing potential wishing to get pregnant, and HCV-infected healthcare workers who perform exposure-prone procedures [37].

An accurate assessment of fibrosis is critical in assessing the urgency for treatment [38]. The degree of hepatic fibrosis is one of the most robust prognostic factors used to predict disease progression and clinical outcomes. Those with substantial fibrosis defined at Metavir F2 or higher should be given priority for therapy in an effort to decrease the risk of clinical consequences such as cirrhosis, liver failure, and hepatocellular cancer. Although liver biopsy is the diagnostic standard, sampling error and observer variability limit test performance. Noninvasive tests to stage the degree of fibrosis in patients with HCV infection include models incorporating indirect serum biomarkers and vibration controlled transient liver elastography. No single method is recognized to have high accuracy alone and each test must be interpreted carefully [38,39].

When therapy is deferred, it is very important to monitor liver disease in these patients. Among individuals with less advanced stages of fibrosis, fibrosis progression over time will help determine the urgency of subsequent antiviral therapy. Fibrosis progression varies markedly between individuals based on host, environmental, and viral factors. Host factors associated with more rapid fibrosis progression include male sex, longer duration of infection, and an older age at the time of infection. Many hepatitis C patients have concomitant nonalcoholic fatty liver disease, the presence of hepatic stenosis with or without steatohepatitis on liver biopsy, as well as elevated body mass index and insulin resistance. Chronic alcohol consumption is an important risk factor because alcohol consumption has been associated with more rapid fibrosis progression [38,39]. The level of virus in serum or plasma is not highly correlated with the stage of disease (degree of inflammation or fibrosis). Available data suggest that fibrosis progression occurs most rapidly in patients with genotype 3 HCV infection. Aside from coinfections with HBV or HIV, no other viral factors are consistently associated with disease progression [38].

Patients who are cured of their HCV infection experience numerous health benefits, including decreased liver inflammation as reflected by improved liver function and a reduction in the rate of progression of liver fibrosis. In addition, these patients have decreased symptoms for severe extrahepatic manifestations including cryoglobulinemic vasculitis, a condition affecting 10–15% of HCV-infected patients [38].

INITIAL TREATMENT OF HCV INFECTION

HCV Genotype 1

Three highly potent direct-acting antiviral or DAA oral combination regimens are recommended for HCV genotype 1 infected individuals, although there are differences in the recommended regimens based on the HCV subtype [39]. Patients with HCV genotype 1a tend to have higher relapse rates than patients with genotype 1b with certain regimens. Genotype 1 HCV infection that cannot be subtyped should be treated as a genotype 1a infection [38,40–58].

For HCV genotype 1a infected, treatment-naïve patients or patients who do not have cirrhosis, or in whom prior INF and RBV treatment has failed, there are three regimens of comparable efficacy: (1) daily fixed-dose combination of Ledipasvir/Sofosbuvir for 12 weeks, (2) daily fixed-dose combination of Paritaprevir/Ritonavir/Ombitasvir plus twice daily dose Dasabuvir and weight-based RBV for 12 weeks for patients without cirrhosis or 24 weeks for patients with cirrhosis, and (3) daily Sofosbuvir plus Simeprevir with or without weight-based RBV for 12 weeks for patients without cirrhosis or 24 weeks for patients with cirrhosis. More importantly, the safety profiles of all the recommended regimens listed are excellent. Across numerous phase III trials, less than 1% of patients without cirrhosis discontinued treatment early and adverse events were mild. Most adverse events occurred in RBV-containing arms. Discontinuation rates were higher for patients with cirrhosis [40–47].

For genotype 1b infected, treatment-naïve patients and patients who do not have cirrhosis, in whom prior INF/RBV treatment has failed, there are three regimens of comparable efficacy: (1) daily fixed combination of Ledipasvir for 12 weeks, (2) daily fixed-dose combination of Paritaprevir/Ritonavir/Ombitasvir plus twice daily dosed Dasabuvir for 12 weeks (the addition of weight-based RBV is recommended), and (3) daily Sofosbuvir plus Simeprevir for 12 weeks for patients without cirrhosis or 24 weeks for patients with cirrhosis. For patients with prior treatment failure

the addition or not of weight-based RBV is an option. Also discontinuation rates were higher for patients with cirrhosis [40–45].

Although regimens of Sofosbuvir and RBV or INF and RBV plus Sofosbuvir, Simeprevir, TVR, or BOC for 12–18 weeks (some using response-guided therapy (RGT)) are also FDA approved, they are inferior to the current recommended regimens. Most of the IFN-containing regimens are associated with higher rates of serious side adverse events (eg, anemia and rash), longer treatment duration, high pill burden, numerous drug–drug interactions, more frequent dosing, and higher intensity of monitoring for continuation and stopping of therapy, and the requirements to be taken with food or with high fats meals. Although clinical trials for Sofosbuvir reported the highest SVR rate of 89% for an INF-containing regimen in combination with INF and weight-based RBV in HCV genotype 1 infection and limited exposure to INF to just 12 weeks, the safety and tolerability profile limits its usefulness in the setting of FDA-approved, highly efficacious oral DAA combinations. INF and RBV for 48 weeks for treatment-naïve patients with HCV genotype 1 has been replaced by treatments incorporating DAA [38].

Recommendations for patients infected with HCV genotype 1a or genotype 1b with compensated cirrhosis, in whom prior INF and RBV treatment has failed are: (1) daily fixed-dose combination of Ledipasvir/Sofosbuvir for 24 weeks, (2) daily fixed dose of Ledipasvir/Sofosbuvir plus weight-based RBV for 12 weeks, (3) daily fixed dose of Paritaprevir/Ritonavir/Ombitasvir plus twice daily dosed Dasabuvir and weight-based RBV for 24 weeks for HCV genotype 1a and 12 weeks for HCV genotype 1b, and (4) daily Sofosbuvir plus Simeprevir with or without weight-based RBV for 24 weeks [40–45].

Recommended treatment for patients without cirrhosis who have HCV genotype 1 infection, regardless of subtype, in whom a prior INF, RBV, and HCV protease inhibitor regimen has failed is daily dose of Ledipasvir/Sofosbuvir for 12 weeks. In addition, there are two options of similar efficacy for patients with cirrhosis regardless of subtype, in whom INF and RBV, and HCV protease inhibitor regimens have failed. Daily fixed-dose of Ledipasvir/Sofosbuvir for 24 weeks or daily fixed-dose combination of Ledipasvir/Sofosbuvir plus weight-based RBV for 12 weeks [44,45].

HCV Genotype 2

The recommended treatment regimen for treatment-naïve patients or patients in whom prior INF and RBV treatment has failed with HCV genotype 2 infection is daily Sofosbuvir and weight-based RBV for 12 weeks, and extending to 16 weeks for patients with cirrhosis. Although there are no other alternative regimens listed for patients infected for HCV genotype 2, several available DAA have activity in vitro and in vivo against HCV genotype 2. Simeprevir has moderate potency against HCV genotype 2 but has not formally been tested in combination with Sofosbuvir [38,40,48,49].

Because of its poor activity in vitro and in vivo, BOC should not be used as therapy for patients with HCV genotype 2 infection. Although TVR plus INF/RBV has antiviral activity against HCV genotype 2, additional adverse effects and longer duration of therapy required do not support use of these regimens [40].

HCV Genotype 3

For individuals with HCV genotype 3 infection, the recommended regimen for treatment-naïve patients and patients in whom prior INF/RBV treatment has failed is daily Sofosbuvir and weight-based RBV for 24 weeks. The alternative regimen for treatment is daily Sofosbuvir and weight-based RBV plus INF treatment for 12 weeks. The combination of Sofosbuvir plus INF and RBV was evaluated in several clinical trials. Treatment-naïve patients infected with HCV genotype 3 achieved SVR during 4–12 weeks of therapy and RBV. For many patients, the adverse events and increased monitoring requirements of INF make the recommended regimen of Sofosbuvir plus weight-based RBV less desirable. However, the shortened period of treatment may be of interest to some. Due to limited in vitro and in vivo activity against HCV genotype 3, BOC, TVR, and Simeprivir should not be used as therapy for patients with HCV genotype 3 infection. The same is true for regimens including INF and RBV for 24–48 weeks, or monotherapy with INF/RBV, or a DAA [40,50–52].

HCV Genotype 4

There are three recommended therapeutic options with similar efficacy for treatment-naïve HCV genotype 4 infected patients and patients in whom prior INF and RBV treatment has failed: (1) daily fixed-dose combination of Ledipasvir plus Sofosbuvir for 12 weeks, (2) daily fixed-dose combination of Paritaprevir and Ritonavir plus Ombitasvir and weight-based RBV for 12 weeks, and (3) daily Sofosbuvir and weight-based RBV for 24 weeks. Alternative treatment regimens for HCV genotype 4 include daily Sofosbuvir and weight-based RBV plus weekly INF for 12 weeks or daily Sofosbuvir plus Simeprevir with or without weight-based RBV for 12 weeks. An additional recommendation for patients in whom prior INF/RBV

treatment has failed is daily Sofosbuvir and weight-based RBV for 24 weeks. The following therapies are not recommended for HCV genotype 4: INF with or without Simeprevir for 24 or 48 weeks, monotherapy with INF, RBV, or a DAA, or TVR or BOC-based regimens [39].

HCV Genotypes 5 and 6

There is limited data available to help guide decision making for patients infected with HCV genotype 5 or genotype 6. Nonetheless, for these patients for whom immediate treatment is required, the following has been recommended: daily Sofosbuvir and weight-based RBV plus weekly INF for 12 weeks for treatment-naïve patients as well as patients in whom INF and RBV treatment has failed. An alternative treatment for HCV genotype 5 infection is weekly INF plus weight-based RBV for 48 weeks. Recommended treatment for treatment-naïve HCV genotype 6 infected patients is daily fixed-dose combination of Ledipasvir or 12 weeks or daily Sofosbuvir and weight-based RBV plus weekly INF for 12 weeks. It is not recommended for patients infected with either HCV genotype 5 or genotype 6 to utilize monotherapy with INF, RBV, or a DAA, TVR, or BOC-based regimens [39].

MONITORING PATIENTS WHO ARE STARTING HCV TREATMENT, ARE ON TREATMENT, OR HAVE COMPLETED TREATMENT

This section provides guidance on monitoring patients with chronic HCV who are starting treatment, are on treatment, or have completed treatment and addressees pretreatment and ongoing treatment monitoring, and posttreatment follow-up for persons in whom treatment has failed to clear the virus, and posttreatment follow-up for those who achieved SVR (virologic cure) [22,39].

Pretreatment testing assumes that a decision to treat with antiviral medications has been reached and that testing involved in deciding to treat, including testing for HCV genotype and assessment of hepatic fibrosis has already been completed. These tests will determine the best option and regimen for treatment. It is recommended that 12 weeks prior to initiation of treatment patients have a complete blood count, international normalized ratio (or INR), liver function panel, thyroid-stimulating hormone (if INF is used), and calculated glomerular filtration rate (or GFR). In addition, testing for HCV genotype and subtype and quantitative HCV viral load should be done.

During treatment individuals should be followed up at clinically appropriate intervals to ensure medication adherence, assess adverse events and potential drug—drug interactions, and monitor blood test results necessary to assure patient safety. The assessment of HCV viral load at week 4 of therapy is useful to determine initial dose response to therapy and adherence. In phase III clinical trials, almost all patients who did not have cirrhosis had undetectable HCV RNA levels at week 4, those with cirrhosis may require more than 4 weeks of treatment before HCV RNA levels become undetectable. There is not a lot of data on how to use HCV RNA levels during therapy to determine when to stop treatment for futility. The current recommendations to repeat quantitative HCV RNA levels at week 4 of treatment and to discontinue treatment if quantitative HCV RNA levels increase by more than 10-fold ($>1 \log_{10}$ IU/mL) is based on expert opinion. There are no data to support stopping treatment based on detectable HCV RNA results at week 2, 3, or 4 of therapy, or that detectable HCV RNA level at these time point signifies medication nonadherence. Although HCV RNA testing is recommended at week 4 of treatment, the absence of HCV RNA at week 4 does not provide justification reason to discontinue therapy. If quantitative HCV viral load is detectable at week 4 of treatment, repeat quantitative HCV RNA viral load testing is recommended after two additional weeks of therapy (treatment week 6). If quantitative HCV viral load has increased by greater than 10-fold on repeat testing at week 6 or thereafter, then discontinuation of HCV treatment is recommended. The significance of HCV RNA test results at week 4 that remains positive but lower at week 6 or week 8 is unknown. No recommendations to stop therapy or extend therapy are available at this time. Quantitative HCV RNA level testing at the end of treatment will help differentiate viral breakthrough from relapse, if necessary. Quantitative HCV viral load testing can be considered at the end of treatment or 24 weeks or longer following the completion of therapy. Some may choose to forgo end of treatment viral load testing given the high rate of viral response with the newer treatments and focus on the week 12 posttreatment viral load. Virologic relapse is rare at 12 or more weeks after completing treatment. Nevertheless, repeat quantitative HCV RNA testing can be considered at 24 or more weeks after discontinuing treatment for selected patients [22,39].

The availability of INF-free treatment has simplified HCV therapy by allowing shorter duration oral therapy for most patients. However, INF and RBV-based regimens are beneficial for selected patients, and these require specific monitoring for toxic effects associated with this type of therapy.

Patients who do not achieve an SVR, because of failure of the treatment to clear or to maintain clearance of HCV infection with relapse after treatment completion, have ongoing HCV infection and the possibility of continued liver injury and transmission. Such patients should be monitored for progressive liver disease and considered for retreatment when alternative treatments are available. Patients who have undetectable HCV RNA in the serum, when assessed by a sensitive PCR assay, 12 or more weeks after completion of treatment, are deemed to have achieved SVR. In these patients, HCV-related liver injury stops, although the patients remain at risk for non-HCV related liver disease or alcoholic liver disease. Patients with cirrhosis remain at risk for developing hepatocellular carcinoma [22].

HCV RNA Results and Interpretations

Definitions and descriptions of the terms used to describe HCV RNA viral load level are provided in Table 38.2. Major improvements to quantitative HCV viral assays have been developed recently (Table 38.3). It is important to note that if HCV RNA is detected by a PCR test (and lower than the linear range of the test), the result is reported by the software as "HCV RNA detected, less than the lower limit of quantitation (LLOQ)" even if the actual viral load titer is below the sensitivity or limit of detection (LOD) of the test. Being able to detect RNA, even below the LOD may seem counterintuitive since it is typically presumed that if the actual HCV RNA titer is below the LOD then there is nothing to detect. However, the LOD is defined and calculated by the ability of the assay to detect HCV RNA greater than or equal to 95%. This means that even at HCV RNA titers that are half the LOD, the PCR amplification may still detect HCV RNA

TABLE 38.2 Definitions of Key Analytical Performance Terms Used in Defining HCV RNA VL Titer Measurements

Result		Definition
Target not detected		HCV RNA is not detected, no PCR amplification or detection
LLOQ	Lowest limit of quantification	Lowest HCV RNA titer within the test's dynamic range that is quantifiable and accurate
LLOD	Lowest limit of detection	Lowest amount of HCV RNA in a sample that can be detected greater than or equal to 95% of times
ULOQ	Upper limit of quantification	Highest HCV RNA titer result within the test's dynamic range that is quantifiable and accurate

TABLE 38.3 Quantitative Real Time Assays for HCV

Assay	Vendor	Technology (target region)	IVD status	Dynamic range (IU/mL)	LLOQ (IU/mL)	LLOD (IU/mL)
COBAS Ampliprep/COBAS TaqMan v 2.0	Roche Molecular Systems	Real-Time PCR (5′UTR)	FDA, CE	$15-1.00 \times 10^8$	15	15
COBAS TaqMan High Pure System Test, v 2.0	Roche Molecular Systems	Real-Time PCR (5′UTR)	FDA, CE	$25-3.91 \times 10^8$	25	20
Abbott Real Time HCV test	Abbott Diagnostics	Real-Time PCR (5′UTR)	FDA, CE	$12-1.00 \times 10^8$	12	12
Versant HCV RNA test, v 1.0 (kPCR)	Siemens	Real-Time PCR (*pol* gene)	CE	$15-1.00 \times 10^8$	15	15
Artus Hepatitis C Test (QS-RGQ)	Qiagen	Real-Time PCR (target proprietary)	CE	$65-1.00 \times 10^6$	35	21

approximately 50% of the time, in which case, the result will be reported as "HCV RNA detected, below the LLOQ" if the RNA is detected [22,47,59].

Viral Kinetics and RGT

In patients treated with INF and RBV, the best predictor of an SVR was shown to be a rapid on-treatment HCV RNA decline to undetectable levels early in therapy. To this end, a RVR or an undetectable (eg, <50 IU/mL) by 4 weeks of INF and RBV has been used to determine eligibility for shortening therapy (eg, 24 weeks vs 48 weeks for HCV genotype 1 infection). Each real-time assay has its own linear range, with an upper and lower limit of detection (LLOD), but the terminology used to interpret results is the same. It is possible to define (1) LLOQ—the lowest value of HCV RNA that is possible to accurately quantify, HCV RNA is detectable and quantifiable; (2) LOD—the lower amount of HCV RNA that can be detected always or LLOD; (3) HCV RNA that is detectable but not quantifiable, the interpretation is the same as LOD; and (4) target not detected (TND)—no HCV RNA amplification, HCV RNA is undetectable or not detected [59–61]. Timing of sample collection is also assessed by guidelines, depending on the specific futility rules of each drug, as otherwise happens for SOC. However, HCV RNA kinetics induced by DAA treatments exhibits a different trend in comparison to that observed with IFN/RBV bi-phasic therapies.

New Definitions for an Undetectable HCV RNA Viral Load

While the goal of treating chronic HCV patients is to eradicate the infection as measured by an undetectable HCV RNA result, the concept of undetectable HCV levels has evolved alongside the treatment algorithm. For INF/RBV therapy, an undetectable result is any result that is less than 50 IU/mL. In contrast, for INF + BOC or TVR regimens, the term *undetectable* is defined as a target not detected result, that was required for patients to be eligible for shorten therapy, but for SVR assessments, a less than 25 IU/mL HCV RNA detected was acceptable. For the recently approved regimens containing Simeprevir, a stopping rule cutoff of 25 IU/mL is used at 4, 12, or 24 weeks in which all therapies are discontinued if HCV RNA results are above this cutoff. For Sofosbuvir, HCV RNA testing is only recommended following treatment of a fixed duration and to assess SVR. Both regimens use less than 25 IU/mL HCV RNA detected for defining undetectable. Given that the trials used a test with a LLOQ of 25 IU/mL, differences in a test LLOQ are important. These are practical considerations that might cause uncertainty for healthcare providers [59–62].

Under some circumstances, a target not detected result can be used to shorten therapy. With the introduction of BOC and TVR, new RGT rules were introduced which lead to considerable confusion in the terms used to define undetectable and when to apply this interpretation. These rules were assessed on a reanalysis of the BOC and TVR trails data that was published by the FDA where it was concluded that a "HCV RNA detectable, less than LLOQ" result predicted a significantly lower cure rate compared with subjects with an undetectable "target not detected" result. Based on this analysis, it was determined that a confirmed "detectable but below LLOQ" HCV RNA result should not be considered equivalent to an undetectable HCV RNA result (target not detected) for the purpose of RGT [59]. Therefore, the target not detected result at 4 and 12 weeks of INF/RBV/TVR therapy was required to shorten therapy (48 weeks to 24 or 36 weeks of INF and RBV). To further add complexity, stopping rules are also different for BOC and TVR regimens [61,63].

Although all commonly used HCV RNA assays report results in a standardized IU/mL, not all tests have the same performance characteristics. Several reports have demonstrated differences between how assays report results, particularly in detecting low amounts of HCV RNA. Concordance analysis in these studies have shown that HCV RNA differences in reporting results that are "target not detected" versus "HCV RNA detected, less than LLOQ" have become apparent. This is particularly true for a study that examined the results generated from HPS TaqMan HCV Test version 2.0 as part of the phase III clinical trial with Simeprevir plus INF/RBV and compared to it to the Abbott Real Time HCV test. Overall there was concordance between the two assays. However, a large number of samples (26–35%) at week 4 of treatment had detectable HCV RNA level below LLOQ with the Abbott Real Time assay that produced a target not detected result by the HPS TaqMan HCV Test version 2.0. These patients received shorten therapy based on the HPS assay "target not detected" result and high SVR rates were achieved. Thus, if the Abbott Real Time assay results at week 4 had been used to determine treatment duration, these patients may have been over-treated by an additional 6 months [63].

Since these new DDA-containing triple therapies require HCV RNA to be not detected at both 4 and 12 weeks in order to shorten therapy, differences between HCV RNA assays can affect key medical decisions. In this case, differences in testing results lead to a larger portion of patients being treated for longer duration. While BOC and TVR have been replaced by more potent regimens, differences in the performance of HCV RNA viral load tests might be important, particularly if they are not clinically validated.

VIRAL RESISTANCE VARIANTS

The high turnover rate in HCV replication combined with a poor fidelity and high error rate of the RNA-dependent RNA polymerase leads to the continuous production of numerous variants known as HCV-quasispecies. In the natural course of the HCV life cycle, the wild-type virus is predominantly produced. Several isolates within the HCV-quasispecies can carry mutations which confer resistance to DAAs either by direct (binding site) or indirect (functional restoration of the protein) effects. Naturally-occurring resistance-associated variants (RAVs) are selected early in monotherapy with TVR [64] and BOC [65], and an occurrence in treatment-naïve patients could be confirmed [66]. Therefore, selected variants are considered to be preexisting mutations generated during the natural HCV life cycle. The incidence of resistant variants is variable and depends on the binding domain, different populations, and HCV genotypes and subtypes. Deep-sequencing analysis enables detection of viral variants with a sensitivity of approximately 0.5–1%. Using these techniques nearly all described RAVs in the *NS3/4A* gene can be detected [67]. To date, RAVs at very low frequencies have no impact on treatment response. Population-based sequencing showed cumulative frequencies of different protease inhibitor resistant mutations in 10.8% of patients [68]. In this study, treatment-naïve patient with TVR-resistant variants prior to treatment achieved similar SVR rates compared to patients without RAVs [69]. Further analysis of the TVR and BOC phase 3 studies emphasized that treatment response is independent of the presence of preexisting RAVs if there is good responsiveness to the INF/RBV backbone. On the other hand, patients with baseline RAVs who were also poor INF/RBV responders (<1 log decrease in HCV RNA during lead-in-phase) showed lower SVR rates compared to poor INF/RBV responders without baseline RAVs (22% vs 37%). In particular, the presence of V36M, T54S, V55A, or R155K at baseline combined with a poor INF response led to an SVR in only 7% of BOC-treated patients [70]. Prior null-responders with the preexisting variants T54S or R155K treated with TVR in the REALIZE study always had on-treatment virologic failure, whereas patients who had previously relapsed achieved SVR in most cases [71].

The mutational variant Q80K is a preexisting RAV frequently found in HCV genotype 1a (prevalence: South America 9%, Europe 19%, and North America 48%) and is rarely detected in genotype 1b (0.5%) [72]. Resistance is associated with Simeprevir only. As currently shown, SVR and RVR rates were distinctly reduced in patients with prior relapse who exhibit the Q80K mutation at baseline (SVR12 47% vs 79%; RVR 43% vs 76%), which highlights the relevance of Q80K mutation in all patients, including good INF/RBV responders. The NS5B-substitution S282T is the only resistance mutation associated with decreased susceptibility to Sofosbuvir. At baseline of phase 3 studies, none of the 1292 patients exhibited the S282T mutation. Also there was no correlation between other NS5B variants detected prior to treatment and treatment outcome [73].

Viral breakthrough or relapse to HCV treatment is linked to the existence of the RAVs L31M/F and Y93H at baseline. In a Japanese study, 58% of patients infected with HCV genotype 1b with these baseline mutations failed therapy [74]. The natural occurrence of Y93H is reported to be between 4% and 23% and is lower for HCV genotype 1a than for subtype 1b [75]. However, most cases of treatment failure have been observed in genotype 1a without baseline RAVs and many patients with baseline RAVs exhibit SVR [75].

To date, eight amino acid positions within the HCV NS3/4A protease that associate with resistance have been described (major variants: V36A/M, T54S/A, V55A/K, Q80R/K, R155K/T/Q, A156S/D/T/V, D168A/V/T/H, V170A/T). Protease inhibitors, whether linear or macrocyclic, show different profiles of resistance, but are subject to significant cross-resistance. Among all protease inhibitors, quasispecies at position R155 were selected. Second-generation polymerase inhibitors like MK-5172 and ACH-1625 have a broader genotypic coverage and lower levels of resistance. During monotherapy with MK-5172, quasispecies containing resistant variants could be detected, but no virologic breakthrough occurred [73], indicating a higher barrier of resistance. In contrast to other protease inhibitors, MK-5172 interacts with the catalytic triad rather than directly with amino acid sites that confer resistance [76].

Resistance to Simeprevir. During in vivo studies, mutations at three amino acid sites were detected (Q80, R155, D168), whereas in vitro studies showed emerging resistance at positions F43 and A156 [77]. Mutations at position Q80 confer low-level resistance, at positions R155 and F43 moderate-level resistance, and at position D168 high-level resistance. At position A156, the susceptibility to Simeprevir depends on the amino acid change. High-level resistance is observed for A156V and moderate-level resistance for A156G/T [77]. The PROMISE study in patients who relapse, as well as the QUEST trials in treatment-naïve subjects, illustrates the impact of preexistent Q80K in genotype 1 patients on treatment outcome. In QUEST-1, patients with preexisting Q80K variant did not show significantly superior SVR rates compared to INF/RBV [78]. The majority of patients (83.7%) in phase 2b and 3 studies with genotype 1a and baseline Q80K exhibited the emergence of a single R155K variant at time of treatment failure, suggesting that the presence of Q80K alone is not sufficient to explain treatment failure. The median time until loss of mutation was 36 and 24 months for genotype 1a and 1b, respectively. Interestingly, the median time to loss of mutation for the R155K variant without baseline Q80K was 64 months compared to 32 months for patients who had emerging R155K and baseline Q80K [77].

Resistance to Faldaprevir. After FDV treatment failure RAVs are predominantly found at positions R155 and D168 [75]. Most of the RAVs confer moderate- (R155Q, D168G) to high-level resistance (R155K, A156T/V, D168A/V). Although rates of virologic failure were consistently higher in genotype 1a versus genotype 1b, there was no influence of the Q80K mutation on SVR rates. Patients with treatment failure during phase 2 studies predominantly selected R155K variants in genotype 1a and D168V variants in genotype 1b [79].

After treatment with FDV the time until loss of mutations was similar to TVR and BOC (median time 8–11 months) [73].

Resistance to Sofosbuvir. S282T mutation is the primary Sofosbuvir resistance mutation selected in genotype 1–6 in vitro. S282T confers a low- to medium-level resistance. To date S282T variant was only found in few patients after treatment. In an analysis of all patients with treatment failure during phase 3 trials, the S282T substitution was not detected by deep sequencing in any of the 225 patients [80]. In general, Sofosbuvir exhibits a high genetic barrier to resistance. Therefore, together with the low replicative fitness of the S282T variant, to date no viral breakthrough has been observed.

Resistance to Daclatasvir. In general the barrier to resistance for Daclatasvir is low. After 14-day monotherapy with Daclatasvir RAVs at positions M28, Q30, L31, and Y93 for subtype 1a, and L31 and Y93 for subtype 1b were observed in vivo [81]. The major mutations Y93H and L31V confer medium-level resistance to Daclatasvir [81]. The antiviral activity of Daclatasvir is less potent for subtype 1a than subtype 1b and also less potent for genotype 3 than genotype 2 in vitro. In line with these preclinical data, clinical studies demonstrated lower response rates for patients infected with subtype 1a and genotype 3 [61].

Treatment-naïve patients and patients after INF/RBV treatment failure. The frequency of preexisting RAVs in this population is generally low. In fact, successful treatment is independent of the preexistence of RAVs to TVR, BOC, and FDV. Given sufficient adherence to the therapeutic regimen, treatment failure occurs only in combination with other unfavorable factors, primarily unresponsiveness to the INF/RBV backbone. For these rare cases resistance testing is not justified. For Simeprevir the relative high frequency of preexisting Q80K variants in genotype 1a especially in European and North American populations has to be considered (19–48%) as this variant is associated with lower rates of SVR and RVR. Therefore, for genotype 1a patients in whom a treatment with Simeprevir is considered, resistance testing for the detection of Q80K has to be discussed. For Sofosbuvir no preexisting RAVs are known, therefore resistance testing is not indicated in this situation. Although a link between DCV–RAVs at baseline and treatment failure exists, combined with another potent antiviral like Sofosbuvir the rates of breakthrough or virologic relapse are low [82], so resistance testing is not indicated.

In contrast to TVR or BOC, second-wave protease inhibitors represent a significant improvement for dose administration, being administered once daily and are generally better tolerated by the patient. While the first-generation protease inhibitors are most active against

HCV genotype 1, the second generation of protease inhibitors is active against all genotypes with the exception of genotype 3, due to the presence of the natural polymorphism D168Q that confers resistance to available protease inhibitors [77]. Although broad cross-resistance exists between protease inhibitors mainly due to the selection for mutations at positions 155 and 156 (first generation) and 168 (second generation), resistance to first-generation protease inhibitors (TVR and BOC) does not completely overlap with the second generation, such as Simeprevir, ABT-450, Faldaprevir, or Asunaprevir [77,83]. MK-5172 is a second-generation protease inhibitor that is administered as a once a day pill that seems to be very potent with broad HCV genotype coverage. In vitro, MK-5172 is very potent and retains activity against HCV viruses that harbor resistance mutations to other HCV protease inhibitors, such as V36A/M, T54A/S, R155K/Q/T, A156S, V36M + R155K, or T54S + R155K. Moreover, MK-5172 is expected to be broadly active against multiple HCV genotypes [83–85]. The prevalence of natural polymorphisms associated with resistance to HCV protease inhibitors has been evaluated in treatment-naïve patients. Using population-based sequencing, less than 1% of subjects harbored mutations at codons 36, 155, 156, or 168, whereas changes at codons 54 or 55 were seen in 3–7% of patients. However, the polymorphism Q80K is frequently found (19–48%) among NS3 protease sequences from genotype 1a [67,77,80,86]. The overall rates of SVR were lower for genotype 1a than in genotype 1b (63% vs 80%, respectively), similar to that observed for other protease inhibitors. The presence of Q80K did not affect Sovaprevir SVR rates in patients with HCV genotype 1 infection after a 28-day administration with IFN/RBV. No significant effect of Q80K on the SVR at week 12 was recognized during the phase II clinical trial evaluating the efficacy of Faldaprevir in combination with IFN/RBV (75% if wild type vs 82% if Q80K was present) [73,87–89].

The impact of natural polymorphisms at positions involved in DAA resistance may be negligible in the context of combination therapies when other compounds of the regimen retain full activity. Using DAA-based therapies in combination with IFN/RBV the presence of baseline polymorphisms or RAVs might negatively influence the virologic response in poorly INF-responsive patients (ie, genotypes 1a and non-CC IL-28B) [74,90–92]. It is noteworthy that the high prevalence of the Q80K polymorphism among patients infected with HCV genotype 1a (19–48%) negatively impacts virological response to Simeprevir/IFN/RBV. For these reasons, baseline resistance testing for Q80K is strongly recommended for HCV genotype 1a, and alternative treatments to Simeprevir should be considered if this mutation is present.

For INF-free regimens, the presence of baseline polymorphisms and/or RAVs may have a clinically significant impact. The low genetic barrier to resistance for many DAAs, with the exception of nucleoside/ nucleoside analogs, might facilitate the on-therapy emergence of resistance variants in patients harboring baseline polymorphisms and/or resistance mutations. Indeed, baseline polymorphisms associated with resistance to NS5A inhibitors have a nonnegligible prevalence (10–15%) and their presence has been associated with lower rates of virologic response in some Daclatasvir-based regimens [74,81].

Routine monitoring for HCV drug RAVs during therapy is not recommended. Patients in whom antiviral therapy fails to achieve SVR may harbor viruses that are resistant to one or more of the antivirals at the time of virologic breakthrough. However, there is no evidence to date that the presence of RAVs results in more progressive liver injury compared to the damage that would have occurred if the patient was infected with a nonresistant form of the virus. The presence of a baseline RAV does not preclude achieving an SVR with a combination DAA regimen. Furthermore, RAVs are often not detectable with routine methods, or with more sensitive tests of HCV variants, even when patients are followed for several months. Subsequent retreatment with combination antivirals, particularly regimens containing antiviral drugs that have a high barrier to resistance such as NS5B nucleotide polymerase inhibitors, may overcome the presence of resistance to one or more antiviral. The exception is testing for the presence of Q80K polymorphism at baseline in patients with HCV genotype 1a infection before treatment with Simeprevir plus INF/RBV. Testing for RAV before repeat antiviral treatment is not routinely done.

ROLE OF HOST PHARMACOGENETICS AS PREDICTORS OF HCV TREATMENT OUTCOME

The genetic background of the host has an important impact on the natural course of HCV infection. CD8 + T cells are the major effector cells that mediate viral clearance. CD8 + T cells recognize viral peptides bound to HLA class I molecules on virus-infected cells. HLA genes display a high degree of genetic variation among individuals, which is reflected in the variations in binding and presentation of viral epitopes. HLA-B27, HLA-B57, and HLA-A3 alleles are significantly associated with spontaneous clearance of HCV infection. The protective role of these alleles has been linked to viral epitopes which do not allow immune escape mutations because of profound negative effects on viral replication fitness, resulting in a highly crippled virus. Genome-wide association studies (GWAS) have become the standard approach to discovering the genetic basis of human disease. The goal of therapy against HCV is to eradicate infection and to achieve SVR. Given the variability of INF and RBV treatment response in individual patients, and in order to reduce side effects and avoid the heavy medication cost, knowing baseline viral load and host parameters that predict response before the treatment would be quite useful. Several studies have proven the roles of both viral factors (such as HCV genotype, quasispecies diversity, baseline viremia) and host factors (ie, age, sex, ethnicity, grade of liver fibrosis, body mass index, and comorbidities) in predicting the natural course of HCV and response to INF treatment [92]. The observation that the response rate for INF-based regimens in African/American is less than half of that observed in Caucasians suggests additional factors associated with patient genetic background are related to likelihood of SVR and may influence tailored INF-based treatment duration [92,93]. In 2009, there was a major breakthrough in the understanding of host genomics in HCV infection through the discovery of several single-nucleotide polymorphisms (SNPs) upstream of the interleukin 28B locus (IL28B), in particular the SNP rs12979860, which predicts both spontaneous recovery from HCV infection and therapy-induced viral clearance in patients infected with HCV genotype 1. The *IL28B* gene on chromosome 19 encodes the cytokine IFN-lambda3 (INF-λ3), which belongs to the type III INF family (INF-λ). INF-λ is rapidly induced during HCV infection and has antiviral activity against HCV [93,94]. Recent studies have reported an association between the IL28B SNPs and the expression of intrahepatic interferon-stimulated genes (ISG) in liver. Low ISG expression prior to treatment has already been correlated with high response to INF-based therapy and the protective IL28B genotype is associated with ISG expression levels, suggesting that ISG induction in part segregates according to IL28B haplotype. Patients carrying rs12979860 CC genotype had a clearance rate three-fold higher compared with patients carrying the CT or TT genotype. Interestingly, the frequency of the favorable CC genotype differs markedly among ethnic groups, reaching 90% in certain North and Eastern Asian populations, an intermediate frequency in Europe, and the lowest frequencies in Africans [94]. Other SNPs of IL28B (rs8099917, rs12980275, rs8103142, rs81057790, rs11881222, rs28416813, rs4803219, and rs7248668) have been identified in HCV genotype 1 patients [8]. IL28B SNPs rs1299860 (CC) and rs8099917 (TT) genotype are significantly associated with SVR in patients infected with HCV genotype 1 who are treated with INF-based therapies [95,96]. Moreover, a meta-analysis validated the strength of these genotypes as independent predictors of patient response to therapy. Determination of this SNP appears sufficient for predicting response to INF-based therapy and the current consensus for utilization of INF-based therapy suggests that IL28B SNP status is the strongest predictor for response, suggesting the possibility of personalized therapy with significant clinical and pharmacoeconomic implications. Thus, carriers of rs1297860 CC infected with HCV genotype 1 or genotype 4 should be treated with INF-based therapies, whereas carriers of the T allele with HCV genotype 1 and no advanced liver fibrosis may delay therapy and wait for new DAAs [96]. IL28B genotyping is less clinically significant among patients infected with HCV genotype 2 or genotype 3 which are more INF responsive. Understanding the mechanisms associated with HCV control in patients with specific IL28B polymorphisms is still limited. IL28B appears to affect INF-λ3 expression, with the unfavorable genotypes resulting in reduced INF-λ3 expression. Patients with the unfavorable genotypes also had a lower induction of innate immune genes, suggesting that the IL28B polymorphism may regulate innate immune functions [97,98].

In the era of DAAs in combination with INF therapy, an important question is whether and how the determination of IL28B polymorphisms may be useful in predicting patient's likelihood of response and the potential implications for treatment decision making [99,100]. However, the rapid move of HCV therapy toward DAAs without INF has weakened the relevance of IL28B genotyping for prediction and management of chronic HCV infection [101–103].

Hemolytic anemia is a common side effect of RBV-based HCV therapy. While this condition is reversible and dose related, it is cause for dose reduction or premature withdrawal from therapy in approximately 15% of cases [104]. Recently, a GWAS study identified two variants in the *ITPA* gene (rs1127354 and rs7270101) that are responsible for ITPA deficiency and correlated with risk of RBV anemia in the European–American population. More recently, it was confirmed that rs1127354 is strongly associated with protection against anemia in a Japanese cohort [105]. These findings were also seen in a HCV/HIV coinfected patient cohort with all HCV genotypes. Two other SNPs (rs11697186 and rs6139030) are located within and around *DDRGK1* gene on chromosome 20, with reduced platelet count in response to INF/RBV treatment in Japanese HCV patients. Even though these studies support the perspective of pharmacogenetic diagnostic tools for tailoring therapy to minimize drug-induced adverse events, the future of DAAs will likely progressively replace RBV-based treatment [32,106].

References

[1] Webster DP, Klenerman P, Dusheiko GM. Hepatitis C. Lancet 2015;385:1124–35.

[2] Hajarizadeh B, Grebely J, Dore GJ. Epidemiology and natural history of HCV infection. Nat Rev Gastroenterol Hepatol 2013;10:553–62.

[3] Rafiq SM, Banik GR, Khan S, Rashid H, Khandaker G. Current burden of hepatitis C virus infection among injecting drug users: a mini systematic review of prevalence studies. Infect Disord Drug Targets 2014;14:93–100.

[4] Lee MH, Yang HI, Yuan Y, L'Italien G, Chen CJ. Epidemiology and natural history of hepatitis C virus infection. World J Gastroenterol 2014;20:9270–80.

[5] Negro F. Epidemiology of hepatitis C in Europe. Dig Liver Dis 2014;46(Suppl 5):S158–64.

[6] Marinaki S, Boletis JN, Sakellariou S, Delladetsima IK. Hepatitis C in hemodialysis patients. World J Hepatol 2015;7:548–58.

[7] Wiessing L, Ferri M, Grady B, et al. Hepatitis C virus infection epidemiology among people who inject drugs in Europe: a systematic review of data for scaling up treatment and prevention. PLoS One 2014;9:e103345.

[8] Ansaldi F, Orsi A, Sticchi L, Bruzzone B, Icardi G. Hepatitis C virus in the new era: perspectives in epidemiology, prevention, diagnostics and predictors of response to therapy. World J Gastroenterol 2014;20:9633–52.

[9] Sebastiani G, Gkouvatsos K, Pantopoulos K. Chronic hepatitis C and liver fibrosis. World J Gastroenterol 2014;20:11033–53.

[10] Marinho RT, Vitor S, Velosa J. Benefits of curing hepatitis C infection. J Gastrointestin Liver Dis 2014;23:85–90.

[11] Gerold G, Pietschmann T. The HCV life cycle: in vitro tissue culture systems and therapeutic targets. Dig Dis 2014;32:525–37.

[12] Li HC, Ma HC, Yang CH, Lo SY. Production and pathogenicity of hepatitis C virus core gene products. World J Gastroenterol 2014;20:7104–22.

[13] Taylor DR. Evolution of cell culture systems for HCV. Antivir Ther 2013;18:523–30.

[14] Zhu YZ, Qian XJ, Zhao P, Qi ZT. How hepatitis C virus invades hepatocytes: the mystery of viral entry. World J Gastroenterol 2014;20:3457–67.

[15] Dubuisson J, Cosset FL. Virology and cell biology of the hepatitis C virus life cycle—an update. J Hepatol 2014;61:S3–13.

[16] Kim CW, Chang KM. Hepatitis C virus: virology and life cycle. Clin Mol Hepatol 2013;19:17–25.

[17] Gu M, Rice CM. Structures of hepatitis C virus nonstructural proteins required for replicase assembly and function. Curr Opin Virol 2013;3:129–36.

[18] Echeverría N, Moratorio G, Cristina J, Moreno P. Hepatitis C virus genetic variability and evolution. World J Hepatol 2015;7:831–45.

[19] Saludes V, González V, Planas R, Matas L, Ausina V, Martró E. Tools for the diagnosis of hepatitis C virus infection and hepatic fibrosis staging. World J Gastroenterol 2014;20:3431–42.

[20] Marascio N, Torti C, Liberto M, Focà A. Update on different aspects of HCV variability: focus on NS5B polymerase. BMC Infect Dis 2014;14(Suppl 5):S1.

[21] Scheel TK, Rice CM. Understanding the hepatitis C virus life cycle paves the way for highly effective therapies. Nat Med 2013;19:837–49.

[22] Cobb B, Heilek G, Vilchez RA. Molecular diagnostics in the management of chronic hepatitis C: key considerations in the era of new antiviral therapies. BMC Infect Dis 2014;14(Suppl 5):S8.

[23] Devaki P, Jencks D, Yee BE, Nguyen MH. Sustained virologic response to standard interferon or pegylated interferon and ribavirin in patients with hepatitis C virus genotype 5: systematic review and meta-analysis of ten studies and 423 patients. Hepatol Int 2015;9:431–7.

[24] Simmons B, Saleem J, Heath K, Cooke GS, Hill A. Long-term treatment outcomes of patients infected with hepatitis C virus: a systematic review and meta-analysis of the survival benefit of achieving a sustained virological response. Clin Infect Dis 2015; 61:730–40.

[25] Wang CH, Wey KC, Mo LR, Chang KK, Lin RC, Kuo JJ. Current trends and recent advances in diagnosis, therapy, and prevention of hepatocellular carcinoma. Asian Pac J Cancer Prev 2015;16:3595–604.

[26] Welch NM, Jensen DM. Pegylated interferon based therapy with second-wave direct-acting antivirals in genotype 1 chronic hepatitis C. Liver Int 2015;35(Suppl 1):11–17.

[27] Taieb V, Pacou M, Ho S, et al. A network meta-analysis to compare simeprevir with boceprevir and telaprevir in combination with peginterferon-α and ribavirin in patients infected with genotype 1 hepatitis C virus. J Med Econ 2015;26:1–10.

[28] Khalilieh S, Feng HP, Hulskotte EG, Wenning LA, Butterton JR. Clinical pharmacology profile of boceprevir, a hepatitis C virus NS3 protease inhibitor: focus on drug-drug interactions. Clin Pharmacokinet 2015;54:599–614.

[29] Gogela NA, Lin MV, Wisocky JL, Chung RT. Enhancing our understanding of current therapies for hepatitis C virus (HCV). Curr HIV/AIDS Rep 2015;12:68–78.

[30] Kirby BJ, Symonds WT, Kearney BP, Mathias AA. Pharmacokinetic, pharmacodynamic, and drug-interaction profile of the hepatitis C virus NS5B polymerase inhibitor Sofosbuvir. Clin Pharmacokinet 2015;54:677–90.

[31] Janardhan SV, Reau NS. Should NS5A inhibitors serve as the scaffold for all-oral anti-HCV combination therapies? Hepat Med 2015;7:11–20.

[32] Barth H. Hepatitis C virus: is it time to say goodbye yet? Perspectives and challenges for the next decade. World J Hepatol 2015;7:725–37.

[33] Shaheen MA, Idrees M. Evidence-based consensus on the diagnosis, prevention and management of hepatitis C virus disease. World J Hepatol 2015;7:616–27.

[34] Keating GM. Ledipasvir/Sofosbuvir: a review of its use in chronic hepatitis C. Drugs 2015;75:675–85.

[35] Pawlotsky JM. NS5A inhibitors in the treatment of hepatitis C. J Hepatol 2013;59:375–82.

[36] Sharma SA, Feld JJ. Management of HCV in cirrhosis—a rapidly evolving landscape. Curr Gastroenterol Rep 2015;17:443.

[37] Ferenci P. Treatment of hepatitis C in difficult-to-treat patients. Nat Rev Gastroenterol Hepatol 2015;12:284–92.

[38] Deborah Friedman N, Green JH, Weber HM, et al. Hepatitis C virus treatment in the 'real-world': how well do 'real' patients respond? J Clin Exp Hepatol 2014;4:214–20.

[39] Minaei AA, Kowdley KV. ABT-450/ ritonavir and ABT-267 in combination with ABT-333 for the treatment of hepatitis C virus. Expert Opin Pharmacother 2015;16:929–37.

[40] Papastergiou V, Karatapanis S. Current status and emerging challenges in the treatment of hepatitis C virus genotypes 4 to 6. World J Clin Cases 2015;3:210–20.

[41] Kattakuzhy S, Levy R, Kottilil S. Sofosbuvir for treatment of chronic hepatitis C. Hepatol Int 2015;9:161–73.

[42] Kumari R, Nguyen MH. Fixed-dose combination of sofosbuvir and ledipasvir for the treatment of chronic hepatitis C genotype 1. Expert Opin Pharmacother 2015;16:739–48.

[43] Asselah T, Marcellin P. Optimal IFN-free therapy in treatment-naïve patients with HCV genotype 1 infection. Liver Int 2015;35:56–64.

[44] Bunchorntavakul C, Reddy KR. Review article: the efficacy and safety of daclatasvir in the treatment of chronic hepatitis C virus infection. Aliment Pharmacol Ther 2015;42:258–72.

[45] Yokosuka O, Omata M, Kanda T. Faldaprevir for the treatment of hepatitis C. Int J Mol Sci 2015;16:4985–96.

[46] McCormack PL. Daclatasvir: a review of its use in adult patients with chronic hepatitis C virus infection. Drugs 2015;75:515–24.

[47] Klibanov OM, Gale SE, Santevecchi B. Ombitasvir/paritaprevir/ritonavir and dasabuvir tablets for hepatitis C virus genotype 1 infection. Ann Pharmacother 2015;49:566–81.

[48] Rai D, Wang L, Jiang X, et al. The changing face of hepatitis C: recent advances on HCV inhibitors targeting NS5A. Curr Med Chem 2015; [Epub ahead of print].

[49] Au TH, Destache CJ, Vivekanandan R. Hepatitis C therapy: looking toward interferon-sparing regimens. J Am Pharm Assoc (2003 2015;55:e72–84.

[50] Alexopoulou A, Karayiannis P. Interferon-based combination treatment for chronic hepatitis C in the era of direct acting antivirals. Ann Gastroenterol 2015;28:55–65.

[51] Childs-Kean LM, Hand EO. Simeprevir and sofosbuvir for treatment of chronic hepatitis C infection. Clin Ther 2015;37:243–67.

[52] Quigley JM, Bryden PA, Scott DA, Kuwabara H, Cerri K. Relative efficacy and safety of simeprevir and telaprevir in treatment-naïve hepatitis C-infected patients in a Japanese population: a Bayesian network meta-analysis. Hepatol Res 2015;45:E89–98.

[53] Sanford M. Simeprevir: a review of its use in patients with chronic hepatitis C virus infection. Drugs 2015;75:183–96.

[54] Cortez KJ, Kottilil S. Beyond interferon: rationale and prospects for newer treatment paradigms for chronic hepatitis C. Ther Adv Chronic Dis 2015;6:4–14.

[55] Kayali Z, Schmidt WN. Finally sofosbuvir: an oral anti-HCV drug with wide performance capability. Pharmgenomics Pers Med 2014;7:387–98.

[56] Manzano-Robleda Mdel C, Ornelas-Arroyo V, Barrientos-Gutiérrez T, Méndez-Sánchez N, Uribe M, Chávez-Tapia NC. Boceprevir and telaprevir for chronic genotype 1 hepatitis C virus infection. A systematic review and meta-analysis. Ann Hepatol 2015;14:46–57.

[57] Guidelines for the screening, care and treatment of persons with hepatitis C infection. Geneva: World Health Organization; 2014.

[58] Triple therapy for hepatitis C in previous non-responders: a review of the clinical effectiveness and safety. Ottawa (ON): Canadian Agency for Drugs and Technologies in Health; 2014. www.ncbi.nlm.nih.gov/books/NBK264036/.

[59] Maasoumy B, Hunyady B, Calvaruso V, et al. Performance of two HCV RNA assays during protease inhibitor-based triple therapy in patients with advanced liver fibrosis and cirrhosis. PLoS One 2014;9:e110857.

[60] Wiesmann F, Naeth G, Sarrazin C, et al. Variation analysis of six HCV viral load assays using low viremic HCV samples in the range of the clinical decision points for HCV protease inhibitors. Med Microbiol Immunol 2015;204:515–25.

[61] Schønning K. Comparison of the QIAGEN artus HCV QS-RGQ test with the Roche COBAS Ampliprep/COBAS TaqMan HCV test v2.0 for the quantification of HCV-RNA in plasma samples. J Clin Virol 2014;60:323–7.

[62] Taylor N, Haschke-Becher E, Greil R, Strasser M, Oberkofler H. Performance characteristics of the COBAS Ampliprep/COBAS TaqMan v2.0 and the Abbott RealTime hepatitis C assays—implications for response-guided therapy in genotype 1 infections. Antivir Ther 2014;19:449–54.

[63] Vermehren J, Aghemo A, Falconer K, et al. Clinical significance of residual viremia detected by two real-time PCR assays for response-guided therapy of HCV genotype 1 infection. J Hepatol 2014;60:913–19.

[64] Sarrazin C, Kieffer TL, Bartels D, et al. Dynamic hepatitis C virus genotypic and phenotypic changes in patients treated with the protease inhibitor telaprevir. Gastroenterology 2007;132:1767–77.

[65] Susser S, Welsch C, Wang Y, et al. Characterization of resistance to the protease inhibitor boceprevir in hepatitis C virus-infected patients. Hepatology 2009;50:1709–18.

[66] Kuntzen T, Timm J, Berical A, et al. Naturally occurring dominant resistance mutations to hepatitis C virus protease and polymerase inhibitors in treatment-naïve patients. Hepatology 2008;48:1769–78.

[67] Verbinnen T, Van Marck H, Vandenbroucke I, et al. Tracking the evolution of multiple in vitro hepatitis C virus replicon variants under protease inhibitor selection pressure by 454 deep sequencing. J Virol 2010;84:11124–33.

[68] Paolucci S, Fiorina L, Piralla A, et al. Naturally occurring mutations to HCV protease inhibitors in treatment-naïve patients. Virol J 2012;9:245.

[69] Bartels DJ, Sullivan JC, Zhang EZ, et al. Hepatitis C virus variants with decreased sensitivity to direct-acting antivirals (DAAs) were rarely observed in DAA-naive patients prior to treatment. J Virol 2013;87:1544–53.

[70] Barnard RJ, Howe JA, Ogert RA, et al. Analysis of boceprevir resistance associated amino acid variants (RAVs) in two phase 3 boceprevir clinical studies. Virology 2013;444:329–36.

[71] De Meyer S, Dierynck I, Ghys A, et al. Characterization of telaprevir treatment outcomes and resistance in patients with prior treatment failure: results from the REALIZE trial. Hepatology 2012;56:2106–15.

[72] Lenz O, Vijgen L, Berke JM, et al. Virologic response and characterisation of HCV genotype 2–6 in patients receiving TMC435 monotherapy (study TMC435-C202). J Hepatol 2013;58:445–51.

[73] Svarovskaia ES, Martin R, McHutchison JG, Miller MD, Mo H. Abundant drug-resistant NS3 mutants detected by deep sequencing in hepatitis C virus-infected patients undergoing NS3 protease inhibitor monotherapy. J Clin Microbiol 2012;50:3267–74.

[74] McPhee F, Hernandez D, Yu F, et al. Resistance analysis of hepatitis C virus genotype 1 prior treatment null responders receiving daclatasvir and asunaprevir. Hepatology 2013;58:902–11.

[75] Suzuki Y, Ikeda K, Suzuki F, et al. Dual oral therapy with daclatasvir and asunaprevir for patients with HCV genotype 1b infection and limited treatment options. J Hepatol 2013;58:655–62.

[76] Romano KP, Ali A, Aydin C, et al. The molecular basis of drug resistance against hepatitis C virus NS3/4A protease inhibitors. PLoS Pathog 2012;8:e1002832.

[77] Lenz O, Verbinnen T, Lin TI, et al. In vitro resistance profile of the hepatitis C virus NS3/4A protease inhibitor TMC435. Antimicrob Agents Chemother 2010;54:1878–87.

[78] Berger KL, Lagacé L, Triki I, et al. Viral resistance in hepatitis C virus genotype 1-infected patients receiving the NS3 protease inhibitor Faldaprevir (BI 201335) in a phase 1b multiple-rising-dose study. Antimicrob Agents Chemother 2013;57:4928–36.

[79] Sulkowski MS, Asselah T, Lalezari J, et al. Faldaprevir combined with pegylated interferon alfa-2a and ribavirin in treatment-naïve patients with chronic genotype 1 HCV: SILEN-C1 trial. Hepatology 2013;57:2143–54.

[80] Wang C, Sun JH, O'Boyle DR, et al. Persistence of resistant variants in hepatitis C virus-infected patients treated with the NS5A replication complex inhibitor daclatasvir. Antimicrob Agents Chemother 2013;57:2054–65.

[81] Dore GJ. The changing therapeutic landscape for hepatitis C. Med J Aust 2012;196:629–32.

[82] Sulkowski MS, Jacobson IM, Nelson DR. Daclatasvir plus sofosbuvir for HCV infection. N Engl J Med 2014;370:1560–1.

[83] De Nicola S, Aghemo A. Second wave anti-HCV protease inhibitors: too little too late? Liver Int 2014;34:e168–70.

[84] Issur M, Götte M. Resistance patterns associated with HCV NS5A inhibitors provide limited insight into drug binding. Viruses 2014;6:4227–41.

[85] Götte M. Resistance to nucleotide analogue inhibitors of hepatitis C virus NS5B: mechanisms and clinical relevance. Curr Opin Virol 2014;8:104–8.

[86] Shang L, Lin K, Yin Z. Resistance mutations against HCV protease inhibitors and antiviral drug design. Curr Pharm Des 2014;20:694–703.

[87] Nakamoto S, Kanda T, Wu S, Shirasawa H, Yokosuka O. Hepatitis C virus NS5A inhibitors and drug resistance mutations. World J Gastroenterol 2014;20:2902–12.

[88] Ness E, Kowdley KV. Update on hepatitis C: epidemiology, treatment and resistance to antiviral therapies. Minerva Gastroenterol Dietol 2015;61:145–58.

[89] Wyles DL, Gutierrez JA. Importance of HCV genotype 1 subtypes for drug resistance and response to therapy. J Viral Hepat 2014;21:229–40.

[90] Najera I. Resistance to HCV nucleoside analogue inhibitors of hepatitis C virus RNA-dependent RNA polymerase. Curr Opin Virol 2013;3:508–13.

[91] Gao M. Antiviral activity and resistance of HCV NS5A replication complex inhibitors. Curr Opin Virol 2013;3:514–20.

[92] Trinks J, Hulaniuk ML, Redal MA, Flichman D. Clinical utility of pharmacogenomics in the management of hepatitis C. Pharmgenomics Pers Med 2014;7:339–47.

[93] Rosso C, Abate ML, Ciancio A, et al. IL28B polymorphism genotyping as predictor of rapid virologic response during interferon plus ribavirin treatment in hepatitis C virus genotype 1 patients. World J Gastroenterol 2014;20:13146–52.

[94] Kamal SM. Pharmacogenetics of hepatitis C: transition from interferon-based therapies to direct-acting antiviral agents. Hepat Med 2014;6:61–77.

[95] Luo Y, Jin C, Ling Z, Mou X, Zhang Q, Xiang C. Association study of IL28B: rs12979860 and rs8099917 polymorphisms with SVR in patients infected with chronic HCV genotype 1 to PEG-INF/RBV therapy using systematic meta-analysis. Gene 2013;513:292–6.

[96] Vidal F, López-Dupla M, Laguno M, et al. Pharmacogenetics of efficacy and safety of HCV treatment in HCV-HIV coinfected patients: significant associations with IL28B and SOCS3 gene variants. PLoS One 2012;7:e47725.

[97] Venegas M, Brahm J, Villanueva RA. Genomic determinants of hepatitis C virus antiviral therapy outcomes: toward individualized treatment. Ann Hepatol 2012;11:827–37.

[98] Aguirre Valadez J, García Juárez I, Rincón Pedrero R, Torre A. Management of chronic hepatitis C virus infection in patients with end-stage renal disease: a review. Ther Clin Risk Manag 2015;11:329–38.

[99] Matsuura K, Watanabe T, Tanaka Y. Role of IL28B for chronic hepatitis C treatment toward personalized medicine. J Gastroenterol Hepatol 2014;29:241–9.

[100] Zheng H, Li M, Chi B, Wu XX, Wang J, Liu DW. IL28B rs12980275 variant as a predictor of sustained virologic response to pegylated-interferon and ribavirin in chronic hepatitis C patients: a systematic review and meta-analysis. Clin Res Hepatol Gastroenterol 2015;39:576–83.

[101] Berger CT, Kim AY. IL28B polymorphisms as a pretreatment predictor of response to HCV treatment. Infect Dis Clin North Am 2012;26:863–77.

[102] Hsu CS, Kao JH. Genomic variation-guided management in chronic hepatitis C. Expert Rev Gastroenterol Hepatol 2012;6: 497–506.

[103] Hayes CN, Imamura M, Aikata H, Chayama K. Genetics of IL28B and HCV—response to infection and treatment. Nat Rev Gastroenterol Hepatol 2012;9:406–17.

[104] Soriano V, Poveda E, Vispo E, Labarga P, Rallón N, Barreiro P. Pharmacogenetics of hepatitis C. J Antimicrob Chemother 2012;67:523–9.

[105] Olmedo DB, Cader SA, Porto LC. IFN-λ gene polymorphisms as predictive factors in chronic hepatitis C treatment-naive patients without access to protease inhibitors. J Med Virol 2015;87:1702–15.

[106] Bartenschlager R, Lohmann V, Penin F. The molecular and structural basis of advanced antiviral therapy for hepatitis C virus infection. Nat Rev Microbiol 2013;11:482–96.

39

Personalized Medicine in Cancer Treatment

V.M. Pratt[1] and S.A. Scott[2]

[1]Pharmacogenomics Laboratory, Department of Medical and Molecular Genetics, Indiana University School of Medicine, Indianapolis, IN, United States [2]Department of Genetics and Genomic Sciences, Icahn School of Medicine at Mount Sinai, New York, NY, United States

INTRODUCTION

Although the primary focus of molecular oncology research and cancer treatment often centers on somatically mutated genes, germline genetic variation can also influence some cancer treatments. Pharmacogenetics is the study of the genetic determinants of drug response variability and this chapter focuses on germline or constitutional variants in an individual's DNA that can impact cancer treatments. Polymorphic variant alleles often occur in genes that encode drug metabolism enzymes and alter their enzymatic activity. These enzymes can activate a prodrug to its active form or catalyze the inactivation and elimination of a drug or metabolite. Prominent among these enzymes are the cytochrome P450 (CYP450) superfamily, which directly influence the pharmacokinetics of many drugs. In addition, genetic variants in cell receptors can influence drug transport, which often plays a role in pharmacodynamics.

Oncology supportive care is frequently targeted toward mollifying adverse effects of cancer treatment while eradicating the cancer. Germline variation can potentially play an important role in the selection and administration of cancer drugs. Demonstrating clinical utility for cancer pharmacogenetic testing has been challenging and is an ongoing area of research. However, clinical tests are currently available for selected genes where clinical validity has largely been established. Examples of cancer therapies and related genes with germline genetic variants that influence patient response are detailed below.

MOLECULAR TARGET: FLUOROPYRIMIDINES AND *DPYD*

Fluoropyrimidines (ie, 5-fluorouracil (5-FU), capecitabine, tegafur) are widely used for the treatment of several solid tumors, including breast and colorectal cancers, and typically are administered in combination with other antineoplastic agents [1]. Both capecitabine and tegafur are inactive prodrugs that are metabolized to 5-FU. The main mechanism of 5-FU activation is believed to be conversion to fluorodeoxyuridine monophosphate, which inhibits the enzyme thymidylate synthase, an important part of the folate–homocysteine cycle and purine and pyrimidine synthesis. The resultant damage occurs due to increased base excision repair causing DNA fragmentation and ultimately cell death. In addition, the fluorouridine triphosphate metabolite can be incorporated into RNA in place of uridine triphosphate and interfering with RNA processing and protein synthesis. Approximately 10–40% of individuals who receive 5-FU develop severe toxicity such as neutropenia, nausea, vomiting, severe diarrhea, stomatitis, mucositis, hand-foot syndrome, and neuropathy [2].

The rate-limiting step of 5-FU catabolism is dihydropyrimidine dehydrogenase (DPD) conversion of 5-FU to dihydrofluorouracil [3]. Importantly, several germline genetic variants in the *DPYD* gene on chromosome 1p21.3 result in deficient DPD activity, and increased drug half-life that can translate to severe and even fatal 5-FU toxicity [4]. In addition, the FDA label for the fluoropyrimidines indicates that variants in *DPYD* are

© 2017 Elsevier Inc. All rights reserved.

TABLE 39.1 CPIC Recommended Dosing of Fluoropyrimidines by DPYD Genotype/Phenotype

Phenotype (genotype)	Examples of diplotypes	Implications for phenotypic measures	Dosing recommendations
Homozygous wild-type or normal, high DPD activity (two or more functional *1 alleles)	*1/*1	Normal DPD activity and "normal" risk for fluoropyrimidine toxicity	Use label-recommended dosage and administration
Heterozygous or intermediate activity (one functional allele *1, plus one nonfunctional allele)	*1/*2; *1/*13	Decreased DPD activity (leukocyte DPD activity at 30–70% that of the normal population) and increased risk for severe or even fatal drug toxicity when treated with fluoropyrimidine drugs	Start with at least a 50% reduction in starting dose followed by titration of dose based on toxicity or pharmacokinetic test (if available)
Homozygous or compound heterozygous variant, DPD deficiency, at risk for toxicity with drug exposure (two nonfunctional alleles)	*2/*2; *2/*13; *13/*13	Complete DPD deficiency and increased risk for severe or even fatal drug toxicity when treated with fluoropyrimidine drugs	Select alternate drug

Adapted from Caudle KE, Thorn CF, Klein TE, et al. Clinical Pharmacogenetics Implementation Consortium guidelines for dihydropyrimidine dehydrogenase genotype and fluoropyrimidine dosing. Clin Pharmacol Ther 2013;94:640–5.

associated with increased risk for adverse and potentially toxic events, and therefore is contraindicated in patients with known DPD deficiency (note that genetic testing or screening of DPD activity is not mentioned in the drug label). However, although several variants have been associated with low DPD activity and fluoropyrimidine toxicity, the presence of these variants does not always result in toxicity, and associations have not been consistently replicated, which is likely due to inconsistencies in treatment regimens across studies [5]. The available literature on DPYD and fluoropyrimidine response prompted Clinical Pharmacogenetics Implementation Consortium (CPIC) practice guidelines that recommend a 50% reduction in starting dose for patients who are heterozygous for a nonfunctional DPYD variant and an alternate therapy for those with two nonfunctional DPYD variants (ie, homozygous or compound heterozygous) (Table 39.1) [5].

DPD deficiency is an autosomal recessive disorder that is characterized by a wide range of severity, with neurological problems in some individuals and no signs or symptoms in others. In individuals with severe DPD deficiency, the disorder becomes apparent in infancy with recurrent seizures, intellectual disability, microcephaly, hypertonia, delayed development of motor skills such as walking, and autistic behaviors that affect communication and social interaction. Other affected individuals are asymptomatic and may be identified only by laboratory testing. More than 50 mutations in the DPYD gene have been identified in people with DPD deficiency. It is estimated that 3–5% of the Caucasian population has partial DPD deficiency and 0.2% have complete DPD deficiency [6].

Molecular Technologies: DPYD

Genetic testing for DPYD can be performed from DNA extracted from whole blood or other tissues, which typically involves targeted genotyping of the DPYD*2A decreased activity allele [7]. However, many other variant alleles have been identified [8]. Some laboratories may offer full gene DPYD sequencing, which will identify all known and novel sequence variants, many potentially with uncertain clinical significance. Clinical laboratories that offer DPYD genetic testing can be found at the voluntary National Institutes of Health Genetic Testing Registry (http://www.ncbi.nlm.nih.gov/gtr/) [9].

Clinical Utility: DPYD

Evidence for the clinical utility of DPYD genotype directed fluoropyrimidine dosing is based on prospective studies [6,10], retrospective genetic studies, case studies of patients with severe toxicity, and meta-analyses [11]. Together, these data suggest that patients heterozygous for loss-of-function DPYD alleles have significantly reduced 5-FU clearances, ranging from 40% to 80% less than the clearances in patients without these variants. Despite the lack of a prospective randomized clinical trial directly evaluating the utility of DPYD genotyping, the available evidence was utilized to inform the CPIC guidelines on dose reduction among DPYD-variant carriers when genotype data is available [5]. Cost-effectiveness studies on DPYD genotype directed fluoropyrimidine dosing have not been reported. As noted above, patients homozygous for loss-of-function DPYD variants have DPD deficiency, a disease with a variable phenotype that ranges from no symptoms to severe convulsive disorders with motor and mental retardation [12,13].

Limitations of Testing: *DPYD*

Targeted *DPYD* genotyping will not detect any alleles that are not directly interrogated so a negative genotyping result does not rule out the possibility that a patient carries another *DPYD* variant. Full gene sequencing will detect all *DPYD* variants, but rare or novel variants will likely be of uncertain clinical significance. In addition, other genes may influence responses to 5-FU, including *ABCB1*, *MTHFR*, and *TYMS* [1,10], which will not be detected by *DPYD* genetic testing. Alternatives to *DPYD* genotyping that assess DPD enzyme activity directly include dihydrouracil/uracil ratio determination in plasma, the uracil breath test method, and measurement of DPD activity in peripheral mononuclear cells [14].

MOLECULAR TARGET: IRINOTECAN AND *UGT1A1*

Irinotecan is used to treat metastatic colorectal cancer, typically given in combination with other anticancer agents (eg, 5-FU, leucovorin). It is also used in combination with cisplatin for the treatment of extensive small cell lung cancer. Irinotecan works by binding to the topoisomerase I–DNA complex and preventing DNA replication, and thus causes double-strand DNA breakage and cell death. The active form of irinotecan is SN-38, which is glucoronized to SN-38 glucoronic acid (SN-38G) and detoxified in the liver via conjugation by the uridine diphosphate glucuronosyltransferase (UGT) 1A1 enzyme, which releases SN-38G into the intestines for elimination [15]. Notably, impaired elimination of the cytotoxic SN-38 metabolite can result in severe toxicities, including myelosuppresion, diarrhea, and neutropenia, which have all been associated with variation in the *UGT1A1* gene on chromosome 2q37.1 [16].

The most important variant *UGT1A1* alleles is *28, which is a promoter polymorphism comprised of seven thymine-adenine (TA) dinucleotide repeats [(TA)$_7$TAA] [17], compared to the normal *UGT1A1*1* allele with six TA repeats [(TA)$_6$TAA]. Importantly, the length of the TA repeat sequence is inversely correlated with UGT1A1 expression and activity [18]. Consequently, *UGT1A1*28* heterozygotes and homozygotes have an approximate 25% and 70% reduction in enzyme activity, respectively [18], and individuals who are homozygous for *UGT1A1*28* are at increased risk for myelosuppression, diarrhea, and neutropenia due to the buildup of SN-38. Although the US FDA package insert for irinotecan does include information on toxicity risk due to *UGT1A1*28* and the availability of testing, clinical *UGT1A1* genetic testing is not required prior to treatment. Instead it is recommended that a loading dose be administered and subsequent dosing based on symptoms (ie, granulocyte count and treatment-related diarrhea).

The UGT family is responsible for the glucuronidation of hundreds of compounds, including hormones, flavonoids, environmental mutagens, and pharmaceutical drugs. Most of the UGTs are expressed in the liver, as well as in other types of tissues, such as intestinal, stomach, or breast tissues. As such, the *UGT1A1*28* allele and other *UGT1A1* missense variants have been implicated in Gilbert syndrome, which is an autosomal recessive unconjugated hyperbilirubinemia [19]. This mild disorder does not indicate liver damage but affects the metabolism of several substances, which can present with jaundice among affected individuals, with mild abdominal pain or nausea triggered by fasting or infections. Given that patients with Gilbert syndrome have normal liver function tests and typically require no treatment, a correct diagnosis is essential to avoid unnecessary testing.

Molecular Technologies: *UGT1A1*

Genetic testing for *UGT1A1* can be performed from DNA extracted from whole blood or other tissues, which typically involves targeted testing of the *UGT1A1*28* TA repeat polymorphism. The most common assay is a laboratory-developed test involving fluorescent PCR amplification and size separation by capillary electrophoresis. Of note, this assay will also detect the five [(TA)$_5$TAA] and eight [(TA)$_8$TAA] repeat alleles. The five TA repeat allele is assumed to maintain efficient transcription, while the uncommon eight TA repeat allele indicates irinotecan sensitivity similar to the seven TA repeat allele (*28). Clinical laboratories that offer *UGT1A1* genetic testing can be found at the voluntary National Institutes of Health Genetic Testing Registry (http://www.ncbi.nlm.nih.gov/gtr/) [9].

Clinical Utility: *UGT1A1*

Although the Evaluation of Genomic Applications in Practice and Prevention (EGAPP) Working Group found that the evidence was insufficient to recommend for or against the routine use of *UGT1A1* genotyping in patients with metastatic colorectal cancer who are treated with irinotecan [20], the Royal Dutch Pharmacists Association—Pharmacogenetics Working Group (KNMP-PWG) has evaluated irinotecan dosing based on *UGT1A1* genotype and recommends dose reduction for *28 homozygous patients receiving more than 250 mg/m^2 [21]. Cost-effectiveness studies on *UGT1A1* genotype directed irinotecan indicate that *UGT1A1*28* testing may be cost-effective, but only if irinotecan dose reduction in homozygotes does not reduce efficacy [22,23]. Individuals who are homozygous for *UGT1A1*28* have a mild unconjugated hyperbilirubinemia, which is consistent with a diagnosis of Gilbert syndrome.

Limitations of Testing: *UGT1A1*

Targeted testing of the *UGT1A1*28* TA repeat polymorphism will not detect any other coding *UGT1A1* variants that may influence irinotecan metabolism. In addition, other genes may influence irinotecan toxicity risk, including *CYP3A4* [24], which will not be detected by *UGT1A1* genetic testing.

MOLECULAR TARGET: RASBURICASE AND G6PD DEFICIENCY

Rasburicase is a drug approved by the FDA for prophylaxis and treatment of hyperuricemia during chemotherapy in adults and children with lymphoma, leukemia, and solid cancers. When chemotherapy is administered, cancer cells are destroyed, releasing large amounts of uric acid into the blood. Rasburicase is a recombinant urate oxidase enzyme that works by breaking down uric acid to allantoin and hydrogen peroxide, which is eliminated from the body by the kidneys. The pegylated form of urate oxidase, pegloticase, is also FDA approved for the treatment of refractory gout [25]. Notably, both rasburicase and pegloticase carry an FDA boxed warning and are contraindicated for use in patients with known glucose-6-phosphate dehydrogenase (G6PD) deficiency due to mutations in the *G6PD* gene on chromosome Xq28 [26].

G6PD deficiency is an X-linked disorder that affects over 400 million people worldwide and approximately 1 in 10 African-American males in the United States [27]. It occurs most frequently in the malarial endemic regions of Africa, Asia, and the Mediterranean due to the protection it provides against malaria infection [28]. Of note, different racial and ethnic groups have predominant founder mutations such as the G6PD Mediterranean (c.563C > T) variant, which has important implications when considering genetic testing. The G6PD enzyme catalyzes the first step in the pentose phosphate pathway, which produces antioxidants to protect cells against oxidative stress [26]. Triggers that heighten oxidative stress in red blood cells result in hemolytic anemia and symptom onset in patients with G6PD deficiency [27].

Without enough functional G6PD, red blood cells are unable to protect themselves from the damaging effects of reactive oxygen species and subsequent hemolysis. Factors such as infections, certain drugs, and ingesting fava beans can increase the levels of reactive oxygen species, causing red blood cells to undergo hemolysis faster than the body can replace them. The loss of red blood cells causes the signs and symptoms of hemolytic anemia such as dark urine, enlarged spleen, fatigue, rapid heart rate, shortness of breath, and jaundice, which are the characteristic features of G6PD deficiency [27].

G6PD variants that result in enzyme deficiency confer a G6PD-deficient phenotype in hemizygous males and homozygous or compound heterozygous females. It is difficult to diagnose G6PD deficiency in heterozygous females due to random X chromosome inactivation. Targeted *G6PD* genotyping can establish a molecular diagnosis of G6PD deficiency. However, prediction of drug response can be difficult without testing G6PD enzyme activity levels. More than 400 disease causing mutations in *G6PD* have been identified, and most are missense mutations that affect protein stability [29]. All *G6PD* variants are broadly divided into five classes according to the resulting level of enzyme activity, with class I being the most severely dysfunctional and class V having the highest enzyme activity (Table 39.2).

The production of hydrogen peroxide following the oxidation of uric acid and allantoin by rasburicase can result in hemolytic anemia after rasburicase administration among G6PD-deficient patients, as evidenced

TABLE 39.2   Classification of *G6PD* Variants

WHO class	Enzyme activity	Associated phenotype	Variant example
I	Severe deficiency	Congenital non-spherocytic hemolytic anemia (CNSHA)	Tondela, Palermo
II	<10% severely deficient	Risk of acute hemolytic anemia	Mediterranean, Canton, Chatham
III	10−60% moderate deficiency	Risk of acute hemolytic anemia	A- Haplotype, Asahi, Orissa, Kalyan-Kerala
IV	60−150% normal activity	No clinical manifestations	B (wildtype), A, Mira d'Aire, Sao Borja
V	150% enhanced activity		Hektoen

Adapted from McDonagh EM, Thorn CF, Bautista JM, Youngster I, Altman RB, Klein TE. PharmGKB summary: very important pharmacogene information for G6PD. Pharmacogenet Genomics 2012;22:219−28.

TABLE 39.3 CPIC Recommended Dosing of Rasburicase by G6PD Phenotype

Phenotype (genotype)	Examples of diplotypes	Implications for phenotypic measures	Dosing recommendations for rasburicase
Normal. A male carrying a nondeficient (class IV) allele or a female carrying two nondeficient (class IV) alleles	Male: B, Sao Boria. Female: B/B, B/Sao Boria	Low or reduced risk of hemolytic anemia	No reason to withhold rasburicase based on G6PD status
Deficient or deficient with CNSHA. A male carrying a class I, II, or III allele, a female carrying two deficient class I–III alleles	Male: A-, Orissa, Kalyan-Kerala, Mediterranean, Canton, Chatham, Bangkok, Villeurbanne. Female: A-/A-, A-/ Orissa, Orissa/Kalyan-Kerala, Mediterranean/ Mediterranean, Chatham/Mediterranean, Canton/Viangchan, Bangkok/Bangkok, Bangkok/Villeurbanne	At risk of acute hemolytic anemia	Rasburicase is contraindicated; alternatives include allopurinol
Variable. A female carrying one nondeficient (class IV) and one deficient (class I–III variants) allele	B/A-, B/Mediterranean, B/Bangkok	Unknown risk of hemolytic anemia	To ascertain that G6PD status is normal, enzyme activity must be measured; alternatives include allopurinol

CNSHA, chronic non-spherocytic hemolytic anemia.

Adapted from Relling MV, McDonagh EM, Chang T, et al. Clinical Pharmacogenetics Implementation Consortium (CPIC) guidelines for rasburicase therapy in the context of G6PD deficiency genotype. Clin Pharmacol Ther 2014;96:169–74. See original reference for complete footnotes.

by numerous clinical reports [30]. Together, these data prompted drug label warnings in several countries that contraindicate rasburicase in patients with G6PD deficiency, and informed the recent CPIC guidelines that recommend an alternative therapy (eg, allopurinol) for G6PD-deficient patients (Table 39.3) [30].

Molecular Technologies: G6PD

Genetic testing for *G6PD* can be performed from DNA extracted from whole blood or other tissues, which typically involves targeted genotyping of a panel of *G6PD*-deficient alleles. However, many *G6PD* variants have been identified [26]. In addition, the US National Newborn Screening Program routinely tests for G6PD deficiency by genotyping in some states with a panel of five variants followed by confirmatory enzyme activity testing [31]. Clinical laboratories that offer *G6PD* genetic testing can be found at the voluntary National Institutes of Health Genetic Testing Registry (http://www.ncbi.nlm.nih.gov/gtr/) [9].

Clinical Utility: G6PD

Despite the lack of a prospective randomized clinical trial directly evaluating the utility of *G6PD* genotype directed rasburicase therapy, the available evidence was utilized to inform the CPIC guidelines recommending an alternative therapy from rasburicase among *G6PD*-deficient patients when genotype data is available [5].

The limited data reported from cost-effectiveness studies on *G6PD* genotyping suggest that G6PD screening may be cost-effective [32]. A number of adverse reactions, such as drug-induced hemolytic anemia, have been reported for several medications among patients with G6PD deficiency. As such, individuals with a diagnosis of G6PD deficiency should exercise caution when initiating new drug treatments to avoid hemolytic anemia and other adverse phenotypes.

Limitations of Testing: G6PD

Targeted *G6PD* genotyping will not detect any alleles that are not directly interrogated so a negative genotyping result does not rule out the possibility that a patient carries another *G6PD* variant. Full gene sequencing will detect all *G6PD* variants, but rare or novel variants will likely be of uncertain clinical significance. In addition, other genes may influence responses to rasburicase, which will not be detected by *G6PD* genetic testing. Alternatives to *G6PD* genotyping that assess G6PD enzyme activity directly are available and can be used to confirm a diagnosis of G6PD deficiency.

MOLECULAR TARGET: TAMOXIFEN AND CYP2D6

Tamoxifen is used for the treatment and prevention of estrogen receptor positive (ER-positive) breast cancer. It has been shown to decrease disease recurrence and mortality rates by as much as 50% and 30%, respectively, and has also been used as a prophylactic treatment for those at high risk of developing breast cancer [33]. However, tamoxifen response has a high degree of interindividual variability that is likely due, in part, to differences in tamoxifen metabolism [34]. Hot flashes, the most common side effect of tamoxifen, affect up to 80% of treated women. Individuals receiving tamoxifen also seem to have approximately 2.5-fold increased risk of developing endometrial cancer. Additionally, tamoxifen may contribute to an increased risk for thromboembolic events, as well as clinical depression. Selective serotonin reuptake inhibitors (SSRIs) are commonly used to treat both hot flashes and depression. However, the potential drug interaction between some SSRIs and tamoxifen necessitates careful consideration of which SSRI to prescribe for these patients [35].

Tamoxifen competitively inhibits cancerous ER-positive cells from getting the estrogen required for growth. However, its metabolites also act as aromatase inhibitors that decrease the amount of available estrogen in the body [36]. The metabolism of tamoxifen is complex, mostly occurring through the 4-hydroxylation and N-demethylation pathways, both of which result in the very potent secondary metabolite, endoxifen [33]. The 4-hydroxylation pathway contributes approximately 7% of tamoxifen metabolism and the N-demethylation to N-desmethyltamoxifen pathway contributes approximately 92% of tamoxifen metabolism [37]. Endoxifen is formed from N-desmethyltamoxifen primarily through hydroxylation by cytochrome P450-2D6 (CYP2D6) and from 4-hydroxy-tamoxifen through demethylation by CYP3A4 [33]. In addition to ER inhibition, endoxifen also targets ERα (coded for by ESR1 gene) for proteasomal degradation, together indicating that endoxifen is the primary metabolite responsible for the efficacy of tamoxifen treatment [38].

The highly polymorphic nature of the CYP2D6 gene on chromosome 22q13.2 and the central role that CYP2D6 plays in the metabolism of tamoxifen to endoxifen have prompted many studies on the potential pharmacogenetic association between CYP2D6 genotype and tamoxifen response. Some of these studies identified a significant association between loss-of-function CYP2D6 alleles and poor prognoses [39–43]. However, others did not detect any association and concluded that clinical CYP2D6 genotyping is not warranted [44,45]. These conflicting findings have resulted in ongoing debate over the clinical validity of CYP2D6 in tamoxifen response, as well as stimulating a recent meta-analysis by the International Tamoxifen Pharmacogenomics Consortium (ITPC) [46]. Notably, the ITPC confirmed the association between CYP2D6 poor metabolizer status and poorer invasive disease-free survival (hazard ratio = 1.25; 95% confidence interval = 1.06, 1.47; $P = 0.009$). However, this was only when implementing a post hoc strict inclusion criteria (eg, dose, duration of treatment, menopausal status, and genotyping quality) as no effect was detected when applying limited or no exclusion criteria among the heterogeneous study populations [46]. Although not independently conclusive, these data suggest that CYP2D6 is likely one of several factors influencing outcome following adjuvant tamoxifen treatment [47]. Moreover, the KNMP-PWG recommends consideration of aromatase inhibitors for postmenopausal women who are CYP2D6 poor or intermediate metabolizers due to increased risk for relapse of breast cancer when treated with tamoxifen [21].

Molecular Technologies: CYP2D6

Genetic testing for CYP2D6 can be performed from DNA extracted from whole blood or other tissues, which typically involves targeted genotyping of a panel of CYP2D6-variant alleles. Several commercial assays are currently available, including the Luminex [48] and AutoGenomics [49] assays, as well as other laboratory-developed tests. The Tag-IT Luminex platform (Luminex Molecular Diagnostics, Toronto, Canada) is a bead array with oligonucleotides bound to microspheres and genotyping by allele-specific primer extension. The AutoGenomics (Carlsbad, CA) platform is a film-based microarray tested on the INFINITI Analyzer. These assays typically interrogate 15–20 important CYP2D6 variants including the deletion and duplication alleles. Clinical laboratories that offer CYP2D6 genetic testing can be found at the voluntary National Institutes of Health Genetic Testing Registry (http://www.ncbi.nlm.nih.gov/gtr/) [9].

Like other CYP450 genes, CYP2D6 alleles are designated by the common star (*) allele nomenclature system, which often include multiple variants on the same haplotype. The CYP2D6*1 allele is the wild-type haplotype encoding normal enzyme activity. More than 100 CYP2D6-variant alleles have been described (http://www.cypalleles.ki.se/cyp2d6.htm). However, many are rare in the general population. Although the effect of all of these variant alleles on enzyme activity has not been established, many of the commonly interrogated CYP2D6 variants include nonfunctional, reduced, and increased function alleles (Table 39.4). The combination of these alleles in a given genotype results in four predicted phenotype categories: (1) ultrarapid, (2) extensive (normal), (3) intermediate, and (4) poor metabolizers [50].

TABLE 39.4 Commonly Interrogated CYP2D6 Alleles

Predicted activity	CYP2D6 alleles (major nucleotide variants: GenBank accession number M33388)
Increased activity	*1xN, *2xN, *35xN
Functional (normal activity)	*1 (wild type)
	*2 (-1584C > G, 1661G > C, 2850C > T, 4180G > C)
	*35 (-1584C > G, 31G > A, 1661G > C, 2850C > T, 4180G > C)
Reduced function	*9 (2613–2615delAGA)
	*10 (100C > T, 1661G > C, 4180G > C), *10xN
	*17 (1023C > T, 1661G > C, 2850C > T, 4180G > C)
	*29 (1659G > A, 1661G > C, 2850C > T, 3183G > A, 4180G > C)
	*41 (1661G > C, 2850C > T, 2988G > A, 4180G > C), *41xN
Nonfunctional	*3 (2549delA)
	*4 (100C > T, 1661G > C, 1846G > A, 2850C > T, 4180G > C), *4xN
	*5 (gene deletion)
	*6 (1707delT, 4180G > C)
	*7 (2935A > C)
	*8 (1661G > C, 1758G > T, 2850C > T, 4180G > C)
	*11 (883G > C, 1661G > C, 2850C > T, 4180G > C)
	*15 (138insT)

Clinical Utility: CYP2D6

Despite the lack of a prospective randomized clinical trial directly evaluating the utility of CYP2D6 genotyping, the available evidence was utilized to inform the KNMP-PWG guidelines that recommend consideration of using aromatase inhibitors for postmenopausal women who are CYP2D6 poor or intermediate metabolizers due to increased risk for relapse of breast cancer with tamoxifen [21]. The limited data reported from cost-effectiveness studies on CYP2D6 genotype directed tamoxifen therapy have generally concluded that there is not enough evidence available to support or reject routine CYP2D6 testing. However, these studies have underscored the heterogeneity in CYP2D6 genotyping across the published retrospective studies and the need for further analyses of large adjuvant aromatase inhibitor trials to better understand any association between CYP2D6 genotype and tamoxifen outcomes [51,52].

Limitations of Testing: CYP2D6

Targeted CYP2D6 genotyping will not detect any alleles that are not directly interrogated so the wild-type CYP2D6*1 allele is typically assigned in the absence of other detected variants. Consequently, when *1 is reported by targeted genotyping, a rare CYP2D6 star (*) allele not included in the genotyping panel would not be detected, which can only be identified by gene sequencing. Furthermore, in addition to duplicated functional CYP2D6 alleles (eg, *1xN, *2xN), duplicated nonfunctional (eg, *4xN) and reduced function (eg, *10xN) alleles have also been described. As such, determining which CYP2D6 allele is duplicated is important for proper interpretation when a gene duplication is identified in conjunction with a heterozygous genotype [53]. Although laboratory guidelines for CYP2D6 genotyping in relation to tamoxifen therapy have been reported [54], no current professional guidelines detail which alleles should be included in clinical CYP2D6 assays. Therefore, different laboratories may include different CYP2D6 alleles in their testing panels, which can result in conflicting CYP2D6 genotypes and predicted phenotypes between laboratories and studies. Other genes likely influence responses to tamoxifen, including other CYP450s and members of the UGT and SULT families [33], which will not be detected by CYP2D6 genetic testing.

MOLECULAR TARGET: THIOPURINES AND THIOPURINE METHYLTRANSFERASE

Thiopurines (ie, azathioprine, mercaptopurine, and thioguanine) are used for the treatment of childhood acute lymphoblastic leukemia, autoimmune diseases, inflammatory bowel diseases, lupus, and transplantation. Specifically, mercaptopurine and azathioprine are used for nonmalignant immunologic disorders, mercaptopurine for lymphoid malignancies, and thioguanine for myeloid leukemias. Thiopurines are inactive precursors that are metabolized by hypoxanthine guanine phosphoribosyl transferase to active thioguanine nucleotides (TGNs), which are inactivated by thiopurine methyltransferase (TPMT) [55]. These drugs are analogs of the nucleic acid guanine and are incorporated into RNA and DNA by phosphodiester linkages, ultimately inhibiting several metabolic pathways and inducing apoptosis. In addition, mercaptopurines are metabolized to methyl-thioinosine monophosphate, which inhibits de novo purine synthesis and cell proliferation, and adding another mechanism of cytotoxicity. However, approximately 10% of the population

TABLE 39.5 CPIC Recommended Dosing of Thioguanine by TPMT Genotype/Phenotype

Phenotype (genotype)	Examples of diplotypes	Implications for pharmacologic measures after thioguanine	Dosing recommendations for thioguanine
Homozygous wild-type or normal, high activity (two functional *1 alleles)	*1/*1	Lower concentrations of TGN metabolites, but note that TGN after thioguanine are 5–10 × higher than TGN after mercaptopurine or azathioprine	Start with normal starting dose. Adjust doses of thioguanine and of other myelosuppressive therapy without any special emphasis on thioguanine. Allow 2 weeks to reach steady state after each dose adjustment
Heterozygote or intermediate activity (one functional allele—*1, plus one nonfunctional allele)	*1/*2, *1/*3A, *1/*3B, *1/*3C, *1/*4	Moderate to high concentrations of TGN metabolites, but note that TGN after thioguanine are 5–10 × higher than TGN after mercaptopurine or azathioprine	Start with reduced doses (reduce by 30–50%) and adjust doses of thioguanine based on degree of myelosuppression and disease-specific guidelines. Allow 2–4 weeks to reach steady state after each dose adjustment. In setting of myelosuppression, and depending on other therapy, emphasis should be on reducing thioguanine over other agents
Homozygous or compound heterozygous variant, mutant, low, or deficient activity (two nonfunctional alleles)	*3A/*3A, *2/*3A, *3C/*3A, *3C/*4, *3C/*2, *3A/*4	Extremely high concentrations of TGN metabolites; fatal toxicity possible without dose decrease	Start with drastically reduced doses (reduce daily dose by 10-fold and dose thrice weekly instead of daily) and adjust doses of thioguanine based on degree of myelosuppression and disease-specific guidelines. Allow 4–6 weeks to reach steady state after each dose adjustment. In setting of myelosuppression, emphasis should be on reducing thioguanine over other agents. For nonmalignant conditions, consider alternative nonthiopurine immunosuppressant therapy

Adapted from Relling MV, Gardner EE, Sandborn WJ, et al. Clinical Pharmacogenetics Implementation Consortium guidelines for thiopurine methyltransferase genotype and thiopurine dosing. Clin Pharmacol Ther 2011;89:387–91. Note that this CPIC guideline also has TPMT-directed guidelines for mercaptopurine and azathioprine.

have intermediate levels of TPMT activity and 0.3% have low or undetectable enzyme activity, which results in significantly increased risks for TGN toxicity and life-threatening myelosuppression [56,57].

Thirty-one variant alleles of the *TPMT* gene on chromosome 6p22.3 have been identified, many of which are missense mutations associated with decreased in vitro activity [58]. The most commonly studied and tested variant alleles are *TPMT*2, *3A, *3B,* and *3C. TPMT*3A* contains two missense variants in *cis,* p.Ala154Thr and p.Tyr240Cys, and is the most common variant allele associated with low TPMT activity in Caucasians (frequency ~5%) [56]. With conventional doses of thiopurines, individuals who inherit two loss-of-function *TPMT* alleles universally experience severe myelosuppression, a high proportion of heterozygous patients show moderate to severe myelosuppression, and homozygous wild-type patients have lower levels of TGN metabolites and a low risk of myelosuppression [56,59,60]. Taken together, these data prompted CPIC practice guidelines recommending dose reductions and/or alternate therapies among loss-of-function *TPMT* allele carriers (Table 39.5) [56].

Molecular Technologies: TPMT

Genetic testing for *TPMT* can be performed from DNA extracted from whole blood or other tissues, which typically involves targeted genotyping of the *TPMT*2, *3A, *3B,* and *3C* alleles [61,62]. PCR-based assays that interrogate the three common variants detect 80–95% of low and intermediate enzyme activity individuals in the Caucasian, African-American, and Asian populations [63]. Clinical laboratories that offer *TPMT* genetic testing can be found at the voluntary National Institutes of Health Genetic Testing Registry (http://www.ncbi.nlm.nih.gov/gtr/) [9].

Clinical Utility: TPMT

Available data suggest that patients with reduced or nonfunctional *TPMT* alleles are at high risk for bone marrow toxicity and require significant dose reduction [56]. Despite the lack of a prospective randomized clinical trial directly evaluating the utility of *TPMT* genotyping, the available evidence was utilized to inform the CPIC guidelines on dose reduction among

TPMT-variant carriers when genotype data is available [56]. Cost-effectiveness studies on *TPMT* genotype directed thiopurine dosing have been reported. However, they largely have concluded with conflicting results [64–66].

Limitations of Testing: *TPMT*

Targeted *TPMT* genotyping will not detect any alleles that are not directly interrogated so a negative genotyping result does not rule out the possibility that a patient carries another *TPMT*. Full gene sequencing will detect all *TPMT* variants, but rare or novel variants will likely be of uncertain clinical significance. In addition, other genes may influence responses to thiopurines, including *ITPA* [1,10], which will not be detected by *TPMT* genetic testing. Alternatives to *TPMT* genotyping are available and include testing TPMT enzyme activity directly and/or TPMT metabolites levels.

OTHER CONSIDERATIONS FOR PERSONALIZED MEDICINE IN CANCER TREATMENT

While pharmacogenetics can play a direct role in some cancer treatments, it also plays a significant role in supportive care for oncology patients. These patients are often treated with additional medications to manage pain, infections, and psychosocial distress (eg, antidepressants), and some of which have clinically actionable pharmacogenetic gene variant associations.

Codeine and *CYP2D6*

Codeine is an opioid analgesic indicated for the relief of mild to moderately severe pain. The analgesic properties of codeine stem from its conversion to morphine, which is predominately mediated by the polymorphic CYP2D6 enzyme (tamoxifen and CYP2D6) [67]. Importantly, CYP2D6 poor metabolizers are unable to efficiently convert codeine to morphine and as a consequence may not experience pain relief [68]. Conversely, CYP2D6 ultrarapid metabolizers may metabolize codeine too efficiently leading to morphine intoxication and toxicity [69]. These data have resulted in several professional societies, including CPIC [70], to recommend an alternate analgesic for CYP2D6 ultrarapid and poor metabolizer patients.

Antidepressants and *CYP2D6*

Significant evidence exists for a role of *CYP2D6* genotyping for individualized treatment with tricyclic antidepressants, which also has prompted a recent CPIC guideline [71]. For example, *CYP2D6* poor metabolizers treated with amitriptyline and nortriptyline have impaired drug metabolism and increased risks of side effects, whereas ultrarapid metabolizers have elevated risks of reduced drug efficacy due to rapid drug elimination. Similarly, evidence supporting a role for *CYP2D6* in SSRI response variability also exists and a CPIC guideline on *CYP2D6* genotype directed SSRI treatment is currently in development (http://www. pharmgkb.org/page/cpic).

References

[1] Thorn CF, Marsh S, Carrillo MW, McLeod HL, Klein TE, Altman RB. PharmGKB summary: fluoropyrimidine pathways. Pharmacogenet Genomics 2011;21:237–42.

[2] Amstutz U, Farese S, Aebi S, Largiader CR. Dihydropyrimidine dehydrogenase gene variation and severe 5-fluorouracil toxicity: a haplotype assessment. Pharmacogenomics 2009;10:931–44.

[3] van Kuilenburg AB, Meinsma R, Zonnenberg BA, et al. Dihydropyrimidinase deficiency and severe 5-fluorouracil toxicity. Clin Cancer Res 2003;9:4363–7.

[4] Van Kuilenburg AB, Vreken P, Abeling NG, et al. Genotype and phenotype in patients with dihydropyrimidine dehydrogenase deficiency. Hum Genet 1999;104:1–9.

[5] Caudle KE, Thorn CF, Klein TE, et al. Clinical Pharmacogenetics Implementation Consortium guidelines for dihydropyrimidine dehydrogenase genotype and fluoropyrimidine dosing. Clin Pharmacol Ther 2013;94:640–5.

[6] Morel A, Boisdron-Celle M, Fey L, et al. Clinical relevance of different dihydropyrimidine dehydrogenase gene single nucleotide polymorphisms on 5-fluorouracil tolerance. Mol Cancer Ther 2006;5:2895–904.

[7] Saif MW, Ezzeldin H, Vance K, Sellers S, Diasio RB. DPYD*2A mutation: the most common mutation associated with DPD deficiency. Cancer Chemother Pharmacol 2007;60:503–7.

[8] McLeod HL, Collie-Duguid ES, Vreken P, et al. Nomenclature for human DPYD alleles. Pharmacogenetics 1998;8:455–9.

[9] Rubinstein WS, Maglott DR, Lee JM, et al. The NIH genetic testing registry: a new, centralized database of genetic tests to enable access to comprehensive information and improve transparency. Nucleic Acids Res 2013;41:D925–35.

[10] Schwab M, Zanger UM, Marx C, et al. Role of genetic and non-genetic factors for fluorouracil treatment-related severe toxicity: a prospective clinical trial by the German 5-FU Toxicity Study Group. J Clin Oncol 2008;26:2131–8.

[11] Terrazzino S, Cargnin S, Del Re M, Danesi R, Canonico PL, Genazzani AA. DPYD IVS14 + 1G > A and 2846A > T genotyping for the prediction of severe fluoropyrimidine-related toxicity: a meta-analysis. Pharmacogenomics 2013;14:1255–72.

[12] Schmidt C, Hofmann U, Kohlmuller D, et al. Comprehensive analysis of pyrimidine metabolism in 450 children with unspecific neurological symptoms using high-pressure liquid chromatography-electrospray ionization tandem mass spectrometry. J Inherit Metab Dis 2005;28:1109–22.

[13] van Gennip AH, Abeling NG, Vreken P, van Kuilenburg AB. Inborn errors of pyrimidine degradation: clinical, biochemical and molecular aspects. J Inherit Metab Dis 1997;20:203–13.

[14] van Staveren MC, Guchelaar HJ, van Kuilenburg AB, Gelderblom H, Maring JG. Evaluation of predictive tests for screening for dihydropyrimidine dehydrogenase deficiency. Pharmacogenomics J 2013;13:389–95.

[15] Whirl-Carrillo M, McDonagh EM, Hebert JM, et al. Pharmacogenomics knowledge for personalized medicine. Clin Pharmacol Ther 2012;92:414–17.

[16] Marsh S, Hoskins JM. Irinotecan pharmacogenomics. Pharmacogenomics 2010;11:1003–10.

[17] Perera MA, Innocenti F, Ratain MJ. Pharmacogenetic testing for uridine diphosphate glucuronosyltransferase 1A1 polymorphisms: are we there yet? Pharmacotherapy 2008;28:755–68.

[18] Zhang D, Zhang D, Cui D, et al. Characterization of the UDP glucuronosyltransferase activity of human liver microsomes genotyped for the UGT1A1*28 polymorphism. Drug Metab Dispos 2007;35:2270–80.

[19] Strassburg CP. Pharmacogenetics of Gilbert's syndrome. Pharmacogenomics 2008;9:703–15.

[20] Evaluation of Genomic Applications in Practice and Prevention Working Group. Recommendations from the EGAPP Working Group: can UGT1A1 genotyping reduce morbidity and mortality in patients with metastatic colorectal cancer treated with irinotecan? Genet Med 2009;11:15–20.

[21] Swen JJ, Nijenhuis M, de Boer A, et al. Pharmacogenetics: from bench to byte—an update of guidelines. Clin Pharmacol Ther 2011;89:662–73.

[22] Gold HT, Hall MJ, Blinder V, Schackman BR. Cost effectiveness of pharmacogenetic testing for uridine diphosphate glucuronosyltransferase 1A1 before irinotecan administration for metastatic colorectal cancer. Cancer 2009;115:3858–67.

[23] Pichereau S, Le Louarn A, Lecomte T, Blasco H, Le Guellec C, Bourgoin H. Cost-effectiveness of UGT1A1*28 genotyping in preventing severe neutropenia following FOLFIRI therapy in colorectal cancer. J Pharm Sci 2010;13:615–25.

[24] van der Bol JM, Mathijssen RH, Creemers GJ, et al. A CYP3A4 phenotype-based dosing algorithm for individualized treatment of irinotecan. Clin Cancer Res 2010;16:736–42.

[25] Sundy JS, Baraf HS, Yood RA, et al. Efficacy and tolerability of pegloticase for the treatment of chronic gout in patients refractory to conventional treatment: two randomized controlled trials. JAMA 2011;306:711–20.

[26] McDonagh EM, Thorn CF, Bautista JM, Youngster I, Altman RB, Klein TE. PharmGKB summary: very important pharmacogene information for G6PD. Pharmacogenet Genomics 2012;22:219–28.

[27] Frank JE. Diagnosis and management of G6PD deficiency. Am Fam Physician 2005;72:1277–82.

[28] Nkhoma ET, Poole C, Vannappagari V, Hall SA, Beutler E. The global prevalence of glucose-6-phosphate dehydrogenase deficiency: a systematic review and meta-analysis. Blood Cells Mol Dis 2009;42:267–78.

[29] Mason PJ, Bautista JM, Gilsanz F. G6PD deficiency: the genotype-phenotype association. Blood Rev 2007;21:267–83.

[30] Relling MV, McDonagh EM, Chang T, et al. Clinical Pharmacogenetics Implementation Consortium (CPIC) guidelines for rasburicase therapy in the context of G6PD deficiency genotype. Clin Pharmacol Ther 2014;96:169–74.

[31] Lin Z, Fontaine JM, Freer DE, Naylor EW. Alternative DNA-based newborn screening for glucose-6-phosphate dehydrogenase deficiency. Mol Genet Genomics 2005;86:212–19.

[32] Khneisser I, Adib SM, Loiselet J, Megarbane A. Cost-benefit analysis of G6PD screening in Lebanese newborn males. Leban Med J 2007;55:129–32.

[33] Klein DJ, Thorn CF, Desta Z, Flockhart DA, Altman RB, Klein TE. PharmGKB summary: tamoxifen pathway, pharmacokinetics. Pharmacogenet Genomics 2013;23:643–7.

[34] Saladores PH, Precht JC, Schroth W, Brauch H, Schwab M. Impact of metabolizing enzymes on drug response of endocrine therapy in breast cancer. Expert Rev Mol Diagn 2013;13:349–65.

[35] Kelly CM, Juurlink DN, Gomes T, et al. Selective serotonin reuptake inhibitors and breast cancer mortality in women receiving tamoxifen: a population based cohort study. BMJ 2010;340:c693.

[36] Jordan VC. Tamoxifen: a most unlikely pioneering medicine. Nat Rev Drug Discov 2003;2:205–13.

[37] Kiyotani K, Mushiroda T, Nakamura Y, Zembutsu H. Pharmacogenomics of tamoxifen: roles of drug metabolizing enzymes and transporters. Drug Metab Pharmacokinet 2012;27:122–31.

[38] Wu X, Hawse JR, Subramaniam M, Goetz MP, Ingle JN, Spelsberg TC. The tamoxifen metabolite, endoxifen, is a potent antiestrogen that targets estrogen receptor alpha for degradation in breast cancer cells. Cancer Res 2009;69:1722–7.

[39] Goetz MP, Rae JM, Suman VJ, et al. Pharmacogenetics of tamoxifen biotransformation is associated with clinical outcomes of efficacy and hot flashes. J Clin Oncol 2005;23:9312–18.

[40] Goetz MP, Knox SK, Suman VJ, et al. The impact of cytochrome P450 2D6 metabolism in women receiving adjuvant tamoxifen. Breast Cancer Res Treat 2007;101:113–21.

[41] Schroth W, Antoniadou L, Fritz P, et al. Breast cancer treatment outcome with adjuvant tamoxifen relative to patient CYP2D6 and CYP2C19 genotypes. J Clin Oncol 2007;25:5187–93.

[42] Schroth W, Goetz MP, Hamann U, et al. Association between CYP2D6 polymorphisms and outcomes among women with early stage breast cancer treated with tamoxifen. JAMA 2009;302:1429–36.

[43] Kiyotani K, Mushiroda T, Imamura CK, et al. Significant effect of polymorphisms in CYP2D6 and ABCC2 on clinical outcomes of adjuvant tamoxifen therapy for breast cancer patients. J Clin Oncol 2010;28:1287–93.

[44] Rae JM, Drury S, Hayes DF, et al. CYP2D6 and UGT2B7 genotype and risk of recurrence in tamoxifen-treated breast cancer patients. J Natl Cancer Inst 2012;104:452–60.

[45] Regan MM, Leyland-Jones B, Bouzyk M, et al. CYP2D6 genotype and tamoxifen response in postmenopausal women with endocrine-responsive breast cancer: the breast international group 1-98 trial. J Natl Cancer Inst 2012;104:441–51.

[46] Province MA, Goetz MP, Brauch H, et al. CYP2D6 genotype and adjuvant tamoxifen: meta-analysis of heterogeneous study populations. Clin Pharmacol Ther 2014;95:216–27.

[47] Province MA, Altman RB, Klein TE. Interpreting the CYP2D6 results from the International Tamoxifen Pharmacogenetics Consortium. Clin Pharmacol Ther 2014;96:144–6.

[48] Melis R, Lyon E, McMillin GA. Determination of CYP2D6, CYP2C9 and CYP2C19 genotypes with Tag-It mutation detection assays. Expert Rev Mol Diagn 2006;6:811–20.

[49] Savino M, Seripa D, Gallo AP, et al. Effectiveness of a high-throughput genetic analysis in the identification of responders/non-responders to CYP2D6-metabolized drugs. Clin Lab 2011;57:887–93.

[50] Owen RP, Sangkuhl K, Klein TE, Altman RB. Cytochrome P450 2D6. Pharmacogenet Genomics 2009;19:559–62.

[51] Fleeman N, Martin Saborido C, Payne K, et al. The clinical effectiveness and cost-effectiveness of genotyping for CYP2D6 for the management of women with breast cancer treated with tamoxifen: a systematic review. Health Technol Assess 2011;15:1–102.

[52] Woods B, Veenstra D, Hawkins N. Prioritizing pharmacogenetic research: a value of information analysis of CYP2D6 testing to guide breast cancer treatment. Value Health 2011;14:989–1001.

[53] Ramamoorthy A, Skaar TC. Gene copy number variations: it is important to determine which allele is affected. Pharmacogenomics 2011;12:299–301.

[54] Lyon E, Gastier Foster J, Palomaki GE, et al. Laboratory testing of CYP2D6 alleles in relation to tamoxifen therapy. Genet Med 2012;14:990—1000.

[55] Evans WE. Pharmacogenetics of thiopurine S-methyltransferase and thiopurine therapy. Ther Drug Monit 2004;26:186—91.

[56] Relling MV, Gardner EE, Sandborn WJ, et al. Clinical Pharmacogenetics Implementation Consortium guidelines for thiopurine methyltransferase genotype and thiopurine dosing. Clin Pharmacol Ther 2011;89:387—91.

[57] Weinshilboum RM, Sladek SL. Mercaptopurine pharmacogenetics: monogenic inheritance of erythrocyte thiopurine methyltransferase activity. Am J Hum Genet 1980;32:651—62.

[58] Appell ML, Berg J, Duley J, et al. Nomenclature for alleles of the thiopurine methyltransferase gene. Pharmacogenet Genomics 2013;23:242—8.

[59] Black AJ, McLeod HL, Capell HA, et al. Thiopurine methyltransferase genotype predicts therapy-limiting severe toxicity from azathioprine. Ann Intern Med 1998;129:716—18.

[60] Relling MV, Hancock ML, Rivera GK, et al. Mercaptopurine therapy intolerance and heterozygosity at the thiopurine S-methyltransferase gene locus. J Natl Cancer Inst 1999;91:2001—8.

[61] Yates CR, Krynetski EY, Loennechen T, et al. Molecular diagnosis of thiopurine S-methyltransferase deficiency: genetic basis for azathioprine and mercaptopurine intolerance. Ann Intern Med 1997;126:608—14.

[62] Evans WE, Hon YY, Bomgaars L, et al. Preponderance of thiopurine S-methyltransferase deficiency and heterozygosity among patients intolerant to mercaptopurine or azathioprine. J Clin Oncol 2001;19:2293—301.

[63] Zhou S. Clinical pharmacogenomics of thiopurine S-methyltransferase. Curr Clin Pharmacol 2006;1:119—28.

[64] Compagni A, Bartoli S, Buehrlen B, Fattore G, Ibarreta D, de Mesa EG. Avoiding adverse drug reactions by pharmacogenetic testing: a systematic review of the economic evidence in the case of TPMT and AZA-induced side effects. Int J Technol Assess Health Care 2008;24:294—302.

[65] Donnan JR, Ungar WJ, Mathews M, Hancock-Howard RL, Rahman P. A cost effectiveness analysis of thiopurine methyltransferase testing for guiding 6-mercaptopurine dosing in children with acute lymphoblastic leukemia. Pediatric Blood Cancer 2011;57:231—9.

[66] Thompson AJ, Newman WG, Elliott RA, Roberts SA, Tricker K, Payne K. The cost-effectiveness of a pharmacogenetic test: a trial-based evaluation of TPMT genotyping for azathioprine. Value Health 2014;17:22—33.

[67] Thorn CF, Klein TE, Altman RB. Codeine and morphine pathway. Pharmacogenet Genomics 2009;19:556—8.

[68] Desmeules J, Gascon MP, Dayer P, Magistris M. Impact of environmental and genetic factors on codeine analgesia. Eur J Clin Pharmacol 1991;41:23—6.

[69] Gasche Y, Daali Y, Fathi M, et al. Codeine intoxication associated with ultrarapid CYP2D6 metabolism. N Engl J Med 2004;351:2827—31.

[70] Crews KR, Gaedigk A, Dunnenberger HM, et al. Clinical Pharmacogenetics Implementation Consortium guidelines for cytochrome P450 2D6 genotype and codeine therapy: 2014 update. Clin Pharmacol Ther 2014;95:376—82.

[71] Hicks JK, Swen JJ, Thorn CF, et al. Clinical Pharmacogenetics Implementation Consortium guideline for CYP2D6 and CYP2C19 genotypes and dosing of tricyclic antidepressants. Clin Pharmacol Ther 2013;93:402—8.

THE FUTURE OF MOLECULAR TESTING

40

Imaging Mass Spectrometry in Clinical Pathology

J.L. Norris, D.B. Gutierrez and R.M. Caprioli

Mass Spectrometry Research Center, and Department of Biochemistry, Vanderbilt University School of Medicine, Nashville, TN, United States

INTRODUCTION

The diagnosis of disease through the examination of cells and tissues in pathology relies heavily on macroscopic, microscopic, immunologic, and biochemical analyses. This discipline has been standardized over many years and today makes use of modern technologies such as immunohistochemistry (IHC), in situ hybridization, flow cytometry, cytogenetics, electron microscopy, and cytopathology, among others. Information obtained from these protocols and technologies allows pathologists to organize and correlate patient specimens in such a way as to deliver a relatively uniform standard of care and ultimately provide a meaningful diagnostic decision. Nevertheless, as new and more detailed molecular information is gathered from the study of diseases in terms of genomic, proteomic, and metabolomic alterations, it is clear that new technology is desperately needed to provide a way for this important molecular information about an individual patient to impact clinical care. These new technologies will ultimately lead to better diagnosis, prognosis, and long-range outcomes.

The ongoing discovery and increased understanding of the molecular basis of disease are revolutionizing the practice and delivery of health care. Application of this knowledge is most essential in the initial diagnosis of disease by pathologists. Single markers for disease are often found to be either unreliable or insufficient to describe the increasingly complex molecular phenotypes now known even in relatively simple diseases. It is essential to rapidly and accurately assess a plurality of disease biomarkers in a cost-effective and high-throughput manner—a capability that has not yet been realized by current anatomical pathology techniques in the clinical laboratory.

Mass spectrometry (MS) technologies continue to grow and provide central capabilities in the clinical laboratory because of their high molecular specificity, rapid analysis time, and cost-effectiveness in high-volume analyses. Liquid chromatography tandem mass spectrometry (LC-MS/MS) is now a standard analytical technology in high-volume clinical laboratories used for the analysis of body fluids (eg, serum, plasma, and urine), vitamins, biomarkers of endocrine function, therapeutic drugs and drugs of abuse, and toxicants [1–3]. Such assays are performed hundreds of thousands of times daily in these laboratories. Although MS is excellent for the analysis of analytes in solution, to the contrary, it does not currently provide a viable capability in the clinical setting for anatomic pathology where the correlation of molecular data with tissue morphology is a major consideration.

Advanced laser-based imaging MS (IMS) technology has been developed for the rapid analysis of tissues and has matured to the point where the molecular specificity of IMS can now be translated to the anatomic pathology laboratory. A major strength of IMS lies in the unique combination of the molecular specificity and sensitivity of MS with the spatial information inherent in images of tissue sections obtained from various forms of microscopy. It is a powerful tool to elucidate the molecular phenotype of the tissue. Further, it does not require a surrogate marker such as an antibody to detect the protein of interest but rather measures the molecules of interest in their native state from the tissue section. Moreover, MS can scan the entire mass range facilitating

© 2017 Elsevier Inc. All rights reserved.

the multiplex analysis of many analytes, such as those formulated for signatures of disease. In all, the unique performance capabilities of imaging by MS are well matched to the current and future needs of anatomic pathology, providing a molecular platform for the diagnosis, prognosis, and assessment of treatment modalities for patients.

The early work using MS for imaging purposes in the basic science laboratory goes back several decades [4,5] and since then, many different MS technologies, especially ionization methods, have been implemented to perform imaging experiments [6]. The common feature of imaging ion sources is the capability to directly irradiate the sample using a high energy beam of photons, atoms, neutral molecules, or fine solvent droplets with sufficient energy to ablate material from the sample. This ablated material is ionized and then introduced into the analyzer of the mass spectrometer. Over recent years, several detailed reviews have been published that describe these technologies from both a historical and technological perspective, and the reader is directed to these works for a detailed discussion of the relative advantages of each approach [7–14].

This chapter will focus on matrix-assisted laser desorption/ionization (MALDI) IMS for imaging tissue specimens [4] and will review its current and potential use in anatomic pathology. For simplicity, MALDI IMS will be termed IMS from this point forward. This platform has emerged as the most widely applicable of the IMS technologies from the viewpoint of its use for the analysis of biological materials, especially tissue sections. IMS can be applied to a wide range of different types of biomolecules such as proteins [15], peptides

[16], lipids [17–23], metabolites [24,25], drugs [26–32], and even biologically relevant metals [33–36]. The modern IMS instrument has great speed because it employs lasers having repetition rates of up to 10 kHz coupled with high-speed analyzers such as time-of-flight (TOF) analyzers. These instruments have high sensitivity and mass resolution and also high molecular weight (MW) capabilities. The lasers used are typically UV lasers with wavelengths of approximately 350 nm. In most commercial instruments, the laser spot size can be focused to about 20–30 μm, ultimately providing image spatial resolutions at this level [9]. Some MS instruments modified or built in special research laboratories have been reported to be capable of delivering 1–10 μm laser spot sizes [37–43].

IMAGING MASS SPECTROMETRY

IMS technology provides spatial information about specific regions of tissue specimens using one of two distinct but related acquisition modes: imaging and profiling/histology-directed acquisition (Fig. 40.1). The particular mode that is used will depend primarily on the overall goal of the experiment, but also somewhat on the specific instruments and accessories at hand. Imaging acquisition is accomplished by ablating and analyzing material from a uniform matrix coating or an ordered array of spots across the tissue surface, with each spot representing a pixel in the final images generated. Each pixel is associated with a mass spectrum containing thousands of m/z values and their individual intensities. Specific molecular images are produced

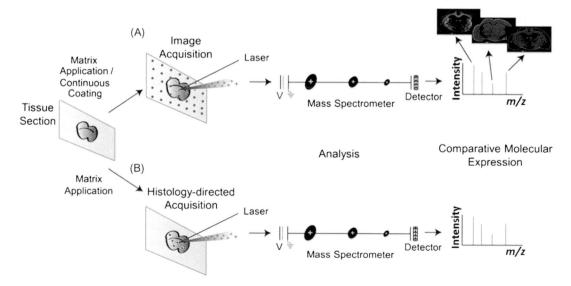

FIGURE 40.1 Sample preparation and data acquisition workflow for MALDI IMS. (A) In a typical IMS workflow, matrix is deposited, either by spray coating or spotting, and data are acquired over the entire tissue section. (B) In a histology-directed workflow, matrix is applied in specific regions of interest annotated by a trained pathologist.

by plotting the intensity of each of the measured peaks over the entire array. Many hundreds of molecular images can be produced from a single raster of the tissue. On the other hand, profiling or histology-directed profiling/acquisition provides similar molecular information from specific spots or regions on the tissue as determined by the investigator from other imaging modalities, such as microscopy. In this way, specific molecular information is spatially placed in the context of a histological image. Often, full imaging and profiling are frequently used together. For example, a few representative samples may be fully imaged at high spatial resolution to complement and validate a larger profiling experiment, which may be performed on hundreds of patient specimens.

Sample preparation protocols for full imaging and histology-directed profiling of a tissue section are similar and can be applied to frozen sections as well as formalin-fixed paraffin-embedded (FFPE) sections. Fresh frozen tissue samples are sectioned on a cryostat to a thickness of approximately 5–20 µm. For protein analyses, the tissues are washed to remove lipids and salts (which interfere with matrix crystallization and may cause ionization suppression) and to dehydrate and fix the proteins while maintaining tissue architecture. This is typically achieved by successive washes in graded ethanol (generally 70%, 90%, 95% for 30 s each) [44–46] or other solvents [47]. Serial sections are often obtained, where one is stained with hematoxylin and eosin (H&E) and used to either guide deposition of matrix on the other section (eg, in a histology-directed fashion) or used to register a full MS image with its histological counterpart. For the analysis of FFPE tissue sections, the overall procedure is much the same except that the last step requires trypsin digestion with the application of a homogeneous spray coating of the enzyme [48] or by individual spots on the tissue [49,50]. This digestion step is needed to release peptides from the protein cross-links induced by formalin. The tryptic peptides released from the sample are subsequently analyzed and identified by MS.

For the MALDI process to occur efficiently and reproducibly over the entire section, the application of matrix must be performed carefully to maintain uniformity and to preserve spatial localization of the endogenous analytes. Typical MALDI matrices are applied in solution to tissues using protocols optimized for analyte extraction, minimal analyte migration, and efficient matrix crystal formation. Sinapinic acid is the matrix primarily used for protein analysis, while α-cyano-4-hydroxycinnamic acid is commonly used for peptides and 2,5-dihydroxybenzoic acid is used for lipids and small molecules. These are typical matrices and many others have been reported, each having some specific advantages for a given kind of analysis [45,51]. Matrix is commonly applied robotically either as individual spots or as a homogeneous spray. Manual application of approximately 0.25–1 µL of matrix solution by pipette also produces excellent MS signals and is often used as a quick check to test and validate instrument performance prior to data acquisition.

For profiling experiments, small accurately placed spots are desired and this can be accomplished through the use of automated robotic spotters [52]. Several commercial instruments are available that are capable of depositing pL volumes of solutions at specified coordinates on the tissue. These result in matrix spots of approximately 100–200 µm on tissue, with multiple passes required for adequate analyte extraction and crystal formation. Placement of individual spots has the advantage of minimizing analyte migration to the diameter of the matrix spot, but these usually are relatively large and cover many cells.

For imaging, obtaining a homogeneous coating of matrix on the tissue section is ideal and is accomplished by spraying the matrix solution over the tissue section using robotic spray devices, electrospray, airbrush, or glass reagent sprayers. The resultant spray forms a thin layer of small crystals when dry. Several commercial instruments are now marketed for this purpose. The best results are obtained, either manually or robotically, when the tissue is sufficiently wetted to allow efficient extraction of analyte and where small crystals are uniformly deposited over the surface of the tissue. Optimal matrix application often takes multiple cycles of spraying and drying to allow a crystal layer to slowly build up on the tissue [45]. In this mode, the primary factor limiting the maximum achievable spatial resolution is the diameter of the laser beam on target, which is approximately 20–80 µm in diameter using the default settings of commercially available instruments.

Data acquisition is performed similarly for profiling and imaging. Most mass spectrometer manufacturers provide software to facilitate experimental design and image acquisition. Precision stepper motors under software control move the sample stage under the laser focus position, and multiple shots are fired. UV lasers are commonly used (N_2 at 335 nm or Nd:YAG at 355 nm) and are capable of firing at repetition rates of up to 10 kHz. Typically approximately 50–400 laser shots are summed for a single average spectrum at each given location. Tissue protein analyses are commonly performed on TOF or TOF/TOF instruments, although commercial hybrid quadrupole TOF and Fourier transform ion cyclotron resonance (FTICR) instruments can also be used. In these experiments, mass spectra range from approximately 2000–50,000 Da. Higher MW proteins over 200 kDa have been measured from tissue [53], although the analysis is not routine and requires optimized instrumental parameters and typically additional acquisition time. The mass accuracy in the protein mass range is approximately 10 ppm for state-of-the-art TOF instruments.

After acquisition, these data are processed, by baseline subtraction, noise reduction, and normalization (eg, total ion current). For profiling features of interest, detected peak areas that meet certain threshold criteria (eg, minimal S/N and prevalence in a certain percentage of sampled spectra) are exported for biostatistical analysis. To accomplish this, packaged software from some MS vendors and third parties are readily available. For example, ClinPro Tools (Bruker Daltonics) enables spectral preprocessing as well as statistical evaluation (eg, average values, standard deviation, and t-test) and classification (eg, via hierarchical clustering, genetic algorithm, and support vector machine) directly from the generated spectra. Image processing may be performed on instrument-associated software or exported into freely available third-party software such as BioMap [54].

Signals, or peaks recorded in a mass spectrum, can be identified in one of several ways. For low MW compounds, up to 4–5 kDa, direct identification can be accomplished using MS/MS-based IMS on the same tissue section. For higher MW proteins (greater than ~5 kDa), more conventional biochemical techniques are commonly used to isolate and identify proteins that correspond to specific m/z signals resulting from histology-directed protein profiling studies. In one approach, tissue samples are suspended in lysis buffer with protease inhibitors, homogenized, and prepared for high-performance liquid chromatography (HPLC) separation [55]. Individual fractions are analyzed by MALDI MS and fractions containing the protein of interest are further separated using one dimension SDS-PAGE [56]. After staining, bands containing the MW region with the m/z signals observed in the MALDI spectra are excised from the gel, reduced, and alkylated. Trypsin is added and the samples are digested. Peptides are extracted from the gel and analyzed by LC-MS/MS. Tandem mass spectra from the LC-MS/MS analyses are then searched against a protein database using commercial software. These data are filtered to a false discovery rate of 5% with a requirement of two or more peptide identifications per protein to obtain minimally acceptable confidence in the identification. Recently, a method was described for top–down protein identification by directly sampling a tissue section via microextraction. This method was able to identify approximately 50–100 proteins from a 1 μL extraction with protein MWs in the range of approximately 5–20 kDa [57]. Another recently published approach to identify the ions observed during IMS experiments introduced the use of trypsin-containing hydrogels to digest and extract proteins directly from small regions (1–4 mm) of tissue [58,59]. The on-tissue hydrogel-mediated digestion is performed subsequent to imaging and provides a similar number of protein identifications as traditional in-solution digests. This technique is able to identify high MW proteins (100–500 kDa) directly from tissue and retains information about spatial localization.

APPLICATIONS TO ANATOMIC PATHOLOGY

Histology-directed molecular analysis is most often used when applying IMS to the analysis of clinically important anatomic pathology investigations. In this workflow, a skilled pathologist would inspect microscopic images of thin tissue sections, suitably stained, and fixed, to visually identify changes in cell morphology and to locate regions indicative of disease. Presently, characterization of disease subtypes often relies on techniques such as IHC, which employs specific tagged antibodies to visualize protein biomarkers at the cellular and subcellular levels. However, IHC requires a suitable antibody directed toward a single antigen in order to be used effectively. Since IMS measures the disease markers directly and can accurately distinguish modified proteins and protein isoforms [60], it is becoming a major technology for the molecular characterization of disease. IMS is able to target small clusters of approximately six to seven cells, as guided by a pathologist's input, and some research instruments are capable of profiling a single cell. This requires a matrix spot diameter of about 50 μm for small cellular clusters and about 10–20 μm for a single cell. By selectively targeting cells in a histology-directed manner based on foreknowledge of putative biomarkers or specific disease traits, the sensitivity and specificity of IMS will increase by minimizing contributions from more numerous unrelated cell types [61]. This task remains a challenge for current commercial MALDI imaging/profiling platforms but promises to be addressed in next-generation instruments to appear in the near future. In its current state, IMS continues to be a powerful tool that advances the understanding of and treatment for significant clinical problems. As examples, this review will summarize the application of IMS to diabetes and cancer.

Diabetes

It is critical to advance our understanding of diabetes in a manner that leads to improved treatment and outcomes. Diabetes is a long-term condition that ultimately leads to a host of complications such as nephropathy, retinopathy, atherosclerosis, etc. Among this spectrum of complications, diabetic nephropathy (DN) is a serious condition that can progress to more acute renal disease affecting approximately one in three patients with diabetes.

Recently, Grove et al. examined lipid distributions using IMS in the glomeruli and tubules of eNOS$^{-/-}$ C57BLKS *db/db* mice, a model for type 2 DN [19]. Mouse kidney sections were imaged in a histology-directed manner at 10 µm spatial resolution in negative ion mode using an *m/z* range of 400−1500. Differences in lipid distributions between control and diabetic mice were observed for four lipid classes: gangliosides, sulfoglycosphingolipids, lysophospholipids,

and phosphatidylethanolamines (PEs). Within the first class, two species had apparent differences— *N*-acetylneuraminic acid (NeuAc)-monosialodihexosy-lganglioside (GM3) and *N*-glycolylneuraminic acid (NeuGc)-GM3, the hydroxylated form of NueAc-GM3. Both species were specific to the glomeruli and NeuAc-GM3 showed similar levels in diabetic, control, and diabetic mice treated with pyridoxamine, a drug that has been shown to slow the progression of early-stage DN. However, NueGc-GM3 was significantly higher in diabetic mice compared to control mice and mice treated with pyridoxamine (Fig. 40.2). Disease-specific expression of lysophospholipids, specifically lysophosphatidylcholine and lysophosphatidic acid

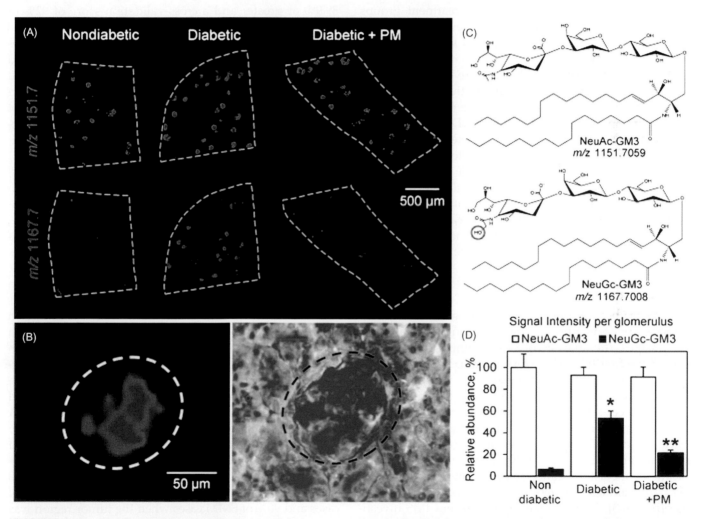

FIGURE 40.2   Gangliosides NeuAc-GM3 and NeuGc-GM3 show distinct changes in diabetic glomeruli. (A) MALDI TOF IMS ion images of *m/z* 1151.7 (NeuAc-GM3) and *m/z* 1167.7 (NeuGc-GM3) in kidneys from nondiabetic control mice, diabetic mice, and diabetic mice treated with pyridoxamine. MALDI IMS was performed at 10 µm spatial resolution and compared with PAS staining of the same section to confirm localization to glomeruli. (B) IMS of the signal at *m/z* 1167.7 and corresponding PAS staining showing the specific localization of NeuGc-GM3 to glomerulus. (C) Structures of gangliosides corresponding to the signals at *m/z* 1151.7 and *m/z* 1167.7 as identified using FTICR MS. The bar graph (D) represents mean ± SEM for three biological replicates per group analyzing 200 glomeruli total. The average signal per glomerulus was determined in ImageJ and data were normalized to nondiabetic NeuAc-GM3. *$P < 0.05$, diabetic versus nondiabetic groups; **$P < 0.05$, diabetic versus diabetic + PM groups. *PAS*, periodic acid-Schiff. *Source: Reprinted with permission from Grove KJ, Voziyan PA, Spraggins JM, et al. Diabetic nephropathy induces alterations in the glomerular and tubule lipid profiles. J Lipid Res 2014;55:1375−85.*

species, was also observed in the glomeruli. These lipids were significantly more abundant in diabetic mice compared to control and pyridoxamine-treated mice. Several sulfoglycolipids were also analyzed and found localized to the tubules. These included sulfogalactoceramide (SM4s), sulfolactoceramide (SM3), gangliotriosylceramide sulfate (SM2a), and gangliotetraosylceramide-*bis*-sulfate (SB1a). SB1a was increased in diabetic mice compared to control mice and diabetic mice treated with pyridoxamine. SM3 and SM4 did not show differences in abundance among the three groups. However, within the tubules these species localized to histologically discrete regions.

PE species unique to the diabetic condition were also detected from these mouse kidney sections. Unmodified PE species did not change among treatment groups. However, glucose-modified PE species were observed in diabetic mouse kidneys, but not in kidneys from the nondiabetic group.

In this study, IMS provided a more complete understanding of the molecular changes that occur in DN and showed where in the tissue these molecular events occur, providing unique insight into DN pathogenesis. IMS is a powerful tool that can characterize the molecular signature of DN, and other diabetes complications, on a large scale (ie, thousands of molecules per experiment) in defined structural and cellular regions, thus reducing confounding and nonspecific signals. This technology offers the potential to reveal key molecular changes that lead to disease progression and to identify important targets for successful treatment.

Cancer

IMS has been applied to a number of cancer investigations, covering a broad range of tissue types and disease stages, the analysis of human glioblastoma being one of the first [62]. This work revealed the power and potential of IMS to characterize and identify disease states, predict outcomes, and advance the understanding of cancer-specific processes [63]. Since then, IMS has been applied to study: (1) the status of human epidermal growth factor receptor 2 in cancer tissue [64,65], (2) bladder cancer [66], (3) the identification of disease origin in metastatic cancers, (4) Wilms tumor [67,68], (5) pancreatic cancer [69], (6) lung cancer [70], as well as others [63]. Here we summarize recent IMS investigations and show how this technology helps drive the understanding of disease at the molecular level.

Skin Cancer. According to the American Cancer Society, in the United States skin cancer has more diagnoses than any other cancer [71]. Two categories of skin cancer include: (1) nonmelanoma skin cancer, which is frequently diagnosed and cured, and (2) melanoma representing less than 2% of the diagnoses, but a disproportionate number of deaths. Approximately 73,000 new melanoma cases will be reported in 2015 [71]. At an early stage, melanoma treatment usually results in a cure. However, as the cancer progresses to stage III the prognosis is variable, with a 5-year survival outcome falling between 24% and 70% [72].

Recently, histology-directed IMS was used to elucidate molecular markers that could subclassify patients with stage III melanoma into groups likely to experience recurrence or survival [72]. In the training set, several proteins were detected at higher levels in the cancerous tissues compared to control lymph node. These proteins were used to establish a classification system for objectively grading stage III melanoma. Four models were developed, each of which had a recognition capability of 89.9% or greater and a cross-validation score of 92% or higher. Seven proteins were correlated with survival and two were associated with recurrence. The intensities of these proteins were used to generate a compound predictor score, which classified a patient with stage III melanoma as poor or favorable for recurrence and survival (Fig. 40.3). This example illustrates the power of IMS to characterize molecular expression in various specific tissue types and provide a diagnostic tool to clinicians for improved patient care.

Laser ablation inductively coupled plasma mass spectrometry was used to study melanoma in human lymph nodes [73]. The study found that ^{31}P readily distinguished the tumor from surrounding lymph tissue, where higher levels of ^{31}P indicated nontumor regions and lower levels of ^{31}P marked areas of tumor presence. Development of this method may provide a diagnostic tool and a means of sampling for metastatic melanoma without surgical removal of biopsies [73].

Histology-directed IMS has been used to distinguish Spitz nevi (SN)—benign melanocytic lesions—from Spitzoid malignant melanomas (SMMs), which has been a longstanding significant challenge in the field of dermatopathology [74]. While there are recognized and verified histopathology standards for distinguishing SN and SMM, these criteria are unable to classify atypical SN. A training set of 26 SN and 25 SMM biopsies was used to establish classification models for both tumor tissue and dermis surrounding the tumor (ie, the tumor microenvironment). The models were applied to a test group of 30 SN and 33 SMM and accurately classified 97% of SN cases and 90% of SMM cases when the tumor region was analyzed. When the surrounding dermis was assessed, the models properly identified 90% of the SMM cases but only 64% of the SN cases. The success of the model for dermis in SMM tissues may be due to higher levels of secreted molecules (eg, cytokines) in the tumor microenvironment compared to dermis in SN tissues [74].

In a case study, a pregnant woman was diagnosed with a malignant melanoma lesion on her arm [75].

(A) Histology directed matrix application

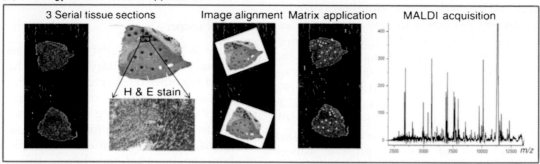

(B) Spectral processing and clinical application

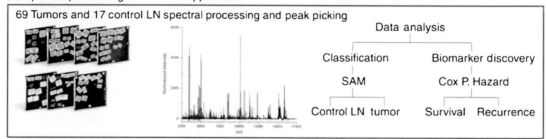

FIGURE 40.3 Evaluation of melanoma by histology-directed IMS. (A) Histology-directed MALDI IMS. Cellular regions are selected by a pathologist to ensure melanoma foci are targeted. (B) The MALDI spectra from these regions were averaged and compared by SAM (control LN vs tumor-positive LN) and by CPH to determine proteins associated with survival and recurrence. *SAM*, significance-analysis-of-microarray; *LN*, lymph node; *CPH*, Cox proportional hazard. *Source: Reprinted with permission from Hardesty WM, Kelley MC, Mi D, Low RL, Caprioli RM. Protein signatures for survival and recurrence in metastatic melanoma. J Proteomics 2011;74:1002−14.*

After delivery of the child, it was apparent that he had melanocytic lesions on his torso, but histological analysis was inconclusive in determining if the lesions were malignant or benign. IMS analysis classified the mother's lesion as malignant and the baby's lesions as benign. Furthermore, genetic analysis of the baby's lesions showed the presence of a Y chromosome, indicating the lesion was indeed a congenital nevi and not a metastasis from the mother. In this case, it is established that the use of IMS technologies on clinical samples has the potential to achieve the ultimate goal of improved patient care and quality of life due to the avoidance of unnecessary surgical procedures.

Cancer Margins. In many cases when cancer is diagnosed, the treatment strategy includes surgical resection, and the extent to which the tumor can be removed influences chances of survival or recurrence [76,77]. It has become evident that the current established methods for assessing tumor margins during surgery do not lead to complete excision of the malignancy in all cases [77,78]. Traditionally, tumor margins are determined visually and tactually during surgery, but intraoperative optical imaging methods are becoming more prominent to visualize the molecular tumor margin, allowing surgeons to more accurately assess tumor boundaries [77,78]. Research that continues to

characterize and define tumor molecular margins will lead to improved markers and outcomes by reducing disease recurrence.

IMS has been used to study the molecular tumor margins of clear-cell renal cell carcinoma (ccRCC) [76]. The results show that histologically normal cells outside of the visual tumor margin are molecularly abnormal, in contrast to the conclusion rendered by morphological analysis alone. As an example, proteins of the electron transport system showed lower abundance in ccRCC tumors compared to normal tissue. Tissue outside of the histologically determined margin also exhibited this lower expression pattern similar to that observed in ccRCC tumors. This effect is postulated to be due to secretions from adjacent tumor cells, infiltration of normal tissue by microfoci of tumor cells, and/or to malignant transformation of cells in the tumor microenvironment to some degree, but not to the extent that the cell is visually affected [76]. This exemplifies the necessity of molecular analyses to help define cancer margins. IMS is an effective tool for such analyses as it can detect thousands of molecules specific to tissue regions with 5−10 μm spatial resolution, providing molecular signatures to discriminate cancer from healthy tissue [76].

Prostate Cancer. Prostate cancer is projected to be the second leading cause of cancer in the male population in 2015 [71]. Chances of survival are nearly 100% for local and regional cases. However, survival drops to 28% among patients with distant metastases. Once prostate cancer is diagnosed, the ability to determine disease aggressiveness is challenging due to the heterogeneity of the disease at each stage [79]. Various imaging modalities, including MALDI IMS [79], have been developed to improve tumor classification and identify biomarkers for prostate cancer [80]. Summarized below are IMS studies that have identified distinguishing features of several prostate cancer stages.

In the first study applying IMS to human prostate tissue, cancerous and noncancerous sections were assessed to determine a proteomic pattern for tissue classification [81]. Proteins were imaged between 1 and 20 kDa, and many were found to display distinct differences in expression between cancerous and noncancerous regions. A classification model was established that distinguished between these two regions with cross-validation, sensitivity, and specificity scores of 88%, 85%, and 91%, respectively. This work did not identify the differentially expressed proteins, but provided a first step and proof-of-concept for the utility of IMS in classifying tissue samples not readily distinguished by classic histology techniques [81]. A similar study was conducted by investigators who identified and confirmed through IHC the ability of a mitogen-activated protein kinase/extracellular signal regulated kinase 2 fragment to distinguish between cancerous and noncancerous prostate tissues [82].

To optimize the distinction of prostate cancer from healthy tissue, investigators combined two techniques, IMS and texture analysis [83]. For the texture analysis, tissue sections were scanned with a high-magnification digital microscope and the textures were assessed. The analysis examined 13 features that discriminated between healthy tissue and prostate cancer regions: 11 gray-level run length matrix features (eg, short and long run emphasis, high and low gray-level run emphasis, and run-length nonuniformity), average pixel value, and variance of the pixel values. IMS profiled peaks from 2 to 45 kDa. Features (ie, texture features and peaks) were selected that best classified the tissue regions into noncancerous or cancerous. The two datasets were computationally combined to optimize a classification model utilizing both techniques. Representative results from one of three experiments showed that the texture analysis alone had a sensitivity and specificity of 87% and 75%, respectively, while the MALDI analysis resulted in 51% and 100% sensitivity and specificity, respectively. When combined, the sensitivity improved compared to the MALDI (80%) and the specificity improved compared to texture analysis (93%) [83].

COMPUTATIONAL APPROACHES IN IMS

Image Fusion

Many IMS experiments utilize or can benefit from additional, complementary images that have been acquired via other modalities, such as microscopy. At a minimum, these ancillary images provide guidance for histology-directed imaging or can be overlaid with ion images to provide an anatomical reference for ion distributions. Beyond this, these complementary sources of data can be used together to provide additional insight by computationally combining the results. This approach, called image fusion, was recently published and the discussion herein comes from that work [84].

When a tissue section is imaged using separate modalities, there is usually an observable set of common features in the patterns acquired. These correlating spatial patterns enable recognition of the same anatomical region within different modalities, even if they are acquired at different spatial resolutions. When these patterns are analyzed and combined via an algorithm, a wealth of new information can be gained.

The field of image fusion captures these cross-correlations in a mathematical model [85,86] in order to relate the patterns observed in one imaging modality (eg, IMS) to correlating measurements in the other modality (eg, microscopy). These correlations facilitate several predictive processes in IMS, such as sharpening ion images to higher spatial resolutions, predicting ion distributions in regions that were never analyzed via IMS but only by another imaging modality, and enhancing biological signals while minimizing artifacts. All of these applications are examples of a new multi-modality concept for tissue exploration whereby mining relationships between different imaging techniques yield a novel imaging modality that combines and surpasses what can be gleaned from the source technologies alone [84].

The concept of sharpening IMS images was inspired by developments in remote sensing and satellite imaging [87]. In the latter case, the concept of pan-sharpening was introduced, where a high-resolution pan-chromatic (gray-level) image and a low-resolution color image were merged into a single high-resolution color image. Such a sharpening procedure provides a mathematically verifiable way of estimating higher spatial resolution versions of IMS datasets and can surpass the physical spatial resolution capabilities of the instrument. However, given the difference in measurement principles and sources of distortion, the algorithms and implementations from the satellite-imaging field cannot be transferred readily to IMS. A custom sharpening algorithm was developed specifically for MS data to facilitate the transformation of an ion image acquired at a set spatial resolution (eg, 100 μm) into a higher resolution estimate (eg, 5 μm) of that same ion image using the information gleaned from an accompanying tissue image, such as a standard H&E stain or a specialized immunohistological stain (Fig. 40.4) [84].

The image fusion workflow can be separated into three broad steps. The first step is to build a data set of locations for which data from both modalities have been recorded. The result is a database of microscopy locations for which both RGB and IMS data are available, and also a list of microscopy locations for which the IMS variables can be estimated since they were not physically measured. The second step is to build a cross-modality model that mathematically links variables from one modality (eg, m/z bins or peak m/z ratios in IMS) to variables in the other modality (eg, red, green, and blue in microscopy). Once a database of IMS-to-microscopy instances has been constructed, building a cross-modality model becomes a task of multivariate regression, a well-documented process in statistics [85]. The general linear model, and subsequently the partial least squares regression were used. In preliminary experiments, these models showed impressive performance and were able to link most of the patterns that showed high contrast in both modalities. Once a model is built it can be treated as a predictor, using microscopy values as inputs and producing estimated IMS values for missing pixels in MS images. The model can then be used to predict ion intensity for various m/z ratios at those pixels for which only microscopy measurements are available. The result is an estimated version of an ion image at a higher spatial resolution. The target resolution can be put anywhere between the IMS and microscopy resolutions, as dictated by the biological application and with the knowledge that a lower upsampling factor means higher reliability [84].

In addition to sharpening the fusion process can also provide additional ways to utilize IMS data. These applications include the capability to predict molecular distributions in tissue regions not measured by IMS but only by microscopy and to enhance molecular discovery through multimodal enrichment and denoise IMS data. A fusion-driven separation of measurements into modality-specific or cross-modality supported variation provides crucial information toward increasing instrumental sensitivity, which is not available in single-technology analysis.

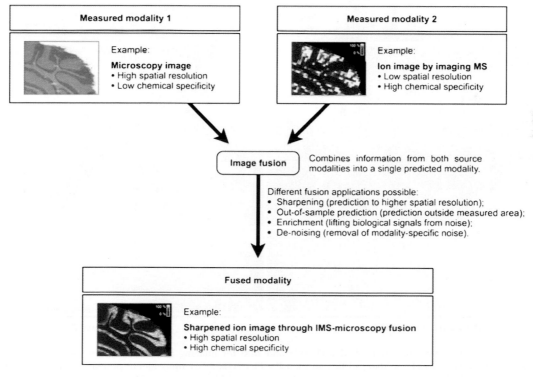

FIGURE 40.4 Concept of image fusion of IMS and microscopy. Image fusion generates a single image from two or more source images, combining the advantages of the different sensor types. The integration of IMS and optical microscopy is given as an example. The IMS—microscopy fusion image is a predictive modality that delivers both the chemical specificity of IMS and the spatial resolution of microscopy in one integrated whole. Each source image measures a different aspect of the content of a tissue sample. The fused image predicts the tissue content as if all aspects were observed concurrently. *Source: Reprinted with permission from Van de Plas R, Yang J, Spraggins J, Caprioli RM. Image fusion of mass spectrometry and microscopy: a multimodality paradigm for molecular tissue mapping. Nat Methods 2015;12:366—72.*

Three-Dimensional Imaging

Recently, images acquired via various MS techniques have been reconstructed into three-dimensional (3D) molecular ion images [88–91]. In a few cases, the resulting 3D images were coregistered with images acquired through magnetic resonance imaging (MRI) to produce overlays for comparing specific molecular information with anatomical features [92–94]. Examples include imaging whole heads from mice containing brain tumors [94], analysis of systemic *Staphylococcus aureus* infection in mice and the effect of linezolid treatment (Fig. 40.5) [92], and imaging of mouse kidney [93]. These data provide valuable insight into clinically significant issues [92,94]. Through the application of image fusion, the analysis of MR and MALDI images can be further refined to include such processes as discussed above—sharpening, denoising and enrichment of specific signals, and even prediction of molecular signals in regions not imaged by MALDI. While these studies are preliminary, they promise a remarkable and informative view of the molecular composition within specific anatomical regions and pathological processes.

CONCLUSIONS

The ability to measure and provide insight into the molecular basis of disease has advanced remarkably in the previous decade. Where classical macroscopic, microscopic, and biochemical investigations are limited, genomic, proteomic, and metabolomic studies can

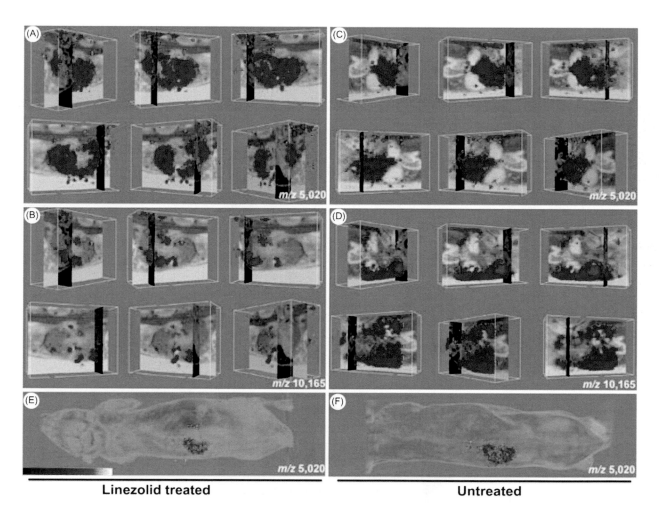

Linezolid treated   **Untreated**

FIGURE 40.5 Three-dimensional integration of MALDI IMS and MRI for imaging the inflammatory response to infection. (A and B) Orthogonal blockface and MRI slice data of linezolid-treated mouse with overlaid (A) alpha-globin protein density (*m/z* 5020) and (B) calgranulin A protein density (*m/z* 10,165) volume renderings. (C and D) Orthogonal blockface and MRI slice data of untreated mouse with overlaid (C) alpha-globin protein density (*m/z* 5020) and (D) calgranulin A protein density (*m/z* 10,165) volume renderings. (E and F) Protein density (*m/z* 5020) from (E) linezolid-treated and (F) untreated mice superimposed on whole mouse image. The data in all panels are presented as arbitrary units of intensity from 0 (dark red) to 1 (white). *Source: Reprinted with permission from Attia AS, Schroeder KA, Seeley EH, et al. Monitoring the inflammatory response to infection through the integration of MALDI IMS and MRI. Cell Host Microbe 2012;11:664–73.*

inform biologists and clinicians of disease diagnosis, prognosis, and treatment. However, despite these advancements there remain numerous diseases and conditions that elude a molecularly specific and confident diagnosis, that could have a better prognosis if the disease were detected at an earlier stage, or for which individual treatment options would vastly improve outcomes. As reviewed here, IMS plays an important role in advancing our knowledge of disease at the molecular level. It can determine and monitor molecular markers that characterize disease stage or disease type where histology is inconclusive, leading to more accurate diagnoses. Further this technology can reveal previously unknown molecular changes associated with a disease stage, providing more reliable prognoses and can aid in the elucidation of disease pathogenesis, leading to effective therapeutics. The ultimate goal is for the use of IMS technology to improve clinical care and patient outcomes.

While IMS technology has developed greatly over the last decade, it is still relatively new and not yet a standard analysis technique in the clinical laboratory. Current technological developments in progress aim to facilitate the integration of IMS as an everyday tool for clinical pathologists. For example, the development of software to automate the process of transferring tissue annotations to regions of interest and registering annotated optical images with the MALDI image is forthcoming. Precoated slides have also been produced [95–98] to improve throughput and reproducibility. Furthermore spatial resolution is improving, with the capability to perform analyses at 5 μm [38,42,43] and even 1–2 μm spatial resolution [39–41]. This provides enhanced molecular specificity for tissues with detailed anatomical structure and in some cases provides cell-specific resolution. Lastly, the combination of IMS data and other imaging outputs through image fusion [84] provides a powerful technique that will be of certain value in the clinical setting. Where MRI or histology can be used to assess morphological and pathological changes, IMS can provide highly specific molecular information.

IMS technology is well suited for clinical applications. It has already proven to aid in better patient care [75]. Future applications should only increase the power of this technique to bring spatially distinct molecular specificity that can inform diagnosis, prognosis, and treatment.

References

[1] Rauh M. LC-MS/MS for protein and peptide quantification in clinical chemistry. J Chromatogr B Analyt Technol Biomed Life Sci 2012;883–884:59–67.

[2] Vogeser M, Seger C. A decade of HPLC-MS/MS in the routine clinical laboratory—goals for further developments. Clin Biochem 2008;41:649–62.

[3] van den Ouweland JM, Kema IP. The role of liquid chromatography-tandem mass spectrometry in the clinical laboratory. J Chromatogr B Analyt Technol Biomed Life Sci 2012;883-884:18–32.

[4] Caprioli RM, Farmer TB, Gile J. Molecular imaging of biological samples: localization of peptides and proteins using MALDI-TOF MS. Anal Chem 1997;69:4751–60.

[5] Galle P. Tissue localization of stable and radioactive nuclides by secondary-ion microscopy. J Nucl Med 1982;23:52–7.

[6] Bhardwaj C, Hanley L. Ion sources for mass spectrometric identification and imaging of molecular species. Nat Prod Rep 2014;31:756–67.

[7] Amstalden van Hove ER, Smith DF, Heeren RM. A concise review of mass spectrometry imaging. J Chromatogr A 2010;1217:3946–54.

[8] Caldwell RL, Caprioli RM. Tissue profiling by mass spectrometry: a review of methodology and applications. Mol Cell Proteomics 2005;4:394–401.

[9] Chughtai K, Heeren RMA. Mass spectrometric imaging for biomedical tissue analysis. Chem Rev 2010;110:3237–77.

[10] Cornett DS, Reyzer ML, Chaurand P, Caprioli RM. MALDI imaging mass spectrometry: molecular snapshots of biochemical systems. Nat Methods 2007;4:828–33.

[11] McDonnell LA, Heeren RMA. Imaging mass spectrometry. Mass Spectrom Rev 2007;26:606–43.

[12] Pacholski ML, Winograd N. Imaging with mass spectrometry. Chem Rev 1999;99:2977–3005.

[13] Schwamborn K, Caprioli RM. Molecular imaging by mass spectrometry—looking beyond classical histology. Nat Rev Cancer 2010;10:639–46.

[14] Walch A, Rauser S, Deininger S-O, Hoefler H. MALDI imaging mass spectrometry for direct tissue analysis: a new frontier for molecular histology. Histochem Cell Biol 2008;130:421–34.

[15] Chaurand P, Norris JL, Cornett DS, Mobley JA, Caprioli RM. New developments in profiling and imaging of proteins from tissue sections by MALDI mass spectrometry. J Proteome Res 2006;5:2889–900.

[16] Groseclose MR, Massion PP, Chaurand P, Caprioli RM. High-throughput proteomic analysis of formalin-fixed paraffin-embedded tissue microarrays using MALDI imaging mass spectrometry. Proteomics 2008;8:3715–24.

[17] Anderson DM, Ablonczy Z, Koutalos Y, et al. High resolution MALDI imaging mass spectrometry of retinal tissue lipids. J Am Soc Mass Spectrom 2014;25:1394–403.

[18] Deeley JM, Hankin JA, Friedrich MG, et al. Sphingolipid distribution changes with age in the human lens. J Lipid Res 2010;51:2753–60.

[19] Grove KJ, Voziyan PA, Spraggins JM, et al. Diabetic nephropathy induces alterations in the glomerular and tubule lipid profiles. J Lipid Res 2014;55:1375–85.

[20] Landgraf RR, Garrett TJ, Conaway MC, Calcutt NA, Stacpoole PW, Yost RA. Considerations for quantification of lipids in nerve tissue using matrix-assisted laser desorption/ionization mass spectrometric imaging. Rapid Commun Mass Spectrom 2011;25:3178–84.

[21] Meriaux C, Franck J, Wisztorski M, Salzet M, Fournier I. Liquid ionic matrixes for MALDI mass spectrometry imaging of lipids. J Proteomics 2010;73:1204–18.

[22] Wang HY, Liu CB, Wu HW. A simple desalting method for direct MALDI mass spectrometry profiling of tissue lipids. J Lipid Res 2011;52:840–9.

[23] Wattacheril J, Seeley EH, Angel P, et al. Differential intrahepatic phospholipid zonation in simple steatosis and nonalcoholic steatohepatitis. PLoS One 2013;8:e57165.

[24] Cha S, Zhang H, Ilarslan HI, et al. Direct profiling and imaging of plant metabolites in intact tissues by using colloidal graphite-assisted laser desorption ionization mass spectrometry. Plant J 2008;55:348−60.

[25] Stoeckli M, Staab D, Schweitzer A. Compound and metabolite distribution measured by MALDI mass spectrometric imaging in whole-body tissue sections. Int J Mass Spectrom 2007;260: 195−202.

[26] Acquadro E, Cabella C, Ghiani S, Miragoli L, Bucci EM, Corpillo D. Matrix-assisted laser desorption ionization imaging mass spectrometry detection of a magnetic resonance imaging contrast agent in mouse liver. Anal Chem 2009;81:2779−84.

[27] Chacon A, Zagol-Ikapitte I, Amarnath V, et al. On-tissue chemical derivatization of 3-methoxysalicylamine for MALDI-imaging mass spectrometry. J Mass Spectrom 2011;46:840−6.

[28] Cornett DS, Frappier SL, Caprioli RM. MALDI-FTICR imaging mass spectrometry of drugs and metabolites in tissue. Anal Chem 2008;80:5648−53.

[29] Fehniger TE, Vegvari A, Rezeli M, et al. Direct dmonstration of tissue uptake of an inhaled drug: proof-of-principle study using matrix-assisted laser desorption ionization mass spectrometry imaging. Anal Chem 2011;83:8329−36.

[30] Greer T, Sturm R, Li L. Mass spectrometry imaging for drugs and metabolites. J Proteomics 2011;74:2617−31.

[31] Reyzer ML, Hsieh Y, Ng K, Korfmacher WA, Caprioli RM. Direct analysis of drug candidates in tissue by matrix-assisted laser desorption/ionization mass spectrometry. J Mass Spectrom 2003;38:1081−92.

[32] Shahidi-Latham SK, Dutta SM, Prieto-Conaway MC, Rudewicz PJ. Evaluation of an accurate mass approach for the simultaneous detection of drug and metabolite distributions via whole-body mass spectrometric imaging. Anal Chem 2012;84:7158−65.

[33] Hood MI, Mortensen BL, Moore JL, et al. Identification of an *Acinetobacter baumannii* zinc acquisition system that facilitates resistance to calprotectin-mediated zinc sequestration. PLoS Pathog 2012;8:e1003068.

[34] Kehl-Fie TE, Zhang Y, Moore JL, et al. MntABC and MntH contribute to systemic *Staphylococcus aureus* infection by competing with calprotectin for nutrient manganese. Infect Immun 2013;81:3395−405.

[35] Lear J, Hare DJ, Fryer F, Adlard PA, Finkelstein DI, Doble PA. High-resolution elemental bioimaging of Ca, Mn, Fe, Co, Cu, and Zn employing LA-ICP-MS and hydrogen reaction gas. Anal Chem 2012;84:6707−14.

[36] Becker JS, Su J, Zoriya MV, Dobrowolska J, Matusch A. Imaging mass spectrometry in biological tissues by laser ablation inductively coupled plasma mass spectrometry. Eur J Mass Spectrom 2007;13:1−6.

[37] Kettling H, Vens-Cappell S, Soltwisch J, et al. MALDI mass spectrometry imaging of bioactive lipids in mouse brain with a Synapt G2-S mass spectrometer operated at elevated pressure: improving the analytical sensitivity and the lateral resolution to ten micrometers. Anal Chem 2014;86:7798−805.

[38] Korte AR, Yandeau-Nelson MD, Nikolau BJ, Lee YJ. Subcellular-level resolution MALDI-MS imaging of maize leaf metabolites by MALDI-linear ion trap-orbitrap mass spectrometer. Anal Bioanal Chem 2015;407:2301−9.

[39] Thiery-Lavenant G, Zavalin AI, Caprioli RM. Targeted multiplex imaging mass spectrometry in transmission geometry for subcellular spatial resolution. J Am Soc Mass Spectrom 2013;24:609−14.

[40] Zavalin A, Todd EM, Rawhouser PD, Yang J, Norris JL, Caprioli RM. Direct imaging of single cells and tissue at subcellular spatial resolution using transmission geometry MALDI MS. J Mass Spectrom 2012;47:1473−81.

[41] Zavalin A, Yang J, Hayden K, Vestal M, Caprioli RM. Tissue protein imaging at 1 μm laser spot diameter for high spatial resolution and high imaging speed using transmission geometry MALDI TOF MS. Anal Bioanal Chem 2015;407:2337−42.

[42] Zavalin A, Yang J, Caprioli R. Laser beam filtration for high spatial resolution MALDI imaging mass spectrometry. J Am Soc Mass Spectrom 2013;24:1153−6.

[43] Zavalin A, Yang J, Haase A, Holle A, Caprioli R. Implementation of a Gaussian beam laser and aspheric optics for high spatial resolution MALDI imaging MS. J Am Soc Mass Spectrom 2014;25:1079−82.

[44] Moore JL, Becker KW, Nicklay JJ, Boyd KL, Skaar EP, Caprioli RM. Imaging mass spectrometry for assessing temporal proteomics: analysis of calprotectin in *Acinetobacter baumannii* pulmonary infection. Proteomics 2014;14:820−8.

[45] Schwartz SA, Reyzer ML, Caprioli RM. Direct tissue analysis using matrix-assisted laser desorption/ionization mass spectrometry: practical aspects of sample preparation. J Mass Spectrom 2003;38:699−708.

[46] Schwamborn K, Caprioli RM. MALDI imaging mass spectrometry—painting molecular pictures. Mol Oncol 2010;4:529−38.

[47] Seeley EH, Oppenheimer SR, Mi D, Chaurand P, Caprioli RM. Enhancement of protein sensitivity for MALDI imaging mass spectrometry after chemical treatment of tissue sections. J Am Soc Mass Spectrom 2008;19:1069−77.

[48] Wenke JL, Schey KL. Microwave-assisted enzymatic digestion on-tissue for membrane protein analysis with MALDI imaging mass spectrometry. In: 61st American Society for mass spectrometry conference on mass spectrometry and allied topics. Minneapolis, MN; 2013.

[49] Aoki Y, Toyama A, Shimada T, et al. A novel method for analyzing formalin-fixed paraffin embedded (FFPE) tissue sections by mass spectrometry imaging. Proc Jpn Acad Ser B Phys Biol Sci 2007;83:205−14.

[50] Casadonte R, Caprioli RM. Proteomic analysis of formalin-fixed paraffin-embedded tissue by MALDI imaging mass spectrometry. Nat Protoc 2011;6:1695−709.

[51] Dreisewerd K. Recent methodological advances in MALDI mass spectrometry. Anal Bioanal Chem 2014;406:2261−78.

[52] Aerni H-R, Cornett DS, Caprioli RM. Automated acoustic matrix deposition for MALDI sample preparation. Anal Chem 2006;78:827−34.

[53] Chaurand P, Caprioli RM. Direct profiling and imaging of peptides and proteins from mammalian cells and tissue sections by mass spectrometry. Electrophoresis 2002; 23:3125−35.

[54] BioMap. MALDI-MSI Interest Group. <http://maldi-msi.org/index.php?option = com_content&task = view&id = 14&Itemid = 39>; 2011 [accessed 27.05.15].

[55] Reyzer ML, Caldwell RL, Dugger TC, et al. Early changes in protein expression detected by mass spectrometry predict tumor response to molecular therapeutics. Cancer Res 2004;64: 9093−100.

[56] Caldwell RL, Gonzalez A, Oppenheimer SR, Schwartz HS, Caprioli RM. Molecular assessment of the tumor protein microenvironment using imaging mass spectrometry. Cancer Genomics and Proteomics 2006;3:279−87.

[57] Schey KL, Anderson DM, Rose KL. Spatially-directed protein identification from tissue sections by top-down LC-MS/MS with electron transfer dissociation. Anal Chem 2013;85: 6767−74.

[58] Harris GA, Nicklay JJ, Caprioli RM. Localized in situ hydrogel-mediated protein digestion and extraction technique for on-tissue analysis. Anal Chem 2013;85:2717−23.

[59] Taverna D, Norris JL, Caprioli RM. Histology-directed microwave assisted enzymatic protein digestion for MALDI MS analysis of mammalian tissue. Anal Chem 2015;87:670−6.

[60] Alomari AK, Klump V, Neumeister V, Ariyan S, Narayan D, Lazova R. Comparison of the expression of vimentin and actin in spitz nevi and spitzoid malignant melanomas. Am J Dermatopathol 2015;37:46−51.

[61] Cornett DS, Mobley JA, Dias EC, et al. A novel histology-directed strategy for MALDI-MS tissue profiling that improves throughput and cellular specificity in human breast cancer. Mol Cell Proteomics 2006;5:1975−83.

[62] Stoeckli M, Chaurand P, Hallahan DE, Caprioli RM. Imaging mass spectrometry: a new technology for the analysis of protein expression in mammalian tissues. Nat Med 2001;7:493−6.

[63] Norris JL, Caprioli RM. Analysis of tissue specimens by matrix-assisted laser desorption/ionization imaging mass spectrometry in biological and clinical research. Chem Rev 2013;113:2309−42.

[64] Balluff B, Elsner M, Kowarsch A, et al. Classification of HER2/neu status in gastric cancer using a breast-cancer derived proteome classifier. J Proteome Res 2010;9:6317−22.

[65] Rauser S, Marquardt C, Balluff B, et al. Classification of HER2 receptor status in breast cancer tissues by MALDI imaging mass spectrometry. J Proteome Res 2010;9:1854−63.

[66] Oezdemir RF, Gaisa NT, Lindemann-Docter K, et al. Proteomic tissue profiling for the improvement of grading of noninvasive papillary urothelial neoplasia. Clin Biochem 2012;45:7−11.

[67] Axt J, Murphy AJ, Seeley EH, et al. Race disparities in Wilms tumor incidence and biology. J Surg Res 2011;170:112−19.

[68] Murphy AJ, Axt JR, de Caestecker C, et al. Molecular characterization of Wilms' tumor from a resource-constrained region of sub-Saharan Africa. Int J Cancer 2012;131:E983−94.

[69] Djidja M-C, Claude E, Snel MF, et al. Novel molecular tumour classification using MALDI-mass spectrometry imaging of tissue micro-array. Anal Bioanal Chem 2010;397:587−601.

[70] Marko-Varga G, Fehniger TE, Rezeli M, Doeme B, Laurell T, Vegvari A. Drug localization in different lung cancer phenotypes by MALDI mass spectrometry imaging. J Proteomics 2011;74:982−92.

[71] Cancer Facts & Figures 2015. American Cancer Society. <http://www.cancer.org/acs/groups/content/@editorial/documents/document/acspc-044552.pdf>; 2015.

[72] Hardesty WM, Kelley MC, Mi D, Low RL, Caprioli RM. Protein signatures for survival and recurrence in metastatic melanoma. J Proteomics 2011;74:1002−14.

[73] Hare D, Burger F, Austin C, et al. Elemental bio-imaging of melanoma in lymph node biopsies. Analyst 2009;134:450−3.

[74] Lazova R, Seeley EH, Keenan M, Gueorguieva R, Caprioli RM. Imaging mass spectrometry—a new and promising method to differentiate Spitz nevi from Spitzoid malignant melanomas. Am J Dermatopathol 2012;34:82−90.

[75] Alomari A, Glusac EJ, Choi J, et al. Congenital melanocytic nevi versus metastatic melanoma in a newborn to a mother with melanoma—diagnosis supported by sex chromosome analysis and imaging mass spectrometry. J Cutan Pathol 2015;42:757−64.

[76] Oppenheimer SR, Mi D, Sanders ME, Caprioli RM. Molecular analysis of tumor margins by MALDI mass spectrometry in renal carcinoma. J Proteome Res 2010;9:2182−90.

[77] Rosenthal EL, Warram JM, Bland KI, Zinn KR. The status of contemporary image-guided modalities in oncologic surgery. Ann Surg 2015;261:46−55.

[78] de Boer E, Harlaar NJ, Taruttis A, et al. Optical innovations in surgery. Br J Surg 2015;102:e56−72.

[79] Spur EM, Decelle EA, Cheng LL. Metabolomic imaging of prostate cancer with magnetic resonance spectroscopy and mass spectrometry. Eur J Nucl Med Mol Imaging 2013;40:S60−71.

[80] Flatley B, Malone P, Cramer R. MALDI mass spectrometry in prostate cancer biomarker discovery. Biochim Biophys Acta 2014;1844:940−9.

[81] Schwamborn K, Krieg RC, Reska M, Jakse G, Knuechel R, Wellmann A. Identifying prostate carcinoma by MALDI-Imaging. Int J Mol Med 2007;20:155−9.

[82] Cazares LH, Troyer D, Mendrinos S, et al. Imaging mass spectrometry of a specific fragment of mitogen-activated protein kinase/extracellular signal-regulated kinase kinase 2 discriminates cancer from uninvolved prostate tissue. Clin Cancer Res 2009;15:5541−51.

[83] Chuang S-H, Li J, Sun X, et al. Prostate cancer region prediction by fusing results from MALDI spectra-processing and texture analysis. Simulation 2012;88:1247−59.

[84] Van de Plas R, Yang J, Spraggins J, Caprioli RM. Image fusion of mass spectrometry and microscopy: a multimodality paradigm for molecular tissue mapping. Nat Methods 2015;12:366−72.

[85] Izenman AJ. Modern multivariate statistical techniques: regression, classification, and manifold learning. Springer texts in statistics. New York, NY: Springer-Verlag; 2008.

[86] Mitchell HB. Image fusion: theories, techniques and applications. Berlin: Springer-Verlag; 2010.

[87] Richards JA, Jia X. Remote sensing digital image analysis: an introduction. 4th ed. Berlin: Springer; 2006.

[88] Andersson M, Groseclose MR, Deutch AY, Caprioli RM. Imaging mass spectrometry of proteins and peptides: 3D volume reconstruction. Nat Methods 2008;5:101−8.

[89] Lanekoff I, Burnum-Johnson K, Thomas M, et al. Three-dimensional imaging of lipids and metabolites in tissues by nanospray desorption electrospray ionization mass spectrometry. Anal Bioanal Chem 2015;407:2063−71.

[90] Seeley EH, Caprioli RM. 3D imaging by mass spectrometry: a new frontier. Anal Chem 2012;84:2105−10.

[91] Trede D, Schiffler S, Becker M, et al. Exploring three-dimensional matrix-assisted laser desorption/ionization imaging mass spectrometry data: three-dimensional spatial segmentation of mouse kidney. Anal Chem 2012;84:6079−87.

[92] Attia AS, Schroeder KA, Seeley EH, et al. Monitoring the inflammatory response to infection through the integration of MALDI IMS and MRI. Cell Host Microbe 2012;11:664−73.

[93] Oetjen J, Aichler M, Trede D, et al. MRI-compatible pipeline for three-dimensional MALDI imaging mass spectrometry using PAXgene fixation. J Proteomics 2013;90:52−60.

[94] Sinha TK, Khatib-Shahidi S, Yankeelov TE, et al. Integrating spatially resolved three-dimensional MALDI IMS with in vivo magnetic resonance imaging. Nat Methods 2008;5:57−9.

[95] Grove KJ, Frappier SL, Caprioli RM. Matrix pre-coated MALDI MS targets for small molecule imaging in tissues. J Am Soc Mass Spectrom 2011;22:192−5.

[96] Manier ML, Spraggins JM, Reyzer ML, Norris JL, Caprioli RM. A derivatization and validation strategy for determining the spatial localization of endogenous amine metabolites in tissues using MALDI imaging mass spectrometry. J Mass Spectrom 2014;49:665−73.

[97] Yang J, Caprioli RM. Matrix precoated targets for direct lipid analysis and imaging of tissue. Anal Chem 2013;85:2907−12.

[98] Manier ML, Reyzer ML, Goh A, et al. Reagent precoated targets for rapid in-tissue derivatization of the anti-tuberculosis drug isoniazid followed by MALDI imaging mass spectrometry. J Am Soc Mass Spectrom 2011;22:1409−19.

41

Whole Genome Sequencing in the Molecular Pathology Laboratory

G.T. Haskell[1] and J.S. Berg[2]

[1]Department of Genetics, Duke University, Durham, NC, United States [2]Department of Genetics, University of North Carolina School of Medicine, Chapel Hill, NC, United States

INTRODUCTION

Genome-scale sequencing technologies have transformed clinical diagnostics and will continue to propel the field of molecular pathology forward over the next several years. Since their demonstrated success in genetic diagnosis a few years ago [1,2], whole exome sequencing (WES) and whole genome sequencing (WGS) have been adopted as a clinical test by several diagnostic laboratories. Genome sequencing is likely to significantly change the way that physicians, laboratories, and genetic counselors handle genetic testing. While this technology has made the opportunity for genetic testing more widely available to patients, it has increased the number of uncertain results. In addition, guidelines for best practices are currently evolving in order to deal with important issues brought up by the use of genome-scale sequencing as a clinical diagnostic test, including informed consent and reporting of incidental findings. The potential for WGS to uncover the full spectrum of genomic pathogenic contribution to disease—both coding and noncoding—ensures this technology will continue to be developed for clinical use, but the challenge will be storing, interpreting, and reporting the wealth of information produced by this type of testing.

The methodology, time, and cost involved in sequencing a whole human genome have changed drastically over the last decade [3] (Fig. 41.1). An increasing number of academic and private laboratories certified by the College of American Pathologists and Clinical Laboratory Improvement Amendments are offering genome-scale sequencing tests, in order to aid diagnosis of rare heritable diseases or for detecting mutations in cancer. The indications for genome-wide testing are expanding and span the spectrum between neonatal emergencies and adult mystery conditions where the diagnosis cannot be made based solely on clinical symptoms [4,5]. In this chapter, we discuss how genomic sequencing is currently being used in the molecular diagnostic laboratory and discuss the ways in which it is likely to be applied in the future.

MOLECULAR TECHNOLOGY

Both WGS and WES are considered next-generation sequencing (NGS) or massively parallel sequencing techniques, aimed at generating base level coverage across the coding regions of the genome and beyond, and collectively can be referred to as clinical genomic sequencing (CGS). Although the majority of disease-causing variants that we know about reside within coding regions of the genome, WES only captures 1–2% of the genome. Thus, the capability of WGS to capture the full potential spectrum of genomic pathogenic contribution to human disease—both coding and noncoding—will require continued optimization to improve clinical utility.

Prior to NGS-based genomic testing, physicians evaluating a patient might develop a differential diagnosis that included the possibility of mutations in many different genes that could potentially cause their patient's phenotype. The genetic testing process might sequentially evaluate one condition at a time using a combination of molecular and biochemical tests. CGS offers a more comprehensive and efficient approach to

© 2017 Elsevier Inc. All rights reserved.

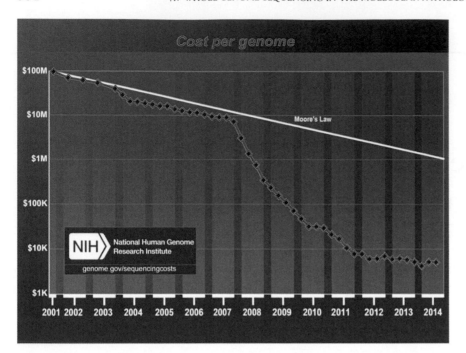

FIGURE 41.1 Cost of sequencing. The cost of sequencing per genome is depicted over time. Since the initiation of the Human Genome Project, when genome sequencing was a multimillion-dollar proposition, the contributions of federally funded researchers and for-profit biotechnology companies have enabled the development of massively parallel sequencing platforms capable of generating sequence data at a fraction of the cost. These advancements have driven the cost of sequencing a human genome below $10,000 and will almost certainly continue to push costs toward the $1000 genome milestone. It should be noted that these costs only reflect the technical side of sequence generation and do not address the interpretive costs.

testing all potentially causative genes at the same time. Thus, CGS exists at the end of a spectrum comprised of single gene tests, gene panels, and genome-scale sequencing tests. For those conditions caused by mutations in only one or a few genes, CGS may not be practical, and instead its application is best suited to genomically heterogeneous conditions. Depending on the number of genes that are known to cause the condition, and the extent to which the laboratory will further pursue novel findings, the use of CGS over NGS-based gene panels, which provide great coverage of the targeted genes, will have to be weighed. Even between WES and WGS, the technological differences between these two comprehensive approaches results in relative advantages and disadvantages which must be weighed in order to choose the right CGS test given the clinical application.

Although WES targets the majority of known disease-causing mutations, WGS may provide more consistent coverage of all genomic regions, including the exome, in a technically efficient manner that can increase turnaround time and limit biases introduced by target enrichment or amplification. On the other hand, WGS typically results in a 100 GB data file (compared to 25 GB for WES), so the need for computational and storage capabilities is increased. Both technologies are prone to mapping and alignment errors in homologous or repetitive genomic regions, and full coverage of clinically relevant genes may be lacking [6]. CGS is indicated for those individuals in whom a genetic etiology is strongly indicated. The test can detect substitutions, small insertions and deletions, inversions, and rearrangements. Given the large

number of potential findings in each patient, the standards for reporting variants are still being developed. Laboratories have developed different reporting thresholds for variants of uncertain significance, with respect to the possibility that they may play a role in the patient's condition.

Limitations of CGS

The impulse to order CGS in those patients who are difficult to diagnose, must be tempered by an understanding of the test limitations. Sequencing is simply not able to detect many clinically relevant genomic changes that for some conditions comprise the bulk of causal pathogenic changes. For example, the majority of hereditary spinal motor neuropathy is due to deletion of exon 7 in *SMN1* and is modified by duplication of *SMN2* [7]. As CGS is limited in its ability to detect copy number changes, it is not an appropriate first-line test for these patients. CGS is also unable to reliably detect low-level mosaicism, imprinting, or uniparental disomy, although optimizing CGS for some of these applications is an ongoing area of research. For detecting rare causes of Mendelian disease, the laboratory may routinely apply an allele frequency filter to sequence variants. In some cases, the laboratory must customize their bioinformatics pipeline to confidently detect particular types of pathogenic variants. Two examples are the Factor V Leiden thrombophilia-associated variant, R506Q, which is the reference allele, and the common *CFTR* Phe508del variant, which can be misaligned due to shifting of reads with the deletion, or might be inadvertently filtered out due to its high frequency in the population [8,9]. In these cases, it may be necessary for the laboratory to manually search for these variants or customize their pipeline to consistently identify known pathogenic (KP) variants. Subsequently, the laboratory must validate the sensitivity and specificity of their detection. If any targeted regions are not well covered, as is often the case for the first exon of many genes, these areas should be highlighted in the reported results.

While the costs of CGS are still formidable (~$10,000 for WGS and ~$5000 for WES), comprehensive genomic testing can make sense for genetically heterogeneous Mendelian conditions and may result in time and cost savings compared to the traditional gene-by-gene approach. Instead, it can be argued that the greatest challenge in the clinical application of WGS is interpreting the large number of variants identified by this type of testing. The initial automated variant analysis typically involves annotation of key metrics, including genomic position of the variant, frequency within the general population, and predicted effect, as well as quality and depth of the sequencing run. Ultimately, variants of interest must be manually investigated through review of the primary literature, use of genome browsers, and online variant databases. Over 100 filtered variants may need to be evaluated in any one individual [10]. As a result, the manual variant curation step has necessitated an expansion in laboratory personnel and is generally one of the most time-consuming aspects of WGS.

In order to limit and prioritize the number of variants that need to be manually analyzed, many laboratories first filter for variants present within a list of genes known to be related to the patient's presenting phenotype. If no pathogenic variants are identified in this initial diagnostic list, the laboratory can subsequently reflex into analyzing variants present on a broader diagnostic gene list. If the clinical diagnosis is unclear, analyzing a broader diagnostic list encompassing all of the possible causal genes might make sense to analyze initially, but increases the workload in terms of the number and breadth of variants that need to be evaluated.

Interpreting Variants in CGS

CGS-identified variants are classified as KP, likely pathogenic (LP), known benign (KB), likely benign (LB), or a variant of unknown significance (VUS). The American College of Medical Genetics and Genomics (ACMG) has recommended guidelines regarding the implementation of this classification scheme for molecular laboratories [11]. Variants generally start as VUS and are upgraded or downgraded depending on the

supporting evidence. Published genetic and functional data, frequency of the variant in the general population, its conservation among species, the incidence of the disease, the demonstrated mode of inheritance, expression, and penetrance of the condition, and the number of independent reports demonstrating the segregation of the variant with disease, are used as supporting evidence (Fig. 41.2). In silico prediction programs designed to score the effect of a missense change do not provide strong evidence on their own, but can be used as mild supportive evidence if all models agree. While truncating variants are presumed to have a deleterious effect on the protein, they may not necessarily be pathogenic, particularly if they occur at the very 3' end of the gene where nonsense-mediated decay may not occur. In such cases, the known mutation spectrum of the disease can help determine whether a novel truncating variant is LP or a VUS. The VUS category is appropriate for those variants in which the evidence does not clearly support classification in the pathogenic or benign categories.

CGS Returns Many Variants of Uncertain Significance

For KP or LP variants with strong supporting evidence, the laboratory may recommend targeted testing of other family members and appropriate medical management changes for those individuals carrying the variant. Many of the variants currently being returned to patients in CGS reports are VUS, and these would not be reported with any recommendations for targeted testing or change in medical management, as the pathogenicity of the variant is unclear. In these cases, the laboratory can offer family testing if that had not been done at the outset in order to evaluate the presence or absence of the VUS in affected and unaffected family members. Presence of the VUS in affected family members, particularly first-degree relatives who have a 50% chance of having inherited the proband's variant, does not prove the variant is causal. However, failure of the variant to segregate with affected family members or the presence of the VUS in unaffected family members can provide moderate to strong evidence against pathogenicity, depending upon the penetrance of the condition. For all reported variants, the laboratory can recommend clinical testing via a separate modality that would be meaningful to the interpretation of a particular variant. In some cases, follow-up clinical testing will reveal a histological, radiological, or biochemical result that either supports or is inconsistent with the original molecular findings. In this case, it is useful for physicians to report this back to the laboratory so that they can more accurately adjudicate the pathogenicity of that variant in the future.

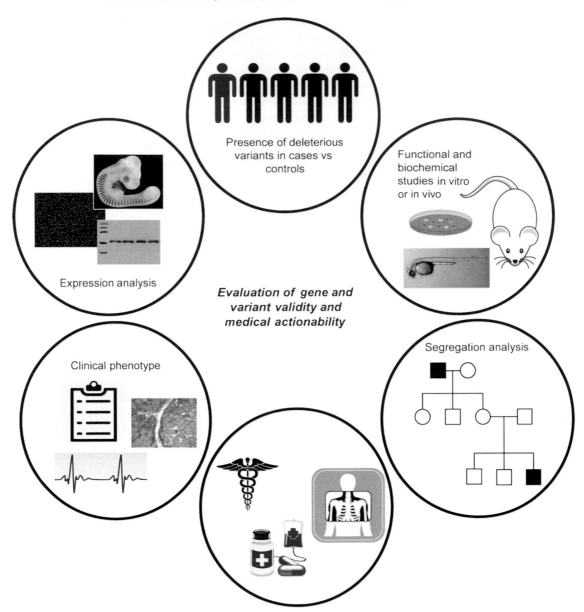

FIGURE 41.2 Validation of discoveries from genome sequencing. In genetics research aimed at the discovery of new gene–disease associations, generation of sequence variant data is now far more straightforward than demonstrating causality or defining mechanisms of disease. This has led to an explosion of studies implicating genes as candidates for various monogenic disorders. Clinical phenotyping, family segregation studies, in vivo animal models, in vitro studies, and other evidence types must complement genetic variant data. The challenge of demonstrating causality for very rare disorders will also impact the ability of variants to be interpreted in the molecular diagnostic laboratory. Diagnosticians must therefore be able to evaluate the evidence supporting a gene–disease association and determine whether a particular finding meets a sufficient threshold of clinical validity to be returned in the setting of clinical molecular diagnostic testing.

For most genomic variants, there is a significant lack of genetic, functional, and clinical evidence, and their current classification as VUS is likely to evolve. As more people are sequenced clinically, it will help define the true frequency of variants in cases and controls and will facilitate genotype–phenotype correlations. Variants previously deemed pathogenic, which may have relied on small internal control populations, have since had their pathogenicity downgraded simply by noting that their allele frequency is much too common in the general population to be considered pathogenic for a rare disease [12]. The exome aggregation consortium has frequency information for over 60,000 individuals, so it is now possible to obtain more precise allele frequencies for many variants than we had previously, at least for those present in the exome [13], although phenotypic details for this dataset are not yet publically available.

The use of trios at the outset can facilitate CGS analysis, particularly for those conditions that are known to be recessive (because phase can be determined) and those conditions for which a large number of pathogenic variants are generated de novo (because variants not present in either parent can be quickly identified), such as intellectual disability/autism [14], or certain seizure conditions [15,16]. Although some laboratories are reporting variants in novel genes, simply relying on trio-based testing to report de novo variants in novel genes, or genes with a weak disease association as pathogenic may result in false positives. The distinction between reporting variants in a known disease gene versus novel candidate genes will result in differences in the diagnostic rate reported for WGS among laboratories. Currently, the average diagnostic yield for CGS is around 15–40%, depending upon the condition being tested, and whether trios are tested simultaneously [10,17].

Incidental Findings and the Ethical, Legal, and Social Implications of CGS

CGS will inevitably uncover incidental findings unrelated to the patient's primary condition. A small number of these variants will be pathogenic for relatively penetrant conditions, for which medical interventions exist. The ACMG has issued guidelines regarding the return of incidental findings [18], and the current consensus is that pathogenic variants in a list of 56 medically actionable genes should be reported if the patient chooses to receive them. Studies are currently looking at how often these secondary findings are encountered, how best to report them, as well as understanding how people differ in their preference to find out about different categories of secondary results. Although the finding of a mutation that highly predisposes to disease in an asymptomatic individual does not provide a medical diagnosis, it can still produce a great deal of anxiety and worry. For many, CGS will be viewed as the opening of Pandora's box. Parents may learn things about their children or about their ex-spouse. Individuals may learn about predispositions to develop disease that they cannot do anything about. These issues warrant a better understanding of the ethical, legal, and social aspects of applying genomic testing clinically.

Given how impactful comprehensive genomic testing information is, genetic counselors must include in their discussions the potential return of secondary findings during the enrollment and return of results process. People are not uniform in terms of what they want—some people may want to learn all of their results at the same time, but there will always be some people who do not want to know everything [19]. This may be particularly true for conditions that are severe and we cannot do anything about, such as Huntington's, but is also highly influenced by people's personal histories. Thus, patient preferences should be considered when developing guidelines regarding how consenters, laboratories, and reporting clinicians handle WGS test results. This awareness has been reflected in part by the recent update of the ACMG's guidelines on incidental findings to include an opt-out for receipt of incidental findings. It will be particularly important for practitioners to be aware of the wishes and privacy of minors—will we allow their parents to receive WGS results that are entered into their medical record? Will parents know of their children's carrier status? If this sort of information is going to be clinically utilized, protections on health information must be sought in parallel, and this is currently an area of legislation [20].

DOES WGS ADD CLINICAL UTILITY?

The ability of WES to provide a molecular diagnosis in a significant fraction of previously undiagnosed patients suspected to have a monogenic condition is clear. However, WGS is not yet routinely used in clinical care. While the computing and data storage issues continue to pose a challenge for routine clinical use of WGS, it is also unclear whether WGS will add any clinical benefit to patients. Because of more consistently even coverage across the genome, WGS may allow for more accurate detection of copy number variants (CNVs) that play a significant role in many diseases. A recent report demonstrated the ability of WGS to significantly increase the diagnostic yield for some patients, even after microarray and WES, and many of these were CNVs [17]. In addition to its potential to call CNVs more reliably, WGS does not solely target the exome, and therefore is able to identify variants in the rest of the genome. The defined noncoding genomic regions that could theoretically be a source of pathogenic variants include enhancers and repressors, insulators, locus control regions, and repetitive regions (Fig. 41.3). The challenge resides in being able to ascertain that a WGS-identified variant indeed falls within one of these important regions, and having enough information on the variant to report it as pathogenic. While coverage of more of the genome is achievable with WGS, the clinical relevance of these regions has not yet been established, and targeted approaches may achieve higher average coverage of the most clinically relevant variants (Fig. 41.4).

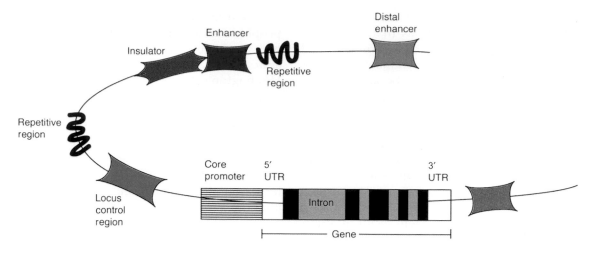

FIGURE 41.3   Model of long-range genomic interactions. To date, most of the variants that cause rare monogenic disorders have been found in the coding regions of genes. In rare cases, disease-causing variants have been demonstrated in the 5'- or 3'-untranslated regions (UTRs). This is one reason why WES has emerged as the preferred assay for research and clinical applications. However, despite the intense focus on the genes as the basic elements encoding the functional proteins, it is well known that noncoding elements, some of them a great distance away from the gene itself, can have important roles in controlling gene expression in a tissue-specific or context-specific manner. These additional genomic elements may provide an important target for the discovery of disease-causing variants in patients whose previous gene-centered testing has been unrevealing.

The Clinical Relevance of Noncoding Regions of the Genome Is Unclear

A number of molecular genetic techniques have recently been developed to define and characterize the landscape of genomic regulatory regions [21,22]. These include ChipSeq, DnaseSeq, FaireSeq, chromatin conformation techniques, cap analysis of gene expression, as well as bioinformatics-based approaches and are used to identify transcription factor binding motifs, ultra-conserved elements, physically interacting regions of the genome, and regions of open chromatin associated with transcription control. Many other noncoding genomic regions have been defined, including mini, micro, or satellite repeats, SINE and LINE elements, and DNA and LTR retrotransposons [23]. As the function of these noncoding regions are validated, and as variants are identified in them that are present in cases and not controls, one could imagine that in the future, CGS may expand to include the return of variants in these more distal genomic regions if the clinical context fits with the genes or pathways predicted to be disrupted by the testing. Currently though, this sort of result would be considered a research finding and would have somewhat questionable clinical value. Until more genomes are sequenced, and functional studies are done on candidate variants, demonstrating the relevance of most of these noncoding variants to clinical phenotypes will be impossible.

It is presumed that many of the disease-associated variants lying outside of coding regions are located within functional enhancers or other regulatory regions of the genome. Enhancers are distal regulatory elements that often reside 10,000–100,000 nucleotides from their target gene. Variants located within transcriptional enhancers (or repressors) can disrupt sequence motifs required for sequence-specific binding of transcription factors, chromatin regulators, and nucleosome positioning signals. The role of distal enhancers in human disease has been suggested for some time, based upon the identification of many patients with Mendelian disorders for which some patients had translocations or structural variants far from the promoter [24,25]. If WGS can pick up some fraction of these clinically relevant noncoding point mutations, small insertions or deletions, or structural variants, it could increase the diagnostic yield for this subset of patients. Structural variants affecting distal enhancers can disrupt their regulatory activity by moving them away from their targets, altering local chromatin conformation, or creating interactions with insulators or repressors that can hinder their action [26]. Although it is thought that looping interactions that facilitate contact with target promoter regions mediate the functional effects of enhancers, the molecular details of enhancer-gene targeting and regulatory mechanism of action are incompletely understood. Although this lack of clarity related to mechanism makes the current interpretation of a genome sequencing identified variant in a putative enhancer region challenging, the clinical relevance of at least some types of noncoding variants has been demonstrated.

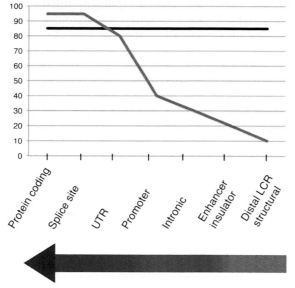

WGS (black line) detects more vaiants than WES (green line) but the clinical relevance of most non-coding genomic variants has not yet been demonstrated

Demonstrated clinical relevance of variant class

FIGURE 41.4 Trade-offs between WGS versus WES. This schematic figure represents important trade-offs between different approaches for genome-scale sequencing. Since by definition any targeted approach (such as WES) enriches for specific parts of the genome, it will have reduced coverage for regions of the genome not included in the target region that is captured. The decision about which platform to use for a given assay will depend on the goals of the test, the spectrum of disease-causing variants, and the cost of generating, storing, and analyzing the sequence data. Thus, WES largely covers the coding regions and nearby intronic sequences relevant for the canonical splice sites, but typically has reduced coverage, or none at all, for deep intronic sequences or more distal regulatory regions. This limitation, when considered in combination with the costs of sequencing, is generally deemed acceptable since the vast majority of clinically relevant variants are known to have an impact on the mRNA and translated protein.

Clinically Relevant Noncoding Variants

Enhancers. One of the most recognized examples of a noncoding variant causing a Mendelian human disease is the dysmorphology associated with mutations in the zone of polarizing activity regulatory sequence (ZRS). Several point mutations, as well as copy number changes in the ZRS, which reside in intron 5 of the *LMBR1* gene, have been described in humans. These mutations cause enhanced sonic hedgehog (SHH) activity, ectopic expression of SHH, and a variable phenotype of preaxial polydactyly, triphalangeal thumb, absent digits, and kidney and cardiac defects [27,28]. Mutations in the ZRS are thought to account for approximately 2–3% of patients with congenital limb abnormalities [29,30] and testing is currently available clinically.

Promoters. Several promoter variants have been described that effect the expression of clinically relevant genes, including *APOE* [31], *CCR5* [32], and *HO1* [33], although clinical testing for these types of variants is not routinely offered as their effects are not consistent with causing monogenic disease, but rather elevating risk to disease. For oncology patients, clinical testing for mutations in a number of gene promoters, including *TERT* for gliomas, thyroid cancer, and melanoma, can clarify diagnosis, inform prognosis, and guide entry into clinical trials [34,35]. Identification of recurrent *TERT* promoter mutations in melanoma suggests that somatic mutations in noncoding gene regulatory regions may represent an important mechanism in tumorigenesis [36]. As the numbers of clinically relevant oncological mutations are identified in noncoding regions of the genome, the utility of a comprehensive test like CGS becomes increasingly relevant to medical management of these patients.

Intronic Region and UTR. Comprehensive gene sequencing that includes intronic regions is clinically available for well-known disease genes with previously described clinically relevant noncoding variants. Cystic fibrosis (CF) represents a significant clinical entity in genetic testing referral, and several intronic variants have been reported to contribute to disease outcome. The poly T tract, a string of thymidine bases located in intron 8 of the *CFTR* gene, can be associated with *CFTR*-related disorders depending on its size. The presence of 5T at this position is considered a variably penetrant mutation, which is thought to decrease the efficiency of intron 8 splicing. Another clinically relevant noncoding region of *CFTR* is the TG tract, which lies just 5′ of the poly T tract and consists of a short string of TG repeats that commonly number 11, 12, or 13. A longer TG tract (12 or 13) in conjunction with a shorter poly T tract (5T) has the strongest adverse effect on proper intron 8 splicing and strongest association with pathogenic phenotype [37,38]. Males with congenital absence of the vas deferens (CAVD) or suspected CAVD, individuals with nonclassic CF, or adult carriers of 5T who wish to further refine their reproductive risks are all appropriate for 5T/TG tract typing. A recent study identified 23 variants in the 6000 bp *CFTR* 5′UTR in phenotypic patients with none or only one *CFTR* coding variant previously identified. Many of these variants led to gene expression changes in vitro, suggesting some of these variants may have functional consequences. The investigators suggested that noncoding variants could be the primary or second hit resulting in this recessive condition for a subset of CF patients [39]. Considering the high frequency of this condition in the population, and how informative diagnostic testing is for these families, there will likely be more research looking into the diagnostic utility of WGS for CF, as its comprehensive nature would allow for concurrent detection of all potential genomic contribution to the disease.

GWAS Variants. In addition to being able to detect monogenic disease-causing variants, WGS can detect variants that may only slightly contribute to disease risk, and the clinical reporting of these variants poses unique challenges. The overwhelming majority of the hundreds of genome-wide association study (GWAS) variants that have now been identified are associated with very modest changes in risk for disease. Nevertheless, the disease association for some of these variants has been consistently replicated, and the functional effect of the variant has been elucidated. If these types of variants are to be returned clinically, clearly their interpretations will need to take into consideration the contribution of other genetic and nongenetic factors that may be working together, to impact the disease outcome [40,41].

WGS for Oncology Applications

Several noncoding variants have been identified that are relevant to the care of subsets of cancer patients, suggesting that WGS may be an attractive comprehensive testing modality for these individuals. However, current WGS read depths average around 30-fold, and thus would likely not be able to detect low levels of genetic mosaicism. Molecular heterogeneity is an important characteristic displayed by many cancers, and surveying the genetic landscape over time, as well as evaluating the relative percentages of genetically distinct clonal subpopulations in tumors can indicate differential sensitivity of the neoplasm to various chemotherapeutic agents [42,43]. This limitation may impair the utility of WGS in oncology testing, until methods are developed to simultaneously evaluate multiple cancer sites at sufficient read depths. In addition, research is examining the use of genome-scale sequencing in conjunction with transcriptome analysis in order to identify pathway dependencies in the tumor [44]. These newer approaches, which look at the entire repertoire of variants in a particular pathway, could identify potentially targetable pathways that may impact how we treat certain malignancies.

One major potential advantage of using CGS in oncology is its capability to detect clinically relevant oncological fusions, copy number changes, inversions, translocations, and other rearrangements that are not currently detectable via WES or chromosomal microarray. How WGS will integrate into the current molecular oncology testing workflow is a continued area of research, and will be determined in large part by how much tissue is available for testing, and the goals of the test. Testing should be prioritized such that those results offering the most clinical utility are evaluated first, such as diagnostic information or results that will impact entry into clinical trials, followed by those results offering less clinical utility (prognostic information or molecular subtype). Since breakpoints can be precisely identified with WGS, incorporating this test might make sense for those cancers where particular fusions are diagnostic or prognostic indicators. While it is theoretically possible that more copy number information can be obtained from WGS than from WES because of covering more of the genomic landscape, laboratories will likely need software packages that are specifically designed for copy number detection, and this is currently an evolving area of research.

APPLICATION OF WGS IN THE FUTURE

Public Health Screening

In the near future, it will likely become more common for patients with suspected genetic conditions to undergo CGS after some modicum of genetic testing has been attempted, or even as a first-line test, particularly for genomically heterogeneous conditions, mystery conditions, or neonates in distress. In addition to its use in diagnostic testing for patients, massively parallel sequencing may be utilized as a public health tool to screen for treatable conditions in children or adults [45]. One could imagine that many of the conditions currently tested for using traditional newborn screening techniques could be augmented by targeted sequencing, by providing genotypic classification of inborn errors of metabolism detected though tandem mass spectrometry, as well as expanding the number of potentially screenable disorders. Some conditions may not be amenable to detection via sequencing, although it will most certainly be able to add to the list of detectable monogenic conditions. The prospect of being able to diagnose children earlier so that they can receive life-saving medicines or be directed into more effective treatment protocols makes genomic sequencing an attractive component of a comprehensive preventative health care strategy. At the same time, we must ensure that testing is focused on those genes that are going to provide the highest clinical utility. While current research is looking into the clinical application of sequencing for newborn screening, concurrent efforts must be undertaken to understand the unique ethical challenges raised by comprehensive sequencing in this population and formulate appropriate protections for minors.

Personalized Medicine and Pharmacogenomics

As the quest for personalized medicine advances, physicians may utilize CGS to screen individuals for the over 500 pharmacogenomic variants that have been associated with a clinical drug response [46]. Pharmaceutical companies that develop treatments and physicians who implement them seek to identify those individuals who are most likely to respond positively and want to be alerted to those individuals who might develop serious side effects. Thus, the increased uptake of genomic testing for companion diagnostics is another area where CGS may be applied in the future to screen for greater numbers of different genomic variants simultaneously. Research is needed to understand how to optimize integrating pharmacogenomic information into the current clinical workflow, in order to ensure maximum clinical utility. Most importantly, outcomes research is needed in order to justify the routine and widespread use of pharmacogenomics in the clinical setting.

The emergence of private companies offering personalized genomic test results portends a patient-driven medical future where individuals inquire about their genetic status regarding a number of different Mendelian and complex diseases, pharmacogenomics variants, and genetic risk factors. Laboratories will need to decide which types of results they offer and at which times during the testing workflow. Physicians will likely be in a position where they will see patients who have received genome-scale sequencing results and are now looking for guidance or clarification on various aspects of those results. The clinical laboratory must therefore be capable of serving as an interpretive and consulting resource for physicians throughout the CGS ordering and interpretation lifecycle.

OPTIMISM SURROUNDING WGS

The ability to perform WGS has provided a great deal of optimism to both patients and physicians. Many physicians view genome-scale sequencing as an opportunity to provide testing to patients who previously would not have qualified for genetic testing— perhaps they did not have a clinical diagnosis, there was no clinical test available for their suspected causal gene, or the causative gene was simply not known. Although better phenotyping at the outset may increase the likelihood of obtaining relevant results, with the advent of this comprehensive testing, a clear clinical diagnosis is not absolutely required. In fact, genome-scale sequencing can often lead to finding pathogenic variants in genes that were not suspected, and this can lead to exciting changes in patient treatment plans [47].

The comprehensive nature of the testing can impact patient expectations regarding CGS results. Compared to single-gene tests, there is more excitement that sweeping the genome will lead to a diagnosis and disappointment when a negative result is returned [48]. However, even with a negative test result, physicians and patients understand that the data can be reevaluated in the future as more information regarding the genetic basis of disease is published. This opportunity for subsequent research gives patients hope that CGS may ultimately facilitate a molecular diagnosis.

References

[1] Ng SB, Buckingham KJ, Lee C, et al. Exome sequencing identifies the cause of a Mendelian disorder. Nat Genet 2010;42:30–5.

[2] Lupski JR, Reid JG, Gonzaga-Jauregui C, et al. Whole-genome sequencing in a patient with Charcot-Marie-Tooth neuropathy. N Engl J Med 2010;362:1181–91.

[3] Wetterstrand K.A. DNA sequencing costs: data from the NHGRI Genome Sequencing Program (GSP). Available from: www.genome.gov/sequencingcosts.

[4] Soden SE, Saunders CJ, Willig LK, et al. Effectiveness of exome and genome sequencing guided by acuity of illness for diagnosis of neurodevelopmental disorders. Sci Transl Med 2014;6:265ra168.

[5] Yang Y, Muzny DM, Xia F, et al. Molecular findings among patients referred for clinical whole-exome sequencing. JAMA 2014;312:1870–9.

[6] Dewey FE, Grove ME, Pan C, et al. Clinical interpretation and implications of whole-genome sequencing. JAMA 2014;311: 1035–45.

[7] Frugier T, Nicole S, Cifuentes-Diaz C, Melki J. The molecular bases of spinal muscular atrophy. Curr Opin Genet Dev 2002; 12:294–8.

[8] Bertina RM, Koeleman BP, Koster T, et al. Mutation in blood coagulation factor V associated with resistance to activated protein C. Nature 1994;369:64–7.

[9] Kerem B, Rommens JM, Buchanan JA, et al. Identification of the cystic fibrosis gene: genetic analysis. Science 1989;245: 1073–80.

[10] Lee H, Deignan JL, Dorrani N, et al. Clinical exome sequencing for genetic identification of rare Mendelian disorders. JAMA 2014;312:1880–7.

[11] Richards S, Aziz N, Bale S, et al. Standards and guidelines for the interpretation of sequence variants: a joint consensus recommendation of the American College of Medical Genetics and Genomics and the Association for Molecular Pathology. Genet Med 2015;17:405–23.

[12] Piton A, Redin C, Mandel J-L. XLID-causing mutations and associated genes challenged in light of data from large-scale human exome sequencing. Am J Hum Genet 2013;93:368–83.

[13] Exome Aggregation Consortium (ExAC). Cambridge, MA. http://exac.broadinstitute.org/.

[14] Iossifov I, O'Roak BJ, Sanders SJ, et al. The contribution of de novo coding mutations to autism spectrum disorder. Nature 2014;515:216–21.

[15] O'Roak BJ, Stessman HA, Boyle EA, et al. Recurrent de novo mutations implicate novel genes underlying simplex autism risk. Nat Commun 2014;5:5595.

[16] Muona M, Berkovic SF, Dibbens LM, et al. A recurrent de novo mutation in KCNC1 causes progressive myoclonus epilepsy. Nat Genet 2014;47:39–46.

[17] Gilissen C, Hehir-Kwa JY, Thung DT, et al. Genome sequencing identifies major causes of severe intellectual disability. Nature 2014;511:344–7.

[18] Green RC, Berg JS, Grody WW, et al. ACMG recommendations for reporting of incidental findings in clinical exome and genome sequencing. Genet Med 2013;15:565–74.

[19] Smith LA, Douglas J, Braxton AA, Kramer K. Reporting incidental findings in clinical whole exome sequencing: incorporation of the 2013 ACMG recommendations into current practices of genetic counseling. J Genet Couns 2015;24:654–62.

[20] "H.R. 493—110th Congress: Genetic Information Nondiscrimination Act of 2008." www.GovTrack.us. 2007. February 2, 2015.

[21] ENCODE Project Consortium. The ENCODE (ENCyclopedia Of DNA Elements) project. Science 2004;306:636–40.

[22] Forrest ARR, Kawaji H, Rehli M, et al. A promoter-level mammalian expression atlas. Nature 2014;507:462–70.

[23] Treangen TJ, Salzberg SL. Repetitive DNA and next-generation sequencing: computational challenges and solutions. Nat Rev Genet 2012;13:36–46.

[24] Kleinjan DA, van Heyningen V. Long-range control of gene expression: emerging mechanisms and disruption in disease. Am J Hum Genet 2005;76:8–32.

[25] Noonan JP, McCallion AS. Genomics of long-range regulatory elements. Ann Rev Genomics Hum Genet 2010;11:1–23.

[26] Ward LD, Kellis M. Interpreting noncoding genetic variation in complex traits and human disease. Nat Biotechnol 2012;30:95–106.

[27] Lettice LA. A long-range Shh enhancer regulates expression in the developing limb and fin and is associated with preaxial polydactyly. Hum Mol Genet 2003;12:1725–35.

[28] Al-Qattan MM, Al Abdulkareem I, Al Haidan Y, Al Balwi M. A novel mutation in the SHH long-range regulator (ZRS) is associated with preaxial polydactyly, triphalangeal thumb, and severe radial ray deficiency. Am J Med Genet 2012;158A:2610–15.

[29] Furniss D, Lettice LA, Taylor IB, et al. A variant in the sonic hedgehog regulatory sequence (ZRS) is associated with triphalangeal thumb and deregulates expression in the developing limb. Hum Mol Genet 2008;17:2417–23.

[30] Furniss D, Kan S-H, Taylor IB, et al. Genetic screening of 202 individuals with congenital limb malformations and requiring reconstructive surgery. J Med Genet 2009;46:730–5.

[31] Bray NJ, Jehu L, Moskvina V, et al. Allelic expression of APOE in human brain: effects of epsilon status and promoter haplotypes. Hum Mol Genet 2004;13:2885–92.

[32] Martin MP, Dean M, Smith MW, et al. Genetic acceleration of AIDS progression by a promoter variant of CCR5. Science 1998;282:1907–11.

[33] Exner M, Minar E, Wagner O, Schillinger M. The role of heme oxygenase-1 promoter polymorphisms in human disease. Free Radic Biol Med 2004;37:1097–104.

[34] Killela PJ, Pirozzi CJ, Healy P, et al. Mutations in IDH1, IDH2, and in the TERT promoter define clinically distinct subgroups of adult malignant gliomas. Oncotarget 2014;5:1515–25.

[35] Gandolfi G, Ragazzi M, Frasoldati A, Piana S, Ciarrocchi A, Sancisi V. TERT promoter mutations are associated with distant metastases in papillary thyroid carcinoma. Eur J Endocrinol 2015;172:403–13.

[36] Huang FW, Hodis E, Xu MJ, Kryukov GV, Chin L, Garraway LA. Highly recurrent TERT promoter mutations in human melanoma. Science 2013;339:957–9.

[37] Cuppens H, Lin W, Jaspers M, et al. Polyvariant mutant cystic fibrosis transmembrane conductance regulator genes. The polymorphic (Tg)m locus explains the partial penetrance of the T5 polymorphism as a disease mutation. J Clin Invest 1998;101:487–96.

[38] Groman JD, Hefferon TW, Casals T, et al. Variation in a repeat sequence determines whether a common variant of the cystic fibrosis transmembrane conductance regulator gene is pathogenic or benign. Am J Hum Genet 2004;74:176–9.

[39] Giordano S, Amato F, Elce A, et al. Molecular and functional analysis of the large 5′ promoter region of CFTR gene revealed pathogenic mutations in CF and CFTR-related disorders. J Mol Diagn 2013;15:331–40.

[40] Sakabe NJ, Savic D, Nobrega MA. Transcriptional enhancers in development and disease. Genome Biol 2012;13:238.

[41] Manolio TA. Bringing genome-wide association findings into clinical use. Nat Rev Genet 2013;14:549–58.

[42] Biswas NK, Chandra V, Sarkar-Roy N, et al. Variant allele frequency enrichment analysis in vitro reveals sonic hedgehog pathway to impede sustained temozolomide response in GBM. Sci Rep 2015;5:7915.

[43] Melisi D, Piro G, Tamburrino A, Carbone C, Tortora G. Rationale and clinical use of multitargeting anticancer agents. Curr Opin Pharmacol 2013;13:536–42.

[44] Dhanasekaran SM, Alejandro Balbin O, Chen G, et al. Transcriptome meta-analysis of lung cancer reveals recurrent aberrations in NRG1 and Hippo pathway genes. Nat Commun 2014;5:5893.

[45] Evans JP, Berg JS, Olshan AF, Magnuson T, Rimer BK. We screen newborns, don't we? Realizing the promise of public health genomics. Genet Med 2013;15:332–4.

[46] Whirl-Carrillo M, McDonagh EM, Hebert JM, et al. Pharmacogenomics knowledge for personalized medicine. Clin Pharmacol Ther 2012;92:414–17.

[47] Fan Z, Greenwood R, Felix ACG, et al. GCH1 heterozygous mutation identified by whole-exome sequencing as a treatable condition in a patient presenting with progressive spastic paraplegia. J Neurol 2014;261:622–4.

[48] Khan CM, Rini C, Bernhardt BA, et al. How can psychological science inform research about genetic counseling for clinical genomic sequencing? J Genet Couns 2015;24:193–204.

Index

Note: Page numbers followed by "*f*" and "*t*" refer to figures and tables, respectively.

CPI Antony Rowe
Eastbourne, UK
August 11, 2017